BIOLOGY

the dynamic science

VOLUME 3

Peter J. Russell
Stephen L. Wolfe
Paul E. Hertz
Cecie Starr
Beverly McMillan

THOMSON
BROOKS/COLE™ Australia • Brazil • Canada • Mexico • Singapore • Spain • United Kingdom • United States

THOMSON
BROOKS/COLE

Biology: The Dynamic Science, Volume 3, First Edition
Peter J. Russell, Stephen L. Wolfe, Paul E. Hertz, Cecie Starr, Beverly McMillan

Vice President, Editor in Chief: Michelle Julet

Publisher: Yolanda Cossio

Managing Editor: Peggy Williams

Senior Development Editors: Mary Arbogast, Shelley Parlante

Development Editor: Christopher Delgado

Assistant Editor: Jessica Kuhn

Editorial Assistant: Rose Barlow

Technology Project Managers: Keli Amann, Kristina Razmara, Melinda Newfarmer

Marketing Manager: Kara Kindstrom

Development Project Manager: Terri Mynatt

Production Manager: Shelley Ryan

Creative Director: Rob Hugel

Art Directors: John Walker, Lee Friedman

Art Developers: Steve McEntee, Dragonfly Media Group

Print Buyer: Karen Hunt

Permissions Editor: Sarah D'Stair

Production Service: Graphic World Inc.

Production Service Manager: Suzanne Kastner

Text Designer: Jeanne Calabrese

Photo Researchers: Linda Sykes, Robin Samper

Copy Editor: Christy Goldfinch

Illustrators: Dragonfly Media Group, Steve McEntee, Precision Graphics, Dartmouth Publishing, Inc.

Cover Designer: Jeremy Mendes

Cover Image: © Leonardo Papini/Getty Images®

Cover Printer: Transcontinental Printing/Interglobe

Compositor: Graphic World Inc.

Printer: Transcontinental Printing/Interglobe

Printed in Canada
1 2 3 4 5 6 7 12 11 10 09 08 07

Library of Congress Control Number: 2007931665

Student Edition:
ISBN-13: 978-0-534-24966-3
ISBN-10: 0-534-24966-3
Volume 3:
ISBN-13: 978-0-495-01034-0
ISBN-10: 0-495-01034-0

Thomson Higher Education
10 Davis Drive
Belmont, CA 94002-3098
USA

About the Authors

PETER J. RUSSELL received a B.Sc. in Biology from the University of Sussex, England, in 1968 and a Ph.D. in Genetics from Cornell University in 1972. He has been a member of the Biology faculty of Reed College since 1972; he is currently a Professor of Biology. He teaches a section of the introductory biology course, a genetics course, an advanced molecular genetics course, and a research literature course on molecular virology. In 1987 he received the Burlington Northern Faculty Achievement Award from Reed College in recognition of his excellence in teaching. Since 1986, he has been the author of a successful genetics textbook; current editions are *iGenetics: A Mendelian Approach*, *iGenetics: A Molecular Approach*, and *Essential iGenetics*. He wrote nine of the BioCoach Activities for The Biology Place. Peter Russell's research is in the area of molecular genetics, with a specific interest in characterizing the role of host genes in pathogenic RNA plant virus gene expression; yeast is used as the model host. His research has been funded by agencies including the National Institutes of Health, the National Science Foundation, and the American Cancer Society. He has published his research results in a variety of journals, including *Genetics*, *Journal of Bacteriology*, *Molecular and General Genetics*, *Nucleic Acids Research*, *Plasmid*, and *Molecular and Cellular Biology*. He has a long history of encouraging faculty research involving undergraduates, including cofounding the biology division of the Council on Undergraduate Research (CUR) in 1985. He was Principal Investigator/Program Director of an NSF Award for the Integration of Research and Education (AIRE) to Reed College, 1998–2002.

STEPHEN L. WOLFE received his Ph.D. from Johns Hopkins University and taught general biology and cell biology for many years at the University of California, Davis. He has a remarkable list of successful textbooks, including multiple editions of *Biology of the Cell*, *Biology: The Foundations*, *Cell Ultrastructure*, *Molecular and Cellular Biology*, and *Introduction to Cell and Molecular Biology*.

PAUL E. HERTZ was born and raised in New York City. He received a bachelor's degree in Biology at Stanford University in 1972, a master's degree in Biology at Harvard University in 1973, and a doctorate in Biology at Harvard University in 1977. While completing field research for the doctorate, he served on the Biology faculty of the University of Puerto Rico at Rio Piedras. After spending 2 years as an Isaac Walton Killam Postdoctoral Fellow at Dalhousie University, Hertz accepted a teaching position at Barnard College, where he has taught since 1979. He was named Ann Whitney Olin Professor of Biology in 2000, and he received The Barnard Award for Excellence in Teaching in 2007. In addition to his service on numerous college committees, Professor Hertz was Chair of Barnard's Biology Department for 8 years. He has also been the Program Director of the Hughes Science Pipeline Project at Barnard, an undergraduate curriculum and research program funded by the Howard Hughes Medical Institute, since its inception in 1992. The Pipleline Project includes the Intercollegiate Partnership, a program for local community college students that facilitates their transfer to 4-year colleges and universities. He teaches one semester of the introductory sequence for Biology majors and preprofessional students as well as lecture and laboratory courses in vertebrate zoology and ecology. Professor Hertz is an animal physiological ecologist with a specific research interest in the thermal biology of lizards. He has conducted fieldwork in the West Indies since the mid-1970s, most recently focusing on the lizards of Cuba. His work has been funded by the National Science Foundation, and he has published his research in such prestigious journals as *The American Naturalist*, *Ecology*, *Nature*, and *Oecologia*.

CECIE STARR is the author of best-selling biology textbooks. Her books include multiple editions of *Unity and Diversity of Life*, *Biology: Concepts and Applications*, and *Biology Today and Tomorrow*. Her original dream was to be an architect. She may not be building houses, but with the same care and attention to detail, she builds incredible books: *"I invite students into a chapter through an intriguing story. Once inside, they get the great windows that biologists construct on the world of life. Biology is not just another house. It is a conceptual mansion. I hope to do it justice."*

BEVERLY MCMILLAN has been a science writer for more than 20 years and is coauthor of a college text in human biology, now in its seventh edition. She has worked extensively in educational and commercial publishing, including 8 years in editorial management positions in the college divisions of Random House and McGraw-Hill. In a multifaceted freelance career, Bev also has written or coauthored six trade books and numerous magazine and newspaper articles, as well as story panels for exhibitions at the Science Museum of Virginia and the San Francisco Exploratorium. She has worked as a radio producer and speechwriter for the University of California system and as a media relations advisor for the College of William and Mary. She holds undergraduate and graduate degrees from the University of California, Berkeley.

Preface

Welcome to *Biology: The Dynamic Science*. The title of our book reflects an explosive growth in the knowledge of living systems over the past few decades. Although this rapid pace of discovery makes biology the most exciting of all the natural sciences, it also makes it the most difficult to teach. How can college instructors—and, more important, college students—absorb the ever-growing body of ideas and information? The task is daunting, especially in introductory courses that provide a broad overview of the discipline.

Our primary goal in this text is to convey fundamental concepts while maintaining student interest in biology

In this entirely new textbook, we have applied our collective experience as college teachers, science writers, and researchers to create a readable and understandable introduction to our field. We provide students with straightforward explanations of fundamental concepts presented from the evolutionary viewpoint that binds all of the biological sciences together. Having watched our students struggle to navigate the many arcane details of college-level introductory biology, we have constantly reminded ourselves and each other to "include fewer facts, provide better explanations, and maintain the narrative flow," thereby enabling students to see the big picture. Clarity of presentation, a high level of organization, a seamless flow of topics within chapters, and spectacularly useful illustrations are central to our approach.

One of the main goals in this book is to sustain students' fascination with the living world instead of burying it under a mountain of disconnected facts. As teachers of biology, we encourage students to appreciate the dynamic nature of science by conveying our passion for biological research. We want to amaze students with *what* biologists know about the living world and *how* we know it. We also hope to excite them about the opportunities they will have to expand that knowledge. Inspired by our collective effort as teachers and authors, some of our students will take up the challenge and become biologists themselves, asking important new questions and answering them through their own innovative research. For students who pursue other career paths, we hope that they will leave their introductory—and perhaps only—biology courses armed with the knowledge and intellectual skills that allow them to evaluate future discoveries with a critical eye.

We emphasize that, through research, our understanding of biological systems is alive and constantly changing

In this book, we introduce students to a biologist's "ways of knowing." Scientists constantly integrate new observations, hypotheses, experiments, and insights with existing knowledge and ideas. To do this well, biology instructors must not simply introduce students to the current state of our knowledge. We must also foster an appreciation of the historical context within which that knowledge developed and identify the future directions that biological research is likely to take.

To achieve these goals, we explicitly base our presentation and explanations on the research that established the basic facts and principles of biology. Thus, a substantial proportion of each chapter focuses on studies that define the state of biological knowledge today. We describe recent research in straightforward terms, first identifying the question that inspired the work and relating it to the overall topic under discussion. Our research-oriented theme teaches students, through example, how to ask scientific questions and pose hypotheses, two key elements of the "scientific process."

Because advances in science occur against a background of past research, we also give students a feeling for how biologists of the past uncovered and formulated basic knowledge in the field. By fostering an appreciation of such discoveries, given the information and theories that were available to scientists in their own time, we can help students to better understand the successes and limitations of what we consider cutting edge today. This historical perspective also encourages students to view biology as a dynamic intellectual endeavor, and not just a list of facts and generalities to be memorized.

One of our greatest efforts has been to make the science of biology come alive by describing how biologists formulate hypotheses and evaluate them using hard-won data, how data sometimes tell only part of a story, and how studies often end up posing more questions than they answer. Although students often prefer to read about the "right" answer to a question, they must be encouraged to embrace "the unknown," those gaps in our knowledge that create opportunities for further research. An appreciation of what we *don't* know will draw more students into the field. And by defining *why* we don't understand interesting phenomena, we encourage students to follow paths dictated by their own curiosity. We hope that this approach will encourage students to make biology a part of their daily lives—to have informal discussions about new scientific discoveries, just as they do about politics, sports, or entertainment.

Special features establish a story line in every chapter and describe the process of science

In preparing this book, we developed several special features to help students broaden their understanding of the material presented and of the research process itself.

- The chapter openers, entitled *Why It Matters,* tell the story of how a researcher arrived at a key insight or how biological research solved a major societal problem or shed light on a fundamental process or phenomenon. These engaging, short vignettes are designed to capture students' imagination and whet their appetite for the topic that the chapter addresses.

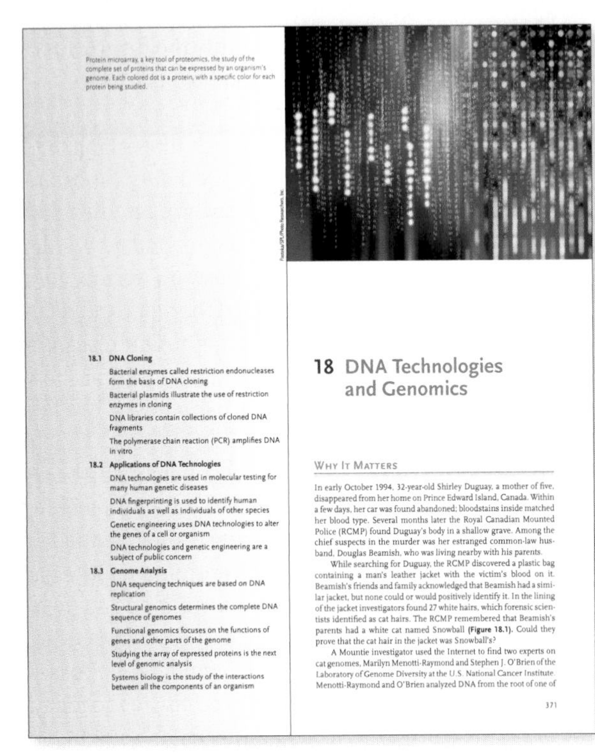

Protein microarray, a key tool of proteomics, the study of the complete set of proteins that can be expressed by an organism's genome. Each colored dot is a protein, with a specific color for each protein being studied.

18.1 DNA Cloning
Bacterial enzymes called restriction endonucleases form the basis of DNA cloning
Bacterial plasmids illustrate the use of restriction enzymes in cloning
DNA libraries contain collections of cloned DNA fragments
The polymerase chain reaction (PCR) amplifies DNA in vitro
18.2 Applications of DNA Technologies
DNA technologies are used in molecular testing for many human genetic diseases
DNA fingerprinting is used to identify human individuals as well as individuals of other species
Genetic engineering uses DNA technologies to alter the genes of a cell or organism
DNA technologies and genetic engineering are a subject of public concern
18.3 Genome Analysis
DNA sequencing techniques are based on DNA replication
Structural genomics determines the complete DNA sequence of genomes
Functional genomics focuses on the functions of genes and other parts of the genome
Studying the array of expressed proteins is the next level of genomic analysis
Systems biology is the study of the interactions between all the components of an organism

18 DNA Technologies and Genomics

WHY IT MATTERS

In early October 1994, 32-year-old Shirley Duguay, a mother of five, disappeared from her home on Prince Edward Island, Canada. Within a few days, her car was found abandoned; bloodstains inside matched her blood type. Several months later the Royal Canadian Mounted Police (RCMP) found Duguay's body in a shallow grave. Among the chief suspects in the murder was her estranged common-law husband, Douglas Beamish, who was living nearby with his parents.

While searching for Duguay, the RCMP discovered a plastic bag containing a man's leather jacket with the victim's blood on it. Beamish's friends and family acknowledged that Beamish had a similar jacket, but none could or would positively identify it. In the lining of the jacket investigators found 27 white hairs, which forensic scientists identified as cat hairs. The RCMP remembered that Beamish's parents had a white cat named Snowball **(Figure 18.1).** Could they prove that the cat hair in the jacket was Snowball's?

A Mountie investigator used the Internet to find two experts on cat genomes, Marilyn Menotti-Raymond and Stephen J. O'Brien of the Laboratory of Genome Diversity at the U.S. National Cancer Institute. Menotti-Raymond and O'Brien analyzed DNA from the root of one of

371

- To complement this historical or practical perspective, each chapter closes with a brief essay, entitled *Unanswered Questions,* often prepared by an expert in the field. These essays identify important unresolved issues relating to the chapter topic and describe cutting-edge research that will advance our knowledge in the future.

UNANSWERED QUESTIONS

Do asexual organisms form species?
As you learned in this chapter, the biological species concept applies only to sexually reproducing organisms because only those organisms can evolve barriers to gene flow (asexual organisms reproduce more or less clonally). Nevertheless, research is starting to show that organisms whose reproduction is almost entirely asexual, such as bacteria, seem to form distinct and discrete clusters in nature. (These clusters could be considered "species.") That is, bacteria and other asexual forms may be as distinct as the species of birds described by Ernst Mayr in New Guinea. Workers are now studying the many species of bacteria in nature (only a small number of which have been discovered) to see if they indeed fall into distinct groups. If they do, then scientists will need a special theory, independent of reproductive isolation, to explain this distinctness. Scientists are now working on theories of whether the existence of discrete ecological niches in nature might explain the possible discreteness of asexual "species."

How often does speciation occur allopatrically versus sympatrically or parapatrically?
Scientists do not know how often speciation occurs between populations that are completely isolated geographically (allopatric speciation) compared with how often it occurs in populations that exchange genes (parapatric or sympatric speciation). The relative frequency of these modes of speciation in nature is an active area of research. The ongoing work includes studies on small isolated islands: if an invading species divides into two or more species in this situation, it probably did so sympatrically or parapatrically, since geographical isolation of populations in small islands is unlikely. In addition, biologists are reconstructing the evolutionary history of speciation using molecular tools and correlating this history with the species' geographical distributions. If this line of research were to show, for example, that the most closely related pairs of species always had geographically isolated distributions, it would imply that speciation was usually allopatric. These lines of research should eventually answer the controversial question of the relative frequency of various forms of speciation.

What are the genetic changes underlying speciation?
Biologists know a great deal about the types of reproductive isolation that prevent gene flow between species, but almost nothing about their genetic bases. Which genes control the difference between flower shape in monkey-flower species? Which genes lead to inviability and sterility of *Drosophila* hybrids? Which genes cause species of ducks to preferentially mate with members of their own species over members of other species? Do the genetic changes that lead to reproductive isolation tend to occur repeatedly at the same genes in a group of organisms, or at different genes? Do the changes occur mostly in protein-coding regions of genes, or in the noncoding regions that control the production of proteins? Were the changes produced by natural selection or by genetic drift? Biologists are now isolating "speciation genes" and sequencing their DNA. With only a handful of such genes known, and all of these causing hybrid sterility or inviability, there will undoubtedly be a lot to learn about the genetics of speciation in the next decade.

Dr. Jerry Coyne conducts research on speciation and teaches at the University of Chicago. To learn more about his research go to http://pondside.uchicago.edu/ecol-evol/faculty/coyne_j.html.

- Each chapter also includes a short boxed essay, entitled *Insights from the Molecular Revolution,* which describes how molecular technologies allow scientists to answer questions that they could not have even posed 20 or 30 years ago. Each *Insight*

INSIGHTS FROM THE MOLECULAR REVOLUTION

A Fragile Connection between DNA Replication and Mental Retardation

One of the most common sources of inherited mental retardation in humans results from breaks that occur in a narrow, constricted region near one end of the X chromosome (see figure). Because the region breaks easily when cultured cells divide, the associated disabilities are called the *fragile X syndrome.* In addition to mental retardation, affected individuals may have an unusually long face and protruding ears; affected males may have oversized testes.

The disorder affects males more frequently than females, as is typical with X-linked traits—about 1 in 1500 male and 1 in 2500 female births are affected by the fragile X syndrome. However, the inheritance pattern of the syndrome also has some unusual characteristics. The disease is passed from a grandfather through his daughter to his grandchild. The grandfather and daughter have apparently normal X chromosomes, although the daughters sometimes have symptoms; however, abnormal X chromosomes and symptoms appear with high frequency in the grandchildren.

Geneticists were baffled by this unusual pattern of inheritance until a partial explanation was supplied by findings in the laboratories of Grant R. Sutherland of Adelaide Children's Hospital in Australia and others. The investigators examined DNA from individuals with fragile X syndrome using "probes"—short, artificially synthesized DNA sequences that are complementary to, and can pair with, DNA sequences that are of interest. They found that probes containing C and G nucleotides in high proportions paired most strongly with DNA in the fragile X region. Sequencing of DNA that paired with those probes showed that the region contains many repeats of the three-nucleotide sequence CCG.

Interestingly, the number of CCG repeats varies in the different groups from 6 to 50 copies in people without the syndrome, 50 to 200 copies in people with mild or no symptoms who transmit the syndrome, and about 230 to 1000 copies in seriously affected people. Somehow the number of CCG repeats increases, which initiates the serious disease symptoms.

The increase in copies occurs by overreplication of the CCG sequence, which begins to occur when the number of copies exceeds about 50. As more CCG copies are added, the region becomes increasingly unstable and the tendency for overreplication also increases. This feature of the process explains why symptoms of the disease often become worse in successive generations.

Scientists still do not understand what causes the overreplication or how it causes fragile X syndrome. However, the increase in CCG copies appears to turn off a nearby gene called *FMR-1,* which is necessary for normal mental development. The increase also inhibits other genes located elsewhere that have the same effect. One hypothesis takes note of the fact that methyl groups are added to cytosines as part of the controls that turn off large blocks of genes in mammals. According to this idea, the extra cytosines of the added CCG groups provide many additional methylation sites, leading to inactivation of the genes near the fragile X region.

The probe that pairs with CCG groups can be used to estimate the number of copies of the sequence in people without the syndrome who may have increased numbers of the sequence in the fragile X region. If the number is elevated above 50 repeats, these individuals can be counseled about the possibility that they could transmit the disease to their offspring several generations down the line.

The constricted region *(arrow)* in the human X chromosome associated with fragile X syndrome. The chromosome is double because it has been duplicated in preparation for cell division.

tein encoded by the gene, which, in turn, may alter how the organism functions. Hence, mutations are highly important to the evolutionary process because they are the ultimate source of the variability in offspring acted on by natural selection.

We now turn from DNA replication and error correction to the arrangements of DNA in eukaryotic and prokaryotic cells. These arrangements organize superstructures that fit the long DNA molecules into the microscopic dimensions of cells and also contribute to the regulation of DNA activity.

STUDY BREAK

Why is a proofreading mechanism important for DNA replication, and what are the mechanisms that correct errors?

294 | UNIT TWO GENETICS

focuses on a single study and includes sufficient detail for its content to stand alone.

- Almost every chapter is further supplemented with one or more short boxed essays that *Focus on Research.* Some of these essays describe seminal studies that provided a new perspective on an important question. Others describe how basic research has solved everyday problems relating to health or the environment. Another set introduces model research organisms—such as *E. coli, Drosophila, Arabidopsis, Caenorhabditis,* and *Anolis*—and explains why they have been selected as subjects for in-depth analysis.

Spectacular illustrations enable students to visualize biological processes, relationships, and structures

Today's students are accustomed to receiving ideas and information visually, making the illustrations and photographs in a textbook more important than ever before. Our illustration program provides an exceptionally clear supplement to the narrative in a style that is consistent throughout the book. Graphs and anatomical drawings are annotated with interpretative explanations that lead students through the major points they convey.

Three types of specially designed Research Figures provide more detailed information about how biologists formulate and test specific hypotheses by gathering and interpreting data.

- *Research Method* figures provide examples of important techniques, such as gel electrophoresis, the use of radioisotopes, and cladistic analysis. Each *Research Method* figure leads a student through the technique's purpose and protocol and

describes how scientists interpret the data it generates.

- *Observational Research* figures describe specific studies in which biologists have tested hypotheses by comparing systems under varying natural circumstances.
- *Experimental Research* figures describe specific studies in which researchers used both experimental and control treatments—either in the laboratory or in the field—to test hypotheses by manipulating the system they study.

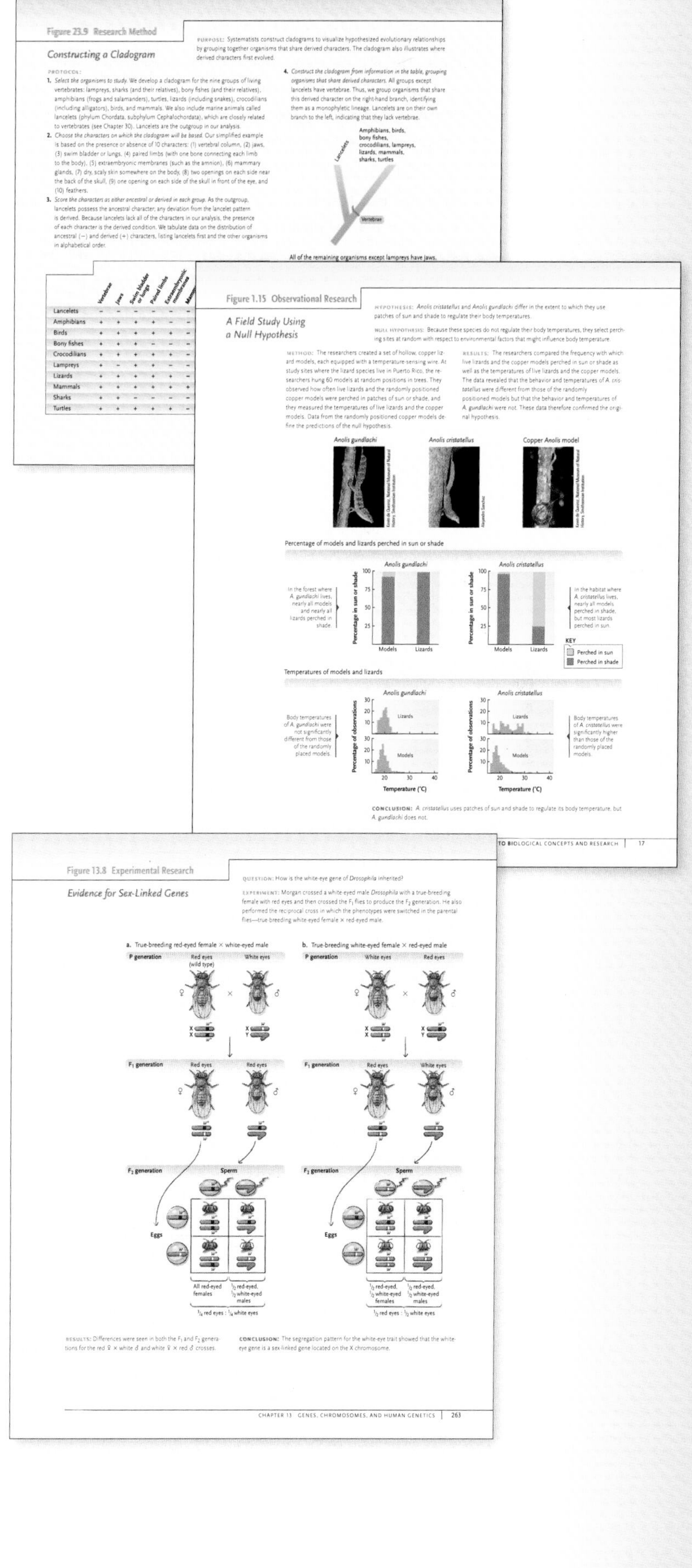

Chapters are structured to emphasize the big picture and the most important concepts

As authors and college teachers, we know how easily students can get lost within a chapter that spans 15 pages or more. When students request advice about how to approach such a large task, we usually suggest that, after reading each section, they pause and quiz themselves on the material they have just encountered. After completing all of the sections in a chapter, they should quiz themselves again, even more rigorously, on the individual sections and, most important, on how the concepts developed in different sections fit together. To assist these efforts, we have adopted a structure for each chapter that will help students review concepts as they learn them.

- The organization within chapters presents material in digestible chunks, building on students' knowledge and understanding as they acquire it. Each major section covers one broad topic. Each subsection, titled with a declarative sentence that summarizes the main idea of its content, explores a narrower range of material.

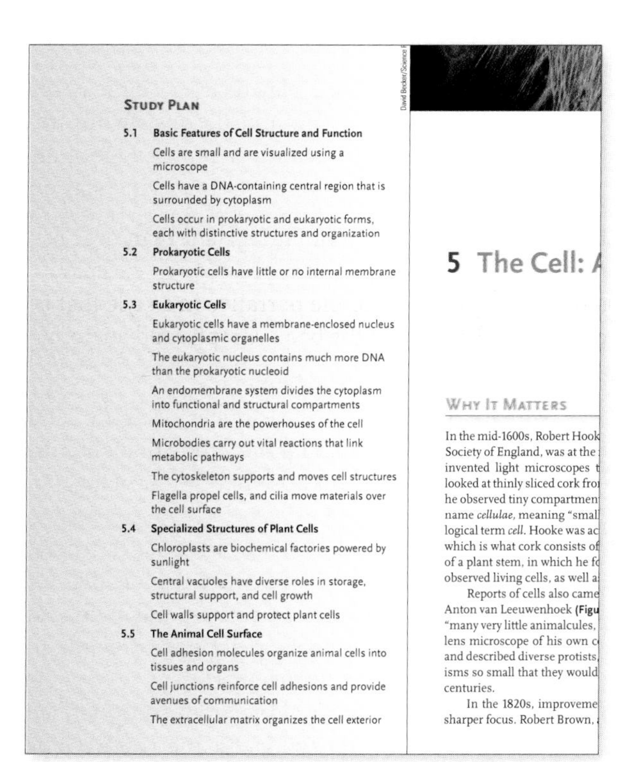

- Whenever possible, we include the derivation of unfamiliar terms so that students will see connections between words that share etymological roots. Mastery of the technical language of biology will allow students to discuss ideas and processes precisely. At the same time, we have minimized the use of unnecessary jargon as much as possible.
- Sets of embedded *Study Break* questions follow every major section. These questions encourage students to pause at the end of a section and review what they have learned before going on to the next topic within the chapter. Short answers to these questions appear in an appendix.

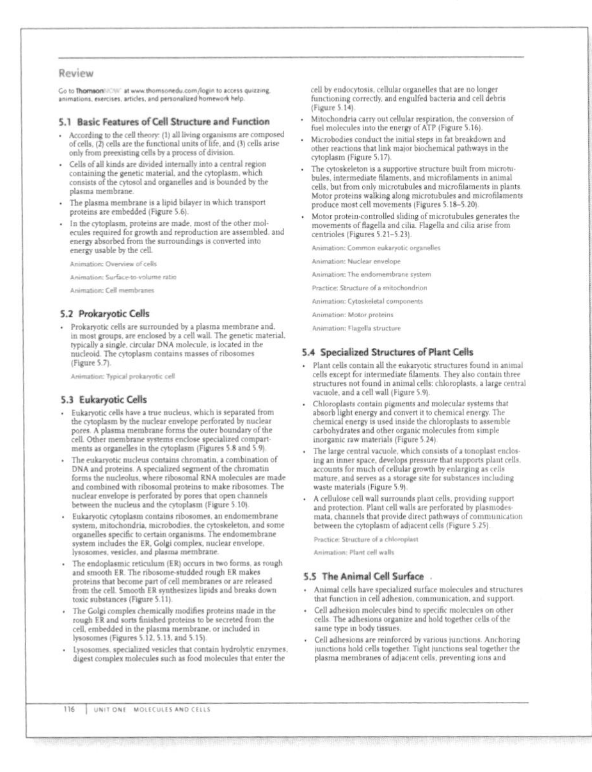

Review

Go to ThomsonNOW at www.thomsonedu.com/login to access quizzing, animations, exercises, articles, and personalized homework help.

5.1 Basic Features of Cell Structure and Function

- According to the cell theory: (1) all living organisms are composed of cells, (2) cells are the functional units of life, and (3) cells arise only from preexisting cells by a process of division.
- Cells of all kinds are divided internally into a central region containing the genetic material, and the cytoplasm, which consists of the cytosol and organelles and is bounded by the plasma membrane.
- The plasma membrane is a lipid bilayer in which transport proteins are embedded (Figure 5.6).
- In the cytoplasm, proteins are made, most of the other molecules required for growth and reproduction are assembled, and energy absorbed from the surroundings is converted into energy usable by the cell.

Animation: Overview of cells

Animation: Surface-to-volume ratio

Animation: Cell membranes

5.2 Prokaryotic Cells

- Prokaryotic cells are surrounded by a plasma membrane and, in most groups, are enclosed by a cell wall. The genetic material, typically a single, circular DNA molecule, is located in the nucleoid. The cytoplasm contains masses of ribosomes (Figure 5.7).

Animation: Typical prokaryotic cell

5.3 Eukaryotic Cells

- Eukaryotic cells have a true nucleus, which is separated from the cytoplasm by the nuclear envelope perforated by nuclear pores. A plasma membrane forms the outer boundary of the cell. Other membrane systems enclose specialized compartments as organelles in the cytoplasm (Figures 5.8 and 5.9).
- The eukaryotic nucleus contains chromatin, a combination of DNA and proteins. A specialized segment of the chromatin forms the nucleolus, where ribosomal RNA molecules are made and combined with ribosomal proteins to make ribosomes. The nuclear envelope is perforated by pores that open channels between the nucleus and the cytoplasm (Figure 5.10).
- Eukaryotic cytoplasm contains ribosomes, an endomembrane system, mitochondria, microbodies, the cytoskeleton, and some organelles specific to certain organisms. The endomembrane system includes the ER, Golgi complex, nuclear envelope, lysosomes, vesicles, and plasma membrane.
- The endoplasmic reticulum (ER) occurs in two forms, as rough and smooth ER. The ribosome-studded rough ER makes proteins that become part of cell membranes or are released from the cell. Smooth ER synthesizes lipids and breaks down toxic substances (Figure 5.11).
- The Golgi complex chemically modifies proteins made in the rough ER and sorts finished proteins to be secreted from the cell, embedded in the plasma membrane, or included in lysosomes (Figures 5.12, 5.13, and 5.15).
- Lysosomes, specialized vesicles that contain hydrolytic enzymes, digest complex molecules such as food molecules that enter the cell by endocytosis, cellular organelles that are no longer functioning correctly, and engulfed bacteria and cell debris (Figure 5.14).
- Mitochondria carry out cellular respiration, the conversion of fuel molecules into the energy of ATP (Figure 5.16).
- Microbodies conduct the initial steps in fat breakdown and other reactions that link major biochemical pathways in the cytoplasm (Figure 5.17).
- The cytoskeleton is a supportive structure built from microtubules, intermediate filaments, and microfilaments in animal cells, but from only microtubules and microfilaments in plants. Motor proteins walking along microtubules and microfilaments produce most cell movements (Figures 5.18–5.20).
- Motor protein-controlled sliding of microtubules generates the movements of flagella and cilia. Flagella and cilia arise from centrioles (Figures 5.21–5.23).

Animation: Common eukaryotic organelles

Animation: Nuclear envelope

Animation: The endomembrane system

Practice: Structure of a mitochondrion

Animation: Cytoskeletal components

Animation: Motor proteins

Animation: Flagella structure

5.4 Specialized Structures of Plant Cells

- Plant cells contain all the eukaryotic structures found in animal cells except for intermediate filaments. They also contain three structures not found in animal cells: chloroplasts, a large central vacuole, and a cell wall (Figure 5.9).
- Chloroplasts contain pigments and molecular systems that absorb light energy and convert it to chemical energy. The chemical energy is used inside the chloroplasts to assemble carbohydrates and other organic molecules from simple inorganic raw materials (Figure 5.24).
- The large central vacuole, which consists of a tonoplast enclosing an inner space, develops pressure that supports plant cells, accounts for much of cellular growth by enlarging as cells mature, and serves as a storage site for substances including waste materials (Figure 5.9).
- A cellulose cell wall surrounds plant cells, providing support and protection. Plant cell walls are perforated by plasmodesmata, channels that provide direct pathways of communication between the cytoplasm of adjacent cells (Figure 5.25).

Practice: Structure of a chloroplast

Animation: Plant cell walls

5.5 The Animal Cell Surface

- Animal cells have specialized surface molecules and structures that function in cell adhesion, communication, and support.
- Cell adhesion molecules bind to specific molecules on other cells. The adhesions organize and hold together cells of the same type in body tissues.
- Cell adhesions are reinforced by various junctions. Anchoring junctions hold cells together. Tight junctions seal together the plasma membranes of adjacent cells, preventing ions and

116 | UNIT ONE MOLECULES AND CELLS

End-of-chapter material encourages students to review concepts, test their knowledge, and think analytically

Supplementary materials at the end of each chapter help students review the material they have learned, assess their understanding, and think analytically as they apply the principles developed in the chapter to novel situations. Many of the end-of-chapter questions also serve as good starting points for class discussions or out-of-class assignments.

- A brief *Review* that references figures and tables in the chapter provides an outline summary of important ideas developed in the chapter. The *Reviews* are much too short to serve as a substitute for reading the chapter. Instead, students may use them as an outline of the material, filling in the details on their own.
- Each chapter also closes with a set of 10 multiple choice *Self-Test* questions that focus on factual material.
- Several open-ended *Questions for Discussion* emphasize concepts, the interpretation of data, and practical applications of the material.
- A question on *Experimental Analysis* asks students to consider how they would develop and test hypotheses about a situation that relates to the chapter's main topic.
- An *Evolution Link* question relates the subject of the chapter to evolutionary biology.
- The *How Would You Vote?* exercise allows students to weigh both sides of an issue by reading pro/con articles, and then making their opinion known through an online voting process.

We hope that, after reading parts of this textbook, you agree that we have developed a clear, fresh, and well-integrated introduction to biology as it is understood by researchers today. Just as important, we hope that our efforts will excite students about the research process and the new discoveries it generates.

Acknowledgments

We are grateful to the many people who have generously fostered the creation of this text

The creation of a new textbook is a colossal undertaking, and we could never have completed the task without the kind assistance of many people.

Jack Carey first conceived of this project and put together the author team. Michelle Julet and Yolanda Cossio have provided the support and encouragement necessary to move it forward to completion. Peggy Williams has served as the extraordinarily able coordinator of the authors, editors, reviewers, contributors, artists, and production team—we like to think of Peggy as the "cat herder."

Developmental Editors play nearly as large a role as the authors, interpreting and deconstructing reviewer comments and constantly making suggestions about how we could tighten the narrative and stay on course. Mary Arbogast has done banner service as a Developmental Editor, patiently working on the project since its inception. Shelley Parlante has provided very helpful guidance as the manuscript matured. Jody Larson and Catherine Murphy have offered useful comments on many of the chapters.

We are grateful to Christopher Delgado and Jessica Kuhn for coordinating the print supplements, and our Editorial Assistant Rose Barlow for managing all our reviewer information.

Many thanks to Keli Amann, Kristina Razmara, and Christopher Delgado, who were responsible for partnering with our technology authors and media advisory board in creating tools to support students in learning and instructors in teaching.

We appreciate the help of the production staff led by Shelley Ryan and Suzanne Kastner at Graphic

World. We thank our Creative Director Rob Hugel, Art Director John Walker.

The outstanding art program is the result of the collaborative talent, hard work, and dedication of a select group of people. The meticulous styling and planning of the program is credited to Steve McEntee and Dragonfly Media Group, led by Craig Durant and Mike Demaray. The DMG group created hundreds of complex, vibrant art pieces. Steve's role was crucial in overseeing the development and consistency of the art program; he was also the illustrator for the unique Research features.

We appreciate Kara Kindstrom, our Marketing Manager, and Terri Mynatt, our Development Project Manager, whose expertise ensured that you would know all about this new book.

Peter Russell thanks Stephen Arch of Reed College for valuable discussions and advice during the writing of the Unit Six chapters on Animal Structure and Function. Paul E. Hertz thanks Hilary Callahan, John Glendinning, and Brian Morton of Barnard College for their generous advice on many phases of this project, and John Alcock of Arizona State University and James Danoff-Burg of Columbia University for their contributions to the discussions of Animal Behavior and Conservation Biology, respectively. Paul would also like to thank Jamie Rauchman, for extraordinary patience and endless support as this book was written, and his thousands of past students, who have taught him at least as much as he has taught them.

We would also like to thank our advisors and contributors:

Media Advisory Board

Scott Bowling, Auburn University
Jennifer Jeffery, Wharton County Junior College
Shannon Lee, California State University, Northridge
Roderick M. Morgan, Grand Valley State University
Debra Pires, University of California, Los Angeles

Art Advisory Board

Lissa Leege, Georgia Southern University
Michael Meighan, University of California, Berkeley
Melissa Michael, University of Illinois at Urbana–Champaign
Craig Peebles, University of Pittsburgh
Laurel Roberts, University of Pittsburgh

Accuracy Checkers

Brent Ewers, University of Wyoming
Richard Falk, University of California, Davis
Michael Meighan, University of California, Berkeley
Michael Palladino, Monmouth University

End-of-Chapter Questions

Patricia Colberg, University of Wyoming
Elizabeth Godrick, Boston University

Student Study Guide

Carolyn Bunde, Idaho State University
William Kroll, Loyola University Chicago
Mark Sheridan, North Dakota State University
Jyoti Wagle, Houston Community College

Instructor's Resource Manual

Benjie Blair, Jacksonville State University
Nancy Boury, Idaho State University
Mark Meade, Jacksonville State University
Debra Pires, University of California, Los Angeles
James Rayburn, Jacksonville State University

Test Bank

Scott Bowling, Auburn University
Laurie Bradley, Hudson Valley Community College
Jose Egremy, Northwest Vista College
Darrel L. Murray, University of Illinois, Chicago
Jacalyn Newman, University of Pittsburgh
Mark Sugalski, Southern Polytechnic State University

Technology Authors

Catherine Black, Idaho State University
David Byres, Florida Community College, Jacksonville
Kevin Dixon, University of Illinois
Albia Dugger, Miami Dade College
Mary Durant, North Harris College
Brent Ewers, University of Wyoming
Debbie Folkerts, Auburn University
Stephen Kilpatrick, University of Pittsburgh
Laurel Roberts, University of Pittsburgh
Thomas Sasek, University of Louisiana, Monroe
Bruce Stallsmith, University of Alabama–Huntsville

Workshop and Focus Group Participants

Karl Aufderheide, *Texas A&M University*

Bob Bailey, *Central Michigan University*

John Bell, *Brigham Young University*

Catherine Black, *Idaho State University*

Hessel Bouma III, *Calvin College*

Scott Bowling, *Auburn University*

Bob Brick, *Blinn College, Bryan*

Randy Brooks, *Florida Atlantic University*

Nancy Burley, *University of California, Irvine*

Genevieve Chung, *Broward Community College*

Allison Cleveland, *University of South Florida*

Patricia Colberg, *University of Wyoming*

Jay Comeaux, *Louisiana State University*

Sehoya Cotner, *University of Minnesota*

Joe Cowles, *Virginia Tech*

Anita Davelos-Baines, *University of Texas, Pan American*

Donald Deters, *Bowling Green State University*

Kevin Dixon, *University of Illinois at Urbana–Champaign*

Jose Egremy, *Northwest Vista College*

Diana Elrod, *University of North Texas*

Zen Faulkes, *University of Texas–Pan American*

Elizabeth Godrick, *Boston University*

Barbara Haas, *Loyola University Chicago*

Julie Harless, *Montgomery College*

Jean Helgeson, *Collin County Community College*

Mark Hunter, *University of Michigan*

Andrew Jarosz, *Michigan State University, Montgomery College*

Jennifer Jeffery, *Wharton County Junior College*

John Jenkin, *Blinn College, Bryan*

Wendy Keenleyside, *University of Guelph*

Steve Kilpatrick, *University of Pittsburgh at Johnstown*

Gary Kuleck, *Loyola Marymount University*

Allen Kurta, *Eastern Michigan University*

Mark Lyford, *University of Wyoming*

Andrew McCubbin, *Washington State University*

Michael Meighan, *University of California, Berkeley*

John Merrill, *Michigan State University*

Richard Merritt, *Houston Community College, Northwest*

Melissa Michael, *University of Illinois at Urbana–Champaign*

James Mickle, *North Carolina State University*

Betsy Morgan, *Kingwood College*

Kenneth Mossman, *Arizona State University*

Darrel Murray, *University of Illinois, Chicago*

Jacalyn Newman, *University of Pittsburgh*

Dennis Nyberg, *University of Illinois–Chicago*

Bruce Ostrow, *Grand Valley State University–Allendale*

Craig Peebles, *University of Pittsburgh*

Nancy Pencoe, *University of West Georgia*

Mitch Price, *Pennsylvania State University*

Kelli Prior, *Finger Lakes Community College*

Laurel Roberts, *University of Pittsburgh*

Ann Rushing, *Baylor University*

Bruce Stallsmith, *University of Alabama–Huntsville*

David Tam, *University of North Texas*

Franklyn Te, *Miami Dade College*

Nanette Van Loon, *Borough of Manhattan Community College*

Alexander Wait, *Missouri State University*

Lisa Webb, *Christopher Newport University*

Larry Williams, *University of Houston*

Michelle Withers, *Louisiana State University*

Denise Woodward, *Pennsylvania State University*

Class Test Participants

Tamarah Adair, *Baylor University*
Idelissa Ayala, *Broward Community College–Central*
Tim Beagley, *Salt Lake Community College*
Catherine Black, *Idaho State University*
Laurie Bradley, *Hudson Valley Community College*
Mirjana Brockett, *Georgia Tech*
Carolyn Bunde, *Idaho State University*
John Cogan, *Ohio State University*
Anne M. Cusic, *University of Alabama–Birmingham*
Ingeborg Eley, *Hudson Valley Community College*
Brent Ewers, *University of Wyoming*
Miriam Ferzli, *North Carolina State University*
Debbie Folkerts, *Auburn University*
Mark Hens, *University of North Carolina, Greensboro*
Anna Hill, *University of Louisiana, Monroe*
Anne Hitt, *Oakland University*
Jennifer Jeffery, *Wharton County Junior College*
David Jones, *Dixie State College*
Wendy Keenleyside, *University of Guelph*
Brian Kinkle, *University of Cincinnati*
Brian Larkins, *University of Arizona*
Shannon Lee, *California State University, Northridge*
Harvey Liftin, *Broward Community College*
Jim Marinaccio, *Raritan Valley Community College*
Monica Marquez-Nelson, *Joliet Junior College*
Kelly Meckling, *University of Guelph*
Richard Merritt, *Houston Community College–Town and Country*
Russ Minton, *University of Louisiana, Monroe*
Necia Nichols, *Calhoun State Community College*
Nancy Rice, *Western Kentucky University*
Laurel Roberts, *University of Pittsburgh*
John Russell, *Calhoun State Community College*
Pramila Sen, *Houston Community College*
Jacquelyn Smith, *Pima County Community College*
Bruce Stallsmith, *University of Alabama–Huntsville*
Joe Steffen, *University of Louisville*
Gail Stewart, *Camden County College*
Mark Sugalski, *Southern Polytechnic State University*
Marsha Turrell, *Houston Community College*
Fil Ventura-Smolenski, *Santa Fe Community College*
Beth Vlad, *College of DuPage*
Alexander Wait, *Missouri State University*
Matthew Wallenfang, *Barnard College*
David Wolfe, *American River College*

Reviewers

Heather Addy, *University of Calgary*
Adrienne Alaie-Petrillo, *Hunter College–CUNY*
Richard Allison, *Michigan State University*
Terry Allison, *University of Texas–Pan American*
Deborah Anderson, *Saint Norbert College*
Robert C. Anderson, *Idaho State University*
Andrew Andres, *University of Nevada–Las Vegas*
Steven M. Aquilani, *Delaware County Community College*
Jonathan W. Armbruster, *Auburn University*
Peter Armstrong, *University of California, Davis*
John N. Aronson, *University of Arizona*
Joe Arruda, *Pittsburg State University*
Karl Aufderheide, *Texas A&M University*
Charles Baer, *University of Florida*
Gary I. Baird, *Brigham Young University*
Aimee Bakken, *University of Washington*
Marica Bakovic, *University of Guelph*
Michael Baranski, *Catawba College*
Michael Barbour, *University of California, Davis*
Edward M. Barrows, *Georgetown University*
Anton Baudoin, *Virginia Tech*
Penelope H. Bauer, *Colorado State University*
Kevin Beach, *University of Tampa*
Mike Beach, *Southern Polytechnic State University*
Ruth Beattie, *University of Kentucky*
Robert Beckmann, *North Carolina State University*

Jane Beiswenger, *University of Wyoming*

Andrew Bendall, *University of Guelph*

Catherine Black, *Idaho State University*

Andrew Blaustein, *Oregon State University*

Anthony H. Bledsoe, *University of Pittsburgh*

Harriette Howard-Lee Block, *Prairie View A&M University*

Dennis Bogyo, *Valdosta State University*

David Bohr, *University of Michigan*

Emily Boone, *University of Richmond*

Hessel Bouma III, *Calvin College*

Nancy Boury, *Iowa State University*

Scott Bowling, *Auburn University*

Laurie Bradley, *Hudson Valley Community College*

William Bradshaw, *Brigham Young University*

J. D. Brammer, *North Dakota State University*

G. L. Brengelmann, *University of Washington*

Randy Brewton, *University of Tennessee–Knoxville*

Bob Brick, *Blinn College, Bryan*

Mirjana Brockett, *Georgia Tech*

William Bromer, *University of Saint Francis*

William Randy Brooks, *Florida Atlantic University–Boca Raton*

Mark Browning, *Purdue University*

Gary Brusca, *Humboldt State University*

Alan H. Brush, *University of Connecticut*

Arthur L. Buikema, Jr., *Virginia Tech*

Carolyn Bunde, *Idaho State University*

E. Robert Burns, *University of Arkansas for Medical Sciences*

Ruth Buskirk, *University of Texas–Austin*

David Byres, *Florida Community College, Jacksonville*

Christopher S. Campbell, *University of Maine*

Angelo Capparella, *Illinois State University*

Marcella D. Carabelli, *Broward Community College–North*

Jeffrey Carmichael, *University of North Dakota*

Bruce Carroll, *North Harris Montgomery Community College*

Robert Carroll, *East Carolina University*

Patrick Carter, *Washington State University*

Christine Case, *Skyline College*

Domenic Castignetti, *Loyola University Chicago–Lakeshore*

Jung H. Choi, *Georgia Tech*

Kent Christensen, *University Michigan School of Medicine*

John Cogan, *Ohio State University*

Linda T. Collins, *University of Tennessee–Chattanooga*

Lewis Coons, *University of Memphis*

Joe Cowles, *Virginia Tech*

George W. Cox, *San Diego State University*

David Crews, *University of Texas*

Paul V. Cupp, Jr., *Eastern Kentucky University*

Karen Curto, *University of Pittsburgh*

Anne M. Cusic, *University of Alabama–Birmingham*

David Dalton, *Reed College*

Frank Damiani, *Monmouth University*

Peter J. Davies, *Cornell University*

Fred Delcomyn, *University of Illinois at Urbana–Champaign*

Jerome Dempsey, *University of Wisconsin–Madison*

Philias Denette, *Delgado Community College–City Park*

Nancy G. Dengler, *University of Toronto*

Jonathan J. Dennis, *University of Alberta*

Daniel DerVartanian, *University of Georgia*

Donald Deters, *Bowling Green State University*

Kathryn Dickson, *CSU Fullerton*

Kevin Dixon, *University of Illinois at Urbana–Champaign*

Gordon Patrick Duffie, *Loyola University Chicago–Lakeshore*

Charles Duggins, *University of South Carolina*

Carolyn S. Dunn, *University North Carolina–Wilmington*

Roland R. Dute, *Auburn University*

Melinda Dwinell, *Medical College of Wisconsin*

Gerald Eck, *University of Washington*

Gordon Edlin, *University of Hawaii*

William Eickmeier, *Vanderbilt University*

Ingeborg Eley, *Hudson Valley Community College*

Paul R. Elliott, *Florida State University*

John A. Endler, *University of Exeter*

Brent Ewers, *University of Wyoming*

Daniel J. Fairbanks, *Brigham Young University*

Piotr G. Fajer, *Florida State University*

Richard H. Falk, *University of California, Davis*

Ibrahim Farah, *Jackson State University*

Jacqueline Fern, *Lane Community College*

Daniel P. Fitzsimons, *University of Wisconsin–Madison*

Daniel Flisser, *Camden County College*

R. G. Foster, *University of Virginia*

Dan Friderici, *Michigan State University*

J. W. Froehlich, *University of New Mexico*

Paul Garcia, *Houston Community College–SW*

Umadevi Garimella, *University of Central Arkansas*

Robert P. George, *University of Wyoming*

Stephen George, *Amherst College*

John Giannini, *St. Olaf College*

Joseph Glass, *Camden County College*

John Glendinning, *Barnard College*

Elizabeth Godrick, *Boston University*

Judith Goodenough, *University of Massachusetts Amherst*

H. Maurice Goodman, *University of Massachusetts Medical School*

Bruce Grant, *College of William and Mary*

Becky Green-Marroquin, *Los Angeles Valley College*

Christopher Gregg, *Louisiana State University*

Katharine B. Gregg, *West Virginia Wesleyan College*

John Griffin, *College of William and Mary*

Samuel Hammer, *Boston University*

Aslam Hassan, *University of Illinois at Urbana–Champaign, Veterinary Medicine*

Albert Herrera, *University of Southern California*

Wilford M. Hess, *Brigham Young University*

Martinez J. Hewlett, *University of Arizona*

Christopher Higgins, *Tarleton State University*

Phyllis C. Hirsch, *East Los Angeles College*

Carl Hoagstrom, *Ohio Northern University*

Stanton F. Hoegerman, *College of William and Mary*

Ronald W. Hoham, *Colgate University*

Margaret Hollyday, *Bryn Mawr College*

John E. Hoover, *Millersville University*

Howard Hosick, *Washington State University*

William Irby, *Georgia Southern*

John Ivy, *Texas A&M University*

Alice Jacklet, *SUNY Albany*

John D. Jackson, *North Hennepin Community College*

Jennifer Jeffery, *Wharton County Junior College*

John Jenkin, *Blinn College, Bryan*

Leonard R. Johnson, *University Tennessee College of Medicine*

Walter Judd, *University of Florida*

Prem S. Kahlon, *Tennessee State University*

Thomas C. Kane, *University of Cincinnati*

Peter Kareiva, *University of Washington*

Gordon I. Kaye, *Albany Medical College*

Greg Keller, *Eastern New Mexico University*

Stephen Kelso, *University of Illinois–Chicago*

Bryce Kendrick, *University of Waterloo*

Bretton Kent, *University of Maryland*

Jack L. Keyes, *Linfield College Portland Campus*

John Kimball, *Tufts University*

Hillar Klandorf, *West Virginia University*

Michael Klymkowsky, *University of Colorado–Boulder*

Loren Knapp, *University of South Carolina*

Ana Koshy, *Houston Community College–NW*

Kari Beth Krieger, *University of Wisconsin–Green Bay*

David T. Krohne, *Wabash College*

William Kroll, *Loyola University Chicago–Lakeshore*

Josepha Kurdziel, *University of Michigan*

Allen Kurta, *Eastern Michigan University*

Howard Kutchai, *University of Virginia*

Paul K. Lago, *University of Mississippi*

John Lammert, *Gustavus Adolphus College*

William L'Amoreaux, *College of Staten Island*

Brian Larkins, *University of Arizona*

William E. Lassiter, *University of North Carolina–Chapel Hill*

Shannon Lee, *California State University, Northridge*

Lissa Leege, *Georgia Southern University*

Matthew Levy, *Case Western Reserve University*

Harvey Liftin, *Broward Community College–Central*

Tom Lonergan, *University of New Orleans*

Lynn Mahaffy, *University of Delaware*

Alan Mann, *University of Pennsylvania*

Kathleen Marrs, *Indiana University Purdue University Indianapolis*

Robert Martinez, *Quinnipiac University*

Joyce B. Maxwell, *California State University, Northridge*

Jeffrey D. May, *Marshall University*

Geri Mayer, *Florida Atlantic University*

Jerry W. McClure, *Miami University*

Andrew G. McCubbin, *Washington State University*

Mark McGinley, *Texas Tech University*

F. M. Anne McNabb, *Virginia Tech*

Mark Meade, *Jacksonvile State University*

Bradley Mehrtens, *University of Illinois at Urbana–Champaign*

Michael Meighan, *University of California, Berkeley*

Catherine Merovich, *West Virginia University*

Richard Merritt, *Houston Community College–Town and Country*

Ralph Meyer, *University of Cincinnati*

James E. "Jim" Mickle, *North Carolina State University*

Hector C. Miranda, Jr., *Texas Southern University*

Jasleen Mishra, *Houston Community College–SW*

David Mohrman, *University of Minnesota Medical School*

John M. Moore, *Taylor University*

David Morton, *Frostburg State University*

Alexander Motten, *Duke University*

Alan Muchlinski, *California State University, Los Angeles*

Michael Muller, *University of Illinois–Chicago*

Richard Murphy, *University of Virginia*

Darrel L. Murray, *University of Illinois–Chicago*

Allan Nelson, *Tarleton State University*

David H. Nelson, *University of South Alabama*

Jacalyn Newman, *University of Pittsburgh*

David O. Norris, *University of Colorado*

Bette Nybakken, *Hartnell College, California*

Tom Oeltmann, *Vanderbilt University*

Diana Oliveras, *University of Colorado–Boulder*

Alexander E. Olvido, *Virginia State University*

Karen Otto, *University of Tampa*

William W. Parson, *University of Washington School of Medicine*

James F. Payne, *University of Memphis*

Craig Peebles, *University of Pittsburgh*

Joe Pelliccia, *Bates College*

Susan Petro, *Rampao College of New Jersey*

Debra Pires, *University of California, Los Angeles*

Thomas Pitzer, *Florida International University*

Roberta Pollock, *Occidental College*

Jerry Purcell, *San Antonio College*

Kim Raun, *Wharton County Junior College*

Tara Reed, *University of Wisconsin–Green Bay*

Lynn Robbins, *Missouri State University*

Carolyn Roberson, *Roane State Community College*

Laurel Roberts, *University of Pittsburgh*

Kenneth Robinson, *Purdue University*

Frank A. Romano, *Jacksonville State University*

Michael R. Rose, *University of California, Irvine*

Michael S. Rosenzweig, *Virginia Tech*

Linda S. Ross, *Ohio University*

Ann Rushing, *Baylor University*

Linda Sabatino, *Suffolk Community College*

Tyson Sacco, *Cornell University*

Peter Sakaris, *Southern Polytechnic State University*

Frank B. Salisbury, *Utah State University*

Mark F. Sanders, *University of California, Davis*

Andrew Scala, *Dutchess Community College*

John Schiefelbein, *University of Michigan*

Deemah Schirf, *University of Texas–San Antonio*

Kathryn J. Schneider, *Hudson Valley Community College*

Jurgen Schnermann, *University Michigan School of Medicine*

Thomas W. Schoener, *University California, Davis*

Brian Shea, *Northwestern University*

Mark Sheridan, *North Dakota State University–Fargo*

Dennis Shevlin, *College of New Jersey*

Richard Showman, *University of South Carolina*

Bill Simcik, *Tomball College*

Robert Simons, *University of California, Los Angeles*

Roger Sloboda, *Dartmouth College*

Jerry W. Smith, *St. Petersburg College*

Nancy Solomon, *Miami University*

Bruce Stallsmith, *University of Alabama–Huntsville*

Karl Sternberg, *Western New England College*

Pat Steubing, *University of Nevada–Las Vegas*

Karen Steudel, *University of Wisconsin–Madison*

Richard D. Storey, *Colorado College*

Michael A. Sulzinski, *University of Scranton*

Marshall Sundberg, *Emporia State University*

David Tam, *University of North Texas*

David Tauck, *Santa Clara University*

Jeffrey Taylor, *Slippery Rock University*

Franklyn Te, *Miami Dade College*

Roger E. Thibault, *Bowling Green State University*

Megan Thomas, *University of Nevada–Las Vegas*

Patrick Thorpe, *Grand Valley State University–Allendale*

Ian Tizard, *Texas A&M University*

Robert Turner, *Western Oregon University*

Joe Vanable, *Purdue University*

Linda H. Vick, *North Park University*

J. Robert Waaland, *University of Washington*

Douglas Walker, *Wharton County Junior College*

James Bruce Walsh, *University of Arizona*

Fred Wasserman, *Boston University*

Edward Weiss, *Christopher Newport University*

Mark Weiss, *Wayne State University*

Adrian M. Wenner, *University of California, Santa Barbara*

Adrienne Williams, *University of California, Irvine*

Mary Wise, *Northern Virginia Community College*

Charles R. Wyttenbach, *University of Kansas*

Robert Yost, *Indiana University Purdue University Indianapolis*

Xinsheng Zhu, *University of Wisconsin–Madison*

Adrienne Zihlman, *University of California–Santa Cruz*

Brief Contents

The chapters listed below are not included in Volume 3

Brief Contents

Contents

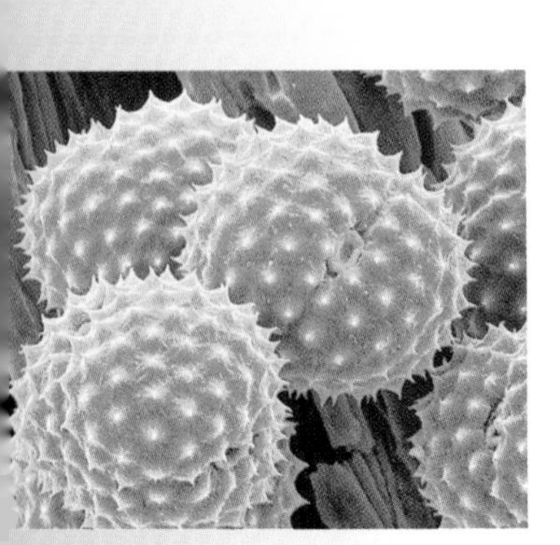

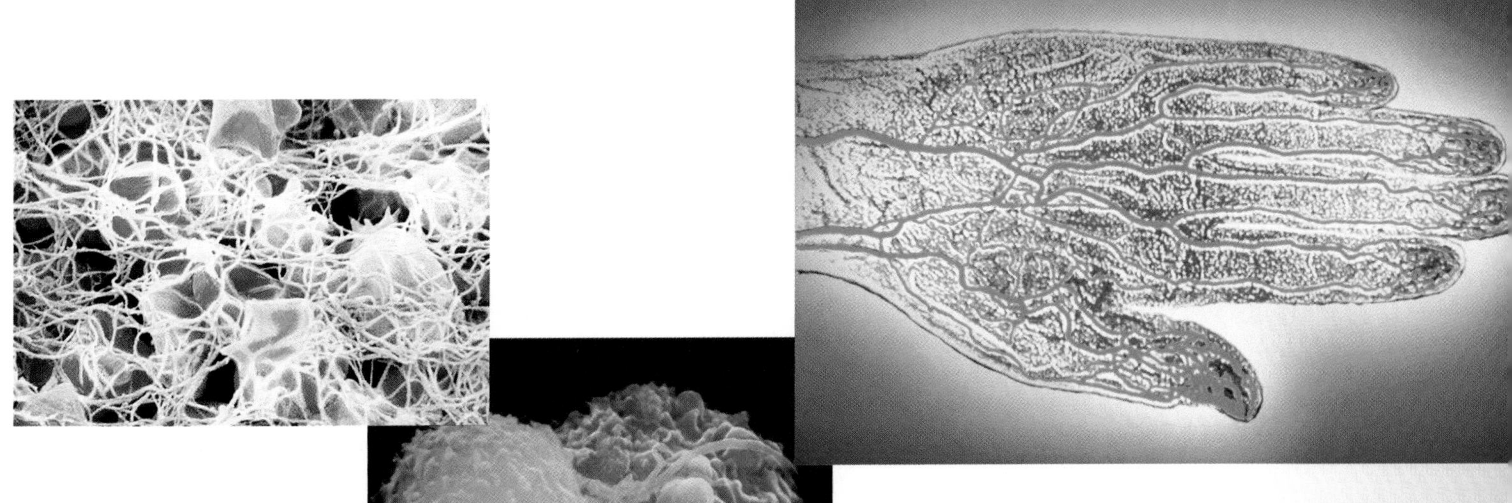

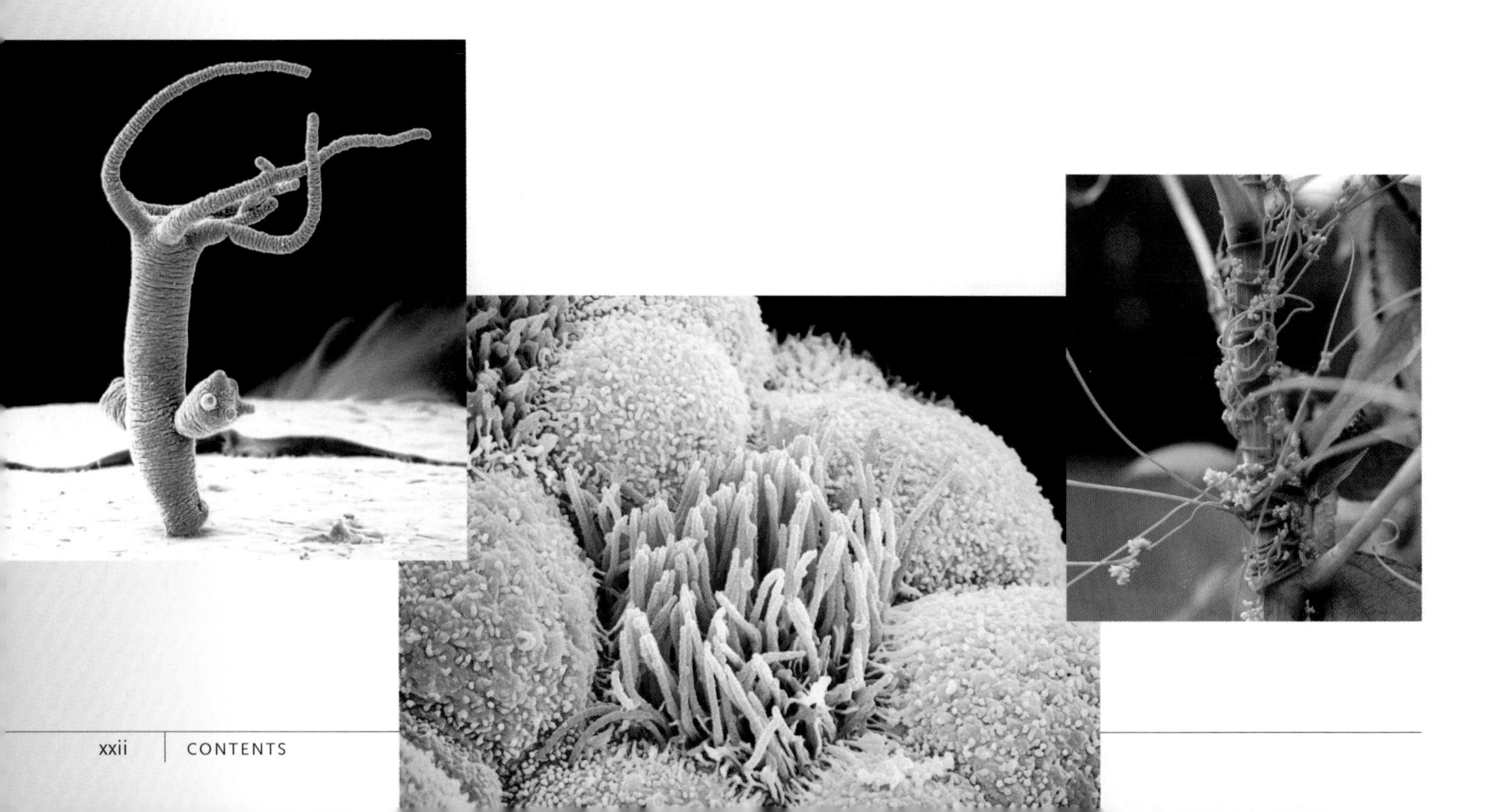

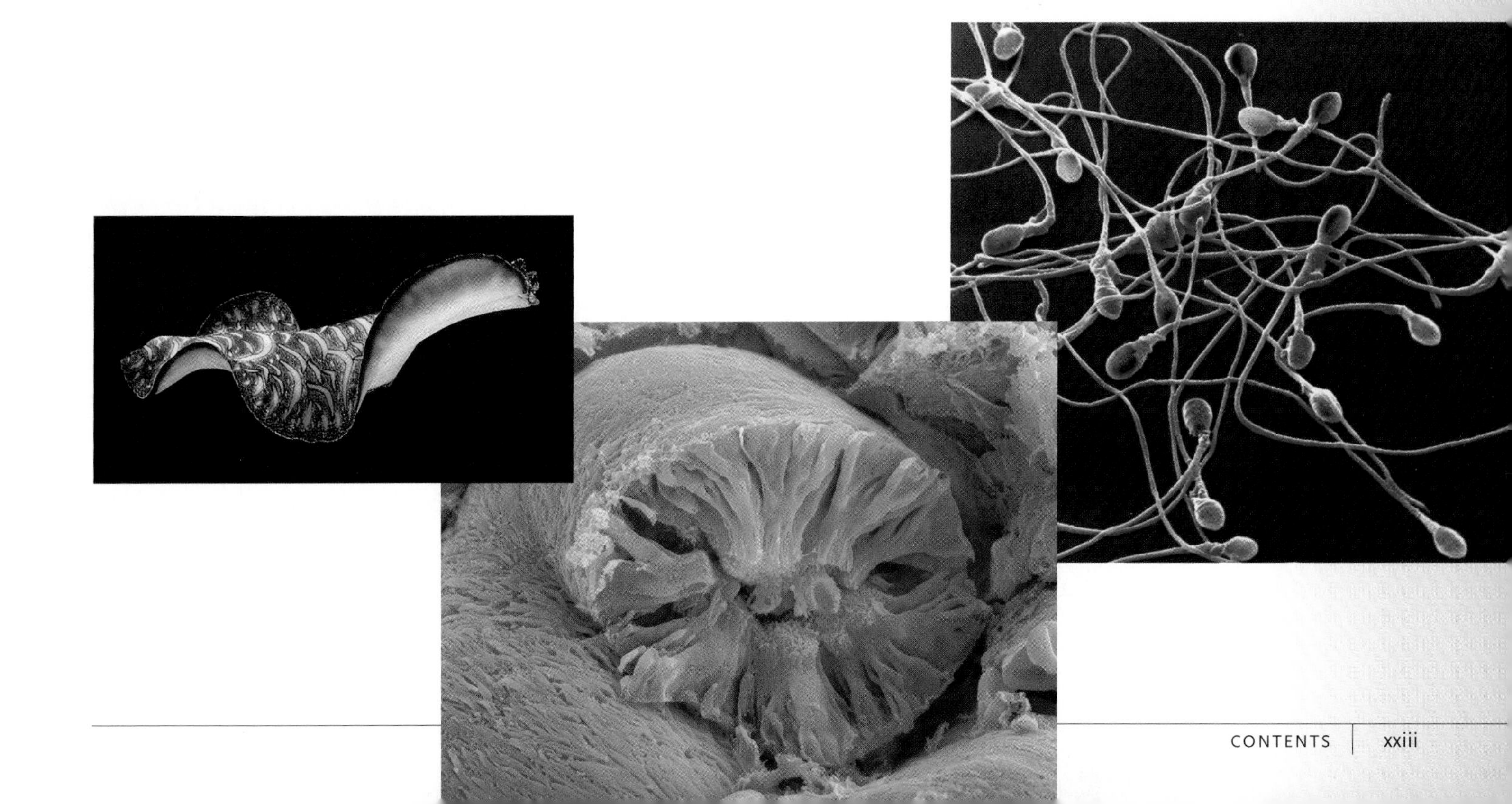

Basic parts of the shoot system of an apple tree *(Malus domestica)*, including leaves, stems, and vividly colored fruits.

31 The Plant Body

Study Plan

Why It Matters

Food, fibers for clothing, wood and other materials for construction, paper and inks, dozens of pharmaceuticals—these and many other essentials of modern human life derive from the parts of plants. In fact, members of the genus *Homo* have been depending on plant parts for their entire history. Fossil teeth discovered in the East African Rift Valley indicate that our early ancestors' diet likely included hard-shelled nuts, dry seeds, soft fruits, and leaves. By about 11,000 years ago humans were domesticating seed plants to provide stable food supplies. Directly or indirectly, leaves, stems, roots, flowers, seeds, and fruits of plants are the basic sources of energy for Earth's human inhabitants and all other animals as well **(Figures 31.1)**.

As you saw in Chapter 27, plants that made the transition from aquatic to terrestrial life did so only as adaptations in form and function helped solve problems posed by the terrestrial environment. These evolutionary adaptations included a shoot system that helps support leaves and other body parts in air, a root system that anchors the plant in soil and provides access to soil nutrients and water, tissues for internal transport of nutrients, and specializations for preventing water loss.

a. Wheat

© Earl Roberge/Photo Researchers, Inc.

b. Antelope feeding on leaves

© W. Perry Conway/Corbis

c. Cedar waxwing consuming berries

© Gregory K. Scott/Photo Researchers, Inc.

Figure 31.1
Examples of plant parts that provide food for animals. **(a)** Mechanized harvesting of *Triticum* seeds, commonly known as wheat grains. **(b)** A pronghorn *(Antilocarpa americana)*, which consumes leaves and grasses. **(c)** Cedar waxwing *(Bombycilla cedrorum)* feeding on plump berries of *Sorbus americana*, the American mountain ash.

This unit surveys the structure and functioning of plants—their morphology, anatomy, and physiology. A plant's *morphology* is its external form, such as the shape of its leaves, and *anatomy* is the structure and arrangement of its internal parts. Plant *physiology* refers to the mechanisms by which the plant's body functions in its environment. Our focus is the plant phylum Anthophyta—angiosperms, or flowering plants—in terms of distribution and sheer numbers of species, the most successful plants on Earth.

31.1 Plant Structure and Growth: An Overview

Plants are photosynthetic autotrophs—"self-feeding" photosynthesizers that need sunlight and the carbon dioxide available in air as well as the water available in soil. In addition, many plants require nutrients that are usually available only in soil, and their aboveground parts may need the physical support of structures anchored in the ground. The evolutionary response to these challenges produced a plant body consisting of two closely linked but quite different components—a photosynthetic *shoot system* extending upward into the air and a nonphotosynthetic *root system* extending downward into the soil. Each system consists of various **organs**—body structures that contain two or more types of tissues and that have a definite form and function. Plant organs include leaves, stems, and roots, among others. A **tissue** is a group of cells and intercellular substances that function together in one or more specialized tasks.

In All Plant Tissues the Cells Share Some General Features

All plant cells share certain features, regardless of the tissue in which they reside. New plant cells develop a primary cell wall around the **protoplast**, the botanical term for the cell's cytoplasm, organelles, and plasma membrane. The primary wall contains cellulose, an insoluble polysaccharide made up of glucose subunits that is embedded in a matrix of other polysaccharides called hemicelluloses. This combination helps make the wall rigid but flexible. Pectin, another polysaccharide, is abundant in the primary wall and in the middle lamella, the layer between the primary walls of neighboring cells that helps bind cells together in tissues. (Plant pectin is often used to congeal jams and jellies.) As a young plant grows, the protoplast of many types of plant cells deposits additional cellulose and other materials inside the primary wall, forming a strong secondary cell wall (see Figure 5.25).

As in animals, all of a plant's cells have the same genes in their nuclei. As each cell matures and *differentiates* (becomes specialized for a particular function), specific genes become active. For the most part, fully differentiated animal cells perform their functions while alive, but some types of plant cells die after differentiating, and their protoplasts disappear. The walls that remain, however, serve key functions, particularly in vascular tissue.

The secondary cell walls of some plants contain lignin, a water-insoluble, inert polymer. **Lignification**, the deposition of lignin in cell walls, anchors the cellulose fibers in the walls, making them stronger and more rigid, and protects the other wall components from physical or chemical damage. Because water cannot penetrate and soften lignified cell walls, lignification also creates a waterproof barrier around the wall's cellulose strands. Many biologists believe that the evolution of large vascular plants became possible when certain cells developed biochemical pathways leading to lignification and could therefore become organized into watertight conducting channels.

Substances pass from one lignified cell to another through various routes. Solutes such as amino acids and sugars move in the plasmodesmata linking adjacent cells (see Figure 5.25). Water moves from cell to cell across *pits,* narrow regions where the secondary wall is absent and the primary wall is thinner and more porous than elsewhere.

Shoot and Root Systems Perform Different but Integrated Functions

A flowering plant's **shoot system** typically consists of stems, leaves, buds, and—during part of the plant's life cycle—reproductive organs known as flowers **(Figure 31.2).** A stem with its attached leaves and buds is a *vegetative* (nonreproductive) shoot; a bud eventually gives rise to an extension of the shoot or to a new, branching shoot. A *reproductive* shoot produces flowers, which later develop fruits containing seeds.

The shoot system is highly adapted for photosynthesis. Leaves greatly increase a plant's surface area and thus its exposure to light. Stems are frameworks for upright growth, which favorably position leaves for light exposure and flowers for pollination. Some parts of the shoot system also store carbohydrates manufactured during photosynthesis.

The **root system** usually grows belowground. It anchors the plant, and sometimes structurally supports its upright parts. It also absorbs water and dissolved minerals from soil and stores carbohydrates. Adaptations in the structure and function of plant cells and tissues were an integral part of the evolution of shoots and roots. For example, vascular tissues specialized to serve as internal pipelines conduct water, minerals, and organic substances throughout the plant. The root hairs sketched in Figure 31.2 are surface cells specialized for absorbing water from soil.

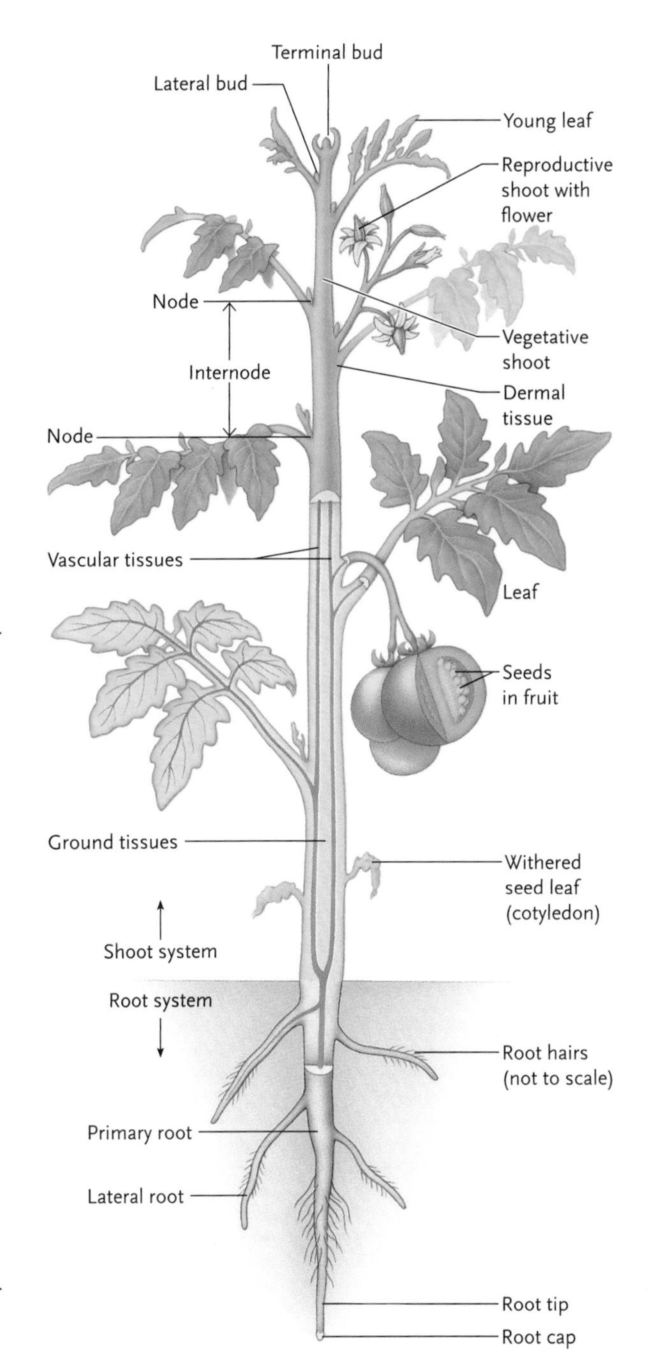

Figure 31.2
Body plan for the commercially grown tomato plant *Solanum lycopersicum*, a typical angiosperm. Vascular tissues (purple) conduct water, dissolved minerals, and organic substances. They thread through ground tissues, which make up most of the plant body. Dermal tissues (epidermis, in this case) cover the surfaces of the root and shoot systems.

Meristems Produce New Tissues Throughout a Plant's Life

As you know from experience, animals generally grow to a certain size, and then their growth slows dramatically or stops. This pattern is called **determinate growth.** In contrast, a plant can grow throughout its life, a pattern called **indeterminate growth.** Individual plant parts such as leaves, flowers, and fruits exhibit determinate growth, but every plant also has self-perpetuating embryonic tissue at the tips of shoots and roots. Under the influence of plant hormones, these **meristems** (*merizein* = to divide) produce new tissues more or less continuously while a plant is alive. A capacity for indeterminate growth gives plants a great deal of flexibility—or what biologists often call *plasticity*—in their possible responses to changes in environmental factors such as light, temperature, water, and nutrients. This plasticity has major adaptive benefits for an organism that cannot move about, as most animals can. For example, if external factors change the direction of incoming light for photosyn-

a. Plants increase in length by cell divisions in apical meristems and by elongation of the daughter cells.

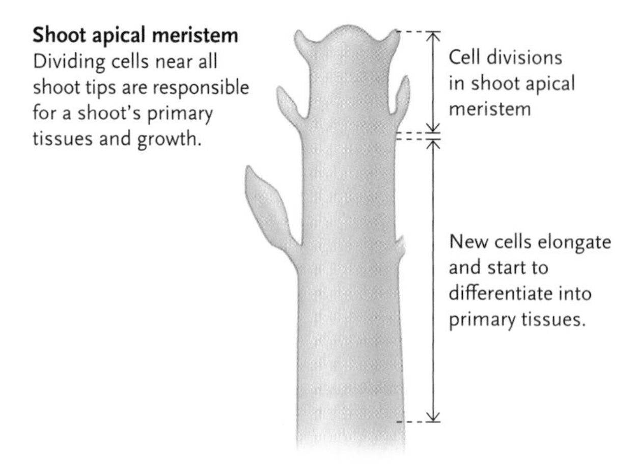

Root apical meristem
Dividing cells near all root tips are responsible for a root's primary tissues and growth.

New cells elongate and start to differentiate into primary tissues.

Cell divisions in root apical meristem

b. Some plants increase in girth by way of cell divisions in lateral meristems.

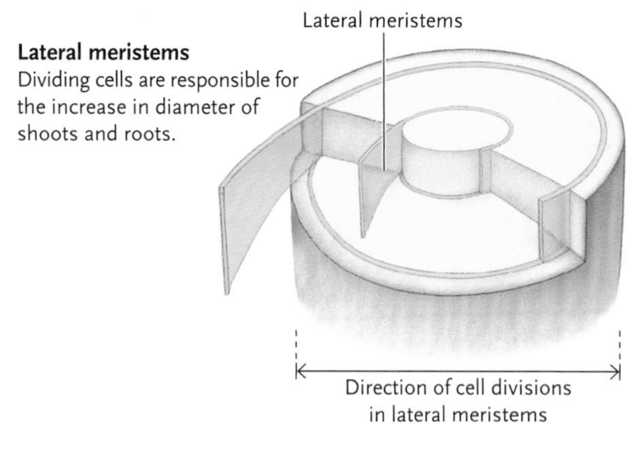

Figure 31.3
Approximate locations of types of meristems that are responsible for increases in the length and diameter of the shoots and roots of a vascular plant.

thesis, stems can "shift gears" and grow in that direction. Likewise, roots can grow outward toward water. These and other plant movements, called tropisms, are a major topic of Chapter 35.

As you know, animals grow mainly by mitosis, which increases the number of body cells. Plants, however, grow by two mechanisms—an increase in the number of cells by mitotic cell division in the meristems, *and* an increase in the size of individual cells. In regions adjacent to the meristems in the tips of shoots and roots, the daughter cells rapidly increase in size—especially in length—for some time after they are produced. In contrast, when animal cells divide mitotically the daughter cells usually increase in size only a little.

Meristems Are Responsible for Growth in Both Height and Girth

Some plants have only one kind of meristem while others have two **(Figure 31.3).** All vascular plants have **apical meristems,** clumps of self-perpetuating tissue at the tips of their buds, stems, and roots (see Figure 31.3a). Tissues that develop from apical meristems are called **primary tissues** and make up the **primary plant body.** Growth of the primary plant body is called **primary growth.**

Some species of plants—grasses and dandelions, for example—show only primary growth, which occurs at the tips of roots and shoots. Others, particularly plants that have a woody body, show **secondary growth,** which originates at self-perpetuating cylinders of tissue called **lateral meristems.** Secondary growth increases the diameter of older roots and stems (see Figure 31.3b). The tissues that develop from lateral meristems, called **secondary tissues,** make up the woody **secondary plant body** we see in trees and shrubs.

Primary and secondary growth can go on simultaneously in a single plant, with primary growth increasing the length of shoot parts and secondary growth adding girth. Each spring, for example, a maple tree undergoes primary growth at each of its root and shoot tips, while secondary growth increases the diameter of its older woody parts. Plant hormones govern these growth processes and other key events that are described in Chapter 35.

Monocots and Eudicots Are the Two General Structural Forms of Flowering Plants

As noted in Chapter 27, several broad categories of body architecture arose as flowering plants evolved. The two major ones are the **monocot** and **eudicot** lineages. Grasses, daylilies, irises, cattails, and palms are examples of monocots. Eudicots include nearly all familiar angiosperm trees and shrubs, as well as many nonwoody (herbaceous) plants. Examples are maples, willows, oaks, cacti, roses, poppies, sunflowers, and garden beans and peas.

Monocots and eudicots, recall, differ in the number of *cotyledons*—the seed leaves associated with plant embryos. Monocot seeds have one cotyledon and eudicot seeds have two. Although monocots and eudicots have similar types of tissues, their body structures differ in distinctive ways **(Table 31.1).** As we discuss the morphology of flowering plants, we will refer frequently to these structural differences.

Flowering Plants Can Be Grouped according to Type of Growth and Lifespan

In evolutionary terms, the distinction between monocot and eudicot flowering plants is most important structurally and developmentally. Yet botanists sometimes use other criteria to distinguish between flowering plants—for example, by whether they have secondary growth. Most monocots and some eudicots are *herbaceous* plants, showing little or no secondary growth during their life cycle. In contrast, many eudicots (and all gymnosperms) are *woody* plants, which do have secondary growth.

We can also distinguish plants by lifespan. **Annuals** are herbaceous plants in which the life cycle is completed in one growing season. With minimal or no secondary growth, annuals typically have only apical meristems. Examples are marigolds (a eudicot) and corn (a monocot). **Biennials** complete the life cycle in two growing seasons, and limited secondary growth occurs in some species. Roots, stems, and leaves form in the first season, then the plant flowers, forms seeds, and dies in the second. Examples are carrots and celery (eudicots). In **perennials,** vegetative growth and reproduction continue year after year. Many perennials, such as trees, shrubs, and some vines, have secondary tissues, although others, such as irises and daffodils, do not.

Study Break

1. Compare and contrast the components and functions of a land plant's shoot and root systems.
2. Explain what meristem tissue is, and name and describe the functions of the basic types of meristems.

Table 31.1 Eudicots and Monocots Compared

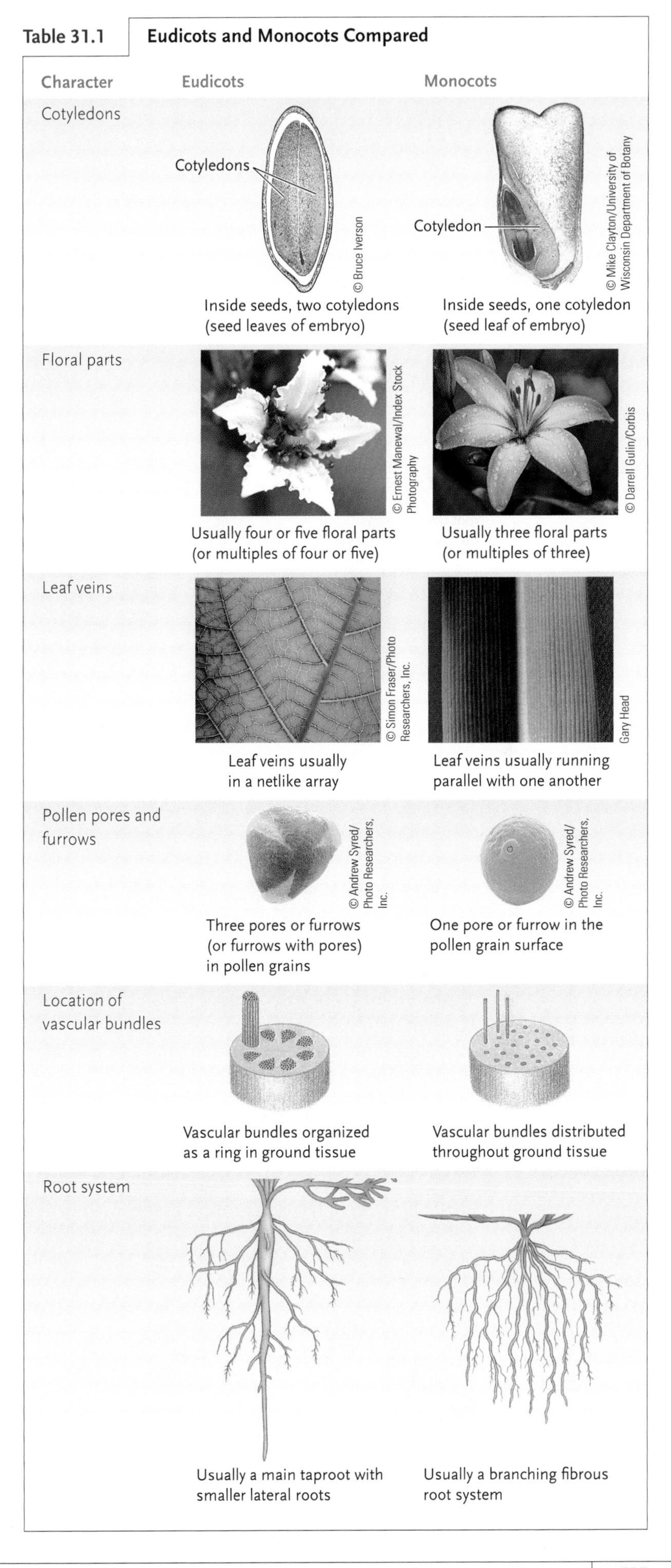

Character	Eudicots	Monocots
Cotyledons	Inside seeds, two cotyledons (seed leaves of embryo)	Inside seeds, one cotyledon (seed leaf of embryo)
Floral parts	Usually four or five floral parts (or multiples of four or five)	Usually three floral parts (or multiples of three)
Leaf veins	Leaf veins usually in a netlike array	Leaf veins usually running parallel with one another
Pollen pores and furrows	Three pores or furrows (or furrows with pores) in pollen grains	One pore or furrow in the pollen grain surface
Location of vascular bundles	Vascular bundles organized as a ring in ground tissue	Vascular bundles distributed throughout ground tissue
Root system	Usually a main taproot with smaller lateral roots	Usually a branching fibrous root system

31.2 The Three Plant Tissue Systems

Plants develop three tissue systems that provide the foundation for the various plant organs. The **ground tissue system,** which makes up most of the plant body, functions in metabolism, storage, and support. The **vascular tissue system** consists of various tubes that transport water and nutrients throughout the plant. The tubes are organized in bundles that are dispersed through the ground tissues. The **dermal tissue system** serves as a skinlike protective covering for the plant body. Figure 31.2 shows the general location of each system in the shoot and root.

Table 31.2	Summary of Flowering Plant Tissues and Their Components		
Tissue System	**Name of Tissue**	**Cell Types in Tissue**	**Tissue Function**
Ground tissue	Parenchyma	Parenchyma cells	Photosynthesis, respiration, storage, secretion
	Collenchyma	Collenchyma cells	Flexible strength for growing plant parts
	Sclerenchyma	Fibers or sclereids	Rigid support, deterring herbivores
Vascular tissue	Xylem	Conducting cells (tracheids, vessel members); parenchyma cells; sclerenchyma cells	Transport of water and dissolved minerals
	Phloem	Conducting cells (sieve tube members); parenchyma cells; sclerenchyma cells	Sugar transport
Dermal tissue	Epidermis	Undifferentiated cells; guard cells and other specialized cells	Control of gas exchange, water loss; protection
	Periderm	Cork; cork cambium; secondary cortex	Protection

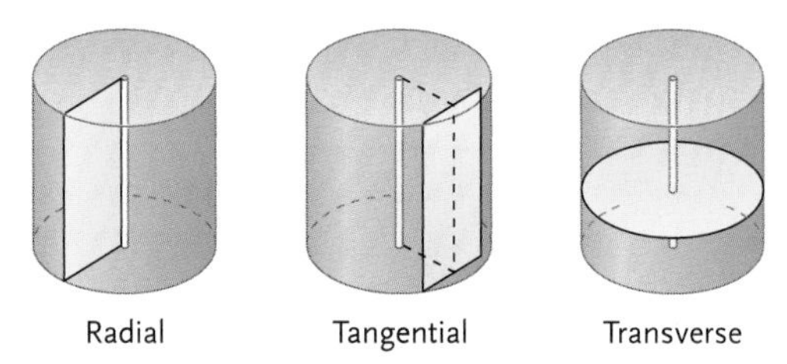

Figure 31.4
Terms that identify how tissue specimens are cut from a plant. Along the radius of a stem or root, longitudinal cuts give radial sections. Cuts at right angles to a root or stem radius give tangential sections. Cuts perpendicular to the long axis of a stem or root give transverse sections (cross sections).

Each tissue system includes several types of tissue, and each tissue is made up of cells with specializations for different functions **(Table 31.2).** *Simple* tissues have only one type of cell. Other tissues are *complex,* with organized arrays of two or more types of cells. **Figure 31.4** will help you interpret micrographs of plant tissues, beginning with the tissues in a transverse section of a stem shown in **Figure 31.5.**

Ground Tissues Are All Structurally Simple, but They Exhibit Important Differences

Plants have three types of ground tissue, each with a distinct structure and function—*parenchyma, collenchyma,* and *sclerenchyma* **(Figure 31.6).** Each type is structurally simple, being composed mainly of one kind of cell. In a very real sense the cells in ground tissues are the "worker bees" of plants, carrying out photosynthesis, storing carbohydrates, providing mechanical support for the plant body, and performing other basic functions. Each kind of cell has a distinctive wall structure, and some have variations in the protoplast as well.

Parenchyma: Soft Primary Tissues. Parenchyma (*para* = around, *chein* = fill in, or pour) makes up the bulk of the soft, moist primary growth of roots, stems, leaves, flowers, and fruits. Most parenchyma cells have only a thin primary wall and so are pliable and permeable to water. Often the cells are spherical or many-sided, although they also can be elongated like a sausage, as in Figure 31.6a. Parenchyma cells sometimes have air spaces between them, especially in leaves (see Figure 31.17). The air spaces may be sizeable in the stems and leaves of aquatic plants, such as water lilies. This adaptation facilitates the movement of oxygen from aerial leaves and stems to submerged parts of the plant, and it also helps the leaves float upward toward the light.

Parenchyma cells may be specialized for tasks as varied as storage, secretion, and photosynthesis. They can occur both as part of parenchyma tissue and as individual cells in other tissues. In many plant species, modified parenchyma cells are specialized for short-distance transport of solutes. Such cells are common in tissues in which water and solutes must be rapidly moved from cell to cell—for example, in vascular tissues and in tissues that secrete nectar. Parenchyma cells usually remain alive and,

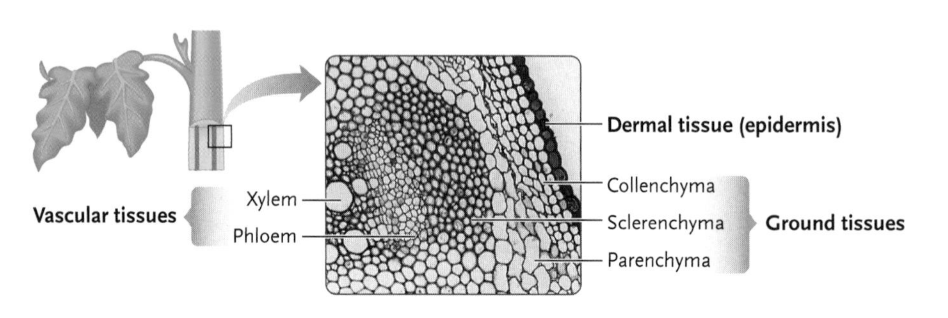

Figure 31.5
Locations of ground, vascular, and dermal tissues in one kind of plant stem, transverse section. Ground tissues are simple tissues while vascular and dermal tissues are complex, containing various types of specialized cells. (Micrograph: James D. Mauseth, Plant Anatomy, Benjamin Cummings, 1988.)

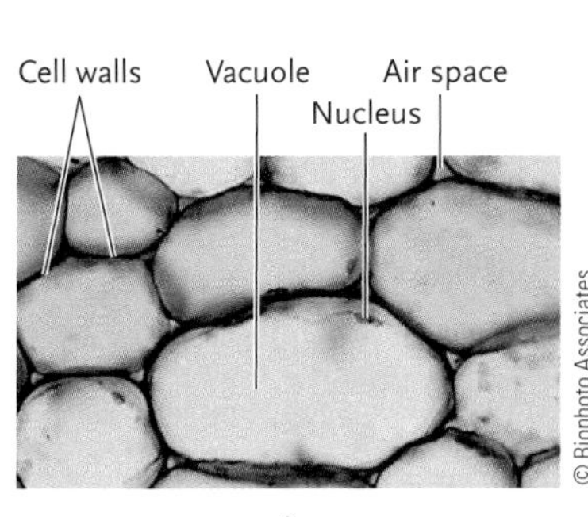

a. Parenchyma tissues consist of soft, living cells specialized for storage, other functions.

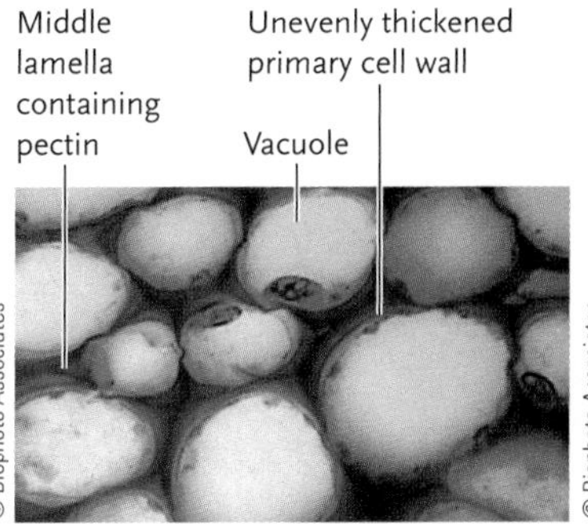

b. Collenchyma tissues provide flexible support.

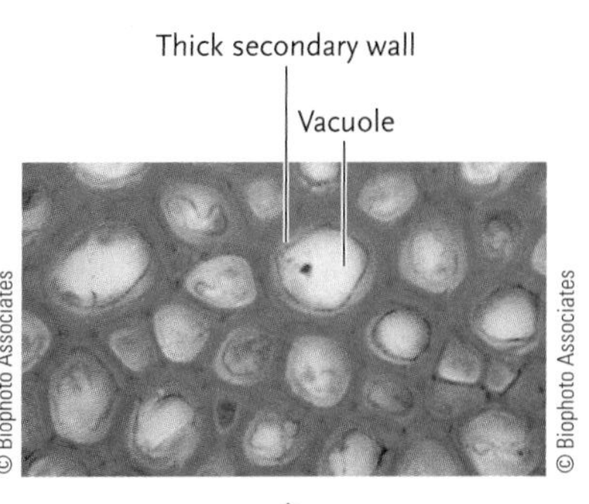

c. Sclerenchyma tissues provide rigid support and protection.

Figure 31.6
Examples of ground tissues from the stem of a sunflower plant *(Helianthus annuus)*.

when mature, retain the capacity to divide; in fact, their mitotic divisions often heal wounds in plant parts.

Collenchyma: Flexible Support. The "strings" in celery are examples of the flexible ground tissue called **collenchyma** (see Figure 31.6b), which helps strengthen plant parts that are still elongating (*kolla* = glue). Collenchyma cells are typically elongated, and collectively they often form strands or a sheathlike cylinder under the dermal tissue of growing shoot regions and the stalklike petioles that attach leaves to stems.

The primary walls of collenchyma cells are built of alternating layers of cellulose and pectin. These walls thicken and stretch as the cell enlarges. Mature collenchyma cells are alive and metabolically active, and they continue to synthesize cellulose and pectin layers as the plant grows.

Sclerenchyma: Rigid Support and Protection. Mature plant parts gain additional mechanical support and protection from **sclerenchyma** (*skleros* = hard). The cells of this ground tissue develop thick secondary walls (see Figure 31.6c), which commonly are lignified and perforated by pits through which water can pass. After mature sclerenchyma cells become encased in lignin, they die because their protoplasts can no longer exchange gases, nutrients, and other materials with the environment. The walls, however, continue to provide protection and support.

The two types of sclerenchyma cells—*sclereids* and *fibers*—differ in their shape and arrangement. Sheetlike arrays of rigid **sclereids** form a protective coat around seeds; examples are the hard casings of a coconut shell or peach pit. Sclereids come in a range of shapes; the gritty texture of a pear comes from the roughly cube-shaped sclereids scattered through its flesh **(Figure 31.7a)**. The long, tapered cells called **fibers** **(Figure 31.7b)** resist stretching, but are more pliable than sclereids. Fibers in the stems of flax plants are massed in parallel; they can flex and twist without stretching and are used to manufacture rope, paper, and linen cloth.

Vascular Tissues Are Specialized for Conducting Fluids

Vascular tissues are complex tissues composed of specialized conducting cells, parenchyma cells, and fibers. *Xylem* and *phloem*, the two kinds of vascular tissues in flowering plants, are organized into bundles of interconnected cells that extend throughout the plant.

Xylem: Transporting Water and Minerals. **Xylem** (*xylon* = wood) conducts water and dissolved minerals absorbed from the soil upward from a plant's roots to the shoot. As you read in Chapter 27, xylem was a key early adaptation allowing plants to make the transition to life on land. Xylem contains two types of conducting cells: *tracheids* and *vessel members*. Both develop thick, lignified secondary cell walls and die at maturity. The empty cell walls of abutting cells serve as pipelines for water and minerals.

Tracheids are elongated, with tapered, overlapping ends **(Figure 31.8a)**. In plants adapted to drier soil conditions, they have strong secondary walls that keep them from collapsing when less water is present. As in sclerenchyma, water can move from cell to cell through

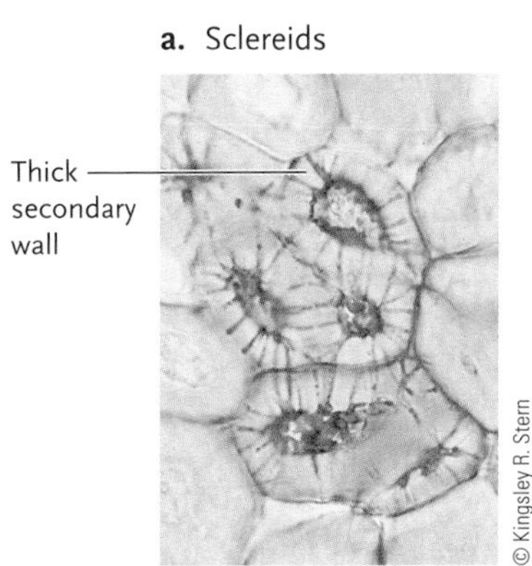

Figure 31.7
Examples of sclerenchyma cells. **(a)** From the flesh of a pear *(Pyrus)*, one type of sclereid: stone cells, each with a thick, lignified wall. **(b)** Strong fibers from stems of a flax plant *(Linum)*.

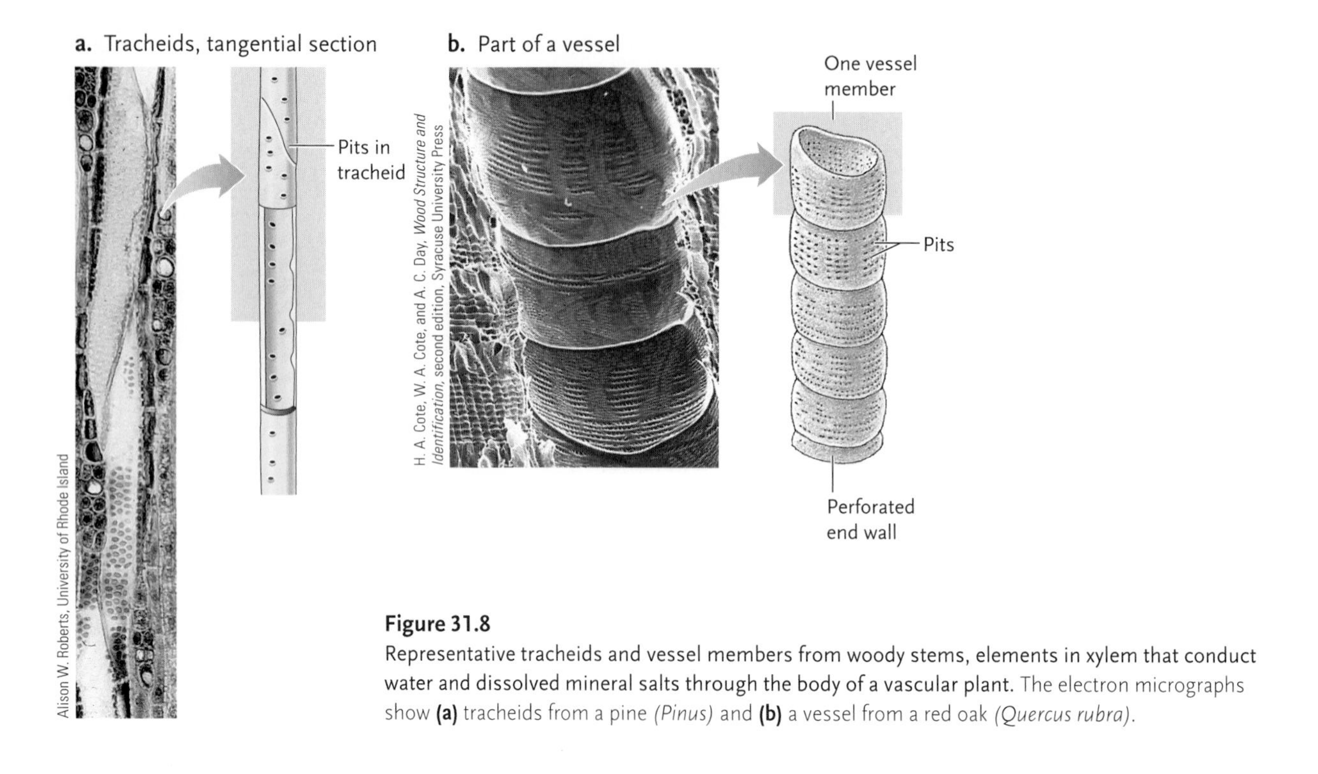

Figure 31.8
Representative tracheids and vessel members from woody stems, elements in xylem that conduct water and dissolved mineral salts through the body of a vascular plant. The electron micrographs show **(a)** tracheids from a pine *(Pinus)* and **(b)** a vessel from a red oak *(Quercus rubra)*.

pits. Usually, a pit in one cell is opposite a pit of an adjacent cell, so water seeps laterally from tracheid to tracheid.

Vessel members (or vessel elements) are shorter cells joined end to end in tubelike columns called vessels **(Figure 31.8b). Vessels** are typically several centimeters long, and in some vines and trees they may be many meters long. Like tracheids, vessel members have pits; however, they also have another adaptation that greatly enhances water flow. As vessel members mature, enzymes break down portions of their end walls, producing perforations. Some vessel members have a single, large perforation, so that the end is completely open (see Figure 31.8b). Others have a cluster of small, round perforations, or ladderlike bars, extending across the open end (see Figure 31.8). The predictability of the perforation patterns suggests that this process is under precise genetic control.

Fossil evidence shows that the forerunners of modern plant species relied solely on tracheids for water transport, and today ferns and most gymnosperms still have only tracheids. Nearly all angiosperms and a few other types of plants have both tracheids and vessel members, however, which confers an adaptive advantage. Flowing water sometimes incorporates air bubbles, which represent a potentially lethal threat to the plant. Water can flow rapidly through vessel members that are linked end to end, but the open channel cannot prevent air bubbles from forming and possibly blocking the flow through the whole vessel. By contrast, even though water moves more slowly in tracheids, the pit membranes are impermeable to air bubbles, and a bubble that forms in one tracheid stays there; water continues to move between other tracheids.

Conducting cells that form after the surrounding tissue has reached its maximum size have complete secondary walls, with pits or perforations. In growing plants, however, only a partial secondary wall forms in cells of xylem, so the cell can elongate as the tissue it services grows. At maturity, tracheids and vessel members die as genetic cues cause their protoplasts to degenerate and lignin to be deposited in cell walls.

Parenchyma cells in xylem participate in the transport of minerals through vessel members and tracheids. Sclerenchyma fibers function like steel cables in concrete, helping keep the tissue fairly rigid and lending structural support to the plant.

Phloem: Transporting Sugars and Other Solutes. The vascular tissue **phloem** (*phloios* = tree bark) transports solutes, notably the sugars made in photosynthesis, throughout the plant body. The main conducting cells of phloem are **sieve tube members (Figure 31.9),** which connect end to end, forming a **sieve tube.** As the name implies, their end walls, called sieve plates, are studded with pores. In flowering plants the phloem is strengthened by fibers and sclereids.

Immature sieve tube members contain the usual plant organelles. Over time, however, the cell nucleus and internal membranes in plastids break down, mitochondria shrink, and the cytoplasm is reduced to a thin layer lining the interior surface of the cell wall. Even without a nucleus, the cell lives up to several years in most plants, and much longer in some trees.

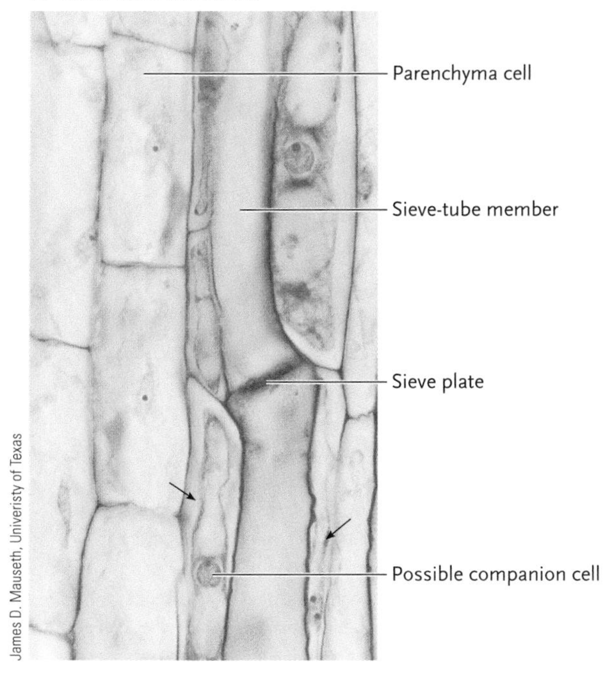

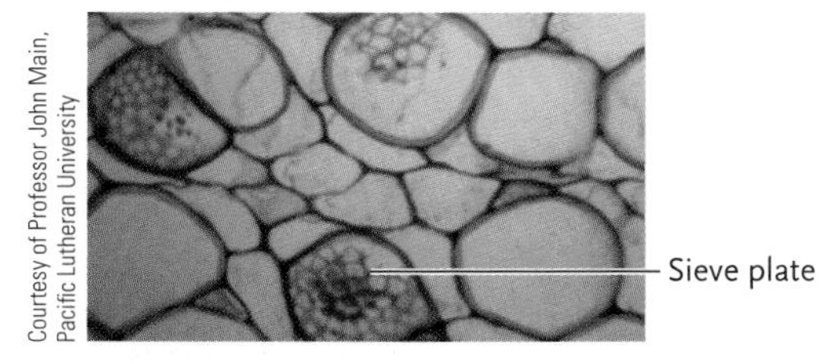

Figure 31.9
Structure of sieve tube members. (a) Micrograph showing sieve tube members in longitudinal section. The arrows point to cells that may be companion cells. Long tubes of sieve tube members conduct sugars and other organic compounds. **(b)** Sieve plate in a cell in phloem, cross section.

In many flowering plants, specialized parenchyma cells known as **companion cells** are connected to mature sieve tube members by plasmodesmata. Unlike sieve tube members, companion cells retain their nucleus when mature. They assist sieve tube members both with the uptake of sugars and with the unloading of sugars in tissues engaged in food storage or growth. They may also help regulate the metabolism of mature sieve tube members. We return to the functions of phloem cells in Chapter 32.

Dermal Tissues Protect Plant Surfaces

A complex tissue called **epidermis** covers the primary plant body in a single continuous layer **(Figure 31.10a)** or sometimes in multiple layers of tightly packed cells. The external surface of epidermal cell walls is coated with waxes that are embedded in cutin, a network of chemically linked fats. Epidermal cells secrete this coating, or **cuticle**, which resists water loss and helps fend off attacks by microbes. A cuticle coats all plant parts except the very tips of the shoot and most absorptive parts of roots; other root regions have an extremely thin cuticle.

Most epidermal cells are relatively unspecialized, but some are modified in ways that represent important adaptations for plants. Young stems, leaves, flower parts, and even some roots have pairs of crescent-shaped **guard cells (Figure 31.10b).** Unlike other cells of the epidermis, guard cells contain chloroplasts and so can carry out photosynthesis. The pore between a pair of guard cells is termed a **stoma** (plural, *stomata*). Water vapor, carbon dioxide, and oxygen move across the epidermis through the stomata, which open and close by way of mechanisms we consider in Chapter 32.

With their exact spacing and vital role in regulating the exchange of gases between a plant and its environment, stomata have captured the interest of

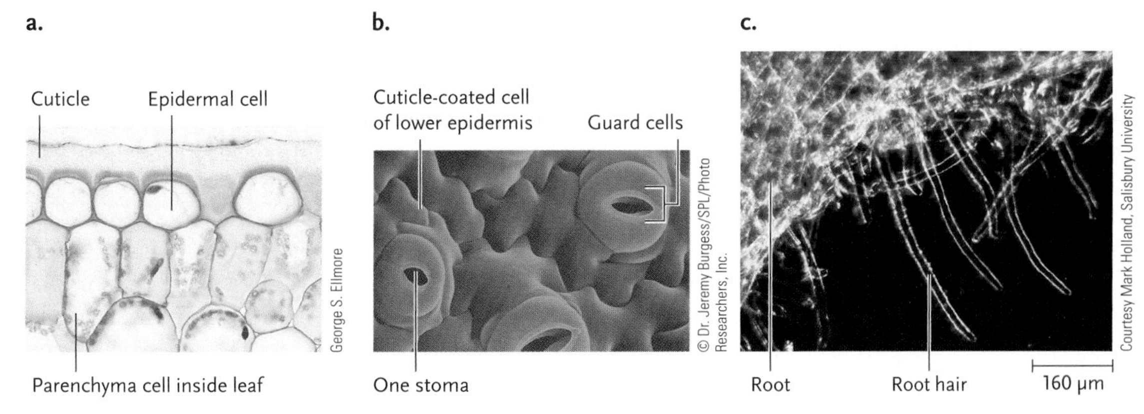

Figure 31.10
Structure and examples of epidermal tissue. (a) Cross section of leaf epidermis from a bush lily *(Clivia miniata)*. **(b)** Scanning electron micrograph of a leaf surface, showing cuticle-covered epidermal cells and stomata. **(c)** Root hairs, an epidermal specialization.

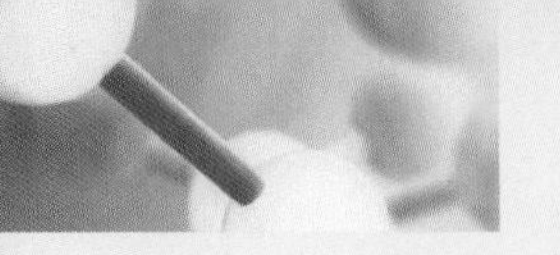

Insights from the Molecular Revolution

Shaping up Flower Color

Different pigments in flowers produce different colors, but are pigments the whole story? A molecular study of flower color in snapdragons *(Antirrhinum majus)* provided a surprising answer. Kenichi Noda of the Nippon Oil Company in Japan and Beverly J. Glover and her colleagues at the John Innes Institute in England were interested in a mutant snapdragon called *mixta,* which produces pale red flower petals with a dull, flat surface rather than the deep red, velvety petals of wild-type plants **(Figure a).**

Through a series of steps, the investigators isolated and cloned the *mixta* gene. Sequencing the gene revealed close similarities to a regulatory gene that activates genes in some other plants. The similarities suggested that the normal snapdragon gene also codes for a regulatory protein that produces normal flower color. When the transposable element inserts in the gene, the regulatory protein is lost.

How does the regulatory protein govern flower color? At first Noda and his colleagues thought it regulated production of anthocyanin, a pigment that gives flowers a red color, and that loss of the protein in mutant plants hampered anthocyanin production. They discarded this hypothesis when both the wild-type and *mixta* plants were found to have normal levels of anthocyanin. However, microscopic examination of flower petals revealed that wild-type and *mixta* epidermal cells are shaped differently. Normal plants have conical epidermal cells, with the tip of the cone pointing outward and giving the petals a velvety appearance. Epidermal cells of *mixta* mutants have a flat, irregular surface that produces a dull appearance. **Figure b** shows the surface of a variegated flower petal, which has both conical and flat cells. This structural difference suggested that in wild-type petals, the cone tips act as prisms that make the red pigment clearly visible, while the irregular surface of the mutant cells scatters light and masks the pigment color.

Figure a
Wild-type snapdragon, which has flowers with deep red petals.

Figure b
Scanning electron micrograph of the surface of a variegated snapdragon petal, showing conical cells in the bright red colored areas and flat cells in the pale colored areas. The genetic events that produce variegated petals were clues that helped the research team identify the *mixta* gene.

As a test, the research team removed the cell walls from the epidermal petal cells, a step that eliminated differences in cell shape. Both the normal and mutant cells had the same intense, red color. On this basis, the researchers proposed that the regulatory protein encoded in the normal *mixta* gene activates other genes whose protein products in some way produce the conical cell shape.

geneticists probing the molecular underpinnings of plant development. Working with *Arabidopsis thaliana* (thale cress) plants, researchers have identified an enzyme—encoded by the gene *YDA*—that appears to ultimately control where and how many stomata form. In mutant plants with a defective enzyme, the epidermis is blanketed with stomata packed side by side. The plants often die early in development or are stunted and appear fuzzy—hence the enzyme's name, YODA, recalling the short, hairy Star Wars character. In nonmutated wild-type plants, unequal divisions of precursor cells produce one smaller and one larger daughter cell, and the smaller one gives rise to the two guard cells of a stoma. (The larger cell either divides again or becomes an underlying epidermal cell.) YODA comes into play when a series of precursor reactions phosphorylate it. The activated enzyme then triggers a cascade of reactions that, by some as-yet-unknown mechanism, either promote or restrict these asymmetric divisions.

Other epidermal specializations are the single-celled or multicellular outgrowths collectively called **trichomes**, which give the stems or leaves of some plants a hairy appearance. Some trichomes exude sugars that attract insect pollinators. Leaf trichomes of *Urtica,* the stinging nettle, provide protection by injecting an irritating toxin into the skin of animals that brush against the plant or try to eat it. **Root hairs,** which develop as extensions in the outer wall of root epidermal cells **(Figure 31.10c),** are also trichomes. Root hairs absorb much of a plant's water and minerals from the soil.

The epidermal cells of flower petals (which are modified leaves) synthesize pigments that are partly responsible for a blossom's colors. However, molecular studies have revealed that flower colors and their intensity or brightness also depend on the shape of the epidermal cells, as described in *Insights from the Molecular Revolution*.

Study Break

1. Describe the defining features, cellular components, and functions of the ground tissue system.
2. What are the functions of xylem and phloem?
3. What are the cellular components and functions of the dermal tissue system?

31.3 Primary Shoot Systems

A young flowering plant's shoot system consists of the main stem, leaves, and buds as well as flowers and fruits. Chapter 34 looks more closely at flowers and fruits; here we focus on the growth and organization of stems, buds, and leaves of the primary shoot system.

Stems Are Adapted to Provide Support, Routes for Vascular Tissues, Storage, and New Growth

Stems are structurally adapted for four main functions. First, they provide mechanical support, generally along a vertical (upright) axis, for body parts involved in growth, photosynthesis, and reproduction. These parts include meristematic tissues, leaves, and flowers. Second, they house the vascular tissues (xylem and phloem), which transport products of photosynthesis, water and dissolved minerals, hormones, and other substances throughout the plant. Third, they often are modified to store water and food. And finally, buds and specific stem regions contain meristematic tissue that gives rise to new cells of the shoot.

The Modular Organization of a Stem. A plant stem develops in a pattern that divides the stem into modules, each consisting of a *node* and an *internode*. A **node** is a place on the stem where one or more leaves are attached; the area between two nodes is thus an **internode.** The upper angle between the stem and an attached leaf is an **axil.** New primary growth occurs in buds—a **terminal bud** at the apex of the main shoot, and **lateral buds,** which produce branches (lateral shoots), in the leaf axils. Meristematic tissue in buds gives rise to leaves, flowers, or both **(Figure 31.11).**

In eudicots, most growth in a stem's length occurs directly below the apical meristem, as internode cells divide and elongate. Internode cells nearest the apex are most active, so the most visible new growth occurs at the ends of stems. In grasses and some other monocots, by contrast, the upper cells of an internode stop dividing as the internode elongates, and cell divisions are limited to a meristematic region at the base of the internode. The stems of bamboo and other grasses elongate as the internodes are "pushed up" by the growth of such meristems. This adaptation allows grasses to grow back readily after grazing by herbivores (or being chopped off by a lawnmower), because the meristem is not removed.

Terminal buds release a hormone that inhibits the growth of nearby lateral buds, a phenomenon called **apical dominance.** Gardeners who want a bushier plant can stimulate lateral bud growth by periodically cutting off the terminal bud. The flow of hormone signals then dwindles to a level low enough that lateral buds begin to grow. In nature, apical dominance is an adaptation that directs the plant's resources into growing up toward the light.

a. Location of nodes and buds

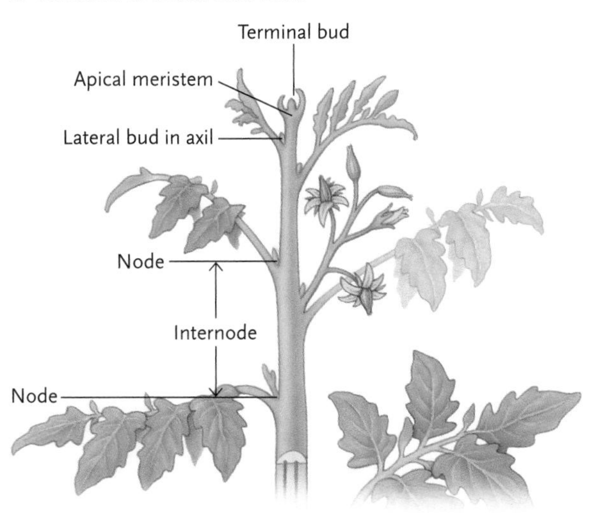

b. Leaves at a terminal bud

Figure 31.11
Modular structure of a stem. **(a)** The arrangement of nodes and buds on a plant stem. **(b)** Formation of leaves at a terminal bud of a dogwood (genus *Cornus*).

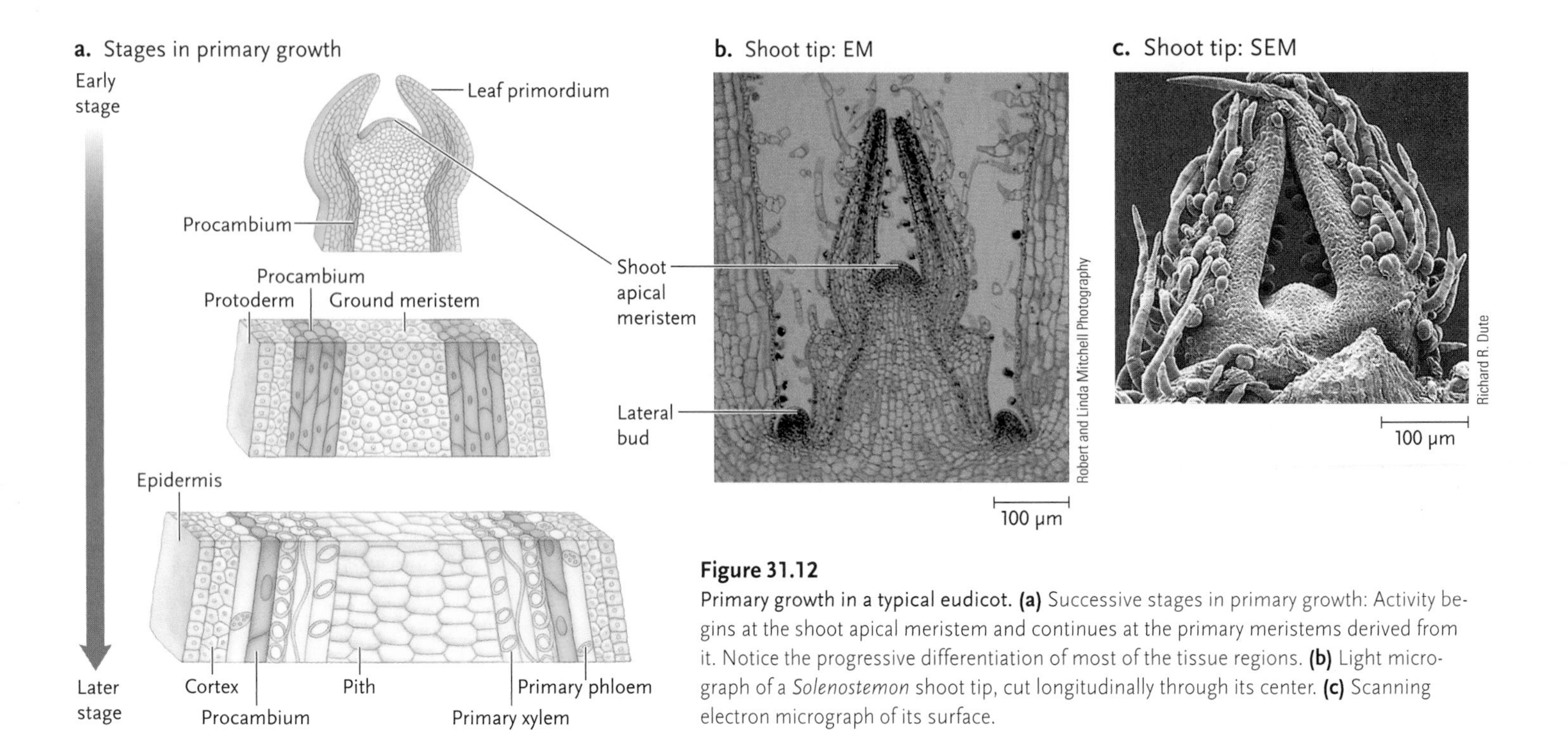

Figure 31.12
Primary growth in a typical eudicot. **(a)** Successive stages in primary growth: Activity begins at the shoot apical meristem and continues at the primary meristems derived from it. Notice the progressive differentiation of most of the tissue regions. **(b)** Light micrograph of a *Solenostemon* shoot tip, cut longitudinally through its center. **(c)** Scanning electron micrograph of its surface.

Primary Growth and Structure of a Stem. Primary growth, the cell divisions and enlargement that produce the primary plant body, begins in the shoot and root apical meristems. The sequence of events is similar in roots and shoots; it is shown for a eudicot shoot in **Figure 31.12.**

The shoot apical meristem is a dome-shaped mass of cells. When one cell divides, one of its daughter cells becomes an **initial,** a cell that remains as part of the meristem. The other daughter cell becomes a **derivative.** The derivative typically divides once or twice and then enters on the path to differentiation. When initials divide, they replenish the supply of derivatives in the meristem.

As derivatives differentiate, they give rise to three **primary meristems:** *protoderm, procambium,* and *ground meristem* (see Figure 31.12). These primary meristems are relatively unspecialized tissues with cells that differentiate in turn into specialized cells and tissues. In eudicots, the primary meristems are also responsible for elongation of the plant body.

How do the genetically identical cells of an apical meristem give rise to three types of primary meristem cells, and ultimately to all the specialized cells of the plant? *Focus on Research* describes some experiments that are probing the genetic mechanisms underlying meristem activity.

Each primary meristem occupies a different position in the shoot tip, as shown in Figure 31.12a. Outermost is **protoderm,** a meristem that will produce the stem's epidermis. While protoderm cells divide and the resulting derivatives are maturing, the shoot tip continues to grow. Eventually, the protoderm cells differentiate into specific types of epidermal cells, including guard cells and trichomes. Some monocots, such as palms, have a primary thickening meristem just under the protoderm; this tissue contributes to both lateral growth and elongation of the stem.

Inward from the protoderm is the **ground meristem,** which will give rise to ground tissue, most of it parenchyma. **Procambium,** which produces the primary vascular tissues, is sandwiched between ground meristem layers. Procambial cells are long and thin, and their spatial orientation foreshadows the future function of the tissues they produce. In most plants, inner procambial cells give rise to xylem and outer procambial cells to phloem. In plants with secondary growth, a thin region of procambium between the primary xylem and phloem remains undifferentiated. Later on it will give rise to the lateral meristems.

The developing vascular tissues become organized into **vascular bundles,** multistranded cords of primary xylem and phloem that are wrapped in sclerenchyma and thread lengthwise through the parenchyma. In the stems and roots of most eudicots and some conifers, the vascular bundles form a **stele** (Greek *stele* = pillar), also known as a *vascular cylinder,* that vertically divides the column of ground tissue into an outer **cortex** and an inner **pith (Figure 31.13a).** Both cortex and pith consist mainly of parenchyma; in some plant species the pith parenchyma stores starch reserves. In the stems of most monocots, vascular bundles are dispersed through the ground tissue **(Figure 31.13b),** so separate cortical and pith regions do not form. In some monocots, including bamboo, the pith breaks down, leaving the stem with a hollow core. The hollow stems of certain hard-walled bamboo species are used to make bamboo flutes.

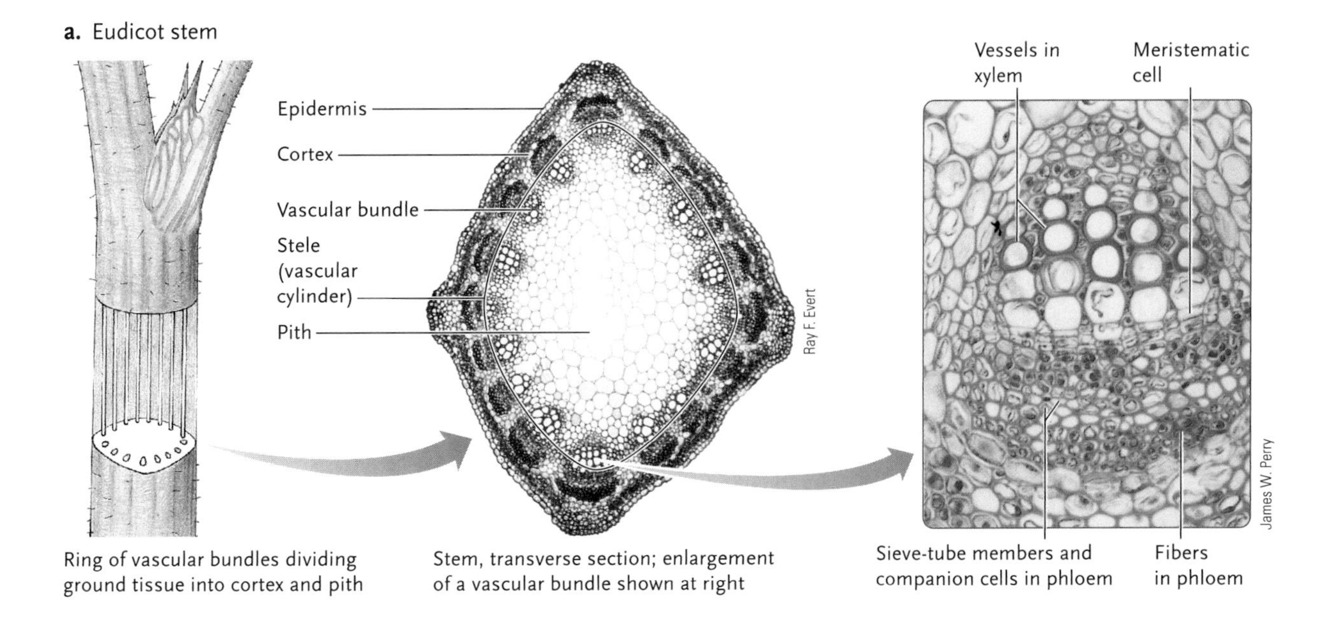

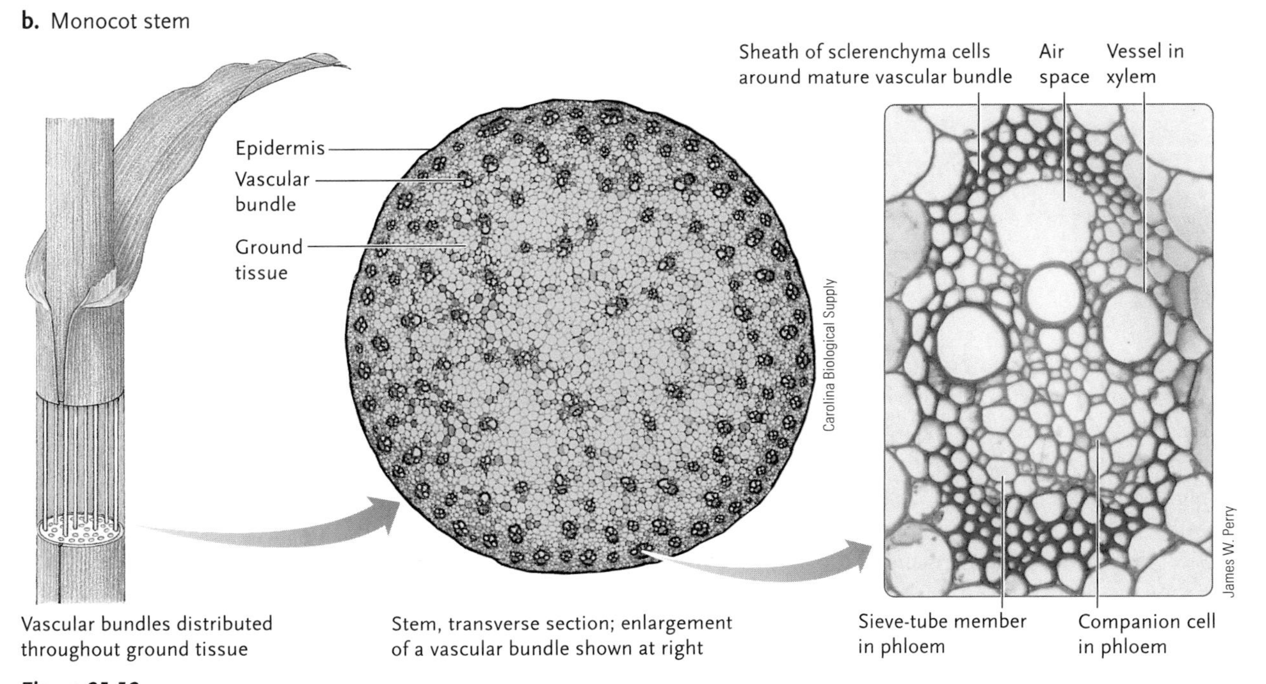

Figure 31.13

Organization of cells and tissues inside the stem of a eudicot and a monocot. (a) Part of a stem from alfalfa *(Medicago)*, a eudicot. In many species of eudicots and conifers, the vascular bundles develop in a more or less ringlike array in the ground tissue system, as shown here. **(b)** Part of a stem from corn *(Zea mays)*, a monocot. In most monocots and some herbaceous eudicots, vascular bundles are scattered through the ground tissue, as shown here.

As leaves and buds appear along a stem, some vascular bundles in the stem branch off into these developing tissues. The arrangement of vascular bundles in a plant ultimately depends on the number of branch points to leaves and buds and on the number and distribution of leaves.

Stem Modifications. Evolution has produced a range of stem specializations, including structures modified for reproduction, food storage, or both **(Figure 31.14)**. An onion or a garlic head is a *bulb,* a modified shoot that consists of a bud with fleshy leaves. *Tubers* are stem regions enlarged by the presence of starch-storing parenchyma cells; examples of plants that form tubers are the potato and the cassava (the source of tapioca). The "eyes" of a potato are buds at nodes of the modified stem, and the regions between eyes are internodes. Many grasses, such as Bermuda grass, and some weeds are difficult to eradicate because they have *rhizomes*—long underground stems that can

FOCUS ON RESEARCH

Basic Research: Homeobox Genes: How the Meristem Gives Its Marching Orders

How do descendents of some dividing cells in a shoot apical meristem (SAM) "know" to become stem tissues, while others embark on the developmental path that produces leaves or other shoot parts? Although the full answer to this question is not yet known, research teams at several laboratories around the world have found evidence of a genetic mechanism in plant meristem cells that appears to guide the process.

Working with SAM tissue from maize (*Zea mays*, generally known in North America as corn), investigators have identified more than a dozen regulatory genes whose protein products activate groups of other genes in differentiating cells. Some genes that act in this way to guide development along a particular path are called *homeotic genes*, because they contain a nucleotide sequence called the homeobox. The homeobox (see Chapter 48) binds to a specific promoter region shared by all of the genes that a homeotic gene controls. Interaction with a homeobox sequence turns the affected genes on or off. Homeobox genes were first discovered in studies of how legs, antennae, and other structures develop in *Drosophila*, the common fruit fly.

Researcher Sarah Hake of the Plant Gene Expression Center (U.S. Department of Agriculture) was curious about the action of a homeotic gene in maize known as *knotted-1 (KN-1)*. Normally the *KN-1* gene is expressed in apical meristems, where it maintains the meristem in an undifferentiated state. When a mutated form, *kn-1*, is expressed, however, the mutation causes abnormal knobby growths on leaves—hence the gene's name. Hake's research helped establish that *KN-1* defines developmental pathways that unfold in meristems. For example, when Hake cloned the *KN-1* gene and inserted it into tobacco leaf cells, the cells *de*differentiated and began acting like meristem cells. As they divided, they produced lines that could differentiate into leaves and stems.

Subsequent studies of *KN-1* in species as diverse as sunflowers and garden peas have led to the identification of the family of what are now called knotted-1-like genes, all of which encode regulatory proteins that influence developmental pathways. As in maize, some are typically expressed in SAM tissue. In sunflower, tomato, and perhaps other species, knotted-1-like genes also appear to be expressed in differentiated plant parts including leaves, flowers, stems, and even roots. The early work on SAM tissue and homeobox genes in maize has blossomed into a wide-ranging investigation of the molecular signals that shape plant architecture.

extend as much as half a meter deep into the soil and rapidly produce new shoots when existing ones are pulled out. The pungent, starchy "root" of ginger is a rhizome also. Crocuses and some other ornamental plants develop elongated, fleshy underground stems called *corms,* another starch-storage adaptation. Tubers, rhizomes, and corms all have meristematic tissue at nodes from which new plants can be propagated—a vegetative (asexual) reproductive mode. Other plants, including the strawberry, reproduce vegetatively via slender stems called *stolons,* which grow along the soil surface. New plants arise at nodes along the stolon.

Leaves Carry Out Photosynthesis and Gas Exchange

Each spring a mature maple tree heralds the new season by unfurling roughly 100,000 leaves. Some other tree species produce leaves by the millions. For these

a. Onion bulb

Mike Hill/Getty Images Inc.

b. Potato tuber

Wally Eberhart/Visuals Unlimited

c. Ginger rhizome

Joerg Boethling/Peter Arnold, Inc.

d. Crocus corm

Alan & Linda Detrick/Photo Researchers, Inc.

e. Strawberry stolons

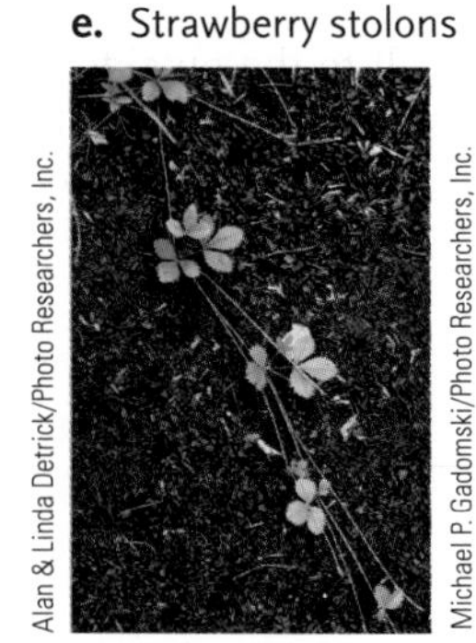

Michael P. Gadomski/Photo Researchers, Inc.

Figure 31.14
A selection of modified stems. (a) The fleshy bulbs of onions *(Allium cepa)* are modified shoots in which the plant stores starch. **(b)** A potato *(Solanum tuberosum)*, a tuber. **(c)** Ginger "root," the pungent, starchy rhizome of the ginger plant *(Zingiber officinale)*. **(d)** Crocus plants (genus *Crocus*) typically grow from a corm. **(e)** A strawberry plant *(Fragaria ananassa)* and stolon.

and most other plants, leaves are the main organs of photosynthesis and gas exchange.

Leaf Morphology and Anatomy. In both eudicots and monocots, the leaf **blade** provides a large surface area for absorbing sunlight and carbon dioxide **(Figure 31.15a).** Studies show that in general, leaves of flowering plants are oriented on the stem axis so that they can capture the maximum amount of sunlight; the stems and leaves of some plants follow the sun's movement during the course of a day by changing position (this phenomenon is described in Chapter 35).

Many eudicot leaves, such as those of maples, have a broad, flat blade attached to the stem by a stalklike **petiole.** Depending on the species, the petiole can be long, short, or in between. A celery stalk is a fleshy petiole. Unless a petiole is very short, it holds a leaf away from the stem and helps prevent individual leaves from shading one another. In many plant species petioles allow leaves to move in the breeze. This helps circulate air around the leaf, replenishing the supply of carbon dioxide for photosynthesis. In most monocot leaves, such as those of rye grass or corn, the blade is longer and narrower and its base simply forms a sheath around the stem.

Leaf Modifications. Leaf forms are based on two basic patterns: simple leaves, which have a single blade **(Figure 31.15b),** and compound leaves, in which the leaf blade is divided into smaller leaflets **(Figure 31.15c).** As with other plant parts, there is huge variety in the morphology of leaves. For instance, leaf edges or margins may be smooth, toothed, or lobed. Some leaves are modified as spines **(Figure 31.16a),** while others have trichomes that take the form of hairs or hooks—all possibly adaptations for defense against herbivores. Leaves or parts of leaves also may be modified into tendrils, like those of the sweet pea **(Figure 31.16b),** or other structures. Epidermal cells on the leaves of the saltbush *Atriplex spongiosa* form balloonlike structures **(Figure 31.16c)** that contain concentrated Na^+ and Cl^- taken up from the salty soil. Eventually, the salt-filled epidermal cells burst or fall off the leaf, releasing the salt to the outside. This adaptation helps control the salt concentration in the plant's tissues—another example of the link between structure, function, and the environment in which a plant lives.

Leaf Primary Growth and Internal Structure. In both angiosperms and gymnosperms, leaves develop on the sides of the shoot apical meristem. Initially, meristem cells near the apex divide and their derivatives elongate. The resulting bulge enlarges into a thin, rudimentary leaf, or **leaf primordium** (see Figure 31.12). As the plant grows and internodes elongate, the leaves that form from leaf primordia become spaced at intervals along the length of the stem or its branches.

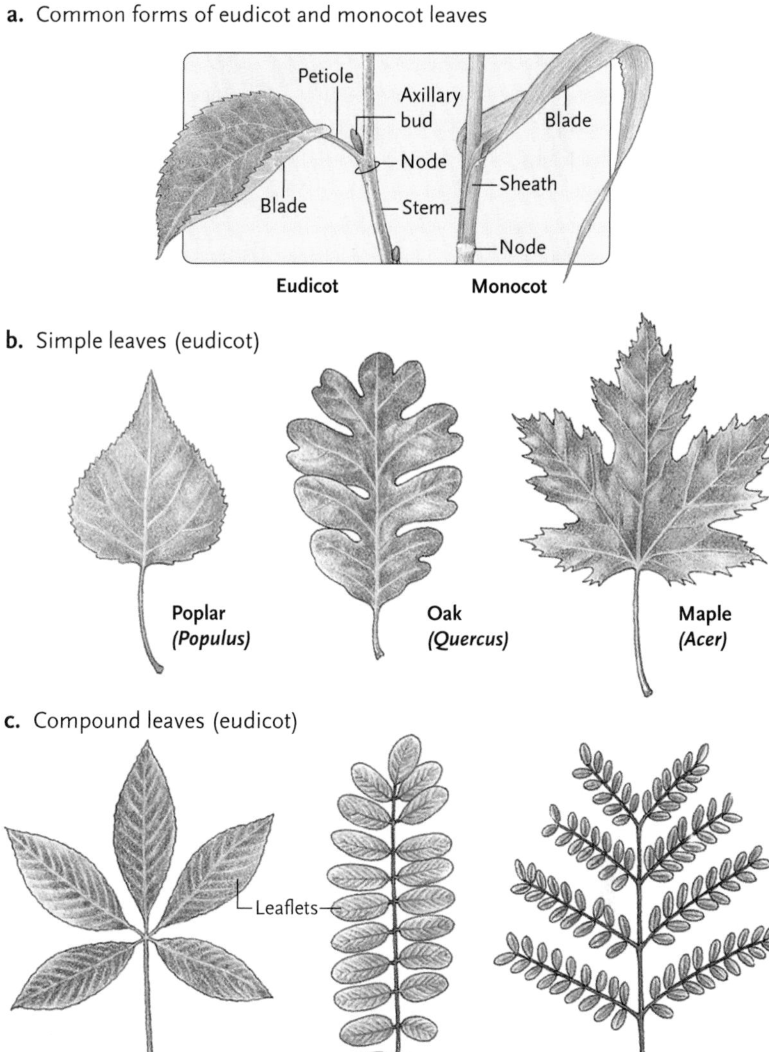

Figure 31.15
Leaf forms. **(a)** Common forms of eudicot and monocot leaves. **(b)** Examples of simple eudicot leaves. **(c)** Examples of compound eudicot leaves.

Leaf tissues typically form several layers **(Figure 31.17).** Uppermost is epidermis, with cuticle covering its outer surface. Just beneath the epidermis is **mesophyll** (*mesos* = middle; *phyllon* = leaf), ground tissue composed of loosely packed parenchyma cells that contain chloroplasts. The leaves of many plants, especially eudicots, contain two layers of mesophyll. *Palisade mesophyll* cells contain more chloroplasts and are arranged in compact columns with smaller air spaces between them, typically toward the upper leaf surface. *Spongy mesophyll,* which tends to be located toward the underside of a leaf, consists of irregularly arranged cells with a conspicuous network of air spaces that gives it a spongy appearance. Air spaces between mesophyll cells enhance the uptake of carbon dioxide and release of oxygen during photosynthesis and account for 15% to 50% of a leaf's volume. Mesophyll also contains collenchyma and sclerenchyma cells, which support the photosynthetic cells.

Below the mesophyll is another cuticle-covered epidermal layer. Except in grasses and a few other

a. Cactus spines

b. Tendrils

c. Salt bladders, a form of trichome

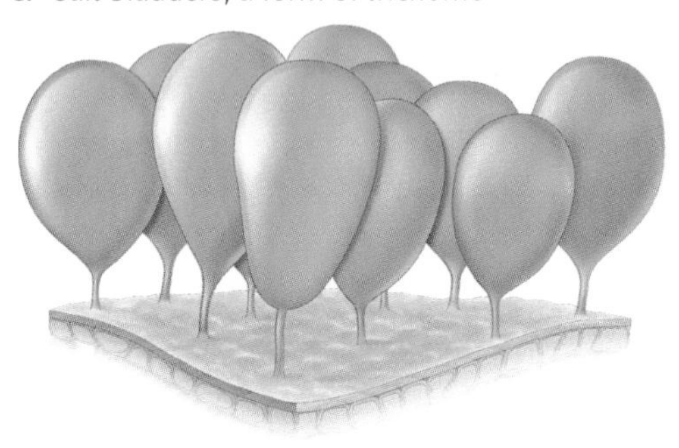

Figure 31.16
A few adaptations of leaves. **(a)** Spines on a barrel cactus *(Ferrocactus covillei)* thwart browsing herbivores and limit the surface area from which water is lost in the plant's arid environment. **(b)** The tendrils of a sweet pea *(Lathyrus odoratus)* help to support the climbing plant's stem. **(c)** SEM of salt bladders on the leaf of a saltbush plant *(Atriplex spongiosa)*. The "bladders" are trichomes, specialized outgrowths of the leaf epidermis in which excess salt from the plant's tissue fluid accumulates. The salt-laden trichomes eventually burst or slough off.

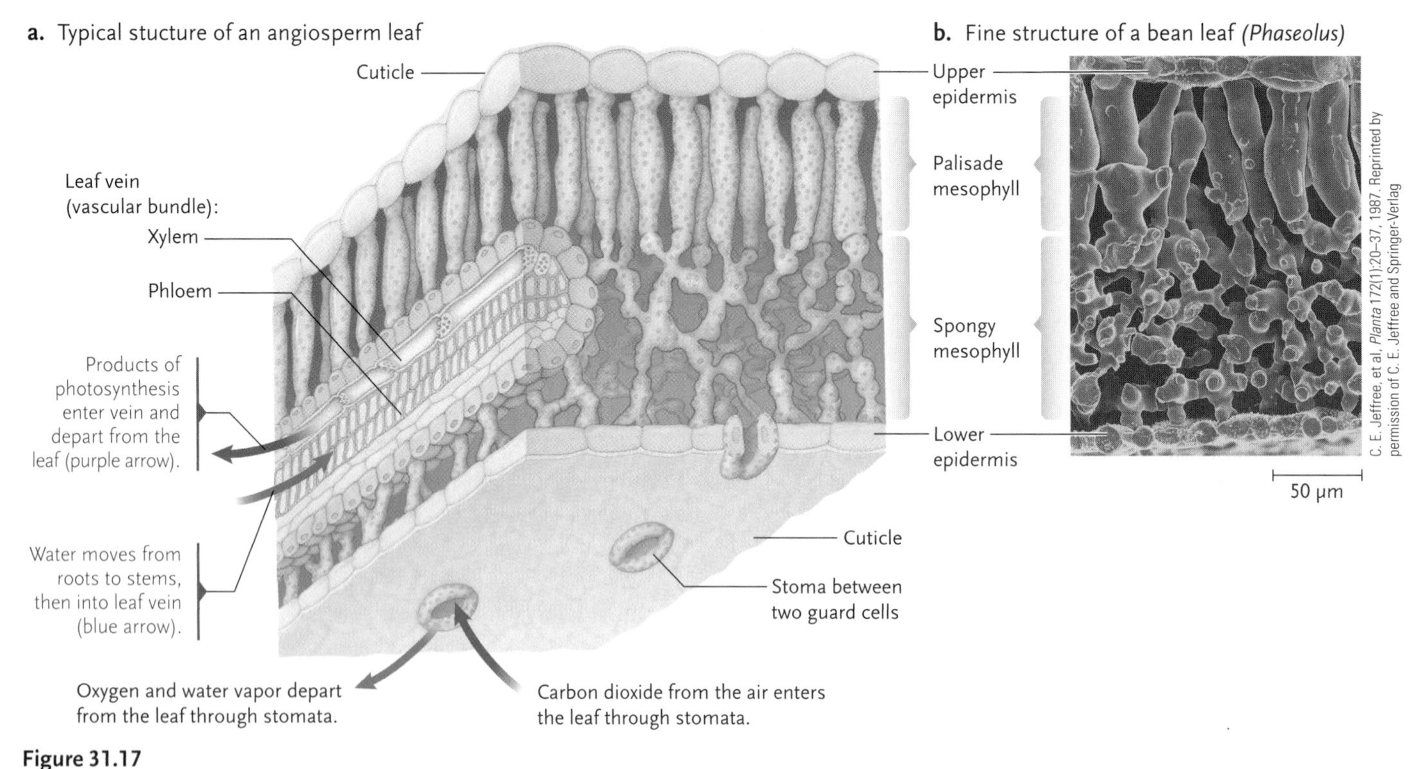

Figure 31.17
Internal structure of a leaf. **(a)** Diagram of typical leaf structure for many kinds of flowering plants. **(b)** Scanning electron micrograph of tissue from the leaf of a kidney bean plant *(Phaseolus)*, transverse section. Notice the compact organization of epidermal cells. See Figure 31.10b for a scanning electron micrograph of stomata.

plants, this layer contains most of the stomata through which water vapor exits the leaf and gas exchange occurs. For example, the upper surface of an apple leaf has no stomata, while a square centimeter of the lower surface has more than 20,000. A square centimeter of the upper epidermis of a tomato leaf has about 1200 stomata, whereas the same area of the lower epidermis has 13,000. The positioning of stomata on the side of the leaf that faces away from the sun may be an adaptation limiting water loss by evaporation through stomatal openings.

Vascular bundles form a lacy network of **veins** throughout the leaf. Eudicot leaves typically have a branching vein pattern; in monocot leaves, veins tend to run in parallel arrays.

In temperate regions, most leaves are temporary structures. In deciduous (*deciduus* = falling off, shedding) species such as birches and maples, hormonal signals cause the leaves to drop from the stem as days shorten in autumn. Other temperate species, such as camellias or hollies, as well as conifers, also drop leaves, but they appear "evergreen" because the leaves may persist for several years and do not all drop at the same time.

Plant Shoots May Have Juvenile and Adult Forms

Leaf shape and other shoot characteristics can mirror the progress of a long-lived plant through its life cycle. Plants that live many years may spend part of their lives in a juvenile phase, then shift to a mature or adult phase. The differences between juveniles and adults often are reflected in leaf size and shape, in the arrangement of leaves on the stem, or in a change from vegetative growth to a reproductive stage—or sometimes all three. For example, oak saplings (genus *Quercus*) have fewer leaves than mature oaks do, but the leaves are considerably larger—an adaptation that probably provides saplings with increased leaf surface area for taking in carbon (in carbon dioxide). Young English ivy plants (*Hedera helix*) grow as vines, have leaves with multiple lobes arranged on the stem in an alternating pattern, and do not flower **(Figure 31.18a)**. By contrast, mature English ivy is a flowering shrub with oval leaves that arise on the stem in a spiral pattern **(Figure 31.18b)**. A magnolia tree (*Magnolia grandiflora*) doesn't flower until its juvenile phase ends, which can be 20 years or more from the time the *Magnolia* seed sprouts. Most woody plants must attain a certain size before their meristem tissue can respond to the hormonal signals that govern flower development, a topic we consider in Chapter 35.

Phase changes provide more examples of the plasticity that characterizes plant development. They almost certainly are associated with changes in the expression of genes that control the development of stem nodes, leaf and flower buds, and other basic aspects of plant growth.

Study Break

1. Describe the functions of stems and stem structure, and list the basic steps in primary growth of stems.
2. Explain the general function of leaves and how leaf anatomy supports this role in eudicots and monocots.
3. Describe the steps in primary growth of a leaf and the structures that result from the process.
4. Describe two examples of the life phases of long-lived plant species.

31.4 Root Systems

Plants must absorb enough water and dissolved minerals to sustain growth and routine cellular maintenance, a task that requires a tremendous root surface. In one study, measurements of the root system of a rye plant (*Secale cereale*) that have been growing for only 4 months may have a surface area of more than 700 m^2—about 130 times greater than the surface area of its shoot system. The roots of carrots, sugar beets, and most other plants also store nutrients produced in photosynthesis, some to be used by root cells and some to be transported later to cells of the shoot. As a root system penetrates downward and spreads out, it also anchors the aboveground parts.

a. Young English ivy

Thomas L. Rost

b. Mature English ivy

Thomas L. Rost

Figure 31.18 Age-related phase changes in English ivy *(Hedera helix)*. **(a)** The juvenile, vine-type growth habit. **(b)** The mature shrub.

Taproot and Fibrous Root Systems Are Specialized for Particular Functions

Most eudicots have a **taproot system**—a single main root, or taproot, that is adapted for storage and smaller branching roots called **lateral roots (Figure 31.19a)**. As the main root grows downward, its diameter increases, and the lateral roots emerge along the length of its older, differentiated regions. The youngest lateral roots are near the root tip. Carrots and dandelions have a taproot system, as do pines and many other conifers. A pine's taproot system can penetrate 6 m or more into the soil.

Grasses and many other monocots develop a **fibrous root system** in which several main roots branch to form a dense mass of smaller roots **(Figure 31.19b)**. Fibrous root systems are adapted to absorb water and nutrients from the upper layers of soil, and tend to spread out laterally from the base of the stem. Fibrous roots are important ecologically because dense root networks help hold topsoil in place and prevent erosion. During the 1930s, drought, overgrazing by livestock, and intensive farming in the North American Midwest destroyed hundreds of thousands of acres of native prairie grasses, contributing to soil erosion on a massive scale. Swirling clouds of soil particles prompted journalists to name the area the Dust Bowl.

In some plants, **adventitious roots** arise from the stem of the young plant. "Adventitious" refers to any structure arising at an unusual location, such as roots that grow from stems or leaves. Adventitious roots and their branchings all are about the same length and diameter. Those of English ivy and some other climbing plants produce a gluelike substance (from trichomes) that allows them to cling to vertical surfaces. The *prop roots* of a corn plant are adventitious roots that develop from the shoot node nearest the soil surface; they both support the plant and absorb water and nutrients. Mangroves and other trees that grow in marshy habitats often have huge prop roots, which develop from branches as well as from the main stem **(Figure 31.19c)**.

a. Taproot system

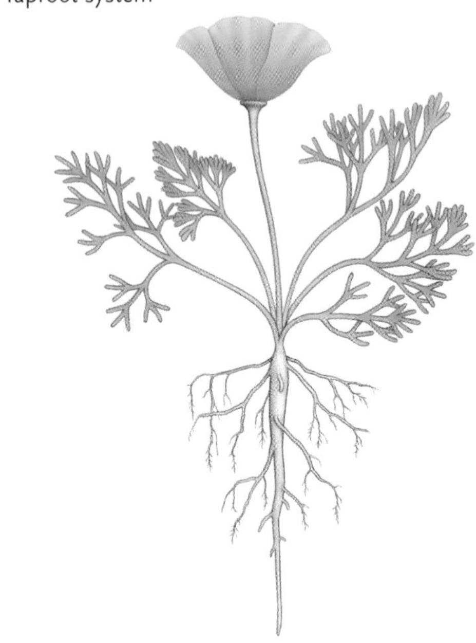

b. Fibrous root system

c. Adventitious roots

Figure 31.19 Types of roots. **(a)** Taproot system of a California poppy *(Eschscholzia californica)*. **(b)** Fibrous root system of a grass plant. **(c)** The prop roots of red mangrove trees *(Rhizophora)*, examples of adventitious roots.

Root Structure Is Specialized for Underground Growth

Like shoots, roots have distinct anatomical parts, each with a specific function. In most plants, primary growth of roots begins when an embryonic root (called a *radicle*) emerges from a germinating seed and its meristems become active. **Figure 31.20** shows the structure of a root tip. Notice that the root apical meristem terminates in a dome-shaped cell mass, the **root cap**. The meristem produces the cap, which in turn surrounds and protects the meristem as the root elongates through the soil. Certain cells in the cap respond to gravity, which guides the root tip downward. Cap cells also secrete a polysaccharide-rich substance that lubricates the tip and eases the growing root's passage through the soil. Outer root cap cells are continually abraded off and replaced by new cells at the cap's base.

Zones of Primary Growth in Roots. Primary growth takes place in successive stages, beginning at the root tip and progressing upward. Just inside the root cap some roots have a small clump of apical meristem cells called the **quiescent center.** Unlike other meristematic cells, cells of the quiescent center divide very slowly unless the root cap or the apical meristem is injured; then they become active and can regenerate the damaged part. The quiescent center also may include cells that synthesize plant hormones controlling root development.

The root apical meristem and the actively dividing cells behind it form the **zone of cell division.** As in the stem, cells of the apical meristem segregate into three primary meristems. Cells in the center of the root tip become the procambium; those just outside the procambium become ground meristem; and those on the periphery of the apical meristem become protoderm.

The zone of cell division merges into the **zone of elongation.** Most of the increase in a root's length comes about here as cells become longer as their vacuoles fill with water. This "hydraulic" elongation pushes the root cap and apical meristem through the soil as much as several centimeters a day.

Above the zone of elongation, cells do not increase in length but they may differentiate further and take on specialized roles in the **zone of maturation.** For instance, epidermal cells in this zone give rise to root hairs, and the procambium, ground meristem, and protoderm complete their differentiation in this region.

a.
Fully grown root hair
Endodermis
Pericycle
Cortex
Epidermis
Stele
Xylem
Phloem
Zone of maturation
The tissue systems complete their differentiation and begin to take on their specialized roles. Root hairs begin to form.
Zone of elongation
Most cells stop dividing but increase in length. The primary meristems begin to differentiate into tissue systems; the phloem matures and the xylem starts to form.
Zone of cell division
Rapidly dividing cells of the root apical meristem segregate into three primary meristems.
Quiescent center
Root cap
b.
100 μm
John Limbaugh/Ripon Microslides, Inc.

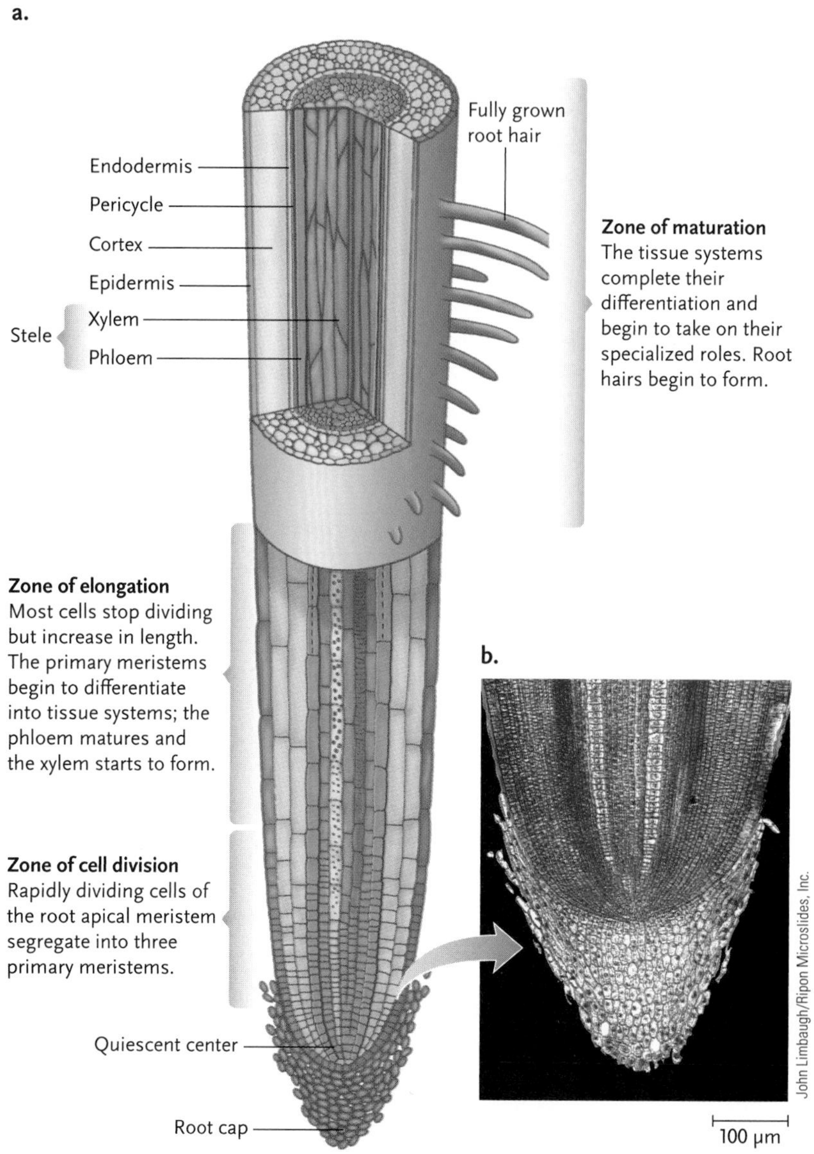

Figure 31.20
Tissues and zones of primary growth in a root tip.
(a) Generalized root tip, longitudinal section. **(b)** Micrograph of a corn root tip, longitudinal section.

Tissues of the Root System. Coupled with primary growth of the shoot, primary root growth produces a unified system of vascular pipelines extending from root tip to shoot tip. The root procambium produces cells that mature into the root's xylem and phloem **(Figure 31.21).** Ground meristem gives rise to the root's cortex, its ground tissue of starch-storing parenchyma cells that surround the stele. In eudicots, the stele runs through the center of the root (see Figure 31.21a). In corn and some other monocots, the stele forms a ring that divides the ground tissue into cortex and pith (see Figure 31.21b).

The cortex contains air spaces that allow oxygen to reach all of the living root cells. Numerous plasmodesmata connect the cytoplasm of adjacent cells of the cortex. In many flowering plants, the outer root cortex cells give rise to an **exodermis,** a thin band of cells that, among other functions, may limit water losses from roots and help regulate the absorption of ions. The innermost layer of the root cortex is the **endodermis,** a thin, selectively permeable barrier that helps control the movement of water and dissolved minerals into the stele. We look in more detail at the roles of exodermis and endodermis in Chapter 32.

Between the stele and the endodermis is the **pericycle,** consisting of one or more layers of parenchyma cells that can still function as meristem. The pericycle gives rise to lateral roots **(Figure 31.22).** In response to chemical growth regulators, **root primordia** (rudimentary roots) arise at specific sites in the pericycle. Gradually, the lateral roots emerge and grow out through the cortex and epidermis, aided by enzymes released by the root primordium that help break down the intervening cells. The distribution and frequency of lateral root formation partly control the overall shape of the root system—and the extent of the soil area it can penetrate.

In some cells in the developing root epidermis the outer surface becomes extended into root hairs (see Figure 31.20). Root hairs can be more than a centimeter long and can form in less than a day. Collectively, the thousands or millions of them on a plant's roots greatly increase the plant's absorptive surface. Root hair structure supports this essential function. Each hair is a slender tube with thin walls made sticky on their surface by a coating of pectin. Soil particles tend to adhere to the wall, providing an intimate association between the hair and the surrounding earth, thus facilitating the uptake of water molecules and

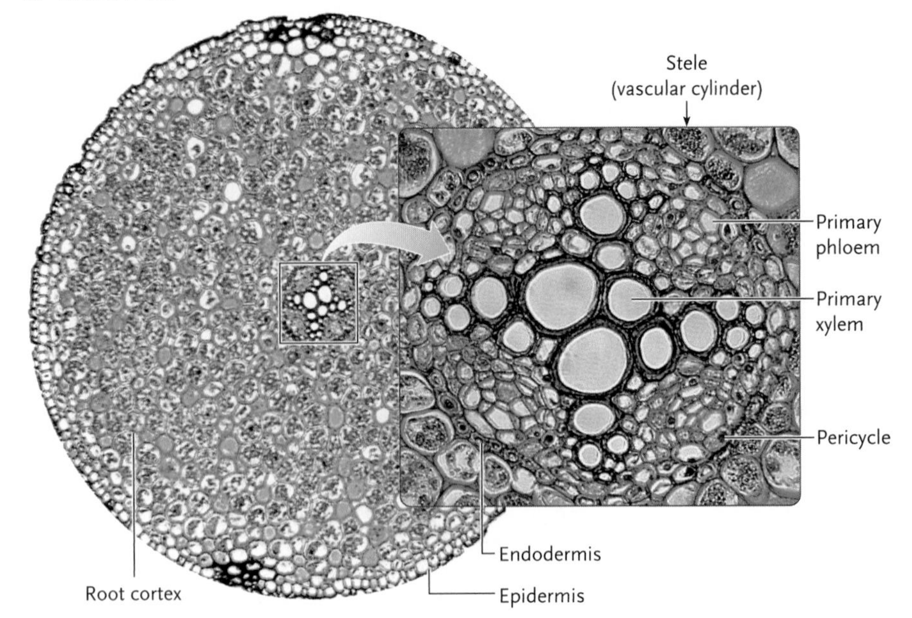

b. Monocot root

Pith
Root cortex
Epidermis
Stele
Primary
xylem
Primary
phloem

Figure 31.21
Stele structure in eudicot and monocot roots compared. **(a)** A young root of the buttercup *Ranunculus*, a eudicot. The close-up shows details of the stele. **(b)** Root of a corn plant *(Zea mays)*, a monocot. Notice how the stele divides the ground tissue into cortex and pith. Both roots are shown in transverse section. (a: Chuck Brown; b: Carolina Biological Supply.)

mineral ions from soil. When plants are transplanted, rough handling can tear off much of the fragile absorptive surface. Unable to take up enough water and minerals, the transplant may die before new root hairs can form.

Study Break

1. Compare the two general types of root systems.
2. Describe the zones of primary growth in roots.
3. Describe the various tissues that arise in a root system and their functions.

31.5 Secondary Growth

All plants undergo primary growth of the root and stem. In addition, some plants have secondary growth processes that add girth to roots and stems over two or more growing seasons. In plant species that have secondary growth, older stems and roots become more massive and woody through the activity of two types of lateral meristems, or *cambia* (singular, cambium). One of these meristems, the **vascular cambium**, produces secondary xylem and phloem. The other, called the **cork cambium**, produces **cork,** a secondary epidermis that is one element of the multilayered structure known as bark. In cells of these tissues, mitosis is periodically reactivated. Hence secondary growth permits woody plants to grow taller and live longer than herbaceous plants.

Vascular Cambium Gives Rise to Secondary Growth in Stems

Recall that after the stem of a woody plant completes its primary growth, each vascular bundle contains a layer of undifferentiated cells between the primary xylem and the primary phloem. These cells, along with parenchyma cells between the bundles, eventually give rise to a cylinder of vascular cambium that wraps around the xylem and pith of the stem **(Figure 31.23).** Vascular cambium consists of two types of cells—*fusiform initials* and *ray initials*—that have different shapes and functions (see Figure 31.23b). Secondary growth takes place as these cells divide. Initials divide at right angles to the stem surface, so their descendants add girth to the stem instead of length. **Fusiform initials,** which are derived from cambium inside the vascular bundles, give rise to secondary xylem and phloem cells. Secondary xylem forms on the inner face of the vascular cambium, and secondary phloem forms on the outer face. **Ray initials** are derived from the parenchyma cells between vascular bundles. As they divide, their descendants form spokelike *rays.* These

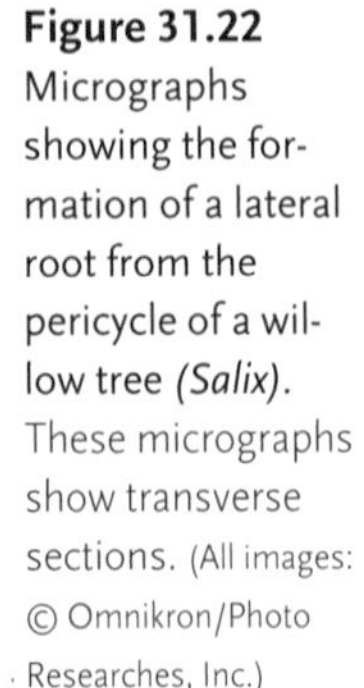

Figure 31.22
Micrographs showing the formation of a lateral root from the pericycle of a willow tree *(Salix)*. These micrographs show transverse sections. (All images: © Omnikron/Photo Researches, Inc.)

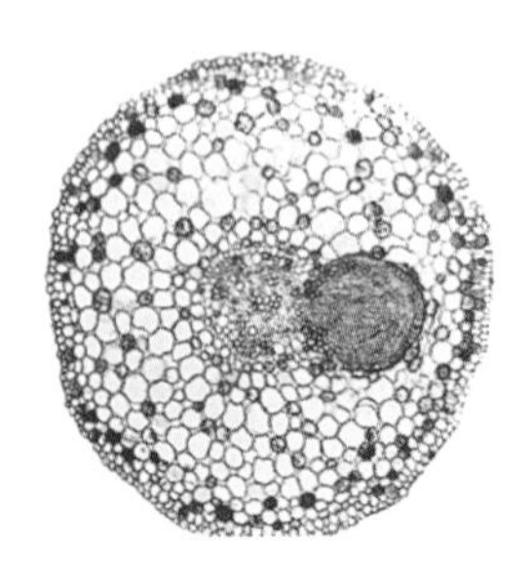

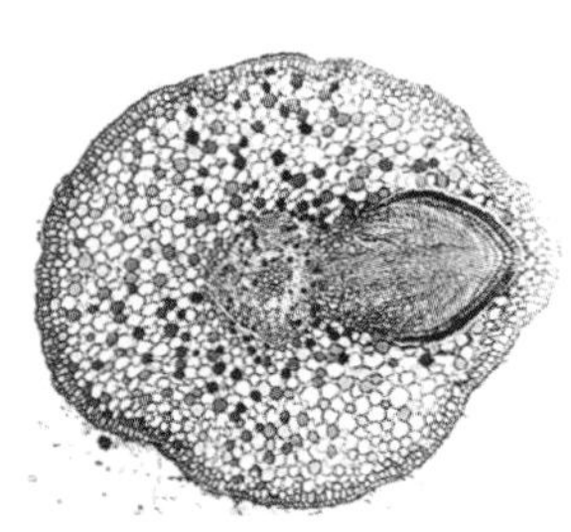

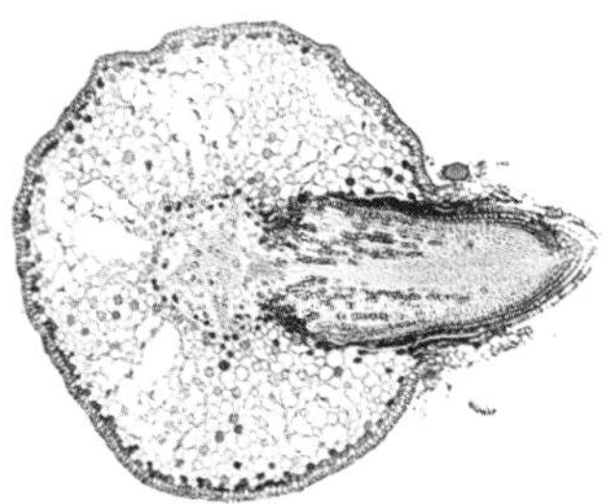

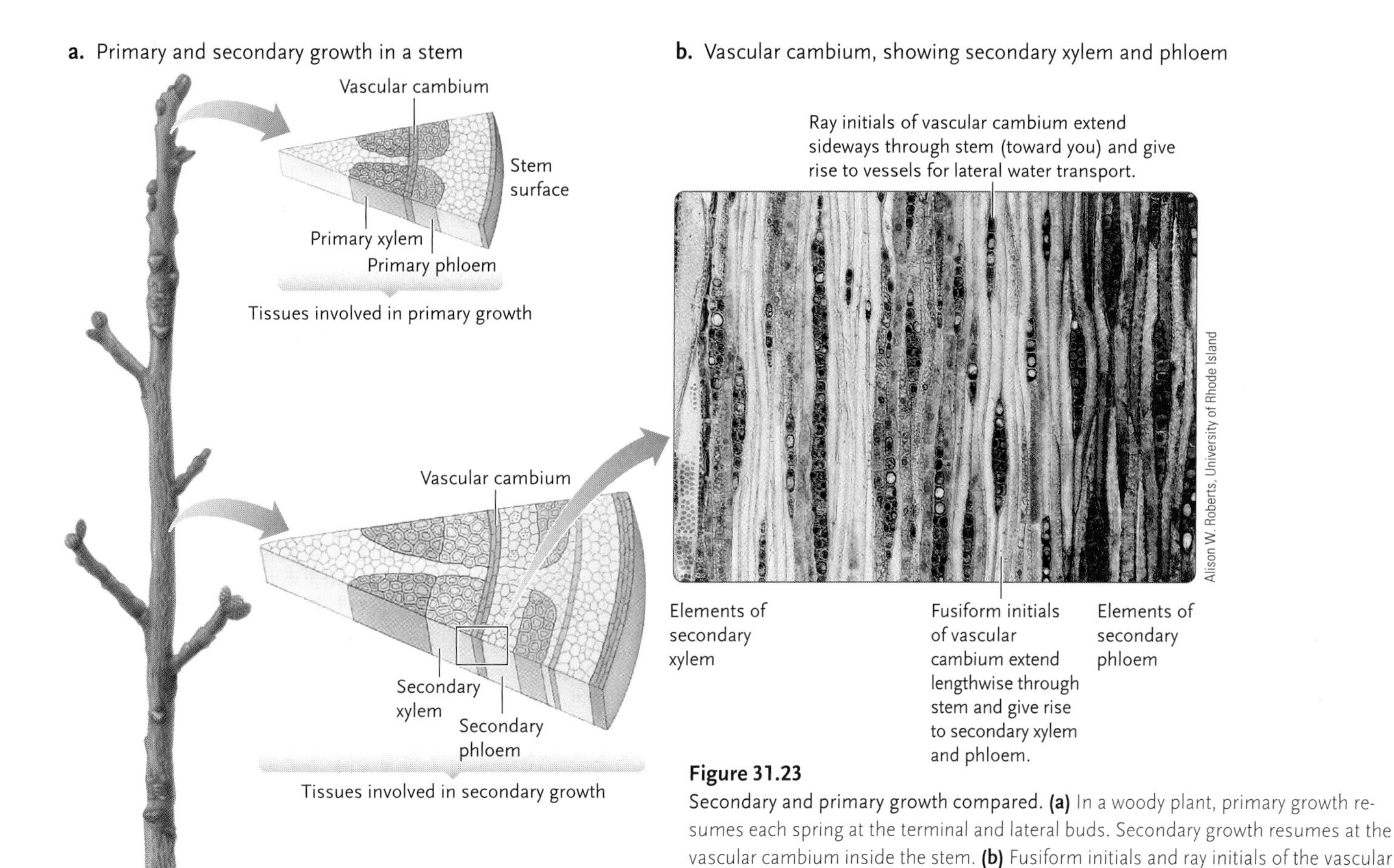
a. Primary and secondary growth in a stem
Vascular cambium
Stem surface
Primary xylem
Primary phloem
Tissues involved in primary growth
Vascular cambium
Secondary xylem
Secondary phloem
Tissues involved in secondary growth
b. Vascular cambium, showing secondary xylem and phloem
Ray initials of vascular cambium extend sideways through stem (toward you) and give rise to vessels for lateral water transport.
Alison W. Roberts, University of Rhode Island
Elements of secondary xylem
Fusiform initials of vascular cambium extend lengthwise through stem and give rise to secondary xylem and phloem.
Elements of secondary phloem

Figure 31.23
Secondary and primary growth compared. **(a)** In a woody plant, primary growth resumes each spring at the terminal and lateral buds. Secondary growth resumes at the vascular cambium inside the stem. **(b)** Fusiform initials and ray initials of the vascular cambium of a walnut tree *(Juglans)*, tangential section.

horizontal channels carry water sideways through the stem, in a radial pattern that resembles a sliced pie. While xylem and phloem mainly conduct fluid lengthwise in the stem, rays ensure that water and solutes also move laterally as stems thicken.

With time, the mass of secondary xylem inside the ring of vascular cambium increases, forming the hard tissue known as **wood.** Outside the vascular cambium, secondary phloem cells also are added each year **(Figure 31.24).** (The primary phloem cells, which have thin walls, are destroyed as they are pushed outward by secondary growth.) As a stem increases in diameter, the growing mass of new tissue eventually causes the cortex, and the secondary phloem beyond

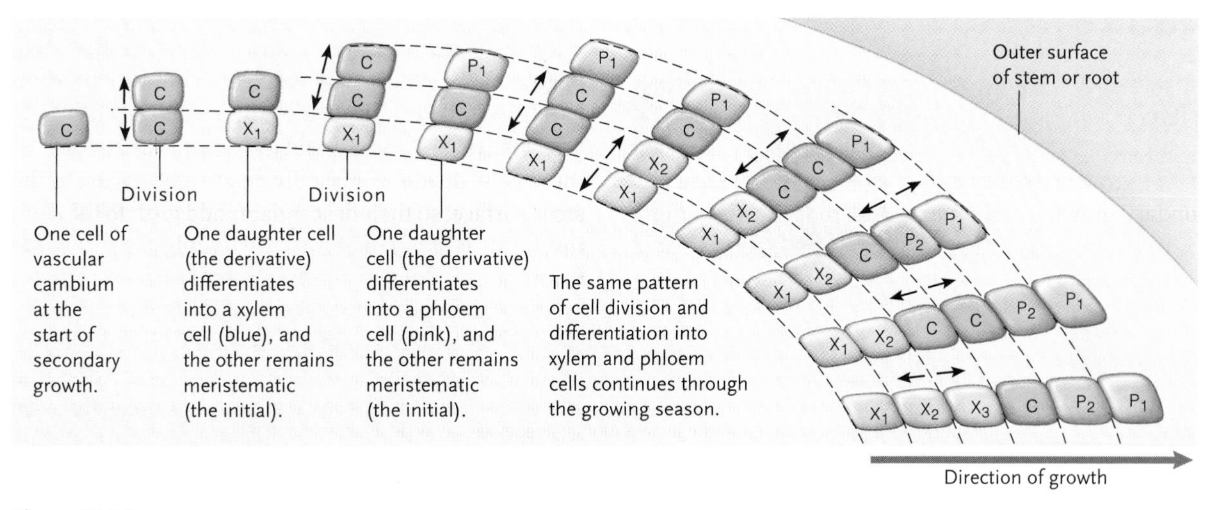
Outer surface of stem or root
Division
Division
One cell of vascular cambium at the start of secondary growth.
One daughter cell (the derivative) differentiates into a xylem cell (blue), and the other remains meristematic (the initial).
One daughter cell (the derivative) differentiates into a phloem cell (pink), and the other remains meristematic (the initial).
The same pattern of cell division and differentiation into xylem and phloem cells continues through the growing season.
Direction of growth

Figure 31.24
Relationship between the vascular cambium and its derivative cells (secondary xylem and phloem). The drawing shows stem growth through successive seasons. Notice how the ongoing divisions displace the cambial cells, moving them steadily outward even as the core of xylem increases the stem or root thickness.

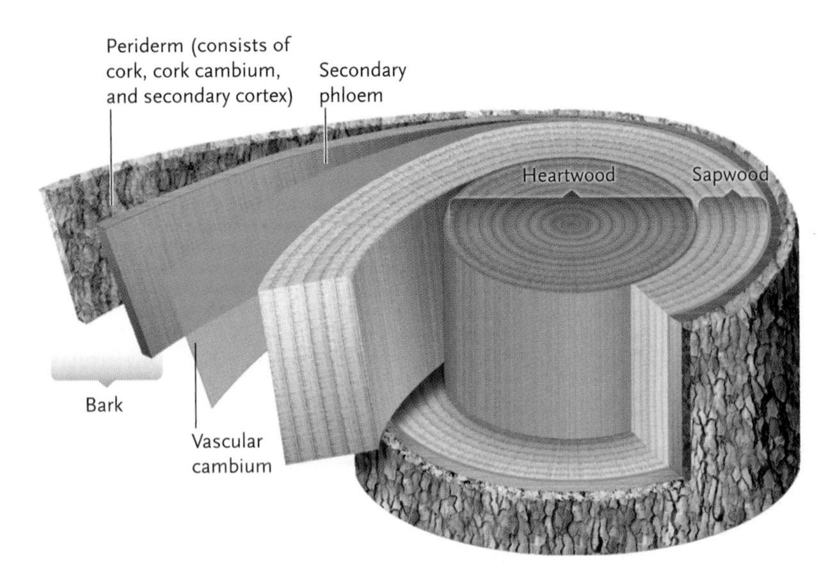

Figure 31.25
Structure of a woody stem showing extensive secondary growth. Heartwood, the mature tree's core, has no living cells. Sapwood, the cylindrical zone of xylem between the heartwood and vascular cambium, contains some living parenchyma cells among the nonliving vessels and tracheids. Everything outside the vascular cambium is bark. Everything inside it is wood.

it, to rupture. Parts of the cortex split away and carry epidermis with them. Cork cambium—produced early in the stem's secondary development by meristem cells in the cortex, epidermis, or secondary phloem—replaces the lost epidermis with cork. The cork cambium produces cork to the outside and secondary cortex to the inside.

Bark encompasses all the living and nonliving tissues between the vascular cambium and the stem surface. It includes the secondary phloem and the **periderm** (*peri* = surrounding; *derma* = skin), the outermost portion of bark that consists of cork, cork cambium, and secondary cortex **(Figure 31.25)**. Girdling a tree by removing a belt of bark around the trunk is lethal because it destroys the secondary phloem layer, and so nutrients from photosynthesis in leaves cannot reach the tree's roots. Natural corks used to seal bottles are manufactured from the especially thick outer bark of the cork oak, *Quercus suber.* Tubular openings called *lenticels* develop in the periderm. They function a bit like snorkels, permitting exchanges of oxygen and carbon dioxide between the living tissues and the outside air.

As a tree ages, changes also unfold in the appearance and function of the wood itself. In the center of its older stems and roots is **heartwood,** dry tissue that no longer transports water and solutes and is a storage depot for some defensive compounds. In time, these substances—including resins, oils, gums, and tannins—clog and fill in the oldest xylem pipelines. Typically they darken heartwood, strengthen it, and make it more aromatic and resistant to decay. **Sapwood** is secondary growth located between heartwood and the vascular cambium. Compared with heartwood, it is wet and not as strong.

In temperate climates, trees produce secondary xylem seasonally, with larger-diameter cells produced in spring and smaller-diameter cells in summer. This "spring wood" and "summer wood" reflect light differently, and it is possible to identify them as alternating light and dark bands. The alternating bands represent annual growth layers, or "tree rings" **(Figure 31.26).**

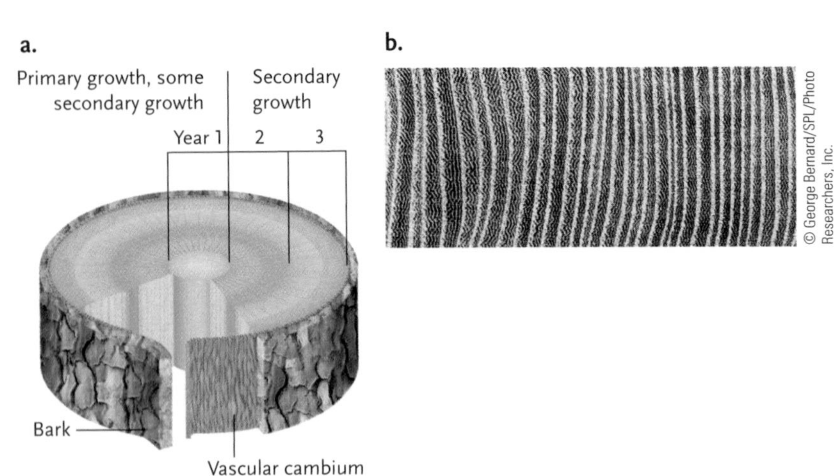

Figure 31.26
Secondary growth and tree ring formation. (a) Radial cut through a woody stem that has three annual rings, corresponding to secondary growth in years 2 through 4. **(b)** Tree rings in an elm *(Ulmas)*. Each ring corresponds to one growing season. Differences in the widths of tree rings correspond to shifts in climate, including the availability of water.

Secondary Growth Can Also Occur in Roots

The roots of grasses, palms, and other monocots are almost always the product of primary growth alone. In plants with roots that have secondary growth, the continuous ring of vascular cambium develops differently than it does in stems. When their primary growth is complete, these roots have a layer of residual procambium between the xylem and phloem of the stele (**Figure 31.27,** step 1). The vascular cambium arises in part from this residual cambium, and in part from the pericycle (step 2). Eventually, the cambial tissues arising from the procambium and those arising from the pericycle merge into a complete cylinder of vascular cambium (step 3). The vascular cambium functions in roots as it does in stems, giving rise to secondary xylem to the inside and secondary phloem to the outside. As secondary xylem accumulates, older roots can become extremely thick and woody. Their ongoing secondary growth is powerful enough to break through concrete sidewalks and even dislodge the foundations of homes.

Cork cambium also forms in roots, where it is produced by the pericycle. In many woody eudicots and in all gymnosperms, most of the root epidermis and cortex fall away, and the surface consists entirely of periderm (step 4).

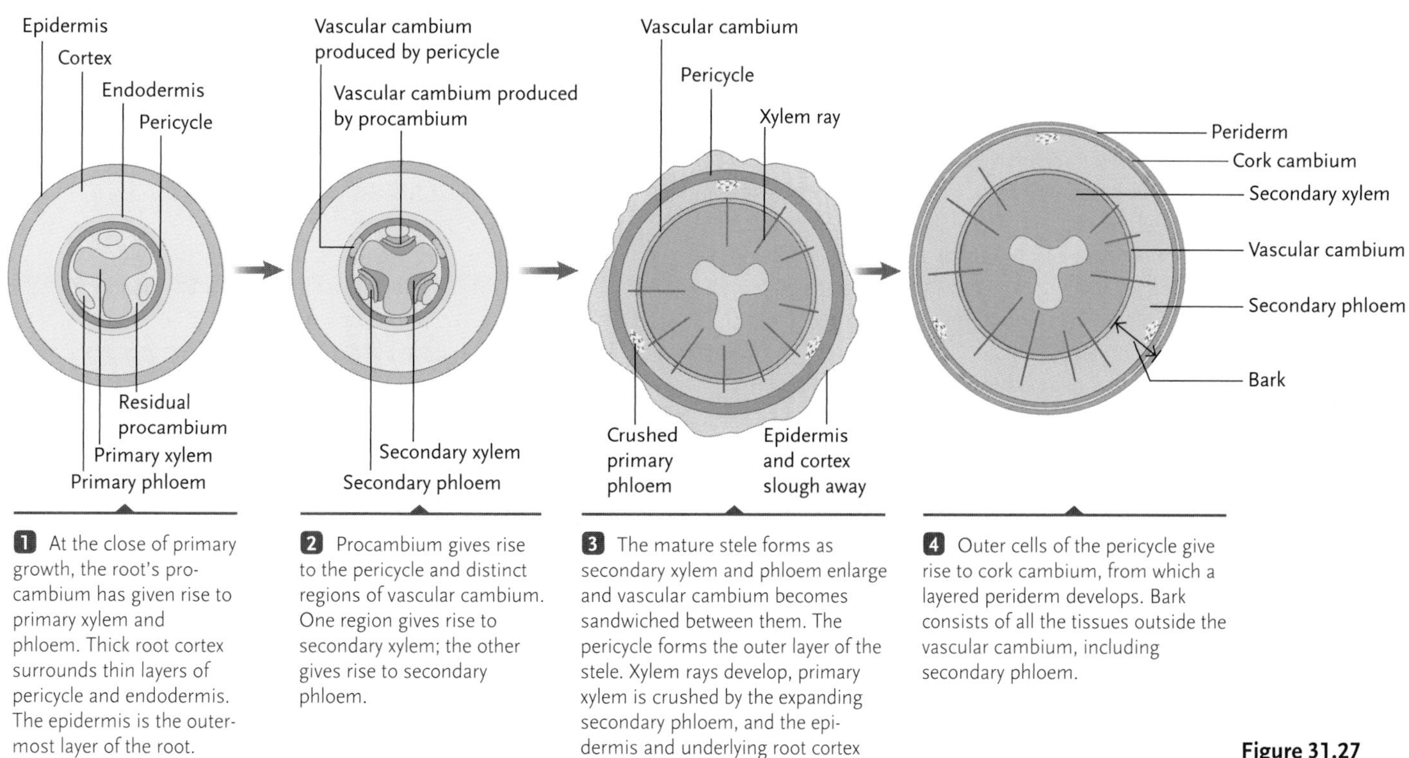

Epidermis
Cortex
Endodermis
Pericycle
Residual procambium
Primary xylem
Primary phloem
Vascular cambium produced by pericycle
Vascular cambium produced by procambium
Secondary xylem
Secondary phloem
Vascular cambium
Pericycle
Xylem ray
Crushed primary phloem
Epidermis and cortex slough away
Periderm
Cork cambium
Secondary xylem
Vascular cambium
Secondary phloem
Bark
1 At the close of primary growth, the root's procambium has given rise to primary xylem and phloem. Thick root cortex surrounds thin layers of pericycle and endodermis. The epidermis is the outermost layer of the root.
2 Procambium gives rise to the pericycle and distinct regions of vascular cambium. One region gives rise to secondary xylem; the other gives rise to secondary phloem.
3 The mature stele forms as secondary xylem and phloem enlarge and vascular cambium becomes sandwiched between them. The pericycle forms the outer layer of the stele. Xylem rays develop, primary xylem is crushed by the expanding secondary phloem, and the epidermis and underlying root cortex begin to slough away.
4 Outer cells of the pericycle give rise to cork cambium, from which a layered periderm develops. Bark consists of all the tissues outside the vascular cambium, including secondary phloem.

Figure 31.27 Secondary growth in the root of one type of woody plant.

Secondary Growth Is an Adaptive Response

Plants, like all living organisms, compete for resources, and woody stems and roots confer some advantages. Plants with taller stems or wider canopies that defy the pull of gravity can intercept more of the light energy from the sun. With a greater energy supply for photosynthesis, they have the metabolic means to increase their root and shoot systems, and thus are better able to acquire resources—and ultimately to reproduce successfully.

In every stage of a plant's growth cycle, growth maintains a balance between the shoot system and root system. Leaves and other photosynthetic parts of the shoot must supply root cells with enough sugars to support their metabolism, and roots must provide shoot structures with water and minerals. As long as a plant is growing, this balance is maintained, even as the complexity of the root and shoot systems increases, whether the plant lives only a few months or—like some bristlecone pines—for 6000 years.

Study Break

1. Explain the nature of secondary growth and where it typically occurs in plants.
2. Describe the components of vascular cambium and their roles in secondary growth in stems, including the development of tissues such as bark, cork, and wood.
3. Compare secondary growth in stems and in roots.

Unanswered Questions

Are plants developmental procrastinators?

It is well established that plants can survive physical insults and exposure to a wide range of environmental fluctuations. What biological resources of plants make them so resilient given their lifestyle constraints? What is the source of phenotypic plasticity that allows a plant's body form to change in response to changes in its habitat? Perhaps the answer lies in the ability of plants to put off making developmental decisions in response to environmental shifts.

Unlike most animal cells, plant cells are pluripotent, retaining their developmental flexibility. Thus, they can behave as stem cells capable of proliferating and producing new structures and even new individuals. Furthermore, many types of plant cells will readily transdifferentiate and assume a new cellular identity even after reaching developmental maturity. In other words, few developmental decisions appear to be final, and many can be tailored to the environmental constraints imposed on the plant. Some plant species appear to have this flexibility even during more global developmental events, such as switching from vegetative growth to reproductive growth or flowering. Why is that the case? Recent findings suggest that by "leaving all options open," plants can quickly adapt to environmental changes and produce progeny, which is the ultimate biological goal for all living organisms.

Research has documented that shifts in environmental context activate genetic changes underlying plants' developmental and phenotypic plasticity. For example, at the whole-organism level some plants, such as *Impatiens balsamina,* can switch between making leaves and making flowers if relative day length changes. In fact, these plants can make leaves that are partial flowers, or flowers that are partial leaves, if light conditions are alternated between short days and long days. Nicholas Battey at the University of Reading in England and his colleagues, who have investigated this phenomenon for many years, have demonstrated that a genetic basis exists for this ability to change body form in response to changing environmental cues. Furthermore, Battey's group suggests that among flowering plants a genetic continuum exists from species that require constant reminders to initiate flowering to species that only require a single signal. For perennial plants such as trees, developmental reprogramming is essential because it allows them to orchestrate seasonally appropriate formation and growth of different organs from the same meristem. Currently, a major effort is under way to understand the genetic basis of developmental evolution as well as how genetic variation may influence phenotypic plasticity. Since plasticity appears to be closely associated with environmental factors, one approach is to study the natural variants of a species from different geographical origins.

Plants respond to environmental variation both spatially and temporally. Perhaps a sort of biological global positioning system (GPS) exists that provides developmentally relevant information in time and space, which the plant translates into a variety of responses. In some species the GPS may be on all the time, while in other species it may only operate at certain times of year, or it may only function once during the plant's life time. What might these genetic GPS devices be? How would we test this idea? Have candidate genes already been identified that might be components within the GPS? Is there a link between a plant's GPS and the genetic basis for its ability to procrastinate developmentally? In short, the answer to all these questions appears to be "maybe," and in all likelihood the full answer will be a complex one. Research conducted by Christopher Cullis at Case Western Reserve University shows that environmentally induced changes in the physical features of flax plants *(Linum usitatissimum)* are accompanied by changes in the entire genome of affected plants, and some of these genetic alterations are heritable. These findings are particularly striking because they demonstrate that in the short term plants can respond to environmental fluctuations not only by altering their developmental output (body form and phenotype) but also by "revising" their genomes. That the very blueprint of life, DNA, is also imbued with significant plasticity is particularly exciting and opens a new realm of inquiry into the mechanisms by which plants may respond to environmental challenges. In some biological contexts being a procrastinator can be advantageous.

Marianne Hopkins is a postdoctoral fellow in the Biology Department at the University of Waterloo in Waterloo, Canada. Her expertise lies in plant genetics and plant molecular biology.

Susan Lolle is an associate professor of biology at the University of Waterloo in Waterloo, Canada. Her research interests include plant development, genetics, and genome biology. To learn more go to http://www.biology.uwaterloo.ca.

Review

Go to **Thomson**NOW™ at www.thomsonedu.com/login to access quizzing, animations, exercises, articles, and personalized homework help.

31.1 Plant Structure and Growth: An Overview

- The vascular plant body consists of an aboveground shoot system with stems, leaves, and flowers, and an underground root system (Figure 31.2).
- Meristems (Figure 31.3) give rise to the plant body and are responsible for a plant's lifelong growth. Each meristem cell divides to produce an initial, which functions as meristem, and a derivative, which may differentiate into a specialized body cell.
- Primary growth of roots and shoots originates at apical meristems at root and shoot tips. Some plants show secondary growth as lateral meristems increase the diameter of stems and roots.
- The two major classes of flowering plants (angiosperms) are monocots and eudicots (Table 31.1).

31.2 The Three Plant Tissue Systems

- Growing plant cells form secondary walls outside the primary walls. Maturing cells become specialized for specific functions, with some functions accomplished by walls of dead cells.

- Plants have three tissue systems (Figure 31.5). Ground tissues make up most of the plant body, vascular tissues serve in transport, and dermal tissue forms a protective cover.
- Of the three types of ground tissues, parenchyma is active in photosynthesis, storage, and other tasks (Figure 31.6); collenchyma and sclerenchyma provide mechanical support.
- Xylem and phloem are the plant vascular tissues. Xylem conducts water and solutes and consists of conducting cells called tracheids and vessel members (Figure 31.8). Phloem, the food-conducting tissue, contains living cells (sieve tube members) joined end to end in sieve tubes (Figure 31.9).
- The dermal tissue, epidermis (Figure 31.10) is coated with a waxy cuticle that restricts water loss. Water vapor and other gases enter and leave the plant through pores called stomata, which are flanked by specialized epidermal cells called guard cells. Epidermal specializations also include trichomes, such as root hairs.

Animation: Tissue systems of a tomato plant

Animation: Apical meristems

Animation: Shoot differentiation

31.3 Primary Shoot Systems

- The primary shoot system consists of the main stem, leaves, and buds, plus any attached flowers and fruits. Stems provide mechanical support, house vascular tissues, and may store food and fluid.
- Stems are organized into modular segments. Nodes are points where leaves and buds are attached, and internodes fall between nodes (Figure 31.11). The terminal bud at a shoot tip consists of shoot apical meristem (Figure 31.12). Lateral buds occur at intervals along the stem. Meristem tissue in buds gives rise to leaves, flowers, or both.
- Derivatives of the apical meristem produce three primary meristems: protoderm makes the stem's epidermis, procambium gives rise to primary xylem and phloem, and ground meristem gives rise to ground tissue.
- Vascular tissues are organized into vascular bundles, with phloem surrounding xylem in each bundle (Figure 31.13).
- Monocot and eudicot leaves have blades of different forms, all providing a large surface area for absorbing sunlight and carbon dioxide (Figure 31.15). Leaf modifications are adaptive responses to environmental selection pressures (Figure 31.16). Leaf characteristics such as shape or arrangement may change over the life cycle of a long-lived plant (Figure 31.18).

Animation: Ground tissues

Animation: Vascular tissues

Animation: Monocot and dicot leaves

Animation: Simple and compound leaves

Animation: Leaf organization

31.4 Root Systems

- Roots absorb water and dissolved minerals and conduct them to aerial plant parts; they anchor and sometimes support the plant and often store food. Root morphologies include taproot systems, fibrous root systems, and adventitious roots (Figure 31.19).
- During primary growth of a root, the primary meristem and actively dividing cells make up the zone of cell division, which merges into the zone of elongation. Past the zone of elongation, cells may differentiate and perform specialized roles in the zone of cell maturation (Figure 31.20).
- A root's vascular tissues (xylem and phloem) usually are arranged as a central stele (Figure 31.21). Parenchyma around the stele forms the root cortex. The root endodermis also wraps around the stele. Inside it is the pericycle, containing parenchyma that can function as meristem. It gives rise to root primordia from which lateral roots emerge (Figure 31.22). Root hairs greatly increase the surface available for absorbing water and solutes.

Animation: Root organization

Animation: Root cross section

Animation: Root systems

31.5 Secondary Growth

- In plants with secondary growth, older stems and roots become more massive and woody via the activity of vascular cambium and cork cambium.
- Vascular cambium consists of two types of cells: fusiform initials, which generate secondary xylem and phloem, and ray initials, which produce horizontal water transport channels called xylem rays (Figures 31.23 and 31.24). Secondary growth takes place as these cells divide.
- Cork cambium gives rise to cork, which replaces epidermis lost when stems increase in diameter. Together, cork cambium and cork make up the periderm (Figure 31.25), the outer portion of bark.
- In root secondary growth, a thin layer of procambium cells between the xylem and phloem differentiates into vascular cambium (Figure 31.27), which gives rise to secondary xylem and phloem. The pericycle produces root cork cambium.

Animation: Secondary growth

Animation: Secondary growth in a root

Animation: Growth in a walnut twig

Animation: Layers in a woody stem

Animation: Annual rings

Questions

Self-Test Questions

1. With respect to growth, plants differ from animals in that:
   a. plant growth involves only an increase in the total number of the organism's cells.
   b. plant cells remain roughly the same size after cell division, whereas animal cells increase in size after they form.
   c. all plants form woody tissues during growth.
   d. plants have indeterminate growth; animals have determinate growth.
   e. plants can grow only when young; animals grow for many years.
2. Identify the correct pairing of a plant tissue and its function.
   a. epidermis: rigid support
   b. xylem: sugar transport
   c. parenchyma: photosynthesis, respiration
   d. phloem: water and mineral transport
   e. periderm: control of gas exchange
3. Identify the correct pairing of a structure and its component(s).
   a. epidermis: companion cells
   b. phloem: sieve tube members
   c. sclerenchyma: nonlignified cell walls

   d. secondary cell wall: cuticle
   e. parenchyma: sclereids
4. Which of the following is *not* part of a stem?
   a. petiole
   b. pith
   c. xylem
   d. procambium
   e. ground meristem
5. Which of the following would be absent in a eudicot leaf?
   a. spongy mesophyll
   b. palisade mesophyll
   c. pericycle
   d. vascular bundles
   e. stoma
6. A student left a carrot in her refrigerator. Three weeks later she noticed slender white fibers growing from its surface. They were not a fungus. Instead they represented:
   a. lateral roots on a taproot.
   b. adventitious roots.
   c. root hairs on a fibrous root.
   d. root hairs on a lateral root.
   e. young prop roots.
7. Which of the following is *not* a structure that results from secondary plant growth?
   a. periderm
   b. sapwood
   c. cork
   d. pith
   e. heartwood
8. Which characteristic do monocots and eudicots share?
   a. the position of the vascular bundles
   b. the pattern of leaf veins
   c. the number of grooves in the pollen grains
   d. the number of cotyledons
   e. the formation of flowers
9. A student forgets to water his plant and the leaves start to droop. The structures first affected by water loss and now not functioning are the:
   a. sieve tubes.
   b. sclereids and fibers.
   c. vessel members and tracheids.
   d. companion cells.
   e. guard cells and stoma.
10. The greatest mitotic activity in a root takes place in the:
   a. zone of maturation.
   b. zone of cell division.
   c. zone of elongation.
   d. root cap.
   e. endodermis.

Questions for Discussion

1. Leaves are modified in diverse ways. Cactus leaves, for example, are transformed into spines. Cacti are adapted to arid habitats in which relatively few other plant species grow. What kinds of selection pressures may have operated to favor the evolution of spinelike cactus leaves?
2. While camping in a national forest you notice a "Do Not Litter" sign nailed onto the trunk of a mature fir tree about 7 feet off the ground. When you return 5 years later, will the sign be at the same height, or will the tree's growth have raised it higher?
3. Peaches, cherries, and other fruits with pits are produced only on secondary branches that are 1 year old. To renew the fruiting wood on a peach tree, how often would you prune it? Where on a branch would you make the cut, and why?
4. African violets and some other flowering plants are propagated commercially using leaf cuttings. Initially, a leaf detached from a parent plant is placed in a growth medium. In time, adventitious shoots and roots develop from the leaf blade, producing a new plant. Which cells in the original leaf tissue are the most likely to give rise to the new structures? What property of the cells makes this propagation method possible?

Experimental Analysis

The sticky cinquefoil (*Potentilla glandulosa*) is a small, deciduous plant with bright yellow flowers that lives in throughout the American West, and its leaf phenotype can vary depending on environmental conditions. Inland, where there are dramatic seasonal temperature swings and unpredictable droughts, plants shed their large "summer leaves" in autumn when the temperature begins to drop. In the spring new leaves are smaller and develop in a compact rosette. This phenotype persists for several months and is thought to be an adaptation that makes the plants less vulnerable to drought (because less water evaporates from reduced leaf surfaces). By contrast, the leaves of *P. glandulosa* plants growing in a coastal climate are always large. In their habitat, seasonal temperature swings are not as great and the annual cycle of winter rain and summer drought is highly predictable. Suppose you decide to explore the hypothesis that the coastal population is genetically capable of exhibiting the same seasonal shift in leaf morphology as the inland plants. Would you need access to a greenhouse where you can control variables, or would it be just as easy to do experiments in the wild? Explain your reasoning and outline your experimental design—including the variable or variables you will test in the first experiment.

Evolution Link

About 90 million years ago flowering plants began their rapid (in a geologic timeframe) rise to dominance in the modern Kingdom Plantae. The first angiosperms may originally have been small, treelike plants in tropical regions, but at some point they began diversifying rapidly into other habitats where early gymnosperms flourished. In the 1990s South African botanist William Bond proposed the "slow seedling" hypothesis to help explain this evolutionary change, and botanists continue to explore and refine it. The hypothesis proposes that angiosperms were able to encroach on and eventually dominate many habitats where ancient gymnosperms lived in part because flowering species increasingly evolved adaptations that made them fast-growing herbaceous plants. Gymnosperms, by contrast grow more slowly. Based on your reading in this chapter and Chapter 27, what are some structural and biochemical features of gymnosperms (such as conifers) that might result in slower growth, putting them at a competitive disadvantage in this scenario?

How Would You Vote?

Large-scale farms and large cities compete for clean, fresh water, which is becoming scarcer as human population growth skyrockets. Should cities restrict urban growth to reduce conflicts over water supplies? Go to www.thomsonedu.com/login to investigate both sides of the issue and then vote.

Cross section of the stem of a geranium *(Pelargonium)* showing parenchyma (pink) wrapping around vessels that transport water and nutrients in plants. In this false-color SEM, large-diameter vessels (xylem) that carry water and minerals appear whitish and bundles of smaller vessels (phloem), which transport sugars, appear pale green.

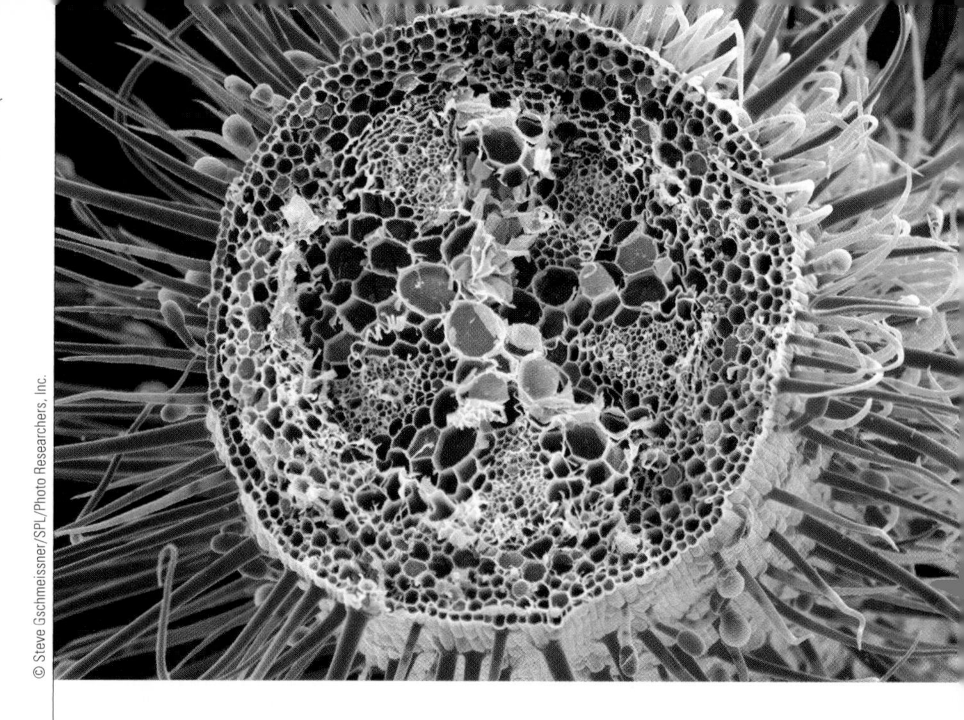

32 Transport in Plants

Study Plan

Why It Matters

The coast redwood, *Sequoia sempervirens* **(Figure 32.1)**, takes life to extremes. Redwood trees can live for more than 2000 years, and they can grow taller than any other organism on Earth. The tallest known specimen, located in Redwood National Park in California, soars 115.5 m, roughly 379 ft, from the dank forest floor. Botanists who have studied these giants estimate that such massive plants consume thousands of liters of water each day to survive. And that water—with its cargo of dissolved nutrients—must be transported the great distances between roots and leaves.

At first, movement of fluids and solutes 100 m or more from a mature redwood's roots to its leafy crown may seem to challenge the laws of physics. Raising water that high above ground in a pipe requires a powerful mechanical pump at the base and substantial energy to counteract the pull of gravity. You also require a pump—your heart—to move fluid over a vertical distance of less than 3 m. Yet a redwood tree has no pump. As you'll learn in this chapter, the evolutionary adaptations that move water and solutes throughout the plant body can move large volumes over great distances by harnessing the cumulative effects of seemingly weak interactions such as cohesion

Owaki-Kulla/Corbis

Figure 32.1
Redwoods *(Sequoia sempervirens)* such as this tree growing in coastal California have reached recorded heights of over 100 m during life spans of more than 2000 years. Such extremely tall trees exemplify the ability of plants to move water and solutes from roots to shoots over amazingly long distances.

and evaporation. Overall, plant transport mechanisms solve a fundamental biological problem—the need to acquire materials from the environment and distribute them throughout the plant body.

Our discussion begins with a brief review of the principles of water and solute movement in plants, a topic introduced in Chapter 6. Then we examine how those principles apply to the movement of water and solutes into and through a plant's vascular pipelines.

32.1 Principles of Water and Solute Movement in Plants

In plants, as in all organisms, the movement of water and solutes begins at the level of individual cells and relies on mechanisms such as osmosis and the operation of transport proteins in the plasma membrane. Once water and nutrients enter a plant's specialized transport systems—the vascular tissues xylem and phloem—other mechanisms carry them between various regions of the plant body in response to changing demands for those substances. Ultimately, these movements of materials result from the integrated activities of the individual cells, tissues, and organs of a single, smoothly functioning organism—the whole plant.

Plant transport mechanisms fall into two general categories—those for short-distance transport and those for long-distance transport. Short-distance transport mechanisms move substances into and between cells across membranes, and also to and from vascular tissues. For example, water, oxygen, and minerals enter roots by crossing the cell membranes of root hairs **(Figure 32.2a),** and nutrients such as carbohydrates from photosynthesis cross plasma membranes to nourish cells of the plant body. Similarly, water and other substances move short distances to and from a plant's xylem and phloem, which are arranged in vascular bundles **(Figure 32.2b).** Long-distance transport mechanisms move substances between roots and shoot parts **(Figure 32.2c).** Thus water and dissolved minerals travel in the xylem from roots to other plant parts, and products of photosynthesis move in the phloem from the leaves and stems into roots and other structures. Carbon dioxide for photosynthesis enters photosynthetic tissues in the shoot.

We consider transport processes in the xylem and phloem later in this chapter. For the moment our focus is on mechanisms that move water and solutes into and out of specific cells in roots, leaves, and stems. Keep in mind that the plant cell wall does not prevent solutes from moving into plant cells. Most solutes can cross the wall by way of the plasmodesmata that connect adjacent cells (see Section 5.4).

Both Passive and Active Mechanisms Move Substances into and out of Plant Cells

Recall from Chapter 6 that in all cells there are two general mechanisms for transporting water and solutes across the plasma membrane into and between cells. In **passive transport,** substances move down a concentration gradient or, if the substance is an ion, down an electrochemical gradient. **Active transport** requires the cell to expend energy in moving substances *against* a gradient, usually by hydrolysis of ATP.

True to its name, simple diffusion is the simplest form of passive transport: as described in Section 6.2, oxygen, carbon dioxide, water, and some other small molecules can readily diffuse across cell plasma membranes, following a concentration gradient. By contrast, in all other types of membrane transport, ions and some larger molecules cross cell membranes assisted by carriers collectively called **transport proteins,** which are embedded in the membrane.

Passive transport of substances down an electrochemical gradient is called *facilitated diffusion* because the transport protein involved "facilitates" the process in some way. Transport proteins called *channel proteins* are configured to form a pore in the plasma membrane. Those called *carrier proteins* change shape in a way that releases the substance to the other side of the membrane.

In active transport, membrane transport proteins use energy to move substances against a concentration gradient or an electrochemical gradient. As you may

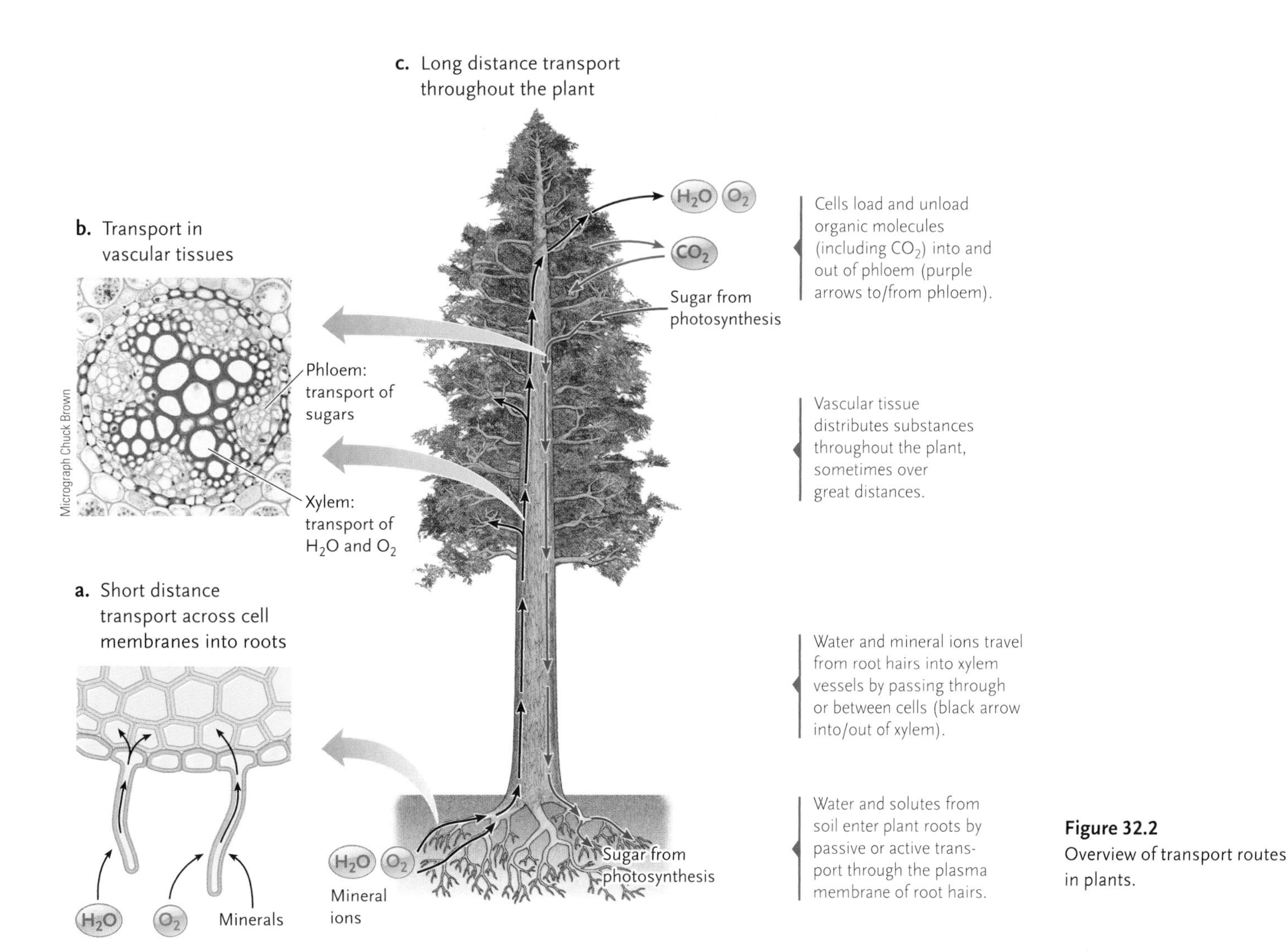

Figure 32.2
Overview of transport routes in plants.

recall from Section 6.4, an electrochemical gradient exists across cell membranes when the concentrations of various ions differ inside or outside the cell. The differences in ion concentration result in a difference in electrical charge across the plasma membrane. In plant cells the cytoplasm is slightly more negative than the fluid outside the cell. This charge difference is measured as an electrical voltage called the **membrane potential.** The word "potential" refers to the fact that the movement of ions across a membrane is a potential source of energy—that is, such ion movements can perform cellular work.

ATP provides the energy for active transport of substances into and out of plant cells. Hydrogen ions (protons), which tend to be more concentrated outside the cell than in the negatively charged cytoplasm, play a central role in the process. First, a proton pump pushes H^+ across the plasma membrane against its electrochemical gradient, from the inside to the outside of the cell **(Figure 32.3a).** As protons accumulate outside the cell, the electrochemical gradient becomes steeper and significant potential energy is available. Crucial solutes such as cations (positively charged ions) often are more concentrated in the extracellular fluid. One result of the increased charge difference created by proton pumping is that cations move into the cell through their membrane channels **(Figure 32.3b).** These cations include mineral ions that have essential roles in plant cell metabolism.

The H^+ gradient also powers *secondary active transport,* a process in which a concentration gradient of an ion is used as the energy source for active transport of another substance. The two secondary mechanisms—*symport* and *antiport*—actively transport ions, sugars, and amino acids into and out of plant cells against their concentration gradient. In **symport,** the potential energy released as H^+ follows its gradient into the cell is coupled to the simultaneous uptake of another ion or molecule **(Figure 32.3c).** In this way, plant cells can take up metabolically important anions such as nitrate (NO_3^-) and potassium (K^+). Nearly all organic substances that enter plant cells move in by symport as well.

In **antiport,** the energy released as H^+ diffuses into the cell powers the active transport of a second molecule, such as Ca^{2+}, in the opposite direction, *out of* the cell **(Figure 32.3d).** One of antiport's key functions is to remove excess Na^+, which readily moves into plant cells by facilitated diffusion through channel proteins. If the Na^+ were not eliminated, it would quickly build up to toxic levels.

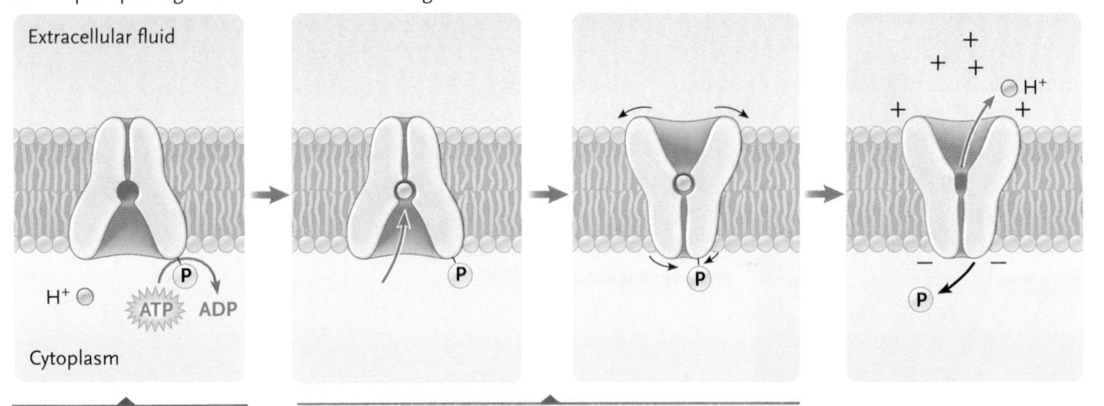

ATP energy pumps hydrogen ions (H^+) out of the cytoplasm, creating an H^+ gradient.

The concentration of H^+ becomes higher outside the membrane than inside. Inward diffusion of H^+ in response to the gradient becomes a source of energy for transporting other ions and neutral molecules such as sugar into the plant cell.

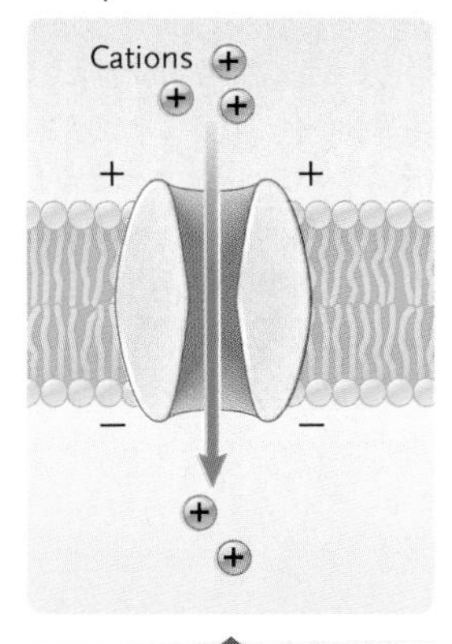

Some cations, such as NH_4^+, enter the cell through selective channel proteins, following the electrochemical gradient created by H^+ pumping.

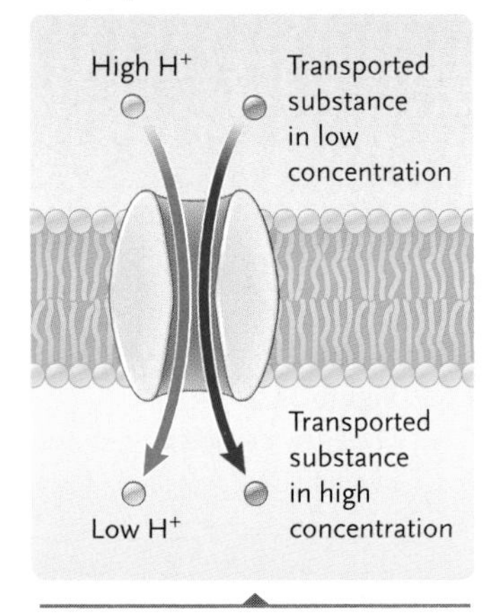

In symport, the inward diffusion of H^+ is coupled with the simultaneous active transport of another substance into the cell.

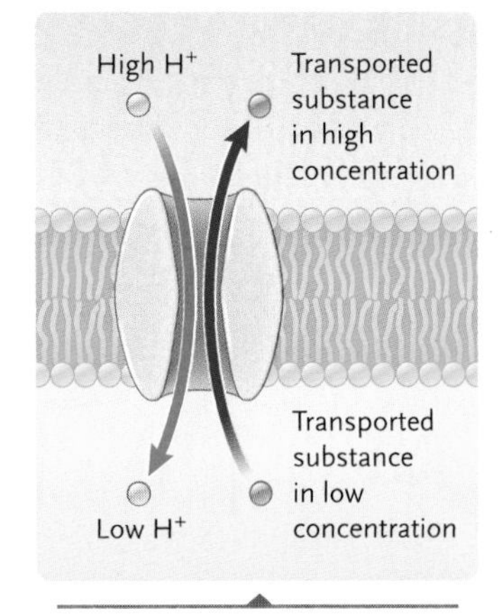

In antiport, H^+ moving into the cell powers the movement of another solute in the opposite direction.

Figure 32.3
Ion transport across the plasma membrane.

Both passive and active transport are selective transport mechanisms that transport specific substances. Two factors govern this specificity. One is the size of the interior channel, which allows only molecules in a particular size range to pass through. The other factor is the distribution of charges along the inside of the channel. A channel that permits cations such as Na^+ to pass through easily may completely bar anions such as Cl^- and vice versa.

Relatively speaking, only small amounts of mineral ions and other solutes move into and out of plant cells. As we see next, H_2O is another matter. Throughout a plant's life large volumes of water enter and exit its cells and tissues by way of osmosis.

Osmosis Governs Water Movement in Plants

One of the most important aspects of plant physiology is how water moves into and through plant cells and tissues. Inside a plant's tubelike vascular tissues, large amounts of water or any other fluid travel by **bulk flow**—the group movement of molecules in response to a difference in pressure between two locations, like water in a closed plumbing system gushing from an open faucet. For example, the dilute solution of water and ions that flows in the xylem, called **xylem sap**, moves by bulk flow from roots to shoot parts. The solution is pulled upward through the plant body in a process that relies on the cohesion of water molecules and that we will consider more fully later in this chapter. Individual cells, however, gain and lose water by **osmosis**, the passive movement of water across a selectively permeable membrane in response to solute concentration gradients, a pressure gradient, or both (see Section 6.3). The driving force for osmosis is energy stored in the water itself. This potential energy, called **water potential**, is symbolized by the Greek letter ψ. By convention, pure water has a ψ value of zero. Two factors that strongly influence this value in living plants are the presence of solutes and physical pressure.

The effect of dissolved solutes on water's tendency to move across a membrane is called *solute potential,* symbolized by ψ_S. In practical terms, water potential is higher where there are more water molecules in a solution relative to the number of solute molecules. Likewise, the water potential is *lower* in a solution with relatively more solutes. The relationship between water potential and solute potential is vital to understanding transport phenomena in plants because water tends to move by osmosis from regions where water potential is higher to regions where it is lower. Solutes are usually more concentrated inside plant cells than in the fluid surrounding them. This means the water potential is higher outside plant cells than inside them, so water tends to enter the cells by osmosis. This in fact is the mechanism that draws soil water into a plant's roots.

The osmotic movement of water into a plant cell cannot continue indefinitely, however, because eventually physical pressure counterbalances it. The wall around a plant cell strictly limits how much the cell can expand. Accordingly, as water moves into plant cells, the pressure inside them increases. This pressure, called **turgor pressure,** rises until it is high enough to prevent more water from entering a cell by osmosis. In effect, when osmotic water movement stops, turgor pressure has increased the water potential inside the cell until it equals the potential of the water outside the cell. The physical pressure required to halt osmotic water movement across a membrane is termed a solution's *pressure potential* and is symbolized as ψ_P.

By convention, plant physiologists measure water potential in units of pressure called **megapascals** (MPa). They use standard atmospheric pressure as a baseline, assigning it a value of zero. Accordingly, the water potential of pure water at standard atmospheric pressure is expressed as 0 MPa. This notation can be used to describe the changing effects under different conditions of solute potential and pressure potential **(Figure 32.4).** Adding pressure increases the MPa while adding solutes reduces it (because the relative concentration of water is lower), and water will flow from a solution of higher MPa to a solution of lower MPa. With these principles in mind, consider now how they operate in living plant cells.

Recall from Section 5.4 that a large **central vacuole** occupies most of the volume of a mature plant cell. The central vacuole, which is surrounded by a vacuolar membrane, or **tonoplast,** contains a dilute solution of sugars, proteins, other organic molecules, and salts. The cell cytoplasm is confined to a thin layer between the tonoplast and the plasma membrane. A major role of the central vacuole is to maintain turgor pressure in

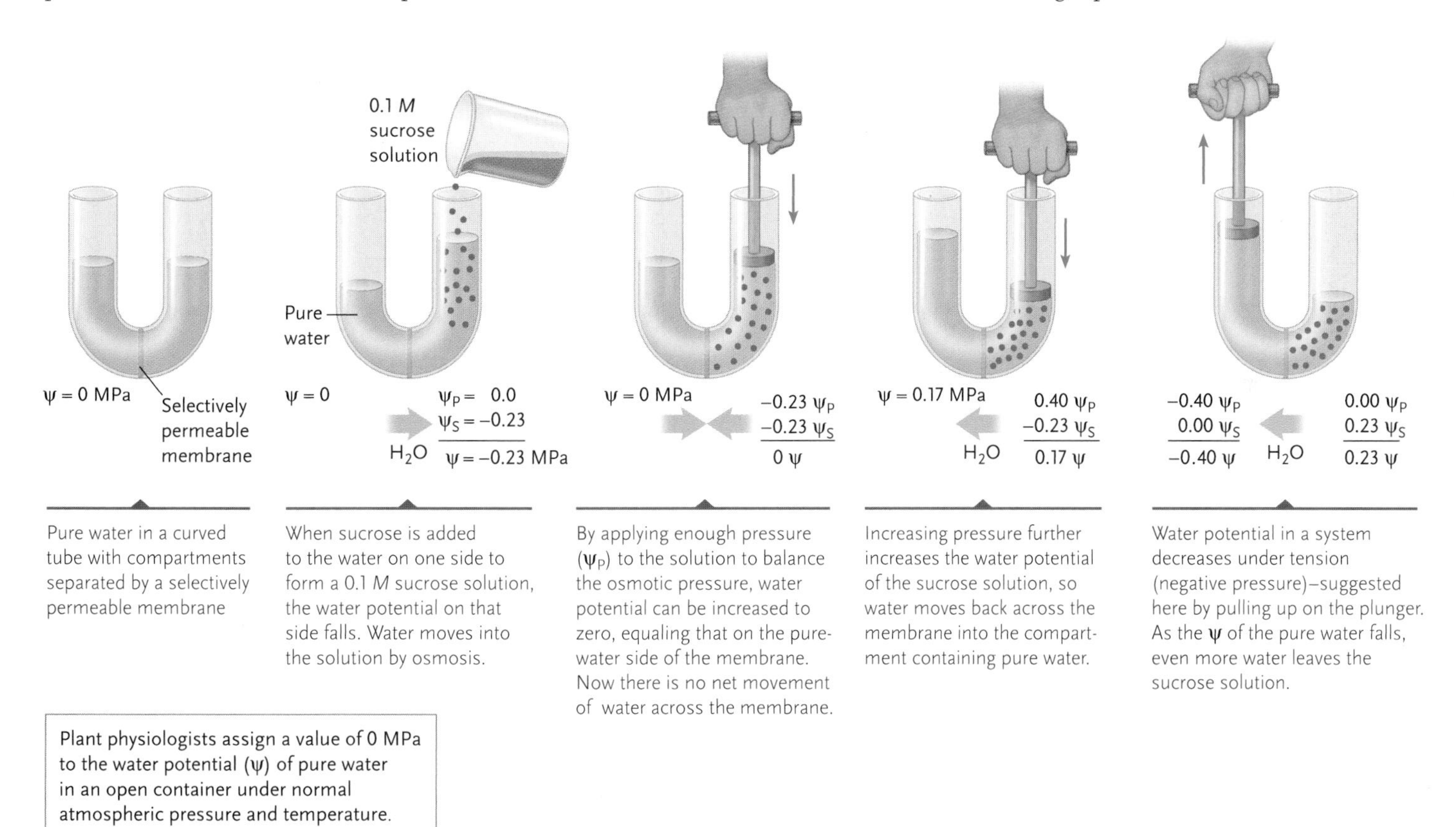

0.1 *M* sucrose solution
Pure water
ψ = 0 MPa
Selectively permeable membrane
ψ = 0
ψP = 0.0
ψS = −0.23
H2O
ψ = −0.23 MPa
ψ = 0 MPa
−0.23 ψP
−0.23 ψS
0 ψ
ψ = 0.17 MPa
0.40 ψP
−0.23 ψS
H2O
0.17 ψ
−0.40 ψP
0.00 ψS
−0.40 ψ
H2O
0.00 ψP
0.23 ψS
0.23 ψ
Pure water in a curved tube with compartments separated by a selectively permeable membrane
When sucrose is added to the water on one side to form a 0.1 *M* sucrose solution, the water potential on that side falls. Water moves into the solution by osmosis.
By applying enough pressure (ψP) to the solution to balance the osmotic pressure, water potential can be increased to zero, equaling that on the pure-water side of the membrane. Now there is no net movement of water across the membrane.
Increasing pressure further increases the water potential of the sucrose solution, so water moves back across the membrane into the compartment containing pure water.
Water potential in a system decreases under tension (negative pressure)—suggested here by pulling up on the plunger. As the ψ of the pure water falls, even more water leaves the sucrose solution.
Plant physiologists assign a value of 0 MPa to the water potential (ψ) of pure water in an open container under normal atmospheric pressure and temperature.

Figure 32.4
The relationship between osmosis and water potential. If the water potential is higher on one side of a selectively permeable membrane, water will cross the membrane to the area of lower water potential. This diagram shows pure water on one side of a selectively permeable membrane and a simple sucrose solution on the other side. In an organism, however, the selectively permeable membranes of cells are rarely if ever in contact with pure water.

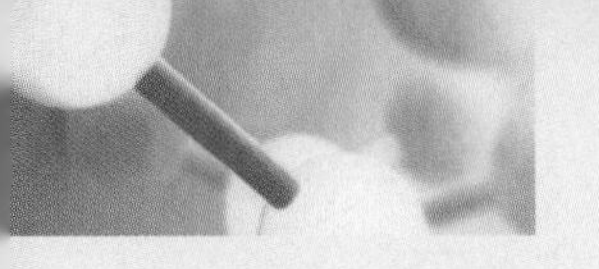

Insights from the Molecular Revolution

A Plant Water Channel Gives Oocytes a Drink

Water moves into or out of the central vacuole of plant cells to compensate for gains or losses of water in the surrounding cytoplasm. Does this water simply diffuse through the lipid part of the tonoplast, or does it move through an aquaporin? Christophe Maurel and his colleagues at the University of California, San Diego, sought to answer this question. They were encouraged by the discovery of aquaporins in the plasma membranes of red blood cells and by the fact that a closely related protein called TIP (tonoplastintrinsic protein) occurs in the tonoplast.

To find out whether TIP functions as the water channel of tonoplasts, the team began by isolating the gene that encodes TIP in *Arabidopsis thaliana* plants. For the later experiments they selected animal cells (oocytes of the frog *Xenopus laevis*) to ensure no other proteins made in plant cells could affect the outcome.

Next they cloned the coding sequence of the TIP gene, inserting it into a bacterial plasmid cloning vector. The vector contained a promoter that allowed in vitro transcription of a cloned coding sequence. In addition, the research team had added to it DNA sequences for 5′ and 3′ UTRs (untranslated regions; see Section 15.3) that function in the processing of the coding sequence in the mRNA transcripts of a *Xenopus* gene. The TIP coding sequence was inserted between the DNA for the UTRs, which ensured that the *Xenopus* oocytes could efficiently translate mRNAs transcribed from the TIP sequence clone. That is, the test-tube transcription of the engineered TIP clone produced mRNAs in a form that could readily be translated into TIP proteins inside *Xenopus* cells.

The test-tube TIP mRNA molecules were then injected into mature *Xenopus* oocytes, which are normally only slightly permeable to water. After 2 to 3 days in an isotonic medium, the oocytes were transferred to a hypotonic medium. Thy swelled and ruptured within 6 minutes. Control oocytes that were not injected with the TIP mRNA, or that were injected only with distilled water, swelled only slightly during the same interval and did not burst when placed in a hypotonic medium. These results supported the conclusion that the TIP protein forms an aquaporin when inserted into a membrane. In this system, the TIP proteins inserted into the oocyte plasma membrane, since animal cells do not have tonoplasts. In its normal location in the tonoplast of plant cells, TIP evidently allows water to move readily in either direction, compensating for water movement between the thin layer of cytoplasm and the extracellular space.

the cell. Many solutes that enter a plant cell are actively transported from the cytoplasm into the central vacuole through channels in the tonoplast. As the solutes accumulate, water follows by osmosis.

The plant cell's relatively small amount of cytoplasm must compensate fairly quickly for water gains or losses caused by changes in osmotic flow. If the medium around a plant cell becomes hypertonic (has a high solute concentration), for example, water flows rapidly out of the cell. Water from the central vacuole replaces it, entering the cytoplasm through water-conducting channel proteins called **aquaporins.** Experiments that identified this channel are the topic of this chapter's *Insights from the Molecular Revolution*.

The water mechanics we have been discussing have major implications for land plants. For instance, the drooping of leaves and stems called **wilting** occurs when environmental conditions cause a plant to lose more water than it gains. Conditions that lead to wilting include dry soil, in which case the water potential in the soil falls below that in the plant. Then the turgor pressure inside the cells falls, and the protoplast shrinks away from the cell wall **(Figure 32.5a)**. By contrast, as long as the ψ of soil is higher than that in root epidermal cells, water will follow the ψ gradient and enter root cells, making them turgid, or firm **(Figure 32.5b)**. As we see in the next section, water and solutes entering roots may move through the plant body by several routes.

Study Break

1. Explain the role(s) of a gradient of protons in moving substances across a plant cell's plasma membrane.
2. How do symport and antiport differ? Give examples of key substances each mechanism transports.
3. What is "water potential," and why is it important with respect to plant cells?

32.2 Transport in Roots

Soil around roots provides a plant's water and minerals, but roots don't simply "soak up" these essential substances. Instead, water and minerals that enter roots first travel laterally through the root cortex to the root xylem. Only then do they begin their journey upward to stems, leaves, and other tissues.

Water Travels to the Root Xylem by Three Pathways

Soil water always enters a root through the root epidermis. Once inside a root, however, water may take one of three routes into the root xylem, traveling either through living cells or in nonliving areas of the root **(Figure 32.6)**.

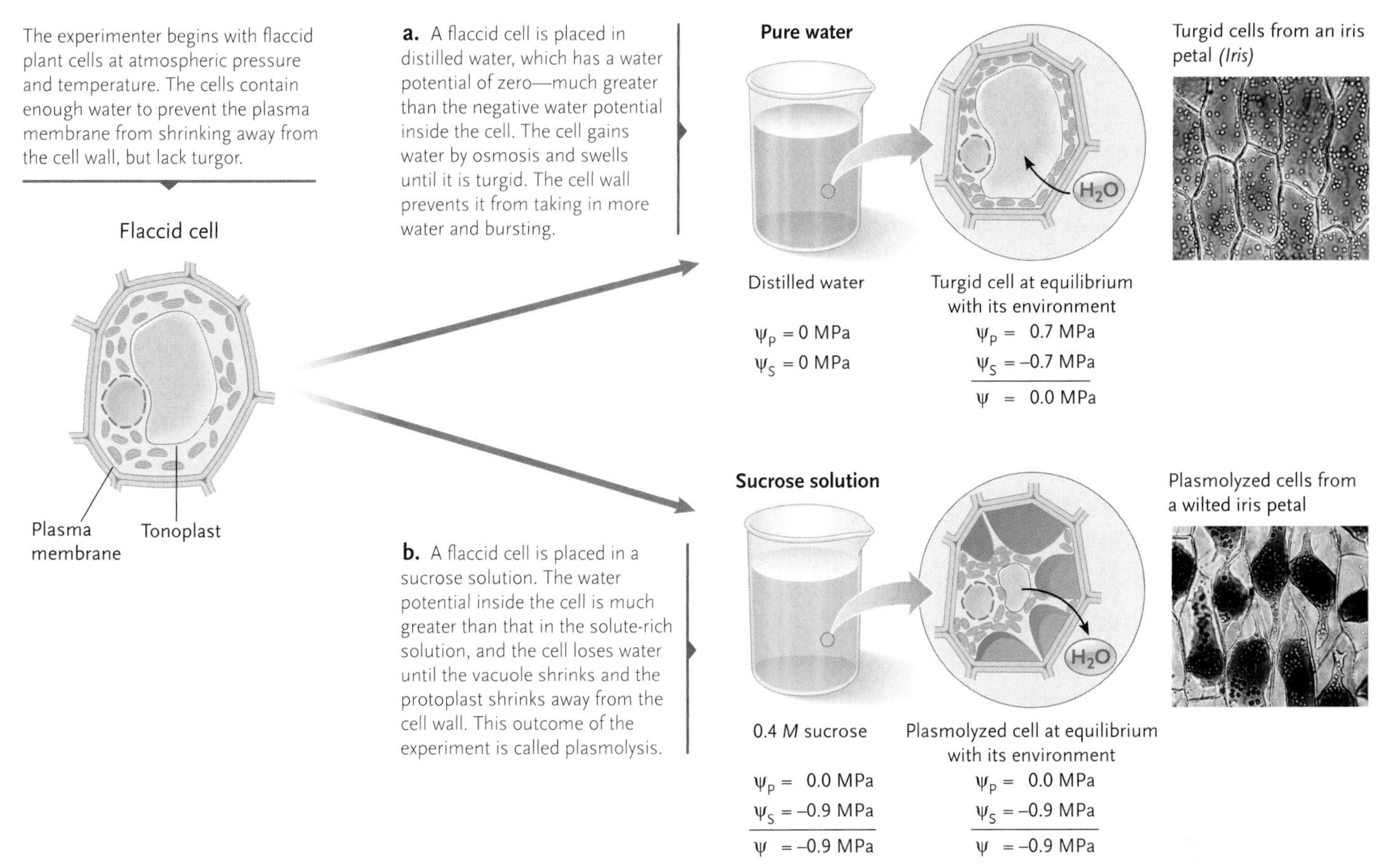

Figure 32.5

An experiment to test the effects of different osmotic environments on plant cells. Notice that in both **(a)** and **(b)** the final condition is the same: the water potential of the plant cell and its environment become equal. (Micrographs: © Claude Nuridsany and Marie Perennou/Science Photo Library/Photo Researchers, Inc.)

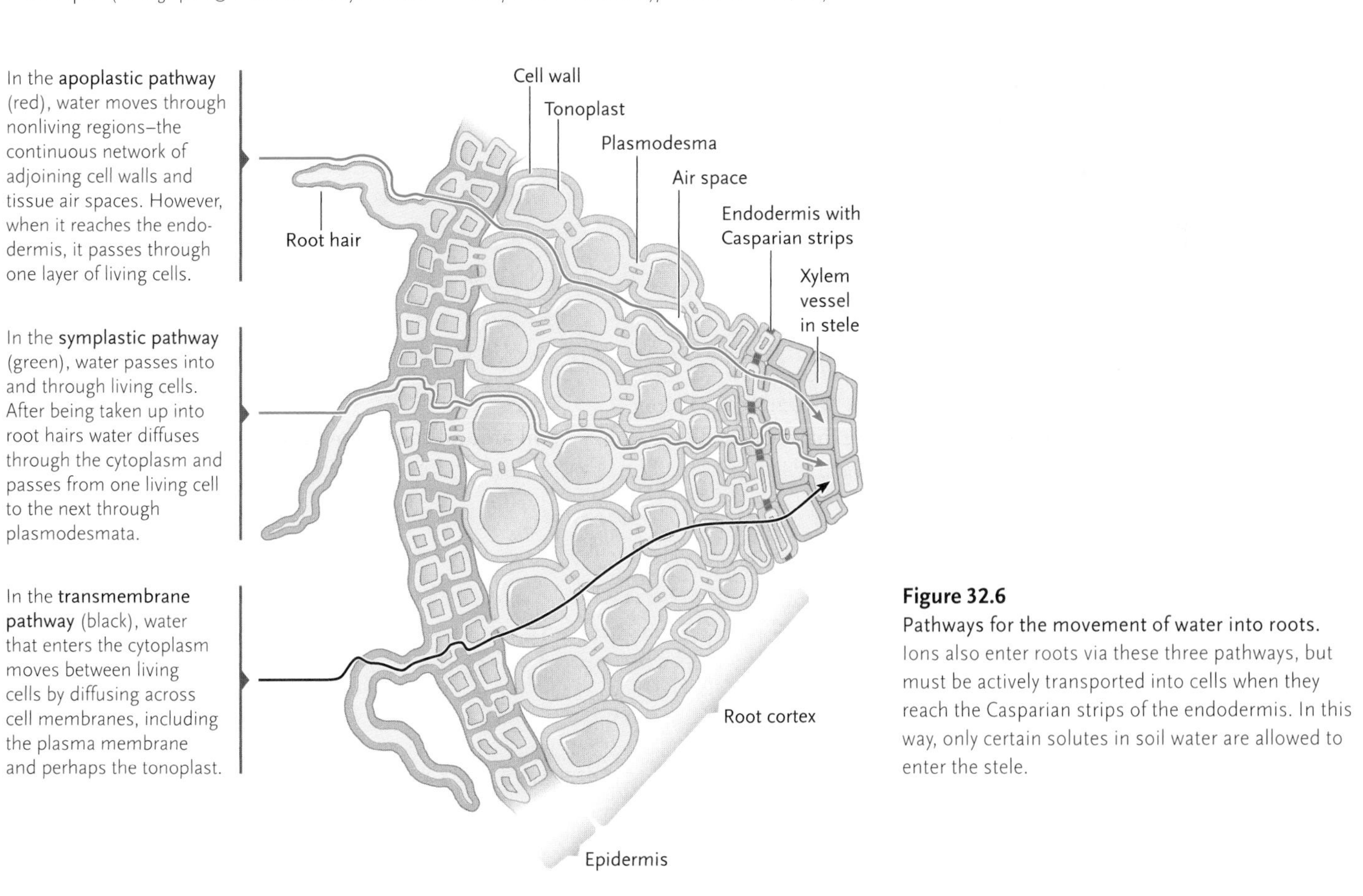

In the apoplastic pathway (red), water moves through nonliving regions–the continuous network of adjoining cell walls and tissue air spaces. However, when it reaches the endodermis, it passes through one layer of living cells.
In the symplastic pathway (green), water passes into and through living cells. After being taken up into root hairs water diffuses through the cytoplasm and passes from one living cell to the next through plasmodesmata.
In the transmembrane pathway (black), water that enters the cytoplasm moves between living cells by diffusing across cell membranes, including the plasma membrane and perhaps the tonoplast.
Cell wall
Tonoplast
Plasmodesma
Air space
Endodermis with Casparian strips
Xylem vessel in stele
Root hair
Root cortex
Epidermis

Figure 32.6

Pathways for the movement of water into roots. Ions also enter roots via these three pathways, but must be actively transported into cells when they reach the Casparian strips of the endodermis. In this way, only certain solutes in soil water are allowed to enter the stele.

Nonliving regions of a plant such as the continuous network of adjoining cell walls and air spaces in root tissue are called the *apoplast*. Thus water follows an **apoplastic pathway** when it moves through the apoplast of roots, a route that does not cross cell membranes. Botanists refer to a plant's living parts as the *symplast*, and water moving through roots in the **symplastic pathway** moves from cell to cell through the open channels of plasmodesmata. Water also can enter root cells across the cell plasma membranes, a **transmembrane pathway.** Water crosses the tonoplast of the central vacuole in this way as well.

When water enters a root, some diffuses into epidermal cells, entering the symplast. But a great deal of the water taken up by plant roots moves into the apoplast, moving along through cell walls and intercellular spaces. This apoplastic water (and any solutes dissolved in it) travels rapidly inward until it encounters the endodermis, the sheetlike single layer of cells that separates the root cortex from the stele. Cells in the root cortex generally have air spaces between them (which helps aerate the tissue), but endodermal cells are tightly packed **(Figure 32.7a).** Each one also has a beltlike **Casparian strip** in its radial and transverse walls, positioned somewhat like a ribbon of packing tape around a rectangular package **(Figure 32.7b–c).** The strip is impregnated with suberin, a waxy substance impermeable to water. Thus the Casparian strip blocks the apoplastic pathway at the endodermis, preventing water and solutes in the apoplast from automatically passing on into the stele. Instead, if molecules are to move into the stele, they must detour across the plasma membranes of endodermal cells, entering the cells (and the symplast) where the wall is not blanketed by a Casparian strip **(Figure 32.7d).** From there water and solutes can pass through plasmodesmata to cells in the outer layer of the stele (the pericycle), then on into the xylem.

Although water molecules can easily cross an endodermal cell's plasma membrane, the semipermeable membrane allows only a subset of the solutes in soil water to cross. Undesirable solutes may be barred, while desirable ones may move into the cell by facilitated diffusion or active transport. Conversely, the endodermis prevents needed substances in the xylem from leaking out, back into the root cortex. In this way the endodermis provides important control over which substances enter and leave a plant's vascular tissue. The roots of most flowering plants also have a second layer of cells with Casparian strips just inside the root

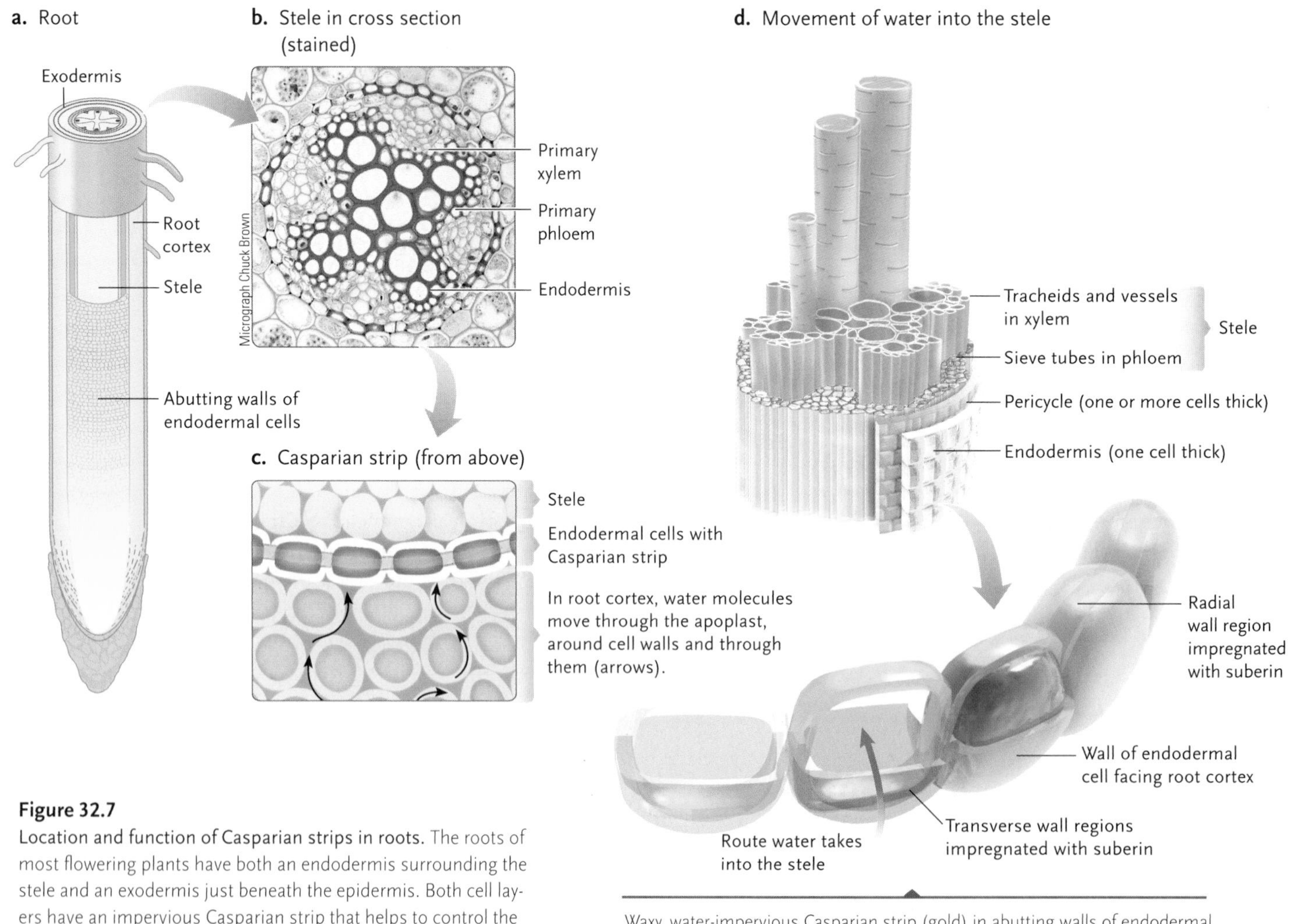

Figure 32.7
Location and function of Casparian strips in roots. The roots of most flowering plants have both an endodermis surrounding the stele and an exodermis just beneath the epidermis. Both cell layers have an impervious Casparian strip that helps to control the uptake of water and dissolved nutrients.

epidermis. This layer, the exodermis, functions like the endodermis.

Roots Take Up Ions by Active Transport

Mineral ions in soil water also enter roots through the epidermis. Some enter the apoplast along with water, but most ions important for plant nutrition tend to be much more concentrated in roots than in the surrounding soil, so they cannot follow a concentration gradient into root epidermal cells. Instead the epidermal cells actively transport ions inward—that is, ions enter the symplast immediately. They travel to the xylem via the symplastic or transmembrane pathways. Other ions can still move inward following the apoplastic pathway until they reach the Casparian strip of the endodermis. If they are to contribute to the plant's nutrition, however, they must be actively transported from the exodermis into cells of the root cortex and, as just described, from the endodermis into the stele. In short, mechanisms that control which solutes will be absorbed by root cells ultimately determine which solutes will be distributed through the plant.

Once an ion reaches the stele, it diffuses from cell to cell until it is "loaded" into the xylem. Experiments to determine whether the loading is passive (by diffusion) or active have been inconclusive, so the details of this final step are not entirely clear. Because the xylem's conducting elements are not living, water and ions in effect reenter the apoplastic pathway when they reach either tracheids or vessels. Once in the xylem, water can move laterally to and from tissues or travel upward in the conducting elements. Minerals are distributed to living cells and taken up by active transport. The following section examines how this "distribution of the wealth" takes place.

Study Break

1. Explain two key differences in how the apoplastic and symplastic pathways route substances laterally in roots.
2. How does an ion enter a root hair and then move to the xylem?

32.3 Transport of Water and Minerals in the Xylem

We return now to the question that opened this chapter: How does the solution of water and minerals called xylem sap move—100 m or more in the tallest trees—from roots to stems, then into leaves? Xylem sap is mostly water, and we know that it moves upward by bulk flow through the tracheids and vessels in xylem. Yet because mature xylem cells are dead, they cannot expend energy to move water into and through the plant shoot. Instead, the driving force for the upward movement of xylem sap from root to shoot is sunlight, which causes water to evaporate from leaves and other aerial parts of land plants. Experiments show that only a small fraction of the water in xylem sap is used in a plant's growth and metabolism. The rest evaporates into the air in a phenomenon called **transpiration.** As described next, transpiration drives the ascent of sap.

The Mechanical Properties of Water Have Key Roles in Its Transport

Chapter 2 introduced several biologically important mechanical properties of water. Two of them interest us here. First, water molecules are strongly *cohesive:* they tend to form hydrogen bonds with one another. Second, water molecules are *adhesive:* they form hydrogen bonds with molecules of other substances, including the carbohydrates in plant cell walls. Water's cohesive and adhesive forces jointly pull water molecules into exceedingly small spaces, such as crevices in cell walls or narrow tubes such as xylem vessels in roots, stems, and leaves. In 1914, plant physiologist Henry Dixon explained the ascent of sap in terms of the relationship between transpiration and water's mechanical properties. His model of xylem transport is now called the **cohesion–tension mechanism of water transport (Figure 32.8).**

According to the cohesion–tension model, water transport begins as water evaporates from the walls of mesophyll cells inside leaves and into the intercellular spaces. This water vapor escapes by transpiration through open stomata, the minute passageways in the leaf surface. As water molecules exit the leaf, they are replaced by others from the mesophyll cell cytoplasm. The water loss gradually reduces the water potential in a transpiring cell below the water potential in the leaf xylem. Now, water from the xylem in the leaf veins follows the gradient into cells, replacing the water lost in transpiration.

In the xylem, water molecules are confined in narrow, tubular xylem cells. The water molecules form a long chain, like a string of weak magnets, held together by hydrogen bonds between individual molecules. When a water molecule moves out of a leaf vein into the mesophyll, its hydrogen bonds with the next molecule in line stretch but don't break. The stretching creates *tension*—a negative pressure gradient—in the column. Adhesion of the water column to xylem vessel walls adds to the tension. Under continuous tension from above, the entire column of water molecules in xylem is drawn upward, in a fashion somewhat analogous to the way water moves up through a drinking straw. Botanists refer to this root-to-shoot flow as the *transpiration stream.*

Transpiration continues regardless of whether evaporating water is replenished by water rapidly taken

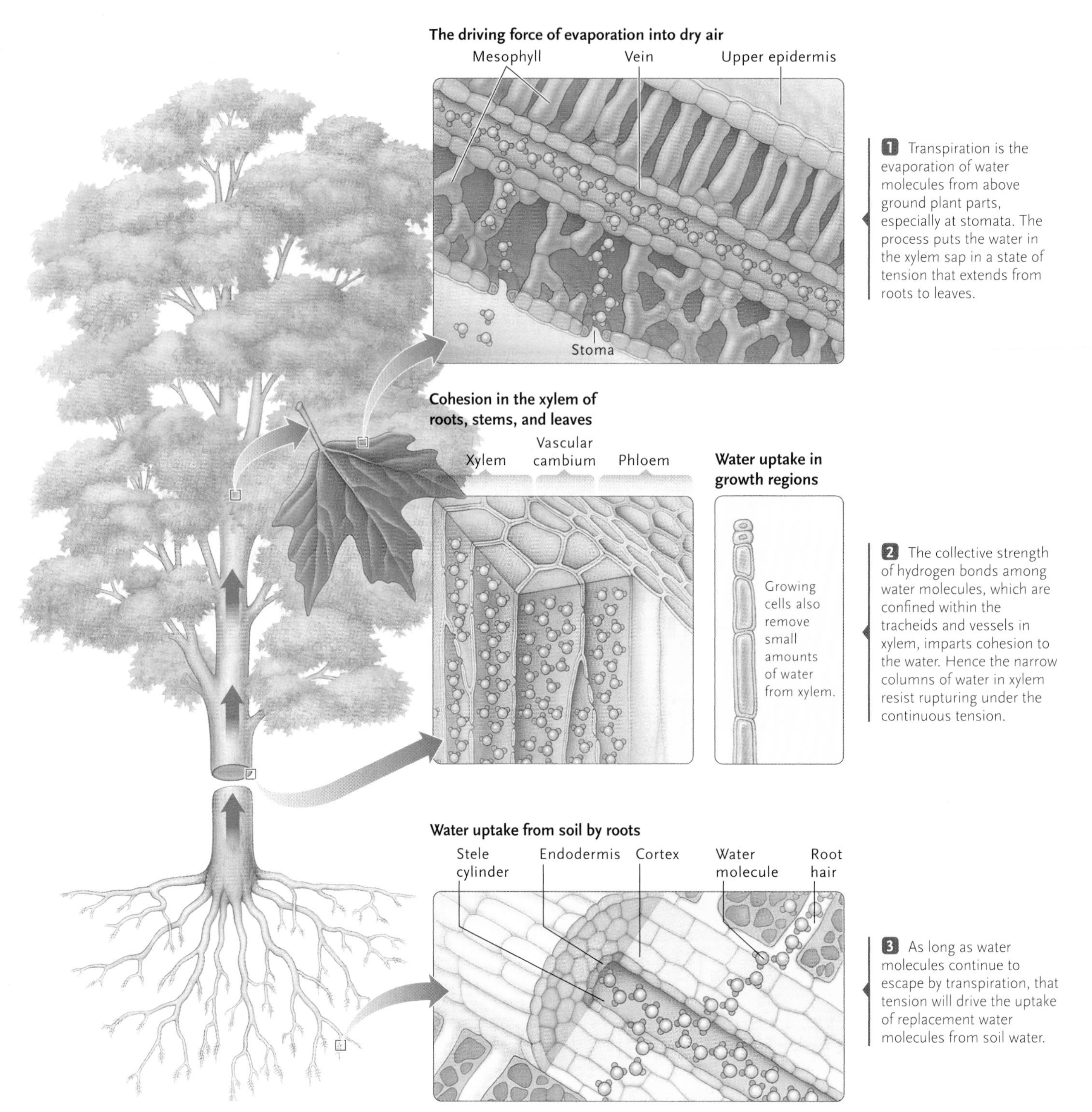
The driving force of evaporation into dry air
Mesophyll
Vein
Upper epidermis
Stoma
1 Transpiration is the evaporation of water molecules from above ground plant parts, especially at stomata. The process puts the water in the xylem sap in a state of tension that extends from roots to leaves.
Cohesion in the xylem of roots, stems, and leaves
Xylem
Vascular cambium
Phloem
Water uptake in growth regions
Growing cells also remove small amounts of water from xylem.
2 The collective strength of hydrogen bonds among water molecules, which are confined within the tracheids and vessels in xylem, imparts cohesion to the water. Hence the narrow columns of water in xylem resist rupturing under the continuous tension.
Water uptake from soil by roots
Stele cylinder
Endodermis
Cortex
Water molecule
Root hair
3 As long as water molecules continue to escape by transpiration, that tension will drive the uptake of replacement water molecules from soil water.

Figure 32.8
Cohesion–tension mechanism of water transport. Transpiration, the evaporation of water from shoot parts, creates tension on the water in xylem sap. This tension, which extends from root to leaf, pulls upward columns of water molecules that are hydrogen-bonded to one another.

up from the soil. Wilting is visible evidence that the water-potential gradient between soil and a plant's shoot parts has shifted. Remember that as soil dries out, the remaining water molecules are held ever more tightly by the soil particles. In effect, the action of soil particles reduces the water potential in the soil surrounding plant roots, and as this happens the roots take up water more slowly. However, because the water that evaporates from the plant's leaves is no longer being fully replaced, the leaves wilt as turgor pressure drops. Reducing the water potential in soil by adding solutes such as NaCl and other salts can have the same wilting effect. When the water potential in the soil finally equals that in leaf cells, a gradient no longer exists. Then movement of water from the soil into roots and up to the leaves comes to a halt.

Leaf Anatomy Contributes to Cohesion–Tension Forces

Leaf anatomy is key to the processes that move water upward into plants. To begin with, as much as two-thirds of a leaf's volume consists of air spaces—thus there is a large internal surface area for evaporation. Leaves also may have thousands to millions of stomata, through which water vapor can escape. Both these factors increase transpiration. Also, every square centimeter of a leaf contains thousands of tiny xylem veins, so that most leaf cells lie within half a millimeter of a vein. This close proximity readily supplies water to cells and the spaces between them, from which the water can readily evaporate.

As water evaporates from a leaf, surface tension at the interface between the water film and the air in the leaf space translates into negative pressure that draws water from the leaf veins. This tension is multiplied many times over in all of the leaves and xylem veins of a plant. It increases further as the plant's metabolically active cells take up xylem sap.

In the Tallest Trees, the Cohesion–Tension Mechanism May Reach Its Physical Limit

A variety of experiments have tested the premises of the cohesion–tension model, and thus far the data strongly support it. For example, the model predicts that xylem sap will begin to move upward at the top of a tree early in the day when water begins to evaporate from leaves. Experiments with several different tree species have confirmed that this is the case. The experiments also showed that sap transport peaks at midday when evaporation is greatest, then tapers off in the evening as evaporative water loss slows.

Other experiments have probed the relationship between xylem transport and tree height. One team of researchers studied eight of the tallest living redwoods, including one that towers nearly 113 m above the forest floor. When the scientists measured the maximum tension exerted in the xylem sap in twigs at the tops of the trees, they discovered that it approached the known physical limit at which the bonds between water molecules in a column of water in a conifer's xylem will rupture. Based on this finding and other evidence, the team has predicted that the maximum height for a healthy redwood tree is 122 to 130 m. Therefore it is possible that the tallest redwoods alive today may grow taller still.

Root Pressure Contributes to Upward Water Movement in Some Plants

The cohesion–tension mechanism accounts for upward water movement in tall trees. In some nonwoody plant species, however—lawn grasses, for instance—a positive pressure can develop in roots and force xylem sap upward. This **root pressure** operates under conditions that reduce transpiration, such as high humidity or low light. In fact, the mechanism that produces root pressure often operates at night, when solar-powered transpiration slows or stops. Then, active transport of ions into the stele sets up a water potential gradient across the endodermis. Because the Casparian strip of the endodermis tends to prevent ions from moving back into the root cortex, the water potential difference becomes quite large. It can move enough water and dissolved solutes into the xylem to produce a relatively high positive pressure. Although not sufficient to force water to the top of a very tall plant, in some smaller plant species root pressure is strong enough to force water out of leaf openings, in a process called **guttation** **(Figure 32.9).** Pushed up and out of vein endings by root pressure, tiny droplets of water that look like dew in the early morning emerge from modified stomata at the margins of leaves.

Dr. John D. Cunningham/Visuals Unlimited

Figure 32.9 Guttation, caused by root pressure. The drops of water appear at the endings of xylem veins along the leaf edges of a strawberry plant (*Fragaria*).

Stomata Regulate the Loss of Water by Transpiration

Three environmental conditions have major effects on the rate of transpiration: relative humidity, air temperature, and air movement. The most important is relative humidity, which is a measure of the amount of water vapor in air. The less water vapor in the air, the more evaporates from leaves (because the water potential is higher in the leaves than in the dry air). The air temperature at the leaf surface also speeds evaporation as it rises. Although evaporation does cool the leaf somewhat, the amount of water lost can double for each 10°C rise in air temperature. Air movement at the leaf surface carries water vapor away from the surface and so makes a steeper gradient. Together these factors explain why on extremely hot, dry, breezy days, the leaves of certain plants must completely replace their water each hour.

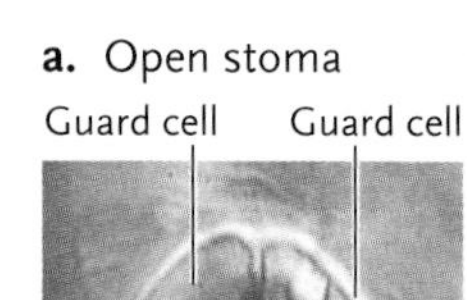

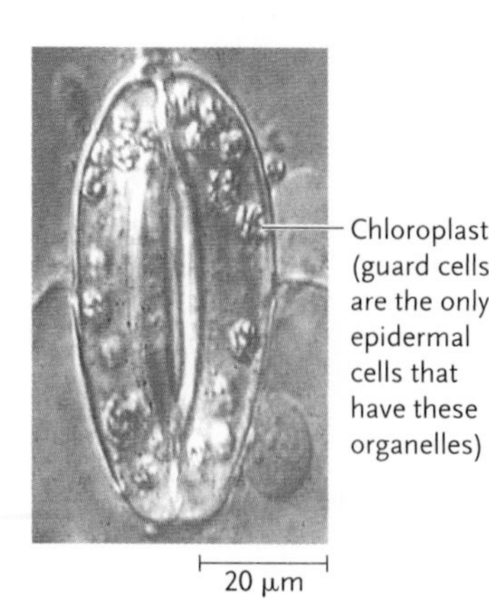

Figure 32.10
Guard cells and stomatal action. **(a)** An open stoma. Water entered collapsed guard cells, which swelled under turgor pressure and moved apart, thus forming the stoma in the needlelike leaf of the rock needlebush *(Hakea gibbosa)*. **(b)** A closed stoma. Water exited the swollen guard cells, which collapsed against each other and closed the stoma.

Even when conditions are not so drastic, more than 90% of the water moving into a leaf can be lost through transpiration. About 2% of the water remaining in the leaf is used in photosynthesis and other activities. These measurements emphasize the need for controls over transpiration, for if water loss exceeds water uptake by roots, the resulting dehydration of plant tissues interferes with normal functioning, and the plant may wilt and die.

The cuticle-covered epidermis of leaves and stems reduces the rate of water loss from aboveground plant parts, but it also limits the rate at which CO_2 for photosynthesis can diffuse into the leaf. The functioning of stomata also affects a plant's water balance. When stomata are open, carbon dioxide can be absorbed, but unless the relative humidity of external air is 100%, water always moves out. However, plants have evolved adaptations that balance water loss with CO_2 uptake. This "transpiration–photosynthesis compromise" involves the regulation of transpiration and gas exchange by opening and closing stomata as environmental conditions change.

Opening and Closing of Stomata. Two guard cells flank each stomatal opening **(Figure 32.10).** Their elastic walls are reinforced by cellulose microfibrils that wrap around the walls like a series of belts. The inward-facing walls are thicker and less elastic than the outer walls.

The opening and closing of stomata are good examples of a symport mechanism (see Figure 32.3c). Stomata open when potassium ions (K^+) flow into the guard cells through ion channels. As a first step, an active transport pump in the plasma membrane begins pumping H^+ ions out of the guard cells. Recall from Section 32.1 that H^+ pumped out of the cell can then follow its concentration gradient back into the cell. This inward flow of H^+ powers the active transport of K^+ into the guard cell. As a result, the K^+ concentration in turgid guard cells may be four to eight times higher than that in flaccid (limp) guard cells **(Figure 32.11).** Water follows inward by osmosis. As turgor pressure builds, the thick inner wall does not expand much, but the outer walls of each guard cell expand lengthwise, so the two cells bend away from each other and create a stoma ("mouth") between them. Stomata close when the H^+ active transport protein stops pumping. K^+ flows passively out of the guard cells, and water follows by osmosis. When the water content of the guard cells dwindles, turgor pressure drops. The guard cells collapse against each other, closing the stomata.

In most plants, stomata open at first light, stay open during daylight, and close at night. Experiments have shown that guard cells respond to a number of environmental and chemical signals, any of which can induce the ion flows that open and close stomata. These signals include light, CO_2 concentration in the air spaces inside leaves, and the amount of water available to the plant.

Light and CO_2 Concentration. Light induces stomata to open through stimulation of blue-light receptors, probably located in the plasma membrane of guard cells. When stimulated, the receptors start the chain of events leading to stomatal opening by triggering activity of the H^+ pumps. Also, as photosynthesis begins in response to light, CO_2 concentration drops in the leaf air spaces as chloroplasts use the gas in carbohydrate production. In some way, this drop in CO_2

a. Open stomata, with potassium mostly in guard cells

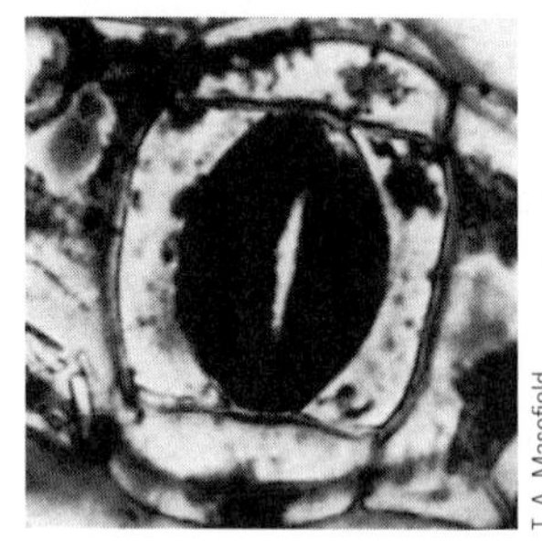

b. Closed stomata, with potassium mostly in epidermal cells

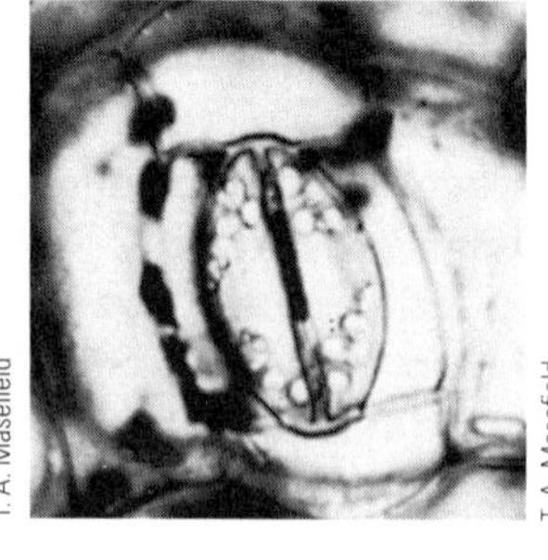

Figure 32.11
Evidence for potassium accumulation in stomatal guard cells undergoing expansion. Strips from the leaf epidermis of a dayflower *(Commelina communis)* were immersed in a solution containing a stain that binds preferentially with potassium ions. **(a)** In leaf samples with open stomata, most of the potassium was concentrated in the guard cells. **(b)** In leaf samples with closed stomata, little potassium was in guard cells; most was present in adjacent epidermal cells.

concentration sets off the series of events increasing the flow of K^+ into guard cells and furthers stomatal opening. The effects of reduced CO_2 concentration have been tested by placing plants in the dark in air containing no CO_2. Even in the absence of light, as the CO_2 concentration falls in leaves, guard cells swell and the stomata open.

Normally, when the sun goes down, a plant's demand for CO_2 drops as photosynthesis comes to a halt. Yet aerobic respiration continues to produce CO_2, which accumulates in leaves. As CO_2 concentration rises, and the blue-light wavelengths that activated the H^+ pumps wane, K^+ is lost from the guard cells and they collapse, closing the stomata. Thus, at night transpiration is reduced and water is conserved.

Water Stress. As long as water is readily available to a plant's roots, the stomata remain open during daylight. However, if water loss stresses a plant, the stomata close or open only slightly, regardless of light intensity or CO_2 concentration. Some simple but elegant experiments have shown that the stress-related closing of stomata depends on a hormone, abscisic acid (ABA), that is released by roots when water is unavailable. Test plants were suspended in containers so that only one-half the root system received water. Even though the roots with access to water could absorb enough water to satisfy the needs of all the plants' leaves, the stomata still closed. Tissue analysis revealed that water-stressed roots rapidly synthesize ABA. Transported through the xylem, this hormone stimulates K^+ loss by guard cells, and water moves out of the cell by osmosis—so the stomata close **(Figure 32.12)**. Mesophyll cells also take up ABA from the xylem and release it, with the same effects on stomata, when their turgor pressure falls due to excessive water loss. ABA can also cause stomata to close when the hormone is added experimentally to leaves.

The Biological Clock. Besides responding to light, CO_2 concentration, and water stress, stomata apparently open and close on a regular daily schedule imposed by a biological clock. Even when plants are placed in continuous darkness, their stomata open and close (for a time) in a cycle that roughly matches the day/night cycle of Earth. Such *circadian rhythms* (*circa* = around; *dies* = day) are also common in animals, and several, including wake/sleep cycles in mammals, are known to be controlled by hormones—a topic pursued in Chapter 40.

In Dry Climates, Plants Exhibit Various Adaptations for Conserving Water

Many plants have other evolutionary adaptations that conserve water, including modifications in structure or physiology **(Figure 32.13)**. Oleanders, for example, have stomata on the underside of the leaf at the bottom of pitlike invaginations of the leaf epidermis lined by hairlike trichomes (see Figure 32.13b). Sunken stomata are less exposed to drying breezes, and trichomes help retain water vapor at the pore opening, so that water evaporates from the leaf much more slowly.

The leaves of *xerophytes*—plants adapted to hot, dry environments in which water stress can be severe—have a thickened cuticle that gives them a leathery feel and provides enhanced protection against evaporative water loss. An example is mesquite (*Prosopis*). In still other plants that inhabit arid landscapes, such as cacti, stems are thick, leaflike pads covered by sharp spines that actually are modified leaves (see Figure 32.13c). These structural alterations reduce the surface area for transpiration.

One intriguing variation on water-conservation mechanisms occurs in CAM plants, including cacti, orchids, and most succulents. As discussed in Section 9.4, **crassulacean acid metabolism** (CAM) is a biochemical variation of photosynthesis that was discovered in a member of the family Crassulaceae. CAM plants generally have fewer stomata than other types of plants, and their stomata follow a reversed schedule. They are closed during the day when temperatures are higher and the relative humidity is lower, and open at night. At night, the plant temporarily fixes carbon dioxide by converting it to malate, an organic acid. In the daytime, the CO_2 is released from malate and diffuses into chloroplasts, so photosynthesis takes place even though a CAM plant's stomata are closed. This adaptation prevents heavy evaporative water losses during the heat of the day.

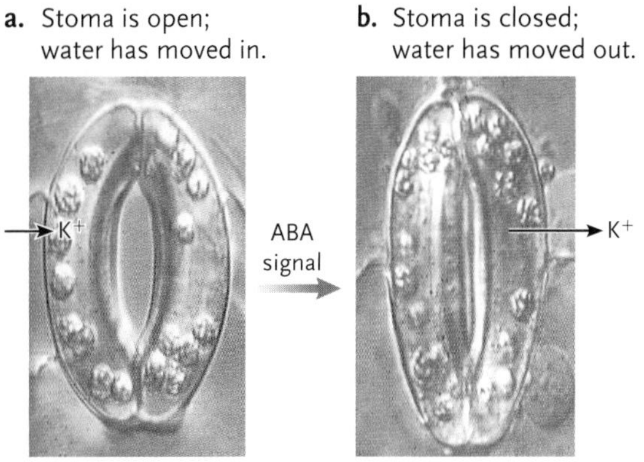

a. Stoma is open; water has moved in.

b. Stoma is closed; water has moved out.

K^+ ABA signal K^+

Figure 32.12
Hormonal control of stomatal closing. **(a)** When a stoma is open, high solute concentrations in the cytoplasm of both guard cells have raised the turgor pressure, keeping the cells swollen open. **(b)** In a water-stressed plant, the hormone abscisic acid (ABA) binds to receptors on the guard cell plasma membrane. Binding activates a signal transduction pathway that lowers solute concentrations inside the cells, which lowers the turgor pressure—so the stoma closes.

a. Oleanders

BIOS Matt Alexander/Peter Arnold, Inc.

b. Oleander leaf

Cuticle

Multilayer epidermis

Recessed stoma

Thomas L. Rost

c. Spines (modified leaves) on a cactus stem

Fritz Polking/Visuals Unlimited

d. CAM plant

Fritz Polking/Visuals Unlimited

Figure 32.13
Some adaptations that enable plants to survive water stress. **(a)** Oleanders *(Nerium oleander)* are adapted to arid conditions. **(b)** As shown in the micrograph, oleander leaves have recessed stomata on their lower surface and a multilayer epidermis covered by a thick cuticle on the upper surface. **(c)** Like many other cacti, the leaves of the Graham dog cactus *(Opuntia grahamii)* are modified into spines that protrude from the underlying stem. Transpiration and photosynthesis occur in the green stems, such as the oval stem in this photograph. **(d)** *Sedum,* a CAM plant, in which the stomata open only at night.

Study Break

1. Explain the key steps in the cohesion–tension mechanism of water transport in a plant.
2. How and when do stomata open and close? In what ways is their functioning important to a plant's ability to manage water loss?

32.4 Transport of Organic Substances in the Phloem

A plant's phloem is another major long-distance transport system, and a superhighway at that: it carries huge amounts of carbohydrates, lesser but vital amounts of amino acids, fatty acids, and other organic compounds, and still other essential substances such as hormones. And unlike the xylem's unidirectional upward flow, the phloem transports substances throughout the plant to wherever they are used or stored. Organic compounds and water in the sieve tubes of phloem are under pressure and driven by concentration gradients.

Organic Compounds Are Stored and Transported in Different Forms

Plants synthesize various kinds of organic compounds, including large amounts of carbohydrates that are stored mainly as starch. Yet regardless of where in a plant a particular compound is destined to be used or stored, starch, protein, and fat molecules cannot leave the cells in which they are formed because all are too large to cross cell membranes. They also may be too insoluble in water to be transported to other regions of the plant body. Consequently, in leaves and other plant parts, specific reactions convert organic compounds to transportable forms. For example, hydrolysis of starch liberates glucose units, which combine with fructose to form sucrose—the main form in which sugars are transported through the phloem of most plants. Proteins are broken down into amino acids, and lipids converted into fatty acids. These forms are also better able to cross cell membranes by passive or active mechanisms.

Organic Solutes Move by Translocation

In plants, the long-distance transport of substances is called **translocation.** Botanists most often use this term to refer to the distribution of sucrose and other organic compounds by phloem, and they understand the mechanism best in flowering plants. The phloem of flowering plants contains interconnecting sieve tubes formed by living sieve tube member cells (see Figure 31.9). Sieve tubes lie end to end within vascular bundles, and they extend through all parts of the plant. Water and organic compounds, collectively called **phloem sap,** flow rapidly through large pores on the sieve tubes' end walls—another example of a structural adaptation that suits a particular function.

Phloem Sap Moves from Source to Sink under Pressure

Over the decades, plant physiologists have proposed several mechanisms of translocation, but it was the tiny aphid, an insect that annoys gardeners, that helped demonstrate that organic compounds flow under pressure in the phloem. An aphid attacks plant leaves and stems, forcing its needlelike stylet (a mouthpart) into sieve tubes to obtain the dissolved sugars and other nutrients inside. Numerous experiments with aphids have shown that in most plant species, sucrose is the main carbohydrate being translocated through the phloem. Studies also verify that the contents of sieve tubes are under high pressure, often five times as much as in an automobile tire. **Figure 32.14** explains a simple and innovative experiment that provided direct confirmation that phloem sap flows under pressure. When a live aphid feeds on phloem sap, this pressure forces the fluid through the aphid's gut and (minus nutrients absorbed) out its anus as "honeydew." If you park your car

Figure 32.14 Experimental Research

Translocation Pressure

HYPOTHESIS: High pressure forces phloem sap to flow through sieve tubes from a source to a sink.

EXPERIMENT: In the late 1970s, John Wright and Donald Fisher at the University of Georgia devised an experiment to directly measure the turgor pressure in sieve tubes of weeping willow saplings *(Salix babylonica)* under nondestructive conditions, using aphids that feed on *S. babylonica* in the wild. Weeping willow saplings were grown in a greenhouse under natural conditions of light and moisture. Aphids were placed on the trees and allowed to begin feeding by inserting their stylets into sieve tubes in the normal fashion. After being anesthetized by exposure to high concentrations of carbon dioxide, the aphids' bodies were cut away and only their stylets were left embedded in the sieve tubes. A tiny pressure-measuring device called a micromanometer then was glued over the end of each stylet. The micromanometer registered the volume and pressure of phloem sap as it was exuded from the stylet over time periods ranging from 30 to 90 minutes.

a. Aphid releasing honeydew

Martin Zimmerman, Science, 1961, 133:73–79, © AAAS

b. Micrograph of aphid stylet in sieve tube

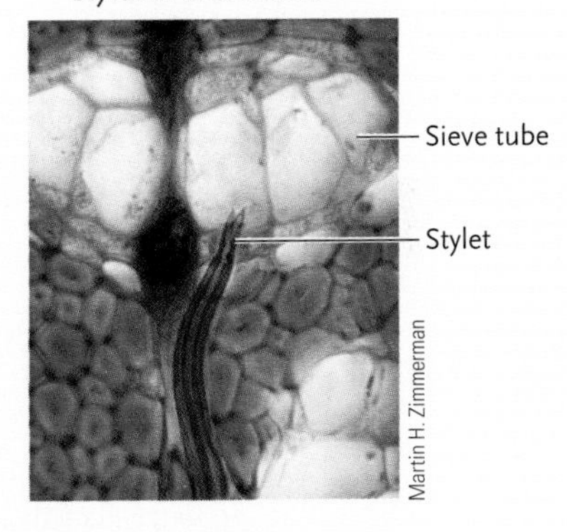

Martin H. Zimmerman

RESULTS: In nearly all cases, a high volume of pressurized sap flowed through the severed stylets into the micromanometer during the test periods.

CONCLUSION: The evidence supports pressure flow as the mechanism that moves phloem sap through sieve tubes.

Other experiments have confirmed that both turgor pressure and the concentration of sucrose are highest in sieve tubes closest to the sap source. Phloem sap also moves most rapidly closest to the source, where pressure is highest.

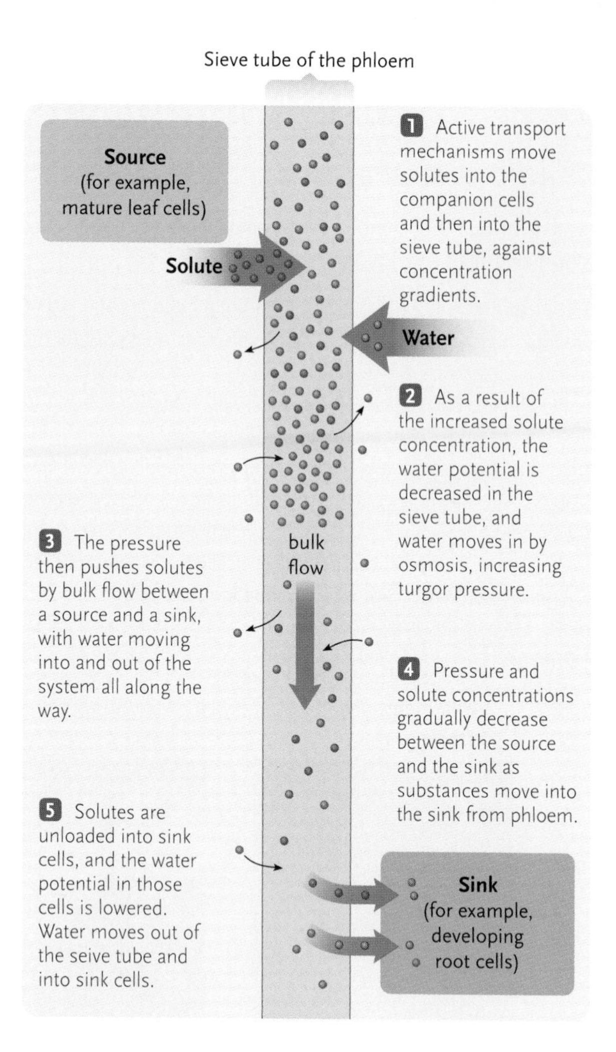

Sieve tube of the phloem
Source
(for example,
mature leaf cells)
Solute
1 Active transport mechanisms move solutes into the companion cells and then into the sieve tube, against concentration gradients.
Water
2 As a result of the increased solute concentration, the water potential is decreased in the sieve tube, and water moves in by osmosis, increasing turgor pressure.
3 The pressure then pushes solutes by bulk flow between a source and a sink, with water moving into and out of the system all along the way.
bulk
flow
4 Pressure and solute concentrations gradually decrease between the source and the sink as substances move into the sink from phloem.
5 Solutes are unloaded into sink cells, and the water potential in those cells is lowered. Water moves out of the seive tube and into sink cells.
Sink
(for example,
developing
root cells)

Figure 32.15 Summary of the pressure flow mechanism in the phloem of flowering plants. Organic solutes are loaded into sieve tubes at a source, such as a leaf, and move by bulk flow toward a sink, such as roots or rapidly growing stem parts.

under a tree being attacked by aphids, it might get spattered with sticky honeydew droplets, thanks to the high fluid pressure in the tree's phloem.

A great deal of what botanists know about the transport of phloem sap has come from studies of sucrose transport in flowering plants. A fundamental discovery is that in flowering plants sucrose-laden phloem sap flows from a starting location, called the *source,* to another site, called the *sink,* along gradients of decreasing solute concentration and pressure. A **source** is any region of the plant where organic substances are being loaded into the phloem's sieve tube system. A **sink** is any region where organic substances are being unloaded from the sieve tube system and used or stored. What causes sucrose and other solutes produced in leaf mesophyll to flow from a source to a sink? In flowering plants, the **pressure flow mechanism** builds up at the source end of a sieve tube system and pushes those solutes by bulk flow toward a sink, where they are removed. **Figure 32.15** summarizes this mechanism.

The site of photosynthesis in mature leaves is an example of a source. Another example is a tulip bulb. In spring, stored food is mobilized for transport upward to growing plant parts, but after the plants bloom, the bulb becomes a sink as sugars manufactured in the tulip plant's leaves are translocated into it for storage. Young leaves, roots, and developing fruits generally start out as sinks, only to become sources when the season changes or the plant enters a new developmental phase. In general, sinks receive organic compounds from sources closest to them. Hence, the lower leaves on a rose bush may supply sucrose to roots, while leaves farther up the shoot supply the shoot tip.

Most substances carried in phloem are loaded into sieve tube members by active transport **(Figure 32.16a).** Sucrose will be our example here. In leaves, sucrose formed inside mesophyll cells is exported and eventually reaches the apoplast (adjoining cell walls and air spaces) next to a small phloem vein. Here, it is actively pumped into companion cells by symport (see Figure 32.3), in which H^+ ions moves into the cell through the same carrier that takes up the sugar molecules. From the companion cells, most sucrose crosses into the living sieve tube members through plasmodesmata. Some sucrose also is loaded into sieve tube members by symport.

In some plants, companion cells become modified into **transfer cells** that facilitate the short-distance transport of organic solutes from the apoplast into the symplast. Transfer cells generally form when large amounts of solutes must be loaded or unloaded into the phloem, and they shunt substances through plasmodesmata to sieve tube members. As a transfer cell is forming, parts of the cell wall grow inward like pleats. This structural feature increases the surface area across which solutes can be taken up. The underlying plasma membrane, packed with transport proteins, then expands to cover the ingrowths. Transfer cells also enhance solute transport between living cells in the xylem, and they occur in glandlike tissues that secrete nectar. Botanists have discovered transfer cells in species from every taxonomic group in the plant kingdom, as well as in fungi and algae. In part because they arise from differentiated cells (instead of from meristem cells like other plant types of plant cells), researchers are working to define the molecular mechanisms that trigger their development.

When sucrose is loaded into sieve tubes its concentration rises inside the tubes. Thus the water potential falls, and water flows into the sieve tubes by osmosis. In fact, the phloem typically carries a great deal of water. As water enters sieve tubes, turgor pressure in the tubes increases, and the sucrose-rich fluid moves by bulk flow into the increasingly larger sieve tubes of larger veins. Eventually, the fluid is pushed out of the leaf into the stem and toward a sink **(Figure 32.16b).** When sucrose is unloaded at the sink, water in the tube "follows solutes," moving by osmosis into the surrounding cells **(Figure 32.16c).** Ultimately, the water enters the xylem and is recirculated.

Sieve tubes are mostly passive conduits for translocation. The system works because companion cells

a. Loading at a source

Upper epidermis
Photosynthetic cell
Sieve tube in phloem
Companion cell
Section from a leaf
Lower epidermis

Photosynthetic cells in leaves are a common source of carbohydrates that must be distributed through a plant. Small, soluble forms of these compounds move from the cells into phloem (in a leaf vein).

b. Translocation along a distribution path

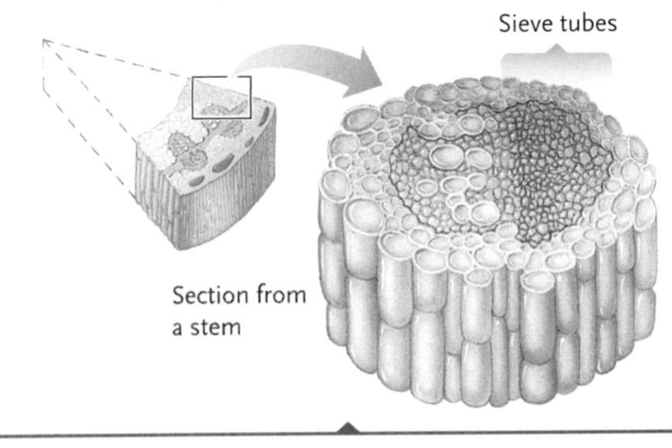

Fluid pressure is greatest inside sieve tubes at the source. It pushes the solute-rich fluid to a sink, which is any region where cells are growing or storing food. There, the pressure is lower because cells are withdrawing solutes from the tubes and water follows the solutes.

c. Unloading at the sink

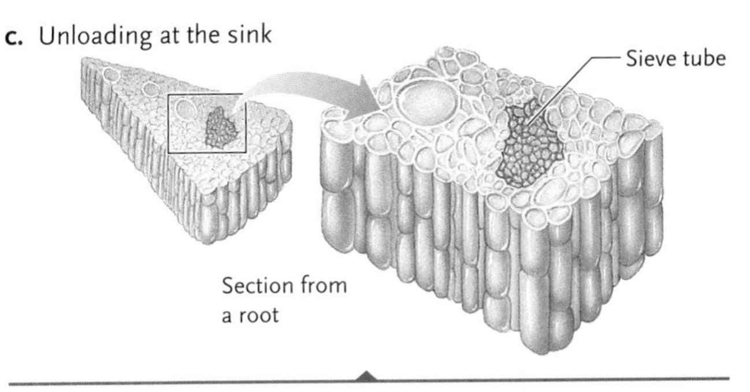

Solutes are unloaded from sieve tubes into cells at the sink; water follows. Translocation continues as long as solute concentration gradients and a pressure gradient exist between the source and the sink.

Figure 32.16
Translocation in the tissues of *Sonchus*, commonly called sow thistle. Research on *Sonchus* provided experimental evidence for the pressure flow mechanism.

UNANSWERED QUESTIONS

What are plasmodesmata made of, and exactly how do they function?

Plasmodesmata, the cytoplasmic channels through plant cell walls, connect plant cells to each other. Yet two fundamental questions about plasmodesmata remain unanswered: Exactly how do plasmodesmata function, and what are their structural components?

As described in this chapter and in Chapter 5, botanists have long assumed that nutrients, water, and small molecules that serve as growth regulators move through plasmodesmata, which form part of the symplastic pathway in plant tissues. Recent studies have demonstrated that larger molecules, including viruses and important proteins involved in plant growth and development, also move from cell to cell through plasmodesmata. For example, Patricia Zambryski and K. M. Crawford at the University of California at Berkeley reported that proteins, including transcription factors, travel via plasmodesmata from the cell that produces the proteins to adjacent cells where the factors promote or inhibit the expression of particular genes.

While the normal functions of plasmodesmata in plant growth and development still are not well understood, ongoing research by Zambryski and other plant scientists has begun to shed light on the workings of these vital channels. For instance, a variety of studies of the processes by which viruses spread through plant tissues have revealed that plasmodesmata are not simply static, open channels. Instead they are dynamic structures with the capacity to close, reopen, widen, and narrow. This capacity for structural change is not triggered by viral infection: rather, it seems that viruses simply take over the plant's natural mechanism for moving molecules from one cell to another.

Plasmodesmata were first observed using electron microscopy several decades ago, and they appear to be lined with proteins as well as membranes. Multiple biochemical approaches have failed to identify the proteins, probably because of the difficulty of purifying proteins that are associated with both a membrane and the cell wall. Genetic screens to identify plasmodesmata proteins, as well as the genes that regulate the functioning of plasmodesmata, are currently under way and may finally reveal details of plasmodesmata structure. As our understanding of the architecture of plasmodesmata and how they function grows, so will insights into the mechanisms of plant development, how plants interact with viral pathogens, and other questions as well.

Beverly McMillan

supply most of the energy that loads sucrose and other solutes at the source, and because solutes are removed at their sinks. As sucrose enters a sink, for example, its concentration in sieve tubes decreases, with a corresponding decrease in pressure. Thus for sucrose and other solutes transported in the phloem, there is always a gradient of concentration from source to sink—and a pressure gradient that keeps the solute moving along.

As noted previously, phloem sap moving through a plant carries a wide variety of substances, including hormones, amino acids, organic acids, and agricultural chemicals. The phloem also transports organic nitrogen compounds and mineral ions that are removed from dying leaves and stored for reuse in root tissue.

The transport functions of xylem and phloem are closely integrated with phenomena discussed later in this unit—reproduction and embryonic development, and the hormone-based regulation of plant growth.

Study Break

1. Compare and contrast translocation and transpiration.
2. Using sucrose as your example, summarize how a substance moves from a source into sieve tubes and then is unloaded at a sink. What is this mechanism called, and why?

Review

Go to ThomsonNOW at www.thomsonedu.com/login to access quizzing, animations, exercises, articles, and personalized homework help.

32.1 Principles of Water and Solute Movement in Plants

- Plants have mechanisms for moving water and solutes (1) into and out of cells, (2) laterally from cell to cell, and (3) long-distance from the root to shoot or vice versa (Figure 32.2).
- Both passive and active mechanisms move substances into and out of plant cells. Solutes generally are transported by carriers (facilitated diffusion), either passively down a concentration or electrochemical gradient (in the case of ions), or actively against a gradient, which requires cellular energy. An H^+ gradient creates the membrane potential that drives the cross-membrane transport of many ions or molecules (Figure 32.3).
- Most organic substances enter plant cells by symport, in which the energy of the H^+ gradient is coupled with uptake of a different solute. Some substances cross the plant cell membrane by antiport, in which energy of the H^+ gradient powers movement of a second solute out of cells.
- Water crosses plant cell membranes by osmosis, which is driven by water potential (ψ). Water tends to move osmotically from regions where water potential is higher to regions where it is lower.
- Water potential reflects a balance between turgor pressure and solute potential. Water potential is measured in megapascals (MPa) (Figures 32.4 and 32.5).
- Water and solutes also move into and out of the cell's central vacuole, transported from the cytoplasm across the tonoplast. Aquaporins across the tonoplast enhance water movement. Water in the central vacuole is vital for maintaining turgor pressure inside a plant cell.
- Bulk flow of fluid occurs when pressure at one point in a system changes with respect to another point in the system.

32.2 Transport in Roots

- Water and mineral ions entering roots travel laterally through the root cortex to the root xylem, following one or more of three major routes: the apoplastic pathway, the symplastic pathway, and the transmembrane pathway (Figure 32.6).
- In the apoplastic pathway, water diffuses into roots between the walls of root epidermal cells. By contrast, water and solutes absorbed by roots can enter either the symplastic or transmembrane pathway, both of which pass through cells.
- Casparian strips form a barrier that forces water and solutes in the apoplastic pathway to pass through cells in order to enter the stele. When an ion reaches the stele, it diffuses from cell to cell to reach the xylem (Figure 32.7). Roots of many flowering plants have a second layer of cells with Casparian strips (exodermis) just inside the root epidermis.

Animation: Water absorption

Animation: Root functioning

32.3 Transport of Water and Minerals in the Xylem

- In the conducting cells of xylem, tension generated by transpiration extends down from leaves to roots. By the cohesion–tension mechanism of water transport, water molecules are pulled upward by tension created as water exits a plant's leaves (Figure 32.8).
- In tall trees, negative pressure generated in the shoot drives bulk flow of xylem sap. In some plants, notably herbaceous species, positive pressure sometimes develops in roots and can force xylem sap upward (Figure 32.9).
- Transpiration and carbon dioxide uptake occur mostly through stomata. Environmental factors such as relative humidity, air temperature, and air movement at the leaf surface affect the transpiration rate.
- Most plants lose water and take up carbon dioxide during the day, when stomata are open. At night, when stomata close, plants conserve water and the inward movement of carbon dioxide falls.
- Stomata open in response to falling levels of carbon dioxide in leaves and also to incoming light wavelengths that activate photoreceptors in guard cells.
- Activation of photoreceptors triggers active transport of K^+ into guard cells. Simultaneous entry of anions such as Cl^- and synthesis of negatively charged organic acids increase the solute concentration, lowering the water potential so that water enters by osmosis. As turgor pressure builds, guard cells swell and draw apart, producing the stomatal opening (Figure 32.10).
- Guard cells close when light wavelengths used for photosynthesis wane. The stomata of water-stressed plants close re-

gardless of light or CO_2 needs, possibly under the influence of the plant hormone ABA. The leaves of species native to arid environments typically have adaptations (such as an especially thick cuticle) that enhance the plant's ability to conserve water (Figures 32.12 and 32.13).

Animation: Stomata

Animation: Transpiration

Animation: Interdependent processes

32.4 Transport of Organic Substances in the Phloem

- In flowering plants, phloem sap is translocated in sieve tube members. Differences in pressure between source and sink regions drive the flow. Sources include mature leaves; sinks include growing tissues and storage regions (such as the tubers of a potato) (Figures 32.14 and 32.15).
- In leaves, the sugar sucrose is actively transported into companion cells adjacent to sieve tube members, then loaded into the sieve tubes through plasmodesmata.
- In some plants, transfer cells take up materials and pass them to sieve tube members. Transfer cells in xylem enhance the transport of solutes between tissues.
- As the sucrose concentration increases in the sieve tubes, water potential decreases. The resulting influx of water causes pressure to build up inside the sieve tubes, so the sucrose-laden fluid flows in bulk toward the sink, where sucrose and water are unloaded and distributed among surrounding cells and tissues (Figure 32.16).

Questions

Self-Test Questions

1. Antiport transport mechanisms:
   a. move dissolved materials by osmosis.
   b. transport molecules in the opposite direction of H^+ transported by proton pumps.
   c. transport molecules in the same direction as H^+ is pumped.
   d. are not affected by the size of molecules to be transported.
   e. are not affected by the charge of molecules to be transported.
2. All the following have roles in transporting materials between plant cells except:
   a. the stele.
   b. symport.
   c. the cell membrane.
   d. stomata.
   e. transport proteins.
3. Turgor pressure is best expressed as the:
   a. movement of water into a cell by osmosis.
   b. driving force for osmotic movement of water (ψ).
   c. group movement of large numbers of molecules due to a difference in pressure between two locations.
   d. equivalent of water potential.
   e. pressure exerted by fluid inside a plant cell against the cell wall.
4. Water potential is:
   a. the driving force for the osmotic movement of water into plant cells.
   b. higher in a solution that has more solute molecules relative to water molecules.
   c. a measure of the physical pressure required to halt osmotic water movement across a membrane.
   d. a measure of the combined effects of a solution's pressure potential and its solute potential.
   e. the functional equivalent of turgor pressure.
5. To regulate the flow of water and minerals in the root, the:
   a. Casparian strip of endodermal cells blocks the apoplastic pathway, forcing water and solutes to cross cell plasma membranes in order to pass into the stele.
   b. apoplastic pathway is expanded, allowing a greater variety of substances to move into the stele.
   c. symplastic pathway is modified in ways that make plasma membranes of root cortex cells more permeable to water and solutes.
   d. symplastic pathway shuts down entirely so that substances can move only through the apoplast.
   e. transmembrane pathway augments transport via the apoplast, shunting substances around cells.
6. An indoor gardener leaving for vacation completely wraps a potted plant with clear plastic. Temperature and light are left at low intensities. The effect of this strategy is to:
   a. halt photosynthesis.
   b. reduce transpiration.
   c. cause guard cells to shrink and stomata to open.
   d. destroy cohesion of water molecules in the xylem.
   e. increase evaporation from leaf mesophyll cells.
7. Stomata open when:
   a. water has moved out of the leaf by osmosis.
   b. K^+ flows out of guard cells.
   c. turgor pressure in the guard cells lessens.
   d. the H^+ active transport protein stops pumping.
   e. outward flow of H^+ sets up a concentration gradient that moves K^+ in via symport.
8. A factor that contributes to the movement of water up a plant stem is:
   a. active transport of water into the root hairs.
   b. an increase in the water potential in the leaf's mesophyll layer.
   c. cohesion of water molecules in stem and leaf xylem.
   d. evaporation of water molecules from the walls of cells in root epidermis and cortex and in the stele.
   e. absorption of raindrops on a leaf's epidermis.
9. In translocation of sucrose-rich phloem sap:
   a. the sap flows toward a source as pressure builds up at a sink.
   b. crassulacean acid metabolism reduces the rate of photosynthesis.
   c. companion cells use energy to load solutes at a source and the solutes then follow their concentration gradients to sinks.
   d. sucrose diffuses into companion cells while H^+ simultaneously leaves the cells by a different route.
   e. companion cells pump sucrose into sieve tube members.
10. In Vermont in early spring, miles of leafless maple trees have buckets hanging from "spigots" tapped into them to capture the fluid raw material for making maple syrup. This fluid flows into the buckets because:
    a. the tap drains phloem sap stored in the heartwood.
    b. phloem sap is moving from its source in maple tree roots to its sink in the developing leaf buds.

c. phloem sap is moving from where it was synthesized to the closest sink.
d. bulk flow results as phloem sap is actively transported from smaller to larger veins.
e. phloem sap is diverted into the tap from transfer cells.

Questions for Discussion

1. Many popular houseplants are native to tropical rain forests. Among other characteristics, many nonwoody species have extraordinarily broad-bladed leaves, some so ample that indigenous people use them as umbrellas. What environmental conditions might make a broad leaf adaptive in tropical regions, and why?
2. Insects such as aphids that prey on plants by feeding on phloem sap generally attack only young shoot parts. Other than the relative ease of piercing less mature tissues, suggest a reason why it may be more adaptive for these animals to focus their feeding effort on younger leaves and stems.
3. So-called systemic insecticides often are mixed with water and applied to the soil in which a plant grows. The chemicals are effective against sucking insects no matter which plant tissue the insects attack, but often don't work as well against chewing insects. Propose a reason for this difference.
4. Concerns about global warming and the greenhouse effect (see Chapter 51) center on rising levels of greenhouse gases, including atmospheric carbon dioxide. Plants use CO_2 for photosynthesis, and laboratory studies suggest that increased CO_2 levels could cause a rise in photosynthetic activity. However, as one environmentalist noted, "What plants do in environmental chambers may not happen in nature, where there are many other interacting variables." Strictly from the standpoint of physiological effects, what are some possible ramifications of a rapid doubling of atmospheric CO_2 on plants in temperate environments? In arid environments?

Experimental Analysis

In an experiment designed to explore possible links between ion uptake by roots and loading of ions into the xylem, a length of root was suspended through an impermeable barrier that separated two compartments—the root tip in one compartment and the cut end of the root in the other. Initially the solutions in the two compartments were identical, except that a known quantity of a radioactive tracer (representing an ion) was added to the one in which the root tip was suspended. The experimenters could then measure the relationship between ion uptake in the root and loading of the ion into the root xylem under different chemical conditions (such as the addition of a hormone or protein synthesis inhibitor). The research has provided evidence that ion uptake in the root is independent of loading of the ion into xylem. How does the experimental design support this kind of testing?

Evolution Link

A variety of structural features of land plants reflect the conflicting demands for conserving water and taking in carbon dioxide for photosynthesis. Identify at least four fundamental structural adaptations that help resolve this dilemma and explain how each one contributes to a land plant's survival.

How Would You Vote?

Phytoremediation using genetically engineered plants can increase the efficiency with which a contaminated site is cleaned up. Do you support planting genetically engineered plants for such projects? Go to www.thomsonedu.com/login to investigate both sides of the issue and then vote.

Lush azaleas *(Rhododendron)* and a stately Southern live oak *(Quercus virginiana)* draped with the unusual flowering plant called Spanish moss *(Tillandsia usneoides)*. The roots of shrubs, trees, and most other plants take up water and minerals from soil, but Spanish moss is an epiphyte—it lives independently on other plants and obtains nutrients via absorptive hairs on its leaves and stems.

33 Plant Nutrition

Study Plan

33.1 Plant Nutritional Requirements

Plants require macronutrients and micronutrients for their metabolism

Nutrient deficiencies cause abnormalities in plant structure and function

33.2 Soil

The components of a soil and the size of the particles determine its properties

The characteristics of soil affect root–soil interactions

33.3 Obtaining and Absorbing Nutrients

Root systems allow plants to locate and absorb essential nutrients

Nutrients move into and through the plant body by several routes

Plants depend on bacterial metabolism to provide them with usable nitrogen

Some plants obtain scarce nutrients in other ways

Why It Matters

Tropical rainforests are remarkable for many reasons, but for biologists the key one may be that they are the most biologically diverse ecosystems on Earth. In addition to containing countless thousands of species of animals, fungi, protists, and prokaryotes, these amazingly lush domains are dense with broadleaved, evergreen trees, some of which soar 40 or 50 m skyward. With rain a near-daily event, it may not seem surprising that the trees' foliage is a luxuriant deep green **(Figure 33.1).** Yet tropical rainforests are demanding places for plants to survive, in large part because the soil is chronically deficient in nutrients, the chemical elements necessary for plant metabolism. This nutrient scarcity is a direct outcome of the incessant rain and the high acidity of tropical rainforest soil. There is ample moisture in the upper layer of soil, but in acid soil mineral nutrients vital to plant metabolism, such as potassium, calcium, magnesium, and phosphorus, are subject to **leaching**—being washed into deeper soil levels that are not as accessible to plant roots. In addition, in the warm, moist environment of a tropical rainforest, bacteria and fungi speedily decompose fallen leaves and other organic remains. Just as rapidly, established trees and vines

Figure 33.1
A lush tropical rain forest growing in Southeast Asia.

take up any nutrients these decomposers have released, leaving few or none to enrich the soil. As falling rain dissolves some atmospheric CO_2, it creates carbonic acid—a type of "acid rain" that exacerbates the leaching problem even more.

Such poor soil and the near perpetual twilight at the forest floor make it extremely difficult for small shrubs and herbaceous plants to survive there. Nearly all such plants climb upward as vines using the tree trunks for mechanical support, or they live attached to the upper branches of taller species, where they can absorb needed minerals from falling dust or from the surfaces of other plants. These intricate adaptations to their particular environment allow the plants to secure energy and raw materials and to utilize both for growth and development.

Tropical rainforests are not unique in posing nutritional challenges for plants. In fact, plants rarely have ready access to a full complement of necessary resources. In a rainforest, the carbon, hydrogen, and oxygen plants need for photosynthesis are relatively easy to come by: plants there usually get enough carbon from the CO_2 in air, and their roots can take up enough water to gain the necessary hydrogen and oxygen. But soils in other environments are frequently dry, making water a limited resource, and almost nowhere in nature do soils hold lavish amounts of dissolved minerals such as nitrogen, calcium, and others that are vital for a plant's survival. In response to the challenge of obtaining nutrients, plants have evolved the range of structural and physiological adaptations that we consider in this chapter.

33.1 Plant Nutritional Requirements

No organism grows normally when deprived of a chemical element essential for its metabolism. In the latter half of the nineteenth century, plant physiologists exploited rapid advances in chemistry to probe both the chemical composition of plants and the essential nutrients plants need to survive. Because plants require some nutrients in only trace amounts, in recent times researchers have brought to bear sophisticated methods in their studies of plant nutrition.

Plants Require Macronutrients and Micronutrients for Their Metabolism

By weight, the tissues of most plants are more than 90% water. Early researchers could obtain a rough idea of the composition of a plant's dry weight by burning the plant and then analyzing the ash. This method typically yielded a long list of elements, but the results were flawed. Chemical reactions during burning can dissipate quantities of some important elements, such as nitrogen. Also, plants take up a variety of ions that they don't use; depending on the minerals present in the soil where a plant grows, a plant's tissues can contain nonnutritive elements such as gold, lead, arsenic, and uranium.

Studying Plant Nutrition Using Hydroponics. In 1860, German plant physiologist Julius von Sachs pioneered an experimental method for identifying the minerals

Figure 33.2 Research Method

Hydroponic Culture

PURPOSE: In studies of plant nutritional requirements, using hydroponic culture allows a researcher to manipulate and precisely define the types and amounts of specific nutrients that are available to test plants.

PROTOCOL: In a typical hydroponic apparatus, many plants are grown in a single solution containing pure water and a defined mix of mineral nutrients. The solution is replaced or refreshed as needed and is aerated with a bubbling system:

a. Basic components of a hydroponic apparatus

A "complete" solution contains all the known and suspected essential plant nutrients. An "incomplete" solution contains all but one of the same nutrients, in the same amounts. For experiments, researchers first grow plants in a complete solution, then transplant some of the plants to an incomplete solution.

b. Procedure for identifying elements essential for proper plant nutrition

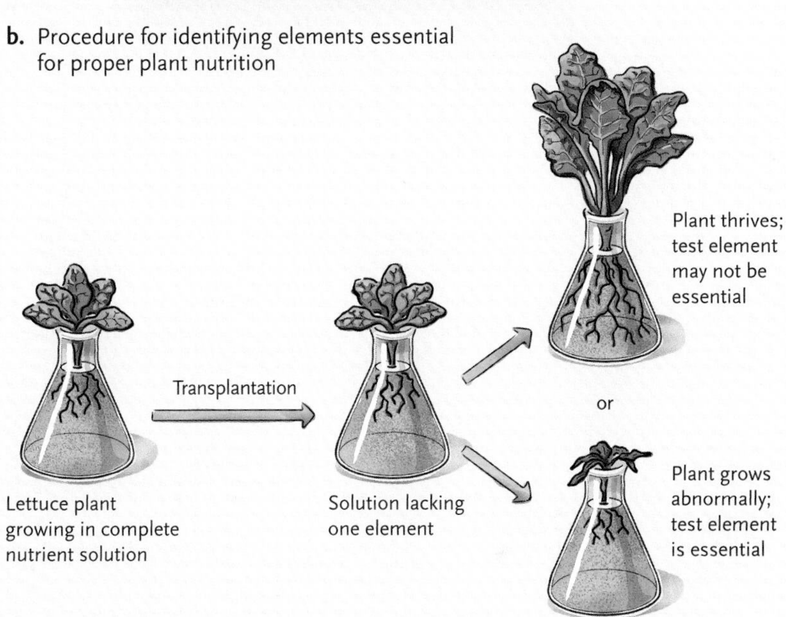

INTERPRETING THE RESULTS: Normal growth of test plants suggests that the missing nutrient is not essential, while abnormal growth is evidence that the missing nutrient may be essential.

absorbed into plant tissues that are essential for plant growth. Sachs carefully measured amounts of compounds containing specific minerals and mixed them in different combinations with pure water. He then grew plants in the solutions, a method called now **hydroponic culture** (*hydro* = water; *ponos* = work). By eliminating one element at a time and observing the results, Sachs deduced a list of six essential plant nutrients, in descending order of the amount required: nitrogen, potassium, calcium, magnesium, phosphorus, and sulfur.

Sachs's innovative research paved the way for decades of increasingly sophisticated studies of plant nutrition. In the spirit of his work, one basic experimental method involves growing a plant in a solution containing a complete spectrum of known and possible essential nutrients **(Figure 33.2a).** The healthy plant is then transferred to a solution that is identical, except that it lacks one element having an unknown nutritional role **(Figure 33.2b).** Abnormal growth of the plant in this solution is evidence that the missing element is essential. If the plant grows normally, the missing element may not be essential; however, only further experimentation can confirm this hypothesis.

In a typical modern hydroponic apparatus, the nutrient solution is refreshed regularly, and air is bubbled into it to supply oxygen to the roots. Without sufficient oxygen for respiration, the plants' roots do not absorb nutrients efficiently. (The same effect occurs in poorly aerated soil.) Variations of this technique are used on a commercial scale to grow some vegetables, such as lettuce and tomatoes.

Essential Macronutrients and Micronutrients. Hydroponics research has revealed that plants generally require 17 essential elements **(Table 33.1)**. By definition, an **essential element** is necessary for normal growth and reproduction, cannot be functionally replaced by a different element, and has one or more roles in plant metabolism. With enough sunlight and the 17 essential elements, plants can synthesize all the compounds they need.

Nine of the essential elements are **macronutrients,** meaning that plants incorporate relatively large amounts

Table 33.1 Essential Plant Nutrients and Their Functions

Element	Commonly Absorbed Forms	Some Known Functions	Some Deficiency Symptoms
Macronutrients			
Carbon*	CO_2	Raw materials for photosynthesis	Rarely deficient
Hydrogen*	H_2O		No symptoms; available from water
Oxygen*	O_2, H_2O, CO_2		No symptoms; available from water and CO_2
Nitrogen	NO_3^-, NH_4^+	Component of proteins, nucleic acids, coenzymes, chlorophylls	Stunted growth; light-green older leaves; older leaves yellow and die (chlorosis)
Phosphorus	$H_2PO_4^-$, HPO_4^{2+}	Component of nucleic acids, phospholipids, ATP, several coenzymes	Purplish veins; stunted growth; fewer seeds, fruits
Potassium	K^+	Activation of enzymes; key role in maintaining water-solute balance and so influences osmosis	Reduced growth; curled, mottled, or spotted older leaves; burned leaf edges; weakened plant
Calcium	Ca^{2+}	Roles in formation and maintenance of cell walls and in membrane permeability; enzyme cofactor	Leaves deformed; terminal buds die; poor root growth
Sulfur	SO_4^{2-}	Component of most proteins, coenzyme A	Light-green or yellowed leaves; reduced growth
Magnesium	Mg^{2+}	Component of chlorophyll; activation of enzymes	Chlorosis; drooping leaves
Micronutrients			
Chlorine	Cl^-	Role in root and shoot growth, and in photosynthesis	Wilting; chlorosis; some leaves die (deficiency not seen in nature)
Iron	Fe^{2+}, Fe^{3+}	Roles in chlorophyll synthesis, electron transport; component of cytochrome	Chlorosis; yellow and green striping in grasses
Boron	H_3BO_3	Roles in germination, flowering, fruiting, cell division, nitrogen metabolism	Terminal buds, lateral branches die; leaves thicken, curl, and become brittle
Manganese	Mn^{2+}	Role in chlorophyll synthesis; coenzyme action	Dark veins, but leaves whiten and fall off
Zinc	Zn^{2+}	Role in formation of auxin, chloroplasts, and starch; enzyme component	Chlorosis; mottled or bronzed leaves; abnormal roots
Copper	Cu^+, Cu^{2+}	Component of several enzymes	Chlorosis; dead spots in leaves; stunted growth
Molybdenum	MoO_4^{2-}	Component of enzyme used in nitrogen metabolism	Pale green, rolled or cupped leaves
Nickel	Ni^{2+}	Component of enzyme required to break down urea generated during nitrogen metabolism	Dead spots on leaf tips (deficiency not seen in nature)

*Carbon, hydrogen, and nitrogen are the nonmineral plant nutrients. All others are minerals.

of them into their tissues. Three of these elements—carbon, hydrogen, and oxygen—account for about 96% of a plant's dry mass. Together, these three elements are the key components of lipids and of carbohydrates such as cellulose; with the addition of nitrogen, they form the basic building blocks of proteins and nucleic acids. Plants also use phosphorus in constructing nucleic acids, ATP, and phospholipids, and they use potassium for functions ranging from enzyme activation to mechanisms that control the opening and closing of stomata. Rounding out the list of macronutrients are calcium, sulfur, and magnesium. Carbon, hydrogen, and oxygen come from the air and water, and are the only plant nutrients that are not considered to be minerals. The other six macronutrients are minerals, inorganic substances available to plants through the soil as ions dissolved in water. Most minerals that serve as nutrients in plants are derived from the weathering of rocks and inorganic particles in the Earth's crust.

The other elements essential to plants are also minerals, and are classed as **micronutrients** because plants require them only in trace amounts. Nevertheless, they are just as vital as macronutrients to a plant's health and survival. For example, 5 metric tons of potatoes contain roughly the amount of copper in a single (copper-plated) penny—yet without it, potato plants are sickly and do not produce normal tubers.

Chlorine was identified as a micronutrient nearly a century after Sachs's experiments. The researchers who discovered its role performed hydroponic culture experiments in a California laboratory near the Pacific Ocean, where the air, like coastal air everywhere, contains sodium chloride. The investigators found that their test plants could obtain tiny but sufficient quantities of chlorine from the air, as well as from sweat (which also contains NaCl) on the researchers' own hands. Great care had to be taken to exclude chlorine from the test plants' growing environment in order to prove that it was essential.

In some cases, plant seeds contain enough of certain trace minerals to sustain the adult plant. For example, nickel (Ni^{2+}) is a component of urease, the enzyme required to hydrolyze urea. Urea is a toxic by-product of the breakdown of nitrogenous compounds, and it will kill cells if it accumulates. In the late 1980s investigators found that barley seeds contain enough nickel to sustain two complete generations of barley plants. Plants grown in the absence of nickel did not begin to show signs of nickel deficiency until the third generation.

Besides the 17 essential elements, some species of plants may require additional micronutrients. Experiments suggest that many, perhaps most, plants adapted to hot, dry conditions require sodium; many plants that photosynthesize by the C_4 pathway (see Section 9.4) appear to be in this group. A few plant species require selenium, which is also an essential micronutrient for animals. Horsetails *(Equisetum)* require silicon, and some grasses (such as wheat) may also need it. Scientists continue to discover additional micronutrients for specific plant groups.

Both micronutrients and macronutrients play vital roles in plant metabolism. Many function as cofactors or coenzymes in protein synthesis, starch synthesis, photosynthesis, and aerobic respiration. As you read in Section 32.1, some also have a role in creating solute concentration gradients across plasma membranes, which are responsible for the osmotic movement of water.

Nutrient Deficiencies Cause Abnormalities in Plant Structure and Function

Plants differ in the quantity of each nutrient they require—the amount of an essential element that is adequate for one plant species may be insufficient for another. Lettuce and other leafy plants require more nitrogen and magnesium than do other plant types, for example, and alfalfa requires significantly more potassium than do lawn grasses. An adequate amount of an essential element for one plant may even be harmful to another. For example, the amount of boron required for normal growth of sugar beets is toxic for soybeans. For these reasons, the nutrient content of soils is an important factor in determining which plants will grow well in a given location.

Plants that are deficient in one or more of the essential elements develop characteristic symptoms (Table 33.1 lists some observable symptoms of nutrient deficiencies). The symptoms give some indication of the metabolic roles the missing elements play. Deficiency symptoms typically include stunted growth, abnormal leaf color, dead spots on leaves, or abnormally formed stems **(Figure 33.3).** For instance, iron is a component of the cytochromes upon which the cellular electron transfer system depends, and it plays a role in reactions that synthesize chlorophyll. Iron deficiency causes **chlorosis,** a yellowing of plant tissues that results from a lack of chlorophyll (see Figure 33.3b). Because ionic iron (Fe^{3+}) is relatively insoluble in water, gardeners often fertilize plants with a soluble iron compound called chelated iron to stave off or cure chlorosis. Similarly, because magnesium is a necessary component of chlorophyll, a plant deficient in this element has fewer chloroplasts than normal in its leaves and other photosynthetic parts. It appears paler green than normal, and its growth is stunted because of reduced photosynthesis (see Figure 33.3c).

Plants that lack adequate nitrogen may also become chlorotic (see Figure 33.3d), with older leaves yellowing first because the nitrogen is preferentially shunted to younger, actively growing plant parts. This adaptation is not surprising, given nitrogen's central role in the synthesis of amino acids, chlorophylls, and other compounds vital to plant metabolism. With some other mineral deficiencies, young leaves are the first to

Figure 33.3
Leaves and stems of tomato plants showing visual symptoms of seven different mineral deficiencies. The plants were grown in the laboratory, where the experimenter could control which nutrients were available.
(Photos by E. Epstein, University of California, Davis.)

show symptoms. These kinds of observations underscore the point that plants utilize different nutrients in specific, often metabolically complex ways.

Soils are more likely to be deficient in nitrogen, phosphorus, potassium, or some other essential mineral than to contain too much, and farmers and gardeners typically add nutrients to suit the types of plants they wish to cultivate. They may observe the deficiency symptoms of plants grown in their locale or have soil tested in a laboratory, then choose a fertilizer with the appropriate balance of nutrients to compensate for the deficiencies. Packages of commercial fertilizers use a numerical shorthand (for example, 15-30-15) to indicate the percentages of nitrogen, phosphorus, and potassium they contain.

Study Break

1. What are the two main categories of the essential elements plants need? Give several examples of each.
2. Do all plants require the same basic nutrients in the same amounts? Explain.

33.2 Soil

Soil anchors plant roots and is the main source of the inorganic nutrients plants require. It also is the source of water for most plants, and of oxygen for respiration in root cells. The physical texture of soil determines whether root systems have access to sufficient water and dissolved oxygen. These characteristics reinforce the conclusion that the physical and chemical properties of soils in different habitats have a major impact on the ability of plant species to grow, survive, and reproduce there.

The Components of a Soil and the Size of the Particles Determine Its Properties

Soil is a complex mix of mineral particles, chemical compounds, ions, decomposing organic matter, air, water, and assorted living organisms. Most soils develop from the physical or chemical weathering of rock (which also liberates mineral ions). The different kinds of soil particles range in size from sand (2.0–0.02 mm) to silt (0.02–0.002 mm) and clay (diameter less than 0.002 mm). These mineral particles usually are mixed with various organic components, including **humus**—decomposing parts of plants and animals, animal droppings, and other organic matter. Dry humus has a loose, crumbly texture. It can absorb a great deal of water and thus contributes to the capacity of soil to hold water. Organic molecules in humus are reservoirs of nutrients, including nitrogen, phosphorus, and sulfur, that are vital to living plants.

The relative proportions of the different sizes of mineral particles give soil its basic texture—gritty if the soil is largely sand, smooth if silt predominates, and dense and heavy if clay is the major component. A soil's texture in turn helps determine the number and volume of pores—air spaces—that it contains. The relative amounts of sand, silt, and clay determine whether a soil is sticky when wet, with few air spaces (mostly clay), or dries quickly and may wash or blow away (mostly sand). Clay soils are more than 30% clay, while sandy soils contain less than 20% clay or silt.

The piles of bagged humus for sale at garden centers each spring reflect the fact that the amount of humus in a soil also affects plant growth. Its plentiful organic material feeds decomposers whose metabolic activities in turn release minerals that plant roots can take up, but that is not its only value in soil. Humus helps retain soil water and, with its loose texture, helps aerate soil as well. Well-aerated soils containing roughly equal proportions of humus, sand, silt, and clay are **loams,** and they are the soils in which most plants do best.

Soil also contains living organisms. Trillions of bacteria, hundreds of millions of fungi, and several million nematodes—not to mention earthworms and insects—are present in every square meter of fertile soil. Together with the roots of living plants, these organisms have a major influence on the composition and characteristics of soil. Bacteria and fungi decompose organic matter; burrowing creatures such as earthworms aerate the soil; and when plant roots die they contribute their organic matter to the soil.

As soils develop naturally, they tend to take on a characteristic vertical profile, with a series of layers or **horizons (Figure 33.4).** Each horizon has a distinct texture and composition that varies with soil type. The top layer of surface litter—twigs and leaves, animal dung, fungi, and similar organic matter—is accordingly called the *O horizon.* The most fertile soil layer, called **topsoil,** occurs just below and forms the *A horizon.* This layer may be less than a centimeter deep on steep slopes to more than a meter deep in grasslands. It consists of humus mixed with mineral particles and usually is fairly loose; it is here that the roots of most herbaceous plants are located. Below the topsoil is the **subsoil** or *B horizon,* a layer of larger soil particles containing relatively little organic matter. Mineral ions, including those that serve as nutrients in plants, tend to accumulate in the B horizon, and mature tree roots generally extend into this layer. Under it is the *C horizon,* a layer of mineral particles and rock fragments that extends down to bedrock.

Regions where the topsoil is deep and rich in humus are ideal for agriculture; the vast grasslands of the North American Midwest and Ukraine are prime examples. Without soil management and intensive irrigation, crops

William Ferguson

O horizon
Fallen leaves and other organic material littering the surface of mineral soil

A horizon
Topsoil, which contains some percentage of decomposed organic material and which is of variable depth; here it extends about 30 cm below the soil surface

B horizon
Subsoil; larger soil particles than the A horizon, not much organic material, but greater accumulation of minerals; here it extends about 60 cm below the A horizon

C horizon
No organic material, but partially weathered fragments and grains of rock from which soil forms; extends to underlying bedrock

Bedrock

Figure 33.4
A representative profile of soil horizons.

usually cannot be grown in deserts due to the lack of rainfall and low humus in the soil. Nor can agriculture flourish for long on land cleared of a tropical rainforest, due to the soil leaching and lack of nutrients described in the chapter introduction.

The Characteristics of Soil Affect Root–Soil Interactions

In different regions, and even in different parts of a local area, the proportions of the types of soil particles can differ dramatically, with corresponding variations in the soil's suitability for plant growth.

Plants have evolved adaptations to many otherwise inhospitable soil environments, as you will see in the following section. First, however, we consider the general ways in which soil composition influences the ability of plant roots to obtain water and minerals.

Water Availability. As water flows into and through soil, gravity pulls much of the water down through the spaces between soil particles into deeper soil layers. This available water is part of the **soil solution (Figure 33.5),** a combination of water and dissolved substances that coats soil particles and partially fills pore spaces. The solution develops through ionic interactions between water molecules and soil particles. Clay particles and the organic components in soil (especially proteins) often bear negatively charged ions on their surfaces. The negative charges attract the polar water molecules, which form hydrogen bonds with the soil particles (see Section 2.4).

Unless a soil is irrigated, the amount of water in the soil solution depends largely on the amount and pattern of precipitation (rain or snow) in a region. How much of this water is actually available to plants depends on the soil's composition—the size of the air spaces in which water can accumulate and the proportions of water-attracting particles of clay and organic matter. By volume, soil is about one-half solid particles and one-half air space.

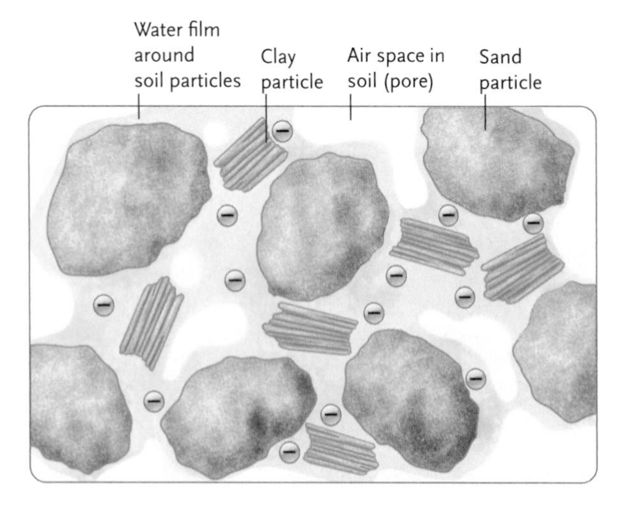

Figure 33.5
Location of the soil solution. Negatively charged ions on the surfaces of soil particles attract water molecules, which coat the particles and fill spaces between them (blue). Hydrogen bonds between water and soil components counteract the pull of gravity and help hold some water in the soil spaces.

The size of the particles in a given soil has a major effect on how well plants will grow there. Sandy soil has relatively large air spaces, so water drains rapidly below the top two soil horizons where most plant roots are located. Soils rich in clay or humus are often high in water content, but in the case of clay, ample water is not necessarily an advantage for plants. Whereas a humus-rich soil contains lots of air spaces, the closely layered particles in clay allow few air spaces—and what spaces there are tend to hold tightly the water that enters them. The lack of air spaces in clay soils also severely limits supplies of oxygen available to roots for cellular respiration, and the plant's metabolic activity suffers. Thus, few plants can grow well in clay soils, even when water content is high. (Overwatered houseplants die because their roots are similarly "smothered" by water.) Plants do not fare much better in drier clay-rich soils, because roots cannot extract the existing water and cannot easily penetrate the densely packed clay. These characteristics explain why good agricultural soils tend to be sandy or silty loams, which contain a mix of humus and coarse and fine particles.

As you learned in Chapter 31, root hairs are specialized extensions of root epidermal cells; they directly contact the soil solution and allow roots to absorb water (and dissolved ions). The soil solution usually contains fewer dissolved solutes than does the water in the cells of plant roots. Accordingly, water tends to diffuse from wet soil into the roots, following the osmotic gradient (see Section 32.2). As roots extract water from the surrounding soil, however, the remaining water molecules are held to the negatively charged clay surfaces with ever-increasing force. Plants start to wilt when the forces that draw water into their root cells equal those holding water in soil. Under these conditions, water no longer diffuses into roots, but it continues to evaporate from leaves and to be used in photosynthesis. Plants that survive in deserts or in salty soils have adaptations that permit their roots to absorb water even when osmotic conditions in soil do not favor water movement into the plant.

Mineral Availability. Some mineral nutrients enter plant roots as cations (positively charged ions) and some as anions (negatively charged ions). Although both cations and anions may be present in soil solutions, they are not equally available to plants.

Cations such as magnesium (Mg^{2+}), calcium (Ca^{2+}), and potassium (K^+) cannot easily enter roots because they are attracted by the net negative charges on the surfaces of soil particles. To varying degrees, they become reversibly bound to negative ions on the surfaces. Attraction in this form is called *adsorption*. The cations are made available to plant roots through

cation exchange, a mechanism in which one cation, usually H^+, replaces a soil cation **(Figure 33.6)**. There are two main sources for the hydrogen ions. Respiring root cells release carbon dioxide, which dissolves in the soil solution, yielding carbonic acid (H_2CO_3). Subsequent reactions ionize H_2CO_3 to produce bicarbonate (HCO_3^-) and hydrogen ions (H^+). Reactions involving organic acids inside roots also produce H^+, which is excreted. As H^+ enters the soil solution, it displaces adsorbed mineral cations attached to clay and humus, freeing them to move into roots. Other types of cations may also participate in this type of exchange, as shown in Figure 33.6.

By contrast, anions in the soil solution, such as nitrate (NO_3^-), sulfate (SO_4^{2-}), and phosphate (PO_4^-), are only weakly bound to soil particles, and so they generally move fairly freely into root hairs. However, because they are so weakly bound compared with cations, anions are more subject to loss from soil by leaching.

The pH of soil also affects the availability of some mineral ions. Soil pH is a function of the balance between cation exchange and other processes that raise or lower the concentration of H^+ in soil. As noted earlier, in areas that receive heavy rainfall, soils tend to become acidic (that is, they have a pH of less than 7). This acidification occurs in part because moisture promotes the rapid decay of organic material in humus; as the material decomposes, the organic acids it contains are released. Acid precipitation, which results from the release of sulfur and nitrogen oxides into the air, also contributes to soil acidification. By contrast, the soil in arid regions, where precipitation is low, often is alkaline (the pH is greater than 7).

Although most plants are not directly sensitive to soil pH, chemical reactions in very acid ($pH < 5.5$) or very alkaline ($pH > 9.5$) soils can have a major impact on whether plant roots take up various mineral cations. For example, experiments have demonstrated that in the presence of OH^- in alkaline soil, calcium and phosphate ions react to form insoluble calcium phosphates. The phosphate captured in these compounds is as unavailable to roots as if it were completely absent from the soil.

For a soil to sustain plant life over long periods, the mineral ions that plants take up must be replenished naturally or artificially. Over the long run, some mineral nutrients enter the soil from the ongoing weathering of rocks and smaller bits of minerals. In the shorter run, minerals, carbon, and some other nutrients are returned to the soil by the decomposition of organisms and their parts or wastes. Other inputs occur when airborne compounds, such as sulfur in volcanic and industrial emissions, become dissolved in rain and fall to earth. Still others, including compounds of nitrogen and phosphorus, may enter soil in fertilizers.

Although the use of commercial fertilizers maintains high crop yields, agricultural chemicals do not add humus to the soil. Their use can also cause serious problems, as when nitrogen-rich runoff from agricultural fields promotes the serious overgrowth of algae in lakes and bays. In many parts of the world, industrial pollutants such as cadmium, lead, and mercury are increasingly serious soil contaminants. The use of plants to remove such materials from soil, called *phytoremediation*, is the topic of this chapter's *Focus on Research*.

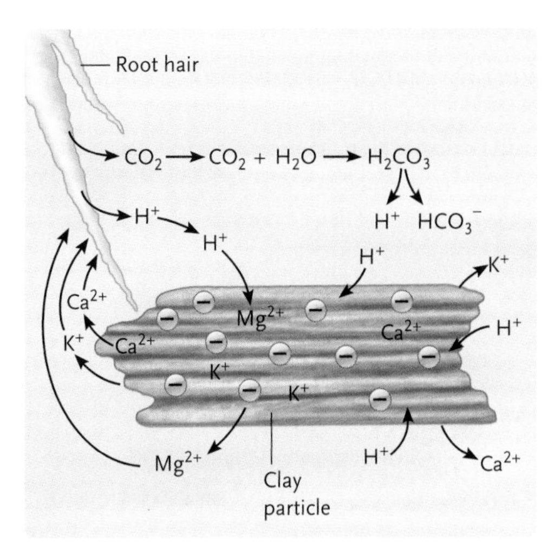

Figure 33.6
Cation exchange on the surface of a clay particle. When cations come into contact with the negatively charged surface of the particle, they become adsorbed. As one type of cation, such as H^+, becomes adsorbed, other ions are liberated and can be taken up by plant roots.

Study Break

1. Why is humus an important component of fertile soil?
2. How does the composition of a soil affect a plant's ability to take up water?
3. What factors affect a plant's ability to absorb minerals from the soil?

33.3 Obtaining and Absorbing Nutrients

Soil managed for agriculture can be plowed, precisely irrigated, and chemically adjusted to provide air, water, and nutrients in optimal quantities for a particular crop. By contrast, in natural habitats there are wide variations in soil minerals, humus, pH, the presence of other organisms, and other factors that influence the availability of essential elements. Although adequate carbon, hydrogen, and oxygen are typically available, other essential elements may not be as abundant. In particular, nitrogen, phosphorus, and potassium are often relatively scarce. The evolutionary solutions to

FOCUS ON RESEARCH

Applied Research: Plants Poised for Environmental Cleanup

For several decades researchers have been searching for efficient modes of *phytoremediation*—the use of plants to remove pollutants from the environment. A high-profile target is the highly toxic organic compound methylmercury (MeHg). This substance is present in coastal soils and wetlands contaminated by industrial wastes that contain an ionic form of the element mercury called Hg (II). Bacteria in contaminated sediments metabolize Hg(II) and generate MeHg as a metabolic by-product. Once MeHg forms, it enters the food web and eventually becomes concentrated in tissues of fishes and other animals. In humans MeHg can lead to degeneration of the nervous system and is the cause of most cases of mercury poisoning due to consuming contaminated fish.

In the 1990s a team of scientists including Scott Bizily and Richard Meagher at the University of Georgia decided to try to modify plants genetically so that they could detoxify mercury-contaminated soil and wetlands. It was already known that bacteria in contaminated sediments possess two genes, *merA* and *merB*, which encode enzymes that convert MeHg into elemental mercury (Hg)—a relatively inert substance that is much less dangerous to organisms. Both these bacterial mercury-resistance genes had already been cloned by others. After modifying the cloned genes so that they could be expressed in plants, the team used a vector (the bacterium *Rhizobium radiobacter*) to introduce each gene into several different sets of *Arabidopsis thaliana* plants (thale cress). They eventually obtained three groups of transgenic plants: some that were *merA* only, some that were *merB* only, and some that were *merA* and *merB*. In a series of experiments, seeds from each group were grown (along with wild-type controls) in five different growth media—one containing no mercury and the other four containing increasing concentrations of methylmercury. Wild-type and *merA* seeds germinated and grew only in the mercury-free growth medium. The *merB* seedlings fared somewhat better: they germinated and grew briefly even at the highest concentrations of MeHg, but soon became chlorotic and died. By contrast, seeds with the *merA/merB* genotype not only germinated, but the resulting seedlings grew into robust plants with healthy root and shoot systems. In later tests *merA/merB* plants were grown in chambers in which the chemical composition of the air was monitored. This study revealed that the doubly transgenic plants also were transpiring large amounts of Hg. The implication of these findings was clear: *A. thaliana* plants having both *merA* and *merB* genes were able to take up the toxic methylmercury with no ill effects and convert it to a harmless form. Meagher and his colleagues now are experimenting with ways of increasing the efficiency of phytoremediating enzymes when plant cells express *merA* and *merB*. They also are studying the mechanisms by which ionic mercury taken up by roots may be transported via the xylem to leaves and other shoot parts. The goal is to engineer plants that accumulate large quantities of mercury in aboveground tissues that can be harvested, leaving the living plant to continue its "work" of detoxifying a contaminated landscape.

these challenges include an array of adaptations in the structure and functioning of plant roots.

Root Systems Allow Plants to Locate and Absorb Essential Nutrients

Immobile organisms such as plants must locate nutrients in their immediate environment, and for plants the adaptive solution to this problem is an extensive root system. Roots make up 20% to 50% of the dry weight of many plants, and even more in species growing where water or nutrients are especially scarce, such as arctic tundra. As long as a plant lives, its root system continues to grow, branching out through the surrounding soil. Roots don't necessarily grow *deeper* as a root system branches out, however. In arid regions, a shallow-but-broad root system may be better positioned to take up water from occasional rains that may never penetrate below the first few inches of soil.

A root system grows most extensively in soil where water and mineral ions are abundant. As described in Section 31.4, roots take up ions in the regions just above the root tips. Over successive growing seasons, long-lived plants such as trees can develop millions, even billions, of root tips, each one a potential absorption site.

Root hairs, the diminutive absorptive structures shown in Figure 31.10c, are another significant adaptation for the uptake of mineral ions and water. In a plant such as a mature red oak *(Quercus rubra)*, which has a vast root system, the total number of root hairs is astronomical. Even in young plants, root hairs greatly increase the root surface area available for absorbing water and ions.

Recall from Chapter 32 that plant cell membranes also have ion-specific transport proteins by which they selectively absorb ions from soil. For example, from studies of plants such as *Arabidopsis thaliana*, a weed that has become a key model organism for plant research, we know that transport channels for potassium ions (K^+) are embedded in the cell membranes of root cortical cells. Such ion transporters absorb more or less of a particular ion depending on chemical conditions in the surrounding soil.

Insights from the Molecular Revolution

Getting to the Roots of Plant Nutrition

One way that mycorrhizal fungi benefit their host plants is by increasing their phosphate uptake from soils. How do the fungi accomplish this beneficial process? A molecular answer to this question came from Maria J. Harrison and Marianne L. van Buuren at the Samuel Roberts Noble Foundation in Ardmore, Oklahoma, who were able to identify a gene in one of these fungi that encodes a phosphate transport protein.

Harrison and van Buuren began with a cDNA library prepared from a plant that had been colonized by the mycorrhizal fungus *Glomus versiforme.* They then probed the plant-derived cDNA with a gene that encodes a phosphate transporter in yeast. The goal was to determine whether any cDNA sequences were similar to the yeast gene. (Recall from Section 18.1 that a cDNA library is a cloned collection of DNA sequences derived from mRNAs isolated from a cell. Hence it represents sequences that encode proteins.) The transport protein encoded by the yeast gene is embedded in the plasma membrane, where it uses an H^+ gradient as an energy source to move phosphate ions into yeast cells by active transport (see Section 6.4).

When mixed with the plant-derived cDNA sequences, the probe did indeed pair with one of the cDNA sequences. Subsequent sequencing of the segment revealed that the cDNA coded for a protein with a structure typical of many eukaryotic and prokaryotic membrane transport proteins.

To eliminate the possibility that the probe was identifying a plant cDNA in the library rather than one from the mycorrhizal fungus, the investigators next used the identified cDNA to probe a preparation containing all the DNA of a plant that had not been colonized by *Glomus.* No pairing occurred with any of the plant DNA fragments, confirming that the cDNA represented a gene came from the fungus. Additional experiments supported this finding.

Harrison and van Buuren carried their investigation further to see whether the fungal gene actually encoded a phosphate transport protein. For this set of experiments, the investigators used a yeast mutant with a nonfunctional phosphate transporter. Because these mutant yeast cells cannot readily take in phosphate, they grow very slowly, even in a culture medium containing a high concentration of phosphate ions.

The researchers added the *Glomus* gene to the mutants under conditions that increased the likelihood that the yeast cells would take and incorporate the DNA. In response, the yeast cells began to grow rapidly and normally, indicating that they could now synthesize a functional phosphate transporter. When radioactive phosphate ions were added to the culture, the cells rapidly became labeled, confirming that they were taking up phosphate ions at a much greater rate than untreated mutants.

Harrison and van Buuren's study was the first to reveal the molecular basis of phosphate transport by the mycorrhizal fungi. More recent studies with potato plants *(Solanum tuberosum)* have identified a gene encoding a phosphate transporter protein that is expressed in parts of potato roots where mycorrhizae form. These lines of research may lead to methods for reducing the amount of phosphate fertilizers added to crop plants by identifying mycorrhizal fungi providing the most efficient phosphate uptake—or by engineering crop plants with an enhanced capacity to take in this essential nutrient.

Mycorrhizae, symbiotic associations between a fungus and the roots of a plant (see Section 28.3) also promote the uptake of water and ions—especially phosphate and nitrogen—in most species of plants. As shown in Figure 28.17, the fungal partner in the association often grows as a network of hyphal filaments around and beyond the plant's roots. Collectively, the hyphae provide a tremendous surface area for absorbing ions from a large volume of soil. As with plant roots, transport proteins shepherd ions into hyphae. Researchers have recently verified experimentally that hyphal transport proteins are encoded by the DNA of the fungus, not that of the plant (as described in *Insights from the Molecular Revolution*). Some of the plant's sugars and nitrogenous compounds nourish the fungus, and as the root grows, it uses some of the minerals that the fungus has secured. In other types of mycorrhizae, the fungus actually lives inside cells of the root cortex. Orchids, for example, depend on this type of mutualistic association. And, as will be described shortly, some other plants gain access to nitrogen by way of mutually beneficial associations with bacteria.

Nutrients Move into and through the Plant Body by Several Routes

Most mineral ions enter plant roots passively along with the water in which they are dissolved. Some enter root cells immediately. Others travel in solution *between* cells until they meet the endodermis sheathing the root's stele (see Figure 32.6). At the endodermis, the ions are actively transported into the endodermal cells and then into the xylem for transport throughout the plant.

Inside cells, most mineral ions enter vacuoles or the cell cytoplasm, where they become available for metabolic reactions. Some nutrients, such as nitrogen-containing ions, move in phloem from site to site in the plant, as dictated by growth and seasonal needs. In plants that shed their leaves in autumn, before the leaves age and fall significant amounts of nitrogen, phosphorus, potassium, and magnesium move out of them and into twigs and branches. This adaptation conserves the nutrients, which will be used in new growth the next season. Likewise, in late summer, mineral ions move to the roots and lower stem tissues of perennial range grasses that typically die back during the winter. These activities are regulated by hormonal signals, which are the topic of Chapter 35.

Plants Depend on Bacterial Metabolism to Provide Them with Usable Nitrogen

A lack of nitrogen is the single most common limit to plant growth. Air contains plenty of gaseous nitrogen—almost 80% by volume—but plants lack the enzyme necessary to break apart the three covalent bonds in each N_2 molecule (N≡N). Some nitrogen from the atmosphere reaches the soil in the form of nitrate, NO_3^-, and ammonium ion, NH_4^+. Plants can absorb both these inorganic nitrogen compounds, but usually there is not nearly enough of them to meet plants' ongoing needs.

Nitrogen also enters the soil in organic compounds as dead organisms and animal wastes decompose. For example, dried blood is about 12% nitrogen by weight and chicken manure is about 5% nitrogen, but the nitrogen is bound up in complex organic molecules such as proteins, and in that form it is unavailable to plants. Instead, the main natural processes that replenish soil nitrogen and convert it to absorbable form are carried out by bacteria. These processes are described later and summarized in **Figure 33.7.** They are part of the *nitrogen cycle,* the global movement of nitrogen in its various chemical forms from the environment to organisms and back to the environment, which is described in Chapter 51.

Production and Assimilation of Ammonium and Nitrate. The incorporation of atmospheric nitrogen into compounds that plants can take up is called **nitrogen fixation.** Metabolic pathways of *nitrogen-fixing bacteria* living in the soil or in mutualistic association with plant roots add hydrogen to atmospheric N_2, producing two molecules of NH_3 (ammonia) and one H_2 for each N_2 molecule. The process requires a substantial input of ATP and is catalyzed by the enzyme nitrogenase. In a final step, H_2O and NH_3 react, forming NH_4^+ (ammonium) and OH^-.

Another bacterial process, called **ammonification,** also produces NH_4^+ when soil bacteria known as *ammonifying bacteria* break down decaying organic matter. In this way, nitrogen already incorporated into plants and other organisms is recycled.

Although plants use NH_4^+ to synthesize organic compounds, most plants absorb nitrogen in the form of nitrate, NO_3^-. Nitrate is produced in soil by **nitrification,**

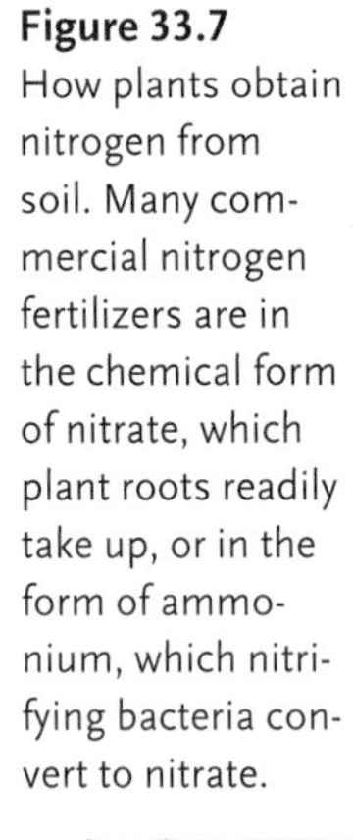
Figure 33.7
How plants obtain nitrogen from soil. Many commercial nitrogen fertilizers are in the chemical form of nitrate, which plant roots readily take up, or in the form of ammonium, which nitrifying bacteria convert to nitrate.

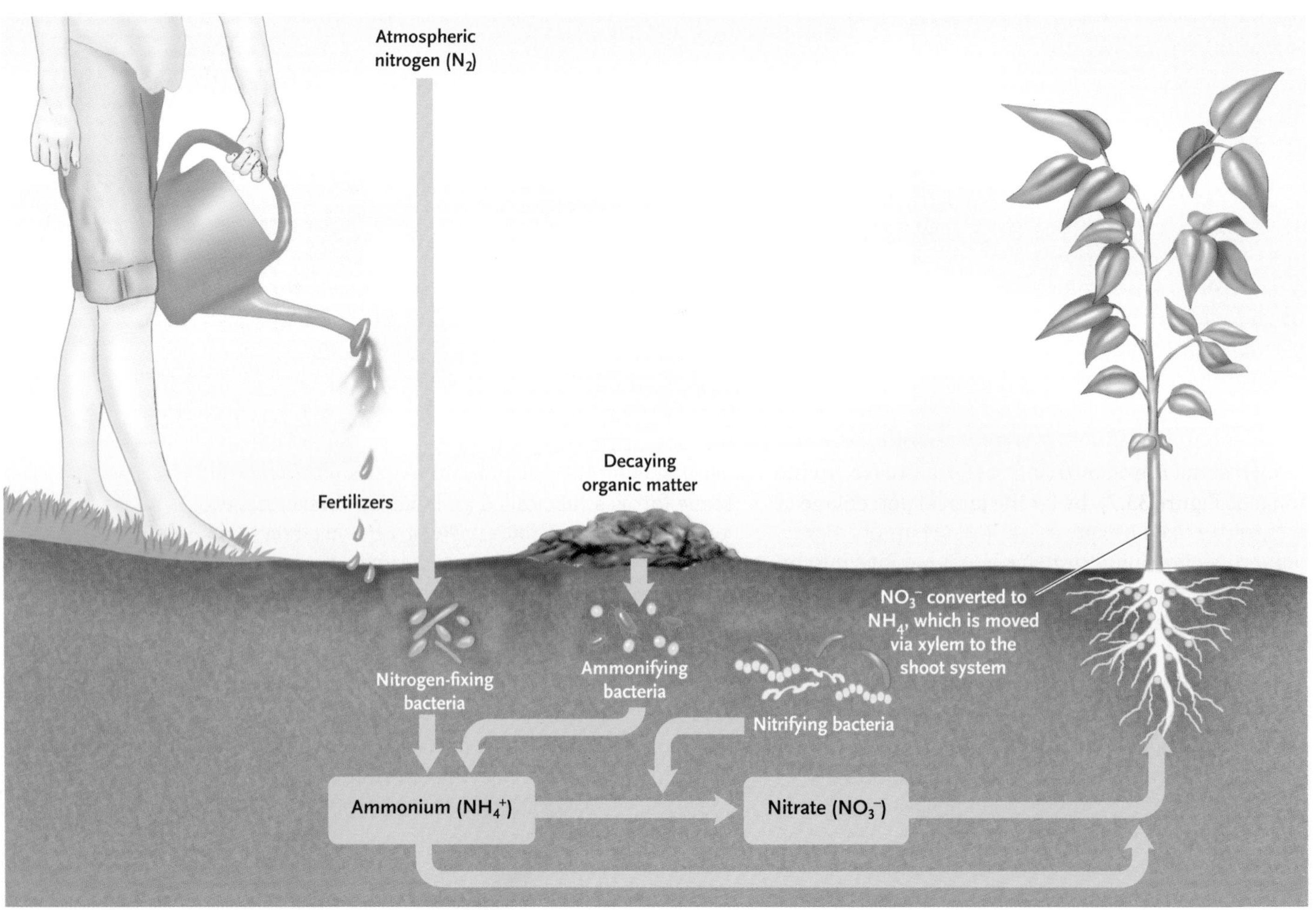

Figure 33.8
The beneficial effect of root nodules. **(a)** Root nodules on a soybean plant *(Glycine max)*. **(b)** Soybean plants growing in nitrogen-poor soil. The plants on the right were inoculated with *Rhizobium* cells and developed root nodules. **(c)** False-color transmission electron micrograph showing membrane-bound bacteroids (red) in a root nodule cell. Membranes that enclose the bacteroids appear blue. The large yellow-green structure is the cell's nucleus.

in which NH_4^+ is oxidized to NO_3^-. Soils generally teem with *nitrifying bacteria,* which carry out this process. Because of ongoing nitrification, nitrate is far more abundant than ammonium in most soils. Usually, the only soils from which plant roots take up ammonium directly are highly acidic, such as in bogs, where the low pH is toxic to nitrifying bacteria.

Nitrogen Assimilation. Once inside root cells, absorbed NO_3^- is converted by a multistep process back to NH_4^+. In this form, nitrogen is rapidly used to synthesize organic molecules, mainly amino acids. These molecules pass into the xylem, which transports them throughout the plant. In some plants, the nitrogen-rich precursors travel in xylem to leaves, where different organic molecules are synthesized. Those molecules travel to other plant cells in the phloem.

Nitrogen Fixation in Plant–Bacteria Associations. Although some nitrogen-fixing bacteria live free in the soil (see Figure 33.7), by far the largest percentage of nitrogen is fixed by species of *Rhizobium* and *Bradyrhizobium,* which form mutualistic associations with the roots of plants in the legume family. The host plant supplies organic molecules that the bacteria use for cellular respiration, and the bacteria supply NH_4^+ that the plant uses to produce proteins and other nitrogenous molecules. In legumes (peas, beans, clover, and alfalfa), the nitrogen-fixing bacteria reside in **root nodules,** localized swellings on roots **(Figure 33.8).** Farmers may exploit root nodules to increase soil nitrogen by rotating crops (for example, planting soybeans and corn in alternating years). When the legume crop is harvested, the root nodules and other tissues remaining in the soil enrich its nitrogen content.

Decades of research have revealed the details of how this remarkable relationship unfolds. Usually, a single species of nitrogen-fixing bacteria colonizes a single legume species, drawn to the plant's roots by chemical attractants—primarily compounds called flavonoids—that the roots secrete. Through a sequence of exchanged molecular signals, bacteria are able to penetrate a root hair and form a colony inside the root cortex.

An association between a soybean plant *(Glycine max)* and *Bradyrhizobium japonicum* illustrates the process. In response to a specific flavonoid released by soybean roots, bacterial genes called *nod* genes (for *nodule*) begin to be expressed **(Figure 33.9a).** Products of the *nod* gene cause the tip of the root hair to curl toward the bacteria and trigger the release of bacterial enzymes that break down the root hair cell wall **(Figure 33.9b).** As bacteria enter the cell and multiply, the plasma membrane forms a tube called an **infection thread** that extends into the root cortex, allowing the bacteria to invade cortex cells **(Figure 33.9c).** The enclosed bacteria, now called **bacteroids,** enlarge and become immobile. Stimulated by still other *nod* gene products, cells of the root cortex begin to divide. This region of proliferating cortex cells forms the root nodule **(Figure 33.9d).** Typically, each cell in a root nodule contains several thousand bacteroids; the plant takes up some of the nitrogen fixed by the bacteroids, and the bacteroids utilize some compounds produced by the plant.

Inside bacteroids, N_2 is reduced to NH_4^+ (ammonium) using ATP produced by cellular respiration. The process is catalyzed by nitrogenase. Ammonium

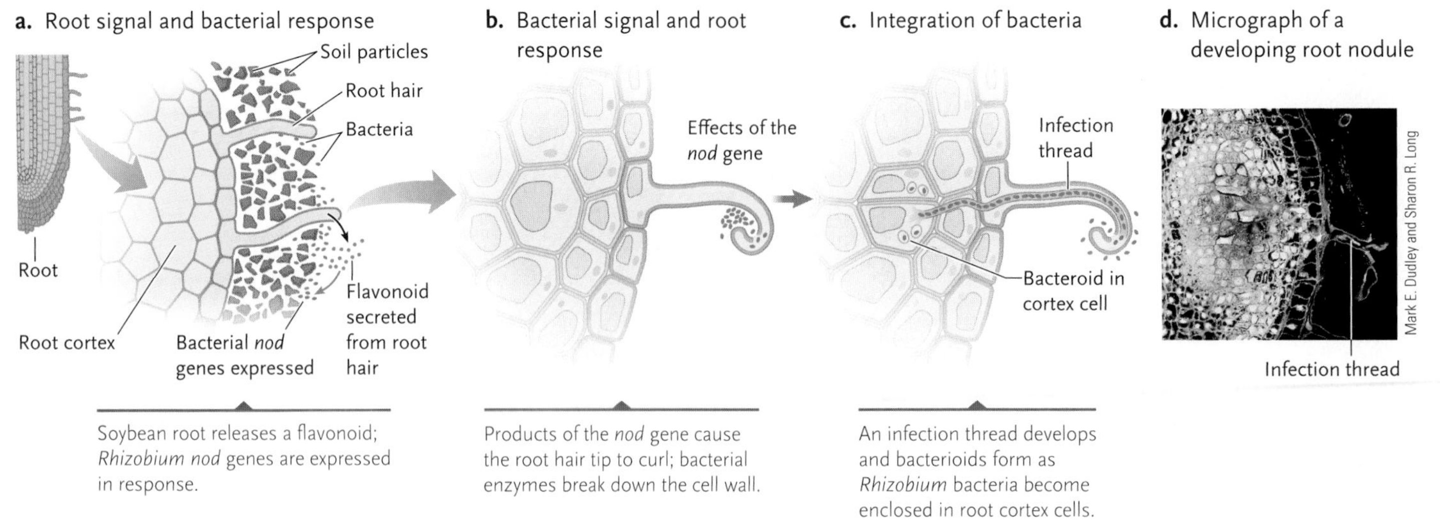

Figure 33.9
Root nodule formation in legumes, which interact mutualistically with the nitrogen-fixing bacteria *Rhizobium* and *Bradyrhizobium*.

is highly toxic to cells if it accumulates, however. Thus, NH_4^+ is moved out of bacteroids into the surrounding nodule cells immediately and converted to other compounds, such as the amino acids glutamine and asparagine.

One factor encoded by the bacterial *nod* genes stimulates plant nodule cells to produce a protein called **leghemoglobin** ("legume hemoglobin"). Like the hemoglobin of animal red blood cells, leghemoglobin contains a reddish, iron-containing heme group that

Unanswered Questions

Is "networking" the key to success for plants in some environments?

Key factors that influence plants' ability to take root, grow, and thrive—notably the availability of water and mineral nutrients—are belowground, in the form of mycorrhizae. As you have read in this chapter, field studies and traditional laboratory analyses established that the symbiotic associations between mycorrhizal fungi and plant roots are crucial elements in the survival of the vast majority of vascular plants. Many researchers, including Peter Kennedy and his colleagues at the University of California at Berkeley, also have wondered about possible broader impacts of mycorrhizal associations, such as the extent of their role, if any, in determining the diversity of plant species in different ecological settings and in determining the particular combinations of species that occur. Now Kennedy and others are harnessing molecular tools to shed light on new kinds of questions about interactions among plants and mycorrhizal fungi.

Researchers' ability to define and amplify fungal DNA sequences has revealed the existence of common mycorrhizal networks (CMNs), in which roots of individual plants of the same or different species all form mycorrhizae with the same individual fungus. This discovery has raised several questions: Do mineral ions or other resources pass between plants in a CMN? Several studies indicate that the answer is yes, but much more research is needed to refine scientific understanding of these interchanges. Does a CMN moderate the effects of competition among plants of different species? Does formation of a CMN improve the survival chances of seedlings, and so help shape the distribution of specific plant species in a given area? Kennedy and his coworkers are exploring these and other questions with respect to CMNs involving two tree species that grow in mixed forests near San Francisco, California—the coast Douglas fir *(Pseudotsuga menziesii)*, a gymnosperm, and the tanbark oak *(Lithocarpus densiflora)*, an angiosperm.

Research efforts by Kennedy and others are examining competition among different species of mycorrhizal fungi, which differ markedly in their resistance to drought and their capacity to take up nutrients. Among other objectives, these studies aim to determine if, or to what extent, the ability of a given plant species to withstand water stress or to gain access to soil nutrients depends on the particular species of fungus with which it forms mycorrhizae. And do the benefits of mycorrhizae increase or decline as environmental conditions change? Answers to such questions will add a new dimension to our understanding of plant nutrition, as well as to our appreciation of what has been called "possibly the most important form of symbiosis in nature."

Beverly McMillan

a. Cobra lily *(Darlingtonia californica)*

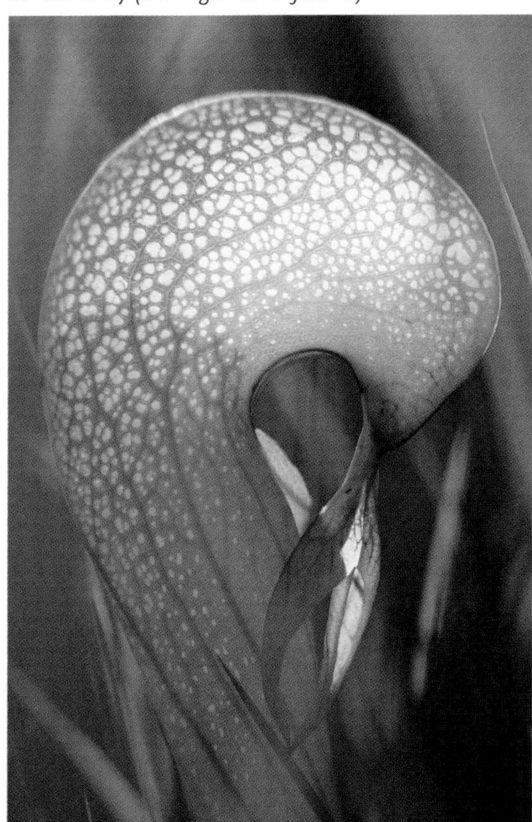

David Cavagnaro/Peter Arnold, Inc.

b. Dodder *(Cuscuta)*

© Grant Heilman Photography

c. Snow plant *(Sarcodes sanguinea)*

Beverly McMillan

d. Lady-of-the-night orchid *(Brassavola nodosa)*

© Prem Subrahmanyan/www.premdesign.com

Figure 33.10
Some plants with unusual adaptations for obtaining nutrients. (a) Cobra lily *(Darlingtonia californica)*, a carnivorous plant. The patterns formed by light shining through the plant's pitcherlike leaves are thought to confuse insects that have entered the pitcher, making an exit more difficult. **(b)** A parasitic dodder, one of the more than 150 *Cuscuta* species. Dodders have slender yellow to orange stems that twine around the host plant before producing haustorial roots that absorb nutrients and water from the host's xylem and phloem. **(c)** Snow plant *(Sarcodes sanguinea)*, which pops up in the deep humus of shady conifer forests after snow has melted in spring. This species lacks chlorophyll and does not photosynthesize. Instead its roots intertwine with hyphae of soil fungi that also form associations with the roots of nearby conifers. Radiocarbon studies have shown that the fungi take up sugars and other nutrients from the trees and pass a portion of this food on to the snow plant. **(d)** The lady-of-the-night orchid *(Brassavola nodosa)*, a tropical epiphyte.

binds oxygen. Its color gives root nodules a pinkish cast (see Figure 33.8). Leghemoglobin picks up oxygen at the cell surface and shuttles it inward to the bacteroids. This method of oxygen delivery is vital, because nitrogenase, the enzyme responsible for nitrogen fixation, is irreversibly inhibited by excess O_2. Leghemoglobin delivers just enough oxygen to maintain bacteroid respiration without shutting down the action of nitrogenase.

Some Plants Obtain Scarce Nutrients in Other Ways

The cobra lily (*Darlingtonia californica;* **Figure 33.10a**) is one of a curious group of plants that obtain nitrogen and other nutrients by trapping and digesting animals. Although such plants are said to be carnivorous (meat eaters), in fact they do not ingest food and digest it, as carnivorous animals do. Rather, they have become adapted to survive in nutrient-deficient environments through elaborate mechanisms for extracellular digestion and absorption. The cobra lily's leaves form a "pitcher" that is partly filled with digestive enzymes. Insects lured in by attractive odors often wander deeper into the pitcher, encountering downward-pointing leaf hairs that have a slick, waxy coating and speed the insect's descent into the pool of enzymes.

Dodders **(Figure 33.10b)** and thousands of other species of flowering plants are parasites that obtain some or all of their nutrients from the tissues of other plants. Parasitic species develop **haustorial roots** (similar to the haustoria of fungi described in Chapter 28) that penetrate deep into the host plant and tap into its vascular tissues. Although some parasitic plants, like mistletoe, contain chlorophyll and thus can photosynthesize, dodders and other nonphotosynthesizers rob the host of sugars as well as water and minerals.

The snow plant (*Sarcodes sanguinea*) shows a variation on this theme. As its deep red color suggests **(Figure 33.10c),** it lacks chlorophyll, but it doesn't have haustorial roots. Instead, the snow plant's roots take up nutrients from mycorrhizae they "share" with the roots of nearby conifers.

Epiphytes, such as the tropical orchid pictured in **Figure 33.10d,** are not parasitic even though they grow on other plants. Some trap falling debris and rainwater among their leaves, while their roots (including mycorrhizae, in the case of the orchid) invade the moist leaf litter and absorb nutrients from it as the litter decomposes. In temperate forests, many mosses and lichens are epiphytes.

These and other strategies plants have evolved for obtaining nutrients and water are only part of the survival equation, however. Plants use nutrients not only for growth and maintenance, but also, of course, for building structures such as pollen, flowers, and seeds used in reproduction—our topic in Chapter 34.

Study Break

1. What is a mycorrhiza, and why are mycorrhizal associations so vital to many plants?
2. Distinguish between nitrogen fixation, ammonification, and nitrification.
3. Summarize the mechanism by which associations with bacteria supply nitrogen to plants such as legumes.

Review

Go to ThomsonNOW™ at www.thomsonedu.com/login to access quizzing, animations, exercises, articles, and personalized homework help.

33.1 Plant Nutritional Requirements

- Plants require carbon, hydrogen, oxygen, nitrogen, and 13 other essential nutrients (Table 33.1). With enough sunlight and these nutrients, plants can synthesize all the compounds they need.
- Nine essential elements are macronutrients, required in relatively large amounts. Four of these elements—carbon, hydrogen, oxygen, and nitrogen—are the main building blocks in the synthesis of carbohydrates, lipids, proteins, and nucleic acids. The essential macronutrients dissolved in the soil solution are nitrogen, potassium, calcium, magnesium, phosphorus, and sulfur.
- Micronutrients are required in only minuscule amounts, but they too are essential. The ones identified to date are chlorine, iron, boron, manganese, zinc, copper, molybdenum, and nickel.
- Each plant species requires specific nutrients in specific amounts. Typical deficiency symptoms are stunted growth, yellowing or other abnormal changes in leaf color, dead spots on leaves, or abnormally formed stems (Figure 33.3).
- Most mineral ions enter plant roots dissolved in water. Inside cells, most mineral ions enter vacuoles or the cell cytoplasm, where they become available for metabolic reactions. Some elements, such as nitrogen and potassium, can move from site to site in phloem as the plant grows.

33.2 Soil

- Soil is composed of sand, silt, and clay particles, usually held together by humus and other organic components. Humus absorbs a great deal of water, and so contributes to the water-holding capacity of soil.
- The relative proportions of various soil mineral particles and humus give soil its basic texture and structure. The best agricultural soils are loams that contain clay, sand, silt, and humus in roughly equal proportions. Topsoil is the most fertile soil layer (Figure 33.4).
- Due to charge differences between soil particles and water molecules, soil particles are thinly coated by the soil solution, a mixture of water and solutes (Figure 33.5). From this solution root hairs and other root epidermal cells absorb water and solutes.
- The amount of water available to plant roots depends mainly on the relative proportions of different soil components. Water moves quickly through sandy soils, while soils rich in clay and humus tend to hold the most water.
- Cations become adsorbed on the negatively charged surfaces of soil particles, potentially limiting their uptake by roots. Cation exchange, in which mineral cations are replaced by H^+, helps make these nutrients available to plants (Figure 33.6). Anions are more weakly bound to soil particles; they move more readily into root hairs but also are more apt to leach out of topsoil. In nature, the soil solution surrounding plant roots generally contains only tiny amounts of essential mineral ions

Animation: Soil profile

33.3 Obtaining and Absorbing Nutrients

- Numerous adaptations help plants solve the problems of obtaining and absorbing essential nutrients. Root systems penetrate the soil towards nutrients and water. Millions or billions of root hairs increase the root's absorptive surface. Ion-specific transporters in root cortical cells adjust the plant's uptake of particular ions. Mycorrhizal associations between fungi and plant roots enhance the absorption of nutrients, particularly phosphorus.
- Nitrogen usually is the scarcest nutrient in soil, and much of the usable soil nitrogen is produced by nitrogen-fixing bacteria. Nitrogen fixation reduces atmospheric N_2 to NH_4^+ (ammonium) in a reaction that requires the enzyme nitrogenase as a catalyst. Nitrifying bacteria rapidly convert NH_4^+ to nitrate, the form in which the roots of most plants absorb nitrogen (Figure 33.7).
- In legumes and a few other species, nitrogen-fixing bacteria reside in root nodules in a mutualistic association (Figure 33.8).
- Bacteria enclosed in a root nodule (bacteroids) reduce N_2 to NH_4^+ (Figure 33.9). The toxic NH_4^+ is moved out of the bacteroids and converted to nitrogen-rich, nontoxic compounds such as amino acids. In plants that do not form root nodules, the ni-

trate directly absorbed by roots is reduced to ammonium, which then is converted to nontoxic forms.

- In many plant species, root cells synthesize amino acids and other organic nitrogenous compounds, and these molecules are transported in xylem throughout the plant. In some plants, the nitrogen-rich precursors travel in xylem to leaves, where different organic molecules are synthesized. Those molecules then travel to other plant cells in phloem.
- A few plant species have evolved alternative mechanisms for obtaining some or all of their nutrients (Figure 33.10). So-called carnivorous plants typically produce insect-attracting secretions that contain enzymes which digest the animal's tissues. The plant then absorbs the released nutrients.
- Some plant species parasitize other plants. The parasite may or may not contain chlorophyll and carry out photosynthesis; species that do not photosynthesize obtain all of their nutrition from the host. Epiphytes grow on other plants but obtain nutrients independently.

Animation: Uptake of nutrients by plants

Questions

Self-Test Questions

1. Which best describes a micronutrient?
   a. It makes up 96% of the plant's dry mass.
   b. It cannot be replaced artificially.
   c. It is early on the periodic chart compared with macronutrients.
   d. It is required in large amounts during sunlight hours.
   e. It is an essential element.
2. Nutrient runoff from fertilizing lush lawns often causes "algal blooms" in nearby lakes, making swimming impossible. The fertilizer components most likely to have caused the blooms are:
   a. iron, magnesium, and nitrogen.
   b. nitrogen, phosphorus, and sulfur.
   c. nitrogen, potassium, and phosphorus.
   d. selenium, magnesium, and potassium.
   e. nitrogen, magnesium, and nickel.
3. Which of the following is/are *not* among the ideal soil conditions for growing crops?
   a. extremely large air spaces
   b. sandy or silty loam
   c. blend of sand and clay
   d. thick top soil
   e. less than 5% humus
4. Which of the following processes contributes to the uptake of mineral ions by plant roots?
   a. chlorosis
   b. osmosis
   c. anion exchange
   d. cation exchange
   e. growth of root hairs
5. Which of the following does *not* influence soil pH?
   a. rainfall
   b. hydroponic growth
   c. release of sulfur and nitrogen oxides into the air
   d. decomposition of organisms
   e. weathering of rock
6. Which of the following is a process that helps plants utilize nitrogen?
   a. nitrogen-fixing bacteria synthesizing nitrate
   b. ammonifying bacteria using ammonium to produce nitrate
   c. nitrifying bacteria converting NH_4^+ to NO_3^-
   d. the absorption of NH_4^+ by root hairs
   e. the absorption of atmospheric N_2 into the xylem
7. The *nod* genes in the bacteria in soybean nodules allow the bacteria to fix nitrogen. Which of the following is *not* a step in this process?
   a. The products of *nod* genes cause cells of the root cortex to divide and become the root nodule in which bacteroids fix nitrogen for the plant.
   b. In the cortex cells bacteria enlarge and become immobile, forming bacteroids.
   c. Bacteria enter the root hair cell and multiply, causing the cell plasma membrane to form an infection thread that extends into the root cortex.
   d. Roots release flavonoid, which turns on the expression of bacterial *nod* genes. Products of *nod* genes cause the tip of the root hair to curl toward the bacteria.
   e. Root hairs trigger release of bacterial enzymes that break down root hair cell walls.
   f. All of the above are steps in the process.
8. Carnivorous plants are deficient in:
   a. oxygen.
   b. phosphorus.
   c. potassium.
   d. nitrogen.
   e. carbon.
9. Haustorial roots are characteristic of plants that are:
   a. parasites.
   b. epiphytes.
   c. nitrate fixers.
   d. leghemoglobin users.
   e. carnivorous.
10. Identify the correct match of a nutrient with its function.
    a. chlorine: component of several enzymes
    b. potassium: component of nucleic acids
    c. phosphorus: component of most proteins
    d. manganese: role in shoot and root growth
    e. calcium: maintenance of cell walls and membrane permeability

Questions for Discussion

1. If you want to study factors that affect plant nutrition in nature, what would be the advantages and disadvantages of using a hydroponic culture method?
2. Gardeners often add a humus-rich "soil conditioner" to garden plots before they plant. Adding the conditioner helps aerate the soil, and the decomposing organic materials in humus provide nutrients. If the plot is for annual plants, it often must be reconditioned year after year, even though the gardener faithfully pulls weeds, fertilizes seedlings, applies chemicals to curtail disease-causing soil microbes, and immediately tosses out the mature plants (along with any plant debris) when they have finished bearing. Suggest some reasons why reconditioning is necessary in this scenario, and some strategies that could help limit the need for it.
3. One effect of acid rain is to dissolve rock, liberating minerals into soil. Accordingly, can a case be made that acid rain confers environmental benefits as well as doing harm? What are some other factors, especially with regard to plant adaptations for gaining nutrients, that bear on this question?
4. Using Table 33.1 as a guide, describe some of the known roles of nitrogen, phosphorus, and potassium in plant function. What are some of the signs that a plant suffers a deficiency in those elements?

Experimental Analysis

A plant in your garden is undersized and develops chlorotic leaves even though you fertilize it with a mixture that contains nitrogen, potassium, and phosphorus. After determining that the plant receives enough sunlight for photosynthesis, you next decide to test whether its mineral nutrition is adequate. What specific hypothesis will your experiment test? How will your experimental design test the hypothesis?

Evolution Link

This chapter's *Focus on Research* discusses phytoremediation, the use of plants to remove environmental pollutants such as heavy metals. Some plant species are "hyperaccumulaters" that take up arsenic and other metallic contaminants and sequester such toxins in shoot parts. How might this activity confer a selective advantage?

The reproductive structures of an ornamental poppy *(Papaver rhoeas)*. Male gametophytes, which produce pollen, surround the female gametophyte, which produces eggs and is the site of fertilization and seed development (photographer's close-up).

Study Plan

34 Reproduction and Development in Flowering Plants

Why It Matters

Seeds of a small flowering tree, *Theobroma cacao,* produce the raw material that modern confectioners turn into chocolate. The tree evolved in the undergrowth of tropical rain forests in Central America, where it was domesticated by the Maya and Aztec peoples. Today cacao trees flourish on vast plantations in the tropical lowlands of Central and South America, the West Indies, West Africa, and New Guinea. Unlike most angiosperms, which produce flowers at the tips of floral shoots, *T. cacao* flowers grow directly from buds on the tree trunk. The flowers are pollinated by insects, primarily midges of the genus *Forcipomyia*. Pollination is the first step toward fertilization of the eggs, and within about 6 months, large, heavy fruits develop from them **(Figure 34.1)**. Each podlike fruit contains from 20 to 60 seeds—the cacao "beans" that chocolate manufacturers process into cocoa, chocolate, and other commercial products.

As in other flowering plants, cacao seeds result from sexual reproduction. Angiosperms have elaborate reproductive systems—housed in flowers—that produce, protect, and nourish sperm, eggs, and developing embryos. As with cacao, the flowers of many species also serve as invitations to animal pollinators, which function in

Courtesy of Caroline Ford, School of Plant Sciences, University of Reading, UK

ZEFA—Rein

Figure 34.1
Flowers and fruits growing from the trunk of a cacao tree *(Theobroma cacao)*, in Central America. Each fruit is the mature ovary of a *T. cacao* flower.

bringing sperm and egg together. Once a new individual forms and begins to grow, finely regulated gene interactions guide the development of flowers and other plant parts. Under certain circumstances, many plants—including cacao—also reproduce asexually, so that individuals of the new generation are clones, genetically identical to their parents.

Sexual reproduction dominates the life cycle of flowering plants, however, and it will be our main focus in the first three sections of this chapter. We then consider asexual reproduction and conclude with a discussion of early plant development. Using methods of molecular biology and a variety of model organisms, plant biologists are beginning to elucidate some of the mechanisms by which plant developmental pathways unfold.

34.1 Overview of Flowering Plant Reproduction

In the living world, sexual reproduction occurs when male and female haploid gametes unite to create a fertilized egg. This fertilized egg—the diploid zygote—then embarks on a developmental course of mitotic cell divisions, cell enlargement, and cell differentiation. In flowering plants, subsequent steps result in distinctive haploid and diploid forms of the individual.

Diploid and Haploid Generations Arise in the Angiosperm Life Cycle

Once an angiosperm zygote has formed, the developmental sequence generates an embryo enclosed within a seed. In a seed, early versions of the basic plant tissue systems are already in place, so the embryo technically is already a **sporophyte**—the diploid, spore-producing body of a plant (see Section 27.1). When most people look at a cherry tree or a rosebush, what they think of as "the plant" is the sporophyte **(Figure 34.2)**.

At some point during one or more seasons of an angiosperm sporophyte's growth and development, one or more of its vegetative shoots undergo changes in structure and function and become *floral shoots*—that is, reproductive shoots that will give rise to a flower or inflorescence (a group of flowers on the same floral shoot). Certain cells in the flowers divide by meiosis. Unlike in animals, however, meiosis in plants does not yield gametes directly. Instead, meiosis gives rise to haploid **spores,** walled cells that develop by mitosis into multicellular haploid **gametophytes.** The gametophytes produce haploid sex cells, the gametes, again by mitosis. Male gametophytes produce sperm cells, the male gametes of flowering plants; female gametophytes produce eggs. This division of a life cycle into a diploid, spore-producing generation and a haploid, gamete-producing one is called **alternation of generations** (a phenomenon described more fully in Chapter 27).

In virtually all plants, the gametophyte and sporophyte are strikingly different from one another in both function and structure. For instance, in bryophytes (mosses and liverworts) the gametophyte is usually larger than the sporophyte; the sporophyte grows out of the gametophyte and is nourished by it (see Section 27.2). In ferns, which are seedless vascular plants, the gametophyte is much smaller than the sporophyte and is free-living for much of its lifespan; in most fern species the gametophyte nourishes itself by photosynthesis. In angiosperms and other seed plants, gametophytes are small structures that are retained *inside* the sporophyte for all or part of their lives. The female gametophyte of a flowering plant usually consists of only seven cells that are embedded in floral tissues, as you will read shortly. Male gametophytes are released into the environment as pollen grains, so small that they

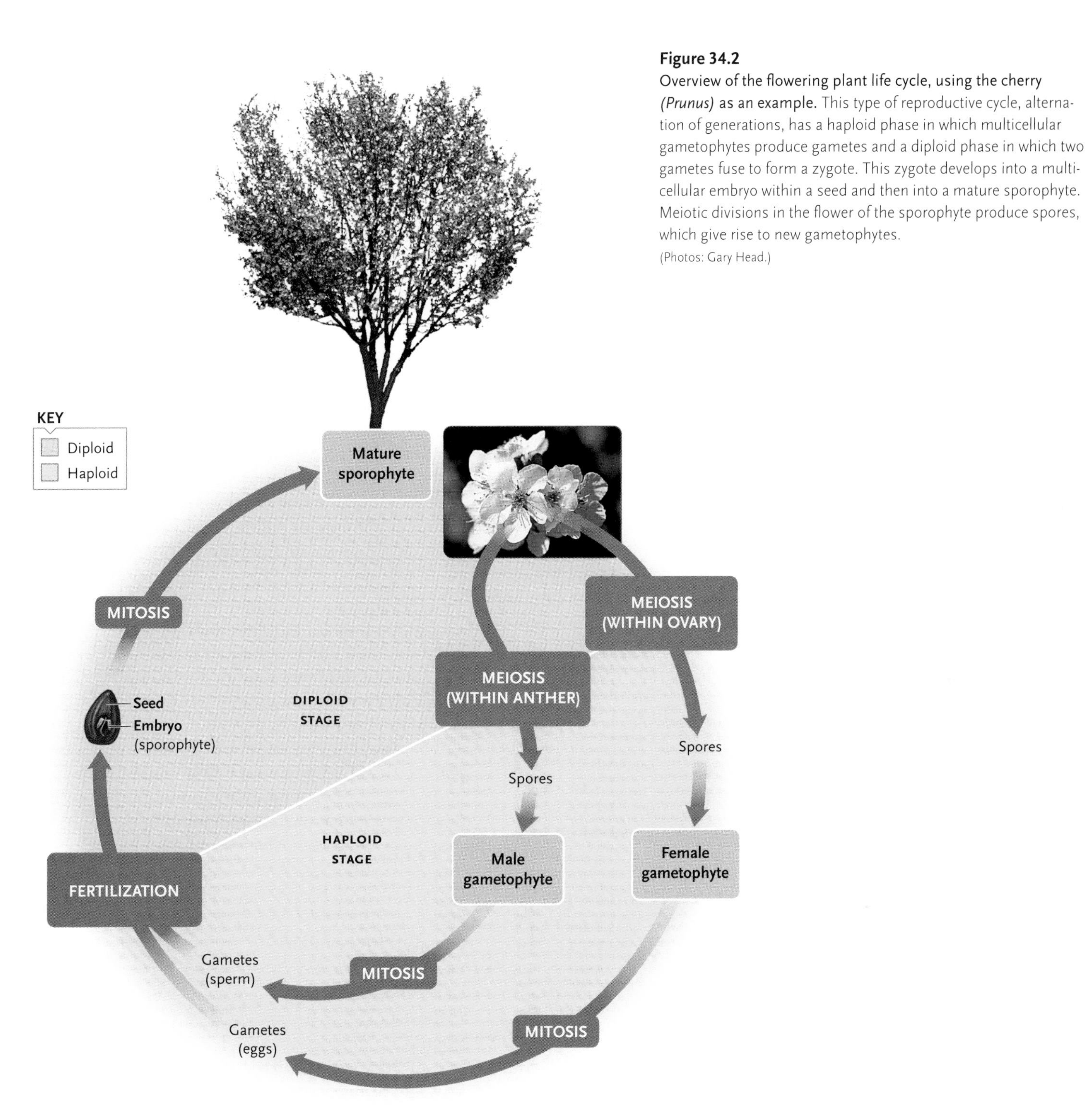

Figure 34.2
Overview of the flowering plant life cycle, using the cherry *(Prunus)* as an example. This type of reproductive cycle, alternation of generations, has a haploid phase in which multicellular gametophytes produce gametes and a diploid phase in which two gametes fuse to form a zygote. This zygote develops into a multicellular embryo within a seed and then into a mature sporophyte. Meiotic divisions in the flower of the sporophyte produce spores, which give rise to new gametophytes.
(Photos: Gary Head.)

are measured in micrometers. The pollen grain matures when it reaches a compatible ovule, resulting in fertilization and production of a new generation of seeds.

Sporophytes may also reproduce asexually. For instance, strawberry plants send out horizontal stolons, and new roots and shoots develop at each node along the stems. Short underground stems of onions and lilies put out buds that grow into new plants. In summer and fall, Bermuda grass produces new plants at nodes along its subterranean rhizomes. Asexual reproduction also can be induced artificially. Whole orchards of genetically identical fruit trees have been grown from cuttings or buds of a single parent tree.

We turn now to our consideration of sexual reproduction in angiosperms, beginning with the crucial step in which flowers develop.

Study Break

1. What are the two "alternating generations" of plants?
2. How do these two life phases differ in structure and function?

34.2 The Formation of Flowers and Gametes

Flowering marks a developmental shift for an angiosperm. Biochemical signals—triggered in part by environmental cues such as day length and temperature—travel to the apical meristem of a shoot and set in motion changes in the activity of cells there. Instead of continuing vegetative growth, the shoot is modified into a floral shoot that will give rise to floral organs.

In Angiosperms, Flowers Contain the Organs for Sexual Reproduction

A flower develops from the end of the floral shoot, called the **receptacle.** Cells in the receptacle differentiate to produce up to four types of concentric tissue regions called *whorls*. The arrangement and number of whorl types varies in different species; **Figure 34.3** shows a typical example in which a flower has one of each of the four whorls. The two outer whorls consist of nonfertile, vegetative structures. The outermost whorl (whorl 1), the **calyx,** is made up of leaflike **sepals.** The calyx is usually green, and, early in the flower's development, it encloses all the other parts, as in an unopened rose bud. The next whorl, the **corolla,** includes the **petals.** Corollas are the "showy" parts of flowers; they have distinctive colors, patterning, and shapes, and these features often function in attracting bees and other animal pollinators.

A flower's two inner whorls are specialized for making gametes. Inside the corolla is the whorl of **stamens** (whorl 3), in which male gametophytes form. In almost all living flowering plant species, a stamen consists of a slender **filament** (stalk) capped by a bilobed **anther.** Each anther contains four **pollen sacs,** in which pollen develops.

The innermost whorl (whorl 4) consists of one or more **carpels,** in which female gametophytes form. The lower part of a carpel is the **ovary.** Inside it is one or more **ovules,** in which an egg develops and fertilization takes place. A seed is a mature ovule. In many flowers that have more than one carpel, the carpels fuse into a single, common ovary containing multiple ovules. Typically, the carpel's slender **style** widens at its upper end, terminating in the **stigma,** which serves as a landing platform for pollen. Fused carpels may share a single stigma and style, or each may retain separate ones. The name angiosperm ("seed vessel") refers to the carpel.

Some species have so-called **complete flowers,** in which all four whorls are present. In other species, flowers lack one or more of the whorls, and thus botanists describe them as **incomplete flowers (Figure 34.4).** Botanists also distinguish flowers on the basis of the sexual parts they contain. Most angiosperms produce **perfect flowers,** which have both kinds of sexual parts—that is, both stamens and carpels. **Imperfect flowers** are a type of incomplete flower that has stamens or carpels, but not both. (Notice that all imperfect flowers are also incomplete.) Species with imperfect flowers are further divided according to whether individual plants produce both sexual types of flowers, or only one. In **monoecious** ("one house") species, such as oaks, each plant has some "male" flowers with only stamens and some "female" flowers with only carpels. In **dioecious** ("two houses") species, such as willows, a given plant produces flowers having only stamens or only carpels. With this basic angiosperm reproductive anatomy in mind, we now turn to the processes by which male and female gametes come into being.

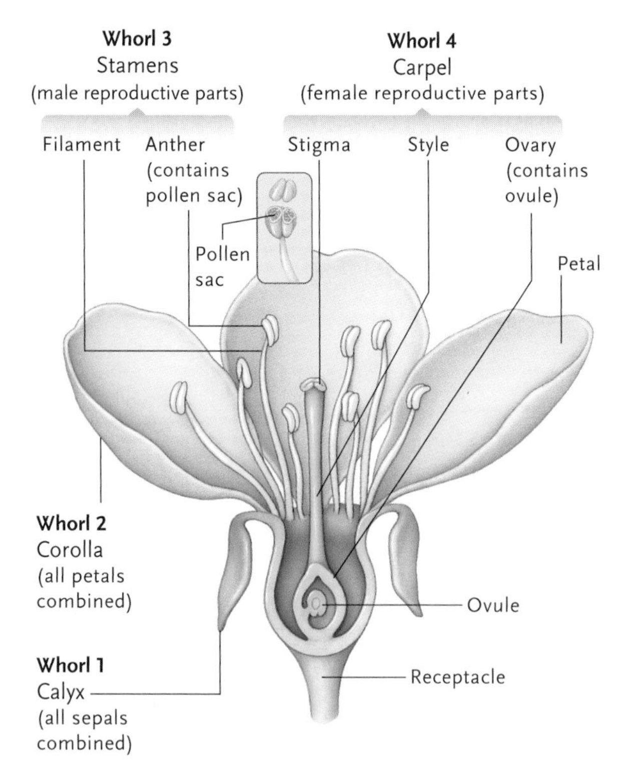

Figure 34.3
Structure of a cherry *(Prunus)* flower, with the four whorls indicated. Like the flowers of many angiosperms, it has a single carpel and several stamens. The anthers of the stamen produce haploid pollen. The stigma of the carpel receives the pollen, and the ovule inside the ovary contains the haploid eggs.

Pollen Grains Arise from Microspores in Anthers

Most of a flowering plant's reproductive life cycle, from production of sperm and eggs to production of a mature seed, takes place within its flowers. **Figure 34.5** shows this cycle as it unfolds in a perfect flower. The spores that give rise to male gametophytes are produced in a flower bud's anthers (see Figure 34.5, left). The pollen sacs inside each anther hold diploid microsporocytes (or *microspore mother cells*); each microsporocyte undergoes meiosis and eventually produces four small haploid **microspores.** Like most plant cells, the

a. Complete flower of an apple tree *(Malus)*

b. Incomplete flower of a Hubbard squash *(Cucurbita)*

Figure 34.4
Examples of complete and incomplete flowers. **(a)** Apple flowers *(Malus)* are complete. Each has many stamens, carpels, and petals, along with petal-like sepals. **(b)** As with other plants in the pumpkin family, the flowers of this Hubbard squash *(Cucurbita maxima)* are both incomplete and imperfect because each has either stamens or carpels, but never both.

microspores are walled, and inside its wall each microspore divides again, this time by mitosis. The result is an immature, haploid male gametophyte—a **pollen grain.**

Of the two nuclei produced by the mitotic division of a microspore, one again divides. After this second round of mitosis the male gametophyte consists of three cells—two sperm cells plus a third cell that controls the development of a **pollen tube.** When pollen lands on a stigma, this tube grows through the tissues of a carpel and carries the sperm cells to the ovary. A mature male gametophyte consists of the pollen tube and sperm cells—the male gametes.

The walls of pollen grains are hardened by the decay-resistant polymer *sporopollenin,* and are tough enough to protect the male gametophyte during the somewhat precarious journey from anther to stigma. These walls are so distinctive that the family to which a plant belongs usually can be identified from pollen alone—based on the size and wall sculpturing of the grains, as well as the number of pores in the wall **(Figure 34.6).** Because they withstand decay, pollen grains fossilize well and can provide revealing clues about the evolution of seed plants and the ecological communities that lived in the past.

Eggs and Other Cells of Female Gametophytes Arise from Megaspores

Meanwhile, in the ovary of a flower, one or more dome-shaped masses form on the inner wall. Each mass becomes an ovule (see Figure 34.5, right), which, if all goes well, develops into a seed. Only one ovule forms in the carpel of some flowers, such as the cherry. Dozens, hundreds, or thousands may form in the carpels of other flowers, such as those of a bell pepper plant *(Capsicum annuum)*. At one end, the ovule has a small opening, called the **micropyle.**

Inside the cell mass, a diploid megasporocyte (or *megaspore mother cell*) divides by meiosis, forming four haploid **megaspores.** In most plants, three of these megaspores disintegrate. The remaining megaspore enlarges and develops into the female gametophyte in a sequence of steps tracked in Figure 34.5.

First, three rounds of mitosis occur *without* cytoplasmic division; the result is a single cell with eight nuclei arranged in two groups of four. Next, one nucleus in each group migrates to the center of the cell; these two **polar nuclei** ("polar" because they migrate from opposite ends of the cell) may fuse or remain separate. The cytoplasm then divides, and a cell wall forms around the two polar nuclei, forming a single large *central cell.* A wall also forms around each of the other nuclei. Three of these walled nuclei become *antipodal cells,* which eventually disintegrate. Three others form a cluster (called the "egg apparatus") near the micropyle; one of them is an **egg cell** that may eventually be fertilized. The other two, called *synergids,* will have a role in pollination. The eventual result of all these events is an **embryo sac** containing seven cells and eight nuclei. This embryo sac is the female gametophyte.

In about a third of flowering plants, biologists have observed variations in the events that produce a female gametophyte. In lilies, for example, changes in the sequence of cell divisions produce several cells with triploid nuclei (see Figure 27.31). The egg cell is not involved, however, so such differences do not affect reproduction. They may have roles in the development and functioning of other embryonic tissues.

As the male and female gametophytes complete their maturation, the stage is set for fertilization and the development of a new individual.

Study Break

1. What is the biological role of flowers, and what fundamental physiological change must occur before an angiosperm can produce a flower?
2. Explain the steps leading to the formation of a mature male gametophyte, beginning with microsporocytes in a flower's anthers. Which structures are diploid and which haploid?
3. Trace the development of a female gametophyte, beginning with the megasporocyte in an ovule of a flower's ovary. Which structures are diploid and which haploid?

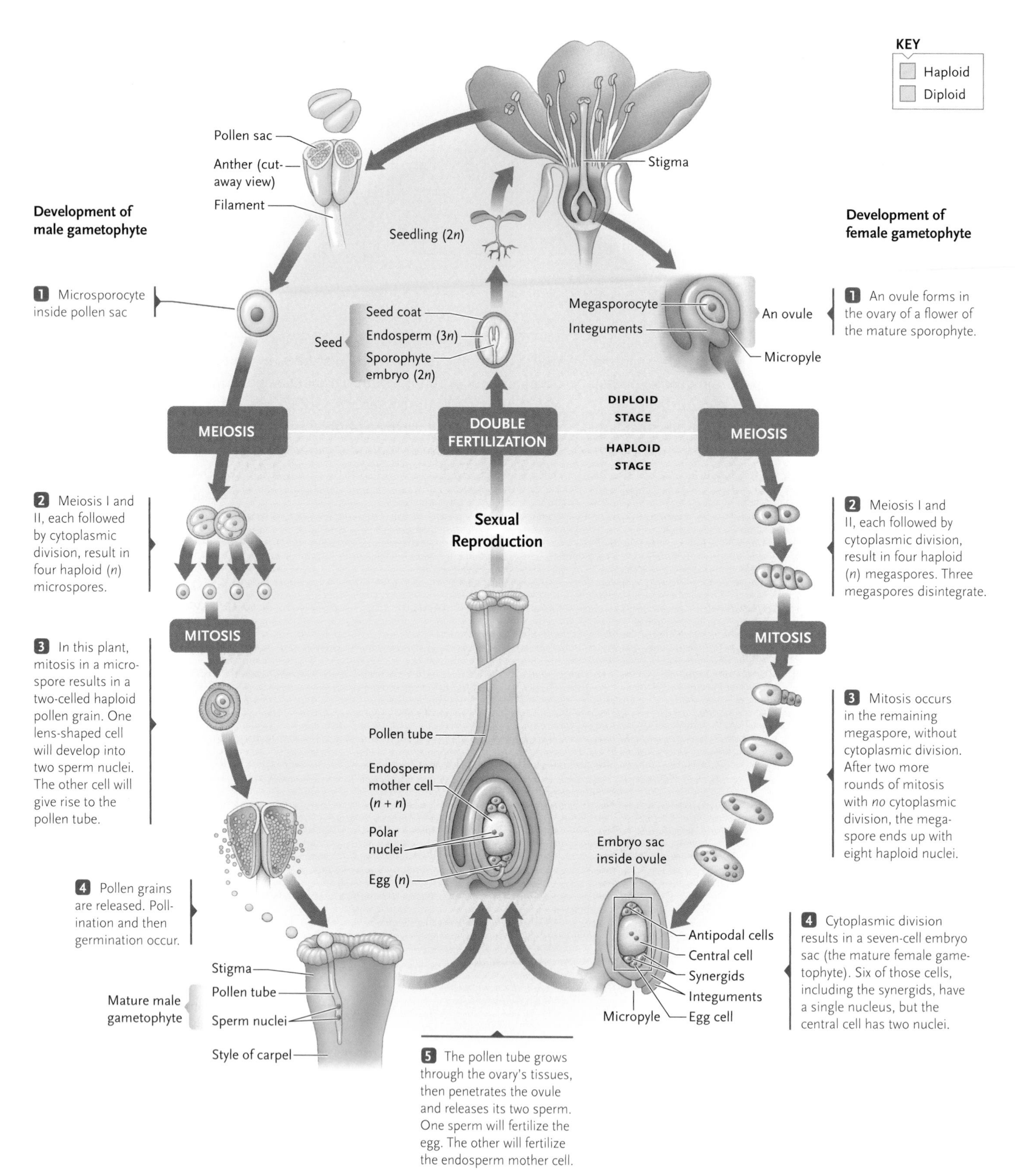
KEY
Haploid
Diploid
Pollen sac
Anther (cut-away view)
Filament
Stigma
Development of male gametophyte
Seedling (2n)
Development of female gametophyte
1 Microsporocyte inside pollen sac
Seed coat
Endosperm (3n)
Seed
Sporophyte embryo (2n)
Megasporocyte
Integuments
An ovule
Micropyle
1 An ovule forms in the ovary of a flower of the mature sporophyte.
MEIOSIS
DOUBLE FERTILIZATION
DIPLOID STAGE
HAPLOID STAGE
MEIOSIS
2 Meiosis I and II, each followed by cytoplasmic division, result in four haploid (n) microspores.
Sexual Reproduction
2 Meiosis I and II, each followed by cytoplasmic division, result in four haploid (n) megaspores. Three megaspores disintegrate.
MITOSIS
MITOSIS
3 In this plant, mitosis in a microspore results in a two-celled haploid pollen grain. One lens-shaped cell will develop into two sperm nuclei. The other cell will give rise to the pollen tube.
Pollen tube
Endosperm mother cell (n + n)
Polar nuclei
Egg (n)
3 Mitosis occurs in the remaining megaspore, without cytoplasmic division. After two more rounds of mitosis with no cytoplasmic division, the megaspore ends up with eight haploid nuclei.
Embryo sac inside ovule
4 Pollen grains are released. Pollination and then germination occur.
Antipodal cells
Central cell
Synergids
Integuments
Egg cell
Micropyle
4 Cytoplasmic division results in a seven-cell embryo sac (the mature female gametophyte). Six of those cells, including the synergids, have a single nucleus, but the central cell has two nuclei.
Stigma
Pollen tube
Mature male gametophyte
Sperm nuclei
Style of carpel
5 The pollen tube grows through the ovary's tissues, then penetrates the ovule and releases its two sperm. One sperm will fertilize the egg. The other will fertilize the endosperm mother cell.

Figure 34.5
Life cycle of cherry *(Prunus)*, a eudicot. Pollen grains develop in pollen sacs within the anthers. An embryo sac forms in the single ovule within the cherry flower's ovary, and an egg forms within the embryo sac. When the pollen grains are released and contact the stigma, double fertilization occurs. An embryo sporophyte and nutritive endosperm develop and become encased in a seed coat.

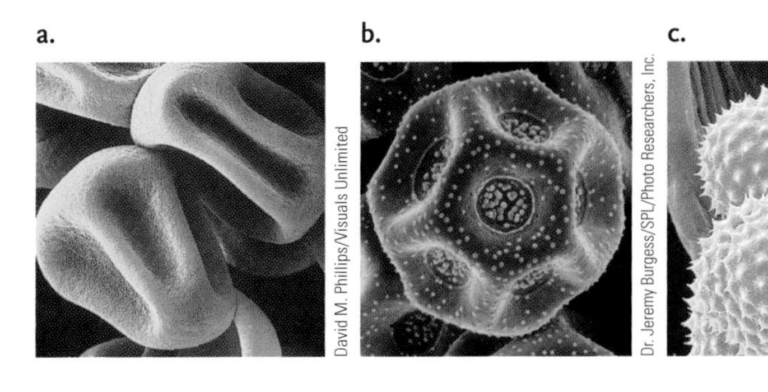

Figure 34.6
Some examples of pollen grain diversity. Scanning electron micrographs of pollen grains from **(a)** a grass, **(b)** chickweed *(Stellaria)*, and **(c)** ragweed *(Ambrosia)* plants.

34.3 Pollination, Fertilization, and Germination

The process by which plants produce seeds—which have the potential to give rise to new individuals—begins with *pollination,* when pollen grains make contact with the stigma of a flower. Air or water currents, birds, bats, insects, or other agents make the transfer. (Section 27.5 discussed the complex relationship between some flowering plants and their animal pollinators.)

Pollination is the first in a series of events leading to *fertilization,* the fusion of an egg and sperm inside the flower's ovary. The resulting embryo and its ovule mature into a seed housing a young sporophyte, and when the seed *germinates,* or sprouts, the sporophyte begins to grow.

Pollination Requires Compatible Pollen and Female Tissues

Even after pollen reaches a stigma, in most cases pollination and fertilization can take place only if the pollen and stigma are compatible. For example, if pollen from one species lands on a stigma from another, chemical incompatibilities usually prevent pollen tubes from developing.

Even when the sperm-bearing pollen and a stigma are from the same species, pollination may not lead to fertilization unless the pollen and stigma belong to genetically distinct individuals. For instance, when pollen from a given plant lands on that plant's own stigma, a pollen tube may begin to develop, but stop before reaching the embryo sac. This **self-incompatibility** is a biochemical recognition and rejection process that prevents self-fertilization, and it apparently results from interactions between proteins encoded by *S* (self) genes.

Research has shown that *S* genes usually have multiple alleles—in some species there may be hundreds—and a common type of incompatibility occurs when pollen and stigma carry an identical *S* allele. The result is a biochemical signal that prevents proper formation of the pollen tube **(Figure 34.7).** For example, studies on plants of the mustard family have revealed that pollen contacting an incompatible stigma produces a protein that prevents the stigma from hydrating the relatively dry pollen grain, an essential step if the pollen tube is to grow. A wide range of self-incompatibility responses has been discovered, however. In cacao, for instance, when incompatible pollen contacts a stigma, a pollen tube grows normally but a hormonal response soon causes the flower to drop off the plant, preventing fertilization.

Self-incompatibility prevents inbreeding and promotes genetic variation, which is the raw material for

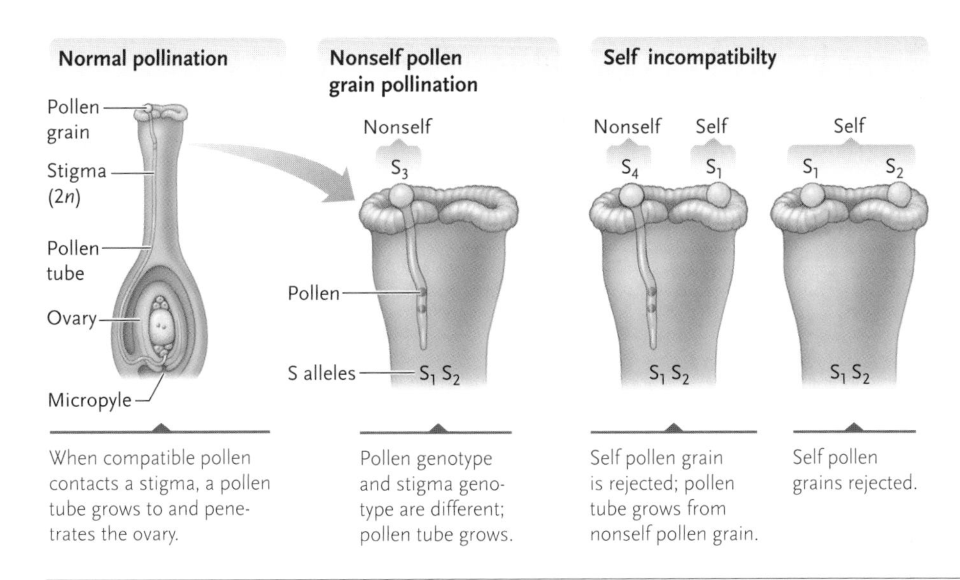

Figure 34.7
Self-incompatibility. When a pollen grain has an *S* allele that matches one in the stigma (which is diploid), the result is a biochemical response that prevents fertilization—in this illustration, by preventing the growth of a pollen tube.

natural selection and adaptation. Even so, many flowering plants do self-pollinate, either partly or exclusively, because that mode, too, has benefits in some circumstances. (Mendel's peas are a classic example.) For instance, "selfing" may help preserve adaptive traits in a population. It also reduces or eliminates a plant's reliance on wind, water, or animals for pollination, and thus ensures that seeds will form when conditions for cross-pollination are unfavorable, as when pollinators or potential mates are scarce.

Double Fertilization Occurs in Flowering Plants

If a pollen grain lands on a compatible stigma, it absorbs moisture and germinates a pollen tube, which burrows through the stigma and style toward an ovule. Chemical cues from the two synergid cells lying close to the egg cell help guide the pollen tube toward its destination. Before or during these events, the pollen grain's haploid sperm-producing cell divides by mitosis, forming two haploid sperm. When the pollen tube reaches the ovule, it enters through the micropyle and an opening forms in its tip. By this time one synergid has begun to die (an example of programmed cell death), and the two sperm are released into the disintegrating cell's cytoplasm. Experiments suggest that elements of the synergid's cytoskeleton guide the sperm onward, one to the egg cell and the other to the central cell.

Next there occurs a remarkable sequence of events called **double fertilization**, which has been observed only in flowering plants and (in a somewhat different version) in the gnetophyte *Ephedra* (see Section 27.4). Typically, one sperm nucleus fuses with the egg to form a diploid ($2n$) zygote. The other sperm nucleus fuses with the central cell, forming a cell with a triploid ($3n$) nucleus. Tissues derived from that $3n$ cell are called **endosperm** ("inside the seed"). They nourish the embryo and, in monocots, the seedling, until its leaves form and photosynthesis has begun.

Embryo-nourishing endosperm forms only in flowering plants, and its evolution coincided with a

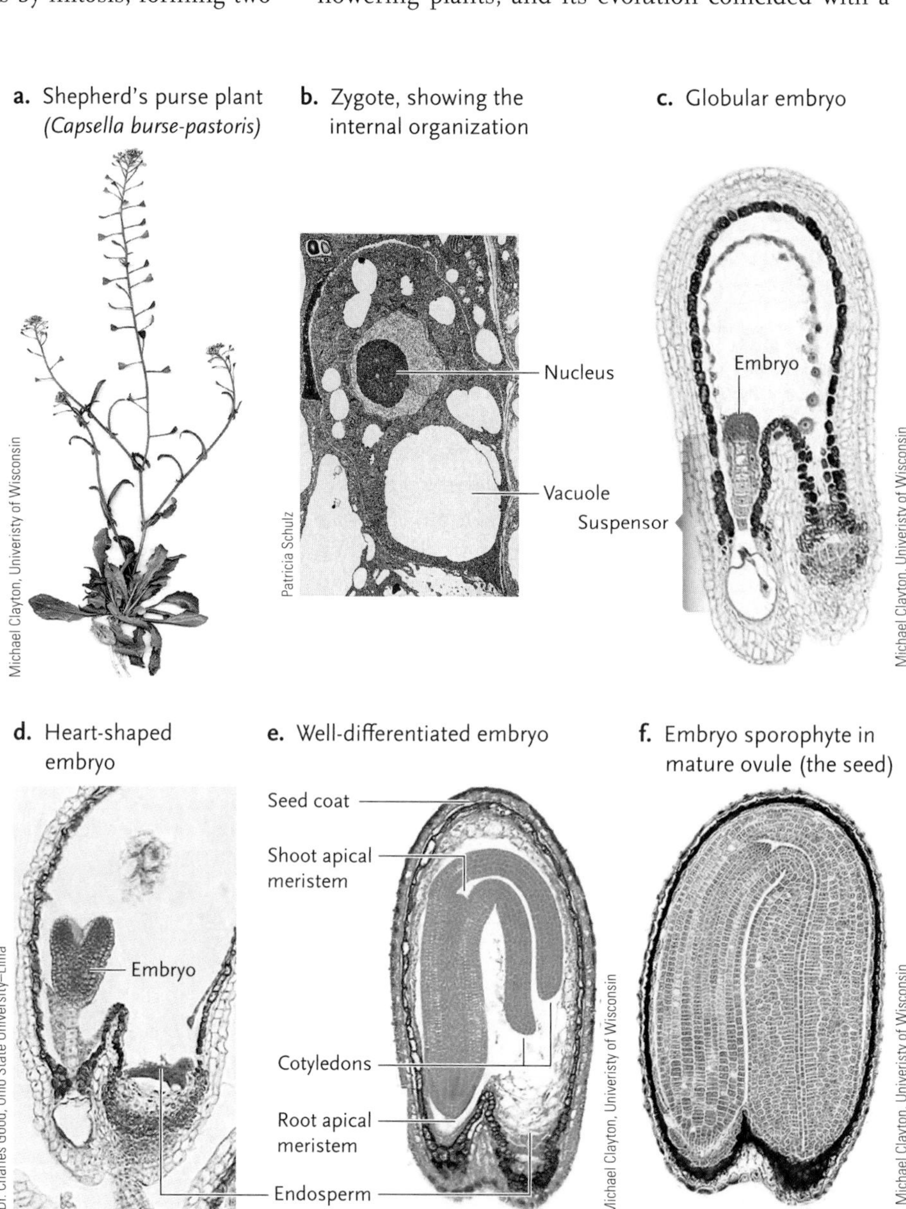

Figure 34.8
Stages in the embryonic development of shepherd's purse *(Capsella bursa-pastoris)*, a eudicot. Figure 34.16 looks in more detail at the development of early plant embryos. The micrographs are not to the same scale.

reduction in the size of the female gametophyte. In other land plants, such as gymnosperms and ferns, the gametophyte itself contains enough stored food to nourish the embryonic sporophytes.

The Embryonic Sporophyte Develops inside a Seed

When the zygote first forms, it starts to develop and elongate even before mitosis begins. For example, in shepherd's purse *(Capsella)*, shown in **Figure 34.8,** most of the organelles in the zygote, including the nucleus, become situated in the top half of the cell, while a vacuole takes up most of the lower half (see Figure 34.8b). The first round of mitosis divides the zygote into an upper *apical cell* and a lower *basal cell*. The apical cell then gives rise to the multicellular embryo, while most descendants of the basal cell form a simple row of cells, the **suspensor,** which transfers nutrients from the parent plant to the embryo (see Figure 34.8c).

The first apical cell divisions produce a globe-shaped structure attached to the suspensor. As they continue to grow, embryos of *Capsella* and other eudicots become heart-shaped (see Figure 34.8d); each lobe of the "heart" is a developing cotyledon (seed leaf), which provides nutrients for growing tissues. Typically, the two cotyledons absorb much of the nutrient-storing endosperm and become plump and fleshy. For instance, mature seeds of a sunflower *(Helianthus annuus)* have no endosperm at all. In some eudicots, however, the cotyledons remain as slender structures; they produce enzymes that digest the seed's ample endosperm and transfer the liberated nutrients to the seedling. Monocots have one, large cotyledon; in many monocot species, especially grasses such as corn and rice, the cotyledon absorbs the endosperm after germination, when the embryo inside the seed begins to grow.

By the time the ovule is mature—that is, a fully developed seed—it has become encased by a protective **seed coat.** Inside the seed, the sheltered embryo has a lengthwise axis with a root apical meristem at one end and a shoot apical meristem at the other (see Figure 34.8e, f).

Figure 34.9a and **Figure 34.9b** illustrate the structure of the seeds of two eudicots, the kidney bean *(Phaseolus vulgaris)* and the castor bean *(Ricinus communis)*. The kidney bean has broad, fleshy cotyledons and the castor bean much thinner ones, but in other ways the embryos are quite similar. The **radicle,** or embryonic root, is located near the micropyle, where the pollen tube entered the ovule prior to fertilization. The radicle

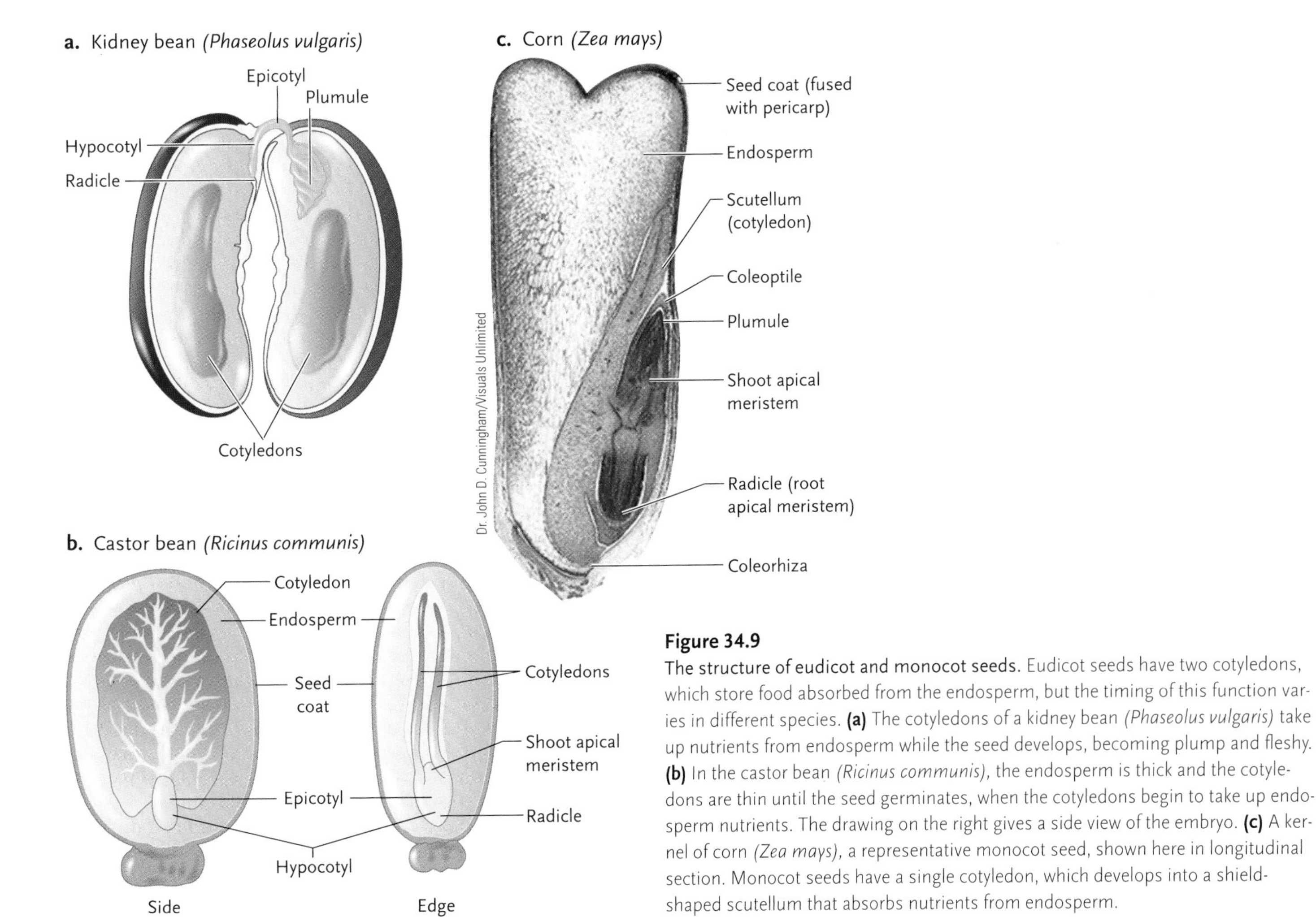

Figure 34.9
The structure of eudicot and monocot seeds. Eudicot seeds have two cotyledons, which store food absorbed from the endosperm, but the timing of this function varies in different species. **(a)** The cotyledons of a kidney bean *(Phaseolus vulgaris)* take up nutrients from endosperm while the seed develops, becoming plump and fleshy. **(b)** In the castor bean *(Ricinus communis)*, the endosperm is thick and the cotyledons are thin until the seed germinates, when the cotyledons begin to take up endosperm nutrients. The drawing on the right gives a side view of the embryo. **(c)** A kernel of corn *(Zea mays)*, a representative monocot seed, shown here in longitudinal section. Monocot seeds have a single cotyledon, which develops into a shield-shaped scutellum that absorbs nutrients from endosperm.

attaches to the cotyledon at a region of cells called the **hypocotyl.** Beyond the hypocotyl is the **epicotyl,** which has the shoot apical meristem at its tip and which often bears a cluster of tiny foliage leaves, the **plumule.** At germination, when the root and shoot first elongate and emerge from the seed, the cotyledons are positioned at the first stem node with the epicotyl above them and the hypocotyl below them.

The embryos of monocots such as corn differ structurally from those of eudicots in several ways **(Figure 34.9c).** They have only one very large cotyledon, called a **scutellum.** In addition, the root and shoot apical meristems of monocots are blanketed by protective tissues. The shoot apical meristem and plumule are covered by a **coleoptile,** a sheath of cells that protects them during upward growth through the soil. A similar covering, the **coleorhiza,** sheathes the radicle until it breaks out of the seed coat and enters the soil as the primary root. The actual embryo of a corn plant is buried deep within the corn "kernel," which technically is called a *grain.* Most of the moist interior of a fresh corn grain is endosperm; the single cotyledon forms a plump, shield-shaped mass that absorbs nutrients from the endosperm.

Fruits Protect Seeds and Aid Seed Dispersal

Most angiosperm seeds are housed inside fruits, which provide protection and often aid seed dispersal. A **fruit** is a matured or ripened ovary. Usually, fruits begin to develop after a flower's ovule or ovules are fertilized by pollen, and the start of ovule growth after pollination is called "fruit set." The fruit wall, called the **pericarp,** develops from the ovary wall and can have several layers. Hormones in pollen grains provide the initial stimulus that turns on the genetic machinery leading to fruit development; additional signals come from hormones produced by the developing seeds.

Fruits are extremely diverse, and biologists classify them into types based on combinations of structural features. A major defining feature is the nature of the pericarp, which may be fleshy (as in peaches) or dry (as in a hazelnut). A fruit also is classified according to the number of ovaries or flowers from which it develops. **Simple fruits,** such as peaches, tomatoes, and the cacao fruits pictured in Figure 34.1, develop from a single ovary, and in many of them at least one layer of the pericarp is fleshy and juicy. Other simple fruits, including grains and nuts, have a thin, dry pericarp, which may be fused to the seed coat. The garden pea *(Pisum sativa)* is a simple fruit, the peas being the seeds and the surrounding shell the pericarp. **Aggregate fruits** are formed from several ovaries in a single flower. Examples are raspberries and strawberries, which develop from clusters of ovaries. Strawberries also qualify as *accessory* fruits, in which floral parts in addition to the ovary become incorporated as the fruit develops. For instance, anatomically, the fleshy part of a strawberry is an expanded receptacle (the end of the floral shoot) and the strawberry fruits are the tiny, dry nubbins (called *achenes*) you see embedded in the fleshy tissue of each berry. **Multiple fruits** develop from several ovaries in multiple flowers. For example, a pineapple is a multiple fruit that develops from the enlarged ovaries of several flowers clustered together in an inflorescence. **Figure 34.10** shows examples of some different types of fruits.

Fruits have two functions: they protect seeds, and they aid seed dispersal in specific environments. For example, the shell of a sunflower seed is a pericarp that protects the seeds within. A pea pod is a pericarp that in nature splits open to disperse the seeds (peas) inside. Maple fruits have winglike extensions for dispersal (see Figure 34.10e). When the fruit drops, the wings cause it to spin sideways and also can carry it away on a breeze. This aerodynamic property propels maple seeds to new locations, where they will not have to compete with the parent tree for water and minerals. Fruits also may have hooks, spines, hairs, or sticky surfaces, and they are ferried to new locations when they adhere to feathers, fur, or blue jeans of animals that brush against them. Fleshy fruits such as blueberries and cherries are nutritious food for many animals, and their seeds are adapted for surviving digestive enzymes in the animal gut. The enzymes remove just enough of the hard seed coats to increase the chance of successful germination when the seeds are expelled from the animal's body in feces.

Seed Germination Continues the Life Cycle

A mature seed is essentially dehydrated. On average, only about 10% of its weight is water—too little for cell expansion or metabolism. After a seed is dispersed and germinates, the embryo inside it becomes hydrated and resumes growth. Ideally, a seed germinates when external conditions favor the survival of the embryo and growth of the new sporophyte. This timing is important, for once germination is underway the embryo loses the protection of the seed coat and other structures that surround it. Overall, the amount of soil moisture and oxygen, the temperature, day length, and other environmental factors influence when germination takes place.

In some species, the life cycle may include a period of seed **dormancy** (*dormire* = to sleep), in which biological activity is suspended. Botanists have described a striking array of variations in the conditions required for dormant seeds to germinate. For instance, seeds may require minimum periods of daylight or darkness, repeated soaking, mechanical abrasion, or exposure to certain enzymes, the high heat of a fire, or a freeze–thaw cycle before they finally break dormancy. In some desert plants, hormones in the seed coat inhibit growth of a seedling until heavy rains flush them away. This adaptation prevents seeds from germinating unless there is

a. Peach *(Prunus)*, a simple fruit

b. Raspberry *(Rubus)*, an aggregate fruit

c. Strawberry *(Fragaria)*, an accessory fruit

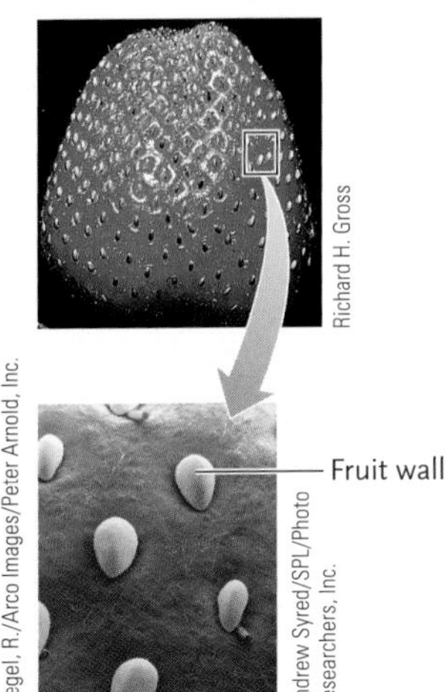

d. Pineapple *(Ananus comosus)*, a multiple fruit

e. Maple *(Acer)* fruit

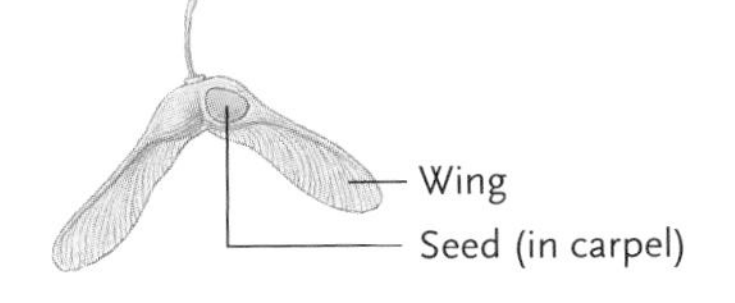

Figure 34.10
Fruits. **(a)** Peach, a fleshy simple fruit. **(b)** Raspberry *(Rubus)*, an aggregate fruit. **(c)** Strawberry *(Fragaria ananassa)*, an accessory fruit that is also an aggregate fruit. **(d)** Pineapple *(Ananas comosus)*, a multiple fruit. **(e)** Winged fruits of maple *(Acer)*.

enough water in the soil to support growth of the plant through the flowering and seed production stages before the soil dries once again. Many desert plants—and plants in harsh environments such as alpine tundra—cycle from germination to growth, flowering, and seed development in the space of a few weeks, and their offspring remain dormant as seeds until conditions once again favor germination and growth.

Seeds of some species appear to remain viable for amazing lengths of time. Thousand-year-old lotus seeds *(Nelumbo lutea)* discovered in a dry lake bed have germinated trouble-free. And in one startling case, seeds of arctic lupine *(Lupinus arcticus)* were discovered in the 10,000-year-old frozen entrails of a lemming. When they were thawed, they readily germinated as well.

Germination begins with **imbibition,** in which water molecules move into the seed, attracted to hydrophilic groups of stored proteins. As water enters, the seed swells, the coat ruptures, and the radicle begins its downward growth into the soil. Within this general framework, however, there are many variations among plants.

Once the seed coat splits, water and oxygen move more easily into the seed. Metabolism switches into high gear as cells divide and elongate to produce the seedling. Stable enzymes that were synthesized before dormancy become active; other enzymes are produced as the genes encoding them begin to be expressed. Among other roles, the increased gene activity and enzyme production mobilize the seed's food reserves in cotyledons or endosperm. Nutrients released by the enzymes sustain the rapidly developing seedling until its root and shoot systems are established.

The events of seed germination have been studied extensively in cereal grains, and **Figure 34.11** illustrates them in barley. Notice that the seed's endosperm is separated from the pericarp by a thin layer of cells called the **aleurone.** As a hydrating seed imbibes water, the embryo produces a *gibberellin,* a hormone that

Figure 34.11
How food reserves are mobilized in a germinated seed of barley *(Hordeum vulgare)*, a monocot.

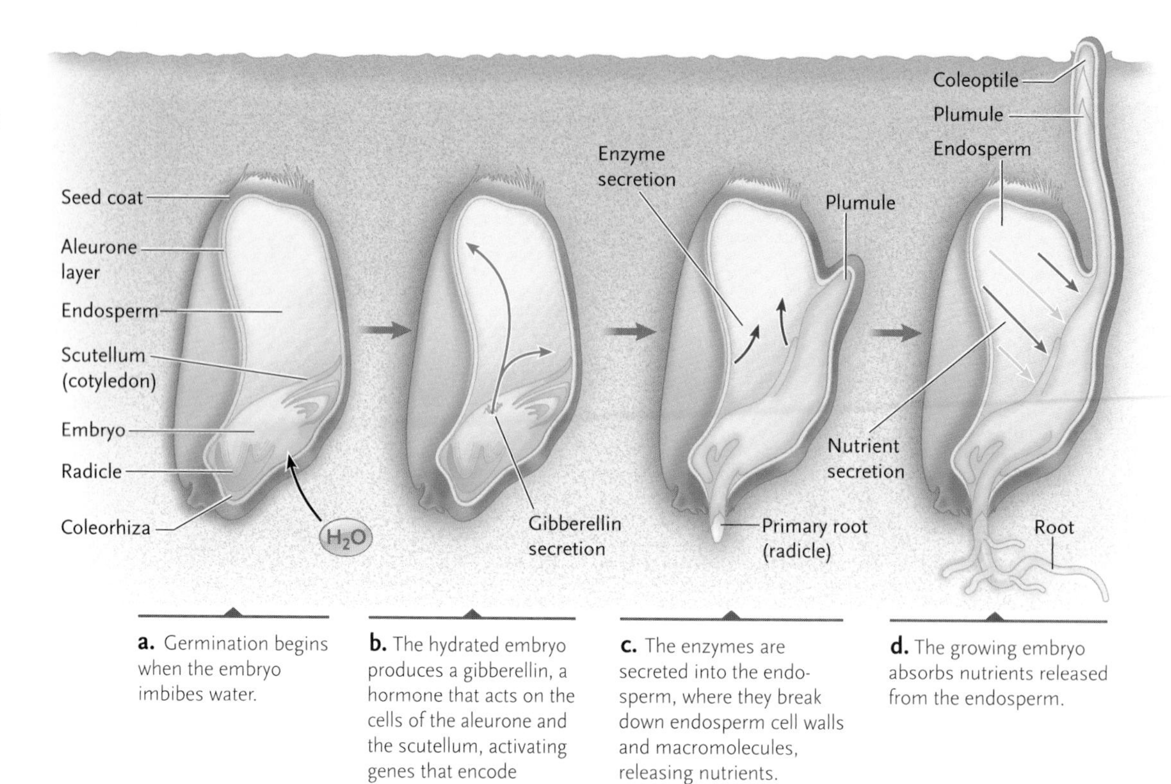

a. Germination begins when the embryo imbibes water.

b. The hydrated embryo produces a gibberellin, a hormone that acts on the cells of the aleurone and the scutellum, activating genes that encode hydrolytic enzymes.

c. The enzymes are secreted into the endosperm, where they break down endosperm cell walls and macromolecules, releasing nutrients.

d. The growing embryo absorbs nutrients released from the endosperm.

Figure 34.12
Stages in the development of a representative eudicot, the kidney bean *(Phaseolus vulgaris)*.

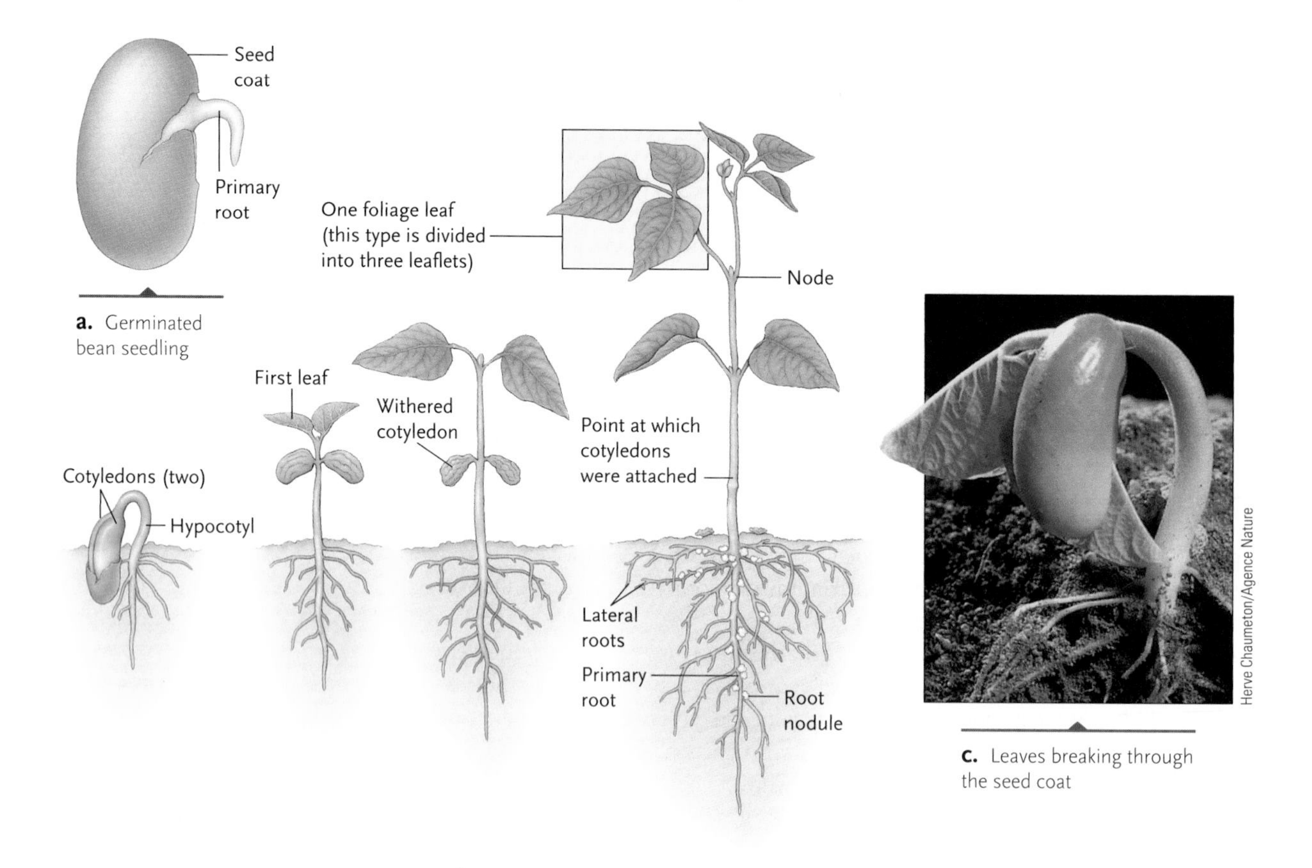

a. Germinated bean seedling

b. Food-storing cotyledons are lifted above the soil surface when cells of the hypocotyl elongate. The hypocotyl becomes hook-shaped and forces a channel through the soil as it grows. At the soil surface, the hook straightens in response to light. For several days, cells of the cotyledons carry out photosynthesis; then the cotyledons wither and drop off. Photosynthesis is taken over by the first leaves that develop along the stem and later by foliage leaves.

c. Leaves breaking through the seed coat

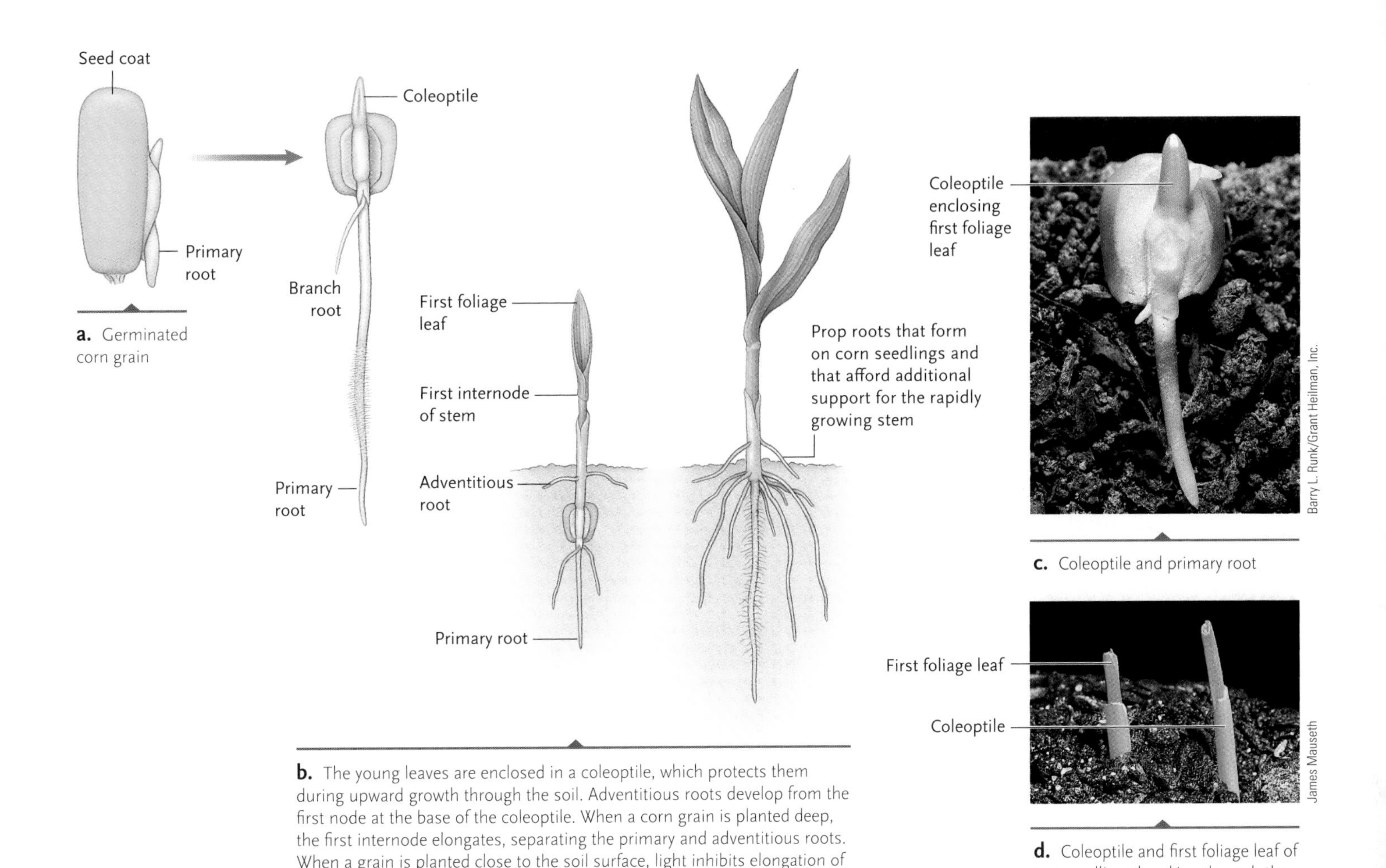

a. Germinated corn grain

b. The young leaves are enclosed in a coleoptile, which protects them during upward growth through the soil. Adventitious roots develop from the first node at the base of the coleoptile. When a corn grain is planted deep, the first internode elongates, separating the primary and adventitious roots. When a grain is planted close to the soil surface, light inhibits elongation of the first internode and the primary and adventitious roots look as if they originate in the same region of the stem.

c. Coleoptile and primary root

d. Coleoptile and first foliage leaf of two seedlings breaking through the soil surface

Figure 34.13 Stages in the development of a representative monocot, the corn plant *(Zea mays)*.

stimulates aleurone cells to manufacture and secrete hydrolytic enzymes. Some of these enzymes digest components of endosperm cell walls; others digest proteins, nucleic acids, and starch of the endosperm, releasing nutrient molecules for use by cells of the young root and shoot. Although it is clear that nutrient reserves are also mobilized by metabolic activity in eudicots and in gymnosperms, the details of the process are not well understood.

Inside a germinating seed, embryonic root cells are generally the first to divide and elongate, giving rise to the radicle. When the radicle emerges from the seed coat as the primary root, germination is complete. **Figure 34.12** and **Figure 34.13** depict the stages of early development in a kidney bean, a eudicot, and in corn, a monocot. As the young plant grows, its development continues to be influenced by interactions of hormones and environmental factors, as you will read in Chapter 35.

Most plants give rise to large numbers of seeds because, in the wild, only a tiny fraction of seeds survive, germinate, and eventually grow into another mature plant. Also, flowers, seeds, and fruits represent major investments of plant resources. Asexual reproduction, discussed next, is a more "economical" means by which many plants can propagate themselves.

Study Break

1. Explain the sequence of events in a flowering plant that begins with formation of a pollen tube and culminates with the formation of a diploid zygote and the $3n$ cell that will give rise to endosperm in a seed.
2. Early angiosperm embryos undergo a series of general changes as a seed matures. Summarize this sequence, then describe the structural differences that develop in the seeds of monocots and eudicots.
3. Germination begins when a seed imbibes water. What are the next key biochemical and developmental events that bring an angiosperm's life cycle full circle?

34.4 Asexual Reproduction of Flowering Plants

As noted in Chapter 31, nodes in the stolons of strawberries and the rhizomes of Bermuda grass each can give rise to new individuals. So can "suckers" that sprout from the roots of blackberry bushes and "eyes" in the tubers of potatoes. All these examples involve asexual or **vegetative reproduction** from a nonreproductive plant part, usually a bit of meristematic tissue in a bud on the root or stem. All of them produce offspring that are clones of the parent. Vegetative reproduction relies on an intriguing property of plants—namely, that many fully differentiated plant cells are **totipotent** ("all powerful"). That is, they have the genetic potential to develop into a whole, fully functional plant. Under appropriate conditions, a totipotent cell can *dedifferentiate*: it returns to an unspecialized embryonic state, and the genetic program that guides the development of a new individual is turned on.

Vegetative Reproduction Is Common in Nature

Various plant species have developed different mechanisms for reproducing asexually. In the type of vegetative reproduction called **fragmentation,** cells in a piece of the parent plant dedifferentiate and then can regenerate missing plant parts. Many gardeners have discovered to their frustration that a chunk of dandelion root left in the soil can rapidly grow into a new dandelion plant in this way.

When a leaf falls or is torn away from a jade plant (*Crassula* species), a new plant can develop from meristematic tissue in the detached leaf adjacent to the wound surface. In the "mother of thousands" plant, *Kalanchoe daigremontiana,* meristematic tissue in notches along the leaf margin gives rise to tiny plantlets **(Figure 34.14)** that eventually fall to the ground, where they can sprout roots and grow to maturity.

Some flowering plants, including some citrus species and the grass variety known as Kentucky blue grass *(Poa pratensis)*, can reproduce asexually through a mechanism called **apomixis.** Typically, a diploid embryo develops from an unfertilized egg or from diploid cells in the ovule tissue around the embryo sac. The resulting seed is said to contain a **somatic embryo,** which is genetically identical to the parent.

In wild plant species, most types of asexual reproduction result in offspring located near the parent. These clonal populations lack the variability provided by sexual reproduction, variation that enhances the odds for survival when environmental conditions change. Yet asexual reproduction offers an advantage in some situations. It usually requires less energy than producing complex reproductive structures such as seeds and showy flowers to attract pollinators. Moreover, clones are likely to be well suited to the environment in which the parent grows.

Ed Reschke/Peter Arnold, Inc.

Figure 34.14 *Kalanchoe daigremontiana*, the mother-of-thousands plant. Each tiny plant growing from the leaf margin can become a new, independent adult plant.

Many Commercial Growers and Gardeners Use Artificial Vegetative Reproduction

For centuries, gardeners and farmers have used asexual plant propagation to grow particular crops and trees and some ornamental plants. They routinely use *cuttings,* pieces of stems or leaves, to generate new plants; placed in water or moist soil, a cutting may sprout roots within days or a few weeks. Trees and wine grapes often are propagated by grafting a bud or branch from a plant with desirable fruit traits—the *scion*—and joining it to a root or stem from a plant with useful root traits—the *stock*. A grafted plant usually produces flowers and fruit identical to those of the scion's parent plant. The scion of a grafted wine grape variety may be chosen for the quality of its fruit and the stock for its hardy, disease-resistant root system. Vegetative propagation can also be used to grow plants from single cells. Rose bushes and fruit trees from nurseries and commercially important fruits and vegetables such as Bartlett pears, McIntosh apples, Thompson seedless grapes, and asparagus come from plants produced vegetatively in tissue culture conditions that cause their cells to dedifferentiate to an embryonic stage.

Vegetative Propagation in Tissue Culture. In groundbreaking experiments in the 1950s, Frederick C. Steward explored the totipotency of plant cells. Together with his coworkers at Cornell University, Steward propagated whole carrot plants in the laboratory by culturing carrot root phloem. Later researchers confirmed that almost any plant cell that has a nucleus and lacks a secondary cell wall may be totipotent.

The method of plant tissue culture Steward pioneered is simple in its general outlines **(Figure 34.15).** Bits of tissue are excised from a plant and grown in a nutrient medium. The procedure disrupts normal interactions between cells in the tissue sample, and the cells dedifferentiate and form an unorganized, white cell mass called a **callus.** When cultured with nutrients and growth hormones, some cells of the callus regain totipotency and develop into plantlets with roots and shoots.

Steward's work laid the foundation for large-scale commercial applications of plant tissue culture, as well as for a whole new field of research on *somatic embryogenesis* in plants. Single cells derived from a callus generated from shoot meristem are placed in a medium containing nutrients and hormones that promote cell differentiation. With some species, totipotent cells in the sample eventually give rise to diploid somatic embryos that can be packaged with nutrients and hormones in artificial "seeds" (see Figure 34.15). Endowed with the same traits as their parent, crop plants grown from somatic embryos are genetically uniform.

However, mutations often occur in the DNA of somatic embryos derived from callus culture. Screening techniques can identify such *somaclonal* mutants with desirable traits—for example, resistance to a disease that attacks wild-type plants of the same species. In plants that are infected with viruses, callus cultures can be restricted to virus-free cells and thus generate virus-free clones. Tissue culture propagation can then produce hundreds or thousands of identical plants from a single specimen. This technique, called **somaclonal selection**, is now a staple tool in efforts to improve major food crops, such as corn, wheat, rice, and soybeans. It is also being used to rapidly increase stocks of hybrid orchids, lilies, and other valued ornamental plants. The yellow and orange tomatoes that have become common in produce markets are the fruits of plants developed by somaclonal selection.

Research on tobacco and some other species has shown that plants can also be regenerated by **protoplast fusion.** In this method, the walls of living cells in solution are first digested away by enzymes, leaving the protoplasts. Then the protoplasts are induced to fuse, either by applying an electric current, a laser beam, or chemical additives to the solution, which briefly "loosen" the plasma membranes. The resulting cell (now $4n$, or tetraploid) is transferred to a solid nutrient medium and allowed to develop into a callus; then individual callus cells are stimulated to develop into embryos. If the fused protoplasts come from somatic cells of a single species, the embryos often grow into fertile plants. It has proven more difficult to grow fertile hybrids from fused protoplasts of different species and genera, probably because there are species-specific signals that govern key physiological functions. Even so, this method has produced the pomato, a cross between a potato and a tomato.

Regardless of how it comes into being, an embryonic sporophyte changes significantly as it begins the developmental journey toward maturity, when it will be capable of reproducing. Next we explore what researchers are learning about these developmental changes.

Study Break

1. Describe three modes of asexual reproduction that occur in flowering plants.
2. What is totipotency, and how do methods of tissue culture exploit this property of plant cells?

34.5 Early Development of Plant Form and Function

Unlike animals, plants have specialized body parts such as leaves and flowers that may arise from meristems throughout an individual's life—sometimes for thousands of years. Accordingly, in plants the biological role of embryonic development is not to generate the tissues and organs of the adult, but to establish a basic body plan—the root–shoot axis and the radial, "outside-to-inside" organization of epidermal, ground, and vascular tissues (see Section 31.1)—and the precursors of the primary meristems. Though they may sound simple, these fundamentals and the stages beyond them all require an intricately orchestrated sequence of molecular events that plant scientists are defining through sophisticated experimentation.

One of the most fruitful approaches has been the study of plants with natural or induced gene mutations that block or otherwise affect steps in development—and accordingly lend insight into the developmental roles of the normal, wild-type versions of those abnormal genes. Some of these genes are **homeotic genes,** regulatory genes in the genome of an organism that encode transcription factors. The transcription factors are proteins that control the expression of other genes, which in turn direct events in development (see *Focus on Research* in Chapter 31). While researchers work with various species to probe the genetic underpinnings of early plant development, the thale cress *(Arabidopsis thaliana)* has become a favorite model organism for plant genetic research (see *Focus on Research*).

Within Hours, an Early Plant Embryo's Basic Body Plan Is Established

The entire *Arabidopsis* genome has been sequenced, providing a powerful molecular "database" for determining how various genes contribute to shaping the plant body. Experimenters' ability to trace the expression of specific genes has shed considerable light on how the root–shoot axis is set and how the three basic plant tissue systems arise.

The Root–Shoot Axis. Shortly after fertilization gives rise to an *Arabidopsis* zygote, the single cell divides. Electron microscopy shows that, as with the *Capsella*

Figure 34.15 Research Method

Plant Cell Culture

PURPOSE: To grow in the laboratory plants that have the same genetic makeup as a parent plant.

PROTOCOL:

1. Typically, bits of somatic tissue are excised, often from root and shoot tips or meristems, because these parts tend to be free of viruses. The excised tissue is cultured in a nutrient medium, under strictly controlled environmental conditions.
2. Within a few days, cells in the excised tissue dedifferentiate and form an unorganized tissue mass called a callus.
3. Individual callus cells can be separated out and cultured in a medium containing growth hormones.
4. Totipotent cells eventually give rise to plantlets with roots and shoots.

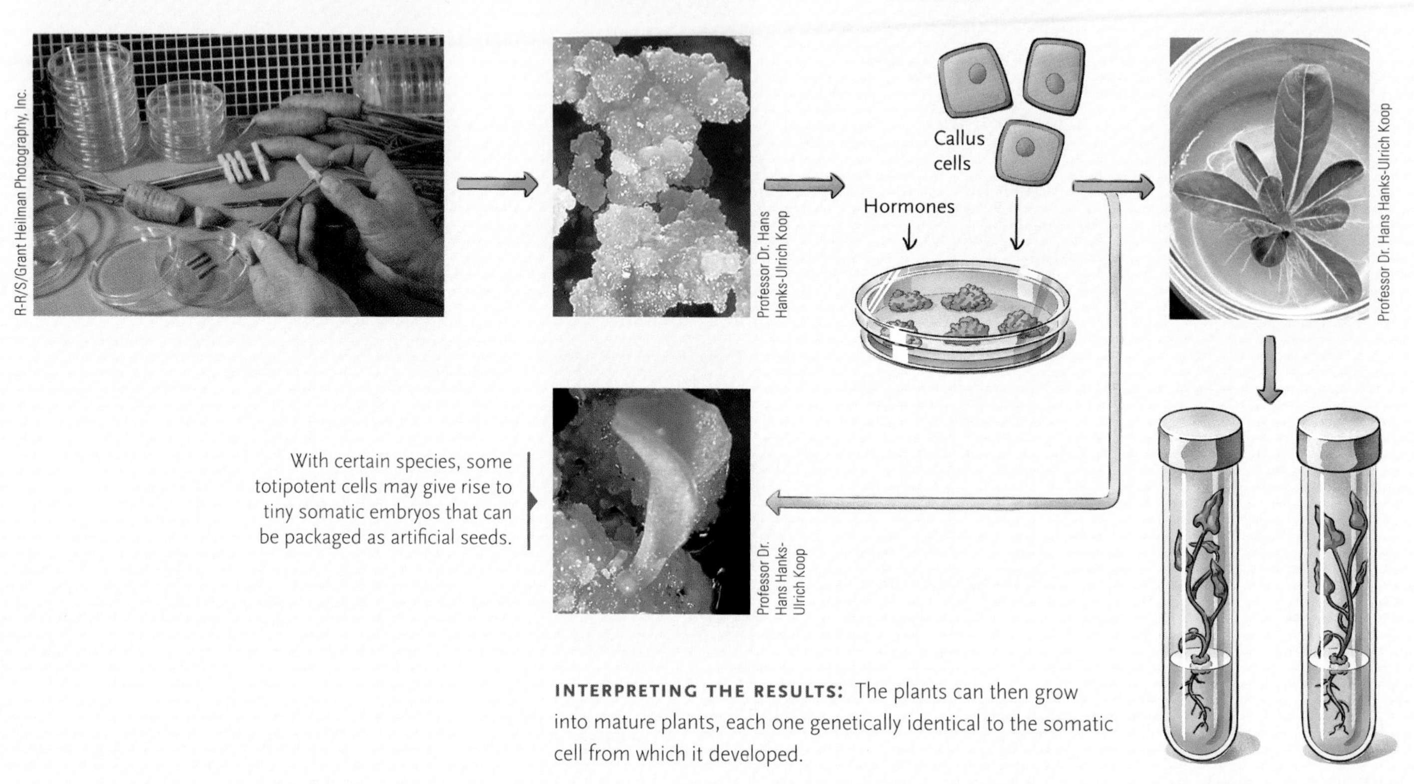

INTERPRETING THE RESULTS: The plants can then grow into mature plants, each one genetically identical to the somatic cell from which it developed.

zygote described earlier, this first round of mitosis produces a small apical cell and a larger basal cell **(Figure 34.16a)**. The apical cell receives the lion's share of the cytoplasm, while the basal cell receives the zygote's large vacuole and less cytoplasm. Researchers have confirmed that this asymmetrical division of the zygote results in the daughter cells receiving different mixes of mRNAs—the gene transcripts that will be translated into proteins.

Translation of differing mRNAs produces proteins that include several transcription factors, and it marks the genetic threshold of the separation of the plant body into root and shoot regions. As transcription factors trigger the expression of differing genes, distinct biochemical pathways unfold in sequence in the two cells. For example, a basal cell initially exports a signaling molecule (a hormone of the auxin family, discussed in Chapter 35) to the apical cell, and this sets in motion steps leading to the development of the various embryonic shoot features. Later, gene expression and the flow of chemical signals shift in ways that promote the development of specific structures from the basal cell, including portions of the root apical meristem.

Several of the genes that influence root–shoot polarity have been identified, and when any of them is disrupted, the result can be a serious defect. For example, when an embryo receives two copies of a mutant gene called *gnom*, the embryo doesn't develop distinct root and shoot regions. Instead it remains a lumpy blob **(Figure 34.16b)**.

Radial Organization of Tissue Layers. A day or so after an *Arabidopsis* egg cell is fertilized, the embryo consists of eight cells. Even at this early stage both the root–shoot axis and the beginnings of tissue systems are present. When an embryo reaches the so-called torpedo stage, cells representing all three basic tissue systems are in place **(Figure 34.16c)**. Again working with mutant plants, investigators have identified

Focus on Research

Model Research Organisms: *Arabidopsis thaliana*

For plant geneticists, the little white-flowered thale cress, *Arabidopsis thaliana*, has attributes that make it a prime subject for genetic research. A tiny member of the mustard family, *Arabidopsis* is revealing answers to some of the biggest questions in plant development and physiology.

Each plant grows only a few centimeters tall, so little laboratory space is required to house a large population. As long as *Arabidopsis* is provided with damp soil containing basic nutrients, it grows easily and rapidly in artificial light. Like Mendel's peas, *Arabidopsis* is self-compatible and self-fertilizing, and the flowers of a single plant can yield thousands of seeds per mating. Seeds grow to mature plants in just over a month and then flower and reproduce themselves in another 3 to 4 weeks. This permits investigators to perform desired genetic crosses and obtain large numbers of offspring having known, desired genotypes with relative ease. Individual *Arabidopsis* cells also grow well in culture.

The *Arabidopsis* genome was the first complete plant genome to be sequenced; at this writing researchers have identified approximately 28,000 genes arranged on five pairs of chromosomes. The genome contains relatively little repetitive DNA, so it is fairly easy to isolate *Arabidopsis* genes, which can then be cloned using genetic engineering techniques. Cloned genes are inserted into bacterial plasmids and the recombinant plasmids transferred to the bacterial species *Agrobacterium tumefaciens*, which readily infects *Arabidopsis* cells. Amplified by the bacteria, the genes and their protein products can be sequenced or studied in other ways.

Typically, researchers use chemical mutagens or recombinant bacteria to introduce changes in the *Arabidopsis* genome. These mutants have become powerful tools for exploring molecular and cellular mechanisms that operate in plant development—for example, elucidation of the homeotic genes responsible for flower development described in this chapter. *Arabidopsis* mutants are also being used to probe fundamental questions such as how plant cells respond to gravity and the role of pigments called phytochromes in plant responses to light.

An ambitious, multinational research effort called the 2010 Project aims to determine the functions of all *Arabidopsis* genes by 2010. The Arabidopsis Information Resource (TAIR) recently estimated the percentages of *A. thaliana* genes in different functional categories **(Figure a).** The goal of Project 2010 is to create a comprehensive genetic portrait of a flowering plant—how each gene affects the functioning of not only individual cells but the plant as a whole.

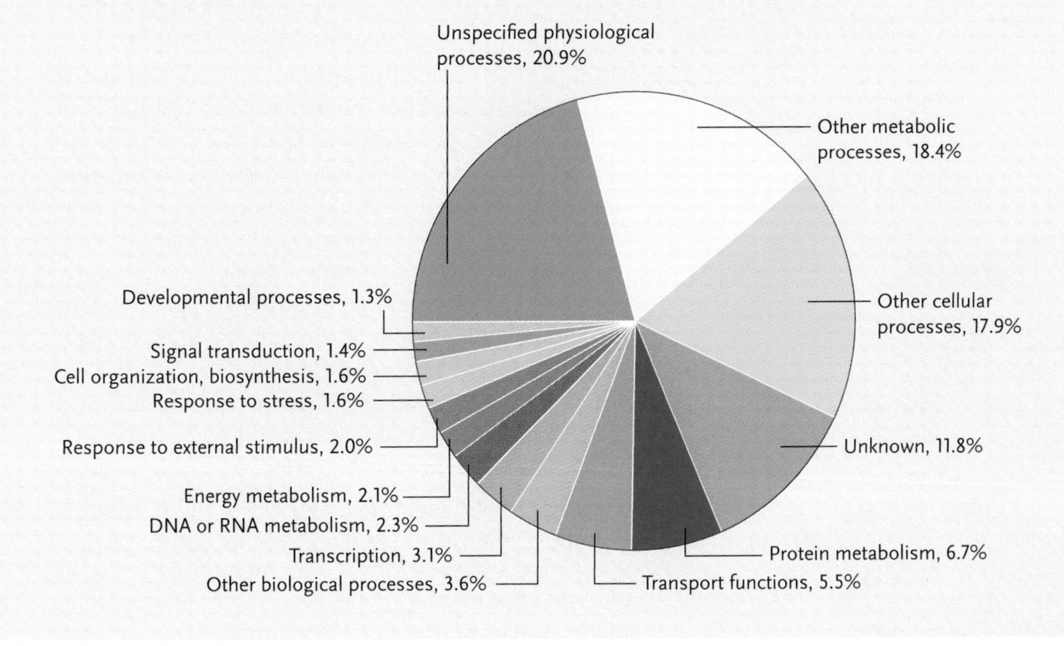

Figure a
The percentages of *A. thaliana* genes that influence different functional categories.
(Courtesy of the Arabidopsis Information Resource, 2005.)

several *Arabidopsis* genes that help govern early development of tissue systems. For example, a gene called *SCR* encodes a protein that apparently regulates mitotic divisions that produce the first cells of a developing root's cortex and endoderm tissue layers (Figure 31.20 shows the locations of these tissues in a mature root). The roots of a mutant *scr* seedling contain cells with jumbled characteristics of both tissue layers.

No matter what the species, nearly all new plant embryos have the general body plan we have been discussing. As development proceeds, cells at different sites become specialized in prescribed ways as a particular set of genes is expressed in each type of cell—a process known as *differentiation*. Differentiated cells in turn are the foundation of specialized tissues and organs, which come about through processes we consider next.

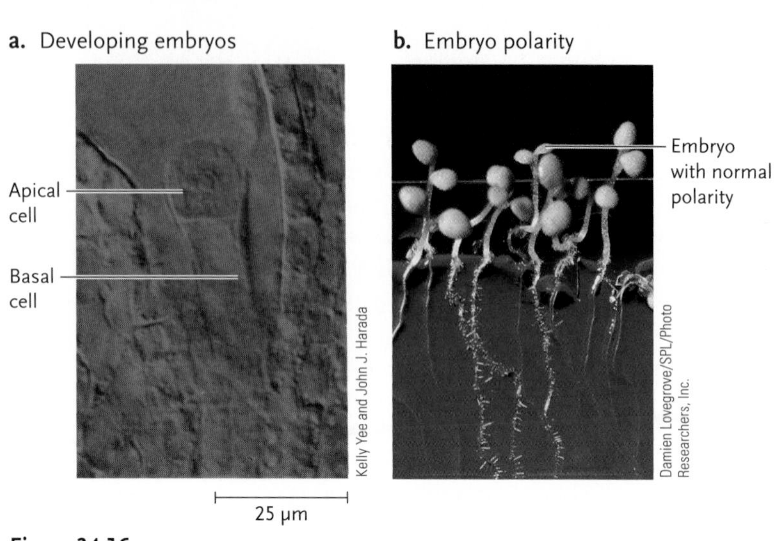

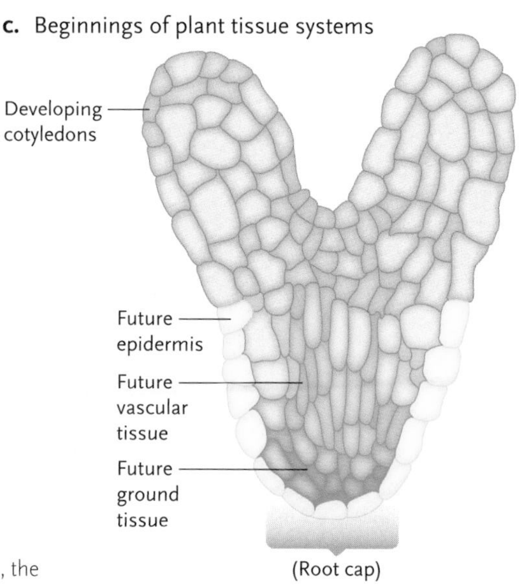

Figure 34.16
Stages in the development of the basic body plan of a plant embryo. (a) After a zygote forms, the first round of cell division produces an embryo with an apical cell that contains much of the zygote's cytoplasm, and a larger basal cell that receives the vacuole and less cytoplasm. The division allots different transcription factors to each cell and establishes the plant's root–shoot axis. **(b)** Normal *A. thaliana* seedlings, in which the root–shoot polarity has become established. **(c)** The approximate locations of early embryonic cells that are the forerunners of epidermal, ground, and vascular tissue systems, respectively.

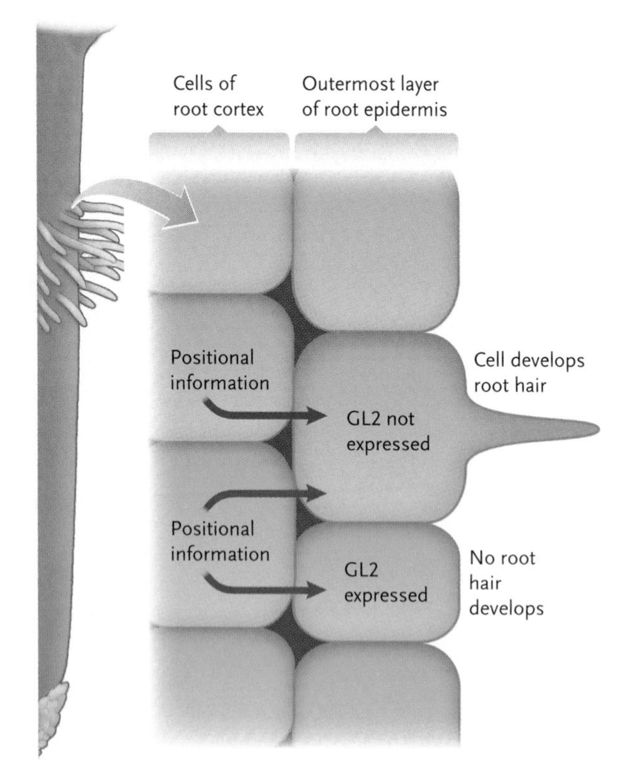

Figure 34.17
One model of how positional information influences the development of root hairs. In this model, the only epidermal cells that develop root hairs are those whose inner wall is in contact with two root cortex cells. Such positioning gives rise to signals that block the expression of the GL2 (GLABRA2) gene. When GL2 is expressed, a root epidermal cell will not develop a root hair.

Key Developmental Cues Are Based on a Cell's Position

Although many of the specifics of development differ in animals and plants, one fundamental holds true for both: Normal development produces ordered spatial arrangements of differentiated tissues. Examples in plants include root and shoot apical meristems at opposite ends of the root–shoot axis, the cotyledons that divide the shoot into an upper epicotyl and a lower hypocotyl, and the nested layers of vascular, ground, and epidermal tissue systems. Developmental biologists call this progressive ordering of parts **pattern formation,** and a wealth of research has shown that it is guided by the position of cells relative to one another. Such *positional information* helps establish a cell's developmental fate: that is, it provides cues that "tell" cells where they are in the developing embryo and thus lay the groundwork for an appropriate genetic response.

Numerous researchers have explored how cells in a developing plant or plant part receive and respond to positional information. Experiments have demonstrated, for example, that only certain cells in the epidermis of an embryonic root will give rise to root hairs, the type of trichomes that take up water and minerals from soil (see Section 31.2). These specialized root epidermal cells all share the same position with respect to the underlying root cortex—each abuts two cortical cells. By contrast, no root hair extension will develop from an epidermal cell that lines up against only one cortical cell. **Figure 34.17** diagrams one model of what happens next. In this scenario, one or more

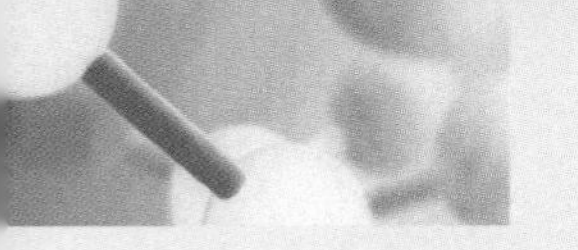

Insights from the Molecular Revolution

Trichomes: Window on Development in a Single Plant Cell

The delicate plant cell extensions called trichomes are helping to illuminate developmental processes that go on in a single plant cell as it differentiates—that is, as it acquires its ultimate specialized structure and function. In *Arabidopsis* each of these minute protuberances consists of a single cell with a branching tripartite pattern **(Figure a).**

A curious feature of trichomes is that as one differentiates, increases in size, and extends branches in different directions, its chromosomes—and the cell's DNA—and duplicated several times over without mitosis (a process called endoreduplication). As a result, the cell has multiple copies of chromosomes. Experiments that isolate the effects of different mutants have helped confirm that the amount of DNA in the cell strongly influences the cell's structure, and that several genes interact to determine it. One of these genes is called *TRY* (for *TRIPTYCHON*); when it is mutated the affected plant's trichomes have a double complement of DNA and develop five branches **(Figure b).** But genes that regulate the cell cycle are only part of the story. Experiments with other mutants show that several other genes also help produce the characteristic three-pronged trichome branching. For example, when a gene called *TUBULIN FOLDING CO-FACTOR C (TFCC)* is affected, the normal organization of microtubules in mutant *tfcc* trichomes is disrupted and the resulting trichome has just two short branches, resembling the oar handles of a canoe. When yet another gene, *STICHEL*, is mutated, *sti* mutants don't develop any branches at all **(Figure c).** The underlying reason for this phenotype is not yet understood.

The examples described here underscore how complex molecular interactions affecting multiple aspects of a cell's functioning ultimately shape a cell's form and function. Because the genes that operate in trichomes are also involved in the development of other types of plant cells, understanding their effects and interactions promises to shed light on processes that operate to generate differentiated cells throughout the plant body.

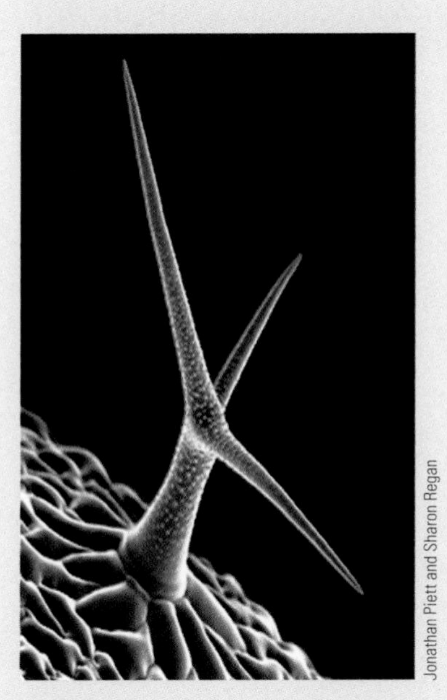

Jonathan Piett and Sharon Regan

Figure a
Normal trichome from the epidermis of a leaf of *Arabidopsis thaliana.*

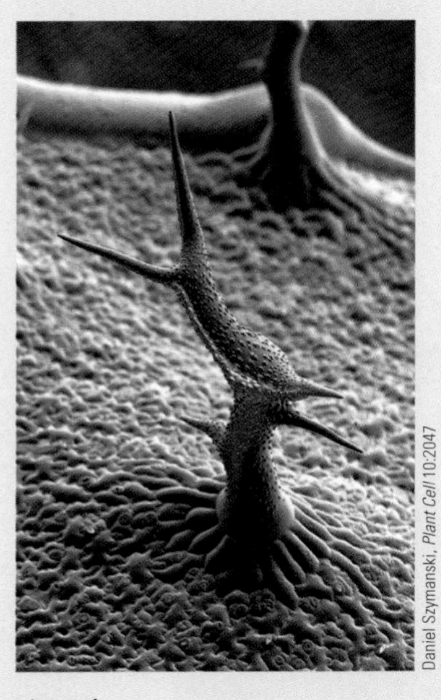

Daniel Szymanski, *Plant Cell* 10:2047

Figure b
Five-pronged trichome from a *try* mutant.

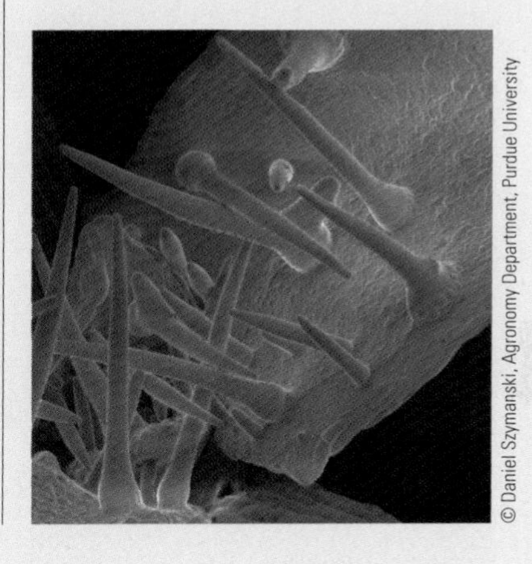

Figure c
The unbranched trichomes of an *sti* mutant.

chemical signals may cross from cortical to epidermal cells by way of plasmodesmata. When an epidermal cell receives signals from a single cortex cell, a series of genes are expressed in a cascade of effects that culminate in the expression of a gene called GL2 (or GLABRA2). The product of GL2 blocks the formation of root hairs. If, on the other hand, an epidermal cell aligns with two cortex cells, it receives signals from both and the cascade of gene effects blocks expression of GL2—and a root hair develops. *Insights from the Molecular Revolution* gives more examples of ways that trichomes such as root hairs have become popular experimental models for studying the differentiation of plant cells.

Morphogenesis Shapes the Plant Body

As a plant embryo grows and tissues of differentiated cells form, the stage is set for different body regions to develop characteristic shapes and structures that correlate with their function. This process, called **morphogenesis**, shapes the new shoot and root parts produced by dividing cells in meristems. In animals, morphogenesis involves localized cell division and growth, as well as migration of cells and entire tissues from one site to another (see Chapter 48). Plant cells, however, are enclosed within thick walls and usually cannot move. Thus morphogenesis in plants relies on mechanisms that don't require mobility. One of these

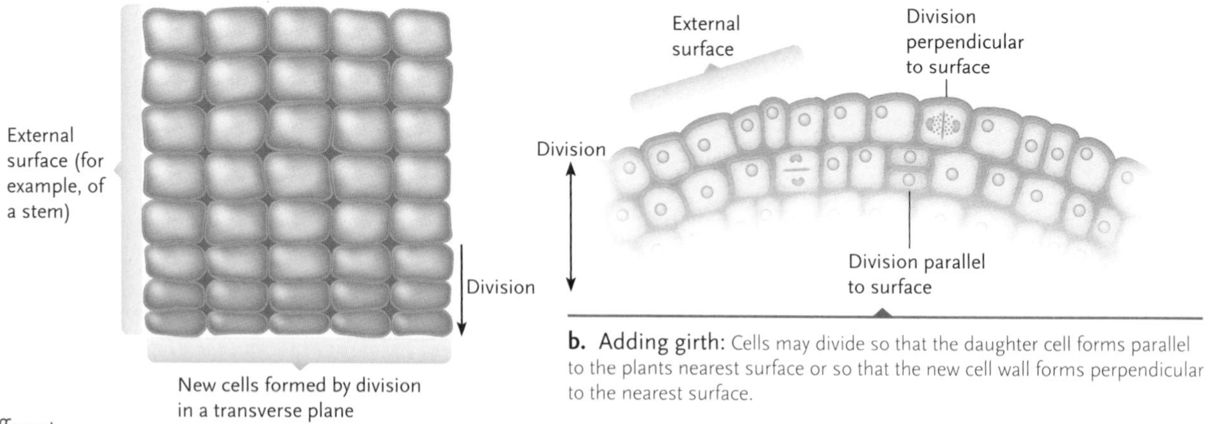

Figure 34.18
Plant cell division in different planes. The external surface nearest the dividing cell is the reference point for establishing division planes.

a. Adding length: Division in a transverse plane parallel to the root-shoot axis produces daughter cells stacked like bricks.

b. Adding girth: Cells may divide so that the daughter cell forms parallel to the plants nearest surface or so that the new cell wall forms perpendicular to the nearest surface.

mechanisms is **oriented cell division**, which establishes the overall shape of a plant organ, and another is **cell expansion**, which enlarges the cells in specific directions in a developing organ.

Oriented Cell Division. As described in Chapter 31, roots and stems grow lengthwise as the division and expansion of cells produce columns of cells parallel to the root–shoot axis. The cell divisions occur in a transverse plane—that is, new cell walls, and then the cell plate, form so that the cells become stacked one atop the other like wooden blocks **(Figure 34.18a).** A plant adds girth—increases in circumference—by way of cell divisions in other planes. For instance, new cell walls

Figure 34.19
How the plane of cell division is determined in a plant cell. This series of micrographs shows events in onion *(Allium cepa)* root tip cells. The arrows mark the eventual location of the new cell wall.
(All: S. M. Wick, *J Cell Biol*, 89:685, 1987, Rockefeller University Press.)

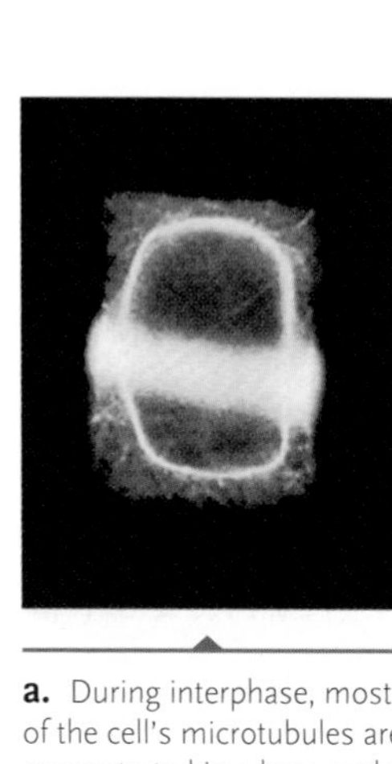

a. During interphase, most of the cell's microtubules are concentrated in a layer under the plasma membrane.

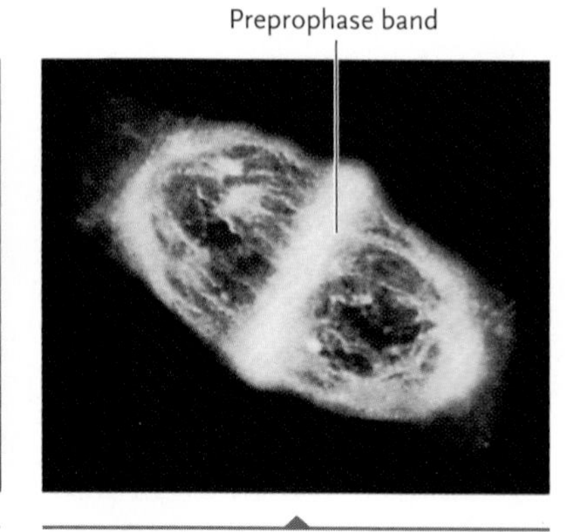

b. In preprophase of mitosis, the microtubules break their links to the plasma membrane and assemble into a thick preprophase band.

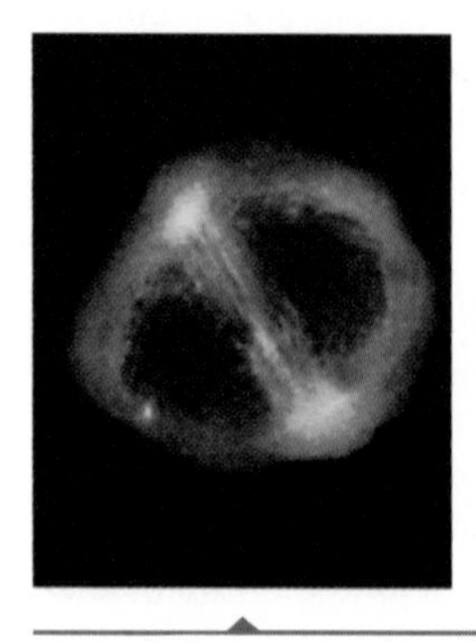

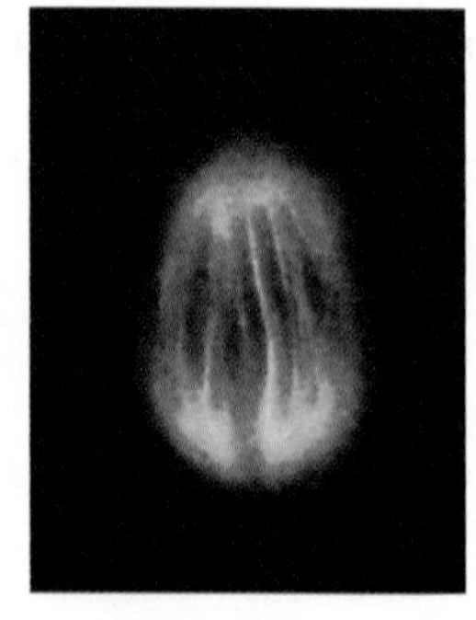

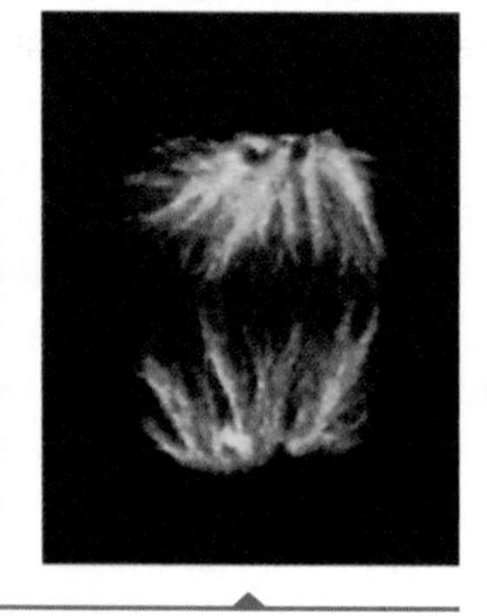

c. As cell division begins, the preprophase band disassembles and microtubules organize into a spindle.

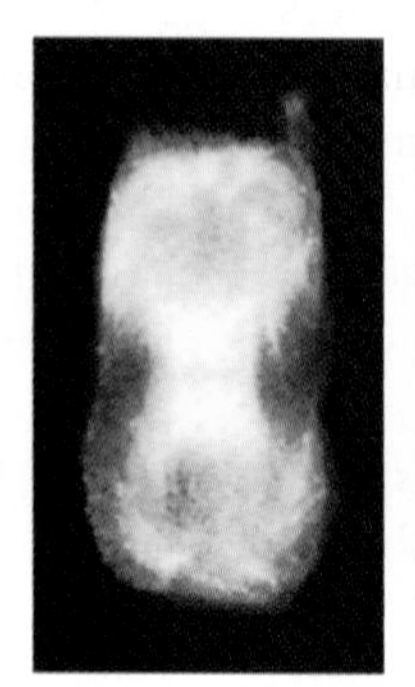

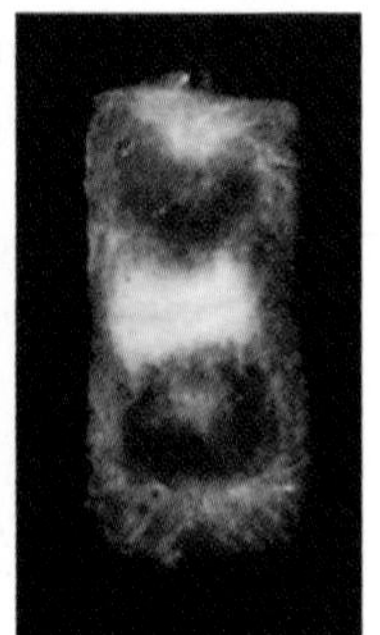

d. After division is complete, the spindle disassembles, leaving a layer of microtubules in the same plane as the preprophase band. The new cell wall and cell plate form there.

may form parallel to the nearest plant surface, or perpendicular both to the nearest surface and to the transverse plane **(Figure 34.18b)**.

You may recall from Chapter 10 that the cell plate forms during the cytokinesis phase of mitosis; it establishes the plane of the middle lamella that will eventually separate the parent and daughter cells. The capacity of dividing plant cells to synthesize a new cell plate in a different plane from the old one underlies morphogenesis in nearly all plant groups. In meristematic tissue, changes in the plane of cell division establish the direction in which structures such as lateral roots, branches, and leaf and flower buds will grow, and so gives the plant body its overall form.

Figure 34.19 shows how the plane of a cell plate is established. While a plant cell destined to divide is still in interphase, hours before mitosis begins, the cell nucleus migrates to a particular location in the cell. The nucleus becomes surrounded by a layer of microtubules and microfilaments that radiate outward from it. Where the layer contacts the cell wall, a belt of microtubules and microfilaments called the *preprophase band* forms briefly, encircling the cell cytoplasm. The band usually disappears as mitosis gets underway in the cell, but its position marks the site where the cell plate forms during cytokinesis. Remnants of microfilaments may guide the edges of the developing cell plate into the proper position against the cell wall.

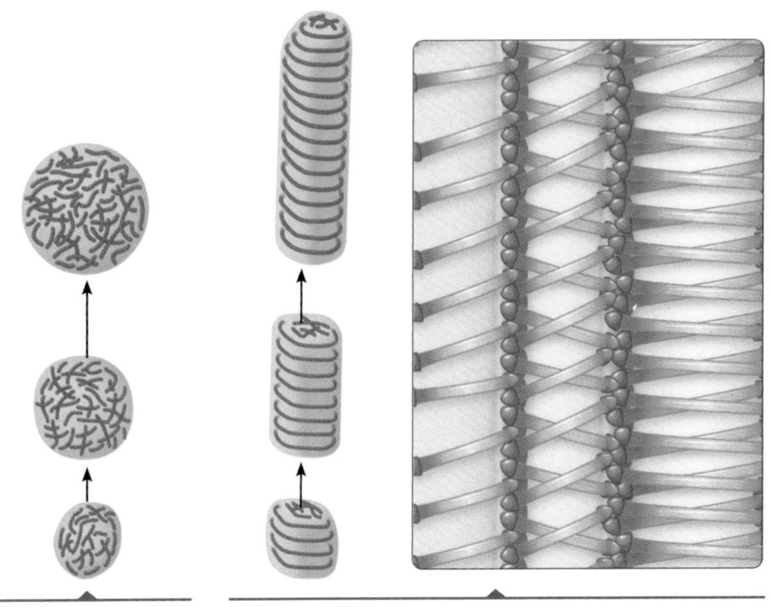

a. When microfibrils are oriented at random, the primary wall is elastic all over, so the cell can grow in all directions.

b. When microfibrils are oriented transversely, the cell can grow only longitudinally.

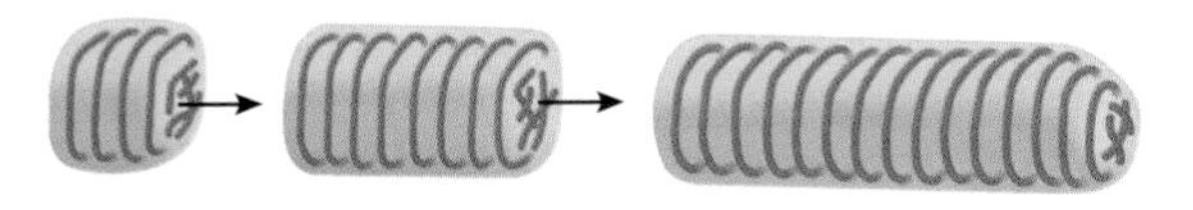

c. When microfibrils are oriented longitudinally, the cell can grow only laterally.

Figure 34.20
Cell expansion and the orientation of cellulose microfibrils. In each cell, microtubules inherited from the parent cell are already oriented in prescribed patterns that govern how cellulose microfibrils will be oriented in the cell wall. Their orientation in turn governs the direction in which a cell can expand.

Cell Expansion. Once a cell has divided, the daughter cells expand to mature size. Yet plant cells are encased in a primary wall of nonliving material. Botanists are beginning to learn how the cell wall expands to accommodate the enlarging cell within.

Primary cell walls consist of a loose mesh of cellulose microfibrils embedded in a gel-like matrix. As plant cells mature, they may elongate to as much as 100 times their embryonic lengths. During this elongation, the cellulose meshwork is first loosened and then stretched. Turgor pressure supplies the force for stretching. The exact mechanism that loosens the wall structure is not known, although experiments indicate that it depends on a dramatic drop in pH. Some researchers suggest that an auxin in the cell cytoplasm may stimulate a plasma membrane proton pump that moves H^+ into the cell wall (see Section 6.4). The acidic wall conditions may activate hydrolytic enzymes that break bonds between wall components, or they may promote loosening in some other way.

During expansion, enzyme complexes in the cell's plasma membrane synthesize new cellulose microfibrils from glucose in the cytoplasm. When each microfibril is fully formed, it is bound in place in the growing wall by pectins and other wall components.

The direction of cell expansion depends on the orientation of the newly formed cellulose microfibrils **(Figure 34.20)**. If the microfibrils are randomly oriented, the cell expands equally in all directions. If they are oriented at right angles to the cell's long axis, the cell expands lengthwise. And if new fibrils are deposited parallel to the long axis of the cell, the cell expands laterally.

Patterns of Cell Division during Early Growth. Like the first mitotic division in an *Arabidopsis* zygote, it's not uncommon for cell divisions in a growing plant to be asymmetrical, so that one daughter cell ends up with more cytoplasm than the other. The unequal distribution of cytoplasm means that the daughter cells differ in their composition and structure, and the differences affect how they interact with their neighbors during growth, even though all cells carry the same genes. Their cytoplasmic differences and interactions with one another trigger selective gene expression. Such events seal the developmental fate of particular cell lineages. Their descendant cells divide in prescribed planes and expand in set directions, producing plant parts with diverse shapes and functions.

Figure 34.21 Experimental Research

Probing the Roles of Floral Organ Identity Genes

QUESTION: What are the genetic mechanisms that govern the formation of the parts of a flower?

EXPERIMENTS: Meyerowitz and his colleagues grew *Arabidopsis thaliana* plants having mutated, inactivated versions of the genes suspected of controlling the proper development of floral organs. They compared the types and arrangements of floral organs in the test plants with the organs present in normal, wild-type *A. thaliana* flowers.

a. Normal *A. thaliana*

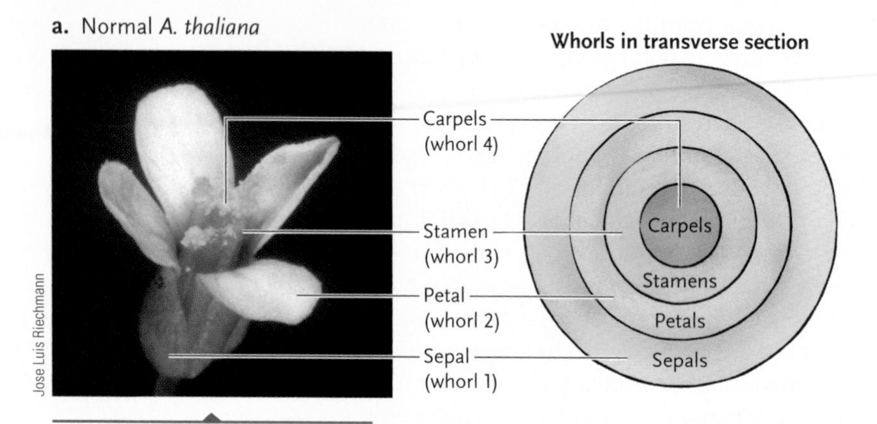

Normal arrangement of organs: carpels in whorl 4, stamens in whorl 3, petals in whorl 2, and sepals in whorl 1

RESULTS: At least three classes of homeotic genes (A, B, and C) regulate different aspects of normal floral organ development.

b. When mutation inactivates the *APETALA2* gene, class A genes are not expressed.

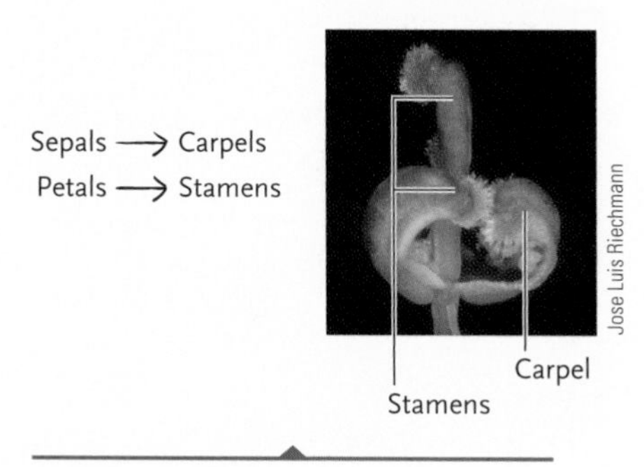

Stamens replace petals and carpels replace sepals. Organ identity in the other whorls does not change.

c. When mutation inactivates the *APETALA3* or *PISTILLATA* gene, class B genes are not expressed.

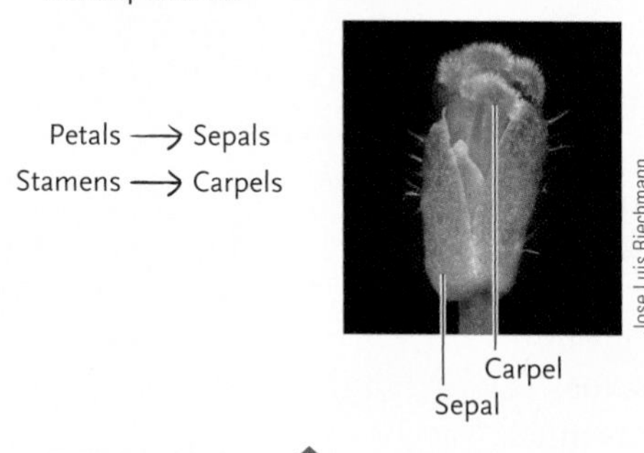

Carpels replace stamens and sepals replace petals. Organ identity in the other whorls does not change.

d. When mutation inactivates the *AGAMOUS* gene, class C genes are not expressed.

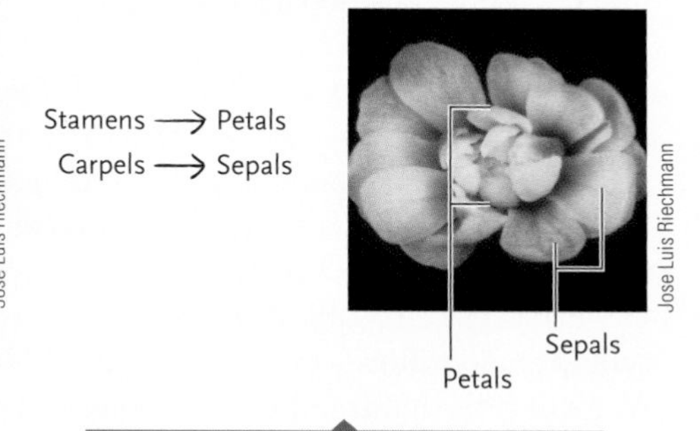

No carpels; instead petals develop in whorl 3 and a version of a floral meristem develops where whorl 4 would normally be. It gives rise to extra petals and sepals. Organ identity in other whorls does not change.

Further experiments with double mutants (inactivation of both A and B, A and C, and B and C) all produce abnormal flowers. For example, only carpels develop in mutants having only an active C class gene, and only sepals develop in plants having only an active B class gene.

e. Overlapping activity fields of floral organ identity genes

Carpels			C	E	**Whorl 4**
Stamens		B	C	E	**Whorl 3**
Petals	A	B		E	**Whorl 2**
Sepals	A				**Whorl 1**

CONCLUSIONS: In *A. thaliana*, A, B, and C activity genes, expressed alone or in pairs, underlie the development of a normal pattern of floral organs. The fields of activity overlap. In addition, A and C activity apparently counteract each other, and if one is absent the other can spread beyond the whorls where it normally appears. Subsequent research revealed that a fourth E class of gene activity is required for proper expression of other organ identity genes.

Regulatory Genes Guide the Development of Floral Organs

Research with several plant species has shed light on the genetic mechanisms that govern the formation of the parts of a flower. For example, experiments with *Arabidopsis* carried out by Elliot Meyerowitz and his colleagues at the California Institute of Technology showed that *floral organ homeotic genes* regulate the development of the sepals, petals, stamens, and carpels in flowers.

The Meyerowitz team studied plants with various mutations in floral organs. By observing the effects of specific mutations on the structure of *Arabidopsis* flowers, the investigators eventually identified three classes of homeotic gene activity—which they named A, B, and C—that regulate different aspects of normal flower development. Subsequent studies by other scientists identified an E class of gene activity that appears to be an essential partner in the functioning of A, B, and C class genes. The effects of the genes overlap, so that A, B, and C class genes are expressed in two whorls, and E class genes in three: Class A genes are expressed in whorls 1 and 2 (sepals and petals), class B genes in whorls 2 and 3 (petals and stamens), class C genes in whorls 3 and 4 (stamens and carpels), and class E genes in whorls 2, 3, and 4 **(Figure 34.21)**.

Abnormal floral patterns such as those in Figure 34.21b–d show how mutations in the floral organ homeotic genes can affect the identity of flower parts. For example, a mutation that deactivates the A-class gene *APETALA2* produces a flower with carpels and stamens in whorl 1 (see Figure 34.21b). Another intriguing finding is that the A and C activity classes normally oppose each other. When no A gene is expressed, C activity spreads into whorls where the A usually occurs, and vice versa. Subsequent studies have examined many other floral homeotic genes, as well as the genes that control the various gene classes.

As the genes governing flower development are isolated, they can be cloned and their nucleotide sequences defined and manipulated. Such cloned genes already are of keen interest in plant genetic engineering, because food grains such as wheat and many other vital agricultural commodities come directly or indirectly from flowers.

Leaves Arise from Leaf Primordia in a Closely Regulated Sequence

A mature leaf may have many millions of differentiated cells organized into tissues such as epidermis and mesophyll. As described in Chapter 31, leaves develop from leaf primordia that arise just behind the tips of shoot apical meristems **(Figure 34.22)**.

Clonal analysis has opened a window on many aspects of plant development, including how leaf primordia originate and give rise to leaves. In this method, the investigator cultures meristematic tissue that contains a mutated embryonic cell having a readily observable trait, such as the absence of normal pigment. (In the laboratory, this kind of mutation can be induced by chemicals or radiation.) The unusual trait then serves as a marker that identifies the mutant cell's clonal descendants, making it possible to map the growing structure. Researchers have used clonal analysis to study leaf development in garden peas, tomatoes, grasses, and tobacco, among others.

Like flowers, leaves arise through a developmental program that begins with gene-regulated activity in meristematic tissue. Hormones or other signals may arrive at target cells via the stem's vascular tissue, activating genes that regulate development. Studies show that small phloem vessels first penetrate a young leaf primordium almost immediately after it begins to bulge out from the underlying meristematic tissue, and xylem soon follows. The early phloem connections are especially vital to the leaf's survival, because the leaf does not begin photosynthesis until it attains one-third of its mature size.

A growing primordium becomes cone-shaped, wider at its base than at its tip. At a certain point, mitosis speeds up in cells along the flanks of the lengthening

a. Leaf primordia of *Coleus*

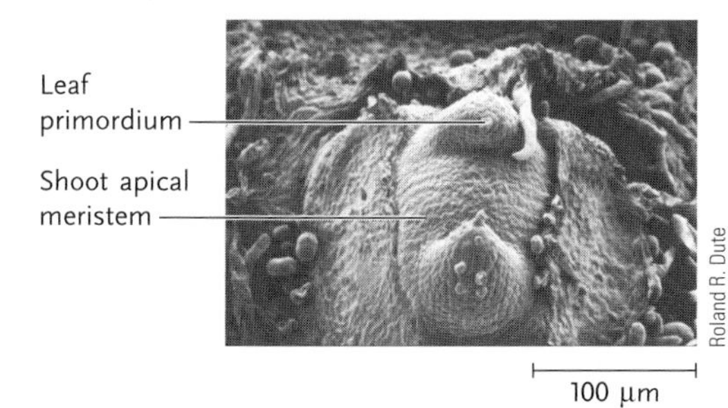

b. Early stages of leaf development

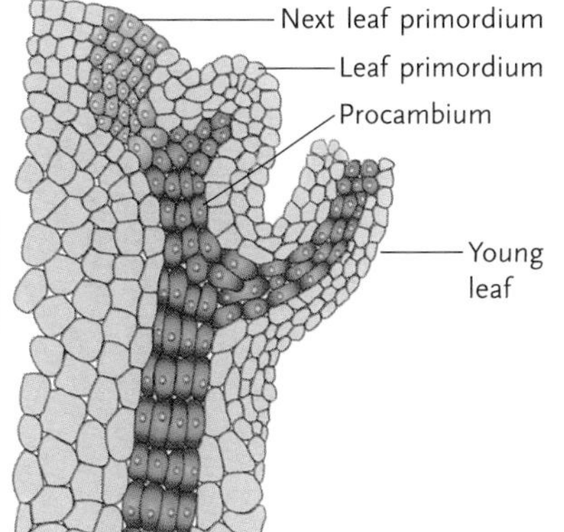

Figure 34.22
Early stages in leaf development. **(a)** Leaf primordia at the shoot tip of *Coleus*, a genus that includes several popular house plants. **(b)** Diagram showing leaf primordia in different stages of development. See also Figure 34.12a, which shows the progression of leaves that form during the early growth and development of a eudicot.

cone. In eudicots, the rapid cell divisions occur perpendicular to the surface and produce the leaf blade. In monocot grasses, which have long, narrow leaves, vertical "files" of cells develop as cells in the meristem at the base of the cone divide in a plane parallel to the surface. The ultimate shape of a leaf depends in part on variations in the plane and rate of these cell divisions.

Cells at the leaf tip, and those of xylem and phloem that service it, are the oldest in the leaf, and it is here that photosynthesis begins. Commonly, leaf tip cells also are the first to stop dividing. By the time a leaf has expanded to its mature size, all mitosis has ended and the leaf is a fully functional photosynthetic organ.

In nature, genes that govern leaf and flower development switch on or off in response to changing environmental conditions. Their signals determine the course of a plant's vegetative growth throughout its life. In many perennials, new leaves begin to develop inside buds in autumn, then become dormant until the following spring, when external conditions favor further growth. Environmental cues stimulate the gene-guided production of hormones that travel through the plant in xylem and phloem, triggering renewed leaf growth and expansion. Leaves and other shoot parts also age, wither, and fall away from the plant as hormonal signals change. The far-reaching effects of plant hormones on growth and development are the subject of Chapter 35.

Study Break

1. What is a homeotic gene? Give at least two examples of plant tissues such genes might govern in a species such as *Arabidopsis*.
2. What are the two basic mechanisms of morphogenesis in plants? Describe the patterns of cell division by which a plant part (a) grows longer and (b) adds girth.
3. Summarize the gene-guided developmental program that gives rise to a leaf.

Unanswered Questions

What are the signaling events that mediate pollen-tube guidance during compatible pollination?

As you learned in this chapter, after pollen lands on the surface of the carpel, it forms a pollen tube, which then invades the carpel, migrates past several different cell types, and enters the micropyle to fertilize an egg and central cell. Recent work has shown that pollen-tube migration is mediated by a series of cell–cell interactions such as attraction, repulsion, and adhesion. However, most of these conclusions have been based on analyzing static images and fixed tissues, and it has been hypothesized that many more subtle, dynamic interactions between the pollen tube and female cells exist to ensure compatible pollination. It has been shown, for example, that pollen tubes gain the competence to successfully find the ovules only when they grow on the pistil tissues, a process that is functionally analogous to the transformation that mammalian spermatozoa undergo after residence for a finite amount of time in the female reproductive tract. It was also recently demonstrated that once a pollen tube penetrates the ovule, additional pollen tubes are prevented from gaining access into the targeted ovule, a process that is functionally analogous to the prevention of polyspermy by a fertilized mammalian egg. These novel signaling events are spurring researchers to take a closer look at pollen-tube guidance to ovules. Only when such signaling events are described can appropriate efforts be taken to identify the cues that mediate these events (see next question).

What kinds of approaches are being taken to identify novel signaling events? Researchers usually first identify a mutant plant that is defective in any of the pollen-tube guidance steps that are essential for successful fertilization. Subsequently, they analyze the defects to learn more about how this process normally happens. Other researchers develop microscopy-based real-time assays to directly observe pollen-tube behavior with a variety of female tissues.

What are the chemical cues from female tissues that facilitate compatible pollination?

As you learned in this chapter, if a pollen tube lands on a compatible stigma, chemical cues produced by the female tissue then guide the pollen tube from the stigma to the embryo sac of an ovule. Recent research has revealed that cues produced by both sporophyte and gametophyte are essential for proper guidance of pollen tubes to ovules. Despite these advances, the identities of these cues remain unknown. What are the hurdles that have hampered efforts to uncover pollen-tube navigation cues of even known signaling events described above? First, the pistil tissue within which the pollen tube elongates to the ovule is comprised of several types of tissue, including stigma, style, and transmitting tract, and these tissues are not readily accessible. Second, analyzing the dynamic responses of pollen tubes is difficult given that pollen-tube navigation occurs well within opaque pistils. Third, it appears that multiple, stage-specific, short-range, and readily labile signals produced in minute quantities mediate pollen-tube guidance. However, recent development of global approaches that are highly sensitive and assays that directly monitor pollen-tube elongation offer hope that guidance cues will be uncovered sooner rather than later.

Ravi Palanivelu is an assistant professor in the Department of Plant Sciences at the University of Arizona. His current research focuses on the isolation and characterization of pollen-tube guidance signals during *Arabidopsis* reproduction, with the long-term goal of understanding the molecular basis of how cells communicate with each other. To learn more about Dr. Palanivelu's research, go to http://www.ag.arizona.edu/research/ravilab.

Review

Go to **Thomson**NOW™ at www.thomsonedu.com/login to access quizzing, animations, exercises, articles, and personalized homework help.

34.1 Overview of Flowering Plant Reproduction

- In most flowering plant life cycles, a multicellular diploid sporophyte (spore-producing plant) stage alternates with a multicellular haploid gametophyte (gamete-producing plant) stage. The sporophyte develops roots, stems, leaves, and, at some point, flowers. The separation of a life cycle into diploid and haploid stages is called alternation of generations (Figure 34.2).

Animation: Flowering plant life cycle

34.2 The Formation of Flowers and Gametes

- A flower develops at the tip of a floral shoot. It can have up to four whorls supported by the receptacle. The calyx and corolla consist of the sepals and petals, respectively. The third whorl consists of stamens, and carpels make up the innermost whorl (Figure 34.3).
- The anther of a stamen contains sacs where pollen grains develop. If compatible pollen lands on the carpel's stigma, which contains an ovary in which eggs develop, fertilization takes place.
- A complete flower has both male and female reproductive parts. In monoecious species each plant has both types of flowers; in dioecious species the "male" and "female" flowers are on different plants (Figure 34.4).
- In pollen sacs, meiosis produces haploid microspores. Mitosis inside each microspore produces a pollen grain, an immature male gametophyte. One of its cells develops into two sperm cells, the male gametes of flowering plants. Another cell produces the pollen tube (Figures 34.5 and 34.6).
- An ovule forms inside a carpel, on the wall of the ovary. Development in the ovule produces a female gametophyte—the embryo sac with egg cell.
- In the ovule, meiosis produces four haploid megaspores. Usually all but one disintegrate. The remaining megaspore undergoes mitosis three times without cytokinesis, producing eight nuclei in a single large cell. Two of these (polar nuclei) migrate to the center of the cell. When cytokinesis occurs, cell walls form around the nuclei, with the two polar nuclei enclosed in a single wall. The result is the seven-celled embryo sac, one cell of which is the haploid egg. The cell containing polar nuclei will help give rise to endosperm.

Animation: Floral structure and function

Animation: Flower parts

Animation: Microspores to pollen

Animation: Apple fruit structure

34.3 Pollination, Fertilization, and Germination

- Upon pollination, the pollen grain resumes growth. A pollen tube develops, and mitosis of the male gametophyte's sperm-producing cell produces two sperm nuclei (Figure 34.7).
- In double fertilization, one sperm nucleus fuses with one egg nucleus to form a diploid ($2n$) zygote. The other sperm nucleus and the two polar nuclei of the remaining cell also fuse, forming a cell that will give rise to triploid ($3n$) endosperm in the seed (Figure 34.8).
- After the endosperm forms, the ovule expands, and the embryonic sporophyte develops. A mature ovule is a seed. Inside the seed, the embryo has a lengthwise axis with a root apical meristem at one end and a shoot apical meristem at the other.
- Eudicot embryos have two cotyledons. The embryonic shoot consists of an upper epicotyl and a lower hypocotyl; also present is an embryonic root, the radicle. The single cotyledon of a monocot forms a scutellum that absorbs nutrients from endosperm. Apical meristems of a monocot embryo are protected by a coleoptile over the shoot tip and a coleorhiza over the radicle (Figure 34.9).
- A fruit is a matured or ripened ovary. Fruits protect seeds and disperse them by animals, wind, or water.
- Fruits are simple, aggregate, or multiple, depending on the number of flowers or ovaries from which they develop. Fruits also vary in the characteristics of their pericarp, which surrounds the seed (Figure 34.10).
- The seeds of most plants remain dormant until external conditions such as moisture, temperature, and day length favor the survival of the embryo and the development of a new sporophyte (Figures 34.11–34.13).

Animation: Bee-attracting flower pattern

34.4 Asexual Reproduction of Flowering Plants

- Many flowering plants also reproduce asexually, as when new plants arise by mitosis at nodes or buds along modified stems of the parent plant. New plants also may arise by vegetative propagation (Figure 34.14).
- Tissue culture methods for developing new plants from a parent plant's somatic (nonreproductive) cells include somatic embryogenesis and protoplast fusion (Figure 34.15).

Animation: Eudicot life cycle

Animation: Eudicot seed development

34.5 Early Development of Plant Form and Function

- In plants that reproduce sexually, development starts at fertilization. Early on, a new embryo acquires its root–shoot axis, and cells in different regions begin to become specialized for particular functions (Figure 34.16). In morphogenesis, body regions develop characteristic shapes and structures that correlate with their function (Figure 34.17).
- Dividing plant cells can synthesize a new cell plate in a different plane from the old one. Such changes establish the direction in which structures such as lateral roots, branches, and leaf and flower buds grow (Figures 34.18 and 34.19).
- Chemical signals that help guide morphogenesis appear to act on certain cells in meristematic tissue, activating homeotic genes that ultimately regulate cell division and differentiation (Figures 34.20 and 34.21).

Animation: ABC model for flowering

Questions

Self-Test Questions

1. An angiosperm life cycle includes:
   a. meiosis within the male gametophyte to produce sperm.
   b. meiosis within the female gametophyte to produce eggs.
   c. meiosis within the ovary to produce megaspores.
   d. fertilization to produce microspores.
   e. fertilization to produce megaspores.
2. In a flower:
   a. the ovary contains the ovule.
   b. the stamens support the petals.
   c. the anther contains the megaspores.
   d. the carpel includes the sepals.
   e. the corolla includes the receptacle.
3. Double fertilization in a flower means:
   a. six sperm fertilize two groups of three eggs each.
   b. one sperm fertilizes the egg; a second sperm fertilizes the $2n$ mother cell.
   c. one microspore becomes a pollen grain; the other microspore becomes a sperm-producing cell.
   d. one pollen grain can make sperm nuclei and a pollen tube.
   e. one sperm can fertilize two endosperm mother cells.
4. A seed is best described as a (an):
   a. epicotyl.
   b. endosperm.
   c. ovary.
   d. mature spore.
   e. mature ovule.
5. The primary root develops from the embryonic:
   a. epicotyl.
   b. hypocotyl.
   c. coleoptile.
   d. radicle.
   e. plumule.
6. Which of the following is *not* a step in the germination of a monocot seed?
   a. Enzymes secreted into the endosperm digest the endosperm cell wall and macromolecules.
   b. The embryo imbibes water and then produces gibberellin.
   c. The embryo absorbs nutrients released from the endosperm.
   d. Endosperm develops as a food reserve.
   e. Gibberellin acts on the cells of the aleurone and scutellum to encode hydrolytic enzymes.
7. A student cuts off a leaflet from a plant and places it in a glass of water. Within a week roots appear on the base of the cutting. A month later she places the growing cutting into soil and it grows to the full size of the "parent" plant. This is an example of:
   a. parthenocarpy.
   b. fragmentation.
   c. grafting.
   d. vegetative reproduction.
   e. tissue culture propagation.
8. Which of the following is *not* an example of pattern formation in developing plants?
   a. an epidermal cell receiving developmental signals from a cortical cell
   b. the loosening of the cell wall to allow the elongation of selected cells to reach mature size
   c. regulation by homeotic genes of the position of different flower parts
   d. oriented cell division that establishes the shape of an organ
   e. cell expansion that directs specific cells to undergo mitosis at a given time and place
9. During the development of a leaf:
   a. mitotic cell divisions occur on planes specific to different plant groups.
   b. xylem vessels are the first to penetrate the leaf primordium.
   c. the growing leaf primordium becomes wider at its base than at its tip.
   d. the leaf primordium bulges from the region behind the shoot apical meristem.
   e. All of the above occur during leaf development.
10. In spring a lone walnut tree in your backyard develops attractive white flowers, and by the end of summer roughly half the flowers have given rise to the shelled fruits we know as walnuts. Walnut trees are self-pollinating. Assuming that pollination was 100% efficient in the case of your tree, which of the following statements best describes your tree's reproductive parts?
    a. Its flowers are in the botanical category of "perfect" flowers.
    b. The tree is monoecious.
    c. The tree is dioecious.
    d. The tree has imperfect, monoecious flowers.
    e. a and b together provide the best description of the tree's flowers.

Questions for Discussion

1. A plant physiologist has succeeded in cloning a gene for pest resistance into petunia cells. How can she use tissue culture to propagate a large number of petunia plants having the gene?
2. A large tree may have tens of thousands of shoot tips, and the cells in each tip can differ genetically from cells in other tips, sometimes substantially. Propose a hypothesis to explain this finding, and speculate about how it might be beneficial to the plant. How might this sort of natural variation be useful to human society?
3. Grocery stores separate displays of fruits and vegetables according to typical uses for these plant foods. For instance, bell peppers, cucumbers, tomatoes, and eggplants are in the vegetable section, while apples, pears, and peaches are displayed with other fruits. How does this practice relate to the biological definition of a fruit?

Experimental Analysis

The developmental genetics of flowers are of keen interest in plant biotechnology, especially with regard to food plants such as wheat and rice. Outline a research program for a crop species that would exploit the genetics of flower development, including the effects of homeotic genes, to engineer a more productive variety.

Evolution Link

Botanists estimate that half or more of angiosperm species may be polyploids that arose initially through hybridization. *Polyploidy*—having more than a diploid set of the parental chromosomes—can result from nondisjunction of homologous chromosomes during meiosis, or when cytokinesis fails to occur in a dividing cell. *Hybridization* is the successful mating of individuals from two different species. Such an interspecific hybrid is likely to be sterile because it has uneven numbers of parental chromosomes, or because the chromosomes are too different to pair during meiosis. A sterile hybrid may reproduce asexually, however, and if by chance its offspring should become polyploid, that plant will be fertile because the original set of chromosomes will have homologs that can pair normally during meiosis. Explain why both the hybrid parent and fertile polyploid offspring may be considered a new species, and describe at least two ways in which this route to speciation differs from speciation in the animal kingdom.

How Would You Vote?

Microencapsulated pesticides are easy to apply and effective for long periods. But they are about the size of pollen grains and are a tempting but toxic threat to certain pollinators. Should we restrict their use? Go to www.thomsonedu.com/login to investigate both sides of the issue and then vote online.

Sunflower plants *(Helianthus)* with flower heads oriented toward the sun's rays—an example of a plant response to the environment.

Study Plan

35 Control of Plant Growth and Development

Why It Matters

In the early 1920s, a researcher in Japan, Eiichi Kurosawa, was studying a rice plant disease that the Japanese called *bakanae*—the "foolish seedling" disease. Stems of rice seedlings that had become infected with the fungus *Gibberella fujikuroi* elongated twice as much as uninfected plants. The lanky stems were weak and eventually toppled over before the plants could produce seeds. Kurosawa discovered that extracts of the fungus also could trigger the disease. Eventually, other investigators purified the fungus's disease-causing substance, naming it gibberellin (GA).

Botanists today recognize more than 100 chemically different gibberellins, the largest class of plant hormones. Gibberellins have been isolated from fungi and from flowering plants, and may exist in other plant groups as well. Like other hormones, gibberellins are intercellular signaling molecules. In flowering plants, gibberellins have major, predictable effects **(Figure 35.1)**, beginning with seed germination.

Basic aspects of plant growth and development are adaptations that promote the survival of organisms that cannot move through their environment. These adaptations range from the triggers for seed germination to the development of a particular body form, the shift

Figure 35.1
Effects of the hormone gibberellin on stem growth of rice plants (Oryza).

from a vegetative phase to a reproductive one, and the timed death of flowers, leaves, and other parts. Although many of the details remain elusive or disputed, ample evidence exists that an elaborate system of molecular signals regulates many of these phenomena. We know, for example, that plant hormones alter patterns of growth, cell metabolism, and morphogenesis in response to changing environmental rhythms, including seasonal changes in day length and temperature and the daily rhythms of light and dark. They also adjust those patterns in response to environmental conditions, such as the amount of sunlight or shade, moisture, soil nutrients, and other factors. Some hormones govern growth responses to directional stimuli, such as light, gravity, or the presence of nearby structures. Often, hormonal effects involve changes in gene expression, although sometimes other mechanisms are at work.

We begin by surveying the different groups of plant hormones and other signaling molecules, and then turn our attention to the remarkable diversity of responses to both internal and environmental signals.

35.1 Plant Hormones

In plants, a **hormone** (*horman* = to stimulate) is a signaling molecule that regulates or helps coordinate some aspect of the plant's growth, metabolism, or development. Plant hormones act in response to two general types of cues: Internal chemical conditions related to growth and development, and circumstances in the external environment that affect plant growth, such as light and the availability of water. Some plant hormones are transported from the tissue that produces them to another plant part, while others exert their effects in the tissue where they are synthesized.

All plant hormones share certain characteristics. They are rather small organic molecules, and all are active in extremely low concentrations. Another shared feature is specificity: each one affects a given tissue in a particular way. Hormones that have effects outside the tissue where they are produced typically are transported to their target sites in vascular tissues, or they diffuse from one plant part to another. Within these general parameters, however, plant hormones vary greatly in their effects. Some stimulate one or more aspects of the plant's growth or development, whereas others have an inhibiting influence. Adding to the potential for confusion, a given hormone can have different effects in different tissues, and the effects also can differ depending on a target tissue's stage of development. And as researchers have increasingly discovered, many physiological responses result from the interaction of two or more hormones.

Biologists recognize at least seven major classes of plant hormones **(Table 35.1):** auxins, gibberellins, cytokinins, ethylene, brassinosteroids, abscisic acid (ABA), and jasmonates. Recent discoveries have added other hormonelike signaling agents to this list of established plant hormones. We now consider each major class of plant hormones and discuss some of the newly discovered signaling molecules as well.

Auxins Promote Growth

Auxins are synthesized primarily in the shoot apical meristem and young stems and leaves. Their main effects are to stimulate plant growth by promoting cell elongation in stems and coleoptiles, and by governing growth responses to light and gravity. Our focus here is indoleacetic acid (IAA), the most important natural auxin. Botanists often use the general term "auxin" to refer to IAA, a practice we follow here.

Experiments Leading to the Discovery of Auxins. Auxins were the first plant hormones identified. The path to their discovery began in the late nineteenth century in the library of Charles Darwin's home in the English countryside (see *Focus on Research* in Chapter 19). Among his many interests, Darwin was fascinated by plant **tropisms**—movements such as the bending of a houseplant toward light. This growth response, triggered by exposure to a directional light source, is an example of a **phototropism.**

Working with his son Francis, Darwin explored phototropisms by germinating seeds of two species of grasses, oat *(Avena sativa)* and canary grass *(Phalaris canariensis)*, in pots on the sill of a sunny window. Re-

Table 35.1 Major Plant Hormones and Signaling Molecules

Hormone/Signaling Compound	Where Synthesized	Tissues Affected	Effects
Auxins	Apical meristems, developing leaves and embryos	Growing tissues, buds, roots, leaves, fruits, vascular tissues	Promote growth and elongation of stems; promote formation of lateral roots and dormancy in lateral buds; promote fruit development; inhibit leaf abscission; orient plants with respect to light, gravity
Gibberellins	Root and shoot tips, young leaves, developing embryos	Stems, developing seeds	Promote cell divisions and growth and elongation of stems; promote seed germination and bolting
Cytokinins	Mainly in root tips	Shoot apical meristems, leaves, buds	Promote cell division; inhibit senescence of leaves; coordinate growth of roots and shoots (with auxin)
Ethylene	Shoot tips, roots, leaf nodes, flowers, fruits	Seeds, buds, seedlings, mature leaves, flowers, fruits	Regulates elongation and division of cells in seedling stems, roots; in mature plants regulates senescence and abscission of leaves, flowers, and fruits
Brassinosteroids	Young seeds; shoots and leaves	Mainly shoot tips, developing embryos	Stimulate cell division and elongation, differentiation of vascular tissue
Abscisic acid	Leaves	Buds, seeds, stomata	Promotes responses to environmental stress, including inhibiting growth/promoting dormancy; stimulates stomata to close in water-stressed plants
Jasmonates	Roots, seeds, probably other tissues	Various tissues, including damaged ones	In defense responses, promote transcription of genes encoding protease inhibitors; possible role in plant responses to nutrient deficiencies
Oligosaccharins	Cell walls	Damaged tissues; possibly active in most plant cells	Promote synthesis of phytoalexins in injured plants; may also have a role in regulating growth
Systemin	Damaged tissues	Damaged tissues	To date known only in tomato; roles in defense, including triggering jasmonate-induced chemical defenses
Salicylic acid	Damaged tissues	Many plant parts	Triggers synthesis of pathogenesis-related (PR) proteins, other general defenses

call from Chapter 34 that the shoot apical meristem and plumule of grass seedlings are sheathed by a protective coleoptile—a structure that is extremely sensitive to light. Darwin did not know this detail, but he observed that as the emerging shoots grew, within a few days they bent toward the light. He hypothesized that the tip of the shoot somehow detected light and communicated that information to the coleoptile. Darwin tested this idea in several ways **(Figure 35.2)** and concluded that when seedlings are illuminated from the side, "some influence is transmitted from the upper to the lower part, causing them to bend."

The Darwins' observations spawned decades of studies—a body of work that illustrates how scientific understanding typically advances step-by-step, as one set of experimental findings stimulates new research. First, scientists in Denmark and Poland showed that the bending of a shoot toward a light source was caused by something that could move through agar (a jellylike culture material derived from certain red algae) but not through a sheet of the mineral mica. This finding prompted experiments establishing that indeed the stimulus was a chemical produced in the shoot tip. Soon afterward, in 1926, experiments by the Dutch plant physiologist Frits Went confirmed that the growth-promoting chemical diffuses downward from the shoot tip to the stem below **(Figure 35.3)**. Using oat seeds, Went first sliced the tips from young shoots that had been grown under normal light conditions. He then placed the tips on agar blocks and left them there long enough for diffusible substances to move into the agar. Meanwhile, the decapitated stems stopped growing, but growth quickly resumed in seedlings that Went "capped" with the agar blocks (see Figure 35.3a). Clearly, a growth-promoting substance in the excised shoot tips had diffused into the agar, and from there into the seedling stems. Went also attached an agar block to one side of a decapitated shoot tip; when the shoot began growing again it bent away from the agar (see Figure 35.3b). Importantly, Went performed his experiments in total darkness, to avoid any "contamination" of his results by the possible effects of light.

Went did not determine the mechanism—differential elongation of cells on the shaded side of a shoot—by which the growth promoter controlled phototropism. However, he did develop a test that correlated specific amounts of the substance, later named auxin (*auxein* = to increase), with particular growth effects. This careful groundwork culminated several

Figure 35.2 Experimental Research

The Darwins' Experiments on Phototropism

QUESTION: Why does a plant stem bend toward the light?

EXPERIMENT 1: The Darwins observed that the first shoot of an emerging grass seedling, which is sheathed by a coleoptile, bends toward sunlight shining through a window. They removed the shoot tip from a seedling and illuminated one side of the seedling.

Original observation

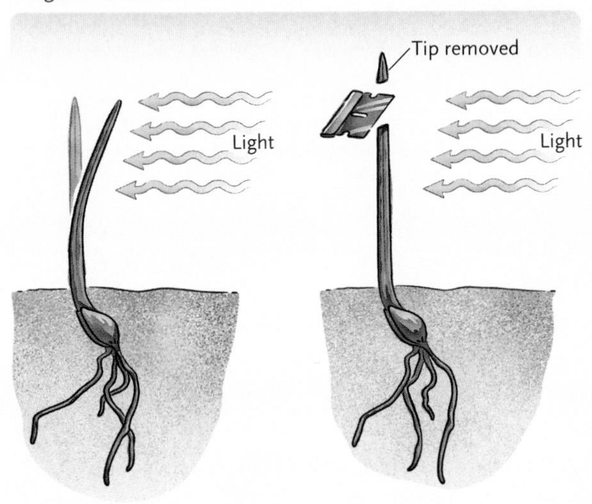

RESULT: The seedling neither grew nor bent.

EXPERIMENT 2: The Darwins divided seedlings into two groups. They covered the shoot tips of one group with an opaque cap and the shoot tips of the other group with a translucent cap. All the seedlings were illuminated from the same side.

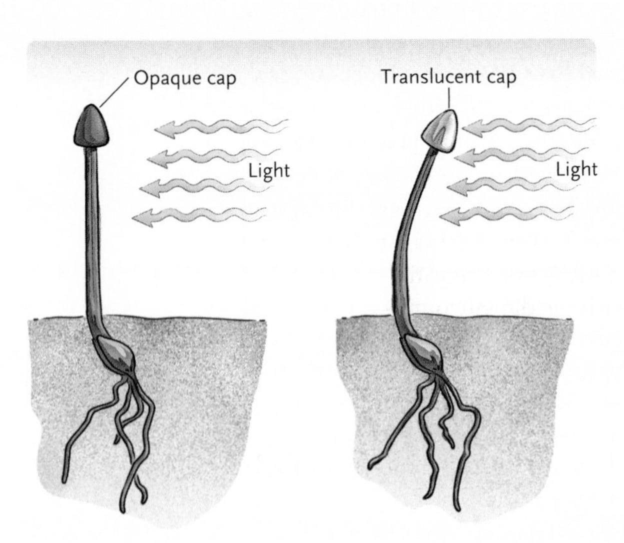

RESULT: The seedlings with opaque caps grew but did not bend. Those with translucent caps both grew *and* bent toward the light.

CONCLUSION: When seedlings are illuminated from one side, an unknown factor transmitted from a seedling's tip to the tissue below causes it to bend toward the light.

years later when other researchers identified auxin as indoleacetic acid (IAA).

Effects of Auxins. As already noted, auxin stimulates aspects of plant growth and development. In fact, recent studies of plant development have revealed that auxin is one of the first chemical signals to help shape the plant body. When the zygote first divides, forming an embryo that consists of a basal cell and an apical cell (see Section 34.3), auxin exported by the basal cell to the apical cell helps guide the development of the various features of the embryonic shoot. As the embryo develops further, IAA is produced mainly by the leaf primordium of the young shoot (see Figure 34.22). While the developing shoot is underground, IAA is actively transported downward, stimulating the primary growth of the stem and root **(Figure 35.4).** Once an elongating shoot breaks through the soil surface, its tip is exposed to sunlight, and the first leaves unfurl and begin photosynthesis. Shortly thereafter the leaf tip stops producing IAA and that task is assumed first by cells at the leaf edges, then by cells at base of the young leaf. Even so, as Section 35.3 discusses more fully, IAA continues to influence a plant's responses to light and plays a role in plant growth responses to gravity as well. IAA also stimulates cell division in the vascular cambium and promotes the formation of secondary xylem, as well as the formation of new root apical meristems, including lateral meristems. Not all of auxin's effects promote growth, however. IAA also maintains apical dominance, which inhibits growth of lateral meristems on shoots and restricts the formation of branches (see Section 31.3). Hence, auxin is a signal that the shoot apical meristem is present and active.

Commercial orchardists spray synthetic IAA on fruit trees because it promotes uniform flowering and helps set the fruit; it also helps prevent premature fruit drop. These effects mean that all the fruit may be picked at the same time, with considerable savings in labor costs.

Some synthetic auxins are used as herbicides, essentially stimulating a target plant to "grow itself to death." An **herbicide** is any compound that, at proper concentration, kills plants. Some herbicides are selective, killing one class of plants and not others. The most widely used herbicide in the world is the synthetic auxin 2,4-D (2,4-dichlorophenoxyacetic acid). This chemical is used extensively to prevent broadleaf weeds (which are eudicots) from growing in fields of cereal crops such as corn (which are monocots). By an unknown mechanism, 2,4-D causes an abnormal burst of growth in which eudicot stems elongate more than 10 times faster than normal—much faster than the plant can support metabolically.

Auxin Transport. To exert their far-reaching effects on plant tissues, auxins must travel away from their main synthesis sites in shoot meristems and young leaves.

a. The procedure showing that IAA promotes elongation of cells below the shoot tip

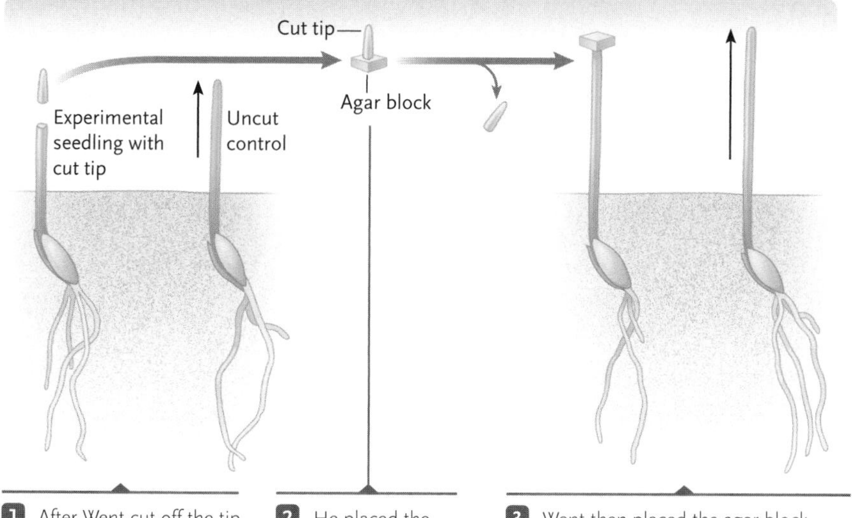

1 After Went cut off the tip of an oat seedling, the shoot stopped elongating, while a control seedling with an intact tip continued to grow.

2 He placed the excised tip on an agar block for 1–4 hours. During that time, IAA diffused into the agar block from the cut tip.

3 Went then placed the agar block containing auxin on another detipped oat shoot, and the shoot resumed elongation, growing about as rapidly as in a control seedling with an intact shoot tip.

b. The procedure showing that cells in contact with IAA grow faster than those farther away

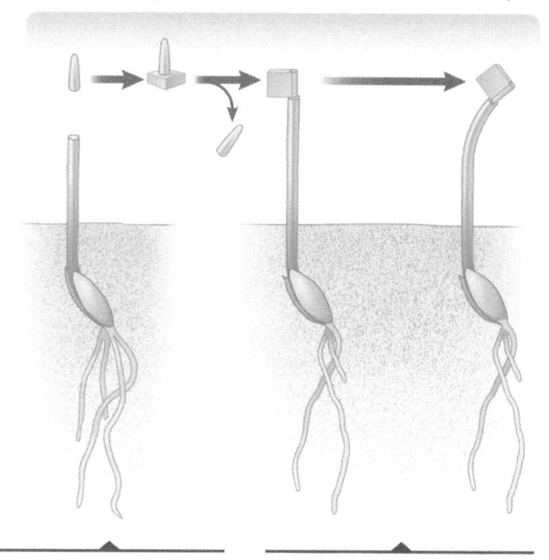

1 Went removed the tip of a seedling and placed it on an agar block.

2 He placed the agar block containing auxin on one side of the shoot tip. Auxin moved into the shoot tip on that side, causing it to bend away from the hormone.

Figure 35.3
Two experiments by Frits Went demonstrating the effect of IAA on an oat coleoptile. Went carried out the experiments in darkness to prevent effects of light from skewing the results.

Yet xylem and phloem sap usually do not contain auxins. Moreover, experiments have shown that while IAA moves through plant tissues slowly—roughly 1 cm/hr—this rate is 10 times faster than could be explained by simple diffusion. How, then, is auxin transported?

Plant physiologists adapted the agar block method pioneered by Went to trace the direction and rate of auxin movements in different kinds of tissues. A research team led by Winslow Briggs at Stanford University determined that the shaded side of a shoot tip contains more IAA than the illuminated side. Hypothesizing that light causes IAA to move laterally from the illuminated to the shaded side of a shoot tip, the team then inserted a vertical barrier (a thin slice of mica) between the shaded and illuminated sides of a shoot tip. IAA could not cross the barrier, and when the shoot tip was illuminated it did not bend. In addition, the concentrations of IAA in the two sides of the shoot tip remained about the same. When the barrier was shortened so that the separated sides of the tip again touched, the IAA concentration in the shaded area increased significantly, and the tip *did* bend. The study confirmed that IAA initially moves laterally in the shoot tip, from the illuminated side to the shaded side, where it triggers the elongation of cells and curving of the tip toward light. Subsequent research showed that IAA then moves downward in a shoot by way of a top-to-bottom mechanism called **polar transport.** That is, IAA in a coleoptile or shoot tip travels from the apex of the tissue to its base, such as from the tip of a developing leaf to the stem. **Figure 35.5** outlines the experimental method that demonstrated polar transport. When IAA reaches roots, it moves toward the root tip.

Inside a stem, IAA appears to be transported via parenchyma cells adjacent to vascular bundles. IAA again moves by polar transport as it travels through and between cells: It enters at one end by diffusing passively through cell walls and exits at the opposite end by active transport across the plasma membrane. The mostly widely accepted explanation for polar IAA transport from cell to cell proposes different mechanisms for moving IAA into and out of plant cells. In this model, IAA enters cells as the result of a high outside/low inside hydrogen ion (H^+) concentration gradient produced by the H^+ pumps in the plasma membrane of all plant cells **(Figure 35.6)**. The movement of H^+ ions out of the cytoplasm into the cell wall also produces an electrochemical gradient. In a neutral pH environment, IAA bears a negative charge (IAA^-), but in acidic surroundings, such as in a cell wall into which H^+ has been pumped, IAA^- reacts with H^+ to form an uncharged molecule, IAAH. The uncharged molecules may then dif-

Figure 35.4
The effect of auxin treatment on a gardenia (*Gardenia*) cutting. Four weeks after an auxin was applied to the base of the cutting on the left, its stem and roots have elongated, but the number of leaves is unchanged. The plant cutting on the right was not treated.

Treated with auxin   Untreated

Kingsley R. Stern

Figure 35.5 Experimental Research

Evidence for the Polar Transport of Auxin in Plant Tissues

QUESTION: Can IAA move both upward and downward in a plant shoot?

EXPERIMENT 1: As a preliminary step, the researchers excised sections of shoot tips from grass seedlings and placed them between blocks of agar containing different concentrations of IAA labeled with radioactive carbon-14. Labeling the IAA allowed them to easily track its movements. The researchers positioned an upright shoot tip section on an untreated agar block, and then placed a second, "donor" agar block containing labeled IAA atop the section.

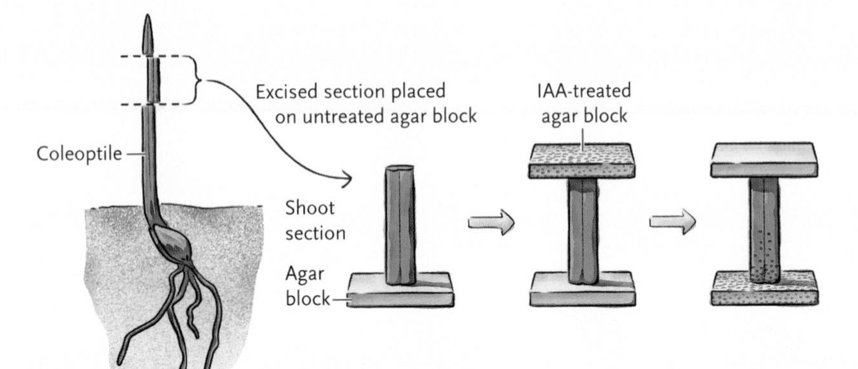

RESULT: IAA traveled from the upper block to the lower one, indicating that the hormone moved downward through the vertical shoot tip.

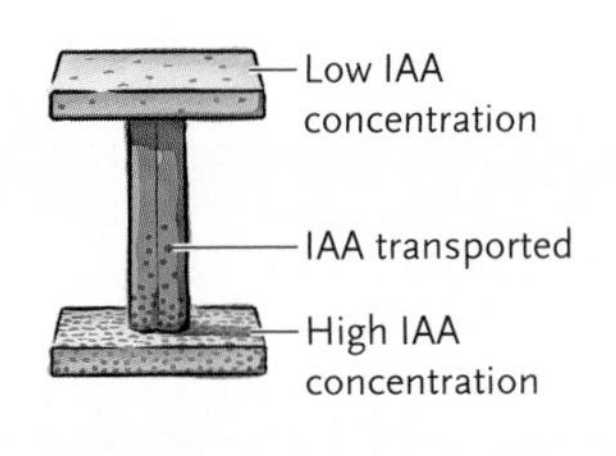

EXPERIMENT 2: The researchers positioned an upright shoot tip section on an agar block containing a high concentration of labeled IAA and placed a donor agar block containing a lower concentration of IAA on top.

RESULT: Even against its concentration gradient, IAA was transported from the upper block to the lower one.

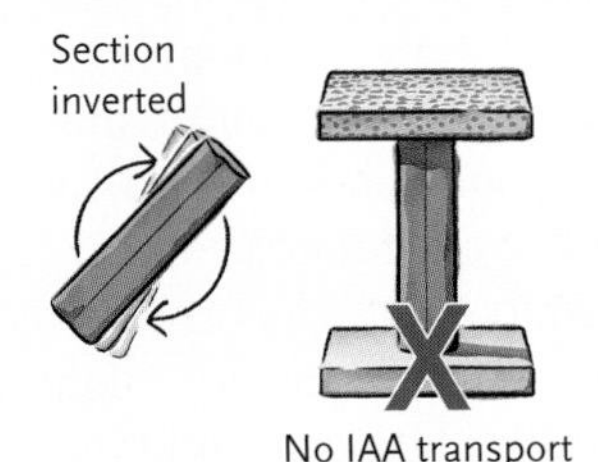

EXPERIMENT 3: The shoot tip from step 2 was inverted (reversing its normal orientation), and the same procedure was performed.

RESULT: No IAA was transported downward.

CONCLUSION: In plant shoot tips, IAA is transported in only one direction, from the shoot tip downward to plant parts below.

fuse across the plasma membrane through membrane transporters called AUX1 proteins (after the gene that encodes them in *Arabidopsis thaliana* plants), or they may enter via cotransport with H^+—or perhaps they move by both means.

A different mechanism moves IAA out of the cell at the opposite pole. Once IAAH reaches the electrically neutral cytoplasm at the apical end of the cell, it dissociates into H^+ and IAA^-. Then, the hormone crosses the cell and diffuses out of it by way of transporters called PIN proteins, which tend to be clustered at the cell's basal end. When the IAA^- diffuses through the transport proteins into the acidic cell wall, it reacts again with H^+, and the process continues with the next cell in line.

There is increasing evidence that auxin also may travel rapidly through plants in the phloem. As this work continues, researchers will undoubtedly gain a clearer understanding of how plants distribute this crucial hormone to their growing parts.

Possible Mechanisms of IAA Action. Ever since auxin was discovered, researchers have actively sought to understand how IAA stimulates cell elongation. You learned in Section 34.5 that as a plant cell elongates, the cellulose meshwork of the cell wall is first loosened

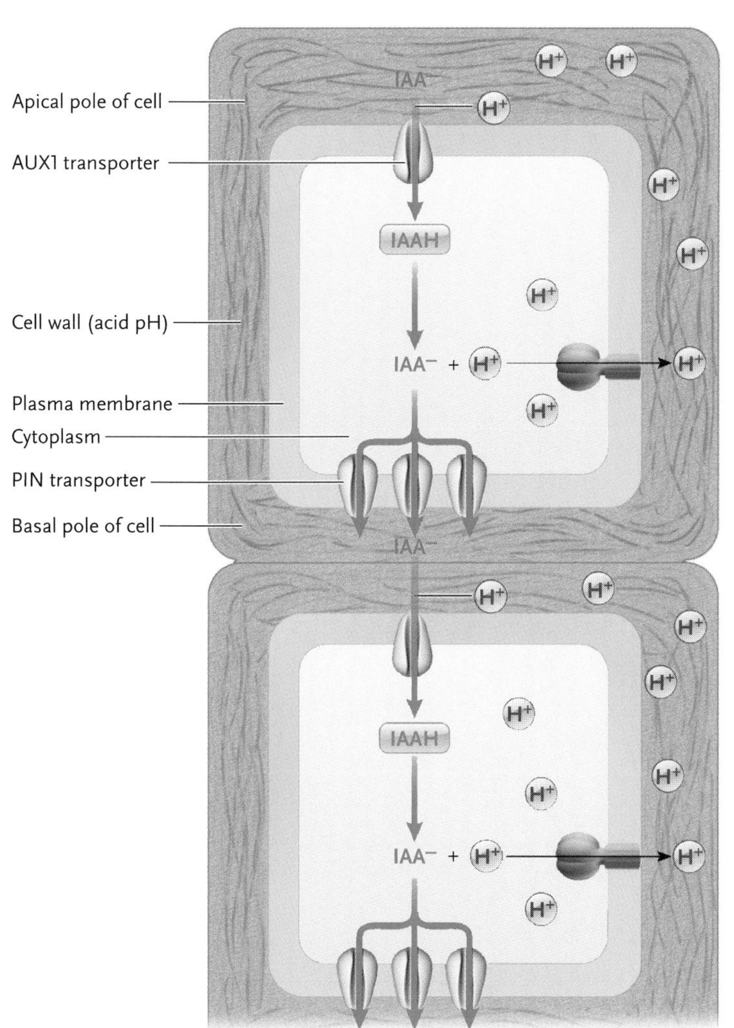

1 As auxin (IAA^-) diffuses through the cell wall, the acid pH makes H^+ bind to it. The resulting nonionized form is IAAH.

2 AUX1 transports IAAH into the cell cytoplasm.

3 In the less acidic cytoplasm, auxin gives up H^+ and reverts to its ionized form.

4 As H^+ is pumped out of the cell, the acidity of the wall increases.

5 Auxin moves out passively through PIN transporters.

6 These steps are repeated in each adjoining parenchyma cell. Thus the auxin transport shows polarity, from auxin's source in a shoot tip and leaves, downward toward the base of the stem.

Figure 35.6
A model for polar auxin transport. A plasma membrane H^+ pump maintains gradients of pH and electrical charge across the membrane, moving H^+ out of the cell using energy from ATP hydrolysis. These gradients are key to transporting IAA from the apical region to the basal region of a cell in a column. Following the gradients, at the basal end of a cell IAA diffuses through transport proteins into the cell wall, then (as IAAH) into the next cell in line.

and then stretched by turgor pressure. Several hormones, and auxin especially, apparently increase the plasticity (irreversible stretching) of the cell wall. Two major hypotheses seek to explain this effect, and both may be correct.

Plant cell walls grow much faster in an acidic environment—that is, when the pH is less than 7. The **acid-growth hypothesis** suggests that auxin causes cells to secrete acid (H^+) into the cell wall by stimulating the plasma membrane H^+ pumps to move hydrogen ions from the cell interior into the cell wall; the increased acidity activates proteins called *expansins,* which penetrate the cell wall and disrupt bonds between cellulose microfibrils in the wall **(Figure 35.7).** In the laboratory, it is easy to measure an increase in the rate at which coleoptiles or stem tissues release acid when they are treated with IAA. Activation of the plasma membrane H^+ pump also produces a membrane potential that pulls K^+ and other cations into the cell; the resulting osmotic gradient draws water into the cell, increasing turgor pressure and helping to stretch the "loosened" cell walls.

A second hypothesis, which also is supported by experimental evidence, suggests that auxin triggers the expression of genes encoding enzymes that play roles in the synthesis of new wall components. Plant cells exposed to IAA don't show increased growth if they are treated with a chemical that inhibits protein synthesis. However, researchers have identified mRNAs that rapidly increase in concentration within 10 to 20 minutes after stem sections have been treated with auxin, although they still do not know exactly which proteins these mRNAs encode.

Gibberellins Also Stimulate Growth, Including the Elongation of Stems

Gibberellins stimulate several aspects of plant growth. Perhaps most apparent to humans is their ability to promote the lengthening of plant stems by stimulating both cell division and cell elongation. Synthesized in shoot and root tips and young leaves, gibberellins, like auxin, modify the properties of plant cell walls in ways that promote expansion (although the gibberellin mechanism does not involve acidification of the cell wall). It may be that the two hormones both affect expansins, or are functionally linked in some other way yet to be discovered. Gibberellins have other known effects as well, such as helping to break the dormancy of seeds and buds.

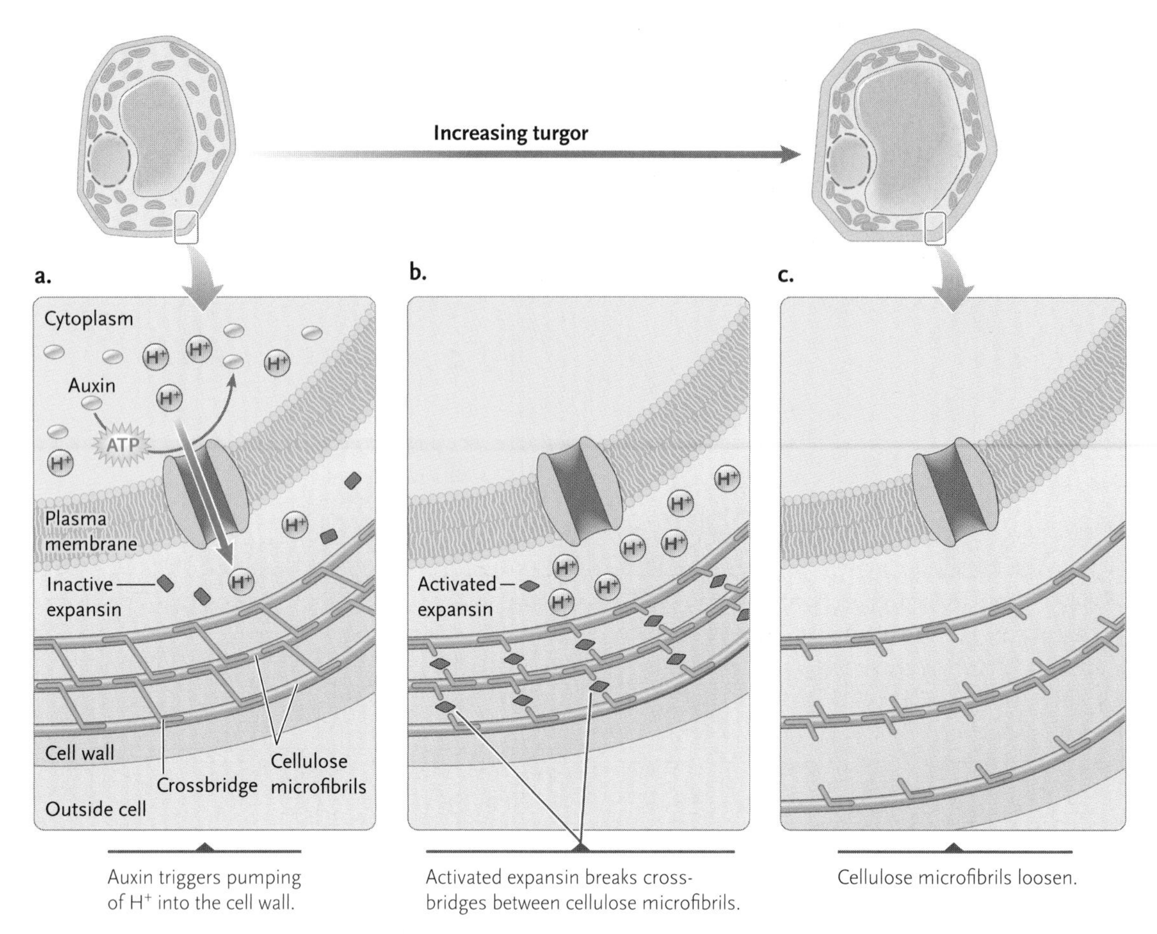

Figure 35.7
How auxin may regulate expansion of plant cells. According to the acid-growth hypothesis, plant cells secrete acid (H^+) when auxin stimulates the plasma membrane H^+ pumps to move hydrogen ions into the cell wall; the increased acidity activates enzymes called *expansins,* which disrupt bonds between cellulose microfibrils in the wall. As a result, the wall becomes extensible and the cell can expand.

Of the 100-plus compounds of the gibberellin family, relatively few are biologically active as hormones. The others are inactive forms or serve as precursors to active forms.

Gibberellins are active in eudicots as well as in a few monocots. In most plant species that have been analyzed, the main controller of stem elongation is the gibberellin called GA_1. Normally, GA_1 is synthesized in small amounts in young leaves and transported throughout the plant in the phloem. When GA_1 synthesis goes awry, the outcome is a dramatic change in the plant's stature. For example, experiments with a dwarf variety of peas *(Pisum sativum)* and some other species show that these plants and their taller relatives differ at a single gene locus. Normal plants make an enzyme required for gibberellin synthesis; dwarf plants of the same species lack the enzyme, and their internodes elongate very little.

Another stark demonstration of the effect gibberellins can have on internode growth is **bolting**, growth of a floral stalk in plants that form vegetative rosettes, such as cabbages *(Brassica oleracea)* and iceberg lettuce *(Lactuca sativa)*. In a rosette plant, stem internodes are so short that the leaves appear to arise from a single node. When these plants flower, however, the stem elongates rapidly and flowers develop on the new stem parts. An experimenter can trigger exaggerated bolting by spraying a plant with gibberellin **(Figure 35.8).** In nature, external cues such as increasing day length or warming after a cold snap stimulate gibberellin synthesis, and bolting occurs soon afterward. This observation supports the hypothesis that in rosette plants and possibly some others, gibberellins switch on internode lengthening when environmental conditions favor a shift from vegetative growth to reproductive growth.

Other experiments using gibberellins have turned up a striking number of additional roles for this hormone family. For example, a gibberellin helps stimulate buds and seeds to break dormancy and resume growth in the spring. Research on barley embryos showed that gibberellin provides signals during germination that lead to the enzymatic breakdown of endosperm, releasing nutrients that nourish the developing seedling (see Section 34.3). In monoecious species, which have flowers of both sexual types on the same plant, applications of gibberellin seem to encourage proportionately more "male" flowers to develop. As a result, there may be more pollen available to pollinate "female" flowers and, eventually, more fruit produced. A gibberellin used by commercial grape growers promotes fruit set and lengthens the stems on which fruits develop, allowing space for individual grapes to grow larger. One result is fruit with greater consumer appeal **(Figure 35.9).**

Cytokinins Enhance Growth and Retard Aging

Cytokinins play a major role in stimulating cell division (hence the name, which refers to cytokinesis). These hormones were first discovered during experiments designed to define the nutrient media required for plant tissue culture. Researchers found that in addition to a carbon source such as sucrose or glucose, minerals, and certain vitamins, cells in culture also required two other substances. One was auxin, which promoted the elongation of plant cells but did not stimulate the cells to divide. The other substance could be coconut milk, which is actually liquid endosperm, or it could be DNA that had been degraded into smaller molecules by boiling. When either was added to a culture medium along with an auxin, the cultured cells would begin dividing and grow normally.

We now know that the active ingredients in both boiled DNA and endosperm are cytokinins, which have a chemical structure similar to that of the nucleic acid base adenine. The most abundant natural cytokinin is zeatin, so-called because it was first isolated from the endosperm of young corn seeds *(Zea mays)*. In endosperm, zeatin probably promotes the burst of cell division that takes place as a fruit matures. As you might expect, cytokinins also are abundant in the rapidly dividing meristem tissues of root and shoot tips. Cytokinins occur not only in flowering plants but also in many conifers, mosses, and ferns. They are also synthesized by many soil-dwelling bacteria and fungi and may be crucial to the growth of mycorrhizae, which help nourish thousands of plant species (see Section 33.3). Conversely, *Agrobacterium* and other microbes that cause plant tumors carry genes that regulate the production of cytokinins.

Cytokinins are synthesized largely (although not only) in root tips and apparently are transported through the plant in xylem sap. Besides promoting cell division, they have a range of effects on plant metabolism and development, probably by regulating protein synthesis. For example, cytokinins promote expansion of young leaves (as leaf cells expand), cause chloroplasts to mature, and retard leaf aging. Another cytokinin effect—coordinating the growth of roots and shoots, in concert with auxin—underscores the point that plant hormones often work together to evoke a particular response. Investigators culturing tobacco tissues found that the relative amounts of auxin and a cytokinin strongly influenced not only growth, but also development **(Figure 35.10).** When the auxin-to-cytokinin ratio is about 10:1, the growing tissue did not differentiate but instead remained as a loose mass of cells, or *callus*. When the relative auxin concentration was increased slightly, the callus produced roots. When the relative concentration of the cytokinin was increased, chloroplasts in the callus cells matured, the callus became green and more compact, and it produced shoots. In nature, the interaction of a cytokinin and auxin may produce the typical balanced growth of roots and shoots, with each region providing the other with key nutrients.

Two untreated cabbages (controls)

Cabbages treated with gibberellins

Sylvan H. Wittwer/Visuals Unlimited

Figure 35.8
A dramatic example of bolting in cabbage *(Brassica oleracea)*, a plant commonly grown as a winter vegetable. The rosette form (left) reflects the plant's growth habit when days are short (and nights are long). Gibberellin was applied to the plants at the right, triggering the rapid stem elongation and subsequent flowering, characteristic of bolting.

Natural cytokinins can prolong the life of stored vegetables. Similar synthetic compounds are already widely used to prolong the shelf life of lettuces and mushrooms and to keep cut flowers fresh.

Ethylene Regulates a Range of Responses, Including Senescence

Most parts of a plant can produce **ethylene,** which is present in fruits, flowers, seeds, leaves, and roots. In different species it helps regulate a wide variety of plant physiological responses, including dormancy of seeds and buds, seedling growth, stem elongation, the ripening of fruit, and the eventual separation of fruits, leaves, and flowers from the plant body. Ethylene is an unusual hormone, in part because it is structurally simple (see Table 35.1) and in part because it is a gas at normal temperature and pressure.

Before a bean or pea seedling emerges from the soil, ethylene simultaneously slows elongation of the stem and stimulates cell divisions that increase stem girth.

Sylvan Wittwer/Visuals Unlimited

Figure 35.9
Effect of gibberellin on seedless grapes *(Vitis vinifera)*. The grapes on the right developed on vines that were treated with a gibberellin.

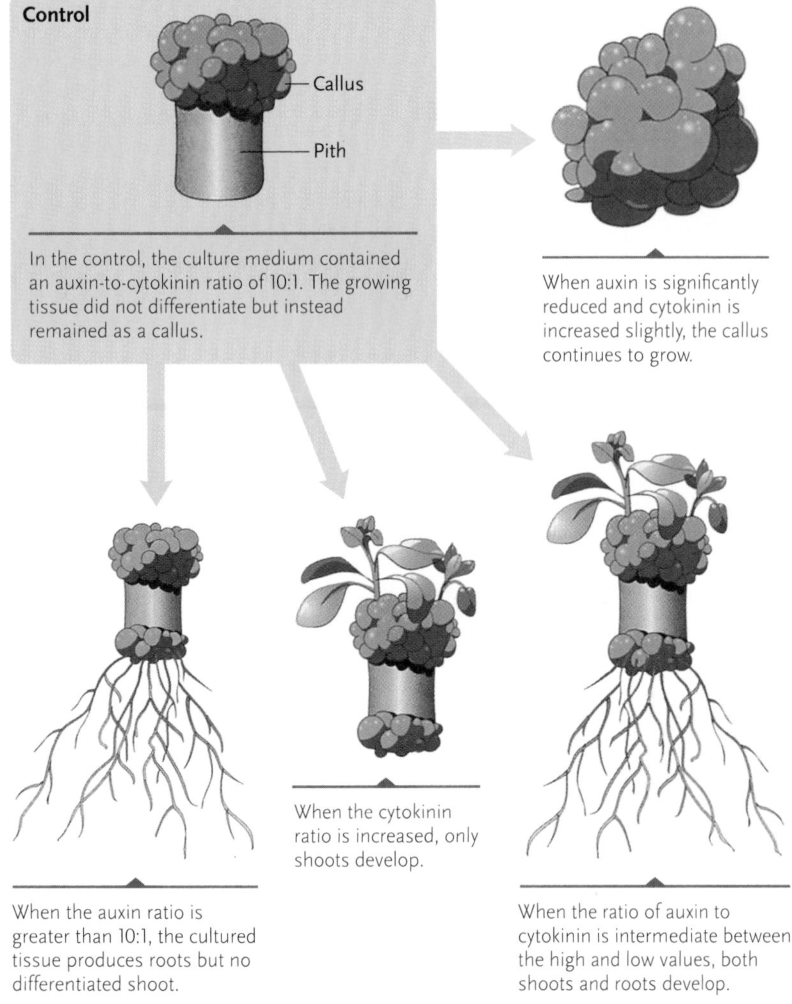

Control
Callus
Pith
In the control, the culture medium contained an auxin-to-cytokinin ratio of 10:1. The growing tissue did not differentiate but instead remained as a callus.
When auxin is significantly reduced and cytokinin is increased slightly, the callus continues to grow.
When the auxin ratio is greater than 10:1, the cultured tissue produces roots but no differentiated shoot.
When the cytokinin ratio is increased, only shoots develop.
When the ratio of auxin to cytokinin is intermediate between the high and low values, both shoots and roots develop.

Figure 35.10
Effects of varying ratios of auxin and cytokinin on tobacco tissues *(Nicotiana tabacum)* grown in culture. The method starts with a block of stem pith, essentially a core of ground tissue removed from the center of a stem. The callus growing on the pith is a disorganized mass of undifferentiated cells.

These alterations push the curved hypocotyl through the soil and into the air (see Figure 34.12). Such ethylene-induced horizontal growth also can help a growing seedling "find its way" into the air if the seed happens to germinate under a pebble or some other barrier.

Ethylene also governs the biologically complex process of aging, or **senescence,** in plants. Senescence is a closely controlled process of deterioration that leads to the death of plant cells. In autumn the leaves of deciduous trees senesce, often turning yellow or red as chlorophyll and proteins break down, allowing other pigments to become more noticeable. Ethylene triggers the expression of genes leading to the synthesis of chlorophyllases and proteases, enzymes that launch the breakdown process. In many plants, senescence is associated with **abscission,** the dropping of flowers, fruits, and leaves in response to environmental signals. In this process, ethylene apparently stimulates the activity of enzymes that digest cell walls in an abscission zone—a localized region at the base of the petiole. The petiole detaches from the stem at that point **(Figure 35.11).**

For some species, the funneling of nutrients into reproductive parts may be a cue for senescence of leaves, stems, and roots. When the drain of nutrients is halted by removing each newly emerging flower or seed pod, a plant's leaves and stems stay green and vigorous much longer **(Figure 35.12).** Gardeners routinely remove flower buds from many plants to maintain vegetative growth. Senescence requires other cues, however. For instance, when a cocklebur is induced to flower under winterlike conditions, its leaves turn yellow regardless of whether the nutrient-demanding young flowers are left on or pinched off. It is as if a "death signal" forms that leads to flowering and senescence when there are fewer hours of daylight (typical of winter days). This observation underscores the general theme that many plant responses to the environment involve the interaction of multiple molecular signals.

Fruit ripening is a special case of senescence. Although the precise mechanisms are not well understood, ripening begins when a fruit starts to synthesize ethylene. The ripening process may involve the conversion of starch or organic acids to sugars, the softening of cell walls, or the rupturing of the cell membrane and loss of cell fluid. The same kinds of events occur in wounded plant tissues, which also synthesize ethylene.

Ethylene from an outside source can stimulate senescence responses, including ripening, when it binds to specific protein receptors on plant cells. The ancient Chinese observed that they could induce picked fruit to ripen faster by burning incense; later, it was found that the incense smoke contains ethylene. Today ethylene gas is widely used to ripen tomatoes, pineapples, bananas, honeydew melons, mangoes, papayas, and other fruit that has been picked and shipped while still green. Ripening fruit itself gives off ethylene, which is why placing a ripe banana in a closed sack of unripe peaches (or some other green fruit) often can cause the fruit to ripen. Oranges and other citrus fruits may be exposed to ethylene to brighten the color of their rind. Conversely, limiting fruit exposure to ethylene can delay ripening. Apples will keep for months without rotting if they are exposed to a chemical that inhibits ethylene production or if they are stored in an environment that inhibits the hormone's effects—including low atmospheric pressure and a high concentration of CO_2, which may bind ethylene receptors.

Brassinosteroids Regulate Plant Growth Responses

The dozens of steroid hormones classed as **brassinosteroids** all appear to be vital for normal growth in plants, for they stimulate cell division and elongation in a wide range of plant cell types. Confirmed as plant hormones in the 1980s, brassinosteroids now are the subject of intense research on their sources and effects. While brassinosteroids have been detected in a wide variety of plant tissues and organs, the highest concentrations are found in shoot tips and in developing seeds and embryos—all examples of young, actively developing parts. In laboratory studies, the hormones have different effects depending on the tissue where they are active. They have promoted cell elongation, differentiation of vascular tissue, and elongation of a pollen tube after a flower is pollinated. By contrast, they inhibit the elongation of roots. First isolated from pollen of a plant in the mustard family, *Brassica napus* (a type of canola), in nature brassinosteroids seem to regulate the expression of genes associated with a plant's growth responses to light. This role was underscored by the outcomes of experiments using mutant *Arabidopsis* plants that were homozygous for a defective genes called *bri1* (for brassinosteroid-insensitive receptor) **(Figure 35.13)**; the results provided convincing evidence that brassinosteroids mediate growth responses to light.

Abscisic Acid Suppresses Growth and Influences Responses to Environmental Stress

Plant scientists ascribe a variety of effects to the hormone **abscisic acid** (ABA), many of which represent evolutionary adaptations to environmental challenges. Plants apparently synthesize ABA from carotenoid pigments inside plastids in leaves and possibly other plant parts. Several ABA receptors have been identified, and in general, we can group its effects into changes in gene expression that result in long-term inhibition of growth, and rapid, short-term physiological changes that are responses to immediate stresses, such as a lack of water, in a plant's surroundings. As its name suggests, at one time ABA was thought to play a major role in abscission. As already described, however, abscission is largely the domain of ethylene.

Suppressing Growth in Buds and Seeds. Operating as a counterpoint to growth-stimulating hormones like gibberellins, ABA inhibits growth in response to environmental cues, such as seasonal changes in temperature and light. This growth suppression can last for many months or even years. For example, one of ABA's major growth-inhibiting effects is apparent in perennial plants, in which the hormone promotes dormancy in leaf buds—an important adaptive advantage in places where winter cold can damage young leaves. If ABA is applied to a growing leaf bud, the bud's normal development stops, and instead protective *bud scales*—modified, nonphotosynthetic leaves that are small, dry, and tough—form around the apical meristem and insulate it from the elements **(Figure 35.14)**. After the scales develop, most cell metabolic activity shuts down and the leaf bud becomes dormant.

In some plants that produce fleshy fruits, such as apples and cherries, abscisic acid is associated with the dormancy of seeds as well. As the seed develops, ABA accumulates in the seed coat, and the embryo does not germinate even if it becomes hydrated. Before such a seed can germinate, it usually will require a long period of cool, wet conditions, which stimulate the breakdown of ABA. The buildup of ABA in developing seeds does more than simply inhibit development, however. As early development draws to a close, ABA stimulates the transcription of certain genes, and large amounts of their protein products are synthesized. These proteins are thought to store nitrogen and other nutrients that the embryo will use when it eventually does germinate. ABA and related growth inhibitors are often applied to plants slated to be shipped to plant nurseries. Dormant plants suffer less shipping damage, and the effects of the inhibitors can be reversed by applying a gibberellin.

Responses to Environmental Stress. ABA also triggers plant responses to various environmental stresses, including cold snaps, high soil salinity, and drought. A

Figure 35.11 Abscission zone in a maple *(Acer)*. This longitudinal section at the left is through the base of the petiole of a leaf.

Figure 35.12 Experimental results showing that the removal of seed pods from a soybean plant *(Glycine max)* delays its senescence.

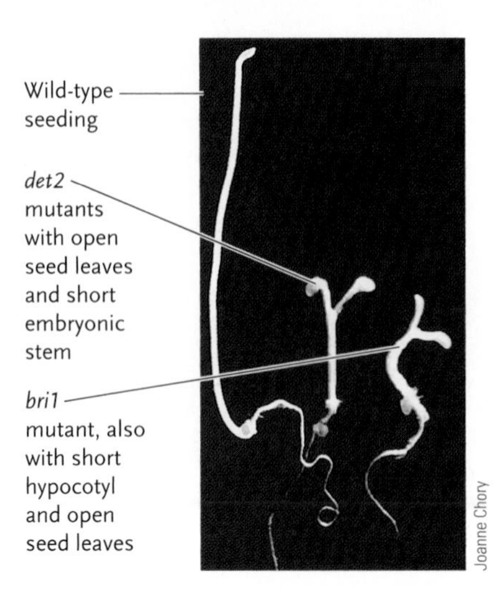

Figure 35.13
Experimental evidence that brassinosteroids can mediate a plant's responses to light by regulating gene expression. In *Arabidopsis*, wild-type seedlings synthesize a protein (encoded by the *DET2* gene) that prevents leaves from developing *(seedling at left)* until photosynthesis is possible, after the seedling breaks out of the dark environment of soil. When the gene is defective, a mutant *det2* plant *(center)* will develop a short hypocotyl (embryonic stem) and open seed leaves (cotyledons) even when there is no light for photosynthesis. Experiments with *bri1* mutants, which lack functioning receptors for a brassinosteroid, resulted in a similar phenotype *(right)*. These findings supported the hypothesis that a brassinosteroid is necessary for normal expression of the *DET2* gene.

great deal of research has focused on how ABA influences plant responses to a lack of water. When a plant is water-stressed, ABA helps prevent excessive water loss by stimulating stomata to close. As described in Section 32.3, flowering plants depend heavily on the proper functioning of stomata. When a lack of water leads to wilting, mesophyll cells in wilted leaves rapidly synthesize and secrete ABA. The hormone diffuses to guard cells, where an ABA receptor binds it. Binding stimulates the release of K^+ and water from the guard cells, and within minutes the stomata close.

Once bound to its receptor, ABA may exert its effects through a cascade of signals that includes phosphorylated proteins. Experiments have shown that an *Arabidopsis* mutant unable to respond to ABA lacks an enzyme that removes phosphate groups from certain proteins. This condition suggests that cleaving phosphates is one step in the ABA response. *Insights from the Molecular Revolution* highlights recent research filling in other steps in the ABA-induced response pathway.

Jasmonates and Oligosaccharins Regulate Growth and Have Roles in Defense

In recent years, studies of plant growth and development have helped define the roles—or revealed the existence—of several other hormonelike compounds in plants. Like the well-established plant hormones just described, these substances are organic molecules and only tiny amounts are required to alter some aspect of a plant's functioning. Some have long been known to exist in plants, but the extent of their signaling roles has only recently become better understood. This group includes **jasmonates** (JA), a family of about 20 compounds derived from fatty acids. Experiments with *Arabidopsis* and other plants have revealed numerous genes that respond to JA, including genes that help regulate root growth and seed germination. JA also appears to help plants "manage" stresses due to deficiencies of certain nutrients (such as K^+). The JA family is best known, however, as part of the plant arsenal to limit damage by pathogens and predators, the topic of the following section.

Figure 35.14
Bud scales, here on the bud of a perennial cornflower *(Centaurea montana)*.

Some other substances also are drawing keen interest from plant scientists, but because their signaling roles are still poorly understood they are not widely accepted as confirmed plant hormones. A case in point involves the complex carbohydrates that are structural elements in the cell walls of plants and some fungi. Several years ago, researchers observed that in some plants, some of these oligosaccharides could serve as signaling molecules. Such compounds were named **oligosaccharins**, and one of their known roles is to defend the plant against pathogens. In addition, oligosaccharins have been proposed as growth regulators that adjust the growth and differentiation of plant cells, possibly by modulating the influences of growth-promoting hormones such as auxin. At this writing, researchers in many laboratories are pursuing a deeper understanding of this curious subset of plant signaling molecules.

Study Break

1. Which plant hormones promote growth and which inhibit it?
2. Give examples of how some hormones have both promoting and inhibiting effects in different parts of the plant at different times of the life cycle.

35.2 Plant Chemical Defenses

Plants don't have immune systems like those that have evolved in animals (the subject of Chapter 43). Even so, over the millennia, in higher plants virtually constant exposure to predation by herbivores and the onslaught

INSIGHTS FROM THE MOLECULAR REVOLUTION

Stressing Out in Plants and People

Unlike people, plants cannot move to more favorable locations when an environmental stress threatens. Instead, to survive stresses plants adjust their responses to environmental factors such as temperature and the availability of water. Recent molecular work shows that responses to stress imposed by drought and cold involve some of the same chemical steps in plants and humans, indicating an ancient link to a common evolutionary ancestor. The research may also point the way to genetic engineering strategies to modify major crop plants for earlier and better responses to stress.

Many plant stress responses are triggered by the hormone abscisic acid (ABA). Although individual steps in the response pathway are unclear, it is known that calcium ions increase in concentration in the cytoplasm when plant cells are exposed to ABA. Soon after the rise in Ca^{2+}, genes are activated that compensate for the stressful situation.

Nam-Hai Chua and his colleagues at the Rockefeller University and the University of Minnesota were interested in piecing together the molecular steps in the plant pathway. One substance they thought might be involved is *cyclic ADP-ribose (cADPR)*, a signaling molecule that was first implicated in calcium release pathways in animal cells.

The Chua team began by injecting two plant genes, *rd29A* and *kin2*, into tomato cells. The two genes are activated by ABA as part of the stress response. Each of the injected genes was linked to an unrelated marker gene that would also be turned on if the gene became active. When ABA was injected into stressed tomato plants grown from the injected cells, the markers were activated, indicating that *rd29A* and *kin2* were turned on by ABA.

The next step was to inject Ca^{2+} and to note whether the injected genes were activated. The injection activated the *rd29A* and *kin2* genes, confirming the role of calcium ions in the pathway. A chemical called EGTA that removes Ca^{2+} from the cytoplasm cancelled the gene activation, as expected if calcium is part of the response pathway.

Then, the investigators injected cADPR to see if it activated *rd29A* and *kin2*. This result was also positive; cADPR had the same effect as either ABA or Ca^{2+}. EGTA blocked the positive response to cADPR, indicating that cADPR lies between ABA and calcium release in the signal pathway.

Another experiment determined whether protein phosphorylation might be part of the pathway. To accomplish this, the investigators injected an inhibitor of protein kinases, the enzymes that phosphorylate proteins as a part of many cellular response pathways (see Section 7.2). After the inhibitor was added, injecting ABA, cADPR, or Ca^{2+} failed to activate *rd29A* and *kin2*, indicating that protein phosphorylation occupies a critical step following these elements in the pathway.

From these results the Chua team was able to reconstruct a major part of the pathway:

ABA → ? → cADPR →
Ca^{2+} → protein kinases →
? → stress gene activation

The question marks indicate one or more unknown steps. Most significant is the specific ABA receptor that carries out the first step in the response pathway. A recently identified ABA receptor that is involved in the events that cause stomata to close (among other effects) may be this "missing link."

In addition to possible benefits for agriculture, the Chua team's research may also shed light on signal pathways in animals in which cADPR plays a part, including one that adjusts the heartbeat and another that regulates insulin release in response to elevated blood glucose. Thus Chua's work may help fill in steps in both plant and animal responses, in pathways inherited from a common ancestor predating both plants and people.

of pathogens have resulted in a striking array of chemical defenses that ward off or reduce damage to plant tissues from infectious bacteria, fungi, worms, or plant-eating insects **(Table 35.2)**. You will discover in this section that as with the defensive strategies of animals, plant defenses include both general responses to any type of attack and specific responses to particular threats. Some get underway almost as soon as an attack begins, while others help promote the plant's long-term survival. And more often than not, multiple chemicals interact as the response unfolds.

Jasmonates and Other Compounds Interact in a General Response to Wounds

When an insect begins feeding on a leaf or some other plant part, the plant may respond to the resulting wound by launching what in effect is a cascade of chemical responses. These complex signaling pathways often rely on interactions among jasmonates, ethylene, or some other plant hormone. As the pathway unfolds it triggers expression of genes leading to chemical and physical defenses at the wound site. For example, in some plants jasmonate induces a response leading to the synthesis of protease inhibitors, which disrupt an insect's capacity to digest proteins in the plant tissue. The protein deficiency in turn hampers the insect's growth and functioning.

A plant's capacity to recognize and respond to the physical damage of a wound apparently has been a strong selection pressure during plant evolution. When a plant is wounded experimentally, numerous defensive chemicals can be detected in its tissues in relatively short order. One of these, **salicylic acid**, or **SA** (a compound similar to aspirin, which is acetylsalicylic acid), seems to have multiple roles in plant defenses,

Table 35.2 Summary of Plant Chemical Defenses

Type of Defense	Effects
General Defenses	
Jasmonate (JA) responses to wounds/injury by pathogens; pathways often include other hormones such as ethylene	Synthesis of defensive chemicals such as protease inhibitors
Hypersensitive response to infectious pathogens (e.g., fungi, bacteria)	Physically isolates infection site by surrounding it with dead cells
PR (pathogenesis-related) proteins	Enzymes, other proteins that degrade cell walls of pathogens
Salicylic acid (SA)	Mobilized during other responses and independently; induces the synthesis of PR proteins, operates in systemic acquired resistance
Systemin (in tomato)	Triggers JA response
Secondary metabolites	
Phytoalexins	Antibiotic
Oligosaccharins	Trigger synthesis of phytoalexins
Systemic acquired resistance (SAR)	Long-lasting protection against some pathogens; components include SA and PR proteins that accumulate in healthy tissues
Specific Defenses	
Gene-for-gene recognition of chemical features of specific pathogens (by binding with receptors coded by R genes)	Triggers defensive response (e.g., hypersensitive response, PR proteins) against pathogens
Other	
Heat-shock responses (encoded by heat-shock genes)	Synthesis of chaperone proteins that reversibly bind other plant proteins and prevent denaturing due to heat stress
"Antifreeze" proteins	In some species, stabilize cell proteins under freezing conditions

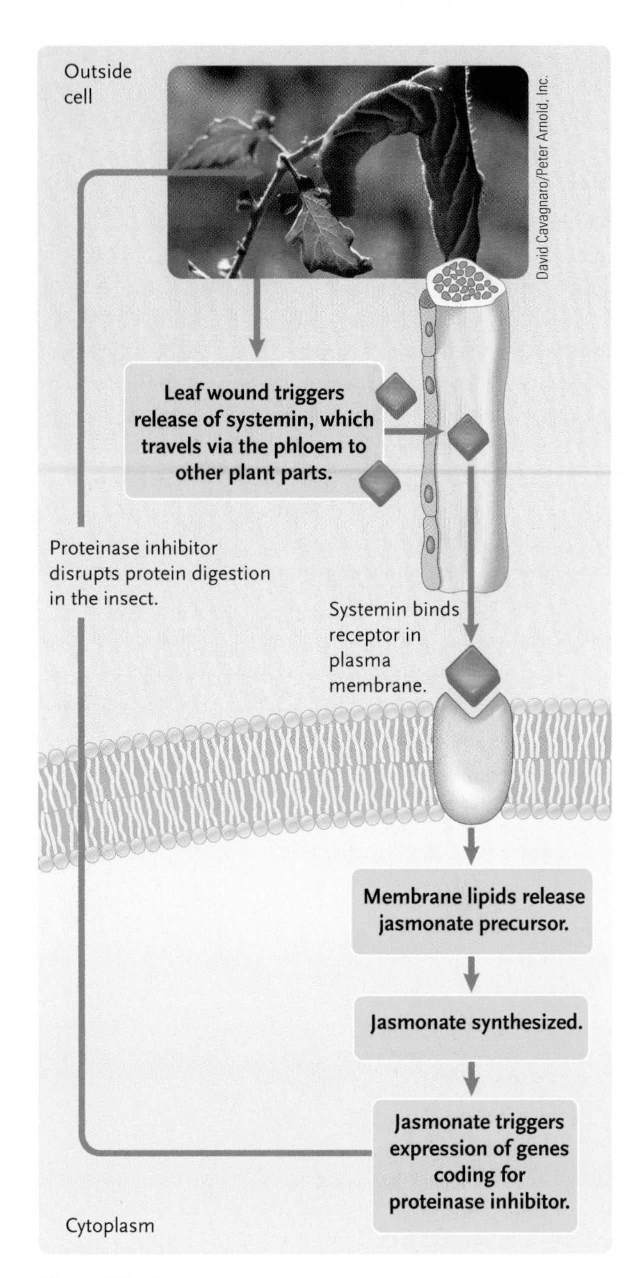

Figure 35.15
The systemin response to wounding. When a plant is wounded, it responds by releasing the protein hormone systemin. Transported through the phloem to other plant parts, in receptive cells systemin sets in motion a sequence of reactions that lead to the expression of genes encoding protease inhibitors—substances that can seriously disrupt an insect predator's capacity to digest protein.

including interacting with jasmonates in signaling cascades.

Researchers are regularly discovering new variations of hormone-induced wound responses in plants. For example, experiments have elucidated some of the steps in an unusual pathway that thus far is known only in tomato *(Lycopersicum esculentum)* and a few other plant species. As diagrammed in **Figure 35.15,** the wounded plant rapidly synthesizes **systemin**, the first peptide hormone to be discovered in plants. (Various animal hormones are peptides, a topic covered in Chapter 40.) Systemin enters the phloem and is transported throughout the plant. Although various details of the signaling pathway have yet to be worked out, when receptive cells bind systemin, their plasma membranes release a lipid that is the chemical precursor of jasmonate. Next jasmonate is synthesized, and it in turn sets in motion the expression of genes that encode protease inhibitors, which protect the plant against attack, even in parts remote from the original wound.

The Hypersensitive Response and PR Proteins Are Other General Defenses

Often, a plant that becomes infected by pathogenic bacteria or fungi counters the attack by way of a **hypersensitive response**—a defense that physically cordons off an infection site by surrounding it with dead cells. Initially, cells near the site respond by producing a burst of highly reactive oxygen-containing compounds (such as hydrogen peroxide, H_2O_2) that can break down nucleic acids, inactivate enzymes, or have other toxic effects on cells. The burst is catalyzed by enzymes in the plant cell's plasma membrane. It may begin the process of killing cells close to the attack site and, as the response advances, programmed cell death may also come into play. In short order, the "sacrificed" dead cells wall off the infected area from the rest of the plant. Thus denied an ongoing supply of nutrients, the invading pathogen dies. A common sign of a successful hypersensitive response is a dead spot surrounded by healthy tissue **(Figure 35.16)**.

While the hypersensitive response is underway, salicylic acid triggers other defensive responses by an infected plant. One of its effects is to induce the synthesis of **pathogenesis-related proteins**, or **PR proteins.** Some PR proteins are hydrolytic enzymes that break down components of a pathogen's cell wall. Examples are chitinases that dismantle the chitin in the cell walls of fungi and so kill the cells. In some cases, plant cell receptors also detect the presence of fragments of the disintegrating wall and set in motion additional defense responses.

Secondary Metabolites Defend against Pathogens and Herbivores

Many plants counter bacteria and fungi by making **phytoalexins**, biochemicals of various types that function as antibiotics. When an infectious agent breaches a plant part, genes encoding phytoalexins begin to be transcribed in the affected tissue. For instance, when a fungus begins to invade plant tissues, the enzymes it secretes may trigger the release of oligosaccharins. In addition to their roles as growth regulators (described in Section 35.1), these substances also can promote the production of phytoalexins, which have toxic effects on a variety of fungi. Plant tissues may also synthesize phytoalexins in response to attacks by viruses.

Phytoalexins are among many *secondary metabolites* produced by plants. Such substances are termed "secondary" because they are not routinely synthesized in all plant cells as part of basic metabolism. A wide range of plant species deploy secondary metabolites as defenses against feeding herbivores. Examples are alkaloids such as caffeine, cocaine, and the poison strychnine (in seeds of the *nux vomica* tree, *Strychnos nux-vomica*), tannins such as those in oak acorns, and various terpenes. The terpene family includes insect-repelling substances in conifer resins and cotton, and essential oils produced by sage and basil plants. Because these terpenes are volatile—they easily diffuse out of the plant into the surrounding air—they also can provide indirect defense to a plant. Released from the wounds created by a munching insect, they attract other insects that prey on the herbivore. Chapter 50 looks in detail at the interactions between plants and herbivores.

Nigel Cattlin/Photo Researchers, Inc.

Figure 35.16 Evidence of the hypersensitive response. The dead spots on these leaves of a strawberry plant *(Fragaria)* are sites where a pathogen invaded, triggering the defensive destruction of the surrounding cells.

Gene-for-Gene Recognition Allows Rapid Responses to Specific Threats

One of the most interesting questions with respect to plant defenses is how plants first sense that an attack is underway. In some instances plants apparently can detect an attack by a specific predator through a mechanism called **gene-for-gene recognition.** This term refers to a matchup between the products of dominant alleles of two types of genes: a so-called **R gene** (for "resistance") in a plant, and an **Avr gene** (for "avirulence") in a particular pathogen. Thousands of R genes have been identified in a wide range of plant species. Dominant R alleles confer enhanced resistance to plant pathogens including bacteria, fungi, and nematode worms that attack roots.

The basic mechanism of gene-for-gene recognition is simple: The dominant R allele encodes a receptor in plasma membranes of a plant's cells, and the dominant pathogen Avr allele encodes a molecule that can bind the receptor. "Avirulence" implies "not virulent," and binding of the Avr gene product triggers an immediate defense response in the plant. Trigger molecules run the gamut from proteins to lipids to carbohydrates that have been secreted by the pathogen or released from its surface **(Figure 35.17)**. Experiments have demonstrated a rapid-fire sequence of early biochemical changes that follow binding of the Avr-encoded molecule; these include changes in ion concentrations inside and outside plant cells and the production of biologically active oxygen compounds that heralds the hypersensitive response. In fact, of the instances of gene-for-gene recognition plant scientists have observed thus far, most trigger the hypersensitive response and the ensuing synthesis of PR proteins, with their antibiotic effects.

Systemic Acquired Resistance Can Provide Long-Term Protection

The defensive response to a microbial invasion may spread throughout a plant, so that the plant's healthy tissues become less vulnerable to infection. This phenomenon is called **systemic acquired resistance,** and experiments using *Arabidopsis* plants have shed light on how it comes about **(Figure 35.18)**. In a key early step, salicylic acid builds up in the affected tissues. By some route, probably through the phloem, the SA passes

Required precondition
A plant has a dominant R gene encoding a receptor that can bind the product of a specfic pathogen dominant Avr gene.

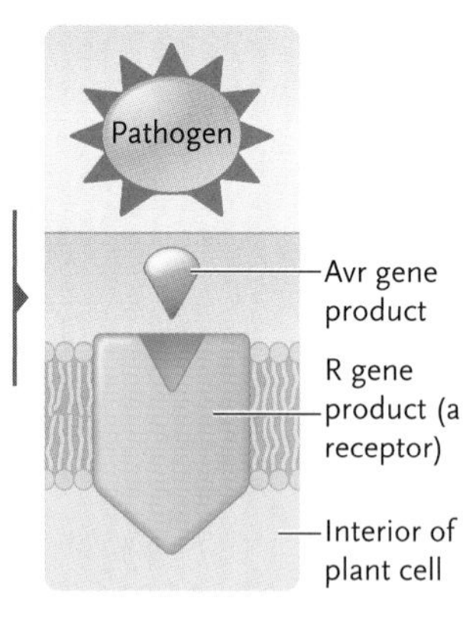

1 When the R-encoded receptor binds its matching Avr product, the binding triggers signaling pathways, leading to various defense responses in the plant.

2 Fluxes of ions and enzyme activity at the plasma membrane contribute to the hypersensitive response. Soon PR proteins, phytoalexins, and salicyic acid (SA) are synthesized. The PR proteins and phytoalexins combat pathogens directly. SA promotes systemic acquired resistance.

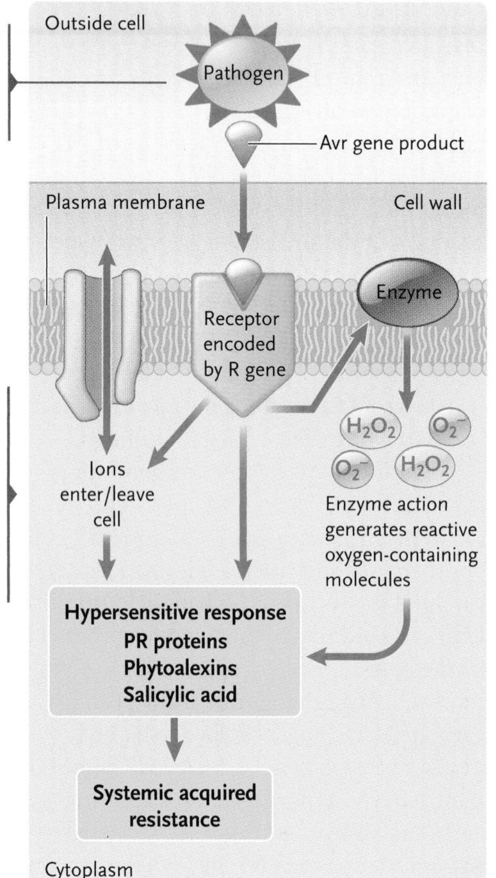

Figure 35.17
Model of how gene-for-gene resistance may operate. For resistance to develop, the plant must have a dominant R gene and the pathogen must have a corresponding dominant Avr gene. Products of such "matching" genes can interact physically, rather like the lock-and-key mechanism of an enzyme and its substrate. Most R genes encode receptors at the plasma membranes of plant cells. As diagrammed in step 1, when one of these receptors binds an Avr gene's product, the initial result may be changes in the movements of specific ions into or out of the cell and the activation of membrane enzymes that catalyze the formation of highly reactive oxygen-containing molecules. Such events help launch other signaling pathways that lead to a variety of defensive responses, including the hypersensitive response (step 2).

from the infected organ to newly forming organs such as leaves, which begin to synthesize PR proteins—again, providing the plant with a "home-grown" antimicrobial arsenal. How does the SA exert this effect? It seems that when enough SA accumulates in a plant cell's cytoplasm, a regulatory protein called NPR-1 (for *n*onexpressor of *p*athogenesis-*r*elated genes) moves from the cytoplasm into the cell nucleus. There it interacts with factors that promote the transcription of genes encoding PR proteins.

In addition to synthesizing SA that will be transported to other tissues by a plant's vascular system, the damaged leaf also synthesizes a chemically similar compound, methyl salicylate. This substance is volatile, and researchers speculate that it may serve as an airborne "harm" signal, promoting defense responses in the plant that synthesized it and possibly in nearby plants as well.

Extremes of Heat and Cold Also Elicit Protective Chemical Responses

Plant cells also contain **heat-shock proteins (HSPs)**, a type of chaperone protein (see Section 3.5) found in cells of many species. HSPs bind and stabilize other proteins, including enzymes, that might otherwise stop functioning if they were to become denatured by rising temperature. Plant cells may rapidly synthesize HSPs in response to a sudden temperature rise. For example, experiments with cells and seedlings of soybean *(Glycine max)* showed that when the temperature rose 10°–15°C, in less than five minutes mRNA transcripts coding for as many as 50 different HSPs were present in cells. When the temperature returns to a normal range, HSPs release bound proteins, which can then resume their usual functions. Further studies have revealed that heat-shock proteins help protect plant cells subjected to other environmental stresses as well, including drought, salinity, and cold.

Like extreme heat, freezing can also be lethal to plants. If ice crystals form in cells they can literally tear the cell apart. In many cold-resistant species, dormancy (discussed in Section 35.4) is the long-term strategy for dealing with cold, but in the short term, such as an unseasonable cold snap, some species also undergo a rapid shift in gene expression that equips cold-stressed cells with so-called antifreeze proteins. Like heat-shock proteins, these molecules are thought to help maintain the structural integrity of other cell proteins.

Study Break

1. Which plant chemical defenses are general responses to attack, and which are specific to a particular pathogen?
2. Why is salicylic acid considered to be a general systemic response to damage?
3. How is the hypersensitive response integrated with other chemical defenses?

35.3 Plant Responses to the Environment: Movements

Although a plant cannot move from place to place as external conditions change, plants do alter the orientation of their body parts in response to environmental stimuli. As noted earlier in the chapter, growth toward or away from a unidirectional stimulus, such as light or gravity, is called a tropism. Tropic movement involves permanent changes in the plant body because cells in particular areas or organs grow differentially in response to the stimulus. Plant physiologists do not fully understand how tropisms occur, but they are fascinating examples of the complex abilities of plants to adjust to their environment. This section will also touch upon two other kinds of movements—developmental responses to physical contact, and changes in the position of plant parts that are not related to the location of the stimulus.

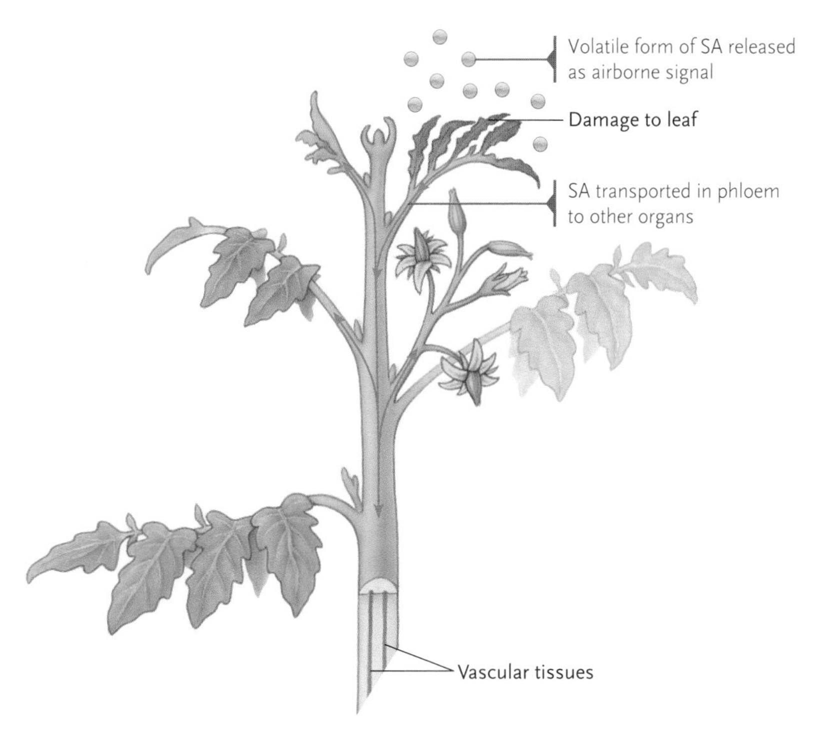

Figure 35.18
A proposed mechanism for systemic acquired resistance. When a plant successfully fends off a pathogen, the defensive chemical salicylic acid (SA) is transported in the phloem to other plant parts, where it may help protect against another attack by stimulating the synthesis of PR proteins. In addition, the plant synthesizes and releases a slightly different, more volatile form of SA. This chemical may serve as an airborne signal to other parts of the plant as well as to neighboring plants.

Phototropisms Are Responses to Light

Light is a key environmental stimulus for many kinds of organisms. Phototropisms, which we have already discussed in the section on auxins, are growth responses to a directional light source. As the Darwins discovered, if light is more intense on one side of a stem, the stem may curve toward the light **(Figure 35.19a)**. Phototropic movements are extremely adaptive for photosynthesizing organisms because they help maximize the exposure of photosynthetic tissues to sunlight.

How do auxins influence phototropic movements? In a coleoptile that is illuminated from one side, IAA moves by polar transport into the cells on the shaded side **(Figure 35.19b–d)**. Phototropic bending occurs because cells on the shaded side elongate more rapidly than do cells on the illuminated side.

The main stimulus for phototropism is light of blue wavelengths. Experiments on corn coleoptiles have shown that a large, yellow pigment molecule

a. Seedlings bend toward light.

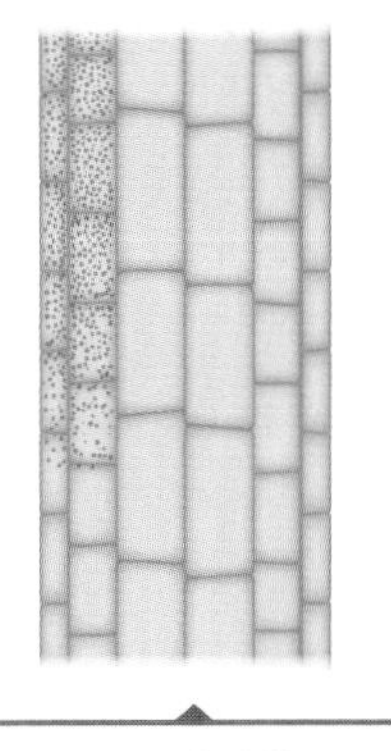

b. Rays from the sun strike one side of a shoot tip.

c. Auxin (red) diffuses down from the shoot tip to cells on its shaded side.

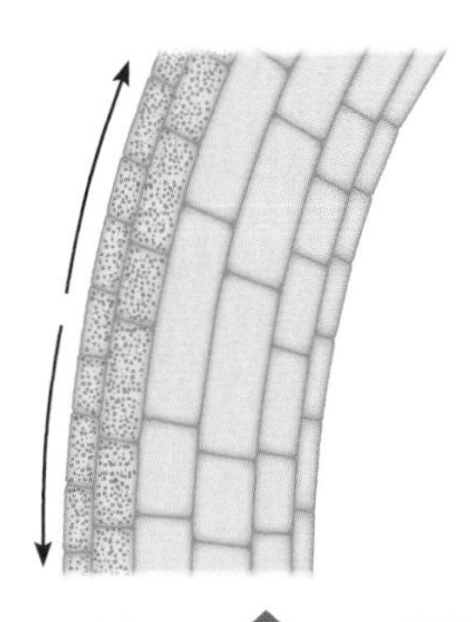

d. The auxin-stimulated cells elongate more quickly, causing the seedling to bend.

Figure 35.19
Phototropism in seedlings. (a) Tomato seedlings grown in darkness; their right side was illuminated for a few hours before they were photographed. **(b–d)** Hormone-mediated differences in the rates of cell elongation bring about the bending toward light. (Auxin is shown in red.)

a. Root oriented vertically

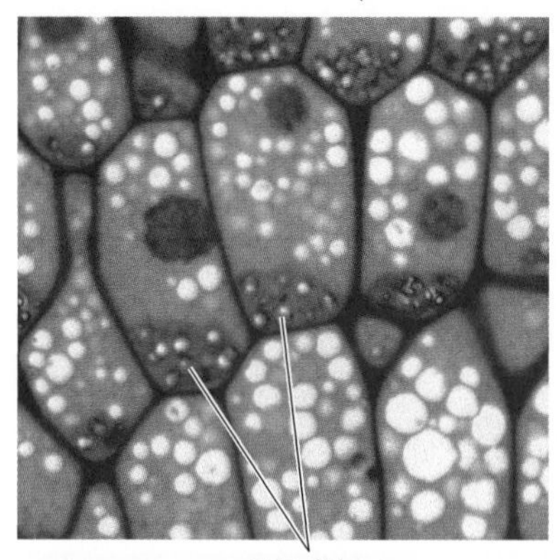

b. Root oriented horizontally

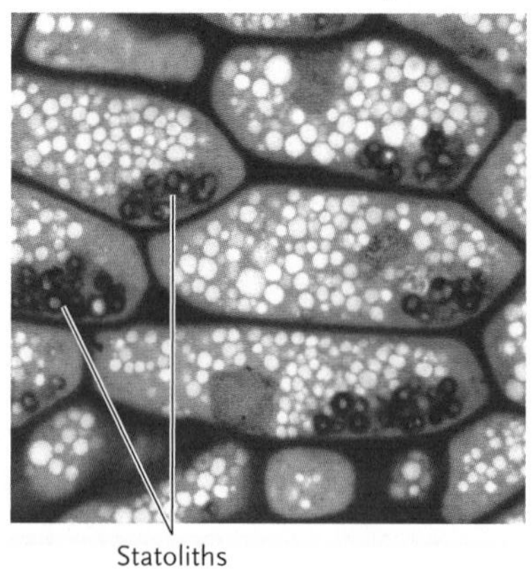

Figure 35.20
Evidence that supports the statolith hypothesis. When a corn root was laid on its side, amyloplasts—statoliths—in cells from the root cap settled to the bottom of the cells within 5 to 10 minutes. Statoliths may be part of a gravity-sensing mechanism that redistributes auxin through a root tip.
(Micrographs courtesy of Randy Moore, from "How Roots Respond to Gravity," M. L. Evans, R. Moore, and K. Hasenstein, *Scientific American*, December 1986.)

called phototropin can absorb blue wavelengths, and it may play a role in stimulating the initial lateral transport of IAA to the dark side of a shoot tip. Studies with *Arabidopsis* suggest there is more than one blue light receptor, however. One is a light-absorbing protein called **cryptochrome**, which is sensitive to blue light and may also be an important early step in the various light-based growth responses. As you will read later, cryptochrome appears to have a role in other plant responses to light as well.

Gravitropism Orients Plant Parts to the Pull of Gravity

Plants show growth responses to Earth's gravitational pull, a phenomenon called **gravitropism.** After a seed germinates, the primary root curves down, toward the "pull" (positive gravitropism), and the shoot curves up (negative gravitropism).

Several hypotheses seek to explain how plants respond to gravity. The most widely accepted hypothesis proposes that plants detect gravity much as animals do—that is, particles called **statoliths** in certain cells move in the direction gravity pulls them. In the semicircular canals of human ears, tiny calcium carbonate crystals serve as statoliths; in most plants the statoliths are amyloplasts, modified plastids that contain starch grains (see Chapter 5). In eudicot angiosperm stems, amyloplasts often are present in one or two layers of cells just outside the vascular bundles. In monocots such as cereal grasses, amyloplasts are located in a region of tissue near the base of the leaf sheath. In roots, amyloplasts occur in the root cap. If the spatial orientation of a plant cell is shifted experimentally, its amyloplasts sink through the cytoplasm until they come to rest at the bottom of the cell **(Figure 35.20).**

How do amyloplast movements translate into an altered growth response? The full explanation appears to be fairly complex, and there is evidence that somewhat different mechanisms operate in stems and in roots. In stems, the sinking of amyloplasts may provide a mechanical stimulus that triggers a gene-guided redistribution of IAA. **Figure 35.21** shows what happens when a potted sunflower seedling is turned on its side in a dark room. Within 15–20 minutes, cell elongation decreases markedly on the upper side of the growing horizontal stem, but increases on the lower side. With the adjusted growth pattern, the stem curves upward, even in the absence of light. Using different types of tests, researchers have been able to document the shifting of IAA from the top to the bottom side of the stem. The changing auxin gradient correlates with the altered pattern of cell elongation.

In roots, a high concentration of auxin has the opposite effect—it inhibits cell elongation. If a root is placed on its side, amyloplasts in the root cap accumulate near the side wall that now is the bottom side of the cap. In some way this stimulates cell elongation in the opposite wall, and within a few hours the root once again curves downward. In root tips of many plants, however, especially eudicots, researchers have not been able to detect a shift in IAA concentration that correlates with the changing position of amyloplasts. One hypothesis is that IAA is redistributed over extremely short distances in root cells, and therefore is difficult to measure. Root cells are much more sensitive to IAA than are cells in stem tissue, and even a tiny shift in IAA distribution could significantly affect their growth.

Along with IAA, calcium ions (Ca^{2+}) appear to play a major role in gravitropism. For example, if Ca^{2+} is added to an otherwise untreated agar block that is then placed on one side of a root cap, the root will bend toward the block. In this way, experimenters have been able to manipulate the direction of growth so that the elongating root forms a loop. Similarly, if an actively bending root is deprived of Ca^{2+}, the gravitropic response abruptly stops. By contrast, the negative gravitropic response of a shoot tip is inhibited when the tissue is exposed to excess calcium.

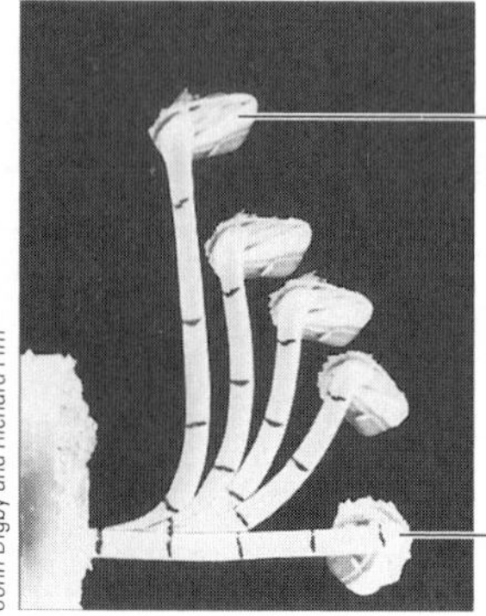

Figure 35.21
Gravitropism in a young shoot. A newly emerged sunflower seedling was grown in the dark for 5 days. Then it was turned on its side and marked at 0.5 cm intervals. Negative gravitropism turned the stem upright in 2 hours.

Just how Ca^{2+} interacts with IAA in gravitropic responses is unknown. One hypothesis posits that calcium functions as an activator. Calcium binds to a small protein called *calmodulin*, activating it in the process. Activated calmodulin in turn can activate a variety of key cell enzymes in many organisms, both plants and animals. One possibility is that calcium-activated calmodulin stimulates cell membrane pumps that enhance the flow of both IAA and calcium through a gravity-stimulated plant tissue.

Some of the most active research in plant biology focuses on the intricate mechanisms of gravitropism. For example, there is increasing evidence that in many plants, cells in different regions of stem tissue are more or less sensitive to IAA, and that gravitropism is linked in some fundamental way to these differences in auxin sensitivity. In a few plants, including some cultivated varieties of corn and radish, the direction of the gravitropic response by a seedling's primary root is influenced by light. Clearly there is much more to be learned.

Thigmotropism and Thigmomorphogenesis Are Responses to Physical Contact

Varieties of peas, grapes, and some other plants demonstrate **thigmotropism** (*thigma* = touch), which is growth in response to contact with a solid object. Thigmotropic plants typically have long, slender stems and cannot grow upright without physical support. They often have *tendrils,* modified stems or leaves that can rapidly curl around a fencepost or the sturdier stem of a neighboring plant. If one side of a grape vine stem grows against a trellis, for example, specialized epidermal cells on that side of the stem tendril shorten while cells on the other side of the tendril rapidly elongate. Within minutes the tendril starts to curl around the trellis, forming tight coils that provide strong support for the vine stem. **Figure 35.22** shows thigmotropic twisting in the passionflower *(Passiflora)*. Auxin and ethylene may be involved in thigmotropism, but most details of the mechanism remain elusive.

The rubbing and bending of plant stems caused by frequent strong winds, rainstorms, grazing animals, and even farm machinery can inhibit the overall growth of plants and can alter their growth patterns. In this phenomenon, called **thigmomorphogenesis,** a stem stops elongating and instead adds girth when it is regularly subjected to mechanical stress. Merely shaking some plants daily for a brief period will inhibit their upward growth **(Figure 35.23).** But although such plants may be shorter, their thickened stems will be stronger. Thigmomorphogenesis helps explain why plants growing outdoors are often shorter, have somewhat thicker stems, and are not as easily blown over as plants of the same species grown indoors. Trees growing near the snowline of windswept mountains show an altered growth pattern that reflects this response to wind stress.

Figure 35.22
Thigmotropism in a passionflower *(Passiflora)* tendril, which is twisted around a support.

Research on the cellular mechanisms of thigmomorphogenesis has begun to yield tantalizing clues. In one study, investigators repeatedly sprayed *Arabidopsis* plants with water and imposed other mechanical stresses, then sampled tissues from the stressed plants. The samples contained as much as double the usual amount of mRNA for at least four genes, which had been activated by the stress. The mRNAs encoded calmodulin and several other proteins that may have roles in altering *Arabidopsis* growth responses. The test plants were also short, generally reaching only half the height of unstressed controls.

Nastic Movements Are Nondirectional

Tropisms are responses to directional stimuli, such as light striking one side of a shoot tip, but many plants also exhibit **nastic movements** (*nastos* = pressed close together)—reversible responses to nondirectional stimuli, such as mechanical pressure or humidity. We see nastic movements in leaves, leaflets, and even flowers. For instance, certain plants exhibit nastic sleep movements, holding their leaves (or flower petals) in roughly horizontal positions during the day but folding

Figure 35.23
Effect of mechanical stress on tomato plants *(Lycopersicon esculentum)*. **(a)** This plant was the control; it was grown in a greenhouse, protected from wind and rain. **(b)** Each day for 28 days this plant was mechanically shaken for 30 seconds at 280 rpm. **(c)** This plant received the same shaking treatment, but twice a day for 28 days.

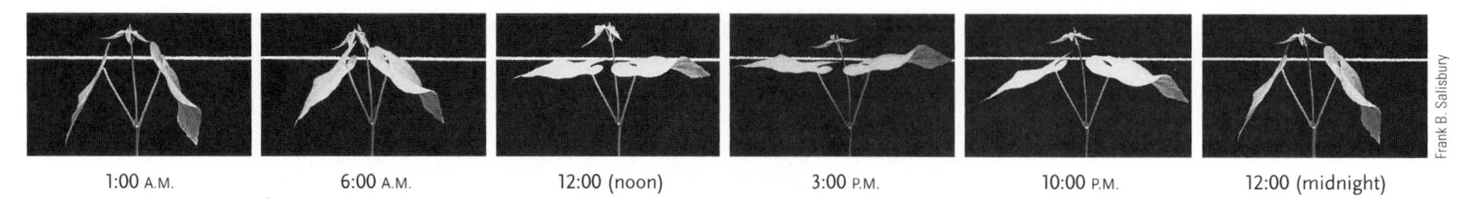

Figure 35.24
Nastic sleep movements in leaves of a bean plant. Although this plant was kept in constant darkness for 23 hours, its sleep movements continued independently of sunrise (6 A.M.) and sunset (6 P.M.). Folding the leaves closer to the stem may prevent phytochrome from being activated by bright moonlight, which could interrupt the dark period necessary to trigger flowering. Or perhaps it helps slow heat loss from leaves otherwise exposed to the cold night air.

them closer to the stem at night **(Figure 35.24).** Tulip flowers "go to sleep" in this way.

Many nastic movements are temporary and result from changes in cell turgor. For example, the daily opening and closing of stomata in response to changing light levels are nastic movements, as is the traplike closing of the lobed leaves of the Venus flytrap when an insect brushes against hairlike sensory structures on the leaves. The leaves of *Mimosa pudica,* the sensitive plant, also close in a nastic response to mechanical pressure. Each *Mimosa* leaf is divided into pairs of leaflets **(Figure 35.25a).** Touching even one leaflet at the leaf tip triggers a chain reaction in which each pair of leaflets closes up within seconds **(Figure 35.25b).**

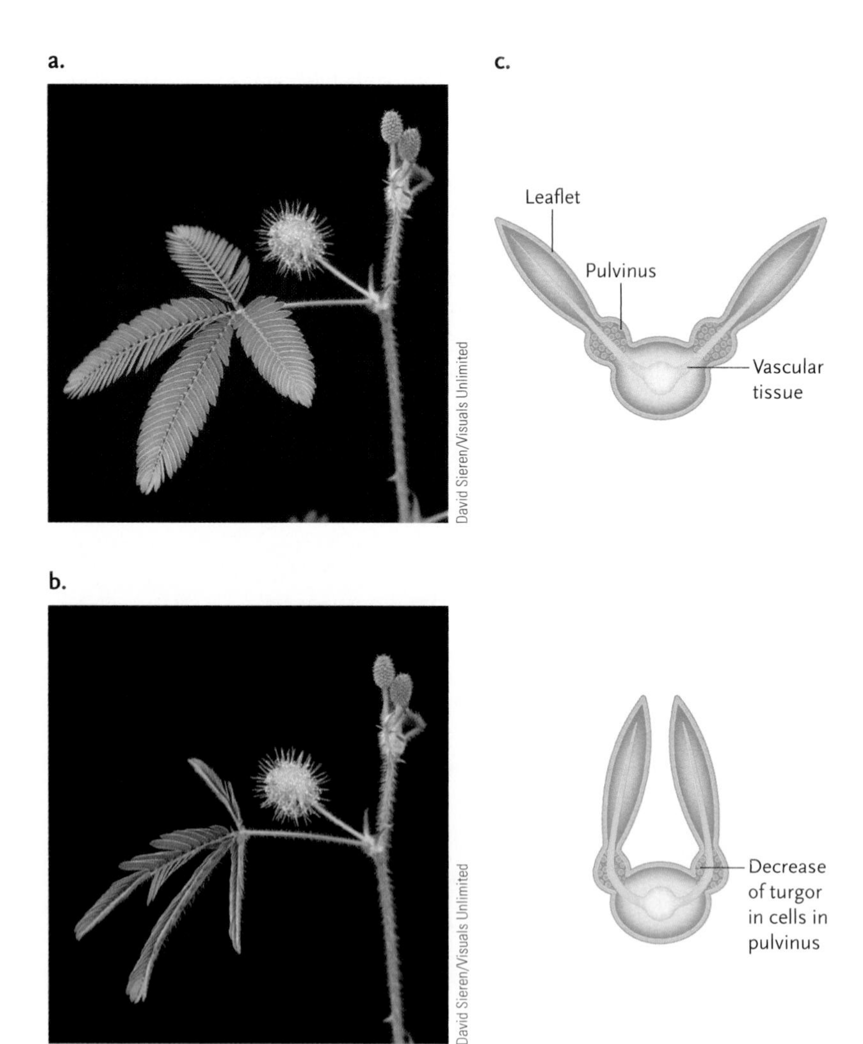

Figure 35.25
Nastic movements in leaflets of *Mimosa pudica*, the sensitive plant. (a) In an undisturbed plant the leaflets are open. If a leaflet near the leaf tip is touched, changes in turgor pressure in pulvini at the base cause the leaf to fold closed. **(b, c).** The diagram sketches this folding movement in cross section. Other leaflets close in sequence as action potentials transmit the stimulus along the leaf.

In many turgor-driven nastic movements, water moves into and out of the cells in **pulvini** (*pulvinus* = cushion), thickened pads of tissue at the base of a leaf or petiole. Stomatal movements depend on changing concentrations of ions within guard cells, and pulvinar cells drive nastic leaf movements in *Mimosa* and numerous other plants by the same mechanism **(Figure 35.25c).**

How is the original stimulus transferred from cells in one part of a leaf to cells elsewhere? The answer lies in the polarity of charge across cell plasma membranes (see Chapter 6). Touching a *Mimosa* leaflet triggers an **action potential**—a brief reversal in the polarity of the membrane charge. When an action potential occurs at the plasma membrane of a pulvinar cell, the change in polarity causes potassium ion (K^+) channels to open, and ions flow out of the cell, setting up an osmotic gradient that draws water out as well. As water leaves by osmosis, turgor pressure falls, pulvinar cells become flaccid, and the leaflets move together. Later, when the process is reversed, the pulvinar cells regain turgor and the leaflets spread apart. Action potentials travel between parenchyma cells in the pulvini via plasmodesmata at the rate of about 2 cm/sec. Animal nerves conduct similar changes in membrane polarity along their plasma membranes (see Chapter 37). These changes in polarity, which are also called action potentials, occur much more rapidly—at velocities between 1 and 100 m/sec.

Stimuli other than touch also can trigger action potentials leading to nastic movements. Cotton, soybean, sunflower, and some other plants display *solar tracking,* nastic movements in which leaf blades are oriented toward the east in the morning, then steadily change their position during the day, following the sun across the sky.

Such movements maximize the amount of time that leaf blades are perpendicular to the sun, which is the angle at which photosynthesis is most efficient.

Study Break

1. What is the direct stimulus for phototropisms? For gravitropism?
2. Explain how nastic movements differ from tropic movements.

35.4 Plant Responses to the Environment: Biological Clocks

Like all eukaryotic organisms, plants have internal time-measuring mechanisms called **biological clocks** that adapt the organism to recurring environmental changes. In plants biological clocks help adjust both daily and seasonal activities.

Circadian Rhythms Are Based on 24-Hour Cycles

Some plant activities occur regularly in cycles of about 24 hours, even when environmental conditions remain constant. These are **circadian rhythms** (*circa* = around, *dies* = day). In Chapter 32, we noted that stomata open and close on a daily cycle, even where plants are kept in total darkness. Nastic sleep movements, described earlier, are another example of a circadian rhythm. Even when such a plant is kept in constant light or darkness for a few days, it folds its leaves into the "sleep" position at roughly 24-hour intervals. In some way, the plant measures time without sunrise (light) and sunset (darkness). Such experiments demonstrate that internal controls, rather than external cues, largely govern circadian rhythms.

Circadian rhythms and other activities regulated by a biological clock help ensure that plants of a single species do the same thing, such as flowering, at the same time. For instance, flowers of the aptly named four-o'clock plant *(Mirabilis jalapa)* open predictably every 24 hours—in nature, in the late afternoon. Such coordination can be crucial for successful pollination. Although some circadian rhythms can proceed without direct stimulus from light, many biological clock mechanisms are influenced by the relative lengths of day and night.

Photoperiodism Involves Seasonal Changes in the Relative Length of Night and Day

Obviously, environmental conditions in a 24-hour period are not the same in summer as they are in winter. In North America, for instance, winter temperatures are cooler and winter day length is shorter. Experimenting with tobacco and soybean plants in the early 1900s, two American botanists, Wightman Garner and Henry Allard, elucidated a phenomenon they called **photoperiodism**, in which plants respond to changes in the relative lengths of light and dark periods in their environment during each 24-hour period. Through photoperiodism, the biological clocks of plants (and animals) make seasonal adjustments in their patterns of growth, development, and reproduction.

In plants, we now know that a blue-green pigment called **phytochrome** often serves as a switching mechanism in the photoperiodic response, signaling the plant to make seasonal changes. Plants synthesize phytochrome in an inactive form, P_r, which absorbs light of red wavelengths. Sunlight contains relatively more red light than far-red light. During daylight hours when red wavelengths dominate, P_r absorbs red light. Absorption of red light triggers the conversion of phytochrome to an active form designated P_{fr}, which absorbs light of far-red wavelengths. At sunset, at night, or even in shade, where far-red wavelengths predominate, P_{fr} reverts to P_r **(Figure 35.26)**.

In nature a high concentration of P_{fr} "tells" a plant that it is exposed to sunlight, an adaptation that is vital given that over time sunlight provides favorable conditions for leaf growth, photosynthesis, and flowering. The exact mechanism of this crucial transfer of environmental information still is not fully understood. Phytochrome activation may stimulate plant cells to take up Ca^{2+} ions, or it may induce certain plant organelles to release them. Either way, when free calcium ions combine with calcium-binding proteins (such as calmodulin), they may initiate at least some responses to light. Botanists suspect that P_{fr} controls the types of

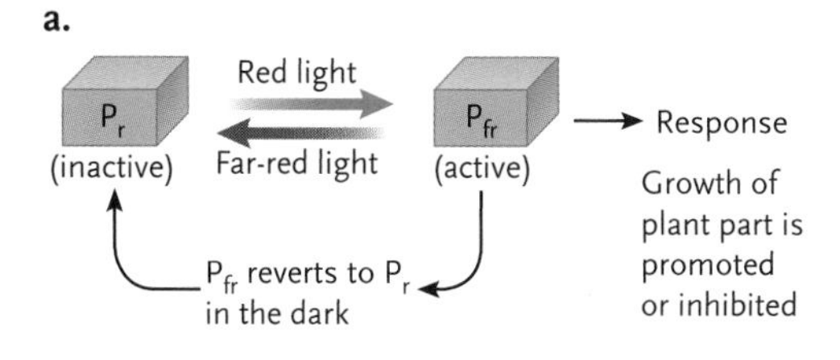

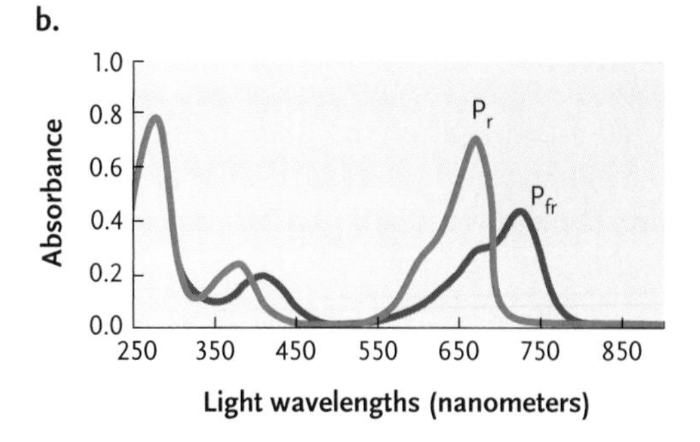

Figure 35.26
The phytochrome switching mechanism, which can promote or inhibit growth of different plant parts. **(a)** Interconversion of phytochrome from the active form (P_{fr}) to the inactive form (P_r). **(b)** The absorption spectra associated with the interconversion of P_r and P_{fr}.

Dwight Kuhn

Figure 35.27
Effects of the absence of light on young bean plants *(Phaseolus)*. The two plants at the right, the control group, were grown in a greenhouse. The other two were grown in darkness for 8 days. Note that the dark-grown plants are yellow; they could form carotenoid pigments but not chlorophyll in darkness. They have longer stems, smaller leaves, and smaller root systems than the controls.

enzymes being produced in particular cells—and different enzymes are required for seed germination, stem elongation and branching, leaf expansion, and the formation of flowers, fruits, and seeds. When plants adapted to full sunlight are grown in darkness, they put more resources into stem elongation and less into leaf expansion or stem branching **(Figure 35.27).**

Cryptochrome—which, recall, is sensitive to blue light and appears to influence light-related growth responses—also interacts with phytochromes in producing circadian responses. Researchers have recently discovered that cryptochrome occurs not only in plants but also in animals such as fruit flies and mice. Does it act as a circadian photoreceptor in both kingdoms? Only further study will provide the answer.

Cycles of Light and Dark Often Influence Flowering

Photoperiodism is especially apparent in the flowering process. Like other plant responses, flowering is often keyed to changes in day length through the year and to the resulting changes in environmental conditions. Corn, soybeans, peas, and other annual plants begin flowering after only a few months of growth. Roses and other perennials typically flower every year or after several years of vegetative growth. Carrots, cabbages, and other biennials typically produce roots, stems, and leaves the first growing season, die back to soil level in autumn, then grow a new flower-forming stem the second season.

In the late 1930s Karl Hamner and James Bonner grew cocklebur plants *(Xanthium strumarium)* in chambers in which the researchers could carefully control environmental conditions, including photoperiod. And they made an unexpected discovery: Flowering occurred only when the test plants were exposed to at least a single night of 8.5 hours of uninterrupted darkness. The length of the "day" in the growth chamber did not matter, but if light interrupted the dark period for even a minute or two, the plant would not flower at all. Subsequent research confirmed that for most angiosperms, it is the length of darkness, not light, that controls flowering.

Kinds of Flowering Responses. The photoperiodic responses of flowering plants are so predictable that botanists have long used them to categorize plants **(Figure 35.28).** The categories, which refer to day length, reflect the fact that scientists recognized the phenomenon of photoperiodic flowering responses long before they understood that darkness, not light, was the cue. **Long-day plants,** such as irises, daffodils, and corn, usually flower in spring when dark periods become shorter and day length becomes longer than some critical value—usually 9–16 hours. **Short-day plants,** including cockleburs, chrysanthemums, and potatoes, flower in late summer or early autumn when dark periods become longer and day length becomes shorter than some critical value. **Intermediate-day plants,** such as sugarcane, flower only when day length falls in between the values for long-day and short-day plants. **Day-neutral plants,** such as dandelions and roses, flower whenever they become mature enough to do so, without regard to photoperiod.

Experiments demonstrate what happens when plants are grown under the "wrong" photoperiod regimes. For instance, spinach, a long-day plant, flowers and produces seeds only if it is exposed to no more than 10 hours of darkness each day for two weeks (see Figure 35.28). **Figure 35.29** illustrates the results of an experiment to test the responses of short-day and long-day plants to night length. In this experiment, bearded iris plants (*Iris* species), which are long-day plants, and chrysanthemums, which are short-day plants, were exposed to a range of light conditions. In each case, when the researchers interrupted a critical dark period

Jan Zeevart

Figure 35.28
Effect of day length on spinach *(Spinacia oleracea)*, a long-day plant.

with a pulse of red light, the light reset the plants' clocks. The experiment provided clear evidence that short-day plants flower only when nights are longer than a critical value—and long-day plants flower only when nights are shorter than a critical value.

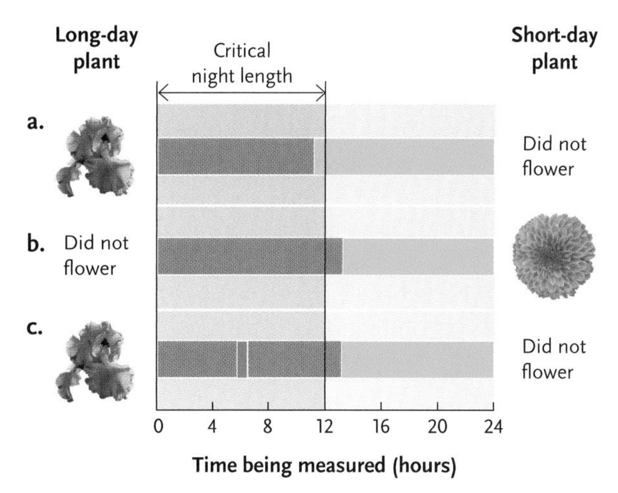

Long-day plant
Critical night length
Short-day plant
a.
Did not flower
b. Did not flower
c.
Did not flower
0 4 8 12 16 20 24
Time being measured (hours)

Figure 35.29
Experiments showing that short-day and long-day plants flower by measuring night length. Each horizontal bar signifies 24 hours. Blue bars represent night, and yellow bars day. **(a)** Long-day plants such as bearded irises flower when the night is shorter than a critical length, while **(b)** short-day plants such as chrysanthemums flower when the night is longer than a critical value. **(c)** When an intense red flash interrupts a long night, both kinds of plants respond as if it were a short night; the irises flowered but the chrysanthemums did not.
(Long-day plant photos: Clay Perry/Corbis; short-day plant photo: Eric Chrichton/Corbis.)

Chemical Signals for Flowering. When photoperiod conditions are right, what sort of chemical message stimulates a plant to develop flowers? In the 1930s botanists began postulating the existence of "florigen," a hypothetical hormone that served as the flowering signal. In a somewhat frustrating scientific quest, researchers spent the rest of the twentieth century seeking this substance in vain. Recently, however, molecular studies using *Arabidopsis* plants have defined a sequence of steps that may collectively provide the internal stimulus for flowering. Here again, we see one of the recurring themes in plant development—major developmental changes guided by several interacting genes.

Figure 35.30 traces the steps of the proposed flowering signal. To begin with, a gene called *CONSTANS* is expressed in a plant's leaves in tune with the daily light/dark cycle, with expression peaking at dusk (step 1). The gene encodes a regulatory protein called CO (not to be confused with carbon monoxide). As days lengthen in spring, the concentration of CO rises in leaves, and as a result a second gene is activated (step 2). The product of this gene, a regulatory protein called FT, travels in

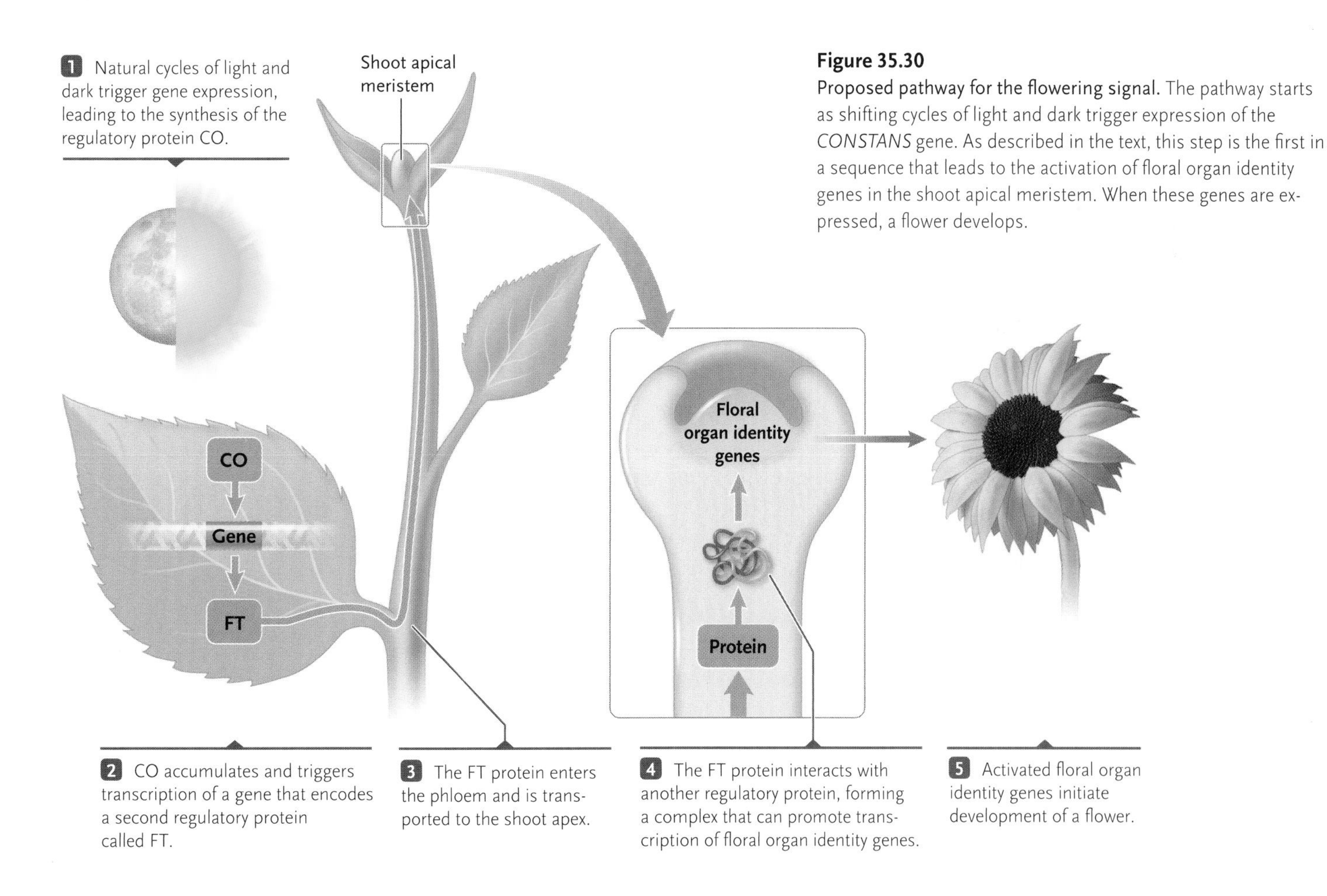

1 Natural cycles of light and dark trigger gene expression, leading to the synthesis of the regulatory protein CO.
Shoot apical meristem
CO
Gene
FT
Floral organ identity genes
Protein
2 CO accumulates and triggers transcription of a gene that encodes a second regulatory protein called FT.
3 The FT protein enters the phloem and is transported to the shoot apex.
4 The FT protein interacts with another regulatory protein, forming a complex that can promote transcription of floral organ identity genes.
5 Activated floral organ identity genes initiate development of a flower.

Figure 35.30
Proposed pathway for the flowering signal. The pathway starts as shifting cycles of light and dark trigger expression of the *CONSTANS* gene. As described in the text, this step is the first in a sequence that leads to the activation of floral organ identity genes in the shoot apical meristem. When these genes are expressed, a flower develops.

the phloem to shoot tips (step 3). Once there, the mRNA is translated into a second regulatory protein (step 4) that in some way interacts with yet a third regulatory protein that is synthesized only in shoot apical meristems (step 5). The encounter apparently sparks the development of a flower (step 6) by promoting the expression of floral organ identity genes in the meristem tissue (see Section 34.5). Key experiments that uncovered this pathway all relied on analysis of DNA microarrays, a technique introduced in Chapter 18 and featured in this chapter's *Focus on Research*.

Vernalization and Flowering. Flowering is more than a response to changing night length. Temperatures also change with the seasons in most parts of the world, and they too influence flowering. For instance, unless buds of some biennials and perennials are exposed to low winter temperatures, flowers do not form on stems in spring. Low-temperature stimulation of flowering is called **vernalization** ("making springlike").

In 1915 the plant physiologist Gustav Gassner demonstrated that it was possible to influence the flowering of cereal plants by controlling the temperature of seeds while they were germinating. In one case, he maintained germinating seeds of winter rye *(Secale cereale)* at just above freezing (1°C) before planting them. In nature, winter rye seeds in soil germinate during the winter, giving rise to a plant that flowers months later, in summer. Plants grown from Gassner's test seeds, however, flowered the same summer even when the seeds were planted in the late spring. Home gardeners can induce flowering of daffodils and tulips by putting the bulbs (technically, *corms*) in a freezer for several weeks before early spring planting. Commercial growers use vernalization to induce millions of plants, such as Easter lilies, to flower just in time for seasonal sales.

Dormancy Is an Adaptation to Seasonal Changes or Stress

As autumn approaches and days grow shorter, growth slows or stops in many plants even if temperatures are still moderate, the sky is bright, and water is plentiful. When a perennial or biennial plant stops growing under conditions that seem (to us) quite suitable for growth, it has entered a state of **dormancy.** Ordinarily, its buds will not resume growth until early spring.

Short days and long nights—conditions typical of winter—are strong cues for dormancy. In one experiment, in which a short period of red light interrupted the long dark period for Douglas firs, the plants responded as if nights were shorter and days were longer; they continued to grow taller **(Figure 35.31).** Conversion of P_r to P_{fr} by red light during the dark period prevented dormancy. In nature, buds may enter dormancy because less P_{fr} can form when day length shortens in late summer. Other environmental cues are at work also. Cold nights, dry soil, and a deficiency of nitrogen apparently also promote dormancy.

The requirement for multiple dormancy cues has adaptive value. For example, if temperature were the only cue, plants might flower and seeds might germinate in warm autumn weather—only to be killed by winter frost.

A dormancy-breaking process is at work between fall and spring. Depending on the species, breaking

Figure 35.31
Effect of the relative length of day and night on the growth of Douglas firs *(Pseudotsuga menziesii)*. The young tree at the left was exposed to alternating periods of 12 hours of light followed by 12 hours of darkness for a year; its buds became dormant because day length was too short. The tree at the right was exposed to a cycle of 20 hours of light and 4 hours of darkness; its buds remained active and growth continued. The middle plant was exposed each day to 12 hours of light and 11 hours of darkness, with a 1-hour light in the middle of the dark period. This light interruption of an otherwise long dark period also prevented buds from going dormant.

Eric Welzel/Fox Hill Nursery, Freeport, Maine
Potted plant grown inside a greenhouse did not flower.
Branch exposed to cold outside air flowered.

Figure 35.32
Effect of cold temperature on dormant buds of a lilac *(Syringa vulgaris)*. In this experiment, a plant was grown in winter inside a warm greenhouse with one branch growing out of a hole. Only the buds on the branch exposed to low outside temperatures resumed growth in spring. This experiment suggests that low-temperature effects are localized.

Focus on Research

Research Methods: Using DNA Microarray Analysis to Track Down "Florigen"

The more plant scientists learn about plant genomes, the more they are relying on DNA microarray assays to elucidate the activity of plant genes.

Recall from Section 18.3 that a DNA microarray, also called a DNA chip, allows an investigator to explore questions such as how the expression of a particular gene differs in different types of cells. To quantify the expression of specific genes in particular types of cells, mRNA transcripts are isolated from the cells; then a cDNA library is created from each mRNA sample, using nucleotides labeled with fluorescent dyes. Probes (nucleotide sequences) representing every gene in the organism's genome are fixed onto a slide; when the labeled cDNAs are added to the slide, each will hybridize to the gene that expressed the mRNA from which it was made. Next, the DNA microarray is scanned with a laser that can detect fluorescence. When a gene is expressed in a cell, the dye fluoresces and gives a color that accords with the degree of its expression. The procedure can be manipulated to reveal the relative amounts of expression of more than one of a cell's genes.

Philip A. Wigge and his colleagues used this method to learn more about the signaling pathway that causes a plant's apical meristem to give rise to flowers. Previous research had established that in leaves, lengthening spring days coincided with rising concentrations of CO, a regulatory protein encoded by the *CONSTANS* gene. But what did CO regulate? Working with *Arabidopsis thaliana,* Wigge's group was able to narrow down the field to four genes, and using microarray analysis of DNA from leaf cells they pinpointed one called FT (for flowering locus T). The researchers found that in leaves, CO causes strong expression of FT: When enough CO is present, FT mRNA is rapidly transcribed, then enters the phloem. (The transport of mRNA in phloem is not unusual.) By contrast, when they tested CO's effects in shoot apex cells, they found that it triggers far less gene expression there. Clearly, CO was not directly triggering the development of flowers. However, FT mRNA moves in the phloem to the shoot apex, where it is translated into protein. Was that protein the direct flowering signal? Other studies had implicated a regulatory protein called FD, which microarray analysis had shown was expressed *only*—but very strongly—in the shoot apex.

To sort out this final piece of the puzzle, the Wigge team examined flowering responses in normal *A. thaliana* plants as well as in mutants having a normal FT protein but a defective *fd,* and vice versa. Flowering was abnormal in both types of mutants, possibly because the mutated "partner" suppressed some aspect of the functioning of the normal protein. On the other hand, in wild-type plants, which had a functioning FD protein, expression of FT triggered a marked increased in the expression of the floral organ gene *APETALA1* **(Figure a).** These results have two major implications. First, they support the hypothesis that FT and FD interact in a normal flowering response. Second, the study suggests that FT, the CO-induced signal from leaves, conveys the environmental signal that it is time for a plant to flower. In that sense, FT may be the long sought "florigen." However, only by interacting with FD does FT "know" where to deliver its flowering signal—in the apical meristems of shoots.

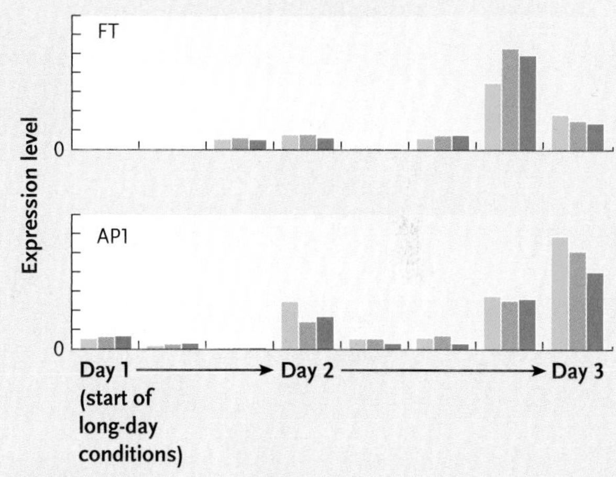

Figure a
Effect of the FT protein on expression of the *APETALA1 (AP1)* floral organ identity gene. In nature, *Arabidopsis thaliana* is a long-day plant, and the experiment was carried out under long-day (that is, short-night) conditions. Three groups of replicates shown here in yellow, orange, and red respectively, were monitored for both AP1 and FT. After a brief delay, the expression of AP1 closely tracked the appearance of the FT regulatory protein, which had been activated by its interaction with the FD protein.

dormancy probably involves gibberellins and abscisic acid, and it requires exposure to low winter temperatures for specific periods **(Figure 35.32).** The temperature needed to break dormancy varies greatly among species. For example, the Delicious variety of apples grown in Utah requires 1230 hours near 43°F (6°C); apricots grown there require only 720 hours at that temperature. Generally, trees growing in the southern United States or in Italy require less cold exposure than those growing in Canada or in Sweden.

Study Break

1. Summarize the switching mechanism that operates in plant responses to changes in photoperiod.
2. Give some examples of how relative lengths of dark and light can influence flowering.
3. Explain why dormancy is an adaptive response to a plant's environment.

35.5 Signal Responses at the Cellular Level

Environmental stimuli such as changing light, temperature, or chemicals on the surface of an attacking pathogen are cues that signal a plant to alter its growth or physiology. For decades plant physiologists have looked avidly for clues about how those signals are converted into a chemical message that produces a change in cell metabolism or growth. As Chapter 7 describes, research on the ways animal cells respond to external signals has revealed some basic mechanisms, and at least some of these mechanisms also apply to plant cells.

Figure 35.33 Signal response pathways in plant cells.

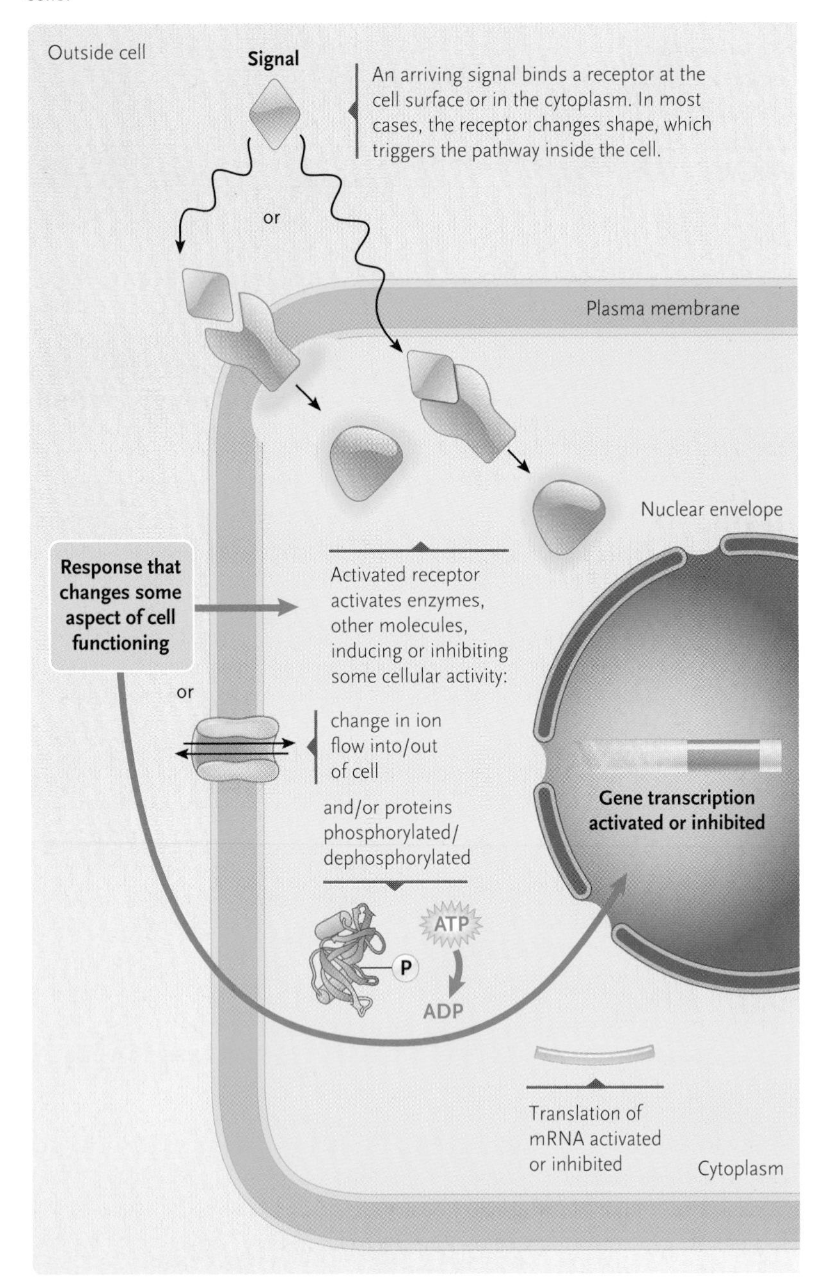

Several Signal Response Pathways Operate in Plants

Hormones and environmental stimuli alter the behavior of target cells, which have receptors to which specific signal molecules can bind and elicit a cellular response. By means of a response pathway, a signal can induce changes in the cell's shape and internal structure or influence the transport of ions and other substances into and out of cells. Some signals cause cells to alter gene activation and the rate of protein synthesis; others set in motion events that modify existing cell proteins. Section 7.1 presents some general features of the signaling process. Here, we'll briefly consider how signal molecules may operate in plants.

Certain hormones and growth factors bind to receptors at the target cell's plasma membrane, on its endoplasmic reticulum (ER), or in the cytoplasm. For example, ethylene receptors are on the ER, and the auxin receptor is a protein in the cytoplasm called TIR1. Research in several laboratories recently confirmed that auxin binds directly to TIR1, setting in motion events that break down protein factors that inhibit transcription. As a result, previously repressed genes are turned on. In many cases (although not with TIR1), binding causes the receptor to change shape. Regardless, binding of a hormone or growth factor triggers a complex pathway that leads to the cell response—the opening of ion channels, activation of transport proteins, or some other event.

Only some cells can respond to a particular signaling molecule, because not all cells have the same types of receptors. For example, particular cells in ripening fruits and developing seeds have ethylene receptors, but few if any cells in stems do. Different signals also may have different effects on a single cell, and may exert those effects by way of different response pathways. One type of signal might stimulate transcription, and another might inhibit it. In addition, as we've seen, some genes controlled by particular receptors encode proteins that regulate still *other* genes.

In plants we know the most about pathways involving auxin, ethylene, salicylic acid, and blue light. **Figure 35.33** diagrams a general model for these response pathways in plant cells. As the figure shows, the response may lead to a change in the cell's structure, its metabolic activity, or both, either directly or by altering the expression of one or more genes.

Second Messenger Systems Enhance the Plant Cell's Response to a Hormone's Signal

We can think of plant hormones and other signaling molecules as external first messengers that deliver the initial physiological signal to a target cell. Often, as with salicylic acid, binding of the signal molecule triggers the synthesis of internal second messengers (introduced in Section 7.4). These intermediary molecules diffuse rapidly through the cytoplasm and provide the main chemical signal that alters cell functioning.

Second messengers usually are synthesized in a sequence of chemical reactions that converts an external signal into internal cell activity. For many years the details of plant second-messenger systems were sketchy and hotly debated. More recently, however, reaction sequences that occur in the cells of animals and some fungi have also been found in plants. The following example describes reactions that close plant stomata in response to a signal from abscisic acid.

As discussed in Section 35.1, abscisic acid helps regulate several responses in plants, including the maturation of seeds and the closing of stomata (see Section 32.3). ABA's role in stomatal closure—triggered by water stress or some other environmental cue—begins when the hormone activates a receptor in the plant cell plasma membrane. Experiments have shown that this binding activates G proteins that in turn activate phospholipase C (see Figure 7.12). This enzyme stimulates the synthesis of second messengers, such as inositol triphosphate (IP_3).

The second messenger diffuses through the cytoplasm and binds with calcium channels in cell structures such as the endoplasmic reticulum, vacuole, and plasma membrane. The bound channels then open, releasing calcium ions that activate protein kinases in the cytoplasm. In turn, the activated enzymes activate their target proteins (by phosphorylating them). Each protein kinase can convert a large number of substrate molecules into activated enzymes, transport proteins, open ion channels, and so forth. Soon the number of molecules representing the final cellular response to the initial signal is enormous.

Recent experimental evidence indicates that, in similar fashion, auxin's hormonal signal is conveyed by cAMP (cyclic adenosine monophosphate), another major second messenger in cells of animals and other organisms.

In addition to the basic pathways described here, other routes may exist that are unique to plant cells.

Unanswered Questions

Do plants have a "backup" copy of their genome?

As described at the beginning of this chapter, land plants manifest adaptations that allow them to survive and reproduce in unfavorable or hostile conditions. Adaptations include changes in growth and development reflecting responses to environmental fluctuations that occur naturally during the normal life cycle of plants. Being physically anchored in one place has driven plant adaptation so that changes in the plant body can facilitate survival. It is also possible, however, that the sessile existence of land plants may have selected for unusual adaptive strategies. Might plants have devised a strategy to utilize previously unknown genetic resources and thereby expand their potential repertoire of adaptive responses? Recent findings demonstrating the existence of a previously unknown mechanism of genetic instability suggest that such a strategy may indeed have been in place during the evolution of land plants. These findings suggest that, at least in *Arabidopsis thaliana*, a "backup" copy of the genome exists that can be accessed under unfavorable conditions.

Why have a backup copy of the genome? Simply put, if the system crashes, it can be restored. By analogy, the genome could be considered the "operating system" stored on the "hard drive" of the organism. If that operating system becomes corrupted, for example by a devastating power surge or a computer virus, a global systems failure might occur. However, if a backup copy were maintained at least in a subset of the population, then, under conditions that might lead to extinction, the backup copy could be used to "restore" the system and increase the chances of survival for that organism or population. In other words, the genome could adapt using the stored information.

An intriguing possibility is that such a backup genome might exist in the form of RNA. The fact that backup copies have not been found using conventional DNA-based detection methods or classical genetic approaches leaves open the exciting possibility that RNA might serve as the storage medium for this information. In a sense, having the information stored in an alternative chemical form (analogous to a different computer language or code) might also make it less susceptible to corruption. Furthermore, it may be that this backup genome is a remnant of an ancestral condition where the genome was RNA-based.

How would you go about testing these different possibilities? First, the findings would have to be independently verified and the existence of a "restoration" mechanism would have to be confirmed by other research groups working on *Arabidopsis* or other plant species. Second, the source and chemical nature of the backup information would need to be identified. Where is it, and is it RNA, DNA, protein, or a combination of these? The question of mechanism would also need to be addressed. How is the system restored, and when does it happen? The question of how widespread this phenomenon is would also need to be considered. Do all plant species maintain a backup copy and do organisms outside of the plant kingdom have a backup genome?

Susan Lolle is associate professor of biology at the University of Waterloo in Waterloo, Canada. Her research interests include plant development, genetics and genome biology. To learn more go to http://www.biology.uwaterloo.ca.

Light is the driving force for photosynthesis, and it may not be farfetched to suppose that plants have evolved other unique light-related biochemical pathways as well. For instance, exciting experiments are extending our knowledge of how plant cells respond to blue light, which, as we have discussed, triggers some photoperiod responses such as the opening and closing of stomata. In all likelihood, much remains to be discovered about this and many other aspects of plant functioning.

Study Break

1. Summarize the various ways that chemical signals reaching plant cells are converted to changes in cell functioning.
2. What basic task does a second messenger accomplish?
3. Thinking back to Chapter 7, can you describe parallels between signal transduction mechanisms in the cells of plants and animals?

Review

Go to ThomsonNOW at www.thomsonedu.com/login to access quizzing, animations, exercises, articles, and personalized homework help.

35.1 Plant Hormones

- At least seven classes of hormones govern flowering plant development, including germination, growth, flowering, fruit set, and senescence (Table 35.1).
- Auxins, mainly IAA, promote elongation of cells in the coleoptile and stem (Figures 35.2–35.7).
- Gibberellins promote stem elongation and help seeds and buds break dormancy (Figures 35.1, 35.8, and 35.9).
- Cytokinins stimulate cell division, promote leaf expansion, and retard leaf aging (Figure 35.10).
- Ethylene promotes fruit ripening and abscission (Figures 35.11 and 35.12).
- Brassinosteroids stimulate cell division and elongation (Figure 35.13).
- Abscisic acid (ABA) promotes stomatal closure and may trigger seed and bud dormancy (Figure 35.14).
- Jasmonates regulate growth and have roles in defense.

Animation: Plant development

Animation: Auxin's effects

35.2 Plant Chemical Defenses

- Plants have diverse chemical defenses that limit damage from bacteria, fungi, worms, or plant-eating insects (Figure 35.15).
- The hypersensitive response isolates an infection site by surrounding it with dead cells (Figure 35.16). During the response salicylic acid (SA) induces the synthesis of PR (pathogenesis-related) proteins.
- Oligosaccharins can trigger the synthesis of phytoalexins, secondary metabolites that function as antibiotics.
- Gene-for-gene recognition enables a plant to chemically recognize a specific pathogen and mount a defense (Figure 35.17).
- Systemic acquired resistance provides long-term protection against some pathogens. Salicylic acid passes from the infected organ to newly forming organs such as leaves, which then synthesize PR proteins (Figure 35.18).
- Heat-shock proteins can reversibly bind enzymes and other proteins in plant cells and prevent them from denaturing when the plant is under heat stress.
- Some plants can synthesize "antifreeze" proteins that stabilize cell proteins when cells are threatened with freezing.

35.3 Plant Responses to the Environment: Movements

- Plants adjust their growth patterns in response to environmental rhythms and unique environmental circumstances. These responses include tropisms.
- Phototropisms are growth responses to a directional light source. Blue light is the main stimulus for phototropism (Figures 35.3 and 35.19).
- Gravitropism is a growth response to Earth's gravitational pull. Stems exhibit negative gravitropism, growing upward, while roots show positive gravitropism (Figures 35.20 and 35.21).
- Some plants or plant parts demonstrate thigmotropism, growth in response to contact with a solid object (Figure 35.22).
- Mechanical stress can cause thigmomorphogenesis, which causes the stem to add girth (Figure 35.31).
- In nastic leaf movements, water enters or exits cells of a pulvinus, a pad of tissue at the base of a leaf or petiole, in response to action potentials (Figures 35.24 and 35.25).

Animation: Gravitropism

Animation: Gravity and statolith distribution

Animation: Phototropism

35.4 Plant Responses to the Environment: Biological Clocks

- Plants have biological clocks, internal time-measuring mechanisms with a biochemical basis. Environmental cues can "reset" the clocks, enabling plants to make seasonal adjustments in growth, development, and reproduction.
- In photoperiodism, plants respond to a change in the relative length of daylight and darkness in a 24-hour period. A switching mechanism involving the pigment phytochrome promotes or inhibits germination, growth, and flowering and fruiting.
- Phytochrome is converted to an active form (P_{fr}) during daylight, when red wavelengths dominate. It reverts to an inactive form (P_r) at sunset, at night, or in shade, when far-red wavelengths predominate. P_{fr} may control the types of metabolic pathways that operate under specific light conditions (Figure 35.26).
- Long-day plants flower in spring or summer, when day length is long relative to night. Short-day plants flower when day length is relatively short, and intermediate-day plants flower when day length falls in between the values for long-day and short-day

plants. Flowering of day-neutral plants is not regulated by light. In vernalization, a period of low temperature stimulates flowering (Figures 35.27–35.29).

- The direct trigger for flowering may begin in leaves, when the regulatory protein CO triggers the expression of the FT gene. The resulting mRNA transcripts move in phloem to apical meristems where translation of the mRNAs yields a second regulatory protein, which in turn interacts with a third. This final interaction activates genes that encode the development of flower parts (Figure 35.30).
- Senescence is the sum of processes leading to the death of a plant or plant structure.
- Dormancy is a state in which a perennial or biennial stops growing even though conditions appear to be suitable for continued growth (Figures 35.31 and 35.32).

Animation: Phytochrome conversions

Animation: Vernalization

Animation: Day length and dormancy

Animation: Flowering response experiments

35.5 Signal Responses at the Cellular Level

- Hormones and environmental stimuli alter the behavior of target cells, which have receptors to which signal molecules can bind. By means of a response pathway that ultimately alters gene expression, a signal can induce changes in the cell's shape or internal structure or influence its metabolism or the transport of substances across the plasma membrane (Figure 35.33).
- Some plant hormones and growth factors may bind to receptors at the target cell's plasma membrane, changing the receptor's shape. This binding often triggers the release of internal second messengers that diffuse through the cytoplasm and provide the main chemical signal that alters gene expression.
- Second messengers usually act by way of a reaction sequence that amplifies the cell's response to a signal. An activated receptor activates a series of proteins, including G proteins and enzymes that stimulate the synthesis of second messengers (such as IP_3) that bind ion channels on endoplasmic reticulum.
- Binding releases calcium ions, which enter the cytoplasm and activate protein kinases, enzymes that activate specific proteins that produce the cell response.

Questions

Self-Test Questions

1. Which of the following plant hormones does *not* stimulate cell division?
   a. auxins
   b. cytokinins
   c. ethylene
   d. gibberellins
   e. abscisic acid
2. Which is the correct pairing of a plant hormone and its function?
   a. salicylic acid: triggers synthesis of general defense proteins
   b. brassinosteroids: promote responses to environmental stress
   c. cytokinins: stimulate stomata to close in water-stressed plants
   d. gibberellins: slow seed germination
   e. ethylene: promotes formation of lateral roots
3. A characteristic of auxin (IAA) transport is:
   a. IAA moves by polar transport from the base of a tissue to its apex.
   b. IAA moves laterally from a shaded to an illuminated side of a plant.
   c. IAA enters a plant cell in the form of IAAH, an uncharged molecule that can diffuse across cell membranes.
   d. IAA exits one cell and enters the next by means of transporter proteins clustered at both the apical and basal ends of the cells.
   e. All of the above are characteristics of auxin transport in different types of cells.
4. Hanging wire fruit baskets have many holes or open spaces. The major advantage of these spaces is that they:
   a. prevent gibberellins from causing bolting or the formation of rosettes on the fruit.
   b. allow the evaporation of ethylene and thus slow ripening of the fruit.
   c. allow oxygen in the air to stimulate the production of ethylene, which hastens the abscission of fruits.
   d. allow oxygen to stimulate brassinosteroids, which hasten the maturation of seeds in/on the fruits.
   e. allow carbon dioxide in the air to stimulate the production of cytokinins, which promotes mitosis in the fruit tissue and hastens ripening.
5. Which of the following is *not* an example of a plant chemical defense?
   a. ABA inhibits leaves from budding if conditions favor attacks by sap-sucking insects.
   b. Jasmonate activates plant genes encoding protease inhibitors that prevent insects from digesting plant proteins.
   c. Acting against fungal infections, the hypersensitive response allows plants to produce highly reactive oxygen compounds that kill selected tissue, thus forming a dead tissue barrier that walls off the infected area from healthy tissues.
   d. Chitinase, a PR hydrolytic protein produced by plants, breaks down chitin in the cell walls of fungi and thus halts the fungal infection.
   e. Attack by fungi or viruses triggers the release of oligosaccharins, which in turn stimulate the production of phytoalexins having antibiotic properties.
6. Which of the following statements about plant responses to the environment is true?
   a. The heat-shock response induces a sudden halt to cellular metabolism when an insect begins feeding on plant tissue.
   b. In gravitropism, amyloplasts sink to the bottom of cells in a plant stem, causing the redistribution of IAA.
   c. The curling of tendrils around a twig is an example of thigmotropism.
   d. Phototropism results when IAA moves first laterally, then downward in a shoot tip when one side of the tip is exposed to light.
   e. Nastic movements, such as the sudden closing of the leaves of a Venus flytrap, are examples of a plant's ability to respond to specific directional stimuli.
7. In nature the poinsettia, a plant native to Mexico, blooms only in or around December. This pattern suggests:
   a. the long daily period of darkness (short day) in December stimulates the flowering.
   b. vernalization stimulates the flowering.
   c. the plant is dormant for the rest of the year.
   d. phytochrome is not affecting the poinsettia flowering cycle.
   e. a circadian rhythm is in effect.

8. Which of the following steps is *not* part of the sequence that is thought to trigger flowering?
   a. Cycles of light and dark stimulate the expression of the *CONSTANS* gene in a plant's leaves.
   b. CO proteins accumulate in the leaves and trigger expression of a second regulatory gene.
   c. mRNA transcribed during expression of a second regulatory gene moves via the phloem to the shoot apical meristem.
   d. Interactions among regulatory proteins promote the expression of floral organ identity genes in meristem tissue.
   e. CO proteins in the floral meristem interact with florigen, a so-called flowering hormone, which provides the final stimulus for expression of floral organ identity genes.
9. Damage from an infectious bacterium, fungus, or worm may trigger a plant defensive response when the pathogen or a substance it produces binds to:
   a. a receptor encoded by the plant's *avirulence (Avr)* gene.
   b. an *R* gene in the plant cell nucleus.
   c. a receptor encoded by a dominant R gene.
   d. PR proteins embedded in the plant cell plasma membrane.
   e. salicylic acid molecules released from the besieged plant cell.
10. In the sequence that unfolds after molecules of a hormone such as ABA bind to receptors at the surface of a target plant cell:
   a. first messenger molecules in the cytoplasm are mobilized, then G proteins carry the signal to second messengers such as protein kinases, which alter the activity of cell proteins such as IP_3.
   b. binding activates G proteins, which in turn activate second messengers such as IP_3; subsequent steps are thought to involve activation of genes that encode protein kinases.
   c. binding activates phospholipase C, which in turn activates G proteins, which then activate molecules of IP_3, a step that leads to the synthesis of protein kinases.
   d. binding stimulates G proteins to activate protein kinases, which then bind calcium channels in ER; the flux of calcium ions activates second messenger molecules that alter the activity of cell proteins or enter the cell nucleus and alter the expression of target genes.
   e. binding activates G proteins, which in turn activate phospholipase C; this substance then stimulates the synthesis of second messenger molecules, the second messengers bind calcium channels in the cell's ER, and finally protein kinases alter the activity of proteins by phosphorylating them.

Questions for Discussion

1. You work for a plant nursery and are asked to design a special horticultural regimen for a particular flowering plant. The plant is native to northern Spain, and in the wild it grows a few long, slender stems that produce flowers each July. Your boss wants the nursery plants to be shorter, with thicker stems and more branches, and she wants them to bloom in early December in time for holiday sales. Outline your detailed plan for altering the plant's growth and reproductive characteristics to meet these specifications.
2. Synthetic auxins such as 2,4-D can be weed killers because they cause an abnormal growth burst that kills the plant within a few days. Suggest reasons why such rapid growth might be lethal to a plant.
3. In some plant species, an endodermis is present in both stems and roots. In experiments, the shoots of mutant plants lacking differentiated endodermis in their root and shoot tissue don't respond normally to gravity, but roots of such plants do respond normally. Explain this finding, based on your reading in this chapter.
4. In *A. thaliana* plants carrying a mutation called *pickle (pkl)*, the primary root meristem retains characteristics of embryonic tissue—it spontaneously regenerates new embryos that can grow into mature plants. However, when the mutant root tissue is exposed to a gibberellin (GA), this abnormal developmental condition is suppressed. Explain why this finding suggests that additional research is needed on the fundamental biological role of GA.

Experimental Analysis

Tiny, thornlike trichomes on leaves are a common plant adaptation to ward off insects. Those trichomes develop very early on, as outgrowths of a seedling's epidermal cells. Biologists have observed, however, that many mature plants develop more leaf trichomes after the fact, as a *response* to insect damage. Researchers at the University of Chicago decided to study this phenomenon, and specifically wanted to determine the effects, if any, of jasmonate, salicylic acid, and gibberellin in stimulating trichome development. Keeping in mind that plant hormones often interact, how many separate experiments, at a minimum, would the research team have had to carry out in order to obtain useful initial data? Do you suppose they used mutant plants for some or all of the tests? Why or why not?

Evolution Link

Cryptochrome occurs in plants and animals. If it was inherited from their shared ancestor, what other major groups of organisms might also have it?

How Would You Vote?

1-Methylcyclopropene, or MCP, is a gas that keeps ethylene from binding to cells in plant tissues. It is used to prolong the shelf life of cut flowers and the storage time for fruits. Should produce that is treated this way be labeled to alert consumers? Go to www.thomsonedu.com/login to investigate both sides of the issue and then vote.

Magnetic resonance imaging (MRI) whole body scans of a man (left), a 9-year-old boy (middle), and a woman. Various organs can be seen in the scans: the whitish skeleton throughout the bodies, the brains within the skulls, lungs (dark) in the chests, lobes of the liver (green and brown ovals) in the abdomens, and bladders (dark ovals) in the lower abdomens.

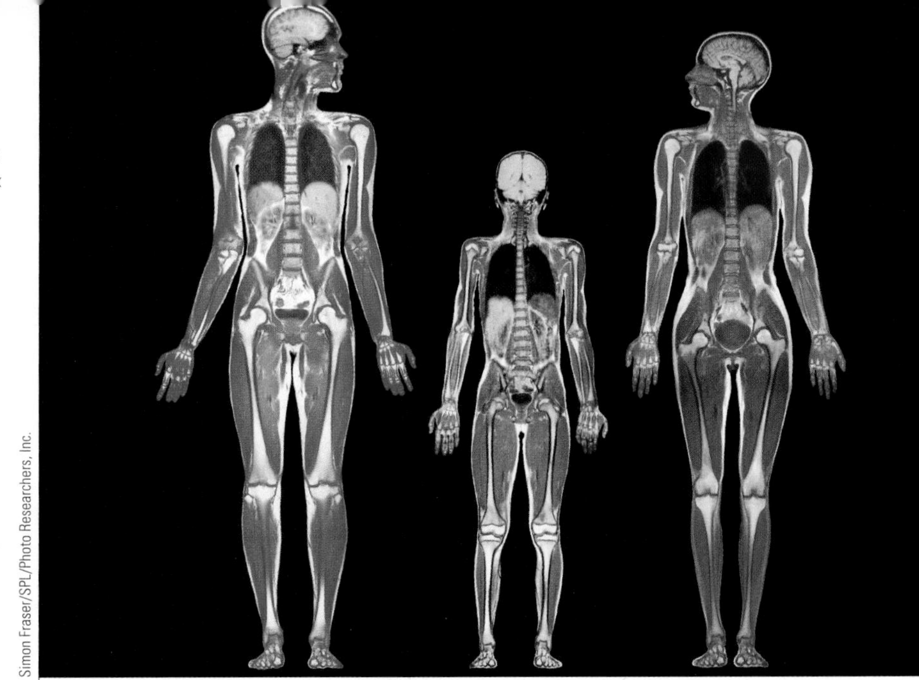

Simon Fraser/SPL/Photo Researchers, Inc.

Study Plan

36 Introduction to Animal Organization and Physiology

Why It Matters

After a cold night in Africa's Kalahari Desert, gray meerkats *(Surricata suricatta)*, a type of mongoose, awaken in their burrows. Although, like all mammals, meerkats regulate their body temperature, their internal temperature falls during cold nights. If the sun is shining in the morning and warms their burrows, the meerkats emerge and stand on their hind legs facing east, warming their bodies in the rays of the sun **(Figure 36.1)**. This sunning behavior helps raise their body temperature.

Once the meerkats warm up, they fan out from their burrows looking for food, mainly insects and an occasional lizard. Their highly integrated body systems allow them to move about, sense the presence of prey, react with speed and precision to capture those prey, and consume them. Within their bodies, the food is broken down into glucose and other nutrient molecules, which are transported throughout the body to provide energy for living. At the same time, balancing mechanisms are constantly at work to maintain the animals' internal environment at a level that keeps body cells functioning. The maintenance of the internal environment in a stable state is called **homeostasis** (*homeo* = the same; *stasis* = standing or stopping). The processes and activi-

Figure 36.1 Meerkats lining up to warm themselves in sunlight.

ties responsible for homeostasis are called **homeostatic mechanisms.** These mechanisms compensate both for the external environmental changes that the meerkats encounter as they explore places with differences in temperature, humidity, and other physical conditions, and for changes in their own body systems.

All animals have body systems for acquiring and digesting nutrients to provide energy for life, growth, reproduction, and movement. Biologists are interested in the structures and functions of these systems. **Anatomy** is the study of the structures of organisms, and **physiology** is the study of their functions—the physico-chemical processes of organisms.

In this chapter we begin with the organization of individual cells into tissues, organs, and organ systems, the major body structures that carry out animal activities. Our discussion continues with a look at how the processes and activities of organ systems coordinate to accomplish homeostasis. The other chapters in this unit discuss the individual organ systems that carry out major body functions such as digestion, movement, and reproduction. Although we emphasize vertebrates throughout the unit, with particular reference to human physiology, we also make comparisons with invertebrates, to keep the structural and functional diversity of the animal kingdom in perspective and to understand the evolution of the structures and processes involved.

36.1 Organization of the Animal Body

In Animals, Specialized Cells Are Organized into Tissues, Tissues into Organs, and Organs into Organ Systems

The individual cells of animals have the same requirements as cells of any kind. They must be surrounded by an aqueous solution that contains ions and molecules required by the cells, including complex organic molecules that can be used as an energy source. The concentrations of these molecules and ions must be balanced to keep cells from shrinking or swelling excessively due to osmotic water movement. Most animal cells also require oxygen to serve as the final acceptor for electrons removed in oxidative reactions. Animal cells must be able to release waste molecules and other by-products of their activities, such as carbon dioxide, to their environment. The physical conditions of the cellular environment, such as temperature, must also remain within tolerable limits.

The evolution of multicellularity (see Section 24.3) made it possible for organisms to create an *internal fluid environment* that supplies all the needs of individual cells, including nutrient supply, waste removal, and osmotic balance. This internal environment allows multicellular organisms to occupy diverse habitats, including dry terrestrial habitats that would be lethal to single cells. Multicellular organisms can also become relatively large because their individual cells remain small enough to exchange ions and molecules with the internal fluid. The fluid occupying the spaces between cells in multicellular animals is called **interstitial fluid,** or **extracellular fluid.**

The evolution of multicellularity also allowed major life functions to be subdivided among specialized groups of cells, with each group concentrating on a single activity. In animals, some groups of cells became specialized for movement, others for food capture, digestion, internal circulation of nutrients, excretion of wastes, reproduction, and other functions. Specialization greatly increases the efficiency by which animals carry out these functions.

In most animals, these specialized groups of cells are organized into tissues, the tissues into organs, and the organs into organ systems **(Figure 36.2).** A **tissue** is a group of cells with the same structure and function, working together as a unit to carry out one or more activities. The tissue lining the inner surface of the intestine, for example, is specialized to absorb nutrients released by digestion of food in the intestinal cavity.

An **organ** integrates two or more different tissues into a structure that carries out a specific function. The eye, liver, and stomach are examples of organs. Thus, the stomach integrates several different tissues into an organ specialized for processing food.

An **organ system** coordinates the activities of two or more organs to carry out a major body function such as movement, digestion, or reproduction. The organ system carrying out digestion, for example, coordinates the activities of organs including the mouth, stomach, pancreas, liver, and small and large intestines. Some organs contribute functions to more than one organ system. For instance, the pancreas forms part of the endocrine system as well as the digestive system.

Study Break

1. What are some advantages for an organism being multicellular?
2. What is the difference between a tissue, an organ, and an organ system?

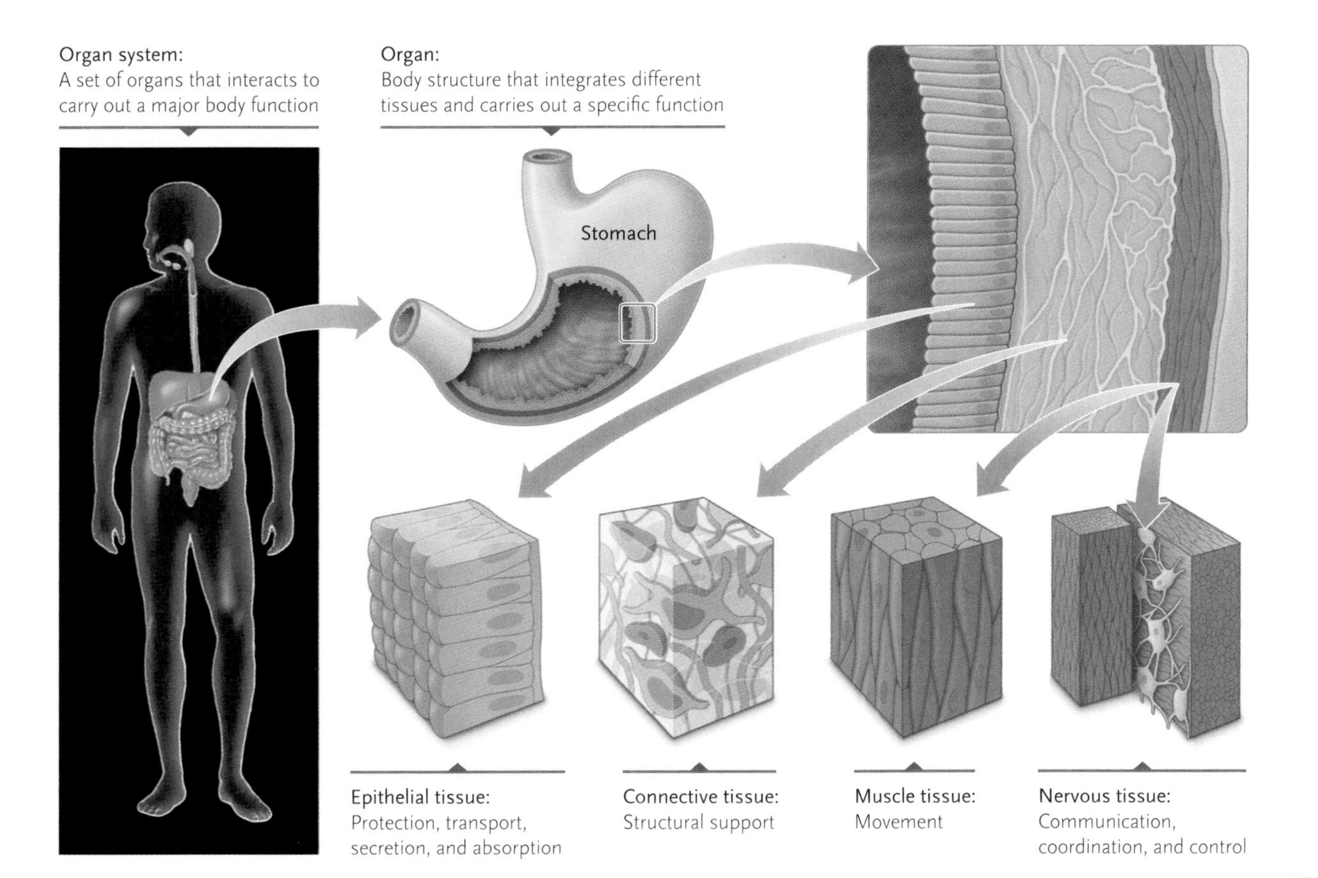

Figure 36.2 Organization of animal cells into tissues, organs, and organ systems.

36.2 Animal Tissues

Although the most complex animals may contain hundreds of distinct cell types, all can be classified into one of four basic tissue groups: *epithelial, connective, muscle,* and *nervous* (see Figure 36.2). Each tissue type is assembled from individual cells. The properties of those cells determine the structure and, therefore, the function of the tissue. More specifically, the structure and integrity of a tissue depend on the structure and organization of the cytoskeleton within the cell, the type and organization of the extracellular matrix (ECM) surrounding the cell, and the junctions holding cells together (see Section 5.5).

Junctions of various kinds link cells into tissues (see Figure 5.27). *Anchoring junctions* form buttonlike spots or belts that weld cells together. They are most abundant in tissues subject to stretching, such as skin and heart muscle. *Tight junctions* seal the spaces between cells, keeping molecules and even ions from leaking between cells. For example, tight junctions in the tissue lining the urinary bladder prevent waste molecules and ions from leaking out of the bladder into other body tissues. *Gap junctions* open channels between cells in the same tissue, allowing ions and small molecules to flow freely from one to another. For example, gap junctions between muscle cells help muscle tissue to function as a unit.

Let us now consider the structural and functional features that distinguish the four types of tissues, with primary emphasis on the forms they take in vertebrates.

Epithelial Tissue Forms Protective, Secretory, and Absorptive Coverings and Linings of Body Structures

Epithelial tissue (*epi* = over; *thele* = covering) consists of sheetlike layers of cells that are usually joined tightly together, with little ECM material between them **(Figure 36.3)**. Also called *epithelia* (singular, *epithelium*), these tissues cover body surfaces and the surfaces of internal organs, as well as line cavities and ducts within the body. They protect body surfaces from invasion by bacteria and viruses, and secrete or absorb substances. For example, the epithelium covering a fish's gill structures serves as a barrier to bacteria and viruses and exchanges oxygen, carbon dioxide, and ions with the aqueous environment. Some epithelia, such as those lining the capillaries of the circulatory system, act as filters, allowing ions and small molecules to leak from the blood into surrounding tissues while barring the passage of blood cells and large molecules such as proteins.

Because epithelia form coverings and linings, they have one free (or outer) surface, which may be exposed to water, air, or fluids within the body. In internal cavities and ducts, the free surface is often covered with *cilia,* which beat like oars to move fluids through the cavity or duct. The epithelium lining the oviducts in mammals, for example, is covered with

Figure 36.3
Structure of epithelial tissues.

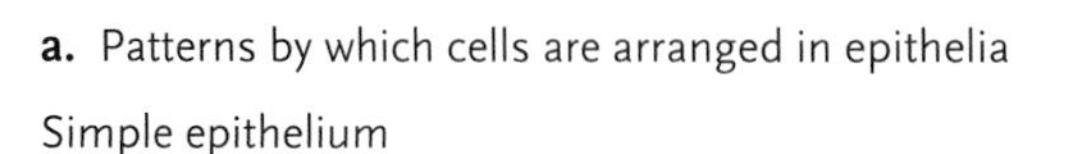

a. Patterns by which cells are arranged in epithelia

Simple epithelium

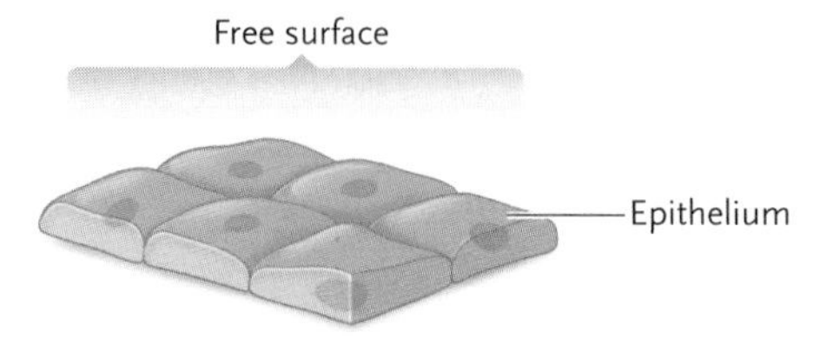

Stratified epithelium

b. The three common shapes of epithelial cells

Squamous epithelium

Ray Simmons/Photo Researchers, Inc.

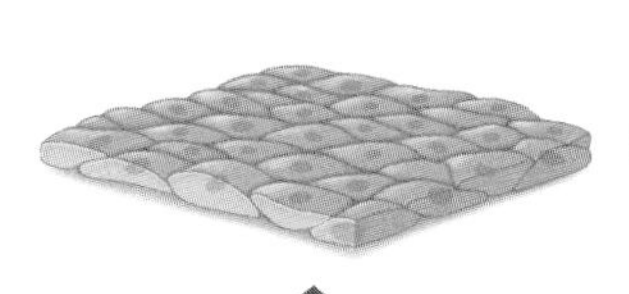

Description: Layer of flattened cells

Common locations: Walls of blood vessels; air sacs of lungs

Function: Diffusion

Cuboidal epithelium

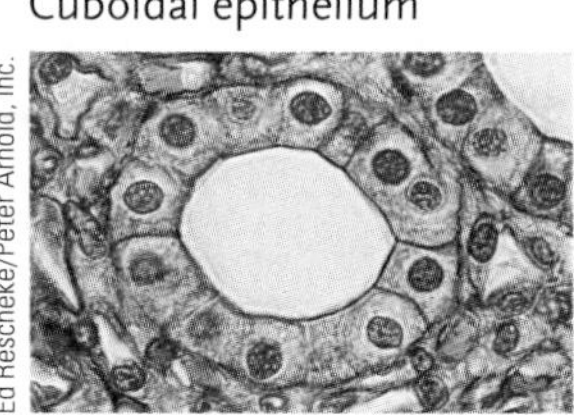

Ed Reschke/Peter Arnold, Inc.

Description: Layer of cubelike cells; free surface may have microvilli

Common locations: Glands and tubular parts of nephrons in kidneys

Function: Secretion, absorption

Columnar epithelium

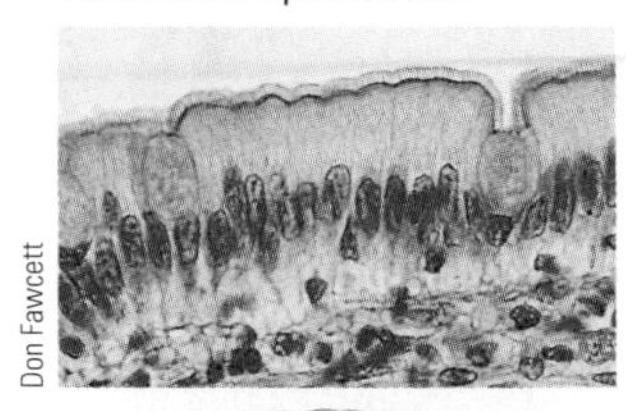

Don Fawcett

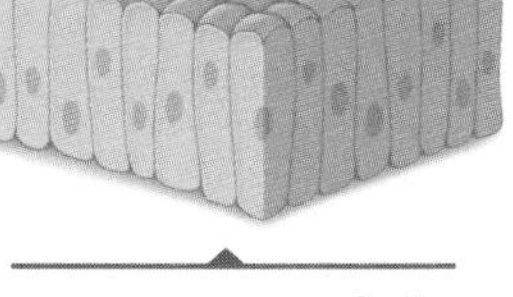

Description: Layer of tall, slender cells; free surface may have microvilli

Common locations: Lining of gut and respiratory tract

Function: Secretion, absorption

cilia that generate fluid currents to move eggs from the ovaries to the uterus. In some epithelia, including the lining of the small intestine, the free surface is crowded with *microvilli,* fingerlike extensions of the plasma membrane that increase the area available for secretion or absorption.

The inner surface of an epithelium adheres to a layer of glycoproteins secreted by the epithelial cells called the **basal lamina,** which fixes the epithelium to underlying tissues. The basal lamina is secreted by connective tissue cells immediately under the epithelium.

Epithelial Cell Structure. Epithelia are classified as *simple*—formed by a single layer of cells—or *stratified*—formed by multiple cell layers (see Figure 36.3a). The shapes of cells within an epithelium may be *squamous* (mosaic, flattened, and spread out), *cuboidal* (shaped roughly like dice or cubes), or *columnar* (elongated, with the long axis perpendicular to the epithelial layer; see Figure 36.3b). For example, the outer epithelium of mammalian skin is stratified and contains columnar, cuboidal, and squamous cells; the epithelium lining blood vessels is simple and squamous; and the intestinal epithelium is simple and columnar.

The cells of some epithelia, such as those forming the skin and the lining of the intestine, divide constantly to replace worn and dying cells. New cells are produced through division of stem cells in the basal (lowest) layer of the skin. *Stem cells* are undifferentiated (unspecialized) cells in the tissue that divide to produce more stem cells as well as cells that differentiate (that is, become specialized into one of the many cell types of the body). Stem cells are found both in adult organisms and in embryos. Besides the skin, adult stem cells are found in tissues of the brain, bone marrow, blood vessels, skeletal muscle, and liver. (*Insights from the Molecular Revolution* describes an effort to culture embryonic stem cells as a source of replacements for damaged tissues and organs.)

Glands Formed by Epithelia. Epithelia typically contain or give rise to cells that are specialized for secretion. Some of these secretory cells are scattered among nonsecretory cells within the epithelium. Others form structures called **glands,** which are derived from pockets of epithelium during embryonic development.

Some glands, called **exocrine glands** (*exo* = external; *crine* = secretion), remain connected to the epithelium by a duct, which empties their secretion at

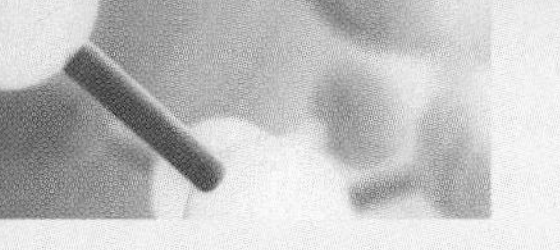

Insights from the Molecular Revolution

Cultured Stem Cells

Stem cells derived from human embryos or fetal tissue have the potential to develop into any tissue, but until recently biomedical researchers had no method for maintaining human stem cells indefinitely in cultures. A reliable supply of cultured stem cells is essential to the growth of tissues for research or for possible use in replacing damaged tissues and organs.

Just a few years ago, James A. Thompson and his coworkers at the University of Wisconsin developed a successful method for culturing stem cells. Their starting point was very early human embryos produced by fertilization of eggs in the test tube. After the embryos had grown for several days, 14 samples of cells were removed and cultured. The challenge was to maintain the cultured cells in the embryonic state and keep them from differentiating into specialized forms.

The researchers' strategy was to place the cells in culture dishes over a bed of mouse fibroblasts. Earlier work had shown that this technique allows mouse and nonhuman primate embryonic cells to survive outside the body, but it had failed to work with some mammalian species. Would it work with human cells?

In 5 of the 14 cultures, the human cells multiplied on the fibroblasts without differentiating as long as the cell masses never contained more than 50 to 100 cells. Larger masses had to be separated and placed in small numbers on a fresh layer of mouse fibroblasts. Using this technique, the five cell cultures were maintained for as long as 8 months in the laboratory with no signs of differentiation or deterioration. They could be frozen, stored, and returned to active cultures at will.

But did they still have full stem-cell function? The investigators ran several tests to answer this question. One experiment tested molecules on the surface of the cultured cells. Stem cells have characteristic surface molecules that change when the cells begin to differentiate into adult tissues. Antibodies against typical stem cell surface molecules all reacted with the cultured cells, indicating that the cells still had the characteristic molecules.

Another experiment tested for telomerase, an enzyme that helps maintain chromosomes at their normal length during rapid cycles of DNA replication and cell division. The enzyme becomes inactive in most cells as they differentiate into adult form. A standard test for the enzyme showed that it was present and remained fully active in the cultured cells.

In the final experiment, the researchers injected samples of the cells into mice to test whether the cultured cells could differentiate into a wide range of tissues. Within these mice, the cells grew into balls of tissue that included skin cells, gut epithelium, cartilage, bone, smooth and striated muscle, and nerve cells. The cultured cells seemed to be able to differentiate into adult tissues when stimulated to do so, and thus to have all the characteristics of stem cells.

The ability to culture stem cells opens many opportunities for future biological and medical discoveries. Observing the differentiation of stem cells into adult types should fill gaps in our knowledge of human development, which until now could be studied only in embryos. These studies may also give clues to the processes that produce birth defects and spontaneous abortion and could indicate means to correct these problems.

Further, if stem cells can be stimulated to differentiate into desired tissues and organs, they may provide an essentially unlimited supply of material for transplants. Many conditions, such as Parkinson disease and juvenile-onset diabetes, result from the death or malfunction of only one or a few cell types. Replacing defective cells from banks of cultured stem cells may be the key to curing the diseases.

These scientific and medical prospects do not provide answers to ethical questions about culturing stem cells derived from embryos. Such questions are the subject of intense scrutiny and debate in the U.S. Congress and among scientists, religious authorities, and the general public. Hopefully, a balance will be found between concerns about the use of human embryonic cells in research and the prospects for significant improvements in human health and scientific knowledge.

the epithelial surface. Exocrine secretions include mucus, saliva, digestive enzymes, sweat, earwax, oils, milk, and venom (**Figure 36.4a** shows an exocrine gland in the skin of a poisonous tree frog). Other glands, called **endocrine glands** (*endo* = inside), become suspended in connective tissue underlying the epithelium, with no ducts leading to the epithelial surface. These ductless glands, such as the pituitary gland, adrenal gland, and thyroid gland **(Figure 36.4b),** release their products—called hormones—directly into the interstitial fluid, to be picked up and distributed by the circulatory system.

Some glands act as both exocrine glands and endocrine glands. The pancreas, for instance, has an exocrine function of secreting pancreatic juice through a duct into the small intestine where it plays an important role in food digestion, and an endocrine function of secreting the hormones insulin and glucagon into the bloodstream to help regulate glucose levels in the blood.

Connective Tissue Supports Other Body Tissues

Most animal body structures contain one or more types of **connective tissue.** Connective tissues support other body tissues, transmit mechanical and other forces, and in some cases act as filters. They consist of cells

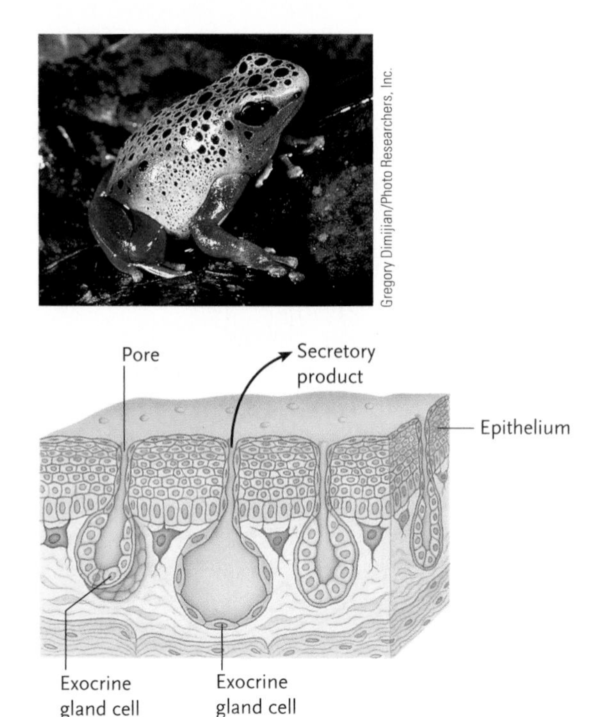

a. Examples of exocrine glands: The mucus- and poison-secreting glands in the skin of a blue poison frog

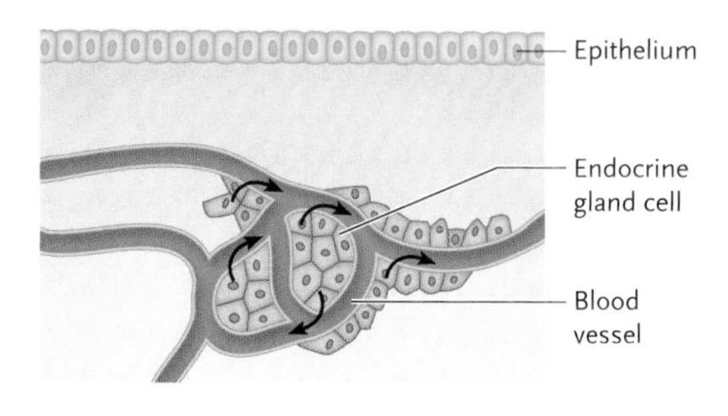

b. Example of an endocrine gland: The thyroid gland, which secretes hormones that regulate the rate of metabolism and other body functions

Figure 36.4
Exocrine and endocrine glands. The poison secreted by the blue poison frog *(Dendrobates azureus)* is one of the most lethal glandular secretions known.

that form networks or layers in and around body structures and that are separated by nonliving material, specifically the ECM secreted by the cells of the tissue (see Section 5.5). Many forms of connective tissue have more ECM material (both by weight and by volume) than cellular material.

The mechanical properties of a connective tissue depend on the type and quantity of its ECM. The consistency of the ECM ranges from fluid (as in blood and lymph), through soft and firm gels (as in tendons), to the hard and crystalline (as in bone). In most connective tissues, the ECM consists primarily of the fibrous glycoprotein **collagen** embedded in a network of proteoglycans—glycoproteins that are very rich in carbohydrates. In bone, the glycoprotein network surrounding the collagen is impregnated with mineral deposits that produce a hard, yet still somewhat elastic, structure. Another class of glycoproteins, **fibronectin,** aids in the attachment of cells to the ECM and helps hold the cells in position.

In some connective tissues another rubbery protein, **elastin,** adds elasticity to the ECM—it is able to return to its original shape after being stretched, bent, or compressed. Elastin fibers, for example, help the skin return to its original shape when pulled or stretched, and give the lungs the elasticity required for their alternating inflation and deflation.

Vertebrates have six major types of connective tissue: *loose connective tissue, fibrous connective tissue, cartilage, bone, adipose tissue,* and *blood.* Each type has a characteristic function correlated with its structure **(Figure 36.5).**

Loose Connective Tissue. **Loose connective tissue** consists of sparsely distributed cells surrounded by a more or less open network of collagen and other glycoprotein fibers (see Figure 36.5a). The cells, called **fibroblasts,** secrete most of the collagen and other proteins in this connective tissue.

Loose connective tissues support epithelia and form a corsetlike band around blood vessels, nerves, and some internal organs; they also reinforce deeper layers of the skin. Sheets of loose connective tissue, covered on both surfaces with epithelial cells, form the **mesenteries,** which hold the abdominal organs in place and provide lubricated, smooth surfaces that prevent chafing or abrasion between adjacent structures as the body moves.

Fibrous Connective Tissue. In **fibrous connective tissue,** fibroblasts are sparsely distributed among dense masses of collagen and elastin fibers that are lined up in highly ordered, parallel bundles (see Figure 36.5b). The parallel arrangement produces maximum tensile strength and elasticity. Examples include **tendons,** which attach muscles to bones, and **ligaments,** which connect bones to each other at a joint. The cornea of the eye is a transparent fibrous connective tissue formed from highly ordered collagen molecules.

Figure 36.5 The six major types of connective tissues.

a. Loose connective tissue

Ed Reschke

Collagen fiber
Fibroblast
Elastin fiber

Description: Fibroblasts and other cells surrounded by collagen and elastin fibers forming a glycoprotein matrix

Common locations: Under the skin and most epithelia

Function: Support, elasticity, diffusion

b. Fibrous connective tissue

Ed Reschke

Collagen fibers
Fibroblast

Description: Long rows of fibroblasts surrounded by collagen and elastin fibers in parallel bundles with a dense extracellular matrix

Common locations: Tendons, ligaments

Function: Strength, elasticity

c. Cartilage

Fred Hossler/Visuals Unlimited

Collagen fibers embedded in an elastic matrix
Chondrocyte

Description: Chondrocytes embedded in a pliable, solid matrix of collagen and chondroitin sulfate

Common locations: Ends of long bones, nose, parts of airways, skeleton of vertebrate embryos

Function: Support, flexibility, low-friction surface for joint movement

d. Bone tissue

Ed Reschke

Fine canals
Central canal containing blood vessel
Osteocytes

Description: Osteocytes in a matrix of collagen and glycoproteins hardened with hydroxyapatite

Common locations: Bones of vertebrate skeleton

Function: Movement, support, protection

e. Adipose tissue

f. Blood

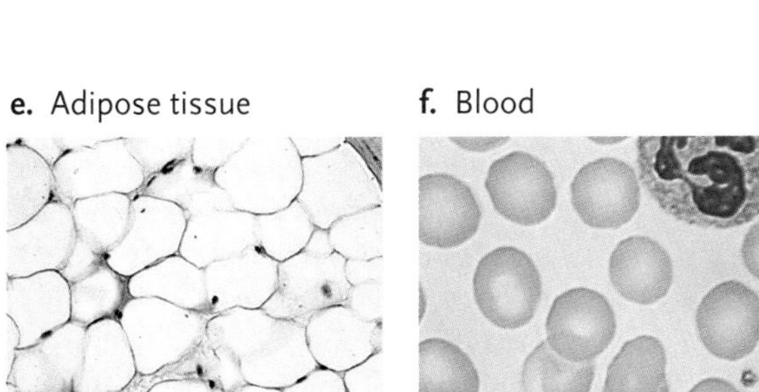

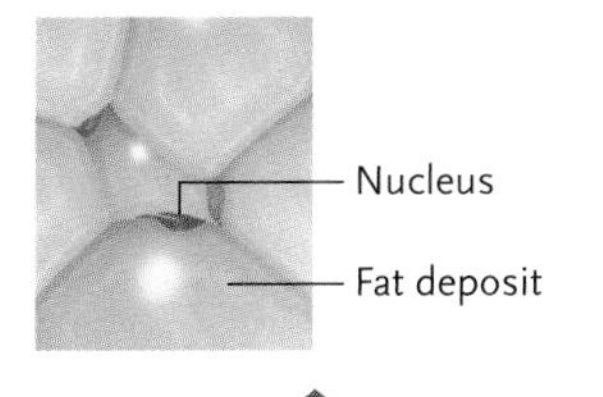

Description: Large, tightly packed adipocytes with little extracellular matrix

Common locations: Under skin; around heart, kidneys

Function: Energy reserves, insulation, padding

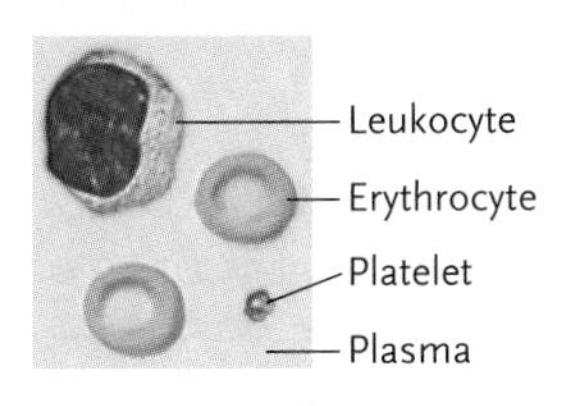

Description: Leukocytes, erythrocytes, and platelets suspended in a plasma matrix

Common locations: Circulatory system

Function: Transport of substances

Cartilage. **Cartilage** consists of sparsely distributed cells called **chondrocytes**, surrounded by networks of collagen fibers embedded in a tough but elastic matrix of the glycoprotein *chondroitin sulfate* (see Figure 36.5c). Elastin is also present in some forms of cartilage.

The elasticity of cartilage allows it to resist compression and stay resilient, like a piece of rubber. Bending your ear or pushing the tip of your nose, which are supported by a core of cartilage, gives a good idea of the flexible nature of this tissue. In humans, cartilage also supports the larynx, trachea, and smaller air passages in the lungs. It forms the disks cushioning the vertebrae in the spinal column and the smooth, slippery capsules around the ends of bones in joints such as the hip and knee. Cartilage also serves as a precursor to bone during embryonic development; in sharks and rays and their relatives, almost the entire skeleton remains as cartilage in adults.

Bone. The densest form of connective tissue, **bone** forms the skeleton, which supports the body, protects softer body structures such as the brain, and contributes to body movements.

Mature bone consists primarily of cells called **osteocytes** (*osteon* = bone) embedded in an ECM containing collagen fibers and glycoproteins impregnated with *hydroxyapatite,* a calcium-phosphate mineral (see Figure 36.5d). The collagen fibers give bone tensile strength and elasticity; the hydroxyapatite resists compression and allows bones to support body weight. Cells called **osteoblasts** (*blast* = bud or sprout) produce the collagen and mineral of bone—as much as 85% of the weight of bone is mineral deposits. Osteocytes, in fact, are osteoblasts that have become trapped and surrounded by the bone materials they themselves produce. **Osteoclasts** (*clast* = break) remove the minerals and recycle them through the bloodstream. Bone is not a stable tissue; it is reshaped continuously by the bone-building osteoblasts and the bone-degrading osteoclasts.

Although bones appear superficially to be solid, they are actually porous structures, consisting of a system of microscopic spaces and canals. The structural unit of bone is the **osteon.** It consists of a minute central canal surrounded by osteocytes embedded in concentric layers of mineral matter (see Figure 36.5d). A blood vessel and extensions of nerve cells run through the central canal, which is connected to the spaces containing cells by very fine, radiating canals filled with interstitial fluid. The blood vessels supply nutrients to the cells with which the bone is built, and the nerve cells hook up the bone cells to the body's nervous system.

Adipose Tissue. The connective tissue called **adipose tissue** mostly contains large, densely clustered cells called *adipocytes* that are specialized for fat storage (see Figure 36.5e). It has little ECM. Adipose tissue also cushions the body and, in mammals, forms an especially important insulating layer under the skin.

The animal body stores limited amounts of carbohydrates, primarily in muscle and liver cells. Excess carbohydrates are converted into the fats stored in adipocytes. The storage of chemical energy as fats offers animals a weight advantage. For example, the average human would weigh about 45 kg (100 pounds) more if the same amount of chemical energy was stored as carbohydrates instead of fats. Adipose tissue is richly supplied with blood vessels, which move fats or their components to and from adipocytes.

Blood. **Blood** (see Figure 36.5f) is considered a connective tissue because its cells are suspended in a fluid ECM, plasma. The straw-colored **plasma** is a solution of proteins, nutrient molecules, ions, and gases.

Blood contains two primary cell types, **erythrocytes** (red blood cells; *erythros* = red) and **leukocytes** (white blood cells; *leukos* = white). Erythrocytes are packed with hemoglobin, a protein that can bind and transport oxygen. There are several types of leukocytes—all help to protect the body against invading viruses, bacteria, and other disease-causing agents. The blood plasma also contains **platelets,** membrane-bound fragments of specialized blood cells, which take part in the reactions that seal wounds with blood clots.

Blood is the major transport vehicle of the body. It carries oxygen and nutrients to body cells, removes wastes and by-products such as carbon dioxide, and maintains the internal fluid environment, including the osmotic balance between cells and the interstitial fluid. Blood also transports hormones and other signal molecules that coordinate body responses. (The components and roles of blood are described in Chapter 42.)

Muscle Tissue Produces the Force for Body Movements

Muscle tissue consists of cells that have the ability to contract (shorten). The contractions, which depend on the interaction of two proteins—*actin* and *myosin*—move body limbs and other structures, pump the blood, and produce a squeezing pressure in organs such as the intestine and uterus. Three types of muscle tissue, *skeletal, cardiac,* and *smooth,* produce body movements in vertebrates **(Figure 36.6).** In all types of muscle tissue, the cells are densely packed, leaving little room for ECM.

Skeletal Muscle. **Skeletal muscle** is so called because most muscles of this type are attached by tendons to the skeleton. Skeletal muscle cells are also called **muscle fibers** because each is an elongated cylinder (see Figure 36.6a). These cells contain many nuclei and are packed with actin and myosin molecules arranged in highly ordered, parallel units that give the tissue a banded or striated appearance when viewed under a microscope. Muscle fibers packed side by side into parallel bundles surrounded by sheaths of connective tissue form many body muscles, such as the biceps.

Skeletal muscle contracts in response to signals carried by the nervous system. The contractions of skeletal muscles, which are characteristically rapid and powerful, move body parts and maintain posture. The contractions also release heat as a by-product of cellular metabolism. This heat helps mammals, birds, and some other vertebrates maintain their body temperatures when environmental temperatures fall. (Skeletal muscle is discussed further in Chapter 41.)

Cardiac Muscle. **Cardiac muscle** is the contractile tissue of the heart (see Figure 36.6b). Cardiac muscle has a striated appearance because it contains actin and myosin molecules arranged like those in skeletal muscle. However, cardiac muscle cells are short and branched, with each cell connecting to several neighboring cells; the joining point between two such cells is called an *intercalated disk.* Cardiac muscle cells thus form an interlinked network, which is stabilized by anchoring junctions and gap junctions. This network makes heart muscle contract in all directions, producing a squeezing or pumping action rather than the

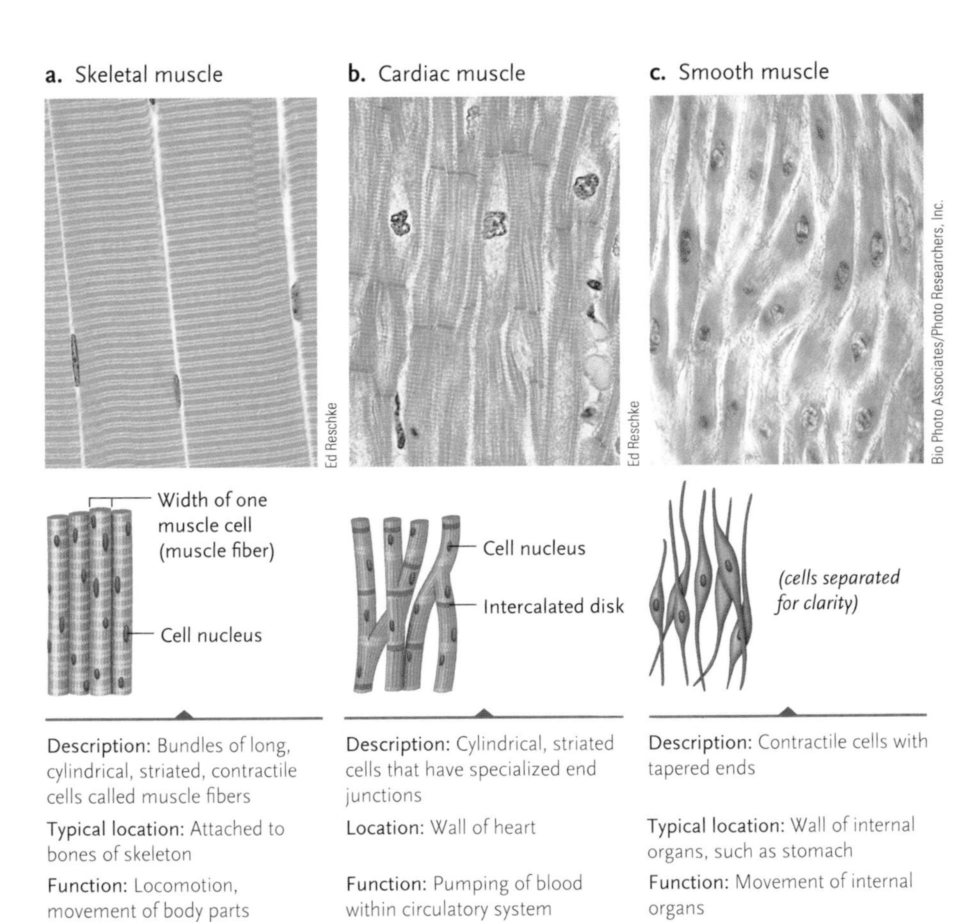
a. Skeletal muscle
b. Cardiac muscle
c. Smooth muscle
Ed Reschke
Ed Reschke
Bio Photo Associates/Photo Researchers, Inc.
Width of one muscle cell (muscle fiber)
Cell nucleus
Cell nucleus
Intercalated disk
(cells separated for clarity)
Description: Bundles of long, cylindrical, striated, contractile cells called muscle fibers
Typical location: Attached to bones of skeleton
Function: Locomotion, movement of body parts
Description: Cylindrical, striated cells that have specialized end junctions
Location: Wall of heart
Function: Pumping of blood within circulatory system
Description: Contractile cells with tapered ends
Typical location: Wall of internal organs, such as stomach
Function: Movement of internal organs

Figure 36.6
Structure of skeletal, cardiac, and smooth muscle.

lengthwise, unidirectional contraction characteristic of skeletal muscle.

Smooth Muscle. **Smooth muscle** is found in the walls of tubes and cavities in the body, including blood vessels, the stomach and intestine, the bladder, and the uterus. Smooth muscle cells are relatively small and spindle-shaped (pointed at both ends), and their actin and myosin molecules are arranged in a loose network rather than in bundles (see Figure 36.6c). This loose network makes the cells appear smooth rather than striated when viewed under a microscope. Smooth muscle cells are connected by gap junctions and enclosed in a mesh of connective tissue. The gap junctions transmit ions that make smooth muscles contract as a unit, typically producing a squeezing motion. Although smooth muscle contracts more slowly than skeletal and cardiac muscle do, its contractions can be maintained at steady levels for a much longer time. These contractions move and mix the stomach and intestinal contents, constrict blood vessels, and push the infant out of the uterus during childbirth.

Nervous Tissue Receives, Integrates, and Transmits Information

Nervous tissue contains cells called **neurons** (also called *nerve cells*) that serve as lines of communication and control between body parts. Billions of neurons are packed into the human brain; others extend throughout the body. Nervous tissue also contains **glial cells** (*glia* = glue), which physically support and provide nutrients to neurons, provide electrical insulation between them, and scavenge cellular debris and foreign matter.

A neuron consists of a *cell body,* which houses the nucleus and organelles, and two types of cell extensions, dendrites and axons **(Figure 36.7).** *Dendrites* receive chemical signals from other neurons or from body cells of other types, and convert them into an electrical signal that is transmitted to the cell body of the receiving neuron. Dendrites are usually highly branched. *Axons* conduct electrical signals away from the cell body to the axon terminals, or endings. At their terminals, axons convert the electrical signal to a chemical signal that stimulates a response in nearby muscle cells, gland cells, or other neurons. Axons are usually unbranched except at their terminals. Depending on the type of neuron and its location in the body, its axon may extend from a few micrometers or millimeters to more than a meter. (Neurons and their organization in body structures are discussed further in Chapters 37, 38, and 39.)

All four major tissue types—epithelial, connective, muscle, and nervous—combine to form the organs and organ systems of animals. The next section depicts the major organs and organ systems of vertebrates, and outlines their main tasks.

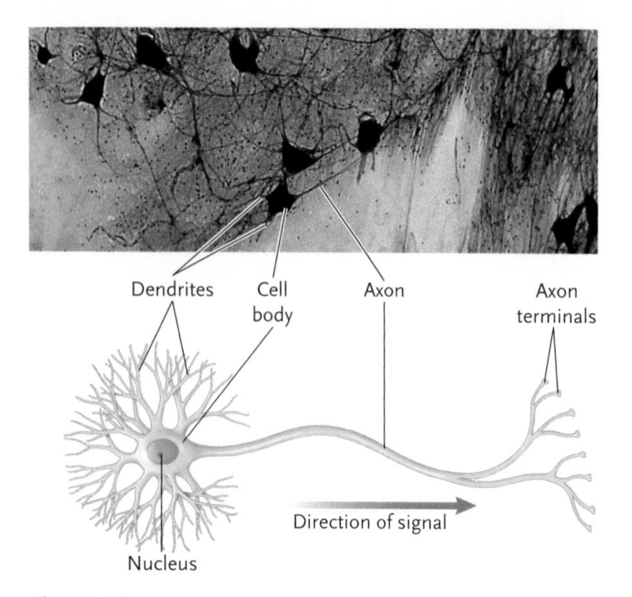

Figure 36.7
Neurons and their structure. The micrograph shows a network of motor neurons, which relay signals from the brain or spinal cord to muscles and glands. (Micrograph: Lennart Nilsson from Behold Man, © 1974 Albert Bonniers Forlag and Little, Brown and Company, Boston.)

STUDY BREAK

1. Distinguish between exocrine and endocrine glands. What is the tissue type of each of these glands?
2. What are the six major types of connective tissue in vertebrates?
3. What three types of muscle tissue produce body movements?

36.3 Coordination of Tissues in Organs and Organ Systems

Organs and Organ Systems Function Together to Enable an Animal to Survive

In the tissues, organs, and organ systems of an animal, each cell engages in the basic metabolic activities that ensure its own survival, and performs one or more functions of the system to which it belongs. All verte-

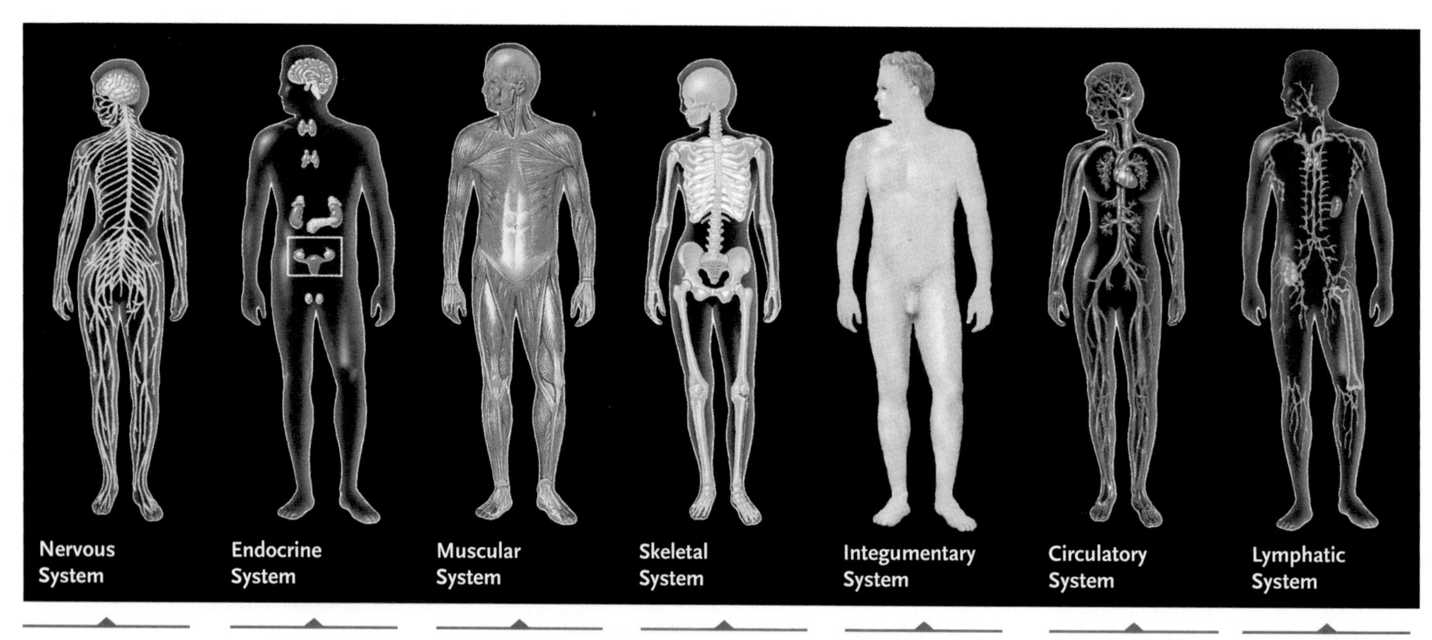

	Nervous System	Endocrine System	Muscular System	Skeletal System	Integumentary System	Circulatory System	Lymphatic System
Main organs:	Brain, spinal cord, peripheral nerves, sensory organs	Pituitary, thyroid, adrenal, pancreas, and other hormone-secreting glands	Skeletal, cardiac, and smooth muscle	Bones, tendons, ligaments, cartilage	Skin, sweat glands, hair, nails	Heart, blood vessels, blood	Lymph nodes, lymph ducts, spleen, thymus
Main functions:	Principal regulatory system; monitors changes in internal and external environments and formulates compensatory responses; coordinates body activities	Regulates and coordinates body activities through secretion of hormones	Moves body parts; helps run bodily functions; generates heat	Supports and protects body parts; provides leverage for body movements	Covers external body surfaces and protects against injury and infection; helps regulate water content and body temperature	Distributes water, nutrients, oxygen, hormones, and other substances throughout body and carries away carbon dioxide and other metabolic wastes; helps stabilize internal temperature and pH	Returns excess fluid to the blood; defends body against invading viruses, bacteria, fungi, and other pathogens as part of immune system

brates (and most invertebrates) have eleven major organ systems, which are summarized in **Figure 36.8,** and discussed in the rest of this unit of the book.

The functions of all these organ systems are coordinated and integrated to accomplish collectively a series of tasks that are vital to all animals, whether a flatworm, a salmon, a meerkat, or a human. These functions include:

1. Acquiring nutrients and other required substances such as oxygen, coordinating their processing, distributing them throughout the body, and disposing of wastes.
2. Synthesizing the protein, carbohydrate, lipid, and nucleic acid molecules required for body structure and function.
3. Sensing and responding to changes in the environment, such as temperature, pH, and ion concentrations.
4. Protecting the body against injury or attack from other animals, and from viruses, bacteria, and other disease-causing agents.
5. Reproducing and, in many instances, nourishing and protecting offspring through their early growth and development.

Together these tasks maintain homeostasis, preserving the internal environment required for survival of the body. Homeostasis is the topic of the next section.

Study Break

What are the major functions of each of the eleven organ systems of the vertebrate body?

36.4 Homeostasis

Homeostasis is the process by which animals maintain their internal environment in a steady state (constant level) or between narrow limits. Homeostasis depends on a number of the body's organ systems, with the

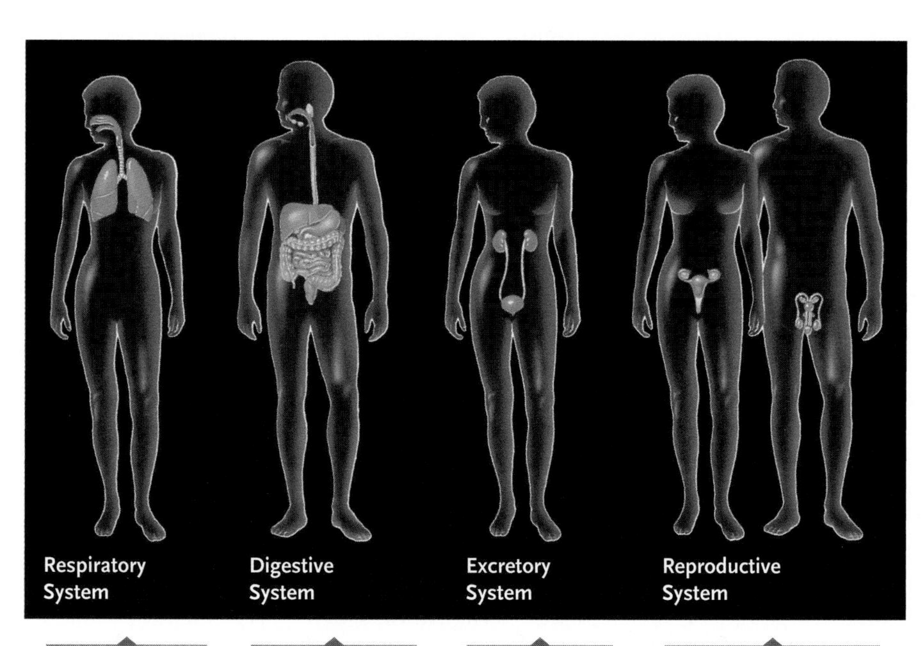

Respiratory System	Digestive System	Excretory System	Reproductive System
Main organs: Lungs, diaphragm, trachea, and other airways	Main organs: Pharynx, esophagus, stomach, intestines, liver, pancreas, rectum, anus	Main organs: Kidneys, bladder, ureter, urethra	Main organs: *Female*: ovaries, oviducts, uterus, vagina, mammary glands *Male*: testes, sperm ducts, accessory glands, penis
Main functions: Exchanges gases with the environment, including uptake of oxygen and release of carbon dioxide	Main functions: Converts ingested matter into molecules and ions that can be absorbed into body; eliminates undigested matter; helps regulate water content	Main functions: Removes and eliminates excess water, ions, and metabolic wastes from body; helps regulate internal osmotic balance and pH	Main functions: Maintains the sexual characteristics and passes on genes to the next generation

Figure 36.8
Organ systems of the human body. The immune system, which is primarily a cellular system, is not shown.

nervous system and endocrine system being the most important. For example, blood pH is controlled by both the nervous and endocrine systems, blood glucose by the endocrine system, internal temperature by the nervous and endocrine systems, and oxygen and carbon dioxide concentrations by the nervous system.

Although the *stasis* part of homeostasis might suggest a static, unchanging process, homeostasis is actually a dynamic process, in which internal adjustments are made continuously to compensate for environmental changes. For example, internal adjustments are needed for homeostasis during exercise or hibernation. The factors controlled by homeostatic mechanisms all require energy that must be constantly acquired from the external environment.

Homeostasis Is Accomplished by Negative Feedback Mechanisms

The primary mechanism of homeostasis is **negative feedback,** in which a stimulus—a change in the external or internal environment—triggers a response that compensates for the environmental change **(Figure 36.9).** Homeostatic mechanisms typically include three elements: a sensor, an integrator, and an effector. The **sensor** consists of tissues or organs that detect a change in external or internal factors such as pH, temperature, or the concentration of a molecule such as glucose. The **integrator** is a control center that compares the detected environmental change with a **set point,** the level at which the condition controlled by the pathway is to be maintained. The **effector** is a system, activated by the integrator, that returns the condition to the set point if it has strayed away. In most animals, the integrator is part of the central nervous system or endocrine system, while effectors may include parts of essentially any body tissue or organ.

The Thermostat as a Negative Feedback Mechanism. The concept of negative feedback may be most familiar in systems designed by human engineers. The thermostat maintaining temperature at a chosen level in a house provides an example. A sensor within the thermostat measures the temperature. If the room temperature changes more than a degree or so from the set point—the temperature that you set in the thermostat—an integrator circuit in the thermostat activates an effector that returns the room temperature to the set point. If the temperature has fallen below the set point, the effector is the furnace, which adds heat to the house until the temperature rises to the set point. If the temperature has risen above the set point, the effector is the air conditioner, which removes heat from the room until the temperature falls to the set point.

Negative Feedback Mechanisms in Animals. Mammals and birds—warm-blooded vertebrates—also have a homeostatic mechanism that maintains body temperature within a relatively narrow range around a set point. The integrator (thermostat) for this mechanism is located in a brain center called the *hypothalamus.* A group of neurons in the hypothalamus detects changes in the temperature of the brain and the rest of the body, and compares it with a set point. For humans, the set point has a relatively narrow range centered at about 37°C.

One or more effectors are activated in humans if the temperature varies beyond the limits of the set point. If the temperature falls below the lower limit, the hypothalamus activates effectors that constrict the blood vessels in the skin. The reduction in blood flow means that less heat is conducted from the blood through the skin to the environment; in short, heat loss from the skin is reduced. Other effectors may induce shivering, a physical mechanism to generate body heat. Also, integrating neurons in the brain, stimulated by signals from the hypothalamus, make us consciously sense a chill, which we may counteract behaviorally by putting on more clothes or moving to a warmer area.

Conversely, if blood temperature rises above the set point, the hypothalamus triggers effectors that dilate the blood vessels in the skin, increasing blood flow and heat loss from the skin. Other effectors induce sweating, which cools the skin and the blood flowing through it as the sweat evaporates. And again, through integrating neurons in the brain, we may consciously sense being overheated, which we may counteract by shedding clothes, moving to a cooler location, or taking a dip in a pool.

Sometimes the temperature set point changes, and the negative feedback mechanisms then operate to maintain body temperature at the new set point. For example, if you become infected by certain viruses and bacteria, the temperature set point increases to a higher level, producing a fever to help overcome the infection. Once the infection is combated, the set point is readjusted down again to its normal level.

All other mammals have similar homeostatic mechanisms that maintain or adjust body temperature. Dogs and birds pant to release heat from their bodies **(Figure 36.10)** and shiver to increase internal heat production. Many terrestrial animals

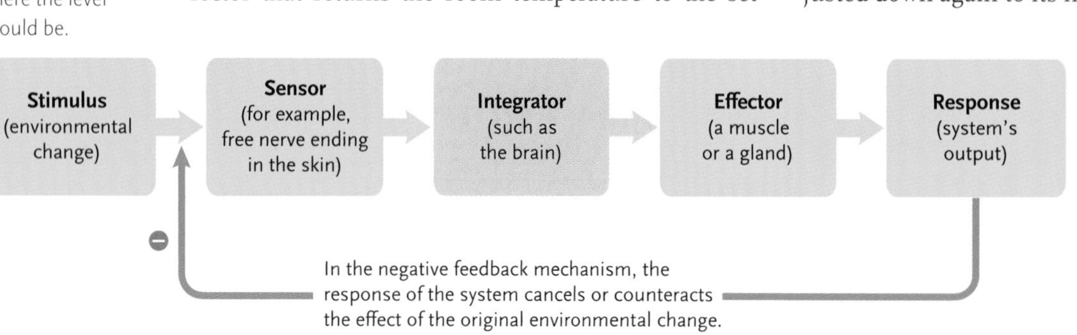

Figure 36.9 Components of a negative feedback mechanism maintaining homeostasis. The integrator coordinates a response by comparing the level of an environmental condition with a set point that indicates where the level should be.

Stimulus
The husky is active on a hot, dry day, and its body surface temperature rises.

Sensors
Neurons in the hypothalamus detect the increase in brain and body temperature.

Integrator
The network of neurons compares brain and body temperature against a set point.

Many **Effectors** carry out specific responses:

Skeletal muscles
Husky starts to pant, increasing heat loss by evaporation of water from lungs, throat, mouth, and tongue.

Smooth muscle in blood vessels
Blood carrying metabolically generated heat circulates through lungs, throat, mouth, and tongue.

Salivary glands
Secretions from glands increase evaporation of water from tongue, mouth, and throat.

Response
Temperature of brain and body decreases.

Fred Bruemmer

Figure 36.10 Homeostatic mechanisms maintaining the body temperature of a husky when environmental temperatures are high.

enter or splash water over their bodies to cool off. Also, recall from the beginning of the chapter how meerkats use behavioral mechanisms to regulate their body temperature.

Whereas mammals regulate their internal body temperature within a narrow range around a set point, certain other vertebrates regulate over a broader range. These vertebrates use other, less precise negative feedback mechanisms for their temperature regulation. Snakes and lizards, for example, respond behaviorally to compensate for variations in environmental temperatures. They may absorb heat by basking on sunny rocks in the cool early morning and move to cooler, shaded spots in the heat of the afternoon. Some fishes, such as the tuna, generate enough heat by contraction of the swimming muscles to maintain body temperature well above the temperature of the surrounding water.

Some invertebrates, such as dragonflies, moths, and butterflies, use muscular contractions equivalent to shivering when their body temperature falls below the level required for flight. The shivering contractions warm the muscles to flying temperature. All of these physiological and behavioral responses depend on negative feedback mechanisms involving sensors, integrators, and effectors.

Animals Also Have Positive Feedback Mechanisms That Do Not Result in Homeostasis

Under certain circumstances, animals respond to a change in internal or external environmental condition by a **positive feedback** mechanism that intensifies or adds to the change. Such mechanisms, with some exceptions, do not result in homeostasis. They operate when the animal is responding to life-threatening conditions (an attack, for instance), or as part of reproductive processes.

The birth process in mammals is a prime example. During human childbirth, initial contractions of the uterus push the head of the fetus against the cervix, the opening of the uterus into the vagina. The pushing causes the cervix to stretch. Sensors that detect the stretching signal the hypothalamus to release a hormone, oxytocin, from the pituitary gland. Oxytocin increases the uterine contractions, intensifying the squeezing pressure on the fetus and further stretching the cervix. The stretching results in more oxytocin release and stronger uterine contraction, repeating the positive feedback circuit and increasing the squeezing pressure until the fetus is pushed entirely out of the uterus.

Because positive feedback mechanisms such as the one triggering childbirth do not result in homeostasis, they occur less commonly than negative feedback in animals. They also operate as part of larger, more inclusive negative feedback mechanisms that ultimately shut off the positive feedback pathway and return conditions to normal limits.

In conclusion, we learned in this chapter about the various tissues and organ systems of the body, and of the involvement of organ systems in homeostasis. Next, we begin a series of chapters describing the organ systems in detail, starting with the nervous system.

Study Break

What are the components of a negative feedback mechanism that results in homeostasis?

Unanswered Questions

Why do so many strokes and heart attacks occur in the morning?

Stroke and heart attack occur most frequently at a particular time of the day—in the morning—exhibiting a profound circadian variation. Circadian rhythms are generated through a discrete set of molecular interactions including the Bmal1, Clock, NPAS2, Cry, and Per proteins. We have recently shown that the biological clock is expressed and oscillating in blood vessels; however, it remains unknown if this "vascular clock" acts to modulate the function of blood vessels. Moreover, if the vascular clock does influence normal vascular function, might a broken clock contribute to the onset of heart attack and stroke? Current research in my laboratory and others is addressing these questions.

How could a vascular clock influence vascular function?

Circadian rhythms are seen in endothelial function, blood pressure, vascular resistance, and blood flow. Could the circadian clock play a role in the regulation of blood vessel homeostasis? What targets within vascular cells might the clock control? One possibility is that the vascular clock may act to regulate production of signaling molecules in endothelial cells, which comprise the inner lining of blood vessels. In addition, direct actions on vascular smooth muscle cells, which contain the nerve and muscle elements critical for constriction and relaxation of blood vessels, may also be under circadian control. To assess these questions, the use of mice with genetic disruption of the circadian clock (knockout or mutant mice) has been invaluable. Garret FitzGerald and colleagues at the University of Pennsylvania demonstrated that *Bmal1* knockout mice and *Clock* mutant mice lack circadian rhythms in blood pressure, in part due to a blunted sympathetic drive. However, the contribution of parasympathetic outflow remains unknown. Might the clock directly induce or inhibit transcription of genes important to vascular function? One approach is to introduce molecular clock components into cultured cells and to assess promoter regulation of a gene of interest. This has already proved useful in identifying the PAI-1 (plasminogen activator inhibitor-1) protein as a target of the molecular clock. PAI-1 inhibits plasminogen activator, an enzyme that breaks up clots. So PAI-1 promotes clot formation. Future studies implementing tissue-specific knockout mice of molecular clock components will determine more directly how and where these targets are regulated by the circadian clock.

Might a dysfunctional vascular clock contribute to chronic vascular disease?

Chronic impairments in blood pressure and blood flow rhythms are sensed by blood vessels, which causes them to respond by changing architecture through a process called vascular remodeling. Vascular remodeling is an extremely intensive and active area of vascular biology research that is important to understanding the progression to blood vessel disease. Using models of blood vessel ligation in mice with a disrupted molecular clock, we are currently assessing the impact of the biological clock on vascular remodeling.

Research to address the role of the vascular clock may ultimately change the way we understand and treat arteriosclerosis, hypertension, and heart attack.

R. Daniel Rudic is an assistant professor in the Department of Pharmacology and Toxicology at the Medical College of Georgia. To learn more about his research on circadian rhythms and vascular biology, go to http://www.mcg.edu/som/phmtox/RudicLab/index.asp.

Review

Go to ThomsonNOW at www.thomsonedu.com/login to access quizzing, animations, exercises, articles, and personalized homework help.

36.1 Organization of the Animal Body

- In most animals, cells are specialized and organized into tissues, tissues into organs, and organs into organ systems. A tissue is a group of cells with the same structure and function, working as a unit to carry out one or more activities. An organ is an assembly of tissues integrated into a structure that carries out a specific function. An organ system is a group of organs that carry out related steps in a major physiological process.

36.2 Animal Tissues

- Animal tissues are classified as epithelial, connective, muscle, or nervous (Figure 36.2). The properties of the cells of these tissues determine the structures and functions of the tissues.
- Various kinds of junctions link cells in a tissue. Anchoring junctions "weld" cells together. Tight junctions seal the cells into a leak-proof layer. Gap junctions form direct avenues of communication between the cytoplasm of adjacent cells in the same tissue.
- Epithelial tissue consists of sheetlike layers of cells that cover body surfaces and the surfaces of internal organs, and line cavities and ducts within the body (Figure 36.3).
- Glands are secretory structures derived from epithelia. They may be exocrine (connected to an epithelium by a duct that empties on the epithelial surface) or endocrine (ductless, with no direct connection to an epithelium) (Figure 36.4).
- Connective tissue consists of cell networks or layers and an extracellular matrix (ECM). It supports other body tissues, transmits mechanical and other forces, and in some cases acts as a filter (Figure 36.5).
- Loose connective tissue consists of sparsely distributed fibroblasts surrounded by an open network of collagen and other glycoproteins. It supports epithelia and organs of the body and forms a covering around blood vessels, nerves, and some internal organs.
- Fibrous connective tissue contains sparsely distributed fibroblasts in a matrix of densely packed, parallel bundles of collagen

and elastin fibers. It forms high tensile-strength structures such as tendons and ligaments.

- Cartilage consists of sparsely distributed chondrocytes surrounded by a network of collagen fibers embedded in a tough but highly elastic matrix of branched glycoproteins. Cartilage provides support, flexibility, and a low-friction surface for joint movement.
- In bone, osteocytes are embedded in a collagen matrix hardened by mineral deposits. Osteoblasts secrete collagen and minerals for the ECM; osteoclasts remove the minerals and recycle them into the bloodstream.
- Adipose tissue consists of cells specialized for fat storage. It also cushions and rounds out the body and provides an insulating layer under the skin.
- Blood consists of a fluid matrix, the plasma, in which erythrocytes and leukocytes are suspended. The erythrocytes carry oxygen to body cells; the leukocytes produce antibodies and initiate the immune response against disease-causing agents.
- Muscle tissue contains cells that have the ability to contract forcibly (Figure 36.6). Skeletal muscle, containing long cells called muscle fibers, moves body parts and maintains posture.
- Cardiac muscle, which contains short contractile cells with a branched structure, forms the heart.
- Smooth muscle consists of spindle-shaped contractile cells that form layers surrounding body cavities and ducts.
- Nervous tissue contains neurons and glial cells. Neurons communicate information between body parts in the form of electrical and chemical signals (Figure 36.7). Glial cells support the neurons or provide electrical insulation between them.

Animation: Cell junctions

Animation: Structure of an epithelium

Animation: Types of simple epithelium

Animation: Soft connective tissues

Animation: Specialized connective tissues

Animation: Muscle tissues

Animation: Functional zones of a motor neuron

Animation: Structure of human skin

Practice: Differences between cell and tissue types

36.3 Coordination of Tissues in Organs and Organ Systems

- Organs and organ systems are coordinated to carry out vital tasks, including maintenance of internal body conditions; nutrient acquisition, processing, and distribution; waste disposal; molecular synthesis; environmental sensing and response; protection against injury and disease; and reproduction.
- In vertebrates and most invertebrates, the major organ systems that accomplish these tasks are the nervous, endocrine, muscular, skeletal, integumentary, circulatory, lymphatic, immune, respiratory, digestive, excretory, and reproductive systems (Figure 36.8).

Animation: Human organ systems

36.4 Homeostasis

- Homeostasis is the process by which animals maintain their internal fluid environment under conditions their cells can tolerate. It is a dynamic state, in which internal adjustments are made continuously to compensate for environmental changes.
- Homeostasis is accomplished by negative feedback mechanisms that include a sensor, which detects a change in an external or internal condition; an integrator, which compares the detected change with a set point; and an effector, which returns the condition to the set point if it has varied (Figure 36.9).
- Animals also have positive feedback mechanisms, in which a change in an internal or external condition triggers a response that intensifies the change, and typically does not result in homeostasis.

Questions

Self-Test Questions

1. Which organ or tissue is an early major defense against viruses and bacteria?
   a. kidneys
   b. skin
   c. stomach
   d. skeletal muscle
   e. heart
2. Which tissue is a constant source of adult stem cells in a mammal?
   a. bone marrow
   b. pancreas
   c. basal lamina
   d. heart muscle
   e. kidneys
3. A flexible, rubbery protein in connective tissue is called ___, whereas a more fibrous, less flexible glycoprotein is called ___.
   a. adipose; cartilage
   b. endocrine; exocrine
   c. sweat; hormones
   d. chondroitin sulfate; hydroxapatite
   e. elastin; collagen
4. Adipose tissue:
   a. gives elasticity under epithelium.
   b. gives strength to tendons.
   c. insulates and is an energy reserve.
   d. provides movement, support, and protection.
   e. supports the nose and airways.
5. The bones of an elderly woman break more easily than those of a younger person. You would surmise that with aging, the cell type that diminishes in activity is the:
   a. osteocyte.
   b. osteoblast.
   c. osteoclast.
   d. chondrocyte.
   e. fibroblast.
6. The enormous mass of weight lifters is due to an increase in the size of:
   a. skeletal muscle.
   b. smooth muscle.
   c. cardiac muscle.
   d. involuntary muscle.
   e. interlinked, branched muscle.
7. Which muscle types appear striated under a microscope?
   a. skeletal muscles only
   b. cardiac muscles only
   c. skeletal muscles and cardiac muscles
   d. smooth muscles only
   e. skeletal muscles and smooth muscles
8. Which of the following is *not* a homeostatic response?
   a. In a contest, a student eats an entire chocolate cake in 10 minutes. Due to hormonal secretions, his blood glucose level does not change dramatically.
   b. The basketball players are dripping sweat at half time.
   c. The pupils in the eyes constrict when looking at a light.

d. Slower breathing in sleep changes carbon dioxide and oxygen blood levels, which affect blood pH.
e. The brain is damaged when a fever rises above 105°F.

8. The pituitary gland secretes a hormone that in turn stimulates the thyroid to secrete hormones. When the thyroid hormones are no longer needed, the pituitary stops or reduces its stimulus. This is an example of:
   a. osmolarity.
   b. environmental sensing.
   c. integration.
   d. positive feedback.
   e. negative feedback.

10. The system that coordinates other organ systems is the:
   a. skeletal system.
   b. reproductive system.
   c. muscular system.
   d. nervous system.
   e. integumentary system.

Questions for Discussion

1. Blood is often described as an atypical connective tissue. If you had to argue that blood is a connective tissue, what reasons would you include? What reasons would you include if you had to argue that blood is not a connective tissue?
2. What effect do you think a program of lifting weights would have on the bones of the skeleton? How would you design an experiment to test your prediction?
3. Positive feedback mechanisms are rare in animals compared with negative feedback mechanisms. Why do you think this is so?
4. Near the time of childbirth, collagen fibers in the connective tissue of the cervix break down, and gap junctions between the smooth muscle cells of the uterus increase in number. What do you think is the significance of these tissue changes?
5. Explain how, when driving, you control the car's speed by a typical negative feedback mechanism.

Experimental Analysis

The regulation of temperature in mammals and birds is an example of homeostasis. Design an experiment to observe and measure processes involved in temperature homeostasis in sedentary versus athletic humans during exercise.

Evolution Link

Steroid hormones are similar in structure and function across a wide array of animal species. For example, estradiol, which plays a critical role in reproductive and sexual functioning, is chemically identical in turtles and humans. What do these observations suggest about the time when steroid hormones evolved?

Section through the cerebellum, a part of the brain that integrates signals coming from particular regions of the body (confocal light micrograph). Neurons, the cells that send and receive signals, are red; glial cells, which provide structural and functional support for neurons, are yellow; and nuclei are purple.

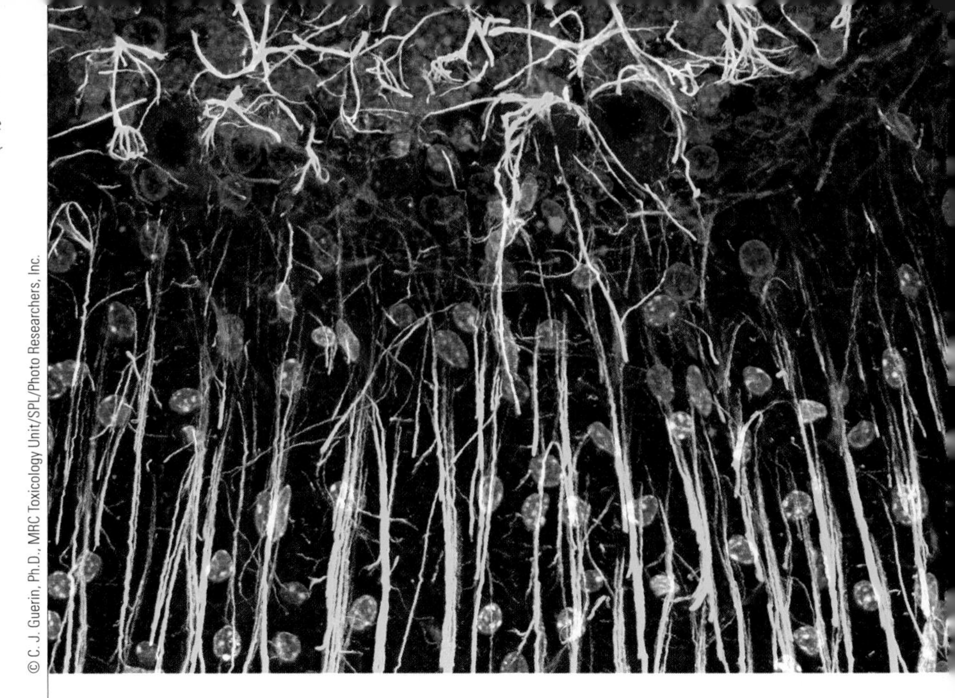

Study Plan

37 Information Flow and the Neuron

Why It Matters

The dog stands alert, muscles tense, motionless except for a wagging tail. His eyes are turned toward his master, a boy poised to throw a Frisbee for him to catch. Even before the Frisbee is released, the dog has anticipated the direction of its flight from the eyes and stance of the boy.

With a snap of the wrist, the boy throws the Frisbee, and the dog springs into action. Legs churning, eyes following the Frisbee, the dog runs beneath its track, closing the distance as the Frisbee reaches the peak of its climb and begins to descend. All this time, parts of the dog's brain have been processing information received through various sensory inputs. The eyes report his travel over the ground and the speed and arc of the Frisbee. Sensors in the inner ears, muscles, and joints detect the position of the dog's body, and his brain sends out signals that keep his movements on track and in balance. Other parts of the brain register inputs from sensors monitoring body temperature and carbon dioxide levels in the blood, and send signals that adjust heart and breathing rate accordingly.

At just the right instant, a burst of signals from the dog's brain causes trunk and leg muscles to contract in a coordinated pattern, and

Figure 37.1 With perfect timing, a dog leaps to catch a Frisbee. The coordinated leap involves processing and integration of information by the dog's nervous system.

the dog leaps to intercept the Frisbee in midair with an assured snap of his jaws **(Figure 37.1).** Now the animal twists, turning his head and eyes toward the ground as his brain calculates the motions required to land on his feet and in balance. The dog makes a perfect landing and trots happily back to his master, ready to repeat the entire performance.

The functions of the dog's nervous system in the chase and capture are astounding in the amount and variety of sensory inputs, the rate and complexity of the brain's analysis and integration of incoming information, and the flurry of signals the brain sends to make compensating adjustments in body activities. Yet they are ordinary in the sense that the same activities take place countless times each day in the nervous system of all but the simplest animals.

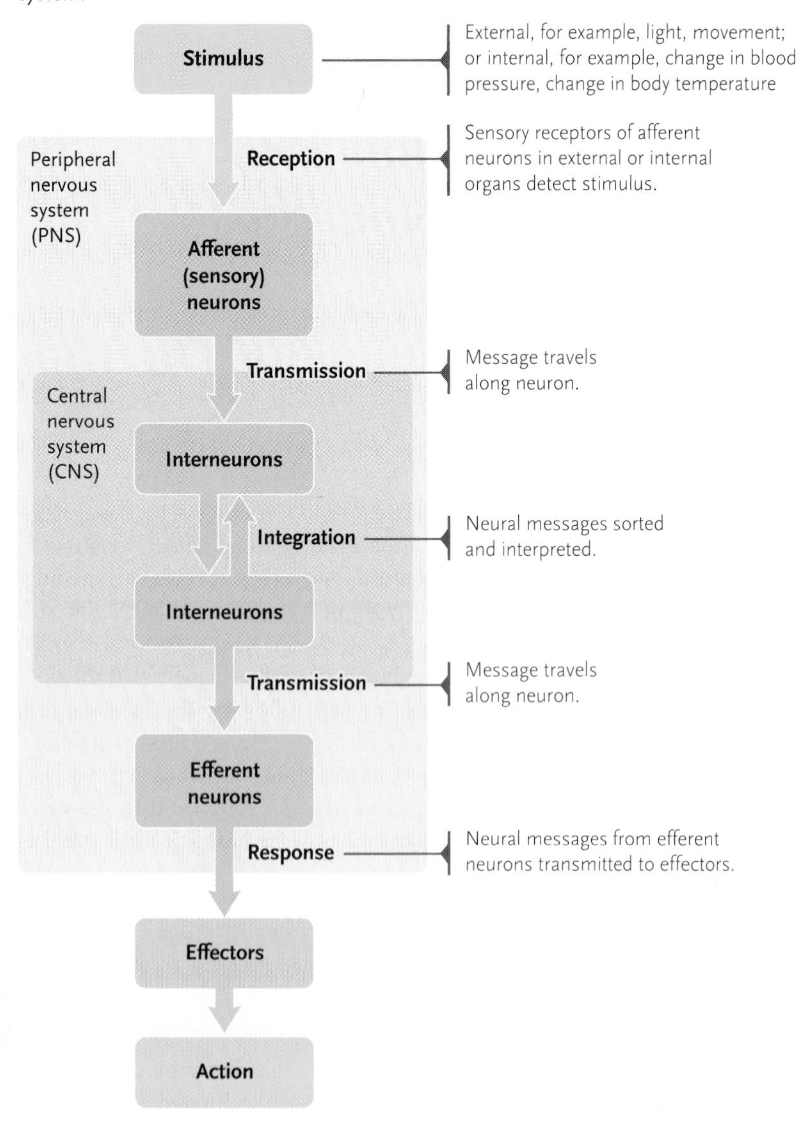

Figure 37.2 Neural signaling: the information-processing steps in the nervous system.

All these activities, no matter how complex, depend on the functions of only two major cell types: *neurons* and *glial cells*. In most animals, these cells are organized into complex networks called *nervous systems*. This chapter describes neuron structure and tells how neurons send and receive signals with the aid and support of glial cells. The next chapter considers the neural networks of the brain and its associated structures. Chapter 39 discusses the sensory receptors that detect environmental changes and convert that information into signals for integration by the nervous system.

37.1 Neurons and Their Organization in Nervous Systems

An animal constantly receives stimuli from both internal and external sources. **Neural signaling**, communication by neurons, is the process by which an animal responds appropriately to a stimulus **(Figure 37.2).** In most animals, the four components of neural signaling are *reception, transmission, integration,* and *response*. **Reception**, the detection of a stimulus, is performed by **neurons**, the cellular components of nervous systems, and by specialized sensory receptors such as those in the eye and skin. **Transmission** is the sending of a message along a neuron, and then to another neuron or to a muscle or gland. **Integration** is the sorting and interpretation of neural messages and the determination of the appropriate response(s). **Response** is the "output" or action resulting from the integration of neural messages. For a dog catching a Frisbee, for instance, sensors in the eye receive light stimuli from the environment, and internal sensors receive stimuli from all the animal's organ systems. The neural messages generated are transmitted through the nervous system and integrated to determine the appropriate response, in this case stimulating the muscles so the dog jumps into the air and catches the Frisbee.

Neurons Are Cells Specialized for the Reception and Transmission of Informational Signals

Neural signaling involves three functional classes of neurons (the blue boxes in Figure 37.2). **Afferent neurons** (also called **sensory neurons**) transmit stimuli collected by their sensory receptors to **interneurons**, which integrate the information to formulate an appropriate response. In humans and some other primates, 99% of neurons are interneurons. **Efferent neurons** carry the signals indicating a response away from the interneuron networks to the **effectors**, the muscles and glands. Efferent neurons that carry signals to skeletal muscle are called **motor neurons.** The information-processing steps in the nervous system can be summarized, therefore, as: (1) sensory receptors on

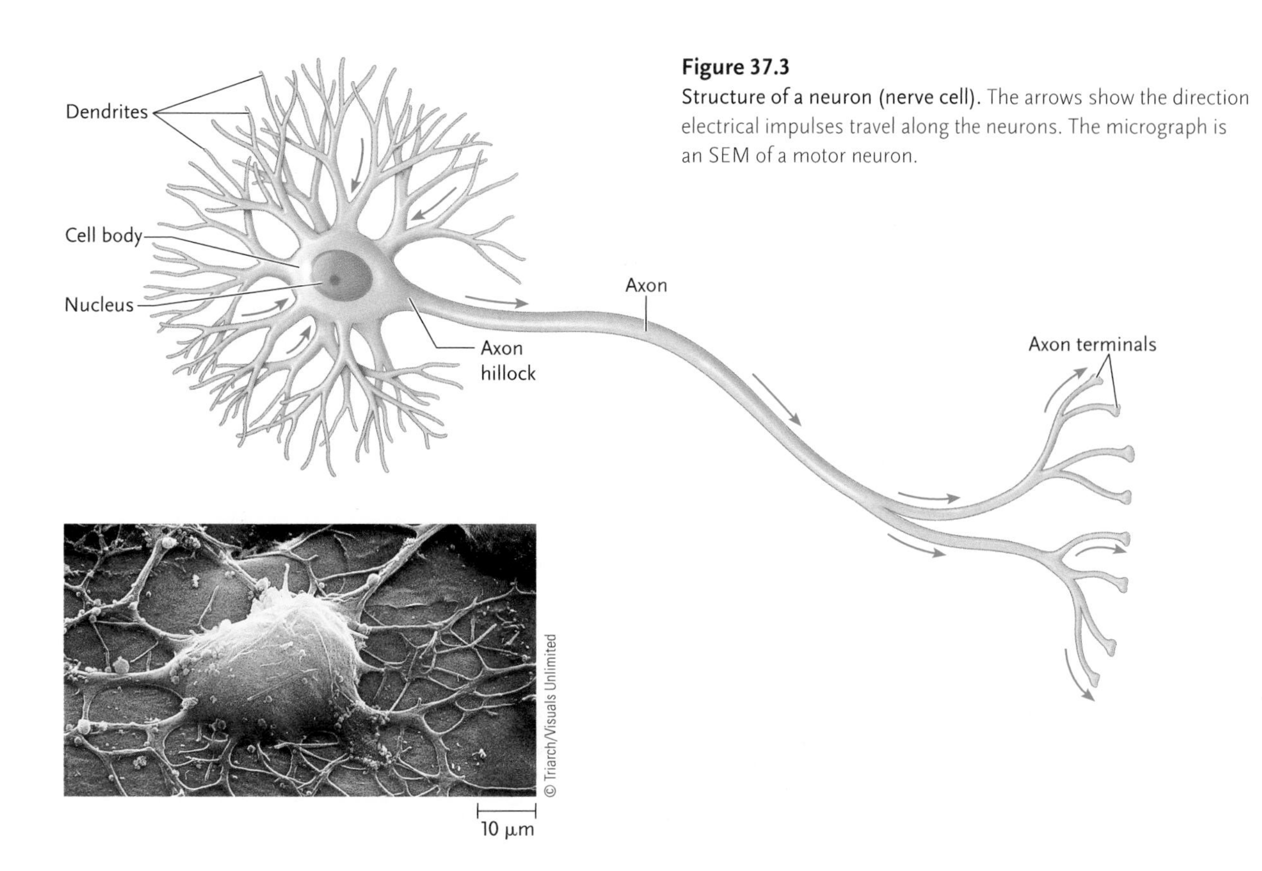

Figure 37.3
Structure of a neuron (nerve cell). The arrows show the direction electrical impulses travel along the neurons. The micrograph is an SEM of a motor neuron.

afferent neurons receive a stimulus; (2) afferent neurons transmit the information to interneurons; (3) interneurons integrate the neural messages; and (4) efferent neurons transmit the neural messages to effectors, which act in a way appropriate to the stimulus.

Neurons vary widely in shape and size. All have an enlarged cell body and two types of extensions or processes, called dendrites and axons **(Figure 37.3).** The **cell body,** which contains the nucleus and the majority of cell organelles, synthesizes most of the proteins, carbohydrates, and lipids of the neuron. Dendrites and axons conduct electrical signals, which are produced by ions flowing down concentration gradients through channels in the plasma membrane of the neuron. **Dendrites** receive the signals and transmit them toward the cell body. Dendrites are generally highly branched, forming a treelike outgrowth at one end of the neuron (*dendros* = tree). **Axons** conduct signals away from the cell body to another neuron or an effector. Neurons typically have a single axon, which arises from a junction with the cell body called an **axon hillock.** The axon has branches at its tip that end as small, buttonlike swellings called **axon terminals.** The more terminals contacting a neuron, the greater its capacity to integrate incoming information.

Connections between axon terminals of one neuron and the dendrites or cell body of a second neuron form **neuronal circuits.** A typical neuronal circuit contains an afferent (sensory) neuron, one or more interneurons, and an efferent neuron. The circuits combine into networks that interconnect the parts of the nervous system. In vertebrates, the afferent neurons and efferent neurons collectively form the *peripheral nervous system (PNS)*. The interneurons form the brain and spinal cord, called the *central nervous system (CNS)*. As depicted in Figure 37.2, afferent (*afferre* = carry toward) information is ultimately transmitted to the CNS where efferent (*efferre* = carry away) information is initiated. The nervous systems of most invertebrates are also divided into central and peripheral divisions.

Neurons Are Supported Structurally and Functionally by Glial Cells

Glial cells are nonneuronal cells that provide nutrition and support to neurons. One type, called **astrocytes** because they are star-shaped **(Figure 37.4),** occurs only in the vertebrate CNS, where they closely cover the surfaces of blood vessels. Astrocytes provide physical support to neurons and help maintain the concentrations of ions in the interstitial fluid surrounding them. Two other types of glial cells—**oligodendrocytes** in the CNS and **Schwann cells** in the PNS—wrap around axons in a jelly roll fashion to form myelin sheaths **(Figure 37.5).** Myelin sheaths have a high lipid content because of the

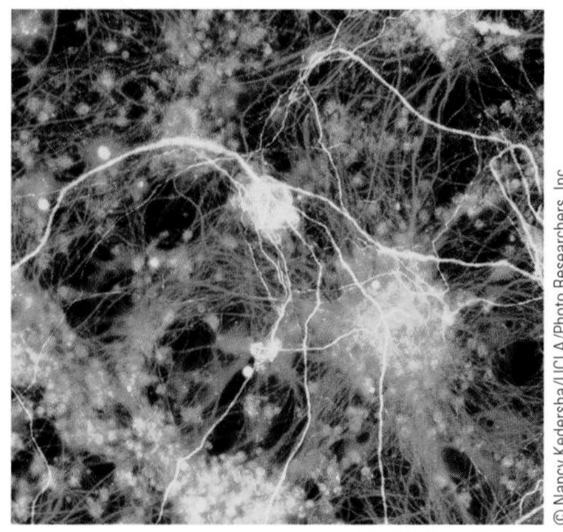

Figure 37.4
Astrocytes (orange), a type of glial cell, and a neuron (yellow) in brain tissue.

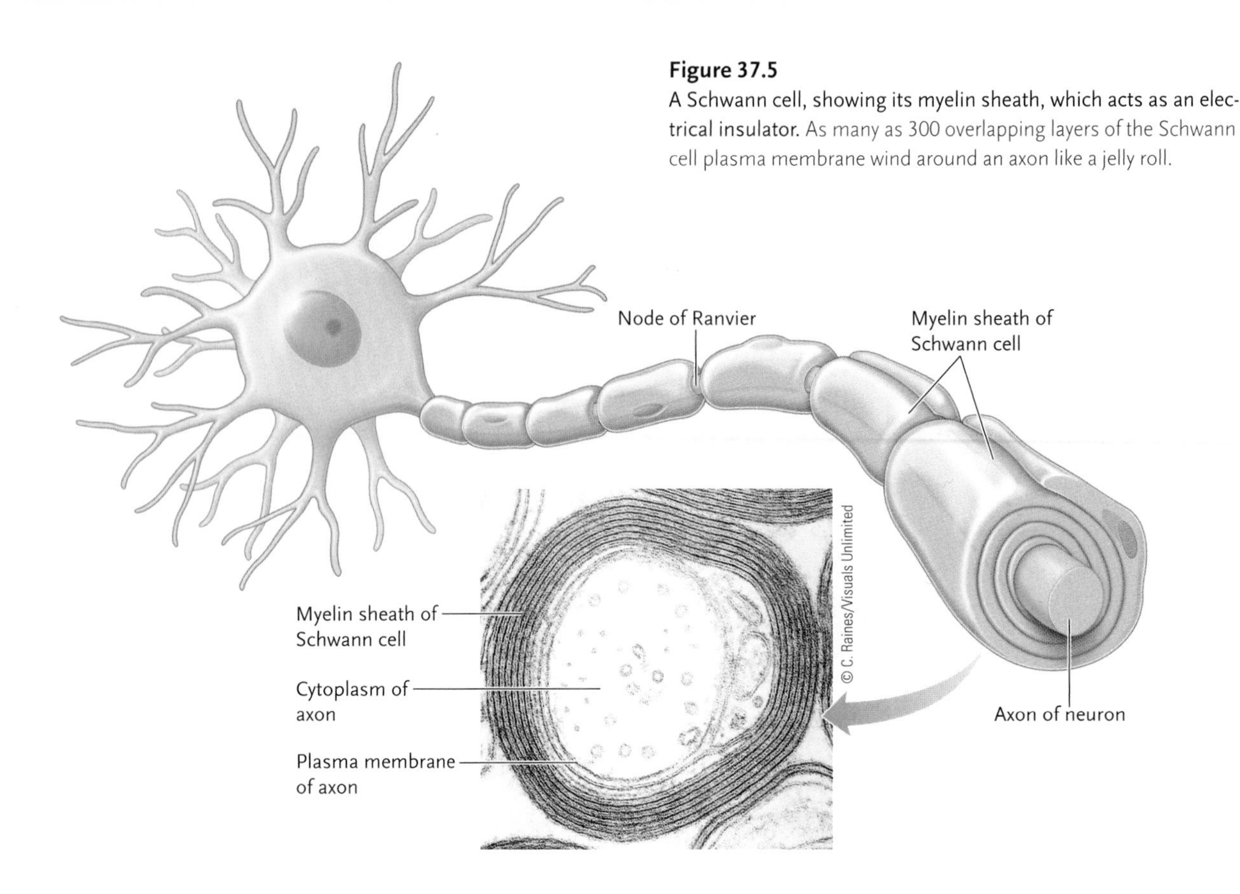

Figure 37.5
A Schwann cell, showing its myelin sheath, which acts as an electrical insulator. As many as 300 overlapping layers of the Schwann cell plasma membrane wind around an axon like a jelly roll.

many layers of plasma membranes of the myelin-forming cells. Because of their high lipid content, the myelin sheaths act as electical insulators. The gaps between Schwann cells, called **nodes of Ranvier,** expose the axon membrane directly to extracellular fluids. This structure speeds the rate at which electrical impulses move along the axons covered by glial cells.

Unlike most neurons, glial cells retain the capacity to divide throughout the life of the animal. This capacity allows glial tissues to replace damaged or dead cells, but also makes them the source of almost all brain tumors, produced when regulation of glial cell division is lost.

Neurons Communicate via Synapses

A **synapse** (*synapsis* = juncture) is a site where a neuron makes a communicating connection with another neuron or with an effector such as a muscle fiber or gland. On one side of the synapse is an axon terminal of the **presynaptic cell**, the neuron that transmits the signal. On the other side is the cell body or a dendrite of the **postsynaptic cell**, the neuron or the surface of an effector that receives the signal. Communication across a synapse may occur by the direct flow of an electrical signal or by means of a **neurotransmitter,** a chemical released by an axon terminal at a synapse. The vast majority of vertebrate neurons communicate by means of neurotransmitters.

In **electrical synapses**, the plasma membranes of the presynaptic and postsynaptic cells are in direct contact **(Figure 37.6a).** When an electrical impulse arrives at the axon terminal, gap junctions (see Section 36.2) allow ions to flow directly between the two cells, leading to unbroken transmission of the electrical signal. Although electrical synapses allow the most rapid conduction of signals, this type of connection is essentially "on" or "off" and unregulated. In humans, electrical synapses occur in locations such as the pulp of a tooth, where they contribute to the almost instant and intense pain we feel if the pulp is disturbed.

In **chemical synapses**, the plasma membranes of the presynaptic and postsynaptic cells are separated by a narrow gap, about 25 nm wide, called the **synaptic cleft (Figure 37.6b).** When an electrical impulse arrives at an axon terminal, it causes the release of a neurotransmitter into the synaptic cleft. The neurotransmitter diffuses across the synaptic cleft and binds to a receptor in the plasma membrane of the postsynaptic cell. If enough neurotransmitter molecules bind to these receptors, the postsynaptic cell generates a new electrical impulse, which travels along its axon to reach a synapse with the next neuron or effector in the circuit. A chemical synapse is more than a simple on-off switch because many factors can influence the generation of a new electrical impulse in the postsynaptic cell, including neurotransmitters that inhibit that cell rather than stimulating it. The balance of stimulatory and inhibitory effects in chemical synapses contributes to the integration of incoming information in a receiving neuron.

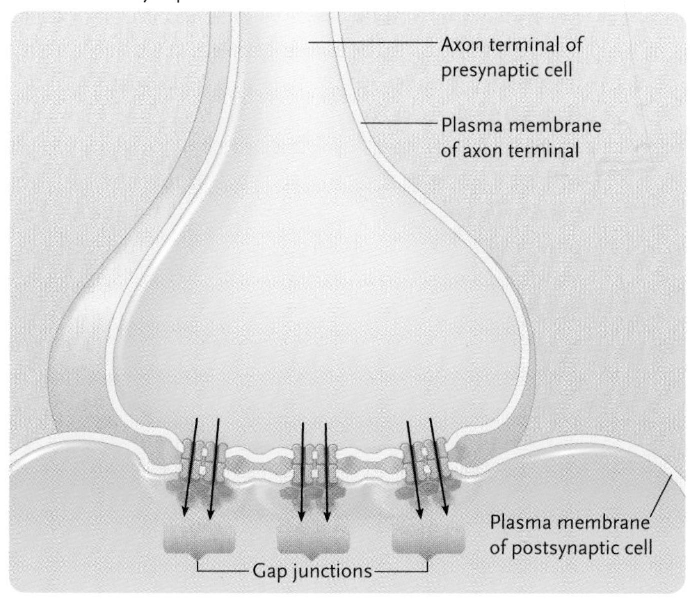

In an electrical synapse, the plasma membranes of the presynaptic and postsynaptic cells make direct contact. Ions flow through gap junctions that connect the two membranes, allowing impulses to pass directly to the postsynaptic cell.

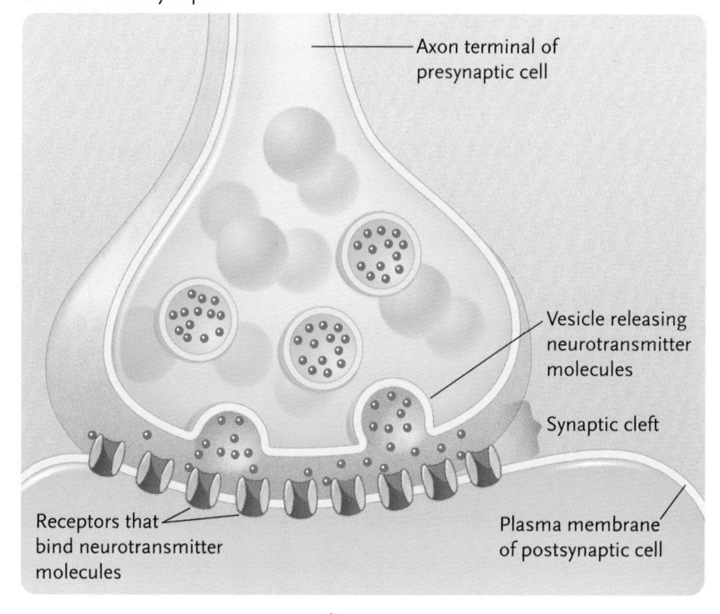

In a chemical synapse, the plasma membranes of the presynaptic and postsynaptic cells are separated by a narrow synaptic cleft. Neurotransmitter molecules diffuse across the cleft and bind to receptors in the plasma membrane of the postsynaptic cell. The binding opens channels to ion flow that may generate an impulse in the postsynaptic cell.

Figure 37.6
The two types of synapses by which neurons communicate with other neurons or effectors.

Study Break

1. Distinguish between a dendrite and an axon.
2. Distinguish between the functions and locations of afferent neurons, efferent neurons, and interneurons.
3. What is the difference between an electrical synapse and a chemical synapse?

37.2 Signal Conduction by Neurons

All cells of an animal have a **membrane potential,** a separation of positive and negative charges across the plasma membrane. Outside the cell the charge is positive, and inside the cell it is negative. This charge separation produces *voltage*—an electrical potential difference—across the plasma membrane.

The membrane potential is caused by the uneven distribution of Na^+ and K^+ inside and outside the cell. As you learned in Chapter 6, plasma membranes are *selectively* permeable in that they allow some ions but not others to move across the membrane through protein channels embedded in the phospholipid bilayer. Plasma membrane-embedded Na^+/K^+ active transport pumps use energy from ATP hydrolysis to pump simultaneously three Na^+ out of the cell for every two K^+ pumped in. This exchange generates a higher Na^+ concentration outside the cell than inside, and a higher K^+ concentration inside the cell than outside, explaining the positive charge outside the cell. The inside of the cell is negatively charged because the cell also contains many negatively charged molecules (anions) such as proteins, amino acids, and nucleic acids.

In most cells, the membrane potential does not change. However, neurons and muscle cells use the membrane potential in a specialized way. That is, in response to electrical, chemical, mechanical, and certain other types of stimuli, their membrane potential changes rapidly and transiently. Cells with this property are said to be *excitable cells.* Excitability, produced by a sudden flow across the plasma membrane, is the basis for nerve impulse generation.

Resting Potential Is the Unchanging Membrane Potential of an Unstimulated Neuron

The membrane of a neuron that is not being stimulated is not conducting an impulse—exhibits a steady negative membrane potential, called the **resting potential** because the neuron is at rest. The resting potential has been measured at between −50 and −60 millivolts (mV) for neurons in the body, and at about −70 mV in isolated neurons **(Figure 37.7).** A neuron exhibiting the resting potential is said to be *polarized.*

The distribution of ions inside and outside an axon that produces the resting potential is shown in **Figure 37.8.** As described earlier in this section, the Na^+/K^+ pump is responsible for creating the imbalance of Na^+ and K^+ inside and outside of the cell, and the concentration of negatively charged molecules within the cell results in the inside being negatively charged and the

Figure 37.7 Research Method

Measuring Membrane Potential

PURPOSE: To determine the membrane potentials of unstimulated and stimulated neurons and muscle cells.

PROTOCOL: Prepare a microelectrode by drawing out a glass capillary tube to a tip with a diameter much smaller than that of a cell and filling it with a salt solution that can conduct an electric current. Under a microscope, use a micromanipulator (mechanical positioning device) to insert the tip of the microelectrode into an axon. Place a reference electrode in the solution outside the cell. Use an oscilloscope or voltmeter to measure the voltage between the microelectrode tip in the axon and the reference electrode outside the cell.

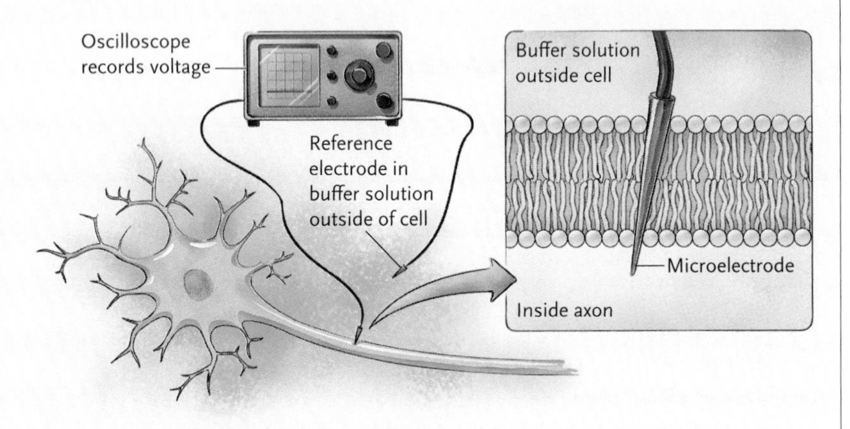

INTERPRETING THE RESULTS: The oscilloscope or voltmeter indicates the membrane potential in volts. Changes in membrane potential caused by stimuli or chemical treatments can be measured and recorded. For an isolated, unstimulated neuron (shown above) the membrane potential is typically about −70 mV.

outside being positively charged. As we will see in the following description of the changes in a neuron that occur when it is stimulated, the *voltage-gated ion channels* for Na^+ and K^+ open and close when the membrane potential changes.

The Membrane Potential Changes from Negative to Positive during an Action Potential

When a neuron conducts an electrical impulse, an abrupt and transient change in membrane potential occurs; this is called the **action potential.** An action potential begins as a stimulus that causes positive charges from outside the neuron to flow inward, making the cytoplasmic side of the membrane less negative **(Figure 37.9).** As the membrane potential becomes less negative, the membrane (which was polarized at rest) becomes **depolarized.** Depolarization proceeds relatively slowly until it reaches a level known as the **threshold potential,** about −50 to −55 mV in isolated neurons. Once the threshold is reached, the action potential fires—and the membrane potential suddenly increases. In less than 1 msec (millisecond, one-thousandth of a second), it rises so high that the inside of the plasma membrane becomes positive due to an influx of positive ions across the cell membrane, momentarily reaching a value of +30 mV or more. The potential then falls again, in many cases dropping to about −80 mV before rising again to the resting potential. When the potential is below the resting value, the membrane is said to be **hyperpolarized.** The entire change, from initiation of the action potential to the return to the resting potential, takes less than 5 msec in the fastest neurons. Action potentials take the same basic form in neurons of all types, with differences in the values of the resting potential and the peak of the action potential, and in the time required to return to the resting potential.

All stimuli cause depolarization of a neuron, but an action potential is produced only if the stimulus is strong enough to cause depolarization to reach the threshold. This is referred to as the **all-or-nothing principle;** once triggered, the changes in membrane potential take place independently of the strength of the stimulus.

Beginning at the peak of an action potential, the membrane enters a **refractory period** of a few milliseconds during which the threshold required for generation of an action potential is much higher than normal. The refractory period lasts until the membrane has stabilized at the resting potential. As we shall see, the refractory period keeps impulses traveling in a one-way direction in neurons.

The Action Potential Is Produced by Ion Movements through the Plasma Membrane

The action potential is produced by movements of Na^+ and K^+ through the plasma membrane. The movements are controlled by specific **voltage-gated ion channels,** membrane-embedded proteins that open and close as the membrane potential changes (see Figure 37.8). Voltage-gated Na^+ channels have two gates, an *activation gate* and an *inactivation gate,* whereas voltage-gated K^+ channels have one gate, an *activation gate.*

How the two voltage-gated ion channels operate to generate an action potential is shown in **Figure 37.10.** When the membrane is at the resting potential, the activation gates of both the Na^+ and K^+ channels are closed. As a depolarizing stimulus raises the membrane potential to the threshold, the activation gate of the Na^+ channels opens, allowing a burst of Na^+ ions to flow into the axon along their concentration gradient. Once above the threshold, more Na^+ channels open, causing a rapid inward flow of positive charges that raises the membrane potential to-

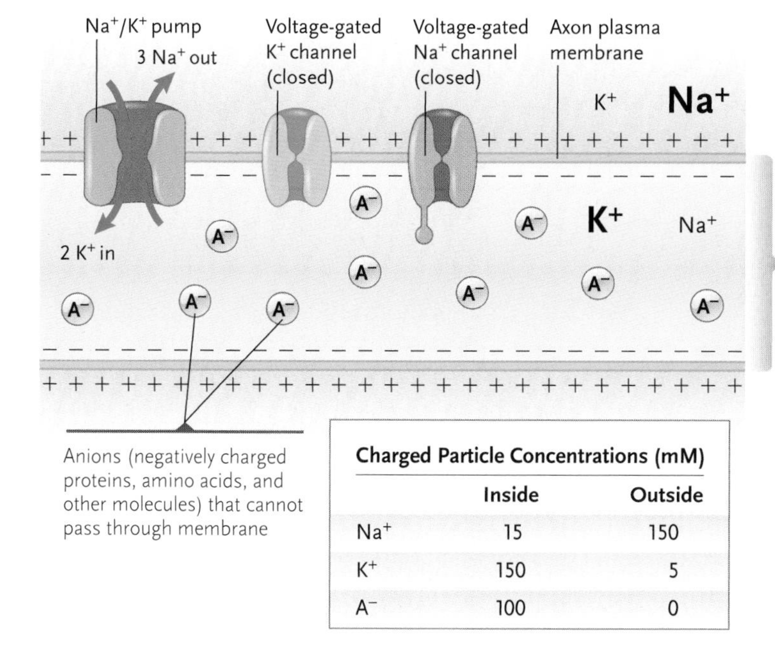

Charged Particle Concentrations (mM)	Inside	Outside
Na^+	15	150
K^+	150	5
A^-	100	0

Figure 37.8
The distribution of ions inside and outside an axon that produces the resting potential, −70 mV. The distribution of ions that do not directly affect the resting potential, such as Cl^-, is not shown. The voltage-gated ion channels open and close when the membrane potential changes.

ward the peak of the action potential. As the action potential peaks, the inactivation gate of the Na^+ channel closes (resembling putting a stopper in the sink), which stops the inward flow of Na^+. The refractory period now begins.

At the same time, the activation gates of the K^+ channels begin to open, allowing K^+ ions to flow rapidly outward in response to their concentration gradient. The K^+ ions contribute to the refractory period and compensate for the inward movement of Na^+ ions, returning the membrane to the resting potential. As the resting potential is reestablished, the activation gates of the K^+ channels close, as do those of the Na^+ channels, and the inactivation gates of the Na^+ channels open. These events end the refractory period and ready the membrane for another action potential.

In some neurons, closure of the gated K^+ channels lags, and K^+ continues to flow outward for a brief time after the membrane returns to the resting potential. This excess outward flow causes the hyperpolarization shown in Figure 37.9, in which the membrane potential dips briefly below the resting potential.

At the end of an action potential, the membrane potential has returned to its resting state, but the ion distribution has changed slightly. That is, some Na^+ ions have entered the cell, and some K^+ ions have left the cell—but not many, relative to the total number of ions, and the distribution is not altered enough to prevent other action potentials from occurring. In the long term, the Na^+/K^+ active transport pumps restore the Na^+ and K^+ to their original locations.

Some of what is known about how ion flow through channels can change membrane potential has come from experiments using the *patch-clamp* technique. In the patch part of the technique, a micropipette with a tip 1 to 3 μm in diameter is touched to the plasma membrane of a neuron (or other cell type). The contact seals the membrane to the micropipette and, when the micropipette is pulled away, a patch of membrane with one or a few ion channels comes with it. The clamp part of the technique refers to a voltage clamp, in which an electronic device holds the membrane potential of the patch at a steady value chosen by the investigator. The investigator can add a stimulus that is expected to open or close ion channels. The amount of current the clamping device needs to keep the voltage constant is directly related to the number and charge of the ions moving through the channels and, hence, measures channel activity.

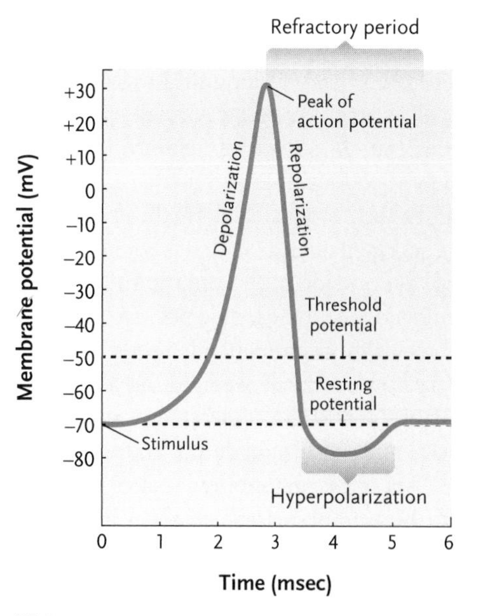

Figure 37.9
Changes in membrane potential during an action potential.

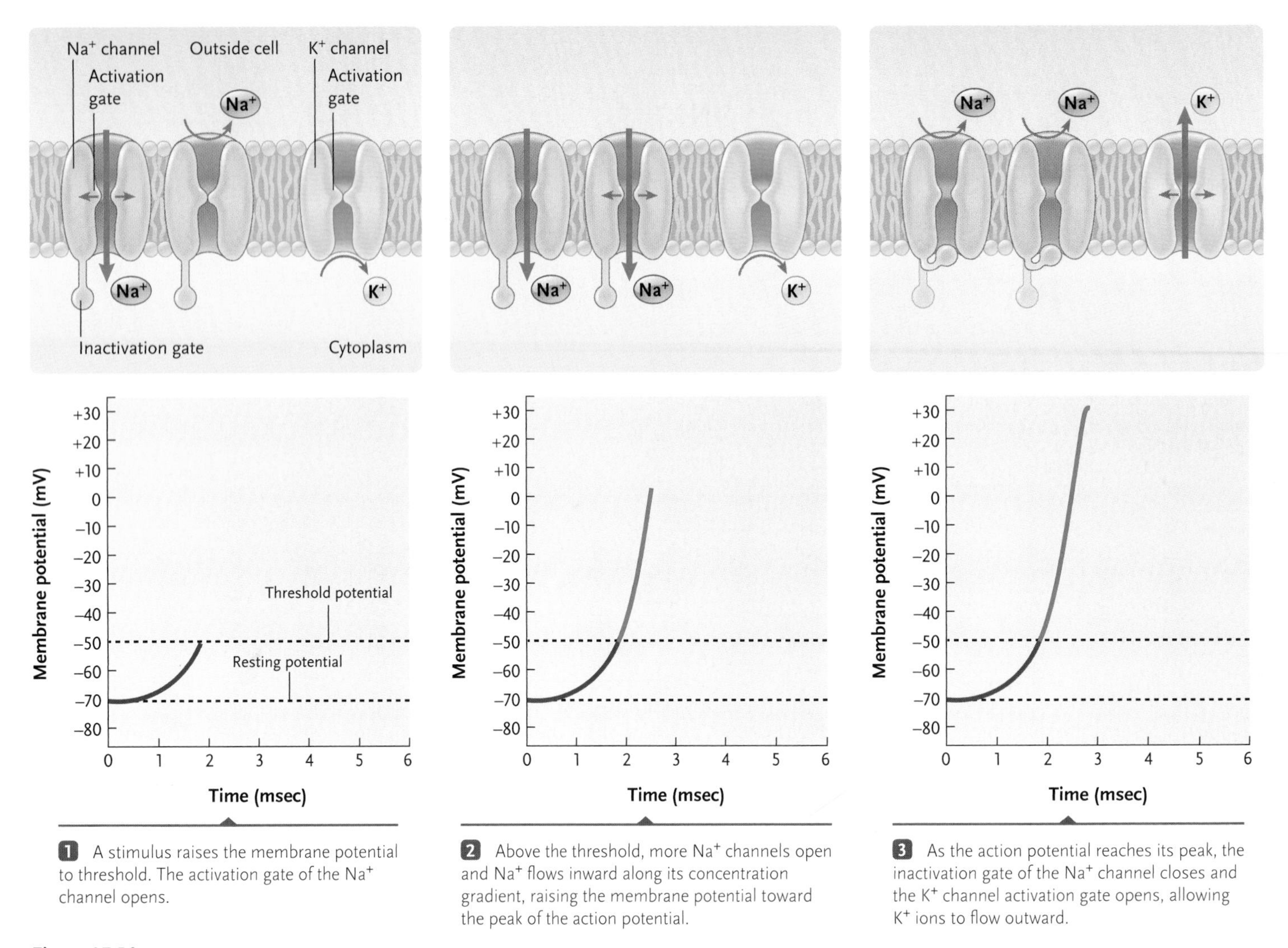

1 A stimulus raises the membrane potential to threshold. The activation gate of the Na^+ channel opens.

2 Above the threshold, more Na^+ channels open and Na^+ flows inward along its concentration gradient, raising the membrane potential toward the peak of the action potential.

3 As the action potential reaches its peak, the inactivation gate of the Na^+ channel closes and the K^+ channel activation gate opens, allowing K^+ ions to flow outward.

Figure 37.10
Changes in voltage-gated Na^+ and K^+ channels that produce the action potential.

Neural Impulses Move by Propagation of Action Potentials

Once an action potential is initiated at the dendrite end of the neuron, it passes along the surface of a nerve or muscle cell as an automatic wave of depolarization traveling away from the stimulation point **(Figure 37.11).** The action potential does not need further trigger events in order for it to be propagated along the axon to the terminals. In a segment of an axon that is generating an action potential, the outside of the membrane becomes temporarily negative and the inside positive. Because opposites attract, as the region outside becomes negative, local current flow occurs between the area undergoing an action potential and the adjacent downstream inactive area both inside and outside the membrane (arrows, Figure 37.11). This current flow makes nearby regions the axon membrane less positive on the outside and more positive on the inside; in other words, they depolarize the membrane.

The depolarization is large enough to push the membrane potential past the threshold, opening the voltage-gated Na^+ and K^+ channels and starting an action potential in the downstream adjacent region. In this way, each segment of the axon stimulates the next segment to fire, and the action potential moves rapidly along the axon as a nerve impulse.

The refractory period keeps an action potential from reversing direction at any point along an axon; only the region in front of the action potential can fire. The refractory period results from the properties of the voltage-gated ion channels. Once they have been opened to their activated state, the upstream voltage-gated ion channels need time to reset to their original positions before they can open again. Therefore, only downstream voltage-gated ion channels are able to open, ensuring the one-way movement of the action potential along the axon toward the axon tips. By the time the refractory period ends in a membrane segment that has just fired an action potential, the action potential has moved too far away to cause a second action potential to develop in the same segment.

The magnitude of an action potential stays the same as it travels along an axon, even where the axon

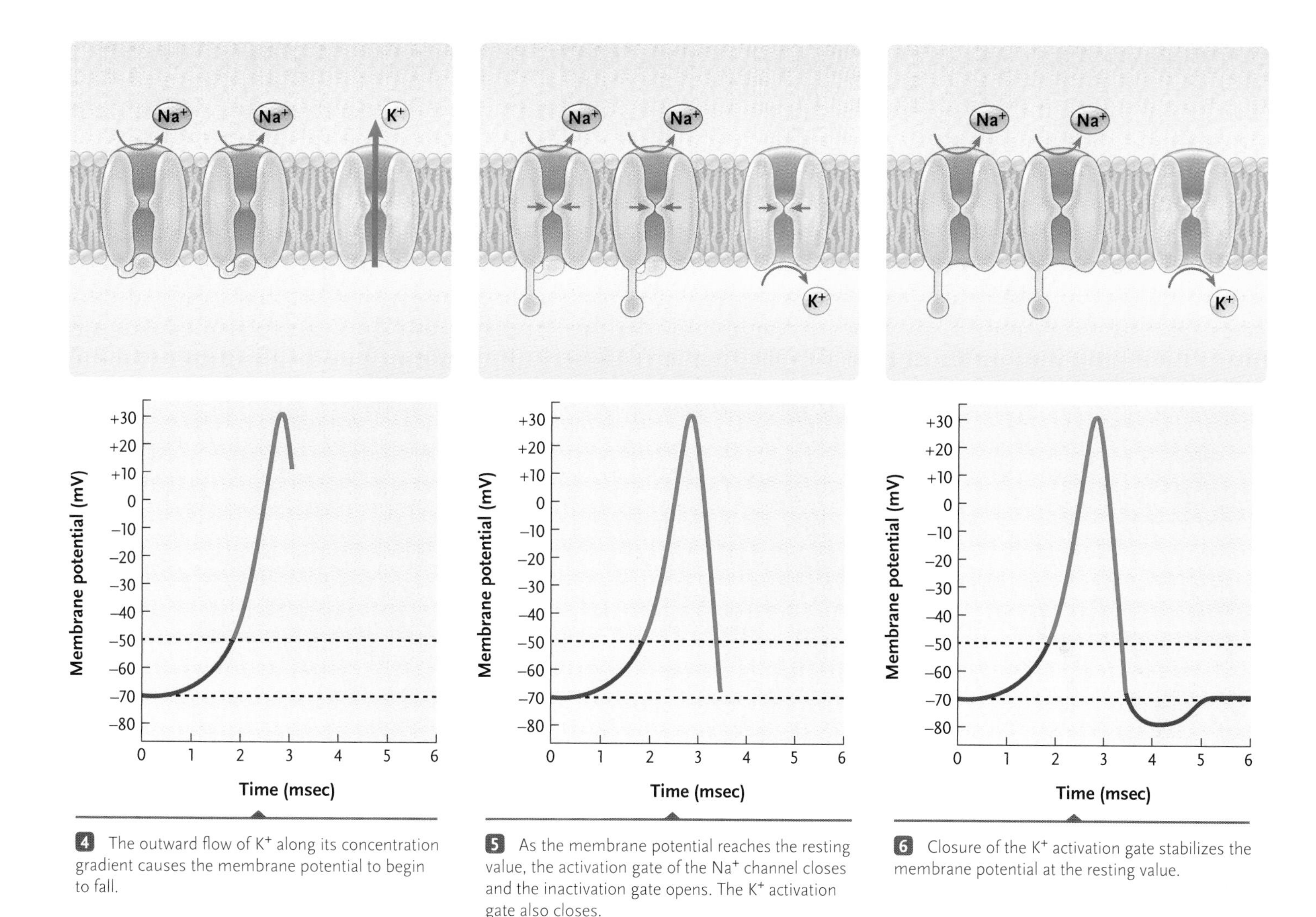

4 The outward flow of K^+ along its concentration gradient causes the membrane potential to begin to fall.

5 As the membrane potential reaches the resting value, the activation gate of the Na^+ channel closes and the inactivation gate opens. The K^+ activation gate also closes.

6 Closure of the K^+ activation gate stabilizes the membrane potential at the resting value.

branches at its tips. Thus the propagation of an action potential resembles a burning fuse, which burns with the same intensity along its length, and along any branches, once it is lit at one end. Unlike a fuse, however, an axon can fire another action potential of the same intensity within a few milliseconds after an action potential passes through.

Due to the all-or-nothing principle of action potential generation, the intensity of a stimulus is reflected in the *frequency* of action potentials—the greater the stimulus, the more action potentials per second, up to a limit depending on the axon type—rather than by the change in membrane potential. For most neuron types, the limit lies between 10 and 100 action potentials per second.

Both natural and synthetic substances target specific parts of the mechanism generating action potentials. Local anesthetics, such as procaine and lidocaine, bind to voltage-gated Na^+ channels and block their ability to transport ions; thus, sensory nerves in the anesthetized region cannot transmit pain signals. The potent poison of the pufferfish, tetrodotoxin, also blocks voltage-gated Na^+ channels in neurons, potentially causing muscle paralysis and death. The pufferfish is highly prized as a delicacy in Japan, eaten after careful preparation to remove organs carrying the tetrodotoxin. A mistake can kill the diners, however, making pufferfish sashimi a kind of culinary Russian roulette.

Saltatory Conduction Increases Propagation Rate in Small-Diameter Axons

In the propagation pattern shown in Figure 37.11, an action potential spreads along every segment of the membrane along the length of the axon. For this type of action potential propagation, the rate of conduction increases with the diameter of the axon. Axons with a very large diameter have evolved in invertebrates such as lobsters, earthworms, and squids as well as a few marine fishes. Giant axons typically carry signals that produce an escape or withdrawal response, such as the sudden flexing of the tail (abdomen) in lobsters that propels the animal backward. The largest known axons, 1.7 mm in diameter, occur in fanworms *(Myxicola)*.

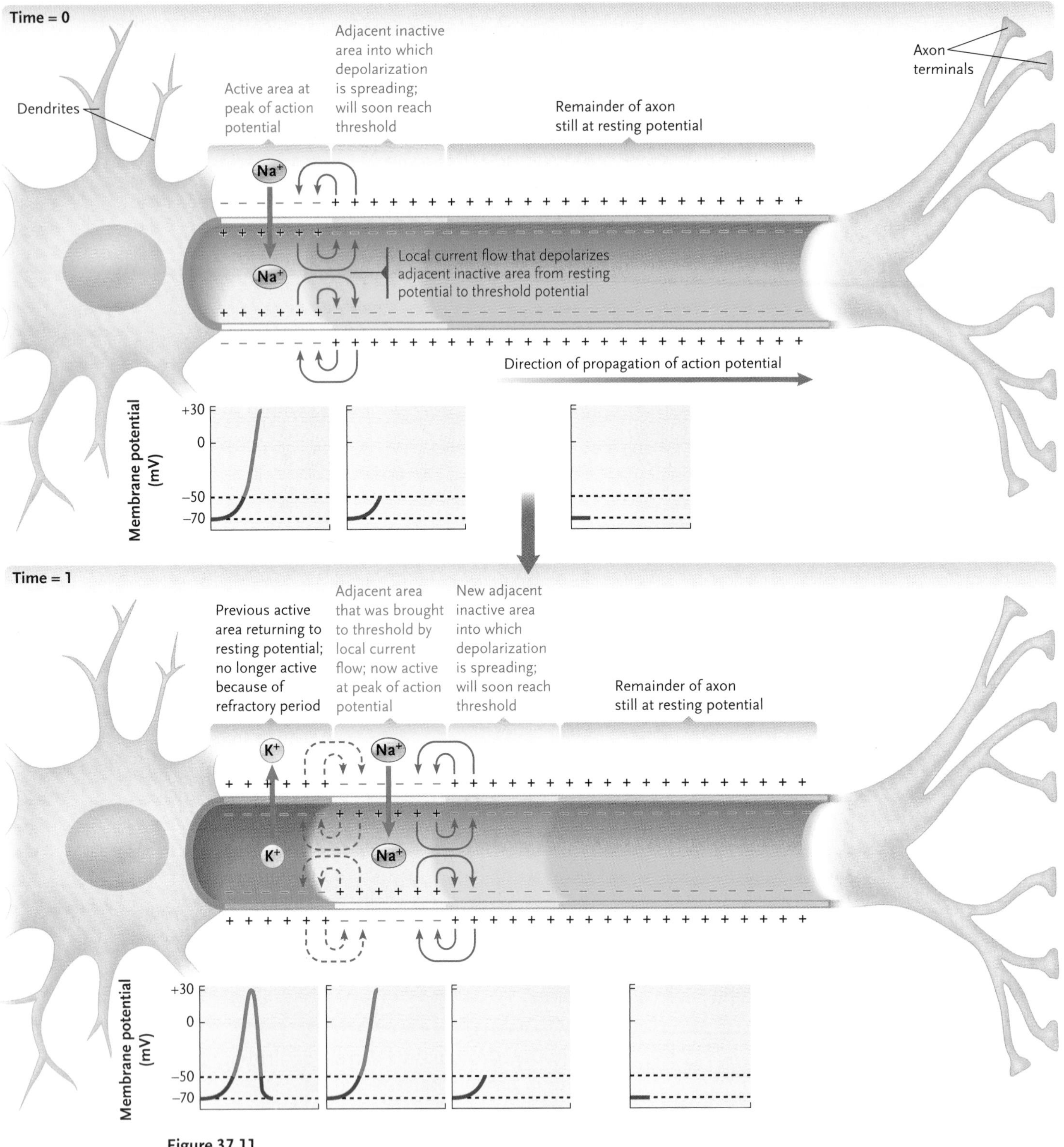

Figure 37.11
Propagation of an action potential along an unmyelinated axon by ion flows between a firing segment and an adjacent unfired region of the axon. Each firing segment induces the next to fire, causing the action potential to move along the axon.

The signals they carry contract a muscle that retracts the fanworm's body into a protective tube when the animal is threatened.

Although large-diameter axons can conduct impulses as rapidly as 25 m/sec (over twice the speed of the world record 100-meter dash), they take up a great deal of space. In complex vertebrates, natural selection has led to a mechanism that allows small-diameter axons to conduct impulses rapidly. The mechanism, called **saltatory conduction** (*saltere* = to leap), allows action potentials to "hop" rapidly along axons instead of burning smoothly like a fuse.

Saltatory conduction depends on the insulating myelin sheath that forms around some axons and in particular on the nodes of Ranvier exposing the axon membrane to extracellular fluids. Voltage-gated Na^+ and K^+ channels crowded into the nodes allow action potentials to develop at these positions **(Figure 37.12)**. The inward movement of Na^+ ions produces depolarization, but the excess positive ions are unable to leave the axon through the membrane regions covered by the myelin sheath. Instead, they diffuse rapidly to the next node where they cause depolarization, inducing an action potential at that node. As this mechanism repeats, the action potential jumps rapidly along the axon from node to node. Saltatory conduction proceeds at rates up to 130 m/sec while an unmyelinated axon of the same diameter conducts action potentials at about 1 m/sec.

Saltatory conduction allows thousands to millions of fast-transmitting axons to be packed into a relatively small diameter. For example, in humans the optic nerve leading from the eye to the brain is only 3 mm in diameter but is packed with more than a million axons. If those axons were unmyelinated, each would have to be about 100 times thicker to conduct impulses at the same velocity, producing an optic nerve about 300 mm (12 inches) in diameter.

The disease *multiple sclerosis* (*sclero* = hard) underscores the importance of myelin sheaths to the operation of the vertebrate nervous system. In this disease, myelin is progressively lost from axons and replaced by hardened scar tissue. The changes block or slow the transmission of action potentials, producing numbness, muscular weakness, faulty coordination of movements, and paralysis that worsens as the disease progresses.

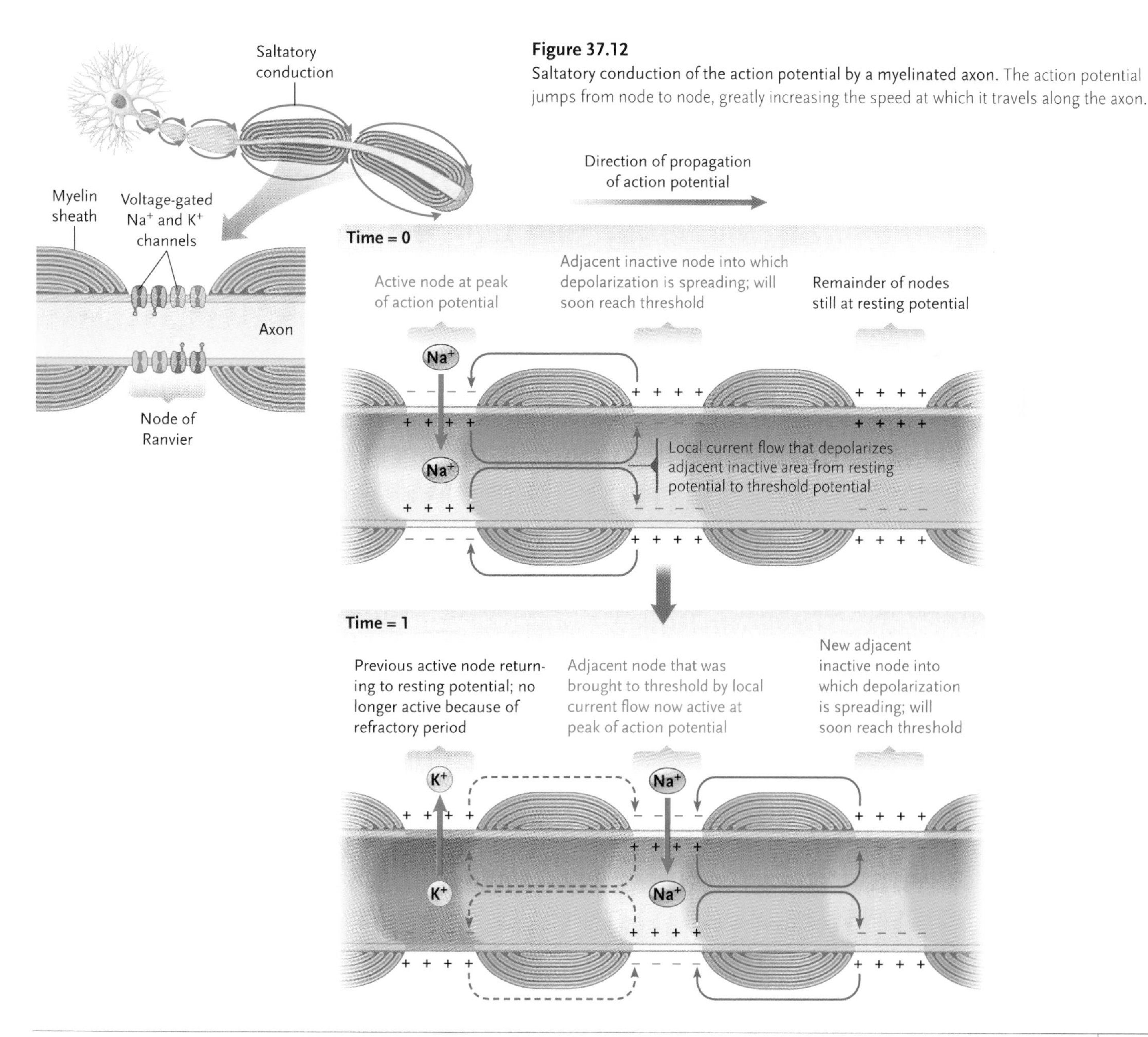

Figure 37.12
Saltatory conduction of the action potential by a myelinated axon. The action potential jumps from node to node, greatly increasing the speed at which it travels along the axon.

STUDY BREAK

1. What mechanism ensures that an electrical impulse in a neuron is conducted in only one direction down the axon?
2. How does having a myelin sheath affect the conduction of impulses in neurons?

37.3 Conduction across Chemical Synapses

Action potentials are transmitted directly across electrical synapses, but they cannot jump across the cleft in a chemical synapse. Instead, the arrival of an action potential causes neurotransmitter molecules—which are synthesized in the cell body of the neuron—to be released by the plasma membrane of the axon terminal, called the **presynaptic membrane (Figure 37.13).** The neurotransmitter diffuses across the cleft and alters ion conduction by activating *ligand-gated ion channels* in the **postsynaptic membrane,** the plasma membrane of the postsynaptic cell. **Ligand-gated ion channels** are channels that open or close when a specific chemical, such as a neurotransmitter, binds to the channel. Neurotransmitter communication from presynaptic to postsynaptic cells is a specialized case of cell-to-cell communication and signal transduction, which you learned about in Chapter 7. Neurobiologists study this phenomenon from the standpoint of understanding the function of neurons, but some of the details being learned are similar in many other types of cells.

Neurotransmitters work in one of two ways. **Direct neurotransmitters** bind directly to a ligand-gated ion channel in the postsynaptic membrane, which opens or closes the channel gate and alters the flow of a specific ion or ions in the postsynaptic cell. The time between arrival of an action potential at an axon terminal and alteration of the membrane potential in the postsynaptic cell may be as little as 0.2 msec.

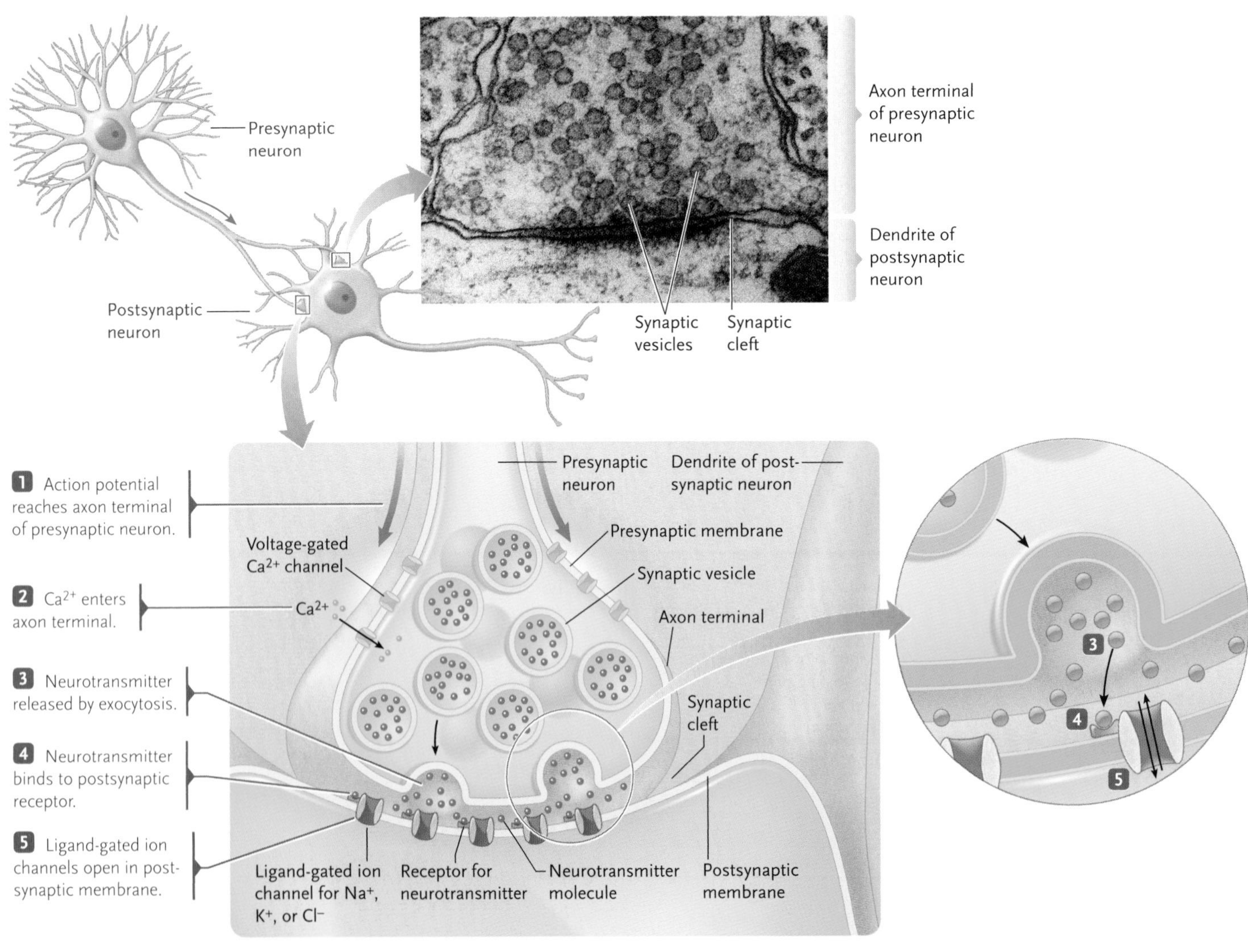

Figure 37.13
Structure and function of chemical synapses.
(Micrograph: © Dennis Kunkel/Visuals Unlimited.)

Indirect neurotransmitters work more slowly (on the order of hundreds of milliseconds). They act as *first messengers,* binding to G-protein-coupled receptors in the postsynaptic membrane, which activates the receptor and triggers generation of a *second messenger* such as cyclic AMP or other processes (see Section 7.4). The cascade of second messenger reactions opens or closes ion-conducting channels in the postsynaptic membrane. Indirect neurotransmitters typically have effects that may last for minutes or hours. Some substances can act as either direct or indirect neurotransmitters, depending on the types of receptors they bind in the receiving cell.

The time required for the release, diffusion, and binding of neurotransmitters across chemical synapses delays transmission as compared with the almost instantaneous transmission of impulses across electrical synapses. However, communication through chemical synapses allows neurons to receive inputs from hundreds to thousands of axon terminals at the same time. Some neurotransmitters have stimulatory effects, while others have inhibitory effects. All of the information received at a postsynaptic membrane is integrated to produce a response. Thus, by analogy, communication by electrical synapses resembles the effect of simply touching one wire to another; communication by direct and indirect neurotransmitters resembles the integration of multiple inputs by a computer chip.

Neurotransmitters Are Released by Exocytosis

Neurotransmitters are stored in secretory vesicles called **synaptic vesicles** in the cytoplasm of an axon terminal. The arrival of an action potential at the terminal releases the neurotransmitters by *exocytosis:* the vesicles fuse with the presynaptic membrane and release the neurotransmitter molecules into the synaptic cleft.

The release of synaptic vesicles depends on voltage-gated Ca^{2+} channels in the plasma membrane of an axon terminal (see Figure 37.13). Ca^{2+} ions are constantly pumped out of all animal cells by an active transport protein in the plasma membrane, keeping their concentration higher outside than inside. As an action potential arrives, the change in membrane potential opens the Ca^{2+} channel gates in the axon terminal, allowing Ca^{2+} to flow back into the cytoplasm. The rise in Ca^{2+} concentration triggers a protein in the membrane of the synaptic vesicle that allows the vesicle to fuse with the plasma membrane, releasing neurotransmitter molecules into the synaptic cleft.

Each action potential arriving at a synapse typically causes approximately the same number of synaptic vesicles to release their neurotransmitter molecules. For example, arrival of an action potential at one type of synapse causes about 300 synaptic vesicles to release a neurotransmitter called acetylcholine. Each vesicle contains about 10,000 molecules of the neurotransmitter, giving a total of some 3 million acetylcholine molecules released into the synaptic cleft by each arriving action potential.

When a stimulus is no longer present, action potentials are no longer generated and a response is no longer needed. In this case a series of events prevents continued transmission of the signal. When action potentials stop arriving at the axon terminal, the voltage-gated Ca^{2+} channels in the axon terminal close and the Ca^{2+} in the axon cytoplasm is quickly pumped to the outside. The drop in cytoplasmic Ca^{2+} stops vesicles from fusing with the presynaptic membrane, and no further neurotransmitter molecules are released. Any free neurotransmitter molecules remaining in the cleft quickly diffuse away, are broken down by enzymes in the cleft, or are pumped back into the axon terminals or into glial cells by active transport. Transmission of impulses across the synaptic cleft ceases within milliseconds after action potentials stop arriving at the axon terminal.

Most Neurotransmitters Alter Ion Flow through Na^+ or K^+ Channels

Neurotransmitters work by opening or closing membrane-embedded ligand-gated ion channels; most of these channels conduct Na^+ or K^+ across the postsynaptic membrane, although some regulate chloride ions (Cl^-). The altered ion flow in the postsynaptic cell that results from the opening or closing of the gates may stimulate or inhibit the generation of action potentials by that cell. For example, if Na^+ channels are opened, the inward Na^+ flow brings the membrane potential of the postsynaptic cell toward the threshold (the membrane becomes depolarized). If K^+ channels are opened, the outward flow of K^+ has the opposite effect (the membrane becomes hyperpolarized). The combined effects of the various stimulatory and inhibitory neurotransmitters at all the chemical synapses of a postsynaptic neuron or muscle cell determine whether the postsynaptic cell triggers an action potential. (*Insights from the Molecular Revolution* describes experiments that worked out the structure and function of an ion channel gated directly by a neurotransmitter.)

Many Different Molecules Act as Neurotransmitters

In all, nearly 100 different substances are now known or suspected to be neurotransmitters. Most of them are relatively small molecules that diffuse rapidly across the synaptic cleft. Some axon terminals release only one type of neurotransmitter while others release several types. Depending on the type of receptor to which it binds, the same neurotransmitter may stimulate or inhibit the generation of action potentials in the post-

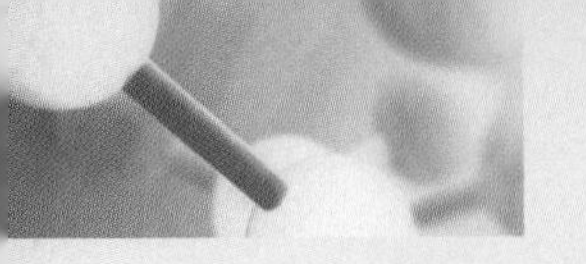

Insights from the Molecular Revolution

Dissecting Neurotransmitter Receptor Functions

Many receptors for direct neurotransmitters are part of an ion channel that is opened or closed by the binding of a neurotransmitter molecule. Each of these receptors has two regions: a large, hydrophilic portion on the outside surface of the plasma membrane that binds the neurotransmitter and a hydrophobic transmembrane portion that anchors the receptor in the plasma membrane and forms the ion-conducting channel.

Jean-Luc Eiselé and his coworkers at the National Center of Scientific Research and the Central Medical University in Switzerland were interested in determining whether the two primary activities of these receptors—binding neurotransmitters and conducting ions—depend on parts of the protein that work independently or reflect an integration of the entire protein structure.

To find out, Eiselé and his colleagues constructed artificial receptors using regions of the receptors for two different neurotransmitters, acetylcholine and serotonin. These two receptors, although related in amino acid sequence and structure, bind different neurotransmitters and react differently to calcium ions. Ion conduction by the acetylcholine receptor is enhanced by Ca^{2+}, while Ca^{2+} ions block the channel of the serotonin receptor and stop ion conduction.

To create the artificial receptors, the investigators broke the genes encoding the acetylcholine and serotonin receptors into two parts. They then reassembled the parts so that in the protein encoded by the composite gene, the part of the acetylcholine receptor located on the membrane surface was joined to the transmembrane channel of the serotonin receptor. Five versions of the artificial gene, encoding proteins in which the two parts were joined at different positions in the amino acid sequence, were then cloned to increase their quantity and injected into oocytes of the clawed frog, *Xenopus laevis.* Once in the oocytes, the genes were translated into the artificial receptor proteins, which were inserted into the oocyte plasma membranes.

Of the five artificial receptors, all were able to bind acetylcholine, but only two were able to conduct ions in response to binding the neurotransmitter, as measured by an increase in the electrical current flowing across the plasma membrane of the oocytes. Agents that inhibit the normal acetylcholine receptor, such as curare, also inhibited the artificial receptors. Serotonin, in contrast, was not bound and did not open the receptor channels, and agents that inhibit the normal serotonin receptor had no effect on the artificial receptors. However, elevated Ca^{2+} concentrations blocked the channel, as in the normal serotonin receptor.

The remarkable research by Eiselé and his coworkers indicates that the parts of a receptor binding a neurotransmitter and conducting ions function independently. Their work also demonstrates the feasibility of constructing composite receptors as a means for dissecting the functions of subregions of the receptors.

synaptic cell. **Figure 37.14** depicts some examples of neurotransmitters.

Acetylcholine acts as a neurotransmitter in both invertebrates and vertebrates. In vertebrates, it acts as a direct neurotransmitter between neurons and muscle cells and as an indirect neurotransmitter between neurons carrying out higher brain functions such as memory, attention, perception, and learning. Acetylcholine-releasing neurons in the brain degenerate in people who develop Alzheimer disease, in which memory, speech, and perceptual abilities decline.

Acetylcholine is the target of many natural and artificial poisons. Curare, a plant extract used as an arrow poison by some indigenous peoples of South America, blocks muscle contraction and produces paralysis by competing directly with acetylcholine for binding sites in synapses that control muscle cells. Atropine, an ingredient of the drops an eye doctor uses to dilate your pupils, is also a plant extract; it relaxes the iris muscles by blocking their acetylcholine receptors. Nicotine also binds to acetylcholine receptors, but acts as a stimulant by turning the receptors on rather than off.

Several amino acids operate as direct neurotransmitters in the CNS of vertebrates and in nerve-muscle synapses of insects and crustaceans. *Glutamate* and *aspartate* stimulate action potentials in postsynaptic cells. They are directly involved in vital brain functions such as memory and learning. *Gamma aminobutyric acid (GABA),* a derivative of glutamate, acts as an inhibitor by opening Cl^- channels in postsynaptic membranes. *Glycine* is also an inhibitor.

Other substances can block the operation of these neurotransmitters. For example, tetanus toxin, released by the bacterium *Clostridium tetani,* blocks GABA release in synapses that control muscle contraction. The body muscles contract so forcibly that the body arches painfully and the teeth become tightly clenched, giving the condition its common name of lockjaw. Once the effects extend to respiratory muscles, the victim quickly dies.

The biogenic amines, which are derived from amino acids, act primarily as indirect neurotransmitters in the CNS. *Norepinephrine, epinephrine,* and *dopamine,* all derived from tyrosine, function as neurotransmitters between interneurons involved in such diverse brain and body functions as consciousness, memory, mood, sensory perception, muscle movements, maintenance of blood pressure, and sleep. Norepinephrine

Figure 37.14
Chemical structures of the major neurotransmitter types.

Acetylcholine

$$H_3C-C(=O)-O-CH_2-CH_2-N^+(CH_3)_3$$

Biogenic amines

Serotonin: HO–(indole ring)–$C-CH_2-CH_2-NH_2$ (ring: CH, N, H)

Norepinephrine: $(HO)_2$–(benzene ring)–$CH(OH)-CH_2-NH_3^+$

Epinephrine: $(HO)_2$–(benzene ring)–$CH(OH)-CH_2-NH-CH_2$

Dopamine: $(HO)_2$–(benzene ring)–$CH_2-CH_2-NH_3^+$

Amino acids

Aspartate: $H_3N^+-CH(COO^-)-CH_2-COO^-$

Glutamate: $H_3N^+-CH(COO^-)-CH_2-CH_2-COO^-$

GABA (gamma aminobutyric acid): $H_3N^+-CH_2-CH_2-CH_2-COO^-$

Glycine: $H_3N^+-CH_2-COO^-$

Neuropeptides

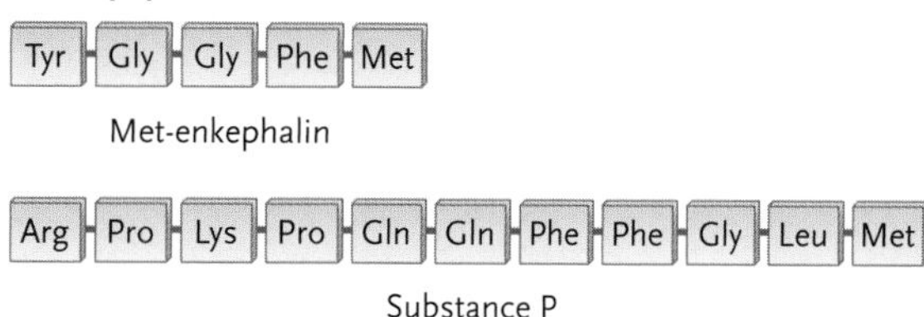

and epinephrine are also released into the general body circulation as hormones. Parkinson disease, in which there is a progressive loss of muscle control, results from degeneration of dopamine-releasing neurons in regions of the brain that coordinate muscular movements. *Serotonin,* which is derived from tryptophan, is released by interneurons in the pathways regulating appetite, reproductive behavior, muscular movements, sleep, and emotional states such as anxiety.

Several drugs enhance or inhibit the action of biogenic amines. For example, cocaine binds to the transporters for active reuptake of certain neurotransmitters such as norepinephrine, dopamine, and serotonin from the synaptic cleft, thereby preventing them from being reabsorbed by the neurons that released them. As a result, the concentrations of the neurotransmitters increase in the synapses, leading to amplification of their natural effects. That is, the affected neurons produce symptoms characteristic of cocaine use, namely high energy from the norepinephrine, euphoria from the dopamine, and feelings of confidence from the serotonin.

Neuropeptides, which are short chains of two or more amino acids, act as indirect neurotransmitters in the central and peripheral nervous systems of both vertebrates and invertebrates. More than 50 neuropeptides are now known. Neuropeptides are also released into the general body circulation as peptide hormones.

Neuropeptides called *endorphins* ("endogenous morphines") are released during periods of pleasurable experience such as eating or sexual intercourse, or physical stress such as childbirth or extended physical exercise. These neurotransmitters have the opiatelike property of reducing pain and inducing euphoria, well known to exercise buffs as a pleasant by-product of their physical efforts. Most endorphins act on the PNS and effectors such as muscles, but *enkephalins,* a subclass of the endorphins, bind to particular receptors in the CNS. Morphine, a potent drug extracted from the opium poppy, blocks the sensation of pain and produces a sensation of well-being by binding to the same receptors in the brain.

Another neuropeptide associated with pain response is *substance P,* which is released by special neurons in the spinal cord. Its effect is to increase messages associated with intense, persistent, or severe pain. For example, suppose you put your hand on a hot barbecue grill. You will snatch your hand away immediately by reflex action, and you will feel the "ouch" of the pain a little later. Why do events occur in this order? The reflex action is driven by rapid nerve impulse conduction along myelinated neurons. The neurons that release substance P are not myelinated, however, so their signal is conducted more slowly and the feeling of pain is delayed. The action of endorphins is antagonistic to substance P, reducing the perception of pain.

In mammals and probably other vertebrates, some neurons synthesize and release dissolved carbon monoxide and nitric oxide as neurotransmitters. For example, in the brain, carbon monoxide regulates the release of hormones from the hypothalamus. Nitric oxide contributes to many nervous system functions such as learning, sensory responses, and muscle movements. By relaxing smooth muscles in the walls of blood vessels, nitric oxide causes the vessels to dilate, increasing the flow of blood. For example, when a male is sexually aroused, neurons release nitric oxide into the erectile tissues in the penis. Relaxation of the muscles increases blood flow into the tissues, causing them to fill with blood and produce an erection. The impotency drug Viagra aids erection by inhibiting an enzyme that normally reduces nitric oxide concentration in the penis.

Study Break

1. What features characterize a substance as a neurotransmitter?
2. Describe how a direct neurotransmitter in a presynaptic neuron controls action potentials in a postsynaptic neuron.

37.4 Integration of Incoming Signals by Neurons

Most neurons receive a multitude of stimulatory and inhibitory signals carried by both direct and indirect neurotransmitters. These signals are integrated by the postsynaptic neuron into a response that reflects their combined effects. The integration depends primarily on the patterns, number, types, and activity of the synapses the postsynaptic neuron makes with presynaptic neurons. Inputs from other sources, such as indirect neurotransmitters and other signal molecules, can modify the integration. The response of the postsynaptic neuron is elucidated by the frequency of action potentials it generates.

Integration at Chemical Synapses Occurs by Summation

As mentioned earlier, depending on the type of receptor to which it binds, a neurotransmitter may stimulate or inhibit the generation of action potentials in the postsynaptic neuron. If a neurotransmitter opens a ligand-gated Na^+ channel, Na^+ enters the cell, causing a depolarization. This change in membrane potential pushes the neuron closer to threshold; that is, it is excitatory and is called an **excitatory postsynaptic potential,** or **EPSP.** On the other hand, if a neurotransmitter opens a ligand-gated ion channel that allows Cl^- to flow into the cell and K^+ to flow out, hyperpolarization occurs. This change in membrane potential pushes the neuron farther from threshold; that is, it is inhibitory and is called an **inhibitory postsynaptic potential,** or **IPSP.** In contrast to the all-or-nothing operation of an action potential, EPSPs and IPSPs are **graded potentials,** in which the membrane potential increases or decreases without necessarily triggering an action potential. And there are no refractory periods for EPSPs and IPSPs.

A neuron typically has hundreds to thousands of chemical synapses formed by axon terminals of presynaptic neurons contacting its dendrites and cell body **(Figure 37.15).** The events that occur at a single synapse produce either an EPSP or an IPSP in that postsynaptic neuron. But how is an action potential produced if a single EPSP is not sufficient to push the postsynaptic neuron to threshold? The answer involves the summation of the inputs received through those many chemical synapses formed by presynaptic neurons. At any given time, some or many of the presynaptic neurons may be firing, producing EPSPs and/or IPSPs in the postsynaptic neuron. The sum of all the EPSPs and IPSPs at a given time determines the total potential in the postsynaptic neuron and, therefore, how that neu-

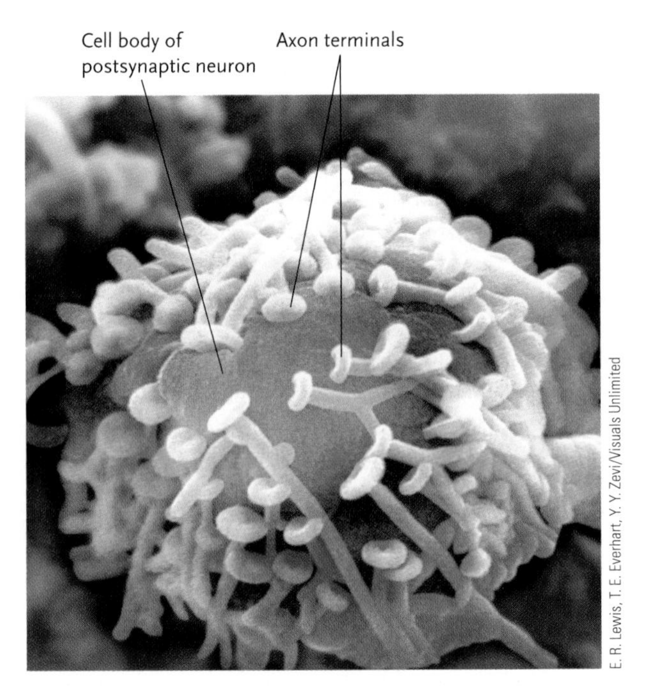

E. R. Lewis, T. E. Everhart, Y. Y. Zevi/Visuals Unlimited

Figure 37.15
The multiple chemical synapses relaying signals to a neuron. The drying process used to prepare the neuron for electron microscopy has toppled the axon terminals and pulled them away from the neuron's surface.

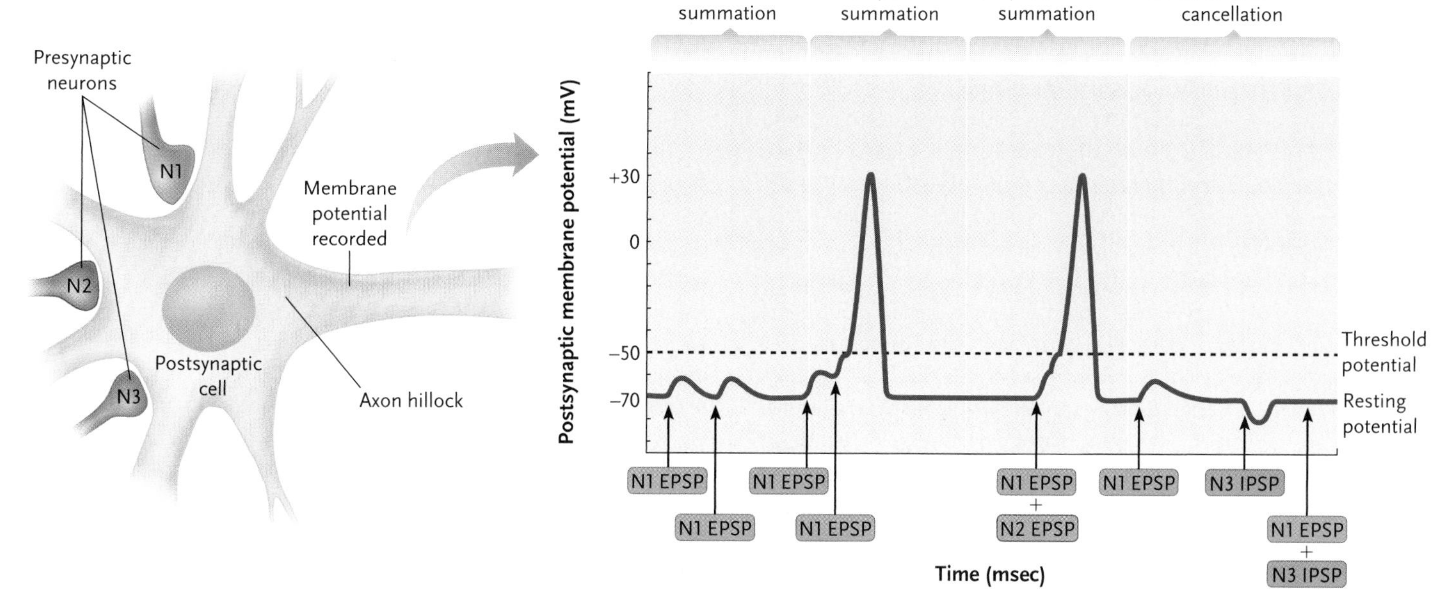

Figure 37.16 Summation of EPSPs and IPSPs by a postsynaptic neuron.

ron responds. **Figure 37.16** shows, in a greatly simplified way, the effects of EPSPs and IPSPs on membrane potential, and how summation of inputs brings a postsynaptic neuron to threshold.

The postsynaptic neuron in the figure has three neurons, N1–N3, forming synapses with it. Suppose that the axon of N1 releases a neurotransmitter, which produces an EPSP in the postsynaptic cell (see Figure 37.16a). The membrane depolarizes, but not enough to reach threshold. If N1 input causes a new EPSP after the first EPSP has died down, it will be of the same magnitude as the first EPSP and no progression toward threshold has taken place—no summation has occurred. If instead, N1 input causes a new EPSP before the first EPSP has died down, the second EPSP will sum with the first and a greater depolarization will have taken place (see Figure 37.16b). This summation of two (or more) EPSPs produced by successive firing of a single presynaptic neuron over a short period of time is called **temporal summation.** If the total depolarization achieved in this way reaches threshold, an action potential will be produced in the postsynaptic neuron. The postsynaptic cell may also be brought to threshold by **spatial summation**, the summation of EPSPs produced by the firing of different presynaptic neurons, such as N1 and N2 (see Figure 37.16c). Lastly, EPSPs and IPSPs can cancel each other out. In the example shown in Figure 37.16d, firing of N1 alone produces an EPSP, firing of N3 alone produces an IPSP, while firing of N1 and N3 simultaneously produces no change in the membrane potential.

The summation point for EPSPs and IPSPs is the axon hillock of the postsynaptic neuron. The greatest density of voltage-gated Na^+ channels occurs in that region, resulting in the lowest threshold potential in the neuron.

The Patterns of Synaptic Connections Contribute to Integration

The total number of connections made by a neuron may be very large—some single interneurons in the human brain, for example, form as many as 100,000 synapses with other neurons. The synapses are not absolutely fixed; they can change through modification, addition, or removal of synaptic connections—or even entire neurons—as animals mature and experience changes in their environments. The combined activities of all the neurons in the nervous system, constantly integrating information from sensory receptors and triggering responses by effectors, control the internal activities of animals and regulate their behavior. This behavior ranges from the simple reflexes of a flatworm to the complex behavior of mammals, including consciousness, emotions, reasoning, and creativity in humans.

Although researchers do not yet understand how processes such as ion flow, synaptic connections, and neural networks produce complex mental activities, they continue to find correspondences between them and the types of neuronal communication described in this chapter. In the next chapter we learn about how nervous systems of animals are organized, and how higher functions such as memory, learning, and consciousness are produced.

Study Break

How does a postsynaptic neuron integrate signals carried by direct and indirect neurotransmitters?

Unanswered Questions

What is the basic wiring diagram of the brain?

In this chapter, you've learned that neurons communicate with each other through synaptic actions that can directly or indirectly cause a change in the membrane potential of a postsynaptic neuron. In order to understand how the brain processes information, it is important first to understand the wiring that handles this information and what forms that communication can take. There are about 100,000,000,000 neurons in the human brain. This is a daunting number, but it can be managed by grouping the neurons into classes or categories. It has been estimated that there are fewer than 10,000 different neuronal classes in the entire brain. Still, each neuron makes and receives synaptic contacts with about 1,000 other neurons on average, making the wiring diagram quite complex. We do not have an adequate means to represent this complexity, nor do we have a means of understanding how information flows through such a complex network. Just as modern sequencing technology allowed a revolution in the field of genomics (the categorization of all of the genes in an organism); similar breakthroughs will need to occur in the field of neuromics (the categorization of all of the neurons and their interactions). These breakthroughs will have to be in data management, computational simulations, and multisite recording techniques.

In our lab, we are working on a way to represent our knowledge of the brain's wiring with an online knowledge base called NeuronBank. To test NeuronBank, we are using the simple nervous systems of sea slugs, especially *Tritonia diomedea*. A sea slug brain has only 10,000 neurons total, many of which are individually identifiable from animal to animal. Eventually, different branches of NeuronBank will represent our knowledge about the basic wiring of the nervous systems of different animals, allowing the neurons and their connections to be compared across species. Having ready access to this information will allow researchers to better design drugs that target specific neurons. This may aid in treatments for neurological conditions ranging from Parkinson's disease to some forms of blindness.

The way information is conveyed between neurons is not yet understood fully. Much of neuroscience has focused on classical neurotransmission. However, the brain uses many other signaling devices. In our lab, we are also using sea slugs to study neuromodulatory signaling by neurons that release the neurotransmitter serotonin, which regulates appetite, reproductive behavior, muscular movements, sleep, and emotional states such as anxiety. We have found that these neurons can change the strength of connections made by other neurons, and that the effects of a serotonin-releasing neuron depend upon the state of the neuron that it is modulating. So, signaling in the nervous system is not a simple matter of summating excitatory and inhibitory inputs; it involves complex, state-dependent actions. Understanding the complexities of neuronal signaling will allow researchers to understand better how the brain processes information. Another unanswered question is simply, "What are all of the different ways that neurons use for communicating information?"

The ultimate question is "How does all of this processing in the brain lead to self-awareness or consciousness?" We do not have an answer for this question as yet. We know that blocking the activity in parts of the brain, through injury, disease, or drugs, can decrease or alter consciousness. But we do not understand how this activity gives rise to the sensation of "being." That is the ultimate question about how the brain works.

Paul Katz is a professor of biology and the director of the Center for Neuromics at Georgia State University. His research interests include neuromics, neuromodulation, and the evolution of neuronal circuits. Learn more about his work at http://www2.gsu.edu/~biopsk.

Review

Go to **Thomson**NOW™ at www.thomsonedu.com/login to access quizzing, animations, exercises, articles, and personalized homework help.

37.1 Neurons and Their Organization in Nervous Systems: An Overview

- The nervous system of an animal (1) receives information about conditions in the internal and external environment, (2) transmits the message along neurons, (3) integrates the information to formulate an appropriate response, and (4) sends out signals to muscles or glands that accomplish the response (Figure 37.2).
- Neurons have dendrites, which receive information and conduct signals toward the cell body, and axons, which conduct signals away from the cell body to another neuron or an effector (Figure 37.3).
- Afferent neurons conduct information from sensory receptors to interneurons, which integrate the information into a response. The response signals are passed to efferent neurons, which activate the effectors carrying out the response (Figure 37.2).
- The combination of an afferent neuron, an interneuron, and an efferent neuron makes up a basic neuronal circuit. The circuits combine into networks that interconnect the peripheral and central nervous systems.
- Glial cells help maintain the balance of ions surrounding neurons and form insulating layers around the axons (Figure 37.5).
- Neurons make connections by two types of synapses. In an electrical synapse, impulses pass directly from the sending to the receiving cell. In a chemical synapse, neurotransmitter molecules released by the presynaptic cell diffuse across a narrow synaptic cleft and bind to receptors in the plasma membrane of the postsynaptic cell (Figure 37.6).

Animation: Neuron structure and function

Animation: Nerve structure

Animation: Impulse travelling through a nerve

37.2 Signal Conduction by Neurons

- The membrane potential of a cell depends on the unequal distribution of positive and negative charges on either side of the membrane, which establishes a potential difference across the membrane.
- Three primary conditions contribute to the resting potential of neurons: (1) an Na^+/K^+ active transport pump that sets up concentration gradients of Na^+ ions (higher outside) and K^+ ions (higher inside); (2) an open channel that allows K^+ to flow out

freely; and (3) negatively charged proteins and other molecules inside the cell that cannot pass through the membrane (Figure 37.8).

- An action potential is generated when a stimulus pushes the resting potential to the threshold value at which voltage-gated Na^+ and K^+ channels open in the plasma membrane. The inward flow of Na^+ changes membrane potential abruptly from negative to a positive peak. The potential falls to the resting value again as the gated K^+ channels allow this ion to flow out (Figure 37.10).
- Action potentials move along an axon as the ion flows generated in one segment depolarize the potential in the next segment (Figure 37.11).
- Action potentials are prevented from reversing direction by a brief refractory period, during which a segment of membrane that has just generated an action potential cannot be stimulated to produce another for a few milliseconds.
- In myelinated axons, ions can flow across the plasma membrane only at nodes where the myelin sheath is interrupted. As a result, action potentials skip rapidly from node to node by saltatory conduction (Figure 37.12).

Animation: Ion concentrations

Animation: Ion flow in myelinated axons

Animation: Action potential propagation

Animation: Measuring membrane potential

Animation: Stretch reflex

37.3 Conduction across Chemical Synapses

- Neurotransmitters released into the synaptic cleft bind to receptors in the plasma membrane of the postsynaptic cell, altering the flow of ions across the plasma membrane of the postsynaptic cell and pushing its membrane potential toward or away from the threshold potential (Figure 37.13).
- A direct neurotransmitter binds to a receptor associated with a ligand-gated ion channel in the postsynaptic membrane; the binding opens or closes the channel.
- An indirect neurotransmitter binds to a receptor in the postsynaptic membrane and triggers generation of a second messenger, which leads to the opening or closing of a gated channel.
- Neurotransmitters are released from synaptic vesicles into the synaptic cleft by exocytosis, which is triggered by entry of Ca^{2+} ions into the cytoplasm of the axon terminal through voltage-gated Ca^{2+} channels opened by the arrival of an action potential.
- Neurotransmitter release stops when action potentials cease arriving at the axon terminal. Neurotransmitters remaining in the synaptic cleft are broken down by enzymes or taken up by the axon terminal or glial cells.
- Types of neurotransmitters include acetylcholine, amino acids, biogenic amines, neuropeptides, and gases such as NO and CO (Figure 37.14). Many of the biogenic amines and neuropeptides are also released into the general body circulation as hormones.

Animation: Chemical synapse

37.4 Integration of Incoming Signals by Neurons

- Neurons carry out integration by summing excitatory postsynaptic potentials (EPSPs) and inhibitory postsynaptic potentials (IPSPs); the summation may push the membrane potential of the postsynaptic cell toward or away from the threshold for an action potential (Figure 37.16).
- The combined effects of summation in all the neurons in the nervous system control behavior in animals and underlie complex mental processes in mammals.

Animation: Synaptic integration

Questions

Self-Test Questions

1. Nerve signals travel in the following manner:
   a. A dendrite of a sensory neuron receives the signal; its cell body transmits the signal to a motor neuron's axon, and the signal is sent to the target.
   b. An axon of a motor neuron receives the signal; its cell body transmits the signal to a sensory neuron's dendrite, and the signal is sent to the target.
   c. Efferent neurons conduct nerve impulses toward the cell body of sensory neurons, which send them on to interneurons and ultimately to afferent motor neurons.
   d. A dendrite of a sensory neuron receives a signal; the cell's axon transmits the signal to an interneuron; the signal is then transmitted to dendrites of a motor neuron and sent forth on its axon to the target.
   e. The axons of oligodendrocytes transmit nerve impulses to the dendrites of astrocytes.
2. Glial cells:
   a. are unable to divide after an animal is born.
   b. in the PNS called Schwann cells form the insulating myelin sheath around axons.
   c. called astrocytes form the nodes of Ranvier in the brain.
   d. called oligodendrocytes are star-shaped cells in the PNS.
   e. are neuronal cells that connect to interneurons.
3. An example of a synapse could be the site where:
   a. neurotransmitters released by an axon travel across a gap and are picked up by receptors on a muscle cell.
   b. an electrical impulse arrives at the end of a dendrite causing ions to flow onto axons of presynaptic neurons.
   c. postsynaptic neurons transmit a signal across a cleft to a presynaptic neuron.
   d. oligodendrocytes contact the dendrites of an afferent neuron directly.
   e. an on-off switch stimulates an electrical impulse in a presynaptic cell to stimulate, not inhibit, other presynaptic cells.
4. The resting potential in neurons requires:
   a. membrane transport channels to be constantly open for Na^+ and K^+ flow.
   b. the inside of neurons to be positive relative to the outside.
   c. a slow movement of K^+ outward with a charge difference in the neural membrane set up by this movement of K^+.
   d. an active Na^+/K^+ pump, which pumps Na^+ and K^+ into the neuron.
   e. three Na^+ ions to be pumped through three Na^+ gates and two K^+ ions to be pumped through two K^+ gates.
5. The major role of the sodium potassium pump is to:
   a. cause a rapid firing of the action potential so the inside of the membrane becomes momentarily positive.
   b. decrease the resting potential to zero.
   c. hyperpolarize the membrane above resting value.
   d. increase a high action potential to enter a refractory period.
   e. maintain the resting potential at a constant negative value.

6. In the propagation of a nerve impulse:
   a. the refractory period begins as the K^+ channel opens, allowing K^+ ions to flow outward with their concentration gradient.
   b. Na^+ ions rush with their concentration gradient out of the axon.
   c. positive charges lower the membrane potential to its lowest action potential.
   d. gated K^+ channels open at the same time as the activation gate of Na^+ channels closes.
   e. the depolarizing stimulus lowers the membrane potential to open the Na^+ gates.
7. Which of the following does not contribute to propagation of action potentials?
   a. As the area outside the membrane becomes negative, it attracts ions from adjacent regions; as the inside of the membrane becomes positive, it attracts negative ions from nearby in the cytoplasm. These events depolarize nearby regions of the axon membrane.
   b. The refractory period allows the impulse to travel in only one direction.
   c. Each segment of the axon prevents the adjacent segments from firing.
   d. The magnitude of the action potential stays the same as it travels down the axon.
   e. Increasing the intensity of the stimulus increases the number of action potentials up to a limit.
8. Which of the following statements best describes saltatory conduction?
   a. It inhibits direct neurotransmitter release.
   b. It transmits the action potential at the nodes of Ranvier and thus speeds up impulses on myelinated axons.
   c. It increases neurotransmitter release at the presynaptic membrane.
   d. It decreases neurotransmitter uptake at chemically gated postsynaptic channels.
   e. It removes neurotransmitters from the synaptic cleft.
9. Transmission of a nerve impulse to its target cell requires:
   a. endocytosis of neurotransmitters by the excitatory presynaptic vesicles.
   b. thousands of molecules of neurotransmitter that had been stored in the postsynaptic cell to be released into the synaptic cleft.
   c. Ca^{2+} ions to diffuse through voltage-gated Ca^{2+} channels.
   d. the fall in Ca^{2+} to trigger a protein that causes the presynaptic vesicle to fuse with the plasma membrane.
   e. an action potential to open the Ca^{2+} gates so that Ca^{2+} ions, in higher concentration outside the axon, can flow back into the cytoplasm of the neuron.
10. Autopsy reports reveal that above a certain threshold, brain size is not related to intelligence. A possible explanation is that the brains of:
   a. gifted people have a much vaster network of neural synapses than do the brains of people with normal intelligence.
   b. people with normal intelligence release far more NO and CO neurotransmitters than do those of the gifted.
   c. people with normal intelligence contain more glutamate and aspartate than do those of the gifted.
   d. gifted people have excessive quantities of gamma aminobutyric acid.
   e. people with normal intelligence contain more glycine than do those of the gifted.

Questions for Discussion

1. In some cases of ADHD (attention deficit hyperactivity disorder) the impulsive, erratic behavior typical of affected people can be calmed with drugs that *stimulate* certain brain neurons. Based on what you have learned about neurotransmitter activity in this chapter, can you suggest a neural basis for this effect?
2. Most sensory neurons form synapses either on interneurons in the spinal cord or on motor neurons. However, in many vertebrates, certain sensory neurons in the nasal epithelium synapse directly on brain neurons that activate behavioral responses to odors. Suggest at least one reason why natural selection might favor such an arrangement.
3. How did evolution of chemical synapses make higher brain functions possible?
4. Use an Internet search engine with the term "Pediatric Neurotransmitter Disease" and, for one such disease, explain how the symptoms relate to neurotransmitter function.

Experimental Analysis

Design an experiment to test whether neurons are connected via electrical or chemical synapses.

Evolution Link

A biologist hypothesized that the mechanism for the propagation of action potentials down a neuron evolved only once. What evidence would you collect from animals living today to support or refute that hypothesis?

Activity in the human brain while reading aloud. The image combines an MRI of a male brain with a PET scan, which shows that blood circulation increases in the language, hearing, and vision areas of the brain, especially in the left hemisphere.

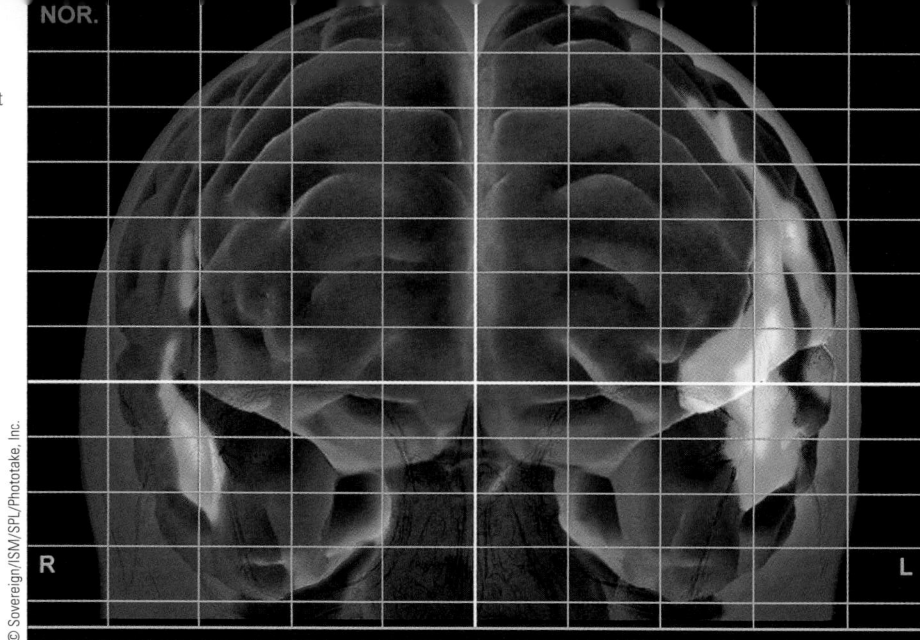

Study Plan

38 Nervous Systems

Why It Matters

The conductor's baton falls and the orchestra plays the first notes of a Mozart symphony. Unaware of the complex interactions of their nervous systems, the musicians translate printed musical notation into melodious sounds played on their instruments. Although their fingers and arms move to produce precise harmonies, the musicians are only vaguely conscious of these movements, learned through years of practice. Their only conscious endeavor is to interpret the music in line with the conductor's directions.

From the back of the hall, a common housefly, *Musca domestica,* moves in random twists and turns that bring it toward the stage. Although far less complex than that of a human, the fly's nervous system contains networks of neurons that work in the same way, in patterns adapted to its lifestyle.

The fly does not register the sounds reverberating through the hall as a significant sensory input. However, some of its receptors are exquisitely sensitive to the presence of potential food molecules, including those in the sweat on the conductor's face. The fly's swoops and turns bring it closer to the conductor; soon it alights on the tip of his nose. When sensory receptors in the fly's footpads detect organic matter on the surface of the nose, they trigger an automated feeding response: the fly's proboscis lowers and its gut begins contractions that suck up the nutrients.

The conductor's eyes notice the insect's approach, and sensory receptors in his skin pinpoint the spot where it lands. Without missing a beat, the conductor's hand flicks toward that exact spot. But his nervous system and effectors, although highly sophisticated, are no match for the escape reflexes of the fly. The fly's sensory receptors detect the motion of the fingers, sending impulses to the fly's leg and wing muscles that launch it into flight long before the fingers reach the nose.

The fly wanders into the orchestra, attracted to potential nutrients on various musicians, who respond with flicking movements that are no more successful than those of the conductor. At last, the fly lands on the left hand of the timpanist, who is listening with pleasure to the music while he awaits his entrance late in the first movement. His right hand holds a mallet. With a skill born of long practice in hitting drums, gongs, and bells with speed and precision, the timpanist deftly swings his mallet and dispatches the fly, ending the latest contest between mammalian and arthropod nervous systems.

The nervous systems underlying these behaviors are one of the features that set animals apart from other organisms. As animals evolved, the need to find food, living space, and mates, and to escape predators and other dangers, provided a powerful selection pressure for increasingly complex and capable nervous systems. Neurons, described in the previous chapter, provide the structural and functional basis for all these systems. We can trace some of the developments along this extended evolutionary pathway by examining the nervous systems of living animals, from invertebrates to mammals, and especially humans.

38.1 Invertebrate and Vertebrate Nervous Systems Compared

The nervous systems of most invertebrates are relatively simple, typically containing fewer neurons, arranged in less complex networks, than vertebrate systems. As animal groups evolved, their nervous systems became more elaborate, providing the ability to integrate more sensory information and to formulate more complex responses. Our comparative survey of nervous systems begins with the simplest invertebrates.

Cnidarians and Echinoderms Have Nerve Nets

Cnidarians and echinoderms are radially symmetrical animals with body parts arranged regularly around a central axis like the spokes of a wheel. Their nervous systems, called **nerve nets,** are loose meshes of neurons organized within that radial symmetry.

The nerve nets of cnidarians such as sea anemones extend into each "spoke" of the body **(Figure 38.1a).** Their neurons lack clearly differentiated dendrites and axons. When part of the animal is stimulated, impulses are conducted through the nerve net in all directions from the point of stimulation. Although there is no cluster of neurons that plays the coordinating role of a brain, nerve cells may be more concentrated in some regions. For example, in scyphozoan jellyfish, which swim by rhythmic contractions of their bells, neurons are denser in a ring around the margin of the bell, in the same area as the contractile cells that produce the swimming movements.

In echinoderms, including sea stars, the nervous system is a modified nerve net, with some neurons organized into **nerves,** bundles of axons enclosed in connective tissue and following the same pathway. A *nerve ring* surrounds the centrally located mouth, and a *radial nerve* that is connected to nerve nets branches throughout each arm **(Figure 38.1b).** If the radial nerve serving an arm is cut, the arm can still move, but not in coordination with the other arms.

More Complex Invertebrates Have Cephalized Nervous Systems

More complex invertebrates have neurons with clearly defined axons and dendrites, and more specialized functions. Some neurons are concentrated into functional clusters called **ganglia** (singular, *ganglion*). A key evolutionary development in invertebrates is a trend toward *cephalization,* the formation of a distinct head region containing both ganglia that constitute a **brain,** the control center of the nervous system, and major sensory structures. One or more solid **nerve cords**—bundles of nerves—extend from the central ganglia to the rest of the body; they are connected to smaller nerves. Another evolutionary trend is toward bilateral symmetry of the body and the nervous system, in which body parts are mirror images on left and right sides. These trends toward cephalization and bilateral symmetry are illustrated here in flatworms, arthropods, and mollusks.

In flatworms, a small brain consisting of a pair of ganglia at the anterior end is connected by two or more longitudinal nerve cords to nerve nets in the rest of the body **(Figure 38.1c).** The brain integrates inputs from sensory receptors, including a pair of anterior eyespots with receptors that respond to light. The brain and longitudinal nerve cords constitute the flatworm's **central nervous system (CNS),** the simplest one known, while the nerves from the CNS to the rest of the body constitute the **peripheral nervous system (PNS).**

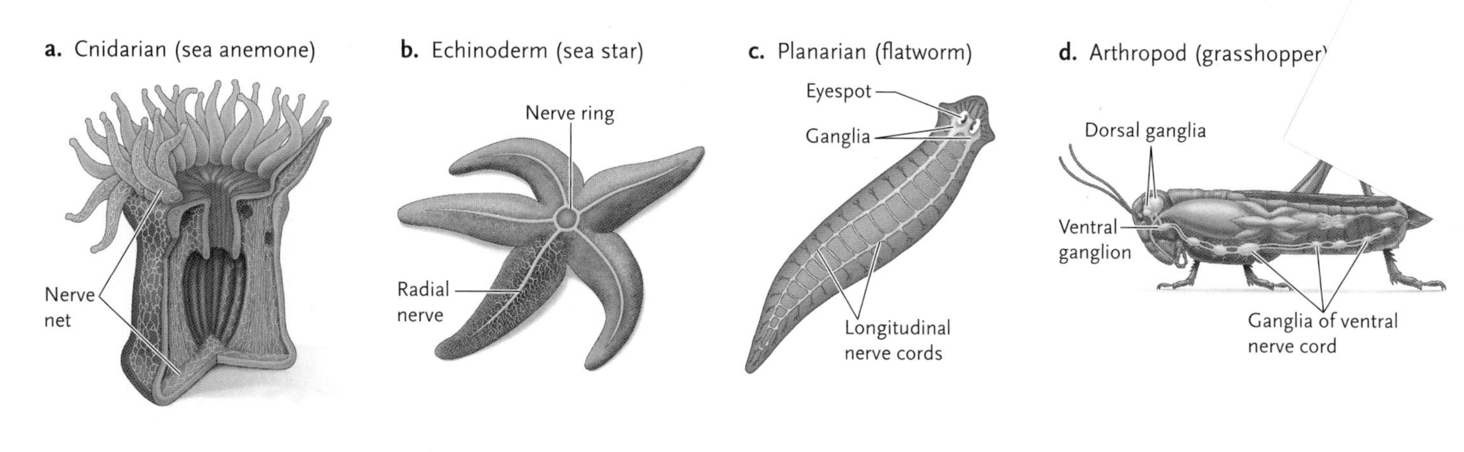

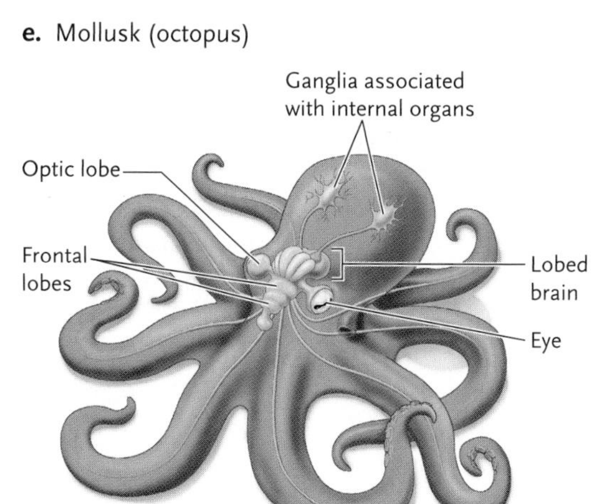

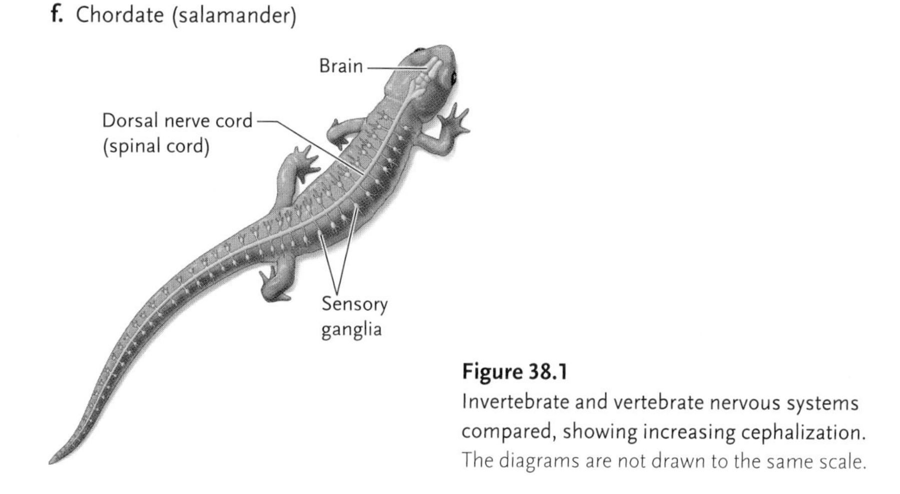

Figure 38.1
Invertebrate and vertebrate nervous systems compared, showing increasing cephalization. The diagrams are not drawn to the same scale.

Arthropods such as insects have a head region that contains a brain consisting of dorsal and ventral pairs of ganglia, and major sensory structures, usually including eyes and antennae **(Figure 38.1d).** The brain exerts centralized control over the remainder of the animal. A ventral nerve cord enlarges into a pair of ganglia in each body segment. In arthropods with fused body segments, as in the thorax of insects, the ganglia are also fused into larger masses forming secondary control centers.

Although different in basic plan from the arthropod system, the nervous systems of mollusks (such as clams, snails, and octopuses) also rely on neurons clustered into paired ganglia and connected by major nerves. Different mollusks have varying degrees of cephalization, with cephalopods having the most pronounced cephalization of any invertebrate group. In the head of an octopus, for example, a cluster of ganglia fuses into a complex, lobed brain with clearly defined sensory and motor regions. Paired nerves link different lobes with muscles and sensory receptors, including prominent optic lobes linked by nerves to large, complex eyes **(Figure 38.1e).** Octopuses are capable of rapid movement to hunt prey and to escape from predators, behaviors that rely on rapid, sophisticated processing of sensory information.

Vertebrates Have the Most Specialized Nervous Systems

In vertebrates, the CNS consists of the brain and spinal cord, and the PNS consists of all the nerves and ganglia that connect the brain and spinal cord to the rest of the body **(Figure 38.1f).** All vertebrate nervous systems are highly cephalized, with major concentrations of neurons in a brain located in the head. In contrast to invertebrate nervous systems, which have solid nerve cords located ventrally, the brain and nerve cord of vertebrates are hollow, fluid-filled structures located dorsally. The head contains specialized sensory organs, which are connected directly to the brain by nerves. Compared with invertebrates, the ganglia are greatly reduced in mass and functional activity except in the gut, which contains extensive interneuron networks.

The structure of the vertebrate nervous system reflects its pattern of development. The nervous system of a vertebrate embryo begins as the hollow **neural tube,** the anterior end of which develops into the brain and the rest into the **spinal cord.** The cavity of the neural tube becomes the fluid-filled **ventricles** of the brain and the **central canal** through the spinal cord. Adjacent tissues give rise to nerves that connect the brain and spinal cord with all body regions.

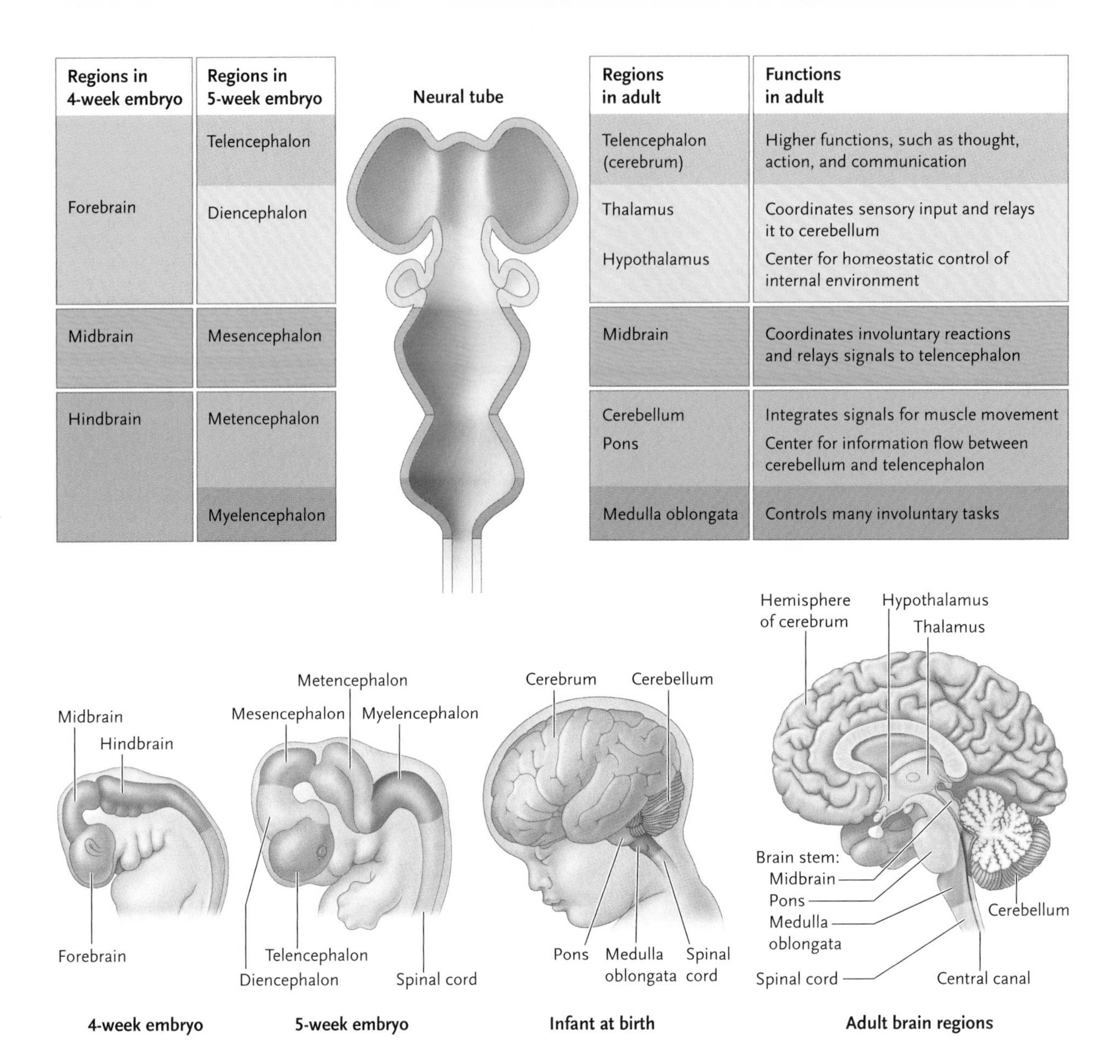

Figure 38.2 Development of the human brain from the anterior end of an embryo's neural tube.

Early in embryonic development, the anterior part of the neural tube enlarges into three distinct regions: the **forebrain, midbrain,** and **hindbrain (Figure 38.2).** A little later, the embryonic hindbrain subdivides into the *metencephalon* and *myelencephalon;* the midbrain develops into the *mesencephalon;* and the forebrain subdivides into the *telencephalon* and *diencephalon.*

The metencephalon gives rise to the *cerebellum,* which integrates sensory signals from the eyes, ears, and muscle spindles with motor signals from the telencephalon, and the *pons,* a major traffic center for information passing between the cerebellum and the higher integrating centers of the adult telencephalon. The myelencephalon gives rise to the *medulla oblongata* (commonly shortened to medulla), which controls many vital involuntary tasks such as respiration and blood circulation. The mesencephalon gives rise to the (adult) midbrain, which with the pons and the medulla constitutes the brain stem. The midbrain has centers for coordinating reflex responses (involuntary reactions) to visual and auditory (hearing) input and relays signals to the telencephalon.

The embryonic telencephalon develops into the *cerebrum* (or adult telencephalon), the largest part of the brain. The cerebrum controls higher functions such as thought, memory, language, and emotions, as well as voluntary movements. The diencephalon gives rise to the *thalamus,* a coordinating center for sensory input and a relay station for input to the cerebellum, and to the *hypothalamus,* the primary center for homeostatic control over the internal environment. In fishes, the cerebrum is little more than a relay station for olfactory (sense of smell) information. In amphibians, reptiles, and birds, it becomes progressively larger and contains greater concentrations of integrative functions. In mammals, the cerebrum is the major integrative structure of the brain.

In the following sections, we examine vertebrate nervous systems, and the human nervous system in particular, beginning with the peripheral nervous system.

Study Break

1. Distinguish between a nerve net, nerves, and nerve cords.
2. What is cephalization?
3. What nervous system structures arise from the embryonic hindbrain, and what are their functions?

38.2 The Peripheral Nervous System

Afferent neurons in the peripheral nervous system transmit signals to the CNS, and signals from the CNS are sent via efferent neurons in the peripheral nervous system to the effectors that carry out responses **(Figure 38.3).** The afferent part of the system includes all the neurons that transmit sensory information from their receptors. The efferent part of the system consists of the axons of neurons that carry signals to the muscles and glands acting as effectors. In mammals, 31 pairs of **spinal nerves** carry signals between the spinal cord and the body trunk and limbs, and 12 pairs of **cranial nerves** connect the brain directly to the head, neck, and body trunk. The efferent part of the PNS is further divided into somatic and autonomic systems (see Figure 38.3).

The Somatic System Controls the Contraction of Skeletal Muscles, Producing Body Movements

The **somatic system** controls body movements that are primarily conscious and voluntary. Its neurons, called motor neurons, carry efferent signals from the CNS to the skeletal muscles. The dendrites and cell bodies of motor neurons are located in the spinal cord; their axons extend from the spinal cord to the skeletal muscle cells they control. As a result, the somatic portions of the cranial and spinal nerves consist only of axons.

Although the somatic system is primarily under conscious, voluntary control, some contractions of skeletal muscles are unconscious and involuntary. These include the reflexes, shivering, and the constant muscle contractions that maintain body posture and balance.

The Autonomic System Is Divided into Sympathetic and Parasympathetic Divisions

The **autonomic nervous system** controls largely involuntary processes including digestion, secretion by sweat glands, circulation of the blood, many functions of the reproductive and excretory systems, and contraction of smooth muscles in all parts of the body. It is organized into *sympathetic* and *parasympathetic* divisions, which

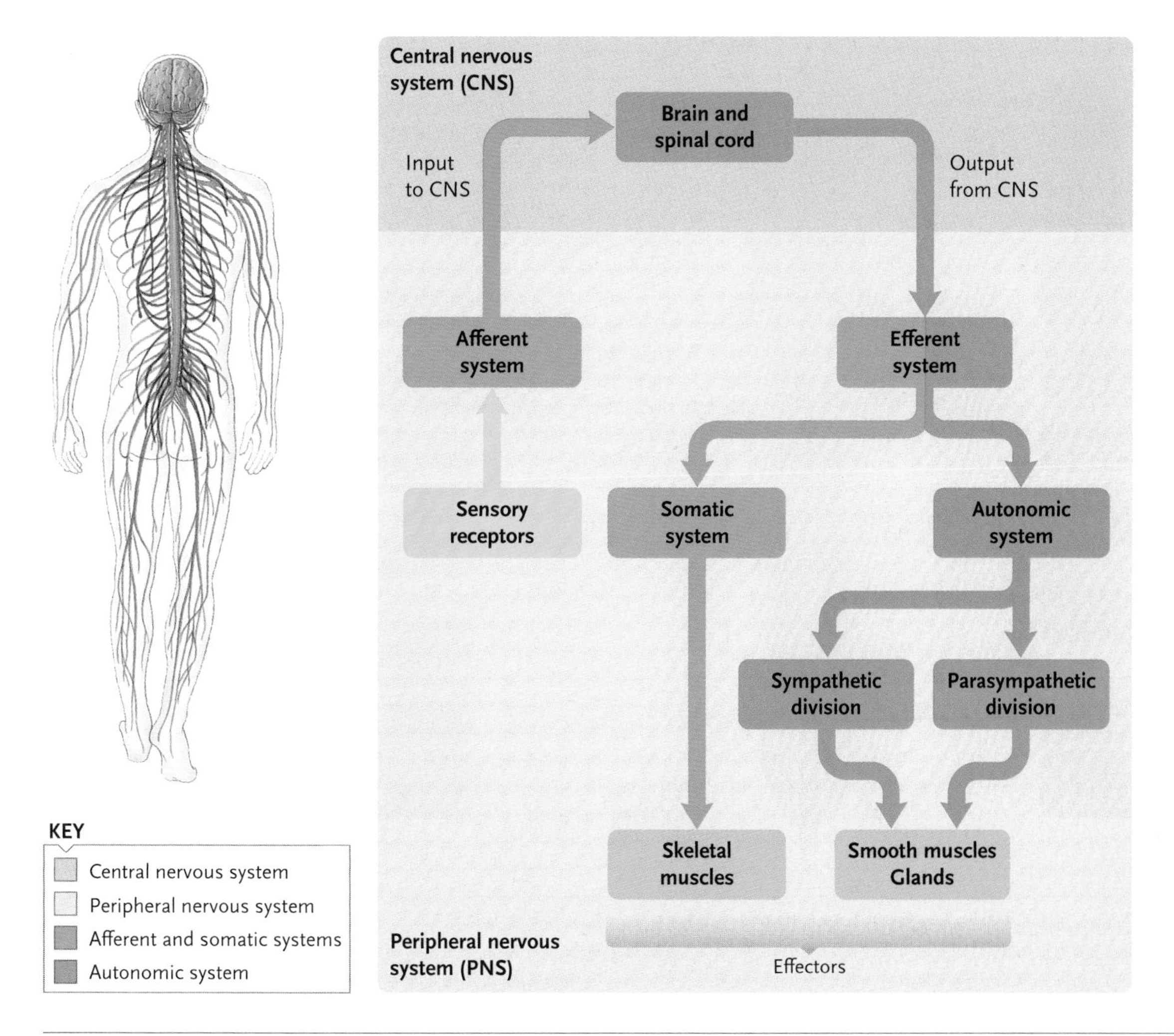

Figure 38.3
The central nervous system (CNS) and peripheral nervous system (PNS), and their subsystems.

are always active, and have opposing effects on the organs they affect, thereby enabling precise control **(Figure 38.4)**. For example, in the circulatory system, sympathetic neurons stimulate the force and rate of the heartbeat, and parasympathetic neurons inhibit these activities. In the digestive system, sympathetic neurons inhibit the smooth muscle contractions that move materials through the small intestine, and parasympathetic neurons stimulate the same activities. These opposing effects control involuntary body functions precisely.

The pathways of the autonomic nervous system include two neurons. The first neuron has its dendrites and cell body in the CNS, and its axon extends to a ganglion outside the CNS in the PNS. There it synapses with the dendrites and cell body of the second neuron in the pathway. The axon of the second neuron extends from the ganglion to the effector carrying out the response.

The **sympathetic division** predominates in situations involving stress, danger, excitement, or strenuous physical activity. Signals from the sympathetic division increase the force and rate of the heartbeat, raise the blood pressure by constricting selected blood vessels, dilate air passages in the lungs, induce sweating, and open the pupils wide. Activities that are less important in an emergency, such as digestion, are suppressed by the sympathetic system. The **parasympathetic division**, in contrast, predominates during quiet, low-stress situations, such as while relaxing. Under its influence the effects of the sympathetic division, such as rapid heartbeat and elevated blood pressure, are reduced and "housekeeping" (maintenance) activities such as digestion predominate.

Study Break

Which of the two autonomic nervous system divisions predominates in the following scenarios? (a) You are hiking on a trail and suddenly a bear appears in your path. (b) It is a hot sunny day. You find a shady tree and sit down. Leaning against its trunk, you feel your eyes becoming heavy.

38.3 The Central Nervous System (CNS) and Its Functions

The central nervous system integrates incoming sensory information from the PNS into compensating responses, thus managing body activities. Our examina-

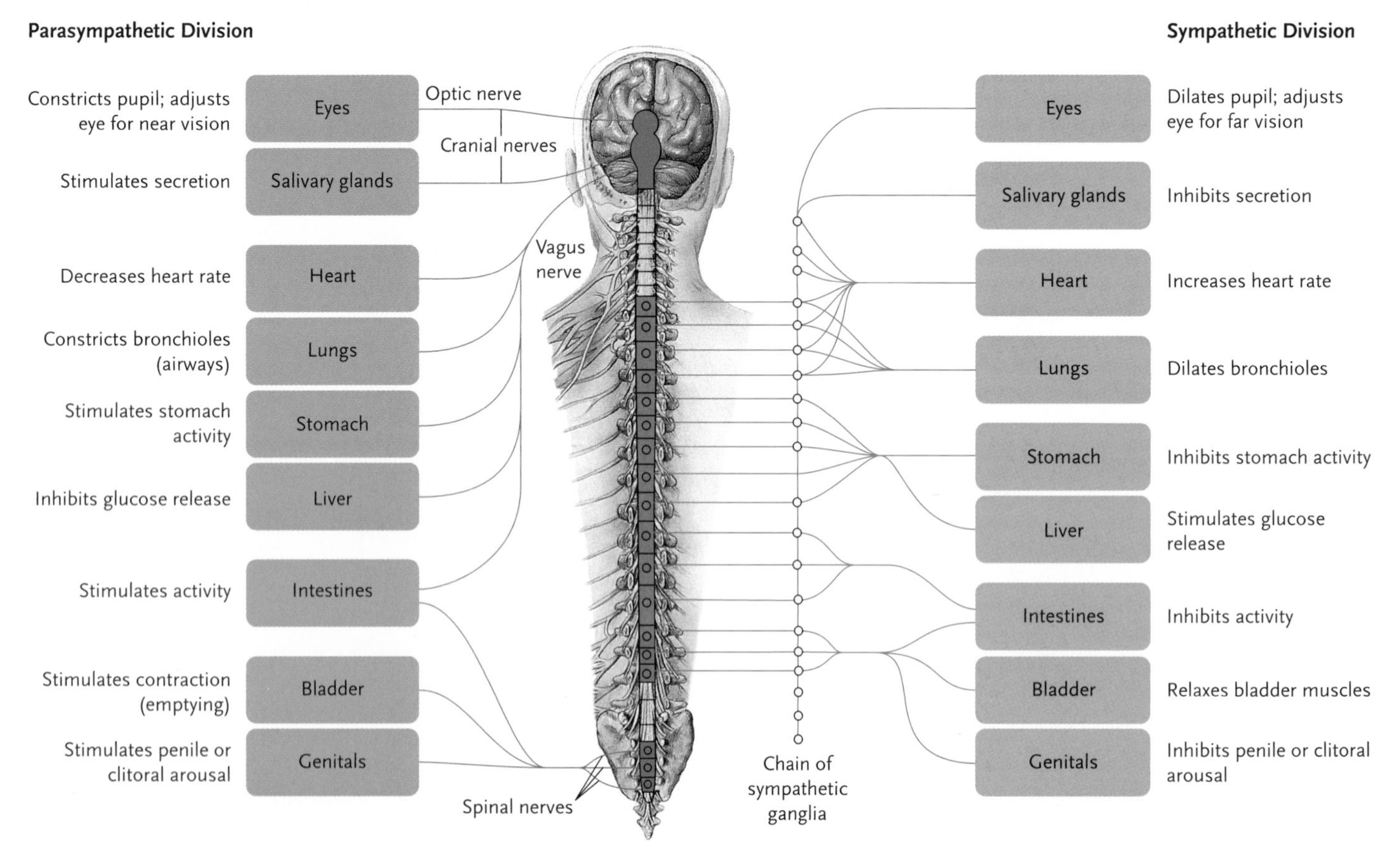

Figure 38.4
Effects of the sympathetic and parasympathetic divisions of the central nervous system on organ and gland function. Only one side of each division is shown; both are duplicated on the left and right sides of the body.

tion of the vertebrate CNS begins with the spinal cord, and then considers the brain and its functions.

The Spinal Cord Relays Signals between the PNS and the Brain and Controls Reflexes

The spinal cord, which extends dorsally from the base of the brain, carries impulses between the brain and the PNS and contains the interneuron circuits that control motor reflexes.

The spinal cord and brain are surrounded and protected by three layers of connective tissue, the **meninges** (*meninga* = membrane), and by **cerebrospinal fluid,** which circulates through the central canal of the spinal cord, through the ventricles of the brain, and between two of the meninges. The fluid cushions the brain and spinal cord from jarring movements and impacts, and it both nourishes the CNS and protects it from toxic substances.

In cross section, the spinal cord has a butterfly-shaped core of **gray matter,** consisting of nerve cell bodies and dendrites. This is surrounded by **white matter,** consisting of axons, many of them surrounded by myelin sheaths. Pairs of spinal nerves connect with the spinal cord at spaces between the vertebrae (**Figure 38.5**).

The afferent axons entering the spinal cord make synapses with interneurons in the gray matter, which send axons upward through the white matter of the spinal cord to the brain. Conversely, axons from interneurons of the brain pass downward through the white matter of the cord and make synapses with the dendrites and cell bodies of efferent neurons in the gray matter of the cord. The axons of these efferent neurons exit the spinal cord through the spinal nerves.

The gray matter of the spinal cord also contains interneurons of the pathways involved in **reflexes,** programmed movements that take place without conscious effort, such as the sudden withdrawal of a hand from a hot surface (see Figure 38.5). When your hand touches the hot surface, the heat stimulates an afferent neuron, which makes connections with at least two interneurons in the spinal cord. One of these interneurons stimulates an efferent neuron, causing the *flexor* muscle of the arm to contract, which bends the arm and withdraws the hand almost instantly from the hot surface. The other interneuron synapses with an efferent neuron connected to an *extensor* muscle, relaxing it so that the flexor can move more quickly. Interneurons connected to the reflex circuits also send signals to the brain, making you aware of the stimulus causing the reflex. You know from experience that when a reflex movement withdraws your hand from a hot surface or other damaging stimulus, you feel the pain shortly *after* the hand is withdrawn. This is the extra time required for impulses to travel from the neurons of the reflex to the brain (see discussion of the neurotransmitter substance P in Section 37.3).

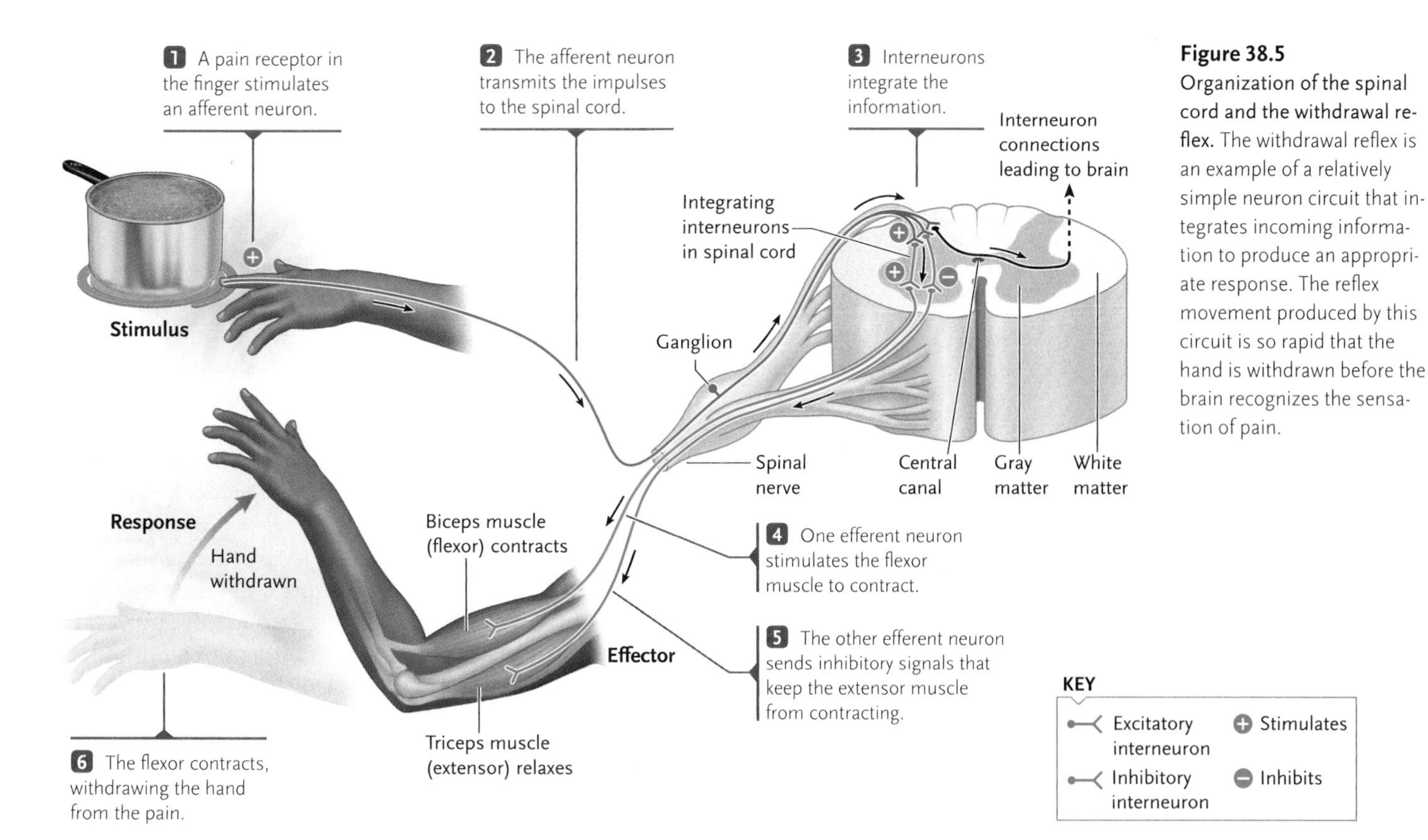
1 A pain receptor in the finger stimulates an afferent neuron.
2 The afferent neuron transmits the impulses to the spinal cord.
3 Interneurons integrate the information.
Interneuron connections leading to brain
Integrating interneurons in spinal cord
Stimulus
Ganglion
Spinal nerve
Central canal
Gray matter
White matter
Response
Hand withdrawn
Biceps muscle (flexor) contracts
4 One efferent neuron stimulates the flexor muscle to contract.
Effector
5 The other efferent neuron sends inhibitory signals that keep the extensor muscle from contracting.
Triceps muscle (extensor) relaxes
6 The flexor contracts, withdrawing the hand from the pain.
KEY
Excitatory interneuron
Inhibitory interneuron
Stimulates
Inhibits

Figure 38.5
Organization of the spinal cord and the withdrawal reflex. The withdrawal reflex is an example of a relatively simple neuron circuit that integrates incoming information to produce an appropriate response. The reflex movement produced by this circuit is so rapid that the hand is withdrawn before the brain recognizes the sensation of pain.

The Brain Integrates Sensory Information and Formulates Compensating Responses

The brain is the major center that receives, integrates, stores, and retrieves information in vertebrates. Its interneuron networks generate responses that provide the basis for our voluntary movements, consciousness, behavior, emotions, learning, reasoning, language, and memory, among many other complex activities.

Major Brain Structures. We have noted that the three major divisions of the embryonic neural tube—forebrain, midbrain, and hindbrain—give rise to the structures of the adult brain. Like the spinal cord, each brain structure contains both gray matter and white matter and is surrounded by meninges and circulating cerebrospinal fluid **(Figure 38.6).**

The hindbrain develops into three major structures in the adult brain: the *cerebellum,* the *pons,* and the *medulla oblongata* (the *medulla*) (see Figure 38.2).

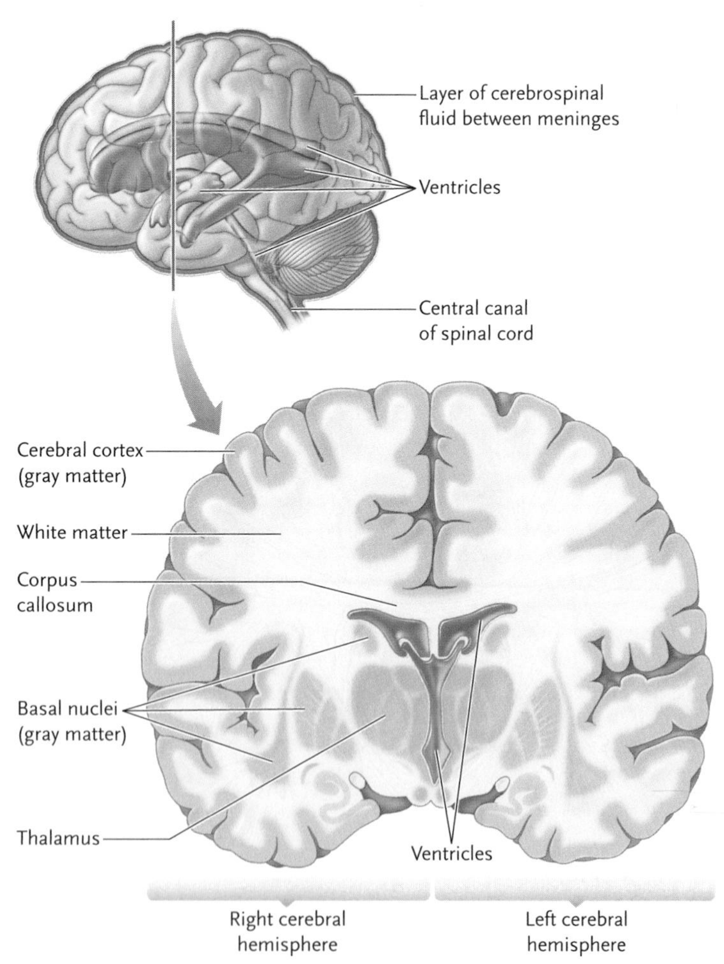

Figure 38.6
The human brain, illustrating the distribution of gray matter and white matter, and the locations of the four ventricles (in blue) with their connection to the central canal of the spinal cord.

The pons and medulla, along with the midbrain, form a stalklike structure known as the **brain stem,** which connects the forebrain with the spinal cord. All but two of the twelve pairs of cranial nerves also originate from the brain stem. The cerebellum, with its deeply folded surface, is an outgrowth of the pons.

The forebrain, which makes up most of the mass of the brain in humans, forms the *telencephalon (cerebrum).* The cerebrum, the largest part of the brain in humans, is organized into the left and right *cerebral hemispheres,* which have many fissures and folds (see Figure 38.6). Each hemisphere consists of **cerebral cortex,** a thin outer shell of gray matter covering a thick core of white matter. The basal nuclei, consisting of several regions of gray matter, are located deep within the white matter.

The Blood-Brain Barrier. Unlike the epithelial cells forming capillary walls elsewhere in the body, which allow small molecules and ions to pass freely from the blood to surrounding fluids, those forming capillaries in the brain are sealed together by tight junctions (see Figure 5.27). The tight junctions set up a **blood-brain barrier** that prevents most substances dissolved in the blood from entering the cerebrospinal fluid and thus protects the brain and spinal cord from viruses, bacteria, and toxic substances that may circulate in the blood.

A few types of molecules and ions, such as oxygen, carbon dioxide, alcohol, and anesthetics, can move directly across the lipid bilayer of the epithelial cell membranes by diffusion. A few other substances—most significantly glucose, the only molecule that brain and spinal cord cells can oxidize for energy—are moved across the plasma membrane by highly selective transport proteins.

The Brain Stem Regulates Many Vital Housekeeping Functions of the Body

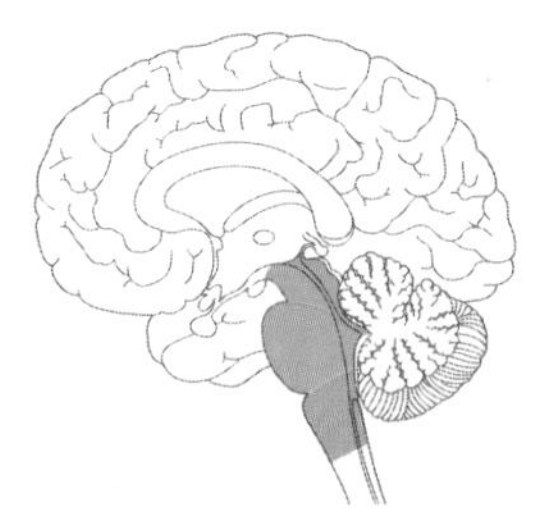

Physicians and scientists have learned much about the functions of various brain regions by studying patients with brain damage from stroke, infection, tumors, or mechanical disturbance. Techniques such as *functional magnetic resonance imaging (fMRI)* and *positron emission tomography (PET)* allow researchers to identify the normal functions of specific brain regions in noninvasive ways. The instruments record a subject's brain activity during various mental and physical tasks by detecting minute increases in blood flow or metabolic activity in specific regions **(Figure 38.7).**

From such medical and experimental analyses, we know that gray-matter centers in the brain stem control

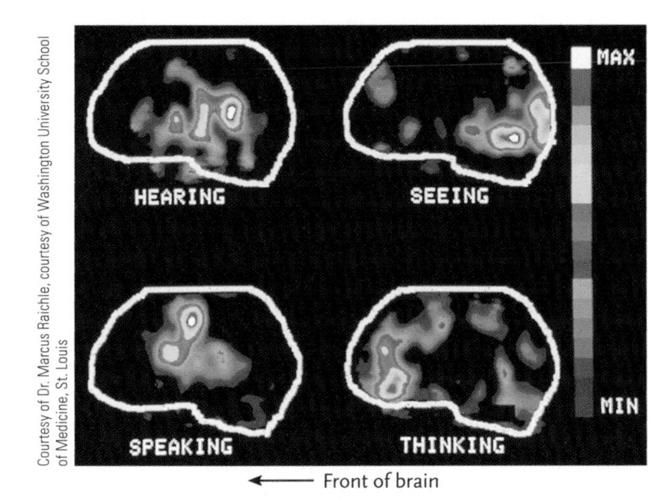

Courtesy of Dr. Marcus Raichle, courtesy of Washington University School of Medicine, St. Louis

Figure 38.7
PET scans showing regions of the brain active when a person performs specific mental tasks. The colors show the relative activity of the sections, with white the most active.

many vital body functions without conscious involvement or control by the cerebrum. Among these functions are the heart and respiration rates, blood pressure, constriction and dilation of blood vessels, coughing, and reflex activities of the digestive system such as vomiting. Damage to the brain stem has serious and sometimes lethal consequences.

A complex network of interconnected neurons known as the **reticular formation** (*reticulum* = netlike structure) runs through the length of the brain stem, connecting to the thalamus at the anterior end and to the spinal cord at the posterior end **(Figure 38.8).** All incoming sensory input goes to the reticular formation, which integrates the information and then sends signals to other parts of the CNS. The reticular formation has two parts. The ascending reticular formation, also called the *reticular activating system,* contains neurons that convey stimulatory signals via the thalamus to arouse and activate the cerebral cortex. It is responsible for the sleep-wake cycle; depending on the level of stimulation of the cortex, various levels of alertness and consciousness are produced. Lesions in this part of the brain stem result in coma. The other part, the descending reticular formation, receives information from the hypothalamus and connects with interneurons in the spinal cord that control skeletal muscle contraction, thereby controlling muscle movement and posture. The reticular formation filters incoming signals, helping to discriminate between important and unimportant ones. Such filtering is necessary because the brain is unable to process every one of the signals from millions of sensory receptors. For example, the action of the reticular formation enables you to sleep through many sounds but waken to specific ones, such as a cat meowing to be let out or a baby crying.

The Cerebellum Integrates Sensory Inputs to Coordinate Body Movements

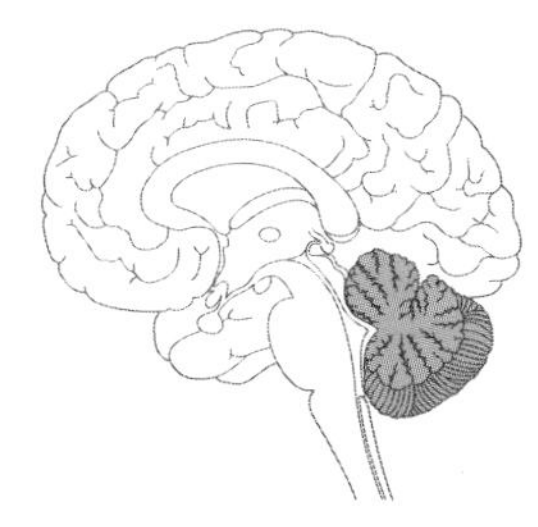

Although the cerebellum is an outgrowth of the pons (see Figure 38.8), it is separate in structure and function from the brain stem. Through its extensive connections with other parts of the brain, the **cerebellum** receives sensory input from receptors in muscles and joints, from balance receptors in the inner ear, and from the receptors of touch, vision, and hearing. These signals convey information about how the body trunk and limbs are positioned, the degree to which different muscles are contracted or relaxed, and the direction in which the body or limbs are moving. The cerebellum integrates these sensory signals and compares them with signals from the cerebrum that control voluntary body movements. Outputs from the cerebellum to the cerebrum, brain stem, and spinal cord modify and fine-tune the movements to keep the body in balance and directed toward targeted positions in space. The cerebellum of all mammals has essentially the same capabilities and works in the same way. The human cerebellum also contributes to the learning and memory of motor skills such as typing.

Gray-Matter Centers Control a Variety of Functions

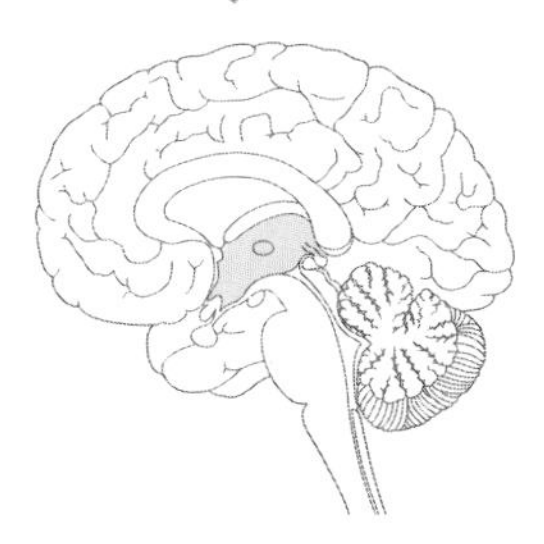

Gray-matter centers derived from the embryonic forebrain include the thalamus, hypothalamus, and basal nuclei **(Figure 38.9).** These centers contribute to the control and integration of voluntary movements, body temperature and glandular secretions, osmotic balance of the blood and extracellular fluids, wakefulness, and the

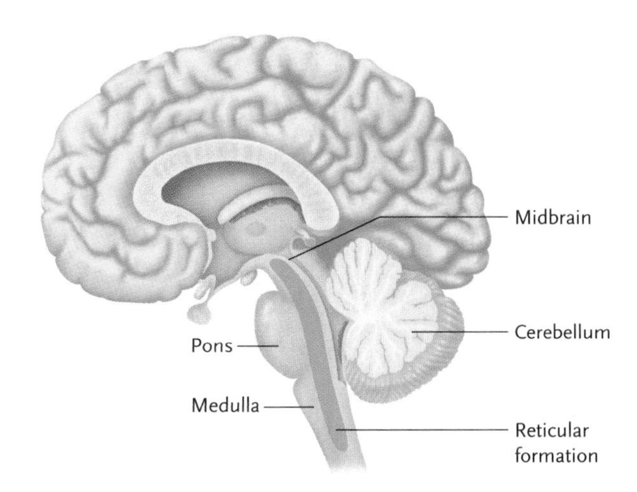

Figure 38.8
Location of the reticular formation (in blue) in the brain stem.

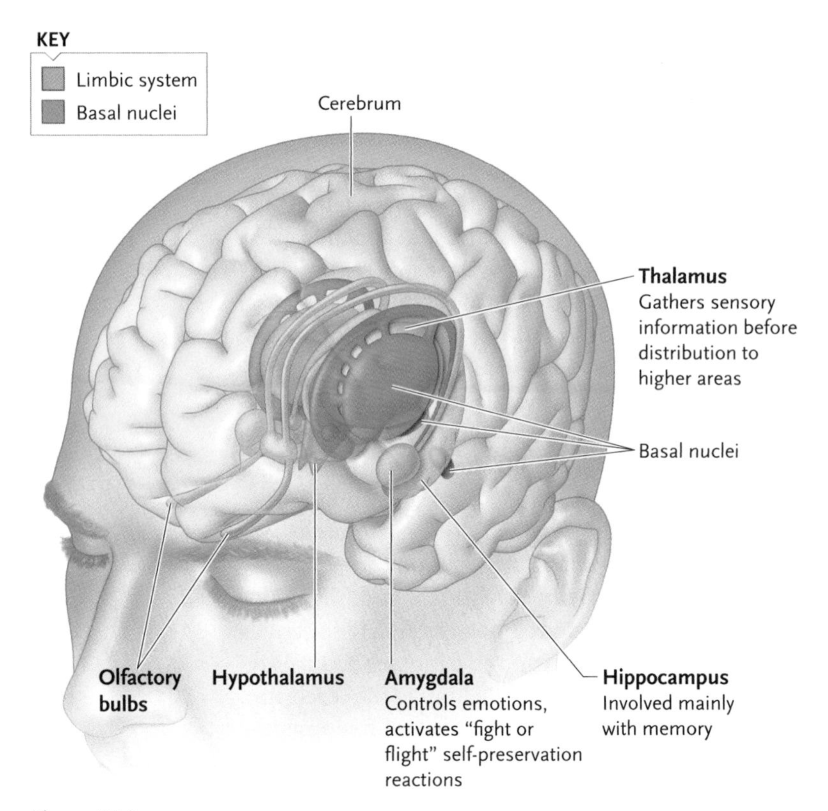

Figure 38.9
Basal nuclei, thalamus, and hypothalamus gray-matter centers. The centers shown in this view are those in the left hemisphere.

emotions, among other functions. Some of the gray-matter centers route information to and from the cerebral cortex, and between the forebrain, brain stem, and cerebellum.

The **thalamus** (see Figure 38.9) forms a major switchboard that receives sensory information and relays it to the regions of the cerebral cortex concerned with motor responses to sensory information of that type. Part of the thalamus near the brain stem cooperates with the reticular formation in alerting the cerebral cortex to full wakefulness, or in inducing drowsiness or sleep.

The **hypothalamus** (see Figure 38.9) contains centers that regulate basic homeostatic functions of the body and contribute to the release of hormones. Some centers set and maintain body temperature by triggering reactions such as shivering or sweating. Others constantly monitor the osmotic balance of the blood by testing its composition of ions and other substances. If departures from normal levels are detected, the hypothalamus triggers responses such as thirst or changes in urine output that restore the osmotic and fluid balance.

The centers of the hypothalamus that detect blood composition and temperature are directly exposed to the bloodstream—they are the only parts of the brain *not* protected by the blood-brain barrier. Parts of the hypothalamus also coordinate responses triggered by the autonomic system, making it an important link in such activities as control of the heartbeat, contraction of smooth muscle cells in the digestive system, and glandular secretion. Some regions of the hypothalamus establish a biological clock that sets up daily metabolic rhythms, such as the regular changes in body temperature that occur on a daily cycle.

The **basal nuclei** are gray-matter centers that surround the thalamus on both sides of the brain (see Figure 38.9). They moderate voluntary movements directed by motor centers in the cerebrum. Damage to the basal nuclei can affect the planning and fine-tuning of movements, leading to stiff, rigid motions of the limbs and unwanted or misdirected motor activity, such as tremors of the hands and inability to start or stop intended movements at the intended place and time. Parkinson disease, in which affected individuals exhibit all of these symptoms, results from degeneration of centers in and near the basal nuclei.

Parts of the thalamus, hypothalamus, and basal nuclei, along with other nearby gray-matter centers—the amygdala, hippocampus, and olfactory bulbs—form a functional network called the **limbic system** (*limbus* = arc), sometimes called our "emotional brain" (see Figure 38.9). The **amygdala** works as a switchboard, routing information about experiences that have an emotional component through the limbic system. The **hippocampus** is involved in sending information to the frontal lobes, and the **olfactory bulbs** relay inputs from odor receptors to both the cerebral cortex and the limbic system. The olfactory connection to the limbic system may explain why certain odors can evoke particular, sometimes startlingly powerful emotional responses.

The limbic system controls emotional behavior and influences the basic body functions regulated by the hypothalamus and brain stem. Stimulation of different parts of the limbic system produces anger, anxiety, fear, satisfaction, pleasure, or sexual arousal. Connections between the limbic system and other brain regions bring about emotional responses such as smiling, blushing, or laughing.

The Cerebral Cortex Carries Out All Higher Brain Functions in Humans

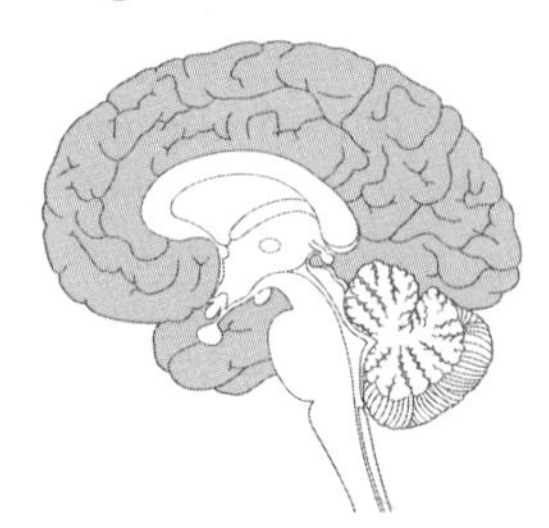

The gray matter of each hemisphere, the cerebral cortex, contains the processing centers for the integration of neural input and the initiation of neural output. The white matter of the cerebral hemispheres, by contrast, contains the neural routes for signal transmission between parts of the cerebral cortex, or from the cerebral cortex to other parts of the CNS. No information processing occurs in the white matter.

Over the course of evolution, the surface area of the cerebral cortex increased by continually folding in

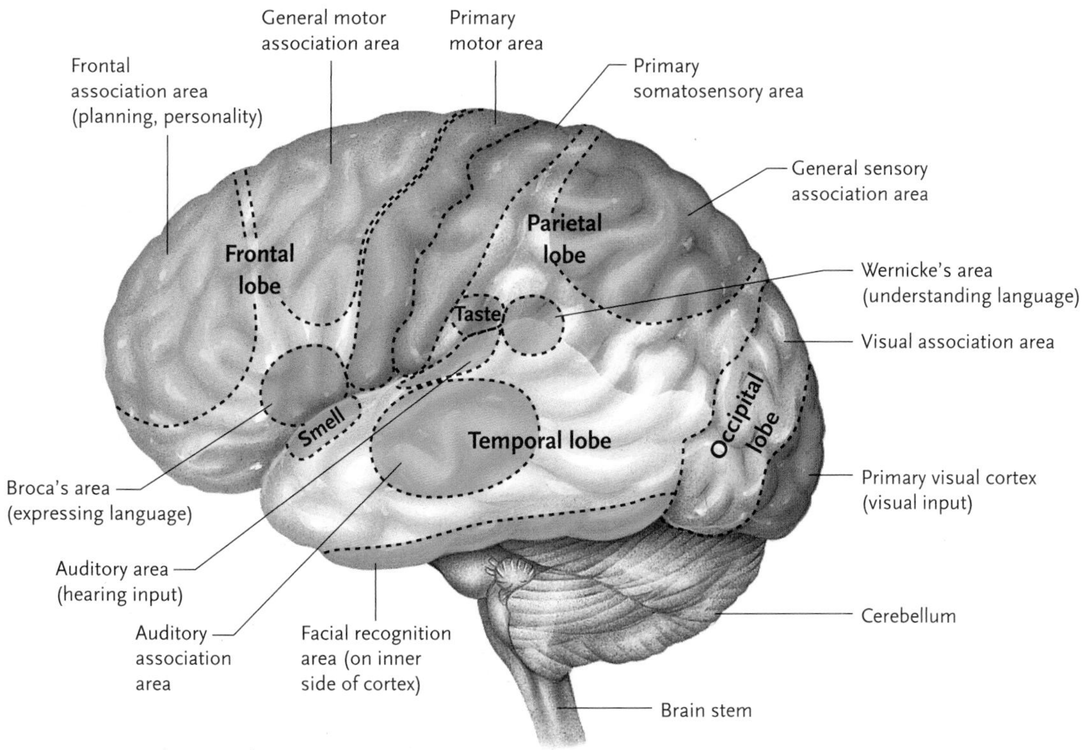

Figure 38.10
The lobes of the cerebrum, showing major regions and association areas of the cerebral cortex.

on itself, thereby expanding the structure into sophisticated information encoding and processing centers. Primates have cerebral cortexes with the largest number of convolutions. In humans, each cerebral hemisphere is divided by surface folds into *frontal, parietal, temporal,* and *occipital* lobes **(Figure 38.10).** Uniquely in mammals, the cerebral cortex of the cerebral hemispheres is organized into six layers of neurons; these layers are the newest part of the cerebral cortex in an evolutionary sense.

The two cerebral hemispheres can function separately, and each has its own communication lines internally and with the rest of the CNS and the body. The left cerebral hemisphere responds primarily to sensory signals from, and controls movements in, the right side of the body. The right hemisphere has the same relationships to the left side of the body. This opposite connection and control reflects the fact that the nerves carrying afferent and efferent signals cross from left to right within the spinal cord or brain stem. Thick axon bundles, forming a structure called the **corpus callosum,** connect the two cerebral hemispheres and coordinate their functions.

Sensory Regions of the Cerebral Cortex. Areas that receive and integrate sensory information are distributed over the cerebral cortex. In each hemisphere, the **primary somatosensory area,** which registers information on touch, pain, temperature, and pressure, runs in a band across the parietal lobes of the brain (see Figure 38.10). Experimental stimulation of this band in one hemisphere causes prickling or tingling sensations in specific parts on the opposite side of the body, beginning with the toes at the top of each hemisphere and running through the legs, trunk, arms, and hands, to the head **(Figure 38.11).**

Other sensory regions of the cerebral cortex have been identified with hearing, vision, smell, and taste (see Figure 38.10). Regions of the temporal lobes on both sides of the brain receive auditory inputs from the ears, while inputs from the eyes are processed in the primary visual cortex in both occipital lobes. Olfactory input from the nose is processed in the olfactory bulbs, located on the ventral side of the temporal lobes. Regions in the parietal lobes receive inputs from taste receptors on the tongue and other locations in the mouth.

Motor Regions of the Cerebral Cortex. The **primary motor area** of the cerebral cortex runs in a band just in front of the primary somatosensory area (see Figure 38.10). Experimental stimulation of points along this band in one hemisphere causes movement of specific body parts on the opposite side of the body, corresponding generally to the parts registering in the primary somatosensory area at the same level (see Figure 38.11). Other areas that integrate and refine motor control are located nearby.

In both the primary somatosensory and primary motor areas, some body parts, such as the lips and fingers, are represented by large regions, and others, such as the arms and legs, are represented by relatively small regions. As shown in Figure 38.11, the relative sizes produce a distorted image of the human body that is quite different from the actual body proportions. The differences are reflected in the precision of

Figure 38.11
The primary somatosensory and motor areas of the cerebrum. The distorted images of the human body show the relative areas of the sensory and motor cortex devoted to different body regions.

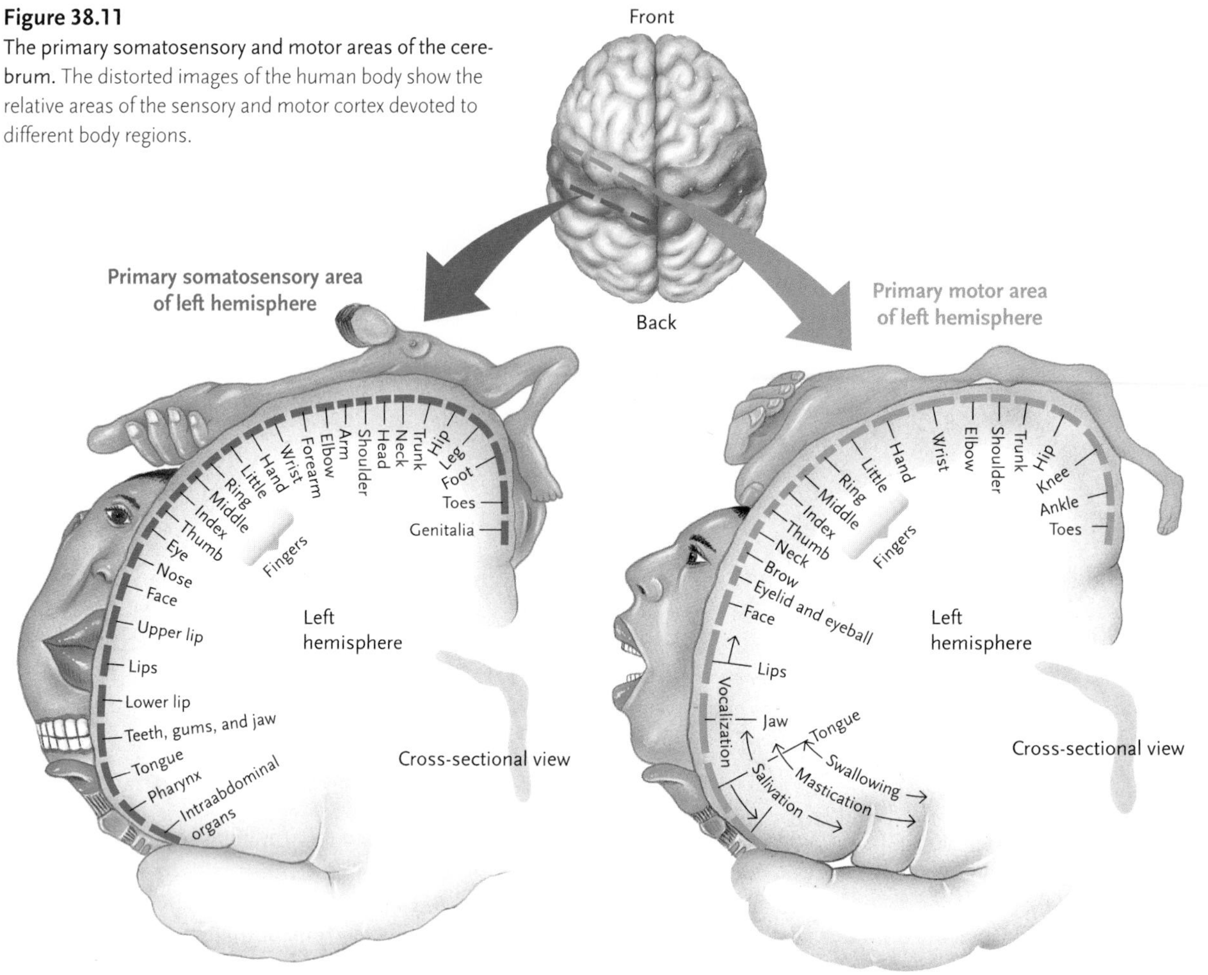

touch and movement in structures such as the lips, tongue, and fingers.

Association Areas. The sensory and motor areas of the cerebral cortex are surrounded by **association areas** (see Figure 38.10), which integrate information from the sensory areas, formulate responses, and pass them on to the primary motor area. Two of the most important association areas are *Wernicke's area* and *Broca's area* (see Figure 38.10), which function in spoken and written language. They are usually present on only one side of the brain—in the left hemisphere in 97% of the human population. Comprehension of spoken and written language depends on Wernicke's area, which coordinates inputs from the visual, auditory, and general sensory association areas. Interneuron connections lead from Wernicke's area to Broca's area, which puts together the motor program for coordination of the lips, tongue, jaws, and other structures producing the sounds of speech, and passes the program to the primary motor area. The brain-scan images in Figure 38.7 dramatically illustrate how these brain regions participate as a person performs different linguistic tasks.

People with damage to Wernicke's area have difficulty comprehending spoken and written words, even though their hearing and vision are unimpaired. Although they can speak, their words usually make no sense. People with damage to Broca's area have normal comprehension of written and spoken language, and know what they want to say, but are unable to speak except for a few slow and poorly pronounced words. Often, such people are also unable to write. Other areas of the brain are also involved in language functions.

Some Higher Functions Are Distributed in Both Cerebral Hemispheres; Others Are Concentrated in One Hemisphere

Most of the other higher functions of the human brain—such as abstract thought and reasoning; spatial recognition; mathematical, musical, and artistic ability; and the associations forming the basis of personality—involve the coordinated participation of many regions of the cerebral cortex. Some of these regions are equally distributed in both cerebral hemispheres, and some are more concentrated in one hemisphere.

Among the functions more or less equally distributed between the two hemispheres is the ability to recognize faces. This function is concentrated along the bottom margins of the occipital and temporal lobes (see Figure 38.10). People with damage to these lobes are often unable to recognize even close relatives by sight but can recognize voices immediately. Functions such as consciousness, the sense of time, and recognizing emotions also seem to be distributed in both hemispheres.

Typically some brain functions are more localized in one of the two hemispheres, a phenomenon called **lateralization.** The unequal distribution of these functions was originally worked out in the 1960s by Roger Sperry and Michael S. Gazzaniga of the California Institute of Technology (Sperry received a Nobel Prize for his research in 1981) in subjects who had had their corpus callosum cut surgically **(Figure 38.12).**

Studies of people with split hemispheres as well as surveys of brain activity by PET and fMRI have confirmed that, for the vast majority of people, the left hemisphere specializes in spoken and written language, abstract reasoning, and precise mathematical calculations. The right hemisphere specializes in nonverbal conceptualizing, intuitive thinking, musical and artistic abilities, and spatial recognition functions such as fitting pieces into a puzzle. The right hemisphere also handles mathematical estimates and approximations that can be made by visual or spatial representations of numbers. Thus the left hemisphere in most people is verbal and mathematical, and the right hemisphere is intuitive, spatial, artistic, and musical.

Study Break

1. Human newborn babies, as well as premature babies, have an incompletely developed blood-brain barrier. Should this condition influence what food and medications are given to them?
2. Distinguish the structure and functions of the cerebellum from those of the cerebral cortex.

38.4 Memory, Learning, and Consciousness

We set memory, learning, and consciousness apart from the other functions because they appear to involve coordination of structures from the brain stem to the cerebral cortex. **Memory** is the storage and retrieval of a sensory or motor experience, or a thought. **Learning** involves a change in the response to a stimulus based on information or experiences stored in memory. **Consciousness** may be defined as awareness of ourselves, our identity, and our surroundings, and an understanding of the significance and likely consequences of events that we experience.

Memory Takes Two Forms, Short Term and Long Term

Psychology research and our everyday experience indicate that humans have at least two types of memory. **Short-term memory** stores information for seconds, minutes, or at most an hour or so. **Long-term memory** stores information from days to years or even for life. Short-term memory, but not long-term memory, is usually erased if a person experiences a disruption such as a sudden fright, a blow, a surprise, or an electrical shock. For example, a person knocked unconscious by an accident typically cannot recall the accident itself or the events just before it, but long-standing memories are not usually disturbed.

To explain these differences, investigators propose that short-term memories depend on transient changes in neurons that can be erased relatively easily, such as changes in the membrane potential of interneurons caused by EPSPs and IPSPs (excitatory and inhibitory postsynaptic potentials) and the action of indirect neurotransmitters that lead to reversible changes in ion transport (see Section 37.3). By contrast, storage of long-term memory is considered to involve more or less permanent molecular, biochemical, or structural changes in interneurons, which establish signal pathways that cannot be switched off easily.

All memories probably register initially in short-term form. They are then either erased and lost, or committed to long-term form. The intensity or vividness of an experience, the attention focused on an event, emotional involvement, or the degree of repetition may all contribute to the conversion from short-term to long-term memory.

The storage pathway typically starts with an input at the somatosensory cortex that then flows to the amygdala, which relays information to the limbic system, and to the hippocampus, which sends information to the frontal lobes, a major site of long-term memory storage. People with injuries to the hippocampus cannot remember information for more than a few minutes; long-term memory is limited to information stored before the injury occurred.

How are neurons and neuron pathways permanently altered to create long-term memory? One change that has been much studied is **long-term potentiation:** a long-lasting increase in the strength of synaptic connections in activated neural pathways following brief periods of repeated stimulation. The synapses become increasingly sensitive over time, so that a constant level of presynaptic stimulation is converted into a larger postsynaptic output that can last hours, weeks, months, or years. (*Insights from the Molecular Revolution* describes experiments investigating the basis of long-term potentiation in neurons of the hippocampus.) Other changes consistently noted as part of long-term memory include more or less permanent alterations in the number and

Figure 38.12 Experimental Research

Investigating the Functions of the Cerebral Hemispheres

QUESTION: Do the two cerebral hemispheres have different functions?

EXPERIMENT: Roger Sperry and Michael Gazzaniga studied split-brain individuals, in whom the corpus callosum connecting the two cerebral hemispheres had been surgically severed to relieve otherwise uncontrollable epileptic convulsions. In one experiment, they tested how subjects perceived words that were projected onto a screen in front of them.

The retinas of the eyes gather visual information and send signals via the optic nerves to the cerebral hemispheres (Figure 38.12a). Light from the *left* half of the visual field reaches light receptors on the *right* sides of the retinas, and parts of the two optic nerves carry signals to the *right* cerebral hemisphere. Light from the *right* half of the visual field reaches light receptors on the *left* sides of the retinas, and signals are sent to the *left* cerebral hemisphere.

The researchers projected words such as COWBOY in such a way that the subjects could see only the left half of the word (COW) with the left eye and the right half of the word (BOY) with the right eye (Figure 38.12b). Sperry asked the subjects to say what word they saw, and he asked them to write the perceived word with the left hand—a hand that was deliberately blocked from the subject's view.

a. Pathway of visual information from eyes to cerebral hemisphere

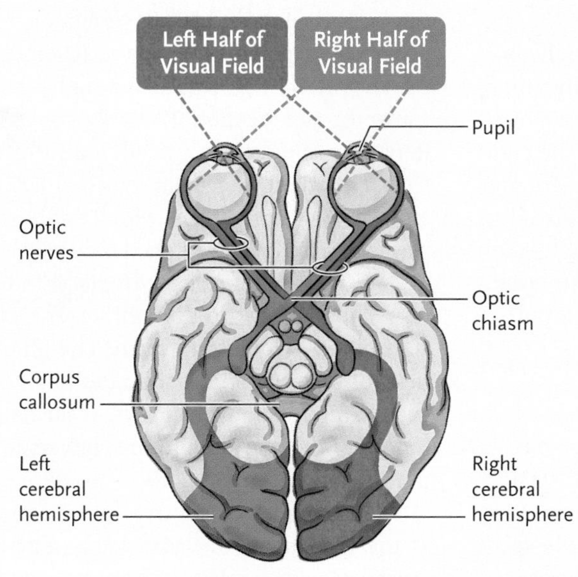

b. Experimental set up—COW seen by right sides of retinas and BOY by left sides

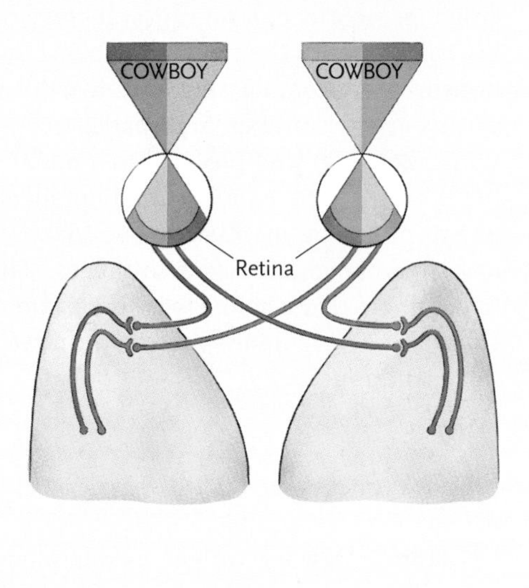

RESULTS: The split-brain subjects said the word in the right half of the visual field (BOY), but wrote the word in the left half of the visual field (COW).

CONCLUSIONS: The studies showed that the left and right hemispheres are specialized in different tasks. The left hemisphere processes language and was able to recognize BOY but received no information about COW. The right hemisphere directs motor activity on the left side of the body and was able to direct the left hand to write COW. However, the subjects could not say what word they wrote. That is, cutting the corpus callosum interrupted communication between the two halves of the cerebrum. In effect, one cerebral hemisphere did not know what the other was doing, and information stored in the memory on one side was not available to the other. In normal individuals, information is shared across the corpus callosum; they would see COWBOY and be able to speak and write the entire word.

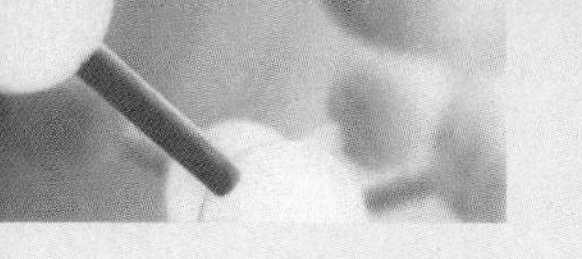

Insights from the Molecular Revolution

Knocked-Out Mice with a Bad Memory

Long-term potentiation (LTP) in neurons of the hippocampus is thought to be central to the conversion of short-term to long-term memory. An indirect neurotransmitter, glutamate, is most often involved in the process. One of the receptors that binds glutamate in postsynaptic membranes of the hippocampus is the *NMDA receptor;* a prolonged series of stimuli causes glutamate to bind to the receptor, which opens an ion channel that forms part of the receptor's structure. Among other effects, Ca^{2+} ions flowing inward through the channel activate a protein kinase in the cytoplasm called *CaMKII.* (Protein kinases are enzymes that, when activated, add phosphate groups to certain proteins; addition of the phosphate groups increases or reduces activity of the proteins.)

One of the target proteins for the activated CaMKII is itself. After it adds phosphate groups to its own structure, CaMKII no longer needs to be activated by calcium—it remains turned on at elevated levels. Its more or less permanent activation makes the neuron more sensitive to incoming signals and increases the number of the signals it sends to other neurons. Researchers hypothesize that this change is a major contributor to LTP in the neuron. Among the supporting evidence is the observation that chemical blockage of the NMDA receptor impairs spatial learning and long-term memory in mice.

A research group at the Massachusetts Institute of Technology led by a Nobel Prize–winning scientist, Susumu Tonegawa, applied molecular techniques to test this hypothesis. Tonegawa's group created a strain of "knockout" mice (see Section 18.2), from which the gene encoding the NMDA receptor has been eliminated.

To test the effects of the knockout on LTP in the neurons, the investigators dissected out brain slices and used microelectrodes to stimulate neurons in the hippocampus. Unlike the response of neurons with an intact NMDA receptor, repeated stimulation was not followed by any potentiation in the knockout neurons.

Would the lack of potentiation in the hippocampal neurons result in the failure of long-term memory storage in the knockout mice? The researchers tested this outcome by placing the knockout mice in a pool in which they had to swim until they could find a submerged platform on which to rest. The knockout mice were much slower than normal mice in finding the platform in initial trials, and, unlike normal mice, were unable to remember its location in later trials.

The results thus support the hypothesis that the NMDA receptor is involved in the conversion from short-term to long-term memory, and that the hippocampus has a central role in this activity. The novel approach used by the Tonegawa group also provides a new, molecular research method by which to analyze higher brain function.

the area of synaptic connections between neurons, in the number and branches of dendrites, and in gene transcription and protein synthesis in interneurons.

Experiments have shown that protein synthesis is critical to long-term memory storage in animals as varied as *Drosophila* and rats. For example, goldfish were trained to avoid an electrical shock by swimming to one end of an aquarium when a light was turned on. The fish could remember the training for about a month under normal conditions; if exposed to a protein synthesis inhibitor while being trained, they forgot the training within a day.

Learning Involves Combining Past and Present Experiences to Modify Responses

As with memory, all animals appear to be capable of learning to some degree. Learning involves three sequential mechanisms, (1) storing memories, (2) scanning memories when a stimulus is encountered, and (3) modifying the response to the stimulus in accordance with the information stored as memory.

One of the simplest forms of memory is **sensitization**—increased responsiveness to mild stimuli after experiencing a strong stimulus. The process was nicely illustrated by Eric Kandel of Columbia University and his associates in experiments with a shell-less marine snail known as the Pacific sea hare, *Aplysia californica,* which is frequently used in research involving reflex behavior, memory, and learning. Many of its neuron circuits have been completely worked out, allowing investigators to follow the reactions of each neuron active in pathways such as learning. The first time the researchers administered a single sharp tap to the siphon (which admits water to the gills), the slug retracted its gills by a reflex movement. However, at the next touch, whether hard or gentle, the siphon retracted much more quickly and vigorously. Sensitization in *Aplysia* has been shown to involve changes in synapses, which become more reactive when more serotonin is released by action potentials. Kandel received the Nobel Prize in 2000 for his research.

Learning skills or procedures, such as tying one's shoes, typing, or playing a musical instrument, involve additional regions of the brain, particularly the cerebellum, where motor activity is coordinated. As we learn such skills, the process gradually becomes automated so that we do not think consciously about each step. (Learning and its relationship to animal behavior are considered further in Chapter 54.)

Consciousness Involves Different States of Awareness

The spectrum of human consciousness ranges from alert wakefulness to daydreaming, dozing, and sleep. Even during sleep there is some degree of awareness, because sleepers can respond to stimuli and waken, unlike someone who is unconscious. Moving between the states of consciousness has been found to involve changes in neural activity over the entire surface of the telencephalon. These changes can be seen using an *electroencephalogram* (EEG), which records voltage changes detected by electrodes placed on the scalp.

When an individual is fully awake, the EEG records a pattern of rapid, irregular *beta waves* **(Figure 38.13).** With mind at rest and eyes closed, the person's EEG pattern changes to slower and more regular *alpha waves.* As drowsiness and light sleep come on, the wave trains gradually become larger, slower, but again less regular; these slower pulsations are called *theta waves.* During the transition from drowsiness to deep sleep, the EEG pattern shifts to even slower *delta waves.* The heart and breathing rates become slower and the skeletal muscles increasingly relaxed, although the sleeper may still change position and move the arms and legs.

Periodically during deep sleep, the delta wave pattern is replaced by the rapid, irregular beta waves characteristic of the waking state. The person's heartbeat and breathing rate increase, the limbs twitch, and the eyes move rapidly behind the closed eyelids, giving this phase its name of **rapid-eye-movement (REM) sleep.** The REM sleep phase occurs about every 1.5 hours while a healthy adult is sleeping, and lasts for 10 to 15 minutes. Sleepers do most of their dreaming during REM sleep, and most research subjects awakened from REM sleep report they were experiencing vivid dreams.

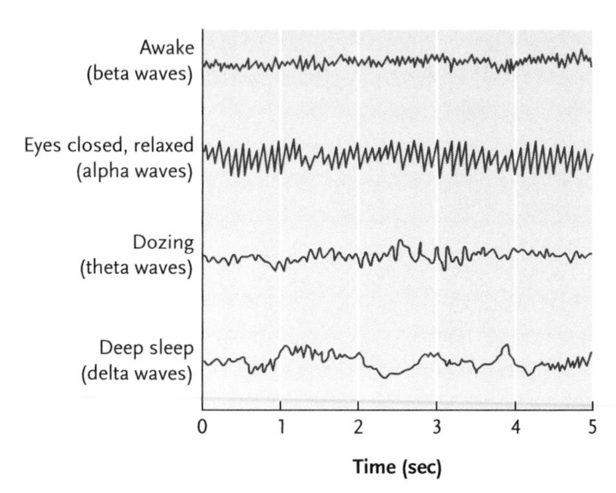

Figure 38.13
Brain waves characteristic of various states of consciousness.

As mentioned earlier, the reticular activating system controls the sleep-wake cycle. It sends signals to the spinal cord, cerebellum, and cerebral cortex, and receives signals from the same locations. The flow of

Unanswered Questions

What gives humans their unique brain capacity?

The human brain is larger relative to our body size than the brains of other mammals and has more functions. How did that come to be the case? At the simplest conceptual level, the answer must lie in the human genome. Researchers led by David Haussler of the University of California, Santa Cruz, have found evidence for unique human DNA that appears to play a central role in giving humans their unique brain capacity. Haussler's group compared the sequences of the human genome with the sequences of the genomes of other primates and other vertebrates. They looked for regions in the human genome that show significantly accelerated rates of base-pair changes since divergence from our common ancestor with the chimpanzee. The investigators found 49 such regions and dubbed them "human accelerated regions," or HARs.

One of the regions—*HAR1*—had dramatic changes that made it stand out from the rest. Haussler's group therefore focused on learning more about the region, looking specifically to see if it contained a gene and, if it did, what that gene encoded. Computer analysis of *HAR1* indicated the presence of what seemed to be a gene, and biochemical analysis confirmed that conclusion by detecting the presence of an RNA encoded by the gene. Interestingly the RNA is not an mRNA, meaning that the gene does not encode a protein. Rather, it is a structural RNA; that is, when it is expressed, it functions in the cell as an RNA. The researchers next investigated where and when the *HAR1* gene was expressed by looking for the RNA product in human embryonic brain tissue samples taken from different stages of development. The results showed that *HAR1* is expressed in a particular type of neuron in the developing human cerebral cortex at 7 to 19 weeks of development. That period of development is crucial for neuron specification in the cerebral cortex.

The researchers conclude that *HAR1* is a highly promising candidate for a gene involved in uniquely human biology. Work is now continuing to determine how the RNA encoded by *HAR1* functions in the cell. Does it interact with a protein, or with another RNA? How does it play a role in neuron specification in the developing cerebral cortex? As the project continues, the investigators hope to be able to use the mouse model to examine key aspects of *HAR1* function, because human experiments are not possible for ethical reasons. Haussler's group is also investigating other HARs to see what genes they might contain and what functions those genes might have in human brain development and function.

Peter J. Russell

signals along these circuits determines whether we are awake or asleep.

Many other animals also alternate periods of wakefulness and sleep or inactivity. Although sleep obviously has restorative effects on mental and physical functions, the physiological basis of these effects remains unknown.

In the previous chapter we learned about neurons, and in this chapter we have discussed the organization of neurons into nervous systems, as well as the structures of the brain and their functions. In the next chapter we consider the sensory systems that provide input for the brain to process.

Study Break

An aging person often experiences a progressive decline in cognitive function. This typically begins with short-term memory loss and the inability to learn new information. What brain changes might be occurring?

Review

Go to **Thomson**NOW™ at www.thomsonedu.com/login to access quizzing, animations, exercises, articles, and personalized homework help.

38.1 Invertebrate and Vertebrate Nervous Systems Compared

- The simplest nervous systems are the nerve nets of cnidarians. Echinoderms have modified nerve nets, with some neurons grouped into nerves (Figure 38.1a–b).
- Flatworms, arthropods, and mollusks have a simple central nervous system (CNS), consisting of ganglia in the head region (a brain), and a peripheral nervous system (PNS), consisting of nerves from the CNS to the rest of the body (Figure 38.1c–e).
- In vertebrates, the CNS consists of a large brain located in the head and a hollow spinal cord, and the PNS consists of all the nerves and ganglia connecting the CNS to the rest of the body (Figure 38.1f).
- In the vertebrate embryo, the anterior end of the hollow neural tube develops into the brain, and the rest develops into the spinal cord. The embryonic brain enlarges into the forebrain, midbrain, and hindbrain, which develop into the adult structures (Figure 38.2).

Animation: Comparisons of animal nervous systems

Animation: Bilateral nervous systems

Animation: Vertebrate nervous system divisions

38.2 The Peripheral Nervous System

- Afferent neurons in the PNS conduct signals to the CNS, and signals from the CNS travel via efferent neurons to the effectors—muscles and glands—that carry out responses (Figure 38.3).
- The somatic system of the PNS controls the skeletal muscles, producing voluntary body movements as well as involuntary muscle contractions that maintain balance, posture, and muscle tone.
- The autonomic system of the PNS, which controls involuntary functions, is organized into the sympathetic division and the parasympathetic division (Figure 38.4).

Animation: Autonomic nerves

38.3 The Central Nervous System (CNS) and Its Functions

- The spinal cord carries signals between the brain and the PNS. Its neuron circuits control reflex muscular movements and some autonomic reflexes (Figure 38.5).
- The medulla, pons, and midbrain form the brain stem, which connects the cerebrum, thalamus, and hypothalamus with the spinal cord.
- The cerebrum is divided into right and left cerebral hemispheres, which are connected by a thick band of nerve fibers, the corpus callosum. Each hemisphere consists of the cerebral cortex, a thin layer of gray matter, covering a thick core of white matter. Other collections of gray matter, the basal nuclei, are deep in the telencephalon (Figure 38.6).
- Cerebrospinal fluid provides nutrients and cushions the CNS (Figure 38.6). A blood-brain barrier allows only selected substances to enter the cerebrospinal fluid.
- Gray-matter centers in the pons and medulla control involuntary functions. Centers in the midbrain coordinate responses to visual and auditory sensory inputs.
- The reticular formation receives sensory inputs from all parts of the body and sends outputs to the cerebral cortex that help maintain balance, posture, and muscle tone. It also regulates states of wakefulness and sleep (Figure 38.8).
- The cerebellum integrates sensory inputs on the positions of muscles and joints, along with visual and auditory information, to coordinate body movements.
- The telencephalon's subcortical gray-matter centers control many functions. The thalamus receives, filters, and relays sensory and motor information to and from regions of the cerebral cortex. The hypothalamus regulates basic homeostatic functions of the body and contributes to the endocrine control of body functions. The basal nuclei affect the planning and fine-tuning of body movements (Figure 38.9).
- The limbic system includes parts of the thalamus, hypothalamus, and basal nuclei, as well as the amygdala and hippocampus. It controls emotions and influences the basic body functions controlled by the hypothalamus and brain stem (Figure 38.9).
- The primary somatosensory areas of the cerebral cortex register incoming information on touch, pain, temperature, and pressure from all parts of the body. In general, the right cerebral hemisphere receives sensory information from the left side of the body and vice versa (Figures 38.10 and 38.11).
- The primary motor areas control voluntary movements of skeletal muscles (Figures 38.10 and 38.11).
- The association areas integrate sensory information and formulate responses that are passed on to the primary motor areas. Wernicke's area integrates visual, auditory, and other sensory information into the comprehension of language; Broca's area coordinates movements of the lips, tongue, jaws, and other structures to produce the sounds of speech (Figure 38.10).
- Long-term memory and consciousness are equally distributed between the two cerebral hemispheres. Spoken and written language, abstract reasoning, and precise mathematical calculations are left hemisphere functions; nonverbal conceptualizing, mathematical estimation, intuitive thinking, spatial recognition,

and artistic and musical abilities are right hemisphere functions (Figure 38.12).

Animation: Organization of the spinal cord

Animation: Regions of the vertebrate brain

Animation: Human brain development

Animation: Sagittal view of a human brain

Animation: Primary motor cortex

Animation: Receiving and integrating areas

Animation: Path to visual cortex

38.4 Memory, Learning, and Consciousness

- Memory is the storage and retrieval of a sensory or motor experience or a thought. Short-term memory involves temporary storage of information, whereas long-term memory is essentially permanent.
- Learning involves modification of a response through comparisons made with information or experiences that are stored in memory.
- Consciousness is the awareness of ourselves, our identity, and our surroundings. It varies through states from full alertness to sleep and is controlled by the reticular activating system (Figure 38.13).

Animation: Structures involved in memory

Questions

Self-Test Questions

1. Ganglia first became enlarged and fused into a lobed brain in the evolution of:
   a. vertebrates.
   b. annelids.
   c. flatworms.
   d. cephalopods.
   e. mammals.
2. The metencephalon develops into the:
   a. spinal cord.
   b. cerebellum.
   c. mesencephalon.
   d. medulla oblongata.
   e. cerebrum.
3. The autonomic nervous system is subdivided into:
   a. afferent and efferent systems.
   b. sympathetic and parasympathetic divisions.
   c. skeletal and smooth muscle innervations.
   d. voluntary and involuntary controls.
   e. peripheral and central systems.
4. People with severe insect-sting allergies carry an *epipen* containing medication that they can inject in an emergency. The medication causes smooth muscles in the lung passages to relax so they can breathe but causes their hearts to pound rapidly. This is an example of stimulation of the:
   a. parasympathetic system.
   b. sympathetic system.
   c. somatic nervous system.
   d. limbic system.
   e. voluntary system.
5. Which one of the following structures participates in a reflex?
   a. the gray matter of the brain
   b. the white matter of the brain
   c. the gray matter of the spinal cord
   d. an interneuron that stimulates an afferent neuron
   e. an interneuron that inhibits an afferent neuron
6. Which of the following statements about the blood-brain barrier is incorrect?
   a. It is formed of capillary walls composed of tight junctions.
   b. It transports glucose to brain cells by means of transport proteins.
   c. It allows alcohol to pass through its lipid bilayer.
   d. It moves oxygen through the lipid bilayer.
   e. It reduces blood supply to brain cells compared with other body cells.
7. A segment of the brain stem that coordinates spinal reflexes with higher brain centers and regulates breathing and wakefulness is the:
   a. reticular formation.
   b. white matter of the pons.
   c. white matter of the medulla.
   d. hypothalamus.
   e. cerebellum.
8. Cushioning and nourishing the brain and spinal cord and filling the ventricles of the brain is (are):
   a. meninges.
   b. myelin.
   c. cerebrospinal fluid.
   d. ganglia.
   e. astrocytes.
9. Which structure and function are correctly paired below?
   a. thalamus: relays emotion signals through the limbic system
   b. basal nuclei: relay inputs from odor receptors to the cerebrum
   c. hypothalamus: releases hormones; sets up daily rhythms
   d. amygdala: relays sensory information to the cerebrum
   e. olfactory bulbs: moderate motor centers in the cerebrum
10. A patient had a tumor in Wernicke's area. It was initially diagnosed when he could not:
    a. understand his morning newspaper.
    b. hear his child crying.
    c. see the traffic light turn red.
    d. speak.
    e. feel if the car heater was on.

Questions for Discussion

1. Meningitis is an inflammation of the meninges, the membranes that cover the brain and spinal cord. Diagnosis involves using a needle to obtain a sample of cerebrospinal fluid to analyze for signs of infection. Why analyze this fluid and not blood?
2. An accident victim arrives at the emergency room with severe damage to the reticular formation. Based on information in this chapter, describe some of the symptoms that the examining physician might discover.
3. In the 1930s and 1940s prefrontal lobotomy, in which neural connections in the frontal lobes of both cerebral hemispheres were severed, was used to treat behavioral conditions such as extreme anxiety and rebelliousness. Although the procedure calmed patients, it had side effects such as apathy and a seriously disrupted personality. In view of the information presented in this chapter, why do you think the operation had these effects?

Experimental Analysis

How would you demonstrate that gene activity in the brain altered with aging in mice?

Evolution Link

How do paleontologists contribute to our understanding of the evolution of the brain?

A greater horseshoe bat *(Rhinolophus ferrumequinum)* hunting a moth. The bat uses its sensory system to pursue prey, and the moth uses its sensory system in attempting to evade capture.

Study Plan

39 Sensory Systems

Why It Matters

An insectivorous bat leaves its cave after a good day's sleep to look for food. As it flies, the bat emits a steady stream of ultrasonic clicking noises. Receptors in the bat's ears detect echoes of the clicks bouncing off objects in the environment and send signals to the brain, where they are integrated into a sound map that the animal uses to avoid trees and other obstacles. This ability to detect objects by *echolocation* is so keenly developed that a bat can detect and avoid a thin wire in the dark.

Besides recognizing obstacles, the bat's sensory system is keenly tuned to the distinctive pattern of echoes reflected by the fluttering wings of its favorite food, a moth. Although the slow-flying moth would seem doomed to become a meal for the foraging bat, natural selection has provided some species of moths with an astoundingly sensitive and efficient auditory sense as well as a programmed escape mechanism. On each side of its abdomen is an "ear," a thin membrane that resonates at the frequencies of the clicks emitted by the bat. The moth's ears register the clicks while the bat is still about 30 m away and initiate a response that turns its flight path directly away from the source of the clicks, giving the moth an early advantage in the nocturnal dance of life and death.

In spite of the moth's evasive turn, the bat's random flight pattern carries it in the direction of its prey. At a distance of about 6 m, when echoes from the moth begin to register in the bat's auditory system, the bat increases the frequency of its clicks, enabling it to pinpoint the moth's position.

The moth has not exhausted its evasive tactics, however. As the bat closes in, the increased frequency of the clicks sets off another programmed response that alters the moth's flight into sudden loops and turns, ending with a closed-wing, vertical fall toward the ground. After dropping a few feet, the moth resumes its fluttering flight.

Although the moth escapes for a moment, the bat also alters its path, turning back toward the moth as its echolocation again locks on the prey. The contest goes on, but the bat's sensory system finally leads it to intercept the moth an instant before its vertical drop, and the bat retires to a branch to eat its meal.

Natural selection has produced highly adaptive sensory receptors in moths, bats, and all other animals. These systems, the subject of this chapter, provide animals with a steady stream of information about their internal and external environments. After integrating the information in the central nervous system (CNS), animals respond in ways that enable them to survive and reproduce. We begin this chapter with a survey of animal sensory systems and the ways in which they work.

39.1 Overview of Sensory Receptors and Pathways

Information about an animal's external and internal environments is picked up by **sensory receptors,** formed by the dendrites of afferent neurons, or by specialized receptor cells making synapses with afferent neurons **(Figure 39.1).** The receptors associated with eyes, ears, skin, and other surface organs detect stimuli from the external environment. Sensory receptors associated with internal organs detect stimuli arising in the body interior.

Sensory receptors respond to stimuli by undergoing a change in membrane potential, caused in most receptors by changes in the rate at which channels conduct positive ions such as Na^{+}, K^{+}, or Ca^{2+} across the plasma membrane. Examples of stimuli are light, heat, sound waves, mechanical stress, and chemicals; the conversion of a stimulus into a change in membrane potential is called **sensory transduction.** The change in membrane potential may generate one or more action potentials, which travel along the axon of an afferent neuron to reach the interneuron networks of the CNS. These interneurons integrate the action potentials, and the brain formulates a compensating response, that is, a response appropriate for the stimulus (see Section 38.3). In animals with complex nervous systems, the

Figure 39.1
Sensory receptors, formed **(a)** by the dendrites of an afferent neuron or **(b)** by a separate cell or structure that communicates with an afferent neuron via a neurotransmitter.

a. Sensory receptor formed by dendrites of an afferent neuron

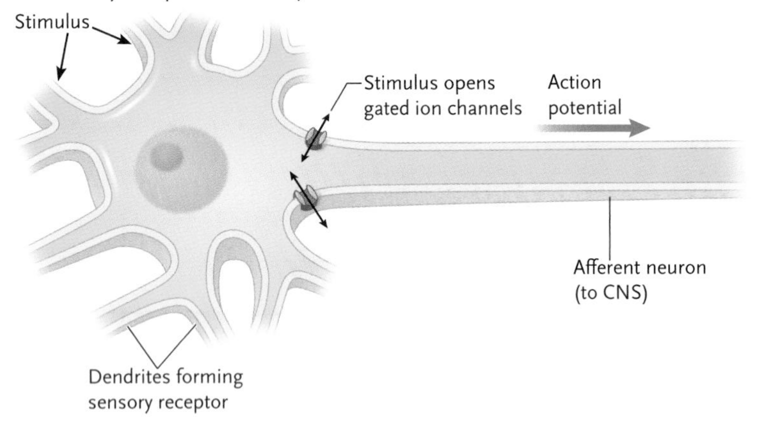

In sensory receptors formed by the dendrites of afferent neurons, a stimulus causes a change in membrane potential that generates action potentials in the axon of the neuron. Temperature and pain receptors are among the receptors of this type.

b. Sensory receptor formed by a cell that synapses with an afferent neuron

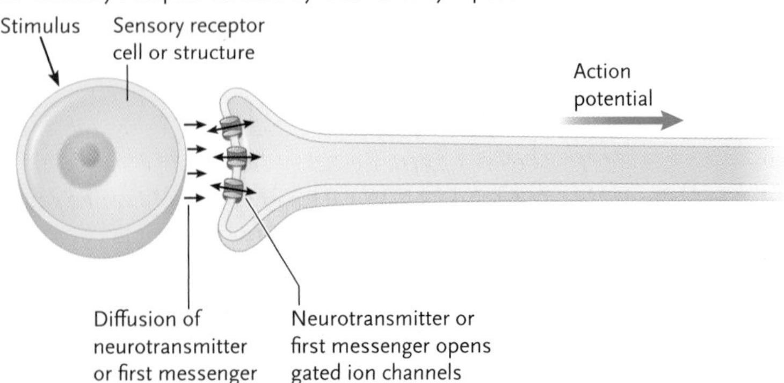

In sensory receptors consisting of separate cells, a stimulus causes a change in membrane potential that releases a neurotransmitter from the cell. The neurotransmitter triggers an action potential in the axon of a nearby afferent neuron. Mechanoreceptors, photoreceptors, and chemoreceptors are examples of receptors of this type.

interneuron networks may also produce an awareness of a stimulus in the form of a conscious sensation or perception.

Five Basic Types of Receptors Are Common to Almost All Animals

Many sensory receptors are positioned individually in body tissues. Others are part of complex sensory organs, such as the eyes or ears, that are specialized for reception of physical or chemical stimuli. Commonly, sensory receptors are classified into five major types, based on the type of stimulus that each detects:

1. **Mechanoreceptors** detect mechanical energy, such as changes in pressure, body position, or acceleration. The auditory receptors in the ears are examples of mechanoreceptors.
2. **Photoreceptors** detect the energy of light. In vertebrates, photoreceptors are mostly located in the retina of the eye.
3. **Chemoreceptors** detect specific molecules, or chemical conditions such as acidity. The taste buds on the tongue are examples of chemoreceptors.
4. **Thermoreceptors** detect the flow of heat energy. Receptors of this type are located in the skin, where they detect changes in the temperature of the body surface.
5. **Nociceptors** detect tissue damage or noxious chemicals; their activity registers as pain. Pain receptors are located in the skin, and also in some internal organs.

In addition to these major types, some animals have receptors that can detect electrical or magnetic fields.

Although humans are traditionally said to have five senses—vision, hearing, taste, smell, and touch—our sensory receptors actually detect more than twice as many kinds of environmental stimuli. Among these are external heat, internal temperature, gravity, acceleration, the positions of muscles and joints, body balance, internal pH, and the internal concentration of substances such as oxygen, carbon dioxide, salts, and glucose.

Afferent Neurons Link Receptors to the CNS

Sensory pathways begin at a sensory receptor and proceed by afferent neurons to the CNS. Because of its wiring, each type of receptor produces a specific kind of response. For example, action potentials arising in the retina of the eye travel along the optic nerve to the visual cortex, where they are interpreted by the brain as differences in the pattern, color, and intensity of light. If you receive a blow to the eye, the stimulus is still interpreted in the visual cortex as differences in the color and intensity of light detected by the eyes—you "see stars"—even though the stimulus is mechanical.

One way in which the intensity and extent of a stimulus is registered is by the frequency (number per unit time) of action potentials traveling along each axon of an afferent pathway. That is, the stronger the stimulus, the more frequently afferent neurons fire action potentials (see Section 37.2). A light touch to the hand, for example, causes action potentials to flow at low frequencies along the axons leading to the primary somatosensory area of the cerebral cortex. As the pressure increases, the number of action potentials per second rises in proportion; in the brain, the increase is interpreted as greater pressure on the hand. Maximum stimulus input is interpreted as pain in the sensory cortex.

The second way in which the intensity and extent of a stimulus is registered is by the number of afferent neurons that the stimulus activates to generate action potentials in the pathway. This way reflects the number of afferent neurons carrying signals from a stimulated region to the brain. The more sensory receptors that are activated, the more axons carry information to the brain. A light touch activates a relatively small number of receptors in a small area near the surface of the finger, for example. As the pressure increases, the resulting indentation of the finger's surface increases in area and depth, activating more receptors. In the appropriate somatosensory area of the brain, the larger number of axons carrying action potentials is interpreted as an increase in pressure spread over a greater area of the finger.

Many Receptor Systems Reduce Their Response When Stimuli Remain Constant

In many systems, the effect of a stimulus is reduced if it continues at a constant level. The reduction, called **sensory adaptation**, reduces the frequency of action potentials generated in afferent neurons when the intensity of a stimulus remains constant. Some receptors adapt quickly and broadly; other receptors adapt only slightly.

For example, when you go to bed, you are initially aware of the touch and pressure of the covers on your skin. Within a few minutes, the sensations lessen or are lost even though your position remains the same. The loss reflects adaptation of mechanoreceptors in your skin. If you move, so that the stimulus changes, the mechanoreceptors again become active. In contrast, nociceptors adapt only slightly, or not at all, to painful stimuli.

In some sensory receptors, biochemical changes in the receptor cell contribute to adaptation. For example, when you move from a dark movie theater into bright sunshine, the photoreceptors of the eye adapt to bright light partly through breakdown of some of the pigments that absorb light.

Sensory adaptation is crucial to animal survival. The adaptation of photoreceptors in our eyes keeps us from being blinded indefinitely as we pass from a darkened room into bright sunlight. Sensory adaptation also increases the sensitivity of receptor systems to *changes* in environmental stimuli, which may be more important to survival than keeping track of environmental factors that remain constant. You may have noticed a cat sitting motionless, focused on its prey, a mouse. As long as the environmental stimuli are constant, the cat's position remains fixed. However, if the mouse moves, the cat will respond rapidly and attempt to capture and kill it.

Many prey animals take advantage of adaptation in predators as a means for concealment or defense. These animals instinctively become motionless when they sense a predator in their environment, which frequently allows them to remain undetected by the adapted senses of their predator.

Nonadapting receptors, such as those detecting pain, are also essential for survival. Pain signals a potential danger to some part of the body, and the signals are maintained until a response by the animal compensates for the stimulus causing the pain.

We now examine the individual receptor types and their characteristics.

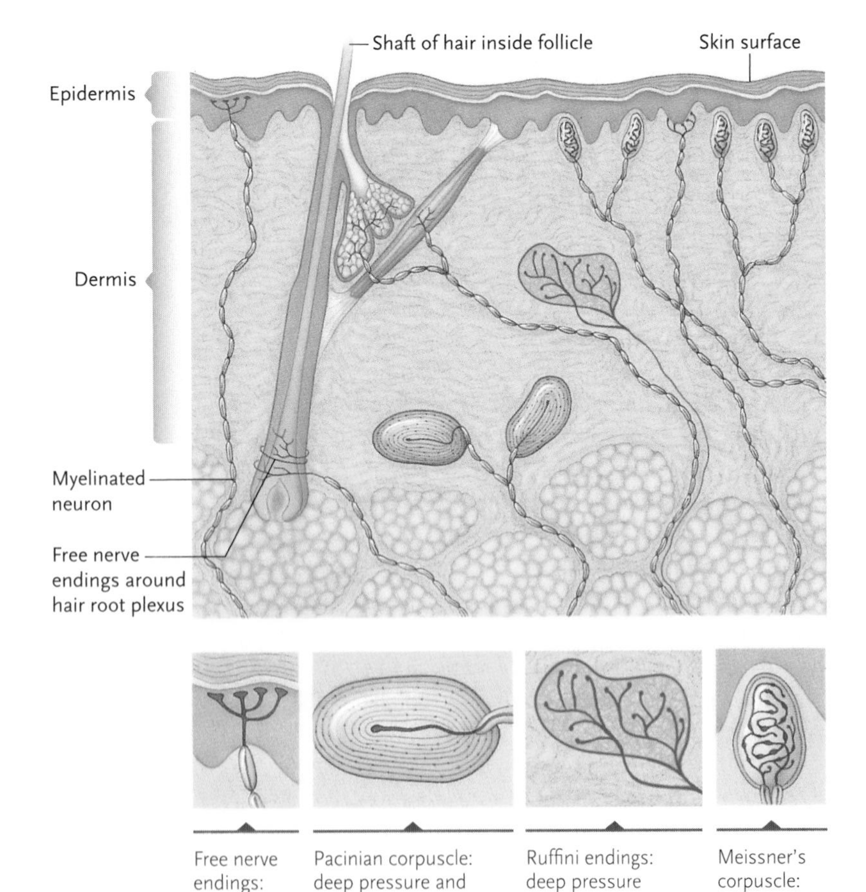

Figure 39.2
Types of mechanoreceptors detecting tactile stimuli in human skin.

Study Break

1. Define sensory transduction.
2. Sensory receptors send signals to the CNS in the same way. How are stimuli of different kinds perceived as being different?

39.2 Mechanoreceptors and the Tactile and Spatial Senses

Mechanical stimuli such as touch and pressure are detected by mechanoreceptors. The mechanical forces of a stimulus distort proteins in the plasma membrane of receptors, altering the flow of ions through the membrane. The changed ion flows generate action potentials in afferent neurons leading to the CNS. Sensory information from the receptors informs the brain of the body's contact with objects in the environment, provides information on the movement, position, and balance of body parts, and underlies the sense of hearing.

Receptors for Touch and Pressure Occur throughout the Body

In vertebrates, mechanoreceptors detecting touch and pressure are embedded in the skin and other surface tissues, in skeletal muscles, in the walls of blood vessels, and in internal organs. In humans, touch receptors in the skin are concentrated in greatest numbers in the fingertips, lips, and tip of the tongue, giving these regions the greatest sensitivity to mechanical stimuli. In other areas, such as the skin of the back, arms, and legs, the receptors are more widely spaced.

You can compare the spacing of receptors by pressing two toothpicks lightly against a fingertip and then against the skin of your arm or leg. On your fingertip, the toothpicks can be quite close together—separated by only a millimeter or so—and still be discerned as two separate points. On your arm or leg, they must be nearly 5 cm (almost 2 inches) apart to be distinguished.

Human skin contains several types of touch and pressure receptors **(Figure 39.2).** Some are free nerve endings, the dendrites of afferent neurons with no specialized structures surrounding them. Others, such as Pacinian corpuscles, have structures surrounding the nerve endings that contribute to reception of stimuli. Free nerve endings wrapped around hair follicles respond when the hair is bent, making you instantly aware, for example, of a spider exploring your arm or leg.

Proprioceptors Provide Information about Movements and Position of the Body

Mechanoreceptors called **proprioceptors** (*proprius* = one's own) detect stimuli that are used in the CNS to maintain body balance and equilibrium and to monitor the position of the head and limbs. The activity of these receptors allows you to touch the tip of your nose with your eyes closed, for example, or reach and scratch an itch on your back precisely. Here we consider examples of proprioceptors found in various animals.

Statocysts in Invertebrates. Many aquatic invertebrates, including jellyfishes, some gastropods, and some arthropods, have organs of equilibrium called **statocysts** (*statos* = standing; *kystis* = bag). Most statocysts are fluid-filled chambers with walls that contain **sensory hair cells** enclosing one or more movable stonelike bodies called **statoliths (Figure 39.3).** For example, lobsters have statoliths consisting of sand grains stuck together by mucus. When the animal moves, the statoliths lag behind the movement, bending the sensory hairs and triggering action potentials in afferent neurons. In this way, the statocysts signal the brain about the body's position and orientation with respect to gravity.

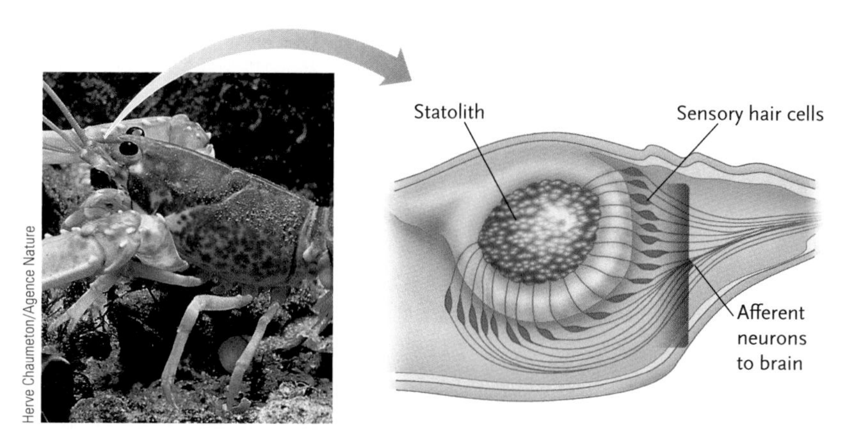

Figure 39.3
A statocyst, an invertebrate organ of equilibrium, and its location at the base of an antenna in a lobster. The statoliths inside are usually formed from fused grains of sand, as they are in the lobster, or from calcium carbonate.

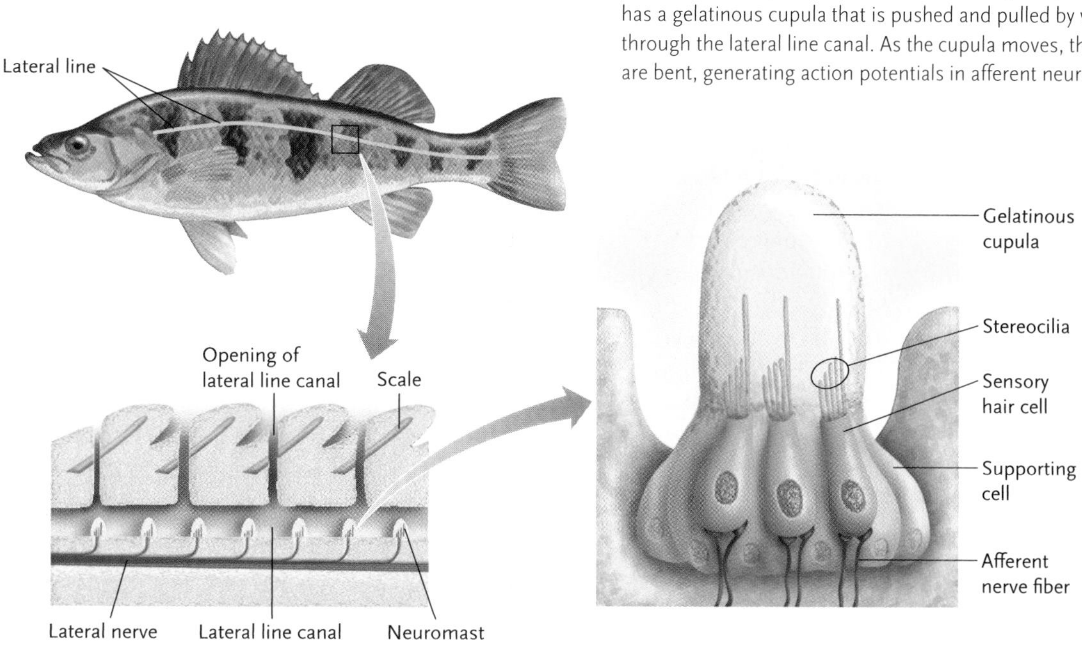

The Lateral Line System in Amphibians and Fish. Fishes and some aquatic amphibians detect vibrations and currents in the water through mechanoreceptors along the length of the body called the **lateral line system (Figure 39.4).** In fish, the mechanoreceptors, known as *neuromasts,* also provide information about the fish's orientation with respect to gravity and its swimming velocity. In some fishes, neuromasts are exposed on the body surface; in others, they are recessed in water-filled canals with porelike openings to the outside (as in Figure 39.4). Each dome-shaped neuromast has sensory hair cells clustered in its base. One surface of the hair cell is covered with **stereocilia,** which are actually microvilli (cell processes reinforced by bundles of microfilaments). The stereocilia extend into a gelatinous structure, the **cupula** (*cupule* = little cup), which moves with pressure changes in the surrounding water. Movement of the cupula bends the stereocilia, which causes the hair cell's plasma membrane to become depolarized and release neurotransmitter molecules; the neurotransmitters then generate action potentials in associated afferent neurons.

Vibrations detected by the lateral line enable fishes to avoid obstacles, orient in a current, and monitor the presence of other moving objects in the water. The system is also responsible for the ability of schools of fish to move in unison, turning and diving in what appears to be a perfectly synchronized aquatic ballet. In actual-

Figure 39.4
The lateral line system of fishes. The sensory receptor of the lateral line, the neuromast, has a gelatinous cupula that is pushed and pulled by vibrations and currents transmitted through the lateral line canal. As the cupula moves, the stereocilia of the sensory hair cells are bent, generating action potentials in afferent neurons that lead to the brain.

Lateral line
Opening of lateral line canal
Scale
Lateral nerve
Lateral line canal
Neuromast
Gelatinous cupula
Stereocilia
Sensory hair cell
Supporting cell
Afferent nerve fiber

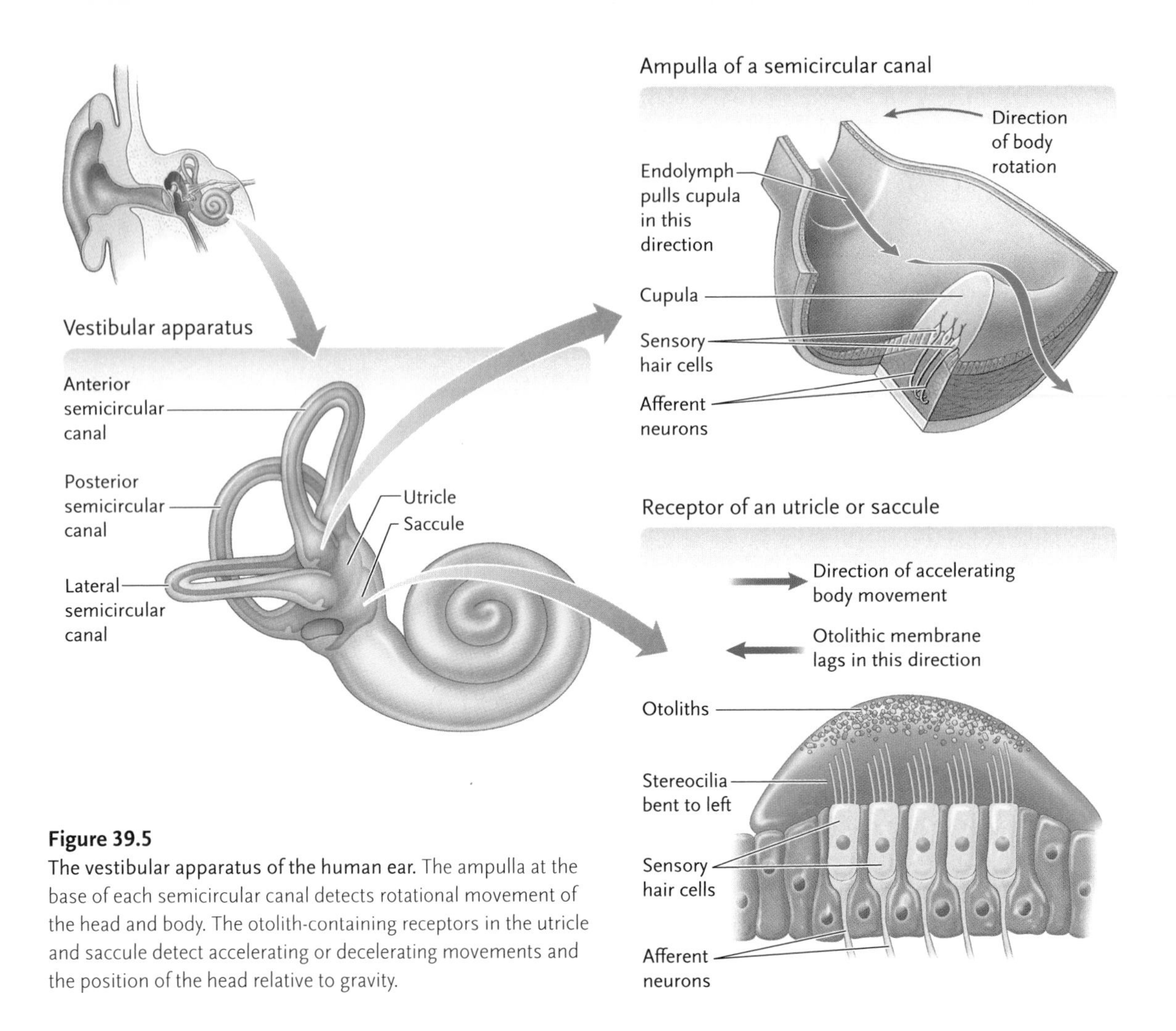

Figure 39.5
The vestibular apparatus of the human ear. The ampulla at the base of each semicircular canal detects rotational movement of the head and body. The otolith-containing receptors in the utricle and saccule detect accelerating or decelerating movements and the position of the head relative to gravity.

ity, the movement of each fish creates a pressure wave in the water that is detected by the lateral line systems of other fishes in the school. Schooling fish can still swim in unison even if blinded, but if the nerves leading from the lateral line system to the brain are severed, the ability to school is lost.

The Vestibular Apparatus in Vertebrates. The inner ear of most terrestrial vertebrates has two specialized sensory structures, the *vestibular apparatus* and the *cochlea*. The **vestibular apparatus** is responsible for perceiving the position and motion of the head and, therefore, is essential for maintaining equilibrium and for coordinating head and body movements. The cochlea is used in hearing, which we discuss later in this chapter.

The vestibular apparatus **(Figure 39.5)** consists of three **semicircular canals** and two chambers, the **utricle** and the **saccule,** filled with a fluid called *endolymph.* The semicircular canals, which are positioned at angles corresponding to the three planes of space, detect rotational (spinning) motions. Each canal has a swelling at its base called an *ampulla,* which is topped with sensory hair cells embedded in a cupula similar to that found in lateral line systems. The cupula protrudes into the endolymph of the canals. When the body or head rotates horizontally, vertically, or diagonally, the endolymph in the semicircular canal corresponding to that direction lags behind, pulling the cupula with it. The displacement of the cupula bends the sensory hair cells and generates action potentials in afferent neurons making synapses with the hair cells.

The utricle and saccule provide information about the position of the head with respect to gravity (up versus down), as well as changes in the rate of linear movement of the body. The ultricle and saccule, which are oriented approximately 30° to each other, each contain sensory hair cells with stereocilia. The hair cells are covered with a gelatinous *otolithic membrane* (which is similar to a cupula) in which **otoliths** (*oto* = ear; *lithos* = stone), small crystals of calcium carbonate, are embedded (see Figure 39.5); otoliths are similar to invertebrate statoliths.

When an animal is upright, the sensory hairs in the utricle are oriented vertically, and those in the saccule are oriented horizontally. When the head is tilted in any direction other than straight up and down, or when there is a change in linear motion of the body, the otolithic membrane of the utricle moves and bends the sensory hairs. Depending on the direction of movement, the hair cells release more or less neurotransmitter, and the brain integrates the signals it receives and generates a perception of the movement. The saccule responds to the tilting of the head away from the horizontal (such as in diving) and to a change in movement up and down (such as jumping up to dunk a basketball). The ultricle and saccule adapt quickly to the body's motion, decreasing their response when there is no change in the rate and direction of movement. In other words, the body adapts to the new position. For instance, when you move your head to the left, that new position becomes the "norm." Then, if you move your head again in any direction, signals from the utricle and saccule tell your brain that your head is moving to a new position.

Stretch Receptors in Vertebrates. In the muscles and tendons of vertebrates, proprioceptors called **stretch receptors** detect the position and movement of the limbs. The stretch receptors in muscles are **muscle spindles,** bundles of small, specialized muscle cells wrapped with the dendrites of afferent neurons and enclosed in connective tissue **(Figure 39.6).** When the muscle stretches, the spindle stretches also, stimulating the dendrites and triggering the production of action potentials. The strength of the response of stretch receptors to stimulation depends on how much and how fast the muscle is stretched. The proprioceptors of tendons, called **Golgi tendon organs,** are dendrites that branch within the fibrous connective tissue of the tendon (see Figure 39.6). These nerve endings measure stretch and compression of the tendon as the muscles move the limbs.

Proprioceptors allow the CNS to monitor the body's position and help keep the body in balance. They also allow muscles to apply constant force under a constant load, and to adjust almost instantly if the load changes. When you hold a cup while someone fills it with coffee, for example, the muscle spindles in your biceps muscle detect the additional stretch as the cup becomes heavier. Signals from the spindles allow you to compensate for the additional weight by increasing the contraction of the muscle, keeping your arm level with no conscious effort on your part. Proprioceptors are typically slow to adapt, so that the body's position and balance are constantly monitored.

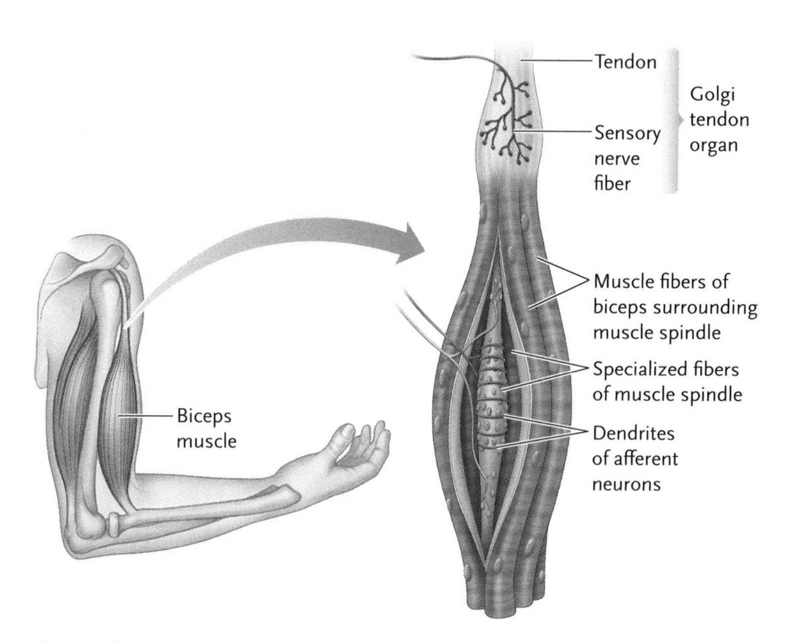

Figure 39.6
Muscle spindles, which detect the stretch and tension of muscles, and Golgi tendon organs, which detect the stretch of tendons.

Study Break

1. What is the function of proprioceptors?
2. What properties qualify proprioceptors as mechanoreceptors?

39.3 Mechanoreceptors and Hearing

In most animals, the receptors that detect sound are closely related to the receptors that detect body movement.

Sounds are vibrations that travel as waves produced by the alternating compression and decompression of the air. Although sound waves travel through air rapidly—at speeds of about 340 meters per second (700 miles per hour) at sea level—the individual air molecules transmitting the waves move back and forth over only a short distance as the wave passes. The vibrations of sound waves travel through water and solids by a similar mechanism.

The loudness, or *intensity,* of a sound depends on the amplitude (height) of the wave. The *pitch* of a sound—whether a musical tone, for example, is a high note or a low note—depends on the frequency of the waves, measured in cycles per second. The more cycles per second, the higher the pitch. Some animals, such as the bat in the introduction to this chapter, can hear sounds well above 100,000 cycles per second. Humans can hear sounds between about 20 and 20,000 cycles per second, which is why we cannot hear the bat's sonar clicks.

Invertebrates Have Varied Vibration-Detecting Systems

Most invertebrates detect sound and other vibrations through mechanoreceptors in their skin or other surface structures. An earthworm, for example, quickly retracts into its burrow at the smallest vibration of the surrounding earth, even though it has no specialized structures serving as ears. Cephalopods such as squids and octopuses have a system of mechanoreceptors on their head and tentacles, similar to the lateral line of fishes, which detects vibrations in the surrounding water. Many insects have sensory receptors in the form

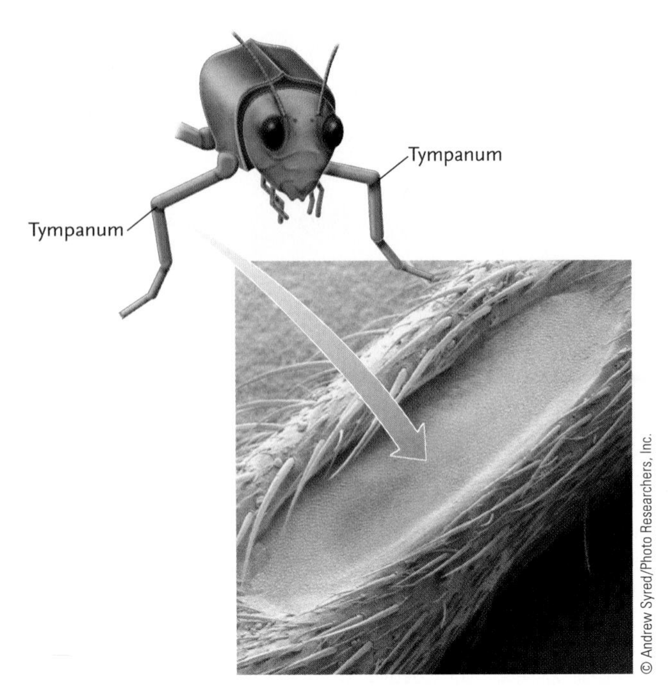

Figure 39.7
The tympanum or eardrum of a cricket, located on the front walking legs.

of hairs or bristles that vibrate in response to sound waves, often at particular frequencies.

Some insects, such as moths, grasshoppers, and crickets, have complex auditory organs on either side of the abdomen or on the first pair of walking legs **(Figure 39.7).** These "ears" consist of a thinned region of the insect's exoskeleton that forms a tympanum (*tympanum* = drum) over a hollow chamber. Sounds reaching the tympanum cause it to vibrate; mechanoreceptors connected to the tympanum translate the vibrations into nerve impulses. Some insect ears respond to sounds only at certain frequencies, such as to the pitch of a cricket's song or, in certain moths, to the frequencies of the echolocation sounds emitted by foraging bats.

Human Ears Are Representative of the Auditory Structures of Mammals

The auditory structures of terrestrial vertebrates transmit the vibrations of sound waves to sensory hair cells, which respond by triggering action potentials. Here, we describe the human ear, which is representative of mammalian auditory structures **(Figure 39.8).**

The **outer ear** has an external structure, the **pinna** (*pinna* = wing or leaf), which concentrates and focuses sound waves. Most mammals can turn the pinnae to help focus sounds, but human pinnae are inefficient compared with the ears of dogs, rabbits, or horses. The sound waves enter the auditory canal, which leads from the exterior, and strike a thin sheet of tissue, the **tympanic membrane,** or eardrum, which vibrates back and forth in response.

Behind the eardrum is the **middle ear,** an air-filled cavity containing three small, interconnected bones: the **malleus** (hammer), the **incus** (anvil), and the **stapes** (stirrup). The stapes is attached to a thin, elastic membrane, the **oval window.** The vibrations of the eardrum, transmitted by the malleus and incus, push the stapes back and forth against the oval window. The levering action of the bones, combined with the much larger size of the eardrum as compared with the oval window, amplifies the vibrations transmitted to the oval window by more than 20 times.

Inside the oval window is the **inner ear.** It contains several fluid-filled compartments, including the semicircular canals, utricle, saccule, and a spiraled tube, the **cochlea** (*kochlias* = snail). The cochlea twists through about two and a half turns; if stretched out flat, it would be about 3.5 cm long. Thin membranes divide the cochlea into three longitudinal chambers, the *vestibular canal* at the top, the *cochlear duct* in the middle, and the *tympanic canal* at the bottom (see Figure 39.8). The vestibular canal and the tympanic canal join at the outer tip of the cochlea, so that the fluid within them is continuous. Within the cochlear duct is the **organ of Corti;** it contains the sensory hair cells that detect sound vibrations transmitted to the inner ear (see Figure 39.8).

The vibrations of the oval window pass through the fluid in the vestibular canal, make the turn at the end, and travel back through the fluid in the tympanic canal. At the end of the tympanic canal, they are transmitted to the **round window,** a thin membrane that faces the middle ear.

The vibrations traveling through the inner ear cause the *basilar membrane* to vibrate in response. The basilar membrane, which forms part of the floor of the cochlear duct, anchors the sensory hair cells in the organ of Corti. The stereocilia of these cells are embedded in the *tectorial membrane,* which extends the length of the cochlear canal. Vibrations of the basilar membrane cause the hair cells to bend, stimulating them to release a neurotransmitter that triggers action potentials in afferent neurons leading from the inner ear.

The basilar membrane is narrowest near the oval window and gradually widens toward the outer end of the cochlear duct. The high-frequency vibrations produced by high-pitched sounds vibrate the basilar membrane most strongly near its narrow end while vibrations of lower frequency vibrate the membrane nearer the outer end. Thus each frequency of sound waves causes hair cells in a different segment of the basilar membrane to initiate action potentials.

More than 15,000 hair cells are distributed in small groups along the basilar membrane. Each group of hairs is connected by synapses to afferent neurons, which in turn are bundled together in the *auditory nerve,* a cranial nerve that leads to the thalamus. From there, the signals are routed to specific regions in the auditory center of the temporal lobe. As sound waves of a particular frequency and intensity stimulate a specific segment of the basilar membrane, the region of

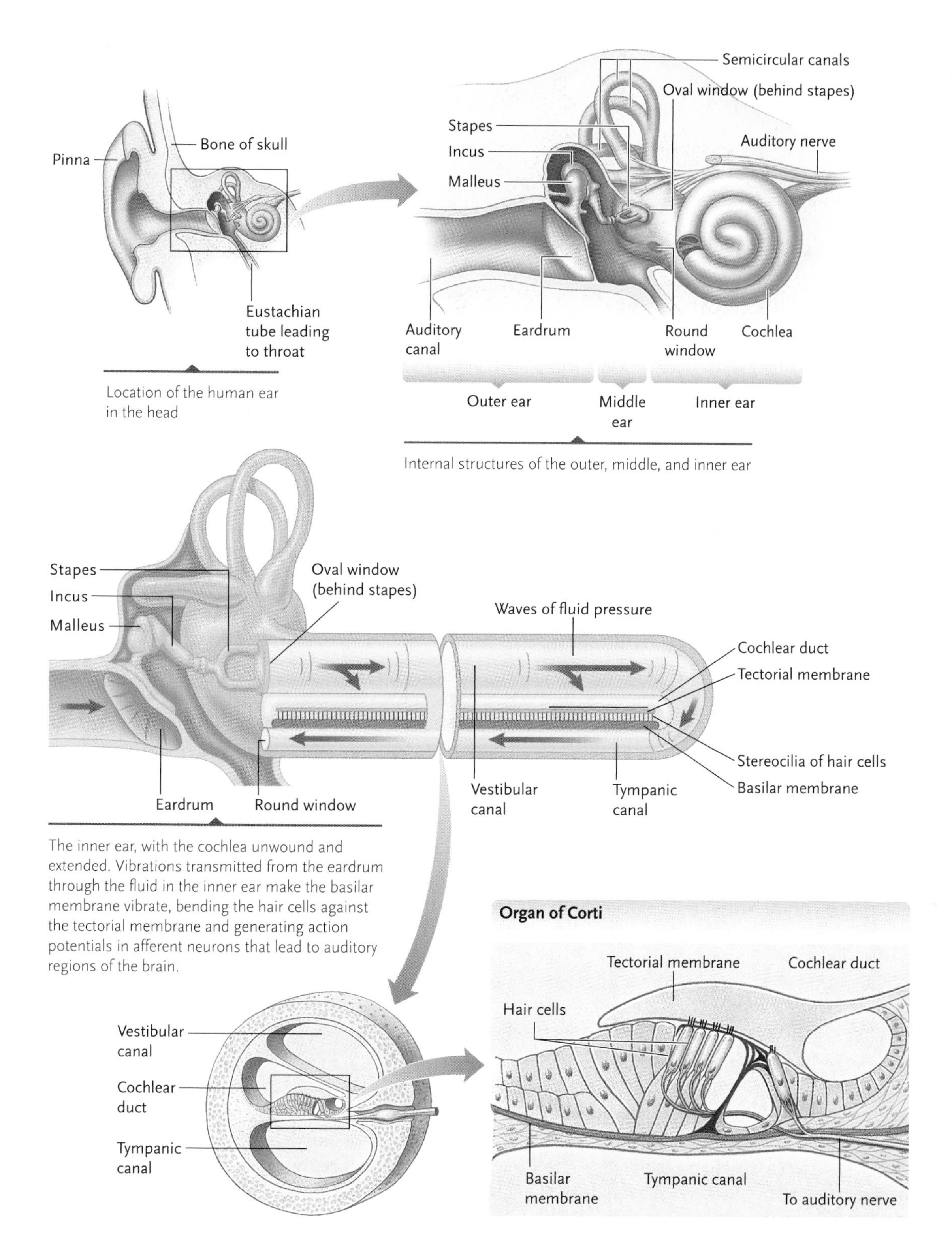

Figure 39.8
Structures of the human ear.

the auditory center to which the signals are sent integrates the information into the perception of sound at a corresponding pitch and loudness.

The sounds we hear are usually a rich combination of vibrations at different frequencies and intensities, which result in hundreds to thousands of different keys being struck simultaneously, with different degrees of force, in the cochlear keyboard. The combination of signals transmitted from the cochlea to the auditory centers is integrated into the perception of a human voice, the song of a sparrow, the roar of a jet plane, or a Mozart symphony.

Another system protects the eardrum from damage by changes in environmental atmospheric pressure. The system depends on the *Eustachian tube,* a duct that leads from the air-filled middle ear to the throat (see Figure 39.8). As we swallow or yawn, the tube opens, allowing air to flow into or out of the middle ear to equalize the pressure on both sides of the eardrum. When swelling or congestion due to infections prevents the tube from admitting air, we complain of having stopped-up ears—we can sense that a pressure difference between the outer and middle ear is bulging the eardrum inward or outward and interfering with the transmission of sounds.

Many Vertebrates Keep Track of Obstacles and Prey by Echolocation

Many vertebrates, like the bat in the introduction to this chapter, locate prey or avoid obstacles by **echolocation**—by making squeaking or clicking noises, and then listening for the echoes that bounce back from objects in their environment. By sensing the direction of the echoes and the time between the squeak or click and the returning echo, the animal can pinpoint the locations of barriers or prey animals.

Porpoises and dolphins locate food fishes in murky water by echolocation, and whales also use echolocation to keep track of the sea bottom and rocky obstacles. Two bird species, the oilbird and the cave swiftlet, use echolocation to avoid obstacles and find their nests in the dark of caves. Even humans can learn to use echolocation—a blind person, for example, listens for the echoes from a tapping cane to avoid posts, walls, and other barriers.

Study Break

1. What vibration-detecting systems are found in cephalopods and insects?
2. How are sounds of particular frequencies distinguished and "heard" by humans?

39.4 Photoreceptors and Vision

Virtually all animals have receptors that can detect and respond to light. As animals evolved and became more complex, the complexity of their visual sensory receptors increased, leading to the highly developed eyes of vertebrates.

We begin our discussion of these remarkable receptors by examining the basic visual structures and their functions in animals. We then compare vision in representative invertebrates and vertebrates.

Vision Involves Detection and Perception of Radiant Energy

Photoreceptors detect light at particular wavelengths, while centers in a brain or central ganglion integrate signals arriving from the receptors into a perception of light. All animals use different forms of a single lipid-like pigment, *retinal* (synthesized from vitamin A), in

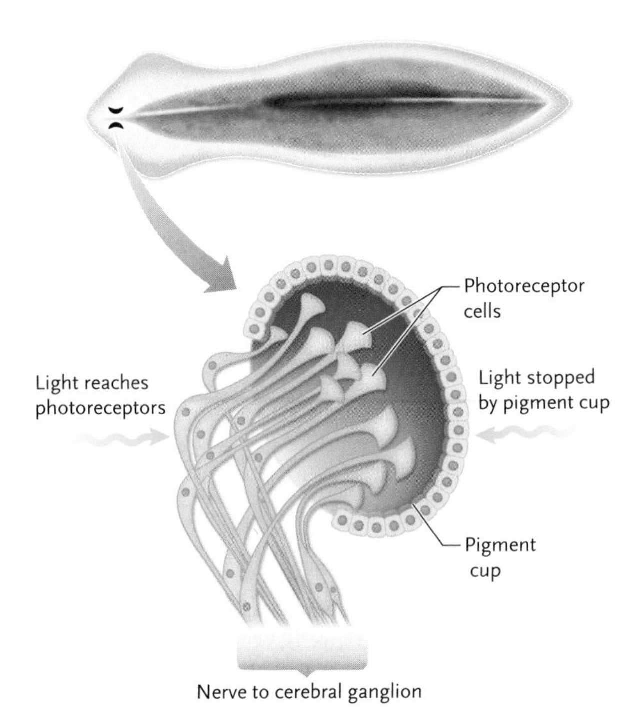

Figure 39.9
The ocellus of *Planaria,* a flatworm, and the arrangement of pigment cells on which its orientation response is based.

the photoreceptors to absorb light energy. The absorbed energy generates action potentials in afferent neurons leading to visual centers in the CNS. The organ of vision that detects light is the *eye.* The simplest eyes are capable only of distinguishing light from dark, while the most complex eyes distinguish shapes and colors and focus an accurate image of objects being viewed onto a layer of photoreceptors. Signals originating from the photoreceptors are integrated in the brain into an accurate, point-by-point perception of the object being viewed.

Invertebrate Eyes Take Many Forms

Some invertebrates, such as earthworms, do not have visual organs; instead, photoreceptors in their skin allow them to sense and respond to light. Earthworms respond negatively to light, as you can easily discover by shining a flashlight on an earthworm outside its burrow at night.

The eyes of other invertebrates are diverse, ranging from collections of photoreceptors with no lens and no image-forming capability to eyes remarkably like those of vertebrates. The photoreceptors of invertebrates are depolarized when they absorb light, and generate action potentials or increase their release of neurotransmitter molecules when they are stimulated. Vertebrate photoreceptors function differently, as we will see.

The simplest eye is the **ocellus** (plural, *ocelli;* also called an *eyespot* or *eyecup*). An ocellus, which detects light but does not form an image, consists of fewer than 100 photoreceptor cells lining a cup or pit. In planarians, for example, photoreceptor cells in a cuplike depression below the epidermis are connected to the dendrites of afferent neurons, which are bundled into nerves that travel from the ocelli to the cerebral ganglion **(Figure 39.9).** Each ocellus is covered on one side by a layer of pigment cells that blocks most of the light rays arriving from the opposite side of the animal. As a result, most of the light received by the pigment cells enters the ocellus from the side that it faces. Through integration of information transmitted to the cerebral ganglion from the eyecups, planarians orient themselves so that the amount of light falling on the two ocelli is equal and diminishes as they swim. This reaction carries them directly away from the source of the light and towards darker areas where the chance of a predator catching them is smaller. Similar ocelli are found in a variety of animals, including a number of insects, arthropods, and mollusks.

Two main types of image-forming eyes have evolved in invertebrates: compound eyes and single-lens eyes. The **compound eye** of insects, crustaceans, and a few annelids and mollusks contains hundreds to thousands of faceted visual units called **ommatidia** (*omma* = eye) fitted closely together **(Figure 39.10).** In insects, light entering an ommatidium is focused by a transparent **cornea** and a *crystalline cone* (just below the cornea) onto a bundle of photoreceptor cells. Microvilli of these cells interdigitate like the fingers of clasped hands, forming a central axis that contains rhodopsin, a **photopigment** (light-absorbing pigment) also found in the rods of vertebrate eyes. Absorption of light by rhodopsin causes action potentials to be generated in afferent neurons connected to the base of the ommatidium. Each ommatidium of a compound eye samples a small part of the visual field. From these signals, the brain receives a mosaic image of the world. Because even the slightest motion is detected simultaneously by many ommatidia, compound eyes are extraordinarily adept at detecting movement—a lesson soon learned by fly-swatting humans.

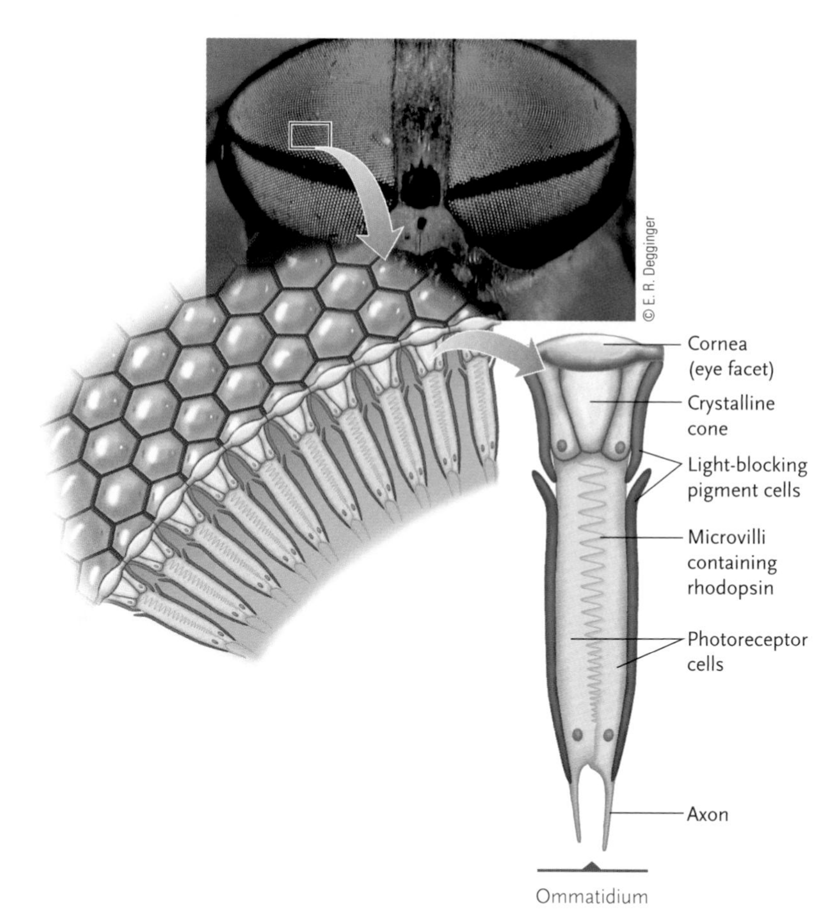

Figure 39.10
The compound eye of a deer fly. Each ommatidium has a cornea that directs light into the crystalline cone; in turn, the cone focuses light on the photoreceptor cells. A light-blocking pigment layer at the sides of the ommatidium prevents light from scattering laterally in the compound eye.

The **single-lens eye** of cephalopods **(Figure 39.11)** resembles a vertebrate eye in that both types operate like a camera. In the cephalopod eye, light enters through the transparent cornea, a **lens** concentrates the light, and a layer of photoreceptors at the back of the eye, the **retina,** records the image. Behind the cornea is the **iris,** which surrounds the **pupil,** the opening through which light enters the eye. Muscles in the iris adjust the size of the pupil to vary the amount of light entering the eye. When the light is bright, circular muscles in the iris contract, shrinking the size of the pupil and reducing the amount of light that enters the eye. In dim light, radial muscles contract and enlarge the pupil, increasing the amount of light that enters the eye. Muscles move the lens forward and back with respect to the retina to focus the image. This is an example of **accommodation,** a process by which the lens changes to enable the eye to focus on objects at different distances.

A neural network lies under the retina, meaning that light rays do not have to pass through the neurons to reach the photoreceptors. The vertebrate eye has the opposite arrangement. This and other differences in structure and function indicate that cephalopod and vertebrate eyes evolved independently.

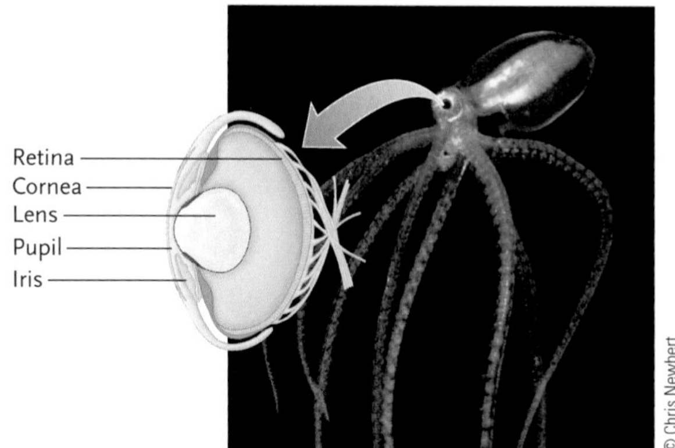

Figure 39.11
The eye of an octopus, a cephalopod mollusk.

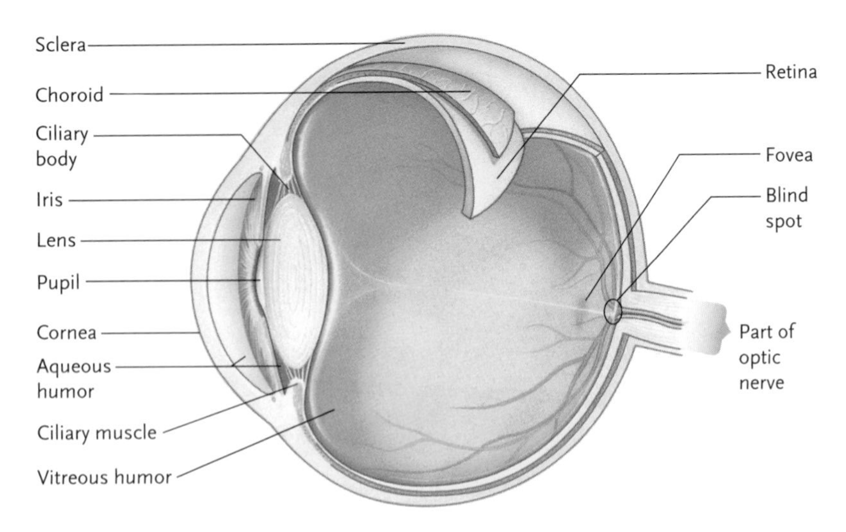

Figure 39.12
Structures of the human eye.

Vertebrate Eyes Have a Complex Structure

The human eye **(Figure 39.12)** has similar structures—cornea, iris, pupil, lens, and retina—to those of the cephalopod eye just described. Light entering the eye through the cornea passes through the iris and then the lens. The lens focuses an image on the retina, and the axons of afferent neurons originating in the retina converge to form the optic nerve leading from the eye to the brain.

A clear fluid called the **aqueous humor** fills the space between the cornea and lens. This fluid carries nutrients to the lens and cornea, which do not contain any blood vessels. The main chamber of the eye, located between the lens and the retina, is filled with the jellylike **vitreous humor** (*vitrum* = glass). The outer wall of the eye contains a tough layer of connective tissue (the *sclera*). Inside it is a darkly pigmented layer (the *choroid*) that prevents light from entering except through the pupil. It also contains the blood vessels nourishing the retina.

Two types of photoreceptors, rods and cones, occur in the retina along with layers of neurons that carry out an initial integration of visual information before it is sent to the brain. The **rods** are specialized for detection of light at low intensities; the **cones** are specialized for detection of different wavelengths (colors).

Accommodation does not occur by forward and back movement of the lens, as described for cephalopods. Rather, the lens of most terrestrial vertebrates is focused by changing its shape. The lens is held in place by fine ligaments that anchor it to a surrounding layer of connective tissue and muscle, the **ciliary body.** These ligaments keep the lens under tension when the ciliary muscle is relaxed. The tension flattens the lens, which is soft and flexible, and focuses light from distant objects on the retina **(Figure 39.13a).** When the ciliary muscles contract, they relieve the tension of the ligaments, allowing the lens to assume a more spherical shape and focusing light from nearby objects on the retina **(Figure 39.13b).**

The Retina of Mammals and Birds Contains Rods and Cones and a Complex Network of Neurons

The retina of a human eye contains about 120 million rods and 6 million cones organized into a densely packed, single layer. Neural networks of the retina are layered on top of the photoreceptor cells, so that light rays focused by the lens on the retina must pass through the neurons before reaching the photoreceptors. The light must also pass through a layer of fine blood vessels that covers the surface of the retina.

In mammals and birds with eyes specialized for daytime vision, cones are concentrated in and around a small region of the retina, the **fovea** (see Figure 39.12). The image focused by the lens is centered on

Figure 39.13
Accommodation in terrestrial vertebrates: the lens changes shape rather than moving forward and back to focus on **(a)** distant and **(b)** near objects.

a. Focusing on distant object

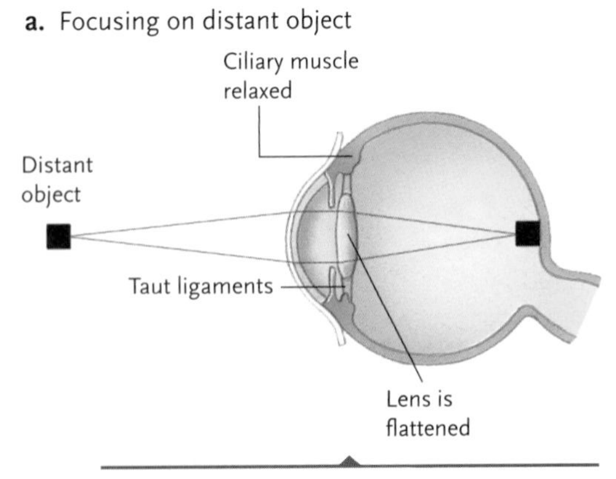

When the eye focuses on a distant object, the ciliary muscles relax, allowing the ligaments that support the lens to tighten. The tightened ligaments flatten the lens, bringing the distant object into focus on the retina.

b. Focusing on near object

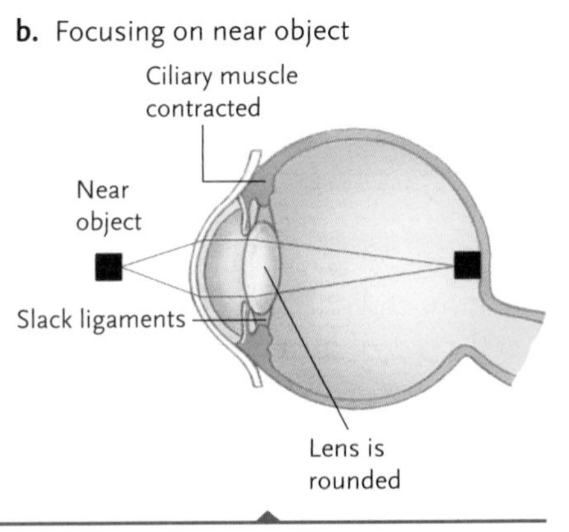

When the eye focuses on a near object, the ciliary muscles contract, loosening the ligaments and allowing the lens to become rounder. The rounded lens focuses a near object on the retina.

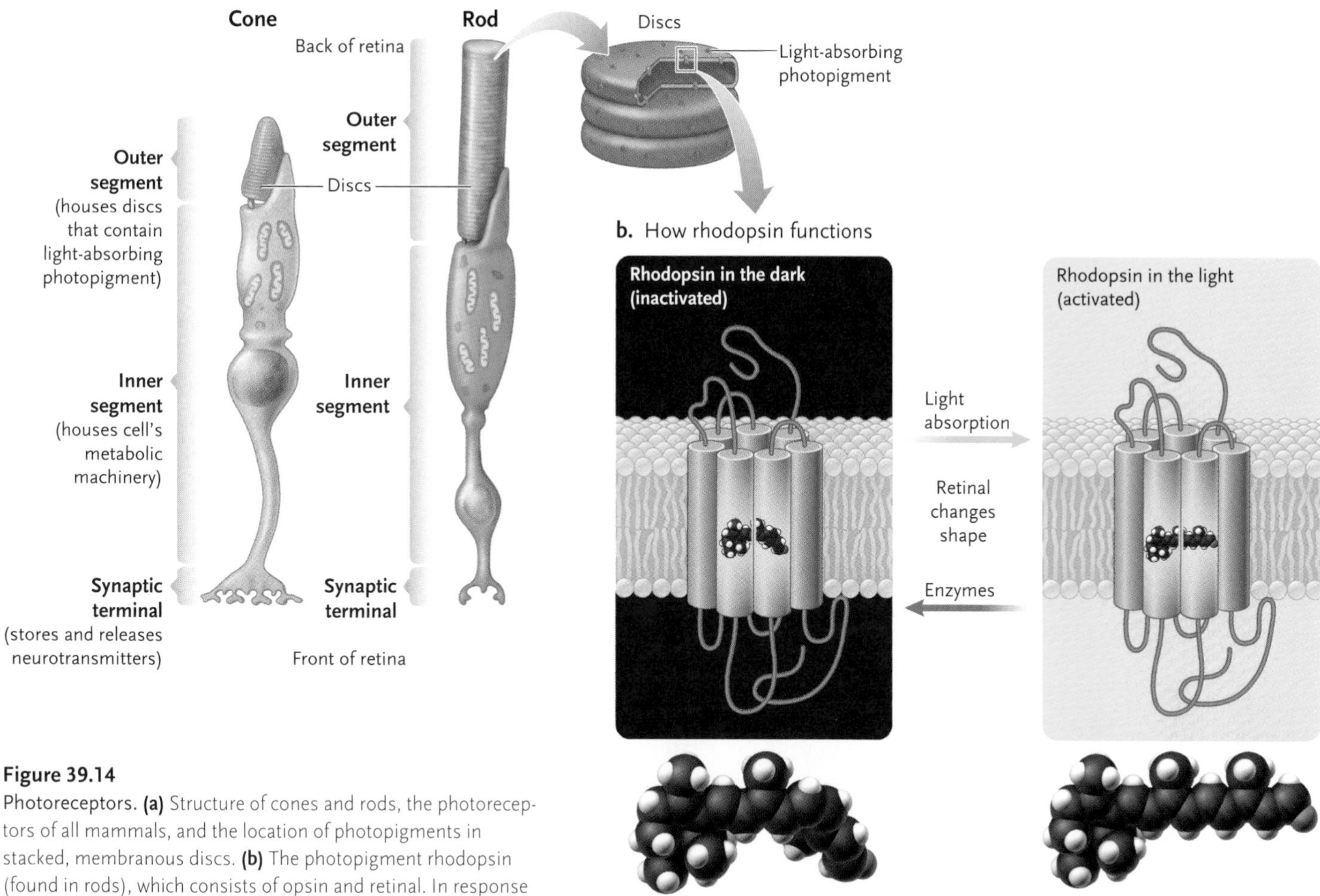

Figure 39.14
Photoreceptors. **(a)** Structure of cones and rods, the photoreceptors of all mammals, and the location of photopigments in stacked, membranous discs. **(b)** The photopigment rhodopsin (found in rods), which consists of opsin and retinal. In response to light, the retinal changes from a bent to a straight structure.

the fovea, which is circular and less than a millimeter in diameter in humans. The rods are spread over the remainder of the retina. We can see distinctly only the image focused on the fovea; the surrounding image is what we term *peripheral vision*. Mammals and birds with eyes specialized for night vision have retinas containing mostly rods, without a clearly defined fovea. Some fishes and many reptiles have cones generally distributed throughout their retina and very few rods.

The rods of mammals are much more sensitive than the cones to light of low intensity; in fact, they can respond to a single photon of light. This is why, in dim light, we can see objects better by looking slightly to the side of the object. This action directs the image away from the cones in the fovea to the highly light-sensitive rods in surrounding regions of the retina.

Sensory Transduction by Rods and Cones. Photoreceptors have three parts: an outer segment consisting of stacked, flattened, membranous discs; an inner segment where the cell's metabolic activities occur; and the synaptic terminal, where neurotransmitter molecules are stored and released **(Figure 39.14a)**. The light-absorbing pigment of rods and cones, retinal, is bonded covalently in the photoreceptors with one of several different proteins called **opsins** to produce **photopigments.** The photopigments are embedded in the membranous discs of the photoreceptors' outer segments **(Figure 39.14b)**. The retinal-opsin photopigment in rods is called **rhodopsin.** Let us see how light stimulating a rod photoreceptor is transduced; the mechanism is essentially the same in the cone photoreceptors.

In the dark, the retinal segment of the unstimulated rhodopsin is in an inactive form known as *cis*-retinal (see Figure 39.14b), and the rods steadily release the neurotransmitter glutamate. When rhodopsin absorbs a photon of light, retinal converts to its active form, *trans*-retinal (see Figure 39.14b), and the rods *decrease* the amount of glutamate they release. This will be discussed in the next section.

Rhodopsin is a membrane-embedded G-protein-coupled receptor (see Section 7.4). Recall that an extracellular signal received by a G-protein-coupled receptor activates the receptor, which triggers a signal transduction pathway within the cell, leading to a cellular response. Here, activated rhodopsin triggers a signal transduction pathway that leads to the closure of Na^+ channels in the plasma membrane **(Figure 39.15)**. Closure of the channels hyperpolarizes the photoreceptor's membrane, thereby decreasing neuro-

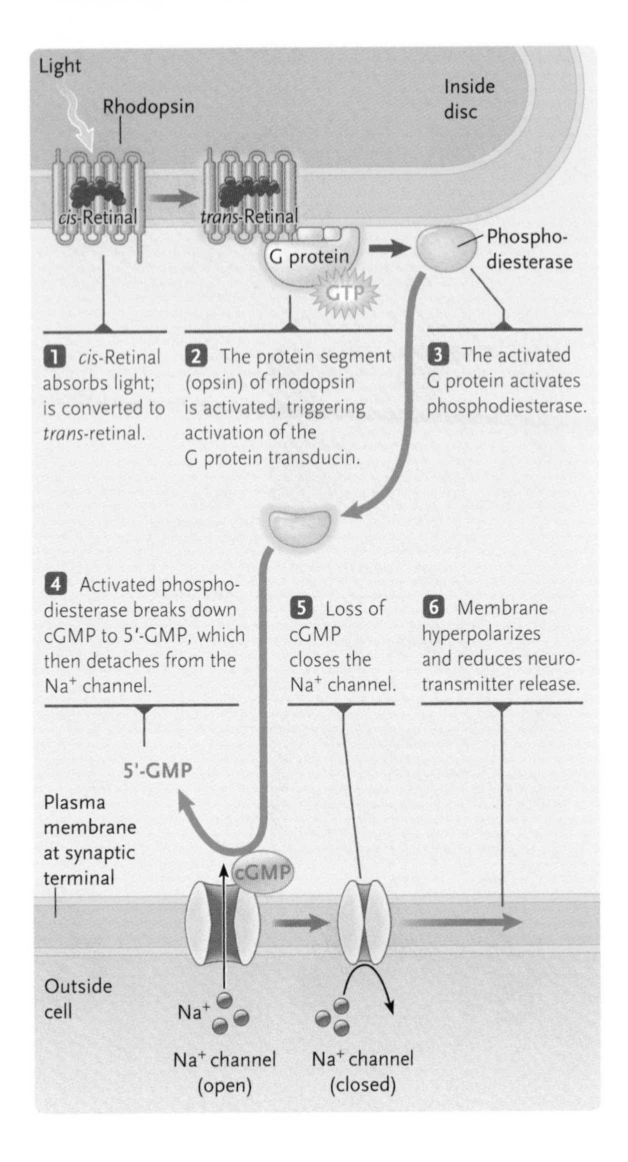
Light
Rhodopsin
Inside disc
cis-Retinal
trans-Retinal
G protein
GTP
Phospho-diesterase
1 cis-Retinal absorbs light; is converted to trans-retinal.
2 The protein segment (opsin) of rhodopsin is activated, triggering activation of the G protein transducin.
3 The activated G protein activates phosphodiesterase.
4 Activated phospho-diesterase breaks down cGMP to 5'-GMP, which then detaches from the Na+ channel.
5 Loss of cGMP closes the Na+ channel.
6 Membrane hyperpolarizes and reduces neuro-transmitter release.
5'-GMP
Plasma membrane at synaptic terminal
cGMP
Outside cell
Na+
Na+ channel (open)
Na+ channel (closed)

Figure 39.15
The signal transduction pathway that closes Na^+ channels in photoreceptor plasma membranes when rhodopsin absorbs light.

transmitter release. The response is graded in the sense that as light absorption by photopigment molecules increases, the amount of neurotransmitter released is reduced proportionately; if light absorption decreases, neurotransmitter release by the photoreceptor increases proportionately. Note that transduction in rods works in the opposite way from most sensory receptors, in which a stimulus increases neurotransmitter release.

Visual Processing in the Retina. In the retina of all vertebrates, the two types of photoreceptors are linked to a network of neurons that carries out initial integration and processing of visual information. The retina of mammals contains four types of neurons **(Figure 39.16).** Just over the rods and cones is a layer of **bipolar cells.** These neurons make synapses with the rods or cones at one end and with a layer of neurons called **ganglion cells** at the other end. The axons of ganglion cells extend over the retina and collect at the back of the eyeball to form the optic nerve, which transmits action potentials to the brain. The point where the optic nerve exits the eye lacks photoreceptors, resulting in a *blind spot* several millimeters in diameter. Two other types of neurons form lateral connections in the retina: **horizontal cells** connect photoreceptor cells, and **amacrine cells** connect bipolar cells and ganglion cells.

In the dark, the steady release of glutamate from rods and cones depolarizes some of the postsynaptic bipolar cells and hyperpolarizes others, depending on the type of receptor those cells have. In the light, the decrease in neurotransmitter release from rods and cones results in the polarized bipolar cells becoming hyperpolarized, and hyperpolarized bipolar cells becoming polarized. These membrane potential changes in response to light are transmitted to the brain for processing.

Signals from the rods and cones may move vertically or laterally in the retina. Signals move vertically from the photoreceptors to bipolar cells and then to ganglion cells. However, while the human retina has over 120 million photoreceptors, it has only about 1 million ganglion cells. This disparity is explained by the fact that each ganglion cell receives signals from a clearly defined set of photoreceptors that constitute the *receptive field* for that cell. Therefore, stimulating numerous photoreceptors in a ganglion cell's receptive field results in only a single message to the brain from that cell. Receptive fields are typically circular and are of different sizes. Smaller receptive fields result in sharper images because they send more precise information to the brain regarding the location in the retina where the light was received.

Signals that move laterally from a rod or cone proceed to a horizontal cell and continue to bipolar cells with which the horizontal cell makes inhibitory connections. To understand this, consider a spot of light falling on the retina. Photoreceptors detect the light and send a signal to bipolar cells and horizontal cells. The horizontal cells inhibit more distant bipolar cells that are outside the spot of light, causing the light spot to appear lighter and its surrounding dark area to appear darker. This type of visual processing is called **lateral inhibition** and serves both to sharpen the edges of objects and enhance contrast in an image.

Three Kinds of Opsin Pigments Underlie Color Vision

Many invertebrates and some species in each class of vertebrates have color vision. Color vision depends on the cones in the retina. Most mammals have only two types of cones, making their color vision limited, while humans and other primates have three types. Each human or primate cone cell contains one of three different photopigments, collectively called **photopsins,** in which retinal is combined with different opsins. The three photopsins absorb light over different, but overlapping, wavelength ranges, with peak absorptions at 445 nm (blue light), 535 nm (green light), and 570 nm

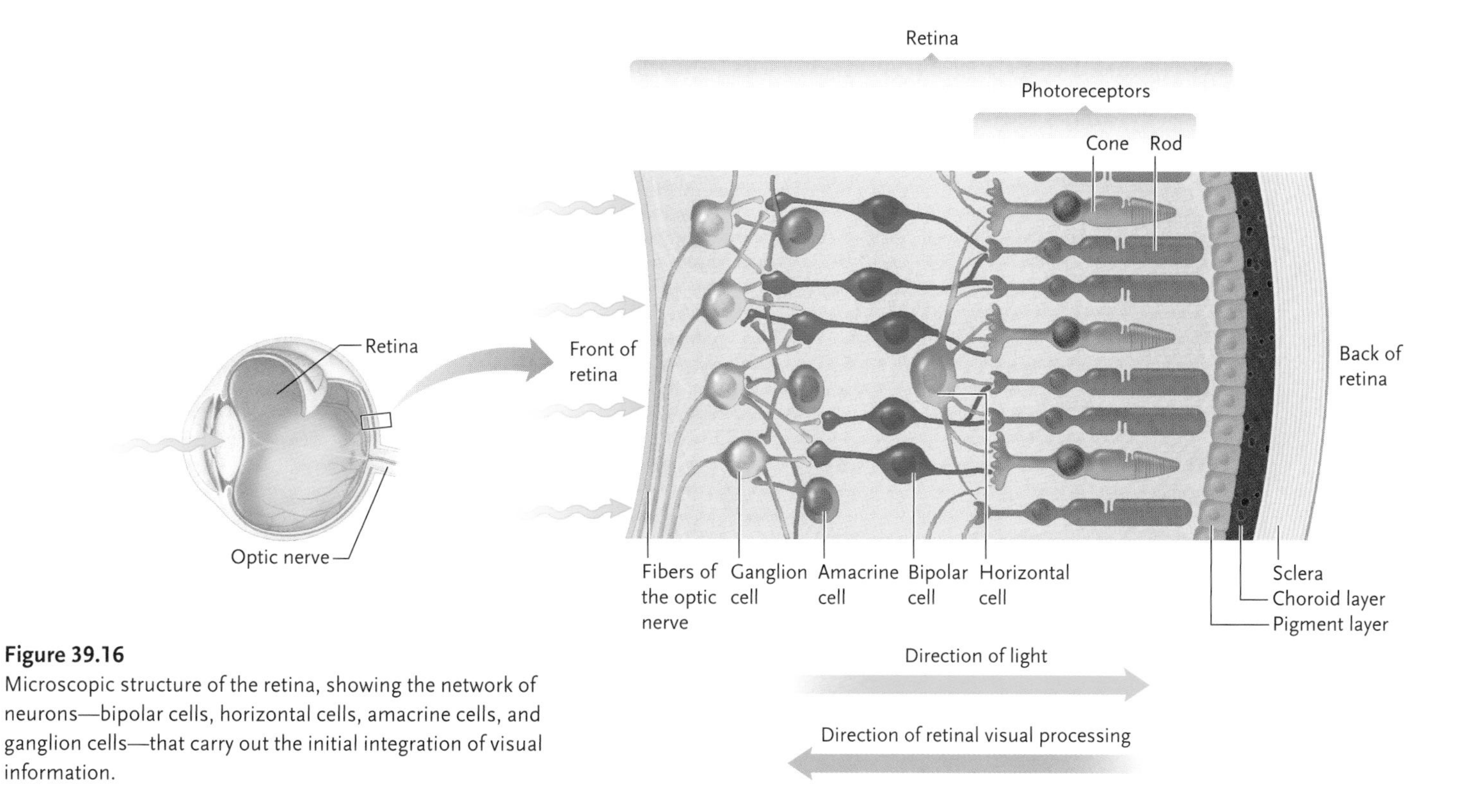

Figure 39.16
Microscopic structure of the retina, showing the network of neurons—bipolar cells, horizontal cells, amacrine cells, and ganglion cells—that carry out the initial integration of visual information.

(red light). The farther a wavelength is from the peak color absorbed, the less strongly the cone responds.

Having overlapping wavelength ranges for the three photoreceptors means that light at any visible wavelength will stimulate at least two of the three types of cones. However, because the maximal absorption of each type of cone is a different wavelength, it is stimulated to a different extent by light at a given wavelength. The differences, relayed to the visual centers of the brain, are integrated into the perception of a color corresponding to the particular wavelength absorbed. Light stimulating all three receptor types equally is seen as white.

The Visual Cortex Processes Visual Information

Just behind the eyes, the optic nerves converge before entering the base of the brain. A portion of each optic nerve crosses over to the opposite side, forming the **optic chiasm** (*chiasma* = crossing place). Most of the axons enter the **lateral geniculate nuclei** in the thalamus, where they make synapses with interneurons leading to the visual cortex **(Figure 39.17)**.

Because of the optic chiasm, the left half of the image seen by both eyes is transmitted to the visual cortex in the right cerebral hemisphere, and the right half of the image is transmitted to the left cerebral hemisphere. The right hemisphere thus sees objects to the left of the center of vision, and the left hemisphere sees objects to the right of the center of vision. Communication between the right and left hemispheres integrates this information into a perception of the entire visual field seen by the two eyes.

If you look at a nearby object with one eye and then the other, you will notice that the point of view is slightly different. Integration of the visual field by the brain creates a single picture with a sense of distance and depth. The greater the difference between the images seen by the two eyes, the closer the object appears to the viewer.

The two optic nerves together contain more than a million axons, more than all other afferent neurons of the body put together. Almost one-third of the cerebral cortex is devoted to visual information. These numbers give some idea of the complexity of the information integrated into the visual image formed by the brain.

Study Break

For vertebrate photoreception, define: (a) photopigment; (b) cone; (c) receptive field.

39.5 Chemoreceptors

Chemoreceptors form the basis of taste (gustation) and smell (olfaction), and measure the levels of internal body molecules such as oxygen, carbon dioxide, and hydrogen ions. All chemoreceptors probably work through membrane receptor proteins that are stimulated when they bind with specific molecules in the

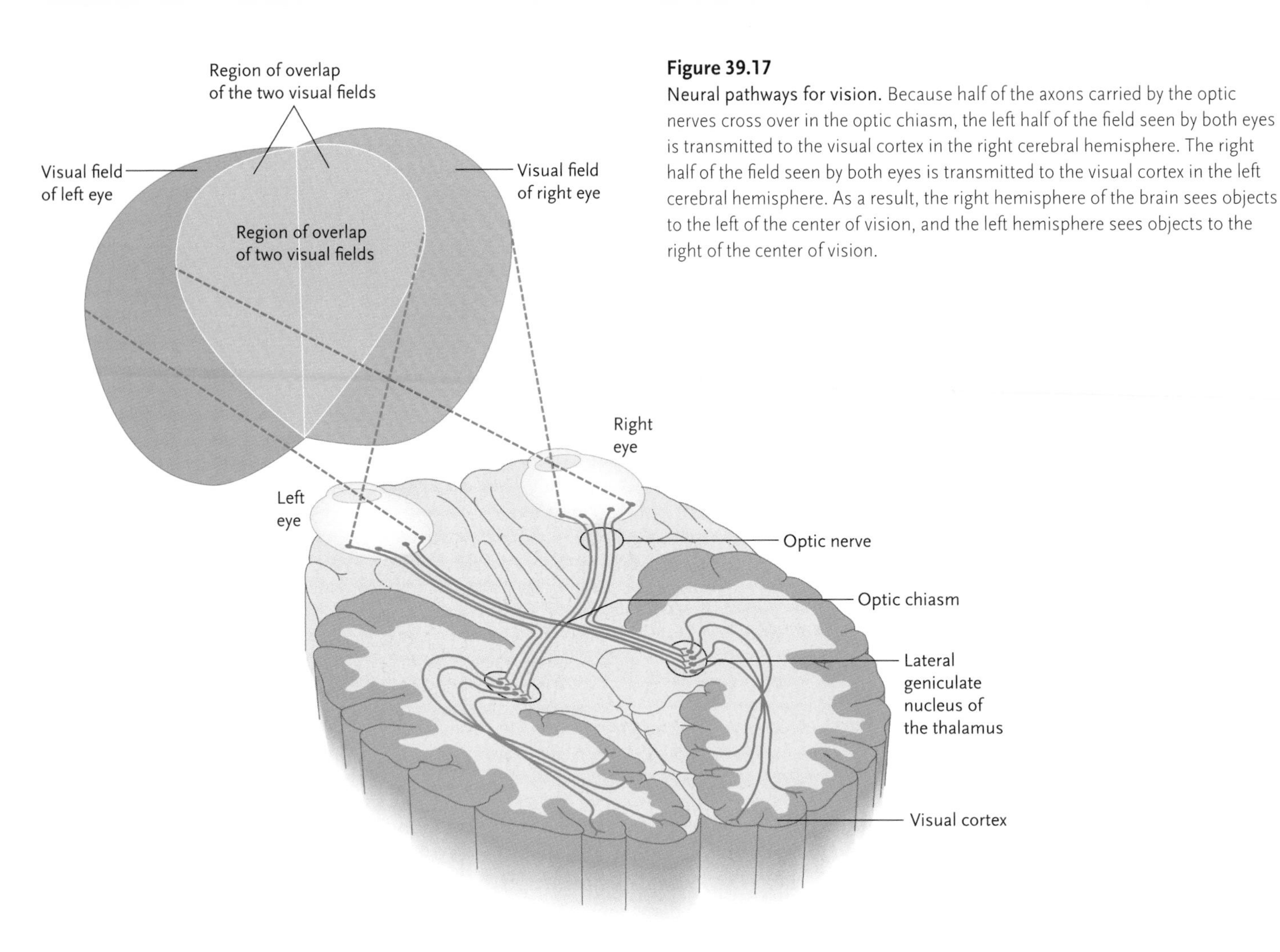

Figure 39.17
Neural pathways for vision. Because half of the axons carried by the optic nerves cross over in the optic chiasm, the left half of the field seen by both eyes is transmitted to the visual cortex in the right cerebral hemisphere. The right half of the field seen by both eyes is transmitted to the visual cortex in the left cerebral hemisphere. As a result, the right hemisphere of the brain sees objects to the left of the center of vision, and the left hemisphere sees objects to the right of the center of vision.

environment, generating action potentials in afferent nerves leading to the CNS.

Invertebrates Have Either the Same or Different Receptors for Taste and Smell

In many invertebrates the same receptors serve for the senses of smell and taste. These receptors may be confined to certain locations or distributed over the body surface. For example, the cnidarian *Hydra* has chemoreceptor cells around its mouth that respond to glutathione, a chemical released from prey organisms ensnared in the cnidarian's tentacles. Stimulation of the chemoreceptors by glutathione causes the tentacles to retract, resulting in ingestion of the prey. By contrast, earthworms have taste/smell receptors distributed over the entire body surface.

Some terrestrial invertebrates, particularly insects, have clearly differentiated taste and smell receptors. In insects, taste receptors occur inside hollow sensory bristles called *sensilla* (singular, *sensillum*), which may be located on the antennae, mouthparts, or feet **(Figure 39.18).** Pores in the sensilla admit molecules from potential food to the chemoreceptors, which are specialized to detect sugars, salts, amino acids, or other chemicals. Many female insects have chemoreceptors on their ovipositors, which allow them to lay their eggs on food appropriate for the hatching larvae.

Insect olfactory receptors detect airborne molecules. Some insects use odor as a means of communication, as with the pheromones released into the air as sexual attractants by female moths. Olfactory receptors in the bristles of male silkworm moth antennae **(Figure 39.19)** have been shown experimentally to be able to detect pheromones released by a female of the same species in concentrations as low as one attractant molecule per 10^{17} air molecules; when as few as 40 of the 20,000 receptor cells on its antennae have been stimulated by pheromone molecules, the male moth responds by fluttering its wings rapidly to attract the female's attention. Ants, bees, and wasps may identify members of the same hive or nest or communicate by means of odor molecules.

Taste and Smell Receptors Are Differentiated in Terrestrial Animals

In terrestrial animals, taste involves the detection of potential food molecules in objects that are touched by a receptor, while smell involves the detection of airborne molecules. Although both taste and smell receptors have hairlike extensions containing the proteins that bind environmental molecules, the hairs of taste receptors are derived from microvilli and contain microfilaments, while the hairs of smell receptors are derived from cilia and contain microtubules. Another significant difference between taste and smell is that information from taste receptors is typically processed in the parietal lobes, while information from smell receptors is processed in the olfactory bulbs and the temporal lobes.

In Vertebrates, Taste Receptors Are Located in Taste Buds

The taste receptors of most vertebrates form part of a structure called a taste bud, a small, pear-shaped capsule with a pore at the top that opens to the exterior **(Figure 39.20).** The sensory hairs of the taste receptors pass through the pore of a taste bud and project to the exterior. The opposite end of the receptor cells forms synapses with dendrites of an afferent neuron.

The taste receptors of terrestrial vertebrates are concentrated in the mouth. Humans have about 10,000 taste buds, each 30 to 40 μm in diameter, scattered over the tongue, roof of the mouth, and throat. Those on the tongue are embedded in outgrowths called *papillae* (*papula* = pimple), which give the surface of the tongue its rough or furry texture.

Taste receptors on the human tongue are thought to respond to five basic tastes: sweet, sour, salty, bitter, and umami (savory). Some of the receptors for umami respond to the amino acid glutamate (familiar as monosodium glutamate or MSG). Recent research indicates that the classes of receptors may all have many subtypes, each binding a specific molecule within that class.

Signals from the taste receptors are relayed to the thalamus. From there, some signals lead to gustatory centers in the cerebral cortex, which integrate them into the perception of taste, while others lead to the brain stem and limbic system, which links tastes to involuntary visceral and emotional responses. Through these connections, a pleasant taste may lead to salivation, secretion of digestive juices, sensations of pleasure, and even sexual arousal, while an unpleasant taste may produce revulsion, nausea, and even vomiting.

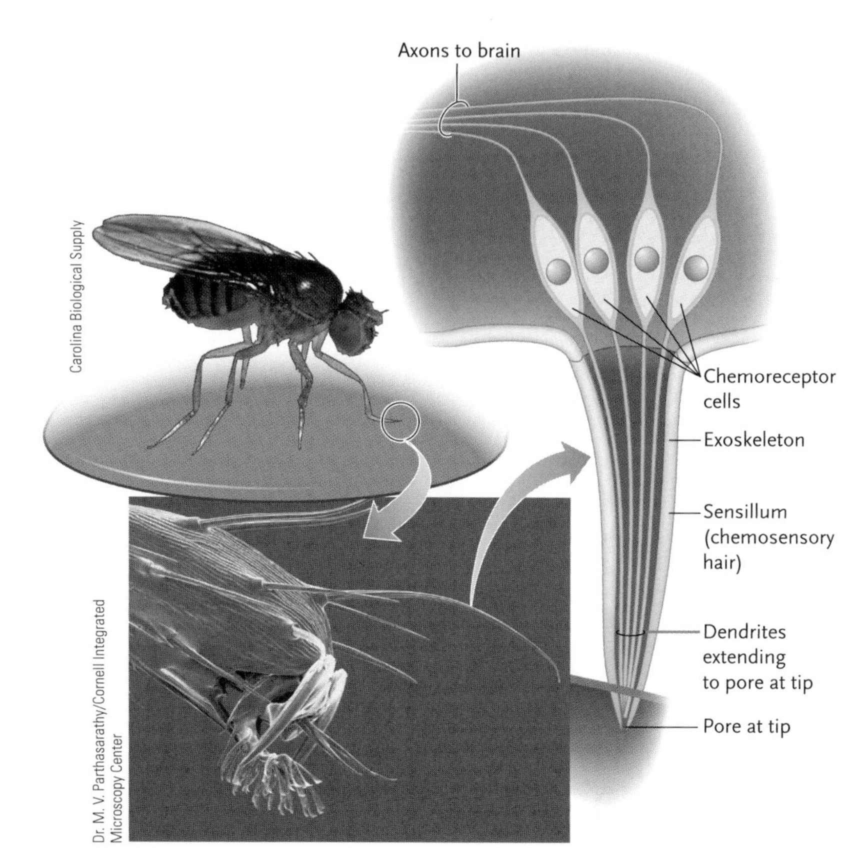

Figure 39.18
Taste receptors on the foot of a fruit fly, *Drosophila*.

Olfactory Receptors Are Concentrated in the Nasal Cavities in Terrestrial Vertebrates

Receptors that detect odors are located in the nasal cavities in terrestrial vertebrates. Bloodhounds have more than 200 million olfactory receptors in patches of olfactory epithelium in the upper nasal passages; humans have about 5 million olfactory receptors.

On one end, each olfactory receptor cell has 10 to 20 sensory hairs that project into a layer of mucus covering the olfactory area in the nose **(Figure 39.21).** To be

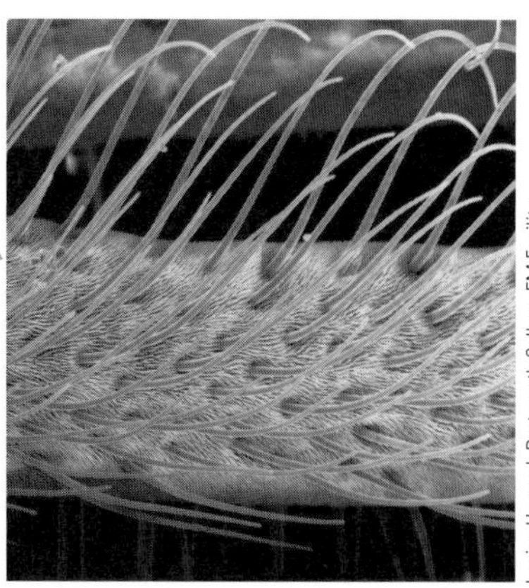

Figure 39.19
The brushlike antennae of a male silkworm moth. Fine sensory bristles containing olfactory receptor cells cover the filaments of the antennae.

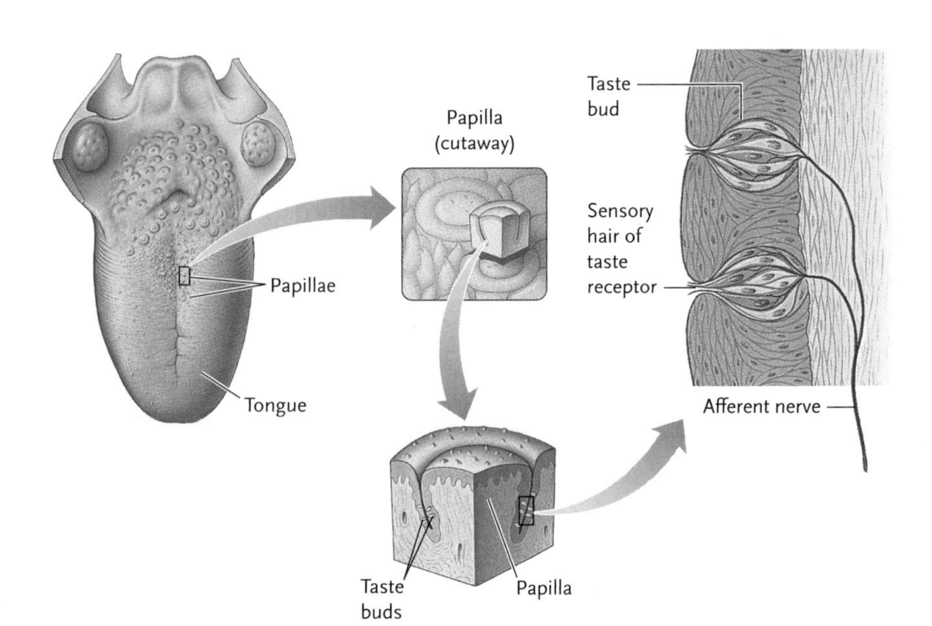

Figure 39.20
Taste receptors in the human tongue. The receptors occur in microscopic taste buds that line the sides of the furry papillae.

detected, airborne molecules must dissolve in the watery mucus solution. On the other end, the olfactory receptors make synapses with interneurons in the olfactory bulbs. Olfactory receptors are the only receptor cells that make direct connections with brain interneurons, rather than via afferent neurons.

From the olfactory bulbs, nerves conduct signals to the olfactory centers of the cerebral cortex, where they are integrated into the perception of tantalizing or unpleasant odors from a rose to a rotten egg. Most odor perceptions arise from a combination of different olfactory receptors. In the early 1990s, Richard Axel and Linda Buck discovered that about 1000 different human genes give rise to an equivalent number of olfactory receptor types, each of which responds to a different class of chemicals. Axel and Buck received the Nobel Prize in 2004 in recognition of their research.

Other connections from the olfactory bulbs lead to the limbic system and brain stem where the signals elicit emotional and visceral responses similar to those caused by pleasant and unpleasant tastes (see Section 38.3). As a result, different odors, like tastes, may give rise to a host of involuntary responses, from salivation to vomiting, as well as conscious responses.

Olfaction contributes to the sense of taste because vaporized molecules from foods are conducted from the throat to the olfactory receptors in the nasal cavities. Olfactory input is the reason why anything that dulls your sense of smell—such as a head cold, or holding your nose—diminishes the apparent flavor of food.

Many mammals use odors as a means of communication. Individuals of the same family or colony are identified by their odor; odors are also used to attract mates and to mark territories and trails. Dogs, for example, use their urine to mark home territories with identifying odors. Humans use the fragrances of perfumes and colognes as artificial sex attractants.

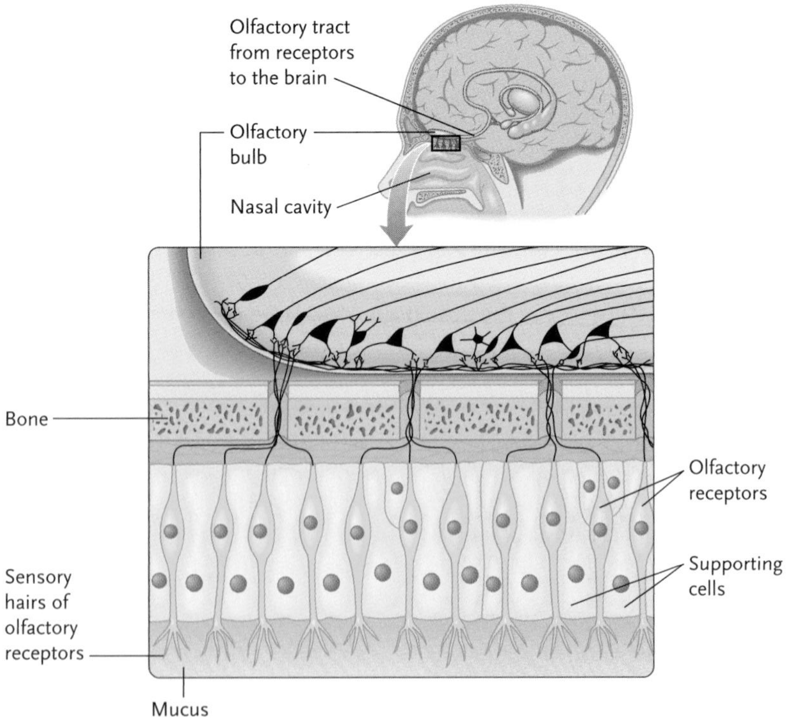

Figure 39.21
Olfactory receptors in the roof of the nasal passages in humans. Axons from these receptors pass through holes in the bone separating the nasal passages from the brain, where they make synapses with interneurons in the olfactory bulbs.

Study Break

1. How do we distinguish different kinds of smells?
2. For terrestrial vertebrates, describe the pathway by which a signal generated by taste receptors leads to a response.

39.6 Thermoreceptors and Nociceptors

Thermoreceptors detect changes in the surrounding temperature. Nociceptors respond to stimuli that may potentially damage their tissues. Both types of receptors consist of free nerve endings formed by the dendrites of afferent neurons, with no specialized receptor structures surrounding them.

Thermoreceptors Can Detect Warm and Cold Temperatures and Temperature Changes

Most animals have thermoreceptors. Some invertebrates such as mosquitoes and ticks use thermoreceptors to locate their warm-blooded prey. Some snakes, including rattlesnakes and pythons, use thermoreceptors called *pit organs* to detect the body heat of warm-blooded prey animals **(Figure 39.22).**

Figure 39.22
The pit organs of an albino Western diamondback rattlesnake *(Crotalus atrox)*, located in depressions on both sides of the head below the eyes. These thermoreceptors detect infrared radiation emitted by warm-bodied prey animals such as mice.

In mammals, distinct thermoreceptors respond to heat and cold. Researchers have shown that three members of the *transient receptor potential* (TRP) gated Ca^{2+} channel family act as heat receptors. One responds when the temperature reaches 33°C and another responds above 43°C, where heat starts to be painful; these two receptors are believed to be involved in thermoregulation. The third receptor responds at 52°C and above, in this case producing a pain response rather than being involved in thermoregulation.

Two cold receptors are known in mammals. One responds between 8 and 28°C, and is thought to be involved in thermoregulation. The second responds to temperatures below 8°C and appears to be associated with pain rather than thermoregulation. The molecular mechanisms that control the opening and closing of heat and cold receptor channels are not currently known.

Some neurons in the hypothalamus of mammals also function as thermoreceptors, an ability that has only recently been investigated. Not only do these neurons sense changes in brain temperature, but they also receive afferent thermal information. These neurons are highly sensitive to shifts from the normal body temperature, and trigger involuntary responses such as sweating, panting, or shivering, which restore normal body temperature.

Nociceptors Protect Animals from Potentially Damaging Stimuli

The signals from nociceptors—receptors in mammals, and possibly other vertebrates, that detect damaging stimuli—are interpreted by the brain as pain. Pain is a protective mechanism; in humans, it prompts us to do something immediately to remove or decrease the damaging stimulus. Often pain elicits a reflex response—such as withdrawing the hand from a hot stove—that proceeds before we are even consciously aware of the sensation.

Various types of stimuli cause pain, including mechanical damage such as a cut, pinprick, or blow to the body, and temperature extremes. Some nociceptors are specific for a particular type of damaging stimulus, while others respond to all kinds.

The axons that transmit pain are part of the somatic system of the PNS (see Section 38.2). They synapse with interneurons in the gray matter of the spinal cord, and activate neural pathways to the CNS by releasing the neurotransmitters glutamate or substance P (see Section 37.3). Glutamate-releasing axons produce sharp, prickling sensations that can be localized to a specific body part—the pain of stepping on a tack, for example. Substance P–releasing axons produce dull, burning, or aching sensations, the location of which may not be easily identified—the pain of tissue damage such as stubbing your toe.

As part of their protective function, pain receptors adapt very little, if at all. Some pain receptors, in fact, gradually intensify the rate at which they send out action potentials if the stimulus continues at a constant level.

The CNS also has a pain-suppressing system. In response to stimuli such as exercise, hypnosis, and stress, the brain releases *endorphins,* natural painkillers that bind to membrane receptors on substance P neurons, reducing the amount of neurotransmitter released.

Nociceptors contribute to the taste of some spicy foods, particularly those that contain hot peppers. In fact, researchers who study pain often use *capsaicin,* the organic compound that gives jalapeños and other peppers their hot taste, to identify nociceptors. To some, the burning sensation from capsaicin is addictive. Here is the reason. Nociceptors in the mouth, nose, and throat immediately transmit pain messages to the brain when they detect capsaicin. The brain re-

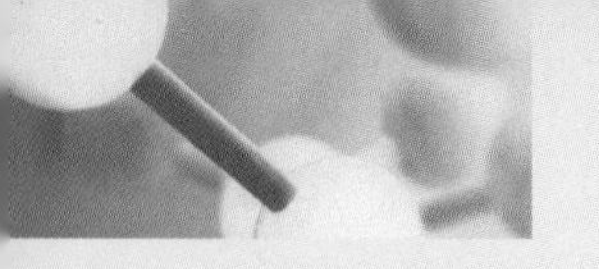

Insights from the Molecular Revolution

Hot News in Taste Research

Biting into a jalapeño pepper (a variety of *Capsicum annuum*) can produce a burning pain in your mouth strong enough to bring tears to your eyes. This painfully hot sensation is due primarily to *capsaicin,* a chemical that probably evolved in pepper plants as a defense against foraging animals. The defense is obviously ineffective against the humans who relish peppers and foods containing capsaicin (such as buffalo wings).

Research by David Julius and his coworkers at the University of California, San Francisco, revealed the molecular basis for detection of capsaicin by nociceptors. They designed their experiments to test the hypothesis that the responding nociceptors have a cell surface receptor that binds capsaicin. Binding the chemical opens a membrane channel in the receptor that admits calcium ions and initiates action potentials interpreted as pain.

The Julius team isolated the total complement of messenger RNAs from nociceptors able to respond to capsaicin and made complementary DNA (cDNA) clones of the mRNAs. The cDNAs, which represented sequences encoding proteins made in the nociceptors, contained thousands of different sequences. The cDNAs were transformed individually into embryonic kidney cells (which do not normally respond to capsaicin), and the transformed cells were screened with capsaicin to identify which took in calcium ions; presumably, these cells had received a cDNA encoding a capsaicin receptor. Messenger RNA transcribed from the identified cDNA clone was injected into both frog oocytes and cultured mammalian cells. Tests showed that both the oocytes and the cultured cells responded to capsaicin by admitting calcium ions, which confirmed that the researchers had found the capsaicin receptor cDNA.

Among the effects noted when the receptor was introduced into oocytes was a response to heat. Increasing the temperature of the solution surrounding the oocytes from 22°C to about 48°C produced a strong calcium inflow. In short, capsaicin and heat produce the same response in cells containing the receptor. Therefore the feeling that your mouth is on fire when you eat a hot pepper probably results from the fact that, as far as your nociceptors and CNS are concerned, it *is* on fire.

sponds by releasing endorphins, which act as a painkiller and create temporary euphoria—a natural high, if you will. *Insights from the Molecular Revolution* describes a series of experiments investigating the molecular basis of the pain caused by capsaicin.

Study Break

What distinguishes thermoreceptors and nociceptors from the other types of sensory receptors discussed previously?

39.7 Magnetoreceptors and Electroreceptors

Some animals have poorly developed visual systems but can gain information about their environment by sensing magnetic or electrical fields. In so doing, they directly sense stimuli that humans can detect only with scientific instruments.

Magnetoreceptors Are Used for Navigation

Some animals that navigate long distances, including migrating butterflies, beluga whales, sea turtles, homing pigeons, and foraging honeybees, have **magnetoreceptors** that allow them to detect and use Earth's magnetic field as a source of directional information (experiments with sea turtles are described in **Figure 39.23**).

The pattern of Earth's magnetic field differs from region to region yet remains almost constant over time, largely unaffected by changing weather and day and night. As a result, animals with magnetic receptors are able to monitor their location reliably. Although little is known about the receptors that detect magnetic fields, they may depend on the fact that moving a conductor, such as an electroreceptor cell, through a magnetic field generates an electric current.

Some magnetoreceptors may depend on the effect of Earth's magnetic field on the mineral *magnetite.* Magnetite is found in the bones or teeth of many vertebrates, including humans, and also in insects—in the abdomen of honeybees and in the heads and abdomens of certain ants, for example.

Other animals, including homing pigeons, which are famous for their ability to find their way back to their nests even when released far from home, navigate by detecting their position with reference to both Earth's magnetic field and the sun. Magnetite is located in the bills of these birds, which is where research indicates magnetoreception likely occurs.

Electroreceptors Are Used for Location of Prey or for Communication

Many sharks and bony fishes, some amphibians, and even some mammals (such as the star-nosed mole and duckbilled platypus) have specialized **electroreceptors**

Figure 39.23 Experimental Research

Demonstration That Magnetoreceptors Play a Key Role in Loggerhead Sea Turtle Migration

QUESTION: Do loggerhead sea turtles use a magnetoreceptor system for migration?

EXPERIMENT: Loggerhead sea turtles *(Caretta caretta)* that hatch along the east coast of Florida spend much of their lives traveling the North Atlantic current system around the Sargasso Sea, a pool of warm water with a unique seaweed ecosystem. Eventually and unerringly, the turtles return to their hatching beach for the mating season. Kenneth Lohmann of the University of North Carolina hypothesized that magnetoreception, likely involving magnetite, plays a central role in loggerhead migration. Lohmann tested his hypothesis using an experimental system in which the direction hatchling turtles swam was analyzed in different magnetic fields.

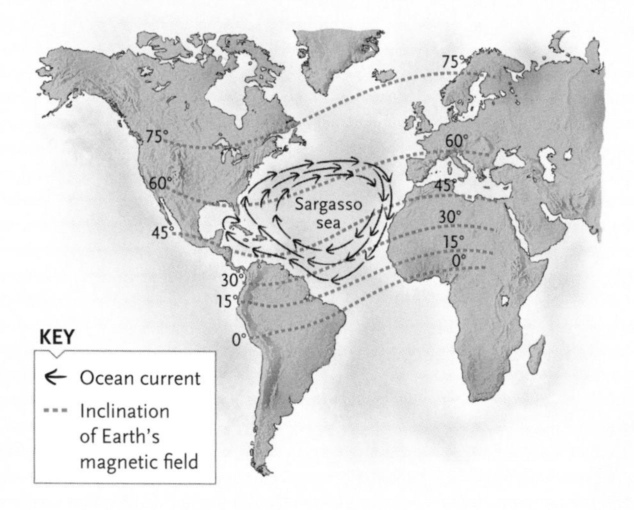

1. Lohmann placed each turtle hatchling he tested in a harness and tethered it to a swiveling, electronic system in the center of a circular pool of water. The pool was surrounded by a large electromagnetic coil system that allowed the researchers to reverse the direction of the magnetic field. The direction the turtle swam was recorded by the tracking system and relayed to a computer.

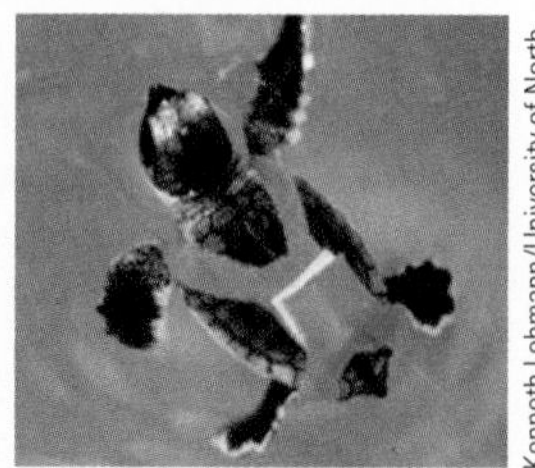

Kenneth Lohmann/University of North Carolina

2. Lohmann allowed the turtles to swim under two experimental conditions: half of the turtles swam in Earth's magnetic field, and the other half swam in a reversed magnetic field.

Coil system to control magnetic field
Turtle
Tether
Lever arm
Digital encoder
Coil system control computer
Output to data analysis computer

RESULTS: The turtle hatchlings tested in Earth's magnetic field swam in an east-to-northeast direction on average, mimicking the direction they follow normally when migrating at sea. The turtle hatchlings tested in the reversed magnetic field on average swam in a direction 180° opposite that of the hatchlings swimming in Earth's magnetic field.

CONCLUSION: The results indicate that loggerhead sea turtle hatchlings have the ability to detect Earth's magnetic field and use it as a way to orient their migration. Their direction of migration, east to northeast, matches the inclination of Earth's magnetic field in the Atlantic Ocean where they migrate (see map figure). Lohmann believes that the magnetoreception system in the turtles involves magnetite.

that detect electrical fields. The plasma membrane of an electroreceptor cell is depolarized by an electrical field, leading to the generation of action potentials. The electrical stimuli detected by the receptors are used to locate prey or navigate around obstacles in muddy water, or, by some fishes, to communicate. Some electroreception systems are passive—they detect electric fields in the environment, not the animal's own electric currents. Passive systems are used mainly to find prey. For example, the electroreceptors of sharks and rays can locate fish buried under the sand from the electrical currents generated by their prey's heartbeat or by the muscle contractions that move water over their gills.

Other electroreception systems are active—the animal emits and receives low voltage electrical signals, either to locate prey or to communicate with members of the same species. The electrical signals

Unanswered Questions

What happens when the senses get scrambled—when listening to music causes you to "see" colors, or when you "taste" certain words?

Synesthesia (joined senses) occurs when two senses, normally separate, are perceived together. For the most part, people with synesthesia are born with it, and it tends to run in families. A recent study by Michael Esterman and his colleagues at the University of California, Berkeley, showed that the posterior parietal cortex, a region of the brain thought to be involved in sensory integration, appears to be crucial to sensory commingling. Some researchers think that this commingling is how the senses function early in development, when the nervous system is still immature. They believe that the senses normally separate from one another around four months after birth. In synesthetes, however, this separation is incomplete and two of their senses remain mingled.

London's Science Museum collaborated with Jamie Ward of University College London in an experiment that paired sounds and music. They wanted to determine if volunteers visiting the museum would prefer combinations of sound and vision as described by synesthetes over combinations randomly generated by a computer. Interestingly, people found the synesthetic combinations more pleasing than the computer-generated ones. Thus, it is possible that everyone may have a built-in understanding of what sounds and colors go together.

In an evolutionary context, which "sense" developed first?

The descriptions of the senses in this chapter focus primarily on vertebrate sensory systems, but the nervous systems of many invertebrates can be quite complex. Indeed, squids, sea hares, leeches, horseshoe crabs, lobsters, and cockroaches have been quite instrumental in helping scientists understand the nervous system. As for senses, squids have sensitive eyes and accentuated smell and taste, and they respond to touch and vibration. Clearly, vertebrates aren't the only multicellular organisms to develop senses.

However, is a nervous system necessary for organisms to have senses? Can single-celled organisms (which, of course, do not have nervous systems) respond to stimuli? Clearly, the answer is yes. Consider *Paramecium tetraurelia,* a single-celled organism covered with cilia that lives in water. It can detect substances in its environment and swim toward certain chemicals while avoiding others. It also responds to solid objects by turning when it runs into one. Thus, it has a "chemical sense" similar to taste or smell, and it responds to a type of "touch." We will probably never know which sense developed first, but some type of touch or chemical sense seems the most likely.

Rona Delay is an associate professor in the Department of Biology at the University of Vermont. Her research centers on understanding how sensory receptors change or transduce information about the external world into a language the brain can understand. The focus of her research is the sense of smell. To learn more go to http://www.uvm.edu/~biology/Faculty/Delay/Delay.html.

are generated by special electric organs. A few species, such as the electric eel and the electric catfish, produce discharges on the order of several hundred volts. These discharges are used to stun or kill prey. The voltage is high enough to stun, but not kill, a human.

Study Break

What are three ways electroreceptors are used in aquatic vertebrates?

Review

Go to ThomsonNOW at www.thomsonedu.com/login to access quizzing, animations, exercises, articles, and personalized homework help.

39.1 Overview of Sensory Receptors and Pathways

- Sensory receptors are formed by the endings of afferent neurons or specialized cells adjacent to the neurons. They detect stimuli such as mechanical pressure, sound waves, light, or specific chemicals. Action potentials generated by the receptors are carried by the axons of afferent neurons to pathways leading to specific parts of the brain, where signals are processed into sensory sensations (Figure 39.1).
- Receptors are specialized as mechanoreceptors, photoreceptors, chemoreceptors, thermoreceptors, and nociceptors. Some animals have receptors that detect electrical or magnetic fields.
- The routing of information from sensory receptors to particular regions of the brain identifies a specific stimulus as a sensation. The intensity of a stimulus is determined by the frequency of action potentials traveling along the neural pathways and the number of afferent neurons carrying action potentials.
- Many sensory systems show sensory adaptation, in which the frequency of action potentials decreases while a stimulus remains constant. Some sensory receptors, such as those related to pain, show little or no sensory adaptation.

Animation: Action potentials

39.2 Mechanoreceptors and the Tactile and Spatial Senses

- Mechanoreceptors detect touch, pressure, acceleration, or vibration. Touch and pressure receptors are free nerve endings or encapsulated nerve endings of sensory neurons (Figure 39.2).
- Mechanoreceptors called proprioceptors detect stimuli used by the CNS to monitor and maintain body and limb positions.
- Proprioceptors based on sensory hair cells generate action potentials when the hairs are moved (Figures 39.3–39.5).
- Receptors in muscles, tendons, and joints of vertebrates detect changes in stretch and tension of body parts (Figure 39.6).

Animation: Dynamic equilibrium

39.3 Mechanoreceptors and Hearing

- Many invertebrates have mechanoreceptors in their skin or other surface structures that detect sound and other vibrations.
- Hearing relies on sensory hair cells in organs that respond to the vibrations of sound waves.
- In terrestrial vertebrates, the ear consists of three parts. The outer ear directs sound to the eardrum. Vibrations of the eardrum are transmitted through one or more bones in the middle ear to the fluid-filled inner ear. In the inner ear, the vibrations are transmitted through membranes that bend the stereocilia of the hair cells, leading to bursts of action potentials that are reflected in the frequency of the sound waves (Figure 39.8).

Animation: Ear structure and function

Animation: Properties of sound

39.4 Photoreceptors and Vision

- Invertebrates possess many forms of eyes, from the simplest, an ocellus, to single-lens eyes that are similar to vertebrate eyes (Figures 39.9–39.11).
- The photoreceptors of all animal eyes contain the pigment retinal, which absorbs the energy of light and uses it to generate changes in membrane potential.
- The transparent cornea admits light into the vertebrate eye. Behind the cornea, the iris controls the diameter of the pupil, regulating the amount of light that strikes the lens. The lens focuses an image on the retina lining the back of the eye, where photoreceptors and neurons carry out the initial integration of information detected by the photoreceptors (Figure 39.12).
- In terrestrial vertebrates, the lens is focused by adjusting its shape (Figure 39.13). The retina contains two types of photoreceptors, rods and cones. Rods are specialized for detecting light of low intensity; cones are specialized for detecting light of different wavelengths, which are perceived as colors.
- The light-absorbing pigment in photoreceptor cells consists of retinal combined with an opsin protein. When it absorbs light, retinal changes form, initiating reactions that alter the amount of neurotransmitter released by the photoreceptor cells (Figures 39.14 and 39.15).
- Rods and cones are linked to neurons in the retina that perform the initial processing of visual information. The processed signal is sent via the optic nerve through the lateral geniculate nuclei to the visual cortex (Figures 39.16 and 39.17).

Animation: Eye structure

Animation: Visual accommodation

Animation: Organization of cells in the retina

Animation: Receptive fields

Animation: Pathway to visual cortex

Animation: Focusing problems

39.5 Chemoreceptors

- Chemoreceptors respond to the presence of specific molecules in the environment. In vertebrates, they form parts of receptor organs for taste (gustation) and smell (olfaction).
- Taste receptors detect molecules from food or other objects that come into direct contact with the receptor and are used primarily to identify foods (Figures 39.18 and 39.20).
- Olfactory receptors detect molecules from distant sources; besides identifying food, they are used to detect predators and prey, identify family and group members, locate trails and territories, and communicate (Figures 39.19 and 39.21).

Animation: Olfactory pathway

Animation: Taste receptors

39.6 Thermoreceptors and Nociceptors

- Thermoreceptors, which consist of free nerve endings located at the body surface and in limited numbers in the body interior, detect changes in body temperature.
- Nociceptors, located on both the body surface and interior, detect stimuli that can damage body tissues. Information from these receptors is integrated in the brain into the sensation of pain.

Animation: Sensory receptors in the human skin

Animation: Referred pain

39.7 Electroreceptors and Magnetoreceptors

- Some vertebrates have electroreceptors that detect electrical currents and fields, or magnetoreceptors that detect magnetic fields (Figure 39.23).

Questions

Self-Test Questions

1. The frequency of a blast from a nearby ambulance siren can cause a dog to howl in pain. Activated under this circumstance are:
   a. thermoreceptors and chemoreceptors.
   b. photoreceptors and nociceptors.
   c. mechanoreceptors and nociceptors.
   d. chemoreceptors and mechanoreceptors.
   e. photoreceptors and chemoreceptors.
2. Two common side effects of Hansen's disease (leprosy) are a permanent numbness in the hands, feet, and buttocks of affected people and a loss of perception of their spatial position. Affected are:
   a. mechanoreceptors.
   b. adapting receptors.
   c. pH change receptors.
   d. the vestibular apparatus.
   e. vibration detecting systems.
3. Neuromasts are best described as:
   a. nonadapting pain receptors.
   b. components of the fish lateral line system.
   c. statoliths that detect motion.
   d. motor axons that activate motion.
   e. cupulas that detect vibrations.
4. Structures are activated by sound waves in the vertebrate ear in the following order:
   a. oval window, tympanum, semicircular canals, Golgi tendon organ, incus, malleus, stapes.
   b. organ of Corti, malleus, incus, stapes, auditory nerve, eardrum.
   c. eustachian tube, round window, vestibular canal, tympanic canal, cochlear canal, oval window, pinna.
   d. basilar membrane, tectorial membrane, otoliths, utricle, saccule, malleus, cochlea.
   e. pinna, tympanic membrane, malleus, incus, stapes, oval window, cochlear duct.

5. The following situation is associated with movement and position in the human body:
   a. Statoliths in statocysts bend sensory hairs and trigger action potentials.
   b. If sensory hairs in the utricle are oriented horizontally and those in the saccule are oriented vertically, the person is lying down.
   c. When the head rotates, the endolymph in the semicircular canal pulls the cupula with it to activate sensory hair cells.
   d. Displacement of the utricle and saccule generates action potentials.
   e. If the body is spinning at a constant rate and direction, the cupula is displaced and action potentials are initiated.
6. The difference between the vertebrate eye and the cephalopod eye is that the vertebrate eye has:
   a. an iris surrounding the pupil, whereas in cephalopods the pupil surrounds the iris.
   b. a lens that changes shape when focusing, whereas in cephalopods the lens moves back and forth to focus.
   c. a retina that moves in the socket when recording the image, whereas in cephalopods the retina changes shape when stimulated.
   d. a pupil that shrinks in size in bright light, whereas cephalopods have a pupil that enlarges in bright light.
   e. retinal synthesized from vitamin A, whereas cephalopods lack retinal.
7. Which of the following events does not occur during light absorption in the vertebrate eye?
   a. The retinal component of rhodopsin changes from *cis* to *trans* form.
   b. Rhodopsin, a G membrane-embedded protein, triggers a signal transduction pathway to close Na^{+} channels in the plasma membrane.
   c. The light stimulus passes from rods and cones to bipolar cells and horizontal cells and then to ganglion cells, whose axons compose the optic nerve.
   d. As light absorption increases, the rhodopsin response causes an increase in the release of neurotransmitters.
   e. When integrating information across the retina, horizontal cells connect the rods and cones, and amacrine cells join with the bipolar cells and ganglion cells.
8. The variety of color seen by humans is directly dependent upon the:
   a. activation of three different photopsins in cones.
   b. transmission of an image to separate brain hemispheres by the optic chiasm.
   c. transmission of impulses from rods across the lateral geniculate nuclei.
   d. lateral inhibition by amacrine cells.
   e. light stimulation of all photoreceptor types equally.
9. In terrestrial animals:
   a. the hairs of taste receptors are derived from cilia and contain microtubules.
   b. the hairs of smell receptors are derived from microvilli and contain microfilaments.
   c. signals from taste receptors are relayed to the temporal lobes.
   d. information from olfactory receptors is processed in the parietal lobes.
   e. connections from the olfactory bulbs lead to the limbic system.
10. In the human response to temperature or pain:
   a. all three transient receptor potential (TRP) gated Ca^{2+} channels act as pain receptors.
   b. cold receptors are activated between 27°C and 37°C.
   c. pain receptors decrease the rate at which they send out action potentials if the pain is constant.
   d. nociceptors, activated by capsaicin in the mouth and nose, can sense pain.
   e. the CNS releases glutamate or substance P to dull the pain sensation.

Questions for Discussion

1. Humans have about 200 million photoreceptors in two eyes, and about 32,000 sensory hair cells in two ears. About 3% of the somatosensory cortex is devoted to hearing, whereas roughly 30% of it is devoted to visual processing. Suggest an explanation for these differences from the perspective of natural selection and adaptation.
2. In owls and many other birds of prey, the fovea is located toward the top of the retina rather than at the center as in humans. This arrangement correlates with the birds' hunting behavior, in which they look down when they fly, scanning the ground for a meal. With this arrangement in mind, why do you think the standing owl in the picture is turning its head upside down?

Chase Smith

3. A patient made an appointment with her doctor because she was experiencing recurrent episodes of dizziness. Her doctor asked questions to distinguish whether she had sensations of lightheadedness, as if she were going to faint, or vertigo, as if she or objects near her were spinning around. Why was this clarification important in the evaluation of her condition?

Experimental Analysis

The fruit fly *Drosophila melanogaster* can distinguish a large repertoire of odors in the environment. Their response may be to move toward food or away from danger. Moreover, particular odors play an important role in their mating behavior. The olfactory organs of a fruit fly are the antennae and an elongated bulge on the head called the maxillary pulp. Because of the ease with which fruit fly genes can be manipulated, identifying and studying their olfactory receptors likely would contribute significantly to our understanding of neural pathways of odor recognition more generally. How could you identify candidate fruit fly genes that encode components of olfactory receptors?

Evolution Link

In 2005, researchers took saliva and blood samples from six cats, including domestic cats, a tiger, and a cheetah, and found that all have a defective gene for one of the two chemoreceptor proteins needed to identify food as sweet. (The scientists conjecture that the lack of a sweet tooth may explain why cats are finicky eaters.) What are the evolutionary implications of the finding?

How Would You Vote?

Noise pollution from commercial shipping and other human activities generates low-frequency sounds that are believed to interfere with the acoustical signals that whales use for navigation, location of food, and communication. To what extent should we limit these activities to protect whales against potential harm? Would you support banning activities that exceeded a certain noise level from U.S. territorial waters? If so, how would you get other nations to do the same? Go to www.thomsonedu.com/login to investigate both sides of the issue and then vote.

Two North American bull elks contesting for cows in Yellowstone National Park. The shorter days of autumn trigger hormone production, battling, and reproductive behavior.

© Mark Wallner

Study Plan

40 The Endocrine System

Why It Matters

Every September, as the days grow shorter and autumn approaches, bull elks *(Cervus canadensis)* begin to strut their stuff. Although they have grazed peacefully together at high mountain elevations from the Yukon to Arizona, they now become testy with each other. They also rasp at tree branches and plow the ground with their antlers. Soon, they descend to lower elevations, where the cow elks have been feeding in large nursery groups with their calves and yearlings.

The bulls move in among the cows and chase away the male yearlings. As part of the mating ritual, the bulls bugle, square off, strut, and circle; then they clash their antlers together, attempting to drive each other from the cows. The winning males claim harems of about 10 females each, a major prize.

After the mating season ends, tranquility returns. The cows again graze in herds; the males form now-friendly bachelor groups that also feed quietly in the meadows. Eating is their major occupation, storing nutrients in preparation for the snowy winter. The young will be born eight to nine months later, when summer returns.

The next year, the shortening days of late summer and fall again trigger the transition to mating behavior. Detected by the eyes and registered in the brain, reduced daylength initiates changes in the secretion of long-distance signaling molecules called **hormones** (*hormaein* = to excite). Hormones are released from one group of cells and are transported through the circulatory system to other cells, their target cells, whose activities they change. Among the changes will be a rise in the concentration of hormones responsible for mating behavior.

We too are driven by our hormones. They control our day-to-day sexual behavior—often as outlandish as that of a bugling bull elk—as well as a host of other functions, from the concentration of salt in our blood, to body growth, to the secretion of digestive juices. Along with the central nervous system, hormones coordinate the activities of multicellular life.

The best-known hormones are secreted by cells of the **endocrine system** (*endo* = within; *krinein* = to separate), although hormones actually are produced by almost all organ systems in the body. The endocrine system, like the nervous system, regulates and coordinates distant organs. The two systems are structurally, chemically, and functionally related, but they control different types of activities. The nervous system, through its high-speed electrical signals, enables an organism to interact rapidly with the external environment, while the endocrine system mainly controls activities that involve slower, longer-acting responses. Typical responses to hormones may persist for hours, weeks, months, or even years.

The mechanisms and functions of the endocrine system are the subjects of this chapter. As in other chapters of this unit, we pay particular attention to the endocrine system of humans and other mammals.

40.1 Hormones and Their Secretion

Cells signal other cells using neurotransmitters, hormones, and local regulators. Recall from Chapters 37 through 39 that a neurotransmitter is a chemical released by an axon terminal at a synapse which affects the activity of a postsynaptic cell. Our focus in this chapter is hormones and local regulators, molecules that act locally rather than over long distances.

The Endocrine System Includes Four Major Types of Cell Signaling

Four types of cell signaling occur in the endocrine system: classical endocrine signaling, neuroendocrine signaling, paracrine regulation, and autocrine regulation. In *classical endocrine signaling,* hormones are secreted into the blood or extracellular fluid by the cells of ductless secretory organs called **endocrine glands** **(Figure 40.1a).** (In contrast, *exocrine glands,* such as the sweat and salivary glands, release their secretions into ducts that lead outside the body or into the cavities of the digestive tract, as described in Section 36.2.) The hormones are circulated throughout the body in the blood and, as a result, most body cells are constantly exposed to a wide variety of hormones. (The cells of the central nervous system are sequestered from the general circulatory system by the blood–brain barrier, described in Section 38.3.) Only *target cells* of a hormone, those with *receptor proteins* recognizing and binding that hormone, respond to it. Through these responses, hormones control such vital functions as digestion, osmotic balance, metabolism, cell division, reproduction, and development. The action of hormones may either speed or inhibit these cellular processes. For example, growth hormone stimulates cell division, whereas glucocorticoids inhibit glucose uptake by most cells in the body.

Hormones are cleared from the body at a steady rate by enzymatic breakdown in their target cells or in either the liver or kidneys. Breakdown products are excreted by the digestive and excretory systems; depending on the hormone, the breakdown takes minutes to days.

In *neuroendocrine signaling,* specialized neurons called **neurosecretory neurons** release a hormone called a *neurohormone* into the circulatory system when appropriately stimulated **(Figure 40.1b).** The neurohormone is distributed by the circulatory system and elicits a response in target cells that have receptors for the hormone. Note that both neurohormones and neurotransmitters are secreted by neurons. Neurohormones are distinguished from neurotransmitters in that neurohormones affect distant target cells, whereas neurotransmitters affect adjacent cells. However, both neurohormones and neurotransmitters function in the same way—they cause cellular responses by interacting with specific receptors on target cells. For instance, gonadotropin-releasing hormone, a neurohormone secreted by the hypothalamus, controls the release of luteinizing hormone from the pituitary.

In *paracrine regulation,* a cell releases a signaling molecule that diffuses through the extracellular fluid and acts on nearby cells—regulation is *local* rather than at a distance, as is the case with hormones and neurohormones **(Figure 40.1c).** In some cases the local regulator acts on the same cells that produced it; this is called *autocrine regulation* **(Figure 40.1d).** For example, many of the growth factors that regulate cell division and differentiation act in both a paracrine and autocrine fashion.

Hormones and Local Regulators Can Be Grouped into Four Classes Based on Their Chemical Structure

More than 60 hormones and local regulators have been identified in humans. Many human hormones are either identical or very similar in structure and function to those in other animals, but other vertebrates as well as invertebrates have hormones not found in humans.

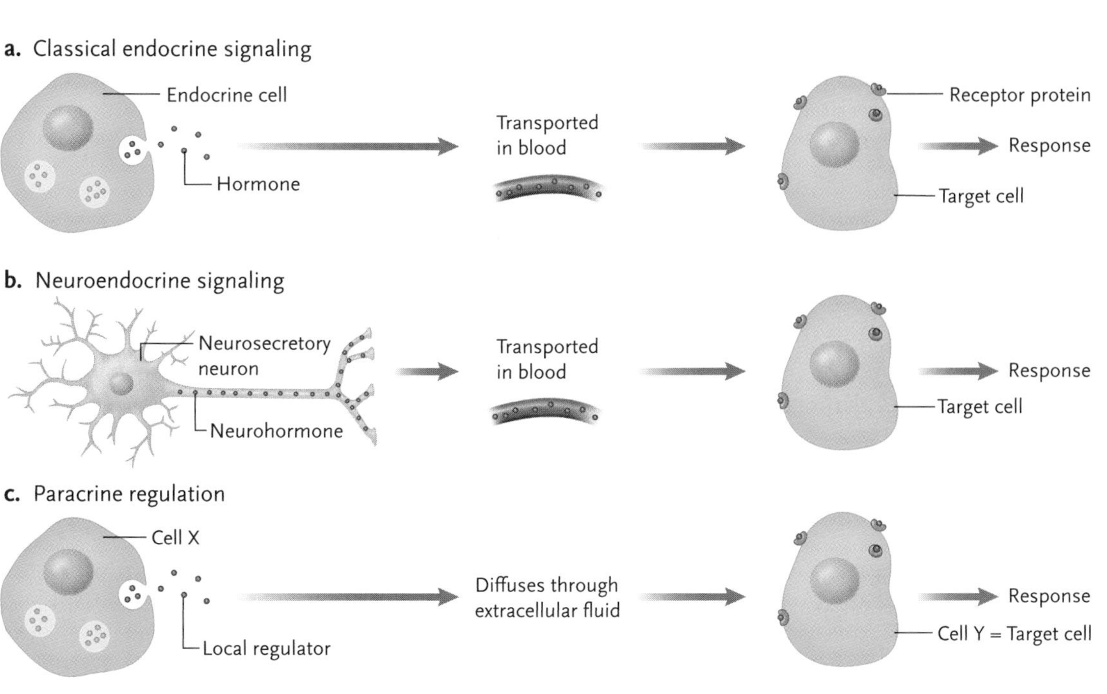

Figure 40.1
The four major types of cell signaling in the endocrine system.

Most of these chemicals can be grouped into four molecular classes: amine, peptide, steroid, and fatty acid–derived molecules.

Amine hormones are involved in classical endocrine signaling and neuroendocrine signaling. Most amine hormones are based on tyrosine. With one major exception, they are hydrophilic molecules, which diffuse readily into the blood and extracellular fluids. On reaching a target cell, they bind to receptors at the cell surface. The amines include epinephrine and norepinephrine, already familiar as neurotransmitters released by some neurons (see Section 37.3). The exception is thyroxine, a hydrophobic amine hormone secreted by the thyroid gland. This hormone, based on a pair of tyrosines, passes freely through the plasma membrane and binds to a receptor inside the target cell, as do steroid hormones (see following discussion and Section 7.5).

The *peptide hormones* consist of amino acid chains, ranging in length from as few as three amino acids to more than 200. Some have carbohydrate groups attached. They are involved in classical endocrine signaling and neuroendocrine signaling. Mostly hydrophilic hormones, peptide hormones are released into the blood or extracellular fluid by exocytosis when cytoplasmic vesicles containing the hormones fuse with the plasma membrane. One large group of peptide hormones, the **growth factors**, regulates the division and differentiation of many cell types in the body. Many growth factors act in both a paracrine and autocrine manner as well as in classical endocrine signaling. Because they can switch cell division on or off, growth factors are an important focus of cancer research.

Steroid hormones are involved in classical endocrine signaling. All are hydrophobic molecules derived from cholesterol and are insoluble in water. They combine with hydrophilic carrier proteins to form water-soluble complexes that can diffuse through extracellular fluids and enter the bloodstream. On contacting a cell, the hormone is released from its carrier protein, passes through the plasma membrane of the target cell, and binds to internal receptors in the nucleus or cytoplasm. Steroid hormones include aldosterone, cortisol, and the sex hormones. Steroid hormones may vary little in structure, but produce very different effects. For example, testosterone and estradiol, two major sex hormones responsible for the development of mammalian male and female characteristics, respectively, differ only in the presence or absence of a methyl group.

Fatty acid–derived molecules are involved in paracrine and autocrine regulation. **Prostaglandins**, for example, are important as local regulators. First discovered in the 1930s in seminal fluid, prostaglandins were so named because they were thought to be secreted by the prostate gland, although actually they are secreted by the seminal vesicles. Scientists later discovered that virtually every cell can secrete prostaglandins, and they are present at essentially all times. In semen, they enhance the transport of sperm through the female reproductive tract by increasing the contractions of smooth muscle cells, particularly in the uterus. During childbirth, prostaglandins secreted by the placenta work with a peptide hormone called oxytocin to stimulate labor contractions. Other prostaglandins induce contraction or relaxation of smooth muscle cells in many parts of the body, including blood vessels and air passages in the lungs. When released as a product of membrane breakdown in injured cells, prostaglandins may also intensify pain and inflammation.

Many Hormones Are Regulated by Feedback Pathways

The secretion of many hormones is regulated by feedback pathways, some of which operate partially or completely independent of neuronal controls. Most of these pathways are controlled by negative feedback—that is, a product of the pathway inhibits an earlier step in the pathway (see Section 36.4). For example, in some mammals, secretion by the thyroid gland is regulated by a negative feedback loop **(Figure 40.2).** Neurosecretory neurons in the hypothalamus secrete thyroid-releasing hormone (TRH) into a vein connecting the hypothalamus to the pituitary gland. In response, the pituitary releases thyroid-stimulating hormone (TSH) into the blood, which stimulates the thyroid gland to release thyroid hormones. As the thyroid hormone concentration in the blood increases, it begins to inhibit TSH secretion by the pituitary; this action is the negative feedback step. As a result, secretion of the thyroid hormones is reduced.

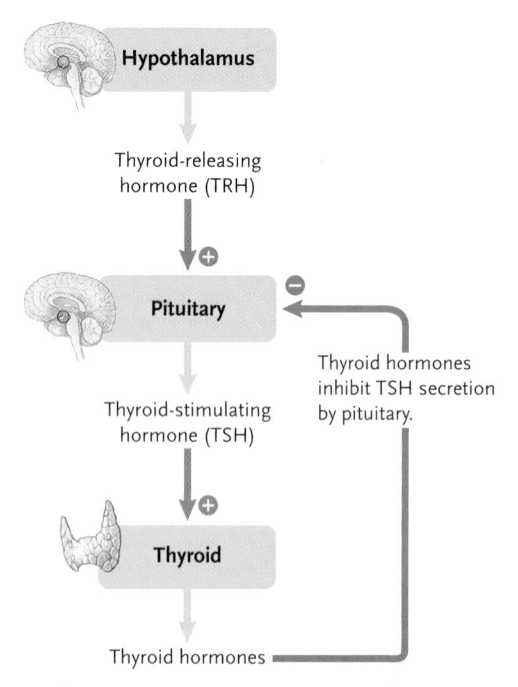

Figure 40.2
A negative feedback loop regulating secretion of the thyroid hormones. As the concentration of thyroid hormones in the blood increases, the hormones inhibit an earlier step in the pathway (indicated by the negative sign).

Body Processes Are Regulated by Coordinated Hormone Secretion

Although we will talk mostly about individual hormones in the remainder of the chapter, body processes are affected by more than one hormone. For example, the blood concentrations of glucose, fatty acids, and ions such as Ca^{2+}, K^+, and Na^+ are regulated by the coordinated activities of several hormones secreted by different glands. Similarly, body processes such as oxidative metabolism, digestion, growth, sexual development, and reactions to stress are all controlled by multiple hormones.

In many of these systems, negative feedback loops adjust the levels of secretion of hormones that act in antagonistic (opposing) ways, creating a balance in their effects that maintains body homeostasis. For example, consider the regulation of fuel molecules such as glucose, fatty acids, and amino acids in the blood. We usually eat three meals a day and fast to some extent between meals. During these periods of eating and fasting, four hormone systems act in coordinated fashion to keep the fuel levels in balance: (1) insulin and glucagon, secreted by the pancreas; (2) growth hormone, secreted by the anterior pituitary; (3) epinephrine and norepinephrine, released by the sympathetic nervous system and the adrenal medulla; and (4) glucocorticoid hormones, released by the adrenal cortex.

The entire system of hormones regulating fuel metabolism resembles the failsafe mechanisms designed by human engineers, in which redundancy, overlapping controls, feedback loops, and multiple safety valves ensure that vital functions are maintained at constant levels in the face of changing and even extreme circumstances.

Study Break

1. Distinguish between a hormone and a neurohormone.
2. Distinguish among the four major types of cell signaling.

40.2 Mechanisms of Hormone Action

Hormones control cell functions by binding to receptor molecules in their target cells. Small quantities of hormones can typically produce profound effects in cells and body functions due to **amplification.** In amplification, an activated receptor activates many proteins, which then activate an even larger number of proteins for the next step in the cellular reaction pathway, and so on, increasing in magnitude for each subsequent step in the pathway.

Hydrophilic Hormones Bind to Surface Receptors, Activating Protein Kinases Inside Cells

Hormones that bind to receptor molecules in the plasma membrane—primarily hydrophilic amine and peptide hormones—produce their responses through signal transduction pathways (see Section 7.2). In brief, when a surface receptor binds a hormone, the receptor is activated and transmits a signal through the plasma membrane. There are two kinds of surface receptors; the cytoplasmic reactions they control when they become activated are described in detail in Sections 7.3 and 7.4. Within the cell, the signal is transduced, changed into a form that causes the cellular response **(Figure 40.3a).** Typically, the reactions of signal transduction pathways involve protein kinases, enzymes that add phosphate groups to proteins. Adding a phosphate group to a protein may activate it or inhibit it, depending on the protein and the reaction. The particular response produced by a hormone depends on the kinds of protein kinases activated, the type of cell that can respond, and the types of target proteins they phosphorylate. Importantly, a small amount of hormone can elicit a large response because of amplification (see Figure 7.6).

The peptide hormone glucagon illustrates the mechanisms triggered by surface receptors. When glucagon binds to surface receptors on liver cells, it triggers the breakdown of glycogen stored in those cells into glucose. The glucose then is released into the circulatory system.

Hydrophobic Hormones Bind to Receptors Inside Cells, Activating or Inhibiting Genetic Regulatory Proteins

After passing through the plasma membrane, the hydrophobic steroid and thyroid hormones bind to internal receptors in the nucleus or cytoplasm **(Figure 40.3b,**

a. Hormone binding to receptor in the plasma membrane

Outside cell
Hydrophilic hormone
Blood vessel
Signal
Reception
Cytoplasmic end of receptor
Activation
Pathway molecule A
Activation
Transduction
Pathway molecule B
Activation
Molecule that brings about response
Pathway molecule C
Response
Change in cell
Cytoplasm

1 Hormone binds to surface receptor and activates it.
2 Activated receptor triggers a signal transduction pathway.
3 Transduction of the signal leads to cellular response.

b. Hormone binding to receptor inside the cell

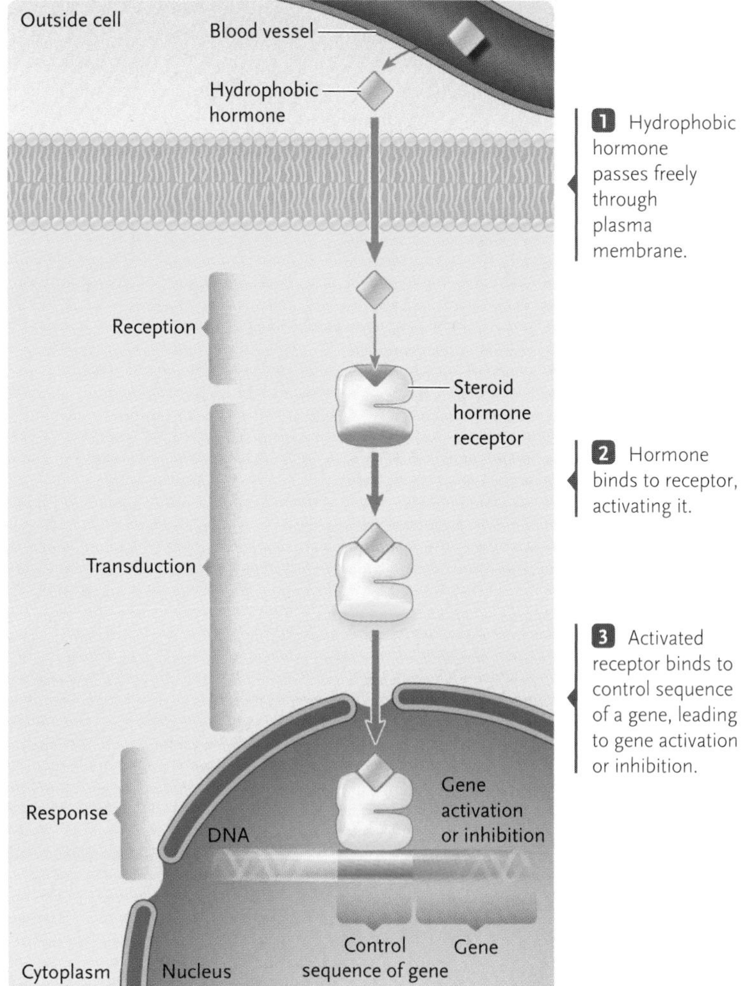

Figure 40.3
The reaction pathways activated by hormones that bind to receptor proteins in the plasma membrane **(a)** or inside cells **(b).** In both mechanisms, the signal—the binding of the hormone to its receptor—is transduced to produce the cellular response.

and described in detail in Section 7.5). Binding of the hormone activates the receptor, which then binds to a control sequence of specific genes. Depending on the gene, binding the control sequence either activates or inhibits its transcription, leading to changes in protein synthesis that accomplish the cellular response. The characteristics of the response depend on the specific genes controlled by the activated receptors, and on the presence of other proteins that modify the activity of the receptor.

One of the actions of the steroid hormone aldosterone illustrates the mechanisms triggered by internal receptors **(Figure 40.4).** If blood pressure falls below optimal levels, aldosterone is secreted by the adrenal glands. The hormone affects only kidney cells that contain the aldosterone receptor in their cytoplasm. When activated by aldosterone, the receptor binds to the control sequence of a gene, leading to the synthesis of proteins that increase reabsorption of Na^+ by the kidney cells. The resulting increase in Na^+ concentration in body fluids increases water retention and, with it, blood volume and pressure.

Target Cells May Respond to More Than One Hormone, and Different Target Cells May Respond Differently to the Same Hormone

A single target cell may have receptors for several hormones and respond differently to each hormone. For example, vertebrate liver cells have receptors for the pancreatic hormones insulin and glucagon. Insulin increases glucose uptake and conversion to glycogen, which decreases blood glucose levels, while glucagon stimulates the breakdown of glycogen into glucose, which increases blood glucose levels.

Conversely, particular hormones interact with different types of receptors in or on a range of target cells. Different responses are then triggered in each target cell type because the receptors trigger different transduction pathways. For example, the amine hormone epinephrine secreted by the adrenal medulla prepares the body for handling stress (including dangerous situations) and physical activity. (Epinephrine is discussed in more detail in Section 40.4.) In mammals, epinephrine can bind to three different plasma membrane-

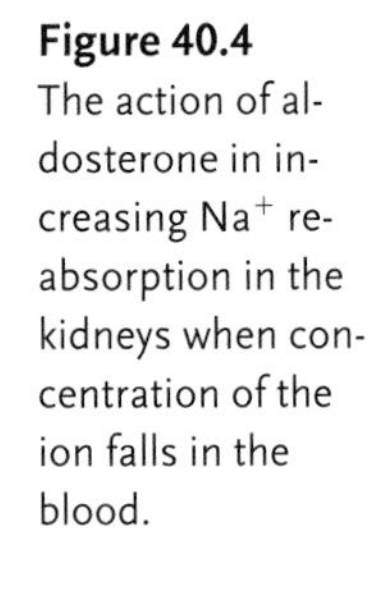

Figure 40.4
The action of aldosterone in increasing Na^+ reabsorption in the kidneys when concentration of the ion falls in the blood.

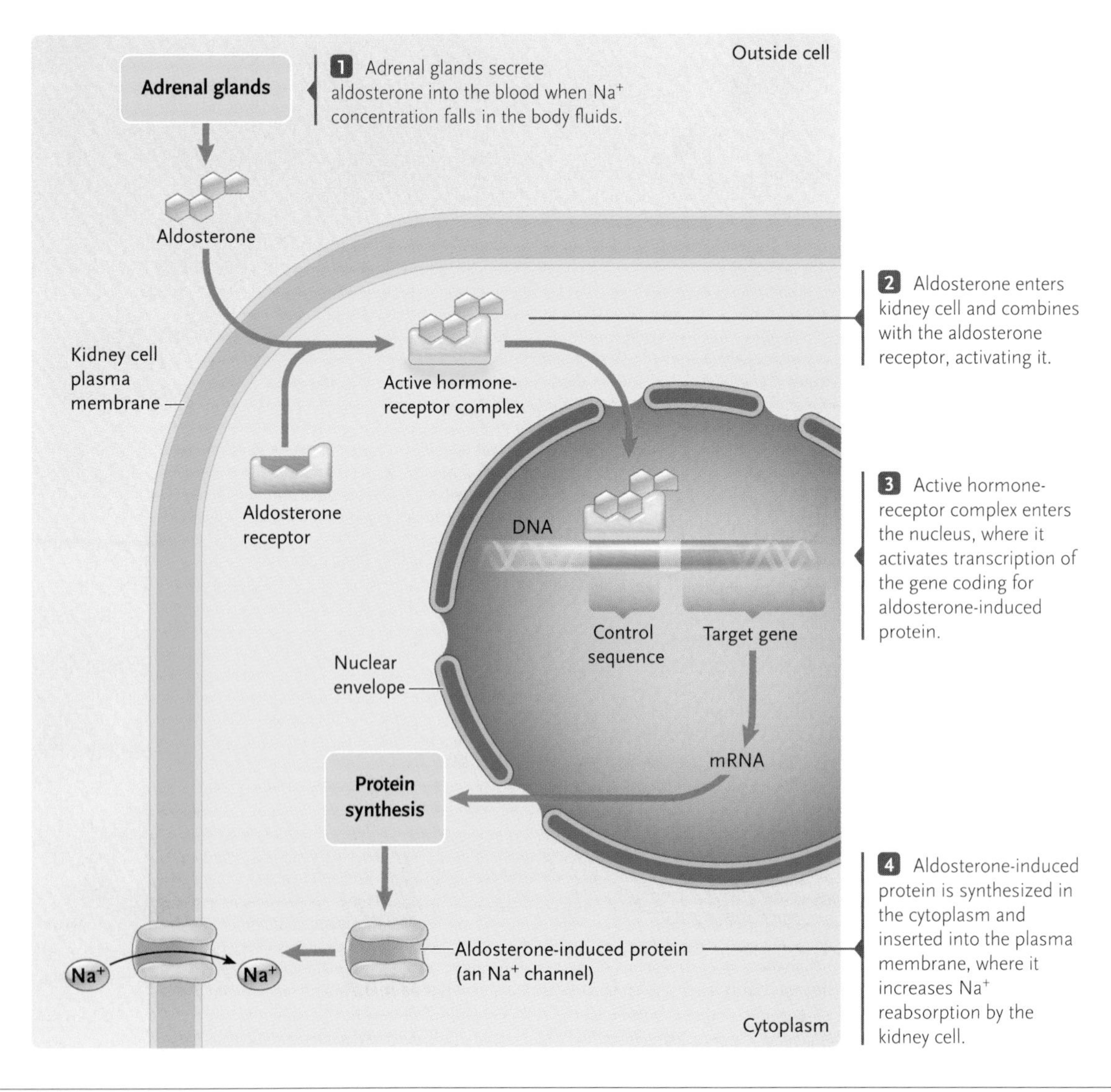

Outside cell
Adrenal glands
1 Adrenal glands secrete aldosterone into the blood when Na+ concentration falls in the body fluids.
Aldosterone
Kidney cell plasma membrane
Active hormone-receptor complex
2 Aldosterone enters kidney cell and combines with the aldosterone receptor, activating it.
Aldosterone receptor
DNA
Control sequence
Target gene
3 Active hormone-receptor complex enters the nucleus, where it activates transcription of the gene coding for aldosterone-induced protein.
Nuclear envelope
mRNA
Protein synthesis
Na+
Na+
Aldosterone-induced protein (an Na+ channel)
Cytoplasm
4 Aldosterone-induced protein is synthesized in the cytoplasm and inserted into the plasma membrane, where it increases Na+ reabsorption by the kidney cell.

Figure 40.5 Experimental Research

Demonstration That Binding of Epinephrine to β Receptors Triggers a Signal Transduction Pathway within Cells

QUESTION: Is binding of epinephrine to β receptors necessary for triggering a signal transduction pathway within cells?

EXPERIMENT: It was known that epinephrine triggers a signal transduction pathway within cells. First, activation of adenylyl cyclase causes the level of the second messenger cAMP to increase, and then cAMP activates protein kinases in a signaling cascade that generates a cellular response (see Section 7.4 for specific details of such pathways). Richard Cerione and his colleagues at Duke University Medical Center performed experiments to show whether the signal transduction pathway is stimulated by binding of epinephrine to β receptors.

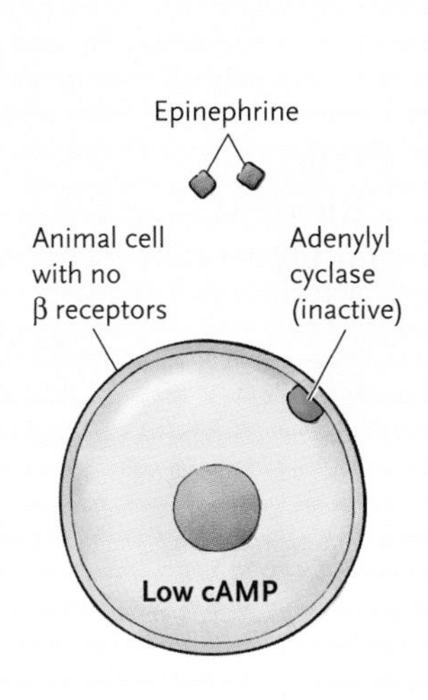

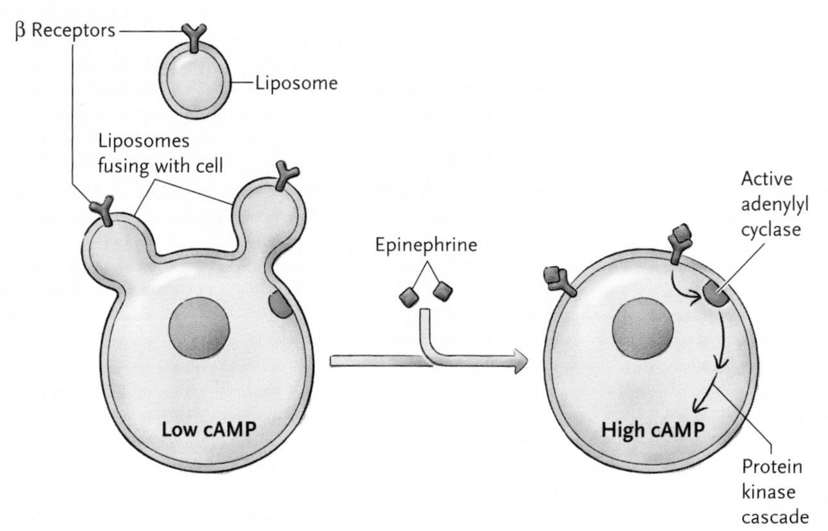

1. Epinephrine was added to animal cells lacking β receptors.

RESULT: No change occurred to the low level of cAMP in those cells. This result demonstrated that epinephrine alone was not able to trigger an increase in cAMP.

2. Liposomes—artificial spherical phospholipid membranes—containing purified β receptors were fused with the animal cells, and then epinephrine was added.

RESULT: When the liposomes fused with the animal cells, β receptors became part of the fused cell's plasma membrane. Then, adding epinephrine triggered synthesis of cAMP, resulting in high levels of cAMP in the cells. This result demonstrated that β receptors must be present in the membrane for epinephrine to trigger an increase in cAMP in the cell. The simplest interpretation was that epinephrine bound to the β receptors, activating adenylyl cyclase within the cell.

CONCLUSION: The cellular response depended upon binding of the hydrophilic hormone to a specific plasma membrane-embedded receptor.

embedded receptors: α, β_1, and β_2 receptors. (The experimental demonstration that the binding of epinephrine to a specific receptor triggers a cellular response is described in **Figure 40.5.**) When epinephrine binds to α receptors on smooth muscle cells, such as those of the blood vessels, it triggers a response pathway that causes the cells to constrict, cutting off circulation to peripheral organs. When epinephrine binds to β_1 receptors on heart muscle cells, the contraction rate of the cells increases, which in turn enhances blood supply. When epinephrine binds to β_2 receptors on liver cells, it stimulates the breakdown of glycogen to glucose, which is released from the cell. The overall effect of these, and a number of other, responses to epinephrine secretion is to supply energy to the major muscles responsible for locomotion—the body is now prepared for handling stress or for physical activity.

Different receptors binding hydrophobic hormones also may generate diverse responses. *Insights from the Molecular Revolution* describes an investigation that tested the cellular responses produced by different receptors binding the same steroid hormone.

In summary, the mechanisms by which hormones work have four major features. First, only the cells that contain surface or internal receptors for a particular hormone respond to that hormone. Second, once

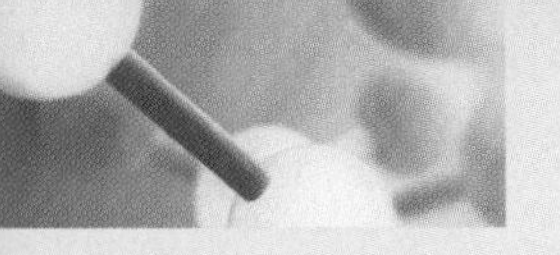

Two Receptors for Estrogens

Estrogens have many effects on female sexual development, behavior, and the menstrual cycle. One negative effect is to stimulate the growth of tumors in breast and uterine cancer. This cancer-enhancing effect can be reduced by administering *antiestrogens,* estrogen-like chemicals that bind competitively to estrogen receptors and block the sites that would normally be bound by the hormone. The antiestrogen *tamoxifen,* for example, inhibits the growth of breast tumors by blocking the activity of estrogen in breast tissue, but patients receiving it are at increased risk of developing uterine cancer.

How can tamoxifen have opposite effects in two different tissues? A group of investigators led by Thomas Scanlan and Peter Kushner at the University of California, San Francisco, joined by others at the Karolinska Institute and the Karo Bio Company in Sweden, had discovered that humans have two highly similar estrogen receptors, ERα and ERβ. Could differences between them account for the opposing effects of tamoxifen in breast and uterine tissues?

To find out, the researchers constructed two pairs of recombinant DNA plasmids. One pair consisted of either the ERα receptor gene or the ERβ receptor gene, adjacent to a promoter for continuous transcription of the receptor gene in human tissue culture cells. The other pair consisted of the firefly luciferase gene, which catalyzes a light-producing reaction, adjacent to one of two gene control sequences, AP1 or ERE, which act in estrogen-regulated systems.

One receptor plasmid and one luciferase plasmid were introduced together into human cell lines that do not normally make estrogen receptors in four possible combinations (see **figure**), with two cell lines making the ERα receptor and two making the ERβ receptor. Estrogen receptors are produced in all four of the resulting cell lines because the gene for the receptor is transcribed from the introduced plasmid.

With this experimental design, the researchers could test whether the ERα or ERβ receptor is activated by binding estrogen, and also whether the activated receptor would bind to the luciferase plasmid. If these two conditions were met, luciferase would be synthesized, and its activity could be measured using a special apparatus. (In this experiment, the luciferase gene acts as a *reporter* for the biological reactions that are occurring—luciferase has nothing to do with estrogen or estrogen activity.)

When the researchers added estrogen to cell lines containing luciferase plasmid with ERE (combinations 1 and 2 in the figure), all the cells produced luciferase. This indicated that both ERα and ERβ could bind the hormone, were activated, and could combine with ERE. When they added tamoxifen alone or along with the estrogen, the cells did not produce luciferase, indicating that tamoxifen acts as an antiestrogen and could combine with either receptor type to block the action of estrogen.

The results were different when the experiment was conducted using cell lines containing luciferase plasmid with AP1 (combinations 3 and 4 in the figure). If the estrogen and tamoxifen were added either separately or together to cells containing gene combination 3, the cells produced luciferase. This result demonstrated that ERα was activated and could combine with the AP1 control sequence, whether it was bound to estrogen or tamoxifen. Thus, in these cells the tamoxifen acted the same as an estrogen.

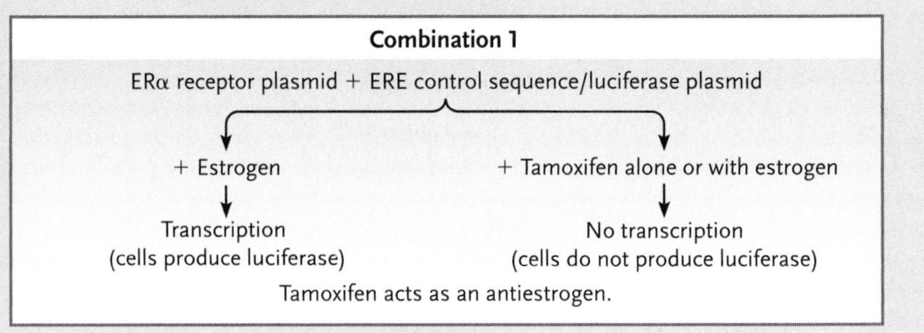

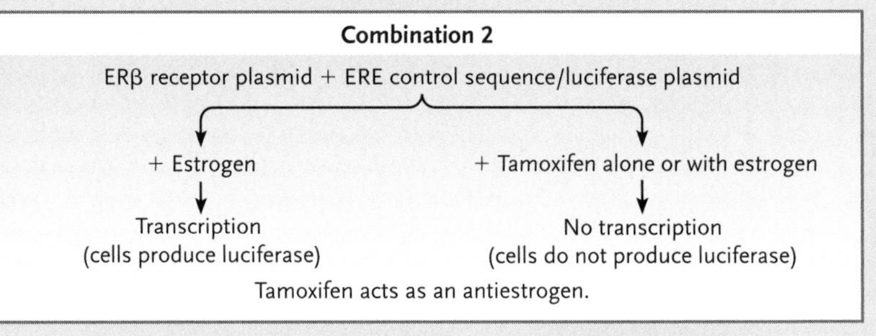

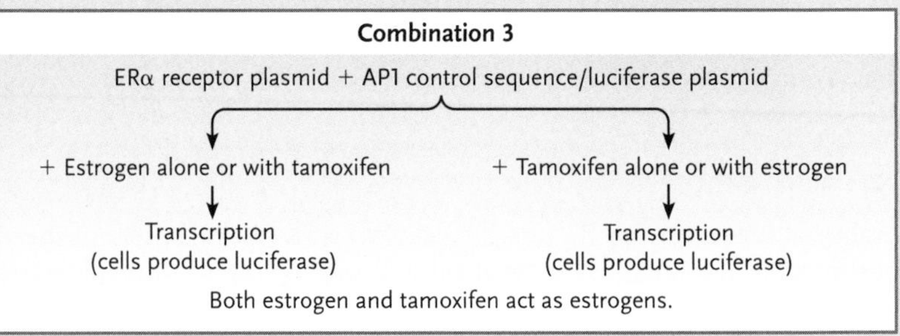

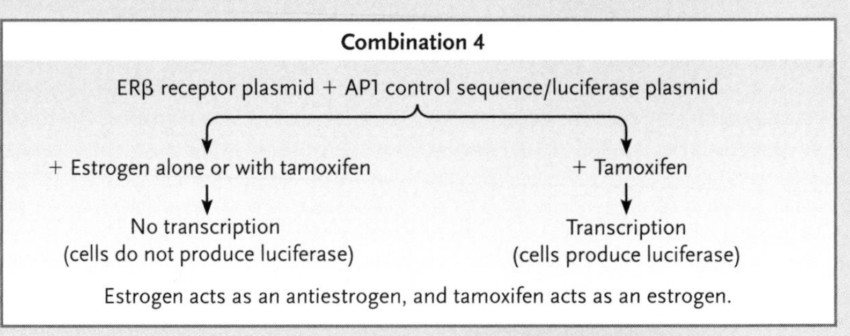

When estrogen was added, alone or with tamoxifen, to cells containing combination 4, no luciferase was produced. However, tamoxifen, if added alone, caused the cells to produce luciferase. These results indicate that ERβ combined with estrogen does not bind and activate AP1. ERβ combined with tamoxifen, however, can bind to AP1 and induce transcription of the luciferase gene. Evidently, estrogen actually acted as an antiestrogen in cells with combination 4—when added along with tamoxifen, the cells did not produce luciferase, indicating that the hormone blocked the action of tamoxifen.

The experiments indicate that the previously baffling and opposing effects of tamoxifen on breast and uterine tissues occur because different estrogen receptors are present, acting on genes controlled by either the ERE or AP1 control sequences. The results emphasize the fact that hormones can have distinct effects in different cell types depending on the types of receptors present. The research also opens the possibility of new cancer treatments that take advantage of the receptor differences.

bound by their receptors, hormones may produce a response that involves stimulation or inhibition of cellular processes through the specific types of internal molecules triggered by the hormone action. Third, because of the amplification that occurs in both the surface and internal receptor mechanisms, hormones are effective in very small concentrations. Fourth, the response to a hormone differs among target organs.

In the next two sections we discuss the major endocrine cells and glands of vertebrates. The locations of these cells and glands in the human body and their functions are summarized in **Figure 40.6** and **Table 40.1.** Pep-

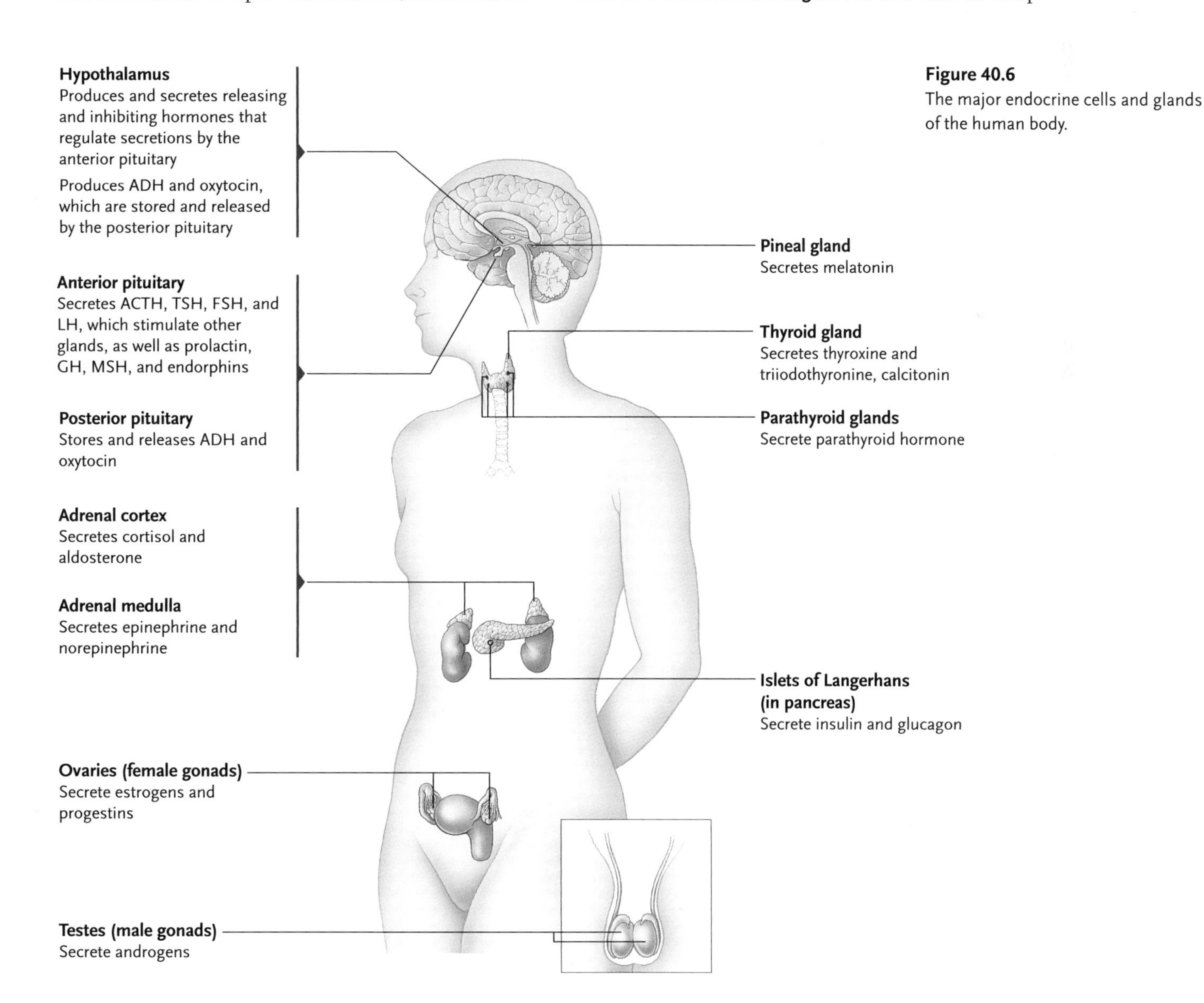

Hypothalamus
Produces and secretes releasing and inhibiting hormones that regulate secretions by the anterior pituitary
Produces ADH and oxytocin, which are stored and released by the posterior pituitary
Anterior pituitary
Secretes ACTH, TSH, FSH, and LH, which stimulate other glands, as well as prolactin, GH, MSH, and endorphins
Posterior pituitary
Stores and releases ADH and oxytocin
Adrenal cortex
Secretes cortisol and aldosterone
Adrenal medulla
Secretes epinephrine and norepinephrine
Ovaries (female gonads)
Secrete estrogens and progestins
Testes (male gonads)
Secrete androgens
Pineal gland
Secretes melatonin
Thyroid gland
Secretes thyroxine and triiodothyronine, calcitonin
Parathyroid glands
Secrete parathyroid hormone
Islets of Langerhans (in pancreas)
Secrete insulin and glucagon

Figure 40.6
The major endocrine cells and glands of the human body.

Table 40.1 The Major Human Endocrine Glands and Hormones

Secretory Tissue or Gland	Hormones	Molecular Class	Target Tissue	Principal Actions
Hypothalamus	Releasing and inhibiting hormones	Peptide	Anterior pituitary	Regulate secretion of anterior pituitary hormones
Anterior pituitary	Thyroid-stimulating hormone (TSH)	Peptide	Thyroid gland	Stimulates secretion of thyroid hormones and growth of thyroid gland
	Adrenocorticotropic hormone (ACTH)	Peptide	Adrenal cortex	Stimulates secretion of glucocorticoids by adrenal cortex
	Follicle-stimulating hormone (FSH)	Peptide	Ovaries in females, testes in males	Stimulates egg growth and development and secretion of sex hormones in females; stimulates sperm production in males
	Luteinizing hormone (LH)	Peptide	Ovaries in females, testes in males	Regulates ovulation in females and secretion of sex hormones in males
	Prolactin (PRL)	Peptide	Mammary glands	Stimulates breast development and milk secretion
	Growth hormone (GH)	Peptide	Bone, soft tissue	Stimulates growth of bones and soft tissues; helps control metabolism of glucose and other fuel molecules
	Melanocyte-stimulating hormone (MSH)	Peptide	Melanocytes in skin of some vertebrates	Promotes darkening of the skin
	Endorphins	Peptide	Pain pathways of PNS	Inhibit perception of pain
Posterior pituitary	Antidiuretic hormone (ADH)	Peptide	Kidneys	Raises blood volume and pressure by increasing water reabsorption in kidneys
	Oxytocin	Peptide	Uterus, mammary glands	Promotes uterine contractions; stimulates milk ejection from breasts
Thyroid gland	Calcitonin	Peptide	Bone	Lowers calcium concentration in blood
	Thyroxine and triiodothyronine	Amine	Most cells	Increase metabolic rate; essential for normal body growth
Parathyroid glands	Parathyroid hormone (PTH)	Peptide	Bone, kidneys, intestine	Raises calcium concentration in blood; stimulates vitamin D activation
Adrenal medulla	Epinephrine and norepinephrine	Amine	Sympathetic receptor sites throughout body	Reinforce sympathetic nervous system; contribute to responses to stress
Adrenal cortex	Aldosterone (mineralocorticoid)	Steroid	Kidney tubules	Helps control body's salt-water balance by increasing Na^+ reabsorption and K^+ excretion in kidneys
	Cortisol (glucocorticoid)	Steroid	Most body cells, particularly muscle, liver, and adipose cells	Increases blood glucose by promoting breakdown of proteins and fats
Testes	Androgens, such as testosterone*	Steroid	Various tissues	Control male reproductive system development and maintenance; most androgens are made by the testes
	Oxytocin	Peptide	Uterus	Promotes uterine contractions when seminal fluid ejaculated into vagina during sexual intercourse
Ovaries	Estrogens, such as estradiol**	Steroid	Breast, uterus, other tissues	Stimulate maturation of sex organs at puberty, and development of secondary sexual characteristics
	Progestins, such as progesterone**	Steroid	Uterus	Prepare and maintain uterus for implantation of fertilized egg and the growth and development of embryo

*Small amounts secreted by ovaries and adrenal cortex.
**Small amounts secreted by testes.

Table 40.1	The Major Human Endocrine Glands and Hormones (Continued)			
Secretory Tissue or Gland	**Hormones**	**Molecular Class**	**Target Tissue**	**Principal Actions**
Pancreas (islets of Langerhans)	Glucagon (alpha cells)	Peptide	Liver cells	Raises glucose concentration in blood; promotes release of glucose from glycogen stores and production from noncarbohydrates
	Insulin (beta cells)	Peptide	Most cells	Lowers glucose concentration in blood; promotes storage of glucose, fatty acids, and amino acids
Pineal gland	Melatonin	Amine	Brain, anterior pituitary, reproductive organs, immune system, possibly others	Helps synchronize body's biological clock with day length; may inhibit gonadotropins and initiation of puberty
Many cell types	Growth factors	Peptide	Most cells	Regulate cell division and differentiation
	Prostaglandins	Fatty acid	Various tissues	Have many diverse roles

tide hormones secreted by other body regions, including the stomach and small intestine, the thymus gland, the kidneys, and the heart will be described in the chapters in which these tissues and organs are discussed.

Study Break

1. Compare and contrast the mechanisms by which glucagon and aldosterone cause their specific responses.
2. Explain how one type of target cell could respond to different hormones, and how the same hormone could produce different effects in different cells.

40.3 The Hypothalamus and Pituitary

The hormones of vertebrates work in coordination with the nervous system. The action of several hormones is closely coordinated by the hypothalamus and its accessory gland, the pituitary.

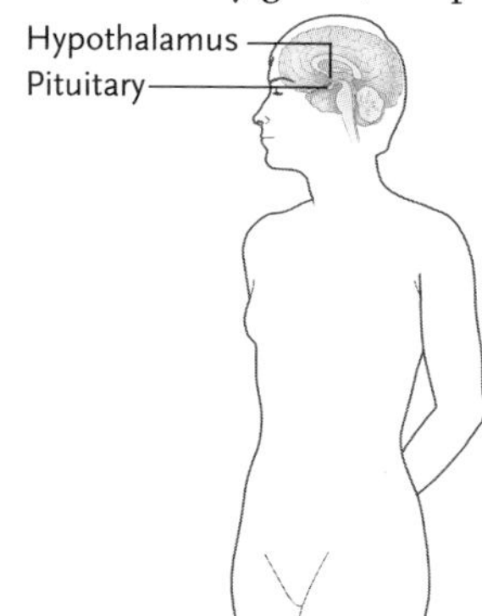

The hypothalamus is a region of the brain located in the floor of the cerebrum (see Section 38.3). The **pituitary gland,** consisting mostly of two fused lobes, is suspended just below it by a slender stalk of tissue that contains both neurons and blood vessels **(Figure 40.7).** The **posterior pituitary** contains axons and nerve endings of neurosecretory neurons that originate in the hypothalamus. The **anterior pituitary** contains nonneuronal endocrine cells that form a distinct gland. The two lobes are separate in structure and embryonic origins.

Under Regulatory Control by the Hypothalamus, the Anterior Pituitary Secretes Eight Hormones

The secretion of hormones from the anterior pituitary is controlled by peptide neurohormones called **releasing hormones (RHs)** and **inhibiting hormones (IHs),** which are released by the hypothalamus. These neurohormones are carried in the blood from the hypothalamus to the anterior pituitary in a *portal vein,* a special vein that connects the capillaries of the two glands. The portal vein provides a critical link between the brain and the endocrine system, ensuring that most of the blood reaching the anterior pituitary first passes through the hypothalamus.

RHs and IHs are **tropic hormones** (*tropic* = stimulating, not to be confused with *trophic,* which means "nourishing"), hormones that regulate hormone secretion by another endocrine gland. RHs and IHs regulate the anterior pituitary's secretion of another group of hormones; those hormones in turn control many other endocrine glands of the body, and also control some body processes directly.

Secretion of hypothalamic RHs is controlled by neurons containing receptors that monitor the blood to detect changes in body chemistry and temperature. For example, TRH, discussed earlier, is secreted in response to a drop in body temperature. Input to the hypothalamus also comes through numerous connections from control centers elsewhere in the brain, including the brain stem and limbic system. Negative feedback pathways regulate secretion of the releasing hormones, such as the pathway regulating TRH secretion.

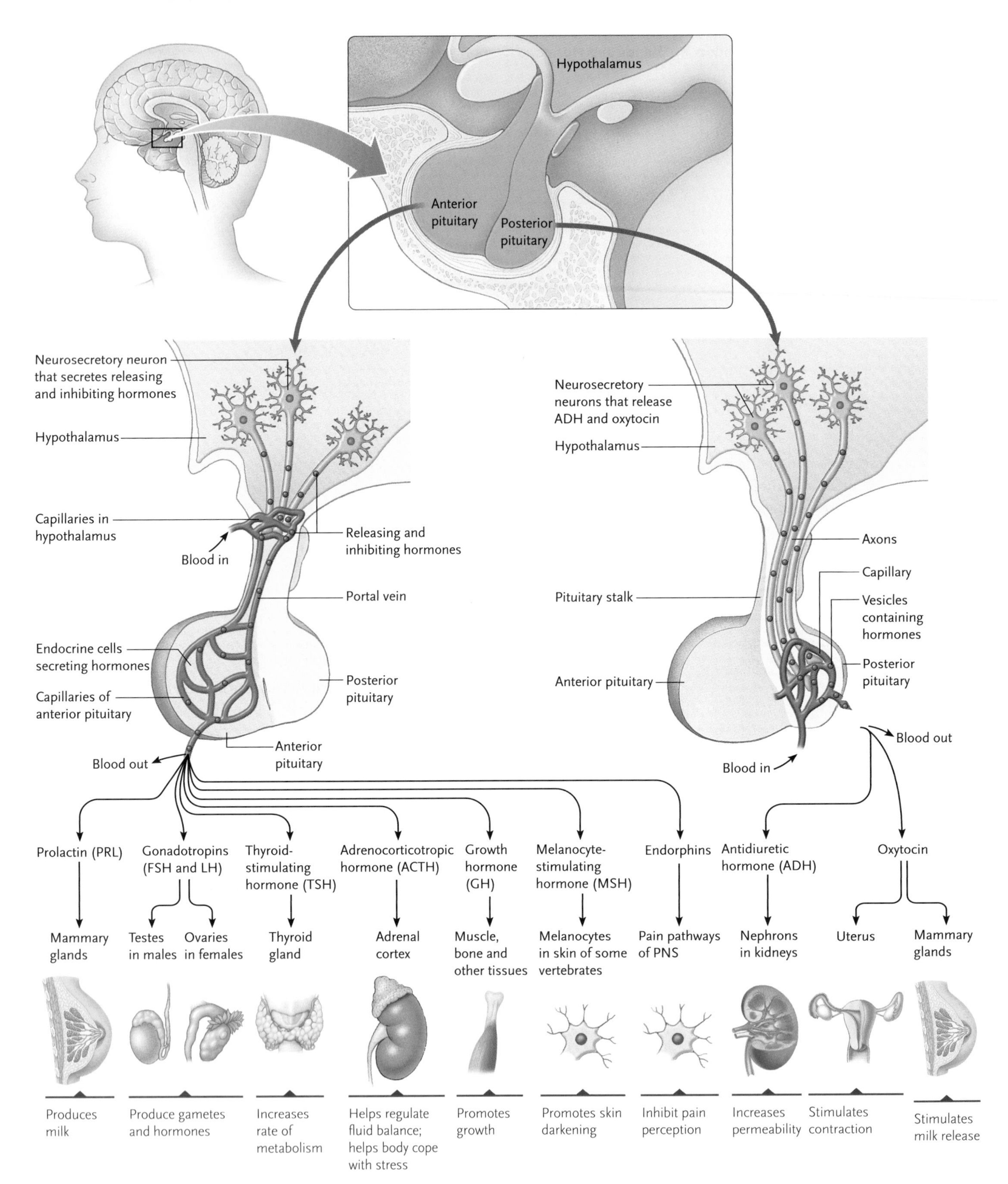

Figure 40.7

The hypothalamus and pituitary. Hormones secreted by the anterior and posterior pituitary are controlled by neurohormones released in the hypothalamus.

Under the control of the hypothalamic RHs, the anterior pituitary secretes six major hormones into the bloodstream: prolactin, growth hormone, thyroid-stimulating hormone, adrenocorticotropic hormone, follicle-stimulating hormone, and luteinizing hormone, and two other hormones, melanocyte-stimulating hormone (MSH) and endorphins. **Prolactin (PRL),** a *nontropic hormone* (a hormone that does not regulate hormone secretion by another endocrine gland), influences reproductive activities and parental care in vertebrates. In mammals, PRL stimulates development of the secretory cells of mammary glands during late pregnancy, and stimulates milk synthesis after a female mammal gives birth. Stimulation of the mammary glands and the nipples, as occurs during suckling, leads to PRL release.

Growth hormone (GH) stimulates cell division, protein synthesis, and bone growth in children and adolescents, thereby causing body growth. GH also stimulates protein synthesis and cell division in adults. For these actions, GH acts as a tropic hormone by binding to target tissues, mostly liver cells, causing them to release insulin-like growth factor **(IGF),** a peptide that directly stimulates growth processes. GH also acts as a nontropic hormone to control a number of major metabolic processes in mammals of all ages, including the conversion of glycogen to glucose and fats to fatty acids as a means of regulating their levels in the blood. In addition, GH stimulates body cells to take up fatty acids and amino acids and limits the rate at which muscle cells take up glucose. These actions help maintain the availability of glucose and fatty acids to tissues and organs between feedings; this is particularly important for the brain. In humans, deficiencies in GH secretion during childhood produce *pituitary dwarfs,* who remain small in stature **(Figure 40.8).** Overproduction of GH during childhood or adolescence, often due to a tumor of the anterior pituitary, produces *pituitary giants,* who may grow above seven feet in height.

Figure 40.8
The results of overproduction and underproduction of growth hormone by the anterior pituitary. The man on the left is of normal height. The man in the center is a pituitary giant, whose pituitary produced excess GH during childhood and adolescence. The man on the right is a pituitary dwarf, whose pituitary produced too little GH.

The other four major hormones secreted by the anterior pituitary are tropic hormones that control endocrine glands elsewhere in the body. **Thyroid-stimulating hormone (TSH)** stimulates the thyroid gland to grow in size and secrete thyroid hormones. **Adrenocorticotropic hormone (ACTH)** triggers hormone secretion by cells in the adrenal cortex. **Follicle-stimulating hormone (FSH)** controls egg development and the secretion of sex hormones in female mammals, and sperm production in males. **Luteinizing hormone (LH)** regulates part of the menstrual cycle in human females and the secretion of sex hormones in males. FSH and LH are grouped together as **gonadotropins** because they regulate the activity of the gonads (ovaries and testes). The roles of the gonadotropins and sex hormones in the reproductive cycle are described in Chapter 47.

Melanocyte-stimulating hormone (MSH) and **endorphins** are nontropic hormones produced by the anterior pituitary. MSH is named because of its effect in some vertebrates on melanocytes, skin cells that contain the black pigment melanin. For example, an increase in secretion of MSH produces a marked darkening of the skin of fishes, amphibians, and reptiles. The darkening is produced by a redistribution of melanin from the centers of the melanocytes throughout the cells. In humans, an increase in MSH secretion also causes skin darkening, although the effect is by no means as obvious as in the other vertebrates mentioned. For example, MSH secretion increases in pregnant women. That, with the effects of increased estrogens, results in increased skin pigmentation; the effects fade after birth of the child.

Endorphins, nontropic peptide hormones produced by the hypothalamus and pituitary, are also released by the anterior pituitary. In the peripheral nervous system (PNS), endorphins act as neurotransmitters in pathways that control pain, thereby inhibiting the perception of pain. Hence, endorphins are often called "natural painkillers."

The Posterior Pituitary Secretes Two Hormones into the Circulatory System

The neurosecretory neurons in the posterior pituitary secrete two nontropic peptide hormones, antidiuretic hormone and oxytocin, directly into the circulatory system (see Figure 40.7).

Antidiuretic hormone (ADH) stimulates kidney cells to absorb more water from urine, thereby increasing the volume of the blood. The hormone is released when sensory receptor cells of the hypothalamus detect an increase in the blood's Na^+ concentration during periods of body dehydration or after a salty meal. Ethyl alcohol and caffeine inhibit ADH secretion, explaining in part why alcoholic drinks and coffee increase the volume of urine excreted. Nicotine and emotional stress, in contrast, stimulate ADH secretion and water retention. After severe stress is relieved, the return to normal ADH secretion often makes a trip to the bathroom among our most pressing needs. The hypothalamus also releases a flood of ADH when an injury results in heavy blood loss or some other event triggers a severe drop in blood pressure. ADH helps maintain blood pressure by reducing water loss and also by causing small blood vessels in some tissues to constrict.

Hormones with structure and action similar to ADH are also secreted in fishes, amphibians, reptiles, and birds. In amphibians, these ADH-like hormones increase the amount of water entering the body through the skin and from the urinary bladder.

We have noted that **oxytocin** stimulates the ejection of milk from the mammary glands of a nursing mother. Stimulation of the nipples in suckling sends neuronal signals to the hypothalamus, and leads to release of oxytocin from the posterior pituitary. The released oxytocin stimulates more oxytocin secretion by a positive feedback mechanism. Oxytocin causes the smooth muscle cells surrounding the mammary glands to contract, forcibly expelling the milk through the nipples. The entire cycle, from the onset of suckling to milk ejection, takes less than a minute in mammals. Oxytocin also plays a key role in childbirth, as we discussed in Section 36.4.

In males, oxytocin is secreted into the seminal fluid by the testes. Like prostaglandins, when the seminal fluid is ejaculated into the vagina during sexual intercourse, oxytocin stimulates contractions of the uterus that aid movement of sperm through the female reproductive tract.

Study Break

1. Summarize the functional interactions between the hypothalamus and the anterior pituitary gland.
2. Distinguish between how tropic hormones and nontropic hormones produce responses.

40.4 Other Major Endocrine Glands of Vertebrates

Besides the hypothalamus and pituitary, the body has seven major endocrine glands or tissues, many of them regulated by the hypothalamus-pituitary connection. These glands are the thyroid gland, parathyroid glands, adrenal medulla, adrenal cortex, gonads, pancreas, and pineal gland (shown in Figure 40.6 and summarized in Table 40.1).

The Thyroid Hormones Stimulate Metabolism, Development, and Maturation

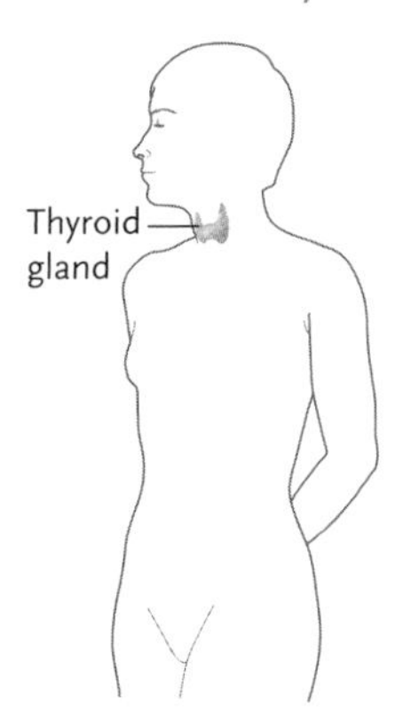

The **thyroid gland,** which is located in the front of the throat in humans, has a shape similar to that of a bowtie. It secretes the same hormones in all vertebrates. The primary thyroid hormone, **thyroxine,** is known as $\mathbf{T_4}$ because it contains four iodine atoms. The thyroid also secretes smaller amounts of a closely related hormone, **triiodothyronine** or $\mathbf{T_3}$, which contains three iodine atoms. A supply of iodine in the diet is necessary for production of these hormones. Normally, their concentrations are kept at finely balanced levels in the blood by negative feedback loops such as that described in Figure 40.2.

Both T_4 and T_3 enter cells; however, once inside, most of the T_4 is converted to T_3, the form that combines with internal receptors. Binding of T_3 to receptors alters gene expression, which brings about the hormone's effects.

The thyroid hormones are vital to growth, development, maturation, and metabolism in all vertebrates. They interact with GH for their effects on growth and development. Thyroid hormones also increase the sensitivity of many body cells to the effects of epinephrine and norepinephrine, hormones released by the adrenal medulla as part of the "fight or flight response" (discussed further later).

In amphibians such as frogs, thyroid hormones trigger **metamorphosis,** or change in body form from tadpole to adult **(Figure 40.9).** Thyroid hormones also contribute to seasonal changes in the plumage of birds and coat color in mammals.

In human adults, low thyroid output, *hypothyroidism,* causes affected individuals to be sluggish mentally and physically; they have a slow heart rate and weak pulse, and often feel confused and depressed. Hypothyroidism in infants and children leads to cretinism, that is, stunted growth and diminished intelligence. Overproduction of thyroid hormones in human adults, *hyperthyroidism,* produces nervousness and emotional instability, irritability, insomnia, weight loss, and a rapid, often irregular heartbeat. The most common form of hyperthyroidism is *Graves' disease,* characterized by inflamed, protruding eyes in addition to the other symptoms mentioned.

Insufficient iodine in the diet can cause *goiter,* enlargement of the thyroid. Without iodine, the thyroid cannot make T_3 and T_4 in response to stimulation by

TSH. Because the thyroid hormone concentration remains low in the blood, TSH continues to be secreted, and the thyroid grows in size. Dietary iodine deficiency has been eliminated in developed regions of the world by the addition of iodine to table salt.

In mammals, the thyroid also has specialized cells that secrete **calcitonin,** a nontropic peptide hormone. The hormone lowers the level of Ca^{2+} in the blood by inhibiting the ongoing dissolution of calcium from bone. Calcitonin secretion is stimulated when Ca^{2+} levels in blood rise above the normal range and inhibited when Ca^{2+} levels fall below the normal range.

The Parathyroid Glands Regulate Ca^{2+} Level in the Blood

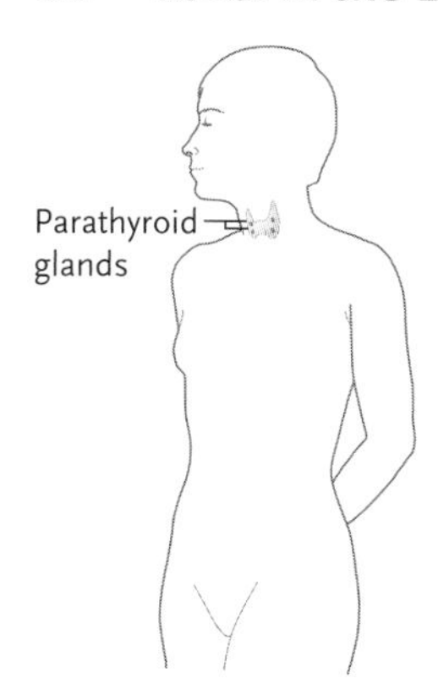

The **parathyroid glands** occur only in tetrapod vertebrates—amphibians, reptiles, birds, and mammals. Each is a spherical structure about the size of a pea. Mammals have four parathyroids located on the posterior surface of the thyroid gland, two on each side. The single hormone they produce, a nontropic hormone called **parathyroid hormone (PTH),** is secreted in response to a fall in blood Ca^{2+} levels. PTH stimulates bone cells to dissolve the mineral matter of bone tissues, releasing both calcium and phosphate ions into the blood. The released Ca^{2+} is available for enzyme activation, conduction of nerve signals across synapses, muscle contraction, blood clotting, and other uses. How blood Ca^{2+} levels control PTH and calcitonin secretion is shown in **Figure 40.10.**

PTH also stimulates enzymes in the kidneys that convert **vitamin D,** a steroidlike molecule, into its fully active form in the body. The activated vitamin D increases the absorption of Ca^{2+} and phosphates from ingested food by promoting the synthesis of a calcium-binding protein in the intestine; it also increases the release of Ca^{2+} from bone in response to PTH.

PTH underproduction causes Ca^{2+} concentration to fall steadily in the blood, disturbing nerve and muscle function—the muscles twitch and contract uncontrollably, and convulsions and cramps occur. Without treatment, the condition is usually fatal, because the severe muscular contractions interfere with breathing. Overproduction of PTH results in the loss of so much calcium from the bones that they become thin and fragile. At the same time, the elevated Ca^{2+} concentration in the blood causes calcium deposits to form in soft tissues, especially in the lungs, arteries, and kidneys (where the deposits form kidney stones).

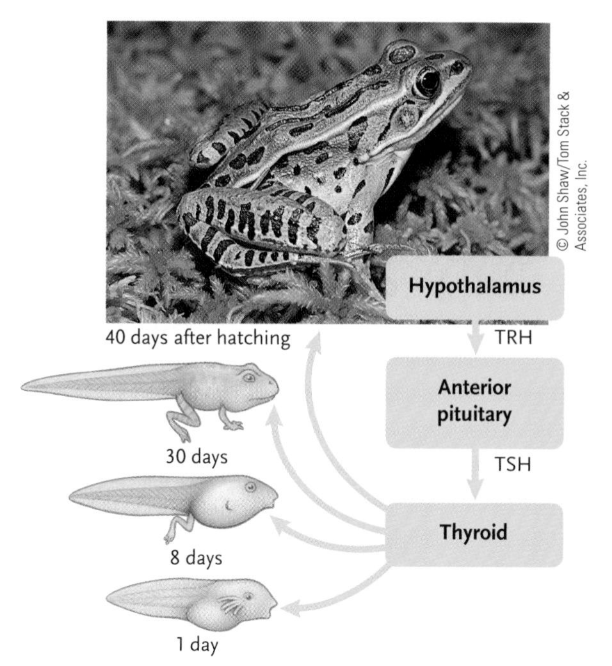

Figure 40.9
Metamorphosis of a tadpole into an adult frog, under the control of thyroid hormones. As a part of the metamorphosis, changes in gene activity lead to a change from an aquatic to a terrestrial habitat. TRH, thyroid-releasing hormone; TSH, thyroid-stimulating hormone.

The Adrenal Medulla Releases Two "Fight or Flight" Hormones

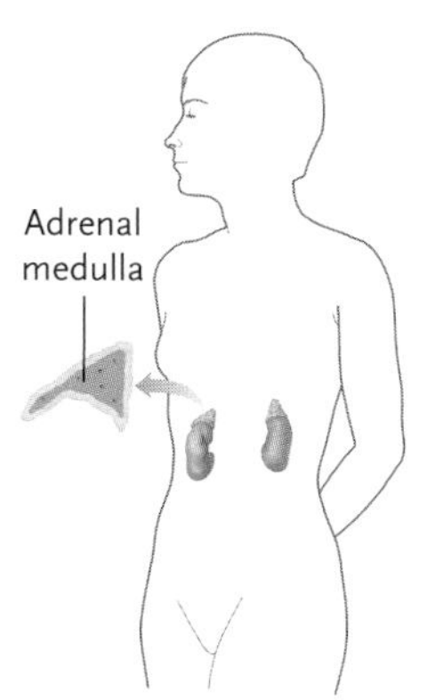

The adrenal glands (*ad* = next to, *renes* = kidneys) of mammals consist of two distinct regions. The central region, the **adrenal medulla,** contains neurosecretory neurons; the tissue surrounding it, the **adrenal cortex,** contains endocrine cells. The two regions secrete hormones with entirely different functions. Nonmammalian vertebrates have glands equivalent to the adrenal medulla and adrenal cortex of mammals, but they are separate. Most of the hormones produced by these glands have essentially the same functions in all vertebrates. The only major exception is aldosterone, which is secreted by the adrenal cortex or its equivalent only in tetrapod vertebrates.

In most species, the adrenal medulla secretes two nontropic amine hormones, **epinephrine** and **norepinephrine,** which are **catecholamines,** chemical compounds derived from the amino acid tyrosine that circulate in the bloodstream. They bind to receptors in the plasma membranes of their target cells. (Epinephrine is also secreted by some cells of the CNS, and norepinephrine is also secreted by some cells of the CNS and neurons of the sympathetic nervous system. In these cases, epinephrine and norepinephrine

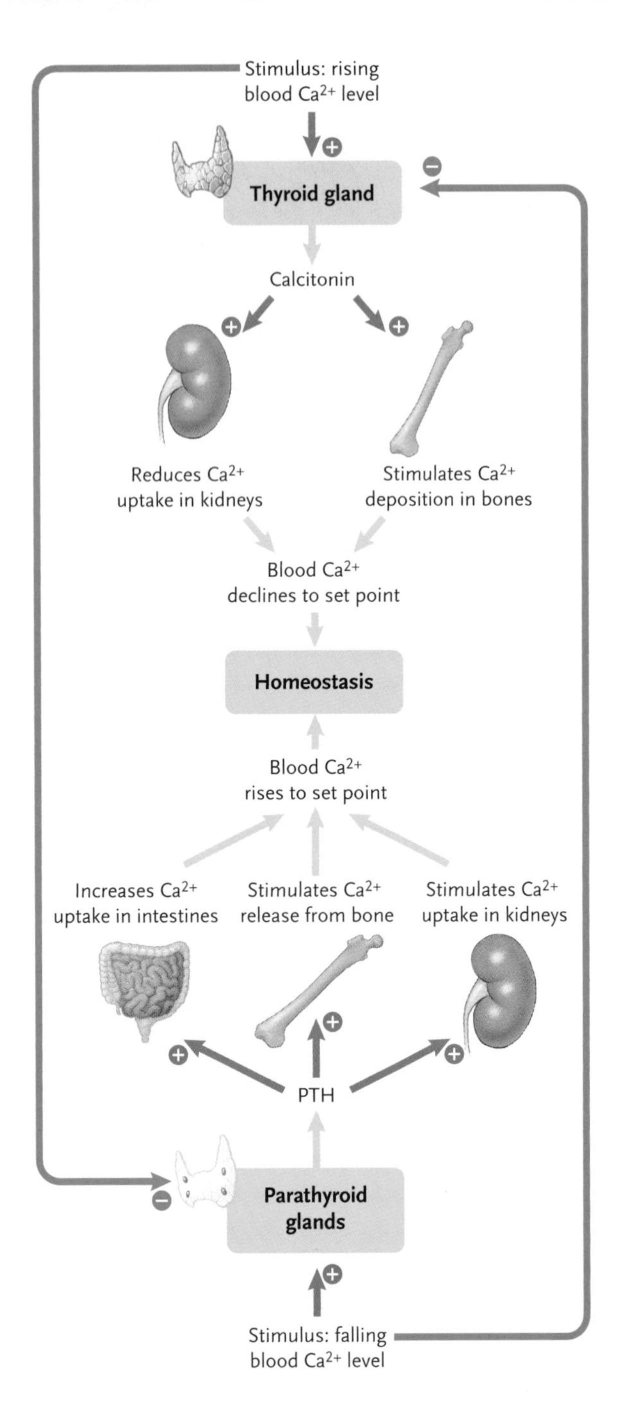

Figure 40.10 Negative feedback control of PTH and calcitonin secretion by blood Ca^{2+} levels.

function as neurotransmitters between interneurons involved in a diversity of brain and body functions; see Section 37.3.)

Epinephrine and norepinephrine, which reinforce the action of the sympathetic nervous system, are secreted when the body encounters stresses such as emotional excitement, danger (fight-or-flight situations), anger, fear, infections, injury, even midterm and final exams. Epinephrine in particular prepares the body for handling stress or physical activity. The heart rate increases. Glycogen and fats break down, releasing glucose and fatty acids into the blood as fuel molecules. In the heart, skeletal muscles, and lungs, the blood vessels dilate to increase blood flow. Elsewhere in the body, the blood vessels constrict, raising blood pressure, reducing blood flow to the intestine and kidneys, and inhibiting smooth muscle contractions, which reduces water loss and slows down the digestive system. Airways in the lungs also dilate, helping to increase the flow of air.

The effects of norepinephrine on heart rate, blood pressure, and blood flow to the heart muscle are similar to those of epinephrine. However, in contrast to epinephrine, norepinephrine causes blood vessels in skeletal muscles to constrict. This antagonistic effect is largely canceled out because epinephrine is secreted in much greater quantities.

No known human diseases are caused by underproduction of the hormones of the adrenal medulla, as long as the sympathetic nervous system is intact. Overproduction of epinephrine and norepinephrine, which can occur if there is a tumor in the adrenal medulla, leads to symptoms duplicating a stress response.

The Adrenal Cortex Secretes Two Groups of Steroid Hormones That Are Essential for Survival

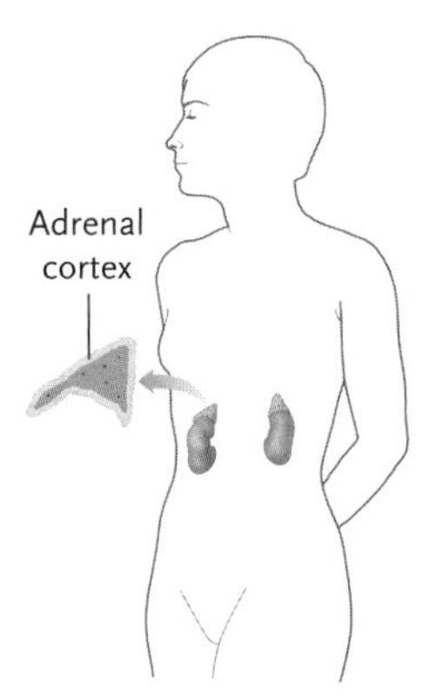

The adrenal cortex of mammals secretes two major types of steroid hormones: **glucocorticoids** help maintain the blood concentration of glucose and other fuel molecules, and **mineralocorticoids** regulate the levels of Na^+ and K^+ ions in the blood and extracellular fluid.

The Glucocorticoids. The glucocorticoids help maintain glucose levels in the blood by three major mechanisms: (1) stimulating the synthesis of glucose from noncarbohydrate sources such as fats and proteins, (2) reducing glucose uptake by body cells except those in the central nervous system, and (3) promoting the breakdown of fats and proteins, which releases fatty acids and amino acids into the blood as alternative fuels when glucose supplies are low. The absence of down-regulation of glucose uptake to the CNS keeps the brain well supplied with glucose between meals and during periods of extended fasting. **Cortisol** is the major glucocorticoid secreted by the adrenal cortex.

Secretion of glucocorticoids is ultimately under control of the hypothalamus **(Figure 40.11).** Low glucose concentrations in the blood, or elevated levels of epinephrine secreted by the adrenal medulla in response to stress, are detected in the hypothalamus, leading to secretion of the tropic hormone ACTH by the anterior pituitary. ACTH promotes the secretion of glucocorticoids by the adrenal cortex.

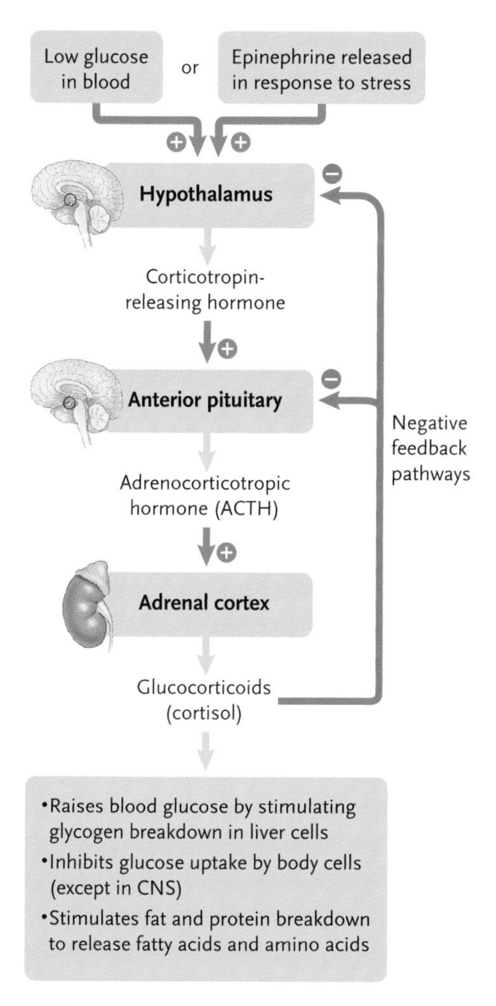

Figure 40.11
Pathways linking secretion of glucocorticoids to low blood sugar and epinephrine secretion in response to stress.

Overproduction of glucocorticoids makes blood glucose rise and increases fat deposition in adipose tissue and protein breakdown in muscles and bones. The loss of proteins from muscles causes weakness and fatigue; loss of proteins from bone, particularly collagens, makes the bones fragile and susceptible to breakage. Underproduction of glucocorticoids causes blood glucose concentration to fall below normal levels in the blood and diminishes tolerance to stress.

Glucocorticoids have anti-inflammatory properties and, consequently, they are used clinically to treat conditions such as arthritis or dermatitis. They also suppress the immune system and are used in the treatment of autoimmune diseases such as rheumatoid arthritis.

The Mineralocorticoids. In tetrapods, the mineralocorticoids, primarily **aldosterone,** increase the amount of Na^+ reabsorbed from the fluids processed by the kidneys and absorbed from foods in the intestine. They also reduce the amount of Na^+ secreted by salivary and sweat glands and increase the rate of K^+ excretion by the kidneys. The net effect is to keep Na^+ and K^+ balanced at the levels required for normal cellular functions, including those of the nervous system. Relatedly, secretion of aldosterone is tightly linked to blood volume and indirectly to blood pressure (see Section 40.2 and Chapter 46).

Moderate overproduction of aldosterone causes excessive water retention in the body, so that tissues swell and blood pressure rises. Conversely, moderate underproduction can lead to excessive water loss and dehydration. Severe underproduction is rapidly fatal unless mineralocorticoids are supplied by injection or other means.

The adrenal cortex also secretes small amounts of androgens, steroid sex hormones responsible for maintenance of male characteristics, which are synthesized primarily by the gonads. These hormones have significant effects only if they are overproduced, as can occur with some tumors in the adrenal cortex. The result is altered development of primary or secondary sex characteristics.

The Gonadal Sex Hormones Regulate the Development of Reproductive Systems, Sexual Characteristics, and Mating Behavior

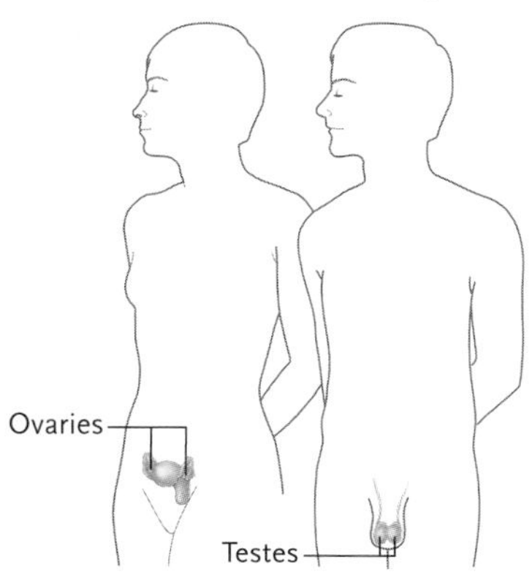

The **gonads**, the testes and ovaries, are the primary source of sex hormones in vertebrates. The steroid hormones they produce, the **androgens**, **estrogens,** and **progestins,** have similar functions in regulating the development of male and female reproductive systems, sexual characteristics, and mating behavior. Both males and females produce all three types of hormones, but in different proportions. Androgen production is predominant in males, while estrogen and progestin production is predominant in females. An outline of the actions of these hormones is presented here, and a more complete picture is given in Chapter 47.

The **testes** of male vertebrates secrete androgens, steroid hormones that stimulate and control the development and maintenance of male reproductive systems. The principal androgen is **testosterone**, the male sex hormone. In young adult males, a jump in testosterone levels stimulates puberty and the development of secondary sexual characteristics, including the growth of facial and body hair, muscle development,

FOCUS ON RESEARCH

Basic Research: Neuroendocrine and Behavioral Effects of Anabolic–Androgenic Steroids in Humans

Anabolic–androgenic steroids (AAS) are synthetic derivatives of the natural steroid hormone testosterone. They were designed to have potent anabolic (tissue building) activity and low androgenic (masculinizing) activity in therapeutic doses. Overall, there are about 60 AAS that vary in chemical structure and, therefore, in their physiological effects.

AAS are used for treating conditions such as delayed puberty and subnormal growth in children, as well as for therapy in chronic conditions such as cancer, AIDS, severe burns, liver and kidney failure, and anemias. AAS are not used exclusively for medical purposes, however. Because of their anabolic effects, which include an increase in muscle mass, strength, and endurance, as well as acceleration of recovery from injuries, AAS are used by athletes such as bodybuilders, weight lifters, baseball players, and football players. This use is actually abuse, because the doses typically administered are far higher than therapeutic doses. AAS abuse is significant: in the early 1990s, about one million Americans had used or were using AAS to increase strength, muscle mass, or athletic ability. While originally limited to elite athletes, use has trickled down to average athletes, including adolescents. It is estimated that perhaps 4% of high school students have used AAS. The greatest increase in AAS abuse over the past decade has been by adolescent girls.

Are AAS harmful at high doses? When researchers gave rodents doses of AAS comparable to those associated with human AAS abuse, they observed significant increases in aggression, anxiety, and sexual behaviors. These changes occur as a result of alterations in the neurotransmitters and other signaling molecules associated with those behaviors. All of these changes have been hypothesized to occur in human AAS abusers.

To study the effect of high doses of AAS on the human endocrine system, R. C. Daly and colleagues at the National Institute of Mental Health, in Bethesda, Maryland, administered the AAS methyltestosterone (MT) to normal (medication-free) human volunteers over a period of time in an inpatient clinic. The subjects were examined for the effects of MT on pituitary–gonadal, pituitary–thyroid, and pituitary–adrenal hormones, and the researchers attempted to correlate endocrine changes with psychological symptoms caused by the MT.

The researchers found, for instance, that high doses of MT caused a significant decrease in the levels of gonadotropins and gonadal steroid hormones in the blood. At the same time, thyroxine and TSH levels increased. No significant increases were seen in pituitary–adrenal hormones.

The decrease in testosterone levels correlated significantly with cognitive problems, such as increased distractibility and forgetfulness. The increase in thyroxine correlated significantly with a rise in aggressive behavior, notably anger, irritability, and violent feelings. There were no changes in activities associated with pituitary–adrenal hormones—energy, disturbed sleep, and sexual arousal—as was expected by the lack of change in those hormones.

In sum, behavioral changes associated with high doses of an AAS suggest that AAS-induced hormonal changes may well contribute to the adverse behavioral and mood changes that occur during AAS abuse. Clearly, there is every reason to believe that taking high doses of AAS for athletic gain alters the normal hormonal balance in humans, as it does in rodents.

changes in vocal cord morphology, and development of normal sex drive. The synthesis and secretion of testosterone by cells in the testes is controlled by the release of luteinizing hormone (LH) from the anterior pituitary, which in turn is controlled by **gonadotropin releasing hormone (GnRH)**, a tropic hormone secreted by the hypothalamus.

Androgens are natural types of **anabolic steroids**, hormones that stimulate muscle development. Natural and synthetic anabolic steroids have been in the news over the years because of their use by bodybuilders and other athletes from sports in which muscular strength is important. *Focus on Research* discusses the potential adverse effects of anabolic–androgenic steroids, synthetic derivatives of testosterone, in humans.

The **ovaries** of females produce estrogens, steroid hormones that stimulate and control the development and maintenance of female reproductive systems. The principal estrogen is **estradiol**, which stimulates maturation of sex organs at puberty and the development of secondary sexual characteristics. Ovaries also produce progestins, principally **progesterone**, the steroid hormone that prepares and maintains the uterus for implantation of a fertilized egg and the subsequent growth and development of an embryo. The synthesis and secretion of progesterone by cells in the ovaries is controlled by the release of follicle-stimulating hormone (FSH) from the anterior pituitary, which in turn is controlled by the same GnRH as in males.

The Pancreatic Islet of Langerhans Hormones Regulate Glucose Metabolism

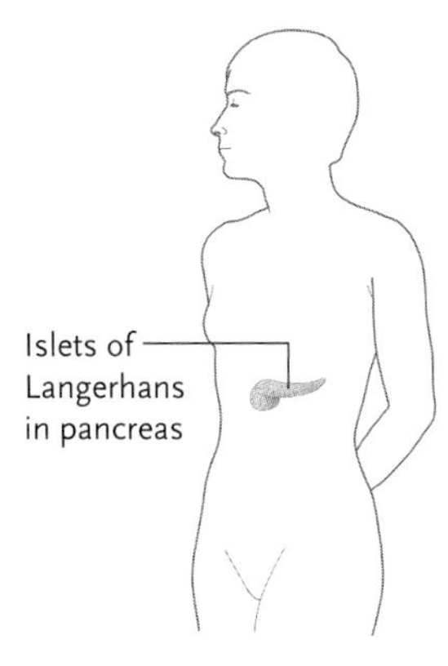

Most of the **pancreas,** a relatively large gland located just behind the stomach, forms an exocrine gland that secretes digestive enzymes into the small intestine (see Chapter 45). However, about 2% of the cells in the pancreas are endocrine cells that form the **islets of Langerhans.** Found in all vertebrates, the islets secrete the peptide hormones insulin and glucagon into the bloodstream.

Insulin and glucagon regulate the metabolism of fuel substances in the body. **Insulin,** secreted by *beta cells* in the islets, acts mainly on cells of nonworking skeletal muscles, liver cells, and adipose tissue (fat). (Brain cells do not require insulin for glucose uptake.) Insulin lowers blood glucose, fatty acid, and amino acid levels and promotes their storage. That is, the actions of insulin include stimulation of glucose transport into cells, glycogen synthesis from glucose, uptake of fatty acids by adipose tissue cells, fat synthesis from fatty acids, and protein synthesis from amino acids. Insulin also inhibits glycogen degradation to glucose, fat degradation to fatty acids, and protein degradation to amino acids.

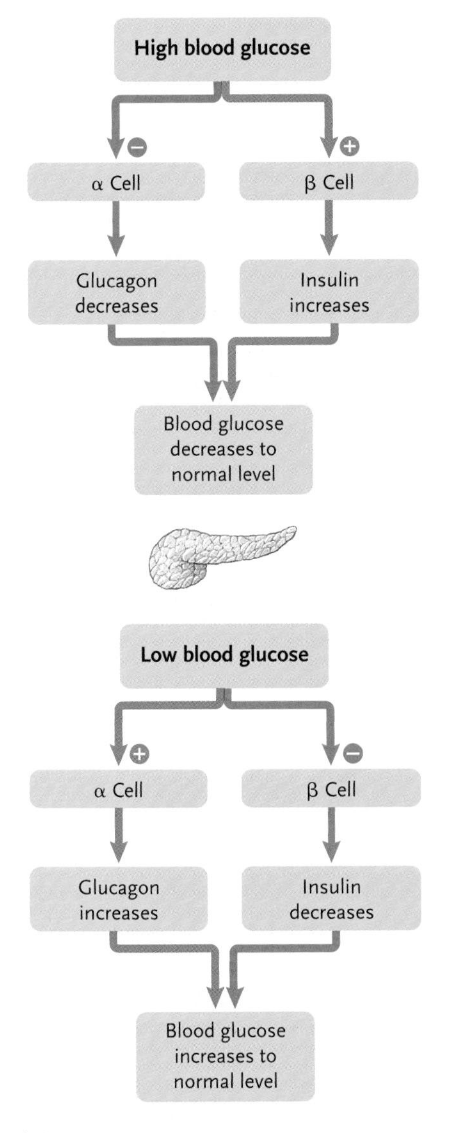

Figure 40.12
The action of insulin and glucagon in maintaining the concentration of blood glucose at an optimal level.

Glucagon, secreted by *alpha cells* in the islets, has effects opposite to those of insulin: it stimulates glycogen, fat, and protein degradation. Glucagon also uses amino acids and other noncarbohydrates as the input for glucose synthesis; this aspect of glucagon function operates during fasting. Negative feedback mechanisms that are keyed to the concentration of glucose in the blood control secretion of both insulin and glucagon to maintain glucose homeostasis **(Figure 40.12).**

Diabetes mellitus, a disease that afflicts more than 14 million people in the United States, results from problems with insulin production or action. The three classic diabetes symptoms are frequent urination, increased thirst (and consequently increased fluid intake), and increased appetite. Frequent urination occurs because without insulin, body cells are not stimulated to take up glucose, leading to abnormally high glucose concentration in the blood; excretion of the excess glucose in the urine requires water to carry it, which causes increased fluid loss and frequent trips to the bathroom. The need to replace the excreted water causes increased thirst. Increased appetite comes about because cells have low glucose levels and, therefore, proteins and fats are broken down as energy sources. Food intake is necessary to offset the negative energy balance or else weight loss will occur. Two of these classic symptoms gave the disease its name: *diabetes* is derived from a Greek word meaning "siphon," referring to the frequent urination, and *mellitus,* a Latin word meaning "sweetened with honey," refers to the sweet taste of a diabetic's urine. (Before modern blood or urine tests were developed, physicians tasted a patient's urine to detect the disease.)

The disease occurs in two major forms called *type 1* and *type 2.* Type 1 diabetes (insulin-dependent diabetes), which occurs in about 10% of diabetics, results from insufficient insulin secretion by the pancreas. This type of diabetes is usually caused by an autoimmune reaction in which an antibody destroys pancreatic beta cells. To survive, type 1 diabetics must receive regular insulin injections (typically, a genetically engineered human insulin called Humulin); careful dieting and exercise also have beneficial effects, because active skeletal muscles do not require insulin to take up and utilize glucose.

In type 2 diabetes (non-insulin-dependent diabetes), insulin is usually secreted at or above normal levels, but target cells have altered receptors that make

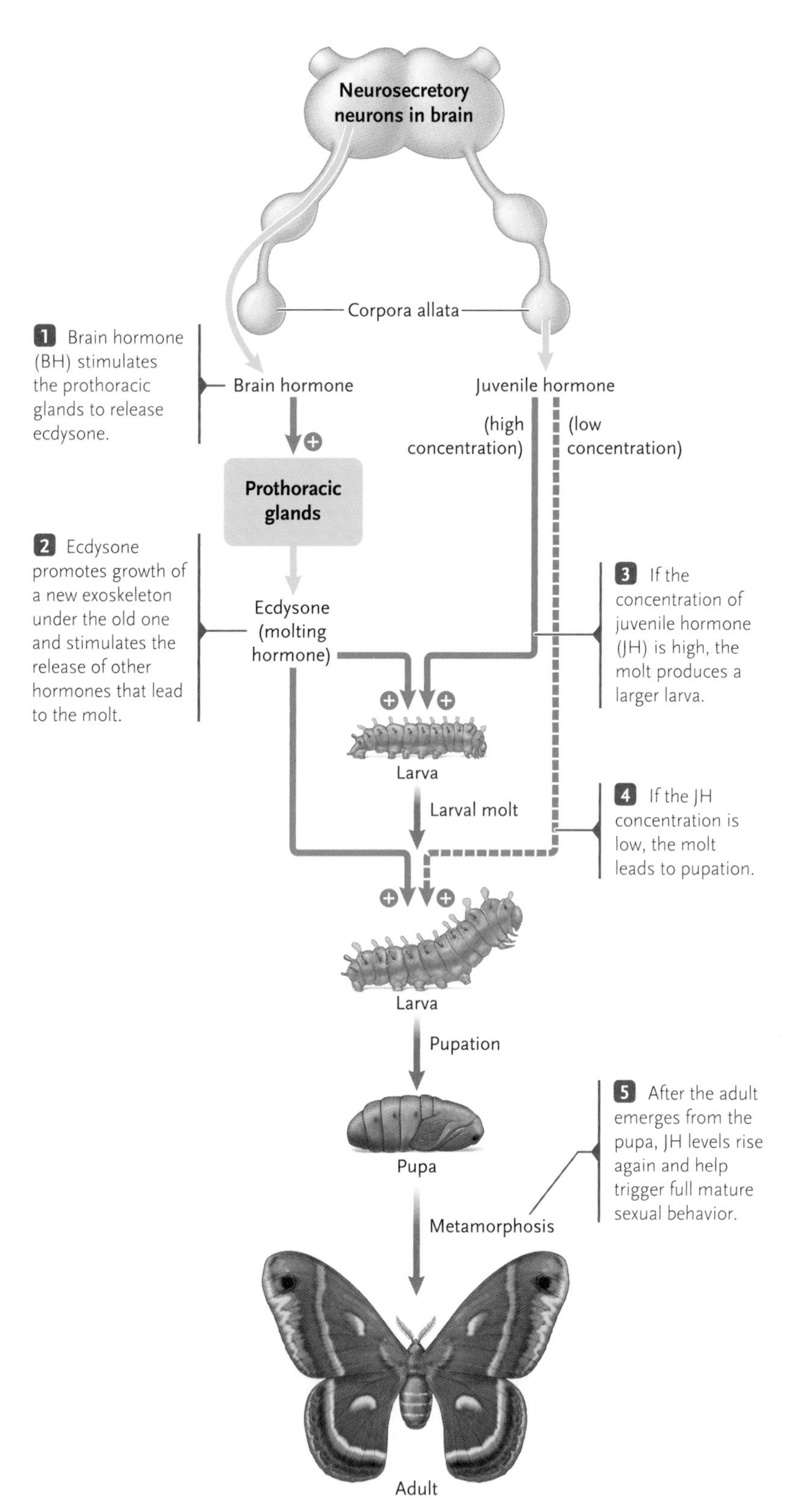

Neurosecretory neurons in brain
Corpora allata
1 Brain hormone (BH) stimulates the prothoracic glands to release ecdysone.
Brain hormone
Juvenile hormone
(high concentration)
(low concentration)
Prothoracic glands
2 Ecdysone promotes growth of a new exoskeleton under the old one and stimulates the release of other hormones that lead to the molt.
Ecdysone (molting hormone)
3 If the concentration of juvenile hormone (JH) is high, the molt produces a larger larva.
Larva
Larval molt
4 If the JH concentration is low, the molt leads to pupation.
Larva
Pupation
5 After the adult emerges from the pupa, JH levels rise again and help trigger full mature sexual behavior.
Pupa
Metamorphosis
Adult

Figure 40.13
The roles of brain hormone, ecdysone, and juvenile hormone in the development of a silkworm moth.

them less responsive to the hormone than cells in normal individuals. About 90% of patients in the developed world who develop type 2 diabetes are obese. A genetic predisposition can also be a factor. Most affected people can lead a normal life by controlling their diet and weight, exercising, and taking drugs that enhance insulin action or secretion.

Diabetes has long-term effects on the body. Its cells, unable to utilize glucose as an energy source, start breaking down proteins and fats to generate energy. The protein breakdown weakens blood vessels throughout the body, particularly in the arms and legs and in critical regions such as the kidneys and retina of the eye. The circulation becomes so poor that tissues degenerate in the arms, legs, and feet. Bleeding in the retina causes blindness at advanced stages of the disease. The breakdown of circulation in the kidneys can lead to kidney failure. In addition, in type 1 diabetes, acidic products of fat breakdown (ketones) are produced in abnormally high quantities and accumulate in the blood. The lowering of blood pH that results can disrupt heart and brain function, leading to coma and death if the disease is untreated.

The Pineal Gland Regulates Some Biological Rhythms

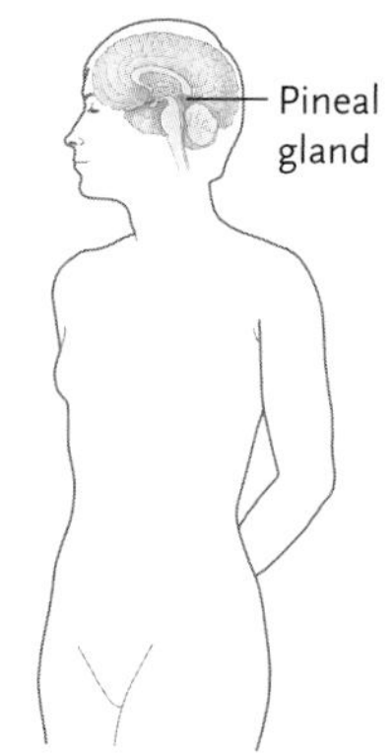

Pineal gland

The **pineal gland** is found at different locations in the brains of vertebrates—for example, in mammals, it is at roughly the center of the brain, while in birds and reptiles, it is on the surface of the brain just under the skull. The pineal gland regulates some biological rhythms.

The earliest vertebrates had a third, light-sensitive eye at the top of the head, and some species, such as lizards and tuataras (New Zealand reptiles), still have an eyelike structure in this location. In most vertebrates, the third eye became modified into a pineal gland, which in many groups retains some degree of photosensitivity. In mammals it is too deeply buried in the brain to be affected directly by light; nonetheless, specialized photoreceptors in the eyes make connections to the pineal gland.

In mammals, the pineal gland secretes a peptide hormone, **melatonin,** which helps to maintain daily biorhythms. Secretion of melatonin is regulated by an inhibitory pathway. Light hitting the eyes generates signals that inhibit melatonin secretion; consequently, the hormone is secreted most actively during periods of darkness. Melatonin targets a part of the hypothalamus called the *suprachiasmatic nucleus,* which is the primary biological clock coordinating body activity to a daily cycle. The nightly release of melatonin may help synchronize the biological clock with daily cycles of light and darkness. The physical and mental discomfort associated with jet lag may reflect the time required for melatonin secretion to reset a traveler's daily biological clock to match the period of daylight in a new time zone.

Melatonin also plays a role in other vertebrates. In some fishes, amphibians, and reptiles, melatonin and other hormones produce changes in skin color through their effects on *melanophores,* the pigment-containing cells of the skin. Skin color may vary with the season, the animal's breeding status, or the color of the background.

Study Break

1. What effect does parathyroid hormone have on the body?
2. What hormones are secreted by the adrenal medulla, and what are their functions?
3. What are the two types of hormones secreted by the adrenal cortex, and what are their functions?
4. To what molecular class of hormones do estradiol and progesterone belong, and what are their functions?

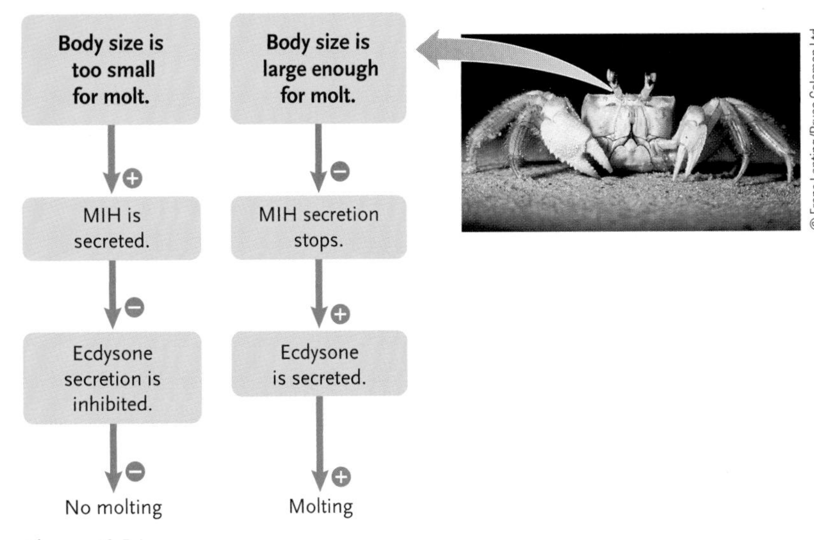

Figure 40.14
Control of molting by molt-inhibiting hormone (MIH), which is secreted by a gland in the eye stalks of crustaceans such as this crab.

40.5 Endocrine Systems in Invertebrates

Invertebrates have fewer hormones, regulating a narrower range of body processes and responses, than vertebrates do. However, in even the simplest animals, such as the cnidarian *Hydra,* hormones produced by neurosecretory neurons control reproduction, growth, and development of some body features. In annelids, arthropods, and mollusks, endocrine cells and glands produce hormones that regulate reproduction, water balance, heart rate, and sugar levels.

Some hormones occur in related forms in invertebrates and vertebrates. For example, both fruit flies and humans have insulin-like hormones and receptors, even though molecular studies suggest that their last common ancestor existed more than 800 million years ago. Both invertebrates and vertebrates secrete peptide and steroid hormones, but most of the hormones have different structures in the two groups, and therefore most have no effect when injected into members of the other group. However, the reaction pathways stimulated by the hormones are the same in both groups, suggesting that these regulatory mechanisms appeared very early in animal evolution.

Hormones Regulate Development in Insects and Crustaceans

Hormones have been studied in detail in only a few invertebrate groups, with the most extensive studies focusing on regulation of metamorphosis in insects. Butterflies, moths, and flies undergo the most dramatic changes as they mature into adults. They hatch from the egg as a caterpillar-like *larva*. During the larval stage, growth is accompanied by one or more *molts,* in which an old exoskeleton is shed and a new one forms. The insects then enter an inactive stage, the *pupa,* in which the body forms a thick, resistant coating, and finally is transformed into an adult.

Three major hormones regulate molting and metamorphosis in insects: **brain hormone** (BH), a peptide hormone secreted by neurosecretory neurons in the brain; **ecdysone** (*ekdysis* = emerging from), a steroid hormone secreted by the *prothoracic glands;* and **juvenile hormone** (JH), a peptide hormone secreted by the *corpora allata,* a pair of glands just behind the brain **(Figure 40.13).** The outcome of the molt depends on the level of JH. If it is high, the molt produces a larger larva; if it is low, the molt leads to pupation and the emergence of the adult.

Hormones that control molting have also been detected in crustaceans, including lobsters, crabs, and crayfish. Before growth reaches the stage at which the exoskeleton is shed, **molt-inhibiting hormone (MIH),** a peptide neurohormone secreted by a gland in the eye stalks, inhibits ecdysone secretion **(Figure 40.14).** As body size increases to the point requiring a molt, MIH secretion is inhibited, ecdysone secretion increases, and the molt is initiated.

In the next chapter we discuss the structure and functions of muscles, and their interactions with the skeletal system to cause movement. Muscle function depends primarily on the action of the nervous system, but the endocrine system plays a role in the control of smooth muscle contraction.

Study Break

How do hormones compare structurally and functionally in invertebrates and vertebrates?

Unanswered Questions

What are the cellular mechanisms for insulin resistance in patients with type 2 diabetes?

Insulin resistance is the condition in which the normal physiological levels of insulin are inadequate to produce a normal insulin response in the body. It plays an important role in the development of type 2 diabetes. Gerald Shulman and his research group at Yale Medical School have a long-term goal of elucidating the cellular mechanisms of insulin resistance. Once the mechanisms are known, therapeutic agents can then be developed to reverse insulin resistance in patients with this type of diabetes. In their research, Shulman's group studies patients with type 2 diabetes as well as transgenic mouse models of insulin resistance.

Recall from the chapter that one of the effects of insulin is the conversion of glucose to glycogen. In one set of experiments, Shulman's group studied the rate of glucose incorporation into muscle glycogen. They discovered that muscle glycogen synthesis plays a major role in causing insulin resistance in patients with type 2 diabetes. More detailed studies showed that defects in insulin-stimulated glucose transport and glucose phosphorylation activity in muscles correlate with the early stages in the onset of type 2 diabetes.

Could the defect in glucose transport and phosphorylation activity be reversed? Shulman's group answered this question in a study of lean offspring of type 2 diabetes parents. The offspring examined were insulin-resistant and synthesized insulin-stimulated muscle glycogen at a level only 50% that of normal individuals but, in contrast with their parents, they showed normal blood glucose levels. The potential for these individuals to develop type 2 diabetes later in life is high; that is, they are considered to be prediabetic. After six weeks of following a four-times-a-week aerobic exercise regime on a StairMaster, their insulin-stimulated muscle glycogen synthesis rates returned to normal due to correction of the glucose transport and glucose phosphorylation defects. Thus, the results suggest that regular aerobic exercise potentially could be useful in reversing insulin resistance in prediabetic individuals such as these offspring and, hence, that it might prevent the development of type 2 diabetes. More research is needed to see if that is the case.

Peter J. Russell

Review

Go to ThomsonNOW™ at www.thomsonedu.com/login to access quizzing, animations, exercises, articles, and personalized homework help.

40.1 Hormones and Their Secretion

- Hormones are substances secreted by cells that control the activities of cells elsewhere in the body. The cells that respond to a hormone are its target cells. The best-known hormones are secreted by the endocrine system.
- The endocrine system includes four major types of cell signaling: classical endocrine signaling, in which endocrine glands secrete hormones; neuroendocrine signaling, in which neurosecretory neurons release neurohormones into the circulatory system; paracrine regulation, in which cells release local regulators that diffuse through the extracellular fluid to regulate nearby cells; and autocrine regulation, in which cells release local regulators that regulate the same cells that produced it (Figure 40.1).
- Most hormones and local regulators fall into one of four molecular classes: amines, peptides, steroids, and fatty acids.
- Many hormones are controlled by negative feedback mechanisms (Figure 40.2).

Animation: Major human endocrine glands

40.2 Mechanisms of Hormone Action

- Hormones typically are effective in very low concentrations in the body fluids because of amplification.
- Hydrophilic hormones bind to receptor proteins embedded in the plasma membrane, activating them. The activated receptors transmit a signal through the plasma membrane, triggering signal transduction pathways that cause a cellular response. Hydrophobic hormones bind to receptors in the cytoplasm or nucleus, activating them. The activated receptors control the expression of specific genes, the products of which cause the cellular response (Figure 40.3).
- As a result of the types of receptors they have, target cells may respond to more than one hormone, or they may respond differently to the same hormone.
- The major endocrine cells and glands of vertebrates are the hypothalamus, pituitary gland, thyroid gland, parathyroid glands, adrenal medulla, adrenal cortex, testes, ovaries, islets of Langerhans of the pancreas, and pineal gland. Hormones are also secreted by endocrine cells in the stomach and intestine, thymus gland, kidneys, and heart. Most body cells are capable of releasing prostaglandins (Figure 40.6).

Animation: Hormones and target cell receptors

40.3 The Hypothalamus and Pituitary

- The hypothalamus and pituitary together regulate many other endocrine cells and glands in the body (Figure 40.7).
- The hypothalamus produces tropic hormones (releasing hormones and inhibiting hormones) that control the secretion of eight hormones by the anterior pituitary: prolactin (PRL), growth hormone (GH), thyroid-stimulating hormone (TSH), adrenocorticotropic hormone (ACTH), follicle-stimulating hormone (FSH), luteinizing hormone (LH), melanocyte-stimulating hormone (MSH), and endorphins.
- The posterior pituitary secretes antidiuretic hormone (ADH), which regulates body water balance, and oxytocin, which stimulates the contraction of smooth muscle in the uterus as a part of childbirth and triggers milk release from the mammary glands during suckling of the young.

Animation: Posterior pituitary function

Animation: Anterior pituitary function

40.4 Other Major Endocrine Glands of Vertebrates

- The thyroid gland secretes the thyroid hormones and, in mammals, calcitonin. The thyroid hormones stimulate the oxidation of carbohydrates and lipids, and coordinate with growth hormone to stimulate body growth and development. Calcitonin lowers the Ca^{2+} level in the blood by inhibiting the release of Ca^{2+} from bone. In amphibians, such as the frog, thyroid hormones trigger metamorphosis (Figure 40.9).
- The parathyroid glands secrete parathyroid hormone, which stimulates bone cells to release Ca^{2+} into the blood. PTH also stimulates the activation of vitamin D, which promotes Ca^{2+} absorption into the blood from the small intestine (Figure 40.10).
- The adrenal medulla secretes epinephrine and norepinephrine, which reinforce the sympathetic nervous system in responding to stress. The adrenal cortex secretes glucocorticoids, which help maintain glucose at normal levels in the blood, and mineralocorticoids, which regulate Na^+ balance and extracellular fluid volume. The adrenal cortex also secretes small amounts of androgens (Figure 40.11).
- The gonadal sex hormones—androgens, estrogen, and progestins—play a major role in regulating the development of reproductive systems, sexual characteristics, and mating behavior.
- The islet of Langerhans cells of the pancreas secrete insulin and glucagon, which together regulate the concentration of fuel substances in the blood. Insulin lowers the concentration of glucose in the blood and inhibits the conversion of noncarbohydrate molecules into glucose. Glucagon raises blood glucose by stimulating glycogen, fat, and protein degradation (Figure 40.12).
- The pineal gland secretes melatonin, which interacts with the hypothalamus to set the body's daily rhythms.

Animation: Parathyroid hormone action

Animation: Hormones and glucose metabolism

40.5 Endocrine Systems in Invertebrates

- Hormones control development and function of the gonads, manage salt and water balance in the body fluids, and control molting in insects and crustaceans.
- Three major hormones—brain hormone (BH), ecdysone, and juvenile hormone (JH)—control molting and metamorphosis in insects. Hormones that control molting are also present in crustaceans (Figures 40.13 and 40.14).

Questions

Self-Test Questions

1. Amine hormones are usually:
   a. hydrophilic when secreted by the thyroid gland.
   b. based on tyrosine.
   c. paracrine but not autocrine.
   d. not transported by the blood.
   e. repelled by the plasma membrane.
2. Prostaglandins would be best described as inducers of:
   a. male and female characteristics.
   b. cell division.
   c. nerve transmission.
   d. smooth muscle contractions.
   e. cell differentiation.
3. When the concentration of thyroid hormone in the blood increases, it:
   a. inhibits TRH secretion by the hypothalamus.
   b. stimulates a secretion by the hypothalamus.
   c. stimulates the pituitary to secrete TRH.
   d. stimulates the pituitary to secrete TSH.
   e. activates a positive feedback loop.
4. Which of the following statements about endocrine targeting and reception is correct?
   a. The idea that one hormone affects one type of tissue is illustrated when epinephrine binds to smooth muscle cells in blood vessels as well as to beta cells in heart muscle.
   b. The idea that one hormone affects one type of tissue is shown when epinephrine cannot activate both the receptors on liver cells and the beta receptors of heart muscle.
   c. The idea that a target cell can respond to more than one hormone is seen when a vertebrate liver cell can respond to insulin and glucagon.
   d. The idea that a minute concentration of hormone can cause widespread effects demonstrates the specificity of cells for certain hormones.
   e. The idea that the response to a hormone is the same among different target cells is shown when different liver cells are activated by insulin.
5. The posterior pituitary secretes:
   a. tropic hormones, which control the hypothalamus.
   b. IGF, which simulates cell division and protein synthesis.
   c. ADH, which increases water absorption by the kidneys.
   d. oxytocin, which controls egg and sperm development.
   e. prolactin, which stimulates milk synthesis.
6. Blood levels of calcium are regulated directly by:
   a. insulin synthesized by the alpha cells of the pancreas.
   b. PTH made by the pituitary.
   c. vitamin D activated in the liver.
   d. prolactin synthesized by the anterior pituitary.
   e. calcitonin secreted by specialized thyroid cells.
7. If the human body is stressed, glucocorticoids:
   a. promote the breakdown of proteins in the muscles and bones.
   b. increase the amount of sodium reabsorbed from urine in the kidneys.
   c. decrease potassium secretion from the kidneys.
   d. decrease glucose uptake by cells in the nervous system.
   e. inhibit the synthesis of glucose from noncarbohydrate sources.
8. When blood glucose rises:
   a. the alpha cells increase glucagon secretion.
   b. the beta cells increase insulin secretion.
   c. in uncontrolled Type I diabetes, urination decreases.
   d. glucagon uses amino acids as an energy source.
   e. target cells decrease their insulin receptors.
9. In mammals:
   a. the suprachiasmatic nucleus of the pineal gland controls both male and female reproductive systems.
   b. estradiol is produced by the hypothalamus to control ovulation.
   c. melatonin controls anabolic steroid production.
   d. GnRH stimulates LH to control testosterone production.
   e. progesterone increases the secretion of LH from the posterior pituitary.

10. Insect development is regulated by:
    a. ecdysone, a peptide secreted by the brain.
    b. juvenile hormone, a peptide secreted by the corpora allata near the brain.
    c. molt-inhibiting hormone, a steroid secreted by the prothoracic glands.
    d. brain hormone, a steroid secreted by the hypothalamus.
    e. melatonin, a peptide secreted by the brain in the larval stage.

Questions for Discussion

1. A physician sees a patient whose symptoms include sluggishness, depression, and intolerance to cold. What disorder do these symptoms suggest?
2. Cushing's syndrome occurs when an individual overproduces cortisol; this rare disorder is also known as hypercortisolism. In children and teenagers, symptoms include extreme weight gain, retarded growth, excess hair growth, acne, high blood pressure, tiredness and weakness, and either very early or late puberty. Adults with the disease may also exhibit extreme weight gain, excess hair growth, and high blood pressure, and in addition may show muscle and bone weakness, moodiness or depression, sleep disorders, and reproductive disorders. Propose some hypotheses for the overproduction of cortisol in individuals with Cushing's syndrome.
3. A 20-year-old woman with a malignant brain tumor has her pineal gland removed. What kinds of side effects might this loss have?
4. In integrated pest management, a farmer uses a variety of tools to combat unwanted insects. These include applications of either hormones or hormone-inhibiting compounds to prevent insects from reproducing successfully. How might each of these hormone-based approaches disrupt reproduction?

Experimental Analysis

The Environmental Protection Agency (EPA) defines endocrine disruptors as chemical substances that can "interfere with the synthesis, secretion, transport, binding, action, or elimination of natural hormones in the body that are responsible for the maintenance of homeostasis (normal cell metabolism), reproduction, development, and/or behavior." The chemicals, sometimes called environmental estrogens, come from both natural and man-made sources. A simple hypothesis is that endocrine disruptors act by mimicking hormones in the body. Many endocrine disruptors affect sex hormone function and, therefore, reproduction.

Examples of endocrine disruptors are the synthetic chemicals DDT (a pesticide) and dioxins, and natural chemicals such as phytoestrogens (estrogen-like molecules in plants), which are found in high levels in soybeans, carrots, oats, onions, beer, and coffee.

Design an experiment to investigate whether a new synthetic chemical (pick your own interesting scenario) is an endocrine disruptor. (Hint: You probably want to work with a model organism.)

Evolution Link

Which endocrine system evolved earlier, endocrine glands or neurosecretory neurons? Support your conclusion with information obtained from online research.

How Would You Vote?

Crop yields that sustain the human population currently depend on agricultural pesticides, some of which may disrupt hormone function in frogs and other untargeted species. Should chemicals that may cause problems remain in use while researchers investigate them? Go to www.thomsonedu.com/login to investigate both sides of the issue and then vote.

Movement in a long-tailed field mouse *(Apodemus sylvaticus)*. Movement of vertebrates occurs as a result of contractions and relaxations of skeletal muscles. When stimulated by the nervous system, actin filaments in the muscles slide over myosin filaments to cause muscle contractions.

G. Delpho/Peter Arnold, Inc.

Study Plan

41.1 Vertebrate Skeletal Muscle: Structure and Function

The striated appearance of skeletal muscle fibers results from a highly organized internal structure

During muscle contraction, thin filaments on each side of a sarcomere slide over thick filaments

The response of a muscle fiber to action potentials ranges from twitches to tetanus

Muscle fibers differ in their rate of contraction and susceptibility to fatigue

Skeletal muscle control is divided among motor units

Invertebrates move using a variety of striated muscles

41.2 Skeletal Systems

A hydrostatic skeleton consists of muscles and fluid

An exoskeleton is a rigid external body covering

An endoskeleton consists of supportive internal body structures such as bones

Bones of the vertebrate endoskeleton are organs with several functions

41.3 Vertebrate Movement: The Interactions between Muscles and Bones

Joints of the vertebrate endoskeleton allow bones to move and rotate

Vertebrates have muscle–bone interactions optimized for specific movements

41 Muscles, Bones, and Body Movements

Why It Matters

A Mexican leaf frog *(Pachymedusa dacnicolor)* sits motionless, its prominent eyes staring into space **(Figure 41.1).** But when the frog detects an approaching cricket, it lunges forward at just the right moment, thrusts out its sticky tongue, and captures the prey. This sequence of events, from the beginning of the movement until the frog's mouth closes, sealing the cricket's fate, requires only 260 milliseconds (ms)—about one quarter of a second. How does the frog move so swiftly, and so surely?

As its prey draws near, neuronal signals travel from the frog's brain to the muscles that extend the frog's hind legs, causing the muscles to contract and propel the frog forward on its forelimbs toward the cricket. Within 50 ms after the jump begins, other signals contract the muscles of the lower jaw, opening the mouth. Then, a muscle on the upper surface of the tongue contracts, which raises the tongue and flips it out of the mouth. As the tongue shoots forward, muscle contractions along the ventral side of the trunk arch the body and direct the head downward toward the prey. Within 80 ms after the lunge begins, the tip of the frog's tongue contacts the cricket. Completion of the lunge folds the tongue—and the cricket—into the frog's mouth, aided by contraction of a muscle

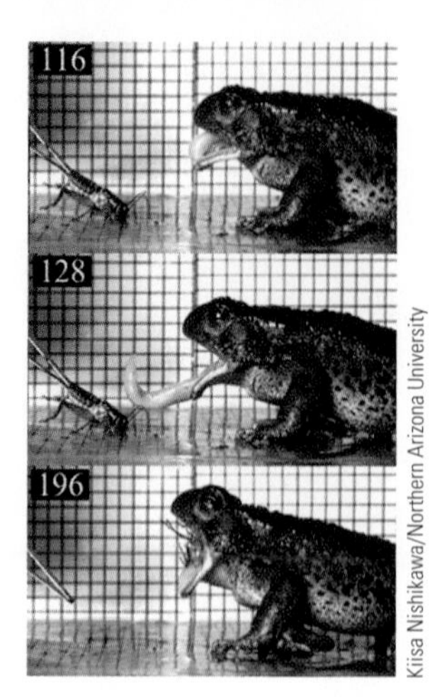

Figure 41.1
A Mexican leaf frog *(Pachymedusa dacnicolor)* capturing a grasshopper.

on the bottom of the tongue. After the mouth closes, further muscle contractions pull the legs forward and fold them under the body.

We know this because Kiisa Nishikawa, Lucie Gray, and James O'Reilly of Northern Arizona University recorded the frog's movements using a high-speed video camera linked to a millisecond timer, with a grid in the background that allowed precise measurement of the distances body parts traveled during the capture. Nishikawa's research group uses the camera's record to study movement in frogs in particular and animals in general.

In Section 36.2 you learned that there are three types of muscle tissue: skeletal, cardiac, and smooth. Skeletal muscle is so named because most muscles of this type are attached by tendons to the skeleton of vertebrates. Cardiac muscle is the contractile muscle of the heart, and smooth muscle is found in the walls of tubes and cavities of the body, including blood vessels and the intestines. In this chapter we describe the structure and function of skeletal muscles, the skeletal systems found in invertebrates and vertebrates, and how muscles bring about movement.

41.1 Vertebrate Skeletal Muscle: Structure and Function

Vertebrate **skeletal muscles** connect to bones of the skeleton. The cells forming skeletal muscles are typically long and cylindrical, and contain many nuclei (shown in Figure 36.6a). Skeletal muscle is controlled by the somatic nervous system.

Most skeletal muscles in humans and other vertebrates are attached at both ends across a joint to bones of the skeleton. (Some, such as those that move the lips, are attached to other muscles or connective tissues under skin.) Depending on its points of attachment, contraction of a single skeletal muscle may extend or bend body parts, or may rotate one body part with respect to another. The human body has more than 600 skeletal muscles, ranging in size from the small muscles that move the eyeballs to the large muscles that move the legs.

Skeletal muscles are attached to bones by cords of connective tissue called *tendons* (see Section 36.2). Tendons vary in length from a few millimeters to some, such as those that connect the muscles of the forearm to the bones of the fingers, that are 20 to 30 cm long.

The Striated Appearance of Skeletal Muscle Fibers Results from a Highly Organized Internal Structure

A skeletal muscle consists of bundles of elongated, cylindrical cells called **muscle fibers,** which are 10 to 100 μm in diameter and run the entire length of the muscle **(Figure 41.2).** Muscle fibers contain many nuclei, reflecting their development by fusion of smaller cells. Some very small muscles, such as some of the muscles of the face, contain only a few hundred muscle fibers; others, such as the larger leg muscles, contain hundreds of thousands. In both cases, the muscle fibers are held in parallel bundles by sheaths of connective tissue that surround them in the muscle and merge with the tendons that connect muscles to bones or other structures. Muscle fibers are richly supplied with nutrients and oxygen by an extensive network of blood vessels that penetrates the muscle tissue.

Muscle fibers are packed with **myofibrils,** cylindrical contractile elements about 1 μm in diameter that run lengthwise inside the cells. Each myofibril consists of a regular arrangement of **thick filaments** (13–18 nm in diameter) and **thin filaments** (5–8 nm in diameter) (see Figure 41.2). The thick and thin filaments alternate with one another in a stacked set.

The thick filaments are parallel bundles of myosin molecules; each myosin molecule consists of two protein subunits that together form a *head* connected to a long double helix forming a *tail.* The head is bent toward the adjacent thin filament to form a *crossbridge.* In vertebrates, each thick filament contains some 200 to 300 myosin molecules and forms as many crossbridges. The thin filaments consist mostly of two linear chains of actin molecules twisted into a double helix, which creates a groove running the length of the molecule. Bound to the actin are *tropomyosin* and *troponin* proteins. Tropomyosin molecules are elongated fibrous proteins that are organized end to end next to the groove of the actin double helix. Troponin is a three-subunit globular protein that binds to tropomyosin at intervals along the thin filaments.

The arrangement of thick and thin filaments forms a pattern of alternating dark bands and light bands, giving skeletal muscle a striated appearance under the microscope (see Figure 41.2). The dark bands, called *A bands,* consist of stacked thick filaments along with the parts of thin filaments that overlap both ends. The lighter-appearing middle region of an A band, which contains only thick filaments, is the *H zone.* In the center of the H zone is a disc of proteins called the *M line,* which holds the stack of thick filaments together. The light bands, called *I bands,* consist of the parts of the thin filaments not in the A band. In the center of each I band is a thin *Z line,* a disc to which the thin filaments are anchored. The region between two adjacent Z lines is a **sarcomere** (*sarco* = flesh; *meros* = segment); sarcomeres are the basic units of contraction in a myofibril.

At each junction of an A band and an I band, the plasma membrane folds into the muscle fiber to form a **T (transverse) tubule (Figure 41.3).** Encircling the sarcomeres is the **sarcoplasmic reticulum,** a complex system of vesicles modified from the smooth endoplasmic reticulum. Segments of the sarcoplasmic re-

Figure 41.2
Skeletal muscle structure. Muscles are composed of bundles of cells called muscle fibers; within each muscle fiber are longitudinal bundles of myofibrils. The unit of contraction within a myofibril, the sarcomere, consists of overlapping myosin thick filaments and actin thin filaments. The myosin molecules in the thick filaments each consist of two subunits organized into a head and a double-helical tail. The actin subunits in the thin filaments form twisted, double helices, with tropomyosin molecules arranged head-to-tail in the groove of the helix and troponin bound to the tropomyosin at intervals along the thin filaments.

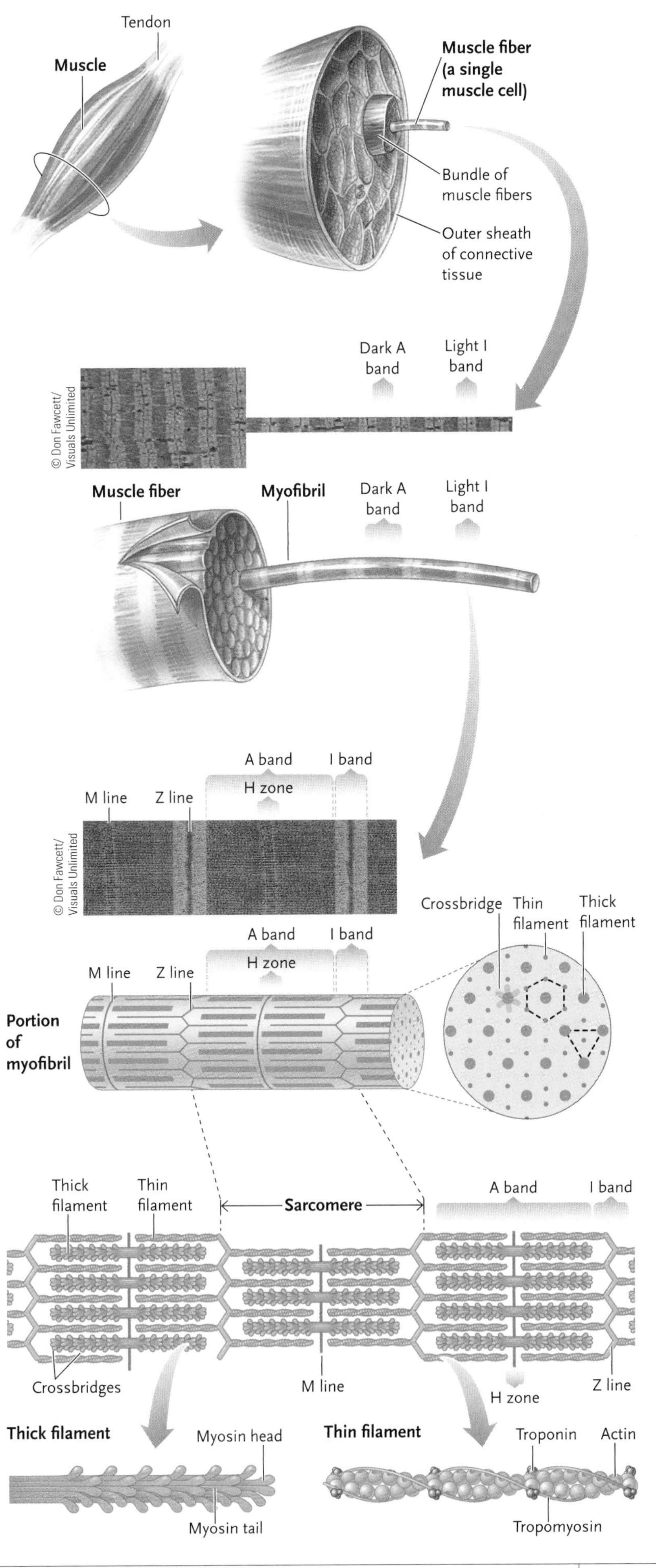

ticulum are wrapped around each A band and I band, and are separated from the T tubules in those regions by small gaps.

An axon of an efferent neuron leads to each muscle fiber. The axon terminal makes a single, broad synapse with a muscle fiber called a **neuromuscular junction** (see Figure 41.3). The neuromuscular junction, T tubules, and sarcoplasmic reticulum are key components in the pathway for stimulating skeletal muscle contraction by neural signals—which starts with action potentials traveling down the efferent neuron—as will be described next.

During Muscle Contraction, Thin Filaments on Each Side of a Sarcomere Slide over Thick Filaments

The precise control of body motions depends on an equally precise control of muscle contraction by a signaling pathway that carries information from nerves to muscle fibers. An action potential arriving at the neuromuscular junction leads to an increase in the concentration of Ca^{2+} in the cytosol of the muscle fiber. The increase in Ca^{2+} triggers a process in which the thin filaments on each side of a sarcomere slide over the thick filaments toward the center of the A band, which brings the Z lines closer together, shortening the sarcomeres and contracting the muscle **(Figure 41.4)**. This *sliding filament mechanism* of muscle contraction depends on dynamic interactions between actin and myosin proteins in the two filament types. That is, the myosin crossbridges make and break contact with actin and pull the thin filaments over the thick filaments—the action is similar to rowing, or a ratcheting process. A model for muscle contraction is shown in **Figure 41.5.**

Conduction of an Action Potential into a Muscle Fiber.

Like neurons, skeletal muscle fibers are *excitable,* meaning that the electrical potential of their plasma membrane can change in response to a stimulus. When an action potential arrives at the neuromuscular junction, the axon terminal releases a neurotransmitter, *acetylcholine,* which triggers an action potential in the muscle fiber (see Figure 41.5, step 1). The action potential travels in all directions over the muscle fiber's

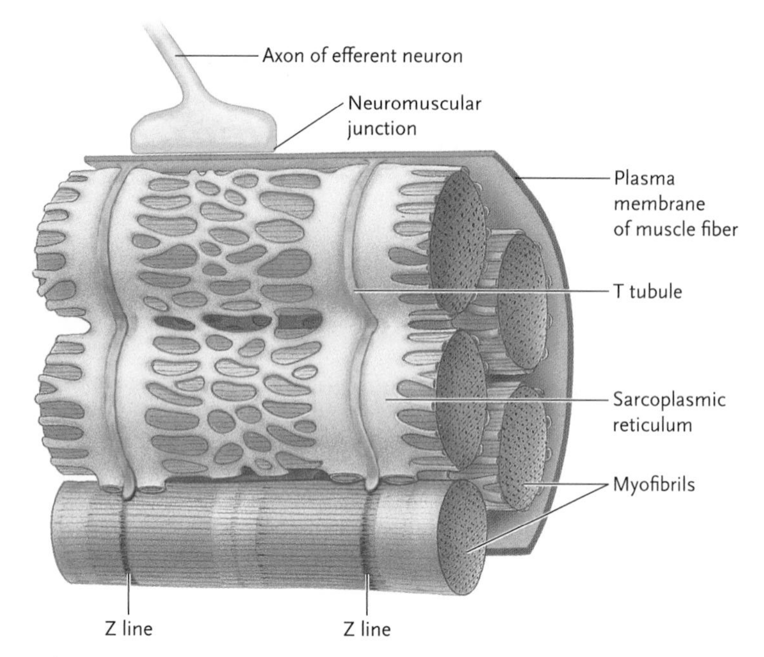

Figure 41.3
Components in the pathway for the stimulation of skeletal muscle contraction by neural signals. T (transverse) tubules are infoldings of the plasma membrane into the muscle fiber originating at each A band–I band junction in a sarcomere. The sarcoplasmic reticulum encircles the sarcomeres and segments of it end in close proximity to the T tubules.

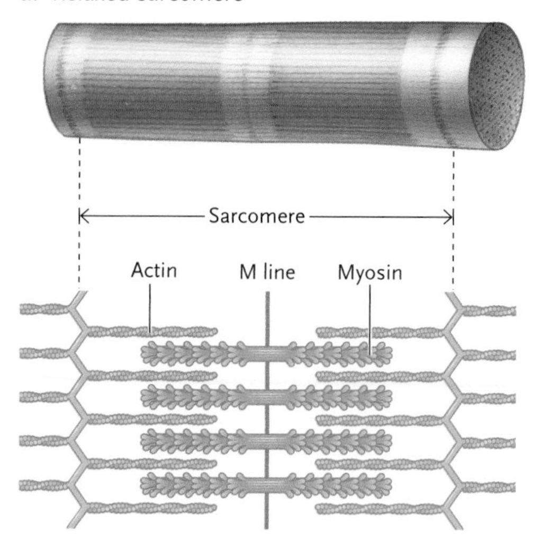

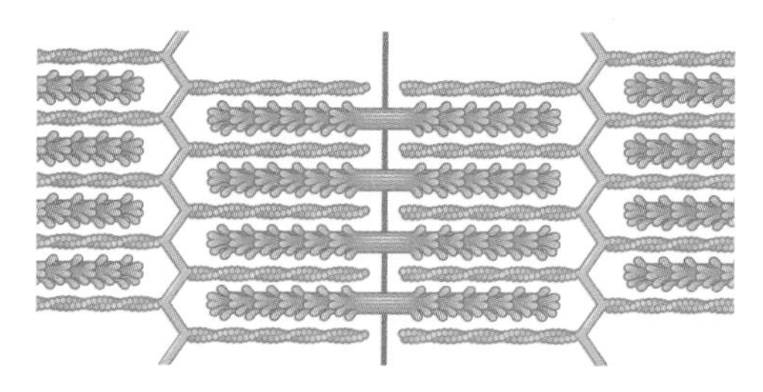

Figure 41.4
Shortening of sarcomeres by the sliding filament mechanism, in which the thin filaments are pulled over the thick filaments.

surface membrane, and also penetrates into the interior of the fiber through the T tubules.

Release of Calcium into the Cytosol of the Muscle Fiber. In the absence of a stimulus, the Ca^{2+} concentration is kept high inside the sarcoplasmic reticulum by active transport proteins that continuously pump Ca^{2+} out of the cytosol and into the sarcoplasmic reticulum. (The active transport proteins are Ca^{2+} pumps, discussed in Section 6.4.) When an action potential reaches the end of a T tubule, it opens ion channels in the sarcoplasmic reticulum that allow Ca^{2+} to flow out into the cytosol (see Figure 41.5, step 2).

When Ca^{2+} flows into the cytosol, the troponin molecules of the thin filament bind the calcium and undergo a conformational change that causes the tropomyosin fibers to slip into the grooves of the actin double helix. The slippage uncovers the actin's binding sites for the myosin crossbridge (see Figure 41.5, step 3). At this point in the process, the myosin crossbridge has a molecule of ATP bound to it, and is not in contact with the thin filament.

The Crossbridge Cycle. Using the energy of ATP hydrolysis, the myosin crossbridge bends away from the tail and binds to a newly exposed myosin crossbridge binding site on an actin molecule (see Figure 41.5, step 4). In effect, this bending compresses a molecular spring in the myosin head. The binding of the crossbridge to actin triggers release of the molecular spring in the crossbridge, which snaps back toward the tail producing the power stroke (motor) that pulls the thin filament over the thick filament (step 5).

The crossbridge now binds another ATP and myosin detaches from actin (see Figure 41.5, step 6). The cycle repeats again, starting with ATP hydrolysis (step 4). Contraction ceases when action potentials stop: Ca^{2+} is pumped back into the sarcoplasmic reticulum, and its effect on troponin is reversed, leading to tropomyosin again blocking myosin crossbridge binding sites on actin. Contraction ceases and the actin thin filaments slide back over the myosin thick filaments to their original relaxed positions (step 7). Crossbridge cycles based on actin and myosin power movements in all living organisms, from cytoplasmic streaming in plant cells and amoebae to muscle contractions in animals.

Although the force produced by a single myosin crossbridge is comparatively small, it is multiplied by the hundreds of crossbridges acting in a single thick filament, and by the billions of thin filaments sliding in a contracting sarcomere. The force, multiplied further by the many sarcomeres and myofibrils in a muscle fiber, is transmitted to the plasma membrane of a muscle fiber by the attachment of myofibrils to elements of the cytoskeleton. From the plasma membrane, it is transmitted to bones and other body parts

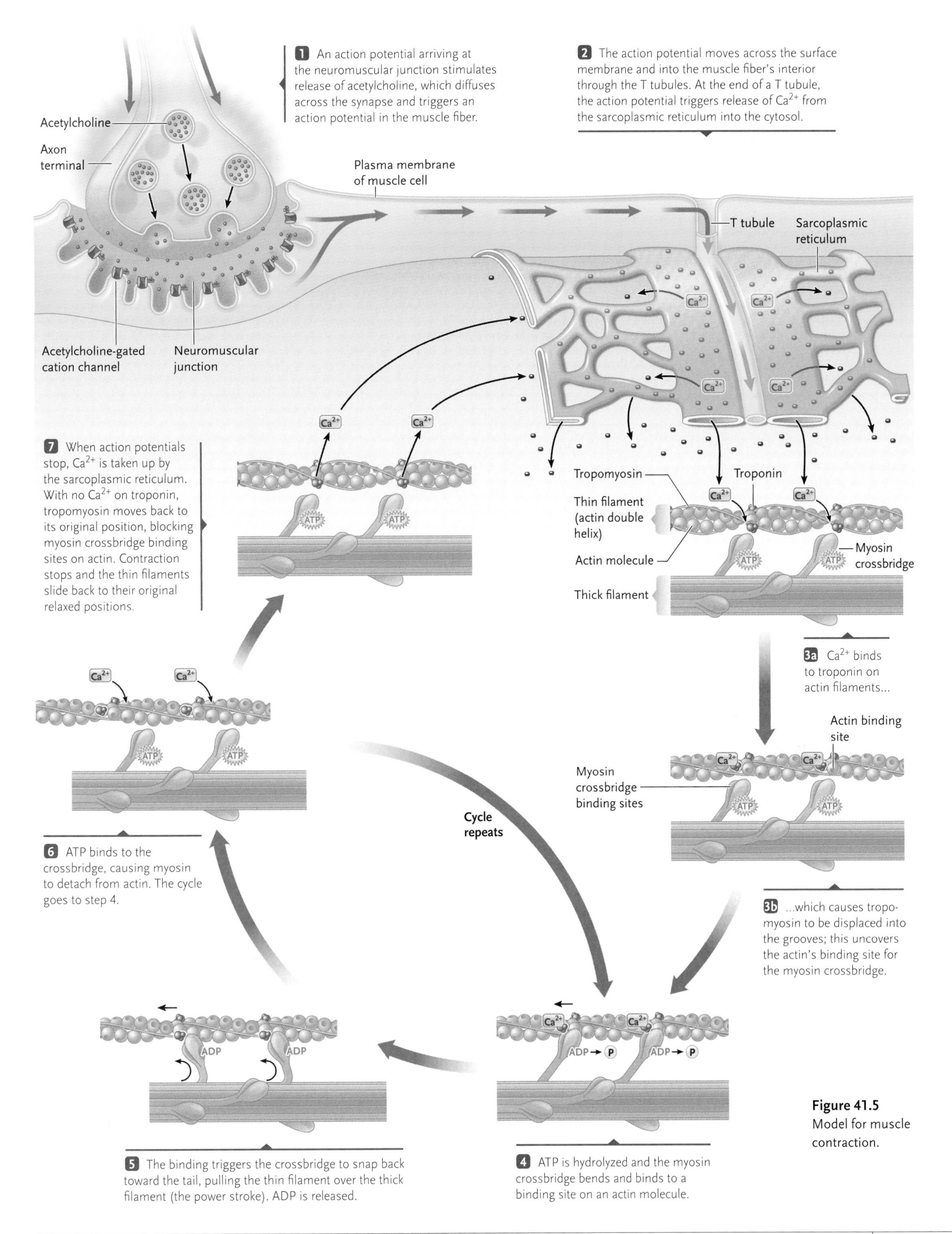

1 An action potential arriving at the neuromuscular junction stimulates release of acetylcholine, which diffuses across the synapse and triggers an action potential in the muscle fiber.
2 The action potential moves across the surface membrane and into the muscle fiber's interior through the T tubules. At the end of a T tubule, the action potential triggers release of Ca^{2+} from the sarcoplasmic reticulum into the cytosol.
Acetylcholine
Axon terminal
Plasma membrane of muscle cell
T tubule
Sarcoplasmic reticulum
Acetylcholine-gated cation channel
Neuromuscular junction
Ca^{2+}
7 When action potentials stop, Ca^{2+} is taken up by the sarcoplasmic reticulum. With no Ca^{2+} on troponin, tropomyosin moves back to its original position, blocking myosin crossbridge binding sites on actin. Contraction stops and the thin filaments slide back to their original relaxed positions.
Tropomyosin
Troponin
Thin filament (actin double helix)
Actin molecule
Myosin crossbridge
Thick filament
ATP
3a Ca^{2+} binds to troponin on actin filaments...
Actin binding site
Myosin crossbridge binding sites
3b ...which causes tropomyosin to be displaced into the grooves; this uncovers the actin's binding site for the myosin crossbridge.
6 ATP binds to the crossbridge, causing myosin to detach from actin. The cycle goes to step 4.
Cycle repeats
ADP
ADP → P
5 The binding triggers the crossbridge to snap back toward the tail, pulling the thin filament over the thick filament (the power stroke). ADP is released.
4 ATP is hydrolyzed and the myosin crossbridge bends and binds to a binding site on an actin molecule.

Figure 41.5
Model for muscle contraction.

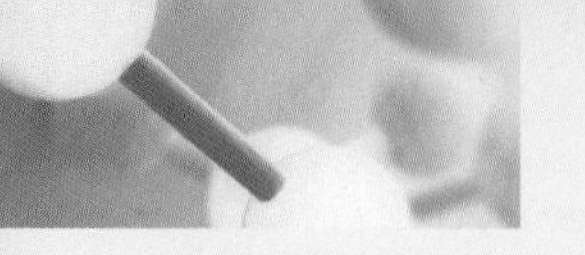

Insights from the Molecular Revolution

A Substitute Player That May Be a Big Winner in Muscular Dystrophy

Duchenne muscular dystrophy (DMD) is an inherited disease, characterized by progressive muscle weakness, that primarily affects males—about 1 out of every 3500 males is born with the disease. When DMD patients are 3 to 5 years old, their muscle tissue begins to break down, and by the time they are in their teens most can walk only with braces. They usually die of complications from degeneration of the heart and diaphragm muscle by their early 20s. Currently, there is no effective treatment for DMD.

The gene that causes DMD, which is located on the X chromosome, was isolated and identified in 1985. In its normal form, the gene encodes the protein *dystrophin*, which anchors a glycoprotein complex in the plasma membrane of a muscle fiber to the underlying actin cytoskeleton (see Section 5.3). In most people with DMD, segments of DNA are missing from the coding sequence of the gene, so the protein cannot function. Without functional dystrophin, the plasma membrane of the muscle fibers is susceptible to tearing during contraction, which leads to muscle destruction. Creatine kinase (CK), an enzyme found predominantly in muscles and in the brain, leaks out of the damaged muscles and accumulates in the blood, which normally contains little CK. Elevated CK in the blood, then, is diagnostic of muscle damage such as that found in DMD.

Many researchers are working to develop a gene-therapy cure for DMD. For example, Kay E. Davies and her colleagues at Oxford University in England have identified a protein that is structurally similar to dystrophin and appears to have a highly similar function. That protein, called *utrophin*, is made in small quantities in muscle fibers and normally functions only in neuromuscular junctions. The utrophin gene and its protein function normally in DMD patients.

The Davies team reasoned that utrophin might be able to substitute for the missing dystrophin in DMD patients if a means could be found to increase its quantity in muscle cells. For their research, they used *mdx* mice, a strain that has the dystrophin gene deleted and is therefore a mouse model of human DMD. First, they introduced an artificial gene (consisting of the mouse utrophin gene under the control of a strong promoter) into fertilized oocytes; the resulting transgenic mice produced much more than the usual amount of utrophin. The researchers were excited to find that CK levels in the blood of the transgenic mice were reduced to 25% of the level in *mdx* mice without the added gene, indicating that muscle damage was markedly decreased. This was confirmed by microscopic examination. Other techniques showed that utrophin, instead of being concentrated in neuromuscular junctions as it is normally, was now distributed throughout the muscle plasma membranes. In short, in these experiments the elevated level of utrophin was able to substitute for dystrophin, and decreased significantly the onset of disease symptoms. Moreover, no deleterious side effects from the overproduction of utrophin could be detected in the genetically engineered mice.

Promising as these results are, germline gene therapy of humans is not allowed, so this approach cannot be used with human patients. Davies's group looked for another way to increase utrophin production, and suggested that upregulating the utrophin gene in all cells of the body could be a strategy to treat DMD. In experiments again using transgenic *mdx* mice, they showed that moderate overproduction of utrophin beginning as late as 10 days after birth caused improvements in muscle appearance compared with controls. Overall, the results show that utrophin overproduction therapy, initiated after birth, can be effective, but that both the timing of therapy and the amount of utrophin expressed are important. Davies's group is now searching for a chemical compound that would increase the levels of utrophin already present in DMD patients. However, much work remains before this can be an effective therapy in humans.

by the connective tissue sheaths surrounding the muscle fibers and by the tendons.

Several mutations affecting muscle and nerve tissues interrupt the transmission of force and cause severe disabilities. Duchenne muscular dystrophy (DMD), for example, is caused by a mutation that weakens the cytoskeleton of the muscle fiber, causing the cells to rupture when contractile forces are generated. *Insights from the Molecular Revolution* describes experiments that may lead to a cure for this debilitating disease.

From Contraction to Relaxation. As long as action potentials continue to arrive at the neuromuscular junction, Ca^{2+} is released in response, and ATP is available, the crossbridge cycle continues to run, shortening the sarcomeres and contracting the muscle fiber.

When action potentials stop, excitation of the T tubules ceases, and the Ca^{2+} release channels in the sarcoplasmic reticulum close. The active transport pumps quickly remove the remaining Ca^{2+} from the cytosol. In response, troponin releases its Ca^{2+} and the tropomyosin fibers are pulled back to cover the myosin binding sites in the thin filaments. The crossbridge cycle stops, and contraction of the muscle fiber ceases. In a muscle fiber that is not contracting, ATP is bound to the myosin head and the crossbridge is not bound to the actin filament (see Figure 41.5, step 7).

Deadly Interruptions of the Crossbridge Cycle. The mechanism controlling vertebrate muscle contraction can be blocked by several toxins and poisons. For example, the bacterium *Clostridium botulinum,* which grows in improperly preserved food, produces a toxin that blocks acetylcholine release in neuromuscular junctions. Many of the body muscles are unable to contract, including the diaphragm, the muscle that is essential for inflating the lungs. As a result, the victim dies from respiratory failure. The toxin is so poisonous that 0.0000001 g is enough to kill a human; 600 g could wipe out the entire human population. This same toxin, under the brand name Botox, is injected in low doses as a cosmetic treatment to remove or reduce wrinkles—if muscles cannot contract, then wrinkles cannot form.

The venom of black widow spiders (genus *Latrodectus*) causes massive release of acetylcholine, leading to convulsive contractions of body muscles; the diaphragm becomes locked in position, causing respiratory failure. Curare, extracted from the bark and sap of some South American trees, blocks acetylcholine from binding to its receptors in muscle fibers. The body muscles, including the diaphragm, become paralyzed and the victim dies of respiratory failure. Some native peoples in South America took advantage of these effects by using curare as an arrow and dart poison.

In a natural process, within a few hours after an animal dies, Ca^{2+} diffuses into the cytoplasm of muscle cells and initiates the crossbridge cycle, producing *rigor mortis,* a strong tension of essentially all the skeletal muscles that stiffens the entire body. As part of rigor mortis, the crossbridges become locked to the thin filaments because ATP production stops (remember that ATP is required to release the crossbridges from actin). The stiffness reverses as actin and myosin are degraded.

The Response of a Muscle Fiber to Action Potentials Ranges from Twitches to Tetanus

A single action potential arriving at a neuromuscular junction usually causes a single, weak contraction of a muscle fiber called a **muscle twitch (Figure 41.6a).** After a muscle twitch begins, the tension of the muscle fiber increases in magnitude for about 30 to 40 ms, and then peaks as the action potential runs its course through the T tubules and the Ca^{2+} channels begin to close. Tension then decreases as the Ca^{2+} ions are pumped back into the sarcoplasmic reticulum, falling to zero in about 50 ms after the peak.

If a muscle fiber is restimulated after it has relaxed completely, a new twitch identical to the first is generated (see Figure 41.6a). However, if a muscle fiber is restimulated before it has relaxed completely, the second twitch is added to the first, producing what is called *twitch summation,* which is basically a summed, stronger contraction **(Figure 41.6b).** And, if action potentials arrive so rapidly (about 25 ms apart) that the fiber cannot relax at all between stimuli, the Ca^{2+} channels remain open continuously and twitch summation produces a peak level of continuous contraction called **tetanus (Figure 41.6c).** (This is not to be confused with the disease of the same name, in which a bacterial toxin causes uncontrolled and con-

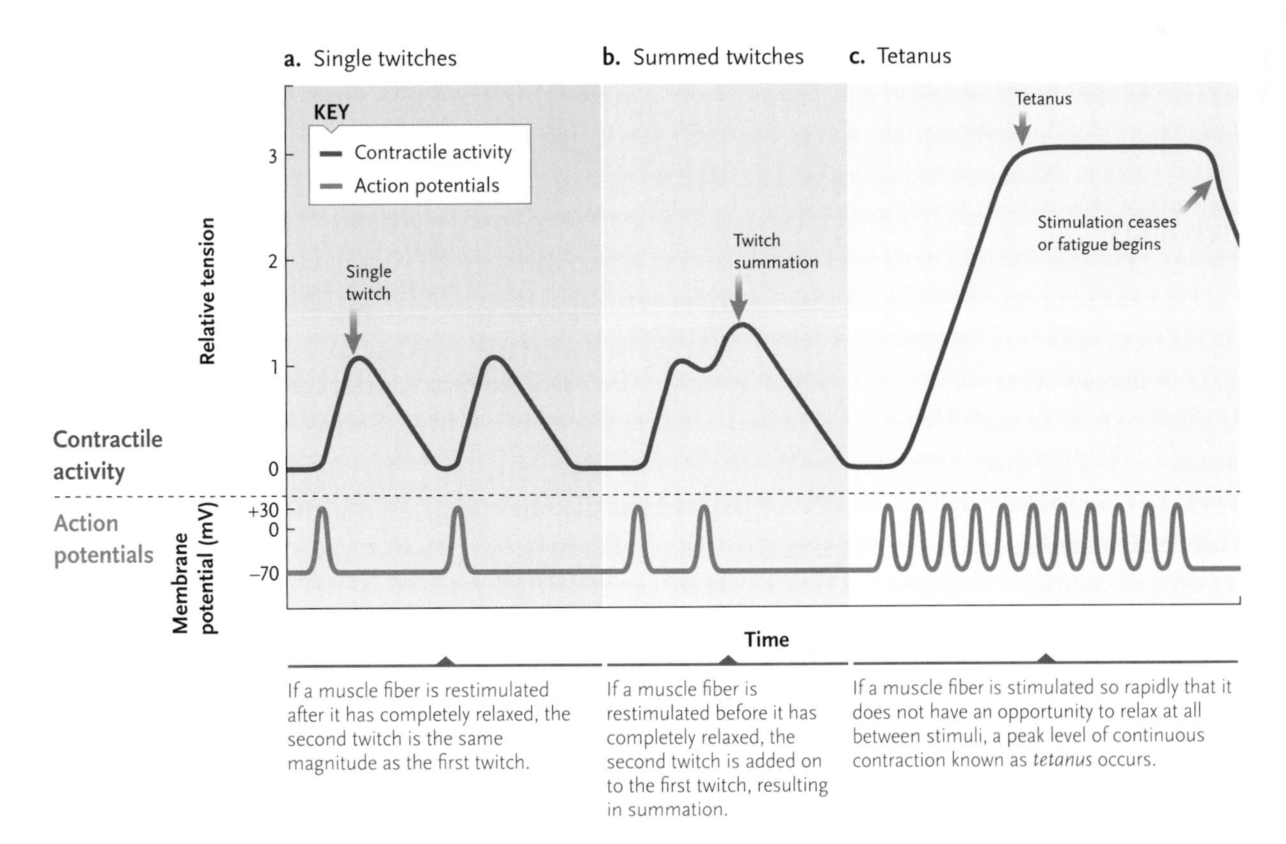
a. Single twitches
b. Summed twitches
c. Tetanus
KEY
Contractile activity
Action potentials
Relative tension
3
2
1
0
Single twitch
Twitch summation
Tetanus
Stimulation ceases or fatigue begins
Contractile activity
Action potentials
Membrane potential (mV)
+30
0
–70
Time
If a muscle fiber is restimulated after it has completely relaxed, the second twitch is the same magnitude as the first twitch.
If a muscle fiber is restimulated before it has completely relaxed, the second twitch is added on to the first twitch, resulting in summation.
If a muscle fiber is stimulated so rapidly that it does not have an opportunity to relax at all between stimuli, a peak level of continuous contraction known as *tetanus* occurs.

Figure 41.6 The relationship of the tension produced in a muscle fiber to the frequency of action potentials.

tinuous muscle contraction.) Contractile activity will then decrease if either the stimuli cease or the muscle fatigues.

Tetanus is an essential part of muscle fiber function. If we lift a moderately heavy weight, for example, many of the muscle fibers in our arms enter tetanus and remain in that state until the weight is released. Even body movements that require relatively little effort, such as standing still but in balance, involve tetanic contractions of some muscle fibers.

Muscle Fibers Differ in Their Rate of Contraction and Susceptibility to Fatigue

Muscle fibers differ in their rate of contraction and resistance to fatigue, and thus can be classified as slow, fast aerobic, and fast anaerobic muscle fibers. Their properties are summarized in **Table 41.1**. The proportions of the three types of muscle fibers tailor the contractile characteristics of each muscle to suit its function within the body.

Slow muscle fibers contract relatively slowly and the intensity of contraction is low because their myosin crossbridges hydrolyze ATP relatively slowly. They can remain contracted for relatively long periods without fatiguing. Slow muscle fibers typically contain many mitochondria and make most of their ATP by oxidative phosphorylation (aerobic respiration). They have a low capacity to make ATP by anaerobic glycolysis. They also contain high concentrations of the oxygen-storing protein **myoglobin**, which greatly enhances their oxygen supplies. Myoglobin is closely related to hemoglobin, the oxygen-carrying protein of red blood cells. Myoglobin gives slow muscle fibers, such as those in the legs of ground birds such as quail, chickens, and ostriches, a deep red color. In sharks and bony fishes, strips of slow muscles concentrated in a band on either side of the body are used for slow, continuous swimming and maintaining body position.

Fast muscle fibers contract relatively quickly and powerfully because their myosin crossbridges hydrolyze ATP faster than those of slow muscle fibers. Fast aerobic fibers have abundant mitochondria, a rich blood supply, and a high concentration of myoglobin, which makes them red in color. They have a high capacity for making ATP by oxidative phosphorylation, and an intermediate capacity for making ATP by anaerobic glycolysis. They fatigue more quickly than slow fibers, but not as quickly as fast anaerobic fibers. Fast aerobic muscle fibers are abundant in the flight muscles of migrating birds such as ducks and geese.

Fast anaerobic fibers typically contain high concentrations of glycogen, relatively few mitochondria, and a more limited blood supply than fast aerobic fibers. They generate ATP mostly by anaerobic respiration (glycolysis) and have a low capacity to produce ATP by oxidative respiration. Fast anaerobic fibers produce especially rapid and powerful contractions but are more susceptible to fatigue. Because their myoglobin supply is limited and they contain few mitochondria, they are pale in color. Some ground birds have flight muscles consisting almost entirely of fast anaerobic muscle fibers. These muscles can produce a short burst of intensive contractions allowing the bird to escape a predator, but they cannot produce sustained flight. Most muscles of lampreys, sharks, fishes, amphibians, and reptiles also contain fast anaerobic muscle fibers, allowing the animals to move quickly to capture prey and avoid danger.

The muscles of humans and other mammals are mixed, and contain different proportions of slow and fast muscle fibers, depending on their functions. Muscles specialized for prolonged, slow contractions, such as the postural muscles of the back, have a high proportion of slow fibers and are a deep red color. The muscles of the forearm that move the fingers have a higher proportion of fast fibers and are a paler red than the back muscles. These muscles can contract rapidly and powerfully, but they fatigue much more rapidly than the back muscles.

The number and proportions of slow and fast muscle fibers in individuals are inherited characteristics. However, particular types of exercise can convert some fast muscle fibers between aerobic and anaerobic types. Endurance training, such as long-distance running, converts fast muscle fibers from the anaerobic to the aerobic type, and regimes such as weight lifting induce the reverse conversion. If the training regimes stop, most of the fast muscle fibers revert to their original types.

Table 41.1 Characteristics of Slow and Fast Muscle Fibers in Skeletal Muscle

	Fiber Type		
Property	**Slow**	**Fast Aerobic**	**Fast Anaerobic**
Contraction speed	Slow	Fast	Fast
Contraction intensity	Low	Intermediate	High
Fatigue resistance	High	Intermediate	Low
Myosin–ATPase activity	Low	High	High
Oxidative phosphorylation capacity	High	High	Low
Enzymes for anaerobic glycolysis	Low	Intermediate	High
Mitochondria	Many	Many	Few
Myoglobin content	High	High	Low
Fiber color	Red	Red	White
Glycogen content	Low	Intermediate	High

Skeletal Muscle Control Is Divided among Motor Units

The control of muscle contraction extends beyond the simple ability to turn the crossbridge cycle on and off. We can adjust a handshake from a gentle squeeze to a strong grasp, or exactly balance a feather or dumbbell in the hand. How are entire muscles controlled in this way? The answer lies in activation of the muscle fibers in blocks called **motor units.**

The muscle fibers in each motor unit are controlled by branches of the axon of a single efferent neuron **(Figure 41.7).** As a result, all those fibers contract each time the neuron fires an action potential. All the muscle fibers in a motor unit are of the same type—either slow, fast aerobic, or fast anaerobic. When a motor unit contracts, its force is distributed throughout the entire muscle because the fibers are dispersed throughout the muscle rather than being concentrated in one segment.

For a delicate movement, only a few efferent neurons carry action potentials to a muscle, and only a few motor units contract. For more powerful movements, more efferent neurons carry action potentials, and more motor units contract.

Muscles that can be precisely and delicately controlled, such as those moving the fingers in humans, have many motor units in a small area, with only a few muscle fibers—about 10 or so—in each unit. Muscles that produce grosser body movements, such as those moving the legs, have fewer motor units in the same volume of muscle but thousands of muscle fibers in each unit. In the calf muscle that raises the heel, for example, most motor units contain nearly 2000 muscle fibers. Other skeletal muscles fall between these extremes, with an average of about 200 muscle fibers per motor unit.

Invertebrates Move Using a Variety of Striated Muscles

Invertebrates also have muscle cells in which actin-based thin filaments and myosin-based thick filaments produce movements by the same sliding mechanism as in vertebrates. Muscles that are clearly striated, which occur in virtually all invertebrates except sponges, have thick and thin filaments arranged in sarcomeres remarkably similar to those of vertebrates, except for variations in sarcomere length and the ratio of thin to thick filaments.

In invertebrates, an entire muscle is typically controlled by one or a few motor neurons. Nevertheless, invertebrate muscles are capable of finely graded contractions because individual neurons make large numbers of synapses with the muscle cells. As action potentials arrive more frequently at the synapses, more Ca^{2+} is released into the cells, and the muscles contract more strongly.

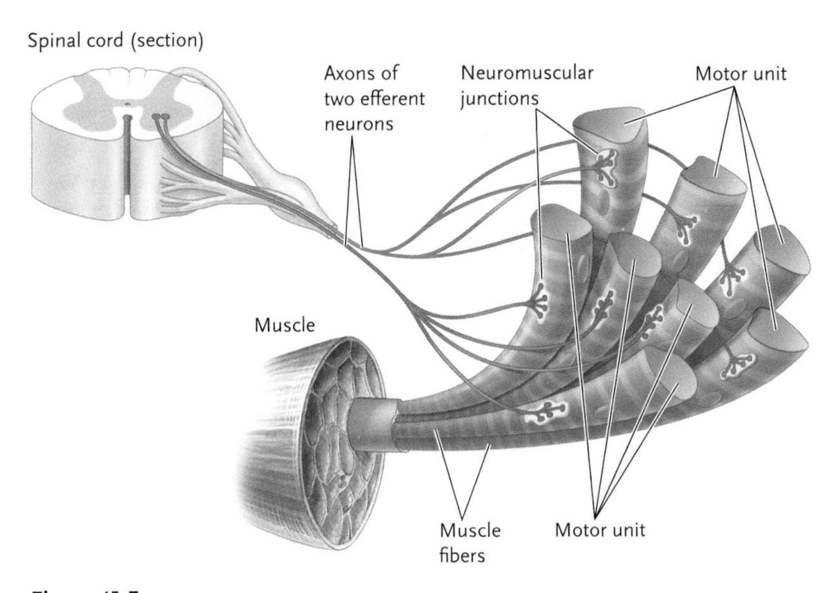

Figure 41.7
Motor units in vertebrate skeletal muscles. Each motor unit consists of groups of muscle fibers activated by branches of a single efferent (motor) neuron.

Study Break

1. Muscle contraction occurs in response to a stimulus from the nervous system. How does this occur?
2. Outline the molecular events that take place in the sliding filament mechanism of muscle contraction.

41.2 Skeletal Systems

Animal skeletal systems provide physical support for the body and protection for the soft tissues. They also act as a framework against which muscles work to move parts of the body or the entire organism. There are three main types of skeletons found in both invertebrates and vertebrates: hydrostatic skeletons, exoskeletons, and endoskeletons.

A Hydrostatic Skeleton Consists of Muscles and Fluid

A **hydrostatic skeleton** (*hydro* = water; *statikos* = causing to stand) is a structure consisting of muscles and fluid that, by themselves, provide support for the animal or part of the animal; no rigid support, like a bone, is involved. A hydrostatic skeleton consists of a body compartment or compartments filled with water or body fluids, which are incompressible liquids. When the muscular walls of the compartment contract, they pressurize the contained fluid. If muscles in one part of the compartment are contracted while muscles in another part are relaxed, the pressurized fluid will

a. Resting position
b. Feeding position
Linda Pitkin/Planet Earth Pictures

Figure 41.8
Sea anemones in (a) the resting and (b) the feeding position. In **(a),** longitudinal muscles in the body wall are contracted, and circular muscles are relaxed. In **(b),** the longitudinal muscles are relaxed, and the circular muscles are contracted. Both sets of muscles work against a hydrostatic skeleton.

move to the relaxed part of the compartment, distending it. In short, the contractions and relaxations of the muscles surrounding the compartments change the shape of the animal.

Hydrostatic skeletons are the primary support systems of cnidarians, flatworms, roundworms, and annelids. In all these animals, compartments containing fluids under pressure make the body semirigid and provide a mechanical support on which muscles act. For example, sea anemones have a hydrostatic skeleton consisting of several fluid-filled body cavities. The body wall contains longitudinal and circular muscles that work against that skeleton. Between meals, longitudinal muscles are contracted (shortened), while the circular ones are relaxed, and the animal looks short and squat **(Figure 41.8a).** It lengthens into its upright feeding position by contracting the circular muscles and relaxing the longitudinal ones **(Figure 41.8b).** In flatworms, roundworms, and annelids, striated muscles in the body wall act on the hydrostatic skeleton to produce creeping, burrowing, or swimming movements. Among these animals, annelids have the most highly developed musculoskeletal systems, with an outer layer of circular muscles surrounding the body, and an inner layer of longitudinal muscles running its length **(Figure 41.9).** Contractions of the circular muscles reduce the diameter of the body and increase the length of the worm; contractions of the longitudinal muscles shorten the body and increase its diameter. Annelids move along a surface or burrow by means of alternating waves of contraction of the two muscle layers that pass along the body, working against the fluid-filled body compartments of the hydrostatic skeleton.

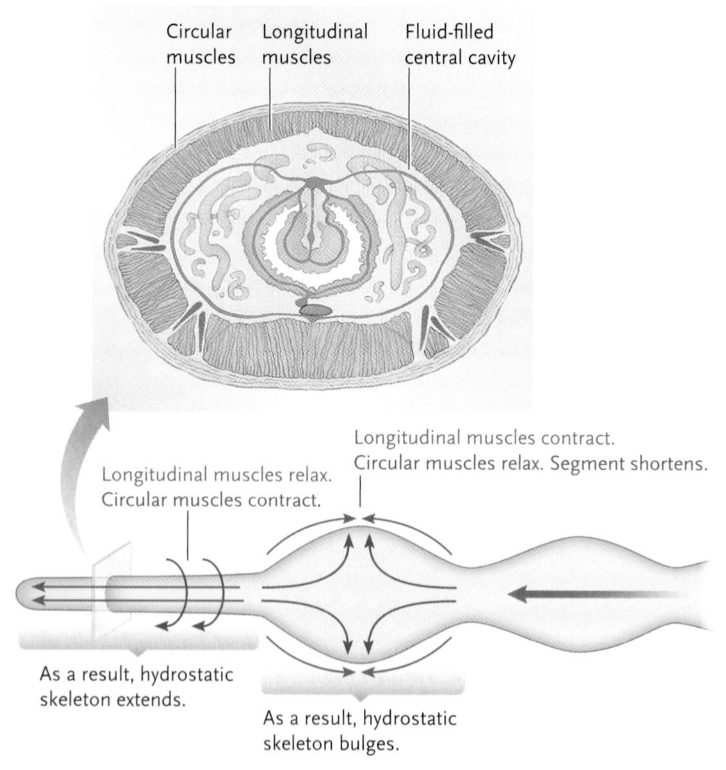
Circular muscles
Longitudinal muscles
Fluid-filled central cavity
Longitudinal muscles relax. Circular muscles contract.
Longitudinal muscles contract. Circular muscles relax. Segment shortens.
As a result, hydrostatic skeleton extends.
As a result, hydrostatic skeleton bulges.

Figure 41.9
Movement of an earthworm, showing how muscles in the body wall act on its hydrostatic skeleton. Contraction of the circular muscles reduce body diameter and increase body length, while contraction of the longitudinal muscles decrease body length and increase body diameter.

Many arthropods have hydrostatic skeletal elements. In the larvae of flying insects, internal fluids held under pressure by the muscular body wall provide some body support. In spiders, the legs are extended from the bent position by muscles exerting pressure against body fluids.

Some structures of echinoderms are supported by hydrostatic skeletons. The tube feet of sea stars and sea urchins, for example, have muscular walls enclosing the fluid of the water vascular system (see Figure 29.46).

In vertebrates, the erectile tissue of the penis is a fluid-filled hydrostatic skeletal structure.

An Exoskeleton Is a Rigid External Body Covering

An **exoskeleton** (*exo* = outside) is a rigid external body covering, such as a shell, that provides support. In an exoskeleton, the force of muscle contraction is applied against that covering. An exoskeleton also protects delicate internal tissues such as the brain and respiratory organs.

Many mollusks, such as clams and oysters, have an exoskeleton consisting of a hard calcium carbonate shell secreted by glands in the mantle. Arthropods, such as insects spiders, and crustaceans, have an external skeleton in the form of a chitinous cuticle, secreted by underlying tissue, that covers the outside surfaces of the animals. Like a suit of armor, the arthropod exoskeleton has movable joints, flexed and extended by muscles that extend across the inside surfaces of the joints **(Figure 41.10).** The exoskeleton protects against dehydration, serves as armor against predators, and provides the levers against which muscles work. In many flying insects, elastic flexing of the exoskeleton contributes to the movements of the wings.

In vertebrates, the shell of a turtle or tortoise is an exoskeletal structure, as are the bony plates, abdominal ribs, collar bones, and most of the skull of the American alligator.

An Endoskeleton Consists of Supportive Internal Body Structures Such as Bones

An **endoskeleton** (*endon* = within) consists of internal body structures, such as bones, that provide support. In an endoskeleton, the force of contraction is applied against those structures. Like exoskeletons, endoskeletons also protect delicate internal tissues such as the brain and respiratory organs.

In mollusks, the mantle of squids and cuttlefish is reinforced by an endoskeletal element commonly called a "pen" (in squid) or the "cuttlebone" in cuttlefish (see Figure 29.22). Squids also have an internal case of cartilage that surrounds and protects the brain; other segments of cartilage support the gills and siphon in squids and octopuses.

Echinoderms have an endoskeleton consisting of *ossicles* (*ossiculum* = little bone), formed from calcium carbonate crystals. The shells of sand dollars and sea urchins are the endoskeletons of these animals.

The endoskeleton is the primary skeletal system of vertebrates. An adult human, for example, has an endoskeleton consisting of 206 bones arranged in two structural groups **(Figure 41.11).** The **axial skeleton,** which includes the skull, vertebral column, sternum, and rib cage, forms the central part of the structure (shaded in red in Figure 41.11). The **appendicular skeleton** (shaded in green) includes the shoulder, hip, leg, and arm bones.

Bones of the Vertebrate Endoskeleton Are Organs with Several Functions

The vertebrate endoskeleton supports and maintains the overall shape of the body and protects key internal organs. In addition, the skeleton is a storehouse for calcium and phosphate ions, releasing them as required to maintain optimal levels of these ions in body fluids. Bones are also sites where new blood cells form.

Bones are complex organs built up from multiple tissues, including bone tissue with cells of several kinds, blood vessels, nerves, and in some, stores of adipose tissue. Bone tissue is distributed between dense, compact bone regions, which have essentially no spaces other than the microscopic canals of the osteons (see Figure 36.5d), and spongy bone regions, which are opened by larger spaces (see Figure 41.11). Compact bone tissue generally forms the outer surfaces of bones, and spongy bone tissue the interior. The interior of some flat bones, such as the hip bones and the ribs, are filled with *red marrow,* a tissue that is the primary source of new red blood cells in mammals and birds. The shaft of long bones such as the femur is opened by a large central canal filled with adipose tissue called *yellow marrow,* which is a source of some white blood cells.

Throughout the life of a vertebrate, calcium and phosphate ions are constantly deposited and withdrawn from bones. Hormonal controls maintain the concentration of Ca^{2+} ions at optimal levels in the blood and extracellular fluids (see Figure 40.10), ensuring that calcium is available for proper functioning of the nervous system, muscular system, and other physiological processes.

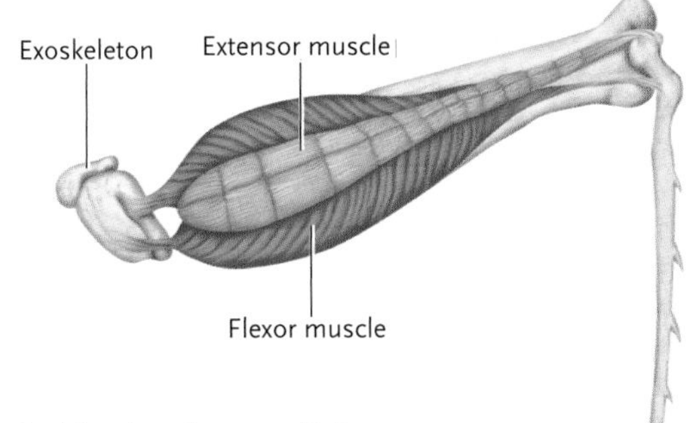

Figure 41.10 Muscles are attached to the inside surfaces of the exoskeleton in a typical insect leg such as this one.

Study Break

1. How do hydrostatic skeletons, exoskeletons, and endoskeletons provide support to the body? Give an example of each of these types in echinoderms and vertebrates.
2. What are the functions of the bones of the vertebrate endoskeleton?

41.3 Vertebrate Movement: The Interactions between Muscles and Bones

The skeletal systems of all animals act as a framework against which muscles work to move parts of the body or the entire organism. In this section, the muscle–bone interactions that are responsible for the movement of vertebrates are described.

Joints of the Vertebrate Endoskeleton Allow Bones to Move and Rotate

The bones of the vertebrate skeleton are connected by joints, many of them movable. The most-movable joints, including those of the shoulders, elbows, wrists, fingers, knees, ankles, and toes, are *synovial joints,* consisting of the ends of two bones enclosed by a fluid-filled capsule of connective tissue **(Figure 41.12a).** Within the joint, the ends of the bones are covered by a smooth layer of cartilage and lubricated by synovial fluid, which makes the bones slide easily as the joint moves. Synovial joints are held together by straps of connective tissue called *ligaments,* which extend across the joints outside the capsule **(Figure 41.12b).** The ligaments restrict the motion of the joint and help prevent it from buckling or twisting under heavy loads.

In other, less movable joints, called *cartilaginous joints,* the ends of bones are covered with layers of cartilage, but have no fluid-filled capsule surrounding them. Fibrous connective tissue covers and connects

Figure 41.11
Major bones of the human body. The inset shows the structure of a limb bone, with the location of red and yellow marrow. The internal spaces lighten the bone's structure. The cartilage layer forms a smooth, slippery cushion between bones in a joint.

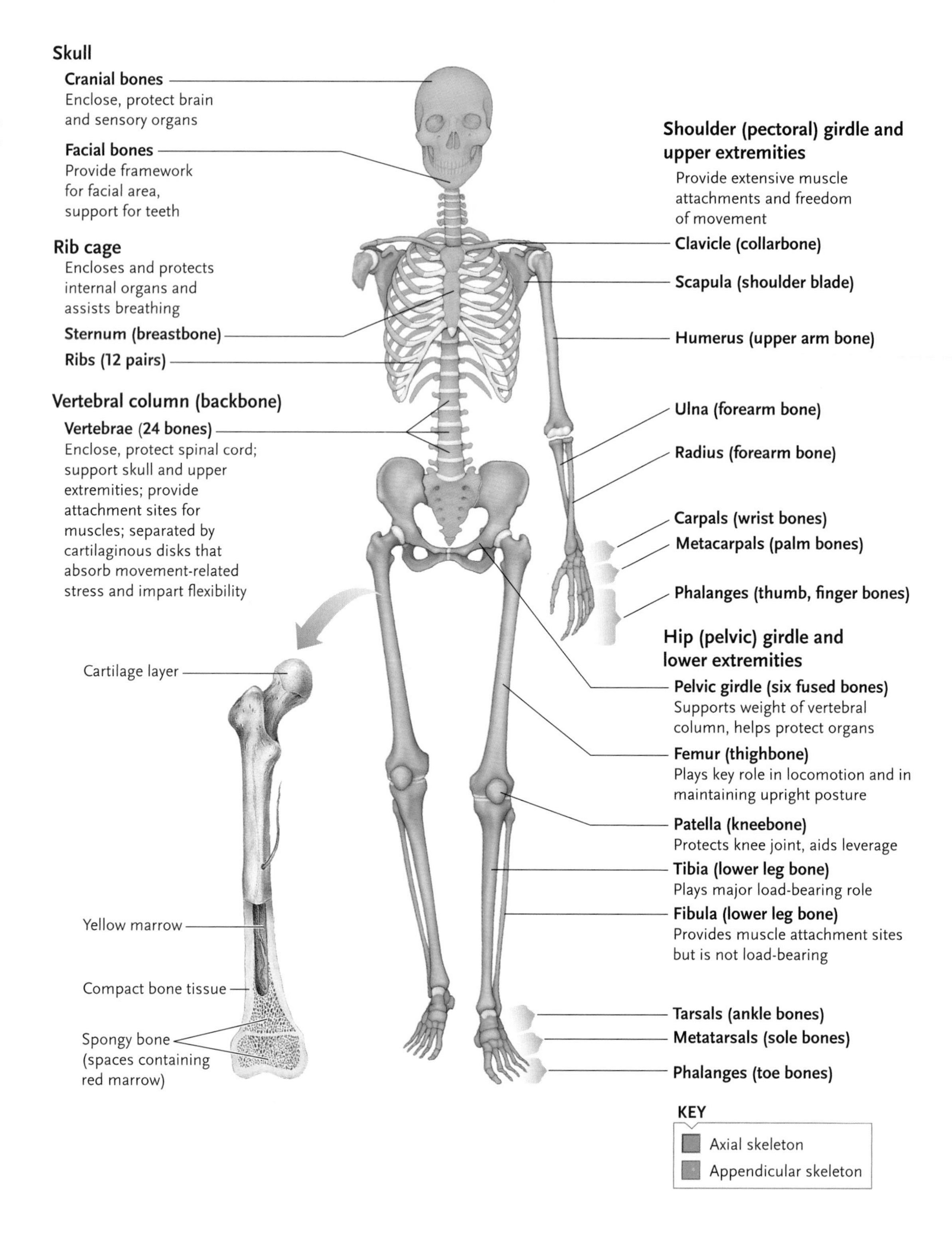

the bones of these joints, which occur between the vertebrae and some rib bones.

In still other joints, called *fibrous joints,* stiff fibers of connective tissue join the bones and allow little or no movement. Fibrous joints occur between the bones of the skull and hold the teeth in their sockets.

The bones connected by movable joints work like levers. A lever is a rigid structure that can move around a pivot point known as a *fulcrum.* Levers differ with respect to where the fulcrum is along the lever and where the force is applied. The most common type of lever system in the body—exemplified by the elbow joint—has the fulcrum at one end, the load at the opposite end, and the force applied at a point between the ends **(Figure 41.13).** For this lever, the force applied must be much greater than the load, but it increases the distance the load moves as compared with the distance over which the force is applied. This allows small muscle movements to produce large body movements, and also allows movements such as running or throwing to be carried out at high speed.

At a joint, a muscle that causes movement in the joint when it contracts is called an **agonist.** In many cases, other muscles that assist the action of an agonist are involved in the movement of a joint. For instance, deltoid and pectoral muscles assist the biceps brachii muscle in lifting a weight.

Most of the bones of vertebrate skeletons are moved by muscles arranged in **antagonistic pairs:** *extensor muscles* extend the joint, meaning increasing the angle between the two bones, while *flexor muscles* do the opposite. (Antagonistic muscles are also used in invertebrates for movement of body parts—for example, the limbs of insects and arthropods.) In humans, one such pair is formed by the biceps brachii muscle at the front of the upper arm and the triceps brachii muscle at the back of the upper arm **(Figure 41.14).** When the biceps muscle contracts, the bone of the lower arm is bent (flexed) around the elbow joint, and the triceps muscle is passively stretched (see Figure 41.14a); when the triceps muscle contracts, the lower arm is straightened (extended) and the biceps muscle is passively stretched (see Figure 41.14b).

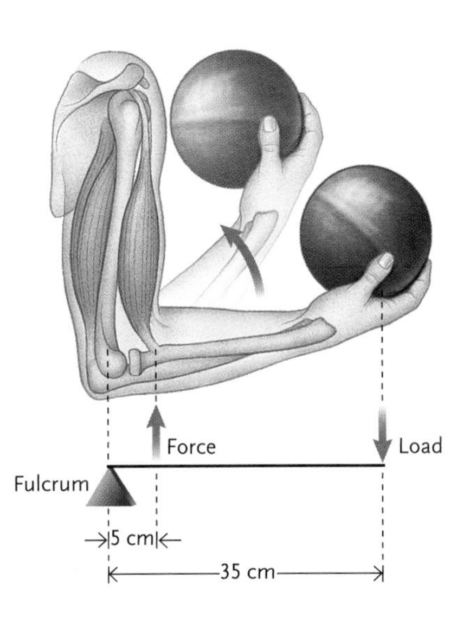

Figure 41.13
A body lever: The lever formed by the bones of the forearm. The fulcrum (the hinge or joint) is at one end of the lever, the load is placed on the opposite end, and the force is exerted at a point on the lever between the fulcrum and the load.

a. Synovial joint cross section

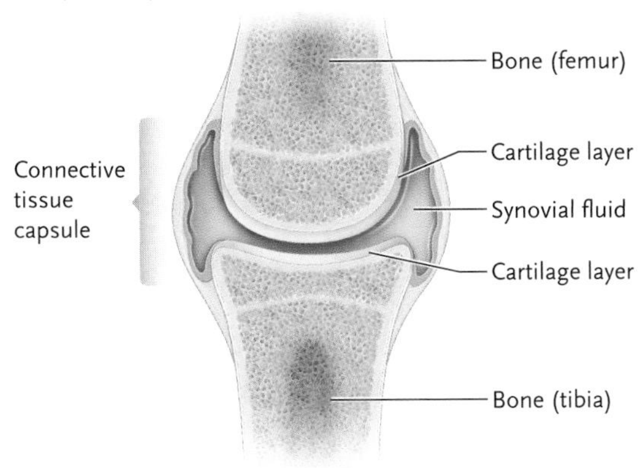

b. Knee joint ligaments

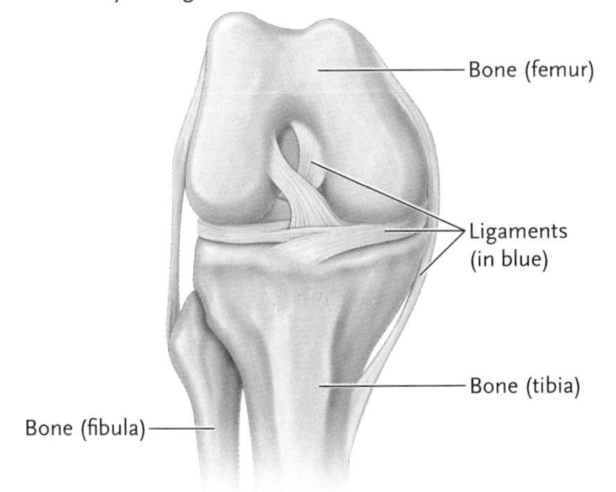

Figure 41.12
A synovial joint. (a) Cross section of a typical synovial joint. **(b)** Ligaments reinforcing the knee joint.

Vertebrates Have Muscle–Bone Interactions Optimized for Specific Movements

Vertebrates differ widely in the patterns by which muscles connect to bones, and in the length and mechanical advantage of the levers produced by these connections. These differences produce limbs and other body parts that are adapted for either power or speed, or the most advantageous compromise between these characteristics. Among burrowing mammals such as the mole, for example, the limb bones are short, thick, and heavy, and the point at which muscles attach produce levers that are slow to move

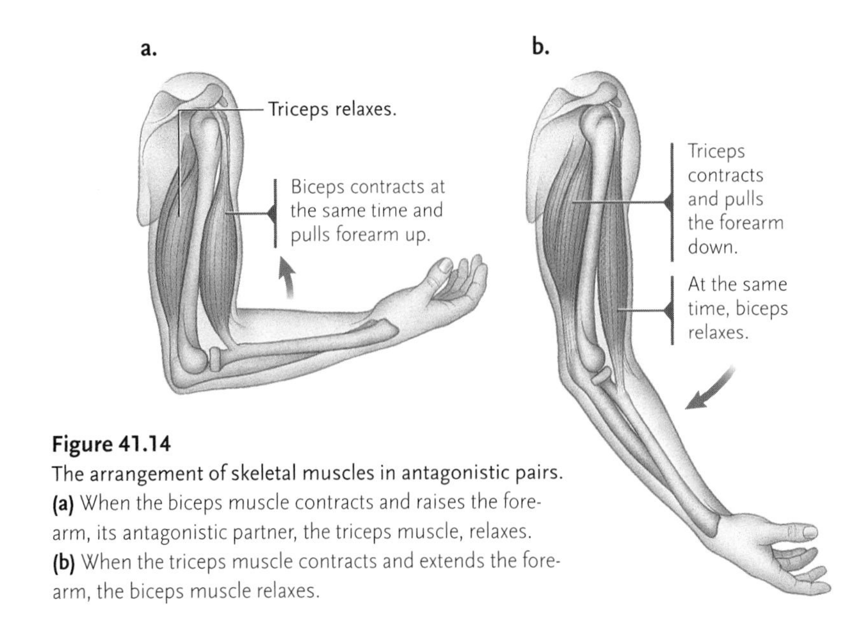

a.
Triceps relaxes.
Biceps contracts at the same time and pulls forearm up.
b.
Triceps contracts and pulls the forearm down.
At the same time, biceps relaxes.

Figure 41.14
The arrangement of skeletal muscles in antagonistic pairs. (a) When the biceps muscle contracts and raises the forearm, its antagonistic partner, the triceps muscle, relaxes. **(b)** When the triceps muscle contracts and extends the forearm, the biceps muscle relaxes.

but that need to apply smaller forces to move a load compared with a human biceps. In contrast, a mammal such as the deer has relatively light and thin bones with muscle attachments producing levers that can produce rapid movement, moving the body easily over the ground.

Study Break

1. Distinguish synovial joints, cartilaginous joints, and fibrous joints.
2. What are antagonistic muscle pairs?

Unanswered Questions

How can muscle growth processes be controlled to improve the clinical treatment of muscular dystrophy and related disorders?
The mechanisms by which different muscle types develop and their impact on organismal metabolism is not well known. However, we do know that one of the proteins produced in muscle cells, myostatin, is a growth factor that inhibits skeletal muscle growth and development. Thus, myostatin is a potential therapeutic target for treating some of the most debilitating types of muscular dystrophy, which is a degenerative and fatal disease associated with the progressive loss of skeletal muscle mass. Animals with mutations that knock out myostatin gene function completely (for example, the Belgian Blue and Piedmontese cattle breeds) or myostatin knockout mice, in which the gene has been removed experimentally (see Section 18.2 and *Focus on Research* in Chapter 43), have significantly enhanced musculature that is commonly referred to as *double muscling*. Such mutations and enhanced skeletal muscle mass have also been described in a racing dog breed, the whippet, and recently in a young boy.

Our laboratory studies focus on developing novel technologies that introduce protein inhibitors of myostatin activity—in essence, inhibiting the inhibitor—and thereby stimulate skeletal muscle growth in both clinical and agricultural settings. We have recently determined that myostatin can also negatively regulate cardiac muscle growth. Thus, disrupting myostatin production or availability may also help heart attack patients. Replacing damaged skeletal and cardiac muscle using adult or embryonic stem cells engineered to match either tissue type is another highly promising technique for treating these disorders; it could be improved by using "antimyostatin" technologies that enhance growth of the transplanted cells.

How much do the metabolic processes of skeletal muscle specifically contribute to energy storage and whole body form?
Complications associated with obesity, particularly type 2 diabetes mellitus, have reached near-epidemic proportions worldwide. Type 2 diabetes differs from type 1 and is caused not by a lack of the pancreatic hormone insulin but rather by insulin resistance, in which an individual's physiological levels of insulin are inadequate to produce a normal insulin response in the tissues. Both types, however, result in the body's inability to properly process and store metabolites, mostly glucose. Type 2 diabetes can be a debilitating and fatal disease if poorly managed and often aggravates other diseases as well. Scientists now recognize that growth and metabolic processes are integrated and controlled by the same hormones, growth factors, and cytokines. Indeed, skeletal muscle is the largest consumer of metabolites and has the greatest potential to impact their circulating levels. Recent studies suggest that increasing muscle mass can significantly reduce fat mass as growing muscle is supported by energy from fat metabolism. Enhancing skeletal muscle growth and/or the ability of the tissue to consume blood metabolites in obese patients with type 2 diabetes could therefore improve treatments for both. The same antimyostatin technologies used to treat muscle growth disorders could also be used to treat severe cases of obesity and type 2 diabetes with the goals of increasing muscle mass, decreasing fat mass, and improving insulin sensitivity.

How do the extremely complex electrical properties of cardiac muscle develop, and how can they be controlled for biomedical purposes?
An ischemic event that blocks blood flow to a region of the heart and ultimately deprives the muscle of oxygen often results in a heart attack and can damage or destroy significant amounts of cardiac muscle. The surviving muscle, however, compensates by increasing the specific force generated by individual myofibers. Scientists have recently determined that these changes are due to the remodeling of electrical properties—changes in the amount and relative distribution as well as the activity of different classes of ion channels—within the surviving muscle itself. Although the specific channels and the mechanisms of regulation are unknown, a better understanding and ultimately control of these processes could help heart attack patients survive.

Buel (Dan) Rodgers is an assistant professor and assistant animal scientist at Washington State University, studying molecular endocrinology and animal genomics, specifically skeletal muscle growth and development. To learn more about his research, visit http://www.ansci.wsu.edu/People/rodgers/faculty.asp.

Review

Go to ThomsonNOW at www.thomsonedu.com/login to access quizzing, animations, exercises, articles, and personalized homework help.

41.1 Vertebrate Skeletal Muscle: Structure and Function

- Skeletal muscles move the joints of the body. They are formed from long, cylindrical cells called muscle fibers, which are packed with myofibrils, contractile elements consisting of myosin thick filaments and actin thin filaments. The two types of filaments are arranged in an overlapping pattern of contractile units called sarcomeres (Figure 41.2).
- Infoldings of the plasma membrane of the muscle fiber form T tubules. The sarcomeres are encircled by the sarcoplasmic reticulum, a system of vesicles with segments separated from T tubules by small gaps (Figure 41.3).

- In the sliding filament mechanism of muscle contraction, the simultaneous sliding of thin filaments on each side of sarcomeres over the thick filaments shortens the sarcomeres and the muscle fibers, producing the force that contracts the muscle (Figure 41.4).
- The sliding motion of thin and thick filaments is produced in response to an action potential arriving at the neuromuscular junction. The action potential causes the release of acetylcholine, which triggers an action potential in the muscle fiber that spreads over its plasma membrane and stimulates the sarcoplasmic reticulum to release Ca^{2+} into the cytosol. The Ca^{2+} combines with troponin, inducing a conformational change that moves tropomyosin away from the myosin-binding sites on thin filaments. Exposure of the sites allows myosin crossbridges to bind and initiate the crossbridge cycle in which the myosin heads of thick filaments attach to a thin filament, pull, and release in cyclic reactions powered by ATP hydrolysis (Figure 41.5).
- When action potentials stop, Ca^{2+} is pumped back into the sarcoplasmic reticulum, leading to Ca^{2+} release from troponin, which allows tropomyosin to cover the myosin-binding sites in the thin filaments, thereby stopping the crossbridge cycle (Figure 41.5).
- A single action potential arriving at a neuromuscular junction causes a muscle twitch. Restimulation of a muscle fiber before it has relaxed completely causes a second twitch, which is added to the first, causing a summed, stronger contraction. Rapid arrival of APs causes the twitches to sum to a peak level of contraction called tetanus. Normally, muscles contract in a tetanic mode (Figure 41.6).
- Muscle fibers occur in three types. Slow muscle fibers contract relatively slowly, but do not fatigue rapidly. Fast aerobic fibers contract relatively quickly and powerfully, and fatigue more quickly than slow fibers. Fast anaerobic fibers can contract more rapidly and powerfully than fast aerobic fibers, but fatigue more rapidly. The fibers differ in their number of mitochondria and capacity to produce ATP (Table 41.1).
- Skeletal muscles are divided into motor units, consisting of a group of muscle fibers activated by branches of a single motor neuron. The total force produced by a skeletal muscle is determined by the number of motor units that are activated (Figure 41.7).
- Invertebrate muscles contain thin and thick filaments arranged in sarcomeres, and contract by the same sliding filament mechanism that operates in vertebrates.

Animation: Structure of skeletal muscle

Animation: Sliding filament model

Animation: Nervous system and muscle contraction

Animation: Troponin and tropomyosin

Animation: Energy sources for contraction

Animation: Types of contractions

41.2 Skeletal Systems

- A hydrostatic skeleton is a structure consisting of a muscle-surrounded compartment or compartments filled with fluid under pressure. Contraction and relaxation of the muscles changes the shape of the animal (Figures 41.8 and 41.9).
- In an exoskeleton, a rigid external covering provides support for the body. The force of muscle contraction is applied against the covering. An exoskeleton can also protect delicate internal tissues (Figure 41.10).
- In an endoskeleton, the body is supported by rigid structures within the body, such as bones. The force of muscle contraction is applied against those structures. Endoskeletons also protect delicate internal tissues. In vertebrates, the endoskeleton is the primary skeletal system. The vertebrate axial skeleton consists of the skull, vertebral column, sternum, and rib cage, while the appendicular skeleton includes the shoulder bones, the forelimbs, the hip bones, and the hind limbs (Figure 41.11).
- Bone tissue is distributed between compact bone, with no spaces except the microscopic canals of the osteons, and spongy bone tissue, which has spaces filled by red or yellow marrow (Figure 41.11).
- Calcium and phosphate ions are constantly exchanged between the blood and bone tissues. The turnover keeps the Ca^{2+} concentration balanced at optimal levels in body fluids.

Animation: Vertebrate skeletons

Animation: Human skeletal system

Animation: Structure of a femur

Animation: Long bone formation

41.3 Vertebrate Movement: The Interactions between Muscles and Bones

- The bones of a skeleton are connected by joints. A synovial joint, the most movable type, consists of a fluid-filled capsule surrounding the ends of the bones forming the joint. A cartilaginous joint, which is less movable, has smooth layers of cartilage between the bones with no surrounding capsule. The bones of a fibrous joint are joined by connective tissue fibers that allow little or no movement (Figure 41.12).
- The bones moved by skeletal muscles act as levers, with a joint at one end forming the fulcrum of the lever, the load at the opposite end, and the force applied by attachment of a muscle at a point between the ends (Figure 41.13).
- At a joint, an agonist muscle, perhaps assisted by other muscles, causes movement. Most skeletal muscles are arranged in antagonistic pairs, in which the members of a pair pull a bone in opposite directions. When one member of the pair contracts, the other member relaxes and is stretched (Figure 41.14).
- Vertebrates have a variety of patterns in which muscles connect to bones, giving different properties to the levers produced. Those properties are specialized for the activities of the animal.

Animation: Opposing muscle action

Animation: Human skeletal muscles

Questions

Self-Test Questions

1. Vertebrate skeletal muscle:
   a. is attached to bone by means of ligaments.
   b. may bend but not extend body parts.
   c. may rotate one body part with respect to another.
   d. is found in the walls of blood vessels and intestines.
   e. is usually attached at each end to the same bone.
2. In a resting muscle fiber:
   a. sarcomeres are regions between two H zones.
   b. discs of M line proteins called the A band separate the thick filaments.
   c. I bands are composed of the same thick filaments seen in the A bands.
   d. Z lines are adjacent to H zones, which attach thick filaments.

   e. dark A bands contain overlapping thick and thin filaments with a central thin H zone composed only of thick filaments.
3. The sliding filament contractile mechanism:
   a. causes thick and thin filaments to slide toward the center of the A band, bringing the Z lines closer together.
   b. is inhibited by the influx of Ca^{2+} into the muscle fiber cytosol.
   c. lengthens the sarcomere to separate the I regions.
   d. depends on the isolation of actin and myosin until a contraction is completed.
   e. uses myosin crossbridges to stimulate delivery of Ca^{2+} to the muscle fiber.
4. During contraction of skeletal muscle:
   a. ATP stimulates Ca^{2+} to move out of the cytosol, which allows tropomyosin to bind myosin causing contraction of the thin filament.
   b. myosin crossbridges use ATP to relax the molecular spring in the myosin head, which pulls the thick filaments away from the thin actin filaments.
   c. actin binds ATP, allowing troponin in the thick filaments to form the myosin crossbridge.
   d. action potentials cause the release of Ca^{2+} into the sarcoplasmic reticulum allowing tropomyosin fibers to uncover the actin binding sites needed for the myosin crossbridge.
   e. botulinum toxin could increase the release of acetylcholine at the contracting muscle site.
5. When a trained marathoner is running, most likely his:
   a. muscles have low concentrations of myoglobin.
   b. slow muscle fibers will do most of the work for the run.
   c. slow muscle fibers will remain in constant tetanus over the length of the run.
   d. fast muscle fibers will be employed in the middle of his run.
   e. slow muscle fibers are using ATP obtained primarily by anaerobic respiration.
6. Which description is characteristic of a motor unit?
   a. A single motor unit's muscle fibers vary among the slow/fast aerobic and slow/fast anaerobic forms.
   b. When receiving an action potential, a motor unit is controlled by a single efferent axon that causes all its fibers to contract.
   c. When a motor unit contracts, certain sections of the muscle as a whole remain relaxed.
   d. If a motor unit controls walking, it is found in large numbers in the same volume of muscle.
   e. If a motor unit controls finger movement, it contains a large number of muscle fibers that are stimulated over a large area.
7. Which of the following is *not* an example of a hydrostatic skeletal structure?
   a. the tube feet of sea urchins
   b. the body wall of annelids
   c. the trunk of an elephant
   d. the body wall of cnidarians
   e. the penis of mammals
8. Endoskeletons:
   a. protect internal organs and provide structures against which the force of muscle contraction can work.
   b. differ from exoskeletons in that endoskeletons do not support the external body.
   c. cannot be found in mollusks and echinoderms.
   d. are composed of appendicular structures that form the skull.
   e. compose the arms and legs, which are part of the axial skeleton.
9. Connecting the vertebrate skeleton are:
   a. nonmovable synovial joints.
   b. ligaments holding together connective tissue of fibrous joints.
   c. cartilaginous joints found in the shoulders and elbows.
   d. synovial joints lubricated by synovial fluid.
   e. fibrous joints that move around a fulcrum allowing small muscle movements to produce large body movements.
10. The movement of vertebrate muscles is:
   a. agonistic when it extends the joint.
   b. antagonistic when it causes movement in the joint.
   c. caused by extensor muscles that flex the joint.
   d. caused by flexor muscles that extend the joint.
   e. most efficient when the biceps and triceps contract simultaneously.

Questions for Discussion

1. A coach must train young athletes for the 100-meter sprint. They need muscles specialized for speed and strength, rather than for endurance. What kinds of muscle characteristics would the training regimen aim to develop? How would it be altered to train marathoners?
2. What kind of exercise program might the coach in question 1 recommend to an older person developing osteoporosis? Why?
3. Based on material in this chapter and in Chapter 40 on endocrine controls, outline some possible causes and physiological effects of calcium deficiency in an active adult.

Experimental Analysis

Design an experiment with rats to determine whether endurance training alters the proportion of slow, fast aerobic, and fast anaerobic muscle fibers.

Evolution Link

What characteristics of vertebrate muscle suggest that the genes for muscle structure were inherited from invertebrate ancestors?

How Would You Vote?

Dietary supplements are largely unregulated. Should they be placed under the jurisdiction of the Food and Drug Administration, which could subject them to more stringent testing for effectiveness and safety? Go to www.thomsonedu.com/login to investigate both sides of the issue and then vote.

Arteries of the human hand (colorized X-ray). Arteries are vessels of the circulatory system that transport molecules, such as oxygen, nutrients, hormones, and wastes, as well as certain cells, from one tissue to another.

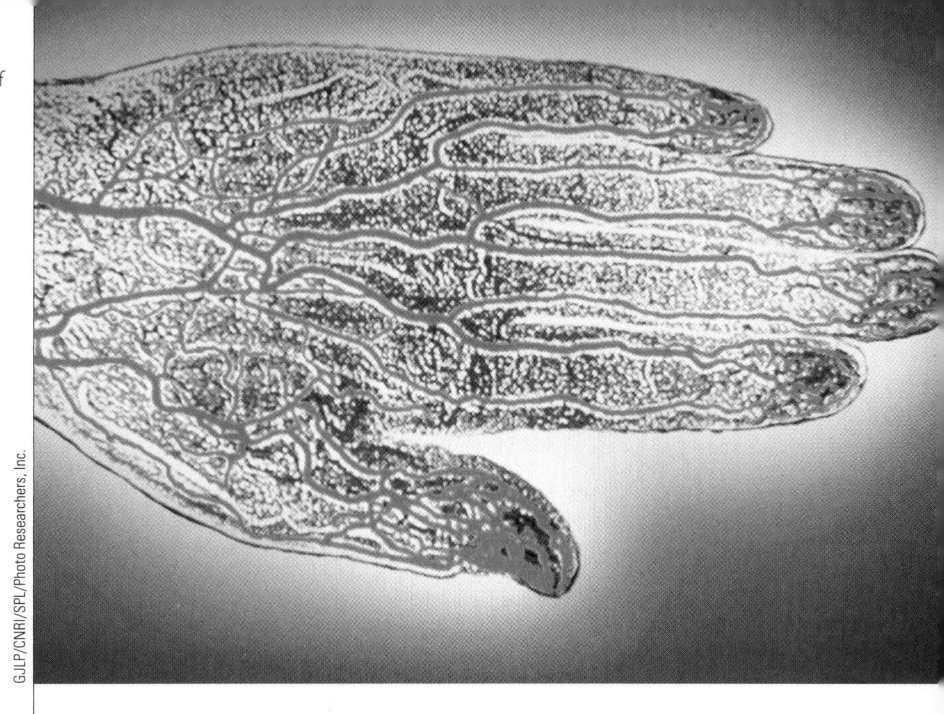

GJLP/CNRI/SPL/Photo Researchers, Inc.

Study Plan

42 The Circulatory System

Why It Matters

Jimmie the bulldog stood on the stage of a demonstration laboratory at a meeting of the Royal Society in London in 1909, with one front paw and one rear paw in laboratory jars containing salt water **(Figure 42.1).** Wires leading from the jars were connected to a galvanometer, a device that can detect electrical currents. Jimmie's master, Dr. Augustus Waller, a physician at St. Mary's Hospital, was relating his experiments in the emerging field of *electrophysiology.* Among other discoveries, Waller had found that his apparatus detected the electrical currents produced each time the dog's heart beat.

Waller originally made his finding by experimenting on himself. He already knew that the heart creates an electrical current as it beats; other scientists had found this out by attaching electrodes directly to the heart of experimental animals. Looking for a painless alternative to that procedure, Waller reasoned that because the human body can conduct electricity, his arms and legs might conduct the currents generated by the heart if they were connected to a galvanometer. Accordingly, Waller set up two metal pans containing salt water and connected wires from the pans to a galvanometer. He put his bare left foot in one of the pans, and his right hand in the other one. The tech-

a. Jimmie the bulldog

b. Electrocardiogram

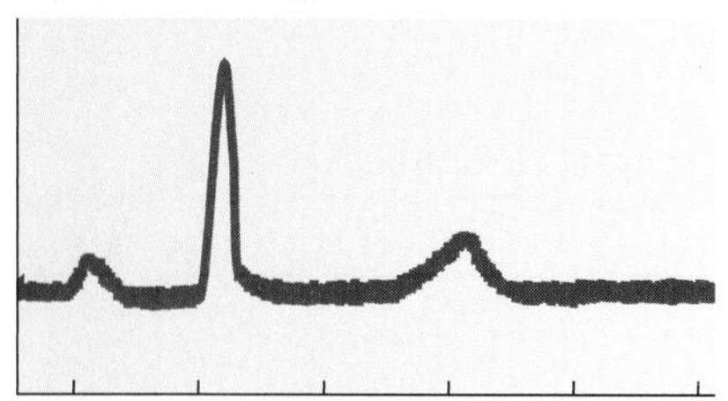

Figure 42.1

The first electrocardiograms. (a) Jimmie the bulldog standing in laboratory jars containing salt water, with wires leading to a galvanometer that recorded the electrical currents produced by his heartbeat. **(b)** One of Waller's early electrocardiograms.

nique worked; the indicator of the galvanometer jumped each time his heart beat. And, it worked with Jimmie.

Waller also invented a method for recording the changes in current, which became the first electrocardiogram (ECG). He constructed a galvanometer by placing a column of mercury in a fine glass tube, with a conducting salt solution layered above the mercury. Changes in the current passing through the tube caused corresponding changes in the surface tension of the mercury, which produced movements that could be detected by reflecting a beam of light from the mercury surface. By placing a moving photographic plate behind the mercury tube, Waller could record the movements of the reflected light on the plate (Figure 42.1b shows one of his records).

The beating of Jimmie's heart, recorded as an electrical trace by Augustus Waller, is part of the actions of the **circulatory system**, an organ system consisting of a fluid, a heart, and vessels for moving important molecules, and often cells, from one tissue to another. Examples of transported molecules are oxygen (O_2), nutrients, hormones, and wastes.

We study these systems in this chapter, with emphasis on the circulatory system of humans and other mammals. We also discuss the **lymphatic system,** an accessory system of vessels and organs that helps balance the fluid content of the blood and surrounding tissues and participates in the body's defenses against invading disease organisms.

42.1 Animal Circulatory Systems: An Introduction

The least complex animals, including sponges, cnidarians, and flatworms, function with no circulatory system. All of these animals are aquatic or, like parasitic flatworms, live surrounded by the body fluids of a host animal. Their bodies are structured as thin sheets of cells that lie close to the fluids of the surrounding environment. Substances diffuse between the cells and the environment through the animal's external surface, or through the surfaces of internal channels and cavities that are open to the environment.

In sponges and cnidarians, flagella or cilia help to circulate the external fluid through the internal spaces. As the water passes over the cells, they pick up O_2 and nutrients, and release wastes and CO_2.

In the sponges, water carrying nutrients and O_2 enters through pores in body walls surrounding a central cavity, passes through the cavity, and leaves through a large exit pore **(Figure 42.2a).** Hydras, jellyfish, sea anemones, and other cnidarians have a central *gastrovascular cavity* with a mouth that opens to the outside and extensions that radiate into the tentacles and all other body regions **(Figure 42.2b).** Water enters and leaves through the mouth, serving both digestion and circulation.

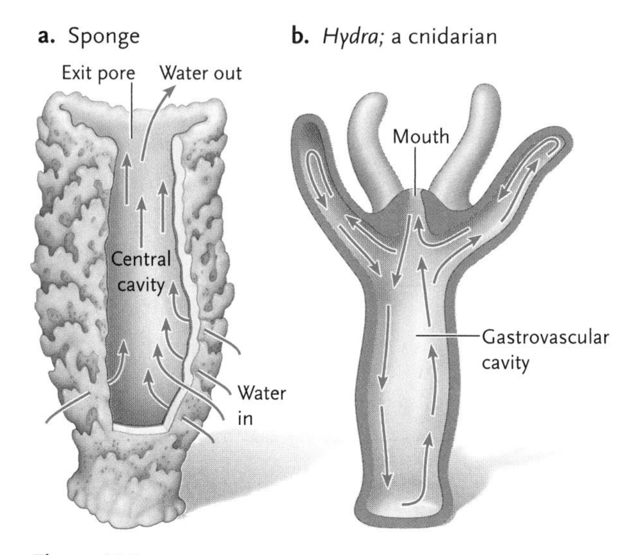

Figure 42.2

Invertebrates with no circulatory system.

Animal Circulatory Systems Share Basic Elements

In larger and more complex animals, most cells lie in cell layers too deep within the body to exchange substances directly with the environment via diffusion. Instead, the animals have a set of tissues and organs—a circulatory system—that conducts O_2, CO_2, nutrients, and wastes between body cells and specialized regions of the animal where substances are exchanged with the external environment. In terrestrial vertebrates, for example, oxygen from the environment is absorbed in the lungs and is carried by the blood to all parts of the body; CO_2 released from body cells is carried by the blood to the lungs, where it is released to the environment. Wastes are conducted from body cells to the kidneys, which remove wastes from the circulation and excrete them into the environment.

The animal circulatory systems carrying out these roles share certain basic features:

1. A specialized fluid medium, such as the blood of mammals and other vertebrates, that carries O_2, CO_2, nutrients, and wastes, and plays a major role in homeostasis (see Section 36.4).
2. A muscular heart that pumps the fluid through the circulatory system.
3. Tubular vessels that distribute the fluid pumped by the heart.

Words associated with the heart often include *cardi(o),* from *kardia,* the Greek word for heart.

Animal circulatory systems take one of two forms, either *open* or *closed* **(Figure 42.3).** In an **open circulatory system,** vessels leaving the heart release bloodlike fluid termed **hemolymph** directly into body spaces, called **sinuses,** that surround organs. Thus there is no distinction between hemolymph and *interstitial fluid,* the fluid immediately surrounding body cells. After flowing through the sinuses, the hemolymph reenters the heart through valves in the heart wall. In a **closed circulatory system,** the fluid—blood—is confined to blood vessels and is distinct from the interstitial fluid. Substances are exchanged between the blood and the interstitial fluid and then between the interstitial fluid and cells.

Most Invertebrates Have Open Circulatory Systems

Arthropods and most mollusks have open circulatory systems with one or more muscular hearts **(Figure 42.4).** In an open system, most of the fluid pressure generated by the heart dissipates when the hemolymph is released into the sinuses. As a result, hemolymph flows relatively slowly. Open systems operate efficiently in these animals because they lead relatively sedentary lives and, as a consequence, their tissues do not require O_2 and nutrients at the rate and quantities required by

a. Open circulatory system: no distinction between hemolymph and interstitial fluid

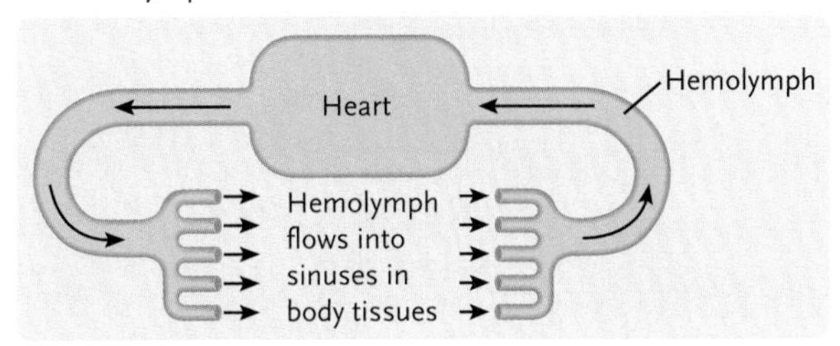

b. Closed circulatory system: blood separated from interstitial fluid

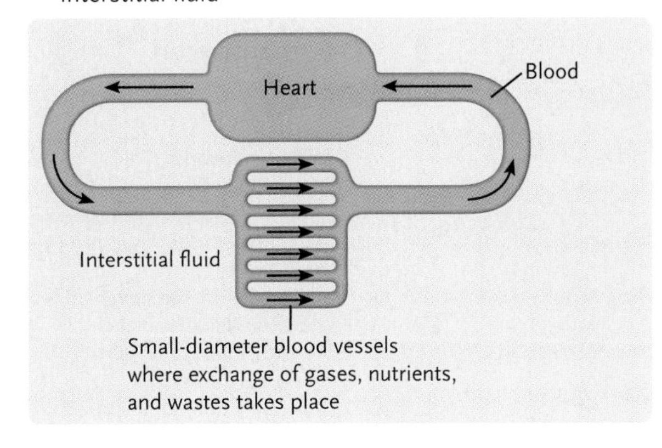

Figure 42.3 Open and closed circulatory systems.

more active species. Among highly mobile and active species with open systems, such as insects and crustaceans, other adaptations compensate for the relatively slow distribution of hemolymph. In insects, for example, O_2 and CO_2 are exchanged efficiently with the environment by specialized air passages that branch throughout the body rather than by the hemolymph (these air passages, called *tracheae,* are discussed in Section 44.2).

Some Invertebrates and All Vertebrates Have Closed Circulatory Systems

Annelids, cephalopod mollusks such as squids and octopuses, and all vertebrates have closed circulatory systems. In these systems, vessels called **arteries** conduct blood away from the heart at relatively high pressure. From the arteries, the blood enters highly branched networks of microscopic, thin-walled vessels called **capillaries** that are well adapted for diffusion of substances. Nutrients and wastes are exchanged between the blood and body tissues as the blood moves through the capillaries. The blood then flows at relatively low

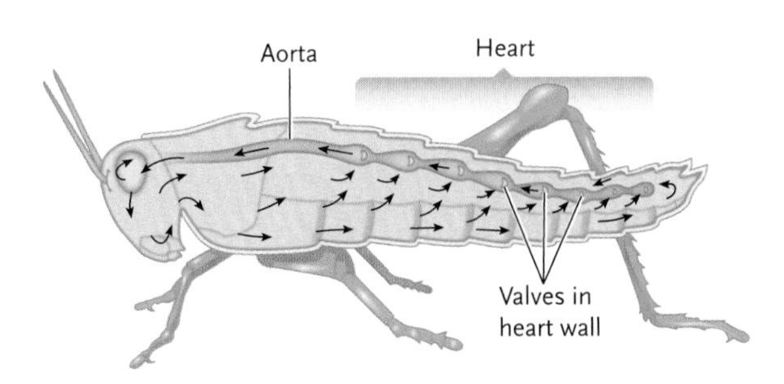

Figure 42.4 Open circulatory system of a grasshopper.

a. Circulatory system of fishes

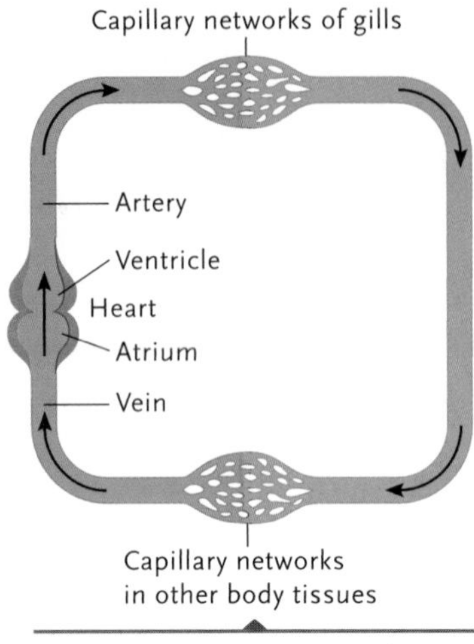

In fishes, a heart consisting of a series of two chambers pumps blood into one circuit. Blood picks up oxygen in the gills and delivers it to the rest of the body. Deoxygenated blood flows back to the heart.

b. Circulatory system of amphibians

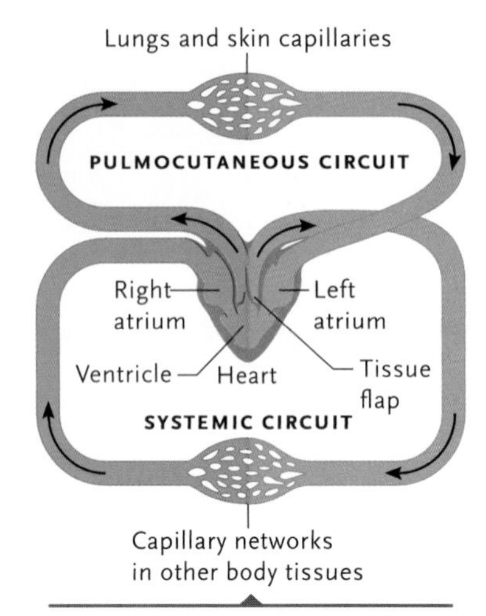

In amphibians, the heart pumps blood through two circuits. Oxygenated blood from the lungs and skin and deoxygenated blood from the rest of the body are kept partially separate by a smooth pattern of flow and a flap of tissue in the large artery leaving the heart.

c. Circulatory system of turtles, lizards, and snakes

Lung capillaries

PULMONARY CIRCUIT

Right atrium

Left atrium

Septum

Ventricles

SYSTEMIC CIRCUIT

Capillary networks in other body tissues

In turtles, lizards, and snakes, a wall of tissue, the septum, improves the separation of oxygenated blood from the lungs and deoxygenated blood from the rest of the body in the single ventricle.

d. Circulatory system of crocodilians, birds, and mammals

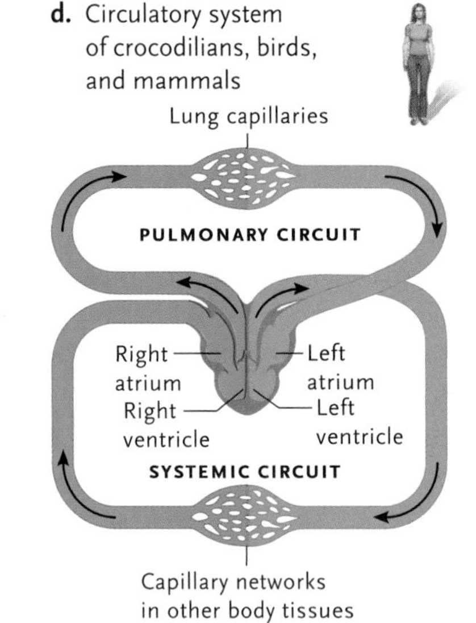

In the four-chambered heart of crocodilians, birds, and mammals, a complete septum forms two ventricles and keeps the flow of oxygenated blood from the lungs and deoxygenated blood entirely separate from the rest of the body.

KEY

Deoxygenated blood

Oxygenated blood

Figure 42.5
Evolutionary developments in the heart and circulatory system of major vertebrate groups.

pressure from the capillaries to larger vessels, the **veins,** which carry the blood back to the heart.

Typically, the blood is maintained at higher pressure and moves more rapidly through the body in closed systems than in open systems. Closed systems are highly efficient in the distribution of O_2 and nutrients and in the clearance of CO_2 and wastes.

In many animals, but particularly the vertebrates, closed systems allow precise control of the distribution and rate of blood flow to different body regions by means of muscles that contract or relax to adjust the diameter of the blood vessels.

Vertebrate Circulatory Systems Have Evolved from Single to Double Blood Circuits

A comparison of the different vertebrate groups reveals several evolutionary trends that accompanied the invasion of terrestrial habitats. Among the most striking is a trend from an effectively single circuit of blood pumped by the heart in sharks and bony fishes to the two completely separate circuits of birds and mammals. As part of this development, the heart evolved from one series of chambers to a double pump acting in two parallel series. In other words, depending on the vertebrate, the heart may consist of one or two **atria** (singular, *atrium*), the chambers that receive blood returning to the heart, and one or two **ventricles,** the chambers that pump blood from the heart.

The Single Blood Circuit of Fishes. In fishes, the heart consists of two chambers arranged in a single line **(Figure 42.5a).** The ventricle of the heart pumps blood into arteries leading to capillary networks in the gills, where the blood releases CO_2 and picks up O_2. The oxygenated blood—now moving more slowly and at lower pressure after passing through the gill capillaries—flows through another series of arteries and is delivered to capillary networks in other body tissues where it delivers O_2 and picks up CO_2. The comparatively slow movement of the blood through this segment of the circulatory system is accelerated by the contractions of skeletal muscles as the animal swims. In some fishes, a vein in the tail is expanded and surrounded by a mass of skeletal muscle that contracts rhythmically, forming an accessory *caudal heart,* which pumps venous blood toward the anterior end of the body. Eventually, the deoxygenated blood enters veins that carry it back to the atrium of the heart, and the single circuit is repeated.

The circulatory system of fishes is highly suited to the environments in which these animals live. Their bodies, supported by water and adapted for swimming, do not use as much energy in locomotion as do animals of an equivalent size moving on the land or in the air. Hence, fishes require less O_2 for their activities than terrestrial vertebrates do, and their relatively simple circulatory systems fully meet their O_2 requirements.

Double Blood Circuits. Vertebrate hearts changed significantly when the first air-breathing fishes such as the lungfish evolved. In these fishes, the lung evolved as a respiratory organ in addition to gills. The lung necessitated a separate circuit because it is an additional organ for oxygenating the blood and, unlike other organs, does not need to receive oxygenated blood from the gills.

In amphibians, the separation into two circuits was accomplished by division of the atrium into two parallel chambers, the left and right atria, to produce a three-chambered heart, and by adaptations that keep oxygenated and deoxygenated blood partially separate as they are pumped by the single ventricle **(Figure 42.5b)**. Amphibians obtain O_2 from gas exchange across their moist skin as well as in their gills or lungs. Oxygenated blood from these organs enters veins that lead to the left atrium, while O_2-depleted blood from the rest of the body enters the right atrium. The atria contract simultaneously, pumping the oxygenated and deoxygenated blood into the single ventricle. The two types of blood remain mostly separated because they differ in density, although some mixing occurs. As the blood is pumped from the ventricle, most of the oxygenated blood enters the branch that leads to the **systemic circuit** of the body, which provides the blood supply for most of the tissues and cells of the body. The deoxygenated blood is directed into the other branch, which leads to the skin and lungs or gills, called the **pulmocutaneous circuit.** Because the blood flows through separate systemic and pulmocutaneous circuits in the amphibians, the blood leaving the heart flows through only one capillary network in each circuit before returning to the heart. This separation greatly increases the blood pressure and flow in the systemic circuit as compared with that of fishes.

Reptiles also have two atria and a single ventricle. The ventricle is divided into right and left halves by a flap of connective tissue called the *septum,* which keeps the flow of oxygenated and deoxygenated blood almost completely separate in a systemic circuit and a **pulmonary circuit,** a branch leading to the lungs **(Figure 42.5c)**. The septum is incomplete in turtles, lizards, and snakes, allowing some mixing of oxygenated and deoxygenated blood. In crocodilians (crocodiles and alligators), the septum is complete **(Figure 42.5d)**.

Birds (which share ancestry with crocodilians) and mammals have a double heart consisting of two atria and two ventricles (as in Figure 42.5d). In effect, each half of the heart operates as a separate pump, restricting the blood circulation to completely separate pulmonary and systemic circuits. Blood is pumped by a ventricle in each circuit, so that both operate at relatively high pressure.

Study Break

1. What are the three basic features of animal circulatory systems?
2. Distinguish between an open circulatory system and a closed circulatory system. Why do you think humans could not function with an open circulatory system?
3. Distinguish among atria, ventricles, arteries, and veins.
4. Distinguish among the systemic, pulmocutaneous, and pulmonary circuits.

42.2 Blood and Its Components

In vertebrates, blood is a complex connective tissue containing blood cells suspended in a liquid matrix called the *plasma.* In addition to transporting molecules, blood helps stabilize the internal pH and salt composition of body fluids, and serves as a highway for cells of the immune system and the antibodies produced by some of these cells. It also helps regulate body temperature by transferring heat between warmer and cooler body regions, and between the body and the external environment (the role of the blood in temperature regulation is discussed in Chapter 46).

For an average-sized adult human, the total blood volume is about 4 to 5 liters—more than a gallon—and makes up about 8% of body weight. The *plasma,* a clear, straw-colored fluid, makes up about 55% of the volume of blood in human males and 58% in human females on average. Suspended in the plasma are three main types of blood cells—*erythrocytes, leukocytes,* and *platelets*—which make up the remainder of the blood volume; on average about 45% and 42% for human males and females, respectively. The typical components of human blood are shown in **Figure 42.6.**

In humans, blood cells develop in red bone marrow primarily in the vertebrae, sternum (breastbone), ribs, and pelvis. Blood cells originate in a single type of cells called *pluripotent* (*plura* = multiple; *potens* = power) *stem cells,* which retain the embryonic capacity to divide **(Figure 42.7)**. Pluripotent stem cells differentiate into two other types of stem cells, myeloid stem cells and lymphoid stem cells. The myeloid stem cells give rise to erythrocytes, platelets, and several types of leukocytes, namely neutrophils, basophils, eosinophils, and monocyte/macrophages. The lymphoid stem cells give rise to other types of leukocytes, namely the natural killer cells, T lymphocytes, and B lymphocytes.

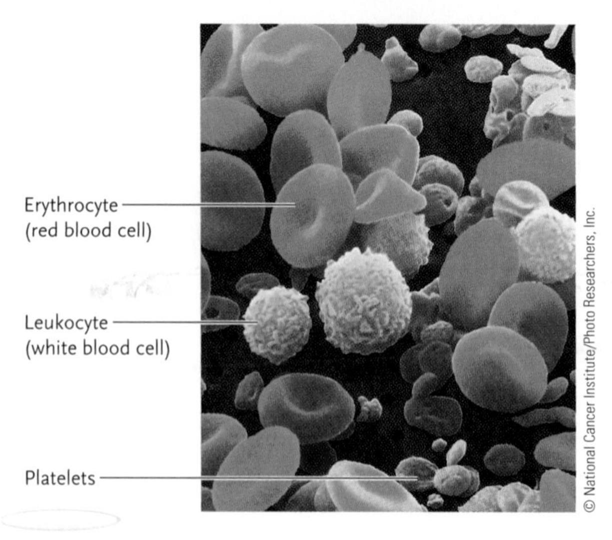

Figure 42.6
Typical components of human blood. The colorized scanning electron micrograph shows the three major cellular components. The sketch of the test tube shows what happens when you centrifuge a blood sample. The blood separates into three layers: a thick layer of straw-colored plasma on top, a thin layer containing leukocytes and platelets, and a thick layer of erythrocytes. The table shows the relative amounts and functions of the various components of blood.

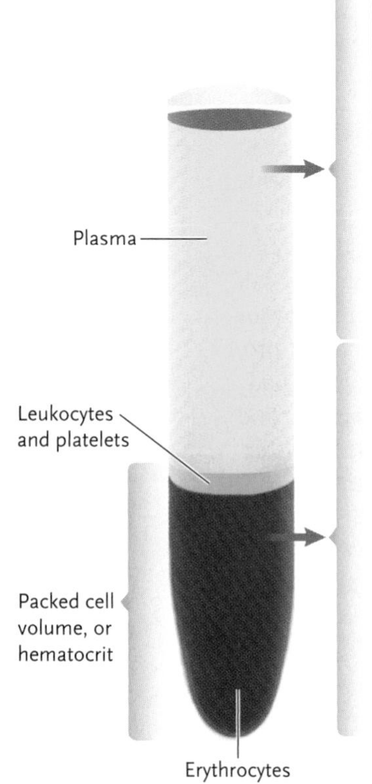

Components	Relative Amounts	Functions
Plasma Portion (55%–58% of total volume):		
1. Water	91%–92% of plasma volume	Solvent
2. Plasma proteins (albumin, globulins, fibrinogen, etc.)	7%–8%	Defense, clotting, lipid transport, roles in extracellular fluid volume, etc.
3. Ions, sugars, lipids, amino acids, hormones, vitamins, dissolved gases	1%–2%	Roles in extracellular fluid volume, pH, etc.
Cellular Portion (45%–42% of total volume):		
1. Erythrocytes (red blood cells)	4,800,000–5,400,000 per microliter	Oxygen, carbon dioxide transport
2. Leukocytes (white blood cells)		
Neutrophils	3,000–6,750	Phagocytosis during inflammation
Lymphocytes	1,000–2,700	Immune response
Monocytes/macrophages	150–720	Phagocytosis in all defense responses
Eosinophils	100–360	Defense against parasitic worms
Basophils	25–90	Secrete substances for inflammatory response and for fat removal from blood
3. Platelets	250,000–300,000	Roles in clotting

Plasma Is an Aqueous Solution of Proteins, Ions, Nutrient Molecules, and Gases

Plasma is so complex that its complete composition is unknown. Among its known components are water (91%–92% of its volume), glucose and other sugars, amino acids, plasma proteins, dissolved gases (mostly O_2, CO_2, and nitrogen), ions, lipids, vitamins, hormones and other signal molecules, and metabolic wastes, including urea and uric acid.

The plasma proteins fall into three classes, the *albumins,* the *globulins,* and *fibrinogen.* The **albumins,** the most abundant proteins of the plasma, are important for osmotic balance and pH buffering. They also transport a wide variety of substances through the circulatory system, including hormones, therapeutic drugs, and metabolic wastes. The **globulins** transport lipids (including cholesterol) and fat-soluble vitamins; a specialized subgroup of globulins, the *immunoglobulins,* constitute antibodies and other molecules contributing to the immune response. Some globulins are also enzymes. **Fibrinogen** plays a central role in the mechanism clotting the blood.

The ions of the plasma include Na^+, K^+, Ca^{2+}, Cl^-, and HCO_3^- (bicarbonate) ions. The Na^+ and Cl^- ions—the components of common table salt—are the most abundant ions. Some of the ions, particularly the bicarbonate ion, help maintain arterial blood at its characteristic pH, which in humans is slightly on the basic side at pH 7.4 (the bicarbonate ion and its role in pH balance are discussed further in Chapter 44).

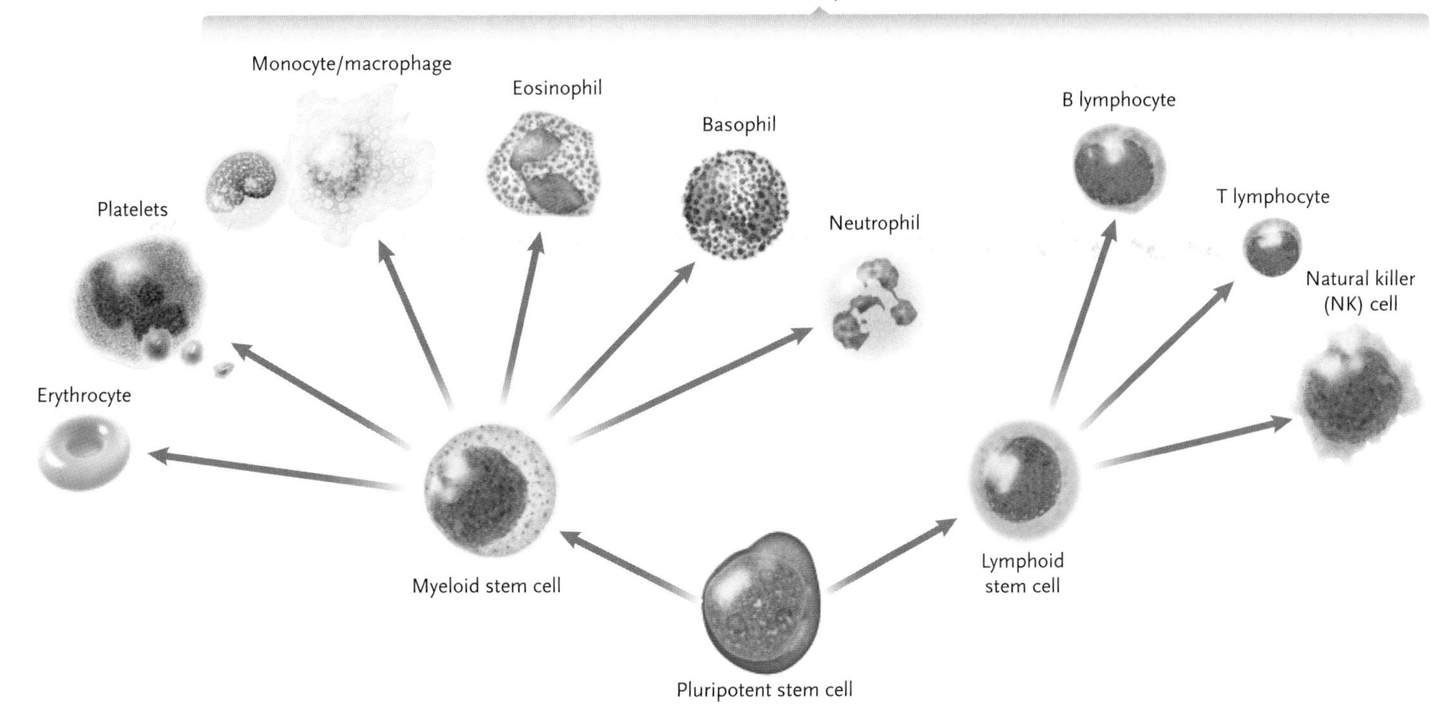

Figure 42.7 Major cellular components of mammalian blood and their origins from stem cells.

Erythrocytes Are the Oxygen Carriers of the Blood

Erythrocytes—the red blood cells—carry O_2 from the lungs to body tissues. Each microliter of human blood normally contains about 5 million erythrocytes, which are small, flattened, and disclike. Indentations on each flattened surface make them *biconcave*—thinner in the middle than at the edges (see Figure 42.6). They measure about 7 μm in diameter and 2 μm in thickness. Erythrocytes are highly flexible cells, able to squeeze through narrow capillaries.

As they mature, mammalian erythrocytes lose their nucleus, cytoplasmic organelles, and ribosomes, thereby limiting their metabolic capabilities and life span. The remaining cytoplasm contains enzymes, which carry out glycolysis, and large quantities of *hemoglobin,* the O_2-carrying protein of the blood. Most of the ATP produced by glycolysis is used to power active transport mechanisms that move ions in and out of erythrocytes.

Hemoglobin, the molecule that gives erythrocytes and the blood its red color, consists of four polypeptides, each linked to a nonprotein *heme* group that contains an iron atom in its center (see Figure 3.18). The iron atom binds O_2 molecules as blood circulates through the lungs and releases the O_2 as blood flows through other body tissues.

Hemoglobin can also combine with carbon monoxide (CO), which is common in the exhaust gases of cars and boats, and may be produced by appliances fueled with natural gas or oil. The combination, which is essentially irreversible, blocks hemoglobin's ability to combine with O_2, which can quickly lead to death if exposure continues. CO is also present in cigarette smoke.

Some 2 million to 3 million erythrocytes are produced in the average human each *second.* The life span of an erythrocyte in the circulatory system is about 120 days. At the end of their useful life, erythrocytes are engulfed and destroyed by *macrophages* (*makros* = big; *phagein* = to eat), a type of large leukocyte, in the spleen, liver, and bone marrow.

A negative feedback mechanism keyed to the blood's O_2 content stabilizes the number of erythrocytes in blood. If the O_2 content drops below the normal level, the kidneys synthesize **erythropoietin** (EPO), a hormone that stimulates stem cells in bone marrow to increase erythrocyte production. Erythropoietin is also secreted after blood loss and when mammals move to higher altitudes. As new red blood cells enter the bloodstream, the O_2-carrying capacity of the blood rises. If the O_2 content of the blood rises above normal levels, erythropoietin production falls in the kidneys and red blood cell production drops. The gene encoding human erythropoietin has been cloned, allowing researchers to produce this protein in large quantities. It can then be injected into the body to stimulate erythrocyte production, for example, in patients with anemia (lower-than-normal hemoglobin levels) caused by kidney failure or chemotherapy. It can also supplement or even replace blood transfusions. Some endurance athletes such as triathletes, bicycle racers, marathon runners, and cross-country skiers have used EPO to increase their erythrocyte levels in order to enhance performance. Such blood doping is deemed illegal by the governing organizations of most endurance sports and, as a result, many athletes have been sanctioned or banned in recent years.

Disorders of erythrocytes are responsible for a number of human disabilities and diseases. The *anemias,* which result from too few or malfunctioning erythrocytes, prevent O_2 from reaching body tissues in sufficient amounts. Shortness of breath, fatigue, and chills are common symptoms of anemia. Anemia can be produced, for example, by blood loss from a wound or bleeding ulcer, or by certain infections.

Leukocytes Provide the Body's Front Line of Defense against Disease

Leukocytes eliminate dead and dying cells from the body, remove cellular debris, and provide the body's first line of defense against invading organisms. They are called white cells because they are colorless, in contrast to the strongly pigmented red blood cells. Also unlike red blood cells, leukocytes retain their nuclei, cytoplasmic organelles, and ribosomes as they mature, and hence are fully functional cells.

Like red blood cells, leukocytes arise from the division of stem cells in red bone marrow (see Figure 42.7). As they mature, they are released into the bloodstream, from which they enter body tissues in large number. Some types of leukocytes are capable of continued division in the blood and body tissues. The specific types of leukocytes and their functions in the immune reaction are discussed in Chapter 43.

Platelets Induce Blood Clots That Seal Breaks in the Circulatory System

Blood **platelets** are oval or rounded cell fragments about 2 to 4 μm in diameter, each enclosed in its own plasma membrane. They are produced in red bone marrow by the division of stem cells (see Figure 42.7).

Platelets contain enzymes and other factors that take part in blood clotting. When blood vessels are damaged, collagen fibers in the extracellular matrix are exposed to the leaking blood. Platelets in the blood then stick to the collagen fibers and release signaling molecules that induce additional platelets to stick to them. The process continues, forming a plug that helps seal off the damaged site. As the plug forms, the platelets release other factors that convert the soluble plasma protein, fibrinogen, into long, insoluble threads of **fibrin.** Cross-links between the fibrin threads form a meshlike network that traps blood cells and platelets and further seals the damaged area **(Figure 42.8).** The entire mass is a blood clot.

Mutations or diseases that interfere with the enzymes and factors taking part in the clotting mechanism can have serious effects and lead to uncontrolled bleeding. In the most common form of hemophilia, for example, a mutation in a single protein (called *clotting factor VIII*) interferes with the clotting reaction. Bleeding is uncontrolled in afflicted individuals; even small cuts and bruises can cause life-threatening blood loss.

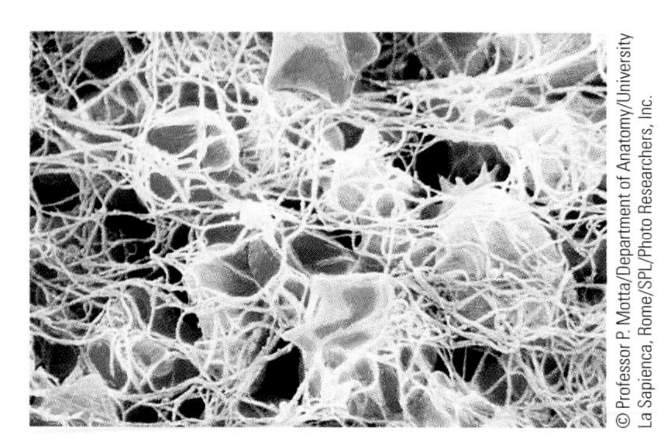

Figure 42.8
Red blood cells caught in a meshlike network of fibrin threads during formation of a blood clot.

Study Break

1. Outline the life cycle of an erythrocyte.
2. How does the body compensate for a lower-than-normal level of oxygen in the blood?
3. What are the roles of leukocytes and platelets?

42.3 The Heart

In mammals, the heart is structured from cardiac muscle cells (see Figure 36.6) forming a four-chambered pump, with two atria at the top of the heart pumping blood into two ventricles at the bottom of the heart **(Figure 42.9).** The powerful contractions of the ventricles push the blood at relatively high pressure into arteries leaving the heart. This arterial pressure is responsible for the blood circulation. Valves between the atria and the ventricles and between the ventricles and the arteries leaving the heart keep the blood from flowing backward.

The heart of mammals pumps the blood through two completely separate circuits of blood vessels: the systemic circuit and the pulmonary circuit **(Figure 42.10).** The right atrium (toward the right side of the body) receives blood returning to the heart in vessels coming from the entire body, except for the lungs: the *superior vena cava* conveys blood from the head and forelimbs, and the *inferior vena cava* conveys blood from the abdominal organs and the hind limbs. This blood is depleted of O_2 and has a high CO_2 content. The right atrium pumps blood into the right ventricle, which contracts to push the blood into the *pulmonary arteries* leading to the lungs. In the capillaries of the lungs, the blood releases CO_2 and picks up O_2. The oxygenated blood completes this pulmonary circuit by returning in *pulmonary veins* to the heart.

Blood returning from the pulmonary circuit enters the left atrium, which pumps the blood into the left ventricle. This ventricle, the most thick-walled and powerful of the heart's chambers, contracts to send the oxygenated blood coursing into a large artery, the **aorta,** which branches into arteries leading to all body regions except the lungs. In the capillary networks of these body regions, the blood releases O_2 and picks up CO_2. The O_2-depleted blood collects in veins, which complete the systemic circuit. The blood from the veins enters the right atrium. The amount of blood pumped by the two halves of the heart is normally balanced so that neither side pumps more than the other.

The heart also has its own circulation, called the *coronary circulation.* The aorta gives off two *coronary arteries* that course over the heart. The coronary arteries branch extensively, leading to dense capillary beds that serve the cardiac muscle cells. The blood from the capillary networks collects into veins that empty into the right atrium.

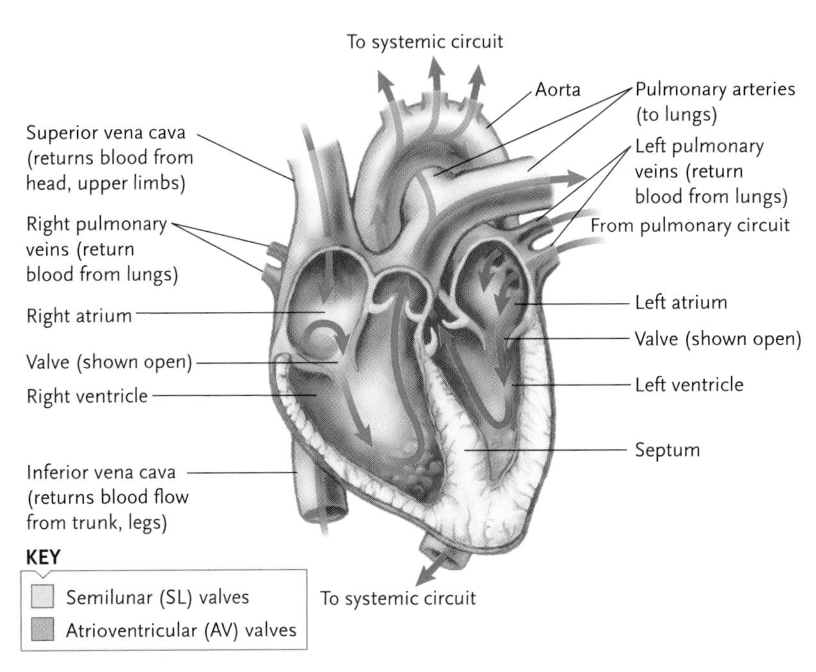

Figure 42.9
Cutaway view of the human heart showing its internal organization.

The Heartbeat Is Produced by a Cycle of Contraction and Relaxation of the Atria and Ventricles

Average heart rates vary among mammals (and among vertebrates generally), depending on body size and the overall level of metabolic activity. A human heart beats 72 times each minute, on average, with each beat lasting about 0.8 second. The heart beat rate of a trained endurance athlete is typically much lower. The heart of a flying bat may beat 1200 times a minute, while that of an elephant beats only 30 times a minute. The period of contraction and emptying of the heart is called the **systole** and the period of relaxation and filling of the heart between contractions is called the **diastole.** The systole-diastole sequence of the heart is called the **cardiac cycle (Figure 42.11).** The following discussion goes through one cardiac cycle.

Starting when both the atria and ventricles are relaxed in diastole, the atria begin filling with blood (see Figure 42.11, step 1). At this point, the **atrioventricular valves (AV valves)** between each atrium and ventricle, and the **semilunar valves (SL valves)** between the ventricles and the aorta and pulmonary arteries, are closed. As the atria fill, the pressure pushes open the AV valves and begins to fill the relaxed ventricles (step 2). When the ventricles are about 80% full, the atria contract and completely fill the ventricles with blood (step 3). Although there are no valves where the veins open into the atria, the atrial contraction compresses the openings, sealing them so that little backflow occurs into the veins.

As the ventricles begin to contract, the rising pressure in the ventricular chambers forces the AV valves shut (step 4). As they continue to contract, the pressure in the ventricular chambers rises above that in the arteries leading away from the heart, forcing open the SL valves. Blood now rushes from the ventricles into the aorta and the pulmonary arteries (step 5).

The completion of the contraction squeezes about two-thirds of the blood in the ventricles into the arteries. Now the ventricles relax, lowering the pressure in the ventricular chambers below that in the arteries. This reversal of the pressure gradient reverses the direction of blood flow in the regions of the SL valves, causing them to close. For about half a second, both the atria and ventricles remain in diastole and blood flows into the atria and ventricles. Then, the blood-filled atria contract, and the cycle repeats.

In an adult human at rest, each ventricle pumps roughly 5 liters of blood per minute—an amount roughly equivalent to the entire volume of blood in the body. At maximum rate and strength the human heart pumps about five times the resting amount, or more than 25 liters per minute.

The heart makes a "lub-dub" sound when it beats, which you can hear by placing an ear against a person's chest or listening to the heart through a stethoscope. The sound is produced mostly by vibrations created when the valves close. The "lub" sound occurs when the AV valves are pushed shut by the contraction of the ventricles; the "dub" sound is made when the SL valves are forced shut as the ventricles relax. *Heart murmurs* are abnormal sounds produced by turbulence created in the blood when one or more of the valves fails to open or close completely and blood flows backward.

The Cardiac Cycle Is Initiated within the Heart

Contraction of cardiac muscle cells is triggered by action potentials that spread across the muscle cell membranes. Some crustaceans, such as crabs and lobsters,

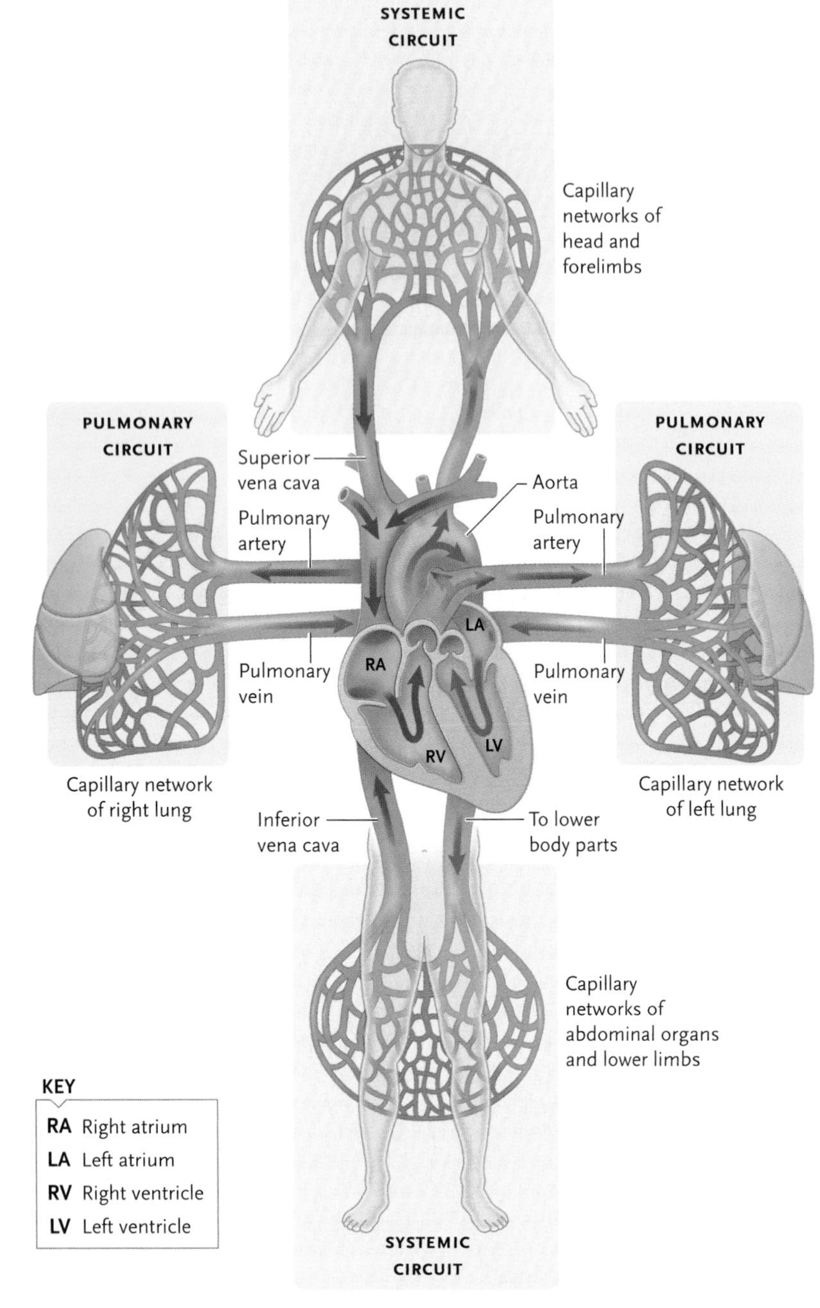

Figure 42.10
The pulmonary and systemic circuits of mammals. The right half of the heart pumps blood into the pulmonary circuit, and the left half of the heart pumps blood into the systemic circuit.

have **neurogenic hearts**, that is, hearts that beat under the control of signals from the nervous system. This type of heart contraction regulation allows for the heart to be stopped completely in these animals, which may not need continuous hemolymph circulation in some situations. Other animals, including all insects and all vertebrates, have **myogenic hearts**, that is, hearts that maintain their contraction rhythm with no requirement for signals from the nervous system. The advantage of a myogenic heart is that blood flow is ensured in the event of serious trauma to the nervous system.

The contraction of individual cardiac muscle cells in mammalian myogenic hearts is coordinated by a region of the heart called the **sinoatrial node (SA node)**, which controls the rate and timing of cardiac muscle cell contraction. The SA node consists of **pacemaker cells**, which are specialized cardiac muscle cells in the upper wall of the right atrium near where the blood enters the heart from the systemic circuit **(Figure 42.12)**. Ion channels in these cells open in a cyclic, self-sustaining pattern that alternately depolarizes and repolarizes their plasma membranes. The regularly timed depolarizations initiate waves of contraction that travel over the heart (step 1). The first effect of a wave is to cause the cells of the atria to contract in unison and fill the ventricles with blood.

A layer of connective tissue separates the atria from the ventricles, in effect placing a layer of electrical insulation between the top and bottom of the heart. The insulating layer keeps a contraction signal from the SA node from spreading directly from the atria to the ventricles (step 2). Instead, the atrial wave of contraction excites cells of the **atrioventricular node (AV node)**, located in the heart wall between the right atrium and right ventricle, just above the insulating layer of connective tissue. The signal produced travels from the AV node to the bottom of the heart via *Purkinje fibers* (step 3). These fibers follow a path downward through the insulating layer to the bottom of the heart, where they branch through the walls of the ventricles. The signal carried by the Purkinje fibers induces a wave of contraction that begins at the bottom of the heart and proceeds upward, squeezing the blood from the ventricles into the aorta and pulmonary arteries (step 4). The transmission of a signal from the AV node to the ventricles takes about 0.1 second; this delay gives the atria time to finish their contraction before the ventricles contract. Cardiac muscle cells have a relatively long refractory period, about 0.25 second, which keeps the signals or contractions from reversing at any point.

As Augustus Waller found, the electrical signals passing through the heart can be detected by attaching electrodes to different points on the surface of the body. The signals change in a regular pattern corresponding to the electrical signals that trigger the cardiac cycle, producing what is known as an **electrocardiogram (ECG**; also **EKG**, from German *elektrocardiogramm*). The highlighted region of the ECG under each stage of the cardiac cycle in Figure 42.12 indicates the electrical activity measured in those stages. Many malfunctions of the heart alter the ECG pattern in characteristic ways, providing clues to the location and type of heart disease.

The spontaneous, rhythmic signal set up by the SA node is the foundation of the normal heartbeat. Because of this internal signaling system, a myogenic heart will continue beating if all the nerves leading to the heart are severed.

The SA node normally dominates the AV node and other conductive regions of the heart, keeping the atria

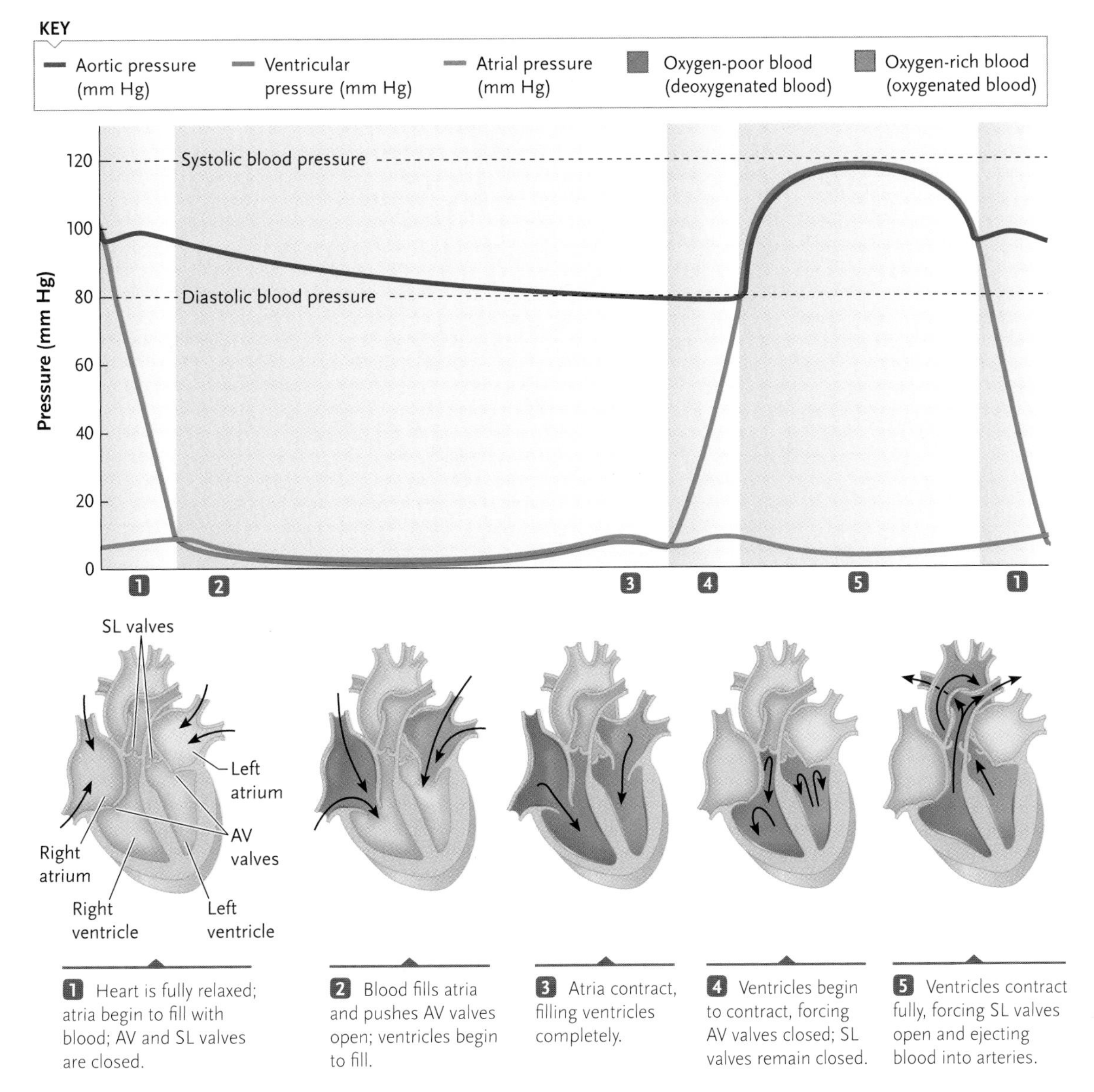

Figure 42.11
The cardiac cycle.

and ventricles beating in a fully coordinated fashion. At times, however, parts of the conductive system outside the SA node may generate signals independently and produce uncoordinated contractions known as *arrhythmias*. Depending on their source and characteristics, arrhythmias range from harmless to life threatening. Most commonly, the ventricles beat prematurely, and then fill more slowly for the next beat, which is proportionately more powerful. This arrhythmia, called a *premature ventricular contraction (PVC)*, feels like a skipped beat but is usually harmless. Consumption of too much caffeine, chocolate, or alcohol can increase the frequency of PVCs. Other arrhythmias, such as those produced when the AV node becomes the dominant pacemaker, can be more dangerous. Some of the more threatening arrhythmias are corrected by surgically implanting an artificial pacemaker that produces an overriding, regular electrical impulse to keep the heart beating at a normal rhythm and rate.

Arterial Blood Pressure Cycles between a High Systolic and a Low Diastolic Pressure

The pressure that a fluid in a confined space exerts is called *hydrostatic pressure*. That is, fluid in a container exerts some pressure on the wall of the container. Blood vessels are essentially tubular containers that are part of a closed system filled with fluid. Hence, the blood in vessels exerts hydrostatic pressure against the walls of the vessels. *Blood pressure* is the measurement of that hydrostatic pressure on the walls of the arteries as the heart pumps blood through the body. Blood pressure is determined by the force and amount of blood pumped by the heart and the size and flexibility of the arteries. In any person, blood pressure changes continually in response to activity, temperature, body position, emotional state, diet, and medications being taken.

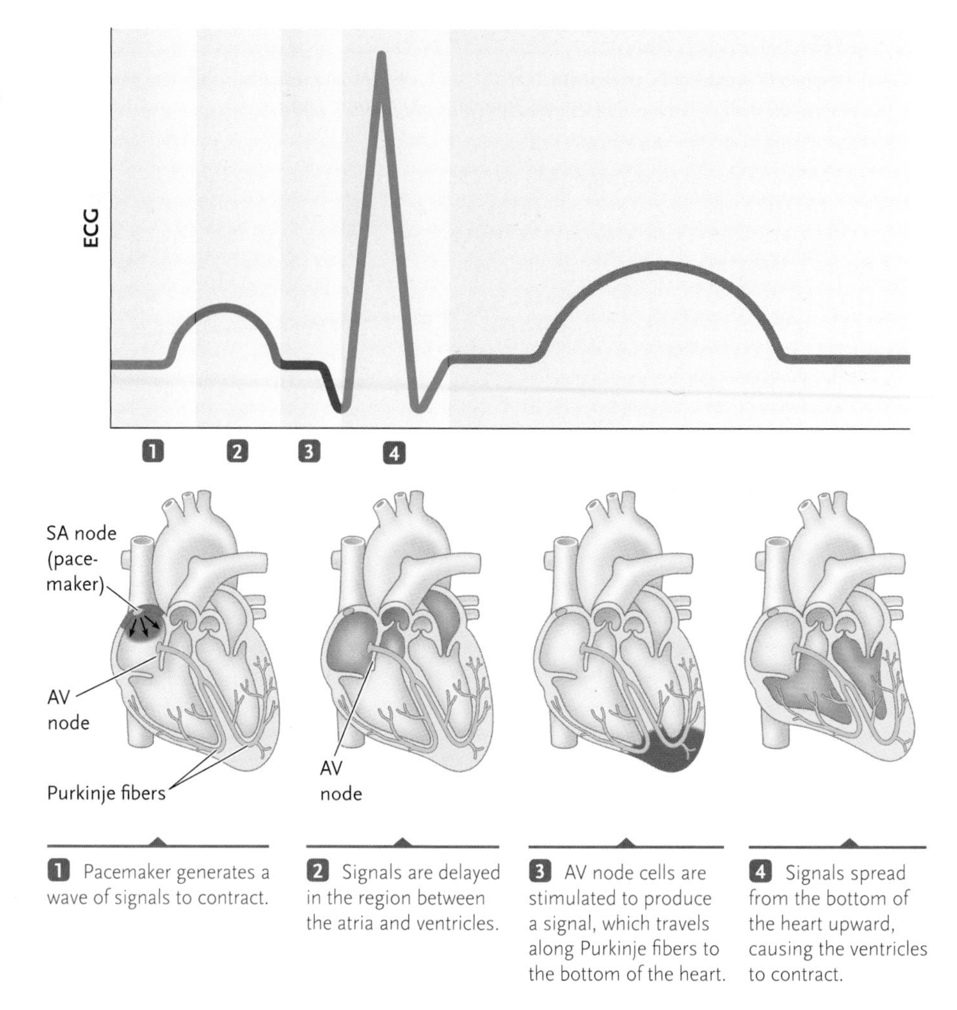

ECG
1
2
3
4
SA node
(pace-
maker)
AV
node
Purkinje fibers
AV
node
1 Pacemaker generates a wave of signals to contract.
2 Signals are delayed in the region between the atria and ventricles.
3 AV node cells are stimulated to produce a signal, which travels along Purkinje fibers to the bottom of the heart.
4 Signals spread from the bottom of the heart upward, causing the ventricles to contract.

Figure 42.12
The electrical control of the cardiac cycle. The top part of the figure shows how a signal originating at the SA node leads to ventricular contraction. The bottom part of the figure shows the electrical activity for each of the stages as seen in an ECG. The colors in the hearts show the location of the signal at each step and correspond to the colors in the ECG.

Two blood pressure measurements are typically made in clinical medicine, representing different parts of the cardiac cycle. As the ventricles contract, a surge of high-pressure blood moves outward through the arteries leading from the heart. This peak of high pressure, called the *systolic blood pressure,* can be felt as a *pulse* by pressing a finger against an artery that lies near the skin, such as the arteries of the neck or the artery that runs along the inside of the wrist. Between ventricular contractions, the arterial blood pressure reaches a low point called the *diastolic blood pressure.*

For most healthy adults at rest, the systolic pressure measured at the upper arm is between 90 and 120 mm Hg, and the diastolic pressure is between 60 and 80 mm Hg (**Figure 42.13** shows how these pressures are measured). The numbers, written in the form 120/80 mm Hg and stated verbally as "120 over 80," refer to the height of a column of mercury in millimeters that would be required to balance the pressure exactly. The systolic and diastolic blood pressures in the pulmonary arteries are typically much lower, about 24/8 mm Hg.

The blood pressure in the systemic and pulmonary circuits is highest in the arteries leaving the heart, and drops as the blood passes from the arteries into the capillaries. By the time the blood returns to the heart, its pressure has dropped to 2 to 5 mm Hg, with no differentiation between systolic and diastolic pressures. The reduction in pressure occurs because the blood encounters resistance as it moves through the vessels, produced primarily by the friction created when blood cells and plasma proteins move over each other and over vessel walls.

Some people have **hypertension,** commonly called high blood pressure, a medical condition in which blood pressure is chronically elevated above normal values; that is, at least 140/90. In some cases, no specific medical cause can be found to explain the hypertension. In other cases, the hypertension results from another medical condition, such as kidney disease, or diseases or cancers affecting the adrenal cortex. Hypertension can also be caused by certain medications, such as ibuprofen and steroids. Age is also a contributor to hypertension because over time the walls of blood vessels become stiffer as more collagen fibers are added, and this causes decreased elasticity of the arteries. During systole, these arteries cannot expand as much as they once could, and this results in a higher arterial blood pressure.

Hypertension is rarely severe enough to cause symptoms. However, in the long term, the increased pressure in the arteries can cause damage to organs. Hence, hypertension is treated because of the correlation with an increased risk for a number of medical conditions, including myocardial infarction (heart attack), cardiovascular accident (stroke), chronic renal failure, and retinal damage. Treatments to reduce hypertension typically involve life style changes, such as weight loss and regular exercise; in the case of moderate to severe hypertension, drugs are also prescribed.

What happens to blood pressure during exercise? The answer depends on the type of exercise. During static exercise involving a sustained contraction of a muscle group or groups, such as weight lifting or Nautilus machine routines, both systolic and diastolic pressure increase. In elite weight lifters, for instance, blood pressure during lifts can reach 300/150 mm Hg. However, during dynamic exercise involving intermittent and rhythmical muscle contractions, such as running, bicycling, and swimming, only the systolic pressure increases.

Study Break

1. Explain the role of each of the four chambers of the mammalian heart in blood circulation.
2. Distinguish systole and diastole.
3. Distinguish neurogenic hearts and myogenic hearts.
4. Describe the electrical events that occur during the cardiac cycle in a mammalian myogenic heart.

42.4 Blood Vessels of the Circulatory System

Both the systemic and pulmonary circuits consist of a continuum of different blood vessel types that begin and end at the heart **(Figure 42.14).** From the heart, large arteries carry blood and branch into progressively smaller arteries, which deliver the blood to the various parts of the body. When a small artery reaches the organ it supplies, it branches into yet smaller vessels, the **arterioles.** Within the organ, arterioles branch into capillaries, the smallest vessels of the circulatory system. The capillaries form a network in the organ that is used to exchange substances between the blood and the surrounding cells. Capillaries rejoin to form small **venules,** which merge into the small veins that leave the organ. The small veins progressively join to form larger veins that eventually become the large veins that enter the heart.

Arteries Transport Blood Rapidly to the Tissues and Serve as a Pressure Reservoir

Arteries have relatively large diameters and, therefore, provide little resistance to blood flow. Structurally, they are adapted to the relatively high pressure of the blood passing through them. The walls of arteries consist of

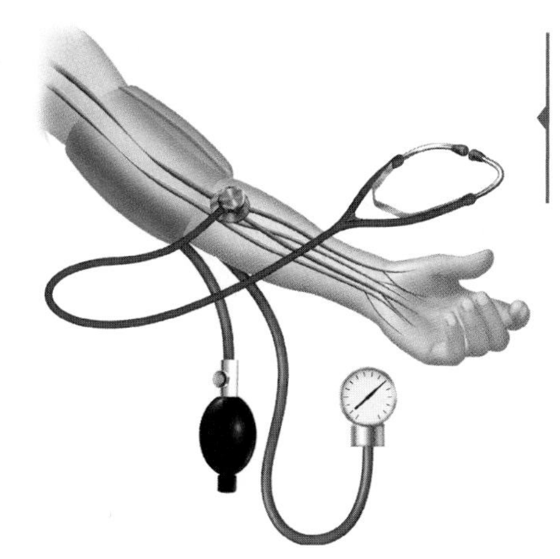

1 The cuff is pumped up until its pressure is higher than the arterial blood pressure, cutting off the flow of blood through the large artery serving the arm.

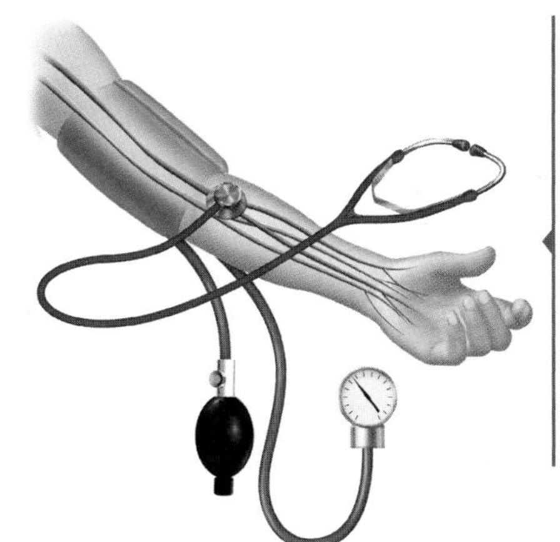

2 A stethoscope is placed over the artery just below the cuff, and the pressure in the cuff is slowly released. When cuff pressure drops to the point that blood just begins to flow through the artery, a faint thumping sound is heard in the stethoscope. The sound is produced by turbulence created as spurts of blood pass through the narrowed artery. The pressure read on the meter when the thumping begins is the systolic pressure.

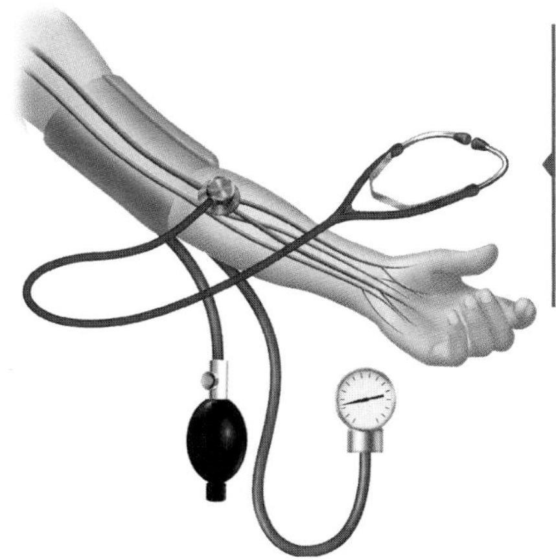

3 The pressure in the cuff is released further until the thumping sound just disappears. At this point, there is no turbulence because the artery is fully open. The pressure read on the meter at this point is the diastolic pressure.

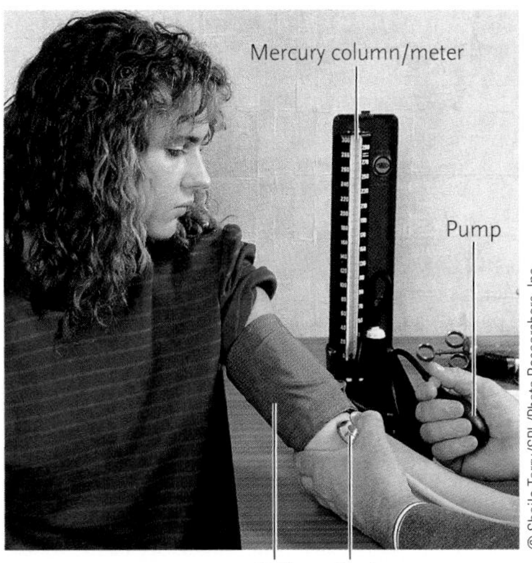

Figure 42.13
Taking blood pressure with a sphygmomanometer. The device consists of a rubber bladder (called a cuff) that is wrapped around an arm or leg and connected to an air pump and a mercury column, which records the pressure inside the bladder in millimeters of mercury (mm Hg).

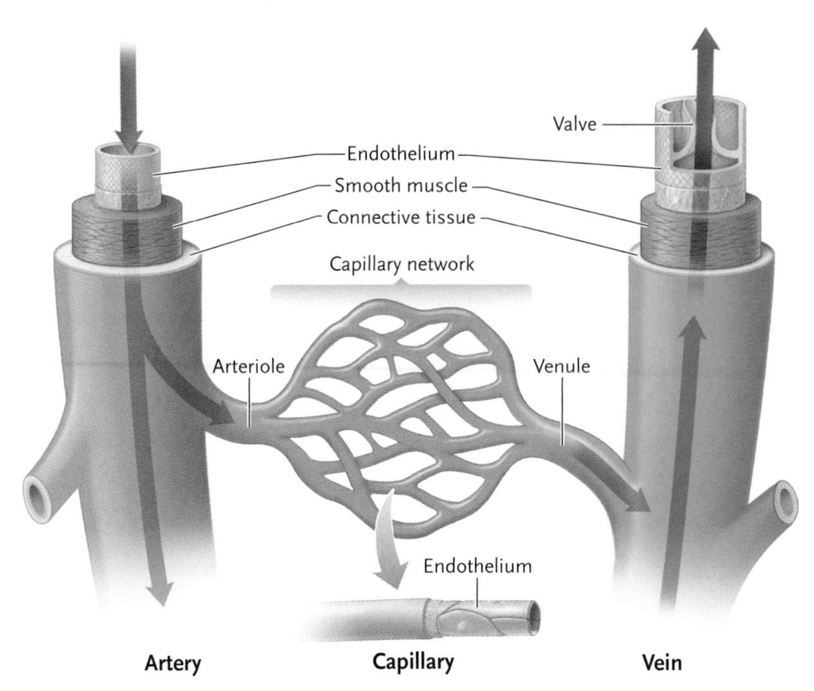

Figure 42.14
The structure of arteries, capillaries, and veins and their relationship in blood circuits.

three major tissue layers: (1) an outer layer of connective tissue containing collagen fibers mixed with fibers of the protein elastin, which gives the vessel recoil ability; (2) a relatively thick middle layer of vascular smooth muscle cells also mixed with elastin fibers; and (3) a one-cell-thick inner layer of flattened cells that forms an *endothelium,* a specialized type of epithelial tissue that lines the entire circulatory system (see Figure 42.14).

In addition to being the conduits for blood traveling to the tissues, arteries also act as a pressure reservoir to generate the force for blood movement when the heart is relaxing. When contraction of the ventricles pumps blood into the arteries, the amount of blood flowing into the arteries is greater than the amount flowing out into the smaller vessels downstream because of the higher resistance to blood flow in those smaller vessels. The arteries can accommodate the excess volume of blood because their elastic walls allow the arteries to expand in diameter. When the heart then relaxes and blood is no longer being pumped into the arteries, the arterial walls recoil passively back to their original state. The recoil pushes the excess blood from the arteries into the smaller downstream vessels. As a result, blood flow to tissues is continuous during systole and diastole.

Capillaries Are the Sites of Exchange between the Blood and Interstitial Fluid

Capillaries thread through nearly every tissue in the body, arranged in networks that bring them within 10 μm of most body cells. They form an estimated 2600 km^2 of total surface area for the exchange of gases, nutrients, and wastes with the interstitial fluid. Capillary walls consist of a single layer of endothelial cells.

Control of Blood Flow through Capillaries. Blood flow through the capillary networks is controlled by contraction of smooth muscle in the arterioles **(Figure 42.15).** The capillaries themselves do not have smooth muscle but, in many cases, a small ring of smooth muscle called a *precapillary sphincter* is present at the junction between an arteriole and a capillary.

When the sphincter is relaxed, blood flows readily through the arterioles and capillary networks (see Figure 42.15a). In the most contracted state, blood flow

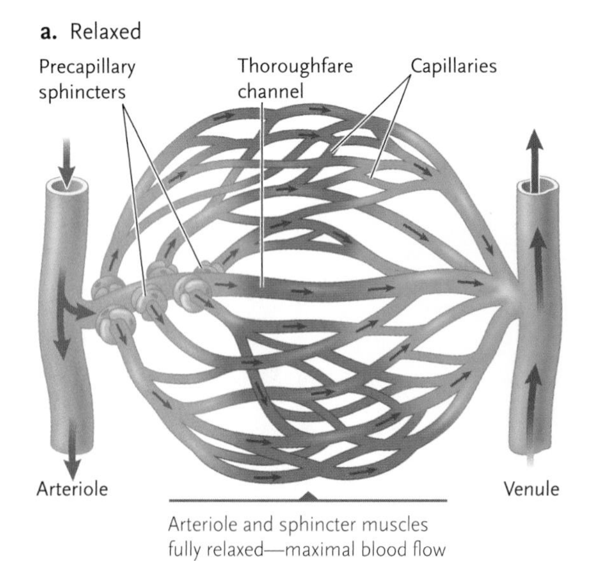

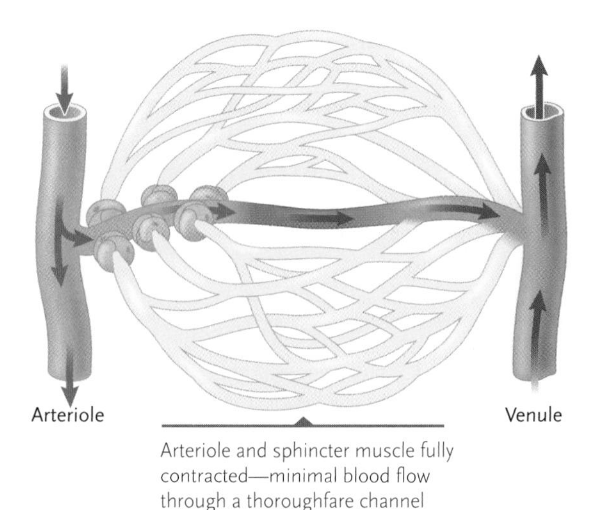

Figure 42.15
Control of blood flow through capillary networks. **(a)** Maximal blood flow when arteriole and sphincter muscles are fully relaxed. **(b)** Minimal blood flow when the arteriole and sphincter muscles are fully contracted.

through the arterioles and capillary networks is limited (see Figure 42.15b). Variation in the contraction of arteriole and sphincter smooth muscles adjusts the rate of flow through the capillary networks between the two flow limits. For example, during exercise, flow of blood through the capillary networks is increased severalfold over the resting state by relaxation of the precapillary sphincters.

Control of Blood Volume to Capillaries by Arterioles. The volume of blood flowing through an organ is adjusted by regulating the internal diameter of the arterioles of the organ. The blood leaving the arterioles enters the capillaries that branch from them. Although their total surface area is astoundingly large, the diameter of individual capillaries is so small that red blood cells must squeeze through most of them in single file **(Figure 42.16).** As a result, each capillary presents a high resistance to blood flow. Yet there are so many billions of capillaries in the networks that their combined diameter is about 1300 times greater than the cross-sectional area of the aorta. As a result, blood slows considerably as it moves through the capillaries **(Figure 42.17).** This is analogous to the slowing that occurs when a narrow, swiftly running streams widens into a broad pool. The slow movement of blood through the capillaries maximizes the time for exchange of substances between blood and tissues. As they leave the tissues, the capillaries rejoin to form veins. Veins have a reduced total cross-sectional area compared with capillaries, so the rate of flow increases as blood returns to the heart (see Figure 42.17).

Exchange of Substances across Capillary Walls. In most body tissues, narrow spaces between the capillary endothelial cells allow water, ions, and small molecules such as glucose to pass freely between the blood and interstitial fluid. Erythrocytes, platelets, and most plasma proteins are too large to pass between the cells and are retained inside the capillaries, except for molecules that are transported through the epithelial cells by specific carriers. Leukocytes, however, are able to squeeze actively between the cells and pass from the blood to the interstitial fluid.

There are exceptions to these general properties in some tissues; in the brain, for example, the capillary endothelial cells are tightly sealed together, preventing essentially all molecules and even ions from passing between them. The tight seals set up the *blood–brain barrier,* which limits the exchange between capillaries and brain tissues to molecules and ions that are specifically transported through the capillary endothelial cells (see Section 38.3 for additional discussion of the blood–brain barrier).

Forces Driving the Exchange. Two major mechanisms drive the exchange of molecules and ions between the capillaries and the interstitial fluid: (1) diffusion along concentration gradients and (2) bulk flow.

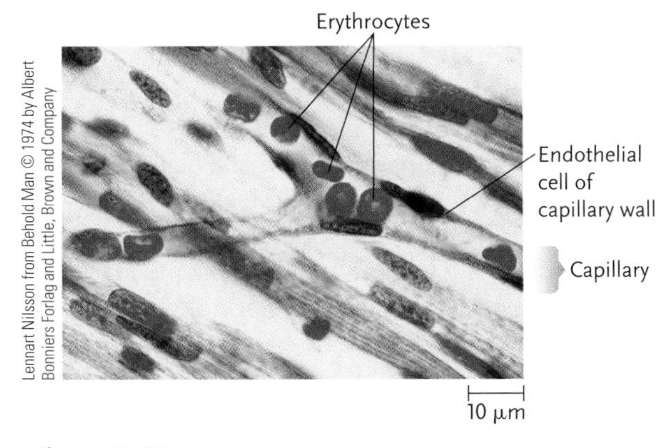

Figure 42.16
Erythrocytes moving through a capillary that is just wide enough to admit the cells in single file.

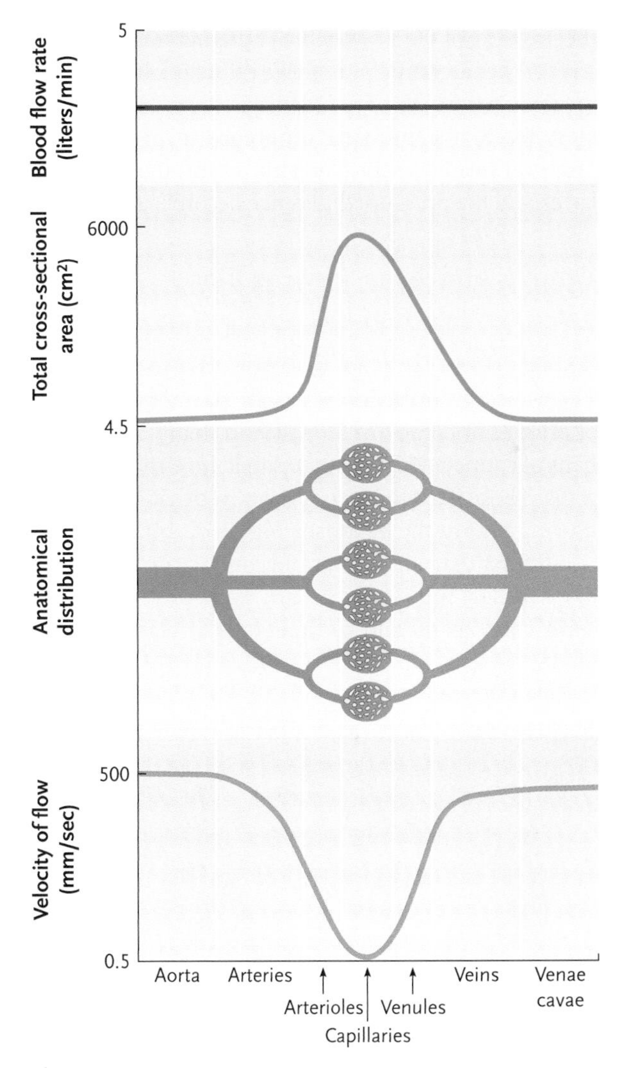

Figure 42.17
Blood flow rate and velocity of flow in relation to total cross-sectional area of the blood vessels. The blood flow rate is identical throughout the circulatory system and is equal to the cardiac output. The velocity of flow in the different types of blood vessels is inversely related to the total cross-sectional area of all the vessels of a particular type: for example, the velocity is highest in the aorta, which has the smallest cross-sectional area, and lowest in the capillaries, which collectively have the largest total cross-sectional area.

Diffusion along concentration gradients occurs both through the spaces between the capillary endothelial cells and through their plasma membranes. Ions or molecules such as O_2 and glucose, which are more concentrated inside the capillaries, diffuse outward along their gradients; other ions or molecules that are more concentrated outside, such as CO_2, diffuse inward. Some molecules, such as O_2 and CO_2, diffuse directly through the lipid bilayer of the endothelial cell plasma membranes; others, such as glucose, pass by facilitated diffusion through transport proteins. The total diffusion is greatest at the ends of capillaries nearest the arterioles, where the concentration differences between blood plasma and interstitial fluid are highest, and lowest at the ends nearest the venules, where the inside/outside concentrations of diffusible substances are almost equal.

Bulk flow, which carries water, ions, and molecules out of the capillaries, occurs through the spaces between capillary endothelial cells. The flow is driven by the pressure of the blood, which is higher than the pressure of the interstitial fluid. Like diffusion, bulk flow is greatest in the ends of the capillaries nearest the arterioles, where the pressure difference is highest, and drops off steadily as the blood moves through the capillaries and the pressure difference becomes smaller.

Venules and Veins Serve as Blood Reservoirs in Addition to Conduits to the Heart

The walls of venules and veins are thinner than those of arteries and contain little elastin. Many veins have flaps of connective tissue that extend inward from their walls. These flaps form one-way valves that keep blood flowing toward the heart (see Figure 42.14).

Rather than stretching and contracting elastically, like arteries, the relatively thin walls of venules and veins can expand and contract over a relatively wide range, allowing them to act as blood reservoirs as well as conduits. At times, the venules and veins may contain from 60% to 80% of the total blood volume of the body. The stored volume is adjusted by skeletal muscle contraction and by the valves, in response to metabolic conditions and signals carried by hormones and neurotransmitters.

Although blood pressure in the venous system is relatively low, several mechanisms assist the movement of blood back to the heart. As skeletal muscles contract, they compress nearby veins, increasing their internal pressure **(Figure 42.18).** The one-way valves in the veins, especially numerous in the larger veins of the limbs, keep the blood from flowing backward when the muscles relax.

When you sit without moving for long periods of time, as you might during an airline flight, the lack of skeletal muscular activity greatly reduces the return of venous blood to the heart. As a result, the blood pools in the veins of the body below the heart, making the hands, legs, and feet swell. The motionless blood can also form clots, particularly in the veins of the legs, a condition called *deep vein thrombosis.* Deep vein thrombosis often does not cause symptoms, but can cause serious medical problems if a clot breaks loose and moves elsewhere in the body, such as to the lungs. Raising the arms and getting up at intervals to exercise or contracting and relaxing the leg muscles as you sit can relieve this condition.

Figure 42.18
How skeletal muscle contraction, and the valves inside veins, help move blood toward the heart.

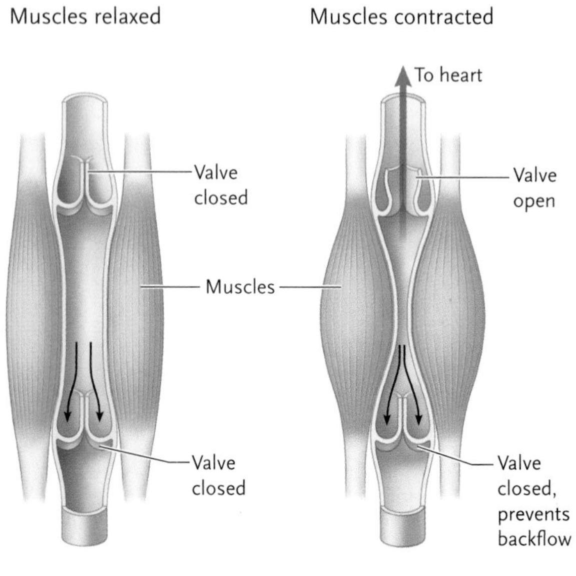

Disorders of the Circulatory System Are Major Sources of Human Disease

The layer of endothelial cells lining the arteries and veins is normally smooth and does not impede blood flow. However, several conditions, including bacterial and viral infections, chronic hypertension, smoking, and a diet high in fats, can damage the endothelial cells, exposing the underlying smooth muscle tissue, which begins a cycle of injury and repair leading to lesions. (*Focus on Research* in Chapter 3 discussed the relationship between fats and cholesterol and coronary artery disease.) Thickened deposits of material called *atherosclerotic plaques* may form at the damaged sites **(Figure 42.19).** The plaques, which consist of cholesterol-rich fatty substances, smooth muscle cells, and collagen deposits, reduce the diameter of the blood vessel and impede blood flow. Worse, the damaged endothelial lining may stimulate platelets to adhere and trigger the formation of blood clots. The clots further reduce the vessel diameter and flow and may break loose, along with segments of plaque material, to block finer vessels in other regions of the body.

Atherosclerosis has its most serious effects in the smaller arteries of the body, particularly in the fine coronary arteries that serve the heart muscle. Here, the plaques and clots reduce or block the flow of blood to the heart muscle cells. Serious blockage can cause a heart attack—the death of cardiac muscle cells deprived of blood flow. The blockage of arteries in the brain by plaque material or blood clots released from atherosclerotic arteries is also a common cause of stroke—a loss of critical brain functions due to the death of nerve cells in the brain. Heart attacks and strokes are the most common causes of death in North America and in Europe.

The risk of heart attacks and stroke can be reduced by avoiding the conditions that damage the blood vessel endothelium. A diet low in fats and cholesterol, avoidance of cigarette smoke, and a program of exercise can reduce epithelial damage and plaque deposition. Medication and exercise, or exercise alone, can also reduce the effects of hypertension. There are good indications that these preventive programs can also reduce the size of existing atherosclerotic plaques.

Study Break

1. How is blood flow through capillary networks controlled?
2. Explain how, in contrast to most body tissues, the brain does not allow exchange of molecules and ions with blood.
3. Describe the two major mechanisms that drive the exchange of molecules and ions between the capillaries and the interstitial fluid.
4. What are atherosclerotic plaques? Indicate, in general, how they form, and provide some examples of factors contributing to their formation.

42.5 Maintaining Blood Flow and Pressure

Arterial blood pressure is the principal force moving blood to the tissues. Blood pressure must be regulated carefully so that the brain and other tissues receive adequate blood flow, but not so high that the heart is overburdened, risking damage to blood vessels. The three main mechanisms for regulating blood pressure are controlling *cardiac output* (the pressure and amount of blood pumped by the left and right ventricles), the degree of constriction of the blood vessels (primarily the arterioles), and the total blood volume. The sympathetic division of the autonomic nervous system and the endocrine system interact to coordinate these mechanisms. The system is effective in counteracting the effects of constantly changing internal and external conditions, such as movement from rest to physical activity or ending a period of fasting by eating a large meal. For example, the heart responds to the initiation of moderate physical activity by increasing blood flow to the heart itself by 367%, to the muscles of the skin by 370%, and to the skeletal muscle by 1066%, while decreasing flow to the digestive tract and liver by 56%, to the kidneys by 45%, and to the bone and most other tissues by 30%. Only the blood flow to the brain remains unchanged.

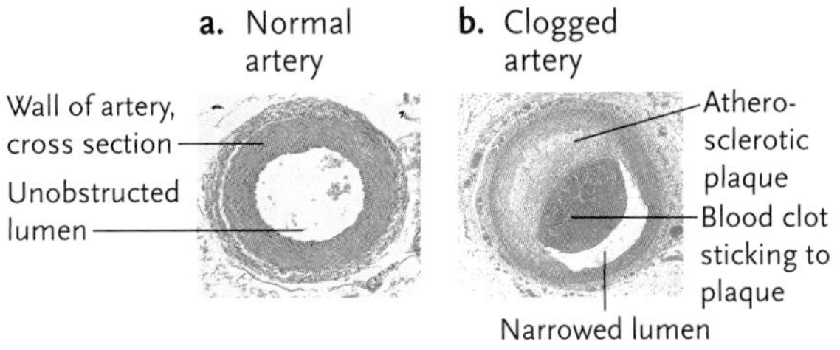

Figure 42.19 Atherosclerosis. **(a)** A normal coronary artery. **(b)** A coronary artery that is partially clogged by an atherosclerotic plaque.
(a: Ed Reschke; b: © Biophoto Associates/Photo Researchers, Inc.)

Cardiac Output Is Controlled by Regulating the Rate and Strength of the Heartbeat

Regulation of the strength and rate of the heartbeat starts at stretch receptors called *baroreceptors* (a type of mechanoreceptor; see Section 39.2), located in the walls of blood vessels. By detecting the amount of stretch of the vessel walls, baroreceptors constantly provide information about blood pressure. The baroreceptors in the cardiac muscle, the aorta, and the carotid arteries (which supply blood to the brain), are the most crucial. Signals sent by the baroreceptors go to the medulla within the brain stem. In response, the brain stem sends signals via the autonomic nervous system that adjust the rate and force of the heartbeat: the heart beats more slowly and contracts less forcefully when arterial pressure is above normal levels, and it beats faster and contracts more forcefully when arterial pressure is below normal levels.

The O_2 content of the blood, detected by chemoreceptors in the aorta and carotid arteries, also influences cardiac output. If the O_2 concentration falls below normal levels, the brain stem integrates this information with the signals sent from baroreceptors and issues signals that increase the rate and force of the heartbeat. Too much O_2 in the blood has the opposite effect, reducing the cardiac output.

Hormones Regulate both Cardiac Output and Arteriole Diameter

Hormones secreted by several glands contribute to the regulation of blood pressure and flow. As part of the stress response, the adrenal medulla reinforces the action of the sympathetic nervous system by secreting epinephrine and norepinephrine into the bloodstream (see Section 40.4). Epinephrine in particular raises the blood pressure by increasing the strength and rate of the heartbeat and stimulating vasoconstriction of arterioles in some parts of the body, including the skin, gut, and kidneys. At the same time, by inducing the vasodilation of arterioles that deliver blood to the heart, skeletal muscles, and lungs, epinephrine increases the blood flow to these structures. *Insights from the Molecular Revolution* shows how gene manipulation experi-

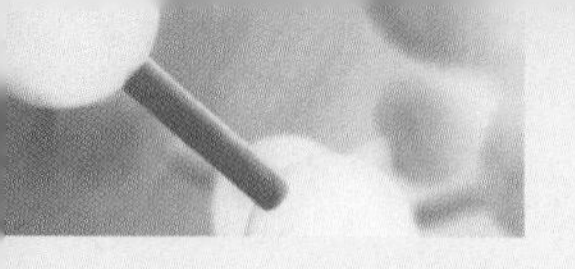

INSIGHTS FROM THE MOLECULAR REVOLUTION

Identifying the Role of a Hormone Receptor in Blood Pressure Regulation Using Knockout Mice

The sympathetic division of the autonomic nervous system brings about a temporary increase in blood pressure when the body experiences stress. That is, the sympathetic nervous system stimulates the adrenal medulla to release the catecholamines epinephrine and norepinephrine. These hormones stimulate the strength and rate of the heartbeat and change arteriole diameter. Epinephrine and norepinephrine exert their effects by binding to specific membrane-embedded receptors on target cells, activating them and triggering a cellular response via a signal transduction pathway. The receptors, which are G-protein–coupled receptors (see Section 7.4), are known as *adrenergic receptors* because of the adrenal origin of the hormones that bind to them.

Different types of adrenergic receptors are responsible for different responses to epinephrine and norepinephrine. With respect to blood pressure regulation, two key receptors are the α_1 receptor and the β_2 receptor. α_1 Receptors are found on cells of all arteriolar smooth muscle, but not in the brain, whereas β_2 receptors are found only on cells of arteriolar smooth muscle in the heart and skeletal muscles. Norepinephrine has strong affinity for the α_1 receptors, while epinephrine has less affinity for this type of receptor. Binding of norepinephrine, and to a lesser extent, epinephrine, to α_1 receptors on arteriolar smooth muscle causes vasoconstriction of arterioles, thereby contributing to an increase in blood pressure.

Researchers have identified three subtypes of α_1-adrenergic receptors: α_{1A}, α_{1B}, and α_{1D}. Tissues that express α_1 receptors can express all subtypes. Each of the subtypes responds to catecholamines, but the contribution of each subtype to the physiological responses caused by the hormones has not been characterized well. For example, despite attempts, scientists have not been able to develop drugs that are completely specific for inhibiting one subtype, making it impossible to use that approach to look at the effects of loss of activity of one subtype. Now, Gozoh Tsujimoto and colleagues at the Tokyo University of Pharmacy and Life Sciences, and the National Children's Medical Research Center, Tokyo, Japan, have used a different approach—making a gene knockout for the gene that encodes the α_{1D} receptor (the technique for making a gene knockout is described in Section 18.2). Using molecular techniques, the researchers showed that, in the knockout mice, no mRNA transcripts of the α_{1D} receptor gene was produced in any of the tissues examined, and there was no change in the expression of the α_{1A} and α_{1B} receptor subtypes. These results indicated that these mice were a highly useful model system for their physiological studies.

Next, the researchers compared the α_{1D} receptor knockout mice with normal mice to determine what had changed in the knockout mice with respect to cardiovascular function. They found that the knockout mice were modestly hypotensive (had slightly lower-than-normal blood pressure) but had a normal heart rate. They also had normal levels of circulating catecholamines. The investigators tested whether norepinephrine would stimulate an increase in blood pressure. They found that increasing doses of norepinephrine progressively increased blood pressure in both knockout and normal mice but that the response was markedly reduced in the knockout mice. Then, to determine whether the α_{1D} receptor is directly involved in vascular smooth muscle contraction, the researchers measured the effect of norepinephrine on contraction of segments of the aortas isolated from knockout and normal mice. Their results showed that norepinephrine induced concentration-dependent contraction of aortal segments from both types of mice but that the contraction response was considerably reduced in the knockout mice. The researchers concluded from their results that the α_{1D}-adrenergic receptor participates directly in sympathetic nervous system–driven regulation of blood pressure by vasoconstriction.

ments illuminated the role of a receptor for these hormones in regulating blood pressure.

The adrenal cortex and the posterior pituitary also release hormones that regulate blood pressure. Those hormones and their effects are described in Chapter 46.

Local Controls Also Regulate Arteriole Diameter

Several automated mechanisms also operate locally to increase the flow of blood to body regions engaged in increased metabolic activity, such as the muscles of your legs during an extended uphill bike ride. Low O_2 and high CO_2 concentrations, produced by the increased oxidation of glucose and other fuels, induce vasodilation of the arterioles serving muscles. The vasodilation increases the flow of blood and the O_2 supply. Nitric oxide (NO) released by arterial endothelial cells in body regions engaged in increased metabolic activity also works as a potent vasodilator. NO is broken down quickly after its release, ensuring that its effects are local.

STUDY BREAK

1. Why is it important to regulate arterial blood pressure?
2. What are the three main mechanisms for regulating blood pressure?
3. How does epinephrine affect blood pressure?

42.6 The Lymphatic System

Under normal conditions, a little more fluid from the blood plasma in the capillaries enters the tissues than is reabsorbed from the interstitial fluid into the plasma. The **lymphatic system** is an extensive network of vessels that collect the excess interstitial fluid and return it to the venous blood **(Figure 42.20).** The interstitial fluid picked up by the lymphatic system is called **lymph.** This system also collects fats that have been absorbed from the small intestine and delivers them to the blood circulation (see Chapter 45). The lymphatic system is also a key component of the immune system.

Vessels of the Lymphatic System Extend throughout Most of the Body

Vessels of the lymphatic system collect the lymph and transport it to *lymph ducts* that empty into veins of the circulatory system. The *lymph capillaries,* the smallest vessels of the lymphatic system, are distributed throughout the body, intermixed intimately with the capillaries of the circulatory system. Although they are several times larger in diameter than the blood capillaries, the walls of lymph capillaries also consist of a single layer of endothelial cells surrounded by a thin network of collagen fibers. Interstitial fluid enters the lymph capillaries—becoming lymph—at sites in their walls where the endothelial cells overlap, forming a flap that is forced open by the higher pressure of the interstitial fluid. The openings are wide enough to admit all components of the interstitial fluid, including infecting bacteria, damaged cells, cellular debris, and lymphocytes.

The lymph capillaries merge into *lymph vessels,* which contain one-way valves that prevent the lymph from flowing backward. The lymph vessels lead to the thoracic duct and the right lymphatic duct (see Figure 42.20), which empty the lymph into a vein beneath the clavicles (collarbones), adding it to the plasma in the vein.

Movements of the skeletal muscles adjacent to the lymph vessels and breathing movements help move the lymph through the vessels, just as they help move the blood through veins. Over a day's time, the lymphatic system returns about 3 to 4 liters of fluid to the bloodstream.

Lymphoid Tissues and Organs Act as Filters and Participate in the Immune Response

The tissues and organs of the lymphatic system include the *lymph nodes,* the *spleen,* the *thymus,* and the *tonsils.* They play primary roles in filtering viruses, bacteria, damaged cells, and cellular debris from the lymph and bloodstream, and in defending the body against infection and cancer. Patches of lymphoid tissue are also scattered in other regions of the body, such as the small intestine and the appendix.

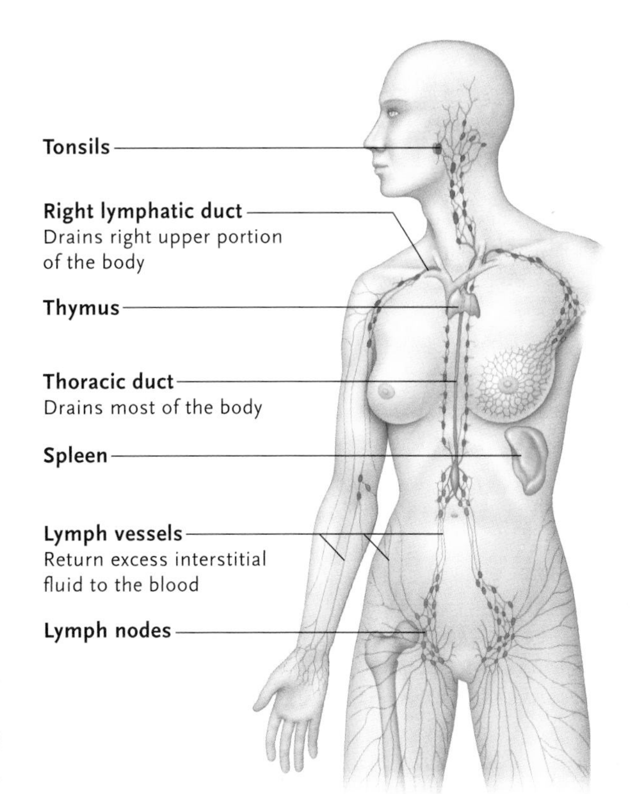

Figure 42.20 The human lymphatic system. Patches of lymphoid tissue in the small intestine and in the appendix also are part of the lymphatic system.

The **lymph nodes** are small, bean-shaped organs spaced along the lymph vessels and clustered along the sides of the neck, in the armpits and groin, and in the center of the abdomen and chest cavity (see Figure 42.20). Spaces in the nodes contain macrophages, a type of leukocyte that engulfs and destroys cellular debris and infecting bacteria and viruses in the lymph. The lymph nodes also contain other leukocytes, which produce antibodies that aid in the destruction of invading pathogens (discussed more in Chapter 43). Cancer cells that lodge in the nodes may be destroyed or may remain to grow and divide, forming new tumors within the nodes. Therefore, to reduce the risk of cancer spread, lymph nodes near a tumor typically are inspected and may be removed during surgery to excise that tumor.

The lymph nodes may become enlarged and painful if large numbers of bacteria or viruses carried by the lymph become trapped inside them. A doctor usually checks for swollen nodes, particularly in the neck, armpits, and groin, as indicators of an infection in the region of the body served by the nodes.

Study Break

What is lymph, and how does it enter the lymph capillaries?

Unanswered Questions

How did vertebrate blood clotting evolve?

Vertebrate blood clotting is a complex process involving about two dozen different proteins found in the blood plasma. The system has the properties of a biochemical amplifier in that the exposure of a tiny amount of tissue initiates a series of proteolytic events, one protease activating another successively, with the climax being a large amount of localized thrombin that transforms fibrinogen into a fibrin clot.

Many years ago when I was a graduate student working in a laboratory devoted to blood proteins, I asked myself the question, How could blood clotting ever have evolved? The process seemed much too complicated to have been concocted in one fell swoop, so, I reasoned, it must have begun in a simpler fashion and gradually become more complicated. Certainly, other people had been asking similar questions about complex organs—the evolution of the eye, for example, had been considered by many scientists, including Charles Darwin—but facts about the evolution of individual proteins were just beginning to emerge.

In particular, Vernon Ingram, a scientist working at MIT, had just determined the amino acid sequences of the alpha and beta chains of hemoglobin and found that these two proteins were about 45% identical. They must be the products of a gene duplication, he reported. Given his observation, it seemed to me that the proteases involved in blood clotting ought also to be the products of gene duplication. At the time, none of their amino acid sequences was known, but several had unique properties that distinguished them from other proteases, so it stood to reason that they were related. Yet, some nonproteolytic proteins were also involved, and they must have been added to the process independently.

So where to start? I decided to compare the blood clotting process in a wide range of animals. As it happens, most animals, invertebrate and vertebrate alike, have a kind of blood (see Section 42.2), and in most cases it can be coagulated by various stressful events. But the vertebrate process looked unique to me and must have evolved independently of the system that occurs in lobsters (Crustacea), for example. Among the vertebrates, the most primitive (early diverging) creature I could get my hands on was the lamprey (see Section 30.4, where its place in the vertebrate lineage is discussed).

Because I was hoping to find a simple, predecessor scheme, I was a little disappointed to find that lampreys have a rather sophisticated coagulation process that involved many of the proteins observed in mammals. Certainly, a small amount of tissue factor provoked a thrombin generation that converted fibrinogen to fibrin, just as in humans, at least in a general way. At the time there was no way to determine whether lampreys use the equivalent of *all* the clotting factors on the way to generating thrombin. Some of the most important proteins in the human system occur in minute amounts, and there was no possibility of ever isolating them from the lamprey, short of collecting several barrels of lamprey blood.

Many years later, after the sequences for many of the clotting factors had been reported for humans, my colleague Da-Fei Feng and I were able to align the sequences with a computer and make a phylogenetic tree. It fit the notion of a series of gene duplications very well. Moreover, it implied that most of the duplications had occurred a long time ago, at the very dawn of the vertebrates.

In 2003, the complete genomic sequence of a modern bony fish, the pufferfish, was determined, and one could scan through it and see what genes it has. It has all but a few of the more peripheral clotting proteins found in humans, the central theme being the same. However, the amount of sequence difference between human and pufferfish proteins compared with the degree of difference observed between the duplicated genes suggested some of the gene duplications had occurred not long before the appearance of bony fish. Why not look at the lamprey genome, which diverged 50 million to 100 million years before the appearance of bony fish, to see if the preduplication genes were there? The reason is that the lamprey genome has not been totally sequenced.

Recently, various genome centers around the world have begun maintaining "trace databases." These are uncurated collections of raw DNA sequences determined by random shotgun methods and robotized sequencers. The data are not assembled in any way, and each entry is at best a fragment of a gene. Several hundred organisms, from bacteria to monkeys, are being logged automatically. Among them is the lamprey! So now one can get a glimpse of what genes the lamprey has. We have been scrutinizing this database, even while recognizing its limitations. The reason is that I am getting on in years and can't wait for some mammoth operation with the resources to assemble all the lamprey DNA fragments into a complete genome.

At this point it looks like the lamprey may lack at least two of the mainline clotting factors (Factors VIII and IX), each of which in other vertebrates is the result of a gene duplication. Although the step-by-step evolution of vertebrate blood clotting factors is still technically an "unanswered question," I'm hoping our efforts with the limited trace database will spur others to go for a more convincing, fully assembled lamprey genome.

Russell Doolittle is professor emeritus at the University of California at San Diego. Finding out how blood clotting evolved is one of his major research interests. To learn more about Dr. Doolittle, go to http://www-biology.ucsd.edu/faculty/doolittle.html.

Review

Go to **Thomson**NOW™ at www.thomsonedu.com/login to access quizzing, animations, exercises, articles, and personalized homework help.

42.1 Animal Circulatory Systems: An Introduction

- Only the simplest invertebrates—the sponges, cnidarians, and flatworms—have no circulatory systems (Figure 42.2).
- Animals with circulatory systems have a muscular heart that pumps a specialized fluid, such as blood, from one body region to another through tubular vessels. The blood carries O_2 and nutrients to body tissues, and carries away CO_2 and wastes.
- Most invertebrates have an open circulatory system, in which the heart pumps hemolymph into vessels that empty into body

spaces called sinuses before returning to the heart. Some invertebrates and all vertebrates have a closed system, in which the blood is confined in blood vessels throughout the body and does not mix directly with the interstitial fluid (Figure 42.3).

- In invertebrates, open circulatory systems occur in arthropods and most mollusks, while closed circulatory systems occur in annelids and in mollusks such as squids and octopuses. In vertebrates, the circulatory system has evolved from a heart with a single series of chambers, pumping blood through a single circuit, to a double heart that pumps blood through separate pulmonary and systemic circuits (Figures 42.4 and 42.5).

Animation: Types of circulatory systems

Animation: Circulatory systems

42.2 Blood and Its Components

- Mammalian blood is a fluid connective tissue consisting of erythrocytes, leukocytes, and platelets, suspended in a fluid matrix, the plasma (Figure 42.6).
- Plasma contains water, ions, dissolved gases, glucose, amino acids, lipids, vitamins, hormones, and plasma proteins. The plasma proteins include albumins, globulins, and fibrinogen.
- Erythrocytes contain hemoglobin, which transports O_2 between the lungs and all body regions (Figure 42.7).
- Leukocytes defend the body against infecting pathogens.
- Platelets are functional cell fragments that trigger clotting reactions at sites of damage to the circulatory system.

Animation: White blood cells

Animation: ABO compatibilities

Animation: Rh factor and pregnancy

Animation: Hemostasis

42.3 The Heart

- The mammalian heart is a four-chambered pump. Two atria at the top of the heart pump the blood into two ventricles at the bottom of the heart, which pump blood into two separate pulmonary and systemic circuits of blood vessels (Figures 42.9 and 42.10).
- In both circuits, the blood leaves the heart in large arteries, which branch into smaller arteries, the arterioles. The arterioles deliver the blood to capillary networks, where substances are exchanged between the blood and the interstitial fluid. Blood is collected from the capillaries in small veins, the venules, which join into larger veins that return the blood to the heart (Figure 42.10).
- Contraction of the ventricles pushes blood into the arteries at a peak (systolic) pressure. Between contractions, the blood pressure in the arteries falls to a minimum (diastolic) pressure. The systole–diastole sequence is the cardiac cycle (Figure 42.11).
- Contraction of the atria and ventricles is initiated by signals from the SA node (pacemaker) of the heart (Figure 42.12).

Animation: Human blood circulation

Animation: Major human blood vessels

Animation: The human heart

Animation: Cardiac cycle

Animation: Cardiac conduction

Animation: Examples of ECGs

42.4 Blood Vessels of the Circulatory System

- Blood is carried from the heart to body tissues in arteries; small branches of arteries, the arterioles, deliver blood to the capillaries, where substances are exchanged with the interstitial fluid. The blood is collected from the capillaries in venules and then returned to the heart in veins (Figure 42.14).
- The walls of arteries consist of an inner endothelial layer, a middle layer of smooth muscle, and an outer layer of elastic fibers. The smallest arteries, the arterioles, constrict and dilate to regulate blood flow and pressure into the capillaries.
- Capillary walls consist of a single layer of endothelial cells. Blood flow through capillaries is controlled by variation in contraction of the smooth muscles of arterioles and precapillary sphincters (Figure 42.15).
- In the capillary networks, the rate of blood flow is considerably slower than that in arteries and veins. This maximizes the time for exchange of substances between blood and tissues. Diffusion along concentration gradients and bulk flow drive the exchange of substances (Figure 42.17).
- Venules and veins have thinner walls than arteries, allowing the vessels to expand and contract over a wide range. As a result, they act as blood reservoirs as well as conduits.
- The return of blood to the heart is aided by pressure exerted on the veins when surrounding skeletal muscles contract and by respiratory movements. One-way valves in the veins prevent the blood from flowing backward (Figure 42.18).

Animation: Capillary forces

Animation: Vessel anatomy

42.5 Maintaining Blood Flow and Pressure

- Blood pressure and flow are regulated by controlling cardiac output, the degree of blood vessel constriction (primarily arterioles), and the total blood volume. The autonomic nervous system and the endocrine system coordinate these mechanisms.
- Regulation of cardiac output starts with baroreceptors, which detect blood pressure changes and send signals to the medulla. In response, the brain stem sends signals via the autonomic nervous system that alter the rate and force of the heartbeat.
- Hormones secreted by several glands contribute to the regulation of blood pressure and flow.
- Local controls respond primarily to O_2 and CO_2 concentrations in tissues. Low O_2 and high CO_2 concentration cause dilation of arteriole walls, increasing the arteriole diameter and blood flow. High O_2 and low CO_2 concentrations have the opposite effects. NO released by arterial endothelial cells acts locally to increase arteriole diameter and blood flow.

Animation: Measuring blood pressure

Animation: Vein function

42.6 The Lymphatic System

- The lymphatic system, a key component of the immune system, is an extensive network of vessels that collect excess interstitial fluid—which becomes lymph—and returns it to the venous blood (Figure 42.20).
- The tissues and organs of the lymphatic system include the lymph nodes, the spleen, the thymus, and the tonsils. They remove viruses, bacteria, damaged cells, and cellular debris from the lymph and bloodstream, and defend the body against infection and cancer.

Animation: Human lymphatic system

Animation: Lymph vascular system

Questions

Self-Test Questions

1. Compared with vertebrates, most invertebrates:
   a. lead more mobile lives.
   b. require a higher level of oxygen.
   c. have more complex layers of cells.
   d. have slower distribution of blood.
   e. require faster delivery and greater quantities of nutrients.
2. Which circulatory system best describes the animal?
   a. Squids and octopuses have open circulatory systems with ventricles that pump blood away from the heart.
   b. Fishes have a single-chambered heart with an atrium that pumps blood through gills for oxygen exchange.
   c. Amphibians have the most oxygenated blood in the pulmocutaneous circuit and the most deoxygenated blood in the systemic circuit.
   d. Amphibians and reptiles use a two-chambered heart to separate oxygenated and deoxygenated blood.
   e. Birds and mammals pump blood to separate pulmonary and systemic systems from two separate ventricles in a four-chambered heart.
3. A healthy student from the coastal city of Boston enrolls at a college in Boulder, Colorado, a mile above sea level. An analysis of her blood in her first months at college would show:
   a. decreased macrophage activity.
   b. increased secretion of erythropoietin by the kidneys.
   c. increased signaling ability of platelets.
   d. anemia caused by malfunctioning erythrocytes.
   e. increased mitosis of leukocytes.
4. A characteristic of blood circulation through or to the heart is that:
   a. the superior vena cava conveys blood to the head.
   b. the inferior vena cava conveys blood to the right atrium.
   c. the pulmonary arteries convey blood from the lungs to the left atrium.
   d. the pulmonary veins convey blood into the left ventricle.
   e. the aorta branches into two coronary arteries that convey blood from heart muscle.
5. The heartbeat includes:
   a. the systole when the heart fills.
   b. the diastole when the heart muscle contracts.
   c. pressure that causes the AV valves to open, filling the ventricles.
   d. rising pressure in the ventricles to open the AV valves and close the SL valves.
   e. the "lub" sound when the SL valves open and the "dub" sound when the AV valves close.
6. Keeping the mammalian cardiac cycle balanced is/are:
   a. an AV node between the right atria and right ventricle, which signals the Purkinje fibers.
   b. pacemaker cells, which compose the AV node and signal the SA node.
   c. an insulating layer that isolates the SA node from the right atrium.
   d. ion channels in pacemaker cells, which close to depolarize their plasma membranes.
   e. neurogenic stimuli from the nervous system.
7. Hydrostatic pressure is best described as:
   a. the uncoordinated contractions that occur during heart attacks.
   b. a premature ventricular contraction that signifies a skipped beat.
   c. a high point of pressure called diastolic blood pressure.
   d. hypertension, which decreases with age.
   e. the pressure of blood on the walls of arteries.
8. Characteristics of veins and venules are:
   a. thick walls.
   b. large muscle mass in walls.
   c. a large quantity of elastin in the walls.
   d. low blood volume compared with arteries.
   e. one-way valves to prevent backflow of blood.
9. When capillaries exchange substances:
   a. red blood cells move through the capillary lumens in double file.
   b. blood flow resistance is lower than it is in arteries and veins.
   c. water, ions, glucose, and erythrocytes pass freely between blood and tissues.
   d. diffusion along a concentration gradient and bulk flow are operating.
   e. diffusion is greatest closest to the arterioles.
10. To increase cardiac output:
    a. the adrenal medulla and sympathetic nervous system secrete epinephrine and norepinephrine.
    b. baroreceptors in the brain signal the sympathetic nerves.
    c. the brain stem signals the baroreceptors, causing the heart to beat faster.
    d. the autonomic nervous system responds to low oxygen on chemoreceptors and decreases the force of the heartbeat.
    e. chemoreceptors, stimulated by excessive blood oxygen, increase the rate of the heartbeat.

Questions for Discussion

1. *Aplastic anemia* develops when certain drugs or radiation destroy red bone marrow, including the stem cells that give rise to erythrocytes, leukocytes, and platelets. Predict some symptoms a person with aplastic anemia would be likely to develop. Include at least one symptom related to each type of blood cell.
2. In addition to the engine exhaust of boats and cars, carbon monoxide is also a component of cigarette smoke. What might be the impact of this phenomenon on a smoker's health?
3. In some people, the pressure of the blood pooling in the legs leads to a condition called *varicose veins*, in which the veins stand out like swollen, purple knots. Explain why this might happen, and why veins closer to the leg surface are more susceptible to the condition than those in deeper leg tissues.

Experimental Analysis

Mice in which the apolipoprotein E gene has been knocked out (deleted) by genetic engineering methods have high levels of plasma cholesterol and readily develop atherosclerosis, particularly on diets high in cholesterol. The immunosuppressant drug rapamycin is being touted also to be a drug that can affect atherosclerosis. Design an experiment to determine whether and at what dose rapamycin is effective in reducing atherosclerosis caused by dietary cholesterol.

Evolution Link

What is the evolutionary advantage of closure of the septum between the two ventricles to create a double circulatory system?

How Would You Vote?

Cardiopulmonary resuscitation (CPR) can make the difference between life and death after a cardiac arrest or a heart attack. Should public high schools in your state require all students to take a course in CPR? Is such a course worth diverting time and resources from the basic curriculum? Go to www.thomsonedu.com/login to investigate both sides of the issue and then vote.

Death of a cancer cell. A cytotoxic T cell (orange) induces a cancer cell (mauve) to undergo apoptosis (programmed cell death). Cytotoxic T cells are part of the body's immune response system programmed to seek out, attach themselves, and kill cancer cells and pathogen-infected host cells.

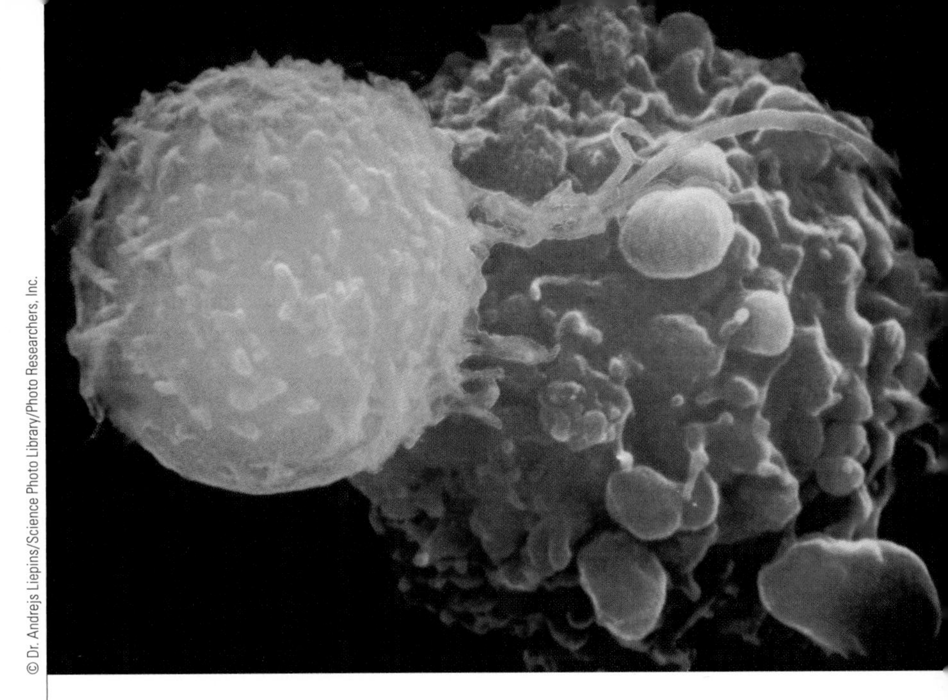

Study Plan

43 Defenses against Disease

Why It Matters

Acquired immune deficiency syndrome (AIDS), which was first identified in the early 1980s, now infects about 40 million people worldwide, and continues to spread. Thousands of health-care workers, physicians, and researchers have joined the effort to control AIDS and develop effective treatments. Their primary aim is to develop an anti-AIDS *vaccine*—a substance that, when swallowed or injected, provides protection against infection by HIV (human immunodeficiency virus), which causes the disease.

The development of vaccines began with efforts to control smallpox, a dangerous and disfiguring viral disease that once infected millions of people worldwide. As early as the twelfth century, healthy individuals in China sought out people who were recovering from mild smallpox infections, ground up scabs from their lesions, and inhaled the powder or pushed it into their skin. Variations on this treatment were effective in protecting many people against smallpox infection.

In 1796, an English country doctor, Edward Jenner, used a more scientific approach. He knew that milkmaids never got smallpox if they had contracted cowpox, a similar but mild disease of cows that can be transmitted to humans. Jenner decided to see if a deliberate infection with cowpox would protect humans from smallpox. He scratched material from a cowpox sore into a boy's arm, and 6 weeks later, after the cowpox infection had subsided, he scratched fluid from human smallpox sores into the boy's skin. (Jenner's use of the boy as an experimental subject would now be considered unethical.) Remarkably, the boy remained free from smallpox. Jenner carried out additional, carefully documented case studies with other patients with the same results. His technique became the basis for worldwide **vaccination** (*vacca* = cow) against smallpox. With improved vaccines, smallpox has now been eradicated from the human population.

Vaccination takes advantage of the **immune system** (*immunis* = exempt), the natural protection that is our main defense against infectious disease. This chapter focuses on the immune system and other defenses against infection, such as the skin. Our description emphasizes human and other mammalian systems, in which most of the scientific discoveries revealing the structure and function of the immune system have been made. At the end of the chapter we compare mammalian systems with the protective systems of nonmammalian vertebrates and invertebrates.

43.1 Three Lines of Defense against Invasion

Every organism is constantly exposed to *pathogens,* disease-causing viruses or organisms such as infectious bacteria, protists, fungi, and parasitic worms. Humans and other mammals have three lines of defense against these threats. The first line of defense involves physical barriers that prevent infection; it is not part of the immune system. The second line of defense is the *innate immunity system,* the inherited mechanisms that protect the body from many kinds of pathogens in a nonspecific way. The third line of defense, the *adaptive immunity system,* involves inherited mechanisms leading to the synthesis of molecules that target pathogens in a specific way. Reaction to an infection takes minutes in the case of the innate immunity system versus days for the adaptive immunity system.

Epithelial Surfaces Are Anatomical Barriers That Help Prevent Infection

An organism's first line of defense is the body surface—the skin covering the body exterior and the epithelial surfaces covering internal body cavities and ducts, such as the lungs and intestinal tract. The body surface forms a barrier of tight junctions between epithelial cells that keeps most pathogens (as well as toxic substances) from entering the body.

Many epithelial surfaces are coated with a mucus layer secreted by the epithelial cells that protects against pathogens as well as toxins and other chemicals. In the respiratory tract, ciliated cells constantly sweep the mucus, with its trapped bacteria and other foreign matter, into the throat, where it is coughed out or swallowed.

Many of the body cavities lined by mucous membranes have environments that are hostile to pathogens. For example, the strongly acidic environment of the stomach kills most bacteria and destroys many viruses that are carried there, including those trapped in swallowed mucus from the respiratory tract. Most of the pathogens that survive the stomach acid are destroyed by the digestive enzymes and bile secreted into the small intestine. The vagina, too, is acidic, which prevents many pathogens from surviving there. The mucus coating in some locations contains the enzyme lysozyme, which was secreted by the epithelial cells. Lysozyme breaks down the walls of some bacteria, causing them to lyse.

Two Immunity Systems Protect the Body from Pathogens That Have Crossed External Barriers

The body's second line of defense is a series of generalized internal chemical, physical, and cellular reactions that attack pathogens that have breached the first line. These defenses include inflammation, which creates internal conditions that inhibit or kill many pathogens, and specialized cells that engulf or kill pathogens or infected body cells.

Innate immunity is the term for this initial response by the body to eliminate cellular pathogens, such as bacteria and viruses, and prevent infection. You are born with an innate immune system. Innate immunity provides an immediate, *nonspecific* response; that is, it targets any invading pathogen and has no memory of prior exposure to the pathogen. It provides some protection against invading pathogens while a more powerful, specific response system is mobilized.

The third and most effective line of defense, **adaptive** (also called **acquired**) **immunity,** is *specific:* it recognizes individual pathogens and mounts an attack that directly neutralizes or eliminates them. It is so named because it is stimulated and shaped by the presence of a specific pathogen or foreign molecule. This mechanism takes several days to become protective. Adaptive immunity is triggered by specific molecules on pathogens that are recognized as being foreign to the body. The body retains a memory of the first exposure to a foreign molecule, enabling it to respond more quickly if the pathogen is encountered again in the future.

Innate immunity and adaptive immunity together constitute the immune system, and the defensive reactions of the system are termed the **immune response.** Functionally, the two components of the immune system interconnect and communicate at the chemical and cellular levels. The immune system is the product of a long evolutionary history of compensating adaptations by both pathogens and their targets. Over millions of years of vertebrate history, the mechanisms by which pathogens attack and invade have become more efficient, but the defenses of animals against the invaders have kept pace.

Study Break

1. What features of epithelial surfaces protect against pathogens?
2. What are the key differences between innate immunity and adaptive immunity?

43.2 Nonspecific Defenses: Innate Immunity

In most cases, the body needs 7 to 10 days to develop a fully effective immune response against a new pathogen, one that is invading the body for the first time. Innate immunity holds off invading pathogens in the meantime, killing or containing them until adaptive immunity comes fully into play. We have already discussed the body's anatomical barriers. Now let us look at the internal mechanisms of innate immunity: secreted molecules and cellular components. As you will see, cellular pathogens (such as bacteria) and viral pathogens elicit different responses.

Innate Immunity Provides an Immediate, General Defense against Invading Cellular Pathogens

Cellular pathogens—typically microorganisms—usually enter the body when injuries break the skin or epithelial surfaces. How does the host body recognize the pathogen as foreign? The answer is that the host has mechanisms to distinguish self from nonself. The innate immune system recognizes particular molecules that are common to many pathogens but absent in the host. An example is the lipopolysaccharide of gram-negative bacteria. The host then responds immediately to combat the pathogen.

Several types of specific cell-surface receptors in the host recognize the various types of molecules on microbial pathogens. The response depends on the receptor. For some receptors, the response is secretion of *antimicrobial peptides,* which kill the microbial pathogen. Other receptors include those that trigger the host cell to engulf, and destroy the pathogen, initiating inflammation, and the soluble receptors of the *complement system.*

Antimicrobial Peptides. All of our epithelial surfaces, namely skin, the lining of the gastrointestinal tract, the lining of the nasal passages and lungs, and the lining of the genitourinary tracts, are protected by antimicrobial peptides called *defensins.* Epithelial cells of those surfaces secrete defensins upon attack by a microbial pathogen. The defensins attack the plasma membranes of the pathogens, eventually disrupting them, thereby killing the cells. In particular, defensins play a significant role in innate immunity of the mammalian intestinal tract.

Inflammation. A tissue's rapid response to injury, including infection by most pathogens, involves **inflammation** (*inflammare* = to set on fire), the heat, pain, redness, and swelling that occur at the site of an infection.

Several interconnecting mechanisms initiate inflammation **(Figure 43.1)**. Let us consider bacteria entering a tissue as a result of a wound. **Monocytes** (a type of leukocyte) enter the damaged tissue from the bloodstream through the endothelial wall of the blood vessel. Once in the damaged tissue, the monocytes differentiate into **macrophages** ("big eaters"), which are phagocytes that are usually the first to recognize pathogens at the cellular level. (**Table 43.1** lists the major types of leukocytes such as macrophages; see also Figure 42.7.) Cell-surface receptors on the macrophages recognize and bind to surface molecules on the pathogen, activating the macrophage to phagocytize (engulf) the pathogen (see Figure 43.1, step 1). Activated macrophages also secrete **cytokines,** molecules that bind to receptors on other host cells and, through signal transduction pathways, trigger a response. Usually, not enough macrophages are present in the area of a bacterial infection to handle all of the bacteria.

The death of cells caused by the pathogen at the infection site activates cells dispersed through connective tissue called **mast cells,** which then release histamine (step 2). This histamine, along with the cytokines from activated macrophages, dilates local blood vessels around the infection site and increases their permeability, which increases blood flow and leakage of fluid from the vessels into body tissues (step 3). The response initiated by cytokines directly causes the heat, redness, and swelling of inflammation.

Cytokines also make the endothelial cells of the blood vessel wall stickier, causing circulating **neutrophils** (another type of phagocytic leukocyte) to attach to it in massive numbers. From there, the neutrophils are attracted to the infection site by **chemokines,** proteins also secreted by activated macrophages (step 4). To get to the infection site, the neutrophils pass between endothelial cells of the blood vessel wall. Neutrophils may also be attracted to the pathogen directly by molecules released from the pathogens themselves. Like macrophages, neutrophils have cell-surface recep-

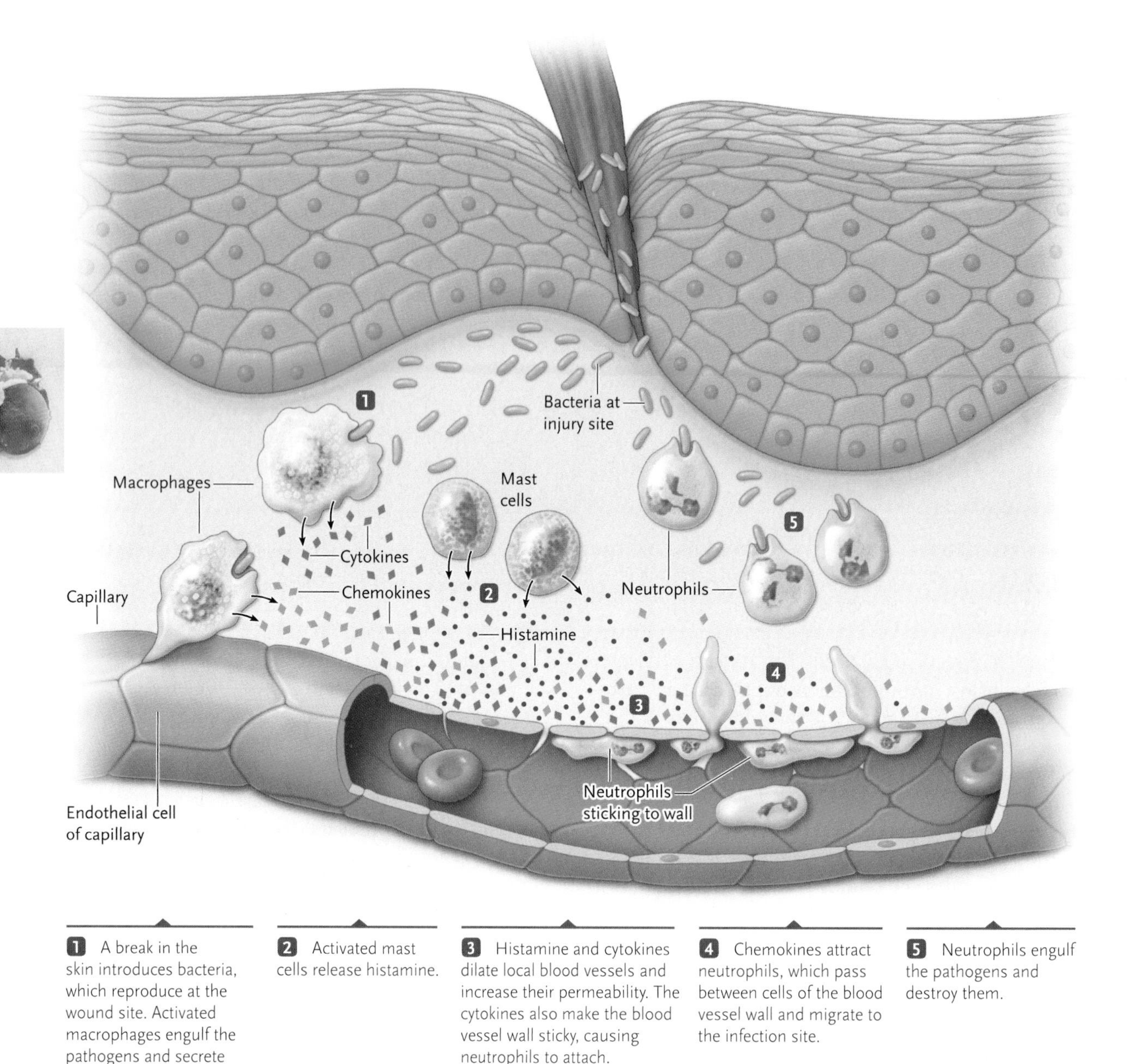

1 A break in the skin introduces bacteria, which reproduce at the wound site. Activated macrophages engulf the pathogens and secrete cytokines and chemokines.

2 Activated mast cells release histamine.

3 Histamine and cytokines dilate local blood vessels and increase their permeability. The cytokines also make the blood vessel wall sticky, causing neutrophils to attach.

4 Chemokines attract neutrophils, which pass between cells of the blood vessel wall and migrate to the infection site.

5 Neutrophils engulf the pathogens and destroy them.

Figure 43.1
The steps producing inflammation. The colorized micrograph on the left shows a macrophage engulfing a yeast cell.

tors that enable them to recognize and engulf pathogens (step 5).

Once a macrophage or neutrophil has engulfed the pathogen, it uses a variety of mechanisms to destroy it. These mechanisms include attacks by enzymes and defensins located in lysosomes and the production of toxic chemicals. The harshness of these attacks usually kills the neutrophils as well, while macrophages usually survive to continue their pathogen-scavenging activities. Dead and dying neutrophils, in fact, are a major component of the pus formed at infection sites. The pain of inflammation is caused by the migration of macrophages and neutrophils to the infection site and their activities there.

Some pathogens, such as parasitic worms, are too large to be engulfed by macrophages or neutrophils. In that case, macrophages, neutrophils, and **eosinophils** (another type of leukocyte) cluster around the pathogen and secrete lysosomal enzymes and defensins in amounts that are often sufficient to kill the pathogen.

The Complement System. Another nonspecific defense mechanism activated by invading pathogens is the **complement system**, a group of more than 30 interacting soluble plasma proteins circulating in the blood and interstitial fluid **(Figure 43.2)**. The proteins are normally inactive; they are activated when they recognize molecules on the surfaces of pathogens. Activated complement proteins participate in a cascade of reactions on pathogen surfaces, producing large numbers of different complement proteins, some of which assemble into **membrane attack complexes.** These complexes insert into the plasma membrane of many types of bacterial cells and create pores that allow ions and small molecules to pass readily through the membrane. As a result, the bacteria can no longer maintain osmotic balance, and they swell and lyse. For the other types of bacterial cells, the cascade of reactions coats the pathogen with fragments of the complement proteins. Cell-surface receptors on phagocytes then recognize these fragments, and engulf and destroy the pathogen.

Several activated proteins in the complement cascade also act individually to enhance the inflammatory response. For example, some of the proteins stimulate mast cells to enhance histamine release, while others cause increased blood vessel permeability.

Combating Viral Pathogens Requires a Different Innate Immune Response

You have learned that specific molecules on cellular pathogens such as bacteria are key to initiating innate immune responses. By contrast, the innate immunity system is unable to distinguish effectively the surface molecules of viral pathogens from those of the host. The host must, therefore, use other strategies to provide some immediate protection against viral infec-

Table 43.1 Major Types of Leukocytes and Their Functions

Type of Leukocyte	Function
Monocyte	Differentiates into a macrophage when released from blood into damaged tissue
Macrophage	Phagocyte that engulfs infected cells, pathogens, and cellular debris in damaged tissues; helps activate lymphocytes carrying out immune response
Neutrophil	Phagocyte that engulfs pathogens and tissue debris in damaged tissues
Eosinophil	Secretes substances that kill eukaryotic parasites such as worms
Lymphocyte	Main subtypes involved in innate and adaptive immunity are natural killer (NK) cells, B cells, plasma cells, helper T cells, and cytotoxic T cells. NK cells function as part of innate immunity to kill virus-infected cells and some cancerous cells of the host. The other cell types function as part of adaptive immunity: they produce antibodies, destroy infected and cancerous body cells, and stimulate macrophages and other leukocyte types to engulf infected cells, pathogens, and cellular debris
Basophil	Located in blood, responds to IgE antibodies in an allergic response by secreting histamine, which stimulates inflammation

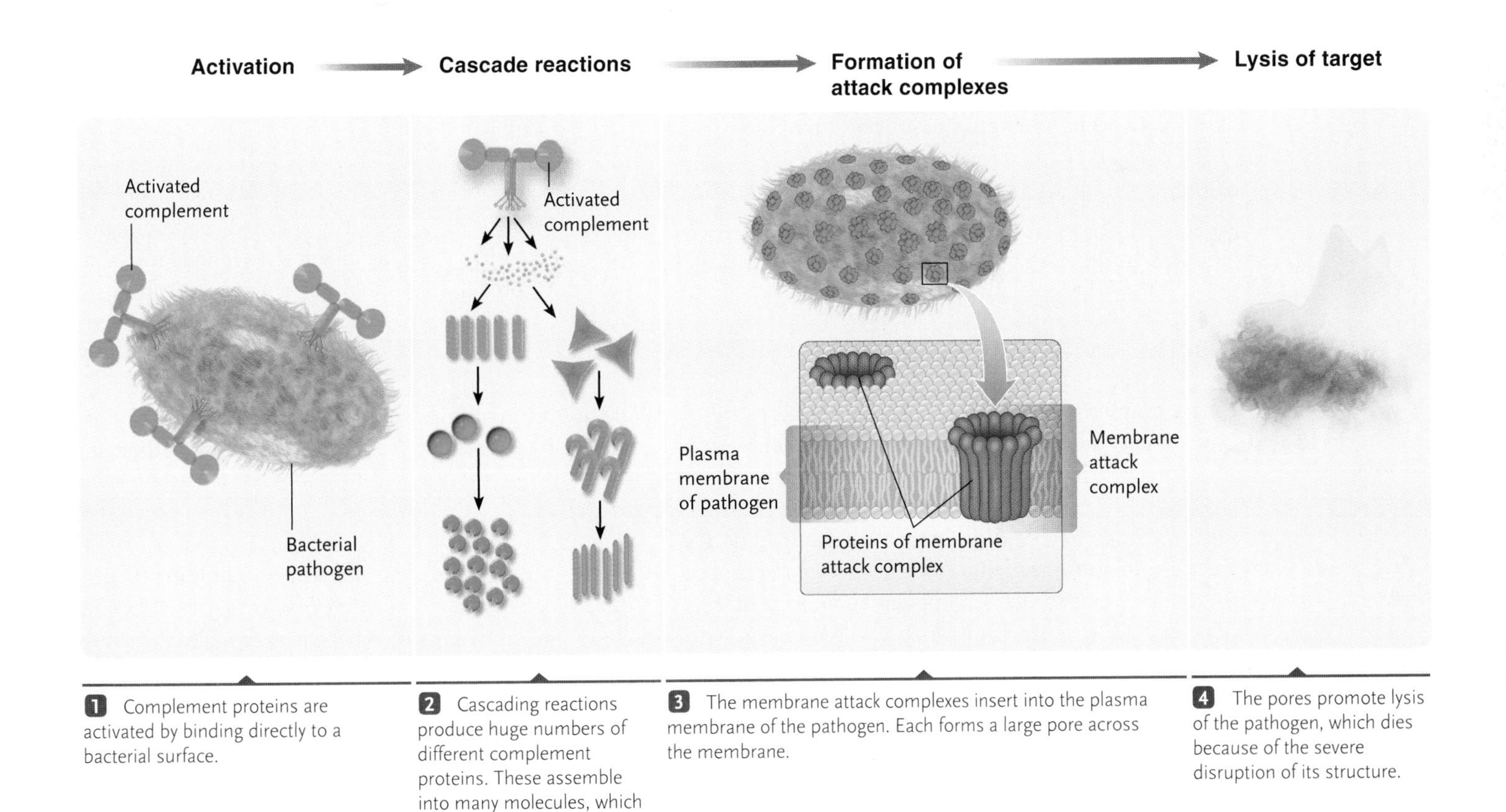

Activation
Cascade reactions
Formation of attack complexes
Lysis of target
Activated complement
Bacterial pathogen
Activated complement
Plasma membrane of pathogen
Membrane attack complex
Proteins of membrane attack complex
1 Complement proteins are activated by binding directly to a bacterial surface.
2 Cascading reactions produce huge numbers of different complement proteins. These assemble into many molecules, which form many membrane attack complexes.
3 The membrane attack complexes insert into the plasma membrane of the pathogen. Each forms a large pore across the membrane.
4 The pores promote lysis of the pathogen, which dies because of the severe disruption of its structure.

Figure 43.2
The role of complement proteins in combating microbial pathogens.

tions until the adaptive immunity system, which can discriminate between viral and host proteins, is effective. Three main strategies involve RNA interference, interferon, and natural killer cells.

RNA Interference. *RNA interference (RNAi)* is a cellular mechanism that is triggered by a virus's double-stranded (ds) RNA molecules (see Section 16.3). Such molecules are a natural part of the life cycles of a number of viruses. The RNAi system destroys dsRNA molecules, thereby inhibiting the virus's life cycle.

Interferon. Viral dsRNA also causes the infected host cell to produce two cytokines, called interferon-α and interferon-β. **Interferons** can be produced by most cells of the body. These proteins act both on the infected cell that produces them, an autocrine effect, and on neighboring uninfected cells, a paracrine effect (see Section 40.1). They work by binding to cell-surface receptors, triggering a signal transduction pathway that changes the gene expression pattern of the cells. The key changes include activation of a ribonuclease enzyme that degrades most cellular RNA and inactivation of a key protein required for protein synthesis, thereby inhibiting most protein synthesis in the cell. These effects on RNA and protein synthesis inhibit replication of the viral genome, while putting the cell in a weakened state from which it often can recover.

Natural Killer Cells. Cells that have been infected with virus must be destroyed. That is the role of **natural killer (NK) cells.** NK cells are a type of **lymphocyte,** a leukocyte that carries out most of its activities in tissues and organs of the lymphatic system (see Figure 42.20). NK cells circulate in the blood and kill target host cells—not only cells that are infected with virus, but also some cells that have become cancerous.

NK cells can be activated by cell-surface receptors or by interferons secreted by virus-infected cells. NK cells are not phagocytes; instead, they secrete granules containing *perforin,* a protein that creates pores in the target cell's membrane. Unregulated diffusion of ions and molecules through the pores causes osmotic imbalance, swelling, and rupture of the infected cell. NK cells also kill target cells indirectly through the secretion of *proteases* (protein-degrading enzymes) that pass through the pores. The proteases trigger **apoptosis,** or programmed cell death (see *Insights from the Molecular Revolution* in Chapter 7). That is, the proteases activate other enzymes that cause the degradation of DNA which, in turn, induces pathways leading to the cell's death.

How does an NK cell distinguish a target cell from a normal cell? The surfaces of most vertebrate cells contain particular *major histocompatibility complex (MHC) proteins.* You will learn about the role of these proteins in adaptive immunity in the next section; for now, just consider them to be tags on the cell surface. NK cells monitor the level of MHC proteins and respond differently depending on their level. An appropriately high level, as on normal cells, inhibits the killing activity of NK cells. Viruses often inhibit the synthesis of MHC proteins in the cells they infect, lowering the levels of those proteins and identifying them to NK cells. Cancer cells also have low or, in some cases, no MHC proteins on their surfaces, which makes them a target for destruction by NK cells as well.

Study Break

1. What are the usual characteristics of the inflammatory response?
2. What processes specifically cause each characteristic of the inflammatory response?
3. What is the complement system?
4. Why does combating viral pathogens require a different response by the innate immunity system than combating bacterial pathogens? What are the three main strategies a host uses to protect against viral infections?

43.3 Specific Defenses: Adaptive Immunity

Adaptive immunity is a defense mechanism that recognizes specific molecules as being foreign and clears those molecules from the body. The foreign molecules recognized may be free, as in the case of toxins, or they may be on the surface of a virus or cell, the latter including pathogenic bacteria, cancer cells, pollen, and cells of transplanted tissues and organs. Adaptive immunity develops only when the body is exposed to the foreign molecules and, hence, takes several days to become effective. This would be a significant problem in the case of invading pathogens were it not for the innate immune system, which combats the invading pathogens in a nonspecific way within minutes after they enter the body. There are two key distinctions between innate and adaptive immunity: innate immunity is nonspecific whereas adaptive immunity is specific, and innate immunity retains no memory of exposure to the pathogen whereas adaptive immunity retains a memory of the foreign molecule that triggered the response, thereby enabling a rapid, more powerful response if that pathogen is encountered again at a later time.

In Adaptive Immunity, Antigens Are Cleared from the Body by B Cells or T Cells

A foreign molecule that triggers an adaptive immunity response is called an **antigen** (meaning "*anti*body *gen*erator"). Antigens are macromolecules; most are large proteins (including glycoproteins and lipoproteins) or polysaccharides (including lipopolysaccharides). Some types of nucleic acids can also act as antigens, as can various large, artificially synthesized molecules.

Antigens may be *exogenous,* meaning they enter the body from the environment, or *endogenous,* meaning they are generated within the body. Exogenous antigens include antigens on pathogens introduced beneath the skin, antigens in vaccinations, and inhaled and ingested macromolecules, such as toxins. Endogenous antigens include proteins encoded by viruses that have infected cells, and altered proteins produced by mutated genes, such as those in cancer cells.

Antigens are recognized in the body by two types of lymphocytes, B cells and T cells. **B cells** differentiate from stem cells in the bone marrow (see Section 42.2). It is easy to remember this as "B for bone." However, the "B" actually refers to the *bursa of Fabricus,* a lymphatic organ found only in birds; B cells were first discovered there. After their differentiation, B cells are released into the blood and carried to capillary beds serving the tissues and organs of the lymphatic system. Like B cells, **T cells** are produced by the division of stem cells in the bone marrow. Then, they are released into the blood and carried to the **thymus,** an organ of the lymphatic system (the "T" in "T cell" refers to the thymus).

The role of lymphocytes in adaptive immunity was demonstrated by experiments in which all of the leukocytes in mice were killed by irradiation with X rays. These mice were then unable to develop adaptive immunity. Injecting lymphocytes from normal mice into the irradiated mice restored the response; other body cells extracted from normal mice could not restore the response. (For more on the use of mice as an experimental organism in biology, see *Focus on Research Organisms.*)

There are two types of adaptive immune responses: **antibody-mediated immunity** (also called *humoral immunity*) and **cell-mediated immunity.** In antibody-mediated immunity, B-cell derivatives called **plasma cells** secrete **antibodies,** highly specific soluble protein molecules that circulate in the blood and lymph recognizing and binding to antigens and clearing them from the body. In cell-mediated immunity, a subclass of T cells becomes activated and, with other cells of the immune system, attacks foreign cells directly and kills them.

The steps involved in the adaptive immune response are similar for antibody-mediated immunity and cell-mediated immunity:

1. Antigen encounter and recognition: Lymphocytes encounter and recognize an antigen.
2. Lymphocyte activation: The lymphocytes are activated by binding to the antigen and proliferate by cell division to produce large clones of identical cells.
3. Antigen clearance: The large clones of activated lymphocytes are responsible for clearing the antigen from the body.
4. Development of immunological memory: Some of the activated lymphocytes differentiate into **memory cells** that circulate in the blood and lymph, ready to initiate a rapid immune response upon subsequent exposure to the same antigen.

These steps will be expanded upon in the following discussions of antibody-mediated immunity and cell-mediated immunity.

Antibody-Mediated Immunity Involves Activation of B Cells, Their Differentiation into Plasma Cells, and the Secretion of Antibodies

An adaptive immune response begins as soon as an antigen is encountered and recognized in the body.

Antigen Encounter and Recognition by Lymphocytes. Exogenous antigens are encountered by lymphocytes in the lymphatic system. As already mentioned, the two key lymphocytes that recognize antigens are B cells and T cells. Each B cell and each T cell is specific for a particular antigen, meaning that the cell can bind to only one particular molecular structure. The binding is so specific because the plasma membrane of each B cell and T cell is studded with thousands of identical receptors for the antigen; in B cells they are called **B-cell receptors (BCRs)** and in T cells they are called **T-cell receptors (TCRs) (Figure 43.3).** Considering the entire populations of B cells and T cells in the body, there are multiple cells that can recognize each antigen but, most importantly, the populations (in normal persons) contain cells capable of recognizing any antigen. For example, each of us has about 10 trillion B cells that collectively have about 100 million different kinds of BCRs. And, these cells are present *before* the body has encountered the antigens.

The binding between antigen and receptor is an interaction between two molecules that fit together like an enzyme and its substrate. A given BCR or TCR typically does not bind to the whole antigen molecule, but to small regions of it called **epitopes** or *antigenic determinants.* Therefore, several different B cells and T cells may bind to the population of a particular antigen encountered in the lymphatic system.

BCRs and TCRs are encoded by different genes and thus have different structures. The BCR on a B cell (see Figure 43.3a) corresponds to the antibody secreted by that particular B cell when it is activated and differentiates into a plasma cell. As you will learn in more detail shortly, an antibody molecule is a protein consisting of four polypeptide chains. At one end, it has two identical *antigen-binding sites,* regions that bind to a specific antigen. In the case of the BCR, at the opposite end of the

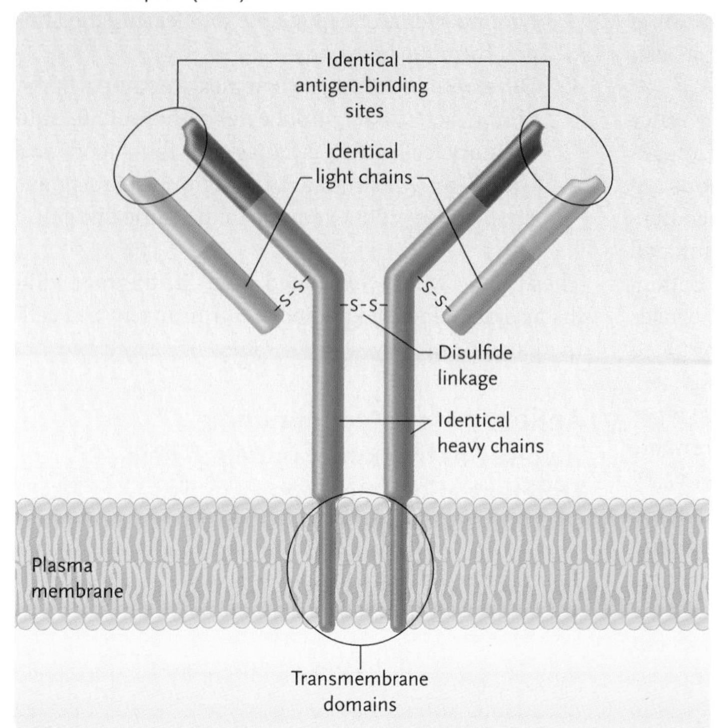

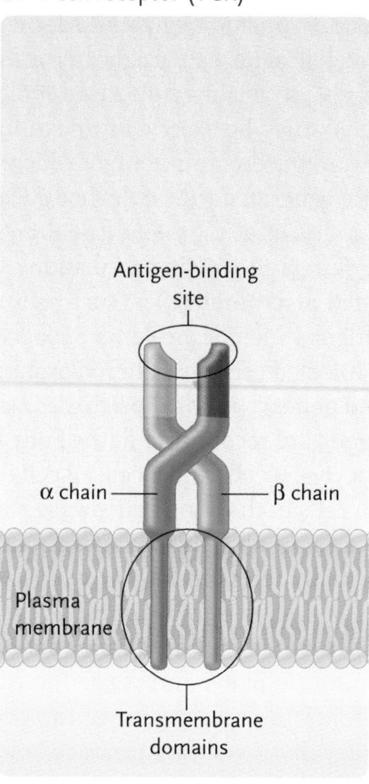

Figure 43.3
Antigen-binding receptors on B cells and T cells.

molecule from the antigen-binding sites are *transmembrane domains,* which embed in the plasma membrane. TCRs are simpler than BCRs, consisting of a protein made up of two different polypeptides (see Figure 43.3b). Like BCRs, TCRs have an antigen-binding site at one end and transmembrane domains at the other end.

Antibodies. Antibodies are the core molecules of antibody-mediated immunity. Antibodies are large, complex proteins that belong to a class of proteins known as **immunoglobulins** (Ig). Each antibody molecule consists of four polypeptide chains: two identical **light chains** and two identical **heavy chains** about twice or more the size of the light chain **(Figure 43.4).** The chains are held together in the complete protein by disulfide (—S—S—) linkages and fold into a Y-shaped structure. The bonds between the two arms of the Y form a hinge that allows the arms to flex independently of one another.

Each polypeptide chain of an antibody molecule has a *constant region* and a *variable region.* The constant region of each antibody type has the same amino acid sequence for that part of the heavy chain, and likewise for that part of the light chain. The variable region of

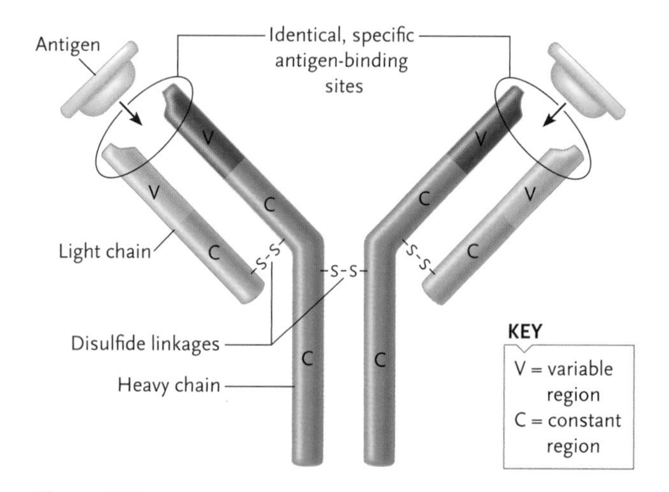

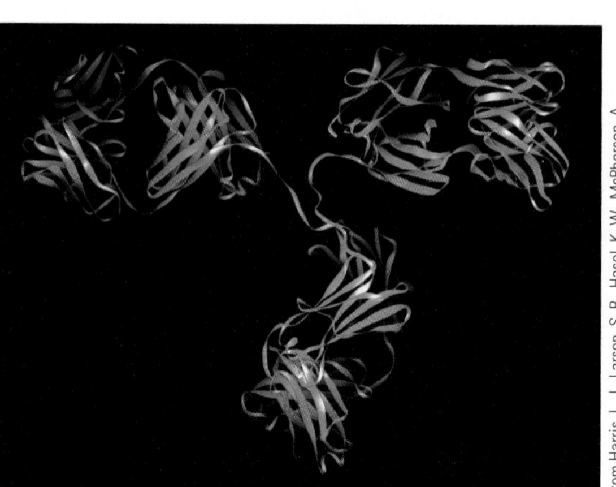

From Harris, L. J., Larson, S. B., Hasel, K. W., McPherson, A., *Biochemistry* 36, p. 1581 (1997). Structure rendered with RIBBONS.

Figure 43.4
The arrangement of light and heavy polypeptide chains in an antibody molecule. As shown, two sites, one at the tip of each arm of the Y, bind the same antigen.

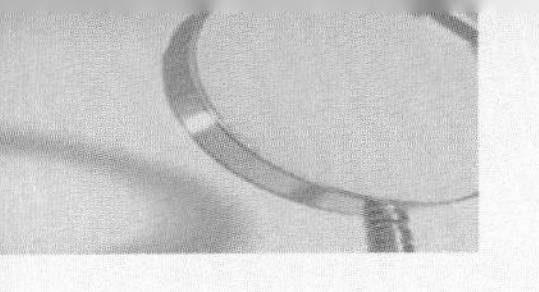

Focus on Research

Research Organisms: The Mighty Mouse

The "wee, sleekit, tim'rous, cowrin' beastie," as the poet Robert Burns called the mouse *(Mus musculus)*, has a much larger stature among scientists. The mouse and its cells have been used to great advantage as models for research on mammalian developmental genetics, immunology, and cancer. The availability of the mouse as a research tool enables scientists to carry out experiments with a mammal that would not be practical or ethical with humans.

Mice are grown by the millions in laboratories all over the world. Its small size makes the mouse relatively inexpensive and easy to maintain in the laboratory, and its short generation time, compared with most other mammals, allows genetic crosses to be carried out within a reasonable time span. Mice can be mated when they are 10 weeks old; within 18 to 22 days the female gives birth to a litter of about 5 to 10 offspring. A female may be rebred within little more than a day after giving birth.

© Peter Skinner/Photo Researchers, Inc.

Mice have a long and highly productive history as experimental animals. Gregor Mendel, the founder of genetics, is known to have kept mice as part of his studies. Toward the end of the nineteenth century, August Weissmann helped disprove an early evolutionary hypothesis, the inheritance of acquired characters, by cutting off the tails of mice for 22 successive generations and finding that it had no effect on tail length. The first example of a lethal allele was also found in mice, and pioneering experiments on the transplantation of tissues between individuals were conducted with mice. During the 1920s, Fred Griffith laid the groundwork for the research showing that DNA is the hereditary molecule in his work with pneumonia-causing bacteria in mice (see Section 14.1). More recently, genetic experiments with mice have revealed more than 500 mutants that cause hereditary diseases, immunological defects, and cancer in mammals including humans.

The mouse has also been the mammal of choice for experiments that introduce and modify genes through genetic engineering. One of the most spectacular results of this research was the production of giant mice by introducing a human growth hormone gene into a line of dwarfed mice that were deficient for this hormone.

Genetic engineering has also produced "knockout" mice, in which a gene of interest is completely nonfunctional (see Section 18.2). The effects of the lack of function of a gene in the knockout mice often allow investigators to determine the role of the normal form of the gene. Some knockout mice have been developed to be defective in genes homologous to human genes that cause serious diseases, such as cystic fibrosis. This allows researchers to study the disease in mice with the goal of developing cures or therapies.

The revelations in developmental genetics from studies with the mouse have been of great interest and importance in their own right. But more and more, as we find that much of what applies to the mouse also applies to humans, the findings in mice have shed new light on human development and opened pathways to the possible cure of human genetic diseases.

In 2002 the sequence of the mouse genome was reported. The sequence is enabling researchers to refine and expand their use of the mouse as a model organism for studies of mammalian biology and mammalian diseases.

both the heavy and light chains, by contrast, has a different amino acid sequence for each antibody molecule in a population. Structurally, the variable regions are the top halves of the polypeptides in the arms of the Y-shaped molecule. The three-dimensional folding of the heavy chain and light chain variable regions of each arm creates the antigen-binding site. The antigen-binding site is identical for the two arms of the same antibody molecule because of the identity of the two heavy chains and the two light chains in a molecule, as mentioned earlier. However, the antigen-binding sites are different from antibody molecule to antibody molecule because of the amino acid differences in the variable regions of the two chain types.

The constant regions of the heavy chains in the tail part of the Y-shaped structure determine the *class* of the antibody. Humans have five different classes of antibodies—*IgM, IgG, IgA, IgE,* and *IgD* **(Table 43.2).** Due to differences in their heavy chain constant regions, they have specific structural and functional differences.

IgM remains bound to the cells that make it due to a region at the end opposite from the antigen-binding end that inserts into the plasma membrane of the cell. BCRs on B cells for antigen recognition are IgM molecules; as we shall see, IgM is also the first type of antibody secreted from plasma cells in the early stages of an antibody-mediated response. (When secreted, they exist as a pentamer.) IgM antibodies activate the complement system when they bind an antigen, and stimulate the phagocytic activity of macrophages.

IgG is the most abundant antibody circulating in the blood and lymphatic system, where it stimulates phagocytosis and activates the complement system when it binds an antigen. IgG is produced in large

Table 43.2	Five Classes of Antibodies		
Class	**Structure (Secreted Form)**	**Location**	**Functions**
IgM		Surfaces of unstimulated B cells (as monomer); free in circulation (as pentamer)	First antibodies to be secreted by B cells in primary response. When bound to antigen, promotes agglutination reaction, activates complement system, and stimulates phagocytic activity of macrophages.
IgG		Blood and lymphatic circulation	Most abundant antibody in primary and secondary responses. Crosses placenta, conferring passive immunity to fetus; stimulates phagocytosis and activates complement system.
IgA		Body secretions such as tears, breast milk, saliva, and mucus	Blocks attachment of pathogens to mucous membranes; confers passive immunity for breastfed infants.
IgE		Skin and tissues lining gastrointestinal and respiratory tracts (secreted by plasma cells)	Stimulates mast cells and basophils to release histamine; triggers allergic responses.
IgD		Surface of unstimulated B cells	Membrane receptor for mature B cells; probably important in B-cell activation (clonal selection).

amounts when the body is exposed a second time to the same antigen.

IgA is found mainly in body secretions such as saliva, tears, breast milk, and the mucus coating of body cavities such as the lungs, digestive tract, and vagina. In these locations, the antibodies bind to surface groups on pathogens and block their attachment to body surfaces. Breast milk transfers IgA antibodies and thus immunity to a nursing infant.

IgE is secreted by plasma cells of the skin and tissues lining the gastrointestinal tract and respiratory tract. IgE binds to basophils and mast cells where it mediates many allergic responses, such as hay fever, asthma, and hives. Binding of a specific antigen to IgE stimulates the cell to which it is bound to release histamine, which triggers an inflammatory response. IgE also contributes to mechanisms that combat infection by parasitic worms.

IgD occurs with IgM as a receptor on the surfaces of B cells; its function is uncertain.

The Generation of Antibody Diversity. The human genome has approximately 20,000–25,000 genes, far fewer than necessary to encode 100 million different antibodies if two genes encoded one antibody, one gene for the heavy chain and one for the light chain. Antibody diversity is generated in a different way from one gene per chain, however, instead involving three rearrangements during B-cell differentiation of DNA segments that encode parts of the light and heavy chains. Let us consider how this process produces light-chain genes for the B-cell receptor **(Figure 43.5)**; the production of heavy-chain genes is similar. The genes for the two different subunits of the TCR undergo similar rearrangements to produce the great diversity in antigen-binding capability of those receptors.

An undifferentiated B cell has three types of light-chain DNA segments: V, J, and C. One of each type is needed to make a complete, functional light-chain gene (see Figure 43.5, top). In humans, there are about 40 different V segments encoding most of the variable region of the chain, 5 different J (joining) segments encoding the rest of the variable region, and only one copy of the segment for the constant (C) part of the chain.

During B-cell differentiation, a DNA rearrangement occurs in which one random V and one random J segment join with the C segment to form a functional light-chain gene (see Figure 43.5, step 1). The rearrangement involves deletion of the DNA between the V and J segments. The positions at which the DNA breaks and rejoins in the V and J joining reaction occur randomly over a distance of several nucleotides, which adds greatly to the variability of the final gene assembly. The DNA between the J segment and the C segment is an intron in the final assembled gene.

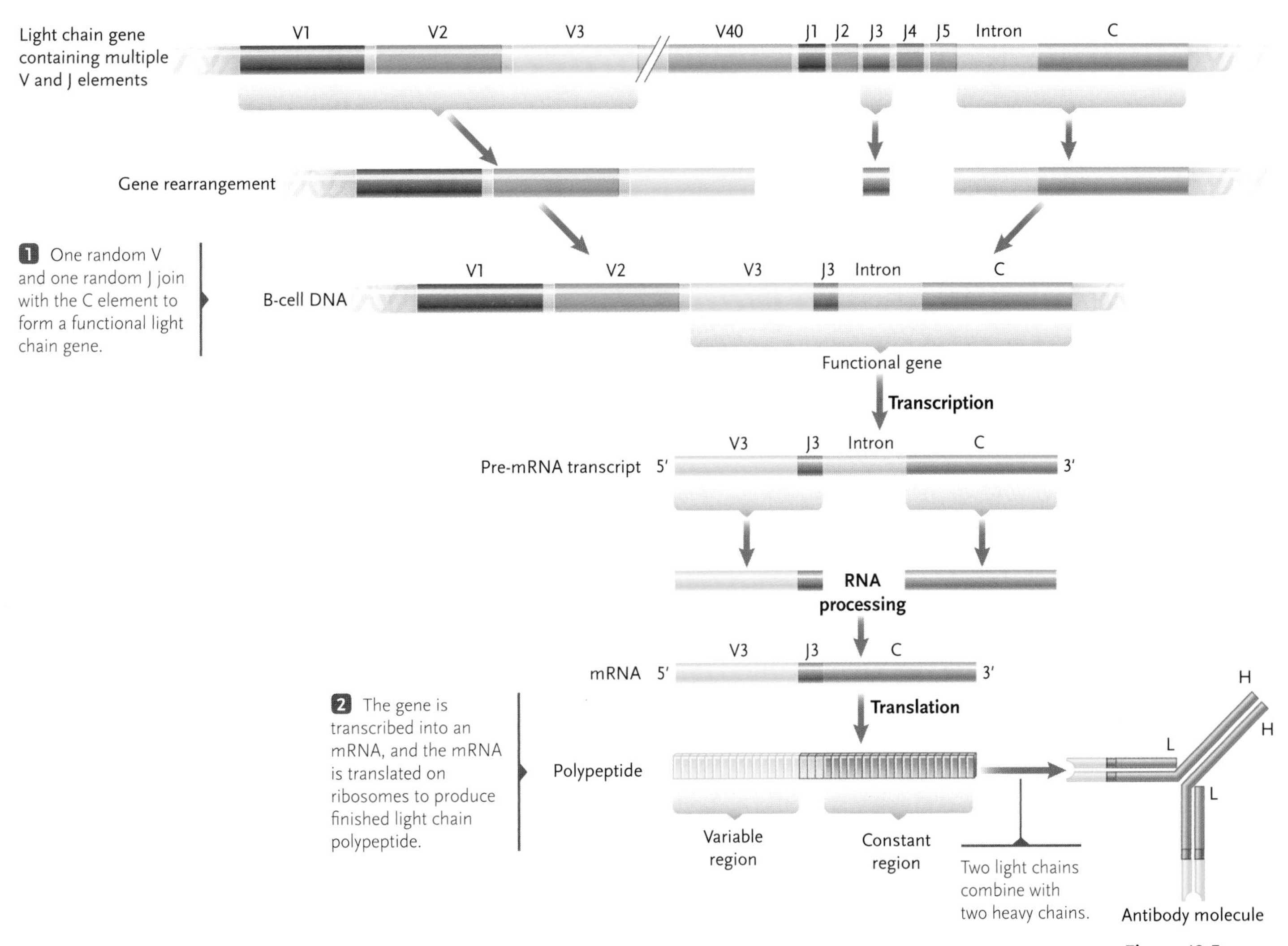

Figure 43.5
The DNA rearrangements producing a functional light-chain gene, in simplified form.

Transcription of this newly assembled gene produces a typical pre-mRNA molecule (see Section 15.3). The intron between the J and C segments is removed during the production of the mRNA by RNA processing. Translation of the mRNA produces the light chain with variable and constant regions (step 2).

As noted earlier, the assembly of functional heavy-chain genes occurs similarly. However, whereas light-chain genes have one C segment, heavy-chain genes have five types of C segments, each of which encodes one of the constant regions of IgM, IgD, IgG, IgE, and IgA. The inclusion of one of the five C segment types in the functional heavy-chain gene therefore specifies the class of antibody that will be made by the cell. Thus, the various DNA rearrangements producing the various light chain and heavy chain genes, along with the various combinations of light and heavy chains, generates the 100 million different antibodies.

T-Cell Activation. Let us now follow the development of an antibody-mediated immune response by linking the recognition of an antigen by lymphocytes, the activation of lymphocytes by antigen binding, and the production of antibodies. Typically, the pathway begins when a type of T cell becomes activated, following the steps outlined in **Figure 43.6.** Let us learn about this pathway by considering the fate of pathogenic bacteria that have been introduced under the skin. Circulating viruses in the blood follow the same pathway.

First, a type of phagocyte called a **dendritic cell** engulfs a bacterium in the infected tissue by phagocytosis (**Figure 43.7,** step 1). Dendritic cells are so named because they have many surface projections resembling dendrites of neurons. They have the same origin as leukocytes, and recognize a bacterium as foreign by the same recognition mechanism used by macrophages in the innate immunity system. In essence, the dendritic cell is part of the innate immunity system, but its primary role is to stimulate the development of an adaptive immune response.

Engulfment of a bacterium activates the dendritic cell; the cell now migrates to a nearby lymph node. Then, within the dendritic cell, the endocytic vesicle containing the engulfed bacterium fuses with a lysosome. In the lysosome, the bacterium's proteins are

Antibody-mediated immune response: T-cell activation

Dendritic cell (a phagocyte) is activated by engulfing a pathogen such as a bacterium.

↓

Pathogen macromolecules are degraded in dendritic cell, producing antigens.

↓

Dendritic cell becomes an antigen-presenting cell (APC) by displaying antigens on surface bound to class II MHC proteins.

↓

APC presents antigen to CD4+ T cell and activates the T cell.

↓

CD4+ T cell proliferates to produce a clone of cells.

↓

Clonal cells differentiate into helper T cells, which aid in effecting the specific immune response to the antigen.

Figure 43.6
An outline of T-cell activation in antibody-mediated immunity.

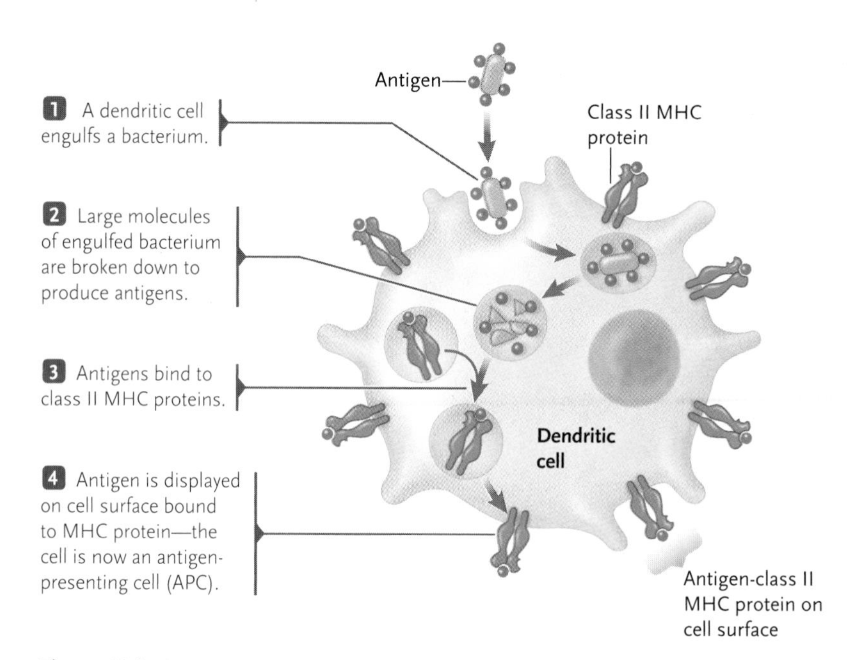
1 A dendritic cell engulfs a bacterium.
2 Large molecules of engulfed bacterium are broken down to produce antigens.
3 Antigens bind to class II MHC proteins.
4 Antigen is displayed on cell surface bound to MHC protein—the cell is now an antigen-presenting cell (APC).
Antigen
Class II MHC protein
Dendritic cell
Antigen-class II MHC protein on cell surface

Figure 43.7
Generation of an antigen-presenting cell when a dendritic cell engulfs a bacterium.

degraded into short peptides, which function as antigens (step 2). The antigens bind intracellularly to **class II major histocompatibility complex (MHC)** proteins (step 3); the interacting molecules then migrate to the cell surface where the antigen is displayed (step 4). These steps, which occur in the dendritic cell after it has migrated to the lymph node, convert the cell into an **antigen-presenting cell (APC)**, ready to present the antigen to T cells in the next step of antibody-mediated immunity. The process is recapped in **Figure 43.8.**

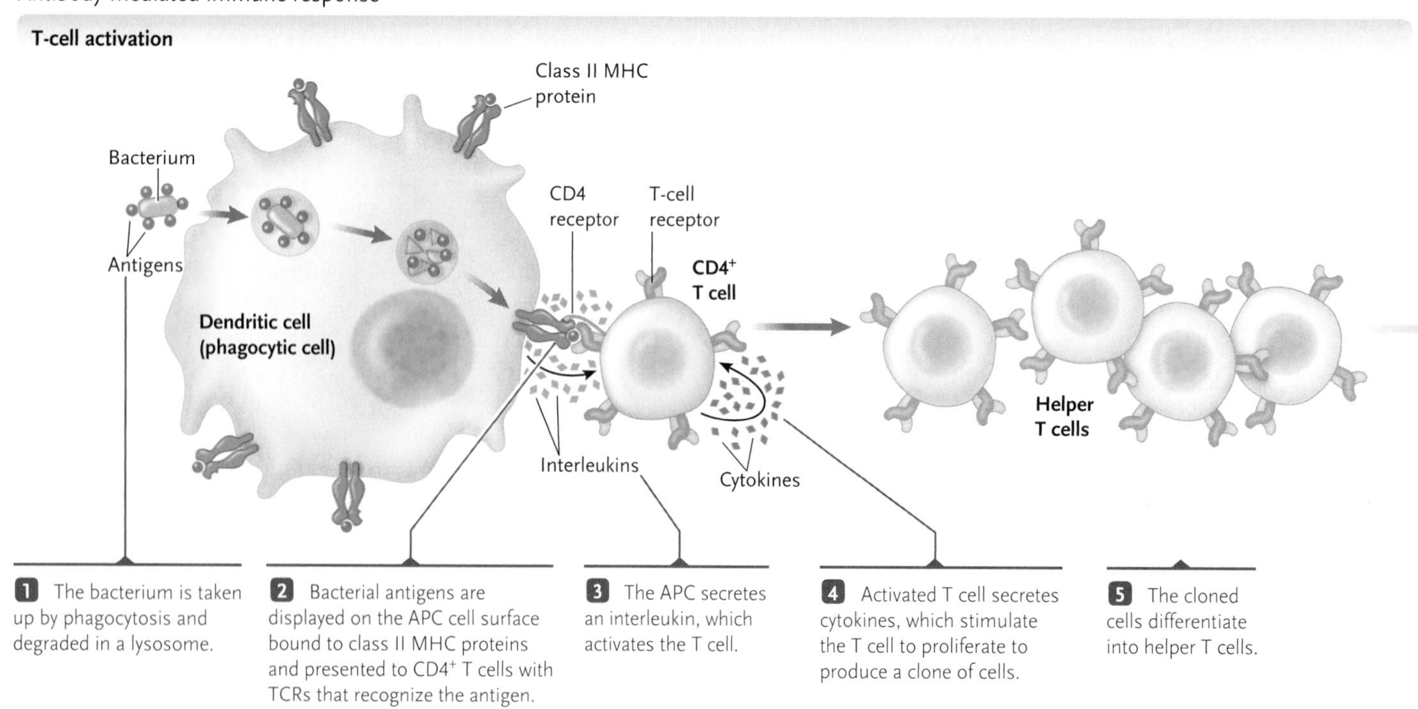
Antibody-mediated immune response
T-cell activation
Class II MHC protein
Bacterium
Antigens
CD4 receptor
T-cell receptor
CD4+ T cell
Dendritic cell (phagocytic cell)
Interleukins
Cytokines
Helper T cells
1 The bacterium is taken up by phagocytosis and degraded in a lysosome.
2 Bacterial antigens are displayed on the APC cell surface bound to class II MHC proteins and presented to CD4+ T cells with TCRs that recognize the antigen.
3 The APC secretes an interleukin, which activates the T cell.
4 Activated T cell secretes cytokines, which stimulate the T cell to proliferate to produce a clone of cells.
5 The cloned cells differentiate into helper T cells.

MHC proteins are named for a large cluster of genes encoding them, called the **major histocompatibility complex.** The complex spans 4 million base pairs and contains 128 genes. Many of these genes play important roles in the immune system. Each individual of each vertebrate species has a unique combination of MHC proteins on almost all body cells, meaning that no two individuals of a species except identical siblings are likely to have exactly the same MHC proteins on their cells. There are two classes of MHC proteins, class I and class II, which have different functions in adaptive immunity, as we will see.

The key function of an APC is to present the antigen to a lymphocyte. In the antibody-mediated immune response, the APC presents the antigen, bound to a class II MHC protein, to a type of T cell in the lymphatic system called a **$CD4^+$ T cell** because it has receptors named CD4 on its surface. A specific $CD4^+$ T cell having a TCR with an antigen-binding site that recognizes the antigen (epitope, actually) binds to the antigen on the APC (see Figure 43.8, step 2). The CD4 receptor on the T cell helps link the two cells together.

When the APC binds to the $CD4^+$ T cell, the APC secretes an *interleukin* (meaning "between leukocytes"), a type of cytokine, which activates the associated T cell (step 3). The activated T cell then secretes cytokines (step 4), which act in an autocrine manner (see Section 40.1) to stimulate **clonal expansion,** the proliferation of the activated $CD4^+$ T cell by cell division to produce a clone of cells. These clonal cells differentiate into **helper T cells,** so named because they assist with the activation of B cells (step 5). A helper T cell is an example of an **effector T cell,** meaning that it is involved in effecting—bringing about—the specific immune response to the antigen.

B-Cell Activation. Antibodies are produced in and secreted by activated B cells. The activation of a B cell requires the B cell to present the antigen on its surface, and then to link with a helper T cell that has differentiated as a result of encountering and recognizing the same antigen. The process is outlined in **Figure 43.9,** and diagrammed in the second phase of Figure 43.8).

The process of antigen presentation on a B-cell surface begins when BCRs on the B cell interact directly with soluble bacterial (in our example) antigens in the blood or lymph. Once the antigen binds to a BCR, the complex is taken into the cell and the antigen is processed in the same way as in dendritic cells, culminating with a presentation of antigen pieces on the B-cell surface in a complex with class II MHC proteins (see Figure 43.8, step 6).

When a helper T cell encounters a B cell displaying the same antigen, usually in a lymph node or in the spleen, the T and B cells become tightly linked together (step 7). The linkage depends on the TCRs, which recognize and bind the antigen fragment displayed by class II MHC molecules on the surface of the B cell, and on CD4, which stabilizes the binding as it did for T-cell binding to the dendritic cell. The linkage between the cells first stimulates the helper T cell to secrete interleukins that activate the B cell and then stimulates the B cell to proliferate, producing a clone of those B cells with identical B-cell receptors (step 8). Some of the cloned cells differentiate into relatively short-lived **plasma cells,** which now secrete the same antibody that

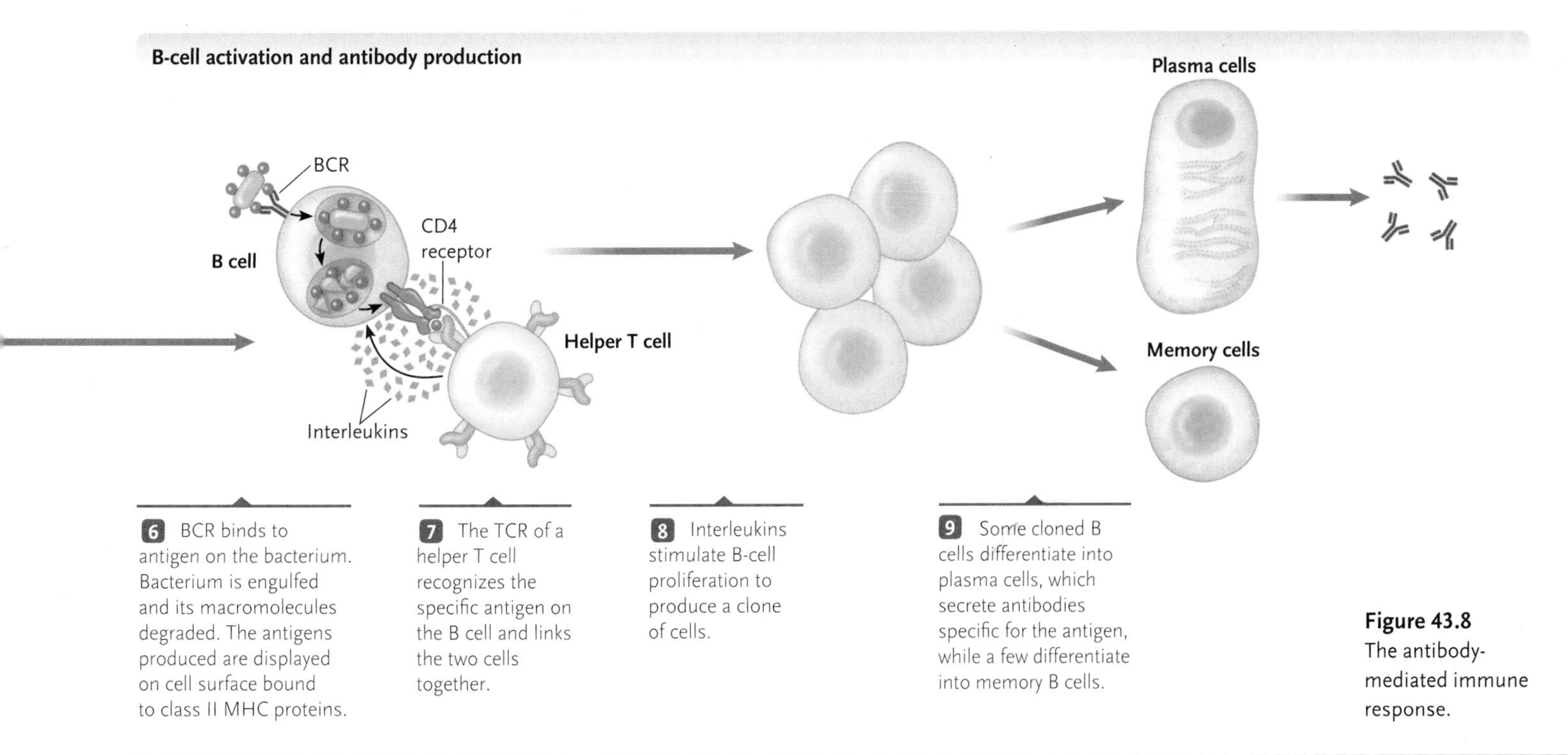

B-cell activation and antibody production
BCR
B cell
CD4 receptor
Helper T cell
Interleukins
Plasma cells
Memory cells
6 BCR binds to antigen on the bacterium. Bacterium is engulfed and its macromolecules degraded. The antigens produced are displayed on cell surface bound to class II MHC proteins.
7 The TCR of a helper T cell recognizes the specific antigen on the B cell and links the two cells together.
8 Interleukins stimulate B-cell proliferation to produce a clone of cells.
9 Some cloned B cells differentiate into plasma cells, which secrete antibodies specific for the antigen, while a few differentiate into memory B cells.

Figure 43.8
The antibody-mediated immune response.

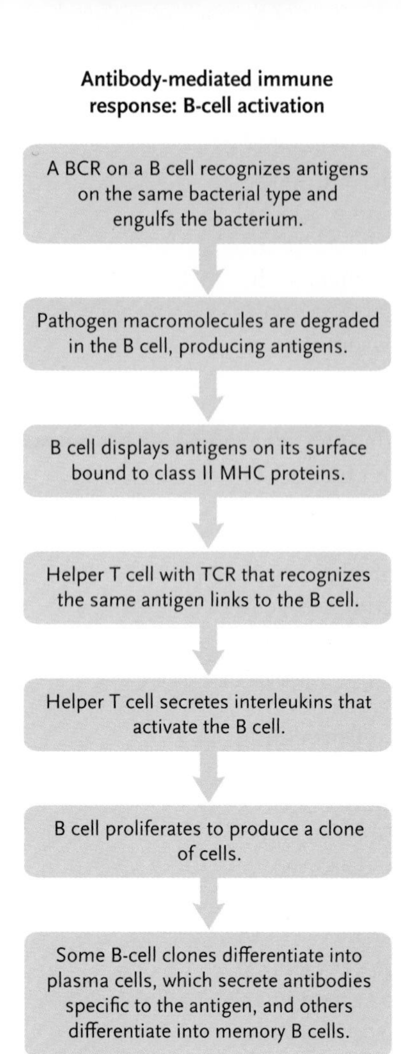

Antibody-mediated immune response: B-cell activation
A BCR on a B cell recognizes antigens on the same bacterial type and engulfs the bacterium.
Pathogen macromolecules are degraded in the B cell, producing antigens.
B cell displays antigens on its surface bound to class II MHC proteins.
Helper T cell with TCR that recognizes the same antigen links to the B cell.
Helper T cell secretes interleukins that activate the B cell.
B cell proliferates to produce a clone of cells.
Some B-cell clones differentiate into plasma cells, which secrete antibodies specific to the antigen, and others differentiate into memory B cells.

Figure 43.9
An outline of B-cell activation in antibody-mediated immunity.

Population of unactivated B cells, each making a specific receptor that is displayed on its surface as a BCR

Binding of antigen and interaction with helper T cell stimulates B cell to divide and produce a clone of cells.

Plasma cells

Memory B cells

Some of the B-cell clones differentiate into plasma cells, which secrete antibodies.

A few B-cell clones differentiate into memory B cells, which respond to a later encounter with the same antigen.

Figure 43.10
Clonal selection. The binding of an antigen to a B cell already displaying a specific antibody to that antigen stimulates the B cell to divide and differentiate into plasma cells, which secrete the antibody, and memory cells, which remain in the circulation ready to mount a response against the antigen at a later time.

was displayed on the parental B cell's surface to circulate in lymph and blood. Others differentiate into **memory B cells,** which are long-lived cells that set the stage for a much more rapid response should the same antigen be encountered later (step 9).

Clonal selection is the process by which a particular lymphocyte is specifically selected for cloning when it recognizes a particular foreign antigen **(Figure 43.10)**. Remember that there is an enormous diversity of randomly generated lymphocytes, each with a particular receptor that may potentially recognize a particular antigen. The process of clonal selection was proposed in the 1950s by several scientists, most notably F. Macfarlane Burnet, Niels Jerne, and David Talmage. Their proposals, made long before the mechanism was understood, described clonal selection as a form of natural selection operating in miniature: antigens select the cells recognizing them, which reproduce and become dominant in the B-cell population. Burnet received the Nobel Prize in 1960 for his research in immunology.

Clearing the Body of Foreign Antigens. How do the antibodies produced in an antibody-mediated immune response clear foreign antigens from the body? Let us consider some examples concerning bacteria and viruses.

Toxins produced by invading bacteria, such as tetanus toxin, can be *neutralized* by antibodies **(Figure 43.11a)**. The antibodies bind to the toxin molecules, preventing them from carrying out their damaging action. For intact bacteria at an infection site or in the circulatory system, antibodies will bind to antigens on their surfaces. Because the two arms of an antibody molecule bind to different copies of the antigen molecule, an antibody molecule may bind to two bacteria with the same antigen. A population of antibodies against the bacterium, then, link many bacteria to-

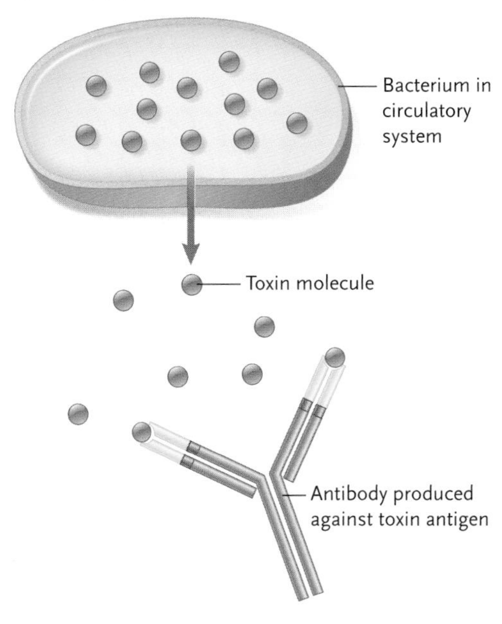

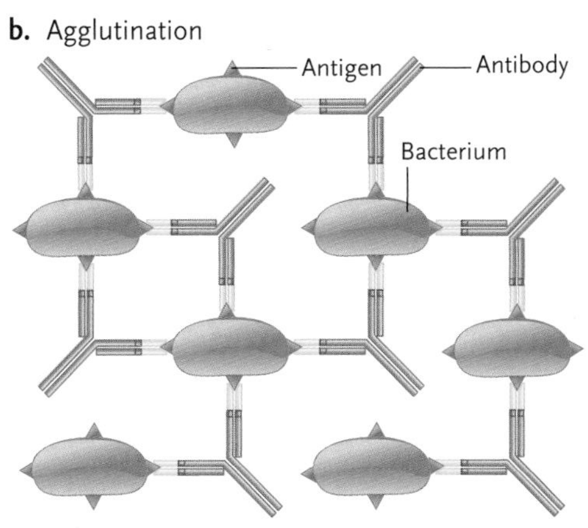

Figure 43.11
Examples of clearing antigens from the body.

gether into a lattice causing *agglutination,* that is, clumping of the bacteria **(Figure 43.11b).** Agglutination immobilizes the bacteria, preventing them from infecting cells. Antibodies can also agglutinate viruses, thereby preventing them from infecting cells.

More importantly, antibodies aid the innate immune response set off by the pathogens. That is, antibodies bound to antigens stimulate the complement system. Membrane attack complexes are formed and insert themselves into the plasma membranes of the bacteria, leading to their lysis and death. In the case of viral infections, membrane attack complexes can insert themselves into the membranes surrounding enveloped viruses, which disrupts the membrane and prevents the viruses from infecting cells.

Antibodies also enhance phagocytosis of bacteria and viruses. Phagocytic cells have receptors on their surfaces that recognize the heavy-chain end of antibodies (the end of the molecule opposite the antigen-binding sites). Antibodies bound to bacteria or viruses therefore bind to phagocytic cells, which then engulf the pathogens and destroy them.

For simplicity, the adaptive immune response has been described here in terms of a single antigen. Pathogens have many different types of antigens on their surfaces, which means that many different B cells are stimulated to proliferate and many different antibodies are produced. Pathogens therefore are attacked by many different types of antibodies, each targeted to one antigen type on the pathogen's surface.

Immunological Memory. Once an immune reaction has run its course and the invading pathogen or toxic molecule has been eliminated from the body, division of the plasma cells and T-cell clones stops. Most or all of the clones die and are eliminated from the bloodstream and other body fluids. However, long-lived memory B cells and **memory helper T cells** (which differentiated from helper T cells), derived from encountering the same antigen, remain in an inactive state in the lymphatic system. Their persistence provides an **immunological memory** of the foreign antigen.

Immunological memory is illustrated in **Figure 43.12.** When exposed to a foreign antigen for the first time, a **primary immune response** results, following the steps already described. The first antibodies appear in the blood in 3 to 14 days and, by week 4, the primary response has essentially gone away. IgM is the main antibody type produced in a primary immune response. The primary immune response curve is followed whenever a new foreign antigen enters the body.

When a foreign antigen enters the body for a second or subsequent time, a **secondary immune response** results, while any new antigen introduced at the same time produces a primary response (see Figure 43.12). The secondary response is more rapid than a primary response because it involves the memory B cells and memory T cells that have been stored in the meantime, rather than having to initiate the clonal selection of a new B cell and T cell. Moreover, less antigen is needed

Figure 43.12
Immunological memory: primary and secondary responses to the same antigen.

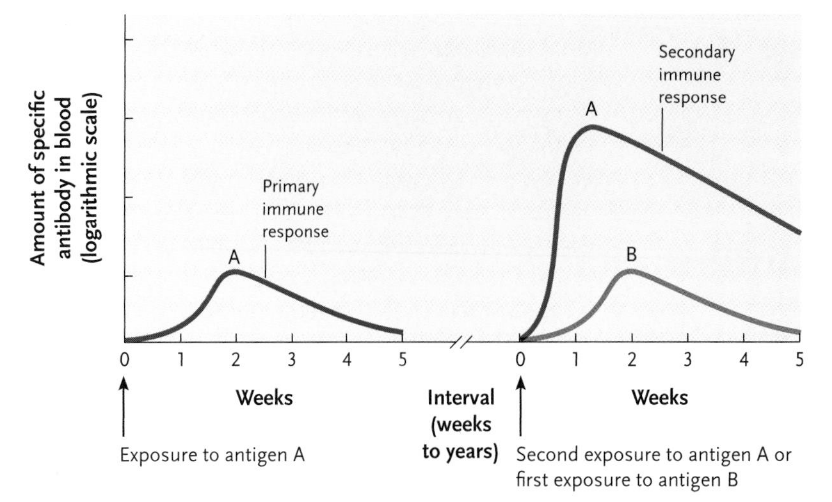

to elicit a secondary response than a primary response, and many more antibodies are produced. The predominant antibody produced in a secondary immune response is IgG; the switch occurs at the gene level in the memory B cells.

Immunological memory forms the basis of vaccinations, in which antigens in the form of living or dead pathogens or antigenic molecules themselves are introduced into the body. After the immune response, memory B cells and memory T cells remaining in the body can mount an immediate and intense immune reaction against similar antigens in the dangerous pathogen. In Edward Jenner's technique, for example, introducing the cowpox virus, a related, less virulent form of the smallpox virus, into healthy individuals initiated a primary immune response. After the response ran its course, a bank of memory B cells and memory T cells remained in the body, able to recognize quickly the similar antigens of the smallpox virus and initiate a secondary immune response. Similarly, a polio vaccine developed by Jonas Salk uses polio viruses that have been inactivated by exposing them to formaldehyde. Although the viruses are inactive, their surface groups can still act as antigens. The antigens trigger an immune response, leaving memory B and T cells able to mount an intense immune response against active polio viruses.

Active and Passive Immunity. **Active immunity** is the production of antibodies in the body in response to exposure to a foreign antigen—the process that has been described up until now. **Passive immunity** is the acquisition of antibodies as a result of direct transfer from another person. This form of immunity provides immediate protection against antigens that the antibodies recognize without the person receiving the antibodies having developed an immune response. Examples of passive immunity include the transfer of IgG antibodies from the mother to the fetus through the placenta and the transfer of IgA antibodies in the first breast milk fed from the mother to the baby. Compared with active immunity, passive immunity is a short-lived phenomenon with no memory, in that the antibodies typically break down within a month. However, in that time, the protection plays an important role. For example, a breast-fed baby is protected until it is able to mount an immune response itself, an ability that is not present until about a month after birth.

Drug Effects on Antibody-Mediated Immunity. Several drugs used to reduce the rejection of transplanted organs target helper T cells. Cyclosporin A, used routinely after organ transplants, blocks the activation of helper T cells and, in turn, the activation of B cells. Unfortunately, cyclosporin and other immunosuppressive drugs also leave the treated individual more susceptible to infection by pathogens.

In Cell-Mediated Immunity, Cytotoxic T Cells Expose "Hidden" Pathogens to Antibodies by Destroying Infected Body Cells

In **cell-mediated immunity**, cytotoxic T cells directly destroy host cells infected by pathogens, particularly those infected by a virus **(Figure 43.13).** The killing pro-

Figure 43.13 The cell-mediated immune response.

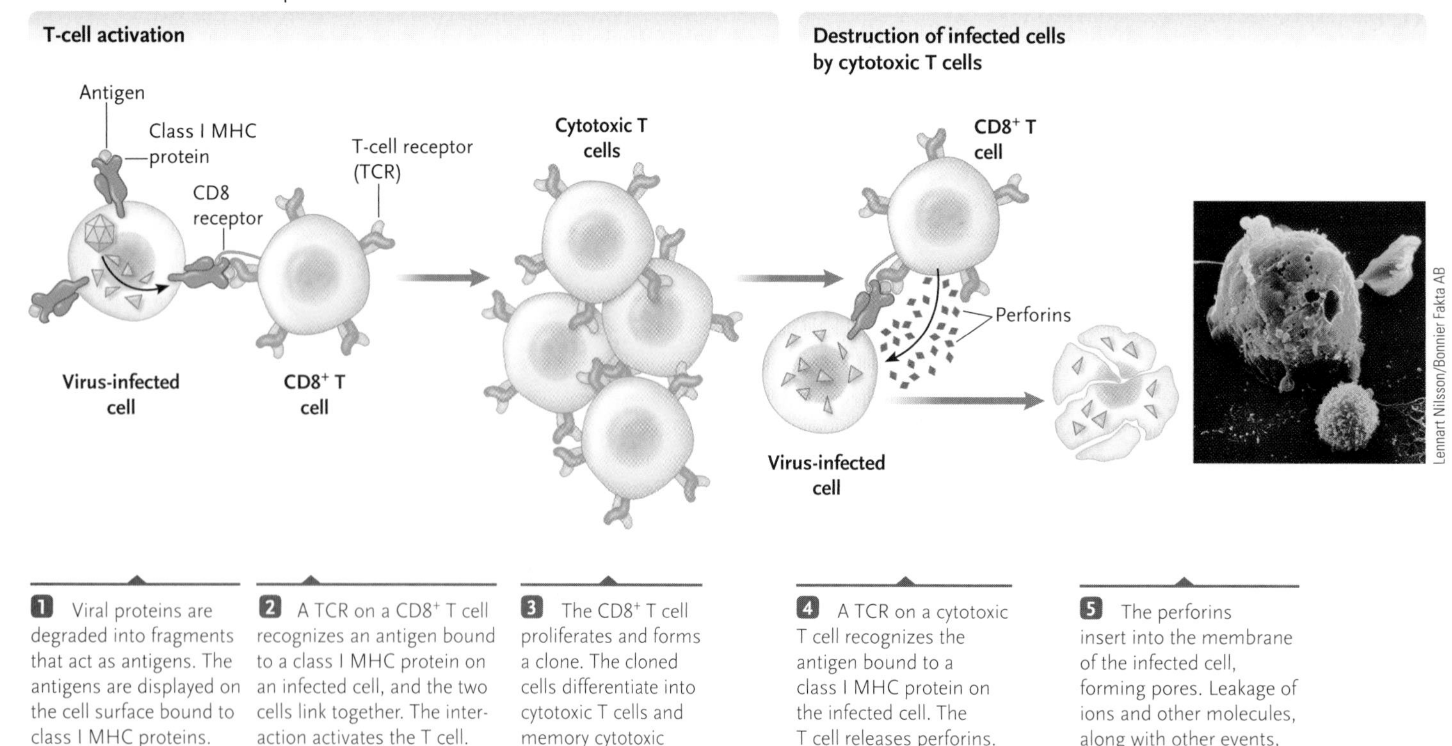

Cell-mediated immune response
T-cell activation
Antigen
Class I MHC protein
CD8 receptor
T-cell receptor (TCR)
Cytotoxic T cells
Virus-infected cell
CD8⁺ T cell
Destruction of infected cells by cytotoxic T cells
CD8⁺ T cell
Perforins
Virus-infected cell
Lennart Nilsson/Bonnier Fakta AB
1 Viral proteins are degraded into fragments that act as antigens. The antigens are displayed on the cell surface bound to class I MHC proteins.
2 A TCR on a CD8⁺ T cell recognizes an antigen bound to a class I MHC protein on an infected cell, and the two cells link together. The interaction activates the T cell.
3 The CD8⁺ T cell proliferates and forms a clone. The cloned cells differentiate into cytotoxic T cells and memory cytotoxic T cells.
4 A TCR on a cytotoxic T cell recognizes the antigen bound to a class I MHC protein on the infected cell. The T cell releases perforins.
5 The perforins insert into the membrane of the infected cell, forming pores. Leakage of ions and other molecules, along with other events, causes the cell to lyse.

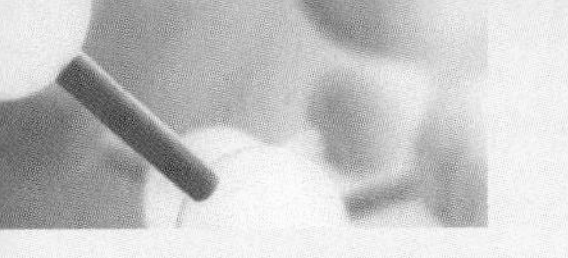

Insights from the Molecular Revolution

Some Cancer Cells Kill Cytotoxic T Cells to Defeat the Immune System

Among the arsenal of weapons employed by cytotoxic T cells to eliminate infected and cancerous body cells is the *Fas–FasL* system. Fas is a receptor that occurs on the surfaces of many body cells; FasL is a signal molecule that is displayed on the surfaces of some cell types, including cytotoxic T cells. If a cell carrying the Fas receptor contacts a cytotoxic T cell with the FasL signal displayed on its surface, the effect for the Fas-bearing cell is something like stepping on a mine. When FasL is bound by the Fas receptor, a cascade of internal reactions initiates apoptosis and kills the cell with the Fas receptor.

Surprisingly, cytotoxic T cells also carry the Fas receptor, so they can kill each other by displaying the FasL signal. This mutual killing plays an important role in reducing the level of an immune reaction after a pathogen has been eliminated. In addition, cells in some regions of the body, such as the eye, the nervous system, and the testis, can be severely damaged by inflammation; these cells can make and display FasL, which kills cytotoxic T cells and reduces the severity of inflammation.

Molecular research by a group of Swiss investigators at the Universities of Lausanne and Geneva now shows that some cancer cells survive elimination by the immune system by making and displaying the FasL signal, and thus killing any cytotoxic T cells attacking the tumor. The researchers were led to their discovery by the observation that many patients with malignant melanoma, a dangerous skin cancer, had a breakdown product associated with FasL in their bloodstream.

The investigators extracted proteins from melanoma cells and tested them with antibodies against FasL. The test was positive, showing that FasL was in the tumor cells. The investigators were also able to detect an mRNA encoding FasL in the tumor cells, showing that the gene encoding FasL was active. As a final confirmation, they tagged antibodies against the FasL protein with a dye molecule to make them visible in the light microscope and added them to sections of melanoma tissue from patients. Intense staining of the cells with the dye showed that FasL was indeed present. Tests for the presence of the Fas receptor were negative, showing that Fas synthesis was turned off in the melanoma cells.

Thus the FasL in melanoma cells kills cytotoxic T cells that invade the tumor. At the same time, the absence of Fas receptors ensures that the tumor cells do not kill each other. The presence of FasL and absence of the Fas receptor may explain why melanomas are rarely destroyed by an immune reaction, and also why many other types of cancer also escape immune destruction.

Melanoma cells originate from pigment cells in the skin called *melanocytes.* Normal melanocytes do not contain FasL, indicating that synthesis of the protein is turned on as a part of the changes transforming normal melanocytes into cancer cells.

The findings of the Swiss group could lead to an effective treatment for cancer using the Fas–FasL system. If melanoma cells could be induced to make Fas as well as FasL, for example, they might eliminate a tumor by killing each other!

cess begins when some of the pathogens break down inside the infected host cells, releasing antigens that are fragmented by enzymes in the cytoplasm. The antigen fragments bind to class I MHC proteins, which are delivered to the cell surface by essentially the same mechanisms as in B cells (step 1). At the surface, the antigen fragments are displayed by the class I MHC protein and the cell then functions as an APC.

In a cell-mediated immune response, the APC presents the antigen fragment to a type of T cell in the lymphatic system called a **CD8$^+$ T cell** because it has receptors named **CD8** on its surface in addition to TCRs. The presence of a CD8 receptor distinguishes this type of T cell from that involved in antibody-mediated immunity. A specific CD8$^+$ T cell having a TCR with an antigen-binding site that recognizes the antigen fragment binds to that fragment on the APC (step 2). The CD8 receptor on the T cell helps the two cells link together.

The interaction between the APC and the CD8$^+$ T cell activates the T cell, which then proliferates to form a clone. Some of the cells differentiate to become **cytotoxic T cells** (step 3), while a few differentiate into *memory cytotoxic T cells.* Cytotoxic T cells are another type of effector T cell. TCRs on the cytotoxic T cells again recognize the antigen fragment bound to class I MHC proteins on the infected cells (the APCs) (step 4). The cytotoxic T cell then destroys the infected cell in mechanisms similar to those used by NK cells. That is, an activated cytotoxic T cell releases perforin, which creates pores in the membrane of the target cell. The leakage of ions and other molecules through the pores causes the infected cell to rupture. The cytotoxic T cell also secretes proteases that enter infected cells through the newly created pores and cause it to self-destruct by apoptosis (step 5, and photo inset). Rupture of dead, infected cells releases the pathogens to the interstitial fluid, where they are open to attack by antibodies and phagocytes.

Cytotoxic T cells can also kill cancer cells if their class I MHC molecules display fragments of altered cellular proteins that do not normally occur in the body. Another mechanism used by cytotoxic T cells to kill cells, and a process used by some cancer cells to defeat the mechanism, is described in *Insights from the Molecular Revolution.*

Figure 43.14 Research Method

Production of Monoclonal Antibodies

PURPOSE: Injecting an antigen into an animal produces a collection of different antibodies that react against different parts of the antigen. Monoclonal antibodies are produced to provide antibodies that all react against the same epitope of a single antigen.

PROTOCOL:

1. Inject antigen into mouse.
2. Extract activated B cells from spleen.
3. Fuse antibody-producing B cells with cancer cells to form fast-growing hybridoma cells.
4. Grow clone from single hybridoma cells; test antibodies produced by clone for reaction against antigen.
5. Grow clone producing antibodies against antigen to large size.
6. Extract and purify antibodies.

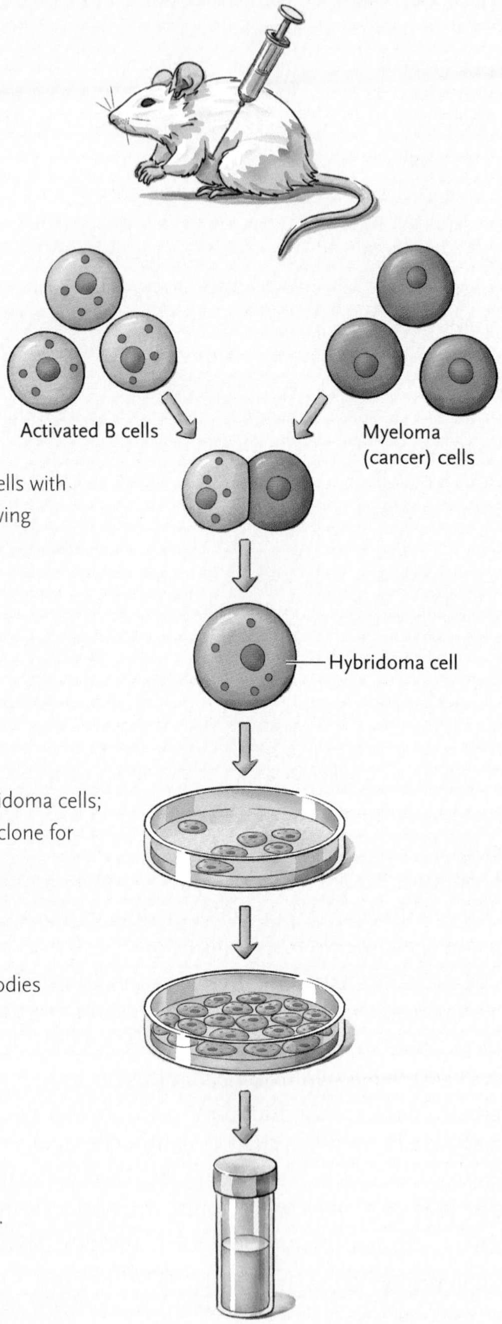

Activated B cells
Myeloma (cancer) cells
Hybridoma cell

Antibodies Have Many Uses in Research

The ability of the antibody-mediated immune system to generate antibodies against essentially any antigen provides an invaluable research tool to scientists, who can use antibodies to identify biological molecules and to determine their locations and functions in cells. To obtain the antibodies, a molecule of interest is injected into a test animal such as a mouse, rabbit, goat, or sheep. In response, the animal develops antibodies capable of binding to the molecule. The antibodies are then extracted and purified from a blood sample.

To identify the cellular location of a molecule, antibodies made against the molecule are combined with a visible marker such as a dye molecule or heavy metal atom. When added to a tissue sample, the marked antibodies can be seen in the light or electron microscope localized to cellular structures such as membranes, ribosomes, or chromosomes, showing that the molecule forms part of the structure.

Antibodies can also be used to "grab" a molecule of interest from a preparation containing a mixture of all kinds of cellular molecules. For such studies, the antibodies are often attached to plastic beads that are packed into a glass column. When the mixture passes through the column, the molecule is trapped by attachment to the antibody and remains in the column. It is then released from the column in purified form by adding a reagent that breaks the antigen-antibody bonds.

Injecting a molecule of interest into a test animal typically produces a wide spectrum of antibodies that react with different parts of the antigen. Some of the antibodies also cross-react with other, similar antigens, producing false results that can complicate the research. These problems have been solved by producing **monoclonal antibodies,** each of which reacts only against the same segment (epitope) of a single antigen.

Georges Kohler and Cesar Milstein pioneered the production of monoclonal antibodies in 1975. In their technique, a test animal (usually a mouse) is injected with a molecule of interest **(Figure 43.14).** After the animal has developed an immune response, fully activated B cells are extracted from the spleen and placed in a cell culture medium. Because B cells normally stop dividing and die within a week when cultured, they are induced to fuse with cancerous lymphocytes called *myeloma cells,* forming single, composite cells called **hybridomas.** Hybridomas combine the desired characteristics of the two cell types—they produce antibodies like fully activated B cells, and they divide continuously and rapidly like the myeloma cells.

Single hybridoma cells are then separated from the culture and used to start clones. Because all the cells of a clone are descended from a single hybridoma cell, they all make the same, highly specific antibody, able to bind the same part of a single antigen. In addition to their use in scientific research, monoclonal antibodies are also widely used in medical applications such as pregnancy tests, screening for prostate cancer, and testing for AIDS and other sexually transmitted diseases.

Study Break

1. How, in general, do the antibody-mediated and cell-mediated immune responses help clear the body of antigens?
2. Describe the general structure of an antibody molecule.
3. What are the principles of the mechanism used for generating antibody diversity?
4. What is clonal selection?
5. How does immunological memory work?

43.4 Malfunctions and Failures of the Immune System

The immune system is highly effective, but it is not foolproof. Some malfunctions of the immune system cause the body to react against its own proteins or cells, producing *autoimmune disease*. In addition, some viruses and other pathogens have evolved means to avoid destruction by the immune system. A number of these pathogens, including the AIDS virus, even use parts of the immune response to promote infection. Another malfunction causes the *allergic reactions* that most of us experience from time to time.

An Individual's Own Molecules Are Normally Protected against Attack by the Immune System

B cells and T cells are involved in the development of **immunological tolerance,** which protects the body's own molecules from attack by the immune system. Although the process is not understood, molecules present in an individual from birth are not recognized as foreign by circulating B and T cells, and do not elicit an immune response. Evidently, during their initial differentiation in the bone marrow and thymus, any B and T cells that are able to react with self molecules carried by MHC molecules are induced to kill themselves by apoptosis, or enter a state in which they remain in the body but are unable to react if they encounter a self molecule. The process of excluding self-reactive B and T cells goes on throughout the life of an individual.

Evidence that immunological tolerance is established early in life comes from experiments with mice. For example, if a foreign protein is injected into a mouse at birth, during the period in which tolerance is established, the mouse will not develop antibodies against the protein if it is injected later in life. Similarly, if mutant mice are produced that lack a given complement protein, so that the protein is absent during embryonic development, they will produce antibodies against that protein if it is injected during adult life. Normal mice do not produce antibodies if the protein is injected.

Autoimmune Disease Occurs When Immunological Tolerance Fails

The mechanisms setting up immunological tolerance sometimes fail, leading to an **autoimmune reaction**—the production of antibodies against molecules of the body. In most cases, the effects of such anti-self antibodies are not serious enough to produce recognizable disease. However, in some individuals, about 5% to 10% of the human population, anti-self antibodies cause serious problems.

For example, type 1 diabetes (see Section 40.4) is an autoimmune reaction against the pancreatic beta cells producing insulin. The anti-self antibodies gradually eliminate the beta cells until the individual is incapable of producing insulin. *Systemic lupus erythematosus (lupus)* is caused by production of a wide variety of anti-self antibodies against blood cells, blood platelets, and internal cell structures and molecules such as mitochondria and proteins associated with DNA in the cell nucleus. People with lupus often become anemic and have problems with blood circulation and kidney function because antibodies, combined with body molecules, accumulate and clog capillaries and the microscopic filtering tubules of the kidneys. Lupus patients may also develop anti-self antibodies against the heart and kidneys. *Rheumatoid arthritis* is caused by a self-attack on connective tissues, particularly in the joints, causing pain and inflammation. *Multiple sclerosis* results from an autoimmune attack against a protein of the myelin sheaths insulating the surfaces of neurons. Multiple sclerosis can seriously disrupt nervous function, producing such symptoms as muscle weakness and paralysis, impaired coordination, and pain.

The causes of most autoimmune diseases are unknown. In some cases, an autoimmune reaction can be traced to injuries that expose body cells or proteins that are normally inaccessible to the immune system, such as the lens protein of the eye, to B and T cells. In other cases, as in type 1 diabetes, an invading virus stimulates the production of antibodies that can also react with self proteins. Antibodies against two viruses, the Epstein-Barr and hepatitis B viruses, can react against myelin basic protein, the protein attacked in multiple sclerosis. Sometimes, environmental chemicals, drugs, or mutations alter body proteins so that they appear foreign to the immune system and come under attack.

Some Pathogens Have Evolved Mechanisms That Defeat the Immune Response

Several pathogens regularly change their surface groups to avoid destruction by the immune system. By the time the immune system has developed antibodies

FOCUS ON RESEARCH

Applied Research: HIV and AIDS

Acquired immune deficiency syndrome (AIDS) is a constellation of disorders that follows infection by the **human immunodeficiency virus, HIV (Figure a).** First reported in various countries in the late 1970s, HIV now infects more than 40 million people worldwide, 64% of them in Africa. AIDS is a potentially lethal disease, although drug therapy has reduced the death rate for HIV-infected individuals in many countries, including the United States.

HIV is transmitted when an infected person's body fluids, especially blood or semen, enter the blood or tissue fluids of another person's body. The entry may occur during vaginal, anal, or oral intercourse, or via contaminated needles shared by intravenous drug users. HIV can also be transmitted from infected mothers to their infants during pregnancy, birth, and nursing. AIDS is rarely transmitted through casual contact, food, or body products such as saliva, tears, urine, or feces.

The primary cellular hosts for HIV are macrophages and helper T cells, which are ultimately destroyed in large numbers by the virus. The infection makes helper T cells unavailable for the stimulation and proliferation of B cells and cytotoxic T cells. The assault on lymphocytes and macrophages cripples the immune system and makes the body highly vulnerable to infections and to development of otherwise rare forms of cancer.

In 1996 researchers confirmed the process by which HIV initially infects its primary target, the helper T cells. First, a glycoprotein of the viral coat, called *gp120,* attaches the virus to a helper T cell by binding to its CD4 receptor. Then, another viral protein triggers fusion of the viral surface membrane with the T-cell plasma membrane, releasing the virus into the cell **(Figure b).** Once inside, a viral enzyme, *reverse transcriptase,* uses the viral RNA as a template for making a DNA copy. (The genetic material of HIV is RNA rather than DNA when it is outside a host cell.) Another viral enzyme, *integrase,* then splices the viral DNA into the host cell's DNA. Once it is part of the host cell DNA, the viral DNA is replicated and passed on as the cell divides. As part of the host cell DNA, the virus is effectively hidden in the helper T cell and protected from attack by the immune system.

The viral DNA typically remains dormant until the helper T cell is stimulated by an antigen. At that point, the viral DNA is copied into new viral RNA molecules, and into mRNAs that direct host cell ribosomes to make viral proteins. The viral RNAs are added to the viral proteins to make infective HIV particles, which are released from the host cell by budding **(Figure c).** The viral particles may infect more body cells or another person. The viral infection also leads uninfected helper T cells to destroy themselves in large numbers by apoptosis, through mechanisms that are still unknown.

At the time of initial infection, many people suffer a mild fever and other symptoms that may be mistaken for the flu or the common cold. The symptoms disappear as antibodies against viral proteins appear in the body, and the number of viral particles

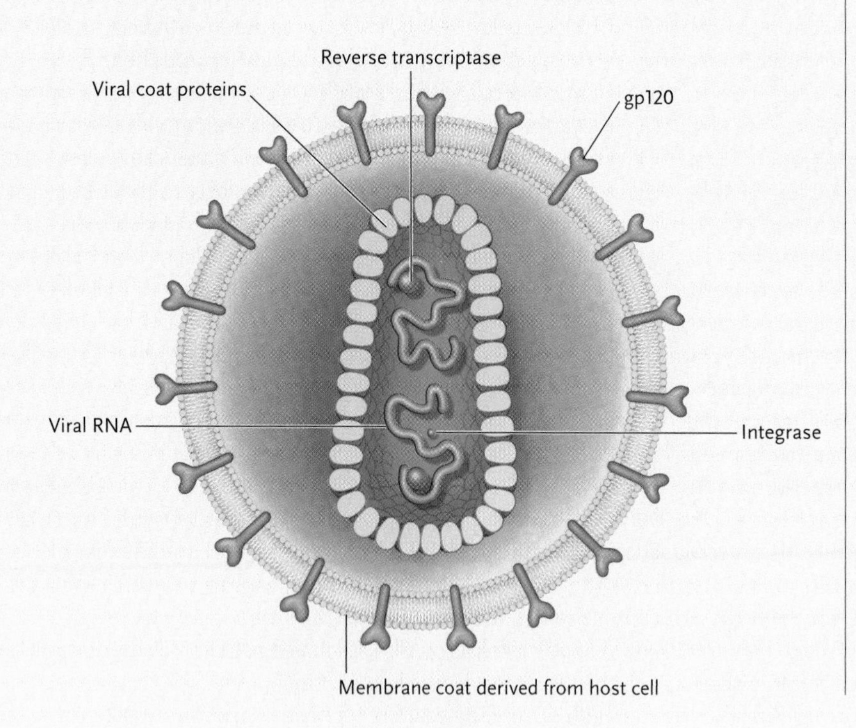

Figure a
Structure of a free HIV viral particle.

against one version of the surface proteins, the pathogens have switched to different surface proteins that the antibodies do not match. These new proteins take another week or so to stimulate the production of specific antibodies; by this time, the surface groups change again. The changes continue indefinitely, always keeping the pathogens one step ahead of the immune system. Pathogens that use these mechanisms to sidestep the immune system include the protozoan causing African sleeping sickness, the bacterium causing gonorrhea, and the viruses causing influenza, the common cold, and AIDS.

Some viruses use parts of the immune system to get a free ride to the cell interior. For example, the

drops in the bloodstream. However, the virus's genome is still present, integrated into the DNA of T cells, and the virus steadily spreads to infect other T cells. An infected person may remain apparently healthy for years, yet can infect others. Both the transmitter and recipient of the virus may be unaware that the disease is present, making it difficult to control the spread of HIV infections.

With time, more and more helper T cells and macrophages are destroyed, eventually wiping out the body's immune response. The infected person becomes susceptible to opportunistic, secondary infections, such as a pneumonia caused by a fungus *(Pneumocystis carinii)*; drug-resistant tuberculosis; persistent yeast *(Candida albicans)* infections of the mouth, throat, rectum, or vagina; and infection by many common bacteria and viruses that rarely infect healthy humans. These secondary infections signal the appearance of full-blown AIDS. Steady debilitation and death typically follow within a period of years in untreated persons.

As yet, there is no cure for HIV infection and no vaccine that can protect against infection. HIV coat proteins mutate constantly, making a vaccine developed against one form of the virus useless when the next form appears. Most of these mutations occur during replication of the virus, when reverse transcriptase makes a DNA copy of the viral RNA.

The development of AIDS can be greatly slowed by drugs that interfere with reverse transcription of the viral genomic RNA into the DNA copy that integrates into the host's genome.

Treatment with a "cocktail" of drugs called *reverse transcriptase inhibitors (RTIs)* inhibits viral reproduction and destruction of helper T cells and extends the lives of people with the AIDS virus. The inhibiting cocktails are not a cure for AIDS, however, because the virus is still present in dormant form in helper T cells. If the therapy is stopped, the virus again replicates and the T-cell population drops.

Presently, the only certain way to avoid HIV infection is to refrain from unprotected sex with people whose HIV status is unknown, and from the use of contaminated needles of the type used to administer drugs intravenously.

Viral RNA genome
Viral DNA
Nucleus
Transcription
Transcription
Viral RNA genome
Viral mRNA
Translation
Viral proteins

1 Viral particle enters cell.

2 Viral reverse transcriptase makes DNA copy of viral RNA genome.

3 DNA copy of viral genome is integrated into the host DNA.

4 Viral DNA is transcribed into viral RNA genomes and into viral mRNAs, which are translated into viral proteins.

5 Viral RNAs and proteins assemble into new viral particles, which bud from cell.

Figure b
The steps in HIV infection of a host cell.

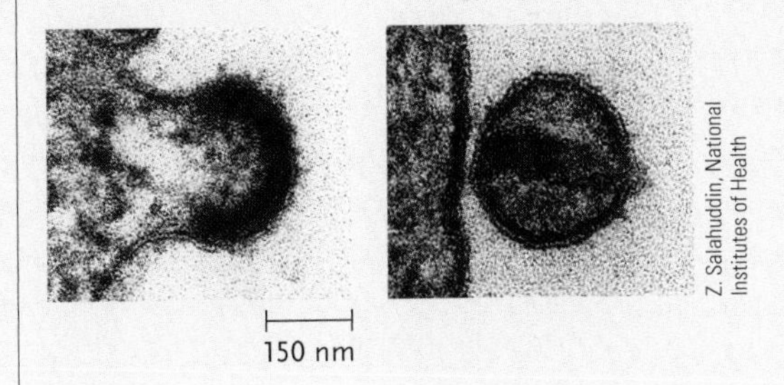

Figure c
An HIV particle budding from a host cell. As it passes from the host cell, it acquires a membrane coat derived from the host cell plasma membrane.

AIDS virus has a surface molecule that is recognized and bound by the CD4 receptor on the surface of helper T cells. Binding to CD4 locks the virus to the cell surface and stimulates the membrane covering the virus to fuse with the plasma membrane of the helper T cell. (The protein coat of the virus is wrapped in a membrane derived from the plasma membrane of the host cell in which it was produced.) The fusion introduces the virus into the cell, initiating the infection and leading to destruction and death of the T cell. (Further details of HIV infection and AIDS are presented in *Focus on Applied Research*.)

a. Initial exposure to allergen

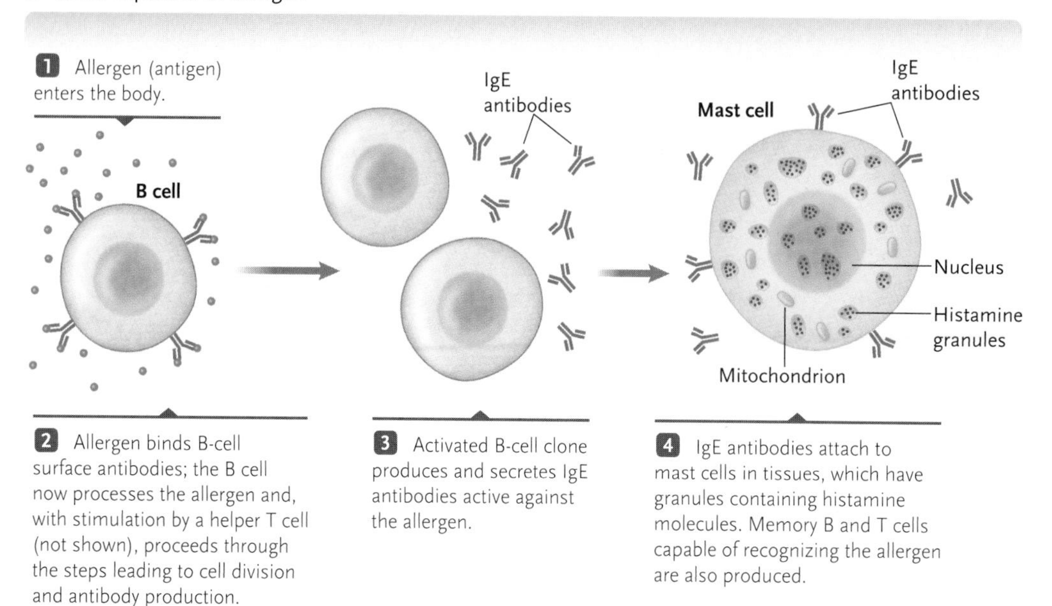

1 Allergen (antigen) enters the body.
B cell
IgE antibodies
Mast cell
IgE antibodies
Nucleus
Histamine granules
Mitochondrion
2 Allergen binds B-cell surface antibodies; the B cell now processes the allergen and, with stimulation by a helper T cell (not shown), proceeds through the steps leading to cell division and antibody production.
3 Activated B-cell clone produces and secretes IgE antibodies active against the allergen.
4 IgE antibodies attach to mast cells in tissues, which have granules containing histamine molecules. Memory B and T cells capable of recognizing the allergen are also produced.

b. Further exposures to allergen

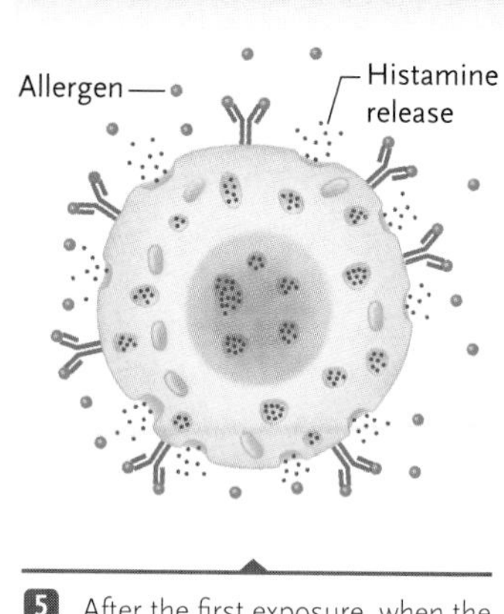

Allergen
Histamine release
5 After the first exposure, when the allergen enters the body, it binds with IgE antibodies on mast cells; binding stimulates the mast cell to release histamine and other substances.

Figure 43.15
The response of the body to allergens. **(a)** The steps in sensitization after initial exposure to an allergen. **(b)** Production of an allergic response by further exposures to the allergen.

Allergies Are Produced by Overactivity of the Immune System

The substances responsible for allergic reactions form a distinct class of antigens called **allergens,** which induce B cells to secrete an overabundance of IgE antibodies **(Figure 43.15).** The IgE antibodies, in turn, bind to receptors on mast cells in connective tissue and on **basophils,** a type of leukocyte in blood (see Table 43.1), inducing them to secrete histamine, which produces a severe inflammation. Most of the inflammation occurs in tissues directly exposed to the allergen, such as the surfaces of the eyes, the lining of the nasal passages, and the air passages of the lungs. Signal molecules released by the activated mast cells also stimulate mucosal cells to secrete floods of mucus and cause smooth muscle in airways to constrict (histamine also causes airway constriction). The resulting allergic reaction can vary in severity from a mild irritation to serious and even life-threatening debilitation. *Asthma* is a severe response to allergens involving constriction of airways in the lungs. Antihistamines (substances that block histamine receptors) are usually effective in countering the effects of the histamine released by mast cells.

An individual is *sensitized* by a first exposure to an allergen, which may produce only mild allergic symptoms or no reaction at all (see Figure 43.15a). However, the sensitization produces memory B and T cells; at the next and subsequent exposures, the system is poised to produce a greatly intensified allergic response (see Figure 43.15b).

In some persons, inflammation stimulated by an allergen is so severe that the reaction brings on a life-threatening condition called **anaphylactic shock.** Extreme swelling of air passages in the lungs interferes with breathing, and massive leakage of fluid from capillaries causes the blood pressure to drop precipitously. Death may result in minutes if the condition is not treated promptly. In persons who have become sensitized to the venom of wasps and bees, for example, a single sting may bring on anaphylactic shock within minutes. Allergies developed against drugs such as penicillin and certain foods can have the same drastic effects. Anaphylactic shock can be controlled by immediate injection of epinephrine (adrenaline), which reverses the condition by constricting blood vessels and dilating air passages in the lungs.

Study Break

1. What is immunological tolerance?
2. Explain how a failure in the immune system can result in an allergy.

43.5 Defenses in Other Animals

Other Vertebrate Groups Have Defenses against Infections

This chapter has emphasized the mammalian immune system, the focus of most immunology research. We know relatively little about defenses against infections in most other vertebrate groups. Yet like all other physiological systems, the mammalian defenses against pathogens are the result of evolution, and evidence of their functions can be seen in other vertebrate groups and also in invertebrates.

For example, molecular studies in sharks and rays have revealed DNA sequences that are clearly related to the sequences coding for antibodies in mammals. If injected with an antigen, sharks produce antibodies, formed from light- and heavy-chain polypeptides, capable of recognizing and binding the antigen. Although embryonic gene segments for the two polypeptides are arranged differently in sharks than they are in mammals, antibody diversity is produced by the same kinds of genetic rearrangements in both. Sharks also mount nonspecific defenses, including the production of a steroid that appears to kill bacteria and neutralize viruses nonspecifically and with high efficiency.

Invertebrates lack specific immune defenses equivalent to antibodies and the activities of B and T cells, so their reactions to invading pathogens most closely resemble the nonspecific defenses of humans and other vertebrates. However, all invertebrates have phagocytic cells, which patrol tissues and engulf pathogens and other invaders. Some of the signal molecules that stimulate phagocytic activity, such as interleukins, appear to be similar in invertebrates and vertebrates.

Antibodies do not occur in invertebrates, but proteins of the immunoglobulin family are widely distributed. In at least some invertebrates, these Ig proteins have a protective function. In moths, for example, an

Unanswered Questions

How does gene expression in a bacterial pathogen change during infection?

Characterizing the genetic events involved in the interactions between pathogens and hosts is important for understanding the development of an infectious disease in the host, and for producing effective therapeutic treatments. You learned in this chapter that pathogenic bacteria are first combated by the innate immunity system. Researchers in James Musser's lab at Baylor College of Medicine in collaboration with researchers at Rocky Mountain Laboratories, National Institute of Allergy and Infectious Diseases in Montana, have now obtained information about changes in gene expression for a bacterial pathogen, group A *Streptococcus* (GAS), during infection of a mammal.

GAS is a Gram-positive bacterium that is the cause of pharyngitis ("strep throat"; 2 million cases annually in the United States) and various other infections, including rheumatic heart disease. The researchers studied how gene expression in GAS changed during an 86-day period following infection that caused pharyngitis in macaque monkeys. They are an excellent model for the study because the progression of pharyngitis caused by GAS in these monkeys is highly similar to that seen in human infections. There are three distinct phases, during which GAS can be detected by culturing throat swabs. In the first phase, colonization, GAS establishes infection of host cells, producing only mild pharyngitis. In the second phase, the acute phase, pharyngitis symptoms peak, as does the number of GAS bacteria. In the third phase, the asymptomatic phase, symptoms decrease and disappear along with a decrease in GAS bacteria.

Experimentally, the researchers analyzed gene expression from the entire genome of GAS in the three phases of pharyngitis using DNA microarrays (see Section 18.3). They found that the pattern of GAS gene expression changed over the course of the disease. Significantly, they saw characteristic gene expression patterns for each of the three phases of pathogen–host interaction. These results indicate that GAS regulates expression of its genes extensively as it establishes an infection, and as the host mounts an innate immune response against it. This work will help direct future research efforts to control infections caused by GAS. More broadly, genomic studies of this kind are likely to provide insights into genetic events contributing to pathogenesis in other pathogen–host interactions.

How does HIV evade the adaptive immunity system?

In cell-mediated immunity, a pathogen-infected antigen-presenting cell (APC) presents an antigen fragment bound to a class I MHC protein to a $CD8^+$ T cell, stimulating the T cell to differentiate into cytotoxic T cells. The cytotoxic T cells then bind to the infected APCs and destroy them. Cytotoxic T cells act particularly against host cells infected by viral pathogens. HIV infects host cells but, rather than being eliminated by the host, this virus establishes a chronic infection that leads to the development of AIDS. That is, HIV evades the adaptive immunity system.

Kathleen Collins and her group at the University of Michigan Medical School have investigated the mechanism of this evasion. They have learned that the virus down-regulates the display of class I MHC proteins on the surfaces of HIV-infected APCs, which thereby limits the presentation of viral antigens by those cells. The down-regulation occurs by the action of the HIV Nef (*ne*gative *f*actor) protein. Nef binds to class I MHC molecules and inhibits them from moving through the Golgi complex to the cell surface. Without class I MHC molecules on the cell surface to present antigens, the immune response is compromised. The action of Nef, therefore, enhances the ability of HIV to induce AIDS. Research now being pursued is directed toward characterizing more completely the role of Nef in disrupting class I MHC movement from the ER through the Golgi complex to the cell surface, identifying and characterizing other proteins that may interact with Nef in an HIV-infected cell, and developing pharmaceutical reagents aimed at blocking Nef action.

Peter J. Russell

Ig-family protein called *hemolin* binds to the surfaces of pathogens and marks them for removal by phagocytes.

Many invertebrates produce antimicrobial proteins such as lysozyme that are able to kill bacteria and other invading cells. Insects, for example, secrete lysozyme in response to bacterial infections.

Study Break

Compare invertebrate and mammalian immune defenses.

Review

Go to **ThomsonNOW** at www.thomsonedu.com/login to access quizzing, animations, exercises, articles, and personalized homework help.

43.1 Three Lines of Defense against Invasion

- Humans and other vertebrates have three lines of defense against pathogens. The first, which is nonspecific, is the barrier set up by the skin and mucous membranes.
- The second line of defense, also nonspecific, is innate immunity, an innate system that defends the body against pathogens and toxins penetrating the first line.
- The third line of defense, adaptive immunity, is specific: it recognizes and eliminates particular pathogens and retains a memory of that exposure so as to respond rapidly if the pathogen is encountered again. The response is carried out by lymphocytes, a specialized group of leukocytes.

43.2 Nonspecific Defenses: Innate Immunity

- In the innate immunity system, molecules on the surfaces of pathogens are recognized as foreign by receptors on host cells. The pathogen is then combated by the inflammation and complement systems.
- Epithelial surfaces secrete defensins, a type of antimicrobial peptide, in response to attack by a microbial pathogen. Defensins disrupt the plasma membranes of pathogens, killing them.
- Inflammation is characterized by heat, pain, redness, and swelling at the infection site. Several interconnecting mechanisms initiate inflammation, including pathogen engulfment, histamine secretion, cytokine release, and local blood vessel dilation and permeability increase. (Figure 43.1).
- Large arrays of complement proteins are activated when they recognize molecules on the surfaces of pathogens. Some complement proteins form membrane attack complexes, which insert into the plasma membrane of many types of bacteria and cause their lysis. Fragments of other complement proteins coat pathogens, stimulating phagocytes to engulf them (Figure 43.2).
- Three nonspecific defenses are used to combat viral pathogens: RNA interference, interferons, and natural killer cells.

Animation: Innate defenses

Animation: Complement proteins

Animation: Inflammatory response

Animation: Immune responses

Animation: Human lymphatic system

43.3 Specific Defenses: Adaptive Immunity

- Adaptive immunity, which is carried out by B and T cells, targets particular pathogens or toxin molecules.
- Antibodies consist of two light and two heavy polypeptide chains, each with variable and constant regions. The variable regions of the chains combine to form the specific antigen-binding site (Figure 43.4).
- Antibodies occur in five different classes: IgM, IgD, IgG, IgA, and IgE. Each class is determined by its constant region (Table 43.2).
- Antibody diversity is produced by genetic rearrangements in developing B cells that combine gene segments into intact genes encoding the light and heavy chains. The rearrangements producing heavy-chain genes and T-cell receptor genes are similar. The light and heavy chain genes are transcribed into precursor mRNAs, which are processed into finished mRNAs, which are translated on ribosomes into the antibody polypeptides (Figure 43.4).
- The antibody-mediated immune response has two general phases: T-cell activation, and B-cell activation and antibody production. T-cell activation begins when a dendritic cell engulfs a pathogen and produces antigens, making the cell an antigen-presenting cell (APC). The APC secretes interleukins, which activate the T cell. The T cell then secretes cytokines, which stimulate the T cell to proliferate, producing a clone of cells. The clonal cells differentiate into helper T cells (Figures 43.3, 43.6, and 43.8).
- B-cell receptors (BCRs) on B cells recognize antigens on a pathogen and engulf it. The B cells then display the antigens. The TCR on a helper T cell activated by the same antigen binds to the antigen on the B cell. Interleukins from the T cell stimulate the B cell to produce a clone of cells with identical BCRs. The clonal cells differentiate into plasma cells, which secrete antibodies specific for the antigen, and memory B cells, which provide immunological memory of the antigen encounter (Figures 43.3, 43.8, and 43.9).
- Clonal expansion is the process of selecting a lymphocyte specifically for cloning when it encounters an antigen from among a randomly generated, large population of lymphocytes with receptors that specifically recognize the antigen (Figure 43.10).
- Antibodies clear the body of antigens by neutralizing or agglutinating them, or by aiding the innate immune response (Figure 43.11).
- In immunological memory, the first encounter of an antigen elicits a primary immune response and later exposure to the same antigen elicits a rapid secondary response with a greater production of antibodies (Figure 43.12).
- Active immunity is the production of antibodies in the body in response to an antigen. Passive immunity is the acquisition of antibodies by direct transfer from another person.
- In cell-mediated immunity, cytotoxic T cells recognize and bind to antigens displayed on the surfaces of infected body cells, or to cancer cells. They then kill the infected body cell (Figure 43.13).
- Antibodies are widely used in research to identify, locate, and determine the functions of molecules in biological systems.
- Monoclonal antibodies are made by isolating fully active B cells from a test animal, fusing them with cancer cells to produce hy-

bridomas, and using single hybridomas to start clones of cells, all of which make highly specific antibodies against the same epitope of an antigen (Figure 43.14).

Animation: Antibody structure

Animation: Gene rearrangements

Animation: Clonal selection of a B cell

Animation: Antibody-mediated response

Animation: Cell-mediated response

43.4 Malfunctions and Failures of the Immune System

- In immunological tolerance, molecules present in an individual at birth normally do not elicit an immune response.
- In some people, the immune system malfunctions and reacts against the body's own proteins or cells, producing autoimmune disease.
- The first exposure to an allergen sensitizes an individual by leading to the production of memory B and T cells, which cause a greatly intensified response at the next and subsequent exposures.
- Most allergies result when antigens act as allergens by stimulating B cells to produce IgE antibodies, which leads to the release of histamine. Histamine produces the symptoms characteristic of allergies (Figure 43.15).

Animation: HIV replication cycle

43.5 Defenses in Other Animals

- Antibodies, complement proteins, and other molecules with defensive functions have been identified in all vertebrates.
- Invertebrates rely on nonspecific defenses, including surface barriers, phagocytes, and antimicrobial molecules.

Questions

Self-Test Questions

1. Which of the following most directly affects a cell harboring a virus?
   a. $CD8^+$ T cells that bind class I MHC proteins holding viral antigen
   b. $CD4^+$ T cells that bind free viruses in the blood
   c. B cells secreting perforin
   d. antibodies that bind the viruses with their constant ends
   e. natural killer cells secreting antiviral antibodies
2. Components of the inflammatory response include all *except*:
   a. macrophages.
   b. neutrophils.
   c. B cells.
   d. mast cells.
   e. eosinophils.
3. When a person resists infection by a pathogen after being vaccinated against it, this is the result of:
   a. innate immunity.
   b. immunological memory.
   c. a response with defensins.
   d. an autoimmune reaction.
   e. an allergy.
4. One characteristic of a B cell is that it:
   a. has the same structure in both invertebrates and vertebrates.
   b. recognizes antigens held on class I MHC proteins.
   c. binds viral infected cells and directly kills them.
   d. makes many different BCRs on its surface.
   e. has a BCR on its surface, which is the IgM molecule.
5. Antibodies:
   a. are each composed of four heavy and four light chains.
   b. display a variable end, which determines the antibody's location in the body.
   c. belonging to the IgE group are the major antibody class in the blood.
   d. found in large numbers in the mucous membranes belong to class IgG.
   e. function primarily to identify and bind antigens free in body fluids.
6. The generation of antibody diversity includes the:
   a. joining of V to C to J segments to make a functional light chain gene.
   b. choice from several different types of C segments to make a functional light chain gene.
   c. deletion of the J segment to make a functional light chain gene.
   d. joining of V to J to C segments to make a functional light chain gene.
   e. initial generation of IgG followed later by IgM on a given cell.
7. An APC:
   a. can be a $CD8^+$ T cell.
   b. derives from a phagocytic cell and is lymphocyte-stimulating.
   c. secretes antibodies.
   d. cannot be a B cell.
   e. cannot stimulate helper T cells.
8. Antibodies function to:
   a. deactivate the complement system.
   b. neutralize natural killer cells.
   c. clump bacteria and viruses for easy phagocytosis by macrophages.
   d. eliminate the chance for a secondary response.
   e. kill viruses inside of cells.
9. After Jen punctured her hand with a muddy nail, in the emergency room she received both a vaccine and someone else's antibodies against tetanus toxin. The immunity conferred here is:
   a. both active and passive.
   b. active only.
   c. passive only.
   d. first active; later passive.
   e. innate.
10. Medicine attempts to enhance the immune response when treating:
    a. organ transplant recipients.
    b. anaphylactic shock.
    c. rheumatoid arthritis.
    d. HIV infection.
    e. Type I diabetes.

Questions for Discussion

1. HIV wreaks havoc with the immune system by attacking helper T cells and macrophages. Would the impact be altered if the virus attacked only macrophages? Explain.

2. Given what you know about how foreign invaders trigger immune responses, explain why mutated forms of viruses, which have altered surface proteins, pose a monitoring problem for memory cells.
3. Cats, dogs, and humans may develop myasthenia gravis, an autoimmune disease in which antibodies develop against acetylcholine receptors in the synapses between neurons and skeletal muscle fibers. Based on what you know of the biochemistry of muscle contraction (see Section 41.1), explain why people with this disease typically experience severe fatigue with even small levels of exertion, drooping of facial muscles, and trouble keeping their eyelids open.

Experimental Analysis

Space, the final frontier! Indeed, but being in space has some problems. Astronauts in space show a decline in their ability to mount an immune response and, consequently, develop a decreased resistance to infection. Two potentially important differences in physiology in space versus on Earth are more fluid flowing to the head and a lack of weight-bearing on the lower limbs. Could they be involved somehow in the deleterious effect on the immune system? Design an experiment to be done on Earth to answer this question.

Evolution Link

Defensins are found in a wide range of organisms, including plants as well as animals. What are the evolutionary implications of this observation?

How Would You Vote?

Drugs are available that can extend the life of patients with AIDS, but their high cost is more than people in most developing countries can afford to pay. Should the federal government offer incentives to companies to discount the drugs for developing countries? What about AIDS patients at home? Who should pay for their drugs? Go to www.thomsonedu.com/login to investigate both sides of the issue and then vote.

Lining of the trachea (windpipe) shown in a colorized SEM, with mucus-secreting cells (white) and epithelial cells with cilia (pink). The trachea is positioned between the larynx and the lungs, providing a conduit for air entering and leaving the body.

Study Plan

44 Gas Exchange: The Respiratory System

Why It Matters

On October 25, 1999, at 9:19 A.M., the captain lined up Learjet N47BA on the runway at Orlando International Airport and opened the throttles. Within seconds, the sleek corporate jet was airborne and climbing; 2 minutes later, at 9:21 EDT, the pilots reported passing through 9500 feet.

As the jet continued its climb, the pressure of the outside air dropped steadily and with it the availability of the oxygen (O_2) that all animal life requires, including the two pilots and three passengers on the jet. Normally, in aircraft, the cabin pressure is maintained at a level equivalent to an altitude of 8000 feet, more than sufficient to keep O_2 available to all on board. But, unknown to the pilots, the pressurization system was not functioning normally.

At 9:27 EDT, the controller at the Jacksonville Control Center instructed the jet to climb to 39,000 feet. The first officer acknowledged the instruction, her voice strong and clear. Her acknowledgment was the last radio transmission anyone was to hear from N47BA.

When humans experience increasingly higher altitudes, each breath brings less O_2 into the body. Of all the cells affected by reduced O_2, the ones most sensitive are those of the eyes and brain. Without an O_2 supply at 25,000 feet, most people progress from fully alert to unconscious in about 3 minutes; at 40,000 feet, the progression takes only 15 seconds.

The jet continued its climb, eventually reaching an altitude of 46,000 feet. When the pilots stopped responding to communications, military jets were sent to investigate. The military pilots could see no movement in the Learjet cabin and there was no response to their transmissions. The forward windshields of the Learjet were frosted over, indicating that warm air from the engines was not ventilating the cabin correctly. Evidently, the aircraft was maintaining its course through the autopilot, without conscious human direction.

Many hours later, at 12:11 P.M. CDT, one of the two engines failed: the aircraft, now unbalanced, rolled over and entered a steep, spiraling descent that ended in a shattering impact in a field near Aberdeen, South Dakota. The subsequent investigation pointed to faulty operation of a single valve controlling cabin pressurization as a likely cause of the accident. This tragic loss of life emphasizes the vital importance of O_2 to the survival of humans and other animals. In this chapter we discuss the respiratory system, the system that allows an animal to exchange CO_2 produced in the body for O_2 from the surroundings. The respiratory systems of animals reflect the environmental conditions under which they live, and this general principle has resulted in a truly remarkable array of adaptations.

44.1 The Function of Gas Exchange

Physiological respiration is the process by which animals exchange gases with their surroundings—how they take in O_2 from the outside environment and deliver it to body cells, and remove CO_2 from body cells and deliver it to the environment **(Figure 44.1)**. The absorbed O_2 is used as the final electron acceptor for the oxidative reactions that produce ATP in mitochondria (see Section 8.4). The CO_2 released to the environment is a product of those oxidative reactions. Because they use O_2 and release CO_2, these ATP-producing reactions are called *cellular respiration.*

Figure 44.1 The relationship between cellular respiration and physiological respiration.

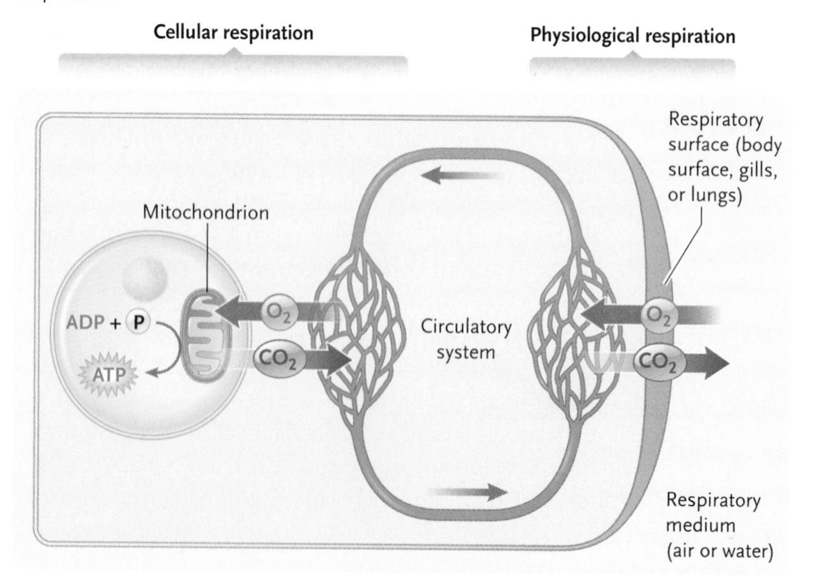

How gas exchange occurs in an animal depends on its respiratory medium—air or water—and the nature of its respiratory surface. The **respiratory medium** is the environmental source of O_2 and the "sink" for released CO_2. For aquatic animals, of course, the respiratory medium is water; for terrestrial animals, it is air. Amphibians and some fishes use both water and air as respiratory media. The exchange of gases with the respiratory medium by animals is called **breathing,** whether the medium is air or water.

The **respiratory surface,** formed by a layer of epithelial cells, provides the interface between the body and the respiratory medium. Oxygen is absorbed across the respiratory surface, and CO_2 is released. In all animals, the exchange of gases across the respiratory surface occurs by simple diffusion, movement of molecules from a region of higher concentration to a region of lower concentration (see Section 6.2).

Generally, the concentration of O_2 is higher in the respiratory medium than on the internal side of the respiratory surface, and thus the net diffusion of O_2 is inward. Carbon dioxide moves in the opposite direction because the CO_2 concentration is higher on the internal side of the respiratory surface than in the respiratory medium.

Respiratory surfaces typically have two structural properties that favor a high rate of diffusion: they are thin, and they have large surface areas. The rate of diffusion is inversely proportional to the square of the distance over which the diffusion occurs; diffusion rates are therefore higher through thin surfaces such as the single layer of epithelial cells forming many respiratory surfaces. And, the rate of diffusion is directly proportional to the surface area across which diffusion occurs, meaning that large surface areas allow for higher rates of gas exchange than small surface areas. In addition, the rate of diffusion becomes higher with larger concentration gradients and with increasing temperature.

In some relatively small animals, such as sponges, ctenophores, roundworms, flatworms, and some annelids, the entire body surface serves as the respiratory surface. All these animals are invertebrates that live in aquatic or moist environments.

In larger animals, specialized structures, *gills* and *lungs,* form the primary respiratory surface for exchanging gases with water and air, respectively. In insects, a **tracheal system,** an extensive system of branching tubes, channels air from the outside to the internal organs and most individual cells of the animal.

Because gases must dissolve in water to enter and leave epithelial cells, the respiratory surface must be wetted to function in gas exchange, either directly by the respiratory medium or by a thin film of water. For this reason, in water-breathing animals, **gills** are *evaginations* of the body: they extend outward into the respiratory medium. In terrestrial animals, **lungs** are typically pockets or *invaginations* of the body surface, buried deeply in the body interior where they are less susceptible to drying out. Also, terrestrial animals have adaptations that moisten dry air before it reaches the respiratory surface. For example, in humans and other mammals, moisture is added to air as it passes through the mouth, nasal passages, throat, and air passages leading to the lungs.

The organ system responsible for gas exchange is termed the **respiratory system.** The respiratory system consists of all the parts of the body involved in exchanging air between the external environment and the blood. In mammals, this includes the airways leading to and into the lungs, the lungs themselves, and the structures of the chest used to move air through the airways into and out of the lungs.

Adaptations That Increase Ventilation and Perfusion of the Respiratory Surface Maximize the Rate of Gas Exchange

Two primary adaptations help animals maintain the difference in concentration between gases outside and inside the respiratory surface, thereby keeping the rate of gas exchange at maximal levels. One is **ventilation,** the flow of the respiratory medium (air or water, depending on the animal) over the external side of the respiratory surface. The second is **perfusion**, the flow of blood or other body fluids on the internal side of the respiratory surface.

Ventilation. As they respire, animals remove O_2 from the respiratory medium and replace it with CO_2. Without ventilation, the concentration of O_2 would fall in the respiratory medium close to the respiratory surface, and the concentration of CO_2 would rise, gradually reducing the concentration gradients and dropping the rate of gas exchange below the minimum level required to sustain life. Examples of ventilation include the one-way flow of water over the gills in fish and many other aquatic animals and the in-and-out flow of air in the lungs of most vertebrates and in the tracheal system of insects.

Perfusion. The constant replacement of blood or another fluid on the internal side of the respiratory surface helps to keep the inside/outside concentration differences of O_2 and CO_2 at a maximum. In animals without a circulatory system, such as roundworms and flatworms, body movements help circulate body fluids beneath the skin. Most animals without a circulatory system are small or have thin, greatly flattened bodies, because all body cells must be located close to the respiratory surface to exchange O_2 and CO_2 adequately. In animals with a circulatory system, the circulatory system brings blood to the internal side of the respiratory surface, transporting CO_2 from all cells of the body—no matter how far they are from the respiratory surface—to exchange for O_2, which is then taken to all cells of the body.

Adaptations That Increase the Area of the Respiratory Surface Maximize the Quantity of Gases Exchanged

Most animals have adaptations that increase the quantity of gases exchanged by increasing the area of the respiratory surface. In animals whose skin serves as the respiratory surface, an elongated or flattened body form increases the area of the respiratory surface **(Figure 44.2a).**

In animals with gills, the respiratory surface is increased by highly branched structures that include many fingerlike or platelike projections **(Figure 44.2b).** Similarly, in animals with lungs or tracheae, the respiratory surface is increased by a multitude of branched tubes, folds, or pockets **(Figure 44.2c).**

Water and Air Have Advantages and Disadvantages as Respiratory Media

Because their respiratory surfaces are exposed directly to the environment, water breathers have no problem keeping the respiratory surface wetted. However, aquatic animals face two main challenges in obtaining O_2 from water compared with terrestrial animals. First, water contains approximately one-thirtieth as much O_2 as air

a. Extended body surface: flatworm

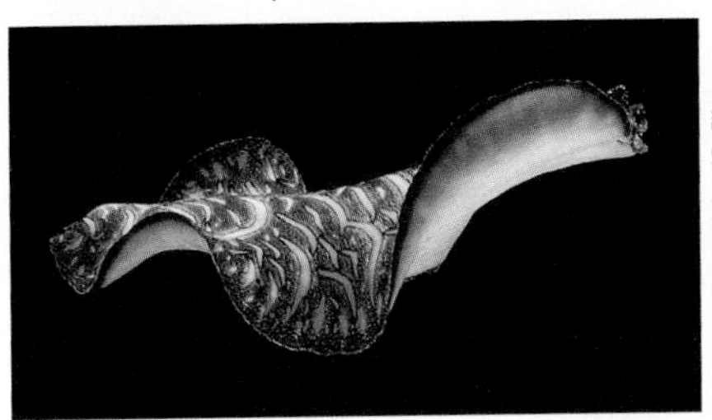

Peter Parks/Oxford Scientific Films

b. External gills: mudpuppy

Jack Dermid/Visuals Unlimited

c. Lungs: human

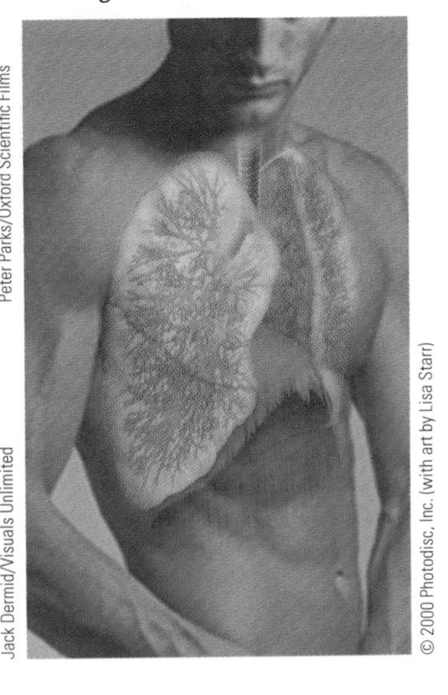

Figure 44.2
Adaptations increasing the area of the respiratory surface. (a) The flattened and elongated body surface of a flatworm. **(b)** The highly branched, feathery structure of the external gills in an amphibian, the mudpuppy *(Necturus)*. **(c)** The many branches and pockets expanding the respiratory surface in the human lung.

does (at 15°C). Therefore, to obtain the same amount of O_2, an aquatic animal must process 30 times as much of its respiratory medium as a terrestrial animal does. Second, water is about 1000 times as dense as air and about 50 times as viscous. Therefore, it takes significantly more energy to move water than air over a respiratory surface. For this reason, ventilation in most aquatic animals takes place in a one-way direction. In bony fishes, for instance, water enters the mouth, flows over the gills, and exits through the gill covers, all in one direction.

In addition, temperature and solutes affect the O_2 content of water. That is, as either the temperature or the amount of solutes increases, the amount of gas that can dissolve in water decreases. Therefore, with respect to obtaining O_2, aquatic animals that live in warm water are at a disadvantage compared with those that live in cold water. And, because levels of solutes (such as sodium chloride) are higher in seawater than in freshwater, aquatic animals living in seawater are at a disadvantage.

The relatively high O_2 content, low density, and low viscosity of air greatly reduce the energy required to ventilate the respiratory surface. These advantages allow animals with lungs to breathe in and out, reversing the direction of flow of the respiratory medium, without a large energy penalty. As you will see later, reversing the direction of flow decreases the efficiency of gas exchange.

Another advantage of breathing air is that gas molecules diffuse nearly 10,000 times faster through air than through water. This increases the rate at which molecules of the gases at the respiratory surface exchange with those located farther away in the air, and reduces the requirement for ventilation as compared with water.

A major disadvantage of air is that it constantly evaporates water from the respiratory surface unless the air is saturated with water vapor. Therefore, except in an environment with 100% humidity, animals lose water by evaporation during breathing and must replace the water to keep the respiratory surface from drying and causing the death of the surface cells.

We next turn to the adaptations that allow water-breathing and air-breathing animals to obtain O_2 and release CO_2 in aquatic and terrestrial environments. These adaptations allow animals to exploit the advantages and circumvent the disadvantages of water and air as respiratory media.

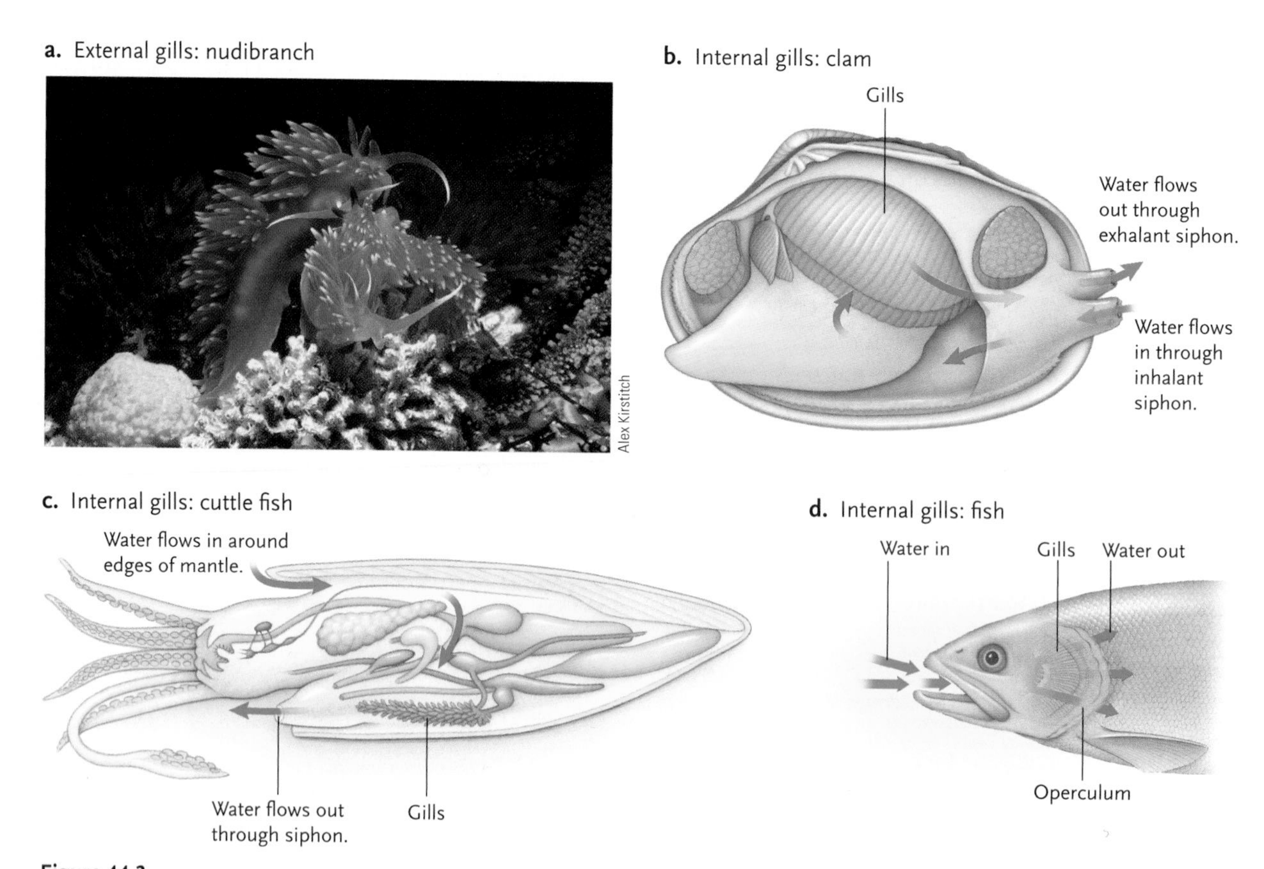

Figure 44.3
External and internal gills. (a) The external gills of a nudibranch *(Flabellina iodinea)*. **(b)** The internal gills in a clam. **(c)** The internal gills in a cuttlefish. **(d)** Internal gills of a bony fish. Water enters through the mouth and passes over the filaments of the gills before exiting through an opening at the edges of the flaplike protective covering, the operculum.

Study Break

1. Distinguish between the roles of the respiratory medium and the respiratory surface in respiratory systems.
2. What is an advantage of water over air as a respiratory medium? What are two key advantages of air over water as a respiratory medium?

44.2 Adaptations for Respiration

Although most animals that live in water exchange gases through the skin or gills, some, such as whales, seals, and dolphins, exchange gases through lungs (which originally evolved in aquatic creatures). And although most animals that live on land exchange gases through lungs, some, such as sow bugs and land crabs, exchange gases through gills, and others, such as insects, exchange gases using a tracheal system.

Aquatic Gill Breathers Exchange Gases More Efficiently Than Skin Breathers

Gills provide water-breathing animals, and a few air-breathers, with more efficient gas exchange than skin breathers have. In combination with the organized circulatory system common to these animals, gills also allow animals to live in more diverse habitats, and to achieve greater body mass, than animals that breathe primarily or exclusively through the skin.

External and Internal Gills. Gills are respiratory surfaces that are branched and folded evaginations of the body surface. **External gills** are gills that do not have protective coverings; they extend out from the body and are in direct contact with the water. **Internal gills,** by contrast, are located within chambers of the body that have a cover providing physical protection for the gills. Water must be brought to internal gills.

Because external gills have no protective coverings, they are exposed to mechanical damage and must be immersed in water to keep them from collapsing or drying. For these reasons, animals with external gills, including some annelids and mollusks **(Figure 44.3a),** aquatic insects, the larval forms of some bony fishes, and some amphibians, are limited to relatively protected aquatic environments.

The coverings of internal gills protect them from mechanical damage and drying. Covered internal gills allow animals to live in highly diverse habitats, ranging from small streams and ponds to rivers, lakes, and the open seas, and even in moist terrestrial habitats. Most crustaceans, mollusks, sharks, and bony fishes have internal gills. Some invertebrates, such as clams and oysters, use beating cilia to circulate water over their internal gills **(Figure 44.3b).** Others, such as the cuttlefish, use contractions of the muscular mantle to pump water over their gills **(Figure 44.3c).** In adult bony fishes, the gills extend into a chamber covered by gill flaps or *opercula* (singular, *operculum* = little lid) on either side of the head. The operculum also serves as part of a one-way pumping system that ventilates the gills **(Figure 44.3d).**

Many Animals with Internal Gills Use Countercurrent Flow to Maximize Gas Exchange

Sharks, fishes, and some crabs take advantage of one-way flow of water over the gills to maximize the amounts of O_2 and CO_2 exchanged with water. In this mechanism, called **countercurrent exchange,** the water flowing over the gills moves in a direction opposite to the flow of blood under the respiratory surface.

Figure 44.4 illustrates countercurrent exchange in the uptake of O_2. At the point where fully oxygenated water first passes over a gill filament in countercurrent flow, the blood flowing beneath it in the opposite direction is also almost fully oxygenated. However, O_2 concentration is still higher in the water than in the blood, and the gas diffuses from the water into the blood, raising the concentration of O_2 in the blood almost to the level of the fully oxygenated water. At the opposite end of the filament, much of the O_2 has been removed from the water, but the blood flowing under the filament, which has just arrived from body tissues and is fully deoxygenated, contains even less O_2. As a result, O_2 also diffuses from the water to the blood at this end of the filament. All along the gill filament, the same relationship exists, so that at any point, the water is more highly oxygenated than the blood, and O_2 diffuses from the water into the blood across the respiratory surface.

The overall effect of countercurrent exchange is the removal of 80% to 90% of the O_2 content of water as it flows over the gills. In comparison, by breathing in and out and constantly reversing the direction of air flow, mammals manage to remove only about 25% of the O_2 content of air. Efficient removal of O_2 from water is important because of the much lower O_2 content of water compared with air.

Insects Use a Tracheal System for Gas Exchange

Insects breathe air by a respiratory system consisting of air-conducting tubes called **tracheae** (singular *trachea,* or "windpipe"). The tracheae are invaginations of the outer epidermis of the animal, reinforced by rings of chitin, the material of the insect exoskeleton. They lead from the body surface and branch so extensively inside the animal that almost every cell is served

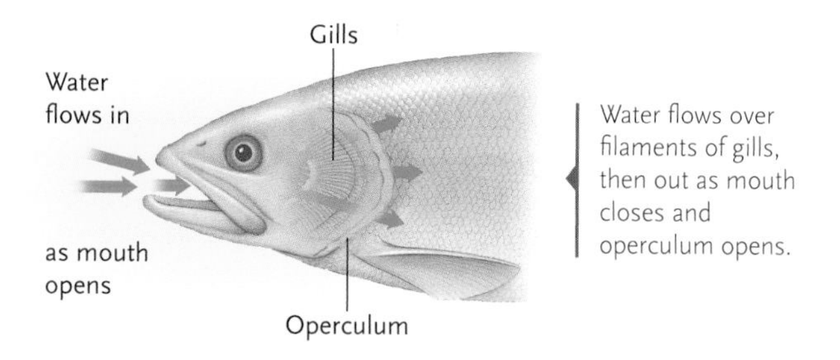

a. The flow of water around the gill filaments

b. Countercurrent flow in fish gills, in which the blood and water move in opposite directions

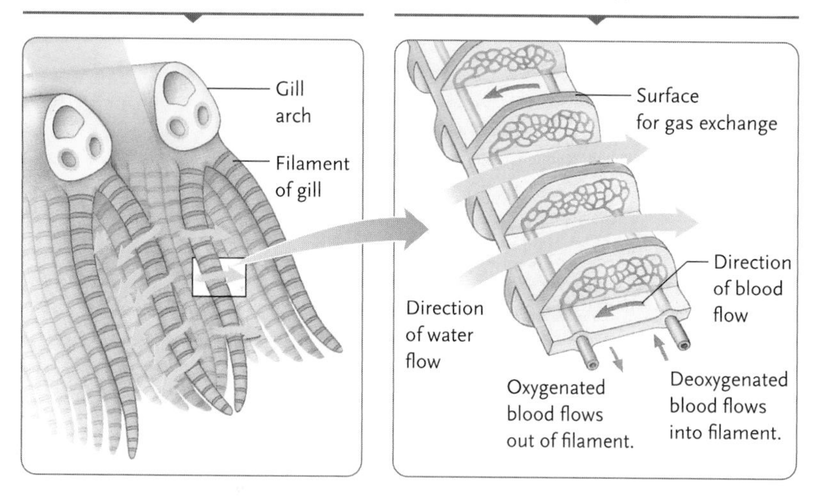

c. In countercurrent exchange, blood leaving the capillaries has the same O_2 content as fully oxygenated water entering the gills

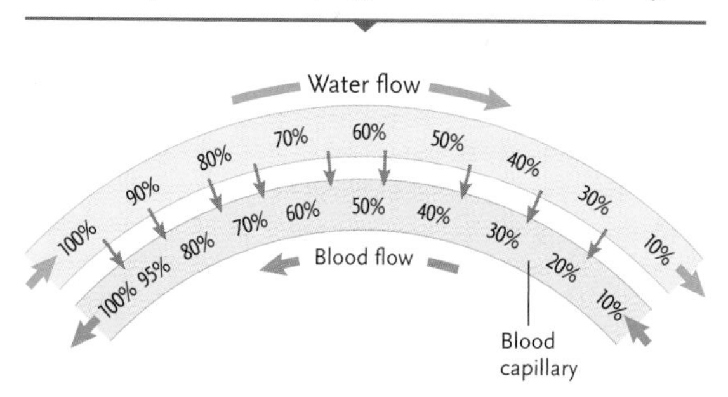

Figure 44.4

Ventilation and countercurrent exchange in bony fishes. (a) Water flows around the gill filaments. **(b)** Water and blood flow in opposite directions through the gill filaments. **(c)** Countercurrent exchange: oxygen from the water diffuses into the blood, raising its oxygen content. The percentages indicate the degree of oxygenation of water (blue) and blood (red).

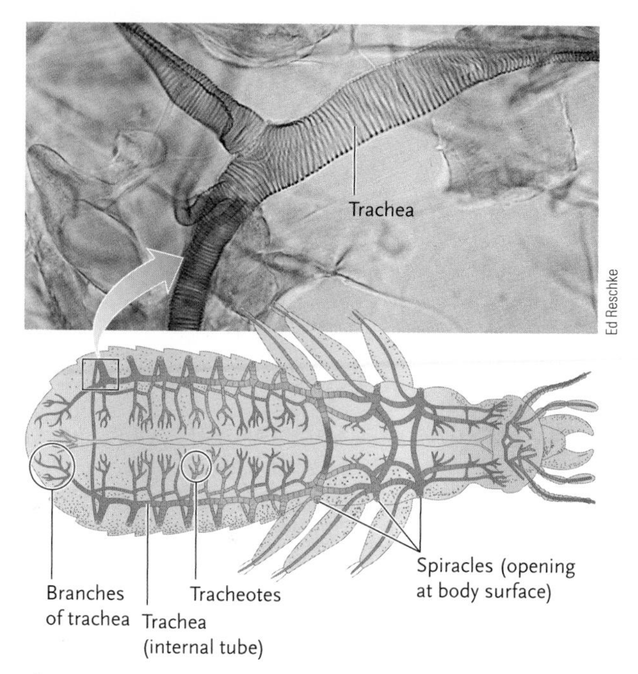

Figure 44.5

The tracheal system of insects. Chitin rings, visible in the photomicrograph, reinforce many of the tracheae.

by a microscopic branch **(Figure 44.5).** Some of the branches even penetrate inside larger cells, such as those of insect flight muscles. The finest branches of the tracheae, called *tracheoles,* form the respiratory surface of the insect system. Tracheoles are dead-end tubes with very small fluid-filled tips that are in contact with cells of the body. Air is transported by the tracheal system to those tips, and gas exchange occurs directly across the plasma membranes of the body cells in contact with the tips. At places within the body, the tracheae expand into internal air sacs that act as reservoirs to increase the volume of air in the system.

Air enters and leaves the tracheal system at openings in the insect's chitinous exoskeleton called **spiracles** (*spiraculum* = airhole). In adult insects, the spiracles are located in a row on either side of the thorax and abdomen. The spiracles open and close in coordination with body movements to compress and expand the air sacs and pump air in and out of the tracheae. During insect flight, alternating compression and expansion of the thorax by the flight muscles also pump air through the tracheal system.

Lungs Allow Animals to Live in Completely Terrestrial Environments

Lungs are one of the primary adaptations that allowed animals to fully invade terrestrial environments. Some fishes and amphibians have lungs, as do all reptiles, birds, and mammals. All lungs are invaginated structures located internally in the body.

In some fishes, such as lungfishes, lungs and air breathing evolved as adaptations to survive in oxygen-poor water or temporarily in air when the water level dropped and exposed them. The lungs of these fishes consist of thin-walled sacs, which branch off from the mouth, pharynx, or parts of the digestive system; air is obtained by **positive pressure breathing,** a gulping or swallowing motion that forces air into the lungs.

The lungs of mature amphibians such as frogs and salamanders are also thin-walled sacs with relatively

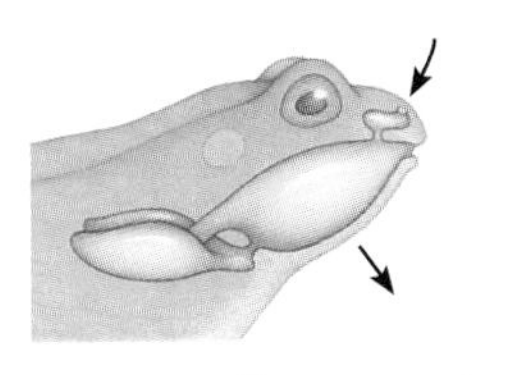

1 The frog lowers the floor of its mouth and inhales through its nostrils.

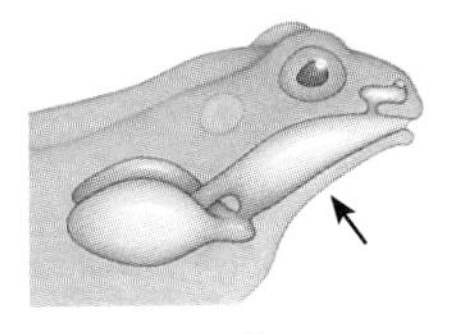

2 It closes its nostrils, opens the glottis, and elevates the floor of the mouth, forcing air into the lungs.

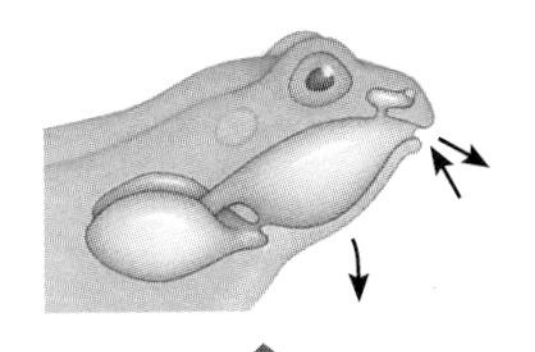

3 Rhythmic ventilation assists in gas exchange.

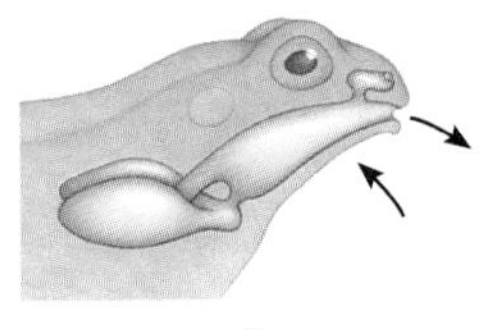

4 Air is forced out when muscles in the body wall above the lungs contract and the lungs recoil elastically.

Figure 44.6
Positive pressure breathing in an amphibian (a frog).

little folding or pocketing. Amphibians also fill their lungs by positive pressure breathing, in this case using a rhythmic motion of the floor of the mouth as the pump, in coordination with opening and closing of the nostrils **(Figure 44.6).**

The lungs of reptiles, birds, and mammals have many pockets and folds that increase the area of the respiratory surface, which contains dense, highly branched capillary networks. Mammalian lungs consist of millions of tiny air pockets, the **alveoli**, each surrounded by dense capillary networks. Reptiles and mammals fill their lungs by **negative pressure breathing**—by muscular contractions that expand the lungs, lowering the pressure of the air in the lungs and causing air to be pulled inward. (Mammalian negative pressure breathing is described in more detail in the next section.)

In birds, a countercurrent exchange system provides the most complex and efficient vertebrate lungs **(Figure 44.7).** In addition to paired lungs, birds have nine pairs of air sacs that branch off the respiratory tract. The air sacs, which collectively contain several times as much air as the lungs, set up a pathway that allows air to flow in one direction through the lungs, rather than in and out as in other vertebrates. Within the lungs, air flows through an array of fine, parallel tubes that are surrounded by a capillary network. The blood flows in the direction opposite to the air flow, setting up a countercurrent exchange. The countercurrent exchange allows bird lungs to extract about one-third of the O_2 from the air as compared with about one-fourth in the lungs of mammals.

Study Break

1. What advantages do gills confer upon a water-breathing animal over skin breathing?
2. What is countercurrent exchange, and how is it beneficial for gas exchange?
3. How does the tracheal system of insects facilitate gas exchange with the cells of the body?
4. Distinguish between positive pressure breathing and negative pressure breathing in animals with lungs.

a. Lungs and air sacs of a bird

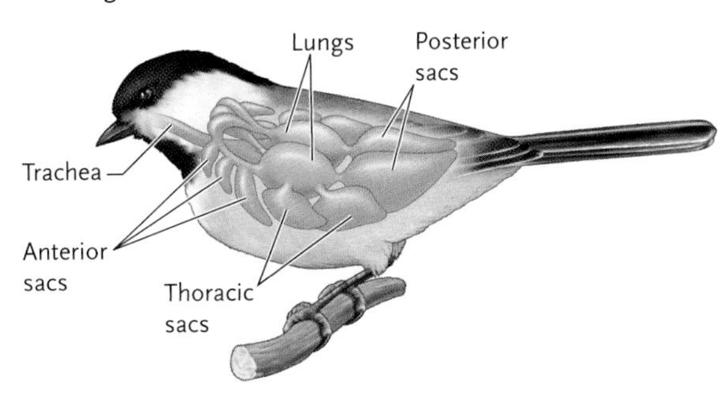

b. Countercurrent exchange

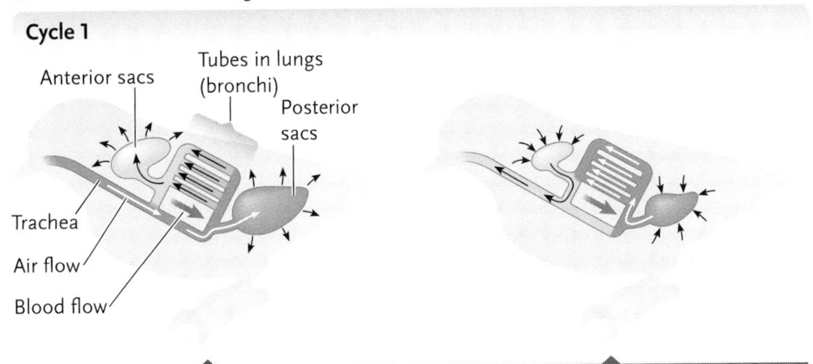

1 During the first inhalation, most of the oxygen flows directly to the posterior air sacs. The anterior air sacs also expand but do not receive any of the newly inhaled oxygen.

2 During the following exhalation, both anterior and posterior air sacs contract. Oxygen from the posterior sacs flows into the gas-exchanging tubes (bronchi) of the lungs.

Cycle 2

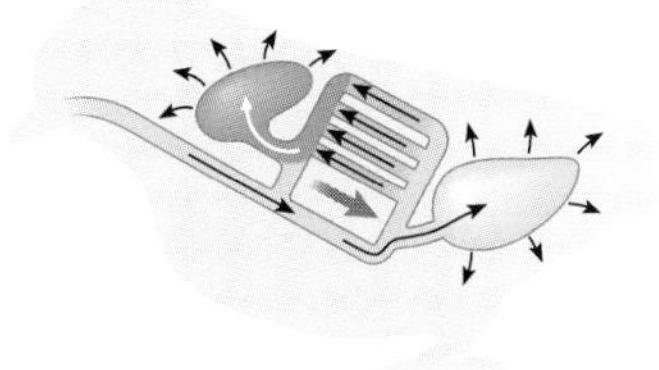

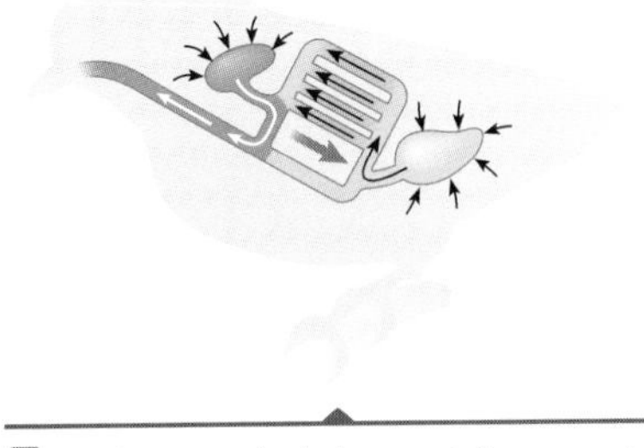

1 During the next inhalation, air from the lung (now deoxygenated) moves into the anterior air sacs.

2 In the second exhalation, air from anterior sacs is expelled to the outside through the trachea.

Figure 44.7
Countercurrent exchange in bird lungs. (a) Unlike mammalian lungs, bird lungs do not expand and contract. Changes in pressure in the expandable air sacs move air in and out. **(b)** Air flows in one direction through the tubes of the lungs; blood flows in the opposite direction in the surrounding capillary network. Two cycles of inhalation and exhalation are needed to move a specific volume of air through the bird respiratory system.

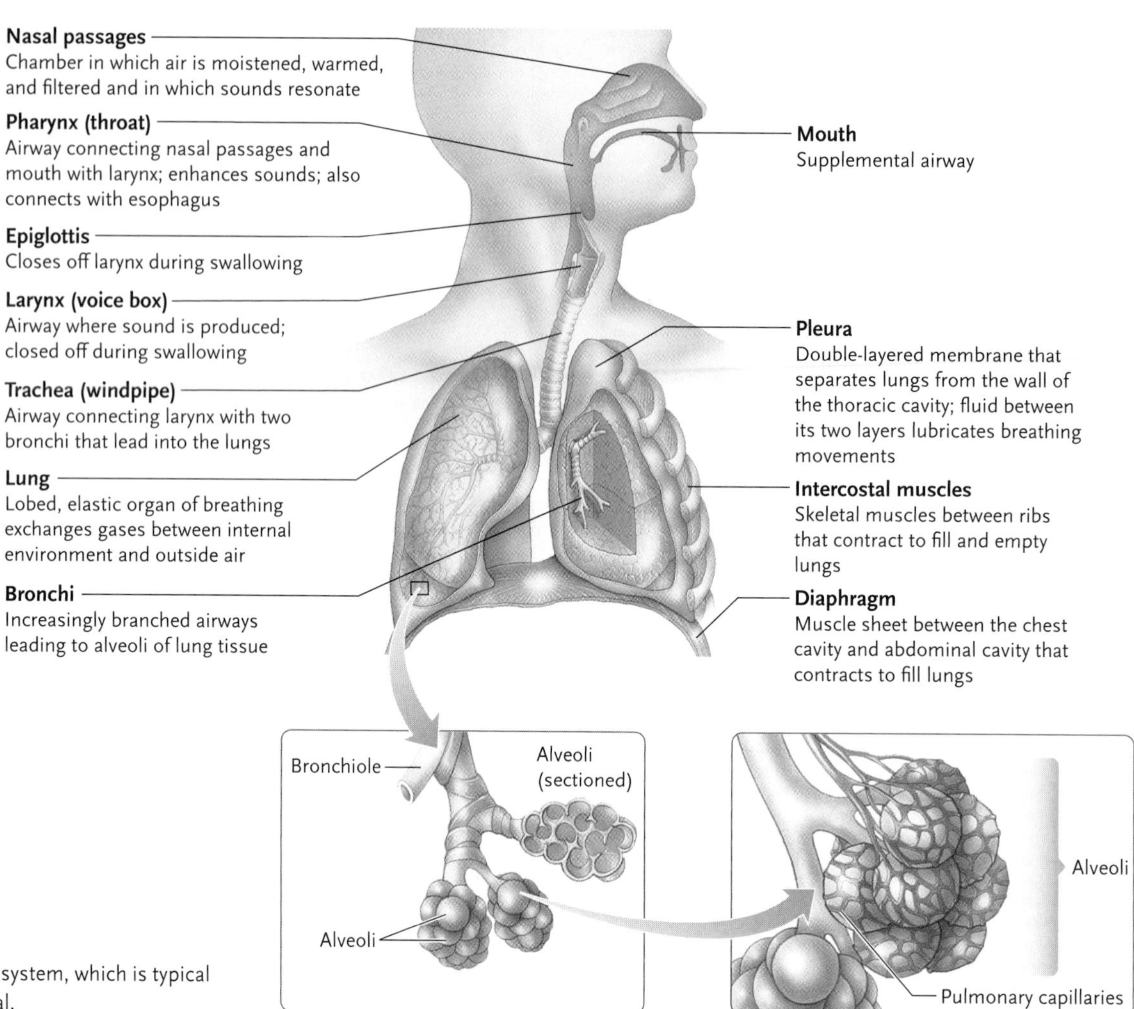
Nasal passages
Chamber in which air is moistened, warmed, and filtered and in which sounds resonate
Pharynx (throat)
Airway connecting nasal passages and mouth with larynx; enhances sounds; also connects with esophagus
Epiglottis
Closes off larynx during swallowing
Larynx (voice box)
Airway where sound is produced; closed off during swallowing
Trachea (windpipe)
Airway connecting larynx with two bronchi that lead into the lungs
Lung
Lobed, elastic organ of breathing exchanges gases between internal environment and outside air
Bronchi
Increasingly branched airways leading to alveoli of lung tissue
Mouth
Supplemental airway
Pleura
Double-layered membrane that separates lungs from the wall of the thoracic cavity; fluid between its two layers lubricates breathing movements
Intercostal muscles
Skeletal muscles between ribs that contract to fill and empty lungs
Diaphragm
Muscle sheet between the chest cavity and abdominal cavity that contracts to fill lungs
Bronchiole
Alveoli (sectioned)
Alveoli
Alveoli
Pulmonary capillaries

Figure 44.8
The human respiratory system, which is typical for a terrestrial mammal.

44.3 The Mammalian Respiratory System

All mammals have a pair of lungs and a diaphragm in the chest cavity that plays an important role in negative pressure breathing. Rapid ventilation of the respiratory surface and perfusion by blood flow through dense capillary networks maximizes gas exchange.

The Airways Leading from the Exterior to the Lungs Filter, Moisten, and Warm the Entering Air

The human respiratory system is typical for a terrestrial mammal **(Figure 44.8).** Air enters and leaves the respiratory system through the nostrils and mouth. Hairs in the nostrils and mucus covering the surface of the airways filter out and trap dust and other large particles. Inhaled air is moistened and warmed as it moves through the mouth and nasal passages.

Next, air moves into the throat, or **pharynx,** which forms a common pathway for air entering the **larynx** (or "voice box") and food entering the esophagus, which leads to the stomach. The airway through the larynx is open except during swallowing.

From the larynx, air moves into the trachea (or "windpipe"), which branches into two airways, the **bronchi** (singular, *bronchus*). The bronchi lead to the two elastic, cone-shaped lungs, one on each side of the chest cavity. Inside the lungs, the bronchi narrow and branch repeatedly, becoming progressively narrower and more numerous. The terminal airways, the **bronchioles,** lead into cup-shaped pockets, the alveoli (singular, *alveolus;* shown in Figure 44.8 insets).

Each of the 150 million alveoli in each lung is surrounded by a dense network of capillaries. By the time inhaled air reaches the alveoli, it has been moistened to the saturation point and brought to body temperature. The many alveoli provide an enormous area for gas exchange. If the alveoli of an adult human were flattened out in a single layer, they would cover an area approaching 100 square meters, about the size of a tennis court!

The larynx, trachea, and larger bronchi are nonmuscular tubes encircled by rings of cartilage that prevent the tubes from compressing. The largest of the rings, which reinforces the larynx, stands out at the front of the throat as the Adam's apple; smaller supporting rings can be felt at the front of the throat just below the larynx. The walls of the smaller bronchi and the bronchioles contain smooth muscle cells that contract or relax to control the diameter of these passages, and with it, the amount of air flowing to and from the alveoli.

The epithelium lining each bronchus contains cilia and mucus-secreting cells. Bacteria and airborne particles such as dust and pollen are trapped in the mucus and then moved upward and into the throat by the beating of the cilia lining the airways. Infection-fighting macrophages also patrol the respiratory epithelium.

Tobacco smoke, by paralyzing the cilia lining the respiratory tract, interferes with the processes that clear bacteria and airborne particles from the lungs. The bacteria and foreign matter persisting in the lungs can cause infections and smoker's cough.

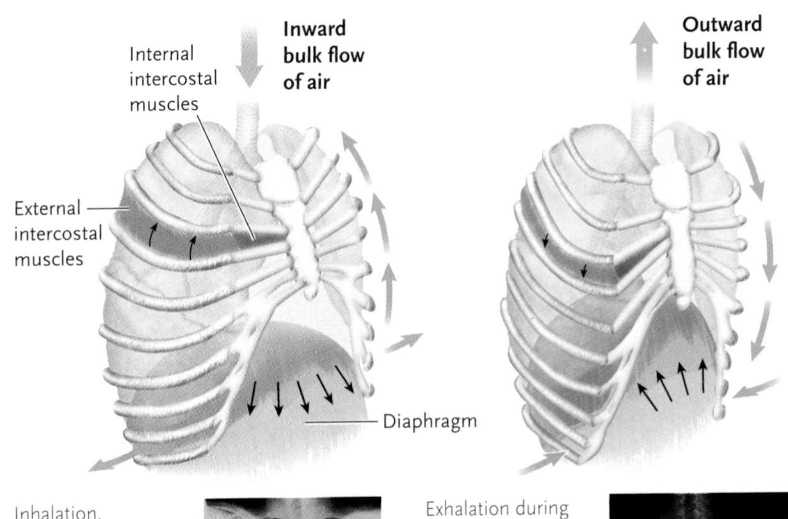

Inhalation. Diaphragm contracts and moves down. The external intercostal muscles contract and lift rib cage upward and outward. The lung volume expands.

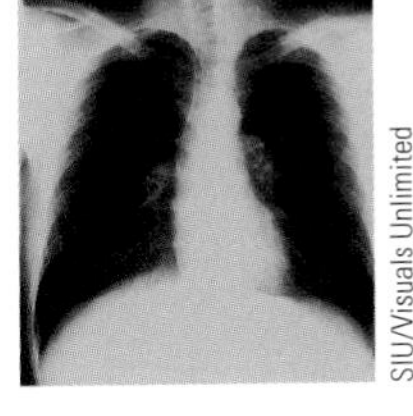

Exhalation during breathing or rest. Diaphragm and external intercostal muscles return to the resting positions. Rib cage moves down. Lungs recoil passively.

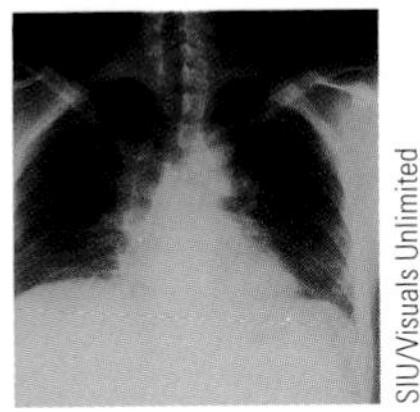

Figure 44.9
The respiratory movements of humans during breathing at rest. The movements of the rib cage and diaphragm fill and empty the lungs. Inhalation is powered by contractions of the external intercostal muscles and diaphragm, and exhalation is passive. During exercise or other activities characterized by deeper and more rapid breathing, contractions of the internal intercostal muscles and the abdominal muscles add force to exhalation. The X-ray images show how the volume of the lungs increases and decreases during inhalation and exhalation.

Contractions of the Diaphragm and Muscles between the Ribs Ventilate the Lungs

The lungs are located in the rib cage above the *diaphragm,* a dome-shaped sheet of skeletal muscle separating the chest cavity from the abdominal cavity. The lungs are covered by a double layer of epithelial tissue called the **pleura.** The inner pleural layer is attached to the surface of the lungs, and the outer layer is attached to the surface of the chest cavity. A narrow space between the inner and outer layers is filled with slippery fluid, which allows the lungs to move within the chest cavity without rubbing or abrasion as they expand and contract.

Contraction of muscles between the ribs and the diaphragm brings air into the lungs by a negative pressure mechanism. As an inhalation begins, the diaphragm contracts and flattens, and one set of muscles between the ribs, the *external intercostal* muscles, contracts, pulling the ribs upward and outward **(Figure 44.9).** These movements expand the chest cavity and lungs, lowering the air pressure in the lungs below that of the atmosphere. As a result, air is drawn into the lungs, expanding and filling them.

The expansion of the lungs is much like filling two rubber balloons. Like balloons, the lungs are elastic, and resist stretching as they are filled. Also like balloons, the stretching stores energy, which can be released to expel air from the lungs. When a person at rest exhales, the diaphragm and muscles between the ribs relax, and the elastic recoil of the lungs expels the air.

When physical activity increases the body's demand for O_2, other muscles help expel the air by forcefully reducing the volume of the chest cavity. Contractions of abdominal wall muscles increase abdominal pressure, exerting an upward-directed force on the diaphragm and thus pushing it upward. Contractions of *internal intercostal* muscles pull the chest wall inward and downward, causing it to flatten. As a result, the dimensions of the chest cavity decrease.

The Volume of Inhaled and Exhaled Air Varies over Wide Limits

The volume of air entering and leaving the lungs during inhalation and exhalation is called the **tidal volume.** In a person at rest, the tidal volume amounts to about 500 mL. As physical activity increases, the tidal volume increases to match the body's needs for O_2; at maximal levels, the tidal volume reaches about 3400 mL in females and 4800 mL in males. The maximum tidal volume is called the **vital capacity** of an individual.

Even after the most forceful exhalation, about 1200 mL of air remains in the lungs in males, and about 1000 mL in females; this is the **residual volume** of the lungs. In fact, the lungs cannot be deflated completely because small airways collapse during forced exhalation, blocking further outflow of air. Because air cannot be removed from the lungs completely, some gas exchange can always occur between blood flowing through the lungs and the air in the alveoli.

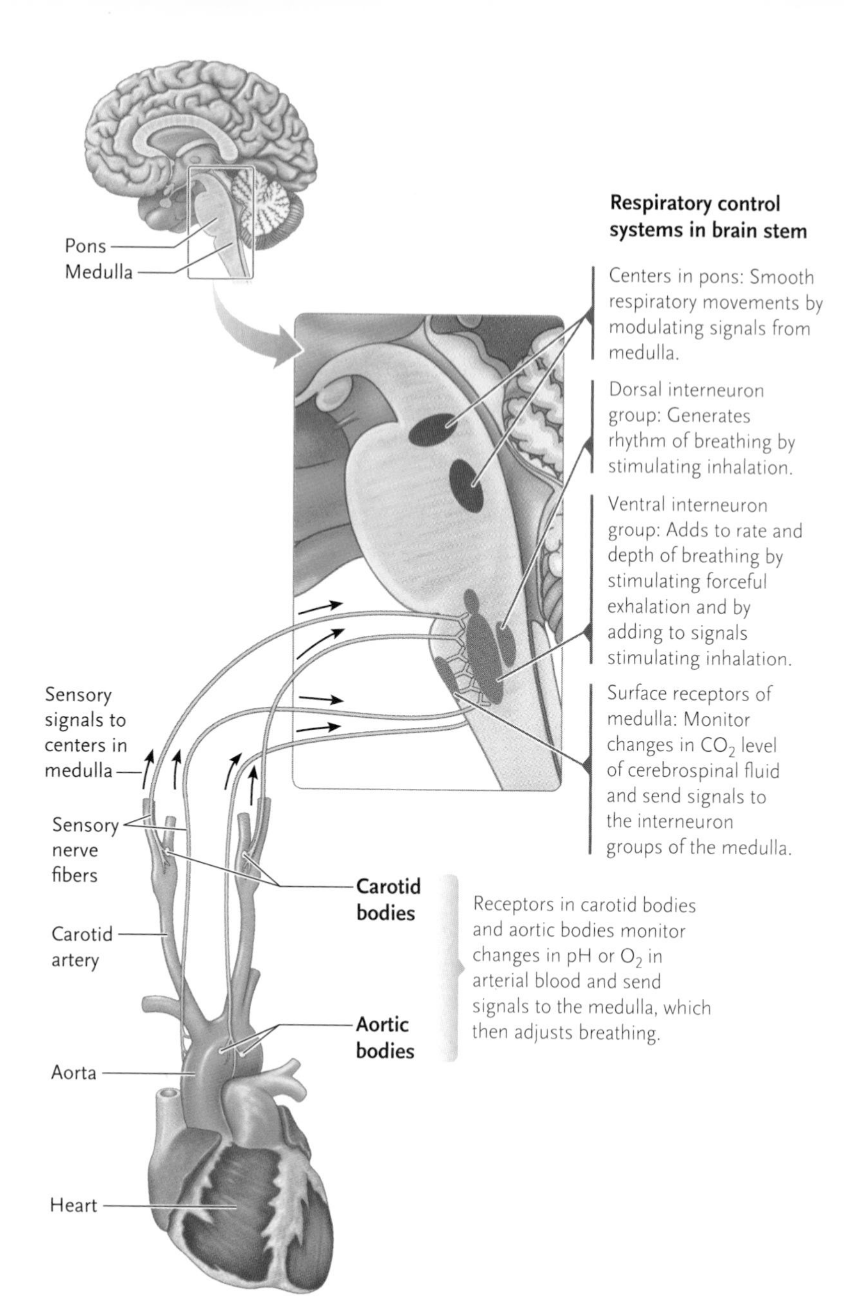
Pons
Medulla
Respiratory control systems in brain stem
Centers in pons: Smooth respiratory movements by modulating signals from medulla.
Dorsal interneuron group: Generates rhythm of breathing by stimulating inhalation.
Ventral interneuron group: Adds to rate and depth of breathing by stimulating forceful exhalation and by adding to signals stimulating inhalation.
Surface receptors of medulla: Monitor changes in CO_2 level of cerebrospinal fluid and send signals to the interneuron groups of the medulla.
Sensory signals to centers in medulla
Sensory nerve fibers
Carotid bodies
Carotid artery
Receptors in carotid bodies and aortic bodies monitor changes in pH or O_2 in arterial blood and send signals to the medulla, which then adjusts breathing.
Aortic bodies
Aorta
Heart

Figure 44.10
Control of breathing. Centers in the pons and medulla control the rhythm, rate, and depth of breathing. Receptors in the carotid arteries and aorta detect changes in the levels of O_2 and CO_2 in blood and body fluids. Signals from these receptors are integrated in the respiratory centers of the medulla and pons.

The Centers That Control Breathing Are Located in the Brain Stem

Breathing is controlled by centers in the medulla and pons, which form part of the brain stem **(Figure 44.10).** Groups of interneurons in the centers regulate the rate and depth of breathing, ranging from shallow, slow breathing when the body is at rest to the deep and rapid breathing of intense physical exercise, excitement, or fear. Over these extremes, the air entering and leaving lungs of a human male varies from as little as 5 to 6 L per minute to (for a brief time only) as much as 150 L per minute.

Interneurons That Regulate Breathing. Signals from interneurons in the medulla carried by efferent (motor) neurons of the autonomic system produce the breathing movements. A set of signals from a dorsal group of interneurons acts as the primary stimulator of inhalation by causing the diaphragm and the external intercostal muscles to contract, which expands the chest cavity and produces an inhalation. In a person at rest, the signal is switched off as the lungs become moderately full: the rib muscles and the diaphragm relax, and a passive exhalation occurs. These signals act as the primary generator of breathing rhythm.

A ventral group of interneurons in the medulla can send signals for both inhalation and exhalation. These neurons become active only during physical exercise, fear, or other situations that require more oxygen when active rather than passive exhalation is needed. In that case, some of the ventral neurons send signals that stimulate the abdominal and internal intercostal muscles to contract, thereby causing active exhalation. Other neurons in the ventral group become stimulated by signals from the dorsal group, and then help increase inhalation activity when faster and deeper breathing is required.

Two interneuron groups in the pons modulate the signals originating from the medulla, fine-tuning and smoothing the muscle contractions so that inhalations and exhalations are gradual and controlled rather than sudden and abrupt. Signals sent from higher brain centers in the cerebrum can override the control of respiratory rate and depth by the brain stem. For example, as we speak or sing, or hold our breath, we can consciously alter or stop breathing to match the demands of these activities. Breathing rate and depth are also modified by emotional states, controlled by centers in the limbic system of the brain (see Section 38.3). Thus breathing is altered as we laugh, gasp, groan, cry, and sigh.

Receptors That Send Information to the Brain Centers. The brain centers controlling the rate and depth of breathing integrate sensory information sent by receptors that monitor O_2 and CO_2 levels in the blood and body fluids. The integration of sensory information serves to match breathing rate to the metabolic demands of the body. These *chemoreceptors* are located centrally on the surface of the medulla, and peripherally in **carotid bodies** in the carotid arteries leading to the brain and in **aortic bodies** in the large arteries leaving the heart (see Figure 44.10).

The receptors of the medulla detect changes in pH in the cerebrospinal fluid; the pH is determined mostly by the CO_2 concentration in the blood. (Remember that pH decreases as CO_2 levels increase.) The receptors in the carotid and aortic bodies detect changes in CO_2 and O_2 concentrations in the blood.

The CO_2 receptors in the medulla have the greatest effects on breathing. If increased body activities cause the CO_2 concentration to rise in the blood, the medulla receptors trigger interneuron groups in the medulla that increase the rate and depth of breathing. If CO_2 concentration falls, the receptors send signals to the medulla that lead to a slowing of the rate and depth of breathing.

The peripheral receptors in the carotid and aortic bodies detect changes in pH or O_2 concentration in arterial blood. When these receptors detect a rise in blood pH they send signals to the medulla that cause the medulla to increase the rate and depth of breathing. Although the receptors in the carotid and aortic bodies also detect the O_2 level in arterial blood, the receptors do not respond until blood O_2 level falls below 60% of normal. This reaction makes the O_2 receptors act as a backup system that comes into play only when blood O_2 concentration falls to critically low levels.

Thus, the level of CO_2 in the blood and body fluids is much more closely monitored, and has a much greater effect on breathing, than the O_2 level. This reflects the fact that small fluctuations in blood pH have much greater effects on the ability of hemoglobin to carry oxygen, and on enzyme activity in the blood and interstitial fluid, than fluctuations in the O_2 level.

Local Controls. Other, automated controls within the lungs match the rates of ventilation and perfusion by responding to O_2 concentrations in the blood. If air flow lags behind capillary blood flow, so that the O_2 level falls in the blood, the reduced O_2 concentration causes smooth muscles in the walls of arterioles in the lungs to contract. This reduces the flow of blood, thereby giving it more time to pick up O_2. Conversely, if blood flow lags behind, the rising blood O_2 concentration causes the smooth muscle cells in arteriole walls to relax, dilating the arterioles and increasing the rate of blood flow through lung capillaries. These local controls, in combination with the neural controls that regulate rate and depth of breathing, ensure that the respiratory system meets the body's varying need to obtain O_2 and release CO_2.

Study Break

1. Explain how inhalation and exhalation occur in a mammal at rest.
2. You can consciously initiate and sustain an exhalation. What is going on muscularly in this case?
3. What is the most important feedback stimulus for breathing?
4. What is the role of the chemoreceptors in the medulla?

44.4 Mechanisms of Gas Exchange and Transport

In both the lungs and body tissues, gas exchange occurs when the gas diffuses from an area of higher concentration to an area of lower concentration. In this section, we consider the mechanics of gas exchange between air and the blood in mammals, and the means by which gases are transported between the lungs and other body tissues. A major part of this story involves hemoglobin, the vertebrate respiratory pigment.

The Proportion of a Gas in a Mixture Determines Its Partial Pressure

For gases, it is often more accurate and convenient to consider concentration differences as differences in pressure. When gases are present in a mixture, the pressure of each individual gas, called its *partial pressure,* is determined by its proportion in the mixture. Air, water, and blood all contain mixtures of gases, including oxygen, carbon dioxide, nitrogen, and other gases, so each gas exerts only a part of the total gas pressure. For example, the proportion of O_2 in dry air is about 21%, or 21/100. In dry air at sea level, the total atmospheric pressure under standard conditions is 760 mm Hg. The partial pressure of O_2, written as P_{O_2}, is equivalent to $760 \times 21/100$, or about 160 mm Hg. The proportion of CO_2 in dry air is about 0.04%, so its partial pressure, P_{CO_2}, is equivalent to $760 \times 0.04/100$, or about 0.3 mm Hg. For O_2 to diffuse inward across a respiratory surface, its partial pressure outside the surface must be greater than inside; for CO_2 to diffuse outward, its partial pressure inside must be greater inside than outside.

In the lungs, even though the P_{O_2} is reduced by mixing with the air in the residual volume, it is still much higher than the P_{O_2} in deoxygenated blood entering the network of capillaries in the lungs **(Figure 44.11).** As a result, O_2 readily diffuses from the alveolar air into the plasma solution in the capillaries.

Hemoglobin Greatly Increases the O_2-Carrying Capacity of the Blood

After entering the plasma, O_2 diffuses into erythrocytes, where it combines with hemoglobin. The combination with hemoglobin removes O_2 from the plasma, lowering the P_{O_2} of the plasma and allowing additional O_2 molecules to diffuse from alveolar air to the blood.

Recall from Section 42.2 that a mammalian hemoglobin molecule has four heme groups, each containing an iron atom that can combine reversibly with an O_2 molecule. A hemoglobin molecule can therefore bind a total of four molecules of O_2. The combination of O_2 with hemoglobin allows blood to carry about

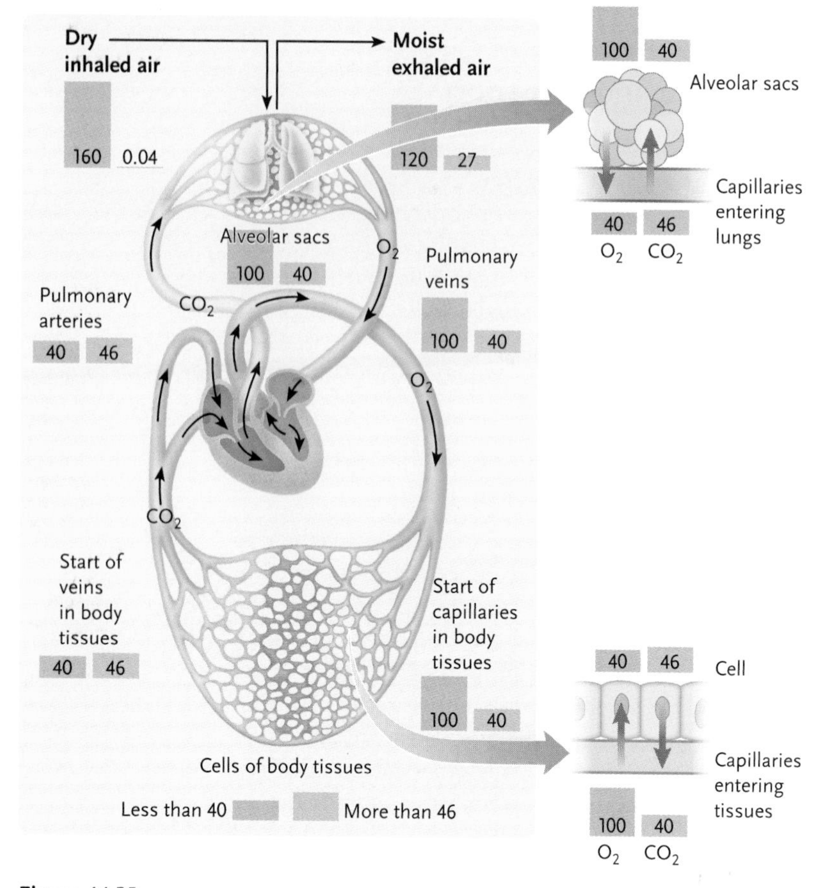

Figure 44.11
The partial pressures of O_2 (pink) and CO_2 (blue) in various locations in the body.

60 times more O_2 (about 200 mL per liter) than it could if the O_2 simply dissolved in the plasma (about 3 mL per liter). About 98.5% of the O_2 in blood is carried by hemoglobin and about 1.5% is carried in solution in the blood plasma.

The reversible combination of hemoglobin with O_2 is related to the partial pressure of O_2 in a pattern shown by the *hemoglobin-O_2* dissociation curve in **Figure 44.12.** (The curve is generated by measuring the amount of hemoglobin saturated at a given P_{O_2}.) The curve is S-shaped with a plateau region, rather than linear. The top, plateau part of the curve above 60 mm Hg is in the blood P_{O_2} range found in the pulmonary capillaries where O_2 is binding to hemoglobin. For this part of the curve, the blood remains highly saturated with O_2 over a relatively large range of P_{O_2}. Even at P_{O_2} levels much higher than shown on the graph (P_{O_2} theoretically can go up to 760 mm Hg), only a small extra amount of O_2 will bind to hemoglobin. The steep part of the curve between 0 and 60 mm Hg is in the blood P_{O_2} range found in the capillaries in the rest of the body. For this part of the curve, small changes in P_{O_2} result in a large change in the amount of O_2 bound to hemoglobin.

Because the partial pressure of O_2 in alveolar air is about 100 mg Hg, most of the hemoglobin molecules are fully saturated in the blood leaving the alveolar networks, meaning that most of the hemoglobin molecules are bound to four O_2 molecules (see Figure 44.12a). The P_{O_2} of the O_2 in solution in the blood plasma has risen to approximately the same level as in the alveolar air, about 100 mm Hg. The blood has also changed color, reflecting the bright red color of oxygenated hemoglobin as compared with the darker red color of deoxygenated hemoglobin.

The oxygenated blood exiting from the alveoli collects in venules, which merge into the pulmonary veins leaving the lungs. These veins carry the blood to the heart, which pumps the blood through the systemic circulation to all parts of the body.

As the oxygenated blood enters the capillary networks of body tissues, it encounters regions in which the P_{O_2} in the interstitial fluid and body cells is lower than that in the blood, ranging from about 40 mm Hg downward to 20 mm Hg or less (see Figure 44.12b). As a result, O_2 diffuses from the blood plasma into the interstitial fluid, and from the fluid into body cells. As O_2 diffuses from the blood plasma into body tissues, it is replaced by O_2 released from hemoglobin.

Several factors contribute to the release of O_2 from hemoglobin, including increased acidity (lower pH) in active tissues. The acidity increases because oxidative reactions release CO_2, which combines with water to form carbonic acid (H_2CO_3). The lowered pH alters hemoglobin's conformation, reducing its affinity for O_2, which is released and used in cellular respiration.

The net diffusion of O_2 from blood to body cells continues until, by the time the blood leaves the capillary networks in the body tissues, much of the O_2 has been removed from hemoglobin. The blood, now with a P_{O_2} of 40 mm Hg or less, returns in veins to the heart, which pumps it through the pulmonary arteries to the lungs for another cycle of oxygenation.

Carbon Dioxide Diffuses down Concentration Gradients from Body Tissues into the Blood and Alveolar Air

The CO_2 produced by cellular oxidations diffuses from active cells into the interstitial fluid, where it reaches a partial pressure of about 46 mm Hg. Because this P_{CO_2} is higher than the 40 mm Hg P_{CO_2} in the blood entering the capillary networks of body tissues (see Figure 44.10), CO_2 diffuses from the interstitial fluid into the blood plasma **(Figure 44.13a)**.

Some of the CO_2 remains in solution as a gas in the plasma. However, most of the CO_2, about 70%, combines with water to produce carbonic acid (H_2CO_3), which dissociates into bicarbonate (HCO_3^-) and H^+ ions. The reaction takes place both in the blood plasma and inside erythrocytes, where an enzyme, *carbonic anhydrase,* greatly speeds the reaction.

Most of the H^+ ions produced by the dissociation of carbonic acid combine with hemoglobin or with proteins in the plasma. The combination, by removing excess H^+ from the blood solution, *buffers* the

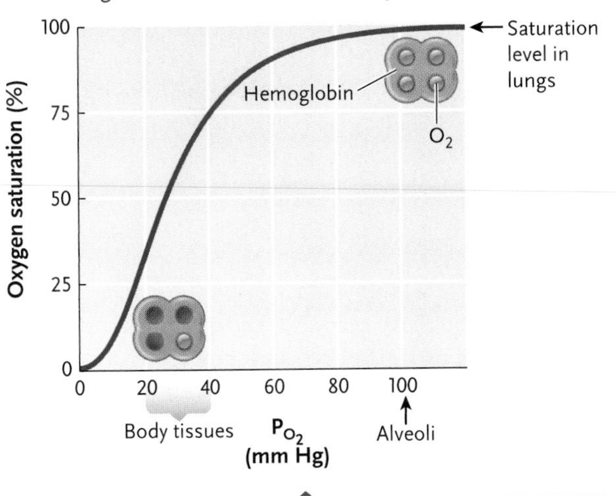

In the alveoli, in which the P_{O_2} is about 100 mm Hg and the pH is 7.4, most hemoglobin molecules are 100% saturated, meaning that almost all have bound four O_2 molecules.

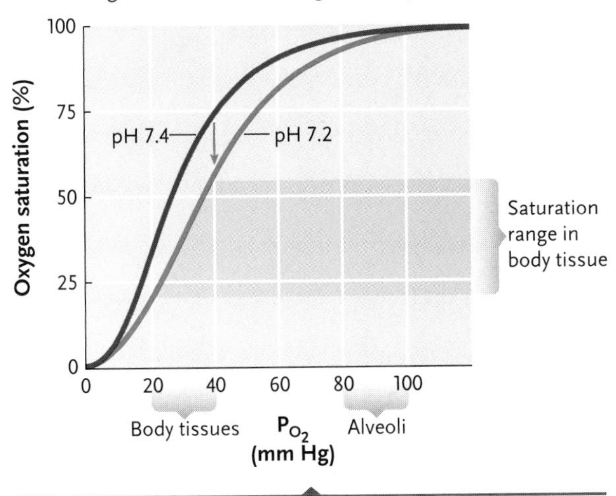

In the capillaries of body tissues, where the P_{O_2} varies between about 20 and 40 mm Hg depending on the level of metabolic activity and the pH is about 7.2, hemoglobin can hold less O_2. As a result, most hemoglobin molecules release two or three of their O_2 molecules to become between 25% and 50% saturated. Note that the drop in pH to 7.2 (red line) in active body tissues reduces the amount of O_2 hemoglobin can hold as a compared with pH 7.4. The reduction in binding affinity at lower pH increases the amount of O_2 released in active tissues.

Figure 44.12
Hemoglobin–O_2 dissociation curves, which show the degree to which hemoglobin is saturated with O_2 at increasing P_{O_2}.

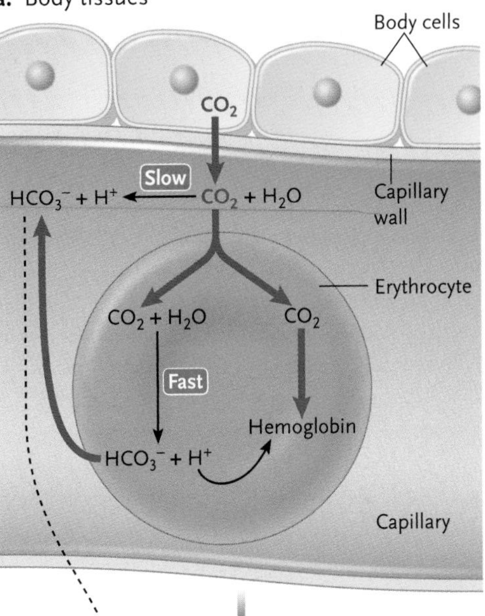

In body tissues, some of the CO_2 released into the blood combines with water in the blood plasma to form HCO_3^- and H^+. However, most of the CO_2 diffuses into erythrocytes, where some combines directly with hemoglobin and some combines with water to form HCO_3^- and H^+. The H^+ formed by this reaction combines with hemoglobin; the HCO_3^- is transported out of erythrocytes to add to the HCO_3^- in the blood plasma.

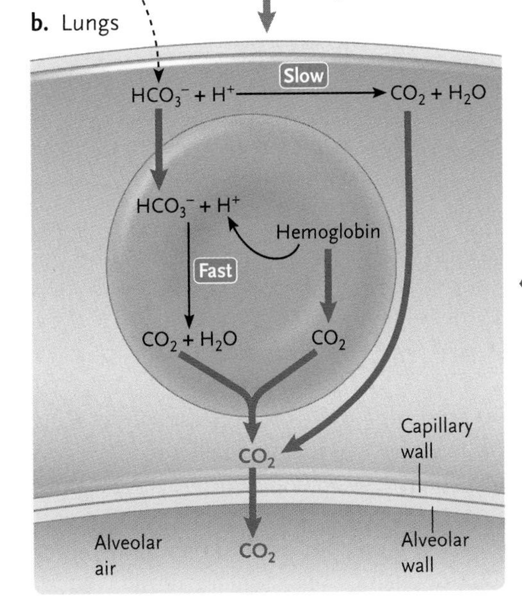

In the lungs, the reactions are reversed. Some of the HCO_3^- in the blood plasma combines with H^+ to form CO_2 and water. However, most of the HCO_3^- is transported into erythrocytes, where it combines with H^+ released from hemoglobin to form CO_2 and water. CO_2 is released from hemoglobin. The CO_2 diffuses from the erythrocytes and, with the CO_2 in the blood plasma, diffuses from the blood into the alveolar air.

Figure 44.13
The reactions occurring during the transfer of CO_2 from body tissues to alveolar air.

blood pH, helping to maintain it at its near-neutral set point of 7.4. (Buffers are discussed in Section 2.5.) The combined pathways absorbing CO_2 in the blood—solution in the plasma, conversion to bicarbonate, and combination with hemoglobin—help maintain the concentration gradient for gaseous CO_2 and keep its diffusion from the interstitial fluid into the blood at optimal levels.

The blood leaving the capillary networks of body tissues is collected in venules and veins and returned to the heart, which pumps it through the pulmonary arteries into the lungs. As the blood enters the capillary networks surrounding the alveoli, the entire process of CO_2 uptake is reversed **(Figure 44.13b).** The P_{CO_2} in the blood, now about 46 mm Hg, is higher than the P_{CO_2} in the alveolar air, about 40 mm Hg (see Figure 44.10). As a result, CO_2 diffuses from the blood into the air. The diminishing CO_2 concentrations in the plasma, along with the lower pH encountered in the lungs, promote the release of CO_2 from hemoglobin. As CO_2 diffuses away, bicarbonate ions in the blood combine with H^+ ions, forming carbonic acid molecules that break down into water and additional CO_2. This CO_2 adds to the quantities diffusing from the blood into the alveolar air. By the time the blood leaves the capillary networks in the lungs, its P_{CO_2} has been reduced to the same level as that of the alveolar air, about 40 mm Hg (see Figure 44.10).

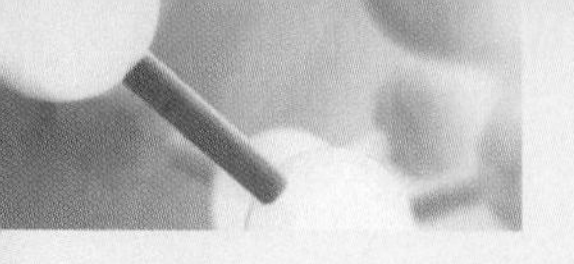

Insights from the Molecular Revolution

Giving Hemoglobin and Myoglobin Air

Carbon monoxide, like O_2, combines directly with the heme group in the proteins myoglobin and hemoglobin. The heme group by itself, unassociated with the polypeptide chain of hemoglobin or myoglobin, has an affinity for CO some 10,000 times greater than for O_2. Combination of the heme group with the proteins reduces its affinity for CO to only 250 times greater than O_2 for hemoglobin and 30 times greater for myoglobin.

How does combination with the proteins reduce the affinity of the heme group for CO so effectively? One hypothesis is that the reduction depends on a stabilizing hydrogen bond between O_2 and an amino acid, histidine, located in a protein pocket formed by a fold in the amino acid chain near the iron atom of the heme group. According to this proposal, the hydrogen bond allows O_2 to displace a water molecule that occupies the pocket when neither O_2 nor CO is bound. Because carbon monoxide cannot form this hydrogen bond, it displaces the water molecule less readily than O_2 does.

Research with myoglobin, conducted by John S. Olson and George Phillips and their colleagues at Rice University, supports the hydrogen-bond hypothesis. For their study, the team used a myoglobin gene isolated from a sperm whale. They chemically altered the DNA of the myoglobin gene so that one of seven other amino acids was substituted for the histidine in the pocket. A highly active bacterial promoter was added to the altered genes, which were then introduced one at a time into *Escherichia coli* bacteria. The bacteria expressed the genes, producing the altered myoglobin molecules in quantity, thereby providing the researchers with seven different forms of myoglobin to test for binding affinity for O_2.

The seven amino acids substituted for the histidine—glycine, alanine, leucine, phenylalanine, threonine, valine, and glutamine—are all nonpolar, or uncharged (histidine is positively charged), and thus unlikely to form a hydrogen bond with O_2. Further, none of these amino acids except glycine and glutamine should be able to hold a water molecule stably in the binding pocket. If the hydrogen-bond hypothesis is correct, the affinity of O_2 for the altered myoglobin should be greatly reduced in the mutant forms of the molecule.

Binding tests showed that the substitutions indeed reduced the affinity for O_2 by a factor that varied between 10 and 100 times. The greatest reduction was produced by the most nonpolar of the amino acids, leucine and phenylalanine. The smallest reduction was observed for glycine and glutamine. Thus the results strongly support the hydrogen-bond hypothesis.

Carbon monoxide (CO), a colorless, odorless gas produced when fuels are incompletely burned, as in automobile exhaust and in faulty furnaces, gas appliances, or space heaters, also binds to hemoglobin if it is inhaled into the lungs. It binds so strongly that it displaces O_2 from hemoglobin and drastically reduces the amount of O_2 carried to body tissues. If CO is inhaled in high quantity for even a few minutes, the reduction in oxygen delivered to the brain can lead to unconsciousness and brain damage. Sustained exposure leads to death by hypoxia (lack of oxygen). Because the brain regulates breathing based on CO_2 levels in blood rather than on O_2 levels, victims breathing CO can die from hypoxia without noticing anything amiss up to the point of unconsciousness. Interestingly, the combination of CO with hemoglobin, carboxyhemoglobin, is bright red. This has led to a myth often seen in textbooks that victims of CO poisoning turn a "classic cherry red" in color. However, this actually occurs in less than 2% of cases. *Insights from the Molecular Revolution* describes recent research testing the molecular basis for the binding of CO to hemoglobin and to *myoglobin,* a muscle protein with structure and properties similar to hemoglobin (myoglobin is discussed in Section 41.1).

Study Break

1. Explain the role of hemoglobin in gas exchange.
2. Why is carbon monoxide potentially lethal?

44.5 Respiration at High Altitudes and in Ocean Depths

This chapter's introduction described some challenges to respiration that arise when humans travel to high altitude. In this concluding section we look more closely at the effects of high altitude on respiration, along with the effects of increased pressures when humans and other mammals dive under water.

High Altitudes Reduce the P_{O_2} of Air Entering the Lungs

As altitude increases, atmospheric pressure decreases, and with it, the P_{O_2} of alveolar air and the concentration gradient of O_2 across the respiratory surface. At 20,000 feet, where most people become unconscious unless they have supplemental O_2, the dry air pressure is about 380 mm Hg and the P_{O_2} is only 380 × 21/100 = about 80 mm Hg, half that at sea level.

Humans who travel from sea level to elevations of 6000 feet or more often experience one or more unpleasant symptoms, including headache, blurred vision, dizziness, nausea, and fatigue. However, after a few weeks at higher elevation, the body adjusts by increasing the number of erythrocytes in the blood. The increase in erythrocyte production is stimulated primarily by a hormone, *erythropoietin* (EPO), which the kidneys secrete in greater quantities in response to a drop in blood O_2. Erythrocyte production slows when people return to lower altitudes. However, the erythrocyte count remains high for several weeks after high-altitude exposure. Athletes often train at high altitudes to increase their erythrocyte count, with the idea that it will improve their stamina and endurance at lower altitudes.

People who live at high altitudes from childhood develop more permanent changes, including an increase in the number of alveoli and more extensive capillary networks in the lungs. These developments are retained if they move to lower altitudes.

Some mammals evolutionarily adapted to high altitudes show genetically determined changes that are present throughout life. For example, llamas, which customarily live at altitudes as high as 4500 m (14,000 feet), have hemoglobin molecules with greater affinity for O_2 than does the hemoglobin of sea level-dwelling mammals. As a result, hemoglobin becomes saturated with O_2 at the lower partial pressures typical of high altitudes. The same adaptation occurs in birds adapted to life at high altitudes, such as the bar-headed goose *(Anser indicus)*. These birds have been observed flying over the peaks of the Great Himalayas, which have altitudes greater than 6000 m.

Diving Mammals Are Adapted to Survive the High Partial Pressures of Gases at Extreme Depths

As a mammal such as a seal or whale dives from the surface, each additional 10 m of depth increases the partial pressure of dissolved gases by about 1 atmo-

Unanswered Questions

Does prenatal nicotine exposure alter development of respiratory neurons in the brainstem?

In this chapter you learned that the muscles of breathing are controlled in the brainstem, by groups of interneurons in the medulla oblongata and pons. The control of respiratory muscles (and thus breathing) by these neurons is called "central ventilatory control." Neonatal mammals that were exposed to nicotine in utero show various breathing abnormalities, such as reduced ventilatory output, increased frequency and duration of apneas (suspension of breathing), and delayed arousal in response to hypoxia (reduced blood oxygen levels) during sleep. One or several of these abnormalities may underlie sudden infant death syndrome (SIDS), also called "crib death" because victims are typically found dead in their crib. Clinical studies have shown that exposure to tobacco smoke is the number one risk factor for SIDS. Accordingly, laboratories, such as our own, are using animal models to examine how nicotine exposure alters development of central ventilatory control.

Several research methods can be used, depending on the particular question being addressed. In all of our experiments, neonatal rodents are exposed to nicotine in utero by implanting a small osmotic pump under the skin of a female rat that is 4 days pregnant. The pump releases nicotine at a prescribed rate, and the developing neonates are exposed as the nicotine passes from mother to neonate via the placenta. When the neonates are born (on the 21st day of pregnancy), we study their breathing responses while they are awake or asleep using a device called a plethysmograph. This device senses the tiny pressure changes that accompany breathing in these small animals, and by adjusting chamber size, animals can be studied from birth to adulthood.

We can also dissect the brainstem, spinal cord, and rib cage from a neonatal animal, and place the preparation in a chamber for in vitro studies. Remarkably, this preparation is able to maintain rhythmic firing of respiratory neurons for up to 6 hours, allowing us to apply drugs and neurotransmitters to brainstem respiratory neurons while recording the electrical activity of neurons and respiratory muscle nerves.

Finally, we can prepare a brainstem slice, containing the most important central respiratory neurons and the hypoglossal nerve (this nerve innervates the tongue muscles, and it contains axons with rhythmic, respiratory-related activity), for detailed electrophysiological studies using the patch clamp technique (see Section 37.2). This preparation allows us to examine how prenatal nicotine exposure influences the membrane potential and firing properties of respiratory neurons.

To date, our studies have demonstrated abnormal breathing in awake neonates, as well as an increase in inhibitory neurotransmission in respiratory neurons. Current and future studies are directed at understanding the detailed cellular mechanisms that lead to the increase in inhibitory neurotransmission caused by prenatal nicotine exposure. Understanding how this occurs will hopefully lead to the development of drugs that can counteract nicotine's impact on the brain, as well as to an increased awareness and acceptance of the link between prenatal nicotine exposure and breathing abnormalities, resulting in more aggressive prevention strategies.

Ralph Fregosi is professor of physiology and neurobiology at the University of Arizona at Tucson. He does research on the neural control of breathing and teaches physiology to undergraduate and graduate students. Learn more about his research at http://www.physiology.arizona.edu/labs/rnlab/.

sphere. Below about 25 m or so, the pressure becomes so great that the lungs collapse and cease to function. Adaptations of diving mammals such as seals and whales allow these animals to survive the extreme pressure and lack of lung function, in some species for over an hour at ocean depths of more than a mile.

Among these adaptations are more blood per unit of body weight and more red blood cells, which are stored in the spleen and released during a dive. In addition, the muscles of these animals contain much greater quantities of the O_2-binding protein myoglobin than the muscles of land-dwelling mammals do. In all, the adaptations pack about twice as much O_2 per kilogram of body weight into a seal, for example, than into a human.

Other adaptations decrease O_2 consumption during a deep and prolonged dive. The heart rate slows by about 80% to 90% and the circulation of blood to internal organs and muscles is cut by as much as 95%, leaving only the brain with its normal blood supply. Even though most of the blood supply to muscles is cut off, the muscles continue to work by shifting to anaerobic oxidation. The lactic acid produced by anaerobic respiration in the muscles is not released into the blood until the animal returns to the surface.

These combined adaptations give seals and whales an amazing ability to dive to great depths and remain under water for extended periods. Although average dives are on the order of 10 to 20 minutes, some sperm whales, tracked by sonar, have reached depths of 2250 m (more than 7000 feet) and remained under water for as long as 82 minutes.

Study Break

List the key adaptations that diving mammals use to survive at significant ocean depths.

Review

Go to ThomsonNOW at www.thomsonedu.com/login to access quizzing, animations, exercises, articles, and personalized homework help.

44.1 The Function of Gas Exchange

- Physiological respiration is the process by which animals exchange O_2 and CO_2 with the environment (Figure 44.1).
- The two primary operating features of gas exchange are the respiratory medium, either air or water, and the respiratory surface, a wetted epithelium over which gas exchange takes place.
- In some invertebrates, the skin serves as the respiratory surface. In other invertebrates and all vertebrates, gills or lungs provide the primary respiratory surface (Figure 44.2).
- Simple diffusion of molecules from regions of higher concentration to regions of lower concentration drives the exchange of gases across the respiratory surface. The area of the respiratory surface determines the total quantity of gases exchanged by diffusion.
- The concentration gradients of O_2 and CO_2 across the respiratory surface are kept at optimal levels by ventilation and perfusion.

Animation: Examples of respiratory surfaces

44.2 Adaptations for Respiration

- Animals breathing water keep the respiratory surface wetted by direct exposure to the environment. The high density and viscosity of water, and its relatively low O_2 content as compared with air, requires water-breathing animals to expend significant energy to keep their respiratory surface ventilated.
- Air is high in O_2 content, allowing air-breathing animals to maintain higher metabolic levels than water breathers. The low density and viscosity of air as compared with water allows air breathers to ventilate the respiratory surface with relatively little energy. To accommodate water loss by evaporation, lungs typically are invaginations of the body surface, allowing air to become saturated with water before it reaches the respiratory surface.
- Gills are evaginations of the body surface. Water moves over the gills by the beating of cilia or is pumped over the gills by contractions of body muscles (Figures 44.2b and 44.3a–d).
- Water moves in a one-way direction over the gills of sharks, bony fishes, and some crabs, allowing these animals to use countercurrent exchange to maximize the exchange of gases over the respiratory surface (Figure 44.4).
- Insects breathe by means of tracheae, air-conducting tubes that lead from the body surface and send branches to essentially every cell in the body. Gas exchange takes place in the fluid-filled tips at the ends of the branches (Figure 44.5).
- Lungs consist of an invaginated system of branches, folds, and pockets. They may be filled by positive pressure breathing, in which air is forced into the lungs by muscle contractions, or by negative pressure breathing, in which muscle contractions expand the lungs, lowering the air pressure inside them and allowing air to be pulled into the lungs (Figures 44.6 and 44.9).

Animation: Bony fish respiration

Animation: Frog respiration

Animation: Vertebrate lungs

Animation: Bird respiration

44.3 The Mammalian Respiratory System

- Air enters the respiratory system through the nose and mouth and passes through the pharynx, larynx, and trachea. The trachea divides into two bronchi, which lead to the lungs. Within the lungs, the bronchi branch into bronchioles, which lead into the alveoli, which are surrounded by dense networks of blood capillaries (Figure 44.8).
- Mammals inhale by a negative pressure mechanism. Air is exhaled passively by relaxation of the diaphragm and the external intercostal muscles between the ribs, and elastic recoil of the lungs. During deep and rapid breathing, the expulsion of air is

forceful, driven by contraction of the internal intercostal muscles (Figure 44.9).

- The tidal volume of the lungs is the air moved in and out of the lungs during an inhalation and exhalation. The vital capacity is the total volume of air a person can inhale and exhale by breathing as deeply as possible. The air remaining in the lungs after as much air as possible is exhaled is the residual volume of the lungs.
- Breathing is controlled by a combination of local chemical controls and regulation by centers in the brain stem. These controls match the rate of air and blood flow in the lungs, and link the rate and depth of breathing to the body's requirements for O_2 uptake and CO_2 release (Figure 44.10).
- The basic rhythm of breathing is produced by interneurons in the medulla. When more rapid breathing is required, another group of interneurons in the medulla sends signals reinforcing inhalation and producing forceful exhalation. Two interneuron groups in the pons smooth and fine-tune breathing by stimulating or inhibiting the inhalation center in the medulla.
- Sensory receptors in the medulla, the carotid bodies, and the aortic bodies detect changes in the levels of O_2 and CO_2 in the blood and body fluids. The control centers in the medulla and pons adjust the rate and depth of breathing to compensate for changes in the blood gases.

Animation: Human respiratory system

Animation: Structure of an alveolus

Animation: Respiratory cycle

Animation: Changes in lung volume and pressure

Animation: Partial pressure gradients

Animation: Pressure-gradient changes during respiration

44.4 Mechanisms of Gas Exchange and Transport

- The partial pressure of O_2 is higher in the alveolar air than in the blood in the capillary networks surrounding the alveoli causing O_2 to diffuse from the alveolar air into the blood. Most of the O_2 entering the blood combines with hemoglobin inside erythrocytes (Figure 44.11).
- A hemoglobin molecule can combine with four O_2 molecules. The large quantities of O_2 that combine with hemoglobin maintain a large gradient in partial pressure between O_2 in the alveolar air and in the blood (Figure 44.12).
- In body tissues outside the lungs, the O_2 concentration in the interstitial fluid and body cells is lower than in the blood plasma. As a result, O_2 diffuses from the blood into the interstitial fluid, and from the fluid into body cells.
- The partial pressure of CO_2 is higher in the tissues than in the blood. About 10% of this CO_2 dissolves in the blood plasma; 70% is converted into H^+ and HCO_3^- (bicarbonate) ions. The remaining 20% combines with hemoglobin (Figures 44.11 and 44.13a).
- In the lungs, the partial pressure of CO_2 is higher in the blood than in the alveolar air. As a result, the reactions packing CO_2 into the blood are reversed, and the CO_2 is released from the blood into the alveolar air (Figure 44.13b).

Animation: Globin and hemoglobin structure

44.5 Respiration at High Altitudes and in Ocean Depths

- In mammals that move to high altitudes, the number of red blood cells and the amount of hemoglobin per cell increase. These changes are reversed if the animals return to lower altitudes.
- Humans living at higher altitudes from birth develop more alveoli and capillary networks in the lungs.
- Some mammals and birds adapted to high altitudes have forms of hemoglobin with greater affinity for O_2, allowing saturation at the lower P_{O_2} typical of high altitudes.
- Marine mammals adapted to deep diving have a greater blood volume per unit of body weight, and their blood contains more red blood cells, with a higher hemoglobin content, than other mammals. Their muscles also contain more myoglobin than those of land mammals, allowing more O_2 to be stored in muscle tissues. During a dive, the heartbeat slows, and circulation is reduced to all parts of the body except the brain.

Questions

Self-Test Questions

1. Which of the following describes a respiratory medium?
   a. In the liver the rate of diffusion is high.
   b. In the brain CO_2 moves from the neurons to the blood.
   c. In the big toe O_2 moves from blood to tissues.
   d. Epithelial cells form thin surfaces in the lungs.
   e. A running brook provides O_2 to fish.
2. Which of the following describes a respiratory surface?
   a. a surface consisting of multiple layers of epithelial cells
   b. the exoskeleton of an insect
   c. the nasal passages of a mammal
   d. a thin surface consisting of a single layer of epithelial cells
   e. the outer membrane of a mitochondrion.
3. At the end of a basketball game, the opposing teams line up and file past each other and shake hands. This efficient exposure of the teams to each other is analogous to:
   a. countercurrent exchange of gases in fish gills and bird lungs.
   b. diffusion of O_2 from blood to cells in shark tissues.
   c. diffusion of CO_2 from cells to blood in crabs.
   d. utilization of O_2 in cells in insects.
   e. excretion of CO_2 from mammalian cells.
4. Tracheal systems are characterized by:
   a. closed circulatory tubes that move gases.
   b. spiracles that move gases between cells and body fluids.
   c. body movements that compress and expand air sacs to pump air.
   d. positive pressure breathing, which swallows air into the body.
   e. negative pressure breathing, which lowers air pressure at the respiratory surfaces.
5. The structures at which one third of O_2 in the atmosphere moves into the blood of humans are:
   a. alveoli.
   b. bronchi.
   c. bronchioles.
   d. tracheae.
   e. pharynges.
6. A speed skater is finishing his last lap. At this time:
   a. the diaphragm and rib muscles contract when he exhales.
   b. positive pressure brings air into his lungs.
   c. his lungs undergo an elastic recoil when he inhales.

d. his tidal volume is at vital capacity.
e. his residual volume momentarily reaches zero.

7. A teenager is frightened when she is about to step onto the stage but then remembers to breathe deeply and slowly as she faces the audience. What is occurring here?
   a. Interneurons in the medulla cause the rib muscles to relax, followed later by stimulation and contraction of the intercostal muscles.
   b. Signals from the pons override the initial brain stem stimuli.
   c. The limbic system stabilized her emotional state, so there is no change in the mechanical movement of air.
   d. The brain signals the aortic bodies in the carotid arteries to adjust the breathing rate.
   e. Initial low CO_2 blood levels causing high pH are followed by increased CO_2 levels that lower pH.
8. Oxygen enters the blood in the lungs because relative to alveolar air:
   a. the CO_2 concentration in the blood is high.
   b. the CO_2 concentration in the blood is low.
   c. the O_2 concentration in the blood is high.
   d. the O_2 concentration in the blood is low.
   e. the process is independent of gas concentrations in the blood.
9. The hemoglobin O_2 dissociation curve:
   a. reflects about 50% saturation of hemoglobin in the alveoli.
   b. shifts to the left when pH rises.
   c. demonstrates that hemoglobin holds less O_2 when the pH is higher.
   d. proves lack of dependence on CO_2 levels.
   e. explains how hemoglobin can bind O_2 at high pH in the lungs and release it at lower pH in the tissues.
10. The majority of CO_2 in the blood:
   a. is in the form of carbonic acid and bicarbonate ions
   b. dissociates to add H^+ to the blood to raise its pH to 7.4.
   c. has a lower P_{CO_2} than the P_{CO_2} in the alveolar air.
   d. increases in the lung capillaries, which have a higher pH than the tissue capillaries.
   e. can be displaced on the hemoglobin molecule by CO if CO is inhaled.

Questions for Discussion

1. Smoking has traditionally been considered to reduce the ability of athletes to run without becoming exhausted. Why might this be true?
2. People are occasionally found unconscious from breathing too much CO_2 (as from a charcoal heater placed indoors) or too much CO (as from auto exhaust in a closed garage). Would it be more advantageous to give pure O_2 to a person breathing too much CO_2 than simply moving the person to fresh air? Why? Which—pure O_2 or fresh air—would be best for a person unconscious from breathing CO? Why?
3. Hyperventilation, or overbreathing, is breathing faster or deeper than necessary to meet the body's needs. Hyperventilation reduces the CO_2 content of blood, but does not significantly increase the amount of O_2 available to tissues. Why might this be so?

Experimental Analysis

Propose a hypothesis for the effect of zero gravity on respiration, and design an experiment to test the hypothesis.

Evolution Link

From what you have learned in this chapter and in Chapter 30, do you think lungs evolved once, or on several occasions? Justify your answer.

How Would You Vote?

Tobacco is a worldwide threat to health and a profitable product for American companies. As tobacco use by its citizens declines, should the United States encourage international efforts to reduce tobacco use around the globe? Go to www.thomsonedu.com/login to investigate both sides of the issue and then vote.

An Alaskan brown bear *(Ursus arctos)* catching a sockeye salmon *(Oncornynchus nerka)*. Animals obtain nutrients by eating other organisms. Their digestive systems break down macromolecules in the food to produce simple organic molecules that are used for fuels and as building blocks for more complex molecules.

Thomas Mangelsen/Minden Pictures

Study Plan

45 Animal Nutrition

Why It Matters

Invisible in the inky darkness, a deep-sea anglerfish (a member of the Order Lophiiformes) lies in wait, its gaping mouth lined with sharp teeth. Just above the mouth dangles a glowing lure suspended from a fishing-rod-like spine that projects from the fish's dorsal fin **(Figure 45.1).** The lure resembles a tiny fish; it even wiggles back and forth in imitation of swimming movements. Its glow is produced by bioluminescent bacteria that live symbiotically in the lure's tissues.

A hapless fish is attracted to the lure. As it comes within range, the anglerfish's mouth expands suddenly, creating a powerful suction that whips the prey in. The backward-angling fangs keep the prey from escaping. The strike takes only 6 ms (milliseconds), among the fastest of any known fishes.

Contractions of throat muscles send the prey to the anglerfish's stomach, which can expand to accommodate a meal as large as the anglerfish itself. In the fish's digestive tract, acids and enzymes dissolve the body of the prey, gradually breaking it into molecules small enough to be absorbed. In this function, the digestive system of the anglerfish is the same that of any other vertebrate, including humans—it provides nutrients that allow the animal to live. And the

David Shale/npl/Minden Pictures

Figure 45.1 A deep sea anglerfish, with its rod and lure lit and ready to attract prey.

anglerfish's adaptations for feeding, although bizarre to human sensibilities, are no more remarkable than those of many other animals.

Animal **nutrition**—which includes the processes by which food is ingested, digested, and absorbed into body cells and fluids—is the subject of this chapter. Our discussion begins with the basic categories of animal foods and **ingestion**, the feeding methods used to take food into the digestive cavity. Then we examine the process of **digestion:** the splitting of carbohydrates, proteins, lipids, and nucleic acids in foods into chemical subunits small enough to be absorbed into an animal's body fluids and cells. The chapter also presents the main structural and functional features of digestive systems, with special emphasis on humans and other mammals. The adaptations animals use to obtain and digest food are among their most strongly defining anatomical and functional characteristics.

45.1 Feeding and Nutrition

All organisms require sources of matter and energy for metabolism, homeostasis (maintaining their internal environment in a stable state; see Section 36.4), growth, and reproduction. For animals, meeting these nutritional requirements involves *feeding,* the uptake of food from the surroundings. Animals employ various feeding methods ranging from the ingestion of molecules in liquid solutions to eating entire organisms in one gulp. Once the food is ingested, digestive processes convert its molecules into absorbable subunits. In this section, we survey animal nutritional requirements and feeding methods as an introduction to animal digestive processes.

Animals Require both Organic and Inorganic Molecules for Nutrition

Plants and other photosynthesizers need only sunlight as an energy source and a supply of simple inorganic precursors such as water, carbon dioxide, and minerals to make all the organic molecules they require. In contrast, animals require a constant diet of organic molecules as a source of both energy and nutrients that they cannot make for themselves.

Animals are classified according to their sources of organic molecules. **Herbivores** such as antelopes, horses, bison, giraffes, kangaroos, manatees, and grasshoppers obtain organic molecules primarily by eating plants. **Carnivores**—cats, Tasmanian devils, penguins, sharks, and spiders, for example—primarily eat other animals. We say "primarily" because many herbivores eat animal matter at times, and a number of carnivores occasionally eat plant material. An antelope will eat insects as it grazes, and a grizzly bear, although primarily carnivorous, also eats berries. **Omnivores,** such as crows, cockroaches, and humans, eat both plants and animals and, in fact, any source of organic matter.

Organic molecules are the basis for two of the most fundamental processes of life: they act as fuels for oxidative reactions supplying energy and as building blocks for making complex biological molecules.

Energy supplies and requirements are usually described in terms of calories. A *calorie* (with a lowercase c) is the amount of heat energy required to raise 1 mL of pure water 1°C, from 14.5°C to 15.5°C. In animal nutrition, calories are usually considered in units of 1000 as kilocalories (kcal; the units listed on food packages in the United States) or Calories (with an uppercase C). One Calorie thus equals 1000 calories. Carbohydrates contain about 4.2 kcal per gram, fats about 9.5 kcal per gram, and proteins about 4.1 kcal per gram. At rest, a human female of average size expends about 1300 to 1500 kcal per day, and a human male about 1600 to 1800 kcal per day. Exercise and physical labor can increase these daily totals.

Carbohydrates and fats are the primary organic molecules used as fuels. Animals whose intake of organic fuels is inadequate, or whose assimilation of such fuels is abnormal, suffer from **undernutrition.** Undernutrition is a form of **malnutrition,** which is a condition resulting from an improper diet. **Overnutrition,** the condition caused by excessive intake of specific nutrients, is another main type of malnutrition.

An animal suffering from undernutrition essentially is starving for one or more nutrients, taking in fewer calories than needed for daily activities. Animals with chronic undernutrition lose weight because they have to use molecules of their own bodies as fuels. Mammals use stored fats and glycogen (animal starch) first. Once those stores have been used up, proteins are metabolized as fuels. The use of proteins as fuels leads to muscle wastage and, in the long term, to organ and brain damage and, therefore, eventually to death.

Organic molecules also serve as building blocks for carbohydrates, lipids, proteins, and nucleic acids. Animals can synthesize many of the organic molecules that they do not obtain directly in the diet by converting one type of building block into another. Typically, however, they cannot make certain amino acids and fatty acids from other organic molecules. These required organic building blocks are called **essential amino acids** and **essential fatty acids** because they must be obtained in the diet. If they are not obtained in the diet over a period of time, there may be serious consequences. For instance, protein synthesis cannot continue unless all 20 amino acids are present. In the absence of essential amino acids in the diet, the animal would have to break down its own proteins to provide them for new protein synthesis.

Animals must also take in **vitamins,** organic molecules required in small quantities that the animal cannot synthesize for itself. Many vitamins are *coenzymes,* nonprotein organic subunits associated with enzymes that assist in enzymatic catalysis (see Section 4.4).

Individual species differ in the vitamins and essential amino acids and fatty acids they require. Various species also have differing dietary requirements for inorganic elements such as calcium, iron, and magnesium. These required inorganic elements are known collectively as **essential minerals.**

The essential amino acids, fatty acids, vitamins, and minerals are known collectively as an animal's **essential nutrients.** The list of essential nutrients differs from animal to animal. For domesticated animals, this means that specific feed formulations must be given to each type of animal. For instance, the essential nutrients for cats and dogs are different, which is why there are specific cat foods and dog foods, and why it is does not make good sense, nutritionally speaking, to feed cats and dogs human food.

Animals Obtain Nutrients in Fluid, Particle, or Bulk Form

All animals display adaptations that allow them to obtain the food they need in particular environments. Although these adaptations are amazingly varied, animals can be classified into one of four groups according to overall feeding methods and the physical state of the organic molecules they consume. These four groups are fluid feeders, suspension feeders, deposit feeders, and bulk feeders **(Figure 45.2).**

a. Fluid feeder

b. Suspension feeder

c. Deposit feeder

d. Bulk feeder

Figure 45.2
Grouping of animals with respect to overall feeding methods and the physical state of the organic molecules they consume. **(a)** Fluid feeders, exemplified by a hummingbird, which obtains nectar from deep within a flower using its long bill and tongue. **(b)** Suspension feeders, exemplified by the northern right whale *(Balaena glacialis)*, which gulps tons of water containing plankton into its mouth, pushes the water out through the sievelike baleen, and swallows the remaining plankton. **(c)** Deposit feeders, exemplified by a fiddler crab (*Uca* species), which sifts edible material from the sediment it takes into its mouth. **(d)** Bulk feeders, exemplified in an extreme way by a python, which ingests its prey (here, a gazelle) whole. Elastic ligaments connecting the jaws allow the snake's mouth to open wide enough to swallow large prey.

Fluid feeders obtain nourishment by ingesting liquids that contain organic molecules in solution. Among the invertebrates, aphids, mosquitoes, leeches, and spiders are examples of fluid feeders. Vertebrate fluid feeders include birds such as hummingbirds (see Figure 45.2a), which feed on flower nectar; parasitic fishes such as lampreys, which feed on body fluids of their hosts; and some bats, which feed on nectar or blood. Many fluid feeders have mouthparts specialized to reach the source of their nourishment. For example, mosquitoes, bedbugs, and aphids have needlelike mouthparts that pierce body surfaces. Nectar-feeding birds and bats have long tongues that can extend deep within flowers. Some fluid feeders use enzymes or other chemicals to liquefy their food or to keep it liquid during feeding. For example, spiders inject digestive enzymes that liquefy tissues inside their victim and then suck up the liquid. The saliva of mosquitoes, leeches, and vampire bats includes an anticoagulant that keeps blood in liquid form during a feeding by inhibiting the clotting reaction.

Suspension feeders ingest small organisms suspended in water, such as bacteria, protozoa, algae, and small crustaceans, or fragments of these organisms. Among the suspension feeders are aquatic invertebrates such as clams, mussels, and barnacles; many fishes; and even some birds and whales (see Figure 45.2b). These animals strain food particles suspended in water through a body structure covered with sticky mucus or through a filtering network of bristles, hairs, or other body parts. The trapped particles are then funneled into the animal's mouth. Bits of organic matter are trapped by the gills of bivalves such as clams and oysters, and plankton is filtered from water by the sievelike fringes of horny fiber hanging in the mouths of baleen whales (see Figure 45.2b).

Deposit feeders pick up or scrape particles of organic matter from solid material they live in or on. Earthworms are deposit feeders that eat their way through soil, taking the soil into their mouth and digesting and absorbing any organic material it contains. Some burrowing mollusks and tube-dwelling polychaete worms use body appendages to gather organic deposits from the sand or mud around them. Mucus on the appendages traps the organic material, and cilia move it to the mouth. The fiddler crab (*Uca* species) is also a deposit feeder (see Figure 45.2c). This animal has claws of markedly different sizes. The small claw picks up sediment and moves it to the mouth where the contents are sifted. The edible parts of the sediment are ingested, and the rest is put back on the sediment as a small ball. The feeding-related movement of the small claw over the larger claw looks like the crab is playing the large claw like a fiddle and hence gives the crab its name.

Bulk feeders are animals that consume sizeable food items whole or in large chunks. Most mammals eat this way, as do reptiles, most birds and fishes, and adult amphibians. Depending on the animal, adaptations for bulk feeding include teeth for tearing or chewing, claws and beaks for holding large food items, and jaws that are hinged or otherwise modified to permit a food mass to enter the mouth (see Figure 45.2d).

We now take up the processes by which animals, having fed, undertake the mechanical and chemical breakdown of food into absorbable molecular subunits.

Study Break

1. What are carnivores, herbivores, and omnivores?
2. What are essential nutrients, and are they the same for all animals?
3. What is the difference between deposit feeders and suspension feeders?

45.2 Digestive Processes

Digestive processes break food molecules into molecular subunits that can be absorbed into body fluids and cells. The breakdown occurs by **enzymatic hydrolysis,** in which chemical bonds are broken by the addition of H^+ and OH^-, the components of a molecule of water (see Section 3.2). Specific enzymes speed these reactions: *amylases* catalyze the hydrolysis of starches, *lipases* break down fats and other lipids, *proteases* hydrolyze proteins, and *nucleases* digest nucleic acids. Depending on the animal, the enzymatic hydrolysis of food molecules may take place inside or outside the body cells.

Intracellular Digestion Takes Place within Cells; Extracellular Digestion Occurs in an Internal Pouch or Tube

In **intracellular digestion,** cells take in food particles by endocytosis (described in Section 6.5). Inside the cell, the endocytic vesicle containing the food particles fuses with a lysosome, a vesicle containing hydrolytic enzymes. The molecular subunits produced by the hydrolysis pass from the vesicle to the cytosol. Any undigested material remaining in the vesicle is released to the outside of the cell by exocytosis (also discussed in Section 6.5). Only a few animals, primarily sponges and some cnidarians, break down food exclusively by intracellular digestion. In sponges, water containing particles of organic matter and microorganisms enter the animal's saclike body through pores in the body wall (see Figure 29.8). In the body cavity, individual *choanocytes* (collar cells) lining the body wall trap the food particles, take them in by endocytosis, and transport them to amoeboid cells, which digest them intracellularly.

Extracellular digestion takes place outside body cells, in a pouch or tube enclosed within the body. Epithelial cells lining the pouch or tube secrete enzymes that digest the food. Processing food in specialized compartments in this way prevents the animal from digesting its own body tissues.

Most invertebrates and all vertebrates digest food primarily by extracellular digestion. From an adaptive standpoint, extracellular digestion greatly expands the range of available food sources by allowing animals to digest much larger food items than single cells can take in. Extracellular digestion also allows animals to eat large batches of food, which can be stored and digested while the animal continues other activities.

Saclike Digestive Systems Have a Single Opening through Which Food Enters and Undigested Matter Exits

Some animals, including flatworms and cnidarians such as hydras, corals, and sea anemones, have a saclike digestive system with a single opening, a mouth, that serves both as the entrance for food and the exit for undigested material. In some of these animals, such as the flatworm *Dugesia,* the digestive cavity is called a **gastrovascular cavity** because it contributes to circulation as well as digestion. Food is brought to the mouth by a protrusible **pharynx** (a throat that can be stuck out) and then enters the gastrovascular cavity **(Figure 45.3)**; glands in the cavity wall secrete enzymes that begin the digestive process. Cells lining the cavity then take up the partially digested material by endocytosis and complete digestion intracellularly. Undigested matter is released to the outside through the pharynx and mouth.

Digestive Tubes Typically Process Nutrients in Five Successive Steps

Most invertebrates and all vertebrates have a tubelike digestive system with two openings that form a separate mouth and anus; the digestive contents move in one direction through specialized regions of the tube, from the mouth to the anus. This type of digestive system is called a **digestive tube,** *gut, alimentary canal, digestive tract,* or *gastrointestinal (GI) tract.* Structurally, the inside of the digestive tube—called the **lumen**—is external to all body tissues. In other words, the lumen is *outside* of the body.

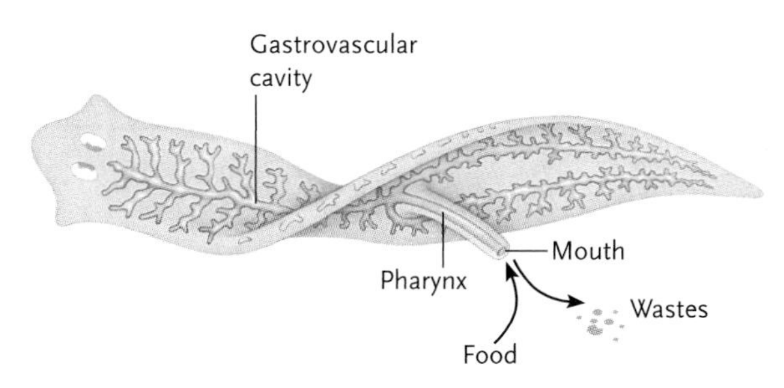

Figure 45.3
The digestive system of the flatworm *Dugesia.* The gastrovascular cavity (in blue) is a blind sac, with one opening to the exterior through which food is ingested and wastes are expelled.

In most animals with a digestive tube, digestion occurs in five successive steps, with each step taking place in a specialized region of the tube. The tube thus acts as a sort of biological disassembly line, with food entering at one end and passing through as many as five areas in which food processing occurs.

1. **Mechanical processing:** Chewing, grinding, and tearing breaks food chunks into smaller pieces, increasing their mobility and the surface area exposed to digestive enzymes.
2. **Secretion of enzymes and other digestive aids:** Enzymes and other substances that aid the process of digestion, such as acids, emulsifiers, and lubricating mucus, are released into the tube.
3. **Enzymatic hydrolysis:** Food molecules are broken down through enzyme-catalyzed reactions into absorbable molecular subunits.
4. **Absorption:** The molecular subunits are absorbed from the digestive contents into body fluids and cells.
5. **Elimination:** Undigested materials are expelled through the anus.

The material being digested is pushed along by muscular contractions of the wall of the digestive tube. During its progress through the tube, the digestive contents may be stored temporarily at one or more locations. The storage allows animals to take in larger quantities of food than they can process immediately, so that feedings can be spaced in time rather than continuous.

Digestion in an Annelid. The earthworm (genus *Lumbricus,* **Figure 45.4a**) is a deposit feeder. As it burrows, it pushes soil particles into its mouth. The particles pass from the mouth through a connecting passage, the **esophagus,** into the **crop,** an enlargement of the digestive tube where the contents are stored and mixed with lubricating mucus. This mixture enters the **gizzard,** which contains grains of sand, and is ground into fine particles by muscular contractions of the wall. The pulverized mixture then enters a long **intestine,** where the organic matter is hydrolyzed by enzymes secreted into the digestive tube. As muscular contractions of the intestinal wall move the mixture along, cells lining the intestine absorb the molecular subunits produced by digestion. The absorptive surface of the intestine is increased by folds of the wall called *typhlosoles.* At the end of the intestine, the undigested residue is expelled through the anus.

Digestion in an Insect. Herbivorous insects such as the grasshopper **(Figure 45.4b)** tear leaves and other plant parts into small particles with hard external

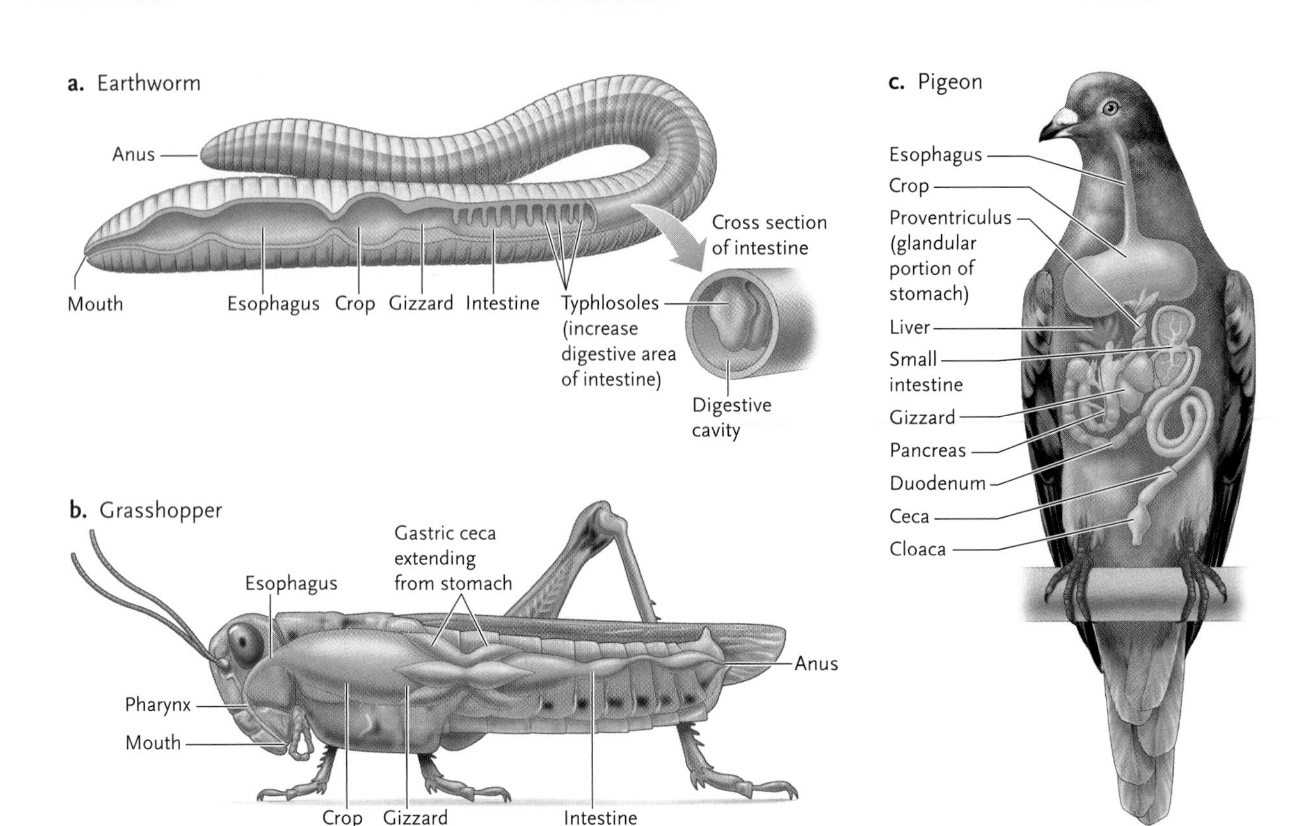

Figure 45.4
The digestive systems of an annelid, the earthworm **(a)**, an insect, the grasshopper **(b)**, and a bird, the pigeon *(Columba)* **(c)**.

mouth parts. From the mouth, the food particles pass through the pharynx, where salivary secretions moisten the mixture before it enters the esophagus and passes into the crop. These secretions begin the process of chemical digestion. From the crop, the food mass enters the gizzard, which grinds it into smaller pieces. These food particles enter the **stomach,** in which food is stored and digestion begins. Insect stomachs have saclike outgrowths, the *gastric ceca* (*caecus* = blind), where enzymes hydrolyze the digestive contents; the products of digestion are absorbed through the walls of the ceca. The undigested contents then move into the intestine for further digestion and absorption. At the end of the intestine, water is absorbed from the undigested matter and the remnants are expelled through the anus. The digestive systems of other arthropods are similar to the insect system.

Digestion in a Bird. A pigeon (*Columba,* **Figure 45.4c**) picks up seeds with its bill. The bird's tongue moves the seeds into its mouth, where they are moistened by mucus-filled saliva and swallowed whole (birds have no teeth). The seeds then pass through the pharynx into the esophagus. (In some cases, birds crack open seeds with their bills and ingest seed kernels in a similar fashion.) The anterior end of the esophagus is tubelike; at the posterior end is the pouchlike crop, in which the bird can store large quantities of food. From the crop, the food passes into the anterior glandular portion of the stomach, called the *proventriculus,* which secretes digestive enzymes and acids. The posterior end is the gizzard, in which the seeds are ground into fine particles, aided by ingested bits of sand and rock. The food particles are released into the intestine, where the liver secretes bile and the pancreas adds digestive enzymes. The molecular subunits produced by enzymatic digestion are absorbed as the mixture passes along the intestine, and the undigested residues are expelled through the anus.

Many of the structures of the pigeon's digestive system, including the mouth, pharynx, esophagus, stomach, intestine, liver, and pancreas, occur in almost all vertebrates.

Study Break

1. Distinguish between extracellular digestion and intracellular digestion.
2. What are the five steps of food processing in a digestive tube?

45.3 Digestion in Humans and Other Mammals

Mammals digest foods using the same five steps as other animals with a digestive tube: mechanical processing, secretion of enzymes and other digestive aids, enzymatic hydrolysis, absorption of molecular subunits, and elimination. The mammalian digestive system is a series of specialized digestive regions that perform these steps, including the mouth, pharynx,

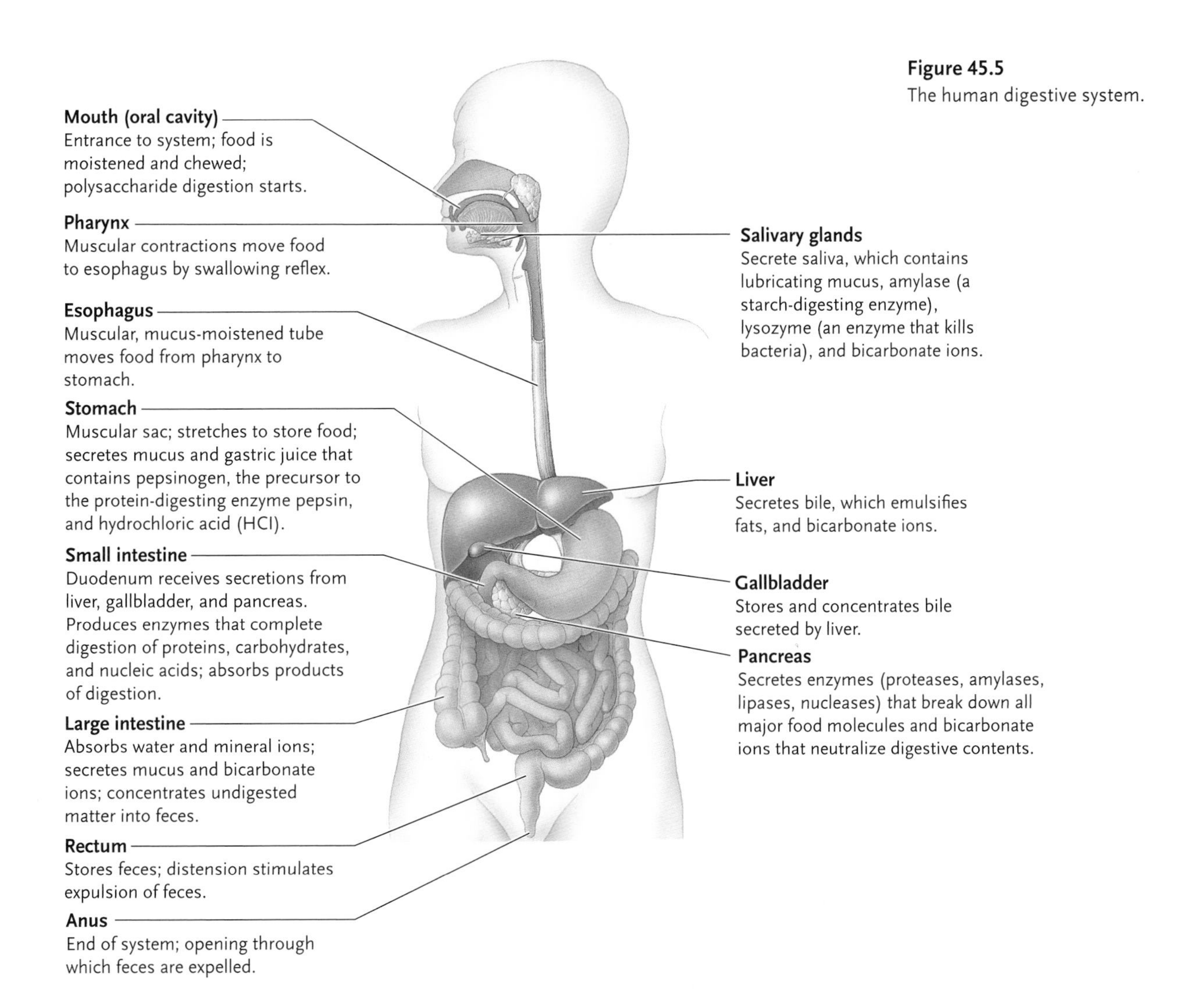

Mouth (oral cavity)
Entrance to system; food is moistened and chewed; polysaccharide digestion starts.
Pharynx
Muscular contractions move food to esophagus by swallowing reflex.
Esophagus
Muscular, mucus-moistened tube moves food from pharynx to stomach.
Stomach
Muscular sac; stretches to store food; secretes mucus and gastric juice that contains pepsinogen, the precursor to the protein-digesting enzyme pepsin, and hydrochloric acid (HCl).
Small intestine
Duodenum receives secretions from liver, gallbladder, and pancreas. Produces enzymes that complete digestion of proteins, carbohydrates, and nucleic acids; absorbs products of digestion.
Large intestine
Absorbs water and mineral ions; secretes mucus and bicarbonate ions; concentrates undigested matter into feces.
Rectum
Stores feces; distension stimulates expulsion of feces.
Anus
End of system; opening through which feces are expelled.
Salivary glands
Secrete saliva, which contains lubricating mucus, amylase (a starch-digesting enzyme), lysozyme (an enzyme that kills bacteria), and bicarbonate ions.
Liver
Secretes bile, which emulsifies fats, and bicarbonate ions.
Gallbladder
Stores and concentrates bile secreted by liver.
Pancreas
Secretes enzymes (proteases, amylases, lipases, nucleases) that break down all major food molecules and bicarbonate ions that neutralize digestive contents.

Figure 45.5
The human digestive system.

esophagus, stomach, small and large intestines, and anus **(Figure 45.5).** These regions are under the control of the nervous and endocrine systems.

Humans Require Specific Essential Amino Acids, Fatty Acids, Vitamins, and Minerals in Their Diet

The human digestive system meets our basic needs for fuel molecules and for a wide range of nutrients, including the molecular building blocks of carbohydrates, lipids, proteins, and nucleic acids. If the diet is adequate, the digestive system also absorbs the essential nutrients—the amino acids, fatty acids, vitamins, and minerals that cannot be synthesized within our bodies.

Essential Amino Acids and Fatty Acids. There are eight essential amino acids for adult humans: lysine, tryptophan, phenylalanine, threonine, valine, methionine, leucine, and isoleucine. Infants and young children also require histidine. The proteins in fish, meat, egg whites, milk, and cheese supply all the essential amino acids, provided those foods are eaten in adequate quantities. In contrast, the proteins of many plants are deficient in one or more of the essential amino acids. Corn, for example, contains inadequate amounts of lysine, and beans contain little methionine. Vegetarians, and especially vegans who eat a diet with no animal-derived nutrients, must choose their foods carefully to obtain all of the essential amino acids **(Figure 45.6).** Such diets typically include combinations of foods, each of which provides some amino acids, and that together contain all of the essential amino acids. An example is including in the diet rice or corn (low in lysine but high in methionine) with legumes such as lentils or with soybeans, perhaps in the form of tofu (low in methionine but high in lysine).

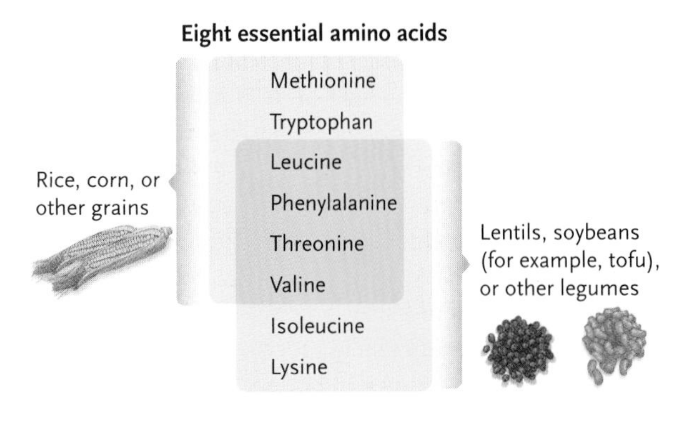

Figure 45.6
Obtaining essential amino acids in a human vegetarian diet.

If the diet lacks one or more essential amino acids, many enzymes and other proteins cannot be synthesized in sufficient quantities. The resulting protein deficiency is most damaging to the young, who must rapidly synthesize proteins for development and growth. Even mild protein starvation during pregnancy or for some months after birth can retard a child's mental and physical development.

Only two fatty acids, linoleic acid and linolenic acid, are essential in the human diet. Both are required for synthesis of phospholipids forming parts of biological membranes and certain hormones. Because almost all foods contain these fatty acids, most people have no problem obtaining them. However, people on a low-fat diet deficient in linoleic acid and linolenic acid are at serious risk for developing coronary heart disease. That is, there is an inverse correlation between the concentration of these essential fatty acids in the diet and the incidence of coronary heart disease. This is illustrated in the case of Hindu vegetarians from India. Their diet consists mainly of low-fat grains and legumes—clearly a low-fat diet—yet their rate of coronary heart disease is higher than that in the United States and Europe, where dietary fat content is higher.

Vitamins. Humans require 13 known vitamins in their diet. Many metabolic reactions depend on vitamins, and the absence of one vitamin can affect the functions of the others. These essential nutrients fall into two classes: **water-soluble** (hydrophilic) **vitamins** and **fat-soluble** (hydrophobic) **vitamins** (summarized in **Table 45.1**). The body stores excess fat-soluble vitamins in adipose tissues, but any amount of water-soluble vitamins above daily nutritional requirements is excreted in the urine. Thus, meeting the daily minimum requirements of water-soluble vitamins is critical. The body can tap its stores of fat-soluble vitamins to meet daily requirements; however, these stores are quickly depleted, so that prolonged deficiencies of the fat-soluble vitamins also become critical to health.

Most of us get all the vitamins we need through a normal and varied diet that includes meats, fish, eggs, cheese, and vegetables. Vitamin supplements are usually necessary only for strict vegetarians, newborns, the elderly, and individuals who are taking medication that affects the body's uptake of nutrients.

Vitamin D (calciferol) differs from other essential vitamins because humans can actually synthesize it themselves, through the action of ultraviolet light on lipids in the skin. However, many people are not exposed to enough sunlight to make sufficient quantities of the vitamin, and so must rely on dietary sources. And, although we cannot make vitamin K, much of our requirement for this vitamin is supplied through the metabolic activity of bacteria living in our large intestine. Vitamin K deficiency, therefore, is exceedingly rare in healthy persons. Vitamin K plays a role in blood clotting, so individuals with vitamin K deficiency will bruise easily and show increased blood clotting times. Vitamin K deficiency can be caused in persons on long-term antibiotic therapy because the antibiotics kill intestinal bacteria.

Other mammals have essentially the same vitamin requirements as humans, with some differences. For example, most other mammals, with the exception of primates, guinea pigs, and fruit bats, can synthesize vitamin C. So far as is known, no animal can synthesize B vitamins, but ruminants such as cattle and deer are supplied with these vitamins by microorganisms that live in the digestive tract (see Section 45.5).

Minerals. Many minerals are essential in the human diet **(Table 45.2)**. Some of them, called **macronutrients**, are required in amounts ranging from 50 mg to more than a gram per day; others, such as zinc, are **micronutrients**, or **trace elements**, required only in small amounts, some less than 1 mg per day. All of the minerals, although listed as elements, are ingested as compounds or as ions in solution.

A normal and varied diet supplies adequate amounts of the essential minerals. Supplements may be required for those on a strict vegetarian diet, the very young, and the aged. Overdoses of some minerals can cause problems; ingesting excess iron, for example, has been linked to liver, heart, and blood vessel damage; too much sodium can lead to elevated blood pressure and excess water retention in tissues.

We now turn to the structures that extract nutrients from ingested foods. We begin with a survey of digestive structures common to all vertebrates.

Four Major Layers of the Gut Each Have Specialized Functions in Digestion

The wall of the gut in mammals and other vertebrates contains four major layers, each with specialized functions. These layers are shown for the stomach in **Figure 45.7.**

1. The **mucosa**, which contains epithelial and glandular cells, lines the inside of the gut. The epithelial cells absorb digested nutrients and seal off the digestive contents from body fluids. The glandular cells secrete enzymes, aids to digestion such as lubricating mucus, and substances that adjust the pH of the digestive contents.
2. The **submucosa** is a thick layer of elastic connective tissue that contains neuron networks and blood and lymph vessels. The neuron networks provide local control of digestive activity and carry signals between the gut and the central nervous system. The lymph vessels carry absorbed lipids to other parts of the body.

Table 45.1	Vitamins: Sources, Functions, and Effects of Deficiencies in Humans		
Vitamin	**Common Sources**	**Main Functions**	**Effects of Chronic Deficiency**
Fat-Soluble Vitamins			
A (retinol)	Yellow fruits, yellow or green leafy vegetables; also in fortified milk, egg yolk, fish liver	Used in synthesis of visual pigments, bone, teeth; maintains epithelial tissues	Dry, scaly skin; lowered resistance to infections; night blindness
D (calciferol)	Fish liver oils, egg yolk, fortified milk; manufactured when body exposed to sunshine	Promotes bone growth and mineralization; enhances calcium absorption from gut	Bone deformities (rickets) in children; bone softening in adults
E (tocopherol)	Whole grains, leafy green vegetables, vegetable oils	Antioxidant; helps maintain cell membrane and red blood cells	Lysis of red blood cells; nerve damage
K (napthoquinone)	Intestinal bacteria; also in green leafy vegetables, cabbage	Promotes synthesis of blood clotting protein by liver	Abnormal blood clotting, severe bleeding (hemorrhaging)
Water-Soluble Vitamins			
B_1 (thiamine)	Whole grains, green leafy vegetables, legumes, lean meats, eggs, nuts	Connective tissue formation; folate utilization; coenzyme forming part of enzyme in oxidative reactions	Beriberi; water retention in tissues; tingling sensations; heart changes; poor coordination
B_2 (riboflavin)	Whole grains, poultry, fish, egg white, milk, lean meat	Coenzyme	Skin lesions
Niacin	Green leafy vegetables, potatoes, peanuts, poultry, fish, pork, beef	Coenzyme of oxidative phosphorylation	Sensitivity to light; contributes to pellagra (damage to skin, gut, nervous system, etc.)
B_6 (pyridoxine)	Spinach, whole grains, tomatoes, potatoes, meats	Coenzyme in amino acid and fatty acid metabolism	Skin, muscle, and nerve damage
Pantothenic acid	In many foods (meats, yeast, egg yolk especially)	Coenzyme in carbohydrate and fat oxidation; fatty acid and steroid synthesis	Fatigue, tingling in hands, headaches, nausea
Folic acid	Dark green vegetables, whole grains, yeast, lean meats; intestinal bacteria produce some folate	Coenzyme in nucleic acid and amino acid metabolism; promotes red blood cell formation	Anemia; inflamed tongue; diarrhea; impaired growth; mental disorders; neural tube defects and low birth weight in newborns
B_{12} (cobalamin)	Poultry, fish, eggs, red meat, dairy foods (not butter)	Coenzyme in nucleic acid metabolism; necessary for red blood cell formation	Pernicious anemia; impaired nerve function
Biotin	Legumes, egg yolk; colon bacteria produce some	Coenzyme in fat and glycogen formation, and amino acid metabolism	Scaly skin (dermatitis), sore tongue, brittle hair, depression, weakness
C (ascorbic acid)	Fruits and vegetables, especially citrus, berries, cantaloupe, cabbage, broccoli, green pepper	Vital for collagen synthesis; antioxidant	Scurvy, delayed wound healing, impaired immunity

3. In most regions of the gut, the **muscularis** is formed by two smooth muscle layers, a *circular layer* that constricts the diameter of the gut when it contracts and a *longitudinal layer* that shortens and widens the gut. The stomach also has an *oblique layer* running diagonally around its wall. The circular and longitudinal muscle layers of the muscularis coordinate their activities to push the digestive contents through the gut **(Figure 45.8).** In this mechanism, called **peristalsis,** the circular muscle layer contracts in a wave that passes along the gut, constricting the gut and pushing the digestive contents onward. Just in front of the advancing constriction, the longitudinal layer contracts, shortening and expanding the tube and making space for the contents to advance.
4. The outermost gut layer, the **serosa,** consists of connective tissue that secretes an aqueous, slippery fluid. The fluid lubricates the areas between the digestive organs and other organs, reducing friction between them as they move together as a result of muscle movement. Along much of the length of the digestive system, the serosa is continuous with the *mesentery,* a tissue that suspends the digestive system from the inner wall of the abdominal cavity.

Table 45.2 Major Minerals: Sources, Functions, and Effects of Deficiencies in Humans

Mineral	Sources	Functions	Effects of Deficiencies
Calcium (Ca)	Dairy products, leafy green vegetables, legumes, whole grains, nuts	Bone, tooth formation; blood clotting; neural and muscle action	Stunted growth; diminished bone mass (osteoporosis)
Chlorine (Cl)	Table salt, meat, eggs, dairy products	HCl formation in stomach, contributes to body's acid-base balance; neural function, water balance	Muscle cramps; impaired growth; poor appetite
Chromium (Cr)*	Meat, liver, cheese, whole grains, brewer's yeast, peanuts	Roles in carbohydrate metabolism	Impaired response to insulin; increases risk of type 2 diabetes mellitus
Cobalt (Co)*	Meat, liver, fish, milk	Constituent of vitamin B_{12} (required for red blood cell maturation)	Same as for vitamin B_{12} (see Table 45.1)
Copper (Cu)*	Nuts, legumes, seafood, drinking water, whole grains, nuts	Used in synthesis of melanin, hemoglobin, and some electron transport chain components in mitochondria	Anemia, changes in bone and blood vessels
Fluorine (F)*	Fluoridated water, tea, seafood	Bone, tooth maintenance	Tooth decay
Iodine (I)*	Marine fish, shellfish, iodized salt	Thyroid hormone formation	Goiter (enlarged thyroid), with metabolic disorders
Iron (Fe)	Liver, whole grains, green leafy vegetables, legumes, nuts, eggs, lean meat, molasses, dried fruit, shellfish	Component of hemoglobin, cytochrome, myoglobin	Iron-deficiency anemia
Magnesium (Mg)	Whole grains, green vegetables, legumes, nuts, dairy products	Required for action of many enzymes; roles in muscle, nerve function	Weak, sore muscles; impaired neural function
Manganese (Mn)*	Whole grains, nuts, legumes, many fruits	Activates many enzymes, including ones with roles in synthesis of urea, fatty acids	Abnormal bone and cartilage
Molybdenum (Mo)*	Dairy products, whole grains, green vegetables, legumes	Component of some enzymes	Impaired nitrogen excretion
Phosphorus (P)	Whole grains, legumes, poultry, red meat, dairy products	Component of bones and teeth, nucleic acids, ATP, phospholipids	Muscular weakness; loss of minerals from bone
Potassium (K)	Meat, milk, many fruits, vegetables	Muscle and neural function; roles in protein synthesis	Muscular weakness
Selenium (Se)*	Meat, seafood, cereal grains, poultry, garlic	Constituent of several enzymes; antioxidant	Muscle pain
Sodium (Na)	Table salt, dairy products, meats, eggs	Acid-base balance, water balance; roles in muscle and neural function	Muscle cramps
Sulfur (S)	Meat, eggs, dairy products	Component of body proteins	Same as protein deficiencies
Zinc (Zn)*	Whole grains, legumes, nuts, meats, seafood	Component of digestive enzymes and transcription factors; roles in normal growth, wound healing, sperm formation, taste and smell	Impaired growth, scaly skin, impaired immune function

*Required in trace amounts in diet

Powerful rings of smooth muscle called **sphincters** form valves between major regions of the digestive tract. By contracting and relaxing, sphincters control the passage of the digestive contents from one region to the next, and ultimately through the anus.

The specialized regions of the gut that perform the sequential processes of digestion in humans allow us to extract nutrients efficiently from the highly varied foods we ingest.

Food Begins Its Travel through the Digestive System in the Mouth, Pharynx, and Esophagus

The human digestive system in its normal contracted state in a living adult is about 4.5 m long. Fully relaxed, it is about twice as long. Food begins its travel through this tract in the mouth, where the teeth cut, tear, and crush food items into small pieces. While chewing is

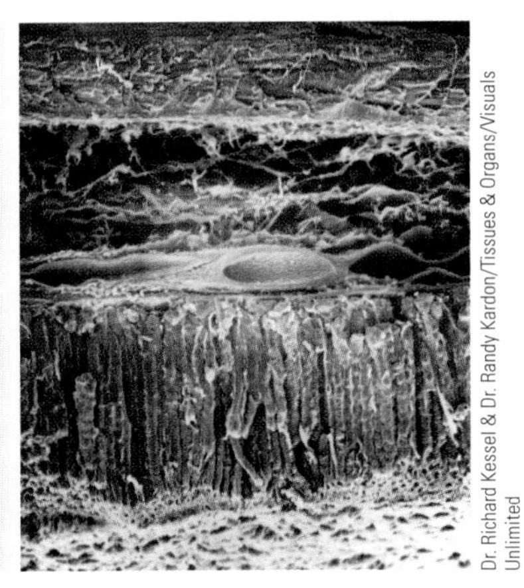

Figure 45.7
Layers of the gut wall in vertebrates, as seen in the stomach wall.

in progress, three pairs of **salivary glands** secrete saliva through ducts that open on the inside of the cheeks and under the tongue.

Saliva, which is more than 99% water, moistens the food. Saliva contains **salivary amylase**, which hydrolyzes starches to the disaccharide maltose. It also contains mucus, which lubricates the food mass, and bicarbonate ions (HCO_3^-), which neutralize acids in the food and keep the pH of the mouth between 6.5 and 7.5, which is the optimal range for salivary amylase to function. Another component of saliva is *lysozyme,* an enzyme that kills bacteria by breaking open their cell walls. Some 1 to 2 L of saliva are secreted into the mouth each day.

After a suitable period of chewing, the food mass, called a **bolus**, is pushed by the tongue to the back of the mouth, where touch receptors detect the pressure and trigger the *swallowing reflex* **(Figure 45.9).** The reflex is an involuntary action produced by contractions of muscles in the walls of the pharynx that direct food into the esophagus. Peristaltic contractions of the esophagus, aided by mucus secreted by the esophagus, propel a bolus towards the stomach. The passage of a bolus down the esophagus stimulates the *gastroesophageal sphincter* at the junction between the esophagus and

Pyloric sphincter
Chyme

1 The circular layer of the muscularis contracts in a wave, constricting the gut and pushing the digestive contents onward.

2 The longitudinal layer contracts, shortening and expanding the gut and making space for the contents to advance.

3 Partially processed food (chyme) enters the small intestine.

Figure 45.8
The waves of peristaltic contractions moving food through the stomach.

Structures of the mouth, pharynx, and esophagus involved in the swallowing reflex

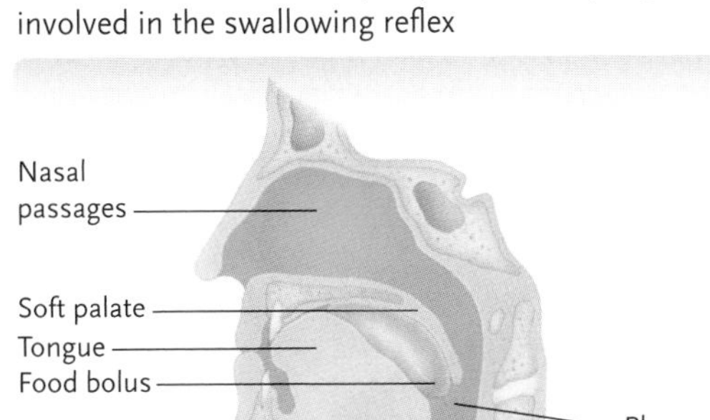

Motions that seal the nasal passages, mouth, and trachea during the swallowing reflex

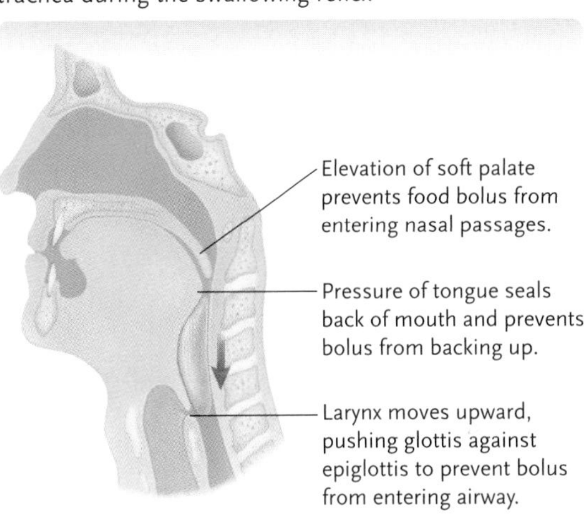

Elevation of soft palate prevents food bolus from entering nasal passages.
Pressure of tongue seals back of mouth and prevents bolus from backing up.
Larynx moves upward, pushing glottis against epiglottis to prevent bolus from entering airway.

Figure 45.9
The swallowing reflex.

the stomach (see Figure 45.7) to open and admit the bolus to the stomach. After the bolus enters the stomach, the sphincter closes tightly. If the closure is imperfect, the acidic stomach contents can enter the esophagus and produce the irritation and pain we recognize as *acid reflux* or heartburn.

We can consciously initiate the swallowing reflex. However, once the swallowing reflex has begun, we cannot voluntarily stop it, as you might have noticed when you get that feeling of a piece of food or a pill being stuck in the throat or chest. This is because the muscles of the pharynx and upper esophagus are skeletal muscles, which you can control, while the muscles below are smooth muscles, which you cannot control.

Involuntary movements of the tongue and soft palate at the back of the mouth prevent food from backing into the mouth or nasal cavities. Entry into the trachea (the airway to the lungs) is blocked by closure of the *glottis* (the space between the vocal cords) and an upward movement of the *larynx* (the voice box) at the top of the trachea, which closes against a flaplike valve, the **epiglottis.** You can feel the larynx and the front of the epiglottis bob upward if you place your hand on your throat while you swallow. If these blocking mechanisms fail, touch receptors in the nasal passages and larynx trigger coughing and sneezing reflexes that clear these passages.

The Stomach Stores Food and Continues Digestion

The stomach is a muscular, elastic sac that stores food and adds secretions that further the process of digestion. The mucosal layer of the stomach is an epithelium covered with tiny *gastric pits* that are entrances to millions of *gastric glands.* These glands extend deep into the stomach wall and contain cells that secrete some of the products needed to digest food.

The entry of food into the stomach activates stretch receptors in its wall. Signals from the stretch receptors stimulate the secretion of **gastric juice (Figure 45.10),** which contains **pepsinogen,** the precursor for the digestive enzyme pepsin, hydrochloric acid (HCl), and lubricating mucus. The stomach secretes about 2 L of gastric juice each day.

Pepsinogen is secreted by *chief cells* in the gastric pits. It is an inactive precursor molecule that is converted to the digestive enzyme **pepsin** by the highly acid conditions of the stomach. Once produced, pepsin itself can catalyze the reaction that converts more pepsinogen to pepsin. Pepsin begins the digestion of proteins by introducing breaks in polypeptide chains. The activation of pepsinogen illustrates a common theme in the digestive system: powerful hydrolytic enzymes that would be dangerous to the cells secreting them are synthesized in the form of inactive precursors and are not converted into active form until they are exposed to the digestive contents.

Parietal cells secrete H^+ and Cl^-, which combine to form HCl in the lumen of the stomach. The HCl lowers the pH of the digestive contents to pH 2 or lower, the level at which pepsin reaches optimal activity. To put this pH in perspective, lemon juice is pH 2.4, and sulfuric acid or battery acid is approximately pH 1. The acidity of the stomach also helps break up food particles and causes proteins in the digestive contents to unfold, exposing their peptide linkages to hydrolysis by pepsin. The acid also kills most of the bacteria that reach the stomach and stops the action of salivary amylase.

A thick coating of alkaline mucus, secreted by *mucous cells,* protects the stomach's mucosal layer from attack by pepsin and HCl. Behind the mucous barrier, tight junctions between cells prevent gastric juice from seeping into the stomach wall. Even so, some break-

Figure 45.10
Cells that secrete mucus, pepsin, and HCl in the stomach lining.

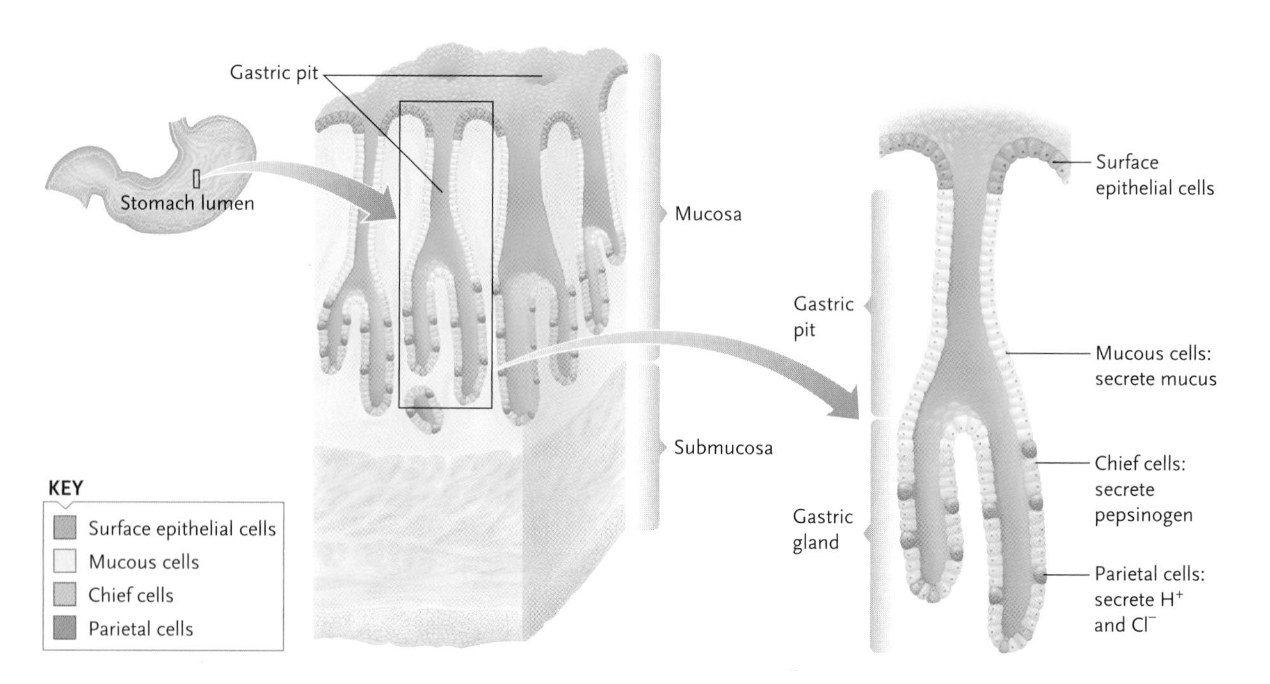

A nephron in a human kidney (colorized SEM). Nephrons are the specialized tubules in kidneys that filter the blood to conserve nutrients and water, balance salts in the body, and concentrate wastes for excretion from the body.

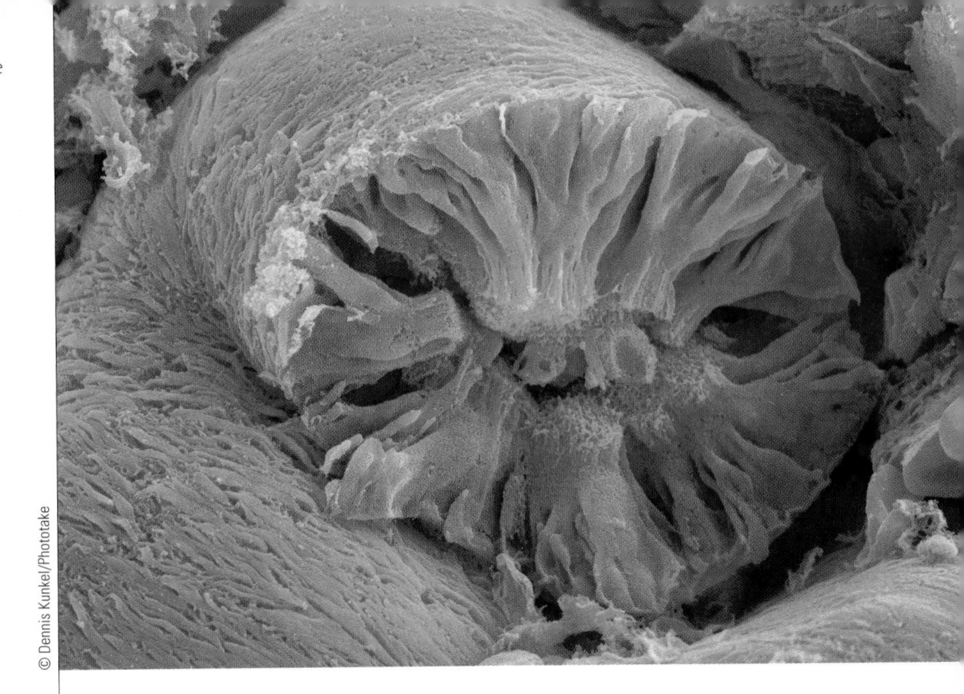

Study Plan

Continued on next page

46 Regulating the Internal Environment

Why It Matters

The crew of the World War II bomber *Lady Be Good* was assigned to fly a night mission to Naples, Italy, from a base on the North African coast on August 4, 1943. But trouble dogged the mission, forcing the crew to turn back before reaching their target. Navigational errors and a cloud layer led them to miss their home base and continue south over the hostile Sahara Desert. Some 440 miles from the coastline, with the fuel running out, the nine crew members parachuted from the aircraft. The bomber remained airborne for a few more minutes and then crashed, leaving its crew miles behind.

The eight men who survived began a northward trek with only half a canteen of water among them, in desert heat that reached 130°F during the day. In a testimony to the physiological mechanisms that conserve water and cool the body, they continued onward for eight days. But then, one by one, they succumbed as the merciless heat and dehydration exceeded their capacity to survive. Rescue teams searched the desert for weeks after their disappearance, but no trace was found of the crew or their airplane.

The fate of the *Lady Be Good* remained unknown until 1958, when an oil exploration team flying over the desert spotted the aircraft, sit-

Continued from previous page

ting largely intact in the desert sands. A 2-year search finally led to the remains of the crew, some of them more than a hundred miles north of the downed bomber. Diaries found among the scattered effects told the poignant story of the flight and the futile struggle against the dehydrating desert environment.

This story illustrates only too clearly the trials of animal life under changing environmental conditions. Water and required nutrients may become more or less abundant. Temperatures may rise or fall. Animals have evolved an astounding capacity to compensate for fluctuating external conditions and to maintain the internal environment of their bodies within the relatively narrow limits that cells can tolerate.

These limits, and the compensating mechanisms that maintain them, are the subjects of this chapter. First we examine **osmoregulation**, the regulation of water and ion balance, and the closely related topic of **excretion**, which helps maintain the body's water and ion balance while ridding the body of metabolic wastes. We then consider **thermoregulation**, the control of body temperature.

46.1 Introduction to Osmoregulation and Excretion

Living cells contain water, are surrounded by water, and constantly exchange water with their environment. For the simplest animals, the water of the external environment directly surrounds cells. For more complex animals, an aqueous extracellular fluid surrounds the cells, and is separated from the external environment by a body covering. In animals with a circulatory system, the extracellular fluid includes both the interstitial fluid immediately surrounding cells and the blood or other circulated fluid; these are commonly called body fluids.

In this section, we review the mechanisms cells use to exchange water and solutes with the surrounding fluid through *osmosis*. We also look at how animals harness osmosis to maintain *water balance,* the equilibrium in inward and outward flow of water.

Osmosis Is a Form of Passive Diffusion

In osmosis (see Section 6.3), water molecules move across a selectively permeable membrane from a region where they are more highly concentrated to a region where they are less highly concentrated. The difference in water concentration is produced by differing numbers of solute molecules or ions on the two sides of the membrane. The side of the membrane with a *lower* solute concentration has a *higher* concentration of water molecules, so water will move osmotically to the other side, where water concentration is *lower.*

Selective permeability is a key factor in osmosis because it helps maintain differences in solute concentration on either side of biological membranes. Proteins are among the most important solutes in establishing the conditions that produce osmosis.

The total solute concentration of a solution, called its **osmolarity,** is measured in *osmoles*—the number of solute molecules and ions (in moles)—per liter of solution. Because the total solute concentration in the body fluids of most animals is less than 1 osmole, osmolarity is usually expressed in thousandths of an osmole, or *milliosmoles* (mOsm). As shown in **Figure 46.1,** the osmolarity of body fluids in humans and other mammals is about 300 mOsm/L; osmolarity in a flounder, a marine teleost (bony fish), is about 330 mOsm/L, and in a goldfish, a freshwater teleost, it is about 290 mOsm/L. By contrast, sharks and many marine invertebrates such as lobsters have osmolarities close to that of seawater, about 1000 mOsm/L, and freshwater invertebrates have an osmolarity of about 225 mOsm/L.

Considering solutions on either side of a selectively permeable membrane, a solution of higher osmolarity is said to be *hyperosmotic* to a solution of lower osmolarity, and a solution of lower osmolarity is said to be *hypoosmotic* to a solution of higher osmolarity. If the solutions on either side of a membrane have the same osmolarity, they are said to be *isoosmotic.* Water moves across the membrane between solutions that differ in osmolarity (see Figure 6.9), whereas when two solutions are isoosmotic, no net water movement occurs.

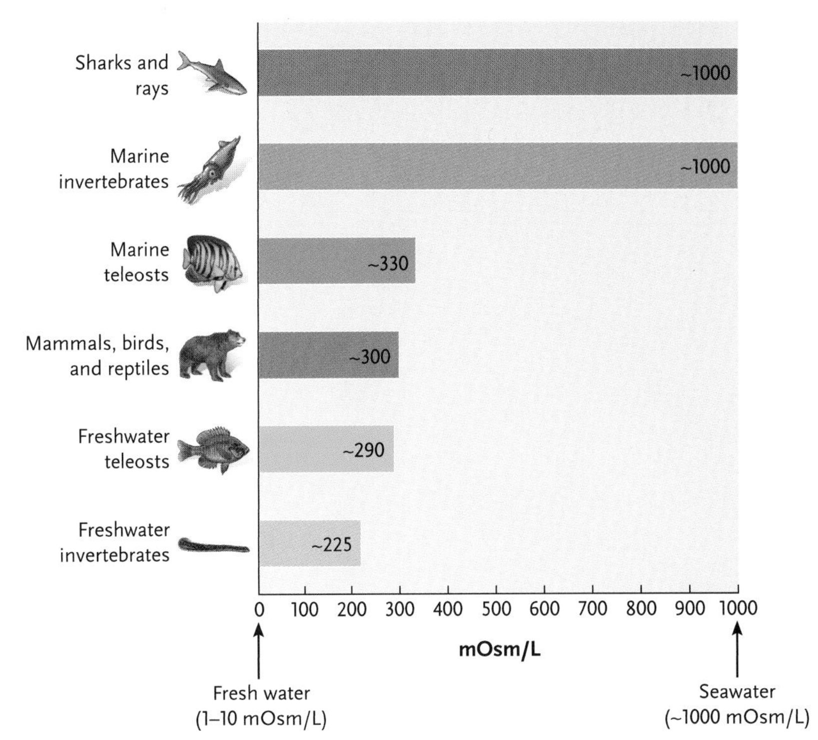

Figure 46.1 Osmolarity of body fluids in some animal groups.

Animals Use Different Approaches to Keep Osmosis from Swelling or Shrinking Their Cells

Because even small differences in osmolarity can cause cells to swell or shrink, animals must keep their cellular and extracellular fluids isoosmotic. In some animals, called **osmoconformers,** the osmolarity of the cellular and extracellular solutions simply matches the osmolarity of the environment. Most marine invertebrates are osmoconformers. Other animals, called **osmoregulators,** use control mechanisms to keep the osmolarity of cellular and extracellular fluids the same, but at levels that may differ from the osmolarity of the surroundings. Most freshwater and terrestrial invertebrates, and almost all vertebrates, are osmoregulators.

For terrestrial animals, one of the greatest challenges to osmoregulation is the limited supply of water in the environment—if the crew of the *Lady Be Good* had had an adequate supply of water, for example, they could probably have reached safety at the North African coast even without food.

Excretion Is Closely Tied to Osmoregulation

Control over osmolarity is partly maintained by removing certain molecules and ions from cells and body fluids and releasing them into the environment; thus, excretion is closely related to osmoregulation. Animals excrete H^+ ions to keep the pH of body fluids near the neutral levels required by cells for survival. They also excrete toxic products of metabolism, such as nitrogenous (nitrogen-containing) compounds resulting from the breakdown of proteins and nucleic acids, and breakdown products of poisons and toxins. Excretion of ions and metabolic products is accompanied by water excretion since water serves as a solvent for those molecules. Animals that take in large amounts of water may also excrete water to maintain osmolarity.

Microscopic Tubules Form the Basis of Excretion in Most Animals

Except in the simplest animals, minute tubular structures carry out osmoregulation and excretion **(Figure 46.2).** The tubules are immersed in body fluids at one end (called the *proximal end* of the tubules), and open directly or indirectly to the body exterior at the other end (called the *distal end* of the tubules). The tubules are formed from a **transport epithelium**—a layer of cells with specialized transport proteins in their plasma membranes. The transport proteins move specific molecules and ions into and out of the tubule by either active or passive transport, depending on the particular substance and its concentration gradient.

Typically, the tubules function in a four-step process:

- **Filtration.** Filtration is the nonselective movement of some water and a number of solutes—ions and small molecules, but not large molecules such as proteins—into the proximal end of the tubules through spaces between cells. In animals with an open circulatory system, the water and solutes come from body fluids, with movement into the tubules driven by the higher pressure of the body fluids compared with the fluid inside the tubule. In animals with a closed circulatory system, such

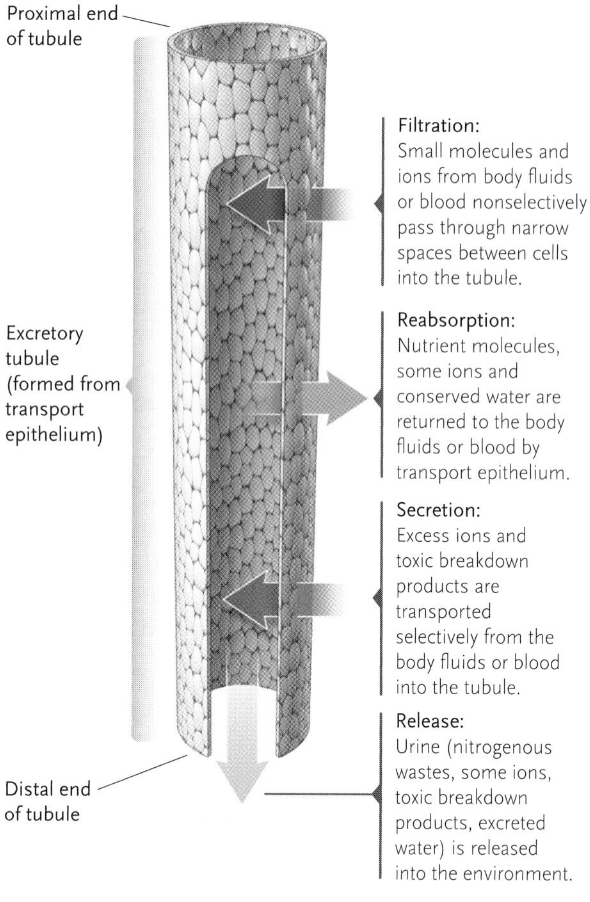

Proximal end of tubule
Excretory tubule (formed from transport epithelium)
Distal end of tubule
Filtration:
Small molecules and ions from body fluids or blood nonselectively pass through narrow spaces between cells into the tubule.
Reabsorption:
Nutrient molecules, some ions and conserved water are returned to the body fluids or blood by transport epithelium.
Secretion:
Excess ions and toxic breakdown products are transported selectively from the body fluids or blood into the tubule.
Release:
Urine (nitrogenous wastes, some ions, toxic breakdown products, excreted water) is released into the environment.

Figure 46.2

Common structures and operations of the tubules carrying out osmoregulation and excretion in animals. The tubules are typically formed from a single layer of cells with transport functions.

as humans, the water and solutes come from the blood in capillaries that surround the tubules, with the movement into the tubules driven by blood pressure. (Open and closed circulatory systems are described in Section 42.1.)

- **Reabsorption.** In reabsorption, some molecules (for example, glucose and amino acids) and ions are transported by the transport epithelium back into the body fluid (animals with open circulatory systems) or into the blood in capillaries surrounding the tubules (animals with closed circulatory systems) as the filtered solution moves through the excretory tubule.
- **Secretion.** Secretion is a selective process in which specific small molecules and ions are transported from the body fluids (animals with open circulatory systems) or blood (animals with closed circulatory systems) into the tubules. Secretion is the second and more important route for eliminating particular substances from the body fluid or blood, filtration being the first. The difference between the two processes is that filtration is nonselective whereas secretion is selective for substances transported. The same substances are transported into the tubule by secretion as in filtration; those substances are added, therefore, to substances already in the tubule as a result of filtration.
- **Release.** The fluid containing waste materials—urine—is released into the environment from the distal end of the tubule. In some animals the fluid is concentrated into a solid or semisolid form.

The tubules may number from hundreds to millions depending on the species. In combination, they expand the transport epithelium to a total surface area large enough to accomplish the osmoregulatory and excretory functions of the animal. In all vertebrates and many invertebrates, the excretory tubules are concentrated in specialized organs, the *kidneys,* which are discussed in later sections.

Animals Excrete Nitrogen Compounds as Metabolic Wastes

The metabolism of ingested food is a source of both energy and molecules for the biosynthetic activities of an animal. Importantly, metabolism of ingested food produces water—called *metabolic water*—that is used in chemical reactions and is involved in physiological processes such as the excretion of wastes.

The proteins, amino acids, and nucleic acids in food are broken down as part of digestion. The same molecules are broken down in body cells as a result of the normal processes of synthesis and replacement. The nitrogenous products of this breakdown are excreted by most animals as *ammonia, urea,* or *uric acid,* or a combination of these substances **(Figure 46.3).** The particular molecule or combination of molecules produced depends on a balance among toxicity, water conservation, and energy requirements.

Ammonia. Ammonia (NH_3) is the result of a series of biochemical steps beginning with the removal of amino groups ($—NH_3^+$) from amino acids as a part of protein breakdown. Ammonia is readily soluble in water, but it is also highly toxic. Therefore, ammonia must either be excreted or be converted to a nontoxic derivative. However, because of its toxicity, ammonia can be excreted from the body only in dilute solutions, making this path possible only in animals with a plentiful supply of water. Those animals include aquatic invertebrates, teleosts, and larval amphibians; ammonia is the primary nitrogenous waste for them. Terrestrial animals, and some aquatic animals, instead detoxify ammonia, converting it either into urea or uric acid.

Urea. All mammals, most amphibians, some reptiles, some marine fishes, and some terrestrial invertebrates combine ammonia with HCO_3^- and convert the product in a series of steps to *urea,* a soluble and relatively nontoxic substance. Although producing urea requires more energy than forming ammonia, excreting urea instead of ammonia requires only about 10% as much water.

Uric Acid. Water is conserved further in some animals, including terrestrial invertebrates, reptiles, and birds, by the formation of uric acid instead of ammonia or urea. Uric acid is nontoxic, and so insoluble that it precipitates in water as a crystal. (The white substance in bird droppings is uric acid.) The embryos of reptiles and birds, which develop within leathery or hard-shelled eggs that are impermeable to liquids, also conserve water by forming uric acid, which is stored as a waste product.

Although making uric acid requires even more energy than urea, molecule for molecule it contains four times as much nitrogen as ammonia. And, because uric acid precipitates from water as a crystal, it can be excreted as a concentrated paste. These factors conserve about 99% of the water that would be required to excrete an equivalent amount of nitrogen as ammonia.

We have now covered the basics of osmoregulation and excretion. In the sections that follow, we look at the specifics of these processes in different animal groups, beginning with the invertebrates.

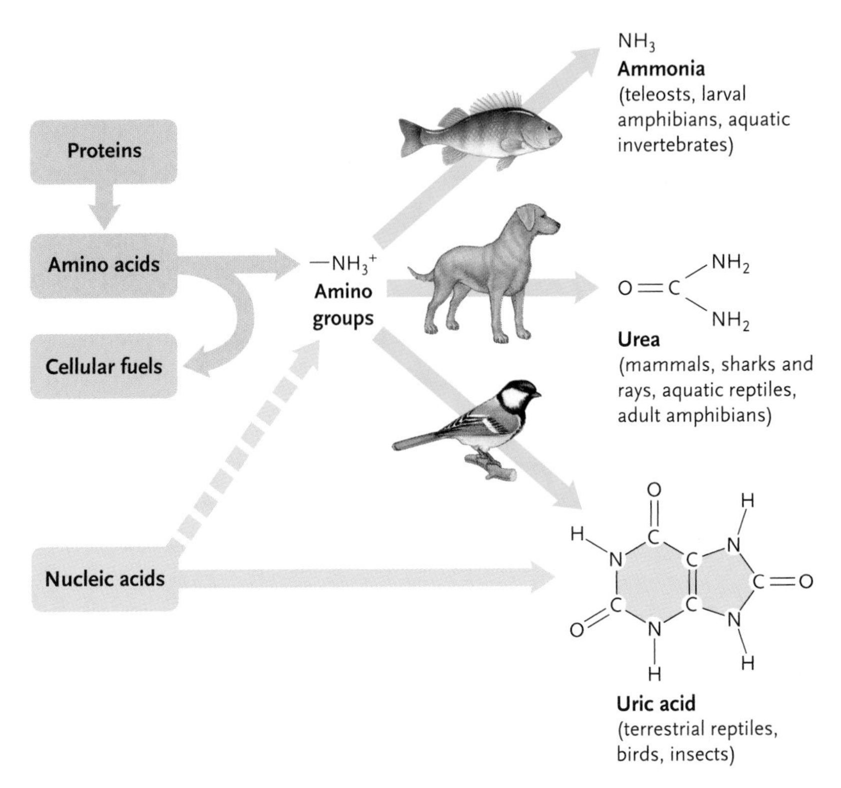

Figure 46.3
Nitrogenous wastes excreted by different animal groups. Although humans and other mammals primarily excrete urea, they also excrete small amounts of ammonia and uric acid.

Study Break

Define the terms osmosis, osmolarity, hypoosmotic, osmoregulator, and transport epithelium.

46.2 Osmoregulation and Excretion in Invertebrates

Both osmoconformers and osmoregulators occur among the invertebrates. Except for the simplest groups, most invertebrates, whether osmoconformers or osmoregulators, carry out excretion by specialized excretory tubules.

Most Marine Invertebrates Are Osmoconformers; All Freshwater and Terrestrial Invertebrates Are Osmoregulators

Most marine invertebrates are osmoconformers. All these animals release water, certain ions, and nitrogenous wastes—usually in the form of ammonia—directly from body cells to the surrounding seawater. The cells of these animals do not swell or shrink because the osmolarity of their intracellular and extracellular fluids and the surrounding seawater is the same, about 1000 mOsm/L. Therefore, they do not have to expend energy to maintain their osmolarity. However, osmoconformers do expend energy to keep some ions, such as Na^+, at lower concentrations inside cells than in the surroundings.

In contrast, all freshwater invertebrates are osmoregulators because their cells could not survive if their internal ion concentrations were reduced to freshwater levels. Terrestrial invertebrates are osmoregulators as well. These animals must expend energy to keep their internal fluids hyperosmotic to their surroundings. Although osmoregulation is energetically expensive, these invertebrates can live in more varied habitats than osmoconformers can.

The internal hyperosmoticity of freshwater osmoregulators such as flatworms and mussels causes water to move constantly from the surroundings into their bodies. This excess water must be excreted, at a considerable cost in energy, to maintain internal hyperosmoticity. These animals must also obtain the salts required to keep their body fluids hyperosmotic to fresh water. The salts are obtained from foods, and by actively transporting salt ions from the water into their bodies (even fresh water contains some dissolved salts). This active ion transport occurs through the skin or gills.

Among terrestrial osmoregulators are annelids (earthworms), arthropods (insects, spiders and mites, millipedes, and centipedes), and mollusks (land snails and slugs). Like their freshwater relatives, these invertebrates must obtain salts from their surroundings, usually in their foods. While they do not have to excrete water entering by osmosis, they must constantly replace water lost from their bodies by evaporation.

In Invertebrate Osmoregulators, Specialized Excretory Tubules Participate in Osmoregulation and Carry Out Excretion

Invertebrate osmoregulators typically use specialized tubules for carrying out excretion. Three common types of these specialized tubules that differ in which body fluids are processed and how are *protonephridia,* found in flatworms and larval mollusks; *metanephridia,* found in annelids and most adult mollusks; and *Malpighian tubules,* found in insects and other arthropods.

Protonephridia. The flatworm *Dugesia* provides an example of the simplest form of invertebrate excretory tubule, the **protonephridium** (*protos* = first; *nephros* = kidney). In *Dugesia,* two branching networks of protonephridia run the length of the body **(Figure 46.4).** The smallest branches of the tubule network end with a large cell containing a bundle of cilia that reach into the tubule and that beat to move fluid through the tubule. This cell is called a *flame cell* because the movement of its cilia resembles a flickering flame. The plasma membrane of the flame cell interdigitates with the plasma membrane of the tubular cell with which it connects. Hemolymph enters the tubule through the membranes in the area where the two membranes interdigitate. As the fluids pass through the protonephridia, some molecules and ions are reabsorbed and others, including nitrogenous wastes, are secreted into the tubules; the urine resulting from this filtration system is released through pores at the ends of the tubules where the tubules reach the body surface.

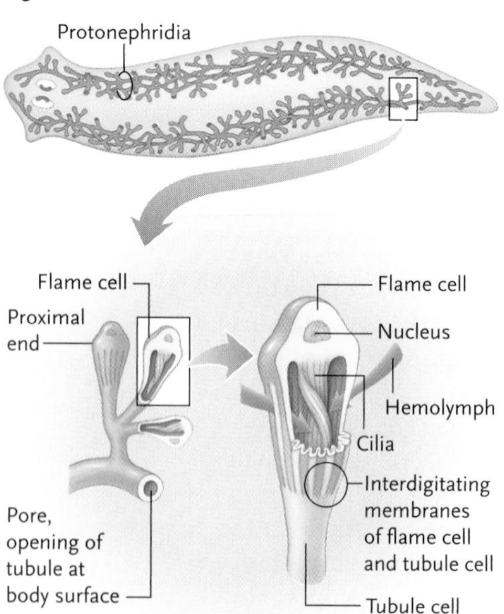

Figure 46.4
The protonephridia of the planarian *Dugesia,* showing a flame cell.

Metanephridia. The excretory tubule of most annelids and adult mollusks, the **metanephridium** (*meta* = between), has a funnel-like proximal end surrounded with cilia that admits hemolymph. Like protonephridia, metanephridia are filtration systems. As hemolymph moves through the tubule, some molecules and ions are reabsorbed, and other ions and nitrogenous wastes are secreted into the tubule and excreted from the body surface.

Figure 46.5 shows the arrangement of metanephridia in an earthworm. The proximal ends of a pair of metanephridia are located in each body segment, one on either side of the animal. Each tubule of the pair extends into the following segment, where it bends and folds into a convoluted arrangement surrounded by a network of blood vessels. Reabsorption and secretion take place in the convoluted section. Urine from the distal end of the tubule collects in a saclike storage organ, the *bladder,* from where it is released through a pore in the surface of the segment. Samples taken with a microneedle from various regions of a metanephridium show that the fluid entering the tubule contains all the smaller molecules and ions of the body fluid; as the fluid moves through the tubules, specific molecules and ions are removed by reabsorption and added by secretion.

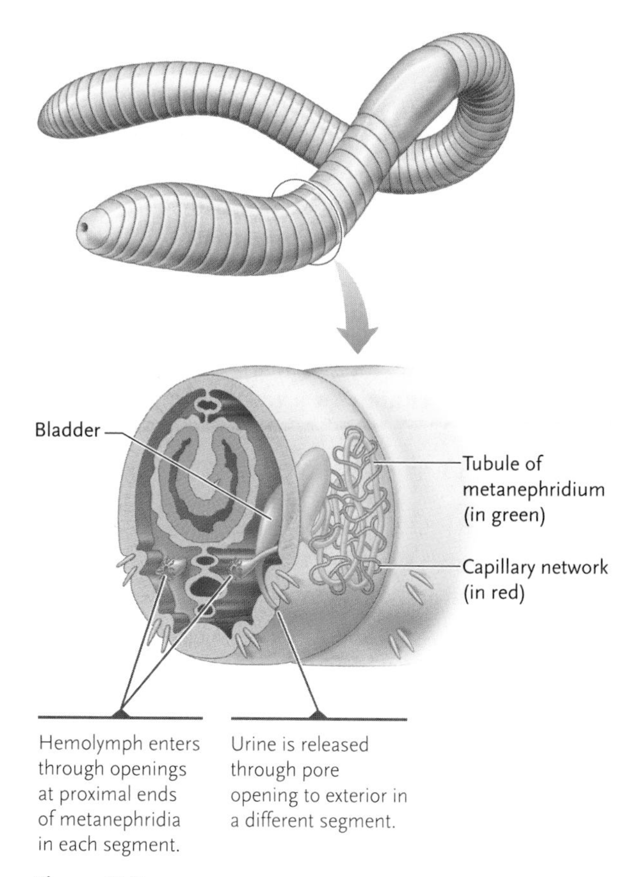

Figure 46.5
The metanephridium of an earthworm.

Malpighian Tubules. The excretory tubule of insects, the **Malpighian tubule,** has a closed proximal end that is immersed in the hemolymph **(Figure 46.6).** The distal ends of the tubules empty into the gut. In contrast to protonephridia and metanephridia, Malpighian tubules do not filter body fluids; instead, they are excretory systems that use secretion to generate the fluid for release from the body. In particular, uric acid and several ions, including Na^+ and K^+, are actively secreted into the tubules. As the concentration of these substances rises, water moves osmotically from the hemolymph into the tubule. The fluid then passes into the hindgut (intestine and rectum) of the insect as dilute urine. Cells in the hindgut wall actively reabsorb most of the Na^+ and K^+ back into the hemolymph; water follows by osmosis. The uric acid left in the gut precipitates into crystals, which mix with the undigested matter in the rectum and are released with the feces.

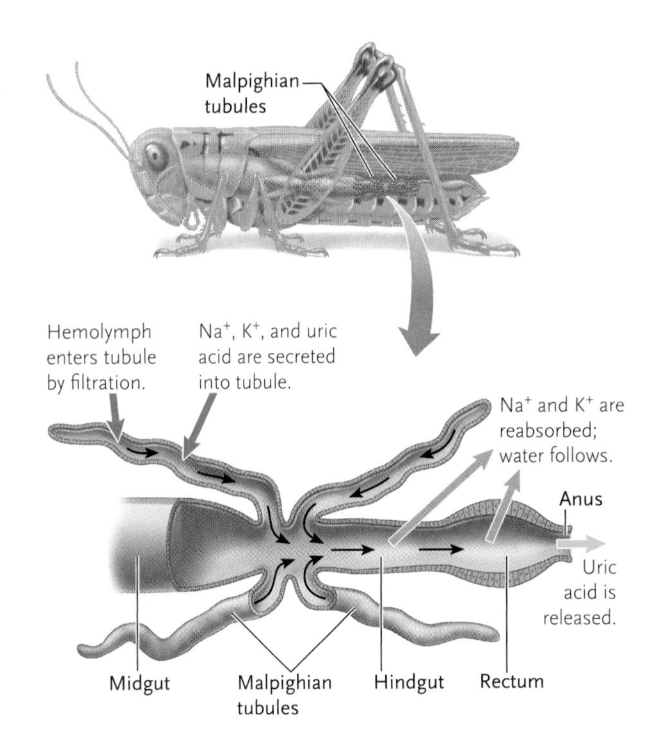

Figure 46.6 Excretion through Malpighian tubules in a grasshopper.

Study Break

Describe protonephridia, metanephridia, and Malpighian tubules. In which animal groups are each of these excretory tubules found?

46.3 Osmoregulation and Excretion in Mammals

In all vertebrates, specialized excretory tubules contribute to osmoregulation and carry out excretion. The excretory tubules, called **nephrons,** are located in a specialized organ, the kidney. We begin our survey of vertebrate osmoregulation and excretion with a description of the structure and function of the mammalian kidney.

The Kidneys and Ureters, the Bladder, and the Urethra Constitute the Urinary System

Mammals have a pair of kidneys, located on either side of the vertebral column at the back of the abdominal cavity **(Figure 46.7).** Internally, the mammalian kidney is divided into an outer **renal cortex** surrounding a central region, the **renal medulla.**

Body fluids are carried in blood through the **renal artery** to a kidney, where metabolic wastes and excess ions are moved into the nephrons and where urine is formed. The filtered blood is routed away from the kidney by the **renal vein.** The urine leaving individual nephrons is processed further in **collecting ducts** and then drains into a central cavity in the kidney called the **renal pelvis.**

From the renal pelvis, the urine flows through a tube called the **ureter** to the **urinary bladder,** a storage sac located outside the kidneys. Urine leaves the bladder through another tube, the **urethra,** which (in most mammals) opens to the outside. In human females, the opening of the urethra is just in front of the vagina; in males, the urethra opens at the tip of the penis. The two kidneys and ureters, the urinary bladder, and the urethra constitute the mammalian urinary system.

Two sphincter muscles control the flow of urine from the bladder to the urethra. In human infants, urination is an autonomic reflex triggered by stretch receptors in the bladder wall. When the bladder becomes full, the sphincters relax, smooth muscles in the bladder wall contract, and the urine is forced to the exterior. At about the age of two years, children learn to override the autonomic reflex by consciously keeping the striated sphincter contracted until urination is convenient.

Mammalian Nephrons Are Differentiated into Regions with Specialized Functions

As in all mammals, human nephrons are differentiated into regions that perform successive steps in excretion. At its proximal end, a human nephron forms the **Bowman's capsule,** an infolded region that cups around a ball of blood capillaries called the **glomerulus (Figure 46.8).** The capsule and glomerulus are located in the renal cortex. Filtration takes place as body fluids are forced into Bowman's capsule from the capillaries of the glomerulus.

Following Bowman's capsule, the nephron forms a **proximal convoluted tubule** in the renal cortex, which descends into the renal medulla in a U-shaped bend called the **loop of Henle** and then ascends again to form a **distal convoluted tubule.** The distal tubule drains the urine into a collecting duct that leads to the renal pelvis. As many as eight nephrons may drain into a single collecting duct. The combined activities of the proximal convoluted tubule, the loop of Henle, the distal convoluted tubule, and the collecting duct convert the filtrate entering the nephron into urine.

Unlike most capillaries in the body, the capillaries in the glomerulus do not lead directly to venules. In-

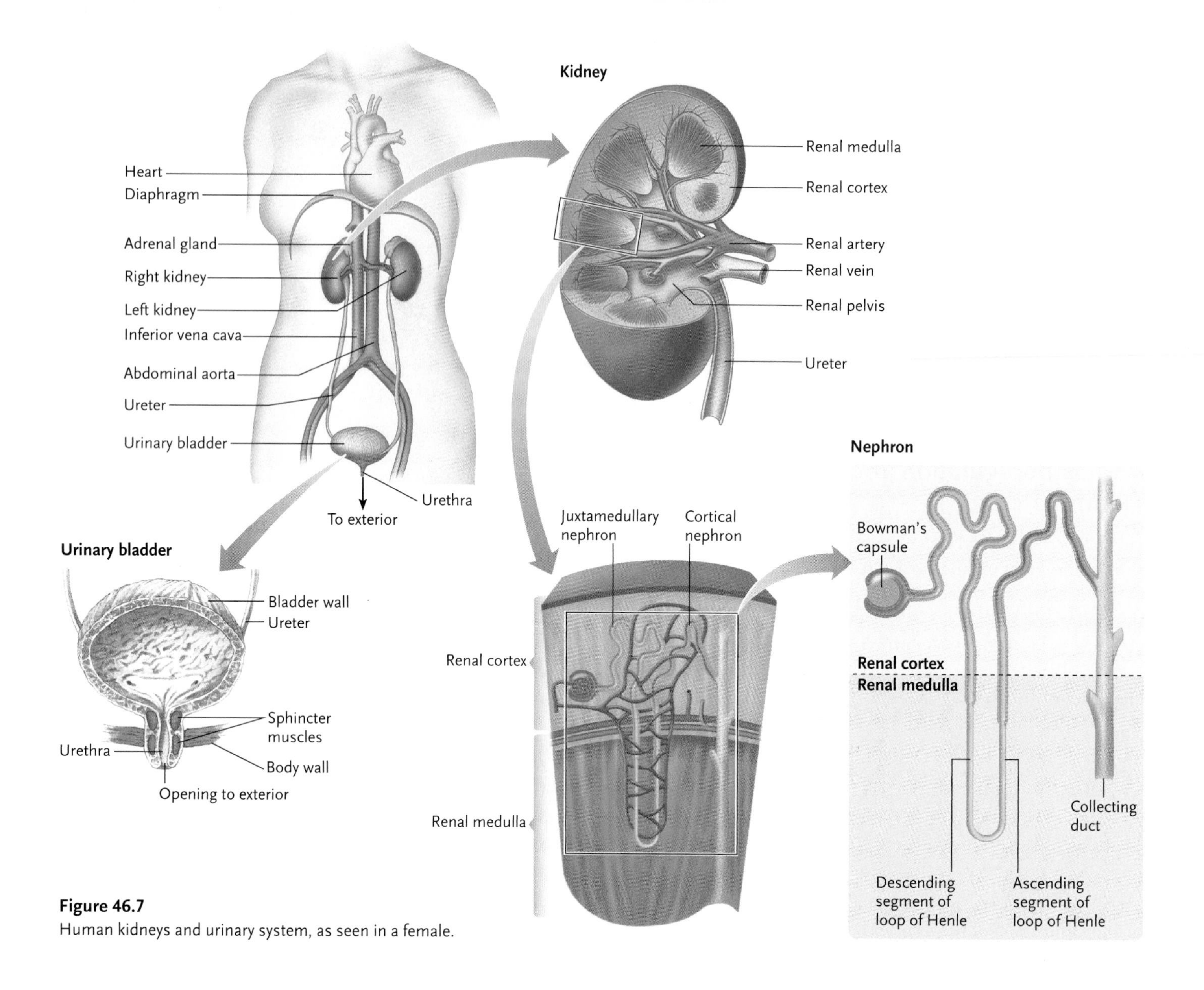

Figure 46.7
Human kidneys and urinary system, as seen in a female.

stead, they form another arteriole that branches into a second capillary network called the **peritubular capillaries.** These capillaries thread around the proximal and distal convoluted tubules and the loop of Henle. Molecules and ions that were reabsorbed during excretion are transferred between the nephron and the peritubular capillaries. However, because the capillaries and the tubules are not in physical contact due to the interstitial fluid between them, this transfer is not direct. Instead, the molecules or ions pass through the wall of the tubule, which is one cell layer thick; diffuse through the interstitial fluid; and then pass into the capillary through its wall, also one cell thick.

Each human kidney has more than a million nephrons. Of these, about 20% (the *juxtamedullary nephrons*) have long loops that descend deeply into the medulla of the kidney. The remaining 80% (the *cortical nephrons*) have shorter loops, most of which are located entirely in the cortex, and the remainder of which extend only partway into the medulla.

Mammalian Nephrons Interact with Surrounding Kidney Structures to Produce Hyperosmotic Urine

In mammals, urine is hyperosmotic to body fluids. All other vertebrates except for a few aquatic bird species produce urine that is hypoosmotic to body fluids, or is at best isoosmotic. Production of hyperosmotic urine is a water-conserving adaptation that is primarily a mammalian characteristic. The production of hyperosmotic urine involves the activities of the mammalian

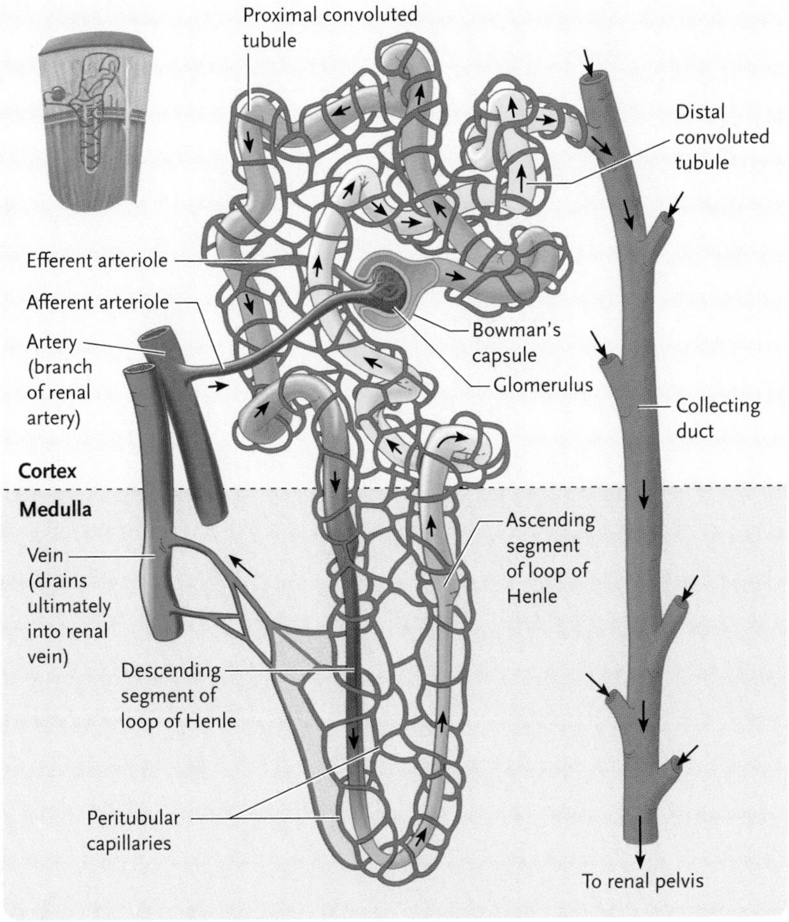

Figure 46.8
A nephron and its blood circulation.

nephron itself and an interaction between nephrons and the highly ordered structure of the mammalian kidney. Three features underlie this interaction:

- The arrangement of the loop of Henle, which descends through the medulla and returns to the cortex again.
- Differences in the permeability of successive regions of the nephron, established by a specific group of membrane transport proteins in each region.
- A gradient in the concentration of molecules and ions in the interstitial fluid of the kidney, which increases gradually from the renal cortex to the deepest levels of the renal medulla.

These features interact to conserve nutrients and water, balance salts, and concentrate wastes for excretion from the body.

Researchers determined the transport activities of specific regions of nephrons by dissecting segments of nephrons out of an animal and experimentally manipulating them in vitro. They placed segments in different buffered solutions and passed solutions containing various components of filtrates through the segment. By labeling specific molecules or ions radioactively, the scientists followed the movements of molecules in the solution surrounding the nephron segment or in the filtrate.

Filtration in Bowman's Capsule Begins the Process of Excretion

The mechanisms of excretion (shown in **Figure 46.9** and summarized in **Table 46.1**) begin in Bowman's capsule. The endothelial cells of the glomerulus capillaries and the cells of the Bowman's capsule are separated by spaces just wide enough to admit water, ions, small nutrient molecules such as glucose and amino acids, and nitrogenous waste molecules, primarily urea. The higher pressure of the blood drives fluid containing these molecules and ions from the capillaries of the glomerulus into the capsule. A thin net of connective tissue between the capillary and Bowman's capsule epithelia contributes to the filtering process. Blood cells and plasma proteins are too large to pass and are retained inside the capillaries.

Two factors help maintain the pressure driving fluid into Bowman's capsule. First, the diameters of

Figure 46.9
The movement of ions, water, and other molecules to and from nephrons and collecting tubules in the human kidney. Nephrons in other mammals and in birds work in similar fashion. The numbers are osmolarity values in mOsm/L.

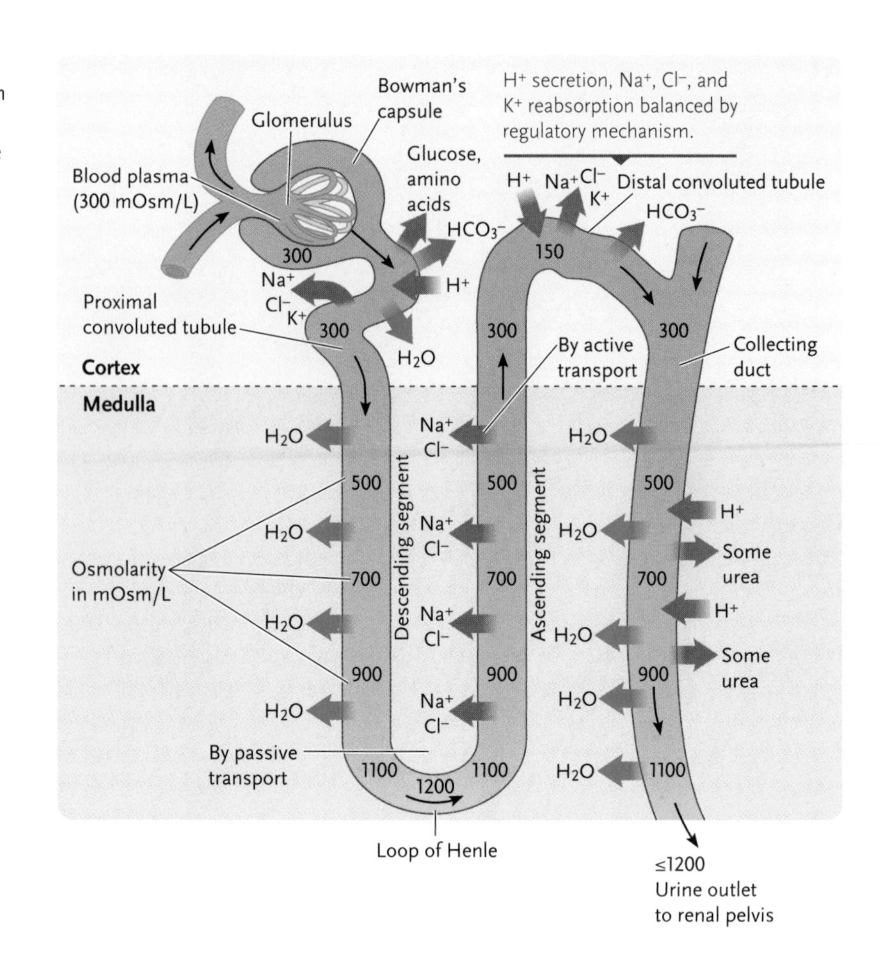

Table 46.1	Filtration, Reabsorption, and Secretion in Nephrons and Collecting Ducts			
Segment	**Location**	**Permeability and Movement**	**Osmolarity of Filtrate and Urine**	**Result of Passage**
Bowman's capsule	Cortex	Water, ions, small nutrients, and nitrogenous wastes move through spaces between epithelia	300 mOsm/L, same as surrounding interstitial fluid	Water and small substances, but not proteins, pass into nephron
Proximal convoluted tubule	Cortex	Na^+ and K^+ actively reabsorbed, Cl^- follows; water leaves through aquaporins; H^+ actively secreted; HCO_3^- reabsorbed into plasma of peritubular capillaries; glucose, amino acids, and other nutrients actively reabsorbed	300 mOsm/L	67% of ions, 65% of water, 50% of urea, and all nutrients return to interstitial fluid; pH maintained
Descending segment of loop of Henle	Cortex into medulla	Water leaves through aquaporins; no movement of ions or urea	From 300 mOsm/L at top to 1200 mOsm/L at bottom of loop	Additional water returned to interstitial fluid
Ascending segment of loop of Henle	Medulla into cortex	Na^+ and Cl^- actively transported out; no entry of water; no movement of urea	From 1200 mOsm/L at bottom to 150 mOsm/L at top of loop	Additional ions returned to interstitial fluid
Distal convoluted tubule	Cortex	K^+ and Na^+ secreted via active transport into urine; Na^+ and Cl^- reabsorbed; water moves into urine through aquaporins; HCO_3^- reabsorbed into plasma of peritubular capillaries	From 150 mOsm/L at beginning to 300 mOsm/L at junction with collecting duct	Ion balance, pH balance
Collecting ducts	Cortex through medulla, empties into renal pelvis	Water moves out via aquaporins; no movement of ions; some urea leaves at bottom of duct	From 300 mOsm/L to 1200 mOsm/L at junction with renal pelvis	More water and some urea returned to interstitial fluid; some H^+ added to urine

the arteriole delivering blood to the glomerulus (called the **afferent arteriole**) and the capillaries of the glomerulus itself are larger than that of arterioles and capillaries elsewhere in the body. The larger diameter maintains blood pressure by presenting less resistance to blood flow. Second, the diameter of the arteriole that receives blood from the glomerulus (called the **efferent arteriole**) is smaller than the diameter of the afferent arteriole, producing a damming effect that backs up the blood in the glomerulus and helps keep the pressure high.

In humans, Bowman's capsules collectively filter about 180 L (47.5 gallons) of fluid each day, from a daily total of 1400 L (369.5 gallons) of blood that pass through the kidneys. The human body contains only about 2.75 L of blood plasma, meaning that the kidneys filter a fluid volume equivalent to 65 times the volume of the blood plasma each day. On average, more than 99% of the filtrate, mostly water, is reabsorbed in the nephrons, leaving about 1.5 L to be excreted daily as urine.

Reabsorption and Secretion Take Place in the Remainder of the Nephron

The fluid filtered into Bowman's capsule contains water, other small molecules, and ions at essentially the same concentrations as the blood plasma. By the time the fluid reaches the distal end of the tubules and passes through the collecting ducts, reabsorption out of the tubules and secretion into them have markedly altered the concentrations of all components of the filtrate.

The Proximal Convoluted Tubule. Reabsorption of water, ions, and nutrients back into the interstitial fluid is the main function of the proximal convoluted tubule. Na^+/K^+ pumps in the epithelium of the proximal convoluted tubule move Na^+ and K^+ from the filtrate into the interstitial fluid surrounding the tubule (see Figure 46.9). The movement of positive charges sets up a voltage gradient that causes Cl^- ions to be reabsorbed from within the tubule with the positive ions. Specific active transport proteins reabsorb essentially all the glucose, amino acids, and other nutrient molecules from the filtrate into the interstitial fluid, making the filtrate hypoosmotic to the interstitial fluid surrounding the tubule. As a result, water moves from the tubule into the interstitial fluid by osmosis. The osmotic movement is aided by *aquaporins,* transport proteins that form passages for water molecules in the transport epithelium of the tubule cells. The nutrients and water that entered the interstitial fluid move into the capillaries of the peritubular network.

Some substances are also secreted into the tubule, however: primarily H^+ ions by active transport and the products of detoxified poisons by passive secretion (detoxification takes place in the liver). The secretion of H^+ ions into the filtrate helps balance the acidity constantly generated in the body by metabolic reactions. H^+ secretion is coupled with HCO_3^- reabsorption from the filtrate in the tubule to the plasma in the peritubular capillaries. Small amounts of ammonia are also secreted into the tubule.

In all, the proximal convoluted tubule reabsorbs about 67% of the Na^+, K^+, and Cl^- ions, 65% of the water, 50% of the urea, and essentially all the glucose, amino acids, and other nutrient molecules in the filtrate. The ions, nutrients, and water reabsorbed by the tubule are transported into the interstitial fluid, and then into capillaries of the peritubular network. Although half of the urea is reabsorbed, the constant flow of filtrate through the tubules keeps the concentration of nitrogenous wastes low in body fluids.

The proximal convoluted tubule has structural specializations that fit its function. The epithelial cells that make up its walls are carpeted on their inner surface by a brush border of microvilli. Like the brush border of epithelial cells in the small intestine (see Section 45.3), these microvilli greatly increase the surface area available for reabsorption and secretion.

The Descending Segment of the Loop of Henle. The filtrate leaving the proximal convoluted tubule enters the descending segment of the loop of Henle, where water is reabsorbed. As this tubule segment descends, it passes through regions of increasingly higher solute concentrations in the interstitial fluid of the medulla (see Figure 46.9). (The generation of this concentration gradient is described later.) As a result, more water moves out of the tubule by osmosis as the fluid travels through the descending segment.

The descending segment has aquaporins, which allow the rapid transport of water. The outward movement of water concentrates the molecules and ions inside the tubule, gradually increasing the osmolarity of the fluid to a peak of about 1200 mOsm/L at the bottom of the loop. This is the same as the osmolarity of the interstitial fluid at the bottom of the medulla.

The Ascending Segment of the Loop of Henle. The fluid then moves into the ascending segment of the loop of Henle, where Na^+ and Cl^- are reabsorbed into the interstitial fluid. As this segment ascends, it passes through regions of gradually lessening osmolarity in the interstitial fluid of the medulla. The ascending segment has membrane proteins that transport salt ions, but no aquaporins. Because water is trapped in the ascending segment, the osmolarity of the urine is reduced as salt ions, primarily Na^+ and Cl^-, move out of the tubule.

In the part of the ascending segment immediately following the loop, the ion concentrations in the tubule filtrate are still high enough to move Na^+ and Cl^- out of the tubule by passive transport. Toward the top of the segment, they are moved out by active transport.

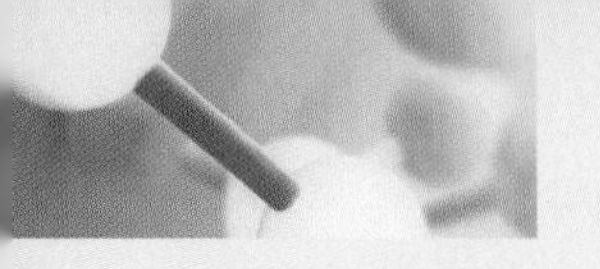

INSIGHTS FROM THE MOLECULAR REVOLUTION

An Ore Spells Relief for Osmotic Stress

Almost all cells respond to osmotic stress—osmotic imbalance with the surroundings—by adjusting the cytoplasmic concentration of small organic molecules called *osmolytes.* When cells are surrounded by a hyperosmotic solution, for example, osmolytes accumulate in the cytoplasm, raising its osmolarity to match that of the surroundings. The almost universal occurrence of osmolytes means that they must have appeared very early in the evolution of life.

In humans, cells in the renal medulla are regularly exposed to high solute concentrations in the interstitial fluid. These cells would quickly die from osmotic water loss if they were not protected by osmolytes. In these cells, as well as in many other types of mammalian cells, one of the primary osmolytes is *sorbitol,* made from glucose in a reaction catalyzed by the enzyme *aldose reductase.* In some unknown way, placing cells in a hyperosmotic medium activates the gene encoding and synthesizing aldose reductase in the cytoplasm.

Joan D. Ferraris and her colleagues at the National Institutes of Health in Bethesda, Maryland, were interested in the molecular steps leading to activation of the aldose reductase gene. To begin their research, the investigators extracted DNA from cells in the renal medulla of a rabbit, and cloned the DNA to increase its quantity. They then probed the DNA with a radioactive DNA segment that could pair with the DNA of aldose reductase genes previously isolated from humans and rats. The probe marked the rabbit version of the gene with radioactivity so that it could be separated from the sample.

The researchers were particularly interested in the promoter sequences controlling the gene, which might contain a region activating transcription in cells exposed to a hypertonic medium. To identify the region, they constructed a composite gene from a segment of the separated DNA containing the promoter and surrounding sequences attached to the coding portion of a gene for luciferase (the firefly enzyme, which catalyzes a cytoplasmic reaction emitting light). The composite gene was then increased in quantity by the polymerase chain reaction.

The composite gene was introduced into cultured renal medulla cells from the rabbit, which were then divided into two groups. One group was maintained in an isotonic medium; the other one was exposed to a medium made hypertonic by added NaCl. The cells exposed to the hypertonic solution glowed with light, showing that the luciferase gene had been turned on by having some part of the promoter segment derived from the aldose reductase gene.

The next step was to isolate the particular control sequence. To accomplish this, the researchers broke the promoter segment into fragments, attached them one at a time to the luciferase gene, and tested them by the same experimental procedure. Eventually, they identified the smallest fragment capable of activating the luciferase gene in cells placed in hypertonic medium; it proved to be the sequence CGGAAAATCAC, beginning 1105 base pairs in advance of the site where transcription begins. The investigators termed the activating sequence an *osmotic response element (ORE).*

Presumably, placing the renal medulla cells under osmotic stress triggers a series of reactions that culminates in synthesis or activation of a nuclear regulatory protein that binds the ORE, leading to activation of the aldose reductase gene. The next step in the investigation is to find the nuclear regulatory protein, and then work backwards one step at a time until the entire series of reactions leading from the first cell receptor to activation of the gene is traced out. If successful, the research will reveal an evolutionarily ancient mechanism that is critical to the survival of virtually all living cells.

Besides reducing the osmolarity of the filtrate in the ascending segment, the reabsorption of salt ions from the tubule into the interstitial fluid helps establish the concentration gradient of the medulla, high near the renal pelvis and low near the renal cortex. The energy required to transport NaCl from higher levels of the ascending segment makes the kidneys one of the major ATP-consuming organs of the body.

By the time the fluid reaches the cortex at the top of the ascending loop, its osmolarity has dropped to about 150 mOsm/L. During the travel of fluid around the entire loop of Henle, water, nutrients, and ions have been conserved and returned to body fluids, and the total volume of the filtrate in the nephron has been greatly reduced. Urea and other nitrogenous wastes have been concentrated in the filtrate. Little secretion occurs in either the descending or ascending segments of the loop of Henle.

The Distal Convoluted Tubule. The transport epithelium of the distal convoluted tubule removes additional water from the fluid in the tubule, and works to balance the salt and bicarbonate concentrations of the tubule fluid against body fluids. In response to hormones triggered by changes in the body's salt concentrations (described in Section 46.4), varying amounts of K^+ and H^+ ions are secreted into the fluid, and varying amounts of Na^+ and Cl^- ions are reabsorbed. Bicarbonate ions are reabsorbed from the filtrate as in the proximal tubule.

In total, more ions move outward than inward and, as a consequence, water moves out of the tubule by osmosis, through aquaporins in the distal tubule. The amounts of urea and other nitrogenous wastes remain the same. By the time the fluid—now urine—enters the collecting ducts at the end of the nephron, its osmolarity is about 300 mOsm/L.

The Collecting Ducts. The collecting ducts concentrate the urine. These ducts, which are permeable to water but not salt ions, descend downward from the cortex through the medulla of the kidney. As the ducts descend, they encounter the gradient of increasing solute concentration in the medulla. This increase makes water move osmotically out of the ducts and greatly increases the concentration of the urine, which can become as high as 1200 mOsm/L at the bottom of the medulla. Near the bottom of the medulla, the walls of the collecting ducts contain passive urea transporters that allow a portion of this nitrogenous waste to pass from the duct into the interstitial fluid. This urea adds significantly to the concentration gradient of solutes in the medulla.

In addition to these mechanisms, H^+ ions are actively secreted into the fluid by the same mechanism as in the proximal and distal convoluted tubules. The balance of the H^+ and bicarbonate ions established in the urine, interstitial fluid, and blood, achieved by secretion of H^+ into the urine by the nephrons and collecting ducts, is important for regulating the pH of blood and body fluids. The kidneys thus provide a safety valve if the acidity of body fluids rises beyond levels that can be controlled by the blood's buffer system (see Section 44.4).

At its maximum value of 1200 mOsm/L, reached when water conservation is at its maximum, the urine at the bottom of the collecting ducts is about four times more concentrated than body fluids. It can also be as low as 50 to 70 mOsm/L, when very dilute urine is produced in response to conditions such as excessive water intake.

The high osmolarity of the interstitial fluid toward the bottom of the medulla would damage the medulla cells if they were not protected against osmotic water loss. The protection comes from high concentrations of otherwise inert organic molecules called *osmolytes* in these cells. The osmolytes, primarily a sugar alcohol called *sorbitol,* raise the osmolarity of the cells to match that of the surrounding interstitial fluid. *Insights from the Molecular Revolution* describes research that identified the genetic controls leading to sorbitol production in kidney medulla cells and other cells subjected to osmotic stress.

Urine flows from the end of the collecting ducts into the renal pelvis, and then through the ureters into the urinary bladder where it is stored. From the bladder, urine exits through the urethra to the outside.

Terrestrial Mammals Have Additional Water-Conserving Adaptations

Terrestrial mammals have other adaptations that complement the water-conserving activities of the kidneys. One is the location of the lungs deep inside the body, which reduces water loss by evaporation during breathing (see Section 44.1). Another is a body covering of keratinized skin. Skin is so impermeable that it almost eliminates water loss by evaporation, except for the controlled loss through evaporation of sweat in mammals with sweat glands.

Among mammals, water-conserving adaptations reach their greatest efficiency in desert rodents such as the kangaroo rat **(Figure 46.10)**. The proportion of nephrons with long loops extending deep into the kidney medulla of kangaroo rats is very high, allowing them to excrete urine that is 20 times more concentrated than body fluids. Further, most of the water in the feces is absorbed in the large intestine and rectum. Lacking sweat glands, kangaroo rats lose little water by evaporation from the body surface. Much of the moisture in their breath is condensed and recycled by specialized passages in the nasal cavities. They stay in burrows during daytime, and come out to feed only at night.

About 90% of the kangaroo rat's daily water supply is generated from oxidative reactions in its cells. (Humans, in contrast, can make up only about 12% of their daily water needs from this source.) The remaining

	Kangaroo Rat	Human
Water gain (milliliters)		
From ingesting food	6.0	850
From drinking liquids	0.0	1400
By metabolism	54.0	350
	60.0	2600
Water loss (milliliters)		
In urine	13.5	1500
In feces	2.6	200
By evaporation	43.9	900
	60.0	2600

David Noble/FPG/Getty Images

Claude Steelman/Tom Stack & Associates

Figure 46.10
A comparison of the sources of water for a human and a kangaroo rat (genus *Dipodymus*). Water conservation in the kangaroo rat is so efficient that the animal never has to drink water.

10% of the kangaroo rat's water comes from its food. These structural and behavioral adaptations are so effective that a kangaroo rat can survive in the desert without ever drinking water.

Marine mammals, including whales, seals, and manatees, eat foods that are high in salt content and never drink fresh water. They are able to survive the high salt intake because they produce urine that is more concentrated than seawater. As a result, they are easily able to excrete all the excess salt they ingest in their diet.

We now turn to the regulatory mechanisms that integrate kidney function with body functions as a whole.

Study Break

1. Describe the structure of a human nephron from the proximal end to the distal end.
2. The urine entering the collecting ducts at the end of the nephron has an osmolarity essentially the same as that of fluids in other parts of the body. How is the urine subsequently made more concentrated?

46.4 Regulation of Mammalian Kidney Function

Mammalian excretory functions are integrated into overall body functions by three primary control systems, which link kidney functions to blood pressure, to the osmolarity and pH of body fluids, and to the body's water balance. An *autoregulation system* located entirely within the kidney keeps glomerular filtration constant during relatively small variations in blood pressure, as when we move from sitting to standing. Two other systems involve hormonal controls that compensate for excessive loss of salt and body fluids and that adjust the rate of water uptake in the kidneys to compensate for excessive water intake or loss. These two hormonal systems regulate interactions between the kidneys and the rest of the body.

Autoregulation Involves Interactions between the Glomerulus and the Nephron

The autoregulation system responds almost instantly to keep the filtration rate constant during small variations in blood pressure. The system depends on signals from receptors in the **juxtaglomerular apparatus** (*juxta* = near) **(Figure 46.11)**, which is located at a point where the distal convoluted tubule contacts the afferent arteriole carrying blood to the glomerulus. The receptors, located in the tubule wall, monitor the pressure and flow of fluid through the distal tubule. If a rise in blood pressure increases the filtration rate, the receptors release chemical signals that trigger constriction of the afferent arteriole. The constriction reduces blood flow through the glomerulus and lowers the filtration rate. A drop in filtration rate due to a fall in blood pressure has the opposite effect: signals from the receptors cause the arterioles to dilate, blood flow increases in the glomerulus, and the filtration rate rises.

The RAAS Responds to Na^+ by Triggering Na^+ Reabsorption

Major changes in blood volume and pressure occur when the body loses or gains Na^+ in excessive amounts. Excessive Na^+ loss may result from prolonged and heavy sweating, repeated vomiting, severe diarrhea, or insufficient Na^+ uptake in the diet. The Na^+ loss reduces the osmolarity of body fluids, which causes less water to be reabsorbed in the kidneys. The water loss reduces the volume of blood and interstitial fluid and causes the blood pressure to drop. Excessive Na^+ intake in salty foods may have the opposite effects. The body must compensate for significant changes in Na^+.

The **renin-angiotensin-aldosterone system (RAAS)** is the most important hormonal system involved in regulating Na^+ (see Figure 46.11). At normal body salt concentrations, the RAAS allows about 10 g of salt to be excreted in the urine each day. If excessive Na^+ is lost in the excreted salt, blood pressure and body fluid volume drop, and the glomerular filtration rate falls below levels that can be restored by the juxtaglomerular apparatus. In response, cells in the juxtaglomerular apparatus secrete the enzyme **renin** into the bloodstream. (The RAAS also is activated to promote renin secretion when blood pressure or blood volume decreases independently of Na^+ levels, as in the case of a hemorrhage.) Renin converts a blood protein into the peptide hormone **angiotensin.** Angiotensin quickly raises blood pressure by constricting arterioles in most parts of the body; it also stimulates the release of the steroid hormone **aldosterone** from the adrenal cortex. Aldosterone increases Na^+ reabsorption in the kidneys, which raises the osmolarity of body fluids. As a result, water moves from the tubules into the extracellular fluid, which conserves water. Angiotensin also stimulates secretion of **antidiuretic hormone (ADH)** by the posterior pituitary (antidiuretic means "against urine output"). ADH increases water absorption in the kidneys. And, angiotensin stimulates thirst so that more water will be brought into the body. Overall, the combined effects of angiotensin act to raise the blood pressure back to normal levels.

In the opposite situation, when salt intake is too high, both body fluid volume and blood pressure rise above normal levels. Under these conditions, renin secretion is inhibited and, as a result, angiotensin production and aldosterone secretion are not stimulated. The reduction in angiotensin lowers blood pressure by al-

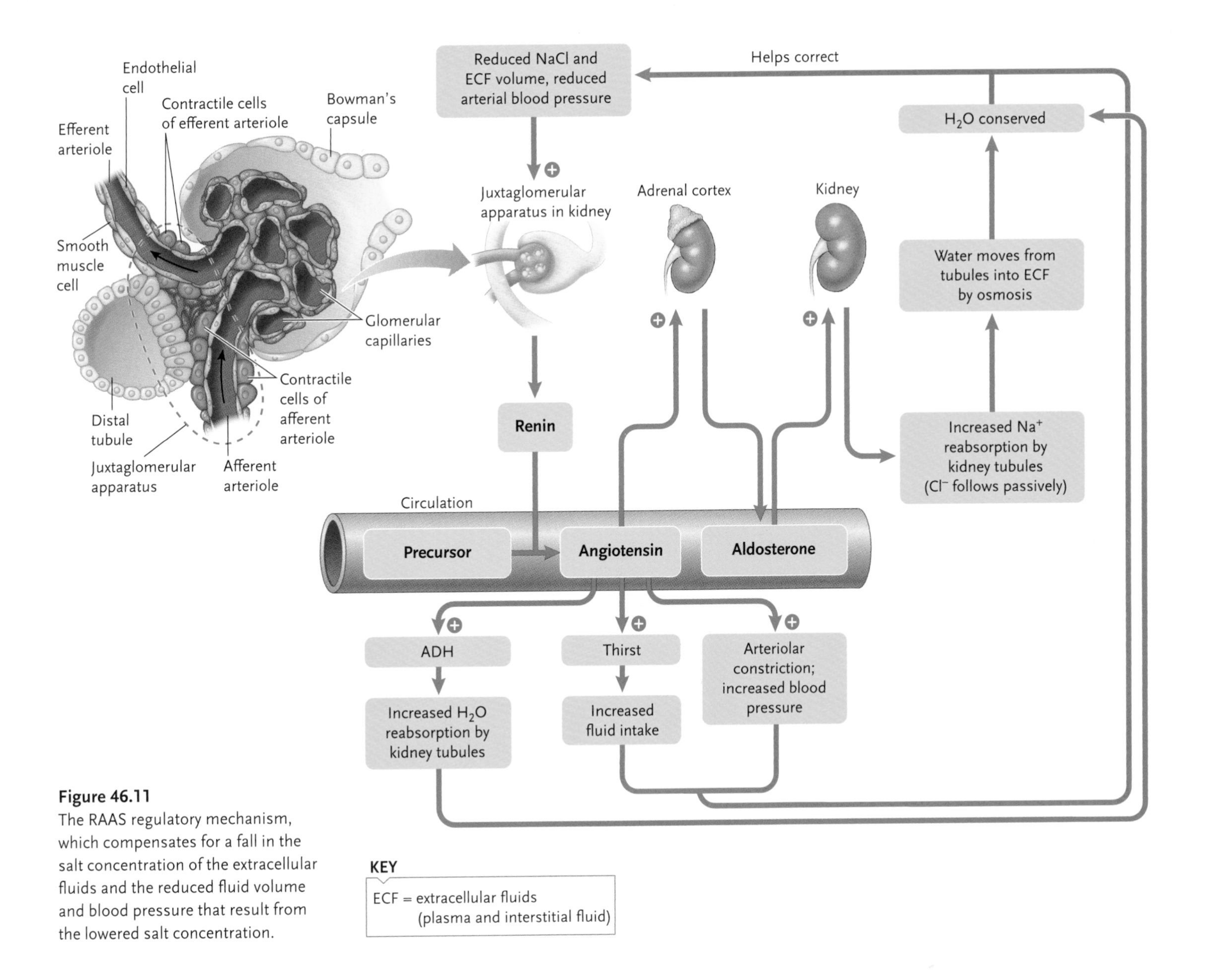

Figure 46.11
The RAAS regulatory mechanism, which compensates for a fall in the salt concentration of the extracellular fluids and the reduced fluid volume and blood pressure that result from the lowered salt concentration.

lowing arterioles to dilate; the reduction in aldosterone increases Na^+ release in the urine by retarding the reabsorption of Na^+ and Cl^- from the kidney tubules.

As a backup to these controls, elevated blood pressure stimulates specialized cells in the heart to release **atrial natriuretic factor (ANF)**, a peptide hormone that also inhibits renin release. ANF also increases the filtration rate by dilating the arterioles that deliver blood to glomeruli and by inhibiting aldosterone release. As less Na^+ is reabsorbed and urine volume increases, both plasma volume and blood pressure fall to normal levels.

The ADH System Also Regulates Osmolarity and Water Balance

You have just learned that the ADH system is stimulated by angiotensin of the RAAS in response to an increase in Na^+. The ADH system regulates osmolarity and water balance—and, therefore, urinary output—by increasing water reabsorption in the kidneys without changing the usual excretion of salt. Independently of being stimulated by angiotensin, the ADH system is triggered by **osmoreceptors**, chemoreceptors in the hypothalamus that respond to changes in the osmolarity of the fluid surrounding them, which reflects the osmolarity generally of the body fluids **(Figure 46.12)**.

When an animal becomes dehydrated, the osmolarity of its body fluids increases and its need for water conservation increases. In this situation, osmoreceptors detect the increase in concentration of salts and other dissolved substances in the extracellular fluid (see Figure 46.12, step 1). Signals from the osmoreceptors are routed to the brain stem, where they trigger thirst (step 2). The resulting increase in water ingestion helps compensate for water loss (step 3).

In addition, neurons of the hypothalamus stimulate the posterior pituitary to secrete ADH (step 4). ADH makes the otherwise impermeable distal convoluted tubules and collecting ducts permeable to water.

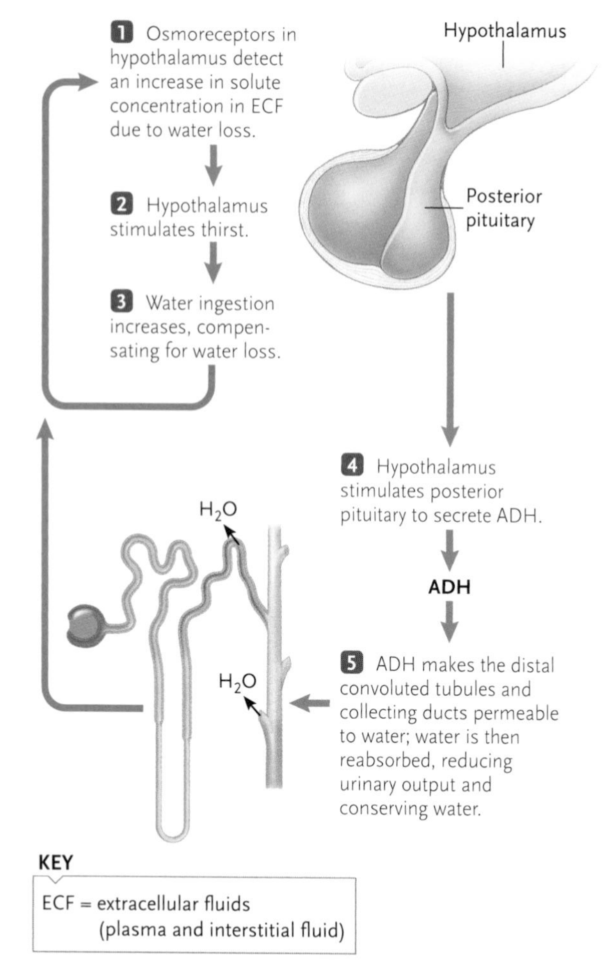

1 Osmoreceptors in hypothalamus detect an increase in solute concentration in ECF due to water loss.
Hypothalamus
2 Hypothalamus stimulates thirst.
Posterior pituitary
3 Water ingestion increases, compensating for water loss.
4 Hypothalamus stimulates posterior pituitary to secrete ADH.
H2O
ADH
H2O
5 ADH makes the distal convoluted tubules and collecting ducts permeable to water; water is then reabsorbed, reducing urinary output and conserving water.
KEY
ECF = extracellular fluids (plasma and interstitial fluid)

Figure 46.12
The ADH regulatory system, which stimulates water reabsorption to compensate for a loss in the fluid volume of the extracellular fluids due to excessive water loss from the body.

As a result, water is reabsorbed into those tubules and ducts so that urinary output is reduced and water is conserved (step 5).

By stimulating thirst and water reabsorption in the kidneys, the body's depleted stock of water is restored. The newly added water dilutes the solutes in the body fluids to normal concentrations.

In the opposite condition, when there is a water excess in extracellular fluids, the osmolarity of those fluids drops below normal levels. Here, there is no stimulation of the osmoreceptors in the hypothalamus. Consequently, there is no sensation of thirst, and no ADH release from the posterior pituitary. (In going from water deficiency to water excess, there is a gradual change in both of these parameters as the body adjusts to match its needs, meaning that the sensation of thirst decreases and ADH release is reduced as water in body fluids increases.) Without ADH, the distal convoluted tubules and collecting ducts again become impermeable to water. The animal excretes large volumes of dilute urine until the osmolarity of the extracellular fluids returns to normal. Alcohol also causes frequent urination by inhibiting ADH release.

Although the RAAS and ADH systems interact to regulate the body's water balance over a wide range of conditions, their regulatory mechanisms cannot compensate for water losses for more than a few days if water is unavailable. Dehydration becomes fatal when water loss amounts to about 12% of the normal fluid volume of the body.

Unlike mammals, most other vertebrates cannot conserve water by producing highly concentrated, hyperosmotic urine. In the next section, we consider some of the adaptations that nonmammalian vertebrates use to maintain the osmolarity of body fluids and water balance while excreting hypoosmotic urine.

Study Break

Outline the roles of the RAAS and ADH system in regulating mammalian kidney function.

46.5 Kidney Function in Nonmammalian Vertebrates

Among nonmammalian vertebrates, only a few species of aquatic birds produce urine that is hyperosmotic to body fluids. The particular adaptations that maintain osmolarity and water balance among these animals vary depending on whether retention of water or salts is the major issue.

Marine Fishes Conserve Water and Excrete Salts

Marine teleosts live in seawater, which is strongly hyperosmotic to their body fluids. As a result, they continually lose water to their environment by osmosis and must replace it by continual drinking. The kidneys of marine teleosts play little role in regulating salt in their body fluids because they cannot produce hyperosmotic urine that would both remove salt and conserve water. Instead, excess Na^+, K^+, and Cl^- ions are eliminated from the body by specialized cells in the gills, called *chloride cells,* which actively transport Cl^- into the surrounding seawater; the Na^+ and K^+ ions are also actively transported to maintain electrical neutrality **(Figure 46.13a).** Certain other ions in the ingested seawater, such as Ca^{2+} and Mg^{2+}, are removed by the kidneys in an isoosmotic urine. On balance, a marine teleost is able to retain most of the water it drinks and eliminate most of the salt, allowing its body fluids to remain hypoosmotic to the surrounding water with no need to secrete hyperosmotic urine. Nitrogenous wastes are released from the gills, primarily as ammonia, by simple diffusion. The kidneys play little role in nitrogenous-waste removal.

Sharks and rays have a different adaptation to seawater—the osmolarity of their body fluids is maintained close to that of seawater by retaining high levels of urea in body fluids, along with another nitrogenous waste, *trimethylamine oxide (TMAO)*. The match in osmolarity keeps sharks and rays from losing water to the surrounding sea by osmosis, and they do not have to drink seawater continually to maintain their water balance. Excess salts ingested with food are excreted in the kidney and by specialized secretory cells in a *rectal salt gland* located near the anal opening.

Freshwater Fishes and Amphibians Excrete Water and Conserve Salts

The body fluids of freshwater fishes and aquatic amphibians (no amphibians live in seawater) are hyperosmotic to the surrounding water, which usually ranges from about 1 to 10 mOsm/L. Water therefore moves osmotically into their tissues. Such animals rarely drink, and they excrete large volumes of dilute urine to get rid of excess water **(Figure 46.13b)**. In freshwater fishes, salt ions lost with the urine are replaced by salt in foods and by active transport of Na^+ and K^+ into the body by the gills; Cl^- follows to maintain electrical neutrality. Aquatic amphibians obtain salt in the diet and by active transport across the skin from the surrounding water. Nitrogenous wastes are excreted from the gills as ammonia in both freshwater fishes and aquatic amphibians.

Terrestrial amphibians must conserve both water and salt, which is obtained primarily in foods. In these animals, the kidneys secrete salt into the urine, causing water to enter the urine by osmosis. In the bladder, the salt is reclaimed by active transport and returned to body fluids. The water remains in the bladder, making the urine very dilute; during times of drought, it is reabsorbed as a water source. Terrestrial amphibians also have behavioral adaptations that help minimize water loss, such as seeking shaded, moist environments and remaining inactive during the day.

Larval amphibians, which are completely aquatic, excrete nitrogenous wastes from their gills as ammonia. Adult amphibians excrete nitrogenous wastes through their kidneys as urea.

Reptiles and Birds Excrete Uric Acid to Conserve Water

Terrestrial reptiles conserve water by secreting nitrogenous wastes in the form of an almost water-free paste of uric acid crystals. Further water conservation occurs as the epithelial cells of the cloaca, the common exit for the digestive and excretory systems, absorb water from feces and urine before those wastes are excreted. Most birds conserve water by the same processes—they excrete nitrogenous wastes as uric acid and absorb water from the urine and feces in the cloaca. In reptiles, the scales covering the skin allow almost no water to escape through the body surface.

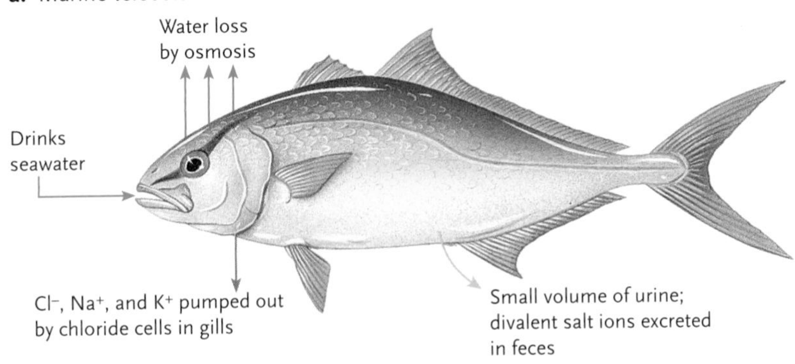

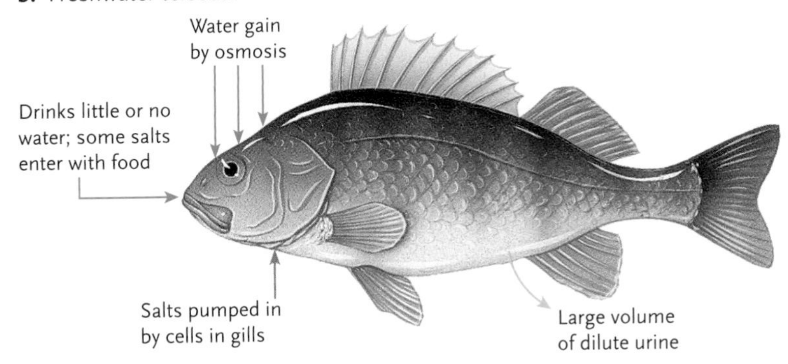

Figure 46.13
The mechanisms balancing the water and salt content of **(a)** marine teleosts and **(b)** freshwater teleosts.

Reptiles and birds that live in or around seawater, including reptiles such as crocodilians, sea snakes, and sea turtles and birds such as seagulls, penguins, and pelicans, take in large quantities of salt with their food and rarely or never drink fresh water. These animals typically excrete excess salt through specialized *salt glands* located in the head **(Figure 46.14)**, which remove salts from the blood by active transport. The salts are secreted to the environment as a water solution in which salts are two to three times more concentrated

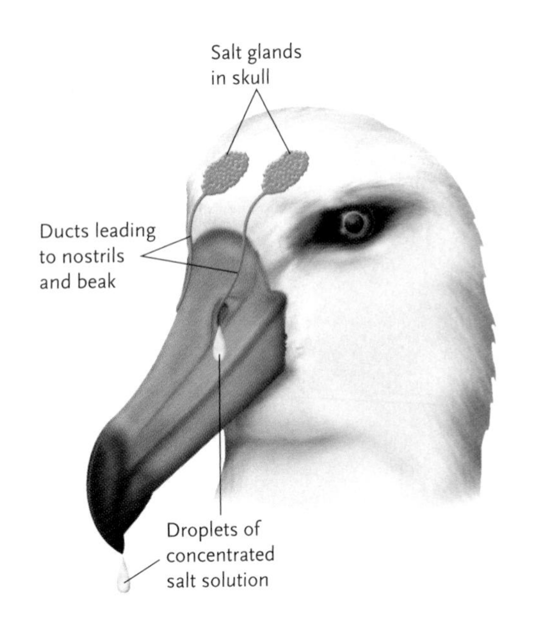

Figure 46.14
Salt glands in a bird living on a seacoast.

than in body fluids. The secretion exits through the nostrils of birds and lizards, through the mouth of marine snakes, and as salty tears from the eye sockets of sea turtles and crocodilians. Neural and hormonal controls, essentially the same as those regulating osmolarity in mammals, control the rate of fluid secretion and its salt concentration.

The adaptations described in this section permit excretion of toxic wastes and allow animals to maintain the concentration of body fluids at levels that keep cells from swelling or shrinking. Animals also have mechanisms that address an equally vital challenge—maintaining their internal environment at temperatures that can be tolerated by body cells. We take up these processes in the next section.

Study Break

1. How do marine and freshwater teleosts differ in water, salt, and nitrogenous-waste regulation?
2. Reptiles and birds excrete nitrogenous wastes in the form of uric acid. Is there an advantage to doing this as opposed to the mammalian process of excreting nitrogenous wastes as urea?

46.6 Introduction to Thermoregulation

Environmental temperatures vary enormously across Earth's surface. However, animal cells can survive only within a temperature range from about 0°C to 45°C (32°F to 113°F). Not far below 0°C, the lipid bilayer of a biological membrane changes from a fluid to a frozen gel, which disrupts vital cell functions, and ice crystals destroy the cell's organelles. At the other extreme, as temperatures approach 45°C, the kinetic motions of molecules become so great that most proteins and nucleic acids unfold from their functional form. Either condition leads quickly to cell death. Animals therefore usually maintain internal body temperatures somewhere within the 0°C to 45°C limits.

Temperature regulation—thermoregulation—is based on negative feedback pathways in which temperature receptors *(thermoreceptors)* detect changes from a temperature *set point*. Signals from the receptors trigger physiological and behavioral responses that return the temperature to the set point (thermoreceptors are discussed in Section 39.6; negative feedback mechanisms and set points are discussed in Section 36.4). All of the responses triggered by negative feedback mechanisms involve adjustments in the rate of heat-generating oxidative reactions within the body, coupled with adjustments in the rate of heat gain or loss at the body surface. The particular adaptations accomplishing these responses vary widely among species, however. And, while body temperature is closely regulated around a set point in all endotherms, the set point itself may vary over the course of a day and between seasons.

In this section, we describe the structures, mechanisms, and behavioral adaptations that enable animals to regulate their temperature.

Thermoregulation Allows Animals to Reach Optimal Physiological Performance

Within the 0°C to 45° range of tolerable temperatures, an animal's *organismal performance*—the rate and efficiency of its biochemical, physiological, and whole-body processes—varies greatly. For example, the speed at which the Middle Eastern lizard *Agama stellio* can sprint is low when the animal's body temperature is cold, rises smoothly with body temperature until it levels to a fairly broad plateau, and then drops off dramatically with further increases in body temperature **(Figure 46.15a)**. Similar patterns of temperature dependence are observed for numerous other body functions **(Figure 46.15b)**. The temperature range that provides good organismal performance varies from one species to another, however.

Animals that maintain body temperature within a fairly narrow optimal range can run quickly, digest food efficiently, and carry out necessary activities and processes rapidly and effectively (see Figure 46.15b). Besides keeping body temperatures within tolerable limits, thermoregulation allows animals to achieve this level of performance.

a. Maximum running speed of a lizard at various body temperatures

Maximum speed (m/s)

2
1
0
20 30 40
Body temperature (°C)

b. Range of optimal physiological performance

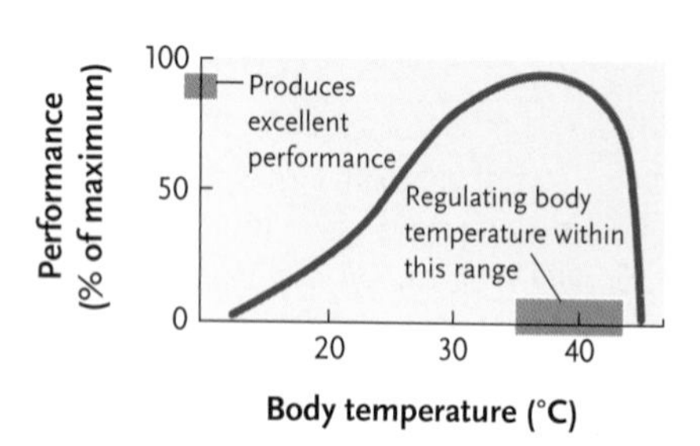

Figure 46.15
Body temperature and organismal performance. **(a)** The maximum sprint speed of a lizard *(Agama stellio)* changes dramatically with body temperature. **(b)** An animal's other behavioral and physiological processes respond to temperature changes in similar ways. The advantage of regulating body temperature within the range indicated by the bar on the horizontal axis is a high level of organismal performance, indicated by the bar on the vertical axis.

Animals Exchange Heat with Their Environments by Conduction, Convection, Radiation, and Evaporation

As part of thermoregulation, animals exchange heat with their environment. Virtually all heat exchange occurs at surfaces where the body meets the external environment. As with all physical bodies, heat flows into animals if they are cooler than their surroundings and flows outward if they are warmer. This heat exchange occurs by four mechanisms: *conduction, convection, radiation,* and *evaporation* **(Figure 46.16).**

Conduction is the flow of heat between atoms or molecules in direct contact. An animal loses heat by conduction when it contacts a cooler object, and gains heat when it contacts an object that is warmer. **Convection** is the transfer of heat from a body to a fluid, such as air or water, that passes over its surface. The movement maximizes heat transfer by replacing fluid that has absorbed or released heat with fluid at the original temperature. **Radiation** is the transfer of heat energy as electromagnetic radiation. Any object warmer than absolute zero (−273°C) radiates heat; as the object's temperature rises, the amount of heat it loses as radiation increases as well. Animals also gain heat through radiation, particularly by absorbing radiation from the sun. **Evaporation** is heat transfer through the energy required to change a liquid to a gas. Evaporation of water from a surface is an efficient way to transfer heat; when the water in sweat evaporates from the body surface, the body cools down because heat is being transferred to the evaporated water in the surrounding air.

All animals gain or lose heat by a combination of these four mechanisms. A marathon runner or a bicycle racer struggling with the heat on a sunny summer day, for example, loses heat by the evaporation of sweat from the skin, by convection as air flows over the skin, and by outward infrared radiation. The runner gains heat from internal biochemical reactions (especially oxidations), by absorbing infrared and solar radiation, and by conduction as the feet contact the hot ground. To maintain a constant body temperature, the heat gained and lost through these pathways must balance.

Figure 46.16
Heat flow to (in red) and from (in blue) a marathon runner on a hot, sunny day. Unlike conduction, convection, and evaporation, which take place through the kinetic movement of molecules, electromagnetic radiation is transmitted through space as waves of energy. (Photo: Rafael Winer/Corbis.)

Ectothermic and Endothermic Animals Rely on Different Heat Sources to Maintain Body Temperature

Different animals use one of two major strategies to balance heat gain and loss. Animals that obtain heat primarily from the external environment are known as **ectotherms** (*ecto* = outside); those obtaining most of their heat from internal physiological sources are called **endotherms** (*endo* = inside). All ectotherms generate at least some heat from internal reactions, however, and endotherms can obtain heat from the environment under some circumstances.

Virtually all invertebrates, fishes, amphibians, and reptiles are ectotherms. Although these animals are popularly described as cold-blooded, the body temperature of some, such as an active lizard, may be as high as or higher than ours on a sunny day. Ectotherms regulate body temperature by controlling the rate of heat exchange with the environment. Through behavioral and physiological mechanisms, they adjust body temperature toward a level that allows optimal physiological performance. However, most ectotherms are unable to maintain optimal body temperature when the temperature of their surroundings departs too far from that optimum, particularly when environmental temperatures fall. As a result, the body temperatures of ectotherms fluctuate with environmental temperatures, and they typically are less active when it is cold. Nevertheless, ectotherms are highly successful, particularly in warm environments.

The endotherms—birds, mammals, some fishes, sea turtles, and some invertebrates—keep their bodies at an optimal temperature by regulating two processes: (1) the amount of heat generated by internal oxidative reactions and (2) the amount of heat exchanged with the environment. Because endotherms use internal heat sources to maintain body temperature at optimal levels, they can remain active over a broader range of environmental temperatures than ectotherms, and they can inhabit a wider range of habitats. However, endotherms require a nearly constant supply of energy to maintain their body temperatures. And because that energy is provided by food, endotherms typically consume much more food than ectotherms of equivalent size.

The difference between ectotherms and endotherms is reflected in their metabolic responses to environmental temperature **(Figure 46.17)**. For example, the metabolic rate of a resting mouse *increases* steadily as the environmental temperature falls from 25°C to 10°C (77°F to 50°F). This increase reflects the fact that in order to maintain a constant body temperature in a colder environment, endotherms must process progressively more food and generate more heat to compensate for their increased rate of heat loss. In this respect, an endotherm can be likened to a house in winter. To maintain a constant internal temperature, the homeowner must burn more oil or gas on a cold day than on a warm day.

By contrast, the metabolic rate of a resting lizard typically *decreases* steadily over the same temperature range. Because ectotherms don't maintain a constant body temperature, their biochemical and physiological functions, including oxidative reactions, slow down as environmental and body temperatures decrease. Thus, an ectotherm consumes and uses less energy when it is cold than when it is warm. This difference between ectotherms and endotherms is so fundamental that even samples of living tissue extracted from an ectotherm consume energy more slowly than equivalent samples from an endotherm.

Ectothermy and endothermy represent different strategies for coping with the variations in environmental temperature that all animals encounter; neither strategy is inherently superior to the other. Endotherms can remain fully active over a wide temperature range. Cold weather does not prevent them from foraging, mating, or escaping from predators, but it does increase their energy and food needs—and, to satisfy their need for food, they may not have the option of staying curled up safely in a warm burrow. Ectotherms do not have the capacity to be active when environmental temperatures drop too low; they move sluggishly and are unable to capture food or escape from predators. However, because their metabolic rates are lower under such circumstances, so are their food needs, and they do not have to actively look for food and expose themselves to danger to the extent that endotherms do.

Having laid the ground rules of heat transfer and weighed the relative advantages and disadvantages of ectothermy and endothermy, we now begin a more detailed examination of how animals actually regulate their body temperatures within these overall strategies.

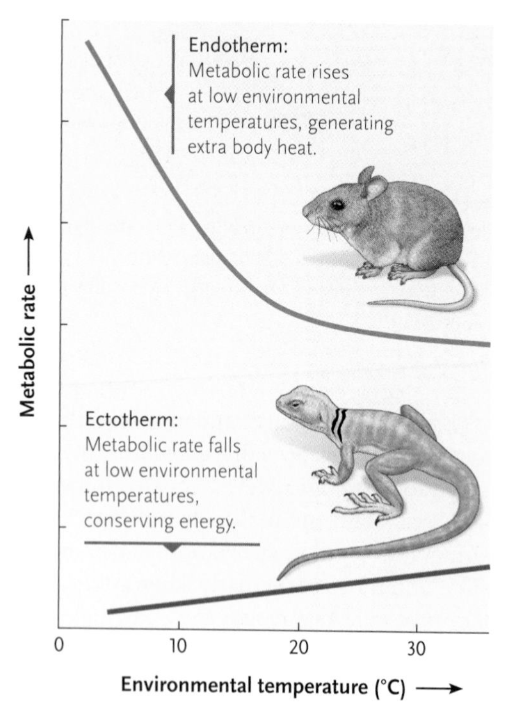

Figure 46.17
Metabolic responses of ectotherms and endotherms to cooling environmental temperatures. At any temperature, the metabolic rates of endotherms are always higher than those of endotherms of comparable size.

Study Break

Distinguish between ectothermy and endothermy. Give one advantage and one disadvantage for each form of thermoregulation.

46.7 Ectothermy

Ectotherms vary widely in their ability to regulate internal body temperatures. For example, most aquatic invertebrates have such limited ability to thermoregulate that their body temperatures closely match those of the surrounding environment. These species live in or seek warm or temperate environments, where temperatures fall within a range that produces optimal physiological performance. Ectotherms with a greater ability to thermoregulate may occupy more varied habitats.

Ectotherms Are Found in All Invertebrate Groups

Most aquatic invertebrates are limited thermoregulators whose body temperature closely follows the temperature of their surroundings. However, even among these animals, some use behavioral responses to regulate body temperature. For example, a South American intertidal mollusk, *Echinolittorina peruviana*, is longer than it is wide. Researchers have shown that this animal orients itself as a means of thermoregulation. On sunny, summer days, it faces the sun, offering a smaller surface area for the sun's rays. On overcast summer days, or during the winter, it orients itself with a lateral side—which has the larger surface area—toward the sun's rays.

Invertebrates living in terrestrial habitats regulate body temperatures more closely. Many also use behavioral responses, such as moving between shaded and sunny regions, to regulate body temperature. Some winged arthropods, including bees, moths, butterflies, and dragonflies, use a combination of behavioral and heat-generating physiological mechanisms for thermoregulation. For example, in cool weather, these animals warm up before taking flight by rapidly vibrating the large flight muscles in the thorax, in a mechanism similar to shivering in humans. The tobacco hawkmoth (*Manduca sexta*) vibrates its flight muscles until its thoracic temperature reaches about 36°C before flying. During flight, metabolic heat generated by the flight muscles sustains the elevated thoracic temperature, so much so that a flying sphinx moth produces more heat per gram of body weight than many mammals.

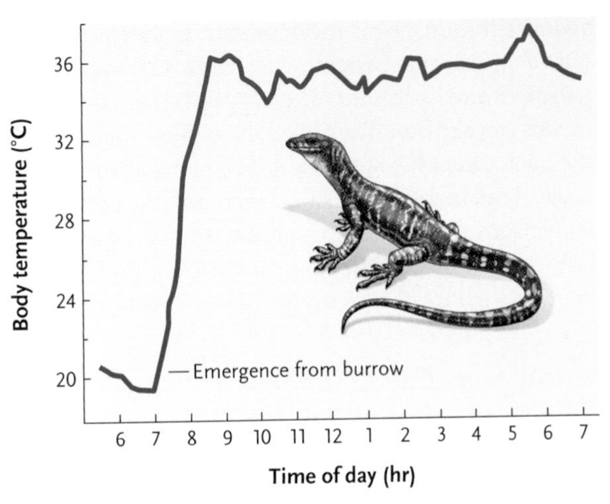

Figure 46.18
An example of excellent thermoregulation in ectotherms. The body temperature of the Australian lizard *Varanus varius* rises quickly after the animal emerges from its burrow and remains relatively stable throughout the day.

Most Fishes, Amphibians, and Reptiles Are Ectotherms

Vertebrate ectotherms—fishes, amphibians, and reptiles—also vary widely in their ability to thermoregulate. Most aquatic species have a more limited thermoregulatory capacity than that found among terrestrial species, particularly the reptiles. Some fishes, however, are highly capable thermoregulators.

Fishes. The body temperatures of most fishes remain within one or two degrees of their aquatic environment. However, many fishes use behavioral mechanisms to keep body temperatures at levels allowing good physiological performance. Freshwater species, for example, may use opportunities provided by the thermal stratification of lakes and ponds (see Figure 52.22). During hot summer days they remain in deep, cool water, moving to the shallows to feed only during early morning and late evening when air and water temperatures fall.

Amphibians and Reptiles. The body temperatures of most amphibians also closely match environmental temperatures. Some, such as the tadpoles of foothill yellow-legged frogs (*Rana boylii*), regulate their body temperature to some degree by changing their location in ponds and lakes to take advantage of temperature differences between deep and shallow water, or between sunny and shaded regions. Some terrestrial amphibians bask in the sun to raise their body temperature, and seek shade to lower body temperature. However, basking can be dangerous to amphibians because they lose water rapidly through their permeable skin. One South American hylid frog (*Phyllomedusa sauvagei*), which often basks in sunlight, avoids this problem by coating itself with waterproofing lipids secreted by glands in its skin.

Thermoregulation is more pronounced among terrestrial reptiles. Some lizard species can maintain temperatures that are nearly as constant as those of endotherms **(Figure 46.18)**. For small lizards, the most common behavioral thermoregulatory mechanism is shuttling between sunny (warmer) and shady (cooler) regions; in the deserts, lizards and other reptiles retreat into burrows during the hottest part of summer days. Some, such as the desert iguana (*Dipsosaurus dorsalis*), lose excess heat by *panting*—rapidly moving air in and out of the airways. The air movement increases heat loss by convection and by evaporation of water from the respiratory tract.

Lizards also frequently adjust their posture to foster heat exchange with the environment, and control the angle of their body relative to the rays of the sun. For example, horned lizards (genus *Phrynosoma*) often warm up by flattening themselves against warm, sunlit rocks to maximize their rate of heat gain by conduction from the rock and radiation from the sun. Snakes and lizards can often be found on large rocks and on roads on chilly nights, taking advantage of the heat retained by the stone or concrete. *Agama savignyi*, a lizard that lives in the Negev Desert in Israel, cools off at midday by climbing into shady bushes, moving away from the hot sand and catching a cooling breeze.

Researchers have demonstrated experimentally that several lizard species couple physiological responses to behavioral mechanisms of thermoregulation. For example, when a Galapagos marine iguana (*Amblyrhynchus cristatus;* see Figure 19.8c) is exposed to heat from infrared radiation, blood flow increases in the heated regions of the skin. The blood absorbs heat rapidly and carries it to critical organs in the core of the body. Conversely, when an area of skin is experimentally cooled, blood flow to it is restricted, thereby preventing the loss of heat to the external environment.

Ectotherms Can Compensate for Seasonal Variations in Environmental Temperature

Many ectotherms undergo physiological changes, called **thermal acclimatization,** in response to seasonal shifts in environmental temperature. These changes allow the animals to attain good physiological performance at both winter and summer temperatures.

For example, in the summer a bullhead catfish (*Ameiurus* species) can survive water temperatures as high as 36°C (97°F), but it cannot tolerate temperatures below 8°C (46°F). In the winter, however, the bullhead cannot survive water temperatures above 28°C (82°F), but can tolerate temperatures near 0°C (32°F). Scientists have hypothesized that the production of different versions of the same enzyme (perhaps encoded by different genes, or produced as a result of alternative splicing; see Section 16.3), each having optimal activity at cooler or warmer temperatures, underlies such acclimatization.

Another acclimatizing change involves the phospholipids of biological membranes (see *Focus on Research* in Chapter 6). For example, membrane phospholipids have higher proportions of double bonds in carp living in colder environments than in carp living in warmer environments. The higher proportion of double bonds makes it harder for the membrane to freeze. A higher proportion of cholesterol also protects membranes from freezing.

When seasonal temperatures fall below 0°C (32°F), some ectotherms add molecules to their body fluids that act as antifreeze molecules to depress their freezing point and retard ice crystal formation. For example, glycerol added to the cellular and extracellular fluids of a parasitic wasp (*Bracon* species) keeps the insect from freezing at temperatures as low as −45°C (−49°F). Similarly, antifreeze proteins allow fishes such as the winter flounder to remain active in seawater as cold as −1.8°C (29°F) (see *Insights from the Molecular Revolution* in Chapter 52).

Ectotherms thus primarily control body temperature by regulating heat exchange with the environment; internal-heat generating mechanisms contribute to the control mechanisms in some species, but are rarely the primary source of body heat. The opposite conditions occur among endotherms: although these animals also regulate heat exchange with the environment, their primary sources of body heat are internal.

Study Break

1. Describe two mechanisms an ectothermic animal can use to regulate its temperature.
2. What is thermal acclimatization?

46.8 Endothermy

Endotherms—mostly birds and mammals—have the most elaborate and extensive thermoregulatory adaptations of all animals. Highly specialized features of body structure interact with both physiological and behavioral mechanisms to keep the body temperature constant within a narrow range. Typically, the body temperatures of fully active individuals are held constant at levels between about 39° to 42°C (102° to 108°F) in birds, and 36° to 39°C (97° to 102°F) in mammals. These internal temperatures are maintained in the face of environmental temperatures that may range over much greater extremes, from as low as −42°C to as high as +48°C (−45°F to +120°F). Some highly specialized endotherms can even survive temperatures beyond these limits.

We begin by describing the basic feedback mechanisms that maintain body temperature, with primary emphasis on the human system. Later sections discuss variations in the responses of other mammals and of birds, and daily and seasonal variations in the temperature set point.

Information from Thermoreceptors Located in the Skin and Internal Structures Is Integrated in the Hypothalamus

Thermoreceptors are found in various locations in the human body, including the **integument** (skin; introduced in Chapter 36), spinal cord, and hypothalamus. Two types of thermoreceptors occur in human skin. One, called a *warm receptor,* sends signals to the hypothalamus as skin temperature rises above 30°C (86°F), and reaches maximum activity when the temperature rises to 40°C (104°F). The other type, the *cold receptor,* sends signals when skin temperature falls below about 35°C (95°F) and reaches maximum activity at 25°C (77°F). By contrast, the highly sensitive thermoreceptors in the hypothalamus send signals when the blood temperature shifts from the set point by as little as 0.01°C (0.02°F).

Signals from the thermoreceptors are integrated in the hypothalamus and other regions of the brain to bring about compensating physiological and behavioral responses **(Figure 46.19).** The responses keep body temperature close to the set point, which varies normally in humans between 35.5° and 37.7°C (96.0° to 99.9°F) for the head and trunk. The appendages typically vary more widely in temperature; in freezing weather, for example, our arms, hands, legs, and feet are typically lower in temperature than the body core—and the ears and nose especially so.

The hypothalamus was identified as a major thermoreceptor and response integrator in mammals by experiments in which various regions of the brain were heated or cooled with a temperature probe. Within the brain, only the hypothalamus produced thermoregulatory responses such as shivering or panting. Later experiments revealed a similar response when regions of the spinal cord were cooled, indicating that thermoreceptors also occur in this location. The hypothalamus is also a major thermoreceptor and response integrator in fishes and reptiles. In birds, thermoreceptors in the spinal cord appear to be most significant in thermoregulation.

Responses When Core Temperature Falls below the Set Point. When thermoreceptors signal a fall in core temperature below the set point, the hypothalamus triggers compensating responses by sending signals through the autonomic nervous system. Among the immediate responses is constriction of the arterioles in the skin (vasoconstriction), which reduces the flow of blood to capillary networks in the skin. The reduced flow cuts down the amount of heat delivered to the skin and lost from the body surface. The reduction in flow is most pronounced in the skin covering the extremities, where blood flow may be reduced by as much as 99% when core temperature falls.

Another immediate response is contraction of the smooth muscles erecting the hair shafts in mammals and feather shafts in birds, which traps air in pockets over the skin, reducing convective heat loss. The response is minimally effective in humans because hair is sparse on most parts of the body—it produces the goose bumps we experience when the weather gets chilly. However, in mammals with fur coats or in birds, erection of the hair or feather shafts significantly increases the thickness of the insulating layer that covers the skin.

Immediate behavioral responses triggered by a reduction in skin temperature also help reduce heat loss from the body. Mammals may reduce heat loss by moving to a warmer locale, curling into a ball, or huddling together. We have all seen puppies huddled together to keep warm; birds such as penguins also keep warm by huddling. We humans may also put on more clothes or slip into a tub of hot water.

If these immediate responses do not return body temperature to the set point, the hypothalamus triggers further responses, most notably the rhythmic tremors of skeletal muscle we know as shivering. The heat released by the muscle contractions and the oxidative reactions powering them can raise the total heat production of the body substantially. At the same time, the hypothalamus triggers secretion of *epinephrine* (from the adrenal medulla) and *thyroid hormone* (see Section 40.4), both of which increase heat production by stimulating the oxidation of fats and other fuels. The generation of heat by oxidative mechanisms in nonmuscle tissue throughout the body is termed **nonshivering thermogenesis.**

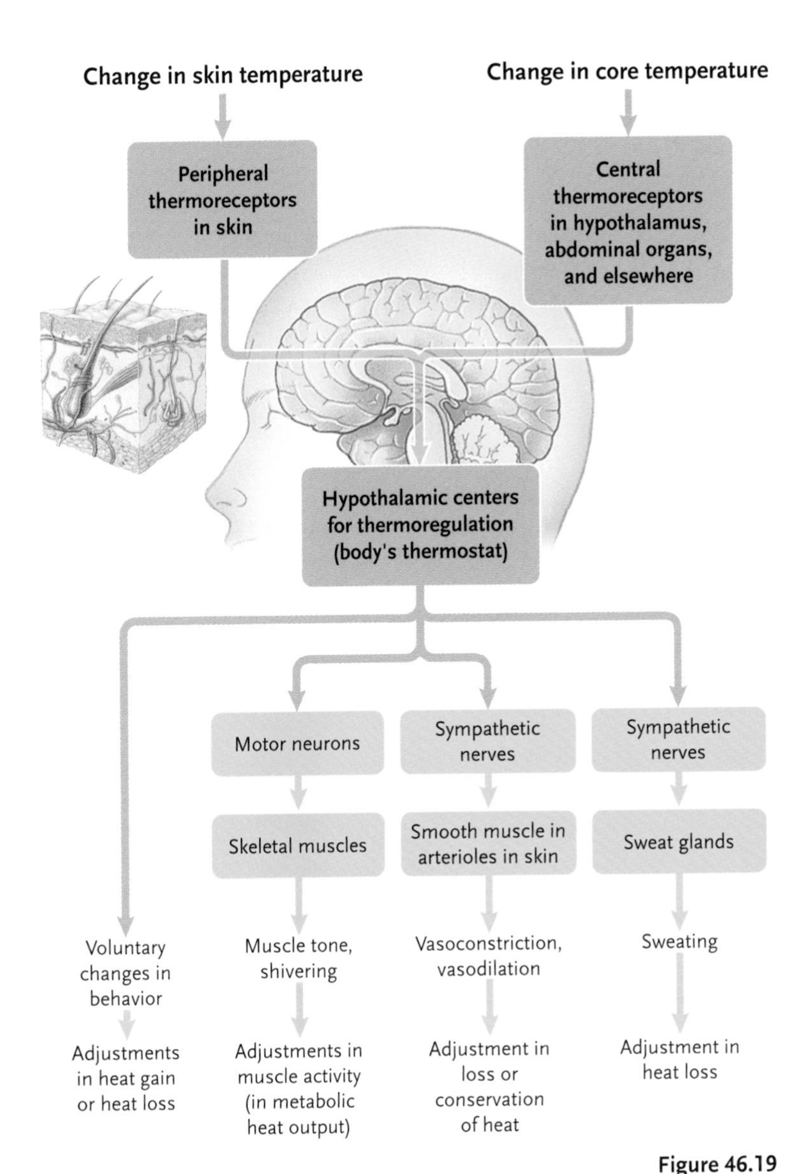

Figure 46.19 The physiological and behavioral responses of humans and other mammals to changes in skin and core temperature.

In human newborn babies and many other mammals, the most intense heat generation by nonshivering thermogenesis takes place in a specialized **brown adipose tissue** (also called **brown fat**) that can produce heat rapidly. Heat is generated by a mechanism that uncouples electron transport from ATP production in mitochondria (see Section 8.4); the heat is transferred throughout the body by the blood. Animals that hibernate or are active in cold regions, as well as the young of many others, contain brown adipose tissue. In most mammals, brown adipose tissue is concentrated between the shoulders in the back and around the neck. In human newborn babies, this tissue accounts for about 5% of body weight. Typically the tissue shrinks as humans age, until it is absent or essentially so in most adults. However, if exposure to cold is ongoing, the tissue remains. For instance, some Japanese and

Korean divers who harvest shellfish in frigid waters, and male Finlanders who work outside during the year, have significant amounts of brown adipose tissue.

If none of these responses succeeds in raising body temperature to the set point, the result is **hypothermia**, a condition in which the core temperature falls below normal for a prolonged period. In humans, a drop in core temperature of only a few degrees affects brain function and leads to confusion; continued hypothermia can lead to coma and death.

Responses When Core Temperature Rises above the Set Point. When core temperature rises above the set point, the hypothalamus sends signals through the autonomic system that trigger responses lowering body temperature. As an immediate response, the signals relax smooth muscles of arterioles in the skin (vasodilation), increasing blood flow and with it, the heat lost from the body surface. In addition, in humans and other mammals with sweat glands, such as antelopes, cows, and horses, signals from the hypothalamus trigger the secretion of sweat, which absorbs heat as it evaporates from the surface of the skin.

Some endotherms, including dogs (which have sweat glands only on their feet) and many birds (which have no sweat glands), use panting as a major way to release heat. These physiological changes are reinforced by behavioral responses such as seeking shade or a cool burrow, plunging into cold water, wallowing in mud, or taking a cold drink. Elephants typically take up water in their trunk and spray it over their body to cool off in hot weather.

When the heat gain of the body is too great to be counteracted by these responses, **hyperthermia** results. An increase of only a few degrees above normal for a prolonged period is enough to disrupt vital biochemical reactions and damage brain cells. Most adult humans become unconscious if their body temperature reaches 41°C (106°F) and die if it goes above 43°C (110°F) for more than a few minutes.

The Skin Is Highly Adapted to Control Heat Transfer with the Environment

Besides its defensive role against infection described in Section 43.1, the skin of birds and mammals is an organ of heat transfer. The arterioles delivering blood to the capillary networks of the skin constrict or dilate to control blood flow and, with it, the amount of heat transferred from the body core to the surface.

The outermost living tissue of human skin, the **epidermis**, consists of cells that grow and divide rapidly **(Figure 46.20)**, becoming packed with fibers of a highly insoluble protein, *keratin* (see Section 5.3). When fully formed, the epidermal cells die and become compacted into a tough, impermeable layer that limits water loss primarily to evaporation of the fluids secreted by the sweat glands.

The sweat glands and hair follicles are embedded in the layer below the epidermis. Called the **dermis**, it is packed with connective tissue fibers such as collagen, which resist compression, tearing, or puncture of the skin. The dermis also contains thermoreceptors and the dense networks of arterioles, capillaries, and venules that transfer heat between the skin and the environment.

The innermost layer of the skin, the **hypodermis**, contains larger blood vessels and additional reinforcing connective tissue. The hypodermis also contains an insulating layer of fatty tissue below the dermal capillary network, which ensures that heat flows between the body core and the surface primarily through the blood. The insulating layer is thickest in mammals that live in cold environments, such as whales, seals, walruses, and polar bears, in which it is known as *blubber*.

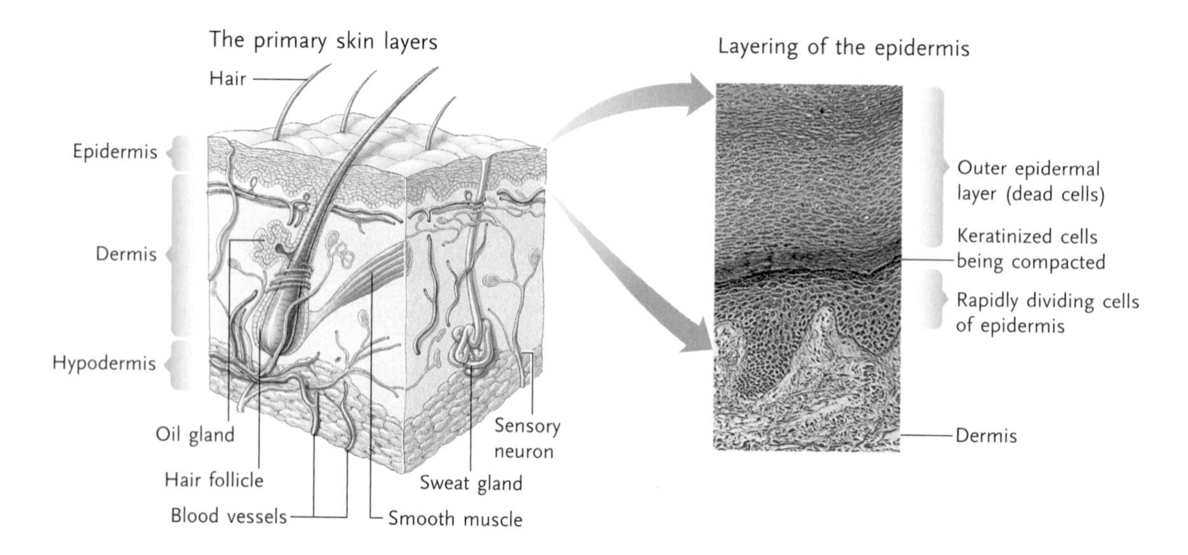

Figure 46.20
The structure of human skin.
(Micrograph: John D. Cunningham/Visuals Unlimited.)

a. Dissipating heat

Joe McDonald/Corbis

b. Conserving heat

Fredrik Broman/Iconica/Getty Images, Inc.

Figure 46.21
Structural and behavioral adaptations controlling heat transfer at the body surface. (a) A jackrabbit *(Lepus californicus)* dissipating heat from its ears on a hot summer day. Notice the dilated blood vessels in its large ears. Both the large surface area of the ears and the extensive network of blood vessels promote the dissipation of heat by convection and radiation. **(b)** A husky *(Canis lupus familiaris)* conserving heat by curling up with the limbs under the body and the tail around the nose.

Many Birds and Mammals Have Additional Thermoregulatory Structures and Responses

The thermoregulatory mechanisms we have described to this point are common to many birds and mammals. Many species also have specialized responses that enhance thermoregulation. In hot weather, for example, many birds fly with their legs extended, so that heat flows from their legs into the passing air. Similarly, penguins expose featherless patches of skin under their wings to cool off on days when the weather is too warm. Jackrabbits **(Figure 46.21a)** and elephants dissipate heat from their large ears, which are richly supplied with blood vessels. In times of significant heat stress, kangaroos and rats spread saliva on their fur to increase heat loss by evaporation; some bats coat their fur with both saliva and urine.

Many mammals have an uneven distribution of fur that aids thermoregulation. In a dog, for example, the fur is thickest over the back and sides of the body and the tail, and thinnest under the legs and over the belly. In cold weather, dogs curl up, pull in their limbs, wrap their tail around the body, and bury their nose in the tail, so that only body surfaces insulated by thick fur are exposed to the air **(Figure 46.21b)**. When the weather is hot, dogs spread their limbs, turn on their side or back, and expose the relatively bare skin of the belly, which acts as a heat radiator. These responses are combined with seeking sun or shade or a warm or cool surface to lie on.

In marine mammals such as whales and seals, heat loss is regulated by adjustments in the blood flow through the thick blubber layer to the skin. In cold water, blood flow is minimized by constriction of the vessels, making the skin temperature close to that of the surrounding water while the body temperature remains constant under the insulating blubber. In warmer water, blood flow to the skin increases, bypassing the blubber and allowing excess heat to be lost from the body surface.

In addition, heat loss in whales and seals is controlled by adjustments of the flow of blood to the flippers, which are not insulated by blubber and act as a heat radiator. When a whale generates excessive internal heat through the muscular activity of swimming, the flow of blood from the body core to the flippers increases. In contrast, when heat must be conserved to maintain core temperature at the set point, blood flow to the flippers is reduced.

As with ectotherms, many mammals also undergo thermal acclimatization to adjust to seasonal temperature change. That is, the development of a thick fur coat in winter, which is shed in summer, enables them to adapt to seasonal temperature changes. Note that the trigger for the coat changes in many cases is day length, rather than temperature; there is a general correlation with temperature and day length over the year. Some arctic and subarctic mammals develop a thicker layer of insulating fat in winter.

The Set Point Varies in Daily and Seasonal Rhythms in Many Birds and Mammals

The temperature set point in many birds and mammals varies in a regular cycle during the day. In some, the daily variations are relatively small and not obviously keyed to changes in environmental temperature. In others, larger variations are correlated with daily or seasonal temperature changes.

Humans are among the endotherms for which daily variations in the temperature set point are small. Normally, human core temperature varies from a minimum of about 35.5°C (95.9°F) in the morning to a maximum of about 37.7°C (99.9°F) in the evening. Women also show a monthly variation keyed to the menstrual cycle, with temperatures rising about 0.5°C (0.9°F) from the time of ovulation until menstruation begins. The physiological significance of these variations is unknown.

Camels undergo a daily variation of as much as 7°C (13°F) in set point temperature. During the day, a camel's set point gradually resets upward, an adaptation that allows its body to absorb a large amount of heat. The heat absorption conserves water that would otherwise be lost by evaporation to keep the body at a lower set point. At night, when the desert is cooler, the thermostat resets again, allowing the body temperature to cool several degrees, releasing the excess heat absorbed during the day.

When the environmental temperature is cool, having a lowered temperature set point greatly reduces the energy required to maintain body temperature. In many animals, the lowered set point is accompanied by reductions in metabolic, nervous, and physical activity (including slower respiration and heartbeat), producing a sleeplike state known as **torpor.**

Entry into **daily torpor**—a period of inactivity keyed to variations in daily temperature—is typical of many small mammals and birds. These animals typically expend more energy per unit of body weight to keep warm than larger animals, because the ratio of body surface to volume increases as body size decreases. Hummingbirds, for example, feed actively during the daytime, when their set point is close to 40°C (104°F). During the cool of night, however, the set point drops to as low as 13°C (55°F), which allows the birds to conserve enough energy to survive overnight without feeding. Some nocturnal animals, including bats and small rodents such as the deer mouse, become torpid in cool locations during daylight hours when they do not actively feed. At night, their temperature set point rises and they become fully active **(Figure 46.22).**

Many animals enter a prolonged state of torpor tied to the seasons, triggered in most cases by a change in day length that signals the transition between summer and winter. The importance of day length has been demonstrated by laboratory experiments in which animals have been induced to enter seasonal torpor by changing the period of artificial light to match the winter or summer day length.

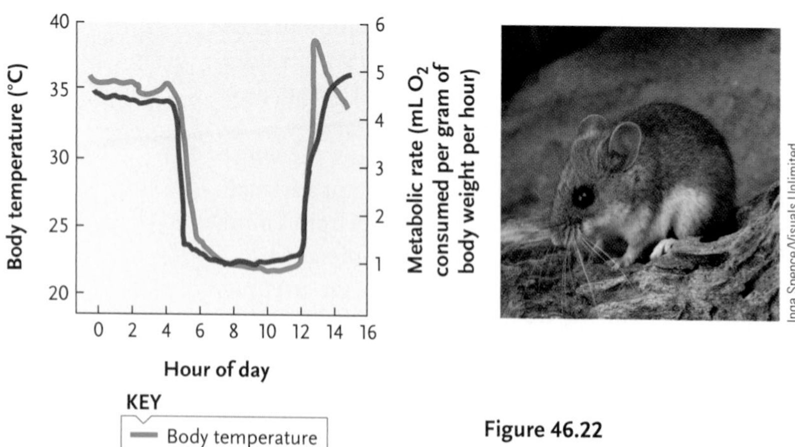

Figure 46.22
Cycle of daily torpor in a deer mouse *(Peromyscus maniculatus).*

Extended torpor during winter, called **hibernation** (*hibernus* = relating to winter), greatly reduces metabolic expenditures when food is unobtainable. Typically, hibernators must store large quantities of fats to serve as energy reserves. The drop in body temperature during hibernation varies with the mammal. In some, such as hedgehogs, woodchucks, and squirrels, body temperature may fall by 20°C (36°F) or more. In certain hedgehogs, for example, body temperature falls from about 38°C (100°F) in the summer to as low as 5° to 6°C (41° to 43°F) during winter hibernation. Body temperature even drops to near 0°C in some small hibernating mammals and, in the Arctic ground squirrel, the body supercools (goes to a below-freezing, unfrozen state) during hibernation, with body temperature dropping to about −3°C. Some ectotherms, including amphibians and reptiles living in northern latitudes and even some insects, also become torpid during winter.

The depth of torpor differs among hibernating mammals. In bears, the core temperature drops only a few degrees. Although sluggish, hibernating bears will waken readily if disturbed. They also waken normally from time to time, as when females wake to give birth during the hibernating season.

Some mammals enter seasonal torpor during summer, called **estivation** (*aestivus* = relating to summer), when environmental temperatures are high and water is scarce. Some ground squirrels, for example, remain inactive in the cooler temperatures of their burrows during extreme summer heat. Many ectotherms, among them land snails, lungfishes, many toads and frogs, and some desert-living lizards, weather such climates by digging into the soil and entering a state of estivation that lasts throughout the hot dry season.

Some Animals Use a Form of Endothermy That Does Not Heat All of Their Cores

In contrast to birds and mammals, some animals exhibit a form of endothermy that does not heat all of their cores. For example, some cold-water marine teleosts (such as tunas and mackerels) and some sharks (such as the great white) use endothermy in their aerobic swimming muscles to maintain a body core temperature as much as 10° to 12°C warmer than their surroundings. These animals have in common the fact that they migrate over long distances, swimming continuously and, therefore, generating constant heat with the swimming muscles. That heat is insufficient to heat the entire body because too much heat is lost at the gill–water interface. These animals have evolved a *countercurrent heat exchanger* system between the swimming muscles and the gills to prevent most of the loss (countercurrent exchange is discussed in Section 44.2).

The system works as follows. Cold blood from the gills is first routed through arteries under the skin,

Unanswered Questions

What is the maximum temperature for life?

As you've read, most nondormant animals cannot live above about 45°C. However, the actual limit remains controversial. Some desert ants, such as *Cataglyphis,* forage on hot Saharan sand with body temperatures exceeding 50°C! In laboratory studies, *Cataglyphis bicolor* remained active at a body temperature of 55°C for short periods. This tolerance for high temperatures may allow it to outcompete others scavengers during the heat of the day.

Most controversial are the alvinellid worms, small tube-building polychaetes that live at hydrothermal vents in the deep sea. Investigators on the submersible *Alvin* (for which the worms were named) found that temperatures in the tubes of *Alvinella* worms were routinely about 60°C, with occasional peaks over 80°C. But other scientists were skeptical, noting that accurate temperature measurements are difficult due to violent currents near the vents, which mix 300°C vent water with the near-freezing water nearby. (At the high pressures of the deep sea, water does not boil at 300°C.) Moreover, laboratory studies of alvinellid enzymes showed that they malfunction above 45°–50°C. To examine tolerance directly, Peter Girguis of Harvard and Raymond Lee of Washington State University placed *Paralvinella* worms (a species related to *Alvinella* and having enzymes with similar thermal limits) in special high-pressure chambers. The chambers have a regulated temperature gradient ranging from 20°C at one end to 61°C at the other. The worms were kept at their natural habitat pressure and observed for seven hours. The animals crawled about and settled around the area at 50°C, where they appeared to behave normally. One worm even survived 55°C water for 15 minutes. These behavioral studies (published in 2006) show that 50°C is *Paralvinella*'s preferred temperature and that it may tolerate higher temperatures briefly. However, the actual limit in nature for these and similar worms remains unknown. What sets the upper limit for animals and other eukaryotes is not known, but is suspected to be fundamental features of gene transcription, RNA processing, and/or translation that cannot be stabilized beyond a certain temperature.

What about limits for archaeans and bacteria? Hyperthermophilic ("high-heat-loving") microbes can be found in abundance at hydrothermal vents. Although none are known to live at the highest temperatures (up to 400°C), one species discovered in 2003 by Derek Lovley and Kazem Kashefi of the University of Massachusetts remained viable in the laboratory after 10 hours at 121°C, a temperature used in autoclaves for sterilizing medical equipment. But the actual upper limit for hyperthermophiles remains speculative. Understanding these aspects of cells is crucial to hypotheses about life's origins and possible life elsewhere in the universe.

How do proteins work in high urea and at high pressure?

You have just read that sharks and their relatives use urea and trimethylamine oxide (TMAO) as osmolytes. Organic osmolytes are often said to be *compatible;* that is, unlike inorganic ions such as Na^+ and Cl^-, they do not perturb proteins even at high concentrations. Thus, organic osmolytes can safely build up in an organism. However, urea is a clear exception. At the urea concentration typical of sharks (300–400 mM), many proteins—including ones in sharks—are perturbed. Indeed, biochemists often use urea to unfold proteins. How, then, can the sharks function with such high urea concentration? In the 1970s, George Somero and I, then of Scripps Institution of Oceanography, found that TMAO is not simply compatible: it actually stabilizes proteins and can *counteract* urea's effect. In mixtures of TMAO and urea at shark levels, stabilizing and destabilizing effects cancel out. Thus, by using two waste products as osmolytes, sharks maintain water balance while not perturbing proteins.

This shark finding led Robert Balaban, Maurice Burg, and coworkers at the National Institutes of Health to realize that a balancing effect could explain how mammalian kidneys survive the high levels of salt and urea in the medulla (where urea can exceed 1 M). In fact, a study led by Serena Bagnasco, now at Johns Hopkins Hospital, found that kidney cells maintain osmotic balance with sorbitol, inositol, glycerophosphorylcholine (GPC), and betaine (and not with salts, as previously believed). Moreover, GPC and betaine are *methylamines,* which, like TMAO, can counteract urea's effects. How methylamines like TMAO actually stabilize proteins remains uncertain. Wayne Bolen and colleagues at the University of Texas Medical Branch at Galveston recently showed that the peptide backbone of proteins is in a sense repelled by solutions of TMAO, making proteins fold up to avoid contact. The physicochemical properties of TMAO responsible for this effect are under investigation.

Stabilizing osmolytes may also help deep-sea organisms cope with high pressure. Low TMAO levels have long been known in bony fish (it is the source of "fishy odor"), but as osmoregulators, all bony fish were thought to have osmotic pressures of 300–400 mOsm/L, with little need for osmolytes. Recently, I and my students at Whitman College found that the deeper a species lives, the more TMAO it has. Indeed, deep-sea species can have osmotic pressures of 600 mOsm/L or more due to TMAO. Why might this be? Laboratory studies showed that TMAO readily counteracts the destabilizing effects that pressure has on protein structure and function.

Such research may be medically useful. William Welch and colleagues at the University of California at San Francisco hypothesize that stabilizing osmolytes might "repair" disease-causing mutant proteins. For example, cystic fibrosis (CF) arises from a chloride channel protein that does not fold properly, leading to symptoms that include impaired production of sweat and digestive secretions. Marybeth Howard in Welch's laboratory and collaborators recently treated cultured CF cells with organic osmolytes, and the mutant protein indeed folded and worked properly. Whether osmolytes can be used in whole mammals is now being studied.

Paul H. Yancey holds the Carl E. Peterson Endowed Chair of Sciences at Whitman College in Walla Walla, Washington. His main research interests are in areas of animal physiology, especially water stress and osmoregulation. To learn more about his research, go to http://marcus.whitman.edu/~yancey/.

which is at the same temperature as the water. The blood enters the body core through small arteries that form the countercurrent heat exchanger along with small veins that bring warm blood from the swimming muscles in the core. Countercurrent exchange warms the arterial blood, which returns to the heart in veins and is then pumped around the body in arteries, including to the core and the gills. The overall result is that heat is retained within the muscles, so that the core body temperature can remain significantly higher than that of the surrounding water.

Study Break

Describe how thermoreceptors and negative feedback pathways achieve temperature regulation in endotherms.

Review

Go to ThomsonNOW™ at www.thomsonedu.com/login to access quizzing, animations, exercises, articles, and personalized homework help.

46.1 Introduction to Osmoregulation and Excretion

- Solute concentration is measured as osmolarity in milliosmoles per liter of solution (mOsm/L). A solution can be comparatively hyperosmotic, hypoosmotic, or isoosmotic to another solution. Water moving from a region of higher osmolarity to a region of lower osmolarity across a selectively permeable membrane is known as osmosis.
- Osmoregulators keep the osmolarity of body fluids different from that of the environment. Osmoconformers allow the osmolarity of their body fluids to match that of the environment.
- Molecules and ions must be removed from the body to keep cellular and extracellular fluids isoosmotic. In most animals, extracellular fluids are filtered through tubules formed from a transport epithelium and released to the exterior of the animal as urine (Figure 46.2).
- Nitrogenous wastes are excreted as ammonia, urea, or uric acid, or as a combination of these substances (Figure 46.3).

Animation: Diffusion, osmosis, and countercurrent systems

Animation: Water and solute balance

46.2 Osmoregulation and Excretion in Invertebrates

- Most marine invertebrates are osmoconformers. Because their body fluids are isoosmotic to seawater, they expend little or no energy on maintaining water balance.
- Freshwater and terrestrial invertebrates are osmoregulators, with body fluids that are hyperosmotic to their surroundings. They must expend energy to excrete water that moves into their cells by osmosis.
- The cells of the simplest marine invertebrates exchange water and solutes directly with the surrounding seawater. More complex invertebrates have specialized excretory tubules (Figures 46.4–46.6).

46.3 Osmoregulation and Excretion in Mammals

- In mammals and other vertebrates, excretory tubules are concentrated in the kidney.
- The mammalian excretory tubule, the nephron, has a proximal end at which filtration takes place, a middle region in which reabsorption and secretion occur, and a distal end that releases urine. A network of capillaries surrounding the nephron takes up ions and water and other molecules absorbed by the nephron. The urine leaving individual nephrons is processed further in collecting ducts and then pools in the renal pelvis. From there it flows through the ureter to the urinary bladder, and through the urethra to the exterior of the animal (Figures 46.7 and 46.8).
- At its proximal end, the nephron forms a cuplike Bowman's capsule around a ball of capillaries, the glomerulus. A filtrate consisting of water, other small molecules, and ions is forced from the glomerulus into Bowman's capsule, from which it travels through the nephron and drains into the collecting ducts and renal pelvis. The proximal convoluted tubule of the nephron secretes H^+ into the filtrate and reabsorbs Na^+, Cl^-, and K^+ along with water, HCO_3^-, and nutrients. In the descending segment of the loop of Henle, water is reabsorbed by osmosis. In the ascending segment of the loop, Na^+ and Cl^- are reabsorbed. In the distal convoluted tubule, the concentrations of H^+ and salts are balanced between the urine and the interstitial fluid surrounding the nephron. In the collecting ducts, additional H^+ is secreted into the urine and water is reabsorbed; some urea is also reabsorbed at the bottom of the ducts (Figure 46.9 and Table 46.1).

Animation: Human urinary system

Animation: Human kidney

Animation: Urine formation

Animation: Tubular reabsorption

46.4 Regulation of Mammalian Kidney Function

- The kidney's autoregulation system is activated by receptors in the juxtaglomerular apparatus. The receptors trigger constriction or dilation of the afferent arteriole to keep blood flow and filtration constant during small variations in blood pressure.
- When blood volume and blood pressure drop, the hormones of the renin-angiotensin-aldosterone system (RAAS) raise blood pressure by stimulating arteriole constriction and increasing NaCl reabsorption in the kidneys (Figure 46.11).
- ADH, which increases water reabsorption and stimulates thirst, is released from the pituitary when osmoreceptors detect an increase in the osmolarity of body fluids (Figure 46.12).

Animation: Structure of the glomerulus

46.5 Kidney Function in Nonmammalian Vertebrates

- Marine teleosts continually drink seawater to replace body water lost by osmosis to their hyperosmotic environment. Excess salts and nitrogenous wastes are excreted by the gills (Figure 46.13a).
- The body fluids of sharks and rays are isoosmotic with seawater. They do not lose water by osmosis, and do not drink seawater. Excess salts are excreted in the kidney and by a rectal salt gland.
- Body fluids of freshwater fishes and amphibians are hyperosmotic to their environment, and these animals must excrete the excess water that enters by osmosis. Body salts are obtained from food and, in fishes, through the gills (Figure 46.13b).

Nitrogenous wastes are excreted from the gills of fishes and larval amphibians as ammonia, and through the kidneys of adult amphibians as urea.

- Reptiles and birds conserve water by secreting nitrogenous wastes as uric acid and by absorbing water from urine and feces in the cloaca.

46.6 Introduction to Thermoregulation

- Animals must maintain body temperature at a level that provides optimal physiological performance. Heat flows between animals and their environment by conduction, convection, radiation, and evaporation (Figures 46.15 and 46.16).
- Ectothermic animals obtain heat energy primarily from the environment; endothermic animals obtain heat energy primarily from internal reactions (Figure 46.17).

Animation: Endotherms and ectotherms

46.7 Ectothermy

- Ectotherms obtain heat energy externally and control body temperature primarily by physiological or behavioral methods of regulating heat exchange with the environment (Figure 46.18).
- Many animals undergo thermal acclimatization, a structural or metabolic change in the limits of tolerable temperatures as the environment alternates between warm and cool seasons.

46.8 Endothermy

- Endotherms obtain heat energy primarily from internal reactions and maintain body temperature over a narrow range by balancing internal heat production against heat loss from the body surface.
- Internal heat production is controlled by negative feedback pathways triggered by thermoreceptors. When deviations from the temperature set point occur, signals from the receptors bring about compensating responses such as changes in blood flow to the body surface, sweating or panting, and behavioral modifications (Figure 46.19).
- The skin of endotherms is water-impermeable, reducing heat lost by direct evaporation of body fluids. The blood vessels of the skin regulate heat loss by constricting or dilating. A layer of insulating fatty tissue under the vessels limits losses to the heat carried by the blood. The hair of mammals and feathers of birds also insulate the skin. Erection of the hair or feathers reduces heat loss by thickening the insulating layer (Figures 46.20 and 46.21).
- The temperature set point in many birds and mammals varies in daily and seasonal patterns. During cooler conditions, a lowered set point is accompanied by torpor (Figure 46.22).
- Some animals exhibit a form of endothermy in which part of their core is maintained at a temperature significantly higher than the surrounding environment.

Animation: Human thermoregulation

Questions

Self-Test Questions

1. Which of the following statements about osmoregulation is true?
   a. In freshwater invertebrates, salts move out of the body into the water because the animal is hypoosmotic to the water.
   b. A marine teleost has to fight gaining water because it is isoosmotic to the sea.
   c. Most land animals are osmoconformers.
   d. Vertebrates are usually osmoregulators.
   e. Terrestrial animals can regulate their osmolarity without expending energy.
2. One role of tubules in excretion is to:
   a. absorb H^+ ions to buffer body fluids.
   b. transport proteins across transport epithelium.
   c. reabsorb glucose and amino acids.
   d. move toxic substances from the filtrate into the cells composing the transport tubules.
   e. filter by maintaining a lower pressure in the fluid outside the tubule than inside it.
3. Products of metabolism in humans, as in:
   a. terrestrial amphibians, can include urea, which requires more energy to produce than ammonia.
   b. birds and reptiles, can include uric acid, which is nontoxic and excreted as a paste.
   c. sharks, are primarily excreted as ammonia.
   d. hydra, must be isoosmotic with the water ingested.
   e. other mammals, cannot be water as water comes only from what they drink.
4. Filtration and/or excretion can be carried out by:
   a. ciliated metanephridia in insects.
   b. protonephridia containing flame cells in flatworms.
   c. a nephron and bladder in insects.
   d. Malpighian tubules on the segments of earthworms.
   e. the hindgut, which reabsorbs Na^+ and K^+ into the hemolymph of earthworms.
5. A mammalian nephron contains the:
   a. Bowman's capsule, which delivers the filtrate to the glomerulus.
   b. Bowman's capsule, which filters fluids, 99% of which will be excreted.
   c. proximal convoluted tubule, which moves Na^+ and K^+ into the filtrate of the interstitial fluids.
   d. proximal convoluted tubule, which reabsorbs K^+, Na^+, Cl^-, H_2O, and urea.
   e. proximal convoluted tubule, which lacks microvilli to ease fluid movement through it.
6. Which of the following correctly describes a part of kidney function?
   a. Collecting ducts dilute urine because they are permeable to salt but not water.
   b. In the ascending loop of Henle, Na^+ and Cl^- move into the tubules because the osmolarity of the filtrate is increased.
   c. The descending loop of Henle receives filtrate from the ascending loop.
   d. The distal convoluted tubule pumps water into the tubule by active transport.
   e. The renal pelvis receives urine from the collecting ducts and carries it to the ureters.
7. Which of the following is an example of autoregulation of kidney function?
   a. The RAAS regulates Na^+ by secreting renin when blood pressure or blood volume decreases.
   b. The ADH system regulates water balance by decreasing water reabsorption and increasing excretion of salt.

c. Receptors in the juxtaglomerular apparatus of the distal convoluted tubule detect drops in blood pressure and cause a higher filtration rate.
d. ANF is released by the kidney to increase renin release.
e. Angiotensin lowers blood pressure by constricting arterioles.

8. Deficient water levels in humans are prevented by:
   a. osmoreceptors on the hypothalamus that detect decreases in salt concentrations.
   b. the hypothalamus stimulating the posterior pituitary to secrete a hormone that allows the collecting ducts and distal convoluted tubules to be permeable to water.
   c. inhibiting ADH, which causes a rise in osmolarity of extracellular fluids.
   d. producing dilute urine.
   e. drinking alcohol, which stimulates aldosterone to raise the osmolarity of body fluids.
9. Which best exemplifies ectothermy?
   a. The metabolic rate increases as the temperature decreases.
   b. Body temperature remains constant when environmental temperatures change.
   c. Food demand increases when temperatures drop.
   d. Virtually all invertebrate groups are ectotherms.
   e. No vertebrate groups are ectotherms.
10. Unique to endotherms is:
   a. torpor.
   b. thermal acclimatization.
   c. a nonchanging body temperature.
   d. response to seasonal temperature changes.
   e. thermoregulation by a hypothalamus.

Questions for Discussion

1. A urinalysis reveals glucose, urea, hemoglobin, and sodium. Which of these substances are abnormal in urine, and why?
2. As a person ages, nephron tubules lose some of their ability to concentrate urine. What is the effect of this change?
3. Shivering increases air movement over the body surface. What effect does this air movement have on heat conservation in the shivering animal?
4. What heat transfer processes might account for the change in body temperature when a mammal's body temperature undergoes daily variations?

Experimental Analysis

Design experiments to demonstrate the role of fluid consumption in thermoregulation during endurance exercise.

Evolution Link

Humans produce urea as an excretion product, whereas reptiles and birds produce uric acid. Indeed, human kidneys are not as efficient as those of reptiles and birds. What does this mean in an evolutionary sense?

How Would You Vote?

Many companies use urine testing to screen for drug and alcohol use among prospective employees. Some people say this is an invasion of privacy. Do you think employers should be allowed to require a person to undergo urine testing before being hired? Go to www.thomsonedu.com/login to investigate both sides of the issue and then vote.

A newly fertilized human egg passing down the oviduct on its way to implantation in the wall of the uterus (colorized SEM).

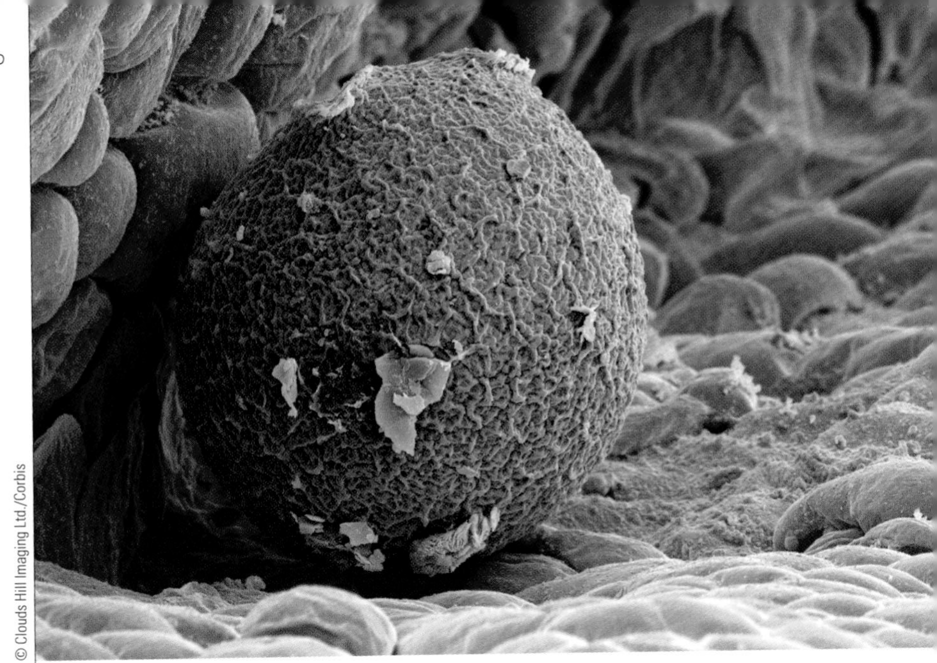

Study Plan

47 Animal Reproduction

Why It Matters

It is 7 days after the October full moon and night is falling. All the inhabitants of the Samoan island of Tutuila who have access to a boat are gathered on the island's large lagoon. Some hold lanterns and look into the water; others have nets at the ready. They are awaiting the palolo worm *(Eunice viridis)*, which has appeared in the water as the moon rises on this same night of the lunar year for as long as the islanders can remember.

The moon peeks over the horizon and the excitement of the crowd rises. Then, there they are, untold thousands of blue and green worms, squirming in the water like animated spaghetti. The boaters scoop up the worms by the netfull and dump them into buckets. When the buckets are full, the islanders glide toward the shore where steaming pots are waiting, for palolo worms are a delicacy that the islanders savor only once a year. For the islanders, a night of feasting, singing, and dancing will follow as they cook and eat the worms. Their flavor has been described by some as similar to that of caviar; by others like, well, palolo worms.

The worms that squirm to the surface to delight the islanders are actually not complete individuals. They are tail sections about 10 to

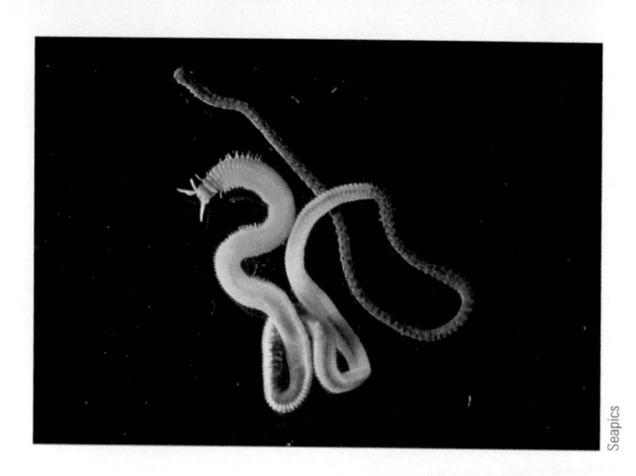

Seapics

Figure 47.1
The palolo worm. Gametes are packed into segments of the tail section (in blue).

20 cm long that break from adults after they become filled with eggs or sperm. The adults are polychaete annelids that live in burrows in coral reefs of the Samoan and Fiji islands **(Figure 47.1).** These annelids develop tail segments once a year, just after the October full moon. On the seventh night following the full moon, the tails break off and swim to the surface, where—if Samoan gourmets do not net them first—they disintegrate and release eggs and sperm by the millions, turning the water of the lagoon milky. The anterior ends of the worms, safe in their burrows, will survive to produce tails for next year's mating frenzy. A biological clock in the worms, timed by periods of moonlight, precisely sets both the appearance of the mating swarm and indirectly, the appearance of the islanders with their boats.

The swarm of the palolo worms is only one of many adaptations that accomplish mating in animals. For animals that reproduce by eggs and sperm, the adaptations are as diverse as the number of species on Earth. This diversity allows individuals of the same species to find each other and unite eggs and sperm. Within the diversity, however, are underlying patterns that are shared by all animals.

Both the underlying patterns and the diversity of animal reproduction are the subjects of this chapter. We also discuss the development of eggs and sperm, and the union of egg and sperm that begins the development of a new individual. The next chapter continues with the events of development after eggs and sperm have united.

47.1 Animal Reproductive Modes: Asexual and Sexual Reproduction

Reproduction is part of a life cycle in which individuals grow, develop, and reproduce according to instructions encoded in DNA. Rather than survival of the individual, reproduction is the means of passing on the individual's genes to new generations of the species. As such, it is among the most vital functions of living organisms.

Two basic modes of reproduction operate in the animal kingdom. In **asexual reproduction,** a single individual gives rise to offspring without fusion of **gametes** (egg and sperm); that is, there is no genetic input from another individual. In **sexual reproduction,** male and female parents produce offspring through the union of egg and sperm generated by meiosis (meiosis is discussed in Chapter 11).

Asexual Reproduction Produces Offspring with Genes from Only One Individual

Many aquatic invertebrates and some terrestrial annelids and insects reproduce asexually. Asexual reproduction is rare among vertebrates. In asexual reproduction, one to many cells of a parent's body develop directly into a new individual. In a few animals that undergo asexual reproduction, the cells taking part are genetically varied products of meiosis, but in most they are products of mitosis. The offspring therefore are genetically identical to one another and to the parent: in other words, they are genetic clones of the parent. For this reason, asexual reproduction of this kind is also called *clonal reproduction.*

Genetic uniformity of offspring can be advantageous in environments that remain stable and uniform. Asexual reproduction tends to preserve gene combinations producing individuals that are successful in such environments. Further, individuals do not have to expend energy to produce gametes or find a mate. Asexual reproduction can also bring reproductive advantages to individuals living in sparsely settled populations, or to sessile animals, which cannot move from place to place.

Asexual reproduction involving mitosis occurs in animals by three basic mechanisms: *fission, budding,* and *fragmentation.* In **fission,** the parent separates into two or more offspring of approximately equal size. Planarians (flatworms), for instance, reproduce asexually by fission; depending on the species, they may divide by transverse or longitudinal fission. In **budding,** a new individual grows and develops while attached to the parent. Sponges, tunicates, and some cnidarians reproduce asexually by this mechanism. The offspring may break free from the parent, or remain attached to form a *colony.* In the cnidarian *Hydra,* for example, an offspring buds and grows from one side of the parent's body and then detaches to become a separate individual **(Figure 47.2).** Among many corals, the buds remain attached when their growth is complete, forming colonies of thousands of interconnected individuals. In **fragmentation,** pieces separate from a parent's body and develop *(regenerate)* into new individuals. Many species of cnidarians, flatworms, annelids, and some echinoderms can reproduce by fragmentation.

Some animals produce offspring by the growth and development of an egg without fertilization. The offspring may be haploid or diploid depending on the species. This form of asexual reproduction is called **parthenogenesis** (*parthenos* = virgin; *genesis* = birth). Because the egg from which a parthenogenetic offspring is produced derives from meiosis in the female parent, the offspring are not genetically identical to the parent or to each other. (How chromosome segregation and genetic recombination during meiosis produces gametes with gene combinations different from the parent is described shortly.)

Parthenogenesis occurs in some invertebrates, including certain aphids, water fleas, bees, and crustaceans. In bees, for instance, haploid male drones are produced parthenogenetically from unfertilized eggs produced by reproductive females (queens) while new queens and sterile workers develop from fertilized eggs. Parthenogenesis also occurs in some vertebrates, for example, in certain fish, salamanders, amphibians, lizards, and turkeys. In these animals, an egg, produced by meiosis, typically doubles its chromosomes to produce a diploid cell that begins development. In single-sex species where females have two identical sex chromosomes, the offspring are female, whereas in single-sex species where males have two identical sex chromosomes, the offspring are male. For instance, all whiptail lizards *(Cnemidophorus)* are females, produced solely by parthenogenesis. Interestingly, these females go through the motions of mating and copulation with each other.

Sexual Reproduction Generates Diversity among Offspring

Animals reproduce sexually by the union of sperm and eggs produced by meiosis. The overriding advantage of sexual reproduction is the generation of genetic diversity among offspring. This diversity increases the chance that, in a changing environment, at least some offspring will grow and reproduce successfully. Diversity also increases the chance that offspring may be able to live and reproduce in environments previously unoccupied by the species.

Two mechanisms that are part of meiosis give rise to the genetic diversity in eggs and sperm: *genetic recombination* (see Section 11.2) and the *independent assortment* of chromosomes of maternal and paternal origin (see Section 12.1). Genetic recombination mixes the alleles of parents into new combinations within chromosomes; independent assortment selects random combinations of maternal and paternal chromosomes to be placed in gamete nuclei. Additional variability is generated at fertilization when eggs and sperm from genetically different individuals fuse together at random to initiate the development of new individuals. To these sources of variability are added random DNA mutations, which are the ultimate source of variability for both sexual and asexual reproduction.

The disadvantages of sexual reproduction include the expenditure of energy and raw materials in producing gametes and finding mates. The need to find mates can also expose animals to predation and takes time from finding food and shelter and caring for existing offspring.

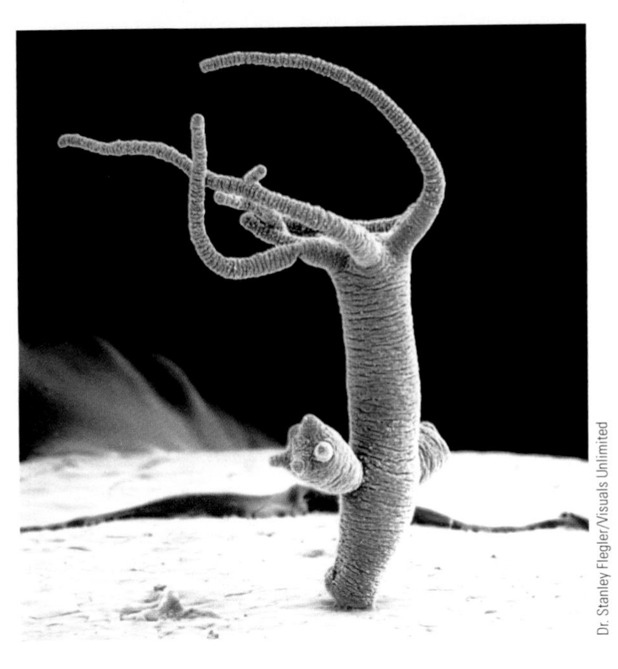
Dr. Stanley Flegler/Visuals Unlimited

Figure 47.2 Asexual reproduction by budding in *Hydra* (colorized SEM).

With these advantages and disadvantages in mind, we now turn to the mechanisms of sexual reproduction, which include both cellular and whole-organism activities. We begin with the cellular mechanisms in the next section.

Study Break

What are the advantages and disadvantages of asexual reproduction? Of sexual reproduction?

47.2 Cellular Mechanisms of Sexual Reproduction

The cellular mechanisms of sexual reproduction are **gametogenesis**, the formation of male and female gametes, and **fertilization**, the union of gametes that initiates development of a new individual. The pairing of a male and a female for the purpose of sexual reproduction is **mating.**

Gametogenesis Involves the Coordinated Events of Meiosis and Sperm and Egg Development

Gametes in most animals form from **germ cells,** a cell line that is set aside early in embryonic development and remains distinct from the other, **somatic cells** of the body. During development, the germ cells collect in specialized gamete-producing organs, the **gonads**—the

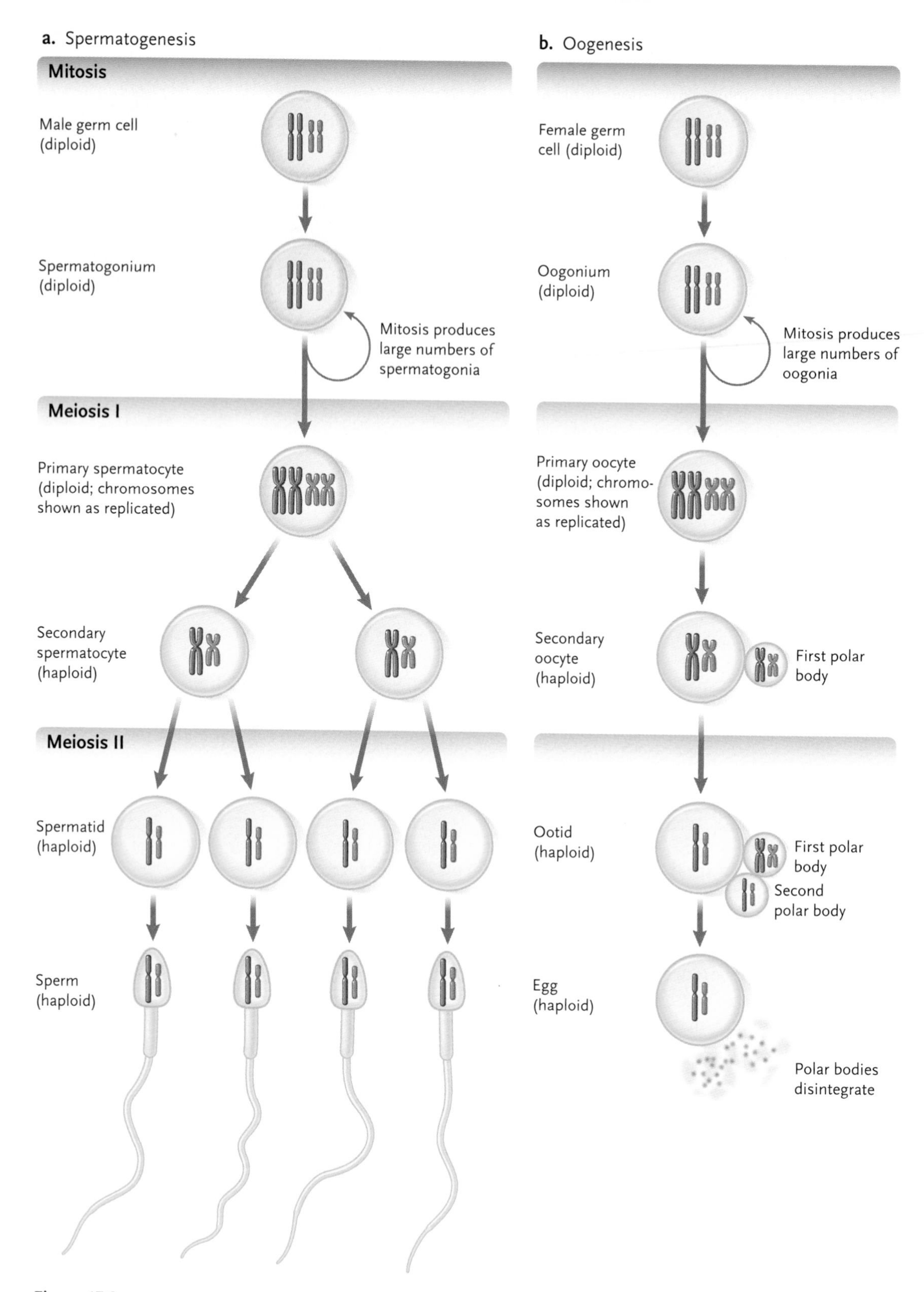

Figure 47.3
The mitotic and meiotic divisions producing eggs and sperm from germ cells. **(a)** Spermatogenesis. **(b)** Oogenesis. The first polar body may or may not divide, depending on the species, so that either two or three polar bodies may be present at the end of meiosis. Two are shown in this diagram.

testes (singular, *testis*) in males and **ovaries** in females. Mitotic divisions of the germ cells produce **spermatogonia** in males and **oogonia** in females; these are the cells that enter meiosis to give rise to gametes **(Figure 47.3).** In some animals, the germ cells also give rise to families of cells that assist gamete development.

Meiosis reduces the number of chromosomes from the diploid level characteristic of somatic cells of the species, in which there are two copies of each chromosome, to the haploid level of gametes, in which there is one copy of each chromosome. The fusion of a haploid sperm and egg during fertilization restores the diploid number of chromosomes and produces a **zygote**, the first cell of a new individual.

During the meiotic divisions, the developing gametes are known as **spermatocytes** or **oocytes;** at the end of meiosis they become *spermatids* or *ootids*. When meiosis is complete, the haploid cells develop into mature sperm cells, also called **spermatozoa** (singular, *spermatozoon*) or simply *sperm;* and egg cells, also called **ova** (singular, *ovum*) or simply *eggs*. The process of producing sperm is called **spermatogenesis** and the process of producing eggs is called **oogenesis.** The sperm of most animal species are motile cells, driven through a watery medium by the whiplike beating of a flagellum that extends from the posterior end of the cell. The eggs of all animals are nonmotile cells, typically much larger than sperm of the same species.

a. Human sperm

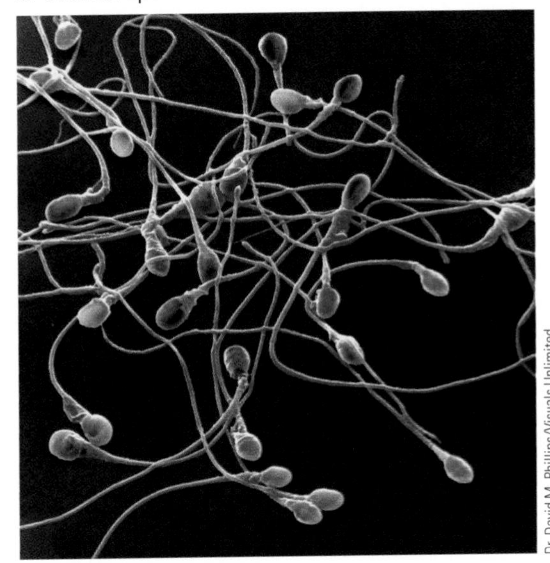

b. Sperm structure

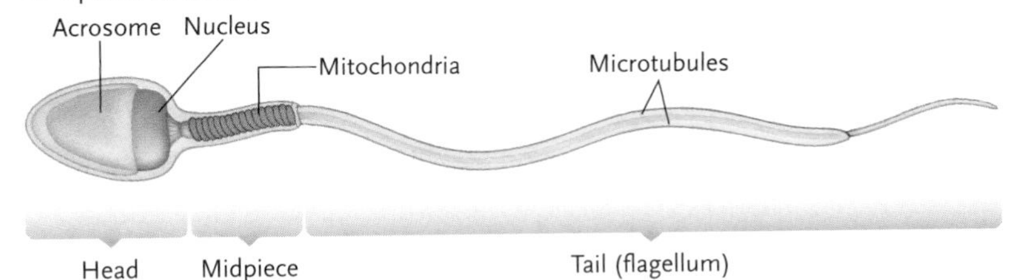

Figure 47.4
Spermatozoa. **(a)** Photomicrograph of human sperm. **(b)** Structure of a sperm.

Spermatogenesis. The events of spermatogenesis produce haploid cells, specialized to deliver their nuclei to eggs of the same species. Two meiotic divisions produce four haploid spermatids (see Figure 47.3a), which develop into mature sperm **(Figure 47.4).** During maturation, most of the cytoplasm is lost, except for mitochondria, which surround the base of a flagellum. These mitochondria produce the ATP used as the energy source for flagellar beating. At the head of the sperm, a specialized secretory vesicle, the **acrosome,** forms a cap over the nucleus. The acrosome contains enzymes and other proteins that help the sperm attach to and penetrate the surface coatings of an egg of the same species.

Oogenesis. In oogenesis, only one of the cell products of meiosis develops into a functional egg, which retains almost all of the parent cell's cytoplasm. The other products form nonfunctional cells called **polar bodies** (see Figure 47.3b). The unequal cytoplasmic divisions concentrate nutrients and other molecules required for development in the egg. In most species, the polar bodies eventually disintegrate and do not contribute to fertilization or embryonic development.

The oocytes of most animals do not actually complete meiosis until fertilization. For example, mammals follow a complex pattern in which oocytes stop developing at the end of the first meiotic prophase, within a few weeks after a female is born. The oocytes remain in the ovary at this stage until the female is sexually mature. In humans, some oocytes may remain in prophase of the first meiotic division for perhaps 50 years.Then, one to several oocytes advance to the metaphase of the second meiotic division and are released from the ovary at intervals ranging from days to months, or at certain seasons, depending on the species. As in other animals, meiosis is completed at fertilization to produce the fully mature egg **(Figure 47.5).** Mature eggs are the largest cell type of an animal species.

An egg typically has specialized features, which include stored nutrients required for at least the early stages of embryonic development; egg coats of one or more kinds, which protect the egg from mechanical injury and infection and, in some species, protect the embryo after fertilization; and mechanisms that prevent the egg from being fertilized by more than one sperm cell (discussed shortly).

Egg coats are surface layers added during oocyte development or fertilization in many species. The **vitelline coat,** called the **zona pellucida** in mammals (see Figure 47.5) is a gel-like matrix of proteins, glycoproteins, or polysaccharides immediately outside of the plasma membrane of the egg cell. Insect eggs have additional outer protein coats that form a hard, water-

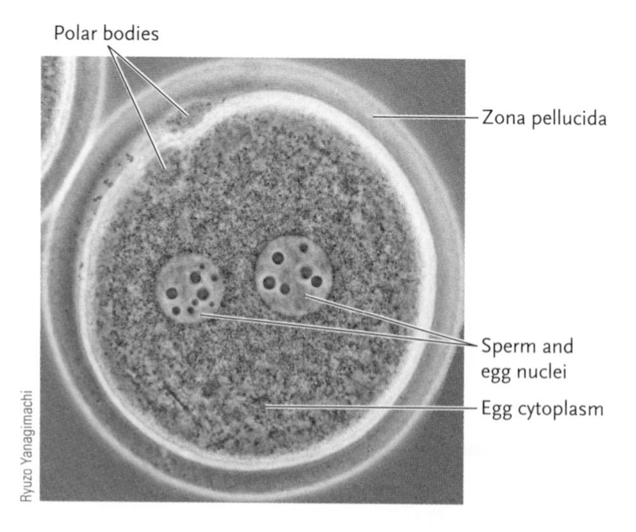

Figure 47.5
A mature hamster egg that has been fertilized.

impermeable layer for preventing desiccation. Amphibians and some echinoderms instead have an additional outer egg jelly layer (see Figure 47.8) that protects the egg from drying.

In birds, reptiles, and one group of egg-laying mammals, the **monotremes,** the egg white, a thick solution of proteins, surrounds the vitelline coat. Outside the white is the *shell* of the egg, flexible and leathery in reptiles and mineralized and brittle in birds. Both the egg white and the shell are added while the egg—fertilized or not—is in transit through the **oviduct,** the tube through which the egg moves from the ovary to the outside of the body. In mammals, the egg is surrounded by **follicle cells** during its development. These cells, which grow from ovarian tissue, nourish the developing egg. They also make up part of the zona pellucida while the egg is in the ovary, and remain as a protective layer after it is released.

The amount of stored nutrients in an egg varies with the animal. Mammalian eggs are microscopic, containing few stored nutrients. In mammals, the embryo develops inside the mother and is supplied with nutrients by the mother's body. In contrast, the relatively huge eggs of birds and reptiles contain all the nutrients required for complete embryonic development: the "yolk" contains the egg cell, and the "white" contains the nutrients. No matter what the size of an animal egg, however, most of the volume is cytoplasm, and the egg nucleus is microscopic or nearly so in all species.

Fertilization Requires an Internal or External Aquatic Medium

Eggs and sperm are delivered from the ovaries and testes to the site of fertilization by oviducts in females and by sperm ducts in males; in many species, external accessory sex organs participate in the delivery. **Figure 47.6** shows examples of invertebrate and vertebrate reproductive systems. The nonmotile eggs move through the oviducts on currents generated by the beating of cilia lining the oviducts, or by contractions of the oviducts or the body wall.

Depending on the species, fertilization may take place externally, in a watery medium outside the body of both parents, or internally, in a watery fluid inside the body of the female. In **external fertilization,** which occurs in most aquatic invertebrates, bony fishes, and amphibians, sperm and eggs are shed into the surrounding water. The sperm swim until they collide with an egg of the same species. The process is helped by synchronization of female and male gamete release, and by the enormous quantities of gametes released. In some animals, such as sea urchins and amphibians, the sperm are attracted to the egg by diffusible attractant molecules released by the egg.

Most amphibians, even terrestrial species such as toads, mate in an aquatic environment. Frogs typically mate by a reflex response called *amplexus,* in which the male clasps the female tightly around the body with his forelimbs **(Figure 47.7).** The embrace stimulates the female to shed a mass of eggs into the water through the *cloaca*—the cavity in reptiles, birds, amphibians, and many fishes into which both the intestinal and genital tracts empty. As the eggs are released, they are fertilized by sperm released by the male.

Internal fertilization takes place in invertebrates such as annelids, some arthropods, and some mollusks, and in vertebrates such as reptiles, birds, mammals, some fishes, and some salamanders. In these animals, the sperm are released by the male close to or inside the entrance of the reproductive tract of the female. The sperm swim through fluids in the reproductive tract until they reach and fertilize each egg. In some species, molecules released by the egg attract the sperm to its outer coats. The physical act involving the introduction of the male's accessory sex organ (for example, penis) into a female's accessory sex organ (for example, vagina) to accomplish internal fertilization is known as **copulation.** Internal fertilization makes terrestrial life possible by providing the aquatic medium required for fertilization inside the female's body without the danger of gametes drying by exposure to the air.

Sharks and rays have evolved a form of internal fertilization in which the male uses a pair of modified pelvic fins as accessory sex organs to channel sperm directly inside the female's cloaca. Male reptiles, birds, and mammals also have accessory sex organs that place sperm directly inside the reproductive tract of females, where fertilization takes place. In reptiles and birds, sperm fertilize eggs as they are released from the ovary and travel through the oviducts, before the shell is added. In mammals, the male's penis delivers sperm into the female's vagina. Unlike the cloaca, which has both sexual and excretory functions, the vagina is specialized for reproduction. Fertilization takes place when sperm swim into the tubular oviducts containing the eggs.

Fertilization Involves Fusion of a Sperm and an Egg, Which Activates the Egg for Development

Once a sperm touches the outer surface of an egg of the same species **(Figure 47.8a),** receptor proteins in the sperm plasma membrane bind the sperm to the vitelline coat or zona pellucida. In most animals, only a sperm from the same species as the egg can recognize and bind to the egg surface.

Species recognition is highly important in animals that carry out external fertilization, because the water surrounding the egg may contain sperm of many different species. It is less important in internal fertilization, because structural adaptations and behavioral patterns of mating usually limit sperm transfer from males to females of the same species.

Fertilization. After the initial attachment of sperm to egg, the events of fertilization proceed in rapid succession **(Figure 47.8b).** The actual attachment event triggers the **acrosome reaction,** in which enzymes contained in the acrosome are released from the sperm and digest a path through the egg coats. The sperm, with its tail still beating, follows the path until its plasma membrane touches and fuses with the plasma membrane of the egg. Fusion introduces the sperm nucleus into the egg cytoplasm and activates the egg to complete meiosis and begin development.

Egg Activation and Blocks to Polyspermy. Two mechanisms can prevent more than one sperm from fertilizing the egg: a *fast block* within seconds of fertilization, and a *slow block* within minutes.

In many invertebrate species, such as the sea urchin, the fusion of egg and sperm opens ion channels in the egg's plasma membrane, spreading a wave of electrical depolarization over the egg surface, much like the nerve impulse traveling along a neuron. The depolarization alters the egg plasma membrane so that it cannot fuse with any additional sperm, thereby eliminating the possibility that more than one set of paternal chromosomes enters the egg. Because it occurs within a few seconds after fertilization, the barrier set up by the wave of depolarization is called the **fast block to polyspermy.**

The fast block depends on a change in the egg's membrane potential from negative to positive. For example, Laurinda Jaffe, of the University of Connecticut Health Center, found that if the membrane potential of a sea urchin egg was artificially kept at a negative value, no fast block was set up, and additional sperm could fuse with the plasma membrane. If the membrane was instead kept positive before sperm contact, fertilization was entirely blocked.

In vertebrates, the wave of membrane depolarization following sperm–egg fusion is not as pronounced, and does not prevent additional sperm from fusing

a. *Drosophila* (fruit fly)

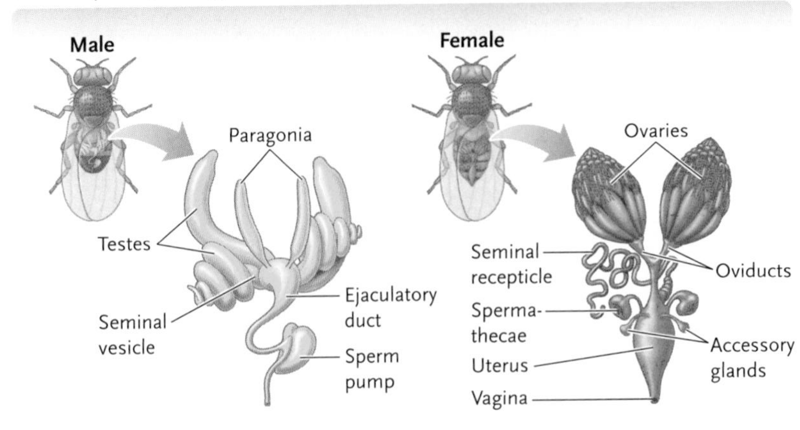

b. Amphibian (frog)

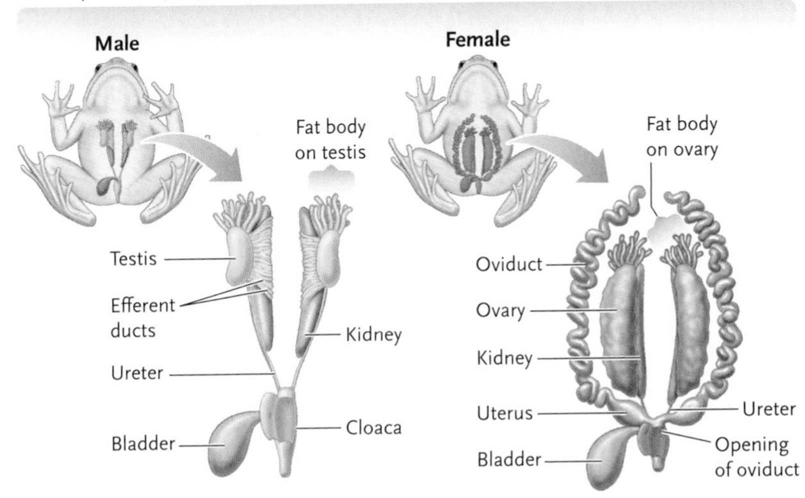

c. Mammal (cat)

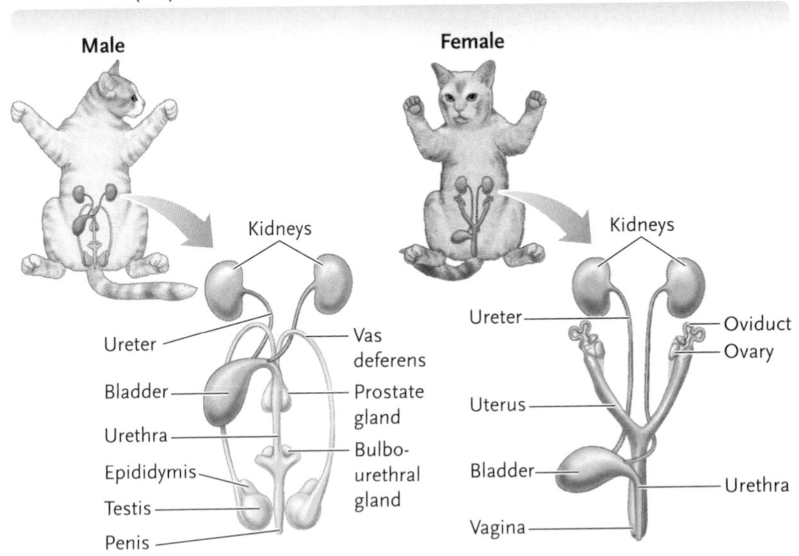

Figure 47.6
Some reproductive systems. **(a)** An insect, *Drosophila* (fruit fly). **(b)** An amphibian, a frog. **(c)** A mammal, a cat. Female systems are shown in blue, and male systems in yellow.

Figure 47.7
A male leopard frog *(Rana pipiens)* clasping a female in a mating embrace known as amplexus. The tight squeeze by the male frog stimulates the female to release her eggs, which can be seen streaming from her body, embedded in a mass of egg jelly. Sperm released by the male fertilize the eggs as they pass from the female.

with the egg. However, any additional sperm nuclei entering the egg cytoplasm usually break down and disappear, so that only the first sperm nucleus to enter fuses with the egg nucleus.

In both invertebrates and vertebrates, fusion of egg and sperm triggers the release of stored calcium (Ca^{2+}) ions from the endoplasmic reticulum into the cytosol. The Ca^{2+} ions activate control proteins and enzymes that initiate intense metabolic activity in the fertilized egg, including a rapid increase in cellular oxidations and synthesis of proteins and other molecules.

The Ca^{2+} ions also trigger the **cortical reaction**, in which **cortical granules**, secretory vesicles just under the plasma membrane, fuse with the egg's plasma membrane and release their contents to the outside (see Figure 47.8b). Enzymes released from the cortical granules alter the egg coats within minutes after fertilization, so that no further sperm can attach and penetrate to the egg. Once this barrier, termed the **slow block to polyspermy**, is set up, no further sperm can reach the egg plasma membrane in any animal species.

The importance of Ca^{2+} to cortical granule release has been demonstrated experimentally: if Ca^{2+} is added to the cytoplasm, the granules are released in

a. Sperm adhering to egg

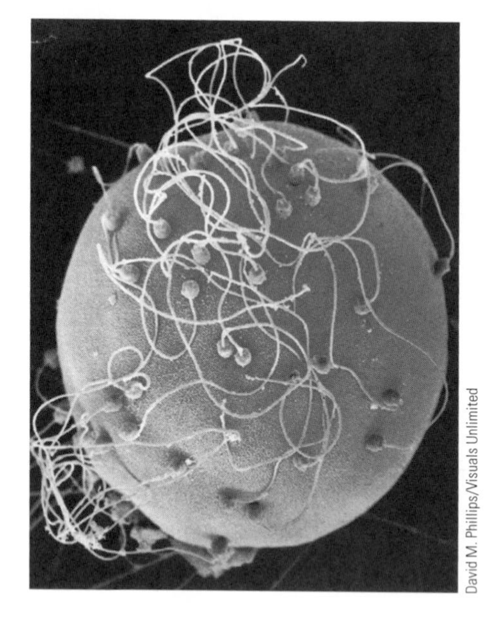

b. Steps in fertilization

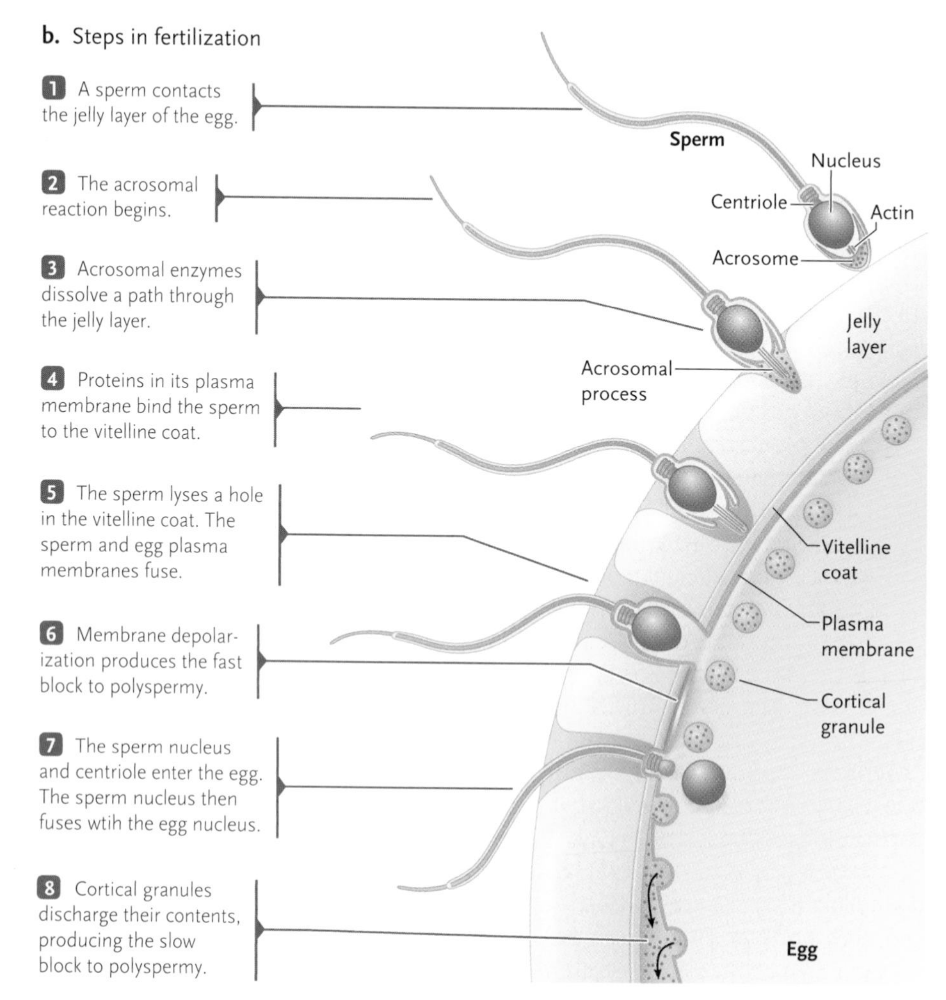

1 A sperm contacts the jelly layer of the egg.
2 The acrosomal reaction begins.
3 Acrosomal enzymes dissolve a path through the jelly layer.
4 Proteins in its plasma membrane bind the sperm to the vitelline coat.
5 The sperm lyses a hole in the vitelline coat. The sperm and egg plasma membranes fuse.
6 Membrane depolarization produces the fast block to polyspermy.
7 The sperm nucleus and centriole enter the egg. The sperm nucleus then fuses wtih the egg nucleus.
8 Cortical granules discharge their contents, producing the slow block to polyspermy.
Sperm
Nucleus
Centriole
Actin
Acrosome
Jelly layer
Acrosomal process
Vitelline coat
Plasma membrane
Cortical granule
Egg

Figure 47.8
Fertilization. **(a)** Sperm adhering to the surface coat of a sea urchin egg. Of the many sperm that may initially adhere to the outer surface of an egg, usually only one accomplishes fertilization. **(b)** Steps of fertilization in a sea urchin.

unfertilized eggs; conversely, if Ca^{2+}-binding chemicals are added to the cytoplasm of unfertilized eggs, so that the Ca^{2+} concentration cannot rise, cortical granule release does not occur after fertilization.

After the sperm nucleus enters the egg cytoplasm, microtubules move the sperm and egg nuclei together in the egg cytoplasm and they fuse. The chromosomes of the egg and sperm nuclei then assemble together and enter mitosis. The subsequent, highly programmed events of embryonic development, which convert the fertilized egg into an individual capable of independent existence, are described in the next chapter.

Of the structures in a sperm cell, only the paternal chromosomes, the microtubule organizing center, and one or two centrioles (see Section 10.3 and Figure 10.11) survive in the egg. Therefore, with the exception of the microtubule organizing center and centrioles, all the cytoplasmic structures of the embryo, and of the new individual, are maternal in origin. The centrioles of the new individual are normally paternal in origin.

Reproductive Systems May Be Oviparous or Viviparous in Animals with Internal Fertilization

In animals with internal fertilization, three major types of support for embryonic development have evolved: *oviparity,* meaning egg laying; *viviparity,* meaning live bearing; and *ovoviparity,* meaning live bearing from eggs that hatch internally. **Oviparous** animals (*ovum* = egg; *parere* = to give birth to) lay eggs that contain the nutrients needed for development of the embryo outside the mother's body. Examples are insects, spiders, most reptiles, and birds. The only oviparous mammals are the *monotremes:* the echidnas and *Ornithorhynchus anatinus* (the duck-billed platypus), both of which inhabit Australia.

Viviparous animals (*vivus* = alive) retain the embryo within the mother's body and nourish it during at least early embryo development. All mammals except the monotremes are viviparous. Viviparity is seen also in all other vertebrate groups except for the crocodiles, turtles, and birds.

In viviparous animals, development of the embryo takes place in a specialized saclike organ, the **uterus** *(womb)*. Among mammals, one group, called the *placental mammals* or *eutherians,* has a specialized temporary structure, the **placenta,** that connects the embryo with the uterus. The placenta facilitates the transfer of nutrients from the mother's blood to the embryo and of wastes in the opposite direction. Humans are placental mammals. The other group of mammals, the *marsupials* or *metatherians,* originally were called nonplacental mammals because of a belief that they lacked a placenta. In fact, they do have a placenta, but it derives from a different tissue than that of eutherians and does not connect the embryo and the uterus. Instead

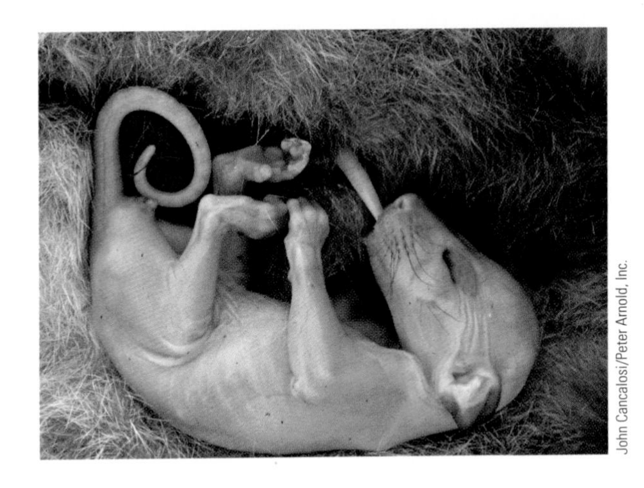

Figure 47.9 Developing offspring of a marsupial mammal, an opossum, attached to nipples in the marsupium (pouch) of the mother.

it provides nutrients to the embryo from an attached membranous sac containing yolk for only the early stages of its development. In many metatherians, the embryo is then born and crawls over the mother's fur to reach the **marsupium,** an abdominal pouch in which it attaches to nipples and continue its development **(Figure 47.9).** Kangaroos, koalas, wombats, and opossums are marsupials.

In some animals, such as some fishes, lizards, and amphibians, many snakes, and many invertebrates, fertilized eggs are retained within the body and the embryo develops using the nutrients provided by the egg. There is no uterus or placenta involved. When development is complete the eggs hatch inside the mother and the young are released to the exterior. Animals showing this form of reproduction are known as **ovoviviparous** animals.

Hermaphroditism Is a Variation on Sexual Reproduction

Some animals have evolved modified mechanisms that they use as their normal sexual reproduction process. One of these mechanisms is **hermaphroditism** (from *Hermes* + *Aphrodite,* a Greek god and goddess), in which both mature egg-producing and mature sperm-producing tissue is present in the same individual. That is, hermaphroditic individuals are able to produce both eggs and sperm. Most flatworms, earthworms, land snails, and numerous other invertebrates are hermaphroditic; in humans and other mammals, hermaphroditism is a rare, abnormal condition.

Most hermaphroditic animals do not fertilize themselves. In those animals, self-fertilization is prevented by anatomical barriers that prevent individuals from introducing sperm into their own body, or by mechanisms in which the egg and sperm mature at different times. The prevention of self-fertilization maintains the genetic variability of sexual reproduction.

Hermaphroditism takes two forms: **simultaneous hermaphroditism,** in which individuals develop functional ovaries and testes at the same time, and

a.

b. Sex organs

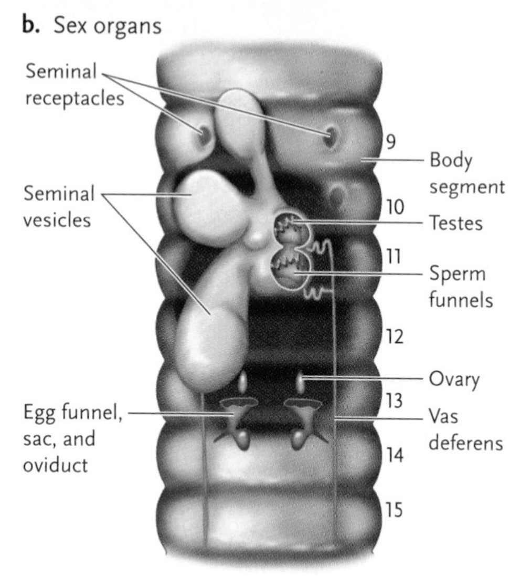

Figure 47.10
Simultaneous hermaphroditism in the earthworm. **(a)** Copulation by a mating pair of earthworms, in which each earthworm releases sperm that fertilizes eggs in its partner. **(b)** Sex organs in the earthworm.

sequential hermaphroditism, in which individuals change from one sex to the other. The two earthworms shown in **Figure 47.10** provide a common example of simultaneous hermaphroditism. The only known vertebrate simultaneous hermaphrodites are hamlets (genus *Hypoplectrus*), a group of predatory sea basses. Sequential hermaphroditism is seen among a number of invertebrates (for example, the gastropod, the slipper shell *Crepidula fornicata*) and some ectothermic vertebrates, notably fishes (for example, the clownfish, genus *Amphiprion*). In some species the initial sex is male (as with the slipper shell and the clownfish), and in others it is female.

Study Break

1. What are egg coats, and what is their function? What egg coats do mammalian and bird eggs have?
2. How is the slow block to polyspermy brought about?

47.3 Sexual Reproduction in Humans

Except for structural details, human reproduction is typical of that of eutherian (placental) mammals. Internally, these mammals have a pair of gonads, either ovaries or testes. The gonads have a dual function in mammals, as they do in all vertebrates: they both produce gametes and secrete hormones responsible for sexual development and mating behavior (see Section 40.4). Males have ducts that carry sperm from the testes to the exterior. Females have an oviduct that leads from each ovary to the uterus, in which fertilized eggs implant and proceed through embryonic development. Nutrients from the mother and wastes from the embryo are exchanged through the placenta. After birth, the newborn offspring is nourished with milk secreted by the mother's mammary glands.

In this section we survey reproductive structures and functions in humans as representative of eutherian mammals. Our story of human development continues in the next chapter, which traces the process from fertilization to birth.

Human Female Sexual Organs Function in Oocyte Production, Fertilization, and Embryonic Development

Human females have a pair of ovaries suspended in the abdominal cavity **(Figure 47.11)**. An oviduct leads from each ovary to the uterus, which is a hollow, saclike organ with walls containing smooth muscle. The uterus is lined by the endometrium, formed by layers of connective tissue with embedded glands and richly supplied with blood vessels. If an egg is fertilized and begins development, it must implant in the endometrium to continue developing. The lower end of the uterus, the **cervix,** opens into a muscular canal, the **vagina,** which leads to the exterior. Sperm enter the female reproductive tract via the vagina and, at birth, the baby passes from the uterus to the outside through the vagina.

At the birth of a female, each ovary contains about 1 million oocytes, arrested at the end of the first meiotic prophase. Of these oocytes, about 200,000 to 400,000 survive until a female becomes sexually mature; about 400 are **ovulated**—released into the oviducts as immature eggs—during a woman's lifetime. The egg is released into the abdominal cavity and pulled into the nearby oviduct by the current produced by the beating of the cilia lining the oviduct. The cilia also propel the egg through the oviduct and into the uterus. Fertilization of the egg occurs in the oviduct.

The external female sex organs, collectively called the **vulva,** surround the opening of the vagina. Two folds of tissue, the **labia minora,** run from front to rear on either side of the opening to the vagina. These folds are partially covered by a pair of fleshy, fat-padded

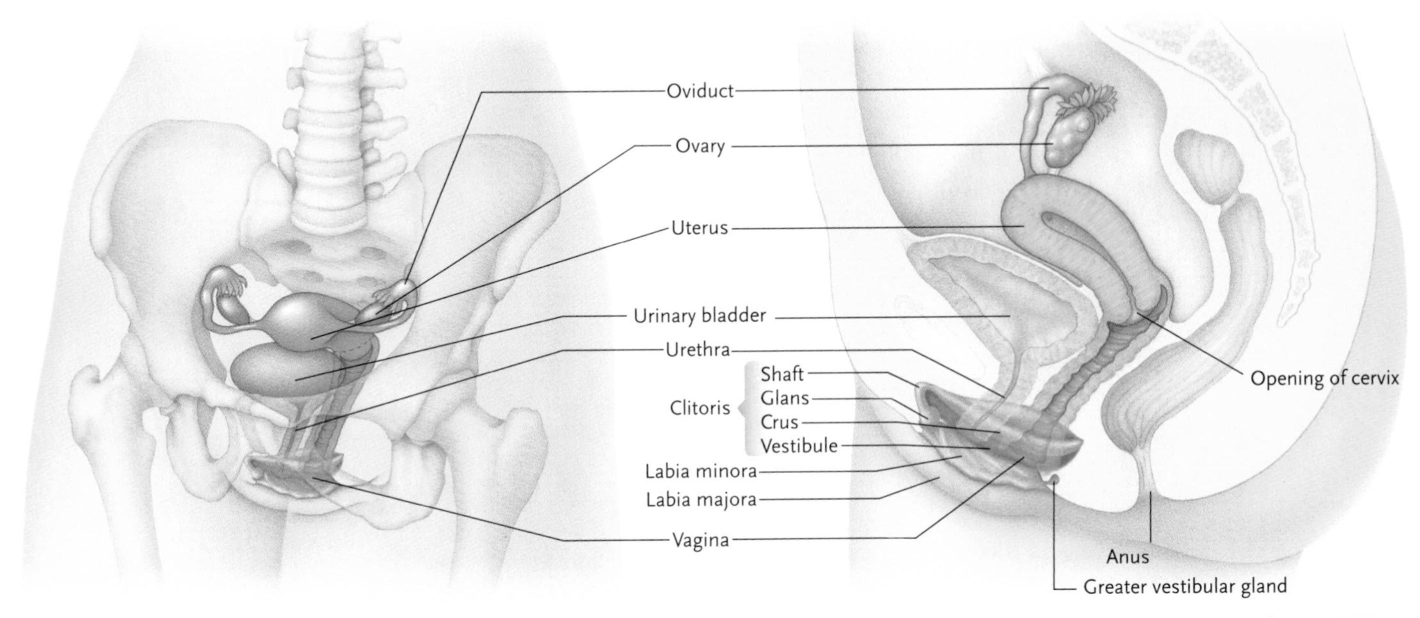

Figure 47.11 The reproductive organs of a human female.

folds, the **labia majora,** which also run from front to rear on either side of the vagina. At the anterior end of the vulva, the labia minora join to partly cover the head of the **clitoris.** The rest of the clitoris is within the body. The clitoris contains erectile tissue and has the same embryonic origins as the penis. A pair of **greater vestibular glands,** with openings near the entrance to the vagina, secretes a mucus-rich fluid that lubricates the vulva. The opening of the urethra, which conducts urine from the bladder, is located between the clitoris and the vaginal opening. Most nerve endings associated with erotic sensations are concentrated in the clitoris, in the labia minora, and around the opening of the vagina. When a human female is born, a thin flap of tissue, the **hymen,** partially covers the opening of the vagina. This membrane, if it has not already been ruptured by physical exercise or other disturbances, is broken by the first sexual intercourse.

Ovulation in Human Females Occurs in a Monthly Cycle

Reproduction in human females is under neuroendocrine control, involving complex interactions between the hypothalamus, pituitary, ovaries, and uterus. Under this control, approximately every 28 days from puberty to menopause, a female releases an egg from one of her ovaries. The cyclic events in the ovary leading to ovulation are known as the **ovarian cycle.** This cycle is coordinated with the **uterine cycle,** or **menstrual cycle** (*menstruus* = monthly), events in the uterus that prepare it to receive the egg if fertilization occurs.

The Ovarian Cycle. The ovarian cycle produces a mature egg **(Figure 47.12).** The starting point for the cycle is a primary oocyte in prophase of meiosis division I. The beginning of the cycle is triggered by an increase in the release of **gonadotropin-releasing hormone (GnRH)** by the hypothalamus. This hormone stimulates the pituitary to release **follicle-stimulating hormone (FSH)** and **luteinizing hormone (LH)** into the bloodstream **(Figure 47.13a).** FSH stimulates 6 to 20 primary oocytes in the ovaries to be released from prophase of meiosis I and continue through the meiotic divisions. As the primary oocytes develop into secondary oocytes—which arrest in metaphase of meiosis II—they become surrounded by cells that form a **follicle** (day 2 of the cycle; **Figure 47.13b**). During this follicular phase, the follicle grows and develops and, at its largest size, becomes filled with fluid and may reach 12 to 15 mm in diameter. Usually only one follicle develops to maturity with release of the egg (secondary oocyte) by ovulation. If two or more follicles develop and their eggs are ovulated, multiple births can result.

As the follicle enlarges, FSH and LH interact to stimulate the follicular cells to secrete **estrogens** (female sex hormones), primarily **estradiol** (see Section 40.4) **(Figure 47.13c).** Initially, the estrogens are secreted in low amounts; at this level, the estrogens have a negative feedback effect on the pituitary, inhibiting its secretion of FSH. As a result, FSH secretion declines briefly. However, estrogen secretion increases steadily, and its level peaks at about 12 days after follicle development begins (day 14 of cycle). The high estrogen level now has a positive feedback effect on the hypothalamus and pituitary, increasing the release of GnRH and stimulating the pituitary to release a burst of FSH and LH. The increased estrogen levels also convert the mucus secreted by the uterus to a thin and watery consistency, making it easier for sperm to swim through the uterus.

The burst in LH secretion stimulates the follicle cells to release enzymes that digest away the wall of the

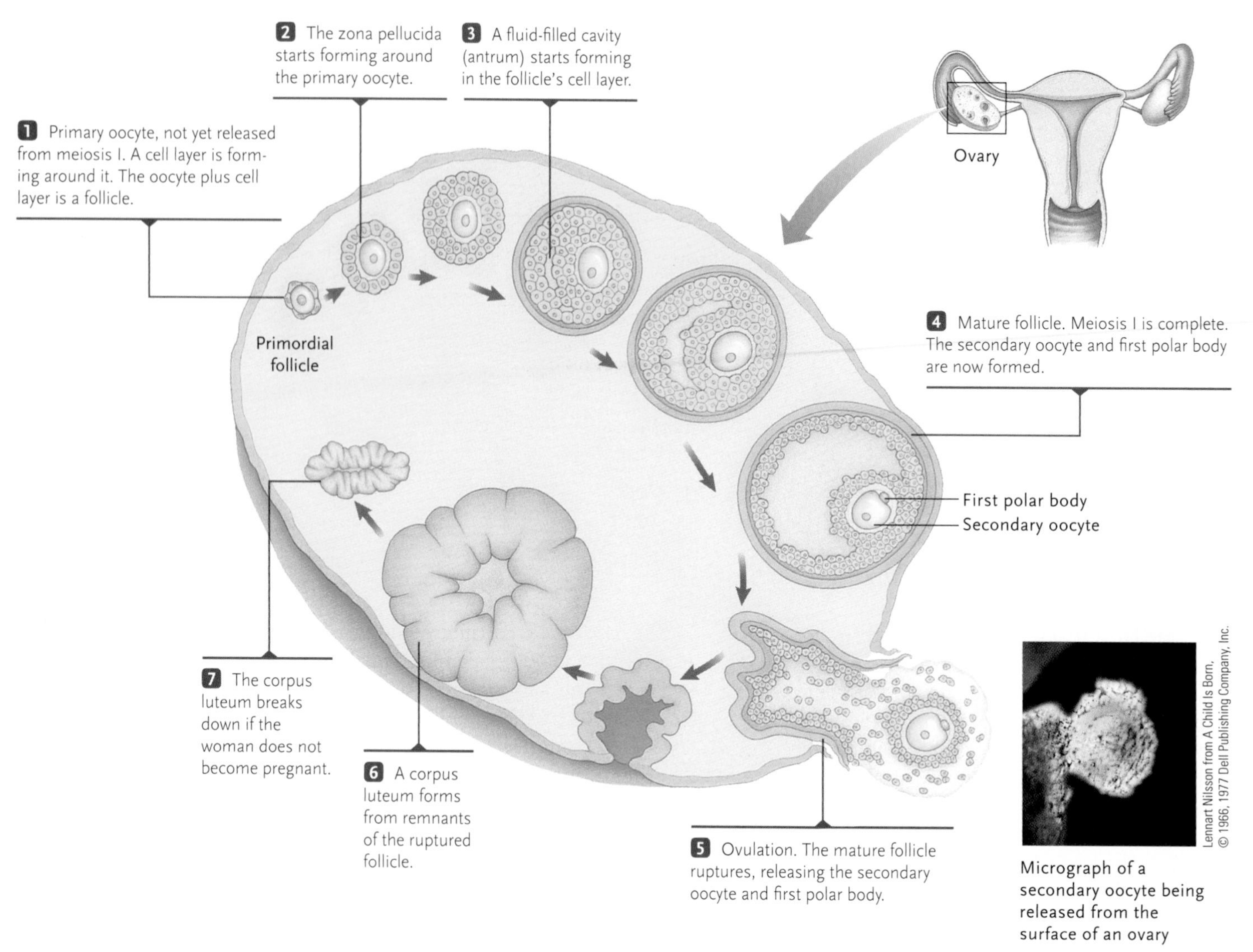

2 The zona pellucida starts forming around the primary oocyte.
3 A fluid-filled cavity (antrum) starts forming in the follicle's cell layer.
1 Primary oocyte, not yet released from meiosis I. A cell layer is forming around it. The oocyte plus cell layer is a follicle.
Ovary
Primordial follicle
4 Mature follicle. Meiosis I is complete. The secondary oocyte and first polar body are now formed.
First polar body
Secondary oocyte
7 The corpus luteum breaks down if the woman does not become pregnant.
6 A corpus luteum forms from remnants of the ruptured follicle.
5 Ovulation. The mature follicle ruptures, releasing the secondary oocyte and first polar body.
Lennart Nilsson from A Child Is Born, © 1966, 1977 Dell Publishing Company, Inc.

Micrograph of a secondary oocyte being released from the surface of an ovary

Figure 47.12
The growth of a follicle, ovulation, and formation of the corpus luteum in a human ovary.

follicle, causing it to burst and release the egg (see Figure 47.12); this is ovulation. LH also initiates the last phase of the menstrual cycle, the *luteal phase*. That is, LH causes the follicle cells remaining at the surface of the ovary to grow into an enlarged, yellowish structure, the **corpus luteum** (*corpus* = body; *luteum* = yellow; see Figure 47.12). Acting as an endocrine gland, the corpus luteum secretes several hormones: estrogens, large quantities of **progesterone,** and **inhibin.** Progesterone, a female sex hormone, stimulates growth of the uterine lining and inhibits contractions of the uterus. Both progesterone and inhibin have a negative feedback effect on the hypothalamus and pituitary. Progesterone inhibits the secretion of GnRH. Without GnRH, the pituitary does not release FSH and LH. FSH secretion from the pituitary is also inhibited directly by inhibin. The fall in FSH and LH levels diminishes the signal for follicular growth, and no new follicles begin to grow in the ovary.

If fertilization does not occur, the corpus luteum gradually degenerates as cells are phagocytized and blood supply is cut off. By about 10 days after ovulation, little tissue remains, meaning that estrogen, progesterone, and inhibin are no longer secreted. In the absence of progesterone, *menstruation* begins (described in the next section). As progesterone and inhibin levels decrease, FSH and LH secretion is no longer inhibited, and a new monthly cycle begins.

The Uterine (Menstrual) Cycle. The hormones that control the ovarian cycle also control the uterine (menstrual) cycle **(Figure 47.13d),** keeping the processes connected physiologically. Day 0 of the monthly cycle in the figure is the beginning of follicular development in the ovary (see Figure 47.13b); in the uterus, this correlates with the time at which menstrual flow begins.

Menstrual flow results from the breakdown of the endometrium, which releases blood and tissue breakdown products from the uterus to the outside through the vagina. When the flow ceases, at day 4 to 5 of the cycle, the endometrium begins to grow again; this is the proliferative phase. As the endometrium gradually thickens, the oocytes in both ovaries begin to develop further, eventually leading to ovulation at about 14 days after the beginning of the cycle, as already described.

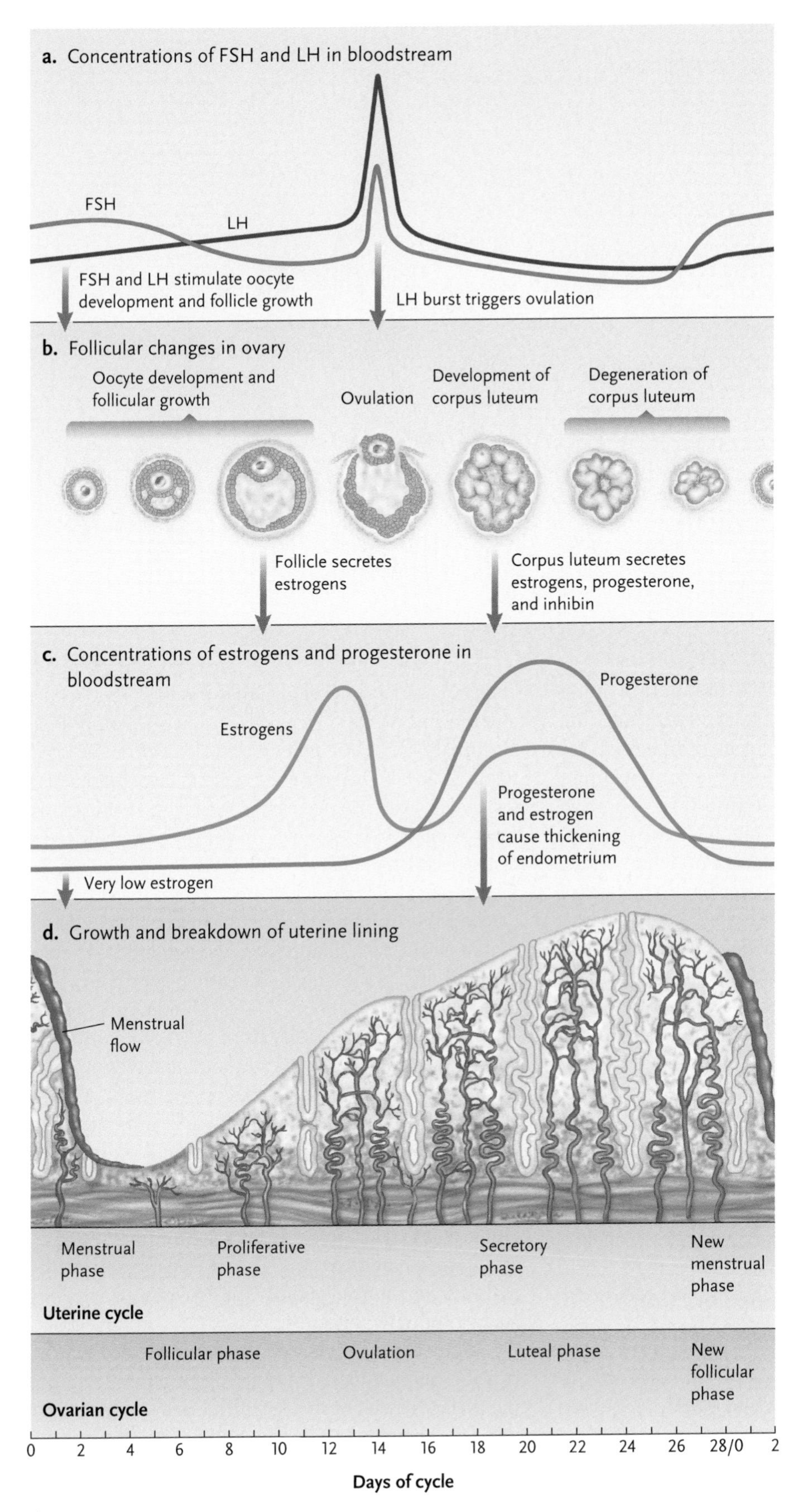

Figure 47.13

The ovarian and uterine (menstrual) cycles of a human female. **(a)** The changing concentrations of FSH and LH in the bloodstream, triggered by GnRH secretion by the hypothalamus. **(b)** The cycle of follicle development, ovulation, and formation of the corpus luteum in the ovary. **(c)** The concentrations of estrogens and progesterone in the bloodstream. **(d)** The growth and breakdown of the uterine lining. The days of the monthly cycle are given in the scale at the bottom of the diagram.

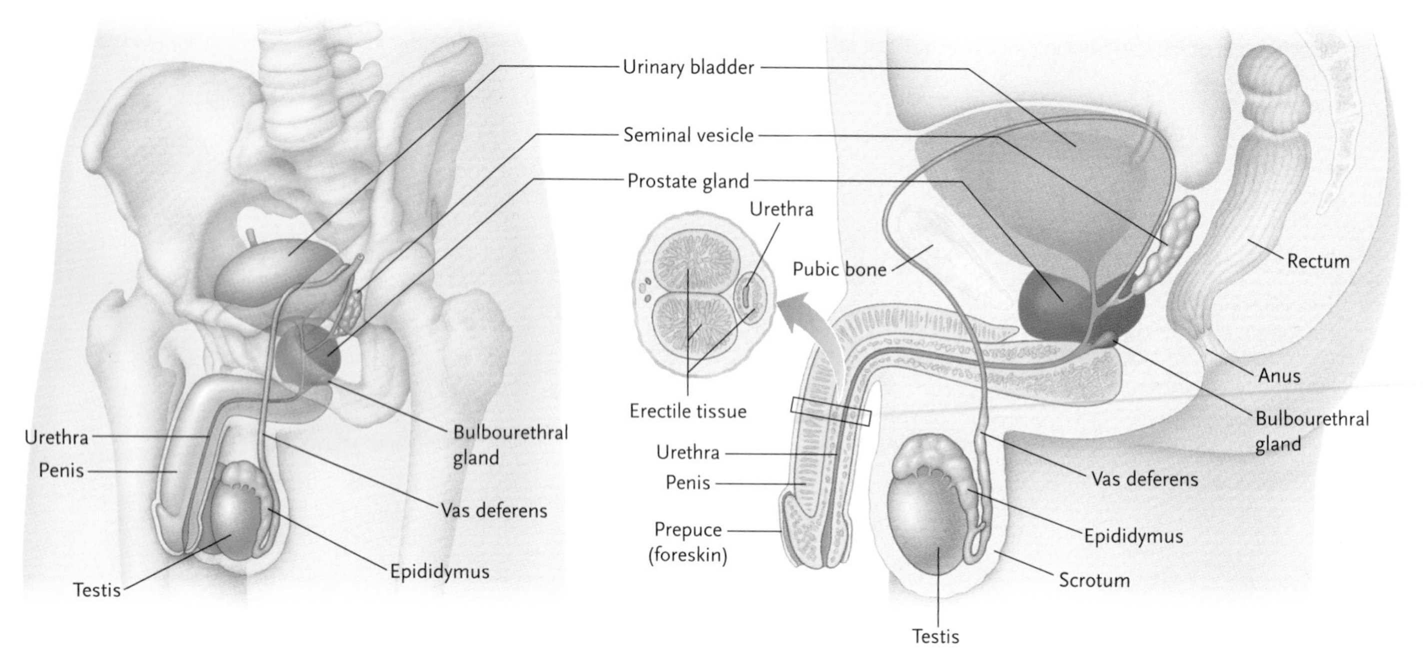

Figure 47.14
The reproductive organs of a human male.

If fertilization does not take place, the uterine lining continues to grow for another 14 days after ovulation; this is the secretory phase. At the end of that time, the absence of progesterone results in the contraction of arteries supplying blood to the uterine lining, shutting down the blood supply and causing the lining to disintegrate. The menstrual flow begins. Contractions of the uterus, no longer inhibited by progesterone, help expel the debris. Prostaglandins released by the degenerating endometrium add to the uterine contractions, making them severe enough to be felt as the pain of "cramps," and also sometimes causing other effects such as nausea, vomiting, and headaches.

Menstruation—the menstrual flow—occurs only in human females and our closest primate relatives, gorillas and chimpanzees. In other mammals, the uterine lining is completely reabsorbed if a fertilized egg does not implant during the period of reproductive activity. The uterine cycle in those mammals is called the *estrous* cycle.

Human Male Sexual Organs Function in Sperm Production and Delivery

Organs that produce and deliver sperm make up the male reproductive system **(Figure 47.14).** The testes are located outside the abdominal cavity; sperm produced by the testes pass through tubules that enter the abdominal cavity and join with the urethra, the duct that carries urine from the bladder to an opening at the tip of the penis.

Male Reproductive Structures. Human males have a pair of testes, suspended in the baglike **scrotum.** Suspension in the scrotum keeps the testes cooler than the body core, at a temperature that provides an optimal environment for sperm development. Some land mammals such as elephants and monotremes have relatively low body temperatures and have internal testes, that is, testes carried within the body. Marine mammals such as whales and dolphins also have internal testes despite relatively high body temperatures. In these animals particular blood vessel networks serve to lower the temperature in the testes to allow for normal function. A testis is packed with about 125 meters of **seminiferous tubules,** in which sperm proceed through all the stages of spermatogenesis **(Figure 47.15).** The entire process, from spermatogonium to sperm, takes about 9 to 10 weeks. The testes produce about 130 million sperm each day.

Supportive cells called **Sertoli cells** completely surround the developing spermatocytes in the seminiferous tubules. They supply nutrients to the spermatocytes and seal them off from the body's blood supply. Other cells located in the tissue surrounding the developing spermatocytes, the **Leydig cells,** produce the male sex hormones, known as **androgens,** particularly **testosterone** (see Figure 47.15).

Mature sperm flow from the seminiferous tubules into the **epididymis,** a coiled storage tubule attached to the surface of each testis. Rhythmic muscular contractions of the epididymis move the sperm into a thick-walled, muscular tube, the **vas deferens** (plural, *vasa deferentia*), which leads into the abdominal cavity. Just below the bladder, the vasa deferentia empty into the urethra. During ejaculation, muscular contractions force the sperm into the urethra and out of the penis. The sperm are activated and become motile as they come in contact with alkaline secretions added to the ejaculated fluid by accessory glands.

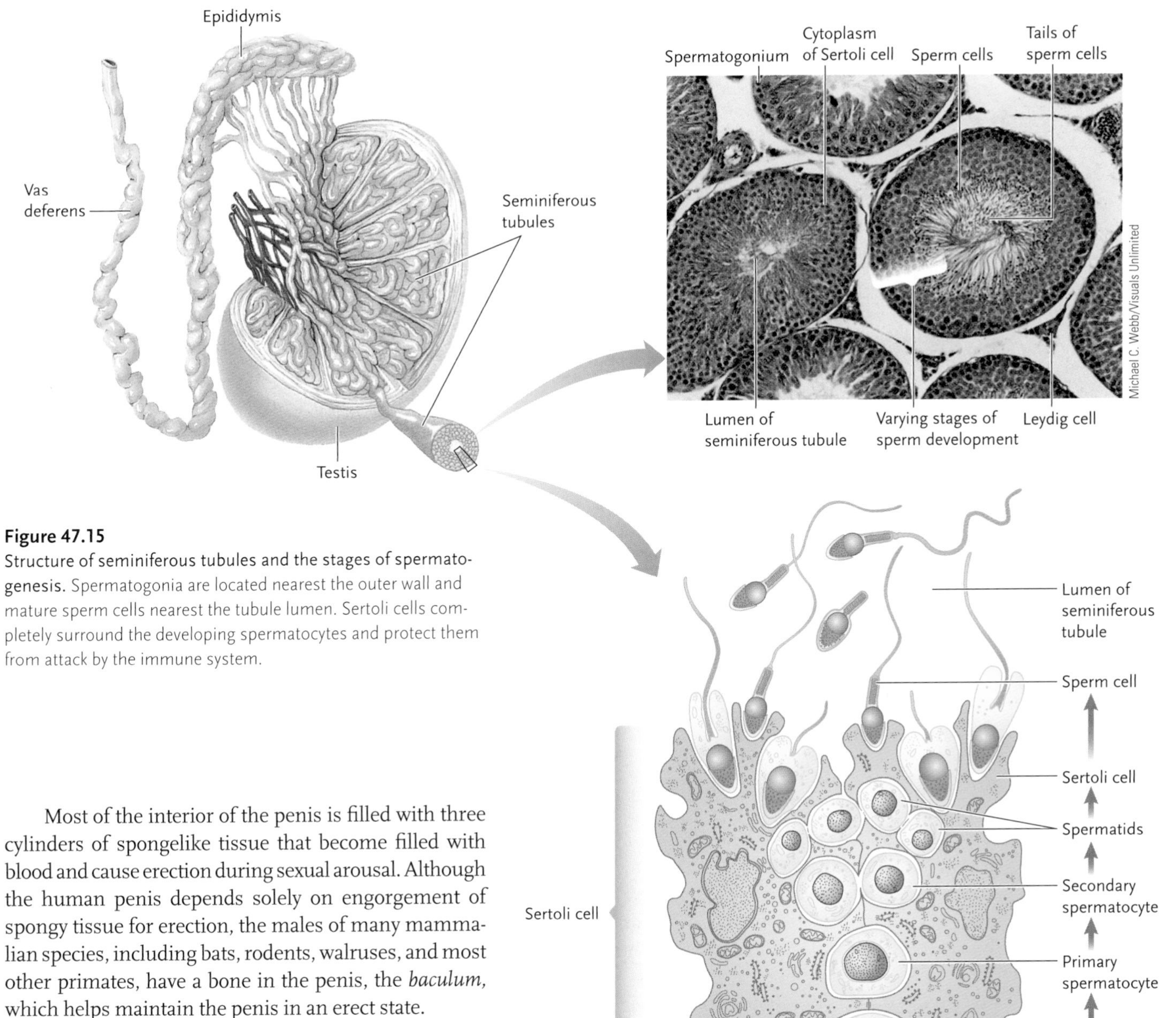

Figure 47.15
Structure of seminiferous tubules and the stages of spermatogenesis. Spermatogonia are located nearest the outer wall and mature sperm cells nearest the tubule lumen. Sertoli cells completely surround the developing spermatocytes and protect them from attack by the immune system.

Most of the interior of the penis is filled with three cylinders of spongelike tissue that become filled with blood and cause erection during sexual arousal. Although the human penis depends solely on engorgement of spongy tissue for erection, the males of many mammalian species, including bats, rodents, walruses, and most other primates, have a bone in the penis, the *baculum,* which helps maintain the penis in an erect state.

The penis ends in a soft, caplike structure, the **glans.** Most of the nerve endings producing erotic sensations are crowded into the glans and the region of the penile shaft just behind the glans. A loose fold of skin, the **prepuce** or **foreskin,** covers the glans (see Figure 47.14). In many cultures the prepuce is removed for hygienic, religious, or other ritualistic reasons by the procedure called **circumcision** ("around cut"). In 2007, the World Health Organization stated that male circumcision is an important strategy to prevent heterosexually acquired HIV infection in males.

Accessory Glands and the Semen. About 150 million to 350 million sperm are released in a single ejaculation. Before they leave the body, these cells are mixed with the secretions of several accessory glands, forming the fluid known as **semen.** In humans, about two-thirds of the volume is produced by a pair of **seminal vesicles,** which secrete a thick, viscous liquid, the **seminal fluid,** into the vasa deferentia near the point where they join with the urethra. The seminal fluid contains prostaglandins that, when ejaculated into the female, trigger contractions of the female reproductive tract that help move the sperm into and through the uterus.

The large **prostate gland,** which surrounds the region where the vasa deferentia empty into the urethra, adds a thin, milky fluid to the semen. The alkaline prostate secretion, which makes up about one-third of the volume of the semen, raises the pH of the semen, and of the vagina, to about pH 6, the level of acidity best tolerated by sperm. The raised pH also activates motility of the sperm. As part of the prostate secretion, a fast-acting enzyme converts the semen to a thick gel when it is first ejaculated. The thickened

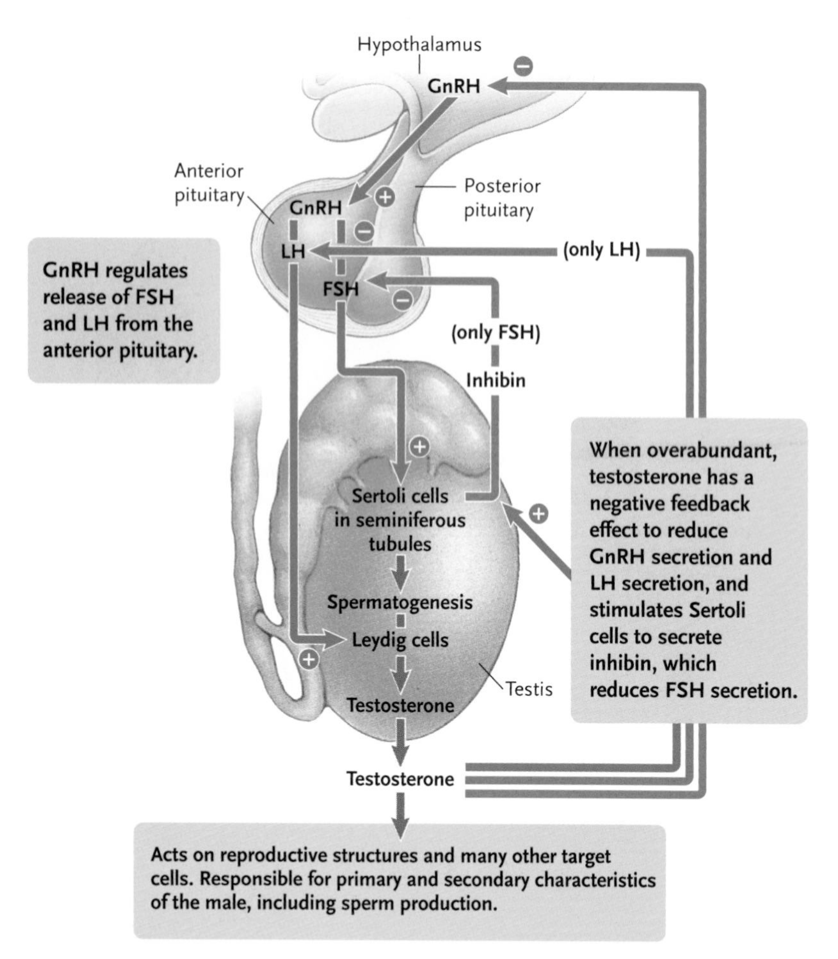
Hypothalamus
GnRH
Anterior pituitary
Posterior pituitary
GnRH
LH
FSH
(only LH)
(only FSH)
Inhibin
GnRH regulates release of FSH and LH from the anterior pituitary.
Sertoli cells in seminiferous tubules
Spermatogenesis
Leydig cells
Testosterone
Testis
When overabundant, testosterone has a negative feedback effect to reduce GnRH secretion and LH secretion, and stimulates Sertoli cells to secrete inhibin, which reduces FSH secretion.
Testosterone
Acts on reproductive structures and many other target cells. Responsible for primary and secondary characteristics of the male, including sperm production.

Figure 47.16
Hormonal regulation of reproduction in the male, and the negative feedback systems controlling hormone levels.

consistency helps keep the semen from draining from the vagina when the penis is withdrawn. A second, slower-acting enzyme in the prostate secretion then gradually breaks down the semen clot and releases the sperm to swim freely in the female reproductive tract.

Finally, a pair of **bulbourethral glands** secretes a clear, mucus-rich fluid into the urethra before and during ejaculation. This fluid lubricates the tip of the penis and neutralizes the acidity of any residual urine in the urethra. In total, the secretions of the accessory glands make up more than 95% of the volume of semen; less than 5% is sperm.

Hormones Also Regulate Male Reproductive Functions

Many of the hormones regulating the menstrual cycle, including GnRH, FSH, LH, and inhibin, also regulate male reproductive functions. Testosterone, secreted by the Leydig cells in the testes, also plays a key role **(Figure 47.16)**.

In sexually mature males, the hypothalamus secretes GnRH in brief pulses every 1 to 2 hours. The GnRH, in turn, stimulates the pituitary to secrete LH and FSH. LH stimulates the Leydig cells to secrete testosterone, which stimulates sperm production and controls the growth and function of male reproductive structures. FSH stimulates Sertoli cells to secrete a protein and other molecules that are required for spermatogenesis.

The concentrations of these hormones are maintained by negative feedback mechanisms. If the concentration of testosterone falls in the bloodstream, the hypothalamus responds by increasing GnRH secretion. If the concentration of testosterone rises too high, the overabundance inhibits GnRH secretion by the hypothalamus and LH secretion by the anterior pituitary. An overabundance of testosterone also stimulates Sertoli cells to secrete inhibin, which inhibits FSH secretion by the anterior pituitary. As a result, testosterone secretion by the Leydig cells drops off, returning the concentration to optimal levels in the bloodstream.

Human Copulation Follows a Typical Mammalian Pattern

When the male is sexually aroused, sphincter muscles controlling the flow of blood to the spongy erectile tissue of the penis relax, allowing the tissue to become engorged with blood. (The penis is a hydrostatic skeleton structure; see Section 41.2.) As the spongy tissue swells, it maintains the pressure by compressing and almost shutting off the veins draining blood from the penis. The engorgement produces an erection in which the penis lengthens, stiffens, and enlarges. During continued sexual arousal, lubricating fluid secreted by the bulbourethral glands may be released from the tip of the penis.

Female sexual arousal results in enlargement and erection of the clitoris, in a process analogous to erection of the penis in males. The labia minora also become engorged with blood and swell in size, and lubricating fluid is secreted onto the surfaces of the vulva by the greater vestibular glands. In addition to these changes, the nipples become erect by contraction of smooth muscle cells, and the breasts swell due to engorgement with blood.

Insertion of the penis into the vagina and the thrusting movements of copulation lead to the reflex actions of ejaculation, including spasmodic contractions of muscles surrounding the vasa deferentia, accessory glands, and urethra. During ejaculation, the sphincter muscles controlling the exit from the bladder close tightly, preventing urine from being released from the bladder and mixing with the ejaculate. Ejaculation is usually accompanied by *orgasm,* a sensation of intense physical pleasure that is the peak—climax—of excitement for sexual intercourse, followed by feelings of relaxation and gratification.

The motions of copulation stretch the vagina and stimulate the clitoris. The stretching and stimulation can also induce orgasm in females. The vaginal stretching also stimulates the hypothalamus to secrete oxyto-

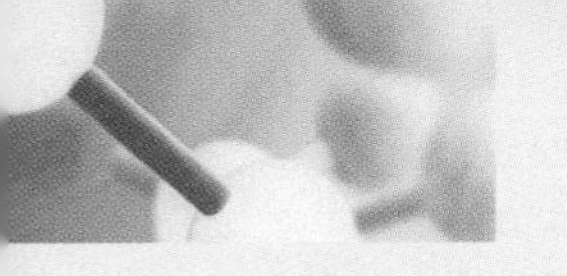

INSIGHTS FROM THE MOLECULAR REVOLUTION

Egging on the Sperm

Whether human eggs release attractants to draw sperm near has long been a subject of speculation and research. Now molecular investigations by Marc Spehr and his colleagues at Ruhr University in Germany indicate that sperm can detect and swim toward attractant chemicals.

Other investigators had found that human sperm cells have receptors able to bind to chemical substances classified as odorants, aroma molecules that can be specifically recognized ("smelled"). In vertebrates, more than a thousand genes encode odorant receptors, most of them olfactory receptors associated with the senses of smell and taste. However, odorant receptors are also located on cell types that do not function in taste and smell, including sperm.

But do the odorant receptors function in sperm–egg attraction? The Spehr team began their investigation of this possibility by testing testicular tissue for odorant-receptor-gene activity, using probes for mRNAs with sequences typical of odorant receptor genes. Only two active odorant receptor genes were found in the testes: hOR17-2 (hOR = human olfactory receptor), which had been discovered by others; and hOR17-4, which was not previously known to be active.

The researchers molecularly cloned (see Section 18.1) the hOR17-4 gene of testicular cells and inserted the gene in a line of cultured human embryonic kidney (HEK) cells. Previous work had shown that when an odorant receptor combines with the chemical it recognizes, it triggers cytoplasmic reactions that lead to Ca^{2+} release in the cytoplasm (the IP_3 pathway, described in Section 7.4). Accordingly, the investigators tested the genetically engineered HEK cells for a Ca^{2+} response to any of the chemicals in a mixture containing 100 different chemicals. Only one of the chemicals, *cyclamal,* caused the HEK cells to respond. HEK cells that did not receive the hOR17-4 gene did not respond to cyclamal. The researchers then tested chemicals closely related in chemical structure to cyclamal and found that they also elicit a Ca^{2+} response in the engineered HEK cells. They used one of these chemicals, *bourgonal,* in further testing because it triggered a stronger response than cyclamal.

Next, the investigators tested human sperm cells to see if they would respond to bourgeonal. The sperm cells responded to the chemical by an increase in cytoplasmic Ca^{2+} concentration, indicating that the hOR17-4 gene was active during spermatogenesis, leading to synthesis and insertion of the hOR17-4 receptor in the sperm plasma membrane.

In a final experiment, human sperm cells were exposed to gradients of bourgeonal solutions in micropipettes. The sperm swam consistently toward the regions of highest concentration, and swam faster and more directly as the concentration increased.

These experiments indicate that human sperm can detect and respond to chemical attractants by swimming toward the source of the attractant. Whether human eggs actually release such attractants remains to be determined. If so, the egg attractant detected by the hOR17-4 receptor is likely to resemble cyclamal and bourgeonal in chemical structure, because odorant receptors are highly specific in their responses.

As part of their research, the Spehr team found that another chemical, *undecanal,* strongly inhibits the binding of the hOR17-4 receptor to cyclamal and bourgeonal. With chemicals at hand that can both stimulate and eliminate sperm attraction, the system might provide a method for either contraception or procreation.

cin, which induces contractions of the uterus. The contractions keep the sperm in suspension and aid their movement through the reproductive tract. Uterine contractions are also induced by the prostaglandins in the semen.

Sperm reach the site of fertilization in the oviducts within 30 minutes after their ejaculation into the vagina. Of the millions of sperm released in a single ejaculation, only a few hundred actually reach the oviducts. After orgasm, the penis, clitoris, and labia minora gradually return to their unstimulated size. Females can experience additional orgasms within minutes or even seconds of a first orgasm, but most males enter a *refractory period* that lasts for 15 minutes or longer before they can regain an erection and have another orgasm.

A Human Egg Can Be Fertilized Only in the Oviduct

A human egg can be fertilized only during its passage through the third of the oviduct nearest the ovary. If the egg is not fertilized during the 12- to 24-hour period that it is in this location, it disintegrates and dies. However, sperm do not swim randomly for a chance encounter with the egg. Rather, they first swim up the cervical canal to reach the oviduct, and then are propelled up the oviduct by contractions of the oviduct's smooth muscles. Further, researchers have found evidence that eggs release chemical attractant molecules that the sperm recognize, causing them to swim directly toward the egg. (*Insights from the Molecular Revolution* describes some of this research.)

To reach the egg, the fertilizing sperm must penetrate the layer of follicle cells surrounding the egg, and then pass through the zona pellucida coating the egg surface **(Figure 47.17).** Enzymes built into the plasma membrane of the sperm cells aid penetration through the follicle cells. Once through the follicle cells, the sperm binds to receptor molecules on the surface of the zona pellucida. The binding triggers the acrosome reaction in which hydrolytic enzymes are released from the acrosome and digest a pathway to the egg. As soon as the first sperm cell reaches the egg through the pathway digested by the released acrosomal enzymes, the sperm and egg plasma membranes fuse, and the sperm cell enters the cytoplasm of the egg. Although only one sperm fertilizes the egg, the combined release of acrosomal enzymes from many sperm greatly increases the chance that a complete channel will be opened through the zona pellucida. Partially for this reason, a low sperm count is often a source of male infertility. Low sperm count has a number of causes, including infection, heat, frequent intercourse, smoking, and excess alcohol consumption.

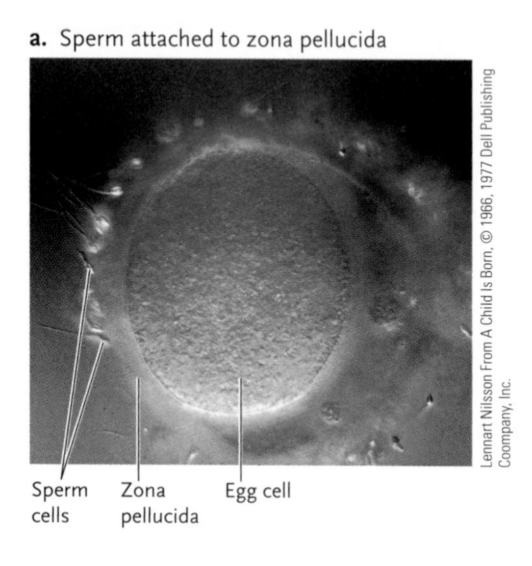
a. Sperm attached to zona pellucida
Sperm cells
Zona pellucida
Egg cell
Lennart Nilsson From A Child Is Born, © 1966, 1977 Dell Publishing Coompany, Inc.

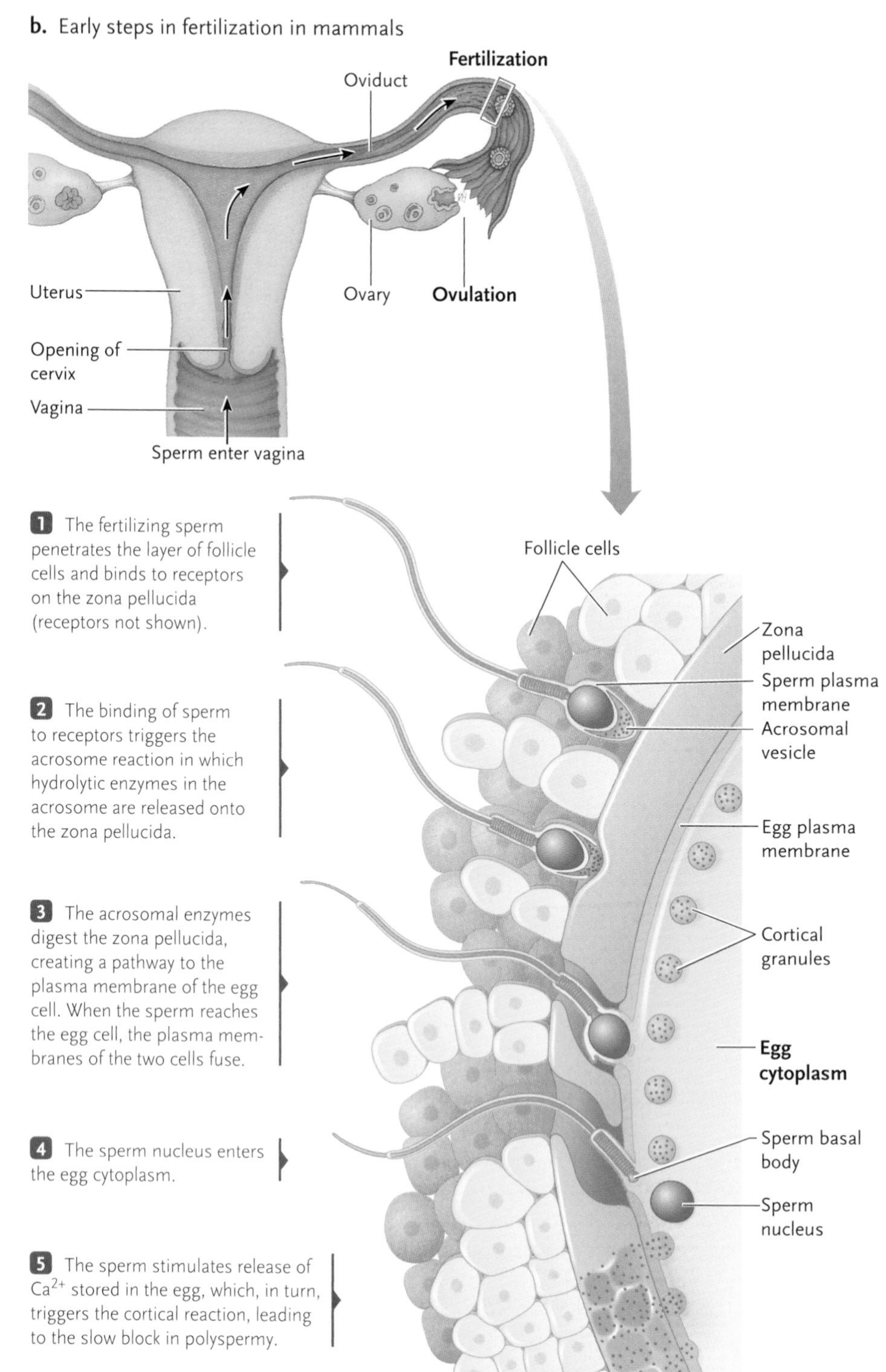
b. Early steps in fertilization in mammals
Fertilization
Oviduct
Uterus
Ovary
Ovulation
Opening of cervix
Vagina
Sperm enter vagina
1 The fertilizing sperm penetrates the layer of follicle cells and binds to receptors on the zona pellucida (receptors not shown).
2 The binding of sperm to receptors triggers the acrosome reaction in which hydrolytic enzymes in the acrosome are released onto the zona pellucida.
3 The acrosomal enzymes digest the zona pellucida, creating a pathway to the plasma membrane of the egg cell. When the sperm reaches the egg cell, the plasma membranes of the two cells fuse.
4 The sperm nucleus enters the egg cytoplasm.
5 The sperm stimulates release of Ca^{2+} stored in the egg, which, in turn, triggers the cortical reaction, leading to the slow block in polyspermy.
Follicle cells
Zona pellucida
Sperm plasma membrane
Acrosomal vesicle
Egg plasma membrane
Cortical granules
Egg cytoplasm
Sperm basal body
Sperm nucleus

Figure 47.17
Fertilization in mammals. **(a)** Sperm attached to the zona pellucida of a human egg cell. **(b)** Early steps in fertilization process.

The membrane fusion activates the egg. The sperm that has entered the egg releases nitric oxide, which stimulates the release of stored Ca^{2+} in the egg. The Ca^{2+} triggers cortical granule release to the outside of the egg. Enzymes from the cortical granules crosslink molecules in the zona pellucida, hardening it and sealing the channels opened by acrosomal enzymes. The enzymes also destroy the receptors that bind sperm to the surface of the zona pellucida. As a result, no further sperm can bind to the zona or reach the plasma membrane of the egg. The Ca^{2+} also triggers the completion of meiosis of the egg (recall that, up to that point, it is a secondary oocyte arrested in metaphase of meiosis II). The sperm and egg nuclei then fuse and the cell is now the zygote. Mitotic divisions of the zygote soon initiate embryonic development.

The first cell divisions of embryonic development take place while the fertilized egg is still in the oviduct. By about 7 days after ovulation, the embryo passes from the oviduct and implants in the uterine lining. During and after implantation, cells associated with the embryo secrete **human chorionic gonadotropin (hCG)**, a hormone that keeps the corpus luteum in the ovary from breaking down. Excess hCG is excreted in the urine; its presence in urine or blood provides the basis of pregnancy tests.

The continued activity of the corpus luteum keeps estrogen and progesterone secretion at high levels, maintaining the uterine lining and preventing menstruation. The high progesterone level also thickens the mucus secreted by the uterus, forming a plug that seals the opening of the cervix from the vagina. The plug keeps bacteria, viruses, and sperm cells from further copulation from entering the uterus.

Later in development, about 10 weeks after implantation, the placenta takes over the secretion of progesterone, hCG secretion drops off, and the corpus luteum regresses. However, the corpus luteum continues to secrete the hormone *relaxin*, which inhibits contraction of the uterus until the time of birth is near.

Study Break

Outline the roles of follicle-stimulating hormone (FSH) and luteinizing hormone (LH) in the ovarian cycle of a human female.

Table 47.1 Pregnancy Rates for Birth Control Methods

Method	Lowest Expected Rate of Pregnancy[a]	Typical-Use Rate of Pregnancy[b]
Rhythm method	1%–9%	25%
Withdrawal	4%	19%
Condom (male)	3%	14%
Condom (female)	5%	21%
Diaphragm and spermicidal jelly	6%	20%
Vasectomy (male sterilization)	0.1%	0.15%
Tubal ligation (female sterilization)	0.5%	0.5%
Contraceptive pill (combination estrogen/progestin)	0.1%	5%
Contraceptive pill (progestin only)	0.5%	5%
Implant (progestin)	0.09%	0.09%
Intrauterine device (IUD) (copper T)	0.6%	0.8%

[a]Rate of pregnancy when the birth control method was used correctly every time.
[b]Rate of pregnancy when the method was used typically, meaning that it may not have been always used correctly every time.
Source: U.S. Food and Drug Administration, http://www.fda.gov/fdac/features/1997/conceptbl.html. Data reported in 1997 for effectiveness of methods in a 1-year period.

47.4 Methods for Preventing Pregnancy: Contraception

In human society, pregnancy can be a blessing or a disaster. An unwanted pregnancy can be inconvenient at the least, or at the worst can have serious physical and social repercussions, particularly for the mother. Many methods exist for achieving contraception—the prevention of pregnancy—some old and others relatively new.

The oldest method of contraception is total abstinence from sex. Unfortunately, millions of years of animal evolution have stacked the cards against total abstinence by making the sex drive among the most powerful of compulsions. Literally millions of unwanted children attest to the failures of this method. Other methods of preventing pregnancy include techniques for (1) preventing the sperm from reaching the site of fertilization, (2) preventing ovulation, or (3) interfering with implantation if fertilization does occur. **Table 47.1** lists the most common contraceptive techniques and their reliability, based on 1 year of use. Two values are given: (1) the lowest expected rate of pregnancy, meaning the rate of pregnancy when the birth control method was used correctly every time; and (2) the typical use rate of pregnancy, meaning the rate of pregnancy when the method was used in a typical manner, meaning that it may not always have been used correctly every time.

Of Methods Preventing Fertilization, Vasectomy and Tubal Ligation Are Most Effective

A natural technique for preventing fertilization is the *rhythm method*, which consists of avoiding intercourse during the time of the month when the egg can be fertilized. Because sperm can survive for as long as 5 days in the female reproductive tract, intercourse should be avoided from 5 days before ovulation and, for safety's sake, for another 4 or 5 days after ovulation. Although conceptually straightforward, the method is difficult to apply because of the unpredictability of the time of ovulation (and the power of the sex drive). The lowest expected rate of pregnancy for this method is 1% to 9%, while the typical rate is 25%.

Another natural method to prevent fertilization is *withdrawal*—starting sexual intercourse, but withdrawing the penis before ejaculation. Unfortunately, once ejaculation begins, it proceeds as a series of reflexes that is extremely difficult to interrupt; in addition, some sperm may be present in lubrication produced prior to ejaculation. The lowest expected rate of pregnancy for this method is 4%, while the typical rate is 19%.

The *condom*, a thin, close-fitting sheath of latex, lambskin, or polyurethane worn over the penis, is one of the traditional methods of preventing ejaculated sperm from entering the vagina. Condoms made from latex may also provide a barrier to the transmission of disease between sexual partners (condoms made from natural skin do not block viruses such as HIV). Pouch-like "female condoms," inserted into the vagina, prevent ejaculated sperm from entering the uterus. The lowest rate of pregnancy for male condoms is 3%, while the typical rate is 14%. The lowest and typical rates for female condoms are 5% and 21%, respectively.

The *diaphragm* is a cuplike rubber device that blocks the cervix in females. (The similar *cervical cap* is smaller and fits more closely over the cervix.) Typically a spermicidal jelly or cream is also used. To be most effective, a diaphragm and the spermicidal jelly must be inserted no more than an hour before intercourse, and left in place for the recommended time afterward. The lowest rate of pregnancy for a diaphragm used with a spermicide is 6%, while the typical rate is 20%.

Fertilization can also be prevented surgically, by cutting and closing off either the vasa deferentia in males or the oviducts in females. In *vasectomy*, the procedure carried out in males, an incision is made in the scrotum and each vas deferens is severed and tied off. After vasectomy, the seminal fluid is still produced and ejaculated, but it does not contain sperm. In *tubal ligation*, the procedure for females, the oviducts are cut and tied off, or seared with heat (cauterized) to close them. The ligation prevents eggs from being fertilized or reaching the uterus. Neither vasectomy nor tubal ligation interferes with the production of sex hormones by the ovaries or testes, or results in any change in sexual behavior. Both operations are highly effective in preventing pregnancy. Although they can be reversed, the procedures are difficult and not always successful. The lowest rate of pregnancy for vasectomy is 0.1%, while the typical rate is 0.15%. The lowest and typical rates for tubal ligation are both 0.5%.

Of Methods Preventing Ovulation, the Oral Contraceptive Pill Is Most Effective

The primary method used to prevent ovulation is the *oral contraceptive pill*, or simply "the pill," containing a combination of estrogen and *progestin* (a synthetic form of progesterone) or progestin alone. In this highly effective method, the pill is taken daily for 20 to 21 days after the end of the menstrual flow and then stopped (actually, placebo pills are taken for the remaining days of the cycle to maintain the routine of pill taking) to allow menstruation; then the next month's course is begun. If pregnancy is desired, the pill is simply not taken after the menstrual flow.

The pill works by inhibiting the secretion of FSH and LH by the pituitary; without these hormones, ovulation does not occur. When the pill is stopped after 20 to 21 days, the resulting drop in progestin concentration causes the uterine lining to break down and initiates the menstrual flow. Since ovulation does not occur, fertilization and pregnancy are not possible.

The lowest rate of pregnancy for the estrogen/progestin pill is 0.1%, while the typical rate is 5%. The lowest and typical rates for the progestin pill are 0.5% and 5%, respectively. Most pregnancies among women taking the pill result from failure to take it on schedule—often simply by forgetting to take the pill for a day or two at the wrong time of the month. Some women, about one in four, experience unpleasant side effects, such as nausea, tenderness of the breasts, irritability, nervousness, or changes in skin color or texture. Modern versions of the pill have almost eliminated the more serious side effects, such as increased incidence of breast cancer and formation of blood clots. However, cigarette smoking significantly increases the risk of heart attacks and strokes for women taking the pill. This risk increases with age and with the number of cigarettes smoked per day.

As an alternative to the pill, progestin is also injected in a time-release form that prevents ovulation throughout the period of release. In one method, plastic tubes containing progestin are implanted under the skin, usually in the upper arm. The tubes release progestin for up to 5 years, making the method effective for women in countries or situations in which obtaining and taking the pill on a daily basis is impractical. The lowest rate and the typical rate of pregnancy for the implant are both 0.09%.

The IUD and the Morning-After Pill Are Effective in Preventing Implantation

A commonly used method for preventing implantation if fertilization occurs is insertion of an *intrauterine device (IUD)*, a small plastic or copper device, into the uterus just inside the cervix. The IUD remains in place as a long-term preventive measure; depending on the type, a single IUD is approved for 5 to 10 years of use. It is not clear how the IUD works; presumably, it causes a mild inflammation of the uterine lining that makes it unreceptive to implantation. The IUD is a refinement of a method used by women since ancient times, in which small pebbles were inserted in the uterus to prevent conception.

The IUD is effective as long as it is not deflected from its correct position in the uterus; unfortunately, this may happen without warning or the user's awareness. A few women also experience unpleasant side effects from the IUD such as cramps, uterine infections, or excessive menstrual bleeding. The lowest rate of pregnancy for the copper T types of IUD is 0.6%, while the typical rate is 0.8%.

Whatever the method of birth control, its effectiveness is improved if sex partners are highly motivated and careful in its use. The effectiveness of condoms, for example, is greatly improved if the penis is withdrawn immediately after ejaculation (before the semen has time to spread under the condom and leak into the vagina) and is not reinserted, with or without a condom, for several hours. Similarly, high motivation in use of the rhythm method, which might require abstaining from intercourse for most of the month except for a few days just after the menstrual flow, considerably improves the percentage of success with this method.

Another method used to prevent pregnancy is the so-called *emergency contraception pill,* commonly referred to as the "morning-after pill." These pills are administered after intercourse has occurred as a means to prevent pregnancy. A high dosage synthetic progestin emergency contraception pill called Plan B is available in the United States without prescription to women who are 18 or older. This pill is highly effective if taken within 72 hours after unprotected sexual intercourse. Pregnancy tests do not work until significantly after this time. Research data show that Plan B works by blocking ovulation; there is no effect of the hormone on implantation of a fertilized egg. But, because sperm can survive in the female reproductive tract for a few days, blocking ovulation can obviously be effective in preventing pregnancy.

Another emergency contraception pill is *mifepristone (RU-486),* which contains a molecule that binds to and blocks progesterone receptors in the uterine lining. The blockage prevents the lining from responding to progesterone and causes it to break down (that is, a menstrual period is initiated), taking with it any embryo that may have implanted. Mifepristone is approved in the United States for terminating pregnancies up to 49 days post-conception; the time period is

Unanswered Questions

Why do male mammals have "female" hormones?

Given the paramount importance of reproduction in biology, there is a surprising lack of knowledge regarding even its basic features. For example, both sexes make the same set of reproductive hormones, but for some of these, we are only beginning to understand their roles in males. For example, oxytocin, released by the pituitary, has long been known to control labor and milk ejection in females. But oxytocin is also found in semen, where it may boost sperm counts and aid sperm transport in the female reproductive tract. And it may also have a role in social behavior. Several studies have shown a role for oxytocin in male–female and mother–child bonding; and recently research groups led by Markus Heinrichs at the University of Zurich, Switzerland, and Ernst Fehr at Collegium Helveticum, Zurich, reported that men who sniffed a nasal spray with oxytocin in the laboratory were much more likely to trust other males in engaging in risky financial transactions. Thus oxytocin has been dubbed the "trust hormone." However, its roles (and how it is regulated) in real social interactions are not known.

Prolactin, also from the pituitary gland, stimulates milk production in females. Tillmann Krüger and colleagues at the Swiss Federal Institute of Technology have found that prolactin levels surge during orgasms in both sexes and that prolactin is associated with a feeling of relaxation and satisfaction. In males, it may also inhibit erection during post-coitus recovery. In 2006, Stuart Brody of the University of Paisley in Scotland and Dr. Krüger found the surge to be much higher after intercourse than masturbation, which may explain why the latter is generally less satisfying. They suggest that excessive prolactin could play a role in male impotence, a possibility now being investigated.

What is the anatomy of the human clitoris?

There is also surprising ignorance on basic reproductive anatomy. Diagrams of the human clitoris are quite misleading, showing it simply as a small external bulb in the female vulva. However, dissections and magnetic resonance images in studies led by Dr. Helen O'Connell, an Australian urology surgeon, revealed the clitoris to be much larger, rivaling the penis. Most of the clitoral tissue consists of large, highly vascular internal bulbs that surround the urethra and vagina. The role of these bulbs is uncertain, but they engorge during intercourse, perhaps to squeeze the urethra closed to prevent infections, to support the vaginal wall during penile penetration, and/or to increase pleasure signaling. Knowledge of this anatomy is crucial for performing safe pelvic surgeries on women.

Paul H. Yancey holds the Carl E. Peterson Endowed Chair of Sciences at Whitman College in Walla Walla, Washington. His main research interests are in the areas of animal physiology, especially water stress and osmoregulation. To learn more about his research, go to http://marcus.whitman.edu/~yancey/.

longer in some foreign countries. It is available only by prescription.

In this chapter we have focused on animal reproduction up to the point of the fertilized egg. In the next chapter, we address the final stage of reproduction in sexually reproducing organisms, the development of a new individual from the fertilized egg.

Study Break

How does the oral contraceptive pill prevent pregnancy?

Review

Go to **Thomson**NOW™ at www.thomsonedu.com/login to access quizzing, animations, exercises, articles, and personalized homework help.

47.1 Animal Reproductive Modes: Asexual and Sexual Reproduction

- In asexual reproduction, a single parent gives rise to offspring without genetic input from another individual. In sexual reproduction, offspring are produced by the union of gametes—eggs and sperm—from two parents.
- Asexual reproduction involving mitosis occurs in animals by fission, budding, or fragmentation (Figure 47.2). In parthenogenesis, a form of asexual reproduction, females produce eggs that develop without being fertilized.
- In sexual reproduction, genetic variability is produced by the meiotic processes of genetic recombination and independent assortment.

47.2 Cellular Mechanisms of Sexual Reproduction

- Sexual reproduction includes two cellular processes, gametogenesis and fertilization, and a whole-organism process, mating. Gametogenesis is the formation of male and female gametes by meiotic cell division, followed by differentiation of the gametes; fertilization is the union of gametes that initiates development of new individuals (Figure 47.3).
- Gametogenesis takes place in the testes of males and in the ovaries of females. Sperm and eggs are delivered to the site of fertilization by sperm ducts in males and oviducts in females. External reproductive structures aid the delivery in many species.
- In male gametogenesis—spermatogenesis—each cell entering meiosis produces four haploid motile sperm cells. In female gametogenesis—oogenesis—each cell entering meiosis produces one haploid egg cell. The meiotic divisions of oogenesis concentrate almost all the cytoplasm in the single egg cell; the other division products are nonfunctional polar bodies (Figures 47.3–47.5).
- The egg contains stored nutrients and information required for at least the early stages of embryonic development. It is covered by one or more protective coats, and it has a mechanism that blocks additional sperm from entering after fertilization (Figure 47.5).
- Fertilization, which follows mating in most animals, may be external or internal. In external fertilization, sperm and eggs are shed into the surrounding water. In internal fertilization, sperm are released close to or inside the female reproductive ducts via copulation (Figure 47.6).
- When a sperm and egg touch during fertilization, their plasma membranes fuse, introducing the sperm nucleus into the egg cytoplasm. The sperm and egg nuclei then fuse to form a diploid zygote nucleus and initiate embryonic development (Figure 47.8).
- Oviparous animals lay eggs in which development of new individuals takes place outside the female's body. In viviparous animals, development takes place inside the female's body. In ovoviviparous animals, fertilized eggs are retained within the body while the embryo develops, the eggs hatch within the mother, and the young are then released from the body.
- In hermaphroditism, single individuals produce both mature egg-producing tissue and mature sperm-producing tissue (Figure 47.10).

Animation: Spermatogenesis

Animation: Fertilization

47.3 Sexual Reproduction in Humans

- In females, eggs released from the ovaries travel through the oviducts to the uterus. The uterus opens into the vagina, the entrance for sperm and the exit for offspring during birth (Figure 47.11).
- The ovarian cycle produces an egg. The cycle begins with the release of GnRH by the hypothalamus, which stimulates the release of FSH and LH from the anterior pituitary. FSH stimulates oocytes in the ovaries to begin meiosis. One oocyte typically develops to maturity surrounded by cells that form a follicle (Figures 47.12 and 47.13).
- The enlarging follicle secretes estrogens, causing a burst in FSH and LH release; at about 14 days, the LH stimulates ovulation, the bursting of the follicle and the release of the egg. The remainder of the follicle forms the corpus luteum, which secretes estrogens, progesterone, and inhibin (Figures 47.12 and 47.13).
- Day 0 of the monthly uterine (menstrual) cycle correlates with the beginning of follicular development in the ovary and the beginning of the menstrual flow. Secretion of estrogen from the developing follicle stimulates the growth of a new endometrium. If fertilization does not occur, progesterone and inhibin maintain the endometrium until the 28th day of the cycle, when the corpus luteum regresses. Without progesterone, the endometrium breaks down and is released as the menstrual flow (Figure 47.13).
- In males, sperm develop in seminiferous tubules in the testes and are released into the epididymis. When a male ejaculates, sperm travel from the epididymis to the vas deferens, and then through the urethra and the penis. The seminal vesicles, prostate gland, and bulbourethral glands add fluids to the sperm traveling to the outside (Figures 47.14 and 47.15).
- Sperm production in males is also controlled by LH and FSH. LH stimulates Leydig cells in the testes to secrete testosterone, which stimulates sperm production. FSH stimulates Sertoli cells in the testes to secrete molecules needed for spermatogenesis (Figure 47.16).
- During copulation, sperm are ejaculated into the vagina of the female. The sperm then swim through the female reproductive tract, aided by contractions of the oviduct and guided by molecules released by the egg. Upon contact with the egg in the oviduct, the acrosomes of sperm release enzymes that digest a path through the egg coats. As the fertilizing sperm contacts the egg, the sperm and egg plasma membranes fuse, releasing the sperm nucleus into the egg cytoplasm and activating the egg.

The egg completes meiosis, and the sperm and egg nuclei fuse, producing the zygote (Figure 47.17).

- As the embryo implants, the hormone hCG sustains the corpus luteum, which continues to secrete estrogen and progesterone at high levels. These hormones maintain the uterine lining and prevent menstruation.

Animation: Male reproductive system

Animation: Route sperm travel

Animation: Hormonal control of sperm production

Animation: Female reproductive system

Animation: Ovarian function

Animation: Hormones and the menstrual cycle

Animation: Menstrual cycle summary

47.4 Methods for Preventing Pregnancy: Contraception

- Methods of contraception work by preventing sperm from reaching the site of fertilization, by preventing ovulation, or by interfering with implantation (Table 47.1).
- Methods for preventing the sperm from reaching the site of fertilization include the condom, the diaphragm or cervical cap, and the rhythm method, as well as vasectomy or tubal ligation.
- The oral contraceptive pill prevents ovulation. It contains a combination of estrogen and the progesterone-like progestin, which inhibiting the secretion of FSH and LH and follicle formation.
- Methods for preventing implantation include the IUD and the morning-after pill.

Questions

Self-Test Questions

1. Asexual reproduction is most successful in:
   a. changing environments.
   b. sessile animals.
   c. densely settled populations.
   d. land animals.
   e. genetically varied individuals.
2. Which of the following processes does *not* increase genetic diversity?
   a. parthenogenesis.
   b. random DNA mutations.
   c. genetic recombination.
   d. independent assortment.
   e. random combinations of paternal and maternal chromosomes.
3. Gametogenesis has parallel stages in egg and sperm formation. The stage in eggs that is equivalent to spermatids is the:
   a. primary oocyte.
   b. oogonium.
   c. ovum.
   d. ootid and polar bodies.
   e. secondary oocyte and polar body.
4. The animal group that exhibits external fertilization is the:
   a. amphibians.
   b. birds.
   c. sharks.
   d. reptiles.
   e. mammals.
5. The slow block to polyspermy:
   a. is caused by a change in membrane potential from negative to positive.
   b. triggers the movement of Ca^{2+} from the cytosol to the endoplasmic reticulum.
   c. triggers a decrease in egg oxidation and protein synthesis.
   d. describes the fusion of egg and sperm nuclei.
   e. includes the fusion of cortical granules with the egg's plasma membrane.
6. Some placental animals provide nutrients to their embryos from an attached membranous yolk-containing sac. They are called:
   a. oviparous animals.
   b. ovoviviparous animals.
   c. metatherians.
   d. eutherians.
   e. mammals.
7. Which activity is a step in the ovarian cycle?
   a. FSH stimulates the pituitary to release GnRH.
   b. When FSH and LH levels fall, the corpus luteum shrinks and the uterine lining breaks down.
   c. Luteinizing hormone stimulates the uterus to make progesterone.
   d. Estrogen levels initially have a positive feedback effect on the pituitary, which is followed by higher estrogen levels causing negative feedback.
   e. A fully developed corpus luteum inhibits uterine lining growth.
8. During spermatogenesis in mammals, sperm travels from the:
   a. Sertoli cells past the epididymis and urethra, through the vas deferens to the prepuce.
   b. seminal vesicles past the prostate gland, through the glans and prepuce to the bulbourethral glands.
   c. vestibular glands past the Leydig cells, through the accessory glands and epididymis to the vas deferens.
   d. labia past the bulbourethral glands, through the vas deferens and urethra to the epididymis.
   e. seminiferous tubules past the Leydig cells, through the epididymis and vas deferens to the urethra.
9. The human egg is fertilized in the:
   a. uterus.
   b. vagina.
   c. oviduct.
   d. cervical canal.
   e. ovary.
10. The most effective method to prevent fertilization is:
    a. the oral contraceptive.
    b. the IUD.
    c. the morning-after pill.
    d. vasectomy and tubal ligation.
    e. the rhythm method.

Questions for Discussion

1. Currently under development is an "anti-pregnancy vaccine" that stimulates a woman's immune system to develop antibodies against human chorionic gonadotropin (hCG). How would this method prevent pregnancy?
2. Men sometimes have reduced fertility because of *testicular varioceles,* varicose veins in the testes in which blood pools. Based on what you now know of the conditions under which sperm develop properly, how do think this condition might impair sperm development?
3. Spermatogenesis produces four sperm for each spermatocyte, but oogenesis produces only one egg for each oocyte. Why might these different outcomes be adaptive?
4. Sertoli cells protect spermatocytes from attack by antibodies during their development in the human male. What structures

might protect the oocyte and egg from attack by antibodies in the human female?

5. Compare the advantages and disadvantages of sexual and asexual reproduction for an aphid and a parasitic worm.
6. It may be possible to develop a birth control drug that would prevent conception by interfering with fertilization. Outline the design for such a drug and explain exactly how it would work. (There may be more than one design that, in theory, would be effective.)

Experimental Analysis

Design experiments to determine if, and at what dose, vitamin E can decrease menstrual cramping significantly.

Evolution Link

The nematode species *Caenorhabditis elegans* and *Caenorhabditis briggsae* are both hermaphroditic. Phylogenetic evidence indicates that the last common ancestor of these two species had a normal male-female mechanism of reproduction. What does this evidence suggest about their hermaphroditism?

How Would You Vote?

Fertility drugs induce multiple ovulations at the same time and increase the likelihood of high-risk multiple pregnancies. Should the use of such drugs be restricted to conditions that limit the number of embryos formed? Go to www.thomsonedu.com/login to investigate both sides of the issue and then vote.

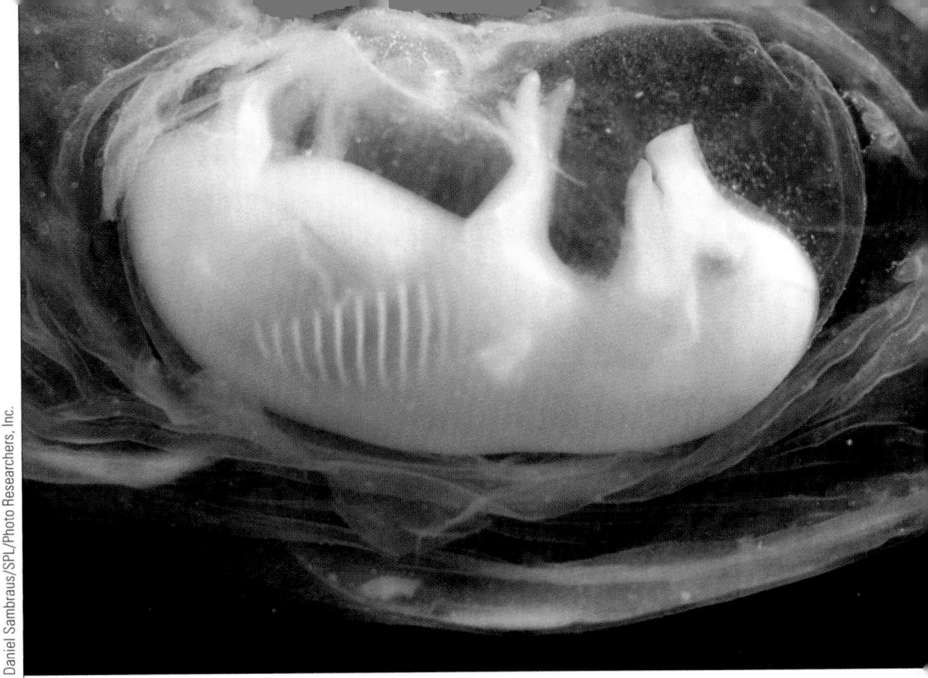

Embryo pig (*Sus scrofa domestica*) after 33 days of development. The embryo is about 16 mm long and is surrounded by several membranous sacs, including the fluid-filled amnion (closest to the embryo), which cushions and protects it.

Daniel Sambraus/SPL/Photo Researchers, Inc.

Study Plan

48 Animal Development

Why It Matters

The uterine contractions announcing birth are taking place at shorter intervals and with greater intensity. The mother-to-be endures the discomfort and apprehension with the knowledge that the child that has been growing in her body will soon come into the world. It began as a fertilized egg, about the size of a period on this page, and grew through a program of cell divisions, complex cell movements, and molecular interactions. She was unaware of these complexities except for movements of the fetus that became apparent about 14 weeks after she became pregnant.

Her baby's development required no conscious attention on her part: human development, like that of all animals, is programmed to proceed inexorably from fertilized egg to free-living offspring. Even childbirth is the result of programmed events that, once started, normally move to conclusion without requiring deliberate input from the mother.

Over the course of its development, the baby's body formed all the organ systems required for independent existence, and at its birth they are already working to sustain its life. Most astonishing, perhaps, is the baby's brain. It began as a tube of nerve tissue that bulged outward and enlarged, continually adding nerve cells and connecting them into circuits until it attained what may well be some of the most complexly organized matter in the universe—all as part of the automated events of development. Still, the human brain is unique only in the degree of its complexity and integrative capacity; the brains of other mammals are basically similar and develop through the same embryonic pathways.

The baby enters the outside world passing head first through the cervix, and then the vagina. Soon the rest of the body slips through, aided and lubricated in its passage by release of the fluid that surrounded and cushioned it in the uterus. In the first indignity of life, the baby is briefly held upside down to drain fluid from its lungs. This action triggers its first breath, followed by a satisfyingly loud cry.

The baby is proudly displayed to the mother, who greets it with love, relief, joy, and realization of the responsibilities the baby will bring. It is a girl, who with further luck and good care will continue developing through childhood, puberty, adult life, and old age, all through programs built into her hereditary molecules. As part of these passages, she may bring her own child into the world.

People have tried since ancient times to understand how development and birth take place. The scientific quest began with Aristotle, who observed chick development and correctly interpreted the functions of the placenta and umbilical cord in humans. The investigators who followed Aristotle concentrated on describing developmental changes in **morphology,** which is the form or shape of an organism, or of a part of an organism. More recently, investigators began to trace the molecular underpinnings of the morphological events.

In this chapter we survey the results of these investigations. We take up the story of animal development where the previous chapter left off, with the fertilized egg. We continue with the early events leading from the fertilized egg to the primary tissues of the embryo, and then trace the development of organs from these tissues. Next, we describe human development as representative of the process in mammals. Then, we survey the cellular and molecular bases of these mechanisms. At the cellular level, the development of an adult animal from a fertilized egg involves cell division, in which more cells are produced by mitosis; **cell differentiation,** in which changes in gene expression establish cells with specialized structure and function; and **morphogenesis** ("form creation"), the generation of the body form of the animal as differentiated cells end up in their appropriate sites. Finally, we discuss the genetic and molecular mechanisms that are largely responsible for directing the course of development.

48.1 Mechanisms of Embryonic Development

Fertilization of an egg by a sperm cell produces a zygote. Embryonic development begins at this point and ultimately produces a free-living individual. All the instructions required for development are packed into the fertilized egg.

Developmental Information Is Located in both the Nucleus and Cytoplasm of the Fertilized Egg

Mitotic divisions of the zygote formed when egg and sperm nuclei fuse are the beginning of developmental activity (see Section 47.2).

Information Storage in the Egg. The information that directs the initiation of development is stored in two locations in the fertilized egg. Part of the information is stored in the zygote nucleus, in the DNA derived from the egg and sperm nuclei. This information directs development as individual genes are activated or turned off in a highly ordered manner. The rest of the information is stored in the egg cytoplasm, in the form of messenger RNA (mRNA) and protein molecules.

Because the fertilizing sperm contributes essentially no cytoplasm, nearly all the cytoplasmic information of the fertilized egg is maternal in origin. The mRNA and proteins stored in the egg cytoplasm are known as **cytoplasmic determinants.** They direct the first stages of animal development, in the period before genes of the zygote become active. Depending on the animal group, the control of early development by cytoplasmic determinants may be limited to the first few divisions of the zygote, as in mammals, or it may last until the actual tissues of the embryo are formed, as in most invertebrates.

Other Components of the Egg. In addition to cytoplasmic determinants, the oocyte cytoplasm also contains ribosomes and other cytoplasmic components required for protein synthesis and the early cell divisions of embryonic development. For example, the egg cytoplasm contains all the tubulin molecules required to form the spindles for early cell divisions. It also contains mitochondria, nutrients stored in granules in the yolk and in lipid droplets, and, in many animals, pigments that color the egg or regions of it.

The **yolk** contains nutrients. When the egg itself supplies all the nutrients for development of the embryo, as in the eggs laid by insects, reptiles, and birds, it contains large amounts of yolk. When the mother supplies most of the nutrients, as in the placental mammals, the egg has a small quantity of yolk that is used only for the earliest stages of development.

Depending on the species, the yolk may be concentrated at one end or in the center of the egg, or distributed evenly throughout the cytoplasm. Its distribution influences the rate and location of cell division during early embryonic development. Typically, cell division proceeds more slowly in the region of the egg containing the yolk. In the large, yolky eggs of birds and reptiles, cell division takes place only in a small, yolk-free patch at the surface of the egg.

Unequal distribution of yolk and other components in a mature egg is termed **polarity.** For example, in most species the egg nucleus is located toward one end of the egg. This end of the egg, called the **animal pole,** typically gives rise to surface structures and the anterior end of the embryo. The opposite end of the egg, the **vegetal pole,** typically gives rise to internal structures such as the gut and the posterior end of the embryo. Yolk, when unequally distributed in the egg cytoplasm, is most frequently concentrated in the vegetal half of the egg. The egg's polarity contributes to the generation of body axes. For example, egg polarity plays a role in setting the three body axes of bilaterally symmetrical animals (such as humans and dogs): the anterior–posterior axis, the dorsal–ventral (back–front) axis, and the left–right axis **(Figure 48.1).**

Cleavage, Gastrulation, and Organogenesis Are Early Events in Development

Soon after fertilization, the zygote begins a series of mitotic **cleavage** divisions, so called because cycles of DNA replication and division occur without the production of new cytoplasm. As a result, the cytoplasm of the egg is partitioned into successively smaller cells without increasing the overall size or mass of the embryo **(Figure 48.2).** These cells are called **blastomeres** (*blastos* = bud or offshoot; *meros* = part or division). In the frog *Xenopus laevis,* for example, a sequence of twelve cleavage divisions produces an embryo of about 4000 cells, which collectively occupy about the same volume and mass as did the original zygote.

Cleavage is the first of three major developmental processes that, with modifications, are common to the early development of most animals (described in detail for particular animals in the next section). Following cleavage, the second major process, **gastrulation,** produces an embryo with three distinct primary tissue layers. Following gastrulation, the development of the major organ systems, called **organogenesis,** gives rise to a free-living individual with the body organization characteristic of its species. Organogenesis involves the same mechanisms used in gastrulation—cell division, cell movements, and cell rearrangements. **Figure 48.3** outlines these stages in the life cycle of a frog.

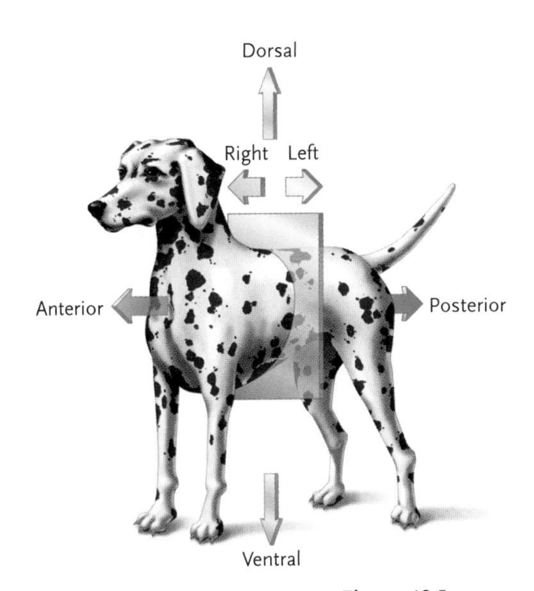

Figure 48.1
Body axes: anterior–posterior, dorsal–ventral, and left–right.

The cleavage divisions lead to three successive developmental stages that are common to the early development of most animals. The first stage, called a **morula** (*morula* = mulberry), is a solid ball or layer of blastomeres. As cleavage divisions continue, the ball or layer hollows out to form the second stage, the **blastula** (*ula* = small), in which the blastomeres enclose a fluid-filled cavity, the **blastocoel** (*koilos* = hollow).

Once cleavage is complete, the cells of the blastula migrate and divide to produce the **gastrula** (*gaster* = gut or belly). This morphogenetic process, gastrulation, dramatically rearranges the cells of the blastula into the three primary cell layers of the embryo: the outer **ectoderm** (*ecto* = outside; *derma* = skin), the inner **endoderm** (*endo* = inside), and the **mesoderm** (*meso* = middle) between the ectoderm and the endoderm. Gastrulation establishes body pattern; that is,

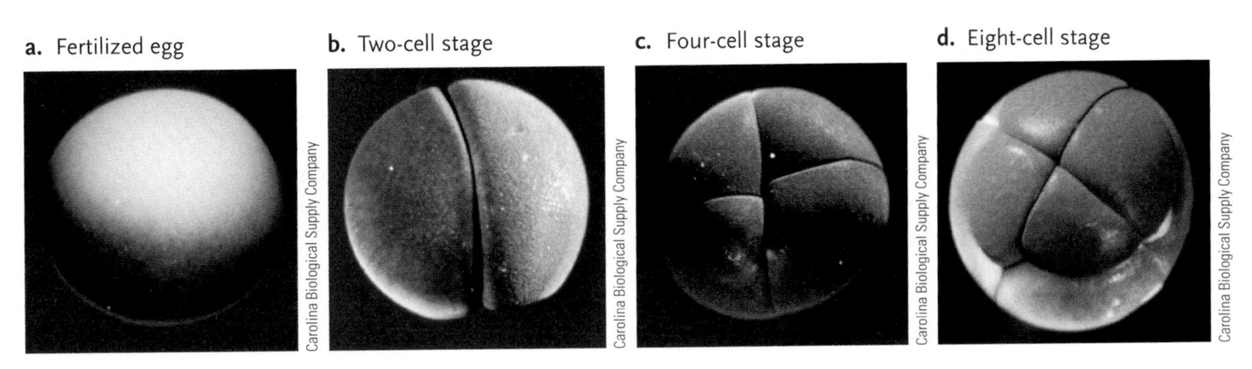

Figure 48.2
The first three cleavage divisions of a frog embryo, which convert the fertilized egg into the eight-cell stage. Note that the cleavage divisions cut the volume of the fertilized egg into successively smaller cells.

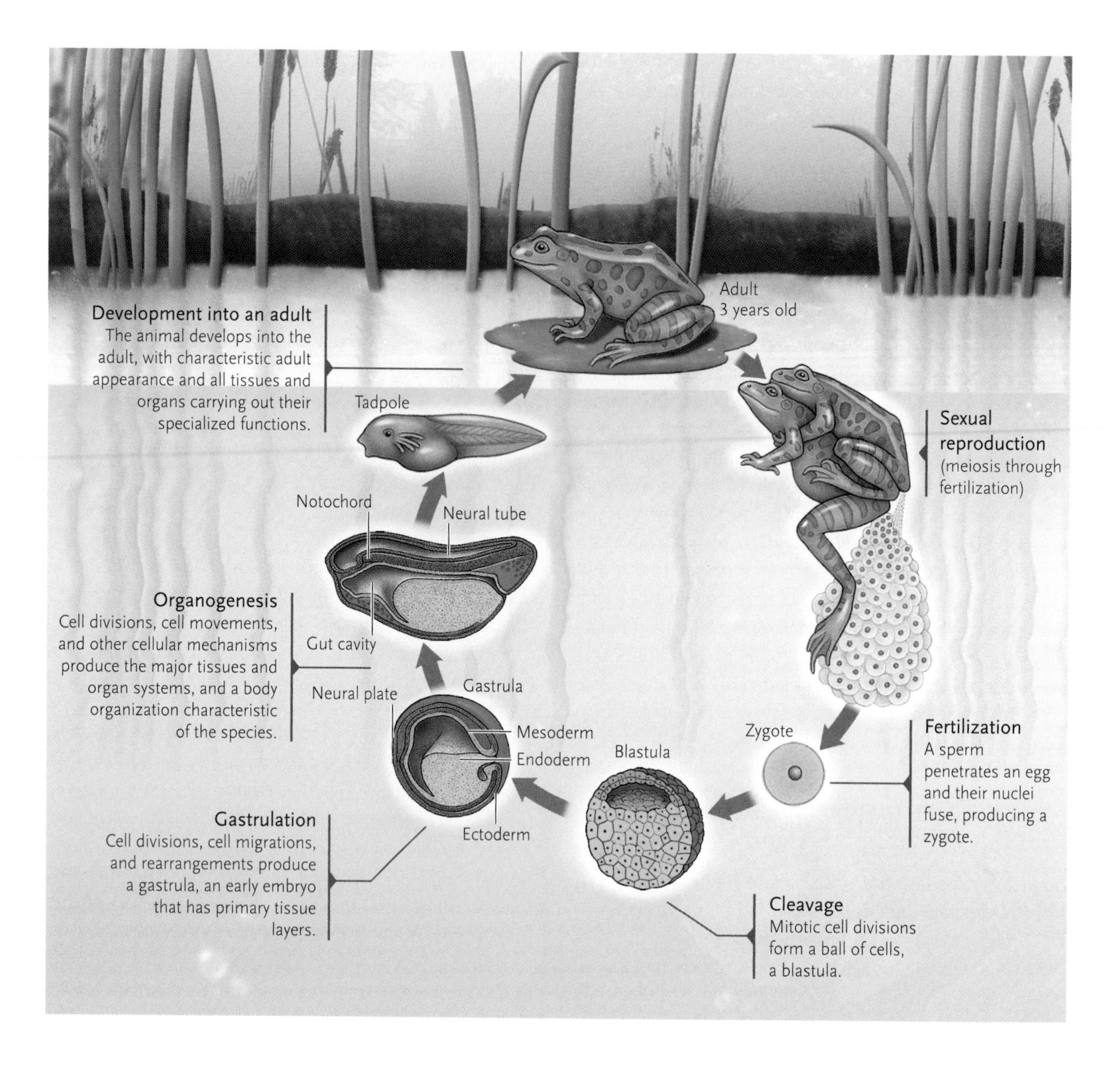

Adult
3 years old
Development into an adult
The animal develops into the adult, with characteristic adult appearance and all tissues and organs carrying out their specialized functions.
Tadpole
Sexual reproduction
(meiosis through fertilization)
Notochord
Neural tube
Organogenesis
Cell divisions, cell movements, and other cellular mechanisms produce the major tissues and organ systems, and a body organization characteristic of the species.
Gut cavity
Neural plate
Gastrula
Mesoderm
Endoderm
Blastula
Zygote
Fertilization
A sperm penetrates an egg and their nuclei fuse, producing a zygote.
Gastrulation
Cell divisions, cell migrations, and rearrangements produce a gastrula, an early embryo that has primary tissue layers.
Ectoderm
Cleavage
Mitotic cell divisions form a ball of cells, a blastula.

Figure 48.3
Stages of animal development shown in a frog.

each tissue and organ of the adult animal originates in one of the three primary cell layers of the gastrula **(Table 48.1)**.

The cell movements also form a new cavity within the embryo, the **archenteron** (*arche* = beginning; *enteron* = intestine or gut), which is lined with endoderm. The archenteron forms the primitive gut of the embryo; an opening at one end, the **blastopore,** gives rise to the anus or mouth of the embryo, depending on the animal group (see Section 29.2). In the protostomes, which include annelids, arthropods, and mollusks, the blastopore develops into the mouth, and the anus forms at the opposite end of the embryonic gut. In the deuterostomes, which include echinoderms and chordates, the blastopore develops into the anus and the mouth forms at the opposite end of the embryonic gut. By the time gastrulation is complete, the embryo has clearly defined anterior, posterior, dorsal, and ventral regions.

As the blastula develops into the gastrula, embryonic cells begin to differentiate: they become recognizably different in biochemistry, structure, and function. The developmental potential of the cells also becomes more limited than that of the fertilized egg from which they originated. For example, although a fertilized egg is capable of developing into a complete embryo, a mesoderm cell may develop into muscle or bone but not normally into outside skin or brain. This restriction of developmental potential does not occur, as was once thought, because the cells have lost all their genes except those for the structure and function of the cell type they will become. Rather, the differentiating cells actually all contain complete genomes of the organism, but each type of cell has a different program of gene expression.

Development in all animals is accomplished by a number of mechanisms that are under genetic control but are influenced to some extent by the environment (for example, temperature affects the rate of cell division). The mechanisms are

1. Mitotic cell divisions.
2. Cell movements.

3. **Selective cell adhesions,** in which cells make and break specific connections to other cells or to the extracellular matrix.
4. **Induction,** in which one group of cells (the inducer cells) causes or influences another nearby group of cells (the responder cells) to follow a particular developmental pathway. The key to induction is that only certain cells can respond to the signal from the inducer cells. Induction typically involves signal transduction events (see Chapter 7). These events are triggered either by direct cell-to-cell contact involving interaction between a membrane-embedded protein on the inducer cell and a receptor protein on the surface of the responder cell, or by a signal molecule released by the inducer cell that interacts with a receptor on the responder cell. (The latter is an example of paracrine signaling; see Section 40.1.)
5. **Determination,** in which the developmental fate of a cell is set. Prior to determination, a cell has the potential to become any cell type of the adult but, after determination, that property is lost as the cell commits to becoming a particular cell type. Typically, determination is the result of induction, but in some cases it results from the asymmetric segregation of cellular determinants.
6. Differentiation, which follows determination, involves the establishment of a cell-specific developmental program in cells. Differentiation results in cell types with clearly defined structures and functions; those features derive from specific patterns of gene expression in cells.

You will see examples of these mechanisms in the examples of development discussed in the following three sections.

Study Break

1. How do cleavage divisions differ from cell division in an adult organism?
2. What are the primary cell layers of the embryo, and what process is responsible for producing them?

48.2 Major Patterns of Cleavage and Gastrulation

With the principles of early embryonic development established, we describe cleavage and gastrulation in three animal groups that have been models in *embryology* (the study of embryos and their development): sea urchins, amphibians, and birds. Later in the chapter, we describe cleavage and gastrulation in humans and other mammals, which resemble the pattern in birds.

Table 48.1 Origins of Adult Tissues and Organs in the Three Primary Tissue Layers

Primary Tissue Layer	Adult Tissues and Organs
Ectoderm	Skin and its elaborations, including hair, feathers, scales, and nails; nervous system, including brain, spinal cord, and peripheral nerves; lens, retina, and cornea of eye; lining of mouth and anus; sweat glands, mammary glands, adrenal medulla, and tooth enamel
Mesoderm	Muscles; most of skeletal system, including bones and cartilage; circulatory system, including heart, blood vessels, and blood cells; internal reproductive organs; kidneys and outer walls of digestive tract
Endoderm	Lining of digestive tract, liver, pancreas, lining of respiratory tract, thyroid gland, lining of urethra, and urinary bladder

Sea Urchin Gastrulation Follows a Symmetrical Pattern That Reflects an Even Distribution of Yolk

Cleavage divisions proceed at approximately the same rate in all regions of a sea urchin embryo **(Figure 48.4,** step 1), reflecting the uniform distribution of yolk in the sea urchin egg. These divisions continue until a blastula containing about a thousand cells is formed (step 2).

Gastrulation begins at the vegetal pole of the blastula. As a result of induction, some cells in the middle of that region become elongated and cylindrical, causing the region to flatten and thicken. Then, some cells break loose and migrate into the blastocoel (see Figure 48.4, step 3). These cells, called *primary mesenchyme cells* (mesenchyme means "middle juice"), move around inside the blastocoel, making and breaking adhesions, until eventually they attach along the ventral sides of the blastocoel. These cells will eventually become the mesoderm (see step 7), which will give rise to skeletal elements of the embryo. Next, the flattened vegetal pole of the blastula invaginates, pushing gradually into the interior (steps 4 and 5). The cells that invaginate are future endoderm cells. The inward movement, in effect much like pushing in the side of a hollow rubber ball, generates a new cavity, the archenteron. The opening of the archenteron is the blastopore.

As the archenteron is forming, the cells of the invaginated cell layer send out extensions that stretch across the blastocoel and contact the inside of the ectoderm (step 6). These extensions make tight adhesions and then contract, pulling the invaginated cell layer inward with them and thereby eliminating most of the blastocoel.

At this point the embryo has two complete cell layers. The outer layer remaining from the original blastula surface makes up the ectoderm of the embryo. The second, inner layer, derived from the cells forming the

Figure 48.4
Cleavage and gastrulation in the sea urchin.

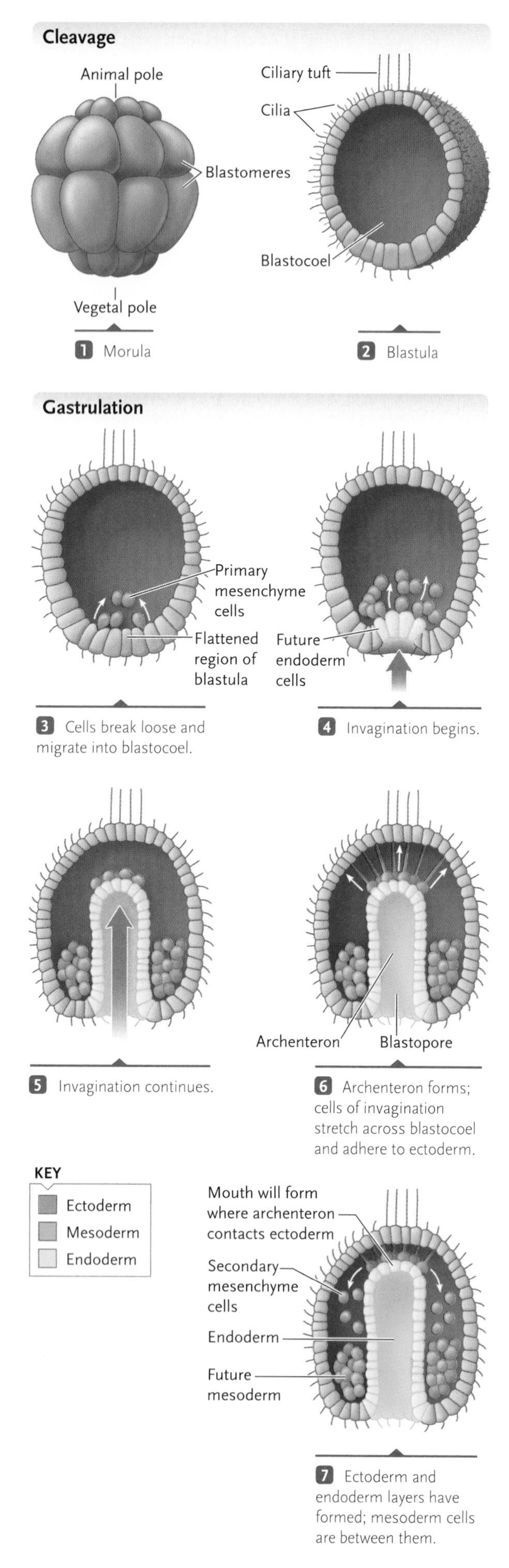

archenteron, makes up the endoderm. Mesodermal cells are also beginning to form a third layer, the mesoderm. Some are derived from the primary mesenchyme cells and others from *secondary mesenchyme cells,* cells that migrated into the space between ectoderm and endoderm (step 7). When the mesoderm layer is complete, the embryo has three complete layers: ectoderm, mesoderm, and endoderm. At this point, cells within each layer begin to differentiate, as evidenced by the synthesis of different proteins in each layer.

As the ectoderm, mesoderm, and endoderm develop, the embryo lengthens into an ellipsoidal shape with the blastopore marking the posterior end of the embryo. From this point on, organ systems differentiate through further cell division, cell movements, selective cell adhesions, induction, and differentiation. The blastopore forms the anus; a mouth will form at the opposite, anterior end of the gut.

Amphibian Cleavage and Gastrulation Are Influenced by an Unequal Distribution of Yolk

In amphibian eggs, such as those of frogs, yolk is concentrated in the vegetal half, which gives it a pale color. The animal half is darkly colored by a layer of pigment granules just below the surface. The sperm normally fertilizes the egg in the animal half (**Figure 48.5,** step 1). After fertilization, the pigmented layer of cytoplasm rotates toward the site of sperm entry, exposing a crescent-shaped region of the underlying cytoplasm at the side opposite the point of sperm entry (step 2). This region, called the **gray crescent,** establishes the dorsal–ventral axis of the embryo, with the gray crescent marking the future dorsal side.

Normally, the first cleavage division runs perpendicular to the long axis of the gray crescent and divides the crescent equally between the resulting cells (step 3). If one of the first two blastomeres does not receive gray crescent material, and the cells are separated experimentally, the cell without gray crescent divides to produce a disordered mass that stops developing. The cell receiving the gray crescent produces a normal embryo. Thus cytoplasmic material localized in the gray crescent is essential to normal development in frog embryos.

As cleavage of a frog embryo continues, cell divisions proceed more rapidly in the animal half, producing smaller and more numerous cells in this region than in the yolky vegetal half. By the time cleavage has produced an embryo with 15,000 cells, the animal half of the embryo has hollowed out, forming the blastula (**Figure 48.6,** step 1, and **Figure 48.7a**).

Gastrulation begins when cells from the animal pole move across the embryo surface and reach the region derived from the gray crescent. This site is marked by a crescent-shaped depression rotated clockwise 90°

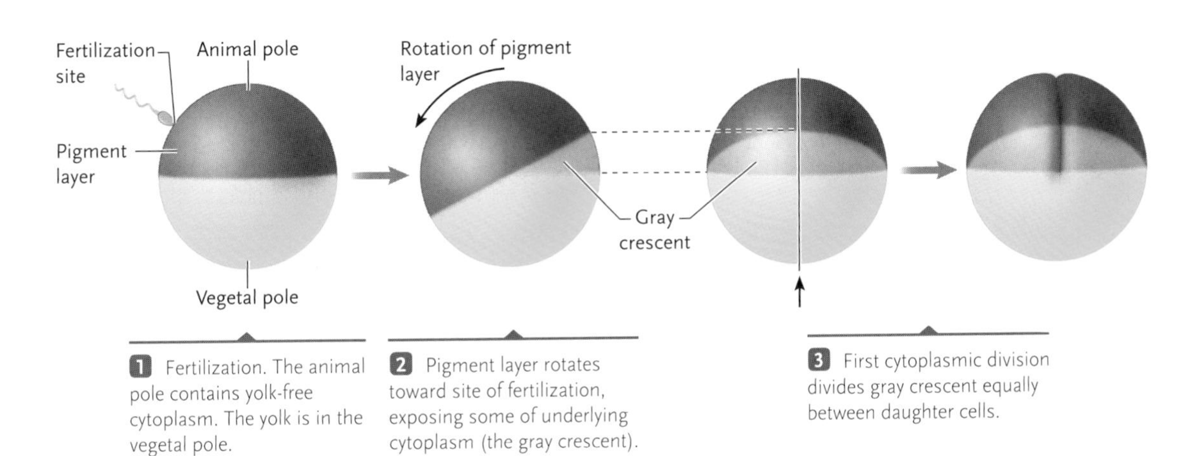
Fertilization site
Animal pole
Pigment layer
Vegetal pole
Rotation of pigment layer
Gray crescent
1 Fertilization. The animal pole contains yolk-free cytoplasm. The yolk is in the vegetal pole.
2 Pigment layer rotates toward site of fertilization, exposing some of underlying cytoplasm (the gray crescent).
3 First cytoplasmic division divides gray crescent equally between daughter cells.

Figure 48.5
Rotation of the pigment layer and development of the gray crescent after fertilization in a frog egg. The gray crescent marks the site where gastrulation of the embryo will begin.

called the **dorsal lip of the blastopore.** Cells changing shape and pushing inward from the surface in a process called **invagination** produce the depression. With continued inward movement of additional cells, the depression eventually forms a complete circle (Figure 48.6, step 2, and **Figure 48.7b**), which is the blastopore.

As cells migrate into the blastopore by a process is called **involution**, the pigmented cell layer of the animal half expands to cover the entire surface of the embryo (Figure 48.6, step 3). The cells of the vegetal half are enclosed by the movement, and show on the outside as a yolk plug in the blastopore (Figure 48.6,

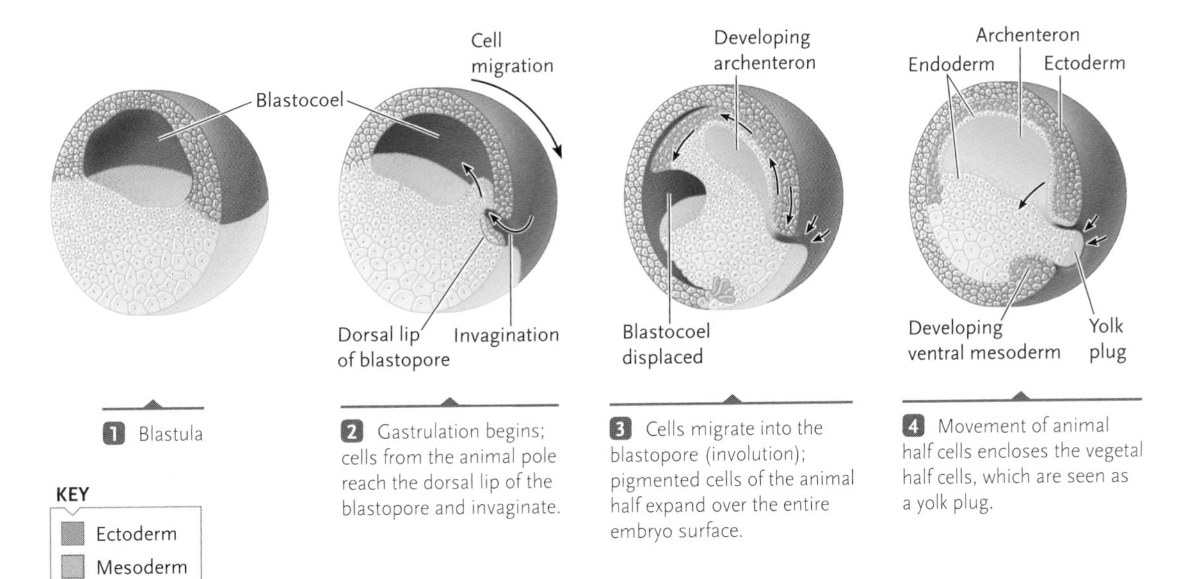
Blastocoel
Cell migration
Dorsal lip of blastopore
Invagination
Developing archenteron
Blastocoel displaced
Archenteron
Endoderm
Ectoderm
Developing ventral mesoderm
Yolk plug
1 Blastula
2 Gastrulation begins; cells from the animal pole reach the dorsal lip of the blastopore and invaginate.
3 Cells migrate into the blastopore (involution); pigmented cells of the animal half expand over the entire embryo surface.
4 Movement of animal half cells encloses the vegetal half cells, which are seen as a yolk plug.
KEY
Ectoderm
Mesoderm
Endoderm

Figure 48.6
Gastrulation in a frog embryo. Yolk cells are shown in darker yellow.

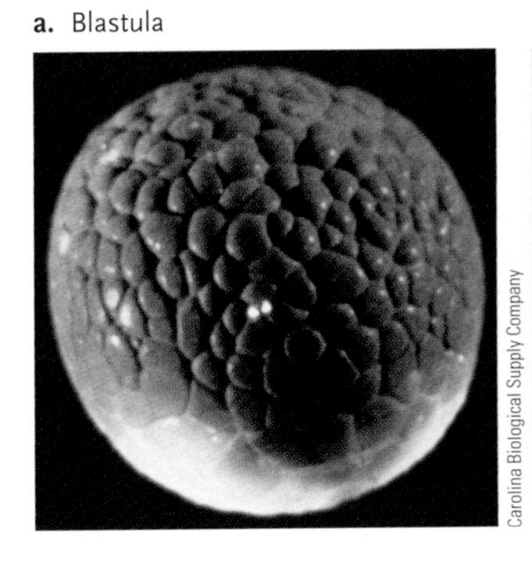
a. Blastula

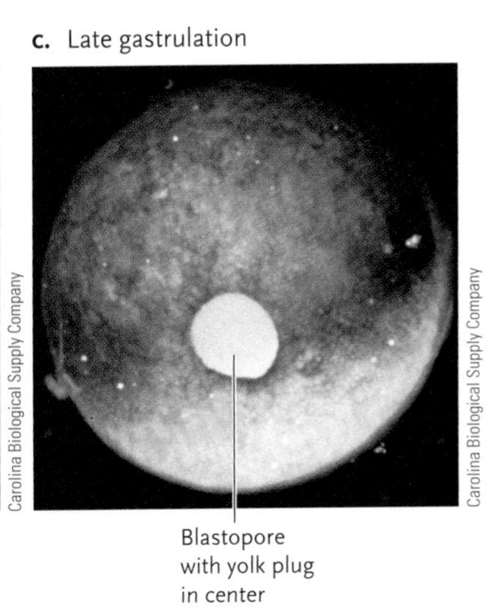
b. Early gastrulation
Dorsal lip of blastopore
c. Late gastrulation
Blastopore with yolk plug in center
Carolina Biological Supply Company

Figure 48.7
Formation of the dorsal lip of the blastopore and the completed blastopore, closed by a yolk plug, in photomicrographs of a frog embryo.

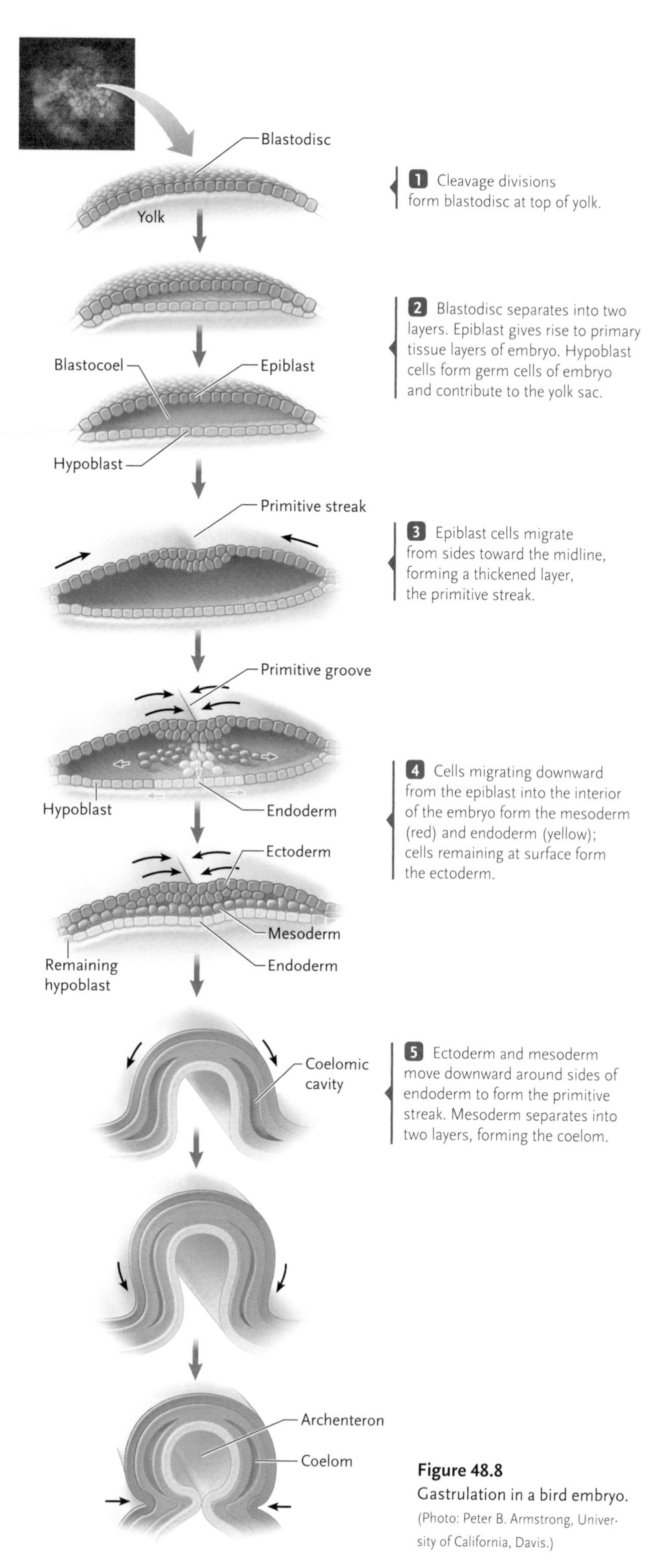

Blastodisc
Yolk
1 Cleavage divisions form blastodisc at top of yolk.
2 Blastodisc separates into two layers. Epiblast gives rise to primary tissue layers of embryo. Hypoblast cells form germ cells of embryo and contribute to the yolk sac.
Blastocoel
Epiblast
Hypoblast
Primitive streak
3 Epiblast cells migrate from sides toward the midline, forming a thickened layer, the primitive streak.
Primitive groove
Hypoblast
Endoderm
4 Cells migrating downward from the epiblast into the interior of the embryo form the mesoderm (red) and endoderm (yellow); cells remaining at surface form the ectoderm.
Ectoderm
Mesoderm
Endoderm
Remaining hypoblast
Coelomic cavity
5 Ectoderm and mesoderm move downward around sides of endoderm to form the primitive streak. Mesoderm separates into two layers, forming the coelom.
Archenteron
Coelom

Figure 48.8
Gastrulation in a bird embryo. (Photo: Peter B. Armstrong, University of California, Davis.)

step 4, and **Figure 48.7c**). As in other vertebrates, the blastopore gives rise to the anus.

Within the embryo, continuing involution moves cells into the interior and upward (see Figures 48.7b and c), forming two layers that line the inside top half of the embryo. The uppermost of these layers is induced to become the dorsal mesoderm (shown in red). The layer beneath it, which contains cells originating from both the outer surface of the embryo and the yolky interior, becomes the endoderm (shown in lighter yellow). The pigmented cells remaining at the surface of the embryo form the ectoderm (shown in blue). The ventral mesoderm begins to be induced near the vegetal pole.

As the mesoderm and endoderm form, the depression created by the inward cell movements gradually deepens and extends inward as the archenteron (see Figures 48.6c and d), which displaces the blastocoel. The cells of the three primary cell layers continue to increase in number by further movements and divisions as development proceeds.

During frog gastrulation, cells of the dorsal lip of the blastopore are inducer cells that control blastopore formation; if the cells in the dorsal lip are removed and transplanted elsewhere in the egg, they cause a second blastopore—and a second embryo—to form in this region (see Section 48.5).

The events of gastrulation in frogs thus include the same developmental mechanisms as in sea urchins—cell divisions, cell movements, selective adhesions, induction, and differentiation.

Gastrulation in Birds Proceeds at One Side of the Yolk

The pattern of gastrulation in birds and reptiles is modified by the distribution of yolk, which occupies almost the entire volume of the egg. (Birds and reptiles, as well as mammals, are all amniotes; see Section 30.7.) The portion of the cytoplasm that divides to give rise to the primary tissues of the embryo is confined to a thin layer at the egg surface. Although mammalian eggs have relatively little yolk, gastrulation follows a similar pattern in mammals, as discussed in Section 48.4.

Cleavage and Gastrulation in Birds. The early cleavage divisions in birds produce a disclike layer of cells at the surface of the yolk called the **blastodisc (Figure 48.8,** step 1). When blastodisc formation is complete, the layer contains about 20,000 cells. The cells of the blastodisc then separate into two layers, called the **epiblast** (top layer) and **hypoblast** (bottom layer). The flattened cavity between them is the blastocoel (step 2).

Gastrulation begins as cells in the epiblast stream toward the midline of the blastodisc, thickening the epiblast in this region. The thickened layer—the **primitive streak**—begins forming in the posterior end of the embryo and extends toward the anterior end as more cells of the epiblast move into it (step 3). The thickening at the anterior end of the primitive streak, called the *primitive knot,* is the functional equivalent of the amphibian dorsal lip of the blastopore. The primitive streak initially defines the axes of the embryo: it extends from posterior to anterior, the region where the streak forms is the dorsal while beneath it is the ventral side, and it defines left and right sides of the embryo.

As the primitive streak forms, its midline sinks, forming the **primitive groove.** The primitive groove is a conduit for migrating cells to move into the blastocoel. The first cells to migrate through the primitive groove are epiblast cells (step 4), which will form the endoderm. Cells migrating laterally between the epiblast and the endoderm form the mesoderm. The epiblast cells left at the surface of the blastodisc form the ectoderm (see step 4). Thus all the primary tissue layers of the chick embryo arise from the epiblast.

Of the cells in the hypoblast, only a few, near the posterior end of the embryo, contribute directly to the embryo. These hypoblast cells form the *germ cells* that, later in development, migrate to the developing gonads and found the cell line leading to eggs and sperm (see Section 47.2).

Initially, the ectoderm, mesoderm, and endoderm are located in three more or less horizontal layers. During gastrulation, the endoderm pushes upward along its midline. At the same time, its left and right sides fold downward, forming a tube oriented parallel to the primitive streak (step 5). The central cavity of the tube is the archenteron, the primitive gut. The mesoderm separates into two layers, forming the coelom, a fluid-filled body cavity (see Section 29.2 and Figure 29.4c). These movements complete formation of the gastrula.

Formation of Extraembryonic Membranes. Each of the primary tissue layers of a bird embryo extends outside the embryo to form four **extraembryonic membranes** **(Figure 48.9),** which conduct nutrients from the yolk to the embryo, exchange gases with the environment outside the egg, and store metabolic wastes removed from the embryo. The **yolk sac** consists of extensions of mesoderm and endoderm that enclose the yolk. Although the yolk sac remains connected to the gut of the embryo by a stalk, yolk does not directly enter the embryo by this route. Instead, it is absorbed by blood vessels in the membrane, which transport the nutrients to the embryo. The **chorion,** produced from ectoderm and mesoderm, is the outermost membrane, which surrounds the embryo and yolk sac completely, and lines the inside of the shell. This membrane exchanges oxygen and carbon dioxide with the environment through the shell of the egg. The **amnion** is the innermost membrane, which closes over the embryo to form the *amniotic cavity.* The cells of the amnion secrete *amniotic fluid* into the cavity, which bathes the embryo and provides an aquatic environment in which it can develop. Reptilian and mammalian embryos are also surrounded by an amnion and amniotic fluid. By providing the embryo with an aquatic environment, this adaptation made possible the development of fully terrestrial vertebrates. The evolutionary importance of the amnion to the fully terrestrial vertebrates is recognized by classifying them together as **amniotes.** A membrane derived from mesoderm and endoderm that has bulged outward from the gut forms a sac called the **allantois.** This sac closely lines the chorion and fills much of the space between the chorion and the yolk sac. The allantois stores nitrogenous wastes (primarily uric acid) removed from the embryo. In addition, the part of the allantoic membrane that lines the chorion forms a rich bed of blood capillaries that is connected to the embryo by arteries and veins. This circulatory system delivers carbon dioxide to the chorion and picks up the oxygen that is absorbed through the shell and chorion.

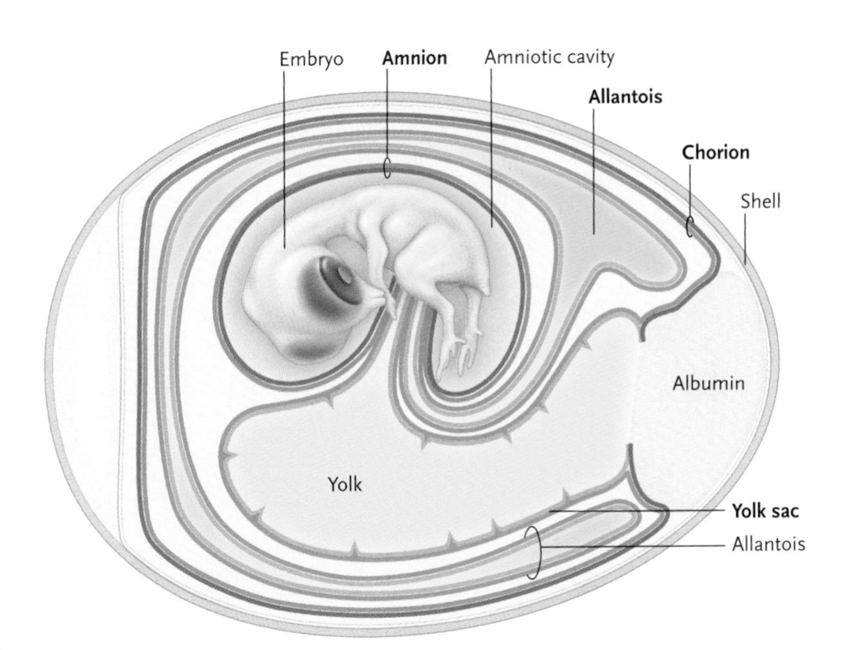

Figure 48.9
The four extraembryonic membranes in a bird embryo (in bold).

Study Break

1. What is the role of the gray crescent in amphibian development?
2. What evidence indicates that cells of the dorsal lip of the blastopore act as inducer cells?
3. What are the extraembryonic membranes in birds, and what are their functions?

48.3 From Gastrulation to Adult Body Structures: Organogenesis

Following gastrulation, organogenesis—the process by which the ectoderm, mesoderm, and endoderm develop into organs—gives rise to an individual with the body organization characteristic of its species. Organo-

Figure 48.10 Development of the neural tube and neural crest cells in vertebrates. Photo is of an amphibian embryo; drawings show steps in a bird embryo.

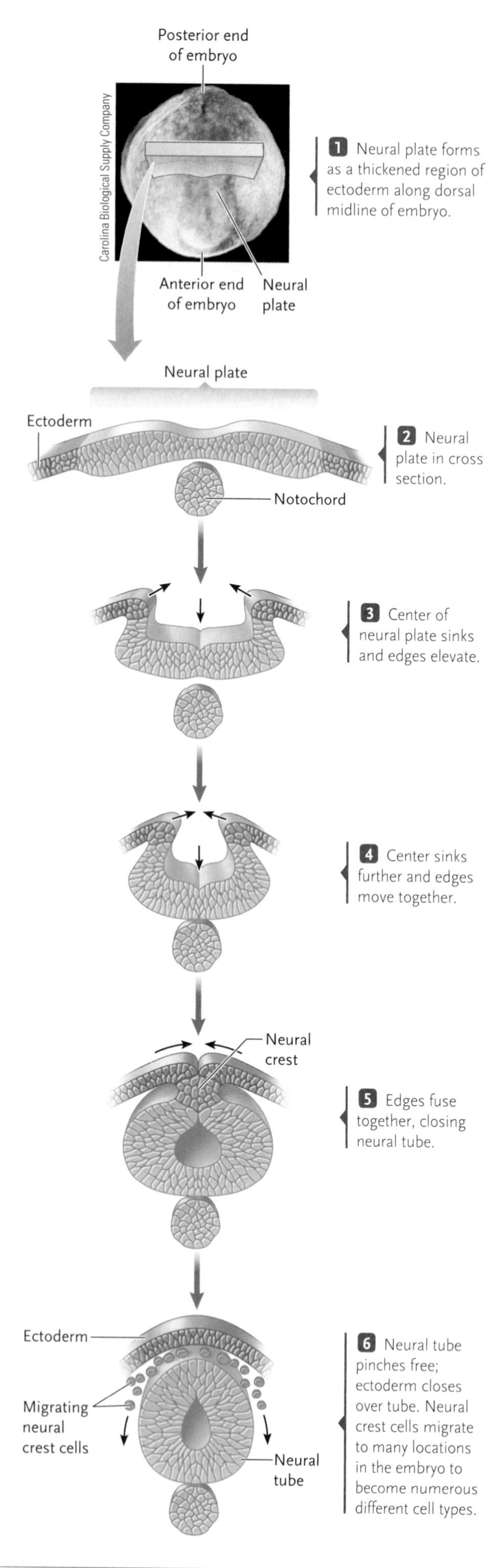
Posterior end of embryo
Carolina Biological Supply Company
1 Neural plate forms as a thickened region of ectoderm along dorsal midline of embryo.
Anterior end of embryo
Neural plate
Neural plate
Ectoderm
2 Neural plate in cross section.
Notochord
3 Center of neural plate sinks and edges elevate.
4 Center sinks further and edges move together.
Neural crest
5 Edges fuse together, closing neural tube.
Ectoderm
Migrating neural crest cells
Neural tube
6 Neural tube pinches free; ectoderm closes over tube. Neural crest cells migrate to many locations in the embryo to become numerous different cell types.

genesis involves the same mechanisms used in gastrulation—cell division, cell movements, selective cell adhesion, induction, and differentiation—plus an additional mechanism, *apoptosis,* in which certain cells are programmed to die (apoptosis is also discussed in Section 43.2). To illustrate how the cellular mechanisms of development interact in organogenesis, we follow the formation of major organ systems in the frog embryo. Then we describe the generation of one organ, the eye, which follows a pathway typical of eye development in all vertebrates.

The Nervous System Develops from Ectoderm

In vertebrates, organogenesis begins with development of the nervous system from ectoderm, a process called **neurulation.** As a preliminary to neurulation, cells of the mesoderm form a solid rod of tissue, the **notochord,** which extends the length of the embryo under the dorsal ectoderm. Notochord cells carry out a major induction, in which they cause the overlying ectoderm to thicken and flatten into a longitudinal band called the **neural plate** (**Figure 48.10,** steps 1 and 2). Experiments have shown that if the notochord is removed, the neural plate does not form.

Once induced, the neural plate sinks downward along its midline (steps 2 and 3), creating a deep longitudinal groove. At the same time, ridges elevate along the sides of the neural plate. The ridges move together and close over the center of the groove (steps 4 and 5), converting the neural plate into a **neural tube** that runs the length of the embryo. The neural tube then pinches off from the overlying ectoderm, which closes over the tube (step 6). The central nervous system, including the brain and spinal cord, develops directly from the neural tube.

During formation of the neural tube, cells of the **neural crest**—the region where the neural tube pinches off from the ectoderm (shown in blue in Figure 48.10)—migrate to many locations in the developing embryo and become numerous different types of cells which contribute to a variety of organ systems. (The neural crest is one of the defining features of vertebrates.) Some cells develop into cranial nerves in the head; others contribute to the bones of the inner ear and skull, the cartilage of facial structures, and the teeth. Yet others form ganglia of the autonomic nervous system, peripheral nerves leading from the spinal cord to body structures, and nerves of the developing gut. Still others move to the skin, where they form pigment cells, or to the adrenal glands, where they form the medulla of these glands. The migration of neural crest cells occurs in the development of all vertebrates.

Other structures differentiate in the embryo while the neural tube is forming. On either side of the notochord, the mesoderm separates into blocks of cells called **somites,** spaced one after the other along both sides of the notochord **(Figure 48.11)**. The somites give rise to the vertebral column, the ribs, the repeating sets of muscles associated with the ribs and vertebral column, and muscles of the limbs. The mesoderm outside the somites, which extends around the primitive gut (lateral mesoderm in Figure 48.11), splits into two layers, one covering the surface of the gut, and the other lining the body wall. The space between the layers is the coelom of the adult (see Section 29.2).

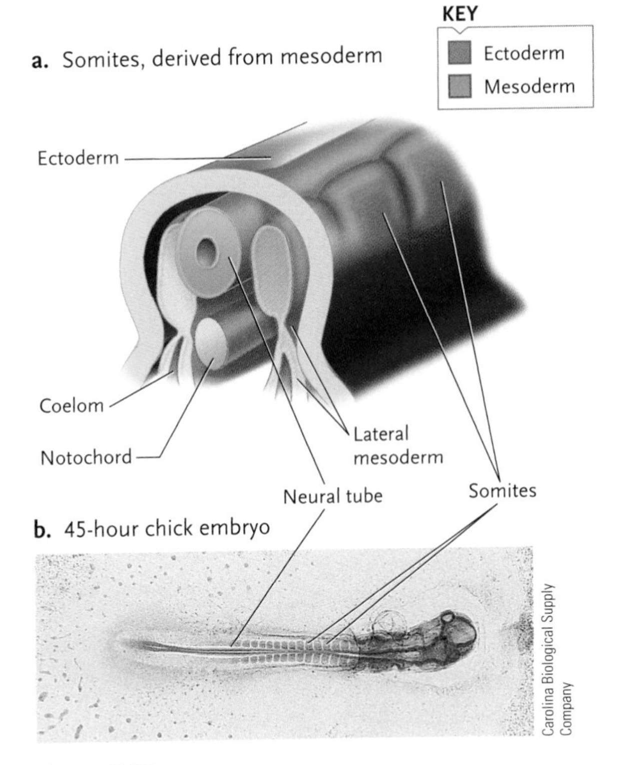

KEY
Ectoderm
Mesoderm
a. Somites, derived from mesoderm
Ectoderm
Coelom
Notochord
Lateral mesoderm
Neural tube
Somites
b. 45-hour chick embryo
Carolina Biological Supply Company

Figure 48.11
Later development of the mesoderm. **(a)** The somites develop into segmented structures such as the vertebrae, the ribs, and the musculature between the ribs. The lateral mesoderm gives rise to other structures, such as the heart and blood vessels and the linings of internal body cavities. **(b)** The somites in a 45-hour chick embryo.

Sequential Inductions and Differentiation Are Central to Eye Development

We now take up the development of the eye, to show how cellular mechanisms interact in organogenesis. Eyes develop by the same pathway in all vertebrates.

The brain forms at the anterior end of the neural tube from a cluster of hollow vesicles that swell outward from the neural tube (**Figure 48.12,** step 1). One paired set of vesicles, the *optic vesicles,* develop into the eyes. The figure depicts the optic vesicles in the brain of a frog embryo; note that the morphology of the forebrain, midbrain, and hindbrain in embryos differs among vertebrates.

The optic vesicles grow outward until they contact the overlying ectoderm, inducing a series of developmental responses in both tissues. The outer surface of the optic vesicle thickens and flattens at the region of contact and then pushes inward, transforming the optic vesicle into a double-walled *optic cup,* which ultimately becomes the retina. The optic cup induces the overlying ectoderm to thicken into a disclike swelling, the *lens placode* (step 2). The center of the lens placode sinks inward toward the optic cup, and its edges eventually fuse together, forming a ball of cells, the *lens vesicle* (step 3).

The developing lens cells begin to synthesize *crystallin,* a fibrous protein that collects into clear, glassy deposits. The lens cells finally lose their nuclei and form the elastic, crystal-clear lens.

As the lens develops, it contacts the overlying ectoderm, which has closed over it. In response, the ectoderm cells lose their pigment granules and become clear, developing into the cornea. Eventually, the developing cornea joins with the edges of the optic cup to complete the primary structures of the eye (step 4). Other cells contribute to accessory structures of the eye; for example, mesoderm and neural crest cells contribute to the reinforcing tissues in the wall of the eye and the muscles that move the eye. Figure 48.12, step 5, shows a fully developed vertebrate eye.

Many experiments have shown that the initial induction by the optic vesicle is necessary for develop-

Figure 48.12
Stages in the formation of the vertebrate eye from the optic vesicle of the brain and the overlying ectoderm.

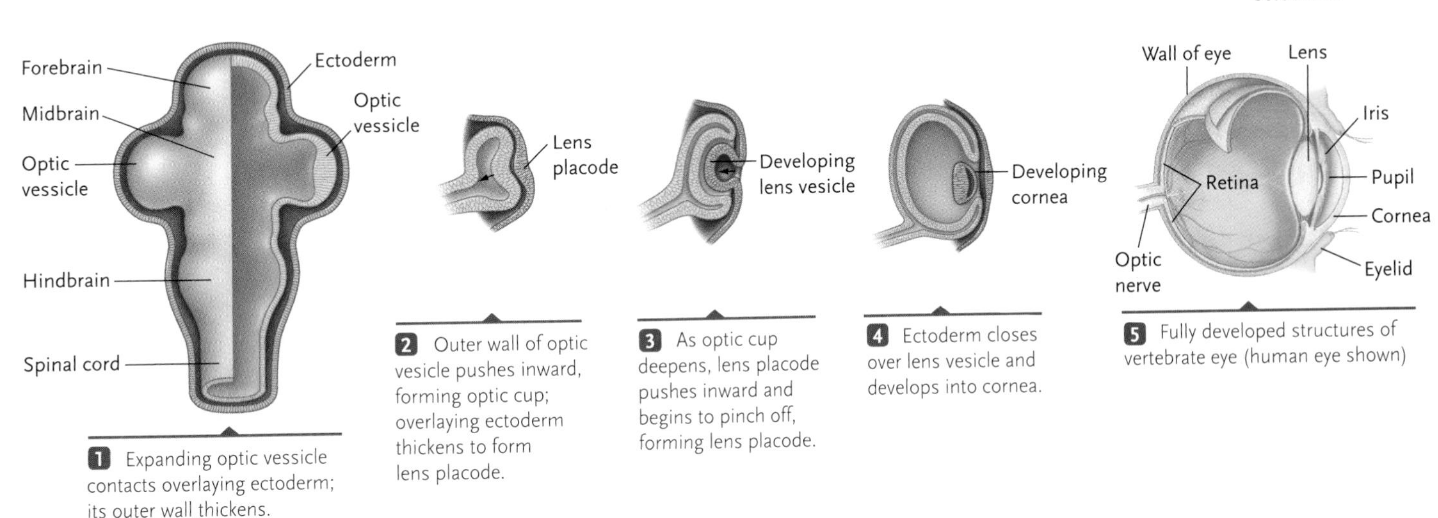

Forebrain
Midbrain
Optic vessicle
Hindbrain
Spinal cord
Ectoderm
Optic vessicle
1 Expanding optic vessicle contacts overlaying ectoderm; its outer wall thickens.
Lens placode
2 Outer wall of optic vesicle pushes inward, forming optic cup; overlaying ectoderm thickens to form lens placode.
Developing lens vesicle
3 As optic cup deepens, lens placode pushes inward and begins to pinch off, forming lens placode.
Developing cornea
4 Ectoderm closes over lens vesicle and develops into cornea.
Wall of eye
Lens
Iris
Retina
Pupil
Cornea
Optic nerve
Eyelid
5 Fully developed structures of vertebrate eye (human eye shown)

ment of the eye. For example, if an optic vesicle is removed before lens formation, the ectoderm fails to develop a lens placode and vesicle. Moreover, placing a removed optic vesicle under the ectoderm in other regions of the head causes a lens to form in the new location. Or, if the ectoderm over an optic vesicle is removed and ectoderm from elsewhere in the embryo is grafted in its place, a normal lens will develop in the grafted ectoderm, even though in its former location it would not differentiate into lens tissue.

Eye development also demonstrates differentiation. Ectoderm cells that are induced to form the lens of the eye synthesize crystallin; in other locations, ectoderm cells typically synthesize a different protein, *keratin,* as their predominant cell product. Keratin is a component of surface structures such as skin, hair, feathers, scales, and horns. In other words, as a response to induction by the optic vesicle, the genes of the ectoderm cells coding for crystallin are activated, while genes coding for keratin are not expressed.

Apoptosis Eliminates Tissues That Are No Longer Required

Induction and differentiation build complex, specialized organs from the three fundamental tissue types. Complementing these processes is *apoptosis,* programmed cell death, which in this case removes tissues present during development but not in the fully formed organ. Apoptosis plays an important role in the development of animals, both invertebrates and vertebrates. The best example of apoptosis in frog development occurs during metamorphosis, in which the tadpole changes into an adult frog. The tail of the tadpole becomes progressively smaller and finally disappears because its cells disintegrate and their components are absorbed and recycled by other cells. Cells that are eliminated by apoptosis, like those of a tadpole's tail, are typically parts of structures required at one stage of development but not for later stages.

In the next section, we describe cleavage, gastrulation, and organogenesis in human and other mammalian embryos.

Figure 48.13
Early stages in the development of the human embryo.

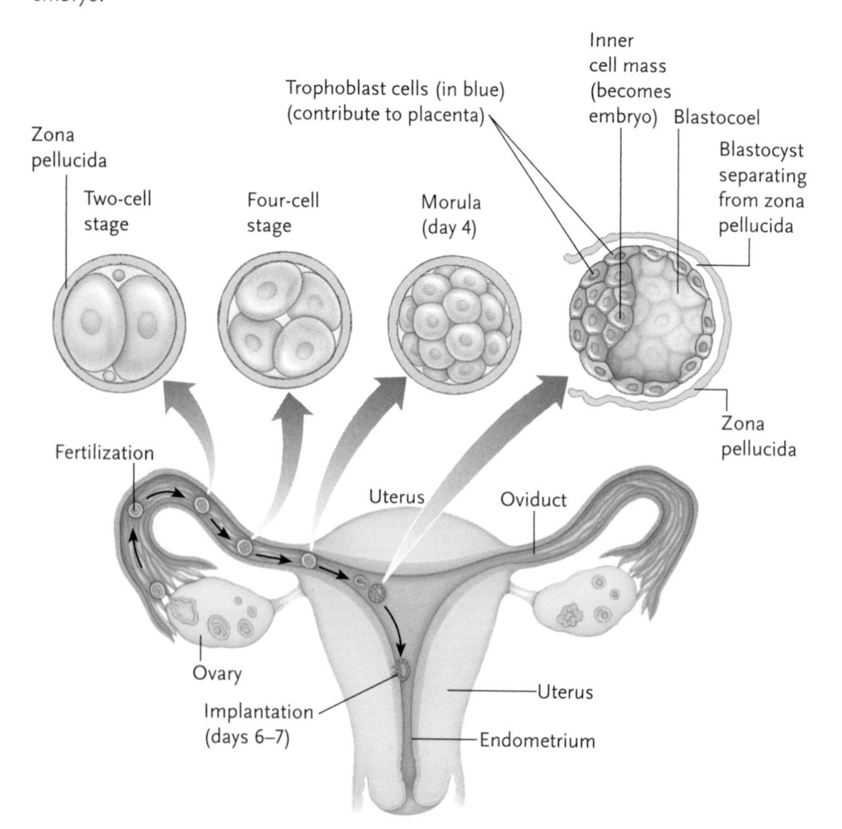

Study Break

1. What is the outcome of organogenesis?
2. What tissues or organs develop from the neural tube and neural crest cells?

48.4 Embryonic Development of Humans and Other Mammals

Human embryonic development is representative of the placental mammals, in which the embryo develops in the uterus of the mother. In the uterus, the embryo is nourished by the placenta, which supplies oxygen and nutrients and carries away carbon dioxide and nitrogenous wastes.

The period of mammalian development that is called **pregnancy** or **gestation** varies in different species. Larger mammals generally have longer gestation periods; for example, gestation takes 600 days in elephants, about 1 year in blue whales, and a mere 21 days in hamsters.

In humans, gestation takes an average of 266 days from the time of fertilization, or about 38 weeks. Because the date of fertilization may be difficult to establish, human gestation is usually calculated from the beginning of the menstrual cycle in which fertilization takes place, giving a period of about 9 months. On this basis, human gestation is divided into three **trimesters,** each 3 months long.

The major developmental events in human gestation—cleavage, gastrulation, and organogenesis—take place during the first trimester. By the fourth week, the embryo's heart is beating, and by the end of the eighth week, the major organs and organ systems have formed. From this point until birth, the developing human is called a **fetus.** Only 5 cm long by the end of the first trimester, the fetus grows during the second and third trimesters to an average length of 50 cm and an average weight of 3.5 kg (or about 19.7 inches and 7.7 pounds). The period of gestation ends with birth.

Cleavage and Implantation Occupy the First 2 Weeks of Development

We noted in Section 47.3 that human fertilization occurs when the egg is in the first third of the oviduct leading from the ovary to the uterus. After fertilization, cleavage divisions take place during passage of the developing embryo down the fallopian tube and while it is still enclosed in the zona pellucida—the original coat of the egg **(Figure 48.13)**.

By day 4, the morula, a 16- or 32-cell ball, has been produced. By the time the endometrium (uterine lining) is ready for implantation (about 7 days after ovulation; see Section 47.3), the morula has reached the uterus and has undergone further cell divisions and differentiation into a blastocyst. At this time, the **blastocyst** is a single-cell-layered hollow ball of about 120 cells with a fluid-filled cavity, the blastocoel, in which a dense mass of cells is localized to one side. This **inner cell mass** will become the embryo itself, while the rest of the blastocyst will become tissues that support the development of the embryo in the uterus. The outer single layer of cells of the blastocyst is the **trophoblast**.

When it is ready to implant, the blastocyst breaks out of the zona pellucida and sticks to the endometrium on its inner cell mass side **(Figure 48.14a)**. Implantation begins when the trophoblast cells overlying the inner cell mass secrete proteases that digest pathways between endometrial cells. Dividing trophoblast cells fill in the digested spaces, appearing like fingerlike projections into the endometrium. These cells continue to digest the nutrient-rich endometrial cells, serving both to produce a hole in the endometrium for the blastocyst and to release nutrients that the developing embryo can use after the small amount of yolk contained in the egg cytoplasm is used up. While the blastocyst is burrowing into the endometrium, the inner cell mass separates into the *embryonic disc,* which consists of two distinct cell layers (see Figure 48.14a). The layer farther from the blastocoel is the epiblast, which gives rise to the embryo proper, and the layer nearer the blastocoel is the hypoblast, which gives rise to part of the extraembryonic membranes. When implantation is complete, the blastocyst has completely burrowed into the endometrium and is covered by a layer of endometrial cells **(Figure 48.14b)**.

Mammalian Gastrulation and Neurulation Resemble the Reptilian–Bird Pattern

Gastrulation proceeds as in birds (see Figure 48.8), with the formation of a primitive streak in the epiblast. Some epiblast cells remain in place, becoming the ectoderm, while others enter the streak to form the endoderm and mesoderm. The ectoderm, mesoderm, and endoderm are located initially in three layers; from this initial arrangement, the endoderm folds to form the primitive gut, and becomes surrounded with ectoderm and mesoderm. Neurulation in human and other mammalian embryos takes place essentially as in birds (see Figure 48.10).

Extraembryonic Membranes Give Rise to the Amnion and Part of the Placenta

Soon after the inner cell mass separates into the epiblast and hypoblast, a layer of cells separates from the epiblast along its top margin (see Figure 48.14b). The fluid-filled space created by the separation becomes the amniotic cavity, and the layer of ectodermal cells forming its roof becomes the amnion, the extraembryonic membrane surrounding the cavity. The amnion expands until eventually it completely surrounds the embryo and suspends it in amniotic fluid. As in birds, the hypoblast develops into the yolk sac. However, in mammals, the mesoderm of the yolk sac gives rise to the blood vessels in the embryonic portion of the placenta.

While the amnion is expanding around the embryonic disc, blood-filled spaces form in maternal tissue, and trophoblast cells grow rapidly around both the embryo and amnion to form the chorion **(Figure 48.14c)**. Next, a connecting stalk forms between the embryonic disc and the chorion, while the chorion begins to grow into the endometrium as fingerlike or treelike extensions called **chorionic villi (Figure 48.14d)**. The chorionic villi greatly increase the surface area of the chorion. Where these villi grow into the endometrium is the area of the future placenta. As the chorion develops, mesodermal cells of the yolk sac grow into it and form a rich network of blood vessels, the embryonic circulation of the placenta. At the same time, the expanding chorion stimulates the blood vessels of the endometrium to grow into the maternal circulation of the placenta **(Figure 48.14e)**.

Within the placenta of humans, apes, monkeys, and rodents, the maternal circulation opens into spaces in which the maternal blood directly bathes the capillaries coming to the placenta from the embryo **(Figure 48.14f)**. (Different types of placentas are found in other mammals.) The embryonic circulation remains closed, however, so that the embryonic and maternal blood do not mix directly. This prevents the mother from developing an immune reaction against cells of the embryo, which may be recognized as foreign by the mother's immune system. Eventually, the placenta and its blood circulation grow to cover about a quarter of the inner surface of the enlarged uterus and reach the size of a dinner plate.

As the embryonic blood circulation develops, this connecting stalk between the embryo and placenta develops into the **umbilical cord**, a long tissue with blood vessels linking the embryo and the placenta. The vessels in the umbilical cord are derived from the extraembryonic membrane, the allantois. They conduct

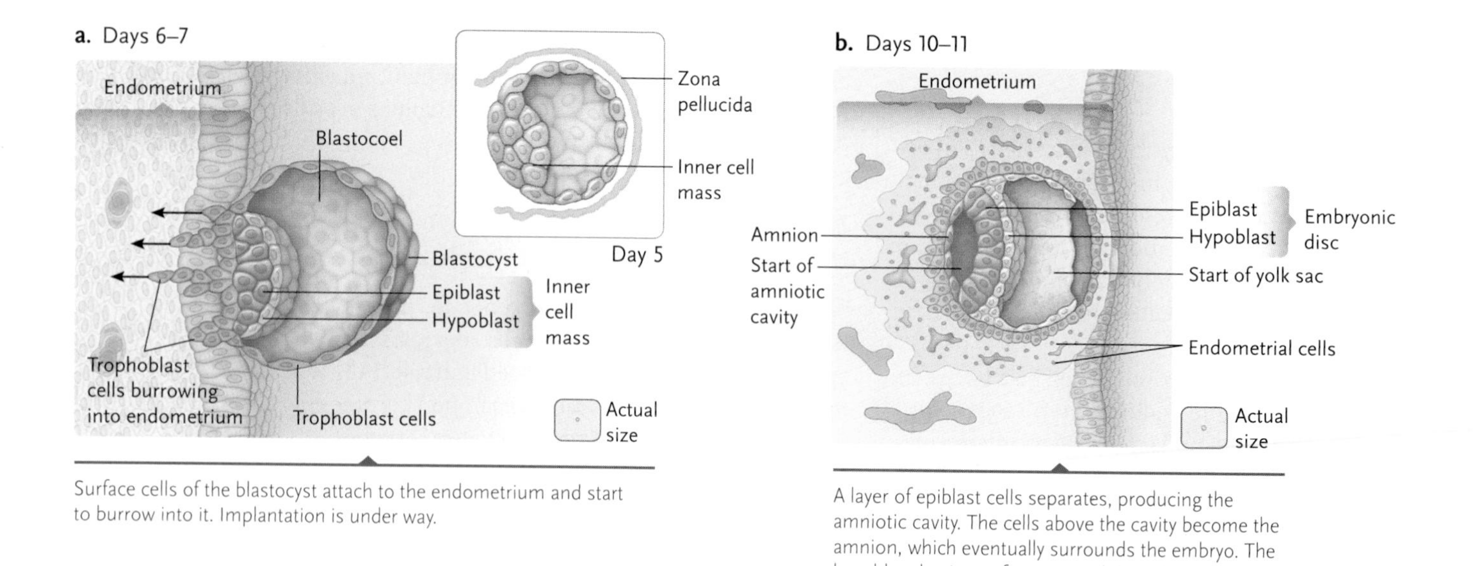

Surface cells of the blastocyst attach to the endometrium and start to burrow into it. Implantation is under way.

A layer of epiblast cells separates, producing the amniotic cavity. The cells above the cavity become the amnion, which eventually surrounds the embryo. The hypoblast begins to form around the yolk sac.

c. Day 12

Blood-filled spaces
Chorion
Yolk sac
Amniotic cavity
Start of chorionic cavity
Actual size

Blood-filled spaces form in maternal tissue. The chorion forms, derived from trophoblast cells, and encloses the chorionic cavity.

d. Day 14

Chorionic villi
Connecting stalk
Chorion
Chorionic cavity
Yolk sac
Actual size

A connecting stalk has formed between the embryonic disk and chorion. Chorionic villi, which will be features of a placenta, start to form.

e. Day 25

Chorion
Chorionic cavity
Embryo
Amniotic cavity
Amnion
Yolk sac
Chorionic villi
Maternal blood

The chorion continues to grow into the endometrium, producing the chorionic villi. The chorion growth stimulates blood vessels of the endometrium to grow into the maternal circulation of the placenta.

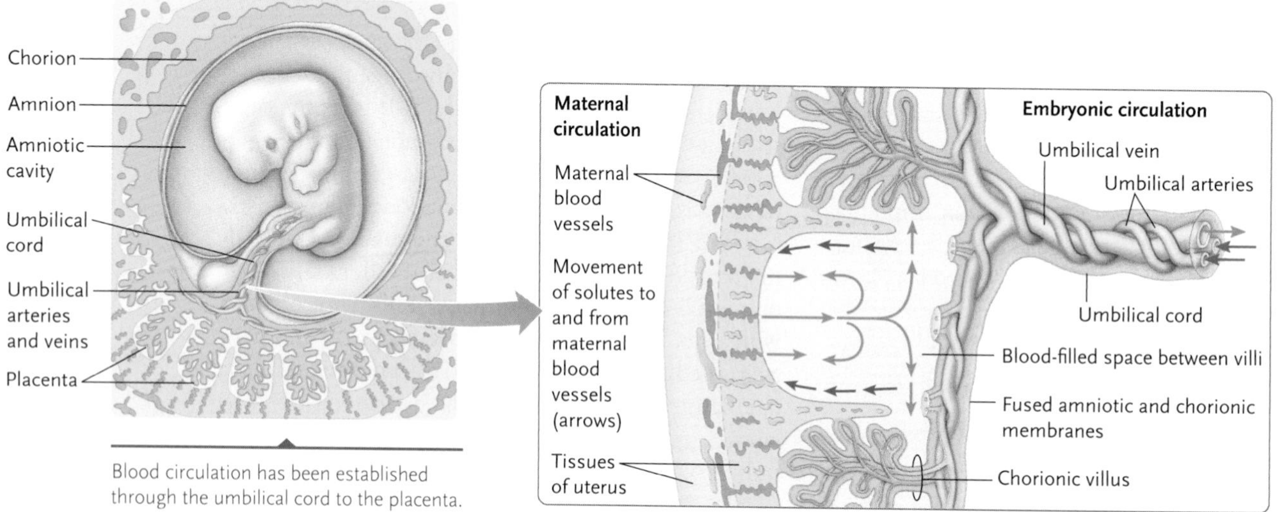

Blood circulation has been established through the umbilical cord to the placenta.

Figure 48.14
Implantation of a human blastocyst in the endometrium of the uterus and the establishment of the placenta.

blood between the embryo and the placenta (shown in the inset for Figure 48.14f).

Within the placenta, nutrients and oxygen pass from the mother's circulation into the circulation of the embryo. Besides nutrients and oxygen, many other substances taken in by the mother—including alcohol, caffeine, drugs, and toxins in cigarette smoke—can pass from mother to embryo. Carbon dioxide and nitrogenous wastes pass from the embryo to the mother, and are disposed of by the mother's lungs and kidneys.

If the presence of a genetic disease such as cystic fibrosis or Down syndrome is suspected, tests can be carried out on cells removed from the embryonic portion of the placenta or from the amniotic fluid, which contains cells derived from the embryo. The test using cells of the placenta is called *chorionic villus sampling;* the test using cells derived from the amniotic fluid is called *amniocentesis* (*centesis* = puncture, referring to the use of a needle, which is pushed through the abdominal wall, to obtain fluid from the amniotic cavity). Chorionic villus sampling can be carried out as early as the eighth week, compared with 14 weeks for amniocentesis. Both tests carry some degree of risk to the embryo.

Further Growth of the Fetus Culminates in Birth

By the end of its fourth week, a human embryo is 3–5 mm long, 250–500 times the size of the zygote **(Figure 48.15a)**. It has a tail and pharyngeal arches, which are embryonic features of all vertebrates (see Section 30.2). The pharyngeal arches contribute to the formation of the face, neck, mouth, nasal cavities, larynx, and pharynx. After 5 to 6 weeks, most of the tail has disappeared and the embryo is beginning to take on recognizable human form **(Figure 48.15b)**. At 8 weeks, the embryo, now a fetus, is about 2.5 cm long **(Figure 48.15c)**. Its organ systems have formed, and its limbs, with fingers or toes at their ends, have developed **(Figure 48.15d)**.

Figure 48.16 shows the hormonal events and associated physical events of birth. As the period of fetal growth comes to a close, the fetus typically turns so that its head is downward, pressed against the cervix. A steep rise in the levels of estrogen secreted by the placenta at this time causes cells of the uterus to express the gene for the receptor of the hormone *oxytocin*. The receptors become inserted into the plasma membranes of those cells. Oxytocin—which is secreted by the pituitary gland—binds to its receptor, triggering the smooth muscle cells of the uterine wall to contract and begin the rhythmic contractions of labor. These contractions mark the beginning of **parturition** (*parturire* = to be in labor), the process of giving birth.

The contractions push the fetus further against the cervix and stretch its walls (see Figure 48.16, step 1). In response, stretch receptors in the walls send nerve signals to the hypothalamus, which responds by stimulating the pituitary to secrete more oxytocin. In turn, the oxytocin stimulates more forceful contractions of the uterus, pressing the fetus more strongly against the cervix, and further stretching its walls. The positive feedback cycle continues, steadily increasing the strength of the uterine contractions.

As the contractions force the head of the fetus through the cervix (step 2), the amniotic membrane bursts, releasing the amniotic fluid. Usually, within 12 to 15 hours after the onset of uterine contractions, the head passes entirely through the cervix. Once the head is through, the rest of the body follows quickly and the entire fetus is forced through the vagina to the exterior, still connected to the placenta by the umbilical cord (step 3).

After the baby takes its first breath the umbilical cord is cut and tied off by the birth attendant. Contractions of the uterus continue, expelling the placenta and any remnants of the umbilical cord and embryonic membranes as the afterbirth, usually within 15 minutes to an hour after the infant's birth. The short length of umbilical cord still attached to the infant dries and shrivels within a few days. Eventually, it separates entirely and leaves a scar, the **umbilicus** or navel, to mark its former site of attachment during embryonic development.

The Mother's Mammary Glands Become Active after Birth

Before birth of the fetus, estrogen and progesterone secreted by the placenta stimulate the growth of the mammary glands in the mother's breasts. However, the high levels of these hormones prevent the mammary glands from responding to *prolactin,* the hormone secreted by the pituitary that stimulates the glands to produce milk. After birth of the fetus and release of the placenta, the levels of estrogen and progesterone fall steeply in the mother's bloodstream, and the breasts begin to produce milk (stimulated by prolactin) and secrete it (stimulated by oxytocin).

Continued milk secretion depends on whether the infant is suckled by the mother. If the infant is suckled, stimulation of the nipples sends nerve impulses to the hypothalamus, which responds by signaling the pituitary to release a burst of prolactin and oxytocin. Hormonal stimulation of milk production and secretion continues as long as the infant is breastfed.

So far, we have followed the development of a generic human, but certain aspects of development differ depending on the offspring's sex. Next we look at the specifics of male and female development.

A Gene on the Y Chromosome Determines the Development of Male or Female Sex Organs

The gonads and their ducts begin to develop during the fourth week of gestation. Until the seventh week, male and female embryos have the same set of inter-

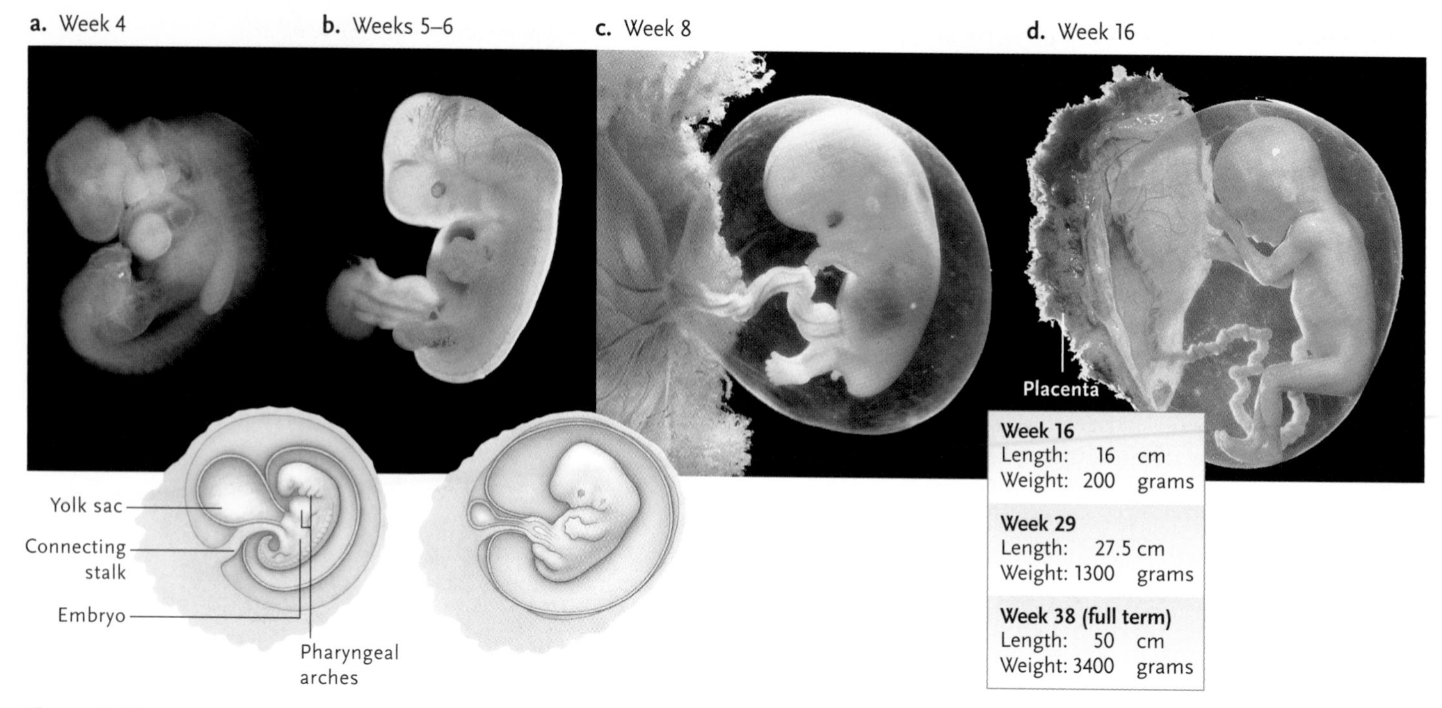

Figure 48.15
The human embryo at various stages of development, beginning at week 4. The chorion has been pulled aside to reveal the embryo in the amnion at week 8 and week 16. By week 16, movements begin as nerves make functional connections with the forming muscles.
(Photos: Lennart Nilsson, A Child Is Born, © 1966, 1977, Dell Publishing Company, Inc.)

nal structures derived from mesoderm, including a pair of gonads **(Figure 48.17a).** Each gonad is associated with two primitive ducts, the **Wolffian duct** and the **Müllerian duct,** which lead to a cloaca. These internal structures are *bipotential:* they can develop into either male or female sexual organs.

The presence or absence of a Y chromosome determines whether the internal structures develop into male or female sexual organs. If the fetus has the XY combination of sex chromosomes, a single gene on the Y chromosome, *SRY (Sex-determining Region of the Y),* becomes active in the seventh week. The protein encoded by the gene sets a molecular switch that causes the primitive gonads to develop into testes. The fetal testes then secrete two hormones, testosterone and the *anti-Müllerian hormone (AMH).* The testosterone stimulates development of the Wolffian ducts into the male reproductive tract, including the epididymis, vas deferens,

Figure 48.16
Birth of the fetus. Hormonal events of birth are at the top, and physical events of birth are at the bottom.

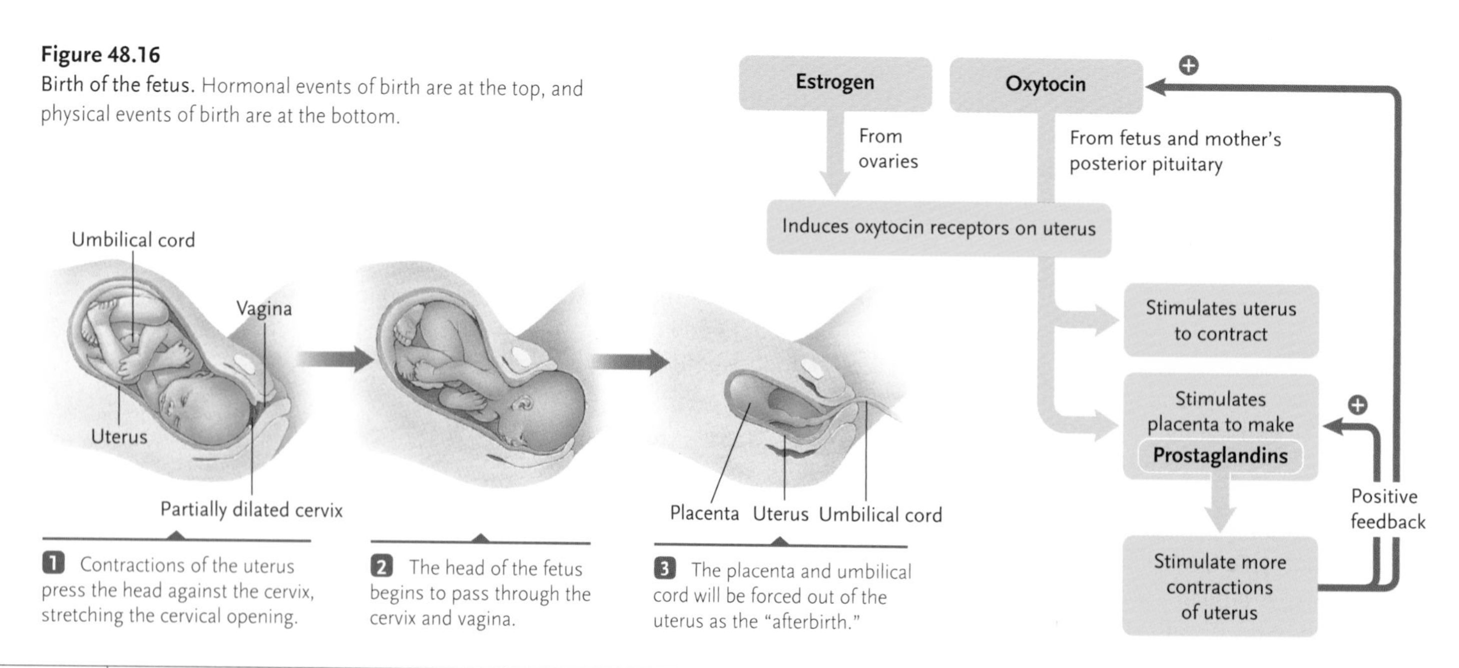

and seminal vesicles **(Figure 48.17b)**. AMH causes the Müllerian ducts to degenerate and disappear. (*Insights from the Molecular Revolution* describes experiments that traced the activity of the *SRY* gene and its encoded protein in male development.) Testosterone additionally stimulates the development of the male genitalia.

If the fetus has the XX combination of chromosomes, no SRY protein is produced and the primitive gonads, under the influence of the estrogens and progesterone secreted by the placenta, develop into ovaries. The Müllerian ducts develop into the oviducts, uterus, and part of the vagina, and the Wolffian ducts degenerate and disappear **(Figure 48.17c)**. The female sex hormones additionally stimulate the development of the female external genitalia.

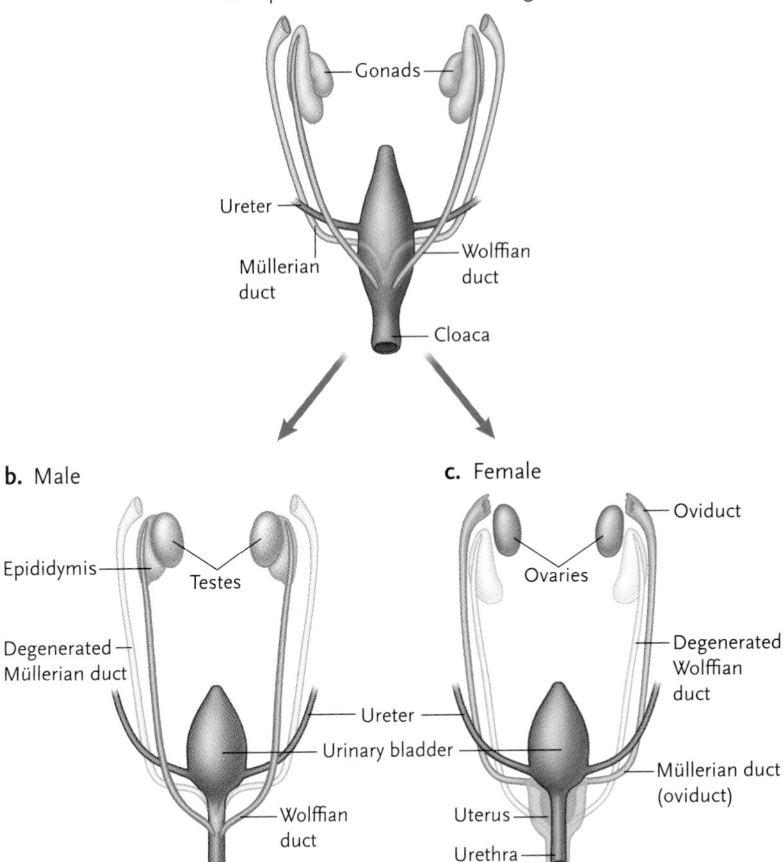

Figure 48.17
Development of the internal sexual organs of males and females from common bipotential origins.

Development Continues after Birth

Once fetal development is over, humans and other mammals, and indeed most other animals, follow a prescribed course of further growth and development that leads to the adult, the sexually mature form of the species. In humans, the internal and external sexual organs mature and secondary sexual characteristics appear at puberty. Similar changes occur in most mammals. There are, in fact, many examples among different animal groups of developmental changes that take place after hatching or birth. In some cases, offspring hatch that are distinctly different in structure from the adult. Examples among invertebrates include insects such as *Drosophila* and butterflies, in which eggs hatch to produce larva that undergo metamorphosis into the adult. Frogs similarly hatch as tadpoles, which undergo metamorphosis to produce the adult.

We have now described embryonic development in animals from a morphological perspective. In the rest of the chapter, we will focus on the cellular and molecular mechanisms that underlie development.

Study Break

1. Distinguish between the roles of the trophoblast and inner cell mass of the blastocyst in mammalian development.
2. What hormone would you use to induce labor in a pregnant woman?

48.5 The Cellular Basis of Development

In the preceding sections, you learned about the processes of development from a mainly structural point of view. Underlying those developmental processes are specific cellular and molecular events. In this section, you will learn about all the cellular events that underlie the stages of development.

Cell Division Varies in Orientation and Rate during Embryonic Development

The *orientation* and *rate* of mitotic cell division have special significance in the development of the shape, size, and location of the organ systems of the embryo. Regulation of these two features of mitotic cell division occurs at all stages of development.

The orientation of cell division refers to the angles at which daughter cells are added to older cells as development proceeds. It is determined by the location of a furrow that separates the cytoplasm after mitotic division of the nucleus (furrowing is discussed in Section 10.2). The furrow forms in alignment with the spindle midpoint. Therefore, when the spindle is centrally positioned in the cell, the furrow leads to symmetric division of the cell. However, when the spindle is displaced to one end of the cell, the furrow leads to asymmetric division of the cell into a smaller and a larger cell. Little is known about how spindle positioning is regulated.

The rate of cell division primarily reflects the time spent in the G_1 period of interphase (see Section 10.2); once DNA replication begins, the rest of the cell cycle is usually of uniform length in all cells of the same species. As an embryo develops and cells differentiate, the time spent in interphase increases and varies in length in different cell types. As a result, different cell types proliferate at various rates, giving rise to tissues and organs with different cell numbers. Some cells, when fully differentiated, remain fixed in interphase and stop replicating their DNA or dividing. Nerve cells in the mammalian brain and spinal cord, for example, stop dividing once the nervous system is fully formed. Ultimately, the rate of cell division is under genetic control.

Frog egg cleavage provides examples of how both changes in orientation and rate of mitotic division affect development. The first two cleavages start at the animal pole and extend to the vegetal pole, producing four equal blastomeres (see Figure 48.2). The third cleavage occurs equatorially. However, because there is yolk in the vegetal region of the embryo, this cleavage furrow forms not at the equator but up higher toward the animal pole. The result is an eight-cell embryo with four small blastomeres in the animal region of the embryo, and four large blastomeres in the vegetal region. The blastomeres in the animal region of the embryo proceed to divide rapidly, while the blastomeres in the vegetal part of the embryo divide more slowly because division is inhibited by yolk. As a result, the morula produced consists of an animal region with many small cells, and a vegetal region with relatively few blastomeres.

Cell-Shape Changes and Cell Movements Depend on Microtubules and Microfilaments

We have seen that embryonic cells undergo changes in shape that generate movements, such as the infolding of surface layers to produce endoderm or mesoderm. Entire cells also move during the embryonic growth of animals, both singly and in groups. Both the shape changes and the whole-cell movements are produced by microtubules, powered by dyneins and kinesins, and microfilaments, powered by myosins (see Section 5.3). Movements are also produced by changes in the rate of growth or by the breakdown of microtubules and microfilaments. Generally speaking, changes in both cell shape and cell movement play important roles in cleavage, gastrulation, and organogenesis.

Changes in Cell Shape. Changes in cell shape typically result from reorganization of the cytoskeleton. For example, during the development of the neural plate in frogs, the ectoderm flattens and thickens; that is, microtubules within cells in the ectoderm layer lengthen and slide farther apart, causing the cells to change from a cubelike to a columnar shape **(Figure 48.18a).**

Once formed, the neural plate sinks downward along its midline. This change is produced by a change in cell shape from columnar to wedgelike **(Figure 48.18b).** As the tops of the cells narrow, the entire cell layer is forced inward—it invaginates. How does this occur? Each wedge-shaped cell contains a group of microfilaments arranged in a circle at its top. Research suggests that the microfilaments slide over each other, tightening the ring like a drawstring and narrowing the top of the cell. This mechanism is supported by experiments in which cytochalasin, a chemical that interferes with microfilament assembly, was added to the cells. As a result, the microfilament circle was dispersed, and no invagination of the ectoderm occurred.

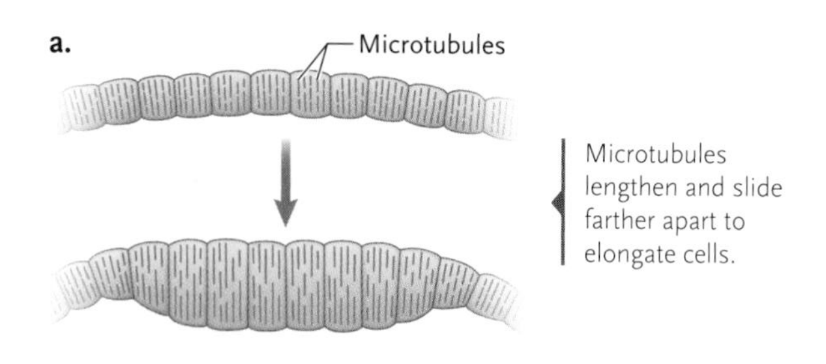
a.
Microtubules
Microtubules lengthen and slide farther apart to elongate cells.

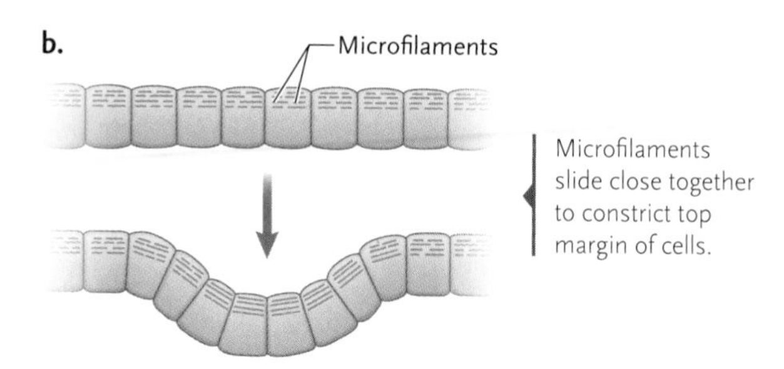
b.
Microfilaments
Microfilaments slide close together to constrict top margin of cells.

Figure 48.18
The roles of microtubules **(a)** and microfilaments **(b)** in the changes in cell shape that produce developmental movements.

Whole-Cell Movements. Among the most striking examples of whole-cell movements in embryonic development are the cell movements during gastrulation and the often long-distance migrations of neural crest cells. These whole-cell movements involve the coordinated activity of microtubules and microfilaments. The typical pattern of movement is a repeating cycle of steps that resemble how an amoeba moves. First, a cell attaches to the substrate (**Figure 48.19,** step 1) and moves forward by elongating from the point of attachment (step 2). The cell next makes a new attachment at the advancing tip (step 3), and then contracts until the rearmost attachment breaks (step 4). The front attachment now serves as the base for another movement.

How do the cells know where to go? Typically, cells migrate over the surface of stationary cells in one of the embryo's layers. In many developmental systems, migrating cells follow tracks formed by molecules of the extracellular matrix (ECM), secreted by the cells along the route over which they travel. An important

Turning On Male Development

The switch to male development in mammalian embryos is triggered by the protein encoded in the *SRY* gene, carried on the Y chromosome. Individuals with a mutation in which *SRY* encodes a faulty, inactive protein develop into females, even though they have the XY combination of sex chromosomes.

Molecular studies revealed that the mutant SRY proteins have changes in single amino acids or have a missing segment. All the single amino acid changes are concentrated in a region of the SRY protein known as the *HMG box,* which can bind to DNA. This discovery suggested that SRY is a regulatory protein that binds to the control regions of genes such as *AMH,* which encodes the anti-Müllerian hormone, and turns them on.

A group of investigators led by Michael Weiss at the Harvard Medical School and Massachusetts General Hospital in Boston carried out molecular studies testing whether SRY directly turns on the *AMH* gene. For their experiments, the researchers attached the control region of the *AMH* gene to the coding portion of a luciferase gene. Luciferase is the firefly enzyme that catalyzes a reaction with the substrate luciferin to produce light. In this experiment, luciferase was used as a reporter; measuring its activity indirectly informed the researchers about the molecular reactions occurring with the *AMH* gene. Luciferase activity is measured by breaking open cells to produce cell extracts, adding the substrate luciferin to samples of the extract, and quantifying the light emitted from the reaction using a special photodetector system. The composite *AMH*-luciferase gene was introduced into embryonic cells removed from the developing gonads of XY rat embryos, taken at the time when differentiation into a testis or ovary would normally begin.

A normal human *SRY* gene was then introduced into the gonad cells containing the artificial gene. High luciferase activity was seen, confirming that the normal SRY protein activates the *AMH* gene. The experiment was repeated with a mutant *SRY* gene isolated from a human patient who had developed into a female even though she had the XY combination of sex chromosomes. In her case, the mutation resulted from a change of a single amino acid in the HMG box of the SRY protein. When her *SRY* gene was added to the embryonic rat gonad cells, there was no luciferase activity, indicating that her altered SRY protein could not turn on the *AMH* gene.

Adding a normal SRY protein to the *AMH* gene in a test tube showed that the protein binds directly to the gene. Tests with DNA-digesting enzymes showed that combination with SRY protects a segment of the control region of the *AMH* gene from attack by the enzymes. This protection indicates that SRY binds in this region, as expected for a regulatory protein.

Current goals include finding the genes activated by SRY that direct development toward the male. The research promises to reveal the complete sequence of molecular events directing male development. It may also lead to treatments for developmental abnormalities produced when the sex-determining system goes awry.

track molecule is *fibronectin,* a fibrous, elongated protein of the ECM. Migrating cells recognize and adhere to the fibronectin; in response, internal changes in the cells trigger movement in a direction based on the alignment of the fibronectin molecules.

Some migrating cells follow concentration gradients instead of molecular tracks. The gradients are created by the diffusion of molecules (often proteins) released by cells in one part of an embryo. Cells with receptors for the diffusing molecule follow the gradient toward its source, or move away from the source.

Selective Cell Adhesions Underlie Cell Movements

Selective cell adhesion, the ability of an embryonic cell to make and break specific connections to other cells, is closely related to cell movement. As development proceeds, many cells break their initial adhesions and move, and then form new adhesions in different locations. Final cell adhesions hold the embryo in its correct shape and form. Junctions of various kinds, including tight, anchoring, and gap junctions, reinforce the final adhesions (see Section 5.5).

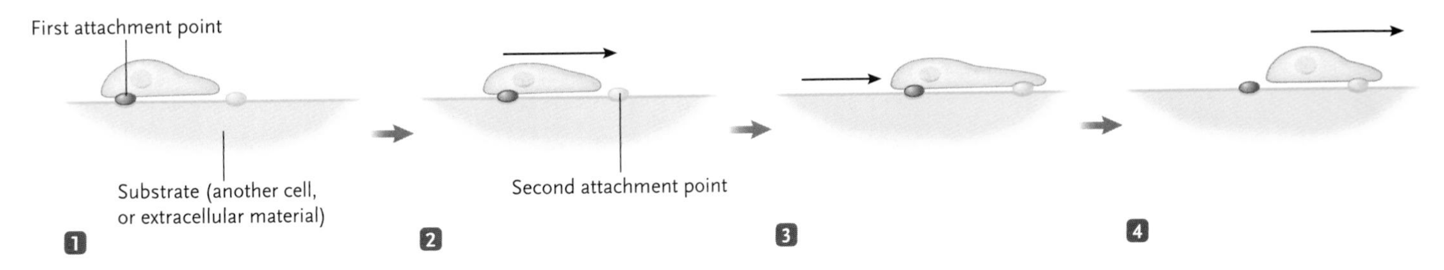

Figure 48.19
The cycle of attachments, stretching, and contraction by which cells move over other cells or extracellular materials in embryos.

The selective nature of cell adhesions was first demonstrated in a classic experiment by Johannes Holtfreter of the University of Rochester and his student P. L. Townes. In this experiment, the researchers removed pieces of ectoderm, mesoderm, and endoderm from living amphibian embryos in the neurulation stage, separated them into individual cells, and added the cells in various combinations to a culture medium. Initially the cells clumped together at random into a ball. After a few hours, they sorted themselves out and moved into arrangements resembling their normal locations in the gastrula **(Figure 48.20).**

Further research has identified many cell surface proteins responsible for selective cell adhesions, including **cell adhesion molecules** (CAMs; see Section 5.5) and **cadherins** *(calcium-dependent adhesion molecules)*. The cadherins are so named because they require calcium ions to set up adhesions. As cells develop, different types of CAMs or cadherins may appear or disappear from their surfaces as they make and break cell adhesions. The changes reflect alterations in gene activity, often in response to molecular signals arriving from other cells. For example, in the neural plate stage of neurulation in the frog, N-cadherin is on neural plate cells, keeping those cells together, while E-cadherin is on the adjacent ectodermal cells, keeping those cells together. The neural tube is produced when the neural plate cells separate from the ectodermal cells, while both cell types retain their respective cadherin type. The neural crest cells have neither cadherin bound to them, so they do not bind to each other and they disperse (as described earlier). However, if N-cadherin is expressed in the ectodermal cells through experimental manipulation, the forming neural tube does not separate from the flanking ectodermal cells because all of the cells are held together by N-cadherin.

Induction Depends on Molecular Signals Made by Inducing Cells

Recall from Section 48.1 that induction is the process in which a group of cells (the inducer cells) causes or influences a nearby group of cells (the responder cells) to follow a particular developmental pathway. Recall also that induction is the major process responsible for determination, in which the developmental fate of a cell is set. Many experiments have shown that induction occurs through the interaction of signal molecules with surface receptors on the responding cells. The signal molecules may be located on the surface of the inducing cells, or they may be released by the inducing cells. The surface receptors are activated by binding the signal molecules; in the activated form, they trigger internal response pathways that produce the developmental changes (surface receptors and their associated signal transduction pathways are discussed in Sections 7.3 and 7.4). Often, the responses include changes in gene activity.

A German scientist, Hans Spemann of the University of Freiburg, carried out the first experiments identifying induction in embryos in the 1920s. He and his doctoral student, Hilde Mangold, found that if the dorsal lip of a newt embryo was removed and grafted into a different position on another newt embryo, on the ventral side for instance, cells moving inward from the dorsal lip induced a neural plate, a neural tube, and eventually an entire embryo to form in the new location **(Figure 48.21).** On the basis of his pioneering research, Spemann proposed that the dorsal lip is an *organizer,* acting on other cells to alter the course of development. This action is now known as *induction,* and the cells responsible for induction are known generally as inducer cells. Spemann received the Nobel Prize in 1935 for his research. (In 1924, the year their research paper was published, Mangold died in an accident when her kitchen gasoline heater exploded. She would likely have also received the Nobel Prize, but they are never awarded posthumously.)

Spemann's findings touched off a search for the inducing molecules that must pass from the inducing cells to the responding cells. It took many years to achieve success. Finally, molecular techniques led to the identification of inducing molecules. For example, in 1992, researchers constructed a DNA library from *Xenopus* gastrulas by isolating and cloning the cellular DNA in gene-size pieces. They made mRNA transcripts of the cloned genes and injected them into early *Xenopus* embryos in which the inducing ability of the mesoderm had been destroyed by exposure to ultraviolet light. Some of the injected mRNAs, translated into proteins in the embryos, were able to induce formation of a neural plate and tube and lead to a normal embryo. More than 10 other proteins acting as inducing molecules have been identified in the *Xenopus* system.

Differentiation Produces Specialized Cells without Loss of Genes

Differentiation is the process by which cells that have committed to a particular developmental fate by the determination process (see Section 48.1) now develop into specialized cell types with distinct structures and functions. As part of differentiation, cells concentrate on the production of molecules characteristic of the specific types. For example, 80% to 90% of the total protein that lens cells synthesize is crystallin.

Research into differentiation confirmed that as cells specialize, they retain all the genes of the original egg cell; except in rare instances, differentiation does not occur through selective gene loss. Several definitive experiments supporting this conclusion were carried out several decades ago by Robert Briggs and Thomas King of Lankenau Hospital Research Institute in Philadelphia (now Fox Chase Cancer Center),

Figure 48.20 Experimental Research

Demonstrating the Selective Adhesion Properties of Cells

QUESTION: Do cells make specific connections to other cells?

EXPERIMENT: Johannes Holtfreter and P. L. Townes demonstrated that cells make specific connections to other cells, that is, that cells have selective adhesion properties.

1. Holtfreter and Townes separated ectoderm, mesoderm, and endoderm tissue from amphibian embryos soon after the neural tube had formed. They used embryos from amphibian species that had cells of different colors and sizes, so they could follow under the microscope where each cell type ended up. (The colors shown here are for illustrative purposes only.)

2. The researchers placed the tissues individually in alkaline solutions, which caused the tissues to break down into single cells.

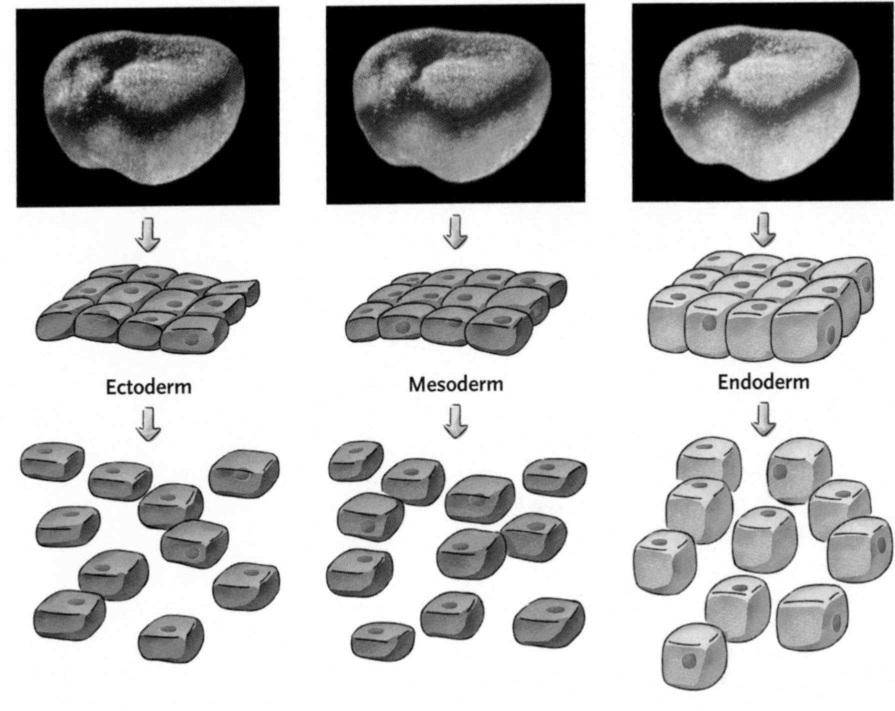

3. Holtfreter and Townes then combined suspensions of single cells in various ways. Shown here are ectoderm + mesoderm, and ectoderm + mesoderm + endoderm. When the pH was returned to neutrality, the cells formed aggregates. Through a microscope, the researchers followed what happened to the aggregates on agar-filled petri dishes.

KEY: Ectoderm | Mesoderm | Endoderm

Ectoderm + Mesoderm

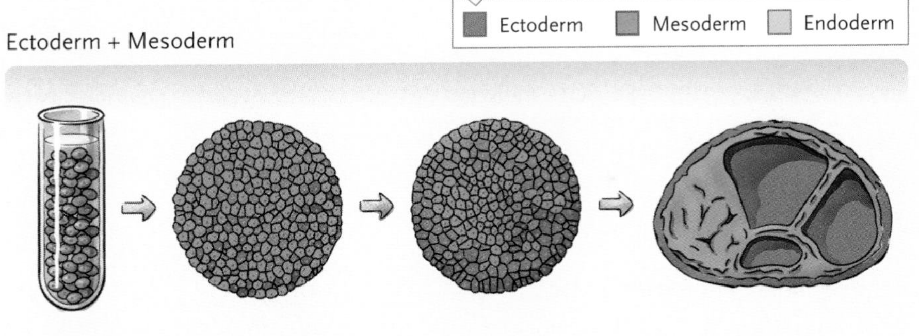

Ectoderm + Mesoderm + Endoderm

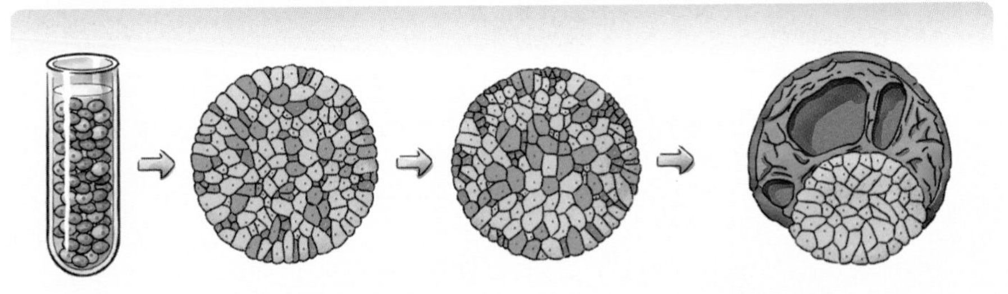

RESULTS: In time the reaggregated cells sorted themselves with respect to cell type; that is, instead of the cell types remaining mixed, each cell type became separated spatially. That is, in the ectoderm + mesoderm mixture, the ectoderm moved to the periphery of the aggregate, surrounding mesoderm cells in the center. In no case did the two cell types remain randomly mixed. The ectoderm + mesoderm + endoderm aggregate showed further that cell sorting in the aggregates generated cell positions reflecting the positions of the cell types in the embryo. That is, the endoderm cells separated from the ectoderm and mesoderm cells and became surrounded by them. In the end, the ectoderm cells were located on the periphery, the endoderm cells were internal, and the mesoderm cells were between the other two cell types.

CONCLUSION: Holtfreter interpreted the results to mean that cells have selective affinity for each other; that is, cells have selective adhesion properties. Specifically, he proposed that ectoderm cells have positive affinity for mesoderm cells but negative affinity for endoderm cells, while mesoderm cells have positive affinity for both ectoderm cells and endoderm cells. In modern terms, these properties result from cell surface molecules that give cells specific adhesion properties.

Figure 48.21 Experimental Research

Spemann and Mangold's Experiment Demonstrating Induction in Embryos

QUESTION: Does induction occur in embryonic development?

EXPERIMENT: Hans Spemann and Hilde Mangold performed transplantation experiments with newt embryos, the results of which demonstrated that specific induction of development occurs in the embryos. The researchers removed the dorsal lip of the blastopore from one newt embryo and grafted it onto a different position—the ventral side—of another embryo. The two embryos were from different newt species that differed in pigmentation, allowing them to follow the fate of the tissue easily. The embryo with the transplant was allowed to develop.

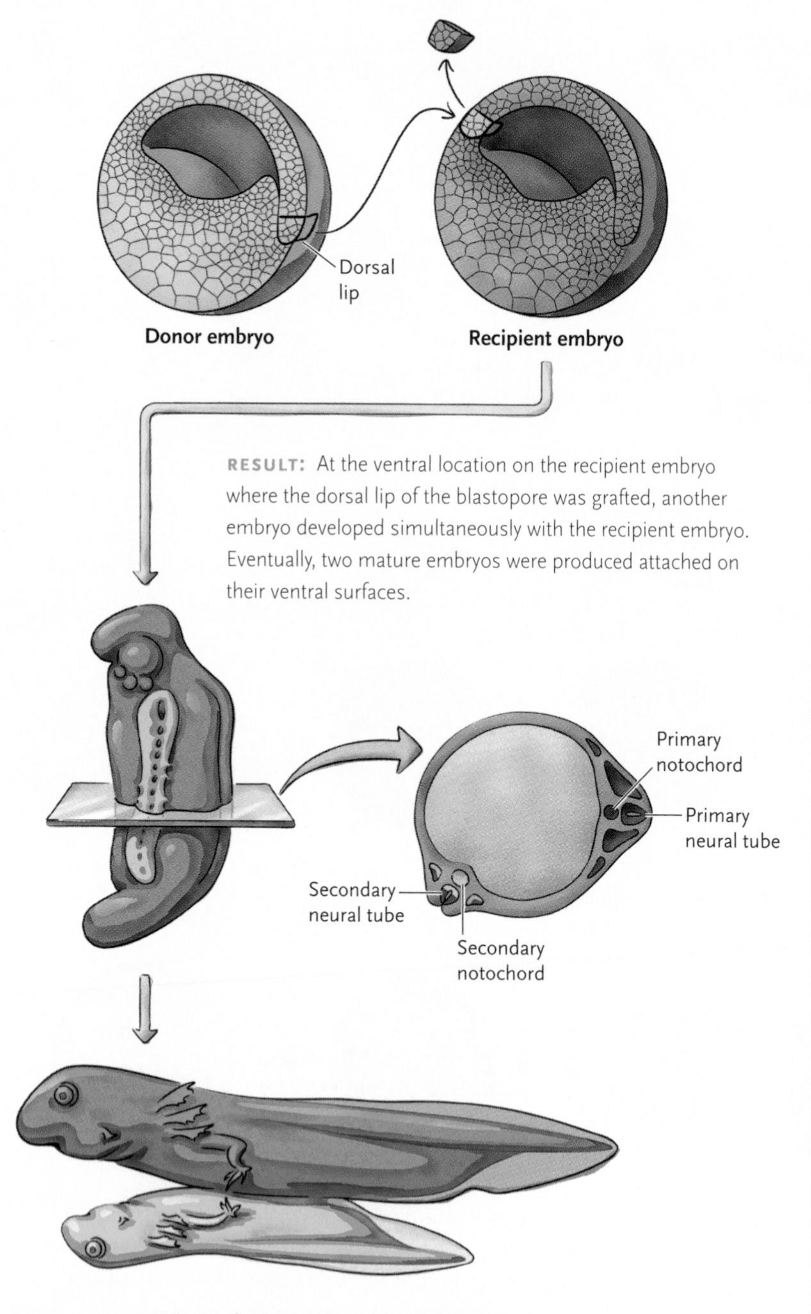

Dorsal lip
Donor embryo
Recipient embryo
RESULT: At the ventral location on the recipient embryo where the dorsal lip of the blastopore was grafted, another embryo developed simultaneously with the recipient embryo. Eventually, two mature embryos were produced attached on their ventral surfaces.
Primary notochord
Primary neural tube
Secondary neural tube
Secondary notochord

CONCLUSION: The grafted dorsal lip of the blastopore induced a second gastrulation and subsequent development in the ventral region of the recipient embryo. The result demonstrated the ability of particular cells to induce the development of other cells.

and extended by John B. Gurdon of the University of Cambridge, United Kingdom. In a typical experiment, the nucleus of a fertilized frog egg was destroyed by ultraviolet light. A micropipette was then used to transfer a nucleus from a fully differentiated tissue, intestinal epithelium, to the enucleated egg. Some of the eggs receiving the transplanted nuclei subsequently developed into normal tadpoles and adult frogs. This outcome was possible only if the differentiated intestinal cells still retained their full complement of genes. This conclusion was extended to mammals in 1997 when Ian Wilmut and his colleagues successfully cloned a sheep—Dolly—starting with an adult cell nucleus. (This experiment is described in Section 18.2.)

Fate Mapping Maps Adult Structures onto Regions of the Embryos from Which They Developed

From the early days of studying development, embryologists have focused on describing not only how embryos form and develop, but exactly how adult tissues and organs are produced from the cells of the embryo. Thus, an important goal in embryology was to trace cell lineages from embryo to adult. For most organisms it is not possible to trace lineages at the individual cell level, primarily because of the complexity of the developmental process and the typical opacity of embryos. However, it has been possible to map adult or larval structures onto the region of the embryo from which each structure developed. This type of study is called *fate mapping*, and the result is called a **fate map**. Experimentally, fate mapping is done by following development of living embryos under the microscope, either using species in which the embryo is transparent, or by marking cells so they can be followed. Cells may be marked with vital dyes (dyes that do not kill cells), fluorescent dyes, or radioactive labels. Fate maps have been produced for a number of organisms, including the chick, *Xenopus*, and *Drosophila*.

In most cases a fate map is not detailed enough to relate how particular cells in the embryo gave rise to cells of the adult. The exception is the fate map of the nematode *Caenorhabditis elegans*, an organism that has a fixed, reproducible developmental pattern. This animal has a transparent body, and scientists have been able to map the fate—trace the **cell lineage**—of every somatic and germ-line cell as the zygote divides and the resulting embryo differentiates into the 959-cell adult hermaphrodite or the 1031-cell adult male **(Figure 48.22)**. They found that all somatic cells of the adult can be traced from five somatic *founder cells* produced during early development. Knowing the cell lineages of *C. elegans* has been a valuable tool for research into the genetic and molecular control of development in this organism, for mutants affecting development have an easily visualized effect.

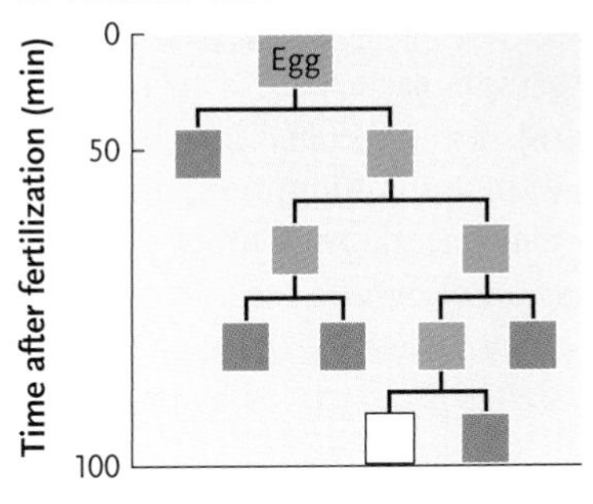

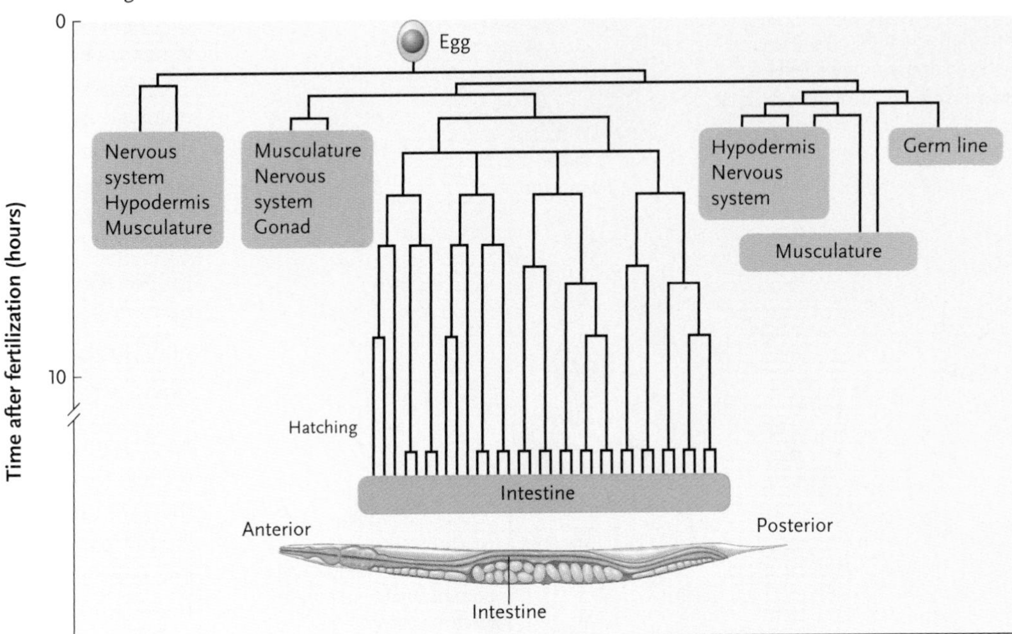

Figure 48.22
Cell lineages of *C. elegans*. **(a)** The founder cells (blue) produced in early cell divisions from which all adult somatic cells are produced. The cell in white gives rise to germ-line cells. **(b)** The cell lineage for cells that form the intestine. The detailed lineages for the other parts of the adult are not shown.

Study Break

1. What are the key cellular events that contribute to morphogenesis in animals?
2. What is induction? What molecules are involved?

48.6 The Genetic and Molecular Control of Development

We have now looked at development at the level of the whole organism, from the fertilized egg to the fully formed individual, and then at the cellular changes and movements that underlie this progression. We now turn to genetic and molecular mechanisms which, to a large extent, determine the course of development. In particular, these include the molecular mechanisms that control gene expression (see Chapter 16).

Developmental biologists are very interested in identifying and characterizing the genes involved in development, and defining how the products of the genes regulate and bring about the elaborate events we see. One productive research approach has been to isolate mutants that affect developmental processes. Researchers can then identify the genes involved, clone these genes, and analyze them in detail to build models for the molecular functions of the gene products in development. A number of model organisms are used for these studies because of the relative ease with which mutants can be made and studied and the ease of performing molecular analyses. These organisms include the fruit fly *(Drosophila melanogaster)* and *C. elegans* among invertebrates, and the zebrafish *(Danio rerio)* and the mouse *(Mus musculus)* among vertebrates. *Focus on Research* describes why the zebrafish is a valuable model organism for genetic and molecular studies of development.

Genes Control Cell Determination and Differentiation

As you have learned, determination, the setting of the developmental fate of a cell, in many cases is the result of induction. The end result of determination is differentiation, which produces cell types of particular kinds, such as skin cells or nerve cells. Both determination and differentiation involve specific, regulated changes in gene expression.

One well-studied example of the genetic control of determination and differentiation is the production of skeletal muscle cells from somites in mammals **(Figure 48.23).** Recall that somites are blocks of mesoderm cells that form along both sides of the notochord (see Figure 48.11). Under genetic control, particular cells of a somite differentiate into skeletal muscle cells. First, paracrine signaling from nearby cells induces those somite cells to express the master regulatory gene, *myoD*. The product of *myoD* is the transcription factor MyoD. By turning on specific muscle-determining genes, the action of MyoD brings about the determination of those cells, converting them to undifferentiated muscle cells known as **myoblasts.** Among the genes that MyoD regulates are the myogenin and MEF genes. These genes are also regulatory genes, expressing transcription factors in the myoblasts that turn on yet another set of genes. The products of those genes—which in-

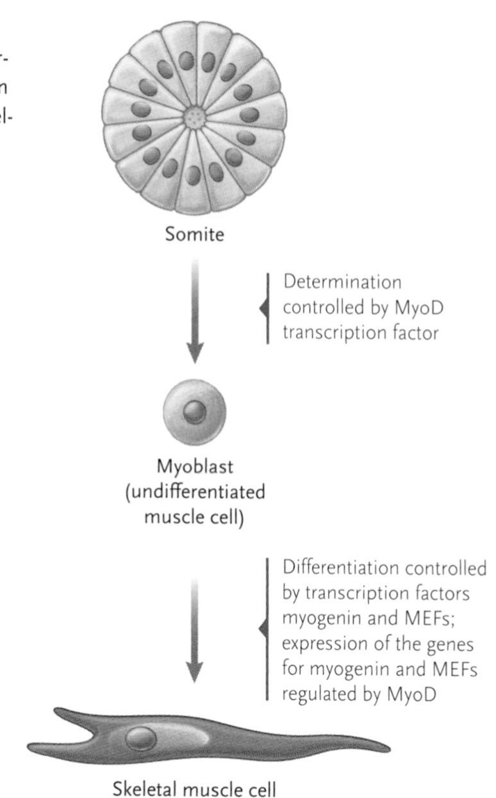

Figure 48.23
The genetic control of determination and differentiation involved in mammalian skeletal muscle cell formation.

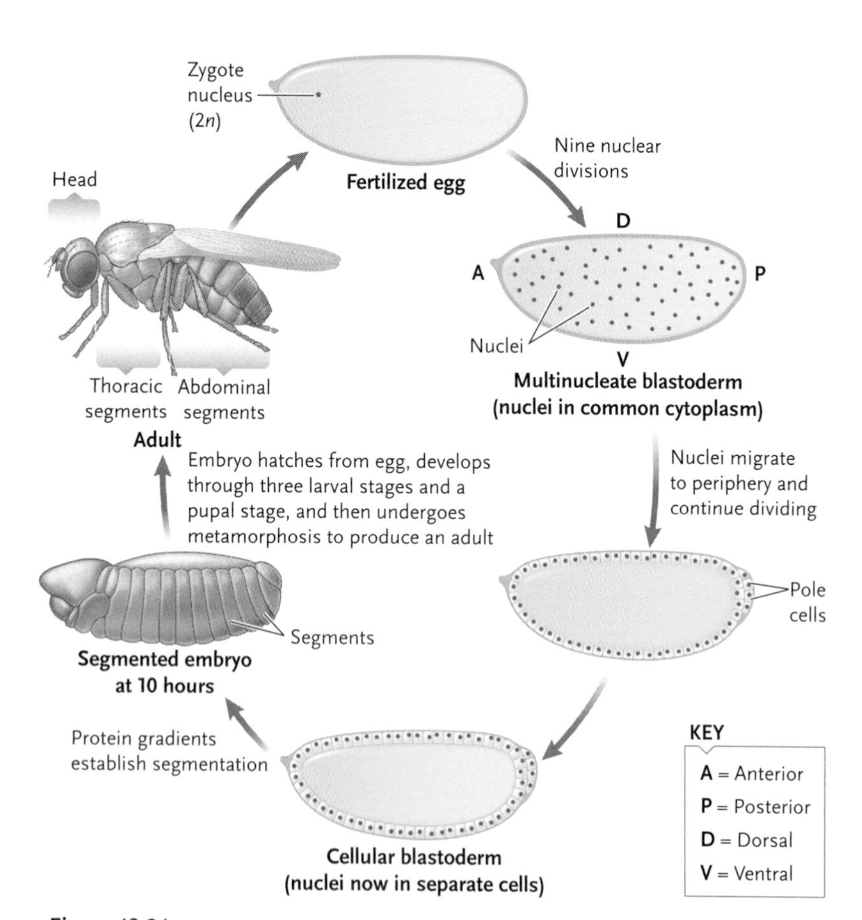

Figure 48.24
Embryogenesis in *Drosophila* and the relationship between segments of the embryo and segments of the adult.

clude myosin, a major protein involved in muscle contraction—are needed for the differentiation of myoblasts into skeletal muscle cells.

Generally speaking, the molecular mechanisms involved in determination and differentiation depend on regulatory genes that encode regulatory proteins controlling the expression of other genes. In essence, the regulatory genes act as master regulators; expression of the regulatory genes is controlled by induction in most cases.

Genes Control Pattern Formation during Development

As a part of the signals guiding differentiation, cells receive positional information that tells them where they are in the embryo. The positional information is vital to **pattern formation:** the arrangement of organs and body structures in their proper three-dimensional relationships. Positional information is laid down primarily in the form of concentration gradients of regulatory molecules produced under genetic control. In most cases, gradients of several different regulatory molecules interact to tell a cell, or a cell nucleus, where it is in the embryo. Below, we describe in brief the results of studies of the genetic control of pattern formation during the development of the fruit fly, *Drosophila melanogaster.* The developmental principles discovered from these studies apply to many other animal species, including humans.

Embryogenesis in *Drosophila*. The production of an adult fruit fly from a fertilized egg occurs in a sequence of genetically controlled development events. Following fertilization, division of the nucleus begins by mitosis, but the cytoplasm does not divide in the early embryo (cytokinesis does not occur) **(Figure 48.24).** The result is a multinucleate blastoderm. At the tenth nuclear division, the nuclei migrate to the periphery of the embryo where, after three more divisions, the 6000 or so nuclei are organized into separate cells. At this stage, the embryo is a *cellular blastoderm,* corresponding to a late blastula stage in the animals we discussed earlier. The cellular blastoderm develops into a segmented embryo (an embryo with distinct segments); at that point, 10 hours have passed since the egg was fertilized. About 24 hours after fertilization, the egg hatches into a larva, which undergoes three molts, then becoming a pupa. The pupa undergoes metamorphosis to produce the adult fly, which emerges about 10–12 days after fertilization. As illustrated by the color usage in Figure 48.24, the segments of the embryo can be mapped to the segments of the adult fly.

Genetic Analysis of *Drosophila* Development. The study of developmental mutants by a large number of researchers has given us important information about

FOCUS ON RESEARCH

Model Research Organisms: The Zebrafish Makes a Big Splash as the Vertebrate Fruit Fly

David M. Parichy

The zebrafish *(Danio rerio)* is a small (3 cm) freshwater fish that gets its name from the black and white stripes running along its body. Native to India, it has spread around the world as a favorite aquarium fish. Beginning about 30 years ago, it began also to be used in scientific laboratories as a model vertebrate organism for studying the roles of genes in development. Its use is now so widespread that it has been dubbed the "vertebrate fruit fly."

The zebrafish brings many advantages as a model research organism. It can be maintained easily in an ordinary aquarium on a simple diet. Although its generation time is relatively long (3 months for the zebrafish as compared with $1\frac{1}{2}$ months for the mouse), a female zebrafish produces about 200 offspring at a time, as compared with an average of 10 for the mouse.

Embryonic development of the zebrafish takes place in eggs released to the outside by the female. The embryos develop rapidly, taking only 3 days from egg laying to hatching. Best of all, the eggs and embryos are transparent, providing an open window that allows researchers to observe developmental stages directly, with little or no disturbance to the embryo. Observational conditions are so favorable that the origin and fate of each cell can be traced from the fertilized egg to the hatchling. Individual nerve cells can be traced, for example, as they grow and make connections in the brain, spinal cord, and peripheral body regions. Removing or transplanting cells and tissues is also relatively easy. Biochemical and molecular studies can be carried out by techniques ranging from the simple addition of reactants to the water surrounding the embryos to injection of chemicals into individual cells.

The zebrafish has some advantages for developmental studies compared with other vertebrate organisms used as developmental models, including the amphibian *Xenopus,* and the mouse. The early developmental stages of a zebrafish are remarkably like those of mammals, and adult structures such as the eye and skeletal system are typically vertebrate. *Xenopus* takes years to become developmentally mature and produce offspring, and it is not readily amenable to genetic analysis. Although the mouse is a mammal with development and anatomy closely related to those of humans, mouse embryos develop inside the mother and can be observed only by removing them from the mother's body; outside the body, they can be maintained only by demanding and elegant experimental techniques. Chemical studies while the embryo is inside the mother are difficult to perform. Additionally, maintaining colonies of mice is expensive.

The advantages of working with the zebrafish have spurred efforts to investigate its genetics, with particular interest in genes that regulate embryonic development. This work has already identified mutants of more than 2000 genes, including more than 400 genes that influence development. Most of the mechanisms controlled by the developmental genes resemble their counterparts in humans and other mammals.

For some of the zebrafish genes, developmental and physiological studies have revealed functions that were previously unknown for their mammalian equivalents. For example, Nancy Hopkins, a developmental geneticist at MIT, found a gene necessary for normal liver and gut development in the zebrafish. The gene is 80% identical in nucleotide sequence to a human gene; identification of the gene's role in zebrafish gave the first clues to its function in mammals. Other zebrafish mutants have been identified that affect development of the brain and spinal cord, the eyes and ears, the skeletal and digestive systems, and the circulatory system, including the heart, blood vessels, and blood cells. The mutants open these systems to biochemical and molecular study and experimentation.

Genetic studies with the zebrafish have been reinforced by a project to obtain the DNA sequence of its entire genome. The sequencing project, which was completed in 2005, will allow all the zebrafish genes to be located, identified, and correlated with their equivalents in humans and other model research organisms, including the mouse, *C. elegans,* and *Drosophila.* Undoubtedly, the zebrafish will continue to be a valuable model in developmental biology research.

Drosophila development. Three researchers performed key, pioneering research with developmental mutants: Edward B. Lewis of the California Institute of Technology, Christiane Nüsslein-Volhard of the Max Planck Institute for Developmental Biology in Tübingen, Germany, and Eric Wieschaus of Princeton University. The three shared a Nobel Prize in 1995 "for their discoveries concerning the genetic control of early embryonic development."

Nüsslein-Volhard and Wieschaus studied early embryogenesis. They searched for *every* gene required for early pattern formation in the embryo. They did this by looking for recessive *embryonic lethal* mutations. These mutations, when homozygous, result in the death of the embryo during development. By examining at what stage of development an embryo died, and how development was disrupted, they gained insights into the role of the particular genes in embryogenesis.

Lewis studied mutants that changed the fates of cells in particular regions in the embryo, producing structures in the adult that normally were produced by other regions. His work was the foundation of research identifying master regulatory genes that control the development of body regions in a wide range of organisms.

Maternal-Effect Genes and Segmentation Genes for Establishing the Body Plan in the Embryo. A number of genes control the establishment of the embryo's body plan. These genes regulate the expression of other genes. There are two classes: *maternal-effect genes,* and *segmentation genes* that work sequentially **(Figure 48.25).**

Many **maternal-effect genes** are expressed by the mother during oogenesis. These genes control the polarity of the egg and, therefore, of the embryo. Some control the formation of the anterior structures of the embryo, others control the formation of the posterior structures, and yet others control the formation of the terminal end.

The *bicoid* gene is the key maternal-effect gene responsible for head and thorax development. The *bicoid* gene is transcribed in the mother during oogenesis, and the resulting mRNAs are deposited in the egg, localizing near the anterior pole **(Figure 48.26).** After the egg is fertilized, translation of the mRNAs produces BICOID protein, which diffuses through the egg to form a gradient with its highest concentration at the anterior end of the egg, and fading to none at the posterior end of the egg. The BICOID protein is a transcription factor that activates some genes and represses others along the anterior–posterior axis of the embryo. Embryos with mutations in the *bicoid* gene have no thoracic structures, but have posterior structures at each end. Researchers concluded, therefore, that the *bicoid* gene in normal embryos is a master regulator gene controlling the expression of genes for the development of anterior structures (head and thorax).

A number of other maternal-effect genes, through the activities of their products in gradients in the embryo, are also involved in axis formation. The *nanos* gene, for instance, is the key maternal-effect gene for the posterior structures. When the *nanos* gene is mutated, embryos lack abdominal segments.

Once the axis of the embryo is set, the expression of at least 24 **segmentation genes** progressively subdivides the embryo into regions, determining the segments of the embryo and the adult (see Figure 48.25). Gradients of BICOID and other proteins encoded by maternal-effect genes regulate expression of the embryo's segmentation genes differentially. That is, each segmentation gene is expressed at a particular time and in a particular location during embryogenesis.

Three sets of segmentation genes are regulated in a cascade of gene activations. The first set to be expressed is the **gap genes,** for example, *hunchback* and *tailless.* These genes are activated based on their positions in the maternally directed anterior–posterior axis of the egg by reading the concentrations of BICOID and other proteins. Gap genes, through their activation of the next genes in the regulatory cascade, control the subdivision of the embryo along the anterior–posterior axis into several broad regions. Mutations in gap genes result in the loss of one or more body segments in the embryo **(Figure 48.27a).**

The products of gap genes are transcription factors that activate **pair-rule genes.** The actions of the prod-

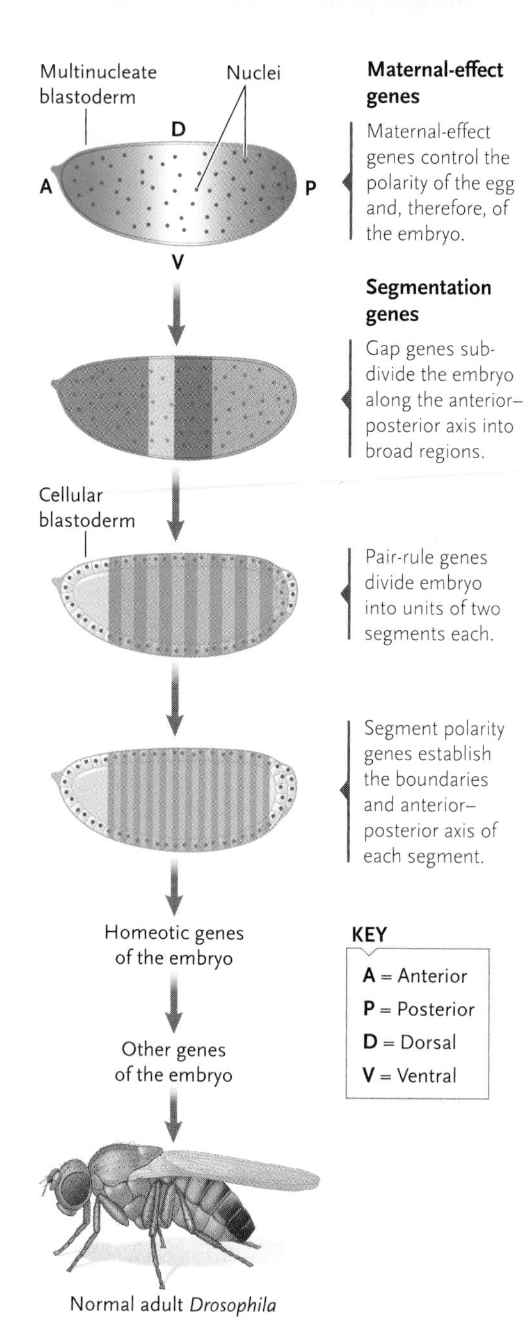

Figure 48.25
Maternal-effect genes and segmentation genes and their role in *Drosophila* embryogenesis.

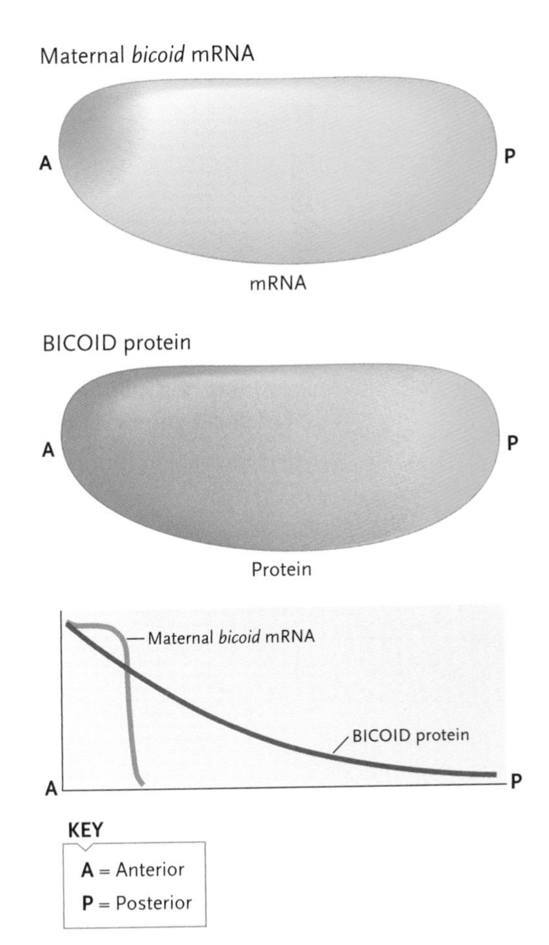

Figure 48.26
Gradients of *bicoid* mRNA and BICOID protein in the *Drosophila* egg.

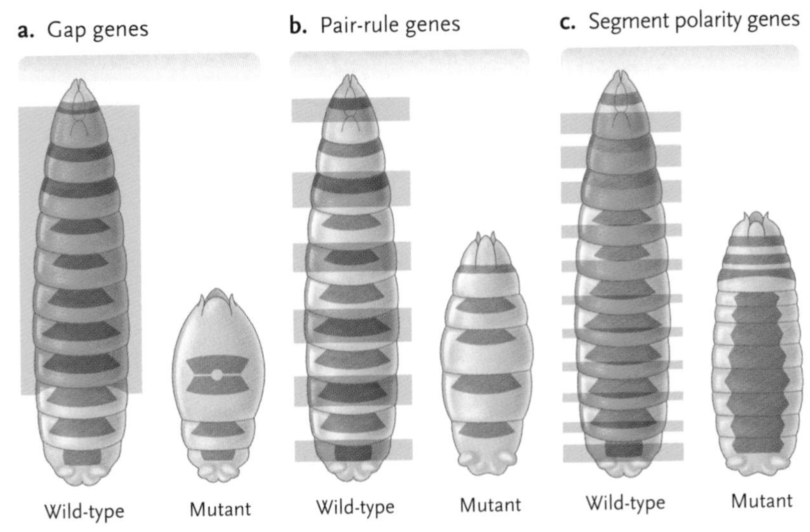

Figure 48.27
Examples of effects of mutations in the different types of segmentation genes of *Drosophila.* Blue highlights indicate wild-type segments that are mutated. **(a)** Gap gene mutants lack one or more segments. **(b)** Pair-rule gene mutants are missing every other segment. **(c)** Segment polarity gene mutants have segments with one part missing and the other part duplicated as a mirror image.

ucts of pair-rule genes divide the embryo into units of two segments each. Mutations in pair-rule genes delete every other segment of the embryo **(Figure 48.27b).**

The products of pair-rule genes are transcription factors that regulate the expression of the last set of genes in the series, the **segment polarity genes.** The actions of the products of segment polarity genes set the boundaries and anterior–posterior axis of each segment in the embryo. Mutations in segment polarity genes produce segments in which one part is missing and the other part is duplicated as a mirror image **(Figure 48.27c).** The products of segment polarity genes are transcription factors and other molecules that regulate other genes involved in laying down the pattern of the embryo.

Homeotic Genes for Specifying the Developmental Fate of Each Segment. Once the segmentation pattern has been set, **homeotic** (structure-determining) **genes** of the embryo specify what that segment will become after metamorphosis. In normal flies, homeotic genes are master regulatory genes that control the development of structures such as eyes, antennae, legs, and wings on particular segments (see Figure 48.24). Researchers discovered the role of homeotic genes from the study of mutations in these genes; such mutations alter the developmental fate of a segment in the embryo in a major way. For example, in flies with a mutation in the *Antennapedia* gene, legs develop in place of antennae **(Figure 48.28).**

How do homeotic genes regulate development? Homeotic genes encode transcription factors that regulate expression of genes responsible for the development of adult structures. Each homeotic gene has a common region called the **homeobox** that is key to its function. A homeobox corresponds to an amino acid section of the encoded transcription factor called the **homeodomain.** The homeodomain of each protein binds to a region in the promoters of the genes whose transcription it regulates.

Homeobox-containing genes are called *Hox* genes. There are eight *Hox* genes in *Drosophila* and, interest-

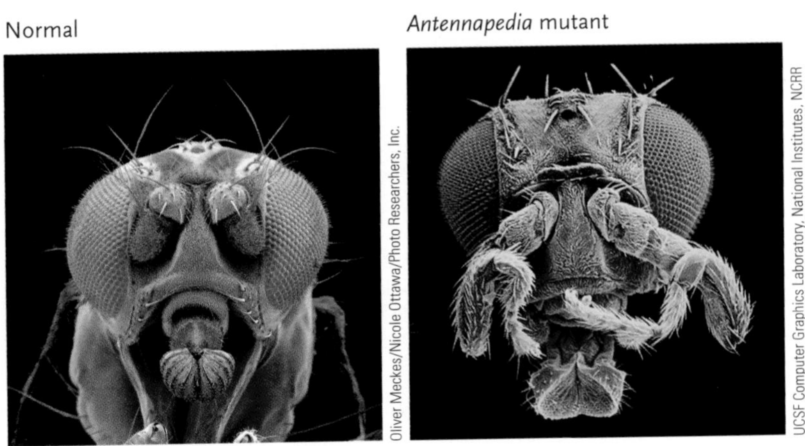

Figure 48.28
Antennapedia, a homeotic mutant of *Drosophila*, in which legs develop in place of antennae.

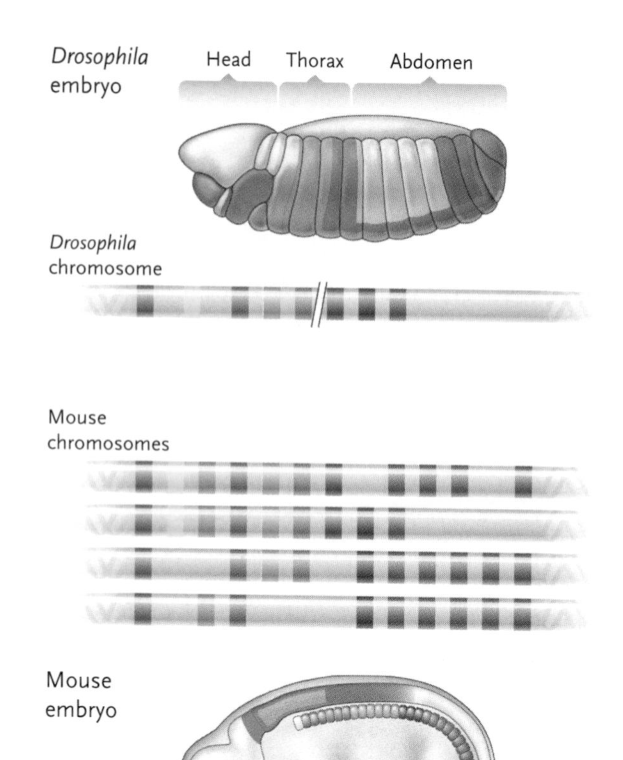
Drosophila embryo
Head
Thorax
Abdomen
Drosophila chromosome
Mouse chromosomes
Mouse embryo

Figure 48.29
The *Hox* genes of the fruit fly and the corresponding regions of the embryo they affect. The mouse has four sets of *Hox* genes on four different chromosomes. Their relationship to the fruit fly genes is shown by the colors.

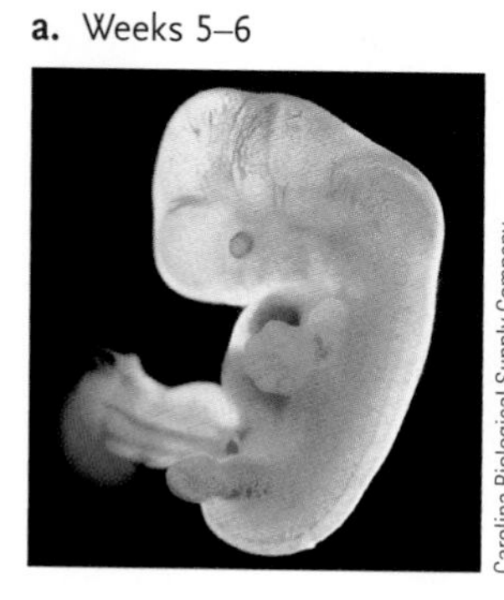
a. Weeks 5–6
Carolina Biological Supply Company

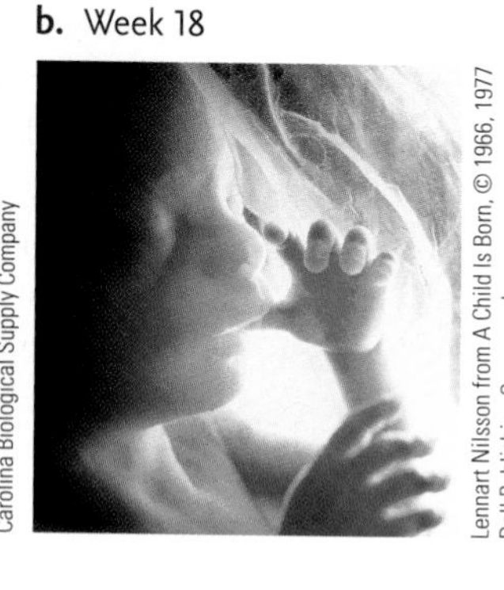
b. Week 18
Lennart Nilsson from A Child Is Born, © 1966, 1977 Dell Publishing Company, Inc.

Figure 48.30
An illustration of apoptosis in humans: the removal of tissue between developing fingers and toes to produce the free fingers and toes later in development.

ingly, they are organized along a chromosome in the same order as they are expressed along the anterior–posterior body axis **(Figure 48.29)**.

The discovery of *Hox* genes in *Drosophila* led to a search for equivalent genes in other organisms. The result of that search has shown that *Hox* genes are present in all major animal phyla. In each case, the genes control the development of the segments/regions of the body and are arranged in order in the genome. (This is a good example of how the results of studies in a model research organism are broadly applicable.) The homeobox sequences in the *Hox* genes are highly conserved,

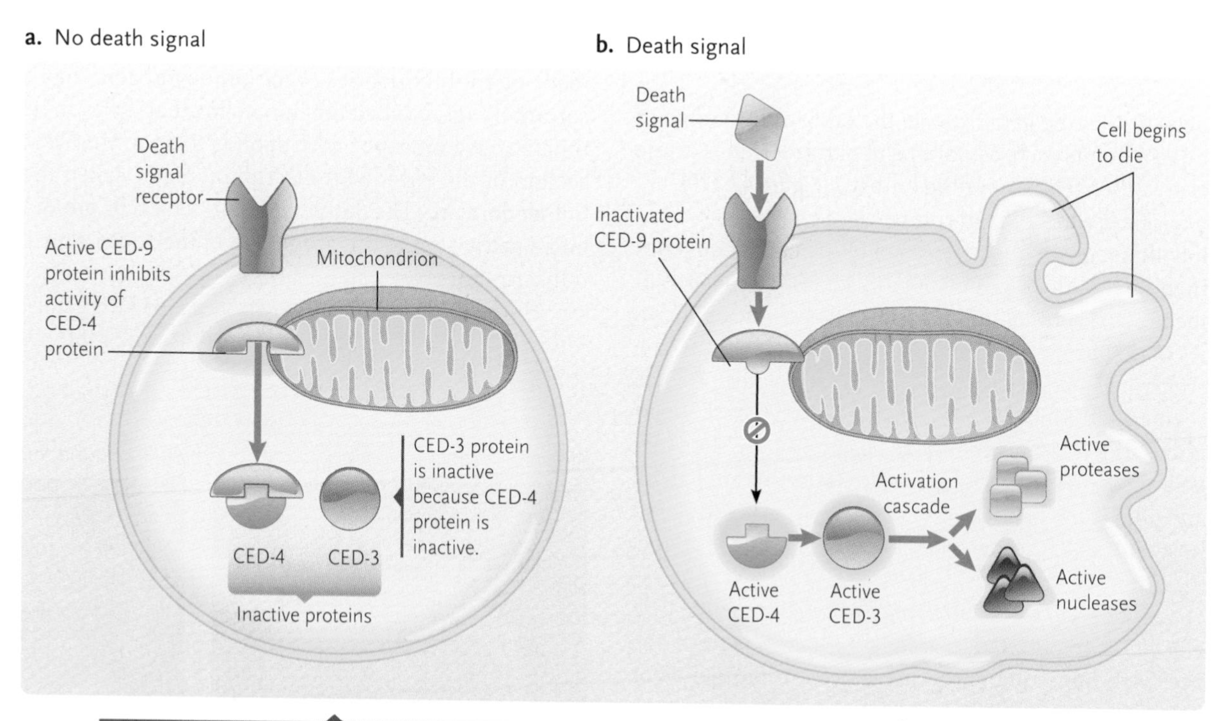
a. No death signal
Death signal receptor
Active CED-9 protein inhibits activity of CED-4 protein
Mitochondrion
CED-3 protein is inactive because CED-4 protein is inactive.
CED-4
CED-3
Inactive proteins
b. Death signal
Death signal
Inactivated CED-9 protein
Cell begins to die
Activation cascade
Active proteases
Active nucleases
Active CED-4
Active CED-3
Apoptosis is inhibited as long as CED-9 protein is active; cell remains alive.
When a death signal binds to the death signal receptor, it activates the receptor, which leads to inactivation of CED-9 protein. As a result, CED-4 protein is no longer inhibited and becomes active, activating CED-3 protein. Active CED-3 triggers a cascade of activations producing active proteases and nucleases, which cause the changes seen in apoptotic cells and eventually to cell death.

Figure 48.31
The molecular basis of apoptosis in *C. elegans.* **(a)** In the absence of a death signal, no apoptosis occurs. **(b)** In the presence of a death signal, activation of CED-4 and CED-3 proteins triggers a pathway that leads to the cell's death.

indicating common function in the wide range of animals in which they are found. For example, the homeobox sequences of mammals are the same or very similar to those of the fruit fly (see Figure 48.29).

Homeotic genes are also found in plants. For example, many homeotic mutations that affect flower development have been identified and analyzed in *Arabidopsis* (see Section 34.5).

Apoptosis Is Triggered by Cell-Death Genes

We have noted the role of apoptosis in the breakdown of a tadpole's tail; there are many other examples of apoptosis in both vertebrate and invertebrate development. In humans, for example, the developing fingers and toes are initially connected by tissue, forming a paddle-shaped structure; later in development, cells of the tissue die by apoptosis resulting in separated fingers and toes **(Figure 48.30).** Like many other mammals, kittens and puppies are born with their eyes sealed shut by an unbroken layer of skin. Just after birth, cells die in a thin line across the middle of each eyelid, freeing the eyelids to open. During the pupation stage that converts a caterpillar to a butterfly, many tissues of the larva break down by apoptosis and are replaced by newly formed adult tissues.

Apoptosis results from gene activation, in response to molecular signals received by receptors at the surfaces of the marked cells. In effect, the signals amount to a death notice, delivered at a specific time during embryonic development. For example, in the nematode *C. elegans,* division of the fertilized egg leads to a total of 1090 cells. Of these, exactly 131 die at prescribed times to produce a total of 959 cells in the adult hermaphrodite.

The molecular basis of apoptosis in *C. elegans* involves a molecule that can be considered a death signal binding to a receptor in the plasma membrane of a target cell for apoptosis. The receptor is activated and this leads to activation of proteins that kill the cell. In the absence of the death signal, the killing proteins remain inactive.

Let's walk through the two situations. In the absence of a death signal, the membrane receptor is inactive **(Figure 48.31a).** This allows a protein associated with the outer mitochondrial membrane, CED-9 (encoded by the *ced-9* *c*ell *d*eath gene) to inhibit CED-4 (encoded

Unanswered Questions

What gene or factor initiates sex determination in birds?

You learned in this chapter that the presence of the *SRY* gene on the Y chromosome determines that a mammal will develop as a male, but the molecular basis for sex determination in birds is unknown. It is known that avian sex determination is chromosomal and that birds have a ZW sex-determination system, which is reversed compared with the mammalian XY system. That is, female birds have two different sex chromosomes—they are ZW—and males have two identical chromosomes—they are ZZ. At least two candidate genes are being investigated for their possible role in avian sex determination: *DMRT1* (Z-linked *d*oublesex- and *m*ab-3-*r*elated *t*ranscription factor gene), which contains a highly conserved DM (cysteine-rich DNA binding) domain and encodes for a protein that is specifically enhanced more in male urogenital development than in female development, and *ASW* (Avian Specific gene on the *W* chromosome, also known as *HINTW, PKC1W,* or *Wpkci*), an altered form of a protein kinase C inhibitor gene.

In mammals, one of the two X chromosomes in females is inactivated through the process of dosage compensation, a genetic regulatory mechanism that equalizes the phenotypic expression determined by X-linked genes so that they are equally expressed in XY males and XX females (as described in Section 13.2). But dosage compensation does not occur in birds; birds lack inactivation of one of the two male Z chromosomes and seem to compensate for a double dose of Z-linked genes by other mechanisms. How do these genes alone or together determine sex in birds? Recent work on tinamou (an ancient group of South American birds) and ratites (flightless birds) by Yayoi Tsuda and Yoichi Matsuda of Hokkaido University in Japan suggests that although all bird sex chromosomes evolved from the same pair of autosomes, the Z and W sex chromosomes have independently diverged from one another several times.

What initiates childbirth?

As discussed in this chapter, oxytocin receptors in the smooth muscle cells of the uterine wall stimulate the muscle to begin the rhythmic contractions of labor marking the beginning of birth or parturition. The initiation of labor is a complex process that has yet to be fully explained, however. What fetal-derived signals lead to the initiation of labor? How do these signals play a role in preterm labor? What roles do signals from the placenta, such as prostaglandins, play in regulating the timing of birth? What roles do the fetal hypothalamus and adrenal glands play in initiating childbirth? What hormones, such as estradiol from the mother or cortisol and corticotropin-releasing factor from the fetus, send the signals that begin the process of parturition? Dr. Louis J. Muglia at Washington University School of Medicine is studying the mechanisms that control normal-term labor and how these mechanisms malfunction to result in preterm labor. Dr. Muglia's research into the timing of birth and the challenges to prevent premature births has demonstrated that prostaglandins are essential for the initiation of parturition, and he has determined that regulation of the *COX-1* gene is important for the initiation of parturition in mice. The *COX-1* gene encodes cyclooxygenase-1, a protein that acts as an enzyme to speed up the production of prostaglandins in the stomach; expression of the *COX-1* gene increases in the uterus during pregnancy. Identifying other regulatory factors may provide critical information into the timing of the onset of parturition.

Laura Carruth, an assistant professor of biology at Georgia State University, studies the genetic and hormonal factors that lead to sex differences in brain development. Her work focuses on model systems in songbirds (the Australian zebra finch) and mice. To learn more about Dr. Carruth's research, go to http://biology.gsu.edu/people/faculty/person.cfm?person=2167.

by the *ced-4* gene) and CED-3 (encoded by the *ced-3* gene), the two proteins that are needed to turn on the cell death program. Cells in this situation, with the *ced-9* gene being expressed and its product CED-9 being active, are those that normally survive to form the adult nematode. If a death signal binds to the receptor, however, the receptor becomes activated and the events that follow are typical of signal transduction pathways (**Figure 48.31b**) (see Sections 7.1–7.4). In this case, the activated receptor leads to inactivation of CED-9. Because CED-9 no longer is inhibiting them, CED-4 is activated and, in turn, activates CED-3. Activated CED-3 triggers a cascade of reactions, including the activation of proteases and nucleases that degrade cell structures and chromosomes as part of the cell death program.

Studies of mutants helped understand the role of the cell death genes in *C. elegans*. In mutants lacking a normal *ced-3* or *ced-4* gene, the 131 marked cells fail to die, producing a highly disorganized embryo. In the nervous system, for example, 103 cells that die by apoptosis in normal embryos live to form neurons in the mutants. These extra neurons, which are inserted at random in the embryo, lead to a disorganized and nonfunctional nervous system.

Genes related to *ced-3* and *ced-4* have been found in all animals that have been tested for their presence. In humans and other mammals, the equivalent of *ced-3* is the *caspase-9* gene, which encodes a protease that degrades cell structures. The *caspase-9* gene becomes active, for example, in the cells that form the webbing between the fingers and toes, and causes the webbing to break down. The equivalent of *ced-4* is the *Apaf* gene (for *A*poptotic *p*rotease-*a*ctivating *f*actor). Mammalian cells are saved from death by the *Bcl* family of genes, which are the equivalent of *ced-9* in *C. elegans*. The genes are so closely related that they retain their effects if they are exchanged between *C. elegans* and human cells.

Study Break

1. In general, how are determination and differentiation controlled?
2. How do the segmentation genes and homeotic genes of *Drosophila* differ in function?

Review

Go to **Thomson**NOW™ at www.thomsonedu.com/login to access quizzing, animations, exercises, articles, and personalized homework help.

48.1 Mechanisms of Embryonic Development

- Developmental information is stored in both the nucleus and cytoplasm of the fertilized egg. The mRNA and protein molecules that direct the first stages of development are the cytoplasmic determinants.
- The unequal distribution of yolk and other components makes eggs polar. The animal pole typically gives rise to surface structures and the anterior end of the embryo, while the vegetal pole typically gives rise to internal structures of the embryo such as the gut (Figure 48.2).
- Following fertilization, cleavage divisions produce the morula. The morula hollows out to form the blastula, which develops into the gastrula, the stage in which rearrangements of cells produce the ectoderm, mesoderm, and endoderm. Gastrulation establishes the body pattern, in that the organs and other structures of embryo arise from these three tissue layers (Figure 48.3; Table 48.1).
- Development proceeds as a result of cell division, cell movements, selective adhesions, induction, determination, and differentiation.

Animation: Where embryos develop

Animation: Stages of development

Animation: Cytoplasmic localization

48.2 Major Patterns of Cleavage and Gastrulation

- In sea urchins eggs, yolk is distributed evenly. As a result, cleavage divisions take place at the same rate in all regions of the embryo, and gastrulation follows a symmetrical pattern (Figure 48.4).
- In amphibian eggs, yolk is distributed unequally, with most in the vegetal pole. As a result, the rate of cell division is more rapid in the animal pole, and gastrulation shows an asymmetric pattern (Figures 48.5 and 48.7).
- In bird and reptile embryos, the cleavage divisions give rise to a flat disc of cells at the top of the yolk, which divides into the epiblast and the hypoblast. In gastrulation, cells of the epiblast migrate to the interior to form the endoderm and the mesoderm. The epiblast cells left at the surface form the ectoderm (Figure 48.8).
- In birds and reptiles, the yolk sac, chorion, amnion, and allantoic membrane form from extensions of the primary tissue layers. These extraembryonic membranes conduct nutrients from the yolk to the embryo, exchange gases with the environment, and store metabolic wastes (Figure 48.9).

Animation: Process of gastrulation

48.3 From Gastrulation to Adult Body Structures: Organogenesis

- In organogenesis, the three primary tissues give rise to the tissues and organs of the embryo. Organogenesis begins with neurulation, the development of the nervous system from ectoderm (Figures 48.10 and 48.11).
- The mesoderm splits into somites, which give rise to the vertebral column and to the muscles of the ribs, vertebral column, and limbs (Figure 48.11).
- Development of the eye from optic vesicles is illustrative of the inductions and differentiations common to organogenesis in vertebrates (Figure 48.12).
- Apoptosis—programmed cell death—plays an important role in development by removing tissues present during development but not in the adult organ.

Animation: Neural tube formation

Animation: Embryonic induction

Animation: AER transplant

Animation: Formation of human fingers

48.4 Embryonic Development of Humans and Other Mammals

- In humans, as in other placental mammals, cleavage divisions produce a morula that differentiates into a blastocyst. The blastocyst implants into the endometrium of the uterus, and its inner cell mass separates into the epiblast and hypoblast. The epiblast produces the ectoderm, mesoderm, and endoderm of the embryo (Figures 48.13 and 48.14a, b).
- Gastrulation, neurulation, differentiation of cell layers, and formation of extraembryonic membranes occur by mechanisms similar to those of bird and reptile embryos. Differentiation of ectoderm, mesoderm, and endoderm into their final tissues and organs also occurs in a similar way to birds and reptiles.
- Extraembryonic membranes form in mammals by processes that are also similar to the reptilian–bird pattern. However, some of the membranes have altered functions, reflecting the minimal amount of yolk in mammalian embryos, and maintenance of the embryo by the placenta (Figure 48.14c–e).
- The placenta is connected to the embryo by the umbilical cord, which conducts blood between the embryo and the placenta (Figure 48.14f).
- Fetal growth proceeds until birth, when the fetus is forced from the uterine cavity and through the vagina by contractions of the uterus, stimulated by oxytocin (Figures 48.15 and 48.16).
- The mother's mammary glands secrete milk once the offspring is born. Suckling by the offspring stimulates prolactin and oxytocin release from the pituitary, which stimulates milk production and secretion from the glands, respectively.
- Embryos develop internal male or female sex organs from the same primitive structures. The presence or absence of a Y chromosome, which carries the key *SRY* gene, determines whether the internal structures develop into male or female sexual organs (Figure 48.17).
- Most animals continue development after hatching or birth, leading to the adult, the sexually mature form of the species.

Animation: Fertilization

Animation: Cleavage and implantation

Animation: First 2 weeks of development

Animation: Weeks 3 to 4 of development

Animation: Proportional changes during development

Animation: Structure of the placenta

Animation: Fetal development

Animation: Birth

Animation: Anatomy of the breast

48.5 The Cellular Basis of Development

- Development in animals involves the regulation of specific cellular events, including cell division, cell movement, and cell adhesion.
- Cell division in development varies in orientation and rate.
- Cell movements in development occur through changes in cell shape or the migrations of entire cells. Shape changes are produced by microtubules or microfilaments. In cell migrations, cells follow tracks in the embryo or move in response to gradients of signal molecules (Figures 48.18 and 48.19).
- Selective cell adhesions, which depend on surface glycoproteins including CAMs and cadherins, underlie many cell movements. The final cell adhesions hold the embryo in its correct shape and form (Figure 48.20).
- Induction results from the effects of signaling molecules of the inducing cells on the responding cells (Figure 48.21).
- In differentiation, cells change from embryonic form to specialized types with distinct structures and functions. Differentiation occurs by differential gene activation.
- For some organisms, the origins of adult or larval structures have been mapped to regions of the embryo from which each structure derived (Figure 48.22).

48.6 The Genetic and Molecular Control of Development

- Pattern formation derives from the positions of cells in the embryo. Typically, positional information is detected by the cells in the form of concentration gradients of regulatory molecules encoded by genes (Figures 48.24–48.27).
- *Hox* genes are evolutionarily conserved regulatory genes that control the development of the segments or regions of the body (Figures 48.28 and 48.29).
- Apoptosis, programmed cell death, typically eliminates structures required for earlier but not later stages of development (Figures 48.30 and 48.31).

Questions

Self-Test Questions

1. Major contributors to the axes of the animal body are the:
   a. sperm and egg cytoplasm.
   b. sperm and egg chromosomes.
   c. ribosomes and mitochondria
   d. egg nucleus and yolk.
   e. pigments.
2. The process by which cells undergo mitosis without a corresponding increase in cytoplasm is called:
   a. polarity.
   b. cleavage.
   c. gastrulation.
   d. organogenesis.
   e. induction.
3. Which of the following mechanisms does *not* contribute to zygote development?
   a. meiosis
   b. mitosis
   c. selective cell adhesions
   d. determination
   e. induction
4. A major event during gastrulation is:
   a. the outward movement of cells at the dorsal lip of the blastopore.
   b. the displacement of the archenteron by the blastocoel.
   c. the formation of the coelom from the endoderm.
   d. the extension of ectoderm and endoderm to form the yolk sac.
   e. the development of ectoderm to form epidermal and neural tissues.
5. To contribute to the formation of a nervous system:
   a. the neural crest develops into motor neurons.
   b. the neural tube is converted into a neural plate.
   c. the notochord induces the overlying ectoderm to become a neural plate.
   d. the roof of the archenteron induces the formation of the neural tube.
   e. somites give rise to the autonomic nervous system.

6. In mammalian development:
   a. the morula develops into a trophoblast.
   b. the chorionic villi allow the blastocyst to move down the oviduct.
   c. the allantois takes over the work of the amnion.
   d. the pharyngeal arches transform into the pharynx, larynx, and nasal cavities.
   e. prolactin stimulates parturition.
7. In the development of the female sex organs:
   a. all ducts in the 7-week embryo become Wolffian ducts.
   b. the *SRY* gene is activated.
   c. anti-Mullerian hormone is secreted.
   d. the Mullerian ducts develop into oviducts.
   e. the mother secretes oxytocin.
8. In the embryonic development of the eye:
   a. the optic vesicle cells permanently adhere to each other to prevent movement, whereas the optic cup cells are very motile.
   b. signals from the optic cup trigger surface receptors on the lens placode.
   c. gradients determine that the ectoderm overlying the lens vesicle develops into the optic vesicle.
   d. microtubules powered by myosins and microfilaments powered by dyneins move the eye components around in the head region.
   e. cadherins function in the presence of calcium to allow the lens placode and optic cup to break apart.
9. In mammals, the nose is located on the anterior end and the heart in the center. These positions are the result of activation of:
   a. *Hox* genes arranged along a number of chromosomes in the same order as they are expressed along the anterior–posterior axis.
   b. maternal-effect genes after somites differentiate into muscle.
   c. *Hox* genes scattered randomly among different chromosomes.
   d. a transcription factor called the homeobox.
   e. a homeodomain that binds ribosomes.
10. During embryonic development in many humans, the cells of the lower earlobe die resulting in an unattached earlobe. For this to occur one would deduce that:
   a. *CED-3* genes are inhibited.
   b. *CED-4* genes are inhibited.
   c. caspase-9 is deactivated.
   d. *CED-9* is actively expressed.
   e. the proteins encoded by *Bcl* genes are inactivated.

Questions for Discussion

1. Experimentally, it is possible to divide an amphibian egg so that the gray crescent is wholly within one of the two cells formed. If the two cells are separated, only the cell with the gray crescent will form an embryo with a long axis, notochord, nerve cord, and back musculature. The other forms a shapeless mass of immature gut and blood cells. Propose an explanation of these outcomes.
2. The renowned developmental biologist Lewis Wolpert once observed that birth, marriage, and death are not the most important events in human life; rather, gastrulation is. In what sense was he correct?
3. Arguably, in sexually reproducing animals development begins when eggs and sperm form in the parents. In a paragraph, explain the rationale for this idea.
4. Investigators discovered a *Drosophila* protein that triggers development of the nerve cord on the ventral side of the embryos. When an mRNA encoding the protein was injected into cells on the ventral side of *Xenopus* embryos, dorsal structures were formed on the ventral side, including incomplete heads. What do these findings suggest about the evolution of embryonic development?

Experimental Analysis

As you have learned in this chapter, embryogenesis in *Drosophila* has been well described. In Chapter 18 you also learned that the genome of *Drosophila* has been completely sequenced, allowing each gene in the genome to be cataloged. Design an experiment to identify all the genes that are activated during embryogenesis, and when they are activated.

Evolution Link

Every one of more than a million species of insects has six legs, a pair on each of the three thoracic segments. By contrast, other arthropods, such as crustaceans, have a variable number of limbs; some species have limbs on every segment in both the thorax and abdomen. Propose a molecular mechanism by which limbs might have been lost during the evolution of the insects.

How Would You Vote?

Sanitation and medical advances have greatly extended the average human life span, especially in developed countries. Some researchers are now looking for ways to extend the human life span even further. Do you think research into life extension should be supported by federal research funding? Go to www.thomsonedu.com/login to investigate both sides of the issue and then vote.

Appendix A
Answers

Chapter 31

Study Break 31.1

1. A land plant's shoot system consists of its photosynthetic tissues and organs—stems, leaves, and buds. Stems are frameworks for upright growth and favorably position leaves for light exposure and flowers for pollination. Leaves increase a plant's surface area and thus its exposure to sunlight. Buds eventually extend the shoot or give rise to a new, branching shoot. The shoot system of a flowering plant also includes flowers and fruits. Parts of the shoot system store carbohydrates manufactured during photosynthesis.

   The root system usually grows belowground. It anchors the plant, and sometimes structurally supports its upright parts. It also absorbs water and dissolved minerals from soil and stores carbohydrates.

2. Meristem tissue is self-perpetuating embryonic tissue. Apical meristems, at the tips of shoots and roots, gives rise to a young plant's stems, buds, roots, and other primary tissues. In plants that show secondary growth, cylinders of lateral meristem tissue give rise to (often woody) secondary tissues that increase the diameter of older stems and roots.

Study Break 31.2

1. The ground tissue system makes up most of the plant body. It includes three types of structurally simple tissues—parenchyma, collenchyma, and sclerenchyma—each of which is composed mainly of one type of cell. Parenchyma makes up most of a plant's primary tissue and typically has air spaces between its cells, which are alive at maturity and can continue to divide. Subgroups of parenchyma cells are specialized for photosynthesis, secretion, and storage (of starch). Collenchyma is flexible ground tissue that contains cellulose. Its cells remain alive and metabolically active at maturity. They provide mechanical support for parenchyma and often collectively form strands or a sheathlike cylinder under the dermal tissue of growing shoot regions and leaf stalks. Cells of sclerenchyma are dead at maturity, but while alive they develop thick secondary walls that typically are lignified and provide additional support and protection in mature plant parts.

2. Xylem and phloem are the tissues of the vascular tissue system. The two types of xylem cells, called tracheids and vessel members, both develop thick, lignified secondary cell walls and die at maturity. The empty cell walls of abutting cells serve as pipelines for water and minerals. The conducting cells of phloem, called sieve tube members, form sieve tubes that conduct solutes, mainly sugars made during photosynthesis, throughout a plant.

3. The dermal tissue system serves as a skinlike protective covering for the plant body. Cells of the epidermis are tightly packed and cover the primary plant body. They secrete a cuticle that coats all plant parts except the very tips of the shoot and most absorptive parts of roots. Some epidermal cells become modified for specialized functions; examples are guard cells, which form stomata; root hairs, which absorb water and minerals; and hairlike trichomes, which function in defense against herbivory or secrete sugars that attract pollinators.

Study Break 31.3

1. Stems have four main functions: (1) they provide mechanical support for body parts involved in growth, photosynthesis, and reproduction; (2) they house the vascular tissues (xylem and phloem), which transport products of photosynthesis, water and dissolved minerals, hormones, and other substances throughout the plant; (3) they often are modified to store water and food; and (4) they have specific stem regions that contain meristematic tissue, which gives rise to new cells of the shoot.

   A plant stem is divided into modules, each consisting of a node, where leaves are attached, and an internode, the space between nodes. New primary growth occurs in buds—a terminal bud at the apex of the main shoot, and lateral buds, which produce branches (lateral shoots), in the leaf axils. Meristem tissue in buds gives rise to leaves, flowers, or both.

   In eudicots, most primary growth in a stem's length occurs directly below the shoot apical meristem. When a meristematic cell divides, one of its daughter cells becomes an initial, a cell that remains as part of the meristem. The other daughter cell becomes a derivative, which typically divides once or twice and then enters on the path to differentiation. As derivatives differentiate, they give rise to three primary meristems: protoderm, procambium, and ground meristem. These primary meristems produce cells that differentiate into specialized cells and tissues. In eudicots, the primary meristems are also responsible for elongation of the plant body. Each primary meristem occupies a different position in the shoot tip. Outermost is protoderm, which gives rise to the stem's epidermis. Inward from the protoderm the ground meristem gives rise to ground tissue (mostly parenchyma). Procambium, which produces the primary vascular tissues, is sandwiched between ground meristem layers. In most plants, inner procambial cells give rise to xylem and outer procambial cells to phloem. The developing vascular tissues become organized into vascular bundles that are wrapped in sclerenchyma and thread lengthwise through the parenchyma. In the stems and roots of most eudicots and some conifers, the vascular bundles form a stele (vascular cylinder) that vertically divides the column of ground tissue into an outer cortex and an inner pith.

2. Leaves are organs specialized for photosynthesis. In both eudicots and monocots, the leaf blade provides a large surface area for absorbing sunlight and carbon dioxide. Many eudicot leaves have a broad, flat blade attached to the stem by a petiole. Unless a petiole is very short, it holds a leaf away from the stem and helps prevent individual leaves from shading one another. In most monocot leaves, such as those of rye grass or corn, the blade is longer and narrower and its base simply forms a sheath around the stem.

3. Leaves develop on the sides of the shoot apical meristem. Initially, meristem cells near the apex divide and their derivatives elongate. The resulting bulge enlarges into a thin, rudimentary leaf, or leaf primordium. As the plant grows and internodes elongate, the leaves become spaced at intervals along the length of the stem or its branches. Leaf tissues typically form several layers. Uppermost is epidermis, with cuticle covering its outer surface. Just beneath the epidermis is mesophyll composed of loosely packed parenchyma cells that contain chloroplasts. Leaves of many plants, especially eudicots, contain two layers of mesophyll. Palisade mesophyll cells contain more chloroplasts and are arranged in compact columns with smaller air spaces between them, typically toward the upper leaf surface. Spongy mesophyll, which tends to be located toward the underside of a leaf, consists of irregularly arranged cells with a network of air spaces that enhance the uptake of carbon dioxide and release of oxygen during photosynthesis and account for 15% to 50% of a leaf's volume. Below the mesophyll is another cuticle-covered epidermal layer. Except in grasses and a few other plants, this layer contains most of

the stomata through which water vapor exits the leaf and gas exchange occurs. Vascular bundles form a network of veins throughout the leaf.

4. Plants that live many years may spend part of their lives in a juvenile phase, then shift to a mature or adult phase. The differences between juveniles and adults often are reflected in leaf size and shape, in the arrangement of leaves on the stem, or in a change from vegetative growth to a reproductive stage. Most woody plants must attain a certain size before their meristem tissue can respond to the hormonal signals that govern flower development.

Study Break 31.4

1. Most eudicots have a taproot system—a single main root, or taproot, that is adapted for storage and smaller branching lateral roots. As the main root grows downward, its diameter increases, and the lateral roots emerge along the length of its older, differentiated regions. Grasses and many other monocots develop a fibrous root system in which several main roots branch to form a dense mass of smaller roots. Fibrous root systems are adapted to absorb water and nutrients from the upper layers of soil, and tend to spread out laterally from the base of the stem.
2. The root apical meristem and the actively dividing cells behind it form the zone of cell division. Cells in the center of the root tip become the procambium; those just outside the procambium become ground meristem; and those on the periphery of the apical meristem become protoderm. The zone of cell division merges into the zone of elongation, most of the increase in a root's length occurs. Above the zone of elongation, cells may differentiate further and take on specialized roles in the zone of maturation.
3. Primary root growth produces a system of vascular pipelines extending from root tip to shoot tip. The root procambium produces cells that mature into the root's xylem and phloem. Ground meristem gives rise to the root's cortex, its ground tissue of starch-storing parenchyma cells that surround the stele. In many flowering plants, the outer root cortex cells give rise to an exodermis, a thin band of cells that may limit water losses from roots and help regulate the absorption of ions. The innermost layer of the root cortex is the thin endodermis, which helps control the movement of water and dissolved minerals into the stele. Between the stele and the endodermis is the pericycle, which gives rise to lateral roots. In some cells in the developing root epidermis the outer surface becomes extended into root hairs.

Study Break 31.5

1. Secondary growth processes add girth to roots and stems over two or more growing seasons. In plant species that have secondary growth, older stems and roots become more massive and woody through the activity of two types of lateral meristems. One of these meristems, the vascular cambium, produces secondary xylem and phloem. The other, the cork cambium, produces cork, a secondary epidermis that is one element of bark.
2. Vascular cambium consists of two types of cells—fusiform initials and ray initials. Fusiform initials are derived from cambium inside the vascular bundles and give rise to secondary xylem and phloem cells. Secondary xylem forms on the inner face of the vascular cambium, and secondary phloem forms on the outer face. Ray initials are derived from the parenchyma cells between vascular bundles. Their descendants form spoke-like rays of parenchyma cells—horizontal channels that carry water sideways through the stem. As the mass of secondary xylem inside the ring of vascular cambium increases, it forms hard tissue known as wood. Bark encompasses all the living and nonliving tissues between the vascular cambium and the stem surface. It includes the secondary phloem and the periderm, the outermost portion of bark that consists of cork, cork cambium, and secondary cortex.
3. Roots of some plant species also undergo secondary growth, but the ring of vascular cambium develops differently than it does in stems. When their primary growth is complete, these roots have a layer of residual procambium between the xylem and phloem of the stele. The vascular cambium arises in part from this residual cambium, and in part from the pericycle. Eventually, the cambial tissues arising from the procambium and those arising from the pericycle merge into a complete cylinder of vascular cambium. As in stems, the vascular cambium gives rise to secondary xylem to the inside and secondary phloem to the outside. As secondary xylem accumulates, older roots can become thick and woody.

Self-Test Questions

1. d 2. c 3. b 4. a 5. c 6. a 7. d 8. e 9. c 10. b

Chapter 32

Study Break 32.1

1. After an H^+ gradient is established (by proton pumping out of the cell), the resulting inward flow of H^+ down its concentration gradient provides the energy to actively transport other substances into the cell.
2. In symport two substances move in the same direction through a cell membrane; in antiport two substances cross the cell membrane in opposite directions. Plants take up organic substances and important anions such as potassium and nitrate by symport. Antiport commonly moves ions of calcium and sodium out of plant cells.
3. Water potential is potential energy stored in water. It is the driving force for osmosis, which in turn is responsible for the moment of water into and out of plant cells, including root cells.

Study Break 32.2

1. In the apoplastic pathway, water and dissolved substances don't pass through living root cells but instead move through the continuous network of adjoining cell walls and air spaces. When apoplastic water (and solutes) reach the endodermis, however, they must detour around the impermeable Casparian strip and pass through cells in order to move into the stele. The symplastic pathway passes through living cells. Water that diffuses into root cells moves in this pathway from cell to cell through plasmodesmata.
2. Epidermal cells of root hairs actively transport most mineral ions into root epidermal cells. These ions travel inward via the transmembrane pathway. Other ions may be dissolved in apoplastic water. They ultimately travel to the xylem in the symplast after crossing into and through endodermal cells of the Casparian strip. Once an ion reaches the stele it enters the xylem.

Study Break 32.3

1. In the cohesion–tension mechanism, water transport begins as water evaporates from the walls of mesophyll cells inside leaves and into the intercellular spaces. This water vapor escapes by transpiration through open stomata. As water molecules exit the leaf, they are replaced by others from the mesophyll cell cytoplasm. The water loss gradually reduces the water potential in a transpiring cell below the water potential in the leaf xylem. Water from the xylem in the leaf veins then follows the gradient into cells, replacing the water lost in transpiration.
2. Stomata open and close in response to changing environmental cues, such as light levels (detected via blue-light receptors), CO_2 concentration in the air spaces inside leaves, and the amount of water available to the plant. Stomata open when hydrogen ions are pumped out of guard cells, setting up the symport of H^+ and K^+ into the guard cells through ion channels. Water then follows by osmosis. Stomata close when H^+ pumping in guard cells ceases and K^+ moves out of guard cells, with water again following by osmosis. Through their ability to open and close, stomata help regulate water loss by plants and the uptake of carbon dioxide for photosynthesis.

Study Break 32.4

1. Translocation is the long-distance transport of substances in plants. The term

generally applies to the transport of organic compounds, mainly sucrose, in phloem. Transpiration is the evaporation of water from a plant's aerial parts, mainly leaves. This water moves from roots upward to aerial parts in the xylem.
2. The mechanism of pressure flow moves sucrose from a source (such as a leaf or stem) into sieve tubes. Pressure builds up at the source end of a sieve tube system as sucrose enters sieve tubes at sources and water follows by osmosis. Under high pressure, sucrose moves by bulk flow toward a sink (plant parts that take up sucrose as metabolic fuel), where the sugar is unloaded.

Self-Test Questions

1. b 2. d 3. a 4. e 5. a 6. b 7. e 8. d 9. c 10. b

Chapter 33

Study Break 33.1

1. Plants require relatively large amounts of macronutrients such as nitrogen, sulfur, potassium, and calcium, and trace amounts of micronutrients such as iron, chlorine, zinc, nickel, and copper.
2. Plants vary in their nutritional requirements. For example, as described in the text, leafy plants require more nitrogen and magnesium than other plant types do, and alfalfa, a grass, requires significantly more potassium than lawn grasses do. An adequate amount of an essential element for one plant also may be toxic for another. For these reasons, the nutrient content of soils is an important factor determining which plants grow well in a given location.

Study Break 33.2

1. Humus is important in soil because it generally contains nutrient-rich organic material and because it absorbs water, which contributes to the water-holding capacity of soil.
2. The amount of water that is available in soil to be taken up by plant roots depends primarily on the relative proportions of different soil components. Water moves quickly through sandy soils, while soils rich in clay and humus tend to hold the most water.
3. A plant's ability to absorb soil minerals depends partly on cation exchange, in which one cation, usually H^+, replaces a soil cation. As H^+ enters the soil solution, it displaces adsorbed mineral cations attached to clay and humus, freeing them to move into roots. Anions in the soil solution, such as nitrate (NO_3^-), sulfate (SO_4^{2-}), and phosphate (PO_4^-), generally move more readily into root hairs. Soil pH also affects the availability of some mineral ions because chemical reactions in very acid ($pH < 5.5$) or very alkaline ($pH > 9.5$) soils can trigger chemical reactions that bind various mineral cations in compounds that are insoluble in soil water.

Study Break 33.3

1. As described in this chapter and in Section 28.3, a mycorrhiza is a symbiotic association between a fungus and plant roots. Most plants form mycorrhizal associations, which facilitate the plant's ability to extract soil nutrients such as nitrogen and phosphorus. As with plant roots, mineral ions enter fungal hyphae by way of transport proteins. Some of the plant's sugars and nitrogenous compounds nourish the fungus, and as the root grows, it takes up a portion of the minerals that the fungus has secured. In some types of mycorrhizae the fungus actually lives inside cells of the root cortex.
2. Nitrogen fixation refers to the incorporation of atmospheric nitrogen into compounds, especially nitrate (NO_3^-), that plants can readily take up. Ammonification is a process in which soil bacteria known as ammonifying bacteria break down decaying organic matter and convert it to ammonium (NH_4^+). In nitrification, nitrifying bacteria oxidize NH_4^+ to NO_3^-. Inside root cells, absorbed NO_3^- is converted by a multistep process back to NH_4^+. In this form, it is rapidly used to synthesize organic molecules, mainly amino acids.
3. Associations with bacteria supply nitrogen to certain types of plants, such as legumes. The host plant provides organic molecules that the bacteria use for cellular respiration, and the bacteria supply NH_4^+ that the plant uses to produce nitrogenous molecules. In legumes the nitrogen-fixing bacteria reside in root nodules. Usually, a single species of nitrogen-fixing bacteria colonizes a single legume species, drawn to the plant's roots by chemical attractants (mainly flavonoids) that the roots secrete. By way of exchanged molecular signals, bacteria then are able to penetrate a root hair and form a colony inside the root cortex. Each cell in a root nodule may contain several thousand bacteria (now called bacteroids). The plant takes up some of the nitrogen fixed by the bacteroids, and the bacteroids utilize some compounds produced by the plant.

Self-Test Questions

1. e 2. c 3. d 4. c 5. b 6. c 7. a 8. d 9. a 10. e

Chapter 34

Study Break 34.1

1. The two alternating generations of plants are the sporophyte (spore-producing) and gametophyte (gamete-producing) generations.
2. Sporophytes produce spores that give rise to gametophytes. Gametophytes then may produce gametes; male gametophytes produce sperms, and female gametophytes produce eggs. In all seed plants the sporophyte is much larger and longer-lived than the gametophyte, and the gametophyte is protected within sporophyte tissues for all or part of its life. Gametophytes also are dependent upon the sporophyte for their nutrition.

Study Break 34.2

1. Flowers are specialized for reproduction. Before an angiosperm can produce a flower, biochemical signals (triggered in part by environmental cues such as day length and temperature) travel to the apical meristem of a shoot. In response, cells there change their activity: Instead of continuing vegetative growth, the shoot is modified into a floral shoot that will give rise to floral organs.
2. Pollen grains are the mature male gametophytes. They arise by the following steps:
   a. Spores that give rise to male gametophytes are produced in a flower bud's anthers.
   b. Diploid microsporocytes inside an anther's pollen sacs undergo meiosis; eventually each one produces four small haploid microspores.
   c. Microspores then divide by mitosis.
   d. One of the two resulting nuclei divides again by mitosis, yielding a three-celled immature gametophyte: two haploid sperm cells and a third cell that will control the development of a pollen tube after pollen lands on a receptive stigma.
   e. A mature male gametophyte consists of the pollen tube and sperm cells.
3. Female gametophytes develop inside ovules in a flower's carpels. They arise by the following steps:
   a. In an ovule, a diploid megasporocyte divides by meiosis, forming four haploid megaspores.
   b. Three (usually) of these megaspores disintegrate.
   c. The remaining megaspore undergoes three rounds of mitosis without cytokinesis. The result is a single large cell with eight nuclei arranged in two groups of four.
   d. One nucleus in each group migrates to the center of the cell
   e. After the cell undergoes cytokinesis a cell wall forms around these two polar nuclei, forming a large "central cell."
   f. A wall also forms around each of the remaining nuclei, and three of them, including an egg cell, cluster near the micropyle.
   g. The result is an embryo sac containing seven cells and eight nuclei. This sac is the mature female gametophyte.

Study Break 34.3

1. A pollen grain that lands on a compatible stigma absorbs moisture and germinates a pollen tube, which burrows through the stigma and style toward an ovule. Chemical cues from the two synergids cells help guide the pollen tube toward the egg. Before or during these events, the pollen grain's haploid sperm-producing cell divides by mitosis, forming two haploid sperm. When the pollen tube reaches the ovule, it enters through the micropyle and an opening forms in its tip. The two sperm are released into the cytoplasm of a disintegrating synergid. Next double fertilization occurs: typically, one sperm nucleus fuses with the egg to form a diploid ($2n$) zygote. The other sperm nucleus fuses with the central cell, forming a cell with a triploid ($3n$) nucleus. Tissues derived from the $3n$ cell are called endosperm.
2. As a seed matures, the embryo inside it develops a root–shoot axis with root apical meristem at one end and shoot apical meristem at the other end. Depending on the plant group, one or two cotyledons also develop. In monocots a single large cotyledon develops and stores endosperm; protective tissues arise around the root and shoot apical meristems. In eudicots, two endosperm-storing cotyledons form. Near the micropyle the radicle (embryonic root) attaches to the cotyledon at a region called the hypocotyl. Beyond the hypocotyl is the epicotyl, which has the shoot apical meristem at its tip and which often bears a cluster of tiny foliage leaves, the plumule. At germination, when the root and shoot first elongate and emerge from the seed, the cotyledons are positioned at the first stem node with the epicotyl above them and the hypocotyl below them.
3. Imbibition causes the seed coat to split, and water and oxygen move more easily into the seed. Metabolism switches into high gear as cells divide and elongate to produce the seedling. Enzymes that were synthesized before dormancy become active; other enzymes are produced as the genes encoding them begin to be expressed. The increased gene activity and enzyme production mobilizes the seed's food reserves in cotyledons or endosperm. Nutrients released by the enzymes sustain the developing seedling sporophyte until its root and shoot systems are established.

Study Break 34.4

1. Flowering plants may reproduce asexually (vegetatively) by fragmentation, in which cells in a piece of the parent plant dedifferentiate and then regenerate a whole plant; by apomixis, in which a diploid embryo develops from an unfertilized egg or from diploid cells in ovule tissue; or by the production of structures such as rhizomes or suckers from a nonreproductive plant part, typically meristem tissues in a bud on a root or stem.
2. Totipotency is the capacity of fully differentiated cells to dedifferentiate, return to an unspecialized embryonic state, and then develop into a fully functional mature plant. Plant tissue culture procedures trigger the development of a mass of dedifferentiated cells (a callus), some of which regain totipotency and develop into plantlets.

Study Break 34.5

1. A homeotic gene is a regulatory gene in the genome of an organism that encodes a transcription factor. In *Arabidopsis*, a eudicot, homeotic genes govern the development of the root and shoot tissue systems, as well as of floral organs.
2. The two basic mechanisms of plant morphogenesis are oriented cell division and cell expansion. Oriented cell division establishes the general shape of a plant organ, and cell expansion enlarges the cells in a developing organ in particular directions. They increase in circumference (girth) when new cell walls form parallel to the nearest plant surface (such as the surface of a stem or tree trunk) or when cell walls form at right angles both to the nearest surface and to the transverse plane.
3. Leaves arise through a developmental program that begins with gene-regulated activity in meristematic tissue. Hormones or other signals may arrive at target cells via the stem's vascular tissue, activating genes that regulate development. Small phloem vessels penetrate a young leaf primordium almost immediately after it begins to bulge out from the underlying meristematic tissue, followed by xylem. A growing primordium becomes cone-shaped (wider at its base than at its tip). Rapid mitosis in a particular plane in cells along the flanks of the cone (perpendicular to the surface in eudicots and parallel to the surface in monocots) produce the leaf blade characteristic of the particular species. Leaf tip cells typically are the first to stop dividing. By the time a leaf has expanded to its mature size, mitosis has ended and the leaf is a fully functional photosynthetic organ.

Self-Test Questions

1. c 2. a 3. b 4. e 5. d 6. c 7. b 8. e 9. c 10. c

Chapter 35

Study Break 35.1

1. Auxins, gibberellins, cytokinins, and brassinosteroids all promote the growth of plant parts, and ethylene stimulates cell division in seedlings. Abscisic acid is the major growth-inhibiting plant hormone.
2. Ethylene is a good example of a hormone that can stimulate or inhibit growth at various stages of the plant life cycle. In seedlings, it simultaneously slows elongation of the stem and stimulates cell divisions that increase stem girth. In mature plants of deciduous species it governs senescence (including fruit ripening) and the abscission of flowers, fruits, and leaves. Studies of brassinosteroids have revealed that this family of steroid hormones have different effects in different tissues—for example, promoting the elongation of vascular tissue and pollen tubes, but inhibiting elongation in roots.

Study Break 35.2

1. General plant responses to attack include mobilization of jasmonates and salicylic acid, systemin (in tomato), the hypersensitive response, PR proteins, and secondary metabolites (phytoalexins). Gene-for-gene recognition is a pathogen-specific response.
2. Salicylic acid (SA) is considered to be a general systemic response to damage because experiments show that when a plant is wounded, soon thereafter SA can be detected in a variety of its tissues.
3. While the hypersensitive response is underway, SA also is synthesized and operates in other defensive chemical pathways in a plant. This effect includes the synthesis of PR (pathogenesis-related) proteins that attack pathogenic cells.

Study Break 35.3

1. Directional light of blue wavelengths is the direct stimulus for phototropism. The most widely accepted scientific explanation for gravitropism is the sinking of amyoplasts in cells surrounding vascular bundles in response to gravity. Sinking amyloplasts may provide a mechanical stimulus that triggers a gene-guided redistribution of IAA. The changing auxin gradient in turn adjusts a plant's growth pattern.
2. Unlike tropisms, nastic movements occur in response to nondirectional stimuli, such as mechanical pressure resulting from an insect brushing against hairlike sensory structures in the leaves of a Venus flytrap plant.

Study Break 35.4

1. Plant responses to changes in photoperiod rely on different chemical forms of the blue-green pigment phytochrome. Daylight converts the inactive phytochrome (P_r) to an active form, (P_{fr}). When light levels fall, P_{fr} reverts to P_r. This switching mechanism helps regulate light-related processes such as photosynthesis.
2. Dormancy is an adaptive response because it attunes a plant's growth to the most favorable environmental conditions for survival.

Study Break 35.5

1. Some response pathways may directly stimulate gene transcription while others might inhibit gene expression. Some pathways have an intermediary step, in which the original signal triggers the synthesis of regulatory proteins that in turn promote or inhibit the expression of still other genes.
2. Second messengers boost cellular responses to a hormonal signal by setting in motion of cascade of activated protein kinases, which in turn activate cell proteins that carry out the cell's response.
3. Although Chapter 7 focuses on signal transduction pathways in animals, in fact the basics of some transduction pathways are similar in animals and plants. In both groups we see examples of one-step signal transduction (in which a signal exerts its effect by binding a cell receptor) and of signaling cascades such as second messenger systems.

Self-Test Questions

1. e 2. a 3. c 4. b 5. b 6. b 7. a 8. d
9. c 10. d

Chapter 36

Study Break 36.1

1. Multicellularity made it possible for animals to create an internal fluid environment for fulfilling the nutrient supply, waste removal, and osmotic balance needs of individual cells. As a result, multicellular organisms could evolve to occupy a variety of habitats, including dry terrestrial environments, in which single cells cannot survive. Multicellularity also allowed major life functions to be distributed among specialized groups of cells, with each group having a single activity. The specialized groups of cells are typically organized into tissues, the tissues into organs, and the organs into organ systems.
2. A tissue is a group of cells with the same structure and function. Cells in the tissue work together to perform one or more activities. An organ integrates two or more different tissues into a structure that carries out a specific function. An organ system coordinates the activities of two or more organs to carry out a major body function.

Study Break 36.2

1. Exocrine and endocrine glands are formed by epithelia. An exocrine gland remains connected to the epithelium by a duct, whereas an endocrine gland is suspended in connective tissue underlying the epithelium, with no ducts leading to the epithelial surface.
2. Loose connective tissue, fibrous connective tissue, cartilage, bone, adipose tissue, and blood.
3. Skeletal, cardiac, and smooth.

Study Break 36.3

See Figure 36.11.

Study Break 36.4

A stimulus—a change in the external or internal environment—starts the homeostatic mechanism. The stimulus is detected by a *sensor,* an *integrator* compares the environmental change with a set point, and the *effector* becomes activated by the integrator and functions to return the environmental parameter to the set point.

Self-Test Questions

1. b 2. a 3. e 4. c 5. b 6. a 7. c 8. e
9. e 10. d

Chapter 37

Study Break 37.1

1. A dendrite receives signals and conducts them toward the cell body. An axon conducts signals away from the cell body toward another neuron or effector.
2. An afferent neuron conducts information from its sensory receptors to interneurons. Efferent neurons conduct signals from interneuron networks to effectors, the muscles and glands that carry out the response. The afferent and efferent neurons constitute the peripheral nervous system (PNS). Interneurons process information received from afferent neurons and send a response to the efferent neurons. Interneurons form the brain and spinal cord, the central nervous system (CNS).
3. In an electrical synapse, the plasma membrane of the axon terminal of the presynaptic cell is in direct contact with the postsynaptic cell, allowing ions to pass directly between the cells when an electrical impulse arrives. In a chemical synapse, the presynaptic and postsynaptic cells are separated by a small gap. Neurotransmitters released from the presynaptic cell diffuse across the gap and bind to receptors in the plasma membrane of the postsynaptic cell. If enough neurotransmitter molecules bind to those receptors, the postsynaptic cell generates a new electrical impulse, which travels along its axon to the next neuron or effector in the circuit.

Study Break 37.2

1. At the peak of an action potential, the plasma membrane of the neuron enters a short refractory period, in which the threshold for generation of an action potential is much higher than normal. Only the region in front of the action potential can fire, meaning that the impulse can only move in one direction, that is, toward the axon tip. The refractory period remains in effect until the membrane again reaches the resting potential. By that time, the action potential has moved too far away to cause a second action potential in the same region.
2. Neurons insulated with myelin sheaths have gaps in the sheaths called nodes of Ranvier, where the axon membrane is exposed to extracellular fluids. The inward movement of sodium ions at a node produces depolarization and an action potential, but the adjacent myelin sheath prevents the excess positive ions from exiting through the membrane. Instead, they diffuse rapidly to the next node where they cause depolarization, inducing an action potential there. Continuation of this process allows the action potential to jump rapidly along the axon from node to node, at a faster rate than a nonmyelinated neuron of the same diameter.

Study Break 37.3

1. A neurotransmitter is synthesized in a neuron, is released into the synaptic cleft from a presynaptic axon terminal, and binds to receptors in the plasma membrane of the postsynaptic cell. Depending on the type of receptor, a neurotransmitter either stimulates or inhibits the generation of action potentials.
2. A direct neurotransmitter, like all neurotransmitters, is stored in synaptic vesicles in the cytoplasm at an axon terminal. When an action potential arrives at the terminal, the change in membrane potential opens voltage-gated Ca^{2+} channels in the axon terminal, allowing Ca^{2+} to flow back into the cytoplasm. The rise in Ca^{2+} concentration triggers the release of the neurotransmitters into the synaptic cleft by exocytosis. Direct neurotransmitters diffuse across the synaptic cleft and open or close ligand-gated ion channels in the postsynaptic neuron's membrane. Most of the channels regulate Na^+ or K^+ movement through the membrane, although some regulate Cl^-. Depending on the ion flow, action potentials are either stimulated or inhibited in the postsynaptic neuron.

Study Break 37.4

One way a postsynaptic neuron integrates signals is through summation of EPSPs and IPSPs that alter the neuron's membrane potential. EPSPs move the membrane potential toward the threshold for an action potential, while IPSPs move the membrane potential away from the threshold for an action potential. The final change in membrane potential depends on the particular array of EPSPs and IPSPs received. The patterns of synaptic connections made by a neuron also contribute to integration.

Self-Test Questions

1. d 2. b 3. a 4. c 5. e 6. a 7. c 8. b
9. e 10. a

Chapter 38

Study Break 38.1

1. A nerve net is a loose meshwork of neurons organized in a radial pattern to reflect the radial symmetry of the animal in which they are found. Nerves are bundles of axons surrounded by connective tissue. Nerve cords are bundles of nerves.
2. Cephalization is the formation of a distinct head region containing ganglia, which form a major central control center or brain, and major sensory structures. Cephalization is found in more complex invertebrates and in all vertebrates.
3. The embryonic hindbrain subdivides into the metencephalon, which gives rise to the cerebellum and the pons, and the myelencephalon, which gives rise to the medulla. The cerebellum integrates sensory signals from eyes, ears, and muscle spindles with motor signals from the forebrain. The pons is a center for information transfer between the cerebellum and the higher integrating centers of the forebrain. The medulla controls vital tasks such as respiration and blood circulation.

Study Break 38.2

(a) Sympathetic nervous system; this is the classic "fight or flight" scenario for this autonomic nervous system division.
(b) Parasympathetic nervous system.

Study Break 38.3

1. The blood-brain barrier prevents most substances dissolved in the blood from entering the cerebrospinal fluid and protects the brain and spinal cord from infectious agents, such as bacteria and viruses, and from toxic substances that may be in the blood. With an incompletely developed blood-brain barrier, it would be important to take precautions that infectious agents, toxic substances, or any chemicals that could affect brain development do not reach the brain.
2. The cerebellum is an outgrowth of the pons. However, it is structurally and functionally separate from the brain stem. The cerebellum has extensive connections with other parts of the brain, through which it receives sensory inputs from receptors in muscles and joints, from balance receptors in the inner ear, and from the receptors of touch, vision, and hearing. The cerebellum integrates the various sensory signals and compares them with signals from the cerebrum that control voluntary body movements. Information flow from the cerebellum to the cerebrum, brain stem, and spinal cord modifies and fine-tunes movements of the body. In humans, the cerebellum also is involved in the learning and memory of motor skills.

   The cerebral cortex is a thin layer of gray matter that forms the surface of the cerebrum. The more evolutionarily advanced an animal is, the more folded its cerebral cortex. The cerebral cortex contains sensory areas that receive and integrate sensory information of many kinds, including touch, pain, temperature, pressure, hearing, vision, smell, and taste. In addition there are motor areas that are involved in controlling body movements and position, and association areas, which integrate information from the sensory areas and send responses to the motor area. Most higher functions of the human brain, including critical thinking, abstract thought, musical ability, and aspects of personality, involve activities of many regions of the cerebral cortex.

Study Break 38.4

Short-term memory involves transient changes in neurons. Short-term memory loss resulting from aging may be caused by a loss of control of the mechanisms involved, or, more likely, by a loss or degeneration of the neurons that constitute the short-term memory system.

Learning involves first storing memories. If the short-term memory system is faulty, perhaps due to loss or degeneration of neurons, then the transfer of short-term memories to the long-term memory system will be impaired. Another possibility is a loss or degeneration of the neurons responsible for long-term memory.

Self-Test Questions

1. d 2. b 3. b 4. b 5. c 6. e 7. a 8. c
9. c 10. a

Chapter 39

Study Break 39.1

1. Sensory transduction is the conversion of a stimulus into a change in membrane potential.
2. Signals from sensory receptors proceed by afferent neurons to the CNS. The afferent neurons from particular receptors are routed to specific regions of the CNS. Processing of the incoming signals by those regions gives the "sense" of the stimulus—a smell, pain, and so forth.

Study Break 39.2

1. Proprioceptors detect stimuli that are processed to provide the animal with information about movements and position of the body.
2. All proprioceptors are stimulated by a mechanical force, hence they are mechanoreceptors. For example, when a vertebrate moves, fluid moving in the vestibular apparatus bends sensory hairs, which generate action potentials in afferent neurons that synapse with the hair cells. The signal is sent to the brain, which integrates the information into a perception of movement.

Study Break 39.3

1. Cephalopods have mechanoreceptors on their head and tentacles, similar to the lateral line of fishes. These mechanoreceptors detect vibrations in the water.

   Many insects have sensory receptors in the form of hairs or bristles that vibrate in response to sound waves, while some insects (for example, moths, grasshoppers, and crickets) have auditory organs on either side of the abdomen or on the first pair of walking legs.
2. Vibrations representing sound frequencies are transmitted into the fluid-filled inner ear. They travel through the inner ear and cause the basilar membrane to vibrate in response, bending the sensory hair cells and stimulating them to release a neurotransmitter that triggers action potentials in afferent neurons leading from the inner ear. The key to "hearing" different frequencies of sound is that the vibrations from a particular sound frequency cause the basilar membrane to vibrate maximally at one particular location. Each frequency of sound waves causes hair cells in a specific region of the basilar membrane to initiate action potentials, and that information is sent to the brain, which integrates it into a perception of the sound stimulus.

Study Break 39.4

(a) A photopigment is an association of retinal with one of several different opsin proteins.
(b) A photoreceptor is a receptor specialized for detection of colors (different wavelengths of light).
(c) A ganglion cell's receptive field is the specific set of photoreceptors that send signals to that cell. Receptive fields are usually circular and vary in size; the smaller the receptive field, the more precise the information sent to the brain, and the sharper the image.

Study Break 39.5

1. Most odor perceptions arise from a combination of different olfactory receptors, which are located in the nasal cavities in humans. We have about 1000 different olfactory receptor types, each of which responds to a different class of chemicals. The stimulated olfactory receptors send signals via the olfactory bulbs to the olfactory centers of the cerebral cortex, where the signals are interpreted as particular smells.
2. Chemicals binds to taste receptors and generate signals. Signals from taste receptors are relayed to the thalamus. From there, some signals go to gustatory centers in the cerebral cortex where they are integrated to produce taste perception. Other signals go to the brain stem and limbic system, which links tastes to involuntary visceral and emotional responses, such as sensations of pleasure or revulsion.

Study Break 39.6

The other sensory receptors involve specialized receptor structures. Afferent neurons synapse with the receptors. When a receptor is stimulated, the change in membrane potential (sensory transduction) is transmitted to the afferent neuron, which transmits the signal to the interneuron networks of the CNS. By contrast, both thermoreceptors and nociceptors consist of free nerve endings formed by the dendrites of afferent neurons, with no specialized receptor structures involved.

Study Break 39.7

Electroreceptors are used in aquatic vertebrates for electrolocation (locating other animals such as prey), electrocommunication (communicating with other members of the same species), and killing prey (involving high voltage discharge).

Self-Test Questions

1. c 2. a 3. b 4. e 5. c 6. b 7. d. 8. a 9. e 10. d

Chapter 40

Study Break 40.1

1. A *hormone* is a signaling molecule secreted by one group of cells and transported through the circulatory system to other, target cells, whose activities they change. Specific target cells react to the hormone because they carry receptors for the hormone. The best-known hormones are secreted by cells of the endocrine system and elicit a response in target cells that have receptors for the hormone. A *neurohormone* is a type of hormone. Specifically, it is a signaling molecule released into the circulatory system from stimulated neurosecretory neurons. Like other hormones, it moves around the body in the circulatory system and causes a response in target cells with receptors for the neurohormone.
2. In classical endocrine signaling, hormones are secreted by endocrine glands into the blood or extracellular fluid and exert their effects on distant target cells.

   In neuroendocrine signaling, neurosecretory neurons respond to and conduct signals, but instead of synapsing with target cells, release neurohormones into the circulatory system. The neurohormones exert their effects on distant target cells.

   In paracrine regulation, a cell releases a signaling molecule—a local regulator—that diffuses through the extracellular fluid and acts on nearby cells, rather than on target cells at a distance.

   Autocrine regulation is similar to paracrine regulation, but the released local regulator acts on the same cells that produced it.

Study Break 40.2

1. Glucagon is a peptide hormone, which triggers a response by binding to a surface receptor. Glucagon binding activates the receptor, which triggers a signal transduction pathway inside the cell, leading to phosphorylation of target proteins. The altered activities of the phosphorylated proteins produce responses in the target cells, in this case the breakdown of glycogen in liver cells to glucose.

   Aldosterone is a steroid hormone, a hydrophobic molecule that passes though the plasma membrane and binds to an internal receptor in the cytoplasm or nucleus, activating it. The hormone-activated receptor complex binds to control sequences in the DNA that the receptor recognizes and either activates or inhibits transcription of the associated target genes. In short, through binding to a receptor, the hormone affects transcription of specific genes in target cells.
2. A target cell could respond to different hormones if it carries receptors for those different hormones. Turning this around, a target cell will respond only to one hormone if it just has the receptor for that hormone. The same hormone could produce different effects in different cells if there are different receptors for that hormone, each triggering a distinct response pathway.

Study Break 40.3

1. The hypothalamus and anterior pituitary gland are connected by nerve and vascular tissues. The hypothalamus releases peptide neurosecretory hormones into the linking blood vessels. These hormones are tropic hormones that regulate the secretion of peptide hormones by the anterior pituitary gland. The pituitary hormones regulate several key body systems.
2. A tropic hormone regulates hormone secretion by another endocrine gland. The hormone or hormones produced by that gland produces the response. A nontropic hormone acts directly on target cells and is directly responsible for the response.

Study Break 40.4

1. Parathyroid hormone stimulates the dissolution of calcium and phosphate ions from bone and their release into the bloodstream.
2. The adrenal medulla secretes epinephrine and norepinephrine. Epinephrine prepares the body for handling stress or physical activity by, among other actions, (1) increasing heart rate; (2) breaking down glycogen and fats, thereby releasing glucose and fatty acids into the blood for fuel; (3) dilating blood vessels in the heart, skeletal muscles, and lungs to increase blood flow; (4) constricting blood vessels elsewhere, thereby raising blood pressure, reducing blood flow to the intestine and kidneys, and inhibiting smooth muscle contraction, which reduces water loss and slows down the digestive system; and (5) dilating airways in the lungs, thereby increasing air flow.

   Norepinephrine has similar effects to epinephrine on heart rate, blood pressure, and blood flow to the heart muscle. In contrast to epinephrine, norepinephrine causes blood vessels in skeletal muscles to contract.
3. Glucocorticoids and mineralocorticoids are the two major classes of steroid hormones secreted by the adrenal cortex. Glucocorticoids help maintain the concentration of glucose and other fuel molecules in the blood, and mineralocorticoids regulate the levels of Na^+ and K^+ in blood and the extracellular fluid.
4. Estradiol is an estrogen and progesterone is a progestin; both are steroid hormones Estradiol produced by the ovaries stimulates maturation of sex organs at puberty and development of secondary sexual characteristics. Progesterone, also produced by the ovaries, prepares and maintains the uterus for implantation of a fertilized egg and the growth and development of an embryo.

Study Break 40.5

In general, invertebrates have fewer hormones, regulating fewer body processes and responses, than vertebrates do. Peptide and steroid hormones are produced in both invertebrates and vertebrates. However, most of those hormones are different in structure and molecular function in the two groups, even though the reaction pathways stimulated by the hormones are the same.

Self-Test Questions

1. b 2. d 3. a 4. c 5. c 6. e 7. a 8. b 9. d 10. b

Chapter 41

Study Break 41.1

1. The tip of an axon of an efferent neuron makes a synapse with a muscle fiber called a neuromuscular junction. When an action potential arrives at that junction, the axon tip releases acetylcholine, which triggers an action potential in the muscle fiber that moves in all directions over its surface, and penetrates to the interior of the fiber through T tubules. When the action potential reaches the end of the T tubules, ion channels are opened in the sarcoplasmic reticulum that allow calcium ions to flow from the sarcoplasmic reticulum into the cytosol. Troponin then binds the ion and undergoes a conformation change that allows tropomyosin to enter the grooves in the actin helix of the thin filaments. As a result, the myosin-binding sites are uncov-

ered, and the crossbridge cycle is turned on, leading to muscle contraction.

2. In the sliding filament mechanism of muscle contraction, the thin filaments on each side of a sarcomere slide over the thick filaments toward the center of the A band, thereby bringing the Z lines closer together. A crossbridge cycle is responsible for the contraction. At the beginning of the cycle, the myosin crossbridge has an ATP bound to it and is not in contact with the actin of the thin filament. The ATP is hydrolyzed, causing the myosin crossbridge to bend away from the tail and bind to a myosin-binding site on the thin filament that was uncovered in response to the release of calcium ions into the cytosol from the sarcoplasmic reticulum. When the myosin crossbridge binds to the actin, the crossbridge snaps back toward the myosin tail to produce the power stroke that pulls the thin filament over the thick filament. ADP and phosphate are released from the crossbridge in this step. A new molecule of ATP now binds to the crossbridge, causing the myosin to detach from the actin. The cycle then repeats.

Study Break 41.2

1. A hydrostatic skeleton provides support to the body or body part through muscles acting on compartments filled with fluid under pressure. In mollusks, the tube feet of sea stars and sea urchins are hydrostatic skeletal structures. In vertebrates, the penis is a hydrostatic skeletal structure.

   An exoskeleton is a rigid external body covering. The force of muscle contraction against the covering provides support to the body or part of the body. Many mollusks, such as clams and oysters, have exoskeletons consisting of a shell secreted by glands in the mantle. In vertebrates, the shell of a turtle or tortoise, and the bony plates in the skin, the abdominal ribs, the collar bones, and most of the bony skull of the American alligator are exoskeletal structures.

   An endoskeleton consists of internal body structures such as bones. The force of muscle contraction against the internal body structures provides support. In mollusks, the mantle of squids and cuttlefish, the case of cartilage surrounding and protecting the squid brain, and segments of cartilage supporting the gills and siphon in squids and octopuses are endoskeletal structures. In vertebrates, the endoskeleton is the primary skeletal system.

2. The bones of the vertebrate endoskeleton provide support for the body and body parts, protect key internal organs, store calcium and phosphate ions, and are the sites where new blood cells form.

Study Break 41.3

1. Synovial joints consist of the ends of two bones enclosed by a fluid-filled capsule of connective tissue. Fluid within the capsule, and a smooth layer of cartilage over the ends of the bones, enable the bones to slide easily as the joint moves. In these joints, ligaments extend across the joints across the capsule. The ligaments serve to confine the motion of the joint and protect it to an extent from deleterious effects of heavy loads.

   Cartilaginous joints have no fluid-filled capsule surrounding them, and the bones involved are covered with cartilage. The bones of the joint are covered by and connected with fibrous connective tissue. Cartilaginous joints are less movable than synovial joints.

   Fibrous joints have stiff fibers of connective tissue joining the bones. As a result, the bones show little or no movement.

2. Antagonistic muscle pairs are muscles arranged so that bones can be extended, flexed, or rotated in opposite directions around a joint. For example, in humans, the biceps and triceps muscles are antagonistic muscle pairs.

Self-Test Questions

1. c 2. e 3. a 4. d 5. b 6. b 7. c 8. a 9. d 10. b

Chapter 42

Study Break 42.1

1. The three basic features of animal circulatory systems are (1) a specialized fluid medium, exemplified by the blood of vertebrates, for transporting molecules; (2) a muscular heart for pumping the fluid; and (3) tubular vessels for distributing the fluid pumped by the heart.

2. In an open circulatory system, there is no distinction between blood and interstitial fluid. Vessels from the heart release hemolymph directly into body spaces and the fluid is subsequently collected and reenters the heart. In a closed circulatory system, blood is channeled in blood vessels leading to and from the heart and is distinct from the interstitial fluid.

   A limitation of an open circulatory system is that most of the fluid pressure generated by the heart dissipates when the blood is released into the body spaces. Consequently, blood flows relatively slowly. Humans could not function as we do with such a system, because we would not be able to distribute oxygen efficiently throughout the body; nor would we be able to eliminate the wastes we produce.

3. Atria are chambers of the heart that receive blood returning to the heart. Ventricles are chambers of the heart that pump blood from the heart. Arteries are vessels of circulatory systems that conduct blood away from the heart at relatively high pressure. Veins are vessels of circulatory systems that carry blood back to the heart.

4. The systemic circuit of the body's circulatory system is the circuit from the heart to most of the tissues and cells of the body and back to the heart. The pulmocutaneous circuit goes from the heart to the skin and lungs or gills in amphibians and back to the heart. The pulmonary circuit goes from the heart to the lungs and back to the heart.

Study Break 42.2

1. Pluripotent stem cells in the red bone marrow are the origin of erythrocytes. In humans and other mammals, erythrocytes lose their nucleus, cytoplasmic organelles, and ribosomes as they mature, becoming essentially a membrane-bound hemoglobin reservoir that is not capable of protein synthesis. At the end of their life span—about 4 months—erythrocytes are engulfed by macrophages, a type of leukocyte, in the spleen, liver, and bone marrow.

2. Low oxygen content triggers a negative feedback mechanism to increase erythrocytes in the blood. The kidneys are stimulated to synthesize the hormone erythropoietin (EPO), which stimulates stem cells in the bone marrow to increase erythrocyte production. EPO synthesis is stopped when the oxygen content of the blood rises above normal levels.

3. Leukocytes are the first line of defense against invading organisms, eliminate dead and dying cells from the body, and remove cellular debris.

   Platelets assist in blood clotting. When blood vessels are damaged, platelets stick to the collagen fibers exposed to leaking blood and recruit other platelets to the site. Eventually a plug forms at the site, sealing off the damaged area.

Study Break 42.3

1. The right atrium receives blood in the systemic circuit returning to the heart in vessels coming from the entire body, with the exception of the lungs. The superior vena cava is the vessel draining blood from the head and forelimbs, and the inferior vena cava is the vessel draining blood from the abdominal organs and the hind limbs. The right atrium pumps blood into the right ventricle.

   The right ventricle receives blood from the right atrium, and pumps it into the pulmonary arteries going to the lungs, beginning the pulmonary circuit.

   The left atrium receives blood returning from the lungs in the pulmonary veins, completing the pulmonary circuit. The left atrium pumps this blood into the left ventricle.

   The left ventricle pumps the blood received from the left atrium into the aorta where the blood begins its path in the systemic circuit.

2. Systole is the period of contraction and emptying of the heart. Diastole is the pe-

riod of relaxation and filling of the heart between contractions.

3. Neurogenic hearts beat under the control of signals from the nervous system. If the signals cease, this type of heart stops beating. Myogenic hearts maintain a contraction rhythm without signals from the nervous system. In the event of a serious trauma to the nervous system, this type of heart keeps beating.
4. Pacemaker cells of the SA node undergo regularly timed depolarizations which initiate waves of contraction that travel over the heart. The waves stop before the they can go to the bottom part of the heart as they encounter an insulating layer of connective tissue. The contraction signals at this point excite cells of the atrioventricular node, which is located in the heart wall between the right atrium and right ventricle. The signal from the AV node passes to the bottom of the heart along Purkinje fibers where it induces a wave of contraction that begins at the bottom of the heart and moves upwards, expelling blood from the ventricles into the aorta and pulmonary arteries.

Study Break 42.4

1. Blood flow through capillary beds is controlled by contraction of smooth muscles in arterioles, and contraction of precapillary sphincters that are at the junctions of capillaries and arterioles.
2. In most body tissues, there are small spaces between capillary endothelial cells, but in the brain the capillary endothelial cells are tightly sealed together, forming a blood–brain barrier.
3. The two major mechanisms are diffusion along concentration gradients and bulk flow. Diffusion along concentration gradients occurs both through the spaces between the capillary endothelial cells and through the plasma membranes. The direction of movement of the molecule or ion depends, then, on the concentration gradient. For instance, oxygen and glucose, which are at higher concentrations in the capillaries, diffuse into the interstitial fluid and then into the cells of the tissues. Carbon dioxide, by contrast, is at a higher concentration in the interstitial and therefore diffuses into the capillaries.

   Bulk flow carries water, ions, and molecules out of the capillaries. Driven by the pressure of the blood, which is higher than the pressure of the interstitial fluid, bulk flow occurs through the spaces between capillary endothelial cells.
4. Atherosclerotic plaques are thickened deposits of cholesterol-rich fatty substances, smooth muscle cells, and collagen that form at damaged sites (lesions) within arteries and veins.

   In general, they form when the endothelial cells lining the blood vessels become damaged, leading to exposure of the underlying smooth muscle tissue. A cycle of injury and repair leads to lesions in the blood vessels which are the potential sites for atherosclerotic plaque formation. Contributing factors to atherosclerotic plaque formation are conditions that cause lesions, such as bacterial and viral infections, hypertension, smoking, and a diet that is high in fats.

Study Break 42.5

1. Arterial blood pressure must be regulated within limits to provide sufficient blood flow for the brain and other tissues, and to prevent damage to blood vessels, tissues, and organs that would occur at high blood pressures.
2. The three main mechanisms for regulating blood pressure are (1) controlling cardiac output (the pressure and amount of blood pumped by the ventricles), (2) controlling the degree of the blood vessels (mostly the arterioles), and (3) controlling the total blood volume.
3. Epinephrine raises blood pressure. The hormone does so by increasing the strength and rate of the heart rate, and stimulating vasoconstriction of arterioles in certain parts of the body.

Study Break 42.6

Lymph is interstitial fluid, an aqueous solution containing molecules, ions, infecting bacteria, damaged cells, cellular debris, and lymphocytes.

Lymph capillaries consist of a single layer of endothelial cells. Interstitial fluid enters these capillaries at sites where the endothelial cells overlap when the pressure of the interstitial forces flaps open. The openings produced are large enough for cells to enter.

Self-Test Questions

1. d 2. e 3. b 4. b 5. c 6. a 7. e 8. e 9. d 10. a

Chapter 43

Study Break 43.1

1. Mucus layers are physical barriers to pathogens. Mucus layers may contain toxins and other chemicals that kill pathogens. The aqueous environment in contact with an epithelial surface may inhibit or kill pathogens by being acidic, or by containing enzymes or bile juices.
2. Innate immunity provides a nonspecific, immediate response to a pathogen. It is the first response system that comes into play when a pathogen is encountered, but it retains no memory of the encounter. The innate immune responses include inflammation and specialized cells that attack pathogens.

   Adaptive immunity, by contrast, provides a specific response to a pathogen, and it retains a memory of exposure to the pathogen so that it can respond more quickly to future attacks. The adaptive immune response takes several days to become protective, in contrast to minutes for an innate immune response.

Study Break 43.2

1. The inflammatory response is characterized by heat, pain, redness, and swelling at the infection site.
2. Heat, redness, and swelling: Inflammation-mediating molecules released at the infection site cause the dilation of local blood vessels and increase their permeability. As a result, blood flow increases and fluid leaks from the blood vessels. Heat, redness, and swelling are direct consequences of these effects.

   Pain: Pain is caused by the migration of macrophages and neutrophils to the infection site, and their activities at the site.
3. The complement system is a nonspecific defense mechanism involving a group of more than 30 interacting soluble plasma proteins. The proteins are activated by molecules on the surface of pathogens. Activated complement system proteins participate in a cascade of reactions, producing membrane attack complexes that insert into the plasma membrane of many types of bacterial cells and coating other types of cells with fragments of the complement proteins. The membrane attack complexes create pores that lead to loss of osmotic balance, which causes the pathogens to lyse. Pathogens coated with complement protein fragments are recognized by phagocytes that then engulf and destroy the pathogens.
4. Viral pathogens do not have surface molecules that are distinct from those of the host. The three main strategies used by the host in innate immunity are RNA interference, interferon, and natural killer cells.

Study Break 43.3

1. The antibody-mediated immune response uses antibodies secreted by plasma cells (differentiated from activated B cells) to target antigens in various body fluids. The cell-mediated immune response uses cytotoxic T cells to target and kill cells infected with a pathogen.
2. An antibody is a multi-subunit protein. It is a member of the immunoglobulin (Ig) family of proteins. It consists of four polypeptides, two of which are identical heavy chains and two of which are identical light chains. Structurally an antibody looks like a Y. Each of the two arms of the Y involves pairing of a light chain with part of the heavy chain. The tail of the Y consists of the remaining parts of the heavy chains paired together. Each heavy chain and each light chain has a variable region and a constant region. The variable regions represent the top half of the arms of the Y and constitute two identical binding sites the antibody has for the antigen with which it can react.
3. DNA rearrangements of genes during differentiation are responsible for gener-

ating diverse light-chain genes and heavy-chain genes for the B-cell receptor and for the T-cell receptor. In brief, there are many different V DNA segments, one C DNA segment, and a few different J (joining) DNA segments. During differentiation, a light-chain gene is assembled from one V segment, one J segment and the C segment, creating a gene that can be transcribed. Transcription of the gene produces a pre-mRNA that is spliced to remove introns and generate the translatable mRNA. The various combinations of segments plus imprecision in the joining at the DNA level of the V and J segments produces many types of light-chain genes. A similar DNA rearrangement process occurs for heavy-chain genes. Together, this results in tremendous variability in the antibody molecules that can be made.

4. Clonal selection is the process by which an antigen stimulates the production of a clone of B-cell-derived plasma cells that secrete antibodies against that antigen. Clonal selection accounts for the rapid response, specific action, and diversity of antibodies seen for the adaptive immune system. It also accounts for immunological memory through the production of memory B cells and memory T cells.
5. Immunological memory is the aspect of adaptive immunity that allows for a rapid, intense immune reaction to develop if the body encounters an antigen it has seen before.

Study Break 43.4

1. Immunological tolerance, a feature of the adaptive immune system, protects the body's own molecules from attack by its immune system. B cells and T cells are involved in the development of immunological tolerance, but the exact process is not understood.
2. Basically, an allergy results from an overreaction of the immune system to a particular antigen. The substances responsible for allergic reactions are a class of antigens known as allergens. Allergens stimulate B cells to secrete IgE antibodies in high amounts. IgE binds to mast cells, which are then induced to oversecrete histamine. Histamine contributes to inflammation, and at high amounts, the histamine-induced inflammation can be severe and even life threatening.

Study Break 43.5

Invertebrates lack the adaptive immunity system seen in mammals; they have no B or T cells, for instance. The invertebrate immune defenses, like the innate immunity system of mammals, are nonspecific. For example, all invertebrates have phagocytic cells that engulf pathogens, and many invertebrates produce antimicrobial proteins such as lysozyme. Many invertebrates produce proteins of the immunoglobulin family. While these are not antibody molecules, in some invertebrates they provide a protective function against some pathogens.

Self-Test Questions

1. a 2. c 3. b 4. e 5. e 6. d 7. b 8. c 9. a 10. d

Chapter 44

Study Break 44.1

1. The respiratory medium is the environmental source of oxygen, and the repository for released carbon dioxide. For aquatic animals, water is the respiratory medium and for terrestrial animals it is air. The respiratory surface is the layer of epithelial cells between the body and the respiratory medium. Gas exchange occurs across the respiratory surface—oxygen in and carbon dioxide out of the body. In certain small animals, the body surface itself is the respiratory surface. Among larger animals, aquatic animals use gills, insects use tracheal systems, and terrestrial animals use lungs as the respiratory surface.
2. The advantage of water over air as a respiratory medium is that it enables the respiratory surface readily to remain wet at all times. Two key advantages of air over water as a respiratory medium are (1) there is much more oxygen in air than in water; and (2) air is much less dense and less viscous than water, so significantly less energy is needed to move air over the respiratory surface than to move water.

Study Break 44.2

1. The advantages of gills to water-breathing animals over skin breathing are greater efficiency of gas exchange, the ability to live in more diverse habitats, and the potential to achieve a greater body mass.
2. In countercurrent exchange, the respiratory medium flows in the opposite direction of the blood flow under the respiratory surface. Examples are the flow of water over the gills in sharks, fishes, and some crabs and the one-way flow of air through the lungs of birds; all these are opposite to the flow of blood. The advantage of this mechanism is that it maximizes the amounts of oxygen and carbon dioxide exchanged with the respiratory medium.
3. In the tracheal system, fine branches called tracheoles end in tips that are in contact with body cells. The tracheole tips are filled with fluid, and gas exchange occurs through the fluid and the plasma membrane of the body cells in contact with the tips. Air enters the tracheal system through spiracles on the body surface. Movement of air through the tracheal system occurs by muscle-driven contractions and expansions of air sacs within the system.
4. In positive pressure breathing, gulping, swallowing, or pumping action forces air into the lungs. In negative pressure breathing, muscular activity expands the lungs, lowering the pressure of air in the lungs and thereby causing air to be pulled inward.

Study Break 44.3

1. Contraction of the diaphragm and the external intercostal muscles pulls the ribs upward and outward, expanding the chest and lungs. By this negative pressure mechanism, the air pressure within the lungs is lower than outside of the body. The higher outside pressure drives air into the lungs, expanding and filling them. Relaxation of the diaphragm and the intercostal muscles reverses the pressure condition, and the elastic recoil of lungs expels the air.
2. For conscious exhalation, you contract the internal intercostal muscles between the ribs to pull the diaphragm down while simultaneously contracting abdominal muscles to compress the abdominal organs and force them upward against the diaphragm to expel air from the lungs.
3. The most important feedback stimulus for breathing is the level of carbon dioxide in the blood.
4. The chemoreceptors in the medulla respond to carbon dioxide levels in the blood. If carbon dioxide levels rise, the medulla responds to trigger an increase in the rate and depth of breathing. The opposite response is triggered if carbon dioxide levels fall.

Study Break 44.4

1. Hemoglobin is present in red blood cells. Oxygen diffuses from alveoli into the plasma solution in the capillaries and then into the red blood cells. In the red blood cells, oxygen binds to hemoglobin, thereby lowering the partial pressure of oxygen in the plasma. This leads to more oxygen molecules diffusing down the oxygen concentration gradient from alveolar air to blood.

   The binding of oxygen to hemoglobin is reversible. Further, the affinity of hemoglobin for oxygen increases as the partial pressure of oxygen increases and vice versa. This property is important for determining the release of oxygen from hemoglobin keyed to tissue requirements.
2. Carbon monoxide has much greater affinity for hemoglobin than oxygen does, and so it displaces oxygen from hemoglobin. This leads to a reduction in the amount of oxygen carried in the blood. Since the brain does not monitor oxygen levels, and other receptors do not respond until blood oxygen levels are critically low, victims can easily lapse into unconsciousness and death.

Study Break 44.5

- More blood per unit body weight than land animals

- Additional red blood cells, many stored in the spleen and released during a dive
- More myoglobin in muscles than land animals
- Slowing of the heart by as much as 90%
- Reduction of blood circulation to internal organs and muscles by up to 95%
- Retention of lactic acid in the muscles with no release into the blood until the animal returns to the surface

Self-Test Questions

1. e 2. d 3. a 4. c 5. a 6. d 7. b 8. d 9. e 10. a

Chapter 45

Study Break 45.1

1. Carnivores primarily eat other animals as their source of organic materials. Herbivores primarily eat plants as their source of organic materials. Omnivores obtain their organic materials from any source.
2. Essential nutrients are the amino acids, fatty acids, vitamins, and minerals that the animal cannot make itself and must obtain from its diet. The list of essential nutrients varies among animal types.
3. Deposit feeders pick up or scrape particles of organic matter from solid material they live in or on. Suspension feeders ingest small organisms that are suspended in water. Depending on the animal, the ingested organisms may be bacteria, protozoa, algae, or small crustaceans, or fragments of those organisms.

Study Break 45.2

1. Extracellular digestion takes place outside body cells, either in a pouch or a tube enclosed by the body, whereas intracellular digestion takes place within cells.
2. (1) Mechanical processing, to break up the food; (2) secretion of enzymes and other substances that aid digestion; (3) enzymatic hydrolysis of food molecules into simpler molecular subunits; (4) absorption of the molecular subunits into body fluids and cells; and (5) elimination of undigested materials.

Study Break 45.3

1. The two classes of vitamins are water-soluble vitamins and fat-soluble vitamins. Fat-soluble vitamins ingested in the diet that are in excess of bodily needs can be stored in adipose tissues. In contrast, excess water-soluble vitamins in the diet are excreted in the urine. Hence, it is critical that humans meet their daily requirements for water-soluble vitamins.
2. The four layers of the gut are the mucosa, the submucosa, the muscularis, and the serosa. The muscularis layer, which is formed from two smooth muscle layers, is responsible for peristalsis.
3. Pepsin is secreted from chief cells into the stomach lumen as an inactive precursor molecule, pepsinogen. The pepsinogen is converted to pepsin by the highly acid conditions of the stomach. Pepsin is a digestive enzyme that begins the digestion of proteins by making breaks in polypeptide chains.
4. In the small intestine, the digestion of macromolecules into their molecular subunits occurs, and those subunits are absorbed. In the large intestine, water and mineral ions are absorbed from the remaining digestive contents; this leaves undigested remnants, the feces, which are expelled from the body.

Study Break 45.4

The hypothalamus has two interneuron centers that play a central role in regulating the digestive processes. One center stimulates appetite and reduces oxidative metabolism. The other center works in opposition to the first center by stimulating the release of a peptide hormone that inhibits appetite.

Study Break 45.5

1. Incisors nip or cut food. Canines bite and pierce food. Premolars and molars crush, grind, and shear food.
2. Symbiotic microorganisms aid digestion in many herbivores by assisting in the breakdown of plant material. The microorganisms synthesize cellulase, an enzyme vertebrates cannot make, which hydrolyzes the cellulose of plant cell walls into glucose subunits.

Self-Test Questions

1. b 2. b 3. a 4. d 5. c 6. e 7. c 8. d 9. e 10. a

Chapter 46

Study Break 46.1

Osmosis is a process in which water molecules move across a selectively permeable membrane from a region where they are more highly concentrated to one where they are less highly concentrated.

Osmolarity is the total solute concentration of a solution. Osmolarity is measured in osmoles per liter of solution, where an osmole is the number of solute molecules and ions (in moles).

A solution that is hypoosmotic has lower osmolarity than the solution on the other side of a selectively permeable membrane. The solution with the higher osmolarity is said to be hyperosmotic.

An osmoregulator is an organism (animal) that uses active control mechanism to keep the osmolarity of cellular and extracellular fluids the same. This osmolarity value may differ from the osmolarity of the surroundings.

Tubules that carry out osmoregulation and excretion are formed from transport epithelium, a layer of cells with specialized transport proteins in their plasma membranes that move specific molecules and ions into and out of the tubule.

Study Break 46.2

Protonephridia are found in flatworms and larval mollusks. They are the simplest invertebrate excretory tubule. Body fluids enter the blind end of a protonephridium, and are propelled through the tube by cilia movement on the flame cell, a large cell at the blind end. As the fluids move through the protonephridium, some molecules and ions are reabsorbed, and others, including nitrogenous wastes, are secreted into the tubules. Excess fluid is released through pores connecting the network of protonephridia to the body surface.

Metanephridia are found in annelids and most adult mollusks. Body fluids enter the funnel-like proximal end, driven by cilia surrounding that end. Some molecules and ions are reabsorbed as the fluids move through the tubule, and other ions and nitrogenous wastes are secreted into the tubule and excreted from the body surface.

Malpighian tubules are found in insects and other arthropods. Body fluids enter the tubules through spaces between the tubule cells. The distal ends of the tubules empty into the gut. Uric acid and several ions, including sodium ions and potassium ions, are actively secreted into the tubules. The concentration of these substances causes water to move from the body fluids into the tubule by osmosis. The fluid then passes into the hindgut as dilute urine. In the hindgut, cells in the gut wall transport most of the ions back into the body fluids; water follows by osmosis. The uric acid remaining is ultimately excreted with the feces.

Study Break 46.3

1. At the proximal end, a human nephron forms a cuplike region called Bowman's capsule. Bowman's capsule surrounds the glomerulus, a complex of blood capillaries. The capsule and the glomerulus are located in the renal cortex.

   Next comes the proximal convoluted tubule in the renal cortex. The tubule descends in a U-shaped bend called the loop of Henle, and ascends again to form the distal convoluted tubule. Up to eight distal convoluted tubules drain into one collecting duct.
2. The collecting ducts are permeable to water but not to salt ions. The ducts begin in the cortex and extend into the medulla of the kidney. As the ducts descend, they become surrounded by an ever-increasing solute concentration in the medulla. As the urine passes down the collecting ducts, water moves osmotically out of the ducts, causing an increase in the concentration of the urine. At the bottom of the collecting ducts, the urine is about four times more concentrated than body fluids.

Study Break 46.4

The RAAS is a regulatory system that compensates for excessive loss of salt and body fluids. The ADH system is a regulatory system that compensates for excessive water intake or loss. Combined, the two systems play an important role in regulating the interactions between the kidneys and the rest of the body.

In brief, the RAAS works as follows: When the sodium ion concentration of body fluids falls, the volume of extracellular fluids and blood pressure also drop. This activates the RAAS. Receptors in the juxtaglomerular apparatus and the heart wall detect the changes and trigger the secretion of renin which, in turn, triggers the production of angiotensin. Angiotensin stimulates constriction of arterioles in many parts of the body, thereby increasing blood pressure. It also stimulates the adrenal cortex to secrete aldosterone, which increases Na^+ reabsorption in the kidneys. The RAAS is suppressed if NaCl concentration in body fluids is higher than normal.

In the ADH system, ADH is released from the pituitary when osmoreceptors in the hypothalamus detect an increase in the osmolarity of body fluids. ADH increases water reabsorption in the kidney; as a result, urinary output is reduced and water is conserved. In the case of a decrease in osmolarity, ADH release from the pituitary is inhibited, thereby decreasing water reabsorption in the kidney.

Study Break 46.5

1. Marine teleosts live in seawater, which is hyperosmotic to their body fluids. These fishes drink seawater continuously to replace water lost to the environment by osmosis. Sodium ions, potassium ions, and chloride ions from the seawater they drink are excreted by the gills. Nitrogenous wastes are released from the gills, primarily as ammonia, by simple diffusion.

   Freshwater fishes live in fresh water, which is hypoosmotic to their body fluids. They take in water by osmosis and excrete excess water. Salts needed for bodily functions are obtained from food and by active transport through the gills from the water. Nitrogenous wastes are excreted from the gills as ammonia.
2. The excretion of urea in mammals involves expelling the urea in solution—urine. Thus, there is loss of water with excretion of nitrogenous waste as urea. Excretion of nitrogenous waste in the form of uric acid conserves water because uric acid crystals are almost water free.

Study Break 46.6

Ectothermy applies to animals that obtain heat energy primarily from the environment, whereas endothermy applies to animals that obtain heat energy primarily from internal reactions.

Generally speaking, ectotherms are highly successful in warm environments, but their bodily functions slow down as the temperature decreases. Endotherms can remain active over a broader range of environmental temperatures than ectotherms do, but they require an almost constant supply of energy to maintain their body temperature.

Study Break 46.7

1. The thermoregulatory responses shown by ectotherms can be physiological, such as regulating blood flow to internal or external body organisms, or behavioral, such as physically moving to a location in the environment suitable for their heat energy needs at the time.
2. Thermal acclimatization refers to the physiological changes that many ectotherms make to compensate for seasonal shifts in environmental temperature.

Study Break 46.8

Temperature regulation in endothermic animals involves mechanisms that balance internal heat production against heat loss from the body. Internal heat production is controlled by negative feedback pathways that are triggered by thermoreceptors in the skin, the hypothalamus, and the spinal cord. When a deviation from the set point occurs, this system operates to return the core temperature to that set point through changes in metabolic processes, behavioral activities, and control of heat loss at the skin surface.

Self-Test Questions

1. d 2. c 3. a 4. b 5. d 6. e 7. a 8. b
9. d 10. e

Chapter 47

Study Break 47.1

Recall that asexual reproduction produces offspring with genes from only one parent. In most animals that undergo asexual reproduction, the offspring are produced by mitosis and, therefore, are genetically identical to one another and to the parent. Such genetic homogeneity can be advantageous in stable and uniform environments. Another advantage is that individuals do not need to use energy to produce gametes or to find and select a mate. Further, for individuals in sparse populations, or for sessile animals, asexual reproduction can be an advantage. A disadvantage is that a genetically homogenous population may not adapt readily to new environments.

By contrast, sexual reproduction always generates genetic diversity among offspring. This provides a population the opportunity to adapt to changing environments, and perhaps to move to and colonize new environments. Disadvantages of sexual reproduction include the expenditure of energy and raw materials to produce gametes and to find and select mates.

Study Break 47.2

1. Egg coats are surface coats around the egg. They are added during oocyte development or fertilization. Egg coats provide protection against mechanical injury, infection by microorganisms, and, in some species, loss of water.

   Mammalian eggs have an egg coat called the zona pellucida immediately surrounding the egg. This gel-like coating is called the vitelline coat in other organisms. In birds, a thick solution of proteins—the egg white—surrounds a vitelline coat. Bounding the egg white is a hard shell.
2. The slow block to polyspermy occurs in many organisms, including mammals. The fusion of egg and sperm triggers an increase in calcium ions in the cytosol. The calcium ions activate proteins that initiate a high level of metabolic activity in the egg. The released calcium ions also cause the cortical granules to fuse with the egg's plasma membrane and release their contents to the outside. Enzymes released from those granules alter the egg coats in a matter of minutes after fertilization, and this blocks further sperm from attaching and penetrating the egg.

Study Break 47.3

- FSH stimulates a number of oocytes to begin meiosis. A follicle forms around each oocyte.
- FSH and LH interact to stimulate follicle cells to secrete estrogens.
- Later in the cycle, an increased level of estrogen leads to a new burst of release of FSH and LH. This new LH stimulates follicle cells to release enzymes that digest away the follicle wall, leading to egg release.
- The new LH also causes the remaining follicle cells to grow into the corpus luteum.
- FSH and LH levels decrease. This removes the stimulatory signal for follicular growth, and no new follicles grow in the ovary.

Study Break 47.4

The oral contraceptive pill inhibits the secretion of FSH and LH by the pituitary. FSH and LH stimulate oocytes in the ovaries to begin meiosis. They also stimulate follicle cells that surround the oocytes to secrete estrogens. Later, LH causes ovulation—the release of an egg. Therefore, without FSH and LH, ovulation cannot occur.

Self-Test Questions

1. b 2. a 3. d 4. a 5. e 6. c 7. b 8. e
9. c 10. d

Chapter 48

Study Break 48.1

1. In cleavage divisions, cycles of DNA replication and division occur without the production of new cytoplasm. Therefore, the cytoplasm of the egg partitions into many cells without increasing in overall size or mass. In cell division in an adult organism, the mitotic cell cycle involves cycles of DNA replication and division interspersed with cell growth. That is the cell grows, then divides, then the two progeny cells grow, then divide and so

on. New cytoplasm is produced during these cell cycles so that overall there is an increase in mass of the cells.

2. The three primary cell layers of the embryo are produced by gastrulation. They are ectoderm, mesoderm, and endoderm.

Study Break 48.2

1. The gray crescent establishes the dorsal–ventral axis, with the gray crescent marking the future dorsal side.
2. If cells of the dorsal lip of the blastopore are transplanted to another location in the egg, they cause a second blastopore, and subsequently a second embryo, to form in that region.
3. The extraembryonic membranes in birds and their functions are as follows:
   - Yolk sac: Surrounds the yolk, which provides nutrients to the embryo.
   - Chorion: Exchanges oxygen and carbon dioxide with the environment through the egg shell.
   - Amnion: Encloses the embryo and secretes amniotic fluid into the space that provides an aquatic environment in which the embryo can develop.
   - Allantois: Stores nitrogenous wastes, primarily in the form of uric acid, that are derived from the embryo. Part of the allantoic membrane forms a bed of capillaries that is connected to the embryo and that delivers carbon dioxide from the embryo to the chorion and picks up oxygen absorbed through the shell.

Study Break 48.3

1. The three primary tissues, ectoderm, mesoderm, and endoderm, give rise to the tissues and organs of the embryo.
2. The central nervous system, notably the brain and spinal cord, develops from the neural tube. Neural crest cells give rise to parts of the nervous system (including cranial nerves, ganglia of the autonomic nervous system, peripheral nerves from the spinal cord to body structures, and nerves of the developing gut) and contribute to a variety of other body structures (for example, bones of the inner ear and skull, cartilage of facial structures, teeth, pigment cells in the skin, and the adrenal medulla).

Study Break 48.4

1. Cells of the trophoblast are responsible for implantation of the blastocyst in the endometrium (uterine lining). When the blastocyst is ready for implantation, trophoblast cells secrete proteases that digest pathways between endometrial cells. Dividing trophoblast cells fill the spaces and, through continued digestion and division, the blastocyst burrows into the endometrium and eventually becomes covered by a layer of endometrial cells. From then on, cells originating in the trophoblast support the development of the embryo/fetus in the uterus through the production of the chorion.

   The inner cell mass becomes the embryo/fetus itself. During implantation, the inner cell mass separates into the epiblast and the hypoblast. The epiblast produces the ectoderm, mesoderm, and endoderm of the developing embryo. The hypoblast gives rise to part of the extraembryonic membranes.
2. Oxytocin. This hormone is responsible for stimulating contractions of the uterus.

Study Break 48.5

1. Cell division, cell movement, and cell adhesion.
2. Induction is the process whereby a region of the embryo acts on other cells to alter the course of development. Induction is brought about by proteins; hence, induction is under genetic control.

Study Break 48.6

1. Determination and differentiation are under molecular control. Regulatory genes encode regulatory proteins that bind to promoters of the genes they control, switching the genes on or off depending on the interaction.
2. The segmentation genes subdivide the embryo progressively into regions, thereby determining the segments of the embryo and the adult. In essence, they organize the embryo into segments. The homeotic genes specify the identity of each segment with respect to the body part it will become.

Self-Test Questions

1. d 2. b 3. a 4. e 5. c 6. d 7. d 8. b 9. a 10. e

Appendix B
Classification System

The classification system presented here is based on a combination of organismal and molecular characters and is a composite of several systems developed by microbiologists, botanists, and zoologists. This classification reflects current trends toward a phylogenetic approach to taxonomy, one that incorporates the ever more detailed information about the relationships of monophyletic lineages provided by new molecular sequence data. In keeping with these trends, we have omitted reference to the traditional taxonomic categories, such as "class" and "order." Instead, we present the major monophyletic lineages in each of the three domains, and we indicate their relationships within a nested hierarchy that parallels that of traditional Linnaean classification.

Although researchers generally agree on the identity of the major monophyletic lineages, the biologists who study different groups have not established universal criteria for identifying the somewhat arbitrary taxonomic categories included in the traditional Linnaean hierarchy. As a result, a "class" or "order" of flowering plants may not be the equivalent of a "class" or "order" of animals. In fact, as described in *Unanswered Questions* at the end of Chapter 23, systematic biologists are shifting toward a more phylogenetic approach to taxonomy and classification, such as the one represented here.

Bear in mind that we include this appendix to introduce the diversity of life and illustrate many of the evolutionary relationships that link monophyletic groups. Like all phylogenetic hypotheses, this classification is open to revision as new information becomes available. Moreover, the classification is incomplete because it includes only those lineages that are described in Unit Four.

Prokaryotes and Eukaryotes

Organisms fall into two groups, prokaryotes and eukaryotes, based on the organization of their cells. Prokaryotes consist of the Domains Bacteria and Archaea and are characterized by a central region, the nucleoid, which has no boundary membrane separating it from the cytoplasm, and by membranes typically limited to the plasma membrane. Most prokaryotes are single-celled, although some are found in simple associations. All other organisms are eukaryotes, which make up the Domain Eukarya. Eukaryotes are characterized by cells with a central, membrane-bound nucleus, and an extensive membrane system. Some eukaryotes are single-celled, while others are multicellular.

Domain Bacteria

The largest and most diverse group of prokaryotes. Includes photoautotrophs, chemoautotrophs, and heterotrophs.

Proteobacteria Purple sulfur bacteria, purple nonsulfur bacteria, and some chemoheterotrophs

Green Bacteria Green sulfur bacteria and green nonsulfur bacteria

Cyanobacteria Photoautotrophic Gram-negative bacteria that use the same chlorophyll as in plants

Gram-Positive Bacteria Chemoheterotrophic bacteria with thick cell walls

Spirochetes Helically spiraled bacteria that move by twisting in a corkscrew pattern

Chlamydias Gram-negative intracellular parasites of animals, with cell walls that lack peptidoglycans

Domain Archaea

Prokaryotes that are evolutionarily between eukaryotic cells and the bacteria. Most are chemoautotrophs. None is photosynthetic. Originally discovered in extreme habitats, they are now known to be widely dispersed. Compared with bacteria, the Archaea have a distinctive cell wall structure and unique membrane lipids, ribosomes, and RNA sequences. Some are symbiotic with animals, but none is known to be pathogenic.

Euryarchaeota Includes methanogens, extreme halophiles, and some extreme thermophiles

Crenarchaeota Includes most of the extreme thermophiles, as well as psychrophiles; mesophilic species comprise a large part of plankton in cool, marine waters

Korarchaeota Known only from DNA isolated from hydrothermal pools. As of this writing, none has been cultured and no species have been named.

Domain Eukarya

Protoctista A collection of single-celled and multicelled lineages, which are almost certainly not a monophyletic group. Some biologists consider the groups listed below to be kingdoms in their own right.

Excavates Single-celled animal parasites that lack mitochondria and move using flagella; most have a hollow, ventral feeding groove

Diplomonadida (diplomonads)—two nuclei; move by multiple free flagella

Parabasala (parabasalids)—move by an undulating membrane and free flagella

Discicristates Mostly single-celled, highly motile cells that swim using flagella, have disc-shaped mitochondrial inner membranes

Euglenoids—free-living photosynthetic autotrophs

Kinetoplastids—nonphotosynthetic, heterotrophs that live as animal parasites

Alveolates Characterized by small membrane-bound vesicles called alveoli in a layer under the plasma membrane

Ciliophora (ciliates)—single-celled heterotrophs; swim by means of cilia

Dinoflagellata (dinoflagellates)—single-celled marine heterotrophs or autotrophs; shell formed from cellulose plates

Apicomplexa (apicomplexans)—nonmotile parasites of animals with apical complex for attachment and invasion of host cells

Heterokonts Characterized by two different flagella

Oomycota (oomycetes)—water molds, white rusts, and mildews

Bacillariophyta (diatoms)—single-celled; covered by a glassy silica shell

Chrysophyta (golden algae)—colonial; each cell of the colony has a pair of flagella and is covered by a glassy shell consisting of plates or scales

Phaeophyta (brown algae)—photoautotrophic protists

Cercozoa Amoebas with stiff, filamentous pseudopodia; some with outer shells

Radiolaria (radiolarians)—heterotrophic; glassy internal skeleton with projecting raylike strands of cytoplasm

Foraminifera (forams)—heterotrophic protists with shells consisting of organic matter reinforced by calcium carbonate

Chlorarachniophyta (chloroarachniophytes)—green, photosynthetic amoebas that also engulf food

Amoebozoa Includes most of the amoebas and the slime molds

Amoebas—single-celled; use non-stiffened pseudopods for locomotion and feeding

Cellular slime molds—heterotrophs; primarily individual cells; move by amoeboid motion, or as a multicellular mass

Plasmodial slime molds—heterotrophs; live as plasmodium, a large composite mass with nuclei in a common cytoplasm, that moves and feeds like a giant amoeba

Opisthokonts A single posterior flagellum at some stage in the life cycle

Choanoflagellata (choanoflagellates)—motile protists with a single flagellum surrounded by collar of closely packed microvilli; likely ancestor of animals and fungi

Archaeplastida Red algae, green algae, and land plants, photosynthesizers with a common evolutionary origin

Rhodophyta (red algae)—marine seaweeds, typically multicellular, reddish in color; with plantlike bodies

Chlorophyta (green algae)—green photosynthetic single-celled, colonial, and multicellular protists that have the same photosynthetic pigments as plants; likely ancestor of land plants

Fungi Heterotrophic, mostly multicellular organisms with cell wall containing chitin and cell nuclei occurring in threadlike hyphae; life cycle typically includes both asexual and sexual phases, with sexual structures used as the basis for phylum-level classification. Single-celled species are known as yeasts.

Zygomycota (zygomycetes)—terrestrial; asexual reproduction via nonmotile haploid spores formed in sporangia; sexual spores (zygospores) form in zygosporangia; aseptate hyphae

Glomeromycota (glomeromycetes)—terrestrial; asexual reproduction via spores at the tips of hyphae; form mycorrhizal associations with plant roots

Ascomycota (ascomycetes/sac fungi)—terrestrial and aquatic; sexual spores form in asci; asexual reproduction occurs via conidia (nonmotile spores); septate hyphae

Basidiomycota (basidiomycetes)—terrestrial; reproduction usually via asexual basidiospores produced by basidia; septate hyphae

Basidiomycetes: mushroom-forming fungi and relatives

Teliomycetes: rusts

Ustomycetes: smuts

Chytridiomycota (chytrids)—mostly aquatic; asexual reproduction by way of motile zoospores; sexual reproduction via gametes produced in gametangia; hyphae mostly aseptate

Conidial fungi—not a true phylum but a convenience grouping of species for which no sexual phase is known

Microsporidia—single-celled sporelike parasites of animals, other groups; phylogeny uncertain

Plantae Multicellular autotrophs, mostly terrestrial, and most of which gain energy via photosynthesis; life cycle characterized by alternation of a gametophyte (gamete-producing) generation and sporophyte (spore-producing) generation

Nonvascular plants (bryophytes)—no vessels for transporting water and nutrients; swimming sperm require liquid water for sexual reproduction

Hepatophyta (liverworts)—leafy or simple flattened thallus with rhizoids; no true leaves, stems, roots, or stomata (porelike openings for gas exchange); spores in capsules

Anthocerophyta (hornworts)—simple flattened thallus, hornlike sporangia

Bryophyta (mosses)—feathery or cushiony thallus; some with hydroids; spores in capsules

Seedless vascular plants—plants in which embryos are not housed inside seeds

Lycophyta (club mosses)—simple leaves, cuticle, stomata, true roots; most species have sporangia on sporophylls; fertilization by swimming sperm

Pterophyta (ferns, whisk ferns, horsetails)—*Ferns*: Finely divided leaves; sporangia in sori. *Whisk ferns*: Branching stem from rhizomes; sporangia on stem scales. *Horsetails*: hollow stem, scalelike leaves, sporangia in strobili.

Seed plants—vascular plants in which embryos develop within seeds

Gymnosperms—seeds born on stems, on leaves, or under scales

Cycadophyta (cycads)—shrubby or treelike with palmlike leaves; male and female strobili on separate plants

Ginkgophyta (ginkgoes)—lineage with a single living species (*Ginkgo biloba*); tree with deciduous, fan-shaped leaves; male, female reproductive structures on separate plants

Gnetophyta (gnetophytes)—shrubs or woody vinelike plants; male and female strobili on separate plants

Coniferophyta (conifers)—predominant extant gymnosperm group; mostly evergreen trees and shrubs with needlelike or scalelike leaves; male and female cones usually on the same plant

Anthophyta (angiosperms/flowering plants)—reproductive structures in flowers

Monocotyledones (monocots)—grasses, palms, lilies, orchids and their relatives; a single cotyledon (seed leaf); pollen grains have one groove

Eudicotyledones (eudicots)—roses, melons, beans, potatoes, most fruit trees, others; two cotyledons; pollen grains have three grooves

Other major angiosperm lineages: magnoliids (magnolias and relatives); water lilies (Family Nymphaeaceae); *Amborella* (Family Amborellaceae)

Animalia Multicellular heterotrophs; nearly all with tissues, organs, and organ systems; motile during at least part of the life cycle; sexual reproduction in most; embryos develop through a series of stages; many with larval and adult stages in life cycle

Parazoa Animals lacking tissues and body symmetry

Porifera (sponges)—multicellular; extract oxygen and particulate food from water drawn into a central cavity

Eumetazoa Animals possessing tissues and either radial or bilateral symmetry

Radiata—acoelomate animals possessing radial symmetry and two tissue layers

Cnidaria (cnidarians)—two tissue layers; single opening into gastrovascular cavity; nerve net; nematocysts for defense and predation; some sessile, some motile; most are predatory, some with photosynthetic endosymbionts; freshwater and marine

Hydrozoa: hydrozoans

Scyphozoa: jellyfishes

Cubozoa: box jellyfishes

Anthozoa: sea anemones, corals

Ctenophora (comb jellies)—two (possibly three) tissue layers; feeding tentacles capture particulate food; beating cilia provide weak locomotion; marine

Bilateria—animals possessing bilateral symmetry and three tissue layers

Protostomia—acoelomate, pseudocoelomate, or schizocoelomate; many with spiral, indeterminate cleavage; blastopore forms mouth; nervous system on ventral side

Lophotrochozoa—many with either a lophophore for feeding and gas exchange or a trochophore larva

Ectoprocta (bryozoans)—coelomate; colonial; secrete hard covering over soft tissues; lophophore; sessile; particulate feeders; marine

Brachiopoda (lamp shells)—coelomate; dorsal and ventral shells; lophophore; sessile; particulate feeders; marine

Phoronida (phoronid worms)—coelomate; secrete tubes around soft tissues; sessile; particulate feeders; lophophore

Platyhelminthes (flatworms)—acoelomate; dorsoventrally flattened; complex reproductive, excretory, and nervous systems; gastrovascular cavity in many; free-living or parasitic, often with multiple hosts; terrestrial, freshwater, and marine

Turbellaria: free-living flatworms

Trematoda: flukes

Cestoda: tapeworms

Rotifera (wheel animals)—pseudocoelomate; microscopic; complete digestive system; well-developed reproductive, excretory, and nervous systems; particulate feeders; major components of marine and freshwater plankton

Nemertea (ribbon worms)—schizocoelomate; proboscis housed within rhynchocoel; complete digestive tract; circulatory system; predatory; mostly marine

Mollusca (mollusks)—schizocoelomate; many with trochophore larva; many with shell secreted by mantle; body divided into head–foot, visceral mass, and mantle; well-developed organ systems; variable locomotion; herbivorous or predatory; terrestrial, freshwater, and marine

Polyplacophora: chitons

Gastropoda: snails, sea slugs, land slugs

Bivalvia: clams, mussels, scallops, oysters

Cephalopoda: squids, octopuses, cuttlefish, nautiluses

Annelida (segmented worms)—schizocoelomate; many with trochophore larva; segmented body and organ systems; well-developed organ systems; many use hydrostatic skeleton for locomotion; some predatory, some particulate feeders, some detritivores; terrestrial, freshwater, and marine

Polychaeta: marine worms

Oligochaeta: freshwater and terrestrial worms

Hirudinea: leeches

Ecdysozoa—cuticle or exoskeleton is shed periodically

Nematoda (roundworms)—pseudocoelomate; body covered with tough cuticle that is shed periodically; well-developed organ systems; thrashing locomotion; many are parasitic on plants or animals; mostly terrestrial

Onychophora (velvet worms)—schizocoelomate; segmented body covered with cuticle; locomotion by many unjointed legs; complex organ systems; predatory; terrestrial

Arthropoda (arthropods)—schizocoelomate; jointed exoskeleton made of chitin; segmented body, some with fusion of segments in head, thorax, or abdomen; complex organ systems; variable modes of locomotion, including flight; specialization of numerous appendages; herbivorous, predatory, or parasitic; terrestrial, freshwater, and marine

Trilobita: trilobites (extinct)

Chelicerata: horseshoe crabs, spiders, scorpions, ticks, mites

Crustacea: shrimps, crayfishes, lobsters, crabs, barnacles, copepods, isopods

Myriapoda: centipedes, millipedes

Hexapoda: springtails and insects

Deuterostomia—enterocoelomate; many with radial, determinate cleavage; blastopore forms anus; nervous system on dorsal side in many

Echinodermata (echinoderms)—secondary radial symmetry, often organized around five radii; hard internal skeleton; unique water vascular system with tube feet; complete digestive system; simple nervous system; no circulatory or respiratory system; generally slow locomotion using tube feet; predatory, herbivorous, particulate feeders, detritivores; exclusively marine

Asteroidea: sea stars

Ophiuroidea: brittle stars

Echinoidea: sea urchins, sand dollars

Holothuroidea: sea cucumbers

Crinoidea: feather stars, sea lilies

Concentricycloidea: sea daisies

Hemichordata (acorn worms)—pharynx perforated with gill slits; proboscis; complex organ systems; tube-dwelling in soft sediments; particulate or deposit feeders; exclusively marine

Chordata (chordates)—notochord; segmental body wall and tail muscles; dorsal hollow nerve chord; perforated pharynx; complex organ systems; variable modes of locomotion; extremely varied diets; terrestrial, freshwater, and marine

- Urochordata: tunicates, sea squirts
- Cephalochordata: lancelets
- Vertebrata: vertebrates
  - Myxinoidea: hagfishes
  - Petromyzontoidea: lampreys
  - Placodermi: placoderms (extinct)
  - Chondrichthyes: sharks, skates, and rays
  - Acanthodii: acanthodians
  - Actinopterygii: ray-finned fishes
  - Sarcopterygii: fleshy-finned fishes
  - Amphibia: salamanders, frogs, caecelians
  - Synapsida: mammals
  - Anapsida: turtles
  - Diapsida: sphenodontids, lizards, snakes, crocodilians, birds

Appendix C

Annotations to a Journal Article

This journal article reports on the movements of a female wolf during the summer of 2002 in northwestern Canada. It also reports on a scientific process of inquiry, observation and interpretation to learn where, how and why the wolf traveled as she did. In some ways, this article reflects the story of "how to do science" told in section 1.4 of this textbook. These notes are intended to help you read and understand how scientists work and how they report on their work.

1 Title of the journal, which reports on science taking place in Arctic regions.

2 Volume number, issue number and date of the journal, and page numbers of the article.

3 Title of the article: a concise but specific description of the subject of study—one episode of long-range travel by a wolf hunting for food on the Arctic tundra.

4 Authors of the article: scientists working at the institutions listed in the footnotes below. Note #2 indicates that P. F. Frame is the corresponding author—the person to contact with questions or comments. His email address is provided.

5 Date on which a draft of the article was received by the journal editor, followed by date on which a revised draft was accepted for publication. Between these dates, the article was reviewed and critiqued by other scientists, a process called peer review. The authors revised the article to make it clearer, according to those reviews.

6 ABSTRACT: A brief description of the study containing all basic elements of this report. First sentence summarizes the background material. Second sentence encapsulates the methods used. The rest of the paragraph sums up the results. Authors introduce the main subject of the study—a female wolf (#388) with pups in a den—and refer to later discussion of possible explanations for her behavior.

7 Key words are listed to help researchers using computer databases. Searching the databases using these key words will yield a list of studies related to this one.

8 RÉSUMÉ: The French translation of the abstract and key words. Many researchers in this field are French Canadian. Some journals provide such translations in French or in other languages.

9 INTRODUCTION: Gives the background for this wolf study. This paragraph tells of known or suspected wolf behavior that is important for this study. Note that (a) major species mentioned are always accompanied by scientific names, and (b) statements of fact or postulations (claims or assumptions about what is likely to be true) are followed by references to studies that established those facts or supported the postulations.

10 This paragraph focuses directly on the wolf behaviors that were studied here.

11 This paragraph starts with a statement of the hypothesis being tested, one that originated in other studies and is supported by this one. The hypothesis is restated more succinctly in the last sentence of this paragraph. This is the inquiry part of the scientific process—asking questions and suggesting possible answers.

1 ARCTIC

2 VOL. 57, NO. 2 (JUNE 2004) P. 196–203

3 Long Foraging Movement of a Denning Tundra Wolf

4 Paul F. Frame,[1,2] David S. Hik,[1] H. Dean Cluff,[3] and Paul C. Paquet[4]

5 *(Received 3 September 2003; accepted in revised form 16 January 2004)*

6 **ABSTRACT** Wolves (*Canis lupus*) on the Canadian barrens are intimately linked to migrating herds of barren-ground caribou (*Rangifer tarandus*). We deployed a Global Positioning System (GPS) radio collar on an adult female wolf to record her movements in response to changing caribou densities near her den during summer. This wolf and two other females were observed nursing a group of 11 pups. She traveled a minimum of 341 km during a 14-day excursion. The straight-line distance from the den to the farthest location was 103 km, and the overall minimum rate of travel was 3.1 km/h. The distance between the wolf and the radio-collared caribou decreased from 242 km one week before the excursion to 8 km four days into the excursion. We discuss several possible explanations for the long foraging bout.

7 *Key words:* wolf, GPS tracking, movements, *Canis lupus*, foraging, caribou, Northwest Territories

8 **RÉSUMÉ** Les loups (*Canis lupus*) dans la toundra canadienne sont étroitement liés aux hardes de caribous des toundras (*Rangifer tarandus*). On a équipé une louve adulte d'un collier émetteur muni d'un système de positionnement mondial (GPS) afin d'enregistrer ses déplacements en réponse au changement de densité du caribou près de sa tanière durant l'été. On a observé cette louve ainsi que deux autres en train d'allaiter un groupe de 11 louveteaux. Elle a parcouru un minimum de 341 km durant une sortie de 14 jours. La distance en ligne droite de la tanière à l'endroit le plus éloigné était de 103 km, et la vitesse minimum durant tout le voyage était de 3,1 km/h. La distance entre la louve et le caribou muni du collier émetteur a diminué de 242 km une semaine avant la sortie à 8 km quatre jours après la sortie. On commente diverses explications possibles pour ce long épisode de recherche de nourriture.

Mots clés: loup, repérage GPS, déplacements, *Canis lupus*, recherche de nourriture, caribou, Territoires du Nord-Ouest

Traduit pour la revue *Arctic* par Nésida Loyer.

9 Introduction

Wolves (*Canis lupus*) that den on the central barrens of mainland Canada follow the seasonal movements of their main prey, migratory barren-ground caribou (*Rangifer tarandus*) (Kuyt, 1962; Kelsall, 1968; Walton et al., 2001). However, most wolves do not den near caribou calving grounds, but select sites farther south, closer to the tree line (Heard and Williams, 1992). Most caribou migrate beyond primary wolf denning areas by mid-June and do not return until mid-to-late July (Heard et al., 1996; Gunn et al., 2001). Consequently, caribou density near dens is low for part of the summer.

10 During this period of spatial separation from the main caribou herds, wolves must either search near the homesite for scarce caribou or alternative prey (or both), travel to where prey are abundant, or use a combination of these strategies.

11 Walton et al. (2001) postulated that the travel of tundra wolves outside their normal summer ranges is a response to low caribou availability rather than a pre-dispersal exploration like that observed in terri torial wolves (Fritts and Mech, 1981; Messier, 1985). The authors postulated this because most such travel was directed toward caribou calving grounds. We report details of such a long-distance excursion by a breeding female tundra wolf wearing a GPS radio collar. We discuss the relationship of the excursion to movements of satellite-collared caribou (Gunn et al., 2001), supporting the hypothesis that tundra wolves make directional, rapid, long-distance movements in response to seasonal prey availability.

[1] Department of Biological Sciences, University of Alberta, Edmonton, Alberta T6G 2E9, Canada
[2] Corresponding author: pframe@ualberta.ca
[3] Department of Resources, Wildlife, and Economic Development, North Slave Region, Government of the Northwest Territories, P.O. Box 2668, 3803 Bretzlaff Dr., Yellowknife, Northwest Territories X1A 2P9, Canada; Dean_Cluff@gov.nt.ca
[4] Faculty of Environmental Design, University of Calgary, Calgary, Alberta T2N 1N4, Canada; current address: P.O. Box 150, Meacham, Saskatchewan S0K 2V0, Canada

![12]

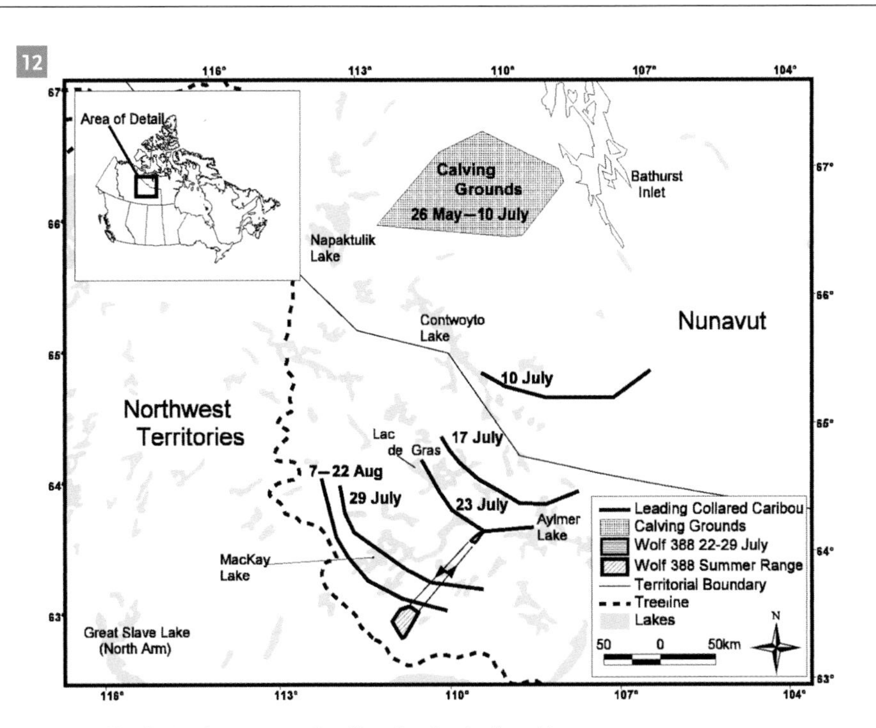

Figure 1. Map showing the movements of satellite radio-collared caribou with respect to female wolf 388's summer range and long foraging movement, in summer 2002.

13 Study Area

Our study took place in the northern boreal forest–low Arctic tundra transition zone (63° 30′ N, 110° 00′ W; Figure 1; Timoney et al., 1992). Permafrost in the area changes from discontinuous to continuous (Harris, 1986). Patches of spruce (*Picea mariana, P. glauca*) occur in the southern portion and give way to open tundra to the northeast. Eskers, kames, and other glacial deposits are scattered throughout the study area. Standing water and exposed bedrock are characteristic of the area.

14 *Details of the Caribou-Wolf System*

The Bathurst caribou herd uses this study area. Most caribou cows have begun migrating by late April, reaching calving grounds by June (Gunn et al., 2001; Figure 1). Calving peaks by 15 June (Gunn et al., 2001), and calves begin to travel with the herd by one week of age (Kelsall, 1968). The movement patterns of bulls are less known, but bulls frequent areas near calving grounds by mid-June (Heard et al., 1996; Gunn et al., 2001). In summer, Bathurst caribou cows generally travel south from their calving grounds and then, parallel to the tree line, to the northwest. The rut usually takes place at the tree line in October (Gunn et al., 2001). The winter range of the Bathurst herd varies among years, ranging through the taiga and along the tree line from south of Great Bear Lake to southeast of Great Slave Lake. Some caribou spend the winter on the tundra (Gunn et al., 2001; Thorpe et al., 2001).

15 In winter, wolves that prey on Bathurst caribou do not behave territorially. Instead, they follow the herd throughout its winter range (Walton et al., 2001; Musiani, 2003). However, during denning (May–

12 This map shows the study area and depicts wolf and caribou locations and movements during one summer. Some of this information is explained below.

13 STUDY AREA: This section sets the stage for the study, locating it precisely with latitude and longitude coordinates and describing the area (illustrated by the map in Figure 1).

14 Here begins the story of how prey (caribou) and predators (wolves) interact on the tundra. Authors describe movements of these nomadic animals throughout the year.

15 We focus on the denning season (summer) and learn how wolves locate their dens and travel according to the movements of caribou herds.

16 Other variables are considered—prey other than caribou and their relative abundance in 2002.

17 METHODS: There is no one scientific method. Procedures for each and every study must be explained carefully.

18 Authors explain when and how they tracked caribou and wolves, including tools used and the exact procedures followed.

19 This important subsection explains what data were calculated (average distance...) and how, including the software used and where it came from. (The calculations are listed in Table 1.) Note that the behavior measured (traveling) is carefully defined.

20 RESULTS: The heart of the report and the observation part of the scientific process. This section is organized parallel to the Methods section.

21 This subsection is broken down by periods of observation. Pre-excursion period covers the time between 388's capture and the start of her long-distance travel. The investigators used visual observations as well as telemetry (measurements taken using the global positioning system (GPS)) to gather data. They looked at how 388 cared for her pups, interacted with other adults, and moved about the den area.

Table 1. Daily distances from wolf 388 and the den to the nearest radio-collared caribou during a long excursion in summer 2002.

Date (2002)	Mean distance from caribou to wolf (km)	Daily distance from closest caribou to den
12 July	242	241
13 July	210	209
14 July	200	199
15 July	186	180
16 July	163	162
17 July	151	148
18 July	144	137
19 July[1]	126	124
20 July	103	130
21 July	73	130
22 July	40	110
23 July[2]	9	104
29 July[3]	16	43
30 July	32	43
31 July	28	44
1 August	29	46
2 August[4]	54	52
3 August	53	53
4 August	74	74
5 August	75	75
6 August	74	75
7 August	72	75
8 August	76	75
9 August	79	79

[1] Excursion starts.
[2] Wolf closest to collared caribou.
[3] Previous five days' caribou locations not available.
[4] Excursion ends.

August, parturition late May to mid-June), wolf movements are limited by the need to return food to the den. To maximize access to migrating caribou, many wolves select den sites closer to the tree line than to caribou calving grounds (Heard and Williams, 1992). Because of caribou movement patterns, tundra denning wolves are separated from the main caribou herds by several hundred kilometers at some time during summer (Williams, 1990:19; Figure 1; Table 1).

16 Muskoxen do not occur in the study area (Fournier and Gunn, 1998), and there are few moose there (H.D. Cluff, pers. obs.). Therefore, alternative prey for wolves includes waterfowl, other ground-nesting birds, their eggs, rodents, and hares (Kuyt, 1972; Williams, 1990:16; H.D. Cluff and P.F. Frame, unpubl. data). During 56 hours of den observations, we saw no ground squirrels or hares, only birds. It appears that the abundance of alternative prey was relatively low in 2002.

17 Methods

Wolf Monitoring

18 We captured female wolf 388 near her den on 22 June 2002, using a helicopter net-gun (Walton et al., 2001). She was fitted with a releasable GPS radio collar (Merrill et al., 1998) programmed to acquire locations at 30-minute intervals. The collar was electronically released (e.g., Mech and Gese, 1992) on 20 August 2002. From 27 June to 3 July 2002, we observed 388's den with a 78 mm spotting scope at a distance of 390 m.

Caribou Monitoring

In spring of 2002, ten female caribou were captured by helicopter net-gun and fitted with satellite radio collars, bringing the total number of collared Bathurst cows to 19. Eight of these spent the summer of 2002 south of Queen Maud Gulf, well east of normal Bathurst caribou range. Therefore, we used 11 caribou for this analysis. The collars provided one location per day during our study, except for five days from 24 to 28 July. Locations of satellite collars were obtained from Service Argos, Inc. (Landover, Maryland).

Data Analysis

19 Location data were analyzed by ArcView GIS software (Environmental Systems Research Institute Inc., Redlands, California). We calculated the average distance from the nearest collared caribou to the wolf and the den for each day of the study.

Wolf foraging bouts were calculated from the time 388 exited a buffer zone (500 m radius around the den) until she re-entered it. We considered her to be traveling when two consecutive locations were spatially separated by more than 100 m. Minimum distance traveled was the sum of distances between each location and the next during the excursion.

We compared pre- and post-excursion data using Analysis of Variance (ANOVA; Zar, 1999). We first tested for homogeneity of variances with Levene's test (Brown and Forsythe, 1974). No transformations of these data were required.

Results 20

Wolf Monitoring

21 **Pre-Excursion Period:** Wolf 388 was lactating when captured on 22 June. We observed her and two other females nursing a group of 11 pups between 27 June and 3 July. During our observations, the pack consisted of at least four adults (3 females and 1 male) and 11 pups. On 30 June, three pups were moved to a location 310 m from the other eight and cared for by an uncollared female. The male was not seen at the den after the evening of 30 June.

Before the excursion, telemetry indicated 18 foraging bouts. The mean distance traveled during these bouts was 25.29 km (± 4.5 SE, range 3.1–82.5 km). Mean greatest distance from the den on foraging

198 *P.F. Frame, et al*

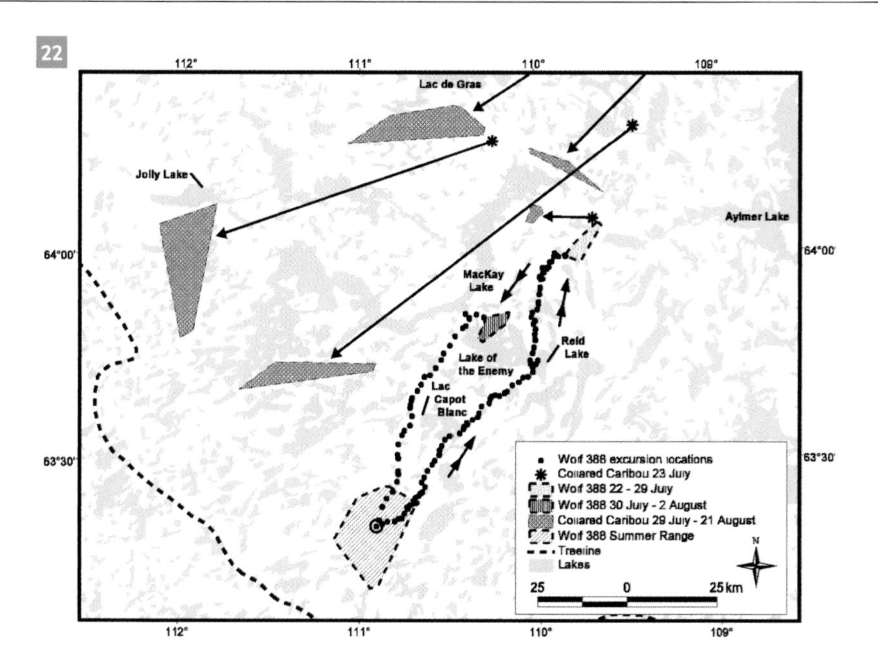

Figure 2. Details of a long foraging movement by female wolf 388 between 19 July and 2 August 2002. Also shown are locations and movements of three satellite radio-collared caribou from 23 July to 21 August 2002. On 23 July, the wolf was 8 km from a collared caribou. The farthest point from the den (103 km distant) was recorded on 27 July. Arrows indicate direction of travel.

bouts was 7.1 km (± 0.9 SE, range 1.7–17.0 km). The average duration of foraging bouts for the period was 20.9 h (± 4.5 SE, range 1–71 h).

The average daily distance between the wolf and the nearest collared caribou decreased from 242 km on 12 July, one week before the excursion period, to 126 km on 19 July, the day the excursion began (Table 1).

23 **Excursion Period:** On 19 July at 2203, after spending 14 h at the den, 388 began moving to the northeast and did not return for 336 h (14 d; Figure 2). Whether she traveled alone or with other wolves is unknown. During the excursion, 476 (71%) of 672 possible locations were recorded. The wolf crossed the southeast end of Lac Capot Blanc on a small land bridge, where she paused for 4.5 h after traveling for 19.5 h (37.5 km). Following this rest, she traveled for 9 h (26.3 km) onto a peninsula in Reid Lake, where she spent 2 h before backtracking and stopping for 8 h just off the peninsula. Her next period of travel lasted 16.5 h (32.7 km), terminating in a pause of 9.5 h just 3.8 km from a concentration of locations at the far end of her excursion, where we presume she encountered caribou. The mean duration of these three movement periods was 15.7 h (± 2.5 SE), and that of the pauses, 7.3 h (± 1.5). The wolf required 72.5 h (3.0 d) to travel a minimum of 95 km from her den to this area near caribou (Figure 2). She remained there (35.5 km2) for 151.5 h (6.3 d) and then moved south to Lake of the Enemy, where she stayed (31.9 km^2) for 74 h (3.1 d) before returning to her den. Her greatest distance from the den, 103 km, was recorded 174.5 h (7.3 d) after the excursion

22 The key in the lower right-hand corner of the map shows areas (shaded) within which the wolves and caribou moved, and the dotted trail of 388 during her excursion. From the results depicted on this map, the investigators tried to determine when and where 388 might have encountered caribou and how their locations affected her traveling behavior.

23 The wolf's excursion (her long trip away from the den area) is the focus of this study. These paragraphs present detailed measurements of daily movements during her two-week trip—how far she traveled, how far she was from collared caribou, her time spent traveling and resting, and her rate of speed. Authors use the phrase "minimum distance traveled" to acknowledge they couldn't track every step but were measuring samples of her movements. They knew that she went at least as far as they measured. This shows how scientists try to be exact when reporting results. Results of this study are depicted graphically in the map in Figure 2.

24 Post-excursion measurements of 388's movements were made to compare with those of the pre-excursion period. In order to compare, scientists often use means, or averages, of a series of measurements—mean distances, mean duration, etc.

25 In the comparison, authors used statistical calculations (F and df) to determine that the differences between pre- and post-excursion measurements were statistically insignificant, or close enough to be considered essentially the same or similar.

26 As with wolf 388, the investigators measured the movements of caribou during the study period. The areas within which the caribou moved are shown in Figure 2 by shaded polygons mentioned in the second paragraph of this subsection.

27 This subsection summarizes how distances separating predators and prey varied during the study period.

28 DISCUSSION: This section is the interpretation part of the scientific process.

29 This subsection reviews observations from other studies and suggests that this study fits with patterns of those observations.

30 Authors discuss a prevailing theory (CBFT) which might explain why a wolf would travel far to meet her own energy needs while taking food caught closer to the den back to her pups. The results of this study seem to fit that pattern.

began, at 0433 on 27 July. She was 8 km from a collared caribou on 23 July, four days after the excursion began (Table 1).

The return trip began at 0403 on 2 August, 318 h (13.2 d) after leaving the den. She followed a relatively direct path for 18 h back to the den, a distance of 75 km.

The minimum distance traveled during the excursionwas 339 km. The estimated overall minimum travel rate was 3.1 km/h, 2.6 km/h away from the den and 4.2 km/h on the return trip.

24 **Post-Excursion Period:** We saw three pups when recovering the collar on 20 August, but others may have been hiding in vegetation.

Telemetry recorded 13 foraging bouts in the post-excursion period. The mean distance traveled during these bouts was 18.3 km (+ 2.7 SE, range 1.2–47.7 km), and mean greatest distance from the den was 7.1 km (+ 0.7 SE, range 1.1–11.0 km). The mean duration of these post-excursion foraging bouts was 10.9 h (+ 2.4 SE, range 1–33 h).

When 388 reached her den on 2 August, the distance to the nearest collared caribou was 54 km. On 9 August, one week after she returned, the distance was 79 km (Table 1).

Pre- and Post-Excursion Comparison

25 We found no differences in the mean distance of foraging bouts before and after the excursion period ($F = 1.5$, $df = 1, 29$, $p = 0.24$). Likewise, the mean greatest distance from the den was similar pre- and post-excursion ($F = 0.004$, $df = 1, 29$, $p = 0.95$). However, the mean duration of 388's foraging bouts decreased by 10.0 h after her long excursion ($F = 3.1$, $df = 1, 29$, $p = 0.09$).

26 *Caribou Monitoring*

Summer Movements: On 10 July, 5 of 11 collared caribou were dispersed over a distance of 10 km, 140 km south of their calving grounds (Figure 1). On the same day, three caribou were still on the calving grounds, two were between the calving grounds and the leaders, and one was missing. One week later (17 July), the leading radio-collared cows were 100 km farther south (Figure 1). Two were within 5 km of each other in front of the rest, who were more dispersed. All radio-collared cows had left the calving grounds by this time. On 23 July, the leading radio-collared caribou had moved 35 km farther south, and all of them were more widely dispersed. The two cows closest to the leader were 26 km and 33 km away, with 37 km between them. On the next location (29 July), the most southerly caribou were 60 km farther south. All of the caribou were now in the areas where they remained for the duration of the study (Figure 2).

A Minimum Convex Polygon (Mohr and Stumpf, 1966) around all caribou locations acquired during the study encompassed 85 119 km^2.

Relative to the Wolf Den: The distance from the nearest collared caribou to the den decreased from 241 km one week before the excursion to 124 km the day it began. The nearest a collared caribou came to the den was 43 km away, on 29 and 30 July. During the study, four collared caribou were located within 100 km of the den. Each of these four was closest to the wolf on at least one day during the period reported. 27

28 Discussion

Prey Abundance

Caribou are the single most important prey of tundra wolves (Clark, 1971; Kuyt, 1972; Stephenson and James, 1982; Williams, 1990). Caribou range over vast areas, and for part of the summer, they are scarce or absent in wolf home ranges (Heard et al., 1996). Both the long distance between radio-collared caribou and the den the week before the excursion and the increased time spent foraging by wolf 388 indicate that caribou availability near the den was low. Observations of the pups' being left alone for up to 18 h, presumably while adults were searching for food, provide additional support for low caribou availability locally. Mean foraging bout duration decreased by 10.0 h after the excursion, when collared caribou were closer to the den, suggesting an increase in caribou availability nearby. 29

Foraging Excursion

One aspect of central place foraging theory (CPFT) deals with the optimality of returning different-sized food loads from varying distances to dependents at a central place (i.e., the den) (Orians and Pearson, 1979). Carlson (1985) tested CPFT and found that the predator usually consumed prey captured far from the central place, while feeding prey captured nearby to dependants. Wolf 388 spent 7.2 days in one area near caribou before moving to a location 23 km back towards the den, where she spent an additional 3.1 days, likely hunting caribou. She began her return trip from this closer location, traveling directly to the den. While away, she may have made one or more successful kills and spent time meeting her own energetic needs before returning to the den. Alternatively, it may have taken several attempts to make a kill, 30 which she then fed on before beginning her return trip. We do not know if she returned food to the pups, but such behavior would be supported by CPFT.

31 Other workers have reported wolves' making long round trips and referred to them as "extraterritorial" or "pre-dispersal" forays (Fritts and Mech, 1981; Messier, 1985; Ballard et al., 1997; Merrill and Mech, 2000). These movements are most often made by young wolves (1–3 years old), in areas where annual territories are maintained and prey are relatively sedentary (Fritts and Mech, 1981; Messier, 1985). The long excursion of 388 differs in that tundra wolves do not maintain annual territories (Walton et al., 2001), and the main prey migrate over vast areas (Gunn et al., 2001).

Another difference between 388's excursion and those reported earlier is that she is a mature, breeding female. No study of territorial wolves has reported reproductive adults making extraterritorial movements in summer (Fritts and Mech, 1981; Messier, 1985; Ballard et al., 1997; Merrill and Mech, 2001). However, Walton et al. (2001) also report that breeding female tundra wolves made excursions.

Direction of Movement

32 Possible explanations for the relatively direct route 388 took to the caribou include landscape influence and experience. Considering the timing of 388's trip and the locations of caribou, had the wolf moved northwest, she might have missed the caribou entirely, or the encounter might have been delayed.

A reasonable possibility is that the land directed 388's route. The barrens are crisscrossed with trails worn into the tundra over centuries by hundreds of thousands of caribou and other animals (Kelsall, 1968; Thorpe et al., 2001). At river crossings, lakes, or narrow peninsulas, trails converge and funnel towards and away from caribou calving grounds and summer range. Wolves use trails for travel (Paquet et al., 1996; Mech and Boitani, 2003; P. Frame, pers. observation). Thus, the landscape may direct an animal's movements and lead it to where cues, such as the odor of caribou on the wind or scent marks of other wolves, may lead it to caribou.

33 Another possibility is that 388 knew where to find caribou in summer. Sexually immature tundra wolves sometimes follow caribou to calving grounds (D. Heard, unpubl. data). Possibly, 388 had made such journeys in previous years and killed caribou. If this were the case, then in times of local prey scarcity she might travel to areas where she had hunted successfully before. Continued monitoring of tundra wolves may answer questions about how their food needs are met in times of low caribou abundance near dens.

34 Caribou often form large groups while moving south to the tree line (Kelsall, 1968). After a large aggregation of caribou moves through an area, its scent can linger for weeks (Thorpe et al., 2001:104). It is conceivable that 388 detected caribou scent on the wind, which was blowing from the northeast on 19–21 July (Environment Canada, 2003), at the same time her excursion began. Many factors, such as odor strength and wind direction and strength, make systematic study of scent detection in wolves difficult under field conditions (Harrington and Asa, 2003). However, humans are able to smell odors such as forest fires or oil refineries more than 100 km away. The olfactory capabilities of dogs, which are similar to wolves, are thought to be 100 to 1 million times that of humans (Harrington and Asa, 2003). Therefore, it is reasonable to think that under the right wind conditions, the scent of many caribou traveling together could be detected by wolves from great distances, thus triggering a long foraging bout.

Rate of Travel

35 Mech (1994) reported the rate of travel of Arctic wolves on barren ground was 8.7 km/h during regular travel and 10.0 km/h when returning to the den, a difference of 1.3 km/h. These rates are based on direct observation and exclude periods when wolves moved slowly or not at all. Our calculated travel rates are assumed to include periods of slow movement or no movement. However, the pattern we report is similar to that reported by Mech (1994), in that homeward travel was faster than regular travel by 1.6 km/h. The faster rate on return may be explained by the need to return food to the den. Pup survival can increase with the number of adults in a pack available to deliver food to pups (Harrington et al., 1983). Therefore, an increased rate of travel on homeward trips could improve a wolf's reproductive fitness by getting food to pups more quickly.

Fate of 388's Pups

36 Wolf 388 was caring for pups during den observations. The pups were estimated to be six weeks old, and were seen ranging as far as 800 m from the den. They received some regurgitated food from two of the females, but were unattended for long periods. The excursion started 16 days after our observations, and it is improbable that the pups could have traveled the distance that 388 moved. If the pups died, this would have removed parental responsibility, allowing the long movement.

Our observations and the locations of radio-collared caribou indicate that prey became scarce in

31 Here our authors note other possible explanations for wolves' excursions presented by other investigators, but this study does not seem to support those ideas.

32 Authors discuss possible reasons for why 388 traveled directly to where caribou were located. They take what they learned from earlier studies and apply it to this case, suggesting that the lay of the land played a role. Note that their description paints a clear picture of the landscape.

33 Authors suggest that 388 may have learned in traveling during previous summers where the caribou were. The last two sentences suggest ideas for future studies.

34 Or maybe 388 followed the scent of the caribou. Authors acknowledge difficulties of proving this, but they suggest another area where future studies might be done.

35 Authors suggest that results of this study support previous studies about how fast wolves travel to and from the den. In the last sentence, they speculate on how these observed patterns would fit into the theory of evolution.

36 Authors also speculate on the fate of 388's pups while she was traveling. This leads to . . .

37 Discussion of cooperative rearing of pups and, in turn, to speculation on how this study and what is known about cooperative rearing might fit into the animal's strategies for survival of the species. Again, the authors approach the broader theory of evolution and how it might explain some of their results.

38 And again, they suggest that this study points to several areas where further study will shed some light.

39 In conclusion, the authors suggest that their study supports the hypothesis being tested here. And they touch on the implications of increased human activity on the tundra predicted by their results.

40 ACKNOWLEDGEMENTS: Authors note the support of institutions, companies and individuals. They thank their reviewers and list permits under which their research was carried out.

41 REFERENCES: List of all studies cited in the report. This may seem tedious, but is a vitally important part of scientific reporting. It is a record of the sources of information on which this study is based. It provides readers with a wealth of resources for further reading on this topic. Much of it will form the foundation of future scientific studies like this one.

the area of the den as summer progressed. Wolf 388 may have abandoned her pups to seek food for herself. However, she returned to the den after the excursion, where she was seen near pups. In fact, she foraged in a similar pattern before and after the excursion, suggesting that she again was providing for pups after her return to the den.

37 A more likely possibility is that one or both of the other lactating females cared for the pups during 388's absence. The three females at this den were not seen with the pups at the same time. However, two weeks earlier, at a different den, we observed three females cooperatively caring for a group of six pups. At that den, the three lactating females were observed providing food for each other and trading places while nursing pups. Such a situation at the den of 388 could have created conditions that allowed one or more of the lactating females to range far from the den for a period, returning to her parental duties afterwards. However, the pups would have been weaned by eight weeks of age (Packard et al., 1992), so nonlactating adults could also have cared for them, as often happens in wolf packs (Packard et al., 1992; Mech et al., 1999).

Cooperative rearing of multiple litters by a pack could create opportunities for long-distance foraging movements by some reproductive wolves during summer periods of local food scarcity. We have recorded multiple lactating females at one or more tundra wolf dens per year since 1997. This reproductive strategy may be an adaptation to temporally and 38 spatially unpredictable food resources. All of these possibilities require further study, but emphasize both the adaptability of wolves living on the barrens and their dependence on caribou.

Long-range wolf movement in response to caribou 39 availability has been suggested by other researchers (Kuyt, 1972; Walton et al., 2001) and traditional ecological knowledge (Thorpe et al., 2001). Our report demonstrates the rapid and extreme response of wolves to caribou distribution and movements in summer. Increased human activity on the tundra (mining, road building, pipelines, ecotourism) may influence caribou movement patterns and change the interactions between wolves and caribou in the region. Continued monitoring of both species will help us to assess whether the association is being affected adversely by anthropogenic change.

40 Acknowledgements

This research was supported by the Department of Resources, Wildlife, and Economic Development, Government of the Northwest Territories; the Department of Biological Sciences at the University of Alberta; the Natural Sciences and Engineering Research Council of Canada; the Department of Indian and Northern Affairs Canada; the Canadian Circumpolar Institute; and DeBeers Canada, Ltd. Lorna Ruechel assisted with den observations. A. Gunn provided caribou location data. We thank Dave Mech for the use of GPS collars. M. Nelson, A. Gunn, and three anonymous reviewers made helpful comments on earlier drafts of the manuscript. This work was done under Wildlife Research Permit – WL002948 issued by the Government of the Northwest Territories, Department of Resources, Wildlife, and Economic Development.

41 References

BALLARD, W.B., AYRES, L.A., KRAUSMAN, P.R., REED, D.J., and FANCY, S.G. 1997. Ecology of wolves in relation to a migratory caribou herd in northwest Alaska. Wildlife Monographs 135. 47 p.

BROWN, M.B., and FORSYTHE, A.B. 1974. Robust tests for the equality of variances. Journal of the American Statistical Association 69:364–367.

CARLSON, A. 1985. Central place foraging in the red-backed shrike (*Lanius collurio* L.): Allocation of prey between forager and sedentary consumer. Animal Behaviour 33:664–666.

CLARK, K.R.F. 1971. Food habits and behavior of the tundra wolf on central Baffin Island. Ph.D. Thesis, University of Toronto, Ontario, Canada.

ENVIRONMENT CANADA. 2003. National climate data information archive. Available online: http://www.climate.weatheroffice.ec.gc.ca/Welcome_e.html

FOURNIER, B., and GUNN, A. 1998. Musk ox numbers and distribution in the NWT, 1997. File Report No. 121. Yellowknife: Department of Resources, Wildlife, and Economic Development, Government of the Northwest Territories. 55 p.

FRITTS, S.H., and MECH, L.D. 1981. Dynamics, movements, and feeding ecology of a newly protected wolf population in northwestern Minnesota. Wildlife Monographs 80. 79 p.

GUNN, A., DRAGON, J., and BOULANGER, J. 2001. Seasonal movements of satellite-collared caribou from the Bathurst herd. Final Report to the West Kitikmeot Slave Study Society, Yellowknife, NWT. 80 p. Available online: http://www.wkss.nt.ca/HTML/08_ProjectsReports/PDF/Seasonal MovementsFinal.pdf

HARRINGTON, F.H., and ASA, C.S. 2003. Wolf communication. In: Mech, L.D., and Boitani, L., eds. Wolves: Behavior, ecology, and conservation. Chicago: University of Chicago Press. 66–103.

HARRINGTON, F.H., MECH, L.D., and FRITTS, S.H. 1983. Pack size and wolf pup survival: Their relationship under varying ecological conditions. Behavioral Ecology and Sociobiology 13:19–26.

HARRIS, S.A. 1986. Permafrost distribution, zonation and stability along the eastern ranges of the cordillera of North America. Arctic 39(1):29–38.

HEARD, D.C., and WILLIAMS, T.M. 1992. Distribution of wolf dens on migratory caribou ranges in the Northwest

Territories, Canada. Canadian Journal of Zoology 70:1504–1510.
HEARD, D.C., WILLIAMS, T.M., and MELTON, D.A. 1996. The relationship between food intake and predation risk in migratory caribou and implication to caribou and wolf population dynamics. Rangifer Special Issue No. 2:37–44.
KELSALL, J.P. 1968. The migratory barren-ground caribou of Canada. Canadian Wildlife Service Monograph Series 3. Ottawa: Queen's Printer. 340 p.
KUYT, E. 1962. Movements of young wolves in the Northwest Territories of Canada. Journal of Mammalogy 43:270–271.
———. 1972. Food habits and ecology of wolves on barren-ground caribou range in the Northwest Territories. Canadian Wildlife Service Report Series 21. Ottawa: Information Canada. 36 p.
MECH, L.D. 1994. Regular and homeward travel speeds of Arctic wolves. Journal of Mammalogy 75:741–742.
MECH, L.D., and BOITANI, L. 2003. Wolf social ecology. In: Mech, L.D., and Boitani, L., eds. Wolves: Behavior, ecology, and conservation. Chicago: University of Chicago Press. 1–34.
MECH, L.D., and GESE, E.M. 1992. Field testing the Wildlink capture collar on wolves. Wildlife Society Bulletin 20:249–256.
MECH, L.D., WOLFE, P., and PACKARD, J.M. 1999. Regurgitative food transfer among wild wolves. Canadian Journal of Zoology 77:1192–1195.
MERRILL, S.B., and MECH, L.D. 2000. Details of extensive movements by Minnesota wolves (*Canis lupus*). American Midland Naturalist 144:428–433.
MERRILL, S.B., ADAMS, L.G., NELSON, M.E., and MECH, L.D. 1998. Testing releasable GPS radiocollars on wolves and white-tailed deer. Wildlife Society Bulletin 26:830–835.
MESSIER, F. 1985. Solitary living and extraterritorial movements of wolves in relation to social status and prey abundance. Canadian Journal of Zoology 63:239–245.
MOHR, C.O., and STUMPF, W.A. 1966. Comparison of methods for calculating areas of animal activity. Journal of Wildlife Management 30:293–304.
MUSIANI, M. 2003. Conservation biology and management of wolves and wolf-human conflicts in western North America. Ph.D. Thesis, University of Calgary, Calgary, Alberta, Canada.
ORIANS, G.H., and PEARSON, N.E. 1979. On the theory of central place foraging. In: Mitchell, R.D., and Stairs, G.F., eds. Analysis of ecological systems. Columbus: Ohio State University Press. 154–177.
PACKARD, J.M., MECH, L.D., and REAM, R.R. 1992. Weaning in an arctic wolf pack: Behavioral mechanisms. Canadian Journal of Zoology 70:1269–1275.
PAQUET, P.C., WIERZCHOWSKI, J., and CALLAGHAN, C. 1996. Summary report on the effects of human activity on gray wolves in the Bow River Valley, Banff National Park, Alberta. In: Green, J., Pacas, C., Bayley, S., and Cornwell, L., eds. A cumulative effects assessment and futures outlook for the Banff Bow Valley. Prepared for the Banff Bow Valley Study. Ottawa: Department of Canadian Heritage.
STEPHENSON, R.O., and JAMES, D. 1982. Wolf movements and food habits in northwest Alaska. In: Harrington, F.H., and Paquet, P.C., eds. Wolves of the world. New Jersey: Noyes Publications. 223–237.
THORPE, N., EYEGETOK, S., HAKONGAK, N., and QITIRMIUT ELDERS. 2001. The Tuktu and Nogak Project: A caribou chronicle. Final Report to the West Kitikmeot/Slave Study Society, Ikaluktuuttiak, NWT. 160 p.
TIMONEY, K.P., LA ROI, G.H., ZOLTAI, S.C., and ROBINSON, A.L. 1992. The high subarctic forest-tundra of northwestern Canada: Position, width, and vegetation gradients in relation to climate. Arctic 45(1):1–9.
WALTON, L.R., CLUFF, H.D., PAQUET, P.C., and RAMSAY, M.A. 2001. Movement patterns of barren-ground wolves in the central Canadian Arctic. Journal of Mammalogy 82:867–876.
WILLIAMS, T.M. 1990. Summer diet and behavior of wolves denning on barren-ground caribou range in the Northwest Territories, Canada. M.Sc. Thesis, University of Alberta, Edmonton, Alberta, Canada.
ZAR, J.H. 1999. Biostatistical analysis. 4th ed. New Jersey: Prentice Hall. 663 p.

Glossary

abscisic acid (ABA) A plant hormone involved in the abscission of leaves, flowers, and fruits, dormancy of buds and seeds, and closing of stomata.

abscission In plants, the dropping of flowers, fruits, and leaves in response to environmental signals.

accommodation A process by which the lens changes to enable the eye to focus on objects at different distances.

acid-growth hypothesis A hypothesis to explain how the hormone auxin promotes growth of plant cells; it suggests that auxin stimulates H^+ pumps in the plasma membrane to move H^+ from the cell interior into the cell wall, which increases wall acidity, making the wall expandable.

acquired immune deficiency syndrome (AIDS) A constellation of disorders that follows infection by the HIV virus.

acrosome A specialized secretory vesicle on the head of an animal sperm, which helps the sperm penetrate the egg.

acrosome reaction The process in which enzymes contained in the acrosome are released from an animal sperm and digest a path through the egg coats.

action potential The abrupt and transient change in membrane potential that occurs when a neuron conducts an electrical impulse.

active immunity The production of antibodies in the body in response to exposure to a foreign antigen.

active transport The mechanism by which ions and molecules move against the concentration gradient across a membrane, from the side with the lower concentration to the side with the higher concentration.

adaptation, sensory *See* sensory adaptation.

adaptive (acquired) immunity A specific line of defense against invasion of the body in which individual pathogens are recognized and attacked to neutralize and eliminate them.

adipose tissue Connective tissue containing large, densely clustered cells called adipocytes that are specialized for fat storage.

adrenal cortex The outer region of the adrenal glands, which contains endocrine cells that secrete two major types of steroid hormones, the glucocorticoids and the mineralocorticoids.

adrenal medulla The central region of the adrenal glands, which contains neurosecretory neurons that secrete the catecholamine hormones epinephrine and norepinephrine.

adrenocorticotropic hormone (ACTH) A hormone that triggers hormone secretion by cells in the adrenal cortex.

adventitious root A root that develops from the stem or leaves of a plant.

afferent arteriole The vessel that delivers blood to the glomerulus of the kidney.

afferent neuron A neuron that transmits stimuli collected by a sensory receptor to an interneuron.

aggregate fruit A fruit that develops from multiple separate carpels of a single flower, such as a raspberry or strawberry.

agonist A muscle that causes movement in a joint when it contracts.

albumin The most abundant protein in blood plasma, important for osmotic balance and pH buffering; also, the portion of an egg that serves as the main source of nutrients and water for the embryo.

aldosterone A mineralocorticoid hormone released from the adrenal cortex that increases the amount of Na^+ reabsorbed from the urine in the kidneys and absorbed from foods in the intestine, reduces the amount of Na^+ secreted by salivary and sweat glands, and increases the rate of K^+ excretion by the kidneys, keeping Na^+ and K^+ balanced at the levels required for normal cellular function.

aleurone The thin layer of cells that separates the endosperm of a seed from the pericarp.

allantois In an amniote egg, an extraembryonic membrane sac that fills much of the space between the chorion and the yolk sac and store's the embryo's nitrogenous wastes.

allergen A type of antigen responsible for allergic reactions, which induces B cells to secrete an overabundance of IgE antibodies.

all-or-nothing principle The principle that an action potential is produced only if the stimulus is strong enough to cause depolarization to reach the threshold.

alternation of generations The regular alternation of mode of reproduction in the life cycle of an organism, such as the alternation between diploid (sporophyte) and haploid (gametophyte) phases in plants.

alveolus (plural, alveoli) One of the millions of tiny air pockets in mammalian lungs, each surrounded by dense capillary networks.

amacrine cell A type of neuron that forms lateral connections in the retina of the eye, connecting bipolar cells and ganglion cells.

ammonification A metabolic process in which bacteria and fungi convert organic nitrogen compounds into ammonia and ammonium ions; part of the nitrogen cycle.

amnion In an amniote egg, an extraembryonic membrane that encloses the embryo, forming the amniotic cavity and secreting amniotic fluid, which provides an aquatic environment in which the embryo develops.

amniota The monophyletic group of vertebrates that have an amnion during embryonic development.

amplification The process in which an activated receptor activates many proteins, which then activate an even larger number of proteins.

amygdala A gray-matter center of the brain that works as a switchboard, routing information about experiences that have an emotional component through the limbic system.

anabolic steroid A steroid hormone that stimulates muscle development.

anaphylactic shock A severe inflammation stimulated by an allergen, involving extreme swelling of air passages in the lungs that interferes with breathing, and massive leakage of fluid from capillaries that causes blood pressure to drop precipitously.

anatomy The study of the structures of organisms.

androgen One of a family of hormones that promote the development and maintenance of sex characteristics.

angiotensin A peptide hormone that raises blood pressure quickly by constricting arterioles in most parts of the body; it also stimulates release of the steroid hormone aldosterone.

animal pole The end of the egg where the egg nucleus is located, which typically gives rise to surface structures and the anterior end of the embryo.

annual An herbaceous plant that completes its life cycle in one growing season and then dies.

antagonistic pair Two skeletal muscles, one of which flexes as the other extends to move joints.

anterior pituitary The glandular part of the pituitary, composed of endocrine cells that synthesize and secrete several tropic and nontropic hormones.

anther The pollen-bearing part of a stamen.

antibody A highly specific soluble protein molecule that circulates in the blood and lymph, recognizing and binding to antigens and clearing them from the body.

antibody-mediated immunity Adaptive immune response in which plasma cells secrete antibodies.

antidiuretic hormone (ADH) A hormone secreted by the posterior pituitary that increases water absorption in the kidneys, thereby increasing the volume of the blood.

antigen A foreign molecule that triggers an adaptive immunity response.

antigen-presenting cell (APC) A cell that presents an antigen to T cells in antibody-mediated immunity and cell-mediated immunity.

antiport A secondary active transport mechanism in which a molecule moves through a membrane channel into a cell and powers the active transport of a second molecule out of the cell.

aorta A large artery from the heart that branches into arteries leading to all body regions except the lungs.

aortic body One of several small clusters of chemoreceptors, baroreceptors, and supporting cells located along the aortic arch, that measures changes in blood pressure and the composition of arterial blood flowing past it.

apical dominance Inhibition of the growth of lateral buds in plants due to auxin diffusing down a shoot tip from the terminal bud.

apical meristem A region of unspecialized dividing cells at shoot tips and root tips of a plant.

apomixis In plants, the production of offspring without meiosis or formation of gametes.

apoplastic pathway The route followed by water moving through plant cell walls and intercellular spaces (the apoplast). *Compare* symplastic pathway.

apoptosis Programmed cell death.

appendicular skeleton The bones comprising the pectoral (shoulder) and pelvic (hip) girdles and limbs of a vertebrate.

appendix A fingerlike sac that extends from the cecum of the large intestine.

aquaporin A specialized protein channel that facilitates diffusion of water through cell membranes.

aqueous humor A clear fluid that fills the space between the cornea and lens of the eye.

archenteron The central endoderm-lined cavity of an embryo at the gastrula stage, which forms the primitive gut.

artery A vessel that conducts blood away from the heart at relatively high pressure.

arteriole A branch from a small artery at the point where it reaches the organ it supplies.

asexual reproduction Any mode of reproduction in which a single individual gives rise to offspring without fusion of gametes; that is, without genetic input from another individual.

association area One of several areas surrounding the sensory and motor areas of the cerebral cortex that integrate information from the sensory areas, formulate responses, and pass them on to the primary motor area.

astrocyte A star-shaped glial cell that provides support to neurons in the vertebrate central nervous system.

atrium (plural, atria) A body cavity or chamber surrounding the perforated pharynx of invertebrate chordates; also, one of the chambers that receive blood returning to the heart.

atrial natriuretic factor (ANF) A peptide hormone that inhibits renin release and increases the filtration rate by dilating the arterioles that deliver blood to glomeruli and by inhibiting aldosterone release.

atrioventricular node (AV node) A region of the heart wall that receives signals from the sinoatrial node and conducts them to the ventricle.

atrioventricular valve (AV valve) A valve composed of endocardium and connective tissue between each atrium and ventricle, which prevents backflow of blood from the ventricle to the atrium during emptying of the heart.

autoimmune reaction The production of antibodies against molecules of the body.

autonomic nervous system A subdivision of the peripheral nervous system that controls largely involuntary processes including digestion, secretion by sweat glands, circulation of the blood, many functions of the reproductive and excretory systems, and contraction of smooth muscles in all parts of the body.

auxin Any of a family of plant hormones that stimulate growth by promoting cell elongation in stems and coleoptiles; inhibit abscission; govern responses to light and gravity, and have other developmental effects.

Avr gene A gene in certain plant pathogens that encodes a product triggering a defensive response in the plant.

axial skeleton The skull, vertebral column, sternum, and rib cage, forming the part of the skeleton along a vertebrate's midline.

axil The upper angle between the stem and an attached leaf.

axon The single elongated extension of a neuron that conducts signals away from the cell body to another neuron or an effector.

axon hillock A junction with the cell body of a neuron from which the axon arises.

axon terminal A branch at the tip of an axon that ends as a small, buttonlike swelling.

B cell A lymphocyte that recognizes antigens in the body.

bacteroid A rod-shaped or branched bacterium in the root nodules of nitrogen-fixing plants.

bark The tough outer covering of woody stems and roots, composed of all the living and nonliving tissues between the vascular cambium and the stem surface.

basal lamina The membrane that fixes the epithelium to underlying tissues (also called the basement membrane).

basal nucleus One of several gray-matter centers that surround the thalamus on both sides of the brain and moderate voluntary movements directed by motor centers in the cerebrum.

basophil A type of leukocyte that is induced to secrete histamine by allergens.

B-cell receptor (BCR) The receptor on B cells that is specific for a particular antigen.

biennial A plant that completes its life cycle in two growing seasons and then dies; limited secondary growth occurs in some biennials.

bile A mixture of substances including bile salts, cholesterol, and bilirubin that is made in the liver, stored in the gallbladder, and used in the digestion of fats.

biological clock An internal time-measuring mechanism that adapts an organism to recurring environmental changes.

bipolar cell A type of neuron in the retina of the eye that connects the rods and cones with the ganglion cells.

blade The expanded part of a leaf that provides a large surface area for absorbing sunlight and carbon dioxide.

blastocoel A fluid filled cavity in the blastula embryo.

blastocyst An embryonic stage in mammals; a single-cell-layered hollow ball of about 120 cells with a fluid-filled blastocoel in which a dense mass of cells is localized to one side.

blastodisc A disclike layer of cells at the surface of the yolk produced by early cleavage divisions.

blastomere A small cell formed during cleavage of the embryo.

blastopore The opening at one end of the archenteron in the gastrula that gives rise to the mouth in protostomes and the anus in deuterostomes.

blastula The hollow ball of cells that is the result of cleavage divisions in an early embryo.

blood A fluid connective tissue composed of blood cells suspended in a fluid extracellular matrix, plasma.

blood-brain barrier A specialize arrangement of capillaries in the brain that prevents most substances dissolved in the blood from entering the cerebrospinal fluid and thus protects the brain and spinal cord from viruses, bacteria, and toxic substances that may circulate in the blood.

bolting Rapid formation of a floral shoot in plant species that form rosettes, such as lettuce.

bolus The food mass after chewing.

bone The densest form of connective tissue, in which living cells secrete the mineralized matrix of collagen and calcium salts that surrounds them; forms the skeleton.

Bowman's capsule An infolded region at the proximal end of a nephron that cups around the glomerulus and collects the water and solutes filtered out of the blood.

brain A single, organized collection of nervous tissue in an organism's head that forms the control center of the nervous system and major sensory structures.

brain hormone (BH) A peptide hormone secreted by neurosecretory neurons in the brain of insects.

brain stem A stalklike structure formed by the pons and medulla, along with the midbrain, which connects the forebrain with the spinal cord.

brassinosteroid Any of a family of plant hormones that stimulate cell division and elongation and differentiation of vascular tissue.

breathing The exchange of gases with the respiratory medium by animals.

bronchus (plural, bronchi) An airway that leads from the trachea to the lungs.

bronchiole One of the small, branching airways in the lungs that lead into the alveoli.

brown adipose tissue A specialized tissue in which the most intense heat generation by nonshivering thermogenesis takes place.

budding A mode of asexual reproduction in which a new individual grows and develops while attached to the parent.

bulbourethral gland One of two pea sized glands on either side of the prostate gland, which secrete a mucous fluid that is added to semen.

bulk feeder An animal that consumes sizeable food items whole or in large chunks.

bulk flow The group movement of molecules in response to a difference in pressure between two locations.

cadherin A cell surface protein responsible for selective cell adhesions that require calcium ions to set up adhesions.

calcitonin A nontropic peptide hormone that lowers the level of Ca^{2+}in the blood by inhibiting the ongoing dissolution of calcium from bone.

callus An undifferentiated tissue that develops on or around a cut plant surface or in tissue culture.

calyx The outermost whorl of a flower, made up of sepals; early in the development of a flower, it encloses all the other parts, as in an unopened bud.

canines Pointed, conical teeth of a mammal, located between the incisors and the first premolars, that are specialized for biting and piercing.

capillary The smallest diameter blood vessel, with a wall that is one cell thick, which forms highly branched networks well adapted for diffusion of substances.

cardiac cycle The systole-diastole sequence of the heart.

cardiac muscle The contractile tissue of the heart.

carnivore An animal that primarily eats other animals.

carotid body A small cluster of chemoreceptors and supporting cells located near the bifurcation of the carotid artery that measures changes in the composition of arterial blood flowing through it.

carpel The reproductive organ of a flower that houses an ovule and its associated structures.

cartilage A tissue composed of sparsely distributed chondrocytes surrounded by networks of collagen fibers embedded in a tough but elastic matrix of the glycoprotein.

Casparian strip A thin, waxy impermeable band that seals abutting cell walls in roots; the strip helps control the type and amount of solutes that enter the stele by blocking the apoplastic pathway at the endodermis and forcing substances to pass through cells (the symplast).

catecholamine Any of a class of compounds derived from the amino acid tyrosine that circulates in the bloodstream, including epinephrine and norepinephrine.

cation exchange Replacement of one cation with another, as on a soil particle.

$CD4^+$ T cell A type of T cell in the lymphatic system that has CD4 receptors on its surface. This type of T cell binds to an antigen-presenting cell in antibody-mediated immunity.

$CD8^+$ T cell A type of T cell in the lymphatic system that has CD8 receptors on its surface. This type of T cell binds to an antigen-presenting cell in cell-mediated immunity.

cecum A a blind pouch formed at the junction of the large and small intestine.

cell adhesion molecule A cell surface protein responsible for selective cell adhesions.

cell body The portion of the neuron containing genetic material and cellular organelles.

cell differentiation A process in which changes in gene expression establish cells with specialized structure and function.

cell expansion A mechanism that enlarges the cells in specific directions in a developing organ.

cell lineage Cell derivation from the undifferentiated tissues of the embryo.

cell-mediated immunity An adaptive immune response in which a subclass of T cells—cytotoxic T cells—becomes activated and, with other cells of the immune system, attacks host cells infected by pathogens, particularly those infected by a virus.

central canal The central portion of the vertebral column in which the spinal cord is found.

central nervous system (CNS) One of the two major divisions of the nervous system containing the brain and spinal cord.

central vacuole A large, water-filled organelle in plant cells that maintains the turgor of the cell and controls movement of molecules between the cytosol and sap.

cerebellum The portion of the brain that receives sensory input from receptors in muscles and joints, from balance receptors in the inner ear, and from the receptors of touch, vision, and hearing.

cerebral cortex A thin outer shell of gray matter covering a thick core of white matter within each hemisphere of the brain; the part of the forebrain responsible for information processing and learning.

cerebrospinal fluid Fluid that circulates through the central canal of the spinal cord and the ventricles of the brain, cushioning the brain and spinal cord from jarring movements and impacts, as well as nourishing the CNS and protecting it from toxic substances.

cervix The lower end of the uterus.

chemical synapse A type of communicating connection between two neurons or a neuron and an effector cell in which an electrical impulse arriving at an axon terminal of the presynaptic cell triggers release of a neurotransmitter that crosses the gap and binds to a receptor on the postsynaptic cell, triggering an electrical impulse in that cell.

chemokine A protein secreted by activated macrophages that attracts other cells, such as neutrophils.

chemoreceptor A sensory receptor that detects specific molecules, or chemical conditions such as acidity.

chlorosis An abnormal yellowing of plant tissues due to lack of chlorophyll; a sign of nutrient deficiency or infection by a pathogen.

chondrocyte A cartilage-producing cell.

chorion In an amniote egg, an extraembryonic membrane that surrounds the embryo and yolk sac completely and exchanges oxygen and carbon dioxide with the environment; becomes part of the placenta in mammals.

chorionic villus (plural, villi) One of many treelike extensions from the chorion, which greatly increase the surface area of the chorion.

chylomicron A small triglyceride droplet covered by a protein coat.

chyme Digested content of the stomach released for further digestion in the small intestine.

ciliary body A fine ligament in the eye that anchors the lens to a surrounding layer of connective tissue and muscle.

circadian rhythm Any biological activity that is repeated in cycles, each about 24 hours long, independently of any shifts in environmental conditions.

circulatory system An organ system consisting of a fluid, a heart, and vessels for moving important molecules, and often cells, from one tissue to another.

circumcision Removal of the prepuce for religious, cultural, or hygienic reasons.

class II major histocompatibility complex (MHC) A collection of proteins that present antigens on the cell surface of an antigen-presenting cell in an antibody-mediated immune response.

cleavage Mitotic cell divisions of the zygote that produce a blastula from a fertilized ovum.

clitoris The structure at the junction of the labia minora in front of the vulva, homologous to the penis in the male.

clonal analysis A method of culturing meristematic tissue that contains a mutated embryonic cell having a readily observable trait, such as the absence of normal pigment.

clonal expansion The proliferation of the activated $CD4^+$ T cell by cell division to produce a clone of cells.

clonal selection The process by which a lymphocyte is specifically selected for cloning when it encounters a foreign antigen from among a randomly generated, enormous diversity of lymphocytes with receptors that specifically recognize the antigen.

closed circulatory system A circulatory system in which the fluid, blood, is confined in blood vessels and is distinct from the interstitial fluid.

cochlea A snail-shaped structure in the inner ear containing the organ of hearing.

cohesion-tension mechanism of water transport A model of how water is transported from roots to leaves in vascular plants; the evaporation of water from leaves pulls water up in xylem by creating a continuous negative pressure (tension) that extends to roots.

coleoptile A protective sheath that covers the shoot apical meristem and plumule of the embryo in monocots, such as grasses, as it pushes up through soil.

coleorhiza A sheath that encloses the radicle of an embryo until it breaks out of the seed coat and enters the soil as the primary root.

collagen Fibrous glycoprotein—very rich in carbohydrates—embedded in a network of proteoglycans.

collecting duct A location where urine leaving individual nephrons is processed further.

collenchyma One of three simple plant tissues. Flexibly supports rapidly growing plant parts. Its elongated cells are alive at maturity and collectively often form strands or a sheathlike cylinder under the dermal tissue of growing shoot regions and leaf stalks.

colon The main part of the large intestine.

companion cell A specialized parenchyma cell that is connected to a mature sieve tube member by plasmodesmata and assists sieve tube members both with the uptake of sugars and with the unloading of sugars in tissues.

complement system A nonspecific defense mechanism activated by invading pathogens, made up of more than 30 interacting soluble plasma proteins circulating in the blood and interstitial fluid.

complete flower A flower in which all four whorls (sepals, petals, stamens, carpels) are present.

compound eye The eye of most insects and some crustaceans, composed of many faceted, light-sensitive units called ommatidia fitted closely together, each having its own refractive system and each forming a portion of an image.

conduction The flow of heat between atoms or molecules in direct contact.

cone In the vertebrate eye, a photoreceptor in the retina that is specialized for detection of different wavelengths (colors); in cone-bearing plants, a cluster of sporophylls.

connective tissue Tissue having cells scattered through an extracellular matrix; forms layers in and around body structures that support other body tissues, transmit mechanical and other forces, and in some cases act as filters.

consciousness Awareness of oneself, identity, and surroundings, with understanding of the significance and likely consequences of events.

convection The transfer of heat from a body to a fluid, such as air or water, that passes over its surface.

copulation The physical act involving the introduction of the accessory sex organ of a male into the accessory sex organ of a female to accomplish internal fertilization.

cork A nonliving, impermeable secondary tissue that is one element of bark.

cork cambium A lateral meristem in plants that forms periderm, which in turn produces cork.

cornea The transparent layer that forms the front wall of the eye, covering the iris.

corolla The structure formed collectively by the petals of a flower.

corpus callosum A structure formed of thick axon bundles that connect the two cerebral hemispheres and coordinate their functions.

corpus luteum Cells remaining at the surface of the ovary during the luteal phase; the structure acts as an endocrine gland, secreting several hormones: estrogens, large quantities of progesterone, and inhibin.

cortex Generally, an outer, rindlike layer. In mammals, the outer layer of the brain, the kidneys, or the adrenal glands. In plants, the outer region of tissue in a root or stem lying between the epidermis and the vascular tissue, composed mainly of parenchyma.

cortical granule A secretory vesicle just under the plasma membrane of an egg cell.

cortical reaction The reaction in which cortical granules fuse with the plasma membrane of the egg and release their contents to the outside.

cortisol The major glucocorticoid steroid hormone secreted by the adrenal cortex, which increases blood glucose by promoting breakdown of proteins and fats.

countercurrent exchange A mechanism in which the water flowing over the gills moves in a direction opposite to the flow of blood under the respiratory surface.

cranial nerve A nerve that connects the brain directly to the head, neck, and body trunk.

crassulacean acid metabolism (CAM) A biochemical variation of photosynthesis that was discovered in a member of the plant family Crassulaceae. Carbon dioxide is taken up and stored during the night to allow the stomata to remain closed during the daytime, decreasing water loss.

crop Of birds, an enlargement of the digestive tube where the digestive contents are stored and mixed with lubricating mucus.

cryptochrome A light-absorbing protein that is sensitive to blue light and that may also be an important early step in various light-based growth responses.

cupula In certain mechanoceptors, a gelatinous structure with stereocilia extending into it that moves with pressure changes in the surrounding water; movement of the cupula bends the stereocilia, which triggers release of neurotransmitters.

cuticle The outer layer of plants and some animals, which helps prevent desiccation by slowing water loss.

cytokine A molecule secreted by one cell type that binds to receptors on other cells and, through signal transduction pathways, triggers a response. In innate immunity, cytokines are secreted by activated macrophages.

cytokinin A hormone that promotes and controls growth responses of plants.

cytoplasmic determinants The mRNA and proteins stored in the egg cytoplasm that direct the first stages of animal development in the period before genes of the zygote become active.

cytotoxic T cell A T lymphocyte that functions in cell-mediated immunity to kill body cells infected by viruses or transformed by cancer.

daily torpor A period of inactivity and lowered metabolic rate that allows an endotherm to conserve energy when environmental temperatures are low.

day-neutral plant A plant that flowers without regard to photoperiod.

dendrite The branched extension of the nerve cell body that receives signals from other nerve cells.

dendritic cell A type of phagocyte, so called because it has many surface projections that resemble dendrites of neurons, which engulfs a bacterium in infected tissue by phagocytosis.

depolarized State of the membrane (which was polarized at rest) as the membrane potential becomes less negative.

deposit feeder An animal that consumes particles of organic matter from the solid substrate on which it lives.

derivative One of the daughter cells produced when a plant cell divides; it typically divides once or twice and then enters on the path to differentiation.

dermal tissue system The plant tissue system that comprises the outer tissues of the plant body, including the epidermis and periderm; it serves as a protective covering for the plant body.

dermis The skin layer below the epidermis; it is packed with connective tissue fibers such as collagen, which resist compression, tearing, or puncture of the skin.

determinate growth The pattern of growth in most animals in which individuals grow to a certain size and then their growth slows dramatically or stops.

determination Mechanism in which the developmental fate of a cell is set.

diabetes mellitus A disease that results from problems with insulin production or action.

diastole The period of relaxation and filling of the heart between contractions.

digestion The splitting of carbohydrates, proteins, lipids, and nucleic acids in foods into chemical subunits small enough to be absorbed into the body fluids and cells of an animal.

digestive tube A tubelike digestive system with two openings that form a separate mouth and anus; the digestive contents move in one direction through specialized regions of the tube, from the mouth to the anus.

dioecious Having male flowers and female flowers on different plants of the same species.

direct neurotransmitter A neurotransmitter that binds directly to a ligand-gated ion channel in the postsynaptic membrane, opening or closing the channel gate and altering the flow of a specific ion or ions in the postsynaptic cell.

distal convoluted tubule The tubule in the human nephron that drains urine into a collecting duct that leads to the renal pelvis.

dormancy A period in the life cycle in which biological activity is suspended.

dorsal lip of the blastopore A crescent-shaped depression rotated clockwise 90° on the embryo surface that marks the region derived from the gray crescent, to which cells from the animal pole move as gastrulation begins.

double fertilization The characteristic feature of sexual reproduction in flowering plants. In the embryo sac, one sperm nucleus unites with the egg to form a diploid zygote from which the embryo develops, and another unites with two polar nuclei to form the primary endosperm nucleus.

duodenum A short region of the small intestine where secretions from the pancreas and liver enter a common duct.

ecdysone A steroid hormone secreted by the prothoracic glands of insects.

echolocation A technique for locating prey by making squeaking or clicking noises, and then listening for the echoes that bounce back from objects in their environment.

ectoderm The outermost of the three primary germ layers of an embryo, which develops into epidermis and nervous tissue.

ectotherm An animal that obtains its body heat primarily from the external environment.

effector In homeostatic feedback, the system that returns the condition to the set point if it has strayed away.

effector T cell A cell involved in effecting—bringing about—the specific immune response to an antigen.

efferent arteriole The arteriole that receives blood from the glomerulus.

efferent neuron A neuron that carries the signals indicating a response away from the interneuron networks to the effectors.

egg cell The female reproductive cell.

elastin A rubbery protein in some connective tissues that adds elasticity to the extracellular matrix—it is able to return to its original shape after being stretched, bent, or compressed.

electrical synapse A mechanical and electrically conductive link between two abutting neurons that is formed at the gap junction.

electrocardiogram (ECG) Graphic representation of the electrical activity within the heart, detected by electrodes placed on the body.

electroreceptor A specialized sensory receptor that detects electrical fields.

embryo sac The female gametophyte of angiosperms, within which the embryo develops; it usually consists of seven cells: an egg cell, an endosperm mother cell, and five other cells with fleeting reproductive roles.

endocrine gland Any of several ductless secretory organs that secrete hormones into the blood or extracellular fluid.

endocrine system The system of glands that release their secretions (hormones) directly into the circulatory system.

endoderm The innermost of the three primary germ layers of an embryo, which develops into the gastrointestinal tract and, in some animals, the respiratory organs.

endodermis The innermost layer of the root cortex; a selectively permeable barrier that helps control the movement of water and dissolved minerals into the stele.

endorphin One of a group of small proteins occurring naturally in the brain and around nerve endings that bind to opiate receptors and thus can raise the pain threshold.

endoskeleton A supportive internal body structure, such as bones, that provides support.

endosperm Nutritive tissue inside the seeds of flowering plants.

endotherm An animal that obtains most of its body heat from internal physiological sources.

enzymatic hydrolysis A process in which chemical bonds are broken by the addition of H^+ and OH^-, the components of a molecule of water.

eosinophil A type of leukocyte that targets extracellular parasites too large for phagocytosis in the inflammatory response.

epiblast The top layer of the blastodisc.

epicotyl The upper part of the axis of an early plant embryo, located between the cotyledons and the first true leaves.

epidermis A complex tissue that covers an organism's body in a single continuous layer or sometimes in multiple layers of tightly packed cells.

epididymis A coiled storage tubule attached to the surface of each testis.

epiglottis A flaplike valve at the top of the trachea.

epinephrine A nontropic amine hormone secreted by the adrenal medulla.

epiphyte A plant that grows independently on other plants and obtains nutrients and water from the air.

epithelial tissue Tissue formed of sheetlike layers of cells that are usually joined tightly together, with little extracellular matrix material between them. They protect body surfaces from invasion by bacteria and viruses, and secrete or absorb substances.

epitope The small region of an antigen molecule to which BCRs or TCRs bind.

erythrocyte A red blood cell, which contains hemoglobin, a protein that transports O_2 in blood.

erythropoietin (EPO) A hormone that stimulates stem cells in bone marrow to increase erythrocyte production.

esophagus A connecting passage of the digestive tube.

essential amino acid Any amino acid that is not made by the human body but must be taken in as part of the diet.

essential element Any of a number of elements required by living organisms to ensure normal reproduction, growth, development, and maintenance.

essential fatty acid Any fatty acid that the body cannot synthesize but needs for normal metabolism.

essential mineral Any inorganic element such as calcium, iron, or magnesium that is required in the diet of an animal.

essential nutrient Any of the essential amino acids, fatty acids, vitamins, and minerals required in the diet of an animal.

estivation Seasonal torpor in an animal that occurs in summer.

estradiol A form of estrogen.

estrogen Any of the group of female sex hormones.

ethylene A plant hormone that helps regulate seedling growth, stem elongation, the ripening of fruit, and the abscission of fruits, leaves, and flowers.

eudicot A plant belonging to the Eudicotyledones, one of the two major classes of angiosperms; their embryos generally have two seed leaves (cotyledons), and their pollen grains have three grooves.

evaporation Heat transfer through the energy required to change a liquid to a gas.

excitatory postsynaptic potential (EPSP) The change in membrane potential caused when a neurotransmitter opens a ligand-gated Na^+ channel and Na^+ enters the cell, making it more likely that the postsynaptic neuron will generate an action potential.

excretion The process that helps maintain the body's water and ion balance while ridding the body of metabolic wastes.

exocrine gland A gland that is connected to the epithelium by a duct and that empties its secretion at the epithelial surface.

exodermis In the roots of some plants, an outer layer of root cortex that may limit water losses from roots and help regulate the absorption of ions.

exoskeleton A hard external covering of an animal's body that blocks the passage of water and provides support and protection.

external fertilization The process in which sperm and eggs are shed into the surrounding water, occurring in most aquatic invertebrates, bony fishes, and amphibians.

external gill A gill that extends out from the body and lacks a protective covering.

extracellular digestion Digestion that takes place outside body cells, in a pouch or tube enclosed within the body.

extracellular fluid The fluid occupying the spaces between cells in multicellular animals.

extraembryonic membrane A primary tissue layer extended outside the embryo that conducts nutrients from the yolk to the embryo, exchanges gases with the environment outside the egg, or stores metabolic wastes removed from the embryo.

fast block to polyspermy The barrier set up by the wave of depolarization triggered when sperm and egg fuse, making it impossible for other sperm to enter the egg.

fast muscle fiber A muscle fiber that contracts relatively quickly and powerfully.

fate map Mapping of adult or larval structures onto the region of the embryo from which each structure developed.

fat-soluble vitamin A vitamin that dissolves in liquid fat or fatty oils, in addition to water.

feces Condensed and compacted digestive contents in the large intestine.

fertilization The union of gametes that initiates development of a new individual.

fetus A developing human from the eighth week of gestation onward, at which point the major organs and organ systems have formed.

fiber In sclerenchyma, an elongated, tapered, thick-walled cell that gives plant tissue its flexible strength.

fibrin A protein necessary for blood clotting; fibrin forms a web-like mesh that traps platelets and red blood cells and holds a clot together.

fibrinogen A plasma protein that plays a central role in the blood-clotting mechanism.

fibroblast The type of cell that secretes most of the collagen and other proteins in the loose connective tissue.

fibronectin A class of glycoproteins that aids in the attachment of cells to the extracellular matrix and helps hold the cells in position.

fibrous connective tissue Tissue in which fibroblasts are sparsely distributed among dense masses of collagen and elastin fibers that are lined up in highly ordered, parallel bundles, producing maximum tensile strength and elasticity.

fibrous root system A root system that consists of branching roots rather than a main taproot; roots tend to spread laterally from the base of the stem.

filament In flowers, the stalk of a stamen, which supports the anther.

filtration The nonselective movement of some water and a number of solutes—ions and small molecules, but not large molecules such as proteins—into the proximal end of the renal tubules through spaces between cells.

blood concentration of glucose and other fuel molecules.

fission The mode of asexual reproduction in which the parent separates into two or more offspring of approximately equal size.

fluid feeder An animal that obtains nourishment by ingesting liquids that contain organic molecules in solution.

follicle cell A cell that grows from ovarian tissue and nourishes the developing egg.

follicle-stimulating hormone (FSH) The pituitary hormone that stimulates oocytes in the ovaries to continue meiosis and become follicles. During follicle enlargement, FSH interacts with luteinizing hormone to stimulate follicular cells to secrete estrogens.

forebrain The largest division of the brain, which includes the cerebral cortex and basal ganglia. It is credited with the highest intellectual functions.

foreskin A loose fold of skin that covers the glans of the penis.

fovea The small region of the retina around which cones are concentrated in mammals and birds with eyes specialized for daytime vision.

fragmentation A type of vegetative reproduction in plants in which cells or a piece of the parent break off, then develop into new individuals.

fruit A mature ovary, often with accessory parts, from a flower.

fusiform initial A cell derived from cambium inside a vascular bundle; gives rise to secondary xylem and phloem cells.

gallbladder The organ that stores bile between meals, when no digestion is occurring.

gamete A haploid cell, an egg or sperm. Haploid cells fuse during sexual reproduction to form a diploid zygote.

gametogenesis The formation of male and female gametes.

gametophyte The phase of the plant life cycle that gives rise to haploid gametes (eggs and sperm).

ganglion A functional concentration of nervous system tissue composed principally of nerve-cell bodies, usually lying outside the central nervous system.

ganglion cell A type of neuron in the retina of the eye that receives visual information from photoreceptors via various intermediate cells such as bipolar cells, amacrine cells, and horizontal cells.

gap gene In *Drosophila* embryonic development, the first activated set of segmentation genes that progressively subdivide the embryo into regions, determining the segments of the embryo and the adult.

gastric juice A substance secreted by the stomach that contains the digestive enzyme pepsin.

gastrovascular cavity A saclike body cavity with a single opening, a mouth, which serves both digestive and circulatory functions.

gastrula The developmental stage resulting when the cells of the blastula migrate and divide once cleavage is complete.

gastrulation The second major process of early development in most animals, which produces an embryo with three distinct primary tissue layers.

gene-for-gene recognition A mechanism in which plants can detect an attack by a specific pathogen; the product of a specific plant gene interacts with the product of a specific pathogen gene, triggering the plant's defensive response.

germ cell An animal cell that is set aside early in embryonic development and gives rise to the gametes.

gestation The period of mammalian development in which the embryo develops in the uterus of the mother.

gibberellin Any of a large family of plant hormones that regulate aspects of growth, including cell elongation.

gill A respiratory organ formed as evagination of the body that extends outward into the respiratory medium.

gizzard The part of the digestive tube that grinds ingested material into fine particles by muscular contractions of the wall.

gland A cell or group of cells that produces and releases substances nearby, in another part of the body, or to the outside.

glans A soft, caplike structure at the end of the penis, containing most of the nerve endings producing erotic sensations.

glial cell A nonneuronal cell contained in the nervous tissue that physically supports and provides nutrients to neurons, provides electrical insulation between them, and scavenges cellular debris and foreign matter.

globulin A plasma protein that transports lipids (including cholesterol) and fat-soluble vitamins; a specialized subgroup of globulins, the immunoglobulins, constitute antibodies and other molecules contributing to the immune response.

glomerulus A ball of blood capillaries surrounded by Bowman's capsule in the human nephron.

glucagon A pancreatic hormone with effects opposite to those of insulin: it stimulates glycogen, fat, and protein degradation.

glucocorticoid A steroid hormone secreted by the adrenal cortex that helps maintain the blood concentration of glucose and other fuel molecules.

Golgi tendon organ A proprioceptor of tendons.

gonad A specialized gamete-producing organ in which the germ cells collect. Gonads are the primary source of sex hormones in vertebrates: ovaries in the female and testes in the male.

gonadotropin A hormone that regulates the activity of the gonads (ovaries and testes).

gonadotropin releasing hormone (GnRH) A tropic hormone secreted by the hypothalamus that causes the pituitary to make luteinizing hormone (LH) and follicle stimulating hormone (FSH).

graded potential A change in membrane potential that does not necessarily trigger an action potential.

gravitropism A directional growth response to Earth's gravitational pull that is induced by mechanical and hormonal influences.

gray crescent A crescent-shaped region of the underlying cytoplasm at the side opposite the point of sperm entry exposed after fertilization when the pigmented layer of cytoplasm rotates toward the site of sperm entry.

gray matter Areas of densely packed nerve cell bodies and dendrites in the brain and spinal cord.

greater vestibular gland One of two glands located slightly below and to the left and right of the opening of the vagina in women. They secrete mucus to provide lubrication, especially when the woman is sexually aroused.

ground meristem The primary meristematic tissue in plants that gives rise to ground tissues, mostly parenchyma.

ground tissue system One of the three basic tissue systems in plants; includes all tissues other than dermal and vascular tissues.

growth factor Any of a large group of peptide hormones that regulates the division and differentiation of many cell types in the body.

growth hormone (GH) A hormone that stimulates cell division, protein synthesis, and bone growth in children and adolescents, thereby causing body growth.

guard cell Either of a pair of specialized crescent-shaped cells that control the opening and closing of stomata in plant tissue.

guttation The exudation of water from leaves as a result of strong root pressure.

haustorium (plural, haustoria) The hyphal tip of a parasitic fungus that penetrates a host plant and absorbs nutrients from it; likewise in parasitic flowering plants, a root that can penetrate a host's tissues and absorb nutrients.

heartwood The inner core of a woody stem; composed of dry tissue and nonliving cells that no longer transport water and solutes and may store resins, tannins, and other defensive compounds.

heat-shock protein (HSP) Any of a group of chaperone proteins that are present in all cells in all life forms. They are induced when a cell undergoes various types of environmental stresses like heat, cold, and oxygen deprivation.

heavy chain The heavier of the two types of polypeptide chains that are found in immunoglobulin and antibody molecules.

helper T cell A clonal cell that assists with the activation of B cells.

hemolymph The circulatory fluid of invertebrates with open circulatory systems, including mollusks and arthropods.

hepatic portal vein The blood vessel that leads to capillary networks in the liver.

herbicide A compound that, at proper concentration, kills plants.

herbivore An animal that obtains energy and nutrients primarily by eating plants.

hermaphroditism The mechanism in which both mature egg-producing and mature sperm-producing tissue are present in the same individual.

hibernation Extended torpor during winter.

hindbrain The lower area of the brain that includes the brain stem, medulla oblongata, and pons.

hippocampus A gray-matter center that is involved in sending information.

homeobox A region of a homeotic gene that corresponds to an amino acid section of the homeodomain.

homeodomain The encoded transcription factor of each protein that binds to a region in the promoters of the genes whose transcription it regulates.

homeostasis The maintenance of the internal environment in a stable state.

homeostatic mechanism Any process or activity responsible for homeostasis.

homeotic gene Any of the family of genes that determines the structure of body parts during embryonic development.

horizon A noticeable layer of soil, such as topsoil, having a distinct texture and composition that varies with soil type.

horizontal cell A type of neuron that forms lateral connections among photoreceptor cells in the retina of the eye.

hormone A signaling molecule secreted by a cell that can alter the activities of any cell with receptors for it.

human chorionic gonadotropin (hCG) A hormone that keeps the corpus luteum in the ovary from breaking down.

human immunodeficiency virus (HIV) A retrovirus that causes acquired immunodeficiency syndrome (AIDS).

humus The organic component of soil remaining after decomposition of plants and animals, animal droppings, and other organic matter.

hybridoma A B cell that has been induced to fuse with a cancerous lymphocyte called a myeloma cell, forming single, composite cell.

hydroponic culture A method of growing plants not in soil but with the roots bathed in a solution that contains water and mineral nutrients.

hydrostatic skeleton A structure consisting of muscles and fluid that, by themselves, provide support for an animal or part of an animal; no rigid support, like a bone, is involved.

hymen A thin flap of tissue that partially covers the opening of the vagina.

hyperpolarized The condition of a neuron when its membrane potential is more negative than the resting value.

hypersensitive response A plant defense that physically cordons off an infection site by surrounding it with dead cells.

hypertension Commonly called high blood pressure, a medical condition in which blood pressure is chronically elevated above normal values.

hyperthermia The condition resulting when the heat gain of the body is too great to be counteracted.

hypoblast The bottom layer of a blastodisc.

hypocotyl The region of a plant embryo's vertical axis between the cotyledons and the radicle.

hypodermis The innermost layer of the skin that contains larger blood vessels and additional reinforcing connective tissue.

hypothalamus The portion of the brain that contains centers regulating basic homeostatic functions of the body and contributing to the release of hormones.

hypothermia A condition in which the core temperature falls below normal for a prolonged period.

imbibition The movement of water into a seed as the water molecules are attracted to hydrophilic groups of stored proteins; the first step in germination.

immune response The defensive reactions of the immune system.

immune system The body's system of defenses against disease, composed of certain white blood cells and antibodies.

immunoglobulin A specific protein substance produced by plasma cells to aid in fighting infection.

immunological memory The capacity of the immune system to respond more rapidly and vigorously to the second contact with a specific antigen than to the primary contact.

immunological tolerance The process that protects the body's own molecules from attack by the immune system.

imperfect flower A type of incomplete flower that has stamens or carpels, but not both.

incisors Flattened, chisel-shaped teeth of mammals, located at the front of the mouth, that are used to nip or cut food.

incomplete flower A flower lacking one or more of the four floral whorls.

incus The second of the three sound-conducting middle ear bones in vertebrates, located between the malleus and the stapes.

indeterminate growth Growth that is not limited by an organism's genetic program, so that the organism grows for as long as it lives; typical of many plants. *Compare* determinate growth.

indirect neurotransmitter A neurotransmitter that acts as a first messenger, binding to a G-protein–coupled receptor in the postsynaptic membrane, which activates the receptor and triggers generation of a second messenger such as cyclic AMP or other processes.

induction A mechanism in which one group of cells (the inducer cells) causes or influences another nearby group of cells (the responder cells) to follow a particular developmental pathway.

infection thread In the formation of root nodules on nitrogen-fixing plants, the tube formed by the plasma membrane of root hair cells as bacteria enter the cell.

inflammation The heat, pain, redness, and swelling that occur at the site of an infection.

ingestion The feeding methods used to take food into the digestive cavity.

inhibin A peptide that, in females, is an inhibitor of FSH secretion from the pituitary thereby diminishing the signal for follicular growth. In males, inhibin inhibits FSH secretion from the pituitary, thereby decreasing spermatogenesis.

inhibiting hormone (IH) A hormone released by the hypothalamus that inhibits the secretion of a particular anterior pituitary hormone.

inhibitory postsynaptic potential (IPSP) A change in membrane potential caused when hyperpolarization occurs, pushing the neuron farther from threshold.

initial A plant cell that remains permanently as part of a meristem and gives rise to daughter cells that differentiate into specialized cell types.

innate immunity A nonspecific line of defense against pathogens that includes inflammation, which creates internal conditions that inhibit or kill many pathogens, and specialized cells that engulf or kill pathogens or infected body cells.

inner cell mass The dense mass of cells within the blastocyst that will become the embryo.

inner ear That part of the ear, particularly the cochlea, that converts mechanical vibrations (sound) into neural messages that are sent to the brain.

insulin A hormone secreted by beta cells in the islets, acting mainly on cells of nonworking skeletal muscles, liver cells, and adipose tissue (fat) to lower blood glucose, fatty acid and amino acids levels, and promote the storage of those molecules.

insulin-like growth factor (IGF) A peptide that directly stimulates growth processes.

integration The sorting and interpretation of neural messages and the determination of the appropriate response(s).

integrator In homeostatic feedback, the control center that compares a detected environmental change with a set point.

integument Skin.

interferon A cytokine produced by infected host cells affected by viral dsRNA, which acts both on the infected cell that produces it, an autocrine effect, and on neighboring uninfected cells, a paracrine effect.

intermediate-day plant A plant that flowers only when daylength falls between the values for long-day and short-day plants.

internal fertilization The process in which sperm are released by the male close to or inside the entrance of the reproductive tract of the female.

internal gill A gill located within the body that has a cover providing physical protection for the gills. Water must be brought to internal gills.

interneuron A neuron that integrates information to formulate an appropriate response.

internode The region between two nodes on a plant stem.

interstitial fluid The fluid occupying the spaces between cells in multicellular animals.

intestinal villus A microscopic, fingerlike extension in the lining of the small intestine.

intestine The portion of digestive system where organic matter is hydrolyzed by enzymes secreted into the digestive tube. As muscular contractions of the intestinal wall move the mixture along, cells lining the intestine absorb the molecular subunits produced by digestion.

intracellular digestion The process in which cells take in food particles by endocytosis.

invagination The process in which cells changing shape and pushing inward from the surface produce an indentation, such as the dorsal lip of the blastopore.

involution The process by which cells migrate into the blastopore.

iris Of the eye, the colored muscular membrane that lies behind the cornea and in front of the lens, which by opening or closing determines the size of the pupil and hence the amount of light entering the eye.

islets of Langerhans Endocrine cells that secrete the peptide hormones insulin and glucagon into the bloodstream.

jasmonate Any of a group of plant hormones that help regulate aspects of growth and responses to stress, including attacks by predators and pathogens.

juvenile hormone (JH) A peptide hormone secreted by the corpora allata, a pair of glands just behind the brain in insects.

juxtaglomerular apparatus A group of receptors that monitor the pressure and flow of fluid through the distal tubule of the kidney.

labia majora A pair of fleshy, fat-padded folds that partially cover the labia minora.

labia minora Two folds of tissue that run from front to rear on either side of the opening to the vagina.

larynx The voice box.

lateral bud A bud on the side of a plant stem from which a branch may grow.

lateral geniculate nuclei Clusters of neurons located in the thalamus that receive visual information from the optic nerves and send it on to the visual cortex.

lateral inhibition Visual processing in which lateral movement of signals from a rod or cone proceeds to a horizontal cell and continues to bipolar cells with which the horizontal cell makes inhibitory connections, serving both to sharpen the edges of objects and enhance contrast in an image.

lateral line system The complex of mechanoreceptors along the sides of some fishes and aquatic amphibians that detect vibrations in the water.

lateral meristem A plant meristem that gives rise to secondary tissue growth. *Compare* primary meristem.

lateral root A root that extends away from the main root (or taproot).

lateralization A phenomenon in which some brain functions are more localized in one of the two hemispheres.

leaching The process by which soluble materials in soil are washed into a lower layer of soil or are dissolved and carried away by water.

leaf primordium A lateral outgrowth from the apical meristem that develops into a young leaf.

leghemoglobin An iron-containing, red-pigmented protein produced in root nodules during the symbiotic association between *Bradyrhizobium* or *Rhizobium* and legumes.

lens The transparent, biconvex intraocular tissue that helps bring rays of light to a focus on the retina.

leukocyte A white blood cell, which eliminates dead and dying cells from the body, removes cellular debris, and participates in defending the body against invading organisms.

Leydig cell A cell that produces the male sex hormones.

ligament A fibrous connective tissue that connects bones to each other at a joint.

ligand-gated ion channel A channel that opens or closes when a specific chemical, such as a neurotransmitter, binds to the channel.

light chain The lighter of the two types of polypeptide chains found in immunoglobulin and antibody molecules.

lignification The deposition of lignin in plant cell walls; it anchors the cellulose fibers in the walls, making them stronger and more rigid, and protects the other wall components from physical or chemical damage.

limbic system A functional network formed by parts of the thalamus, hypothalamus, and basal nuclei, along with other nearby gray-matter centers—the amygdala, hippocampus, and olfactory bulbs—sometimes called the "emotional brain".

liver A large organ whose many functions include aiding in digestion, removing toxins from the body, and regulating the chemicals in the blood.

loam Any well-aerated soil composed of a mixture of sand, clay, silt, and organic matter.

long-day plant A plant that flowers in spring when dark periods become shorter and day length becomes longer.

long-term memory Memory that stores information from days to years or even for life.

long-term potentiation A long-lasting increase in the strength of synaptic connections in activated neural pathways following brief periods of repeated stimulation.

loop of Henle A U-shaped bend of the proximal convoluted tubule.

loose connective tissue A tissue formed of sparsely distributed cells surrounded by a more or less open network of collagen and other glycoprotein fibers.

lumen The inside of the digestive tube.

lung One of a pair of invaginated respiratory surfaces, buried in the body interior where they are less susceptible to drying out; the organs of respiration in mammals, birds, reptiles, and most amphibians.

luteinizing hormone (LH) A hormone secreted by the pituitary that stimulates the growth and maturation of eggs in females and the secretion of testosterone in males.

lymph The interstitial fluid picked up by the lymphatic system.

lymph node One of many small, bean-shaped organs spaced along the lymph vessels that contain macrophages and other leukocytes that attack invading disease organisms.

lymphatic system An accessory system of vessels and organs that helps balance the fluid content of the blood and surrounding tissues and participates in the body's defenses against invading disease organisms.

lymphocyte A leukocyte that carries out most of its activities in tissues and organs of the lymphatic system. Lymphocytes play major roles in immune responses.

macronutrient In humans, a mineral required in amounts ranging from 50 mg to more than 1 gram per day. In plants, a nutrient needed in large amounts for the normal growth and development.

macrophage A phagocyte that takes part in nonspecific defenses and adaptive immunity.

magnetoreceptor A receptor found in some animals that navigate long distances which allows them to detect and use Earth's magnetic field as a source of directional information.

major histocompatibility complex A large cluster of genes encoding the MHC proteins.

malleus The outermost of the sound-conducting bones of the middle ear in vertebrates.

malnutrition A condition resulting from a diet that lacks one or more essential nutrients.

Malpighian tubule The main organ of excretion and osmoregulation in insects, helping them to maintain water and electrolyte balance.

marsupium An external pouch on the abdomen of many female marsupials, containing the mammary glands, and within which the young continue to develop after birth.

mast cell A type of cell dispersed through connective tissue that releases histamine when activated by the death of cells, caused by a pathogen at an infection site.

maternal-effect gene One of a class of genes that regulate the expression of other genes expressed by the mother during oogenesis and that control the polarity of the egg and, therefore, of the embryo.

mating The pairing of a male and a female for the purpose of sexual reproduction.

mechanoreceptor A sensory receptor that detects mechanical energy, such as changes in pressure, body position, or acceleration. The auditory receptors in the ears are examples of mechanoreceptors.

megapascal A unit of pressure used to measure water potential.

megaspore A plant spore that develops into a female gametophyte; usually larger than a microspore.

melanocyte-stimulating hormone (MSH) A hormone secreted by the anterior pituitary that controls the degree of pigmentation in melanocytes.

melatonin A peptide hormone secreted by the pineal gland that helps maintain daily biorhythms.

membrane attack complexes (MAC) An abnormal activation of the complement (protein) portion of the blood, forming a cascade reaction that brings blood proteins together, binds them to the cell wall, and then inserts them through the cell membrane.

membrane potential An electrical voltage that measures the potential inside a cell membrane relative to the fluid just outside; it is negative under resting conditions and becomes positive during an action potential.

memory The storage and retrieval of a sensory or motor experience, or a thought.

memory B cell In antibody-mediated immunity, a long-lived cell expressing an antibody on its surface that can bind to a specific antigen. A memory B cell is activated the next time the antigen is encountered, producing a rapid secondary immune response.

memory cell An activated lymphocyte that circulates in the blood and lymph, ready to initiate a rapid immune response upon subsequent exposure to the same antigen.

memory helper T cell In cell-mediated immunity, a long-lived cell differentiated from a helper T cell, which remains in an inactive state in the lymphatic system after an immune reaction has run its course and ready to be activated upon subsequent exposure to the same antigen.

meninges Three layers of connective tissue that surround and protect the spinal cord and brain.

menstrual cycle A cycle of approximately 1 month in the human female during which an egg is released from an ovary and the uterus is prepared to receive the fertilized egg; if fertilization does not occur, the endometrium breaks down, which releases blood and tissue breakdown products from the uterus to the outside through the vagina.

meristem An undifferentiated, permanently embryonic plant tissue that gives rise to new cells forming tissues and organs.

mesenteries Sheets of loose connective tissue, covered on both surfaces with epithelial cells, which suspend the abdominal organs in the coelom and provide lubricated, smooth surfaces that prevent chafing or abrasion between adjacent structures as the body moves.

mesoderm The middle layer of the three primary germ layers of an animal embryo, from which the muscular, skeletal, vascular, and connective tissues develop.

mesophyll The ground tissue located between the two outer leaf tissues, composed of loosely packed parenchyma cells that contain chloroplasts.

metamorphosis A reorganization of the form of certain animals during postembryonic development.

metanephridium The excretory tubule of most annelids and mollusks.

micronutrient Any mineral required by an organism only in trace amounts.

micropyle A small opening at one end of an ovule through which the pollen tube passes prior to fertilization.

microspore A plant spore from which a male gametophyte develops; usually smaller than a megaspore.

microvilli Fingerlike projections forming a brush border in epithelial cells that cover the villi.

midbrain The uppermost of the three segments of the brainstem, serving primarily as an intermediary between the rest of the brain and the spinal cord.

middle ear The air-filled cavity containing three small, interconnected bones: the malleus, incus, and stapes.

mineralocorticoid A steroid hormone secreted by the adrenal cortex that regulates the levels of Na^+ and K^+ in the blood and extracellular fluid.

molars Posteriormost teeth of mammals, with a broad chewing surface for grinding food.

molt-inhibiting hormone (MIH) A peptide neurohormone secreted by a gland in the eye stalks of crustaceans that inhibits ecdysone secretion.

monoclonal antibody An antibody that reacts only against the same segment (epitope) of a single antigen.

monocot A plant belonging to the Monocotyledones, one of the two major classes of angiosperms; monocot embryos have a single seed leaf (cotyledon) and pollen grains with a single groove.

monocyte A type of leukocyte that enters damaged tissue from the bloodstream through the endothelial wall of the blood vessel.

monoecious Having both "male" flowers (which possess only stamens) and "female" flowers (which possess only carpels).

monotreme A lineage of mammals that lay eggs instead of bearing live young.

morphogenesis Orderly, genetically programmed changes in the size, shape, and proportion of body parts of an organism; the process by which specialized tissues and organs form.

morphology The form or shape of an organism, or of a part of an organism.

morula The first stage of animal development, a solid ball or layer of blastomeres.

motor neuron An efferent neuron that carries signals to skeletal muscle.

motor unit A block of muscle fibers that is controlled by branches of the axon of a single efferent neuron.

mucosa The lining of the gut that contains epithelial and glandular cells.

Müllerian duct The bipotential primitive duct associated with the gonads that leads to a cloaca.

multiple fruit A fruit that develops from several ovaries in multiple flowers; examples are pineapples and mulberries.

muscle fiber A bundle of elongated, cylindrical cells that make up skeletal muscle.

muscle spindle A stretch receptor in muscle; a bundlesof small, specialized muscle cells wrapped with the dendrites of afferent neurons and enclosed in connective tissue.

muscle tissue Cells that have the ability to contract (shorten) forcibly.

muscle twitch A single, weak contraction of a muscle fiber.

muscularis The muscular coat of a hollow organ or tubular structure.

mycorrhiza A mutualistic symbiosis in which fungal hyphae associate intimately with plant roots.

myoblast An undifferentiated muscle cell.

myofibril A cylindrical contractile element about 1 μm in diameter that runs lengthwise inside the muscle fiber cell.

myogenic heart A heart that maintains its contraction rhythm with no requirement for signals from the nervous system.

myoglobin An oxygen-storing protein closely related to hemoglobin.

nastic movement In plants, a reversible response to nondirectional stimuli, such as mechanical pressure or humidity.

natural killer (NK) cell A type of lymphocyte that destroys virus-infected cells.

negative feedback The primary mechanism of homeostasis, in which a stimulus—a change in the external or internal environment—triggers a response that compensates for the environmental change.

negative pressure breathing Muscular contractions that expand the lungs, lowering the pressure of the air in the lungs and causing air to be pulled inward.

nephron A specialized excretory tubule that contributes to osmoregulation and carries out excretion, found in all vertebrates.

nerve A bundle of axons enclosed in connective tissue and all following the same pathway.

nerve cord A bundle of nerves that extends from the central ganglia to the rest of the body, connected to smaller nerves.

nerve net A simple nervous system that coordinates responses to stimuli but has no central control organ, or brain.

nervous tissue Tissue that contains neurons, which serve as lines of communication and control between body parts.

neural crest In development, the region where the neural tube pinches off from the ectoderm.

neural plate Ectoderm thickened and flattened into a longitudinal band, induced by notochord cells.

neural signaling The process by which an animal responds appropriately to a stimulus.

neural tube A hollow tube in vertebrate embryos that develops into the brain, spinal cord, spinal nerves, and spinal column.

neurogenic heart A heart that beats under the control of signals from the nervous system.

neuromuscular junction The junction between a nerve fiber and the muscle it supplies.

neuron An electrically active cell of the nervous system responsible for controlling behavior and body functions.

neuronal circuit The connection between axon terminals of one neuron and the dendrites or cell body of a second neuron.

neurosecretory neuron A neuron that releases a neurohormone into the circulatory system when appropriately stimulated.

neurotransmitter A chemical released by an axon terminal at a chemical synapse.

neurulation The process in vertebrates by which organogenesis begins with development of the nervous system from ectoderm.

neutrophil A type of phagocytic leukocyte that attaches to blood vessel walls in massive numbers when attracted to the infection site by chemokines.

nitrification A metabolic process in which certain soil bacteria convert ammonia or ammonium ions into nitrites that are then converted by other bacteria to nitrates, a form usable by plants.

nitrogen fixation A metabolic process in which certain bacteria and cyanobacteria convert molecular nitrogen into ammonia and ammonium ions, forms usable by plants.

nociceptor A sensory receptor that detects tissue damage or noxious chemicals; their activity registers as pain.

node The point on a stem where one or more leaves are attached.

node of Ranvier The gap between two Schwann cells, which exposes the axon membrane directly to extracellular fluids.

nonshivering thermogenesis The generation of heat by oxidative mechanisms in non-muscle tissue throughout the body.

norepinephrine A nontropic amine hormone secreted by the adrenal medulla.

notochord A flexible rod-like structure, constructed of fluid-filled cells surrounded by tough connective tissue, which supports a chordate embryo from head to tail.

nutrition The processes by which an organism takes in, digests, absorbs, and converts food into organic compounds.

ocellus The simplest eye, which detects light but does not form an image.

olfactory bulb A gray-matter center that relay inputs from odor receptors to both the cerebral cortex and the limbic system.

oligodendrocyte A type of glial cell that populates the CNS and is responsible for producing myelin.

oligosaccharin A complex carbohydrate that in plants serves as a signaling molecule and as a defense against pathogens.

ommatidium (plural, ommatidia) A faceted visual unit of a compound eye.

omnivore An animal that feeds at several trophic levels, consuming plants, animals, and other sources of organic matter.

oocyte A developing gamete that becomes an ootid at the end of meiosis.

oogenesis The process of producing eggs.

oogonium A cell that enters meiosis and gives rise to gametes, produced by mitotic divisions of the germ cells in females.

open circulatory system An arrangement of internal transport in some invertebrates in which the vascular fluid, hemolymph, is released into sinuses, bathing organs directly, and is not always retained within vessels.

opsin One of several different proteins that bond covalently with the light-absorbing pigment of rods and cones (retinal).

optic chiasm Location just behind the eyes where the optic nerves converge before entering the base of the brain, a portion of each optic nerve crossing over to the opposite side.

organ Two or more different tissues integrated into a structure that carries out a specific function.

organ of Corti An organ within the cochlear duct that contains the sensory hair cells detecting sound vibrations transmitted to the inner ear.

organ system The coordinated activities of two or more organs to carry out a major body function such as movement, digestion, or reproduction.

organogenesis The development of the major organ systems, giving rise to a free-living individual with the body organization characteristic of its species.

oriented cell division Cell division in different planes; establishes the overall shape of a plant organ.

osmoconformer An animal in which the osmolarity of the cellular and extracellular solutions matches the osmolarity of the environment.

osmolarity The total solute concentration of a solution, measured in osmoles—the number of solute molecules and ions (in moles)—per liter of solution.

osmoreceptor A chemoreceptor in the hypothalamus that responds to changes in the osmolarity of the fluid surrounding it, which reflects the osmolarity generally of the body fluids.

osmoregulation The regulation of water and ion balance.

osmoregulator An animal that uses control mechanisms to keep the osmolarity of cellular and extracellular fluids the same, but at levels that may differ from the osmolarity of the surroundings.

osmosis The passive transport of water across a selectively permeable membrane in response to solute concentration gradients, a pressure gradient, or both.

osteoblast A cell that produces the collagen and mineral of bone.

osteoclast A cell that removes bone minerals and recycles them through the bloodstream.

osteocyte A mature bone cell.

osteon The structural unit of bone, consisting of a minute central canal surrounded by osteocytes embedded in concentric layers of mineral matter.

otolith One of many small crystals of calcium carbonate embedded in the otolithic membrane of the hair cells.

outer ear The external structure of the ear, consisting of the pinna and meatus.

oval window An opening in the bony wall that separates the middle ear from the inner ear.

ovarian cycle The cyclic events in the ovary leading to ovulation.

ovary In animals, the female gonad, which produces female gametes and reproductive hormones. In flowering plants, the enlarged base of a carpel in which one or more ovules develop into seeds.

overnutrition The condition caused by excessive intake of specific nutrients.

oviduct The tube through which the egg moves from the ovary to the outside of the body.

oviparous Referring to animals that lay eggs containing the nutrients needed for development of the embryo outside the mother's body.

ovoviviparous Referring to animals in which fertilized eggs are retained within the body and the embryo develops using nutrients provided by the egg; eggs hatch inside the mother.

ovulation The process in which oocytes are released into the oviducts as immature eggs.

ovule In plants, the structure in a carpel in which a female gametophyte develops and fertilization takes place.

ovum A female sex cell, or egg.

oxytocin A hormone that stimulates the ejection of milk from the mammary glands of a nursing mother.

pacemaker cell A specialized cardiac muscle cell in the upper wall of the right atrium that sets the rate of contraction in the heart.

pair-rule genes In *Drosophila* embryonic development, the set of segmentation regulatory genes activated by gap genes that divide the embryo into units of two segments each.

pancreas A mixed gland composed of an exocrine portion that secretes digestive enzymes into the small intestine and an endocrine portion, the islets of Langerhans, that secretes insulin and glucagon.

parasympathetic division The division of the autonomic nervous system that predominates during quiet, low-stress situations, such as while relaxing.

parathyroid gland One of a pair of glands that produce parathyroid hormone (PTH) (found only in tetrapod vertebrates).

parathyroid hormone (PTH) The hormone secreted by the parathyroid glands in response to a fall in blood Ca^{2+} levels.

parthenogenesis A mode of asexual reproduction in which animals produce offspring by the growth and development of an egg without fertilization.

parturition The process of giving birth.

passive immunity The acquisition of antibodies as a result of direct transfer from another person.

passive transport The transport of substances across cell membranes without expenditure of energy, as in diffusion.

pathogenesis-related (PR) protein A hydrolytic enzyme that breaks down components of a pathogen's cell wall.

pattern formation The arrangement of organs and body structures in their proper three-dimensional relationships.

pepsin An enzyme made in the stomach that breaks down proteins.

pepsinogen The inactive precursor molecule for pepsin.

perennial A plant in which vegetative growth and reproduction continue year after year.

perfect flower A flower that has both male (stamen) and female (carpel) sexual organs.

perfusion The flow of blood or other body fluids on the internal side of the respiratory surface.

pericarp The fruit wall.

pericycle A tissue of plant roots, located between the endodermis and the phloem, which gives rise to lateral roots.

periderm The outermost portion of bark; consists of cork, cork cambium, and secondary cortex.

peripheral nervous system (PNS) All nerve roots and nerves (motor and sensory) that supply the muscles of the body and transmit information about sensation (including pain) to the central nervous system.

peristalsis The rippling motion of muscles in the intestine or other tubular organs characterized by the alternate contraction and relaxation of the muscles that propel the contents onward.

peritubular capillary A capillary of the network surrounding the glomerulus.

petal Part of the corolla of a flower, often brightly colored. petiole The stalk by which a leaf is attached to a stem.

pharynx The throat. In some invertebrates, a protrusible tube used to bring food into the mouth for passage to the gastrovascular cavity; in mammals, the common pathway for air entering the larynx and food entering the esophagus.

phloem The food-conducting tissue of a vascular plant.

phloem sap The solution of water and organic compounds that flows rapidly through the sieve tubes of flowering plants.

photoperiodism The response of plants to changes in the relative lengths of light and dark periods in their environment during each 24-hour period.

photopigment Light-absorbing pigment.

photopsin One of three photopigments in which retinal is combined with different opsins.

photoreceptor A sensory receptor that detects the energy of light.

phototropism The tendency of a plant shoot to bend toward a source of light.

physiological respiration The process by which animals exchange gases with their surroundings—how they take in oxygen from the outside environment and deliver it to body cells, and remove carbon dioxide from body cells and deliver it to the environment.

physiology The study of the functions of organisms—the physico-chemical processes of organisms.

phytoalexin A biochemical that functions as an antibiotic in plants

phytochrome A blue-green pigmented plant chromoprotein involved in the regulation of light-dependent growth processes.

pineal gland A light-sensitive, melatonin-secreting gland that regulates some biological rhythms.

pinna The external structure of the outer ear, which concentrates and focuses sound waves.

pith The soft, spongelike, central cylinder of the stems of most flowering plants, composed mainly of parenchyma.

pituitary A gland consisting mostly of two fused lobes suspended just below the hypothalamus by a slender stalk of tissue that contains both neurons and blood vessels; it interacts with the hypothalamus to control many physiological functions, including the activity of some other glands.

placenta A specialized temporary organ that connects the embryo and fetus with the uterus in mammals, mediating the delivery of oxygen and nutrients.

plasma The clear, yellowish fluid portion of the blood in which cells are suspended. Plasma consists of water, glucose and other sugars, amino acids, plasma proteins, dissolved gases, ions, lipids, vitamins, hormones and other signal molecules, and metabolic wastes.

plasma cell A large antibody-producing cell that develops from B cells.

platelet An oval or rounded cell fragment enclosed in its own plasma membrane, which is found in the blood; they are produced in red bone marrow by the division of stem cells and contain enzymes and other factors that take part in blood clotting.

pleura The double layer of epithelial tissue covering the lungs.

plumule The rudimentary terminal bud of a plant embryo located at the end of the hypocotyl, consisting of the epicotyl and a cluster of tiny foliage leaves.

polar body A nonfunctional cell produced in oogenesis.

polar nucleus In the embryo sac of a flowering plant, one of two nuclei that migrate into the center of the sac, become housed in a central cell, and eventually give rise to endosperm.

polar transport Unidirectional movement of a substance from one end of a cell (or other structure) to the other.

polarity The unequal distribution of yolk and other components in a mature egg.

pollen grain The male gametophyte of a seed plant.

pollen sac The microsporangium of a seed plant, in which pollen develops.

pollen tube A tube that grows from a germinating pollen grain through the tissues of a carpel and carries the sperm cells to the ovary.

positive feedback A mechanism that intensifies or adds to a change in internal or external environmental condition.

positive pressure breathing A gulping or swallowing motion that forces air into the lungs.

posterior pituitary The neural portion of the pituitary, which stores and releases two hormones made by the hypothalamus, antidiuretic hormone and oxytocin.

postsynaptic cell The neuron or the surface of an effector after a synapse that receives the signal from the presynaptic cell.

postsynaptic membrane The plasma membrane of the postsynaptic cell.

pregnancy The period of mammalian development in which the embryo develops in the uterus of the mother.

premolars Teeth located in pairs on each side of the upper and lower jaws of mammals, positioned behind the canines and in front of the molars.

prepuce Foreskin; a loose fold of skin that covers the glans of the penis.

pressure flow mechanism In vascular plants, pressure that builds up at the source end of a sieve tube system and pushes solutes by bulk flow toward a sink, where they are removed.

presynaptic cell The neuron with an axon terminal on one side of the synapse that transmits the signal across the synapse to the dendrite or cell body of the postsynaptic cell.

presynaptic membrane The plasma membrane of the axon terminal of a presynaptic cell, which releases neurotransmitter molecules into the synapse in response to arrival of an action potential.

primary growth The growth of plant tissues derived from apical meristems. *Compare* secondary growth.

primary immune response The response of the immune system to the first challenge by an antigen.

primary meristem Root and shoot apical meristems, from which a plant's primary tissues develop. *Compare* lateral meristem.

primary motor area The area of the cerebral cortex that runs in a band just in front of the primary somatosensory area and is responsible for voluntary movement.

primary plant body The portion of a plant that is made up of primary tissues.

primary somatosensory area The area of the cerebral cortex that runs in a band across the parietal lobes of the brain and registers information on touch, pain, temperature, and pressure.

primary tissue A plant tissue that develops from an apical meristem.

primitive groove In development of birds, the sunken midline of the primitive streak that acts as a conduit for migrating cells to move into the blastocoel.

primitive streak In development of birds, the thickened region of the embryo produced by cells of the epiblast streaming toward the midline of the blastodisc.

procambium The primary meristem of a plant that develops into primary vascular tissue.

progesterone A female sex hormone that stimulates growth of the uterine lining and inhibits contractions of the uterus.

progestin A class of sex hormones synthesized by the gonads of vertebrates and active predominantly in females.

prolactin (PRL) A peptide hormone secreted by the anterior pituitary that stimulates breast development and milk secretion in mammals.

propagation In animal nervous systems, the concept that the action potential does not need further trigger events to keep going.

proprioceptor A mechanoreceptor that detects stimuli used in the CNS to maintain body balance and equilibrium and to monitor the position of the head and limbs.

prostaglandin One of a group of local regulators derived from fatty acids that are involved in paracrine and autocrine regulation.

prostate gland An accessory sex gland in males that adds a thin, milky fluid to the semen and adjusts the pH of the semen to the level of acidity best tolerated by sperm.

protoderm The primary meristem that will produce stem epidermis.

protonephridium The simplest form of invertebrate excretory tubule.

protoplast The cytoplasm, organelles, and plasma membrane of a plant cell.

protoplast fusion A plant breeding process in which protoplasts are fused into a single cell.

proximal convoluted tubule The tubule between the Bowman's capsule and the loop of Henle in the nephron of the kidney, which carries and processes the filtrate.

pulmocutaneous circuit In amphibians, the branch of a double blood circuit that receives deoxygenated blood and moves it to the skin and lungs or gills.

pulmonary circuit The circuit of the cardiovascular system that supplies the lungs.

pulvinus (plural, pulvini) A jointlike, thickened pad of tissue at the base of a leaf or petiole; flexes when the leaf makes nastic movements

pupil The dark center in the middle of the iris through which light passes to the back of the eye.

quiescent center A region in a root apical meristem where there is no cell division.

R gene A resistance gene in a plant; dominant R alleles confer enhanced resistance to plant pathogens.

radiation The transfer of heat energy as electromagnetic radiation.

radicle The rudimentary root of a plant embryo.

rapid eye movement (REM) sleep The period during deep sleep when the delta wave pattern is replaced by rapid, irregular beta waves characteristic of the waking state. The person's heartbeat and breathing rate increase, the limbs twitch, and the eyes move rapidly behind the closed eyelids.

ray initial A cell in vascular cambium that gives rise to spokelike rays of parenchyma cells.

reabsorption The process in which some molecules (for example, glucose and amino acids) and ions are transported by the transport epithelium back into the body fluid (animals with open circulatory systems) or into the blood in capillaries surrounding the tubules (animals with closed circulatory systems) as the filtered solution moves through the excretory tubule.

receptacle The expanded tip of a flower stalk that bears floral organs.

reception The detection of a stimulus by neurons.

rectum The final segment of the large intestine.

reflex A programmed movement that takes place without conscious effort, such as the sudden withdrawal of a hand from a hot surface.

refractory period A period that begins at the peak of an action potential and lasts a few milliseconds, during which the threshold required for generation of an action potential is much higher than normal.

release The process in which urine is released into the environment from the distal end of the excretory tubule.

releasing hormone (RH) A peptide neurohormone that control the secretion of hormones from the anterior pituitary.

renal artery An artery that carries bodily fluids into the kidney.

renal cortex The outer region of the mammalian kidney that surrounds the renal medulla.

renal medulla The inner region of the mammalian kidney.

renal pelvis The central cavity in the kidney where urine drains from collecting ducts.

renal vein The vein that routes filtered blood away from the kidney.

renin An enzyme secreted by cells in the juxtaglomerular apparatus into the bloodstream that converts a blood protein into the peptide hormone angiotensin.

renin-angiotensin-aldosterone system (RAAS) The most important hormonal system involved in regulation of Na^+ in mammals.

residual volume The air that remains in lungs after exhalation.

respiratory medium The environmental source of O_2 and the "sink" for released CO_2. For aquatic animals, the respiratory medium is water; for terrestrial animals, it is air.

respiratory surface A layer of epithelial cells that provides the interface between the body and the respiratory medium.

respiratory system All the parts of the body involved in exchanging air between the external environment and the blood.

response The "output" or action resulting from the integration of neural messages.

resting potential A steady negative membrane potential exhibited by the membrane of a neuron that is not stimulated—that is, not conducting an impulse.

reticular formation A complex network of interconnected neurons that runs through the length of the brain stem, connecting to the thalamus at the anterior end and to the spinal cord at the posterior end.

retina A light-sensitive membrane lining the posterior part of the inside of the eye.

rhodopsin The retinal-opsin photopigment.

rod In the vertebrate eye, a type of photoreceptor in the retina that is specialized for detection of light at low intensities.

root cap A dome-shaped cell mass that forms a protective covering over the apical meristem in the tip of a plant root.

root hair A tubular outgrowth of the outer wall of a root epidermal cell; root hairs absorb much of a plant's water and minerals from the soil.

root nodule A localized swelling on a root in which symbiotic nitrogen-fixing bacteria reside.

root pressure The pressure that develops in plant roots as the result of osmosis, forcing xylem sap upward and out through leaves. *See also* Guttation.

root primordium A rudimentary root.

root system An underground (or submerged), network of roots with a large surface area that favors the rapid uptake of soil water and dissolved mineral ions.

round window A thin membrane that faces the middle ear.

ruminant An animals that has a complex, four-chambered stomach.

saccule A fluid-filled chamber in the vestibular apparatus that provides information about the position of the head with respect to gravity (up versus down), as well as changes in the rate of linear movement of the body.

salicylic acid (SA) In plants, a chemical synthesized following a wound that has multiple roles in plant defenses, including interaction with jasmonates in signaling cascades.

salivary amylase A substance that hydrolyzes starches to the disaccharide maltose.

salivary gland A gland that secretes saliva through a duct on the inside of the cheek or under the tongue; the saliva lubricates food and begins digestion.

saltatory conduction A mechanism that allows small-diameter axons to conduct impulses rapidly.

sapwood The newly formed outer wood located between heartwood and the vascular cambium. Compared with heartwood, it is wet, lighter in color, and not as strong.

sarcomere The basic unit of contraction in a myofibril.

sarcoplasmic reticulum In vertebrate muscle fibers, a complex system of vesicles modified from the smooth endoplasmic reticulum that encircles the sarcomeres. The sarcoplasmic reticulum is part of the pathway for the stimulation of muscle contraction by neural signals.

Schwann cell A type of glial cell in the PNS that wraps nerve fibers with myelin and also secretes regulatory factors.

sclereid A type of sclerenchyma cell; sclereids typically are short and have thick, lignified walls.

sclerenchyma A ground tissue in which cells develop thick secondary walls, which commonly are lignified and perforated by pits through which water can pass.

scrotum The baglike sac in which the testes are suspended in many mammals.

scutellum The shield-shaped cotyledon of a grass.

Secondary growth Plant growth that originates at lateral meristems and increases the diameter of older roots and stems. *Compare* primary growth.

secondary immune response The rapid immune response that occurs during the second (and subsequent) encounters of the immune system of a mammal with a specific antigen.

secondary plant body The part of a plant made up of tissues that develop from lateral meristems.

secondary tissue In plants, the tissue that develops from lateral meristems.

secretion A selective process in which specific small molecules and ions are transported from the body fluids (in animals with open circulatory systems) or blood (in animals with closed circulatory systems) into the excretory tubules.

seed coat The outer protective covering of a seed.

segment polarity genes In *Drosophila* embryonic development, the set of segmentation regulatory genes activated by pair-rule genes that set the boundaries and anterior-posterior axis of each segment in the embryo.

segmentation genes Genes that work sequentially, progressively subdividing the embryo into regions, determining the segments of the embryo and the adult.

selective cell adhesion A mechanism in which cells make and break specific connections to other cells or to the extracellular matrix.

self-incompatibility In plants, the inability of a plant's pollen to fertilize ovules of the same plant.

semen The secretions of several accessory glands in which sperm are mixed prior to ejaculation.

semicircular canal A part of the vestibular apparatus that detects rotational (spinning) motions.

semilunar valve (SL valve) A flap of endocardium and connective tissue reinforced by fibers that prevent the valve from turning inside out.

seminal fluid Fluid secreted by the seminal vesicles that contains prostaglandins, which when ejaculated into the female trigger contractions of the female reproductive tract that help move the sperm into and through the uterus.

seminal vesicle A vesicle that secretes seminal fluid.

seminiferous tubule One of the tiny tubes in the testes where sperm cells are produced, grow, and mature.

senescence The biologically complex process of aging in mature organisms that leads to the death of cells and eventually the whole organism.

sensitization Increased responsiveness to mild stimuli after experiencing a strong stimulus; one of the simplest forms of memory.

sensor A tissue or organ that detects a change in an external or internal factor such as pH, temperature, or the concentration of a molecule such as glucose.

sensory adaptation A condition in which the effect of a stimulus is reduced if it continues at a constant level.

sensory hair cell A hair cell that send impulses along the auditory nerve to the brain when alternating changes of pressure agitate the basilar membrane on which the organ of Corti rests, moving the hair cells.

sensory neuron A neuron that transmits stimuli collected by their sensory receptors to interneurons.

sensory receptor A receptor formed by the dendrites of afferent neurons, or by specialized receptor cells making synapses with afferent neurons that pick up information about the external and internal environments of the animal.

sensory transduction The conversion of a stimulus into a change in membrane potential.

sepal One of the separate, usually green parts forming the calyx of a flower.

sequential hermaphroditism The form of hermaphroditism in which individuals change from one sex to the other.

serosa The serous membrane: a thin membrane lining the closed cavities of the body; has two layers with a space between that is filled with serous fluid.

Sertoli cell One of the supportive cells that completely surrounds developing spermatocytes in the seminiferous tubules. Follicle-stimulating hormone stimulates Sertoli cells to secrete a protein and other molecules that are required for spermatogenesis.

set point The level at which the condition controlled by a homeostatic pathway is to be maintained.

sexual reproduction The mode of reproduction in which male and female parents produce offspring through the union of egg and sperm generated by meiosis.

shoot system The stems and leaves of a plant.

short-day plant A plant that flowers in late summer or early autumn when dark periods become longer and light periods become shorter.

short-term memory Memory that stores information for seconds.

sieve tube A series of phloem cells joined end to end, forming a long tube through which nutrients are transported; seen mainly in flowering plants.

sieve tube member Any of the main conducting cells of phloem that connect end to end, forming a sieve tube.

simple fruit A fruit that develops from a single ovary; in many of them at least one layer of the pericarp is fleshy and juicy.

simultaneous hermaphroditism A form of hermaphroditism in which individuals develop functional ovaries and testes at the same time.

single-lens eye An eye type that works by changing the amount of light allowed to enter into the eye and by focusing this incoming light with a lens.

sink Any region of a plant where organic substances are being unloaded from the sieve tube system and used or stored.

sinoatrial node (SA node) The region of the heart that controls the rate and timing of cardiac muscle cell contraction.

sinus A body space that surrounds an organ.

skeletal muscle A muscle that connect to bones of the skeleton, typically made up of long and cylindrical cells that contain many nuclei.

slow block to polyspermy The process in which enzymes released from cortical granules alter the egg coats within minutes after fertilization, so that no other sperm can attach and penetrate to the egg.

slow muscle fiber A muscle fiber that contracts relatively slowly and with low intensity.

smooth muscle A relatively small and spindle-shaped muscle cell in which actin and myosin molecules are arranged in a loose network rather than in bundles.

soil solution A combination of water and dissolved substances that coats soil particles and partially fills pore spaces.

somaclonal selection A procedure in which somatic embryos derived from tissue culture are screened to identify those having desired characteristics, such as disease resistance.

somatic cell Any of the cells of an organism's body other than reproductive cells.

somatic embryo A plant embryo that is genetically identical to the parent because it arose through asexual means.

somatic nervous system A subdivision of the peripheral nervous system controlling body movements that are primarily conscious and voluntary.

somites Paired blocks of mesoderm cells along the vertebrate body axis that form during early vertebrate development and differentiate into dermal skin, bone, and muscle.

source In plants, any region (such as a leaf) where organic substances are being loaded into the sieve tube system of phloem.

spatial summation The summation of EPSPs produced by firing of different presynaptic neurons.

spermatocyte A developing gamete that becomes a spermatid at the end of meiosis.

spermatogenesis The process of producing sperm.

spermatogonium (plural, spermatogonia) A cell that enters meiosis and gives rise to gametes, produced by mitotic divisions of the germ cells in males.

spermatozoan Also called sperm; a haploid cell that develops into a mature sperm cell when meiosis is complete.

sphincter A powerful ring of smooth muscle that forms a valve between major regions of the digestive tract.

spinal cord A column of nervous tissue located within the vertebral column and directly connected to the brain.

spinal nerve A nerve that carries signals between the spinal cord and the body trunk and limbs.

spiracle An opening in the chitinous exoskeleton of an insect through which air enters and leaves the tracheal system.

spore A reproductive structure, usually a single cell, that can develop into a new individual without fusing with another cell; found in plants, fungi, and certain protists.

sporophyte The diploid generation in the plant life cycle; it produces haploid spores.

stamen A "male" reproductive organ in flowers, consisting of an anther (pollen producer) and a slender filament.

stapes The smallest of three sound-conducting bones in the middle ear of tetrapod vertebrates.

statocyst A mechanoreceptor in invertebrates that senses gravity and motion using statoliths.

statolith A movable starch- or carbonate-containing stonelike body involved in sensing gravitational pull.

stele The central core of vascular tissue in roots and shoots of vascular plants; it consists of the xylem and phloem together with supporting tissues.

stereocilia Microvilli covering the surface of hair cells clustered in the base of neuromasts.

stigma The receptive end of a carpel where deposited pollen germinates.

stoma (plural, stomata) The opening between a pair of guard cells in the epidermis of a plant leaf or stem, through which gases and water vapor pass.

stomach The portion of the digestive system in which food is stored and digestion begins.

stretch receptor A proprioceptor in the muscles and tendons of vertebrates that detects the position and movement of the limbs.

style The slender stalk of a carpel situated between the ovary and the stigma in plants.

submucosa A thick layer of elastic connective tissue that contains neuron networks and blood and lymph vessels.

subsoil The region of soil beneath topsoil, which contains relatively little organic matter.

suspension feeder An animal that ingests small food items suspended in water.

suspensor In seed plants, a stalklike row of cells that develops from a zygote and helps position the embryo close to the nourishing endosperm.

sympathetic division Division of the autonomic nervous system that predominates in situations involving stress, danger, excitement, or strenuous physical activity.

symplastic pathway The route taken by water that moves through the cytoplasm of plant cells (the symplast). *Compare* apoplastic pathway.

symport The transport of two molecules in the same direction across a membrane.

synapse A site where a neuron makes a communicating connection with another neuron or an effector such as a muscle fiber or gland.

synaptic cleft A narrow gap that separates the plasma membranes of the presynaptic and postsynaptic cells.

synaptic vesicle A secretory vesicle in the cytoplasm of an axon terminal of a neuron, in which neurotransmitters are stored.

systemic acquired resistance A plant defense response to microbial invasion; defensive chemicals including salicylic acid may spread throughout a plant, rendering healthy tissues less vulnerable to infection.

systemic circuit In amphibians, the branch of a double blood circuit that receives oxygenated blood and provides the blood supply for most of the tissues and cells of a body.

systemin A plant peptide hormone that functions in defense responses to wounds.

systole The period of contraction and emptying of the heart.

T (transverse) tubule The tubule that passes in a transverse manner from the sarcolemma across a myofibril of striated muscle.

T cell A lymphocyte produced by the division of stem cells in the bone marrow and then released into the blood and carried to the thymus. T cells participate in adaptive immunity.

taproot system A root system consisting of a single main root from which lateral roots can extend; often stores starch.

T-cell receptor (TCR) A receptor that covers the plasma membrane of a T-cell, specific for a particular antigen.

temporal summation The summation of several EPSPs produced by successive firing of a single presynaptic neuron over a short period of time.

tendon A type of fibrous connective tissue that attaches muscles to bones.

terminal bud A bud that develops at the apex of a shoot.

testis (plural, testes) The male gonad. In male vertebrates, they secrete androgens and steroid hormones that stimulate and control the development and maintenance of male reproductive systems.

testosterone A hormone produced by the testes, responsible for the development of male secondary sex characteristics and the functioning of the male reproductive organs.

tetanus A situation in which a muscle fiber cannot relax at all between stimuli, and twitch summation produces a peak level of continuous contraction.

thalamus A major switchboard of the brain that receives sensory information and relays it to the regions of the cerebral cortex concerned with motor responses to sensory information of that type.

thermal acclimatization A set of physiological changes in ectotherms in response to seasonal shifts in environmental temperature, allowing the animals to attain good physiological performance at both winter and summer temperatures.

thermoreceptor A sensory receptor that detects the flow of heat energy.

thermoregulation The control of body temperature.

thick filament A type of filament in striated muscle composed of myosin molecules; they interact with thin filaments to shorten muscle fibers during contraction.

thigmomorphogenesis A plant response to a mechanical disturbance, such as frequent strong winds; includes inhibition of cellular elongation and production of thick-walled supportive tissue.

thigmotropism Growth in response to contact with a solid object.

thin filament A type of filament in striated muscle composed of actin, tropomyosin, and troponin molecules; they interact with thick filaments to shorten muscle fibers during contraction.

threshold potential In signal conduction by neurons, the membrane potential at which the action potential fires.

thymus An organ of the lymphatic system that plays a role in filtering viruses, bacteria, damaged cells, and cellular debris from the lymph and bloodstream, and in defending the body against infection and cancer.

thyroid gland A gland located beneath the voice box (larynx) that secretes hormones regulating growth and metabolism.

thyroid-stimulating hormone (TSH) A hormone that stimulates the thyroid gland to grow in size and secrete thyroid hormones.

thyroxine (T_4) The main hormone of the thyroid gland, responsible for controlling the rate of metabolism in the body.

tidal volume The volume of air entering and leaving the lungs during inhalation and exhalation.

tissue A group of cells and intercellular substances with the same structure that function as a unit to carry out one or more specialized tasks.

tonoplast The membrane that surrounds a plant cell vacuole.

topsoil The rich upper layer of soil where most plant roots are located; it generally consists of sand, clay particles, and humus.

torpor A sleeplike state produced when a lowered set point greatly reduces the energy required to maintain body temperature, accompanied by reductions in metabolic, nervous, and physical activity.

totipotent Having the capacity to produce cells that can develop into or generate a new organism or body part.

trace element A mineral required by organisms only in small amounts.

trachea In insects, an extensively branched, air-conducting tube formed by invagination of the outer epidermis of the animal, and reinforced by rings of chitin. In vertebrates, the windpipe, which branches into the bronchi.

tracheal system The branching network of tubes that carry air from small openings in the exoskeleton of an insect to tissues throughout its body.

tracheid A conducting cell of xylem, usually elongated and tapered.

transfer cell Any of the specialized cells that form when large amounts of solutes must be loaded or unloaded into the phloem; they facilitate the short-distance transport of organic solutes from the apoplast into the symplast.

translocation The long-distance transport of substances by plant vascular tissues (xylem and phloem).

transmembrane pathway The path followed by water when it enters root cells by crossing across the plasma membrane.

transmission In neural signaling, the sending of a message along a neuron, and then to another neuron or to a muscle or gland.

transpiration The evaporation of water from a plant, principally from the leaves.

transport epithelium A layer of cells with specialized transport proteins in their plasma membranes.

transport protein A protein embedded in the cell membrane that acts as a carrier of ions or some larger molecules across cell membranes.

trichome A single-celled or multicellular outgrowth from the epidermis of a plant that provides protection and shade and often gives the stems or leaves a hairy appearance.

triiodothyronine (T_3) A hormone secreted by the thyroid gland that regulates metabolism.

trimester A division of human gestation, three months in length.

trophoblast The outer single layer of cells of the blastocyst.

tropic hormone A hormone that regulates hormone secretion by another endocrine gland.

tropism The turning or bending of an organism or one of its parts toward or away from an external stimulus, such as light, heat, or gravity.

turgor pressure The normal fullness or tension produced by the fluid content of plant and animal cells.

tympanum A thin membrane in the auditory canal that vibrates back and forth when struck by sound waves.

umbilical cord A long tissue with blood vessels linking the embryo and the placenta.

umbilicus Navel; the scar left when the short length of umbilical cord still attached to the infant after birth dries and shrivels within a few days.

undernutrition A condition in animals in which intake of organic fuels is inadequate, or whose assimilation of such fuels is abnormal.

ureter The tube through which urine flows from the renal pelvis to the urinary bladder.

urethra The tube through which urine leaves the bladder. In most animals, the urethra opens to the outside.

urinary bladder A storage sac located outside the kidneys.

uterine cycle The menstrual cycle.

uterus A specialized saclike organ, in which the embryo develops in viviparous animals.

utricle A fluid-filled chamber of the vestibular apparatus that provides information about the position of the head with respect to gravity (up versus down), as well as changes in the rate of linear movement of the body.

vaccination The process of administering a weakened form of a disease to patients as a means of giving them immunity to a more serious form of the disease.

vagina The muscular canal that leads from the cervix to the exterior.

vas deferens The tube through which sperm travel from the epididymis to the urethra in the male reproductive system.

vascular bundle A cord of plant vascular tissue; often multistranded with both xylem and phloem.

vascular cambium A lateral meristem that produces secondary vascular tissues in plants.

vascular tissue system One of the three tissue systems in plants that provide the foundation for plant organs; it consists of transport tubes for water and nutrients.

vegetal pole The end of the egg opposite the animal pole, which typically gives rise to internal structures such as the gut and the posterior end of the embryo.

vegetative reproduction Asexual reproduction in plants by which new individuals arise (or are created) without seeds or spores; examples include fragmentation from the parent plant or the use of cuttings by gardeners.

vein In a plant, a vascular bundle that forms part of the branching network of conducting and supporting tissues in a leaf or other expanded plant organ. In an animal, a vessel that carries the blood back to the heart.

ventilation The flow of the respiratory medium (air or water, depending on the animal) over the respiratory surface.

ventricle In the brain, an irregularly shaped cavity containing cerebrospinal fluid. In the heart, a chamber that pumps blood out of the heart.

venule A capillary that merges into the small veins leaving an organ.

vernalization The stimulation of flowering by a period of low temperature.

vessel In plants, one of the tubular conducting structures of xylem, typically several centimeters long; most angiosperms and some other vascular plants have xylem vessels.

vessel member Any of the short cells joined end to end in tubelike columns in xylem.

vestibular apparatus The specialized sensory structure of the inner ear of most terrestrial vertebrates that is responsible for perceiving the position and motion of the head and, therefore, for maintaining equilibrium and for coordinating head and body movements.

vital capacity The maximum tidal volume of air that an individual can inhale and exhale.

vitamin An organic molecule required in small quantities that the animal cannot synthesize for itself.

vitamin D A steroidlike molecule that increases the absorption of Ca^{2+} and phosphates from ingested food by promoting the synthesis of a calcium-binding protein in the intestine; it also increases the release of Ca^{2+} from bone in response to PTH.

vitelline coat A gel-like matrix of proteins, glycoproteins, or polysaccharides immediately outside the plasma membrane of an egg cell.

vitreous humor The jellylike substance that fills the main chamber of the eye, between the lens and the retina.

viviparous Referring to animals that retain the embryo within the mother's body and nourish it during at least early embryo development.

voltage-gated ion channel A membrane-embedded protein that opens and closes as the membrane potential changes.

vulva The external female sex organs.

water potential The potential energy of water, representing the difference in free energy between pure water and water in cells and solutions; it is the driving force for osmosis.

water-soluble vitamin A vitamin with a high proportion of oxygen and nitrogen able to form hydrogen bonds with water.

white matter The myelinated axons that surround the gray matter of the central nervous system.

wilting The drooping of leaves and stems caused by a loss of turgor.

Wolffian duct A bipotential primitive duct associated with the gonads that leads to a cloaca.

wood The secondary xylem of trees and shrubs, lying under the bark and consisting largely of cellulose and lignin.

xylem The plant vascular tissue that distributes water and nutrients.

xylem sap The dilute solution of water and solutes that flows in the xylem.

yolk The portion of an egg that serves as the main energy source for the embryo.

yolk sac In an amniote egg, an extraembryonic membrane that encloses the yolk.

zona pellucida A gel-like matrix of proteins, glycoproteins, or polysaccharides immediately outside the plasma membrane of the egg cell.

zone of cell division The region in a growing root that consists of the root apical meristem and the actively dividing cells behind it.

zone of elongation The region in a root where newly formed cells grow and elongate.

zone of maturation The region in a root above the zone of elongation where cells do not increase in length but may differentiate further and take on specialized roles.

zygote A fertilized egg.

Credits

This page constitutes an extension of the copyright page. We have made every effort to trace the ownership of all copyrighted material and to secure permission from copyright holders. In the event of any question arising as to the use of any material, we will be pleased to make the necessary corrections in future printings. Thanks are due to the following authors, publishers, and agents for permission to use the material indicated.

CHAPTER 31 **Page 711** © Mark Bolton/Corbis. **31.1** (a) © Earl Roberge/Photo Researchers, Inc. (b) © W. Percy Conway/Corbis. (c) © Gregory K. Scott/Photo Researchers, Inc. **Table 31.1** (row 1, left) © Bruce Iverson. (row 1, right) © Mike Clayton/University of Wisconsin Department of Botany. (row 2, left) © Ernest Manewal/Index Stock Imagery. (row 2, right) © Darrell Gulin/Corbis. (row 3, left) © Simon Fraser/Photo Researchers, Inc. (row 3, right) Gary Head. (row 4, left and right) © Andrew Syred/Photo Researchers, Inc. **31.5** (right) James D. Mauseth. **31.6** (all) © Biophoto Associates. **31.7** (a) © Kingsley R. Stern. (b) © D. E. Akin and I. L. Rigsby, Richard B. Russel, Agricultural Research Service, U.S. Department of Agriculture, Athens, Georgia. **31.8** (a) Alison W. Roberts, University of Rhode Island. (b) H. A. Cote, W. A. Cote, and A. C. Day, *Wood Structure and Identification,* second edition, Syracuse University Press. **31.9** (a) James D. Mauseth, University of Texas. (b) Courtesy of Professor John Main, Pacific Lutheran University. **31.10** (a) George S. Ellmore. (b) © Dr. Jeremy Burgess/SPL/Photo Researchers, Inc. (c) Courtesy Mark Holland, Salisbury University. **Page 720** (figure a) William E. Ferguson. (figure b) Cathie Martin. **31.11** (b) Jakub Jasinski/Visuals Unlimited. **31.12** (b) Robert and Linda Mitchell Photography. (c) Richard R. Dute. **31.13** (a, center) Ray F. Evert. (a, right) James W. Perry. (b, center) Carolina Biological Supply. (b, right) James W. Perry. **31.14** (a) Mike Hill/Getty Images Inc. (b) Wally Eberhart/Visuals Unlimited. (c) Joerg Boethling/Peter Arnold, Inc. (d) Alan & Linda Detrick/Photo Researchers, Inc. (e) Michael P. Gadomski/Photo Researchers, Inc. **31.16** (a) Joseph Devenney/Getty Images Inc. (b) Maxine Adcock/SPL/Photo Researchers Inc. **31.17** (b) C. E. Jeffree, et al, *Planta* 172(1):20–37, 1987. Reprinted by permission of C. E. Jefree and Springer-Verlag. **31.18** (a, b) Thomas L. Rost. **31.19** (c) © Beth Davidow/Visuals Unlimited. **31.20** (b) John Limbaugh/Ripon Microslides, Inc. **31.21** (a) Chuck Brown. (b) Carolina Biological Supply. **31.22** © Omnikron/Photo Researchers, Inc. **31.23** (b) Alison W. Roberts, University of Rhode Island. **31.26** (b) © George Bernard/SPL/Photo Researchers, Inc. **Page 733** Tony Gibson at the National Science Foundation.

CHAPTER 32 **Page 737** © Steve Gschmeissner/SPL/Photo Researchers, Inc. **32.1** Owaki-Kulla/Corbis. **32.2** Micrograph Chuck Brown. **32.5** (both) © Claude Nuridsany and Marie Perennou/Science Photo Library/Photo Researchers, Inc. **32.7** (b) Micrograph Chuck Brown. **32.9** Dr. John D. Cunningham/Visuals Unlimited. **32.11** (both) T. A. Masefield. **32.13** (a) BIOS Matt Alexander/Peter Arnold, Inc. (b) Thomas L. Rost. (c) Fritz Polking/Visuals Unlimited. (d) Fritz Polking/Visuals Unlimited. **32.14** (a) Martin Zimmerman, ***Science***, 1961, 133: 73–79, © AAAS. (b) Martin H. Zimmermann.

CHAPTER 33 **757** © Ellen McKnight/Alamy. **33.1** Gerry Ellis/The Wildlife Collection. **33.3** (all) E. Epstein, University of California, Davis. **33.4** William Ferguson. **33.8** (a) Adrian P. Davies/Bruce Coleman. (b) NifTAL Project, University of Hawaii, Maui. (c) © Dr. Jeremy Burgess/SPL/Photo Researchers, Inc. **33.9** (d) Mark E. Dudley and Sharon R. Long. **33.10** (a) David Cavagnaro/Peter Arnold, Inc. (b) © Grant Heilman Photography. (c) Beverly McMillan. (d) © Prem Subrahmanyam/www.premdesign.com.

CHAPTER 34 **Page 775** © Ted Kinsman/SPL/Photo Researchers, Inc. **34.1** (left) Courtesy of Caroline Ford, School of Plant Sciences, University of Reading, UK. (right) ZEFA—Rein. **34.2** (both) Gary Head. **34.4** (a) Janet Jones. (b) Karlene V. Schwartz. **34.6** (a) David M. Phillips/Visuals Unlimited. (b) Dr. Jeremy Burgess/SPL/Photo Researchers, Inc. (c) David Scharf/Peter Arnold, Inc. **34.8** (a) Michael Clayton, University of Wisconsin. (b) Patricia Schulz. (c) Michael Clayton, University of Wisconsin. (d) Dr. Charles Good, Ohio State University–Lima. (e, f) Michael Clayton, University of Wisconsin. **34.9** (c) Dr. John D. Cunningham/Visuals Unlimited. **34.10** (b) Siegel, R./Arco Images/Peter Arnold, Inc. (c, top) Richard H. Gross. (c, bottom) Andrew Syred/SPL/Photo Researchers, Inc. (d) Mark Rieger. (e) R. Carr. **34.12** (c) Herve Chaumeton/Agence Nature. **34.13** (c) Barry L. Runk/Grant Heilman, Inc. (c) James Mauseth. **34.14** Ed Reschke/Peter Arnold, Inc. **34.15** (left) R-R/S/Grant Heilman Photography, Inc. (center top and bottom; right) Professor Dr. Hans Hanks-Ulrich Koop. **Page 791** Courtesy of the Arabidopsis Information Resource, 2005. **34.16** (a) Kelly Yee and John J. Harada. (b) Damien Lovegrove/SPL/Photo Researchers, Inc. **Page 793** (left) Jonathan Plett and Sharon Regan. (center) Daniel Szymanski, *Plant Cell* 10:2047. (right) © Dr. Daniel Szymanski, Agronomy Department, Purdue University. **34.19** (all) S. M. Wick, *J Cell Biol*, 89:685, 1981, Rockefeller University Press. **34.21** (all) Jose Luis Riechmann. **34.22** (a) Roland R. Dute.

CHAPTER 35 **Page 801** © Garry Black/Masterfile. **35.1** Nigel Cattlin/Visuals Unlimited. **35.4** Kingsley R. Stern. **35.8** Sylvan H. Wittwer/Visuals Unlimited. **35.9** Sylvan Wittwer/Visuals Unlimited. **35.11** N.R. Lersten. **35.12** Larry D. Nooden. **35.13** Joanne Chory. **35.14** Amanda Darcy/Getty Images Inc. **35.15** David Cavagnaro/Peter Arnold, Inc. **35.16** Nigel Cattlin/Photo Researchers, Inc. **35.19** Cathlyn Melloan/Stone/Getty Images. **35.20** (both) Micrographs courtesy of Randy Moore, from "How Roots Respond to Gravity," M. L. Evans, R. Moore, and K. Hasenstein, *Scientific American,* December 1986. **35.21** (left) Michael Clayton, University of Wisconsin. (right) John Digby and Richard Firn. **35.23** Cary Mitchell. **35.24** (all) Frank B. Salisbury. **35.25** (a, b) David Sieren/Visuals Unlimited. **35.27** Dwight Kuhn. **35.28** Jan Zeevart. **35.29** (left) Clay Perry/Corbis. (right) Eric Chrichton/Corbis. **35.31** R. J. Downs. **35.32** (right) Eric Welzel/Fox Hill Nursery, Freeport, Maine. **Page 827** Tony Gibson at the National Science Foundation.

CHAPTER 36 **Page 831** Simon Fraser/SPL/Photo Researchers, Inc. **36.1** David Macdonald. **36.3** (b, left) Ray Simmons/Photo Researchers, Inc. (b, center) Ed Reschke/Peter Arnold, Inc. (b, right) Don Fawcett. **36.4** (a, top) Gregory Dimijian/Photo Researchers, Inc. **36.5** (a, top) Ed Reschke. (b, top) Ed Reschke. (c, top) Fred Hossler/Visuals Unlimited. (d, top) Ed Reschke. (e, top) Ed Reschke. (f, top) Ed Reschke. **36.6** (a, top) Ed Reschke. (b, top) Ed Reschke. (c, top) BioPhoto Associates/Photo Researchers, Inc. **36.7** Lennart Nilsson from *Behold Man,* © 1974 Albert Bonniers Forlag and Little, Brown and Company, Boston. **36.10** Fred Bruemmer.

CHAPTER 37 **Page 847** © C. J. Guerin, Ph.D., MRC Toxicology Unit/SPL/Photo Researchers, Inc. **37.1** © Gary Gerovac/Masterfile. **37.3** © Triarch/Visuals Unlimited **37.4** © Nancy Kedersha/UCLA/Photo Researchers, Inc. **37.5** © C. Raines/Visuals Unlimited. **37.13** © Dennis Kunkel/Visuals Unlimited. **37.15** E. R. Lewis, T. E. Everhart, Y. Y. Zevi/Visuals Unlimited.

CHAPTER 38 **Page 867** © Sovereign/ISM/SPL/Phototake, Inc. **38.7** Courtesy of Dr. Marcus Raichle, courtesy of Washington University School of Medicine, St. Louis.

CHAPTER 39 **Page 885** © Stephen Dalton/Animals, Animals—Earth Scenes. **39.3** (left) Herve Chaumeton/Agence Nature. **39.7** (bottom) © Andrew Syred/Photo Researchers, Inc. **39.10** (top) © E. R. Degginger. **39.11** © Chris Newbert. **39.18** (top) Carolina Biological Supply. (bottom) Dr. M. V. Parthasarathy/Cornell Integrated Microscopy Center. **39.19** (left) © A. Shay/OSF/Animals Animals—Earth Scenes. (right) Louisa Howard, Dartmouth College EM Facility. **39.22** David Hosking. **39.23** (top right) Kenneth Lohmann/University of North Carolina. **Page 908** Chase Smith.

CHAPTER 40 **Page 909** © Mark Wallner. **40.8** Syndication International Ltd., 1986. **40.9** (top) © John Shaw/Tom Stack & Associates, Inc. **40.14** (right) © Frans Lanting/Bruce Coleman Ltd.

CHAPTER 41 **Page 933** G. Delpho/Peter Arnold, Inc. **41.1** Kiisa Nishikawa/Northern Arizona University. **41.2** (both) © Don Fawcett/Visuals Unlimited. **41.8** Linda Pitkin/Planet Earth Pictures.

CHAPTER 42 **Page 949** GJLP/CNRI/SPL/Photo Researchers, Inc. **42.1** (a) From A. D. Waller, *Physiology, The Servant of Medicine,* Hitchcock Lectures, University of London Press, 1910. **42.6** (top) © National Cancer Institute/Photo Researchers, Inc. **42.8** © Professor P. Motta/Department of Anatomy/University La Sapienca, Rome/SPL/Photo Researchers, Inc. **42.13** (bottom) © Sheila Terry/SPL/Photo Researchers, Inc. **42.16** Lennart Nilsson from *Behold Man* ©1974 published by Albert Bonniers Forlag and Little, Brown and Company. **42.19** (a) Ed Reschke. (b) Biophoto Associates/Photo Researchers, Inc.

CHAPTER 43 **Page 971** © Dr. Andrejs Liepins/Science Photo Library/Photo Researchers, Inc. **43.1** Biology Media/Photo Researchers, Inc. **43.4** (right) From Harris, L. J.; Larson, S. B.; Hasel, K. W.; McPherson, A.; *Biochemistry* 36, p. 1581 (1997). Structure rendered with RIBBONS. **Page 979** © Peter Skinner/Photo Researchers, Inc. **43.13** Lennart Nilsson/Bonnier Fakta AB. **Page 991** (both) Z. Salahuddin, National Institutes of Health.

CHAPTER 44 **Page 997** © Steve Gschmeissner/Science Photo Library/Photo Researchers, Inc. **44.2** (a) Peter Parks/Oxford Scientific Films. (b) Jack Dermid/Visuals Unlimited. (c) © 2000 Photodisc, Inc. (with art by Lisa Starr). **44.3** (a) Alex Kirstitch. **44.5** (top) Ed Reschke. **44.9** (both) SIU/Visuals Unlimited.

CHAPTER 45 **Page 1015** Thomas Mangelsen/Minden Pictures. **45.1** David Shale/npl/Minden Pictures. **45.2** (a) Sanford/Angliolo/Corbis.

(b) James Watt/Animals, Animals—Earth Scenes. (c) Joe McDonald/Corbis. (d) Gunter Ziesler/Bruce Coleman, Inc. **45.7** Dr. Richard Kessel & Dr. Randy Kardon/Tissues & Organs/Visuals Unlimited. **45.11** (left) Mark Nielsen, University of Utah. (right) D.W. Fawcett/Photo Researchers, Inc. **45.18** (left) Fred Bruemmer. (right) Jane Burton/Bruce Coleman, Inc. **45.19** (left) D. Robert Franz/Plant Earth Pictures.

CHAPTER 46 **Page 1039** © Dennis Kunkel/Phototake, Inc. **46.10** (top) David Noble/FPG/Getty Images. (bottom) Claude Steelman/Tom Stack & Associates. **46.16** Rafael Winer/Corbis. **46.20** (right) John D. Cunningham/Visuals Unlimited. **46.21** (a) Joe McDonald/Corbis. (b) Fredrik Broman/Iconica/Getty Images, Inc. **46.22** (right) Inga Spence/Visuals Unlimited.

CHAPTER 47 **Page 1069** © Clouds Hill Imaging Ltd./Corbis. **47.1** Seapics. **47.2** Dr. Stanley Flegler/Visuals Unlimited. **47.4** (a) Dr. David M. Phillips/Visuals Unlimited. **47.5** Ryuzo Yanagimachi. **47.7** Hans Pfletschinger. **47.8** David M. Phillips/Visuals Unlimited. **47.9** John Cancalosi/Peter Arnold, Inc. **47.10** (a) Robin Chittenden/Frank Lane Picture Agency/Corbis. **47.12** Lennart Nilsson from *A Child Is Born,* © 1966, 1977 Dell Publishing Company, Inc. **47.15** (top right) Michael C. Webb/Visuals Unlimited. **47.17** (a) Lennart Nilsson From *A Child Is Born,* © 1966, 1977 Dell Publishing Company, Inc.

CHAPTER 48 **Page 1093** Daniel Sambraus/SPL/Photo Researchers, Inc. **48.2** (all) Carolina Biological Supply Company. **48.7** (all) Carolina Biological Supply Company. **48.8** Peter B. Armstrong, University of California, Davis. **48.10** Carolina Biological Supply Company. **48.11** (b) Carolina Biological Supply Company. **48.15** (all) Lennart Nilsson, *A Child Is Born,* © 1966, 1977, Dell Publishing Company, Inc. **Page 1117** David M. Parichy. **48.28** (left) Oliver Meckes/Nicole Ottawa/Photo Researchers, Inc. (right) UCSF Computer Graphics Laboratory, National Institutes, NCRR Grant 01081. **48.30** (a) Carolina Biological Supply Company. (b) Lennart Nilsson from *A Child Is Born,* © 1966, 1977 Dell Publishing Company, Inc.

Credits

This page constitutes an extension of the copyright page. We have made every effort to trace the ownership of all copyrighted material and to secure permission from copyright holders. In the event of any question arising as to the use of any material, we will be pleased to make the necessary corrections in future printings. Thanks are due to the following authors, publishers, and agents for permission to use the material indicated.

CHAPTER 31 **Page 711** © Mark Bolton/Corbis. **31.1** (a) © Earl Roberge/Photo Researchers, Inc. (b) © W. Percy Conway/Corbis. (c) © Gregory K. Scott/Photo Researchers, Inc. **Table 31.1** (row 1, left) © Bruce Iverson. (row 1, right) © Mike Clayton/University of Wisconsin Department of Botany. (row 2, left) © Ernest Manewal/Index Stock Imagery. (row 2, right) © Darrell Gulin/Corbis. (row 3, left) © Simon Fraser/Photo Researchers, Inc. (row 3, right) Gary Head. (row 4, left and right) © Andrew Syred/Photo Researchers, Inc. **31.5** (right) James D. Mauseth. **31.6** (all) © Biophoto Associates. **31.7** (a) © Kingsley R. Stern. (b) © D. E. Akin and I. L. Rigsby, Richard B. Russel, Agricultural Research Service, U.S. Department of Agriculture, Athens, Georgia. **31.8** (a) Alison W. Roberts, University of Rhode Island. (b) H. A. Cote, W. A. Cote, and A. C. Day, *Wood Structure and Identification*, second edition, Syracuse University Press. **31.9** (a) James D. Mauseth, University of Texas. (b) Courtesy of Professor John Main, Pacific Lutheran University. **31.10** (a) George S. Ellmore. (b) © Dr. Jeremy Burgess/SPL/Photo Researchers, Inc. (c) Courtesy Mark Holland, Salisbury University. **Page 720** (figure a) William E. Ferguson. (figure b) Cathie Martin. **31.11** (b) Jakub Jasinski/Visuals Unlimited. **31.12** (b) Robert and Linda Mitchell Photography. (c) Richard R. Dute. **31.13** (a, center) Ray F. Evert. (a, right) James W. Perry. (b, center) Carolina Biological Supply. (b, right) James W. Perry. **31.14** (a) Mike Hill/Getty Images Inc. (b) Wally Eberhart/Visuals Unlimited. (c) Joerg Boethling/Peter Arnold, Inc. (d) Alan & Linda Detrick/Photo Researchers, Inc. (e) Michael P. Gadomski/Photo Researchers, Inc. **31.16** (a) Joseph Devenney/Getty Images Inc. (b) Maxine Adcock/SPL/Photo Researchers Inc. **31.17** (b) C. E. Jeffree, et al, *Planta* 172(1):20–37, 1987. Reprinted by permission of C. E. Jefree and Springer-Verlag. **31.18** (a, b) Thomas L. Rost. **31.19** (c) © Beth Davidow/Visuals Unlimited. **31.20** (b) John Limbaugh/Ripon Microslides, Inc. **31.21** (a) Chuck Brown. (b) Carolina Biological Supply. **31.22** © Omnikron/Photo Researchers, Inc. **31.23** (b) Alison W. Roberts, University of Rhode Island. **31.26** (b) © George Bernard/SPL/Photo Researchers, Inc. **Page 733** Tony Gibson at the National Science Foundation.

CHAPTER 32 **Page 737** © Steve Gschmeissner/SPL/Photo Researchers, Inc. **32.1** Owaki-Kulla/Corbis. **32.2** Micrograph Chuck Brown. **32.5** (both) © Claude Nuridsany and Marie Perennou/Science Photo Library/Photo Researchers, Inc. **32.7** (b) Micrograph Chuck Brown. **32.9** Dr. John D. Cunningham/Visuals Unlimited. **32.11** (both) T. A. Masefield. **32.13** (a) BIOS Matt Alexander/Peter Arnold, Inc. (b) Thomas L. Rost. (c) Fritz Polking/Visuals Unlimited. (d) Fritz Polking/Visuals Unlimited. **32.14** (a) Martin Zimmerman, *Science*, 1961, 133: 73–79, © AAAS. (b) Martin H. Zimmermann.

CHAPTER 33 **757** © Ellen McKnight/Alamy. **33.1** Gerry Ellis/The Wildlife Collection. **33.3** (all) E. Epstein, University of California, Davis. **33.4** William Ferguson. **33.8** (a) Adrian P. Davies/Bruce Coleman. (b) NifTAL Project, University of Hawaii, Maui. (c) © Dr. Jeremy Burgess/SPL/Photo Researchers, Inc. **33.9** (d) Mark E. Dudley and Sharon R. Long. **33.10** (a) David Cavagnaro/Peter Arnold, Inc. (b) © Grant Heilman Photography. (c) Beverly McMillan. (d) © Prem Subrahmanyam/www.premdesign.com.

CHAPTER 34 **Page 775** © Ted Kinsman/SPL/Photo Researchers, Inc. **34.1** (left) Courtesy of Caroline Ford, School of Plant Sciences, University of Reading, UK. (right) ZEFA—Rein. **34.2** (both) Gary Head. **34.4** (a) Janet Jones. (b) Karlene V. Schwartz. **34.6** (a) David M. Phillips/Visuals Unlimited. (b) Dr. Jeremy Burgess/SPL/Photo Researchers, Inc. (c) David Scharf/Peter Arnold, Inc. **34.8** (a) Michael Clayton, University of Wisconsin. (b) Patricia Schulz. (c) Michael Clayton, University of Wisconsin. (d) Dr. Charles Good, Ohio State University–Lima. (e, f) Michael Clayton, University of Wisconsin. **34.9** (c) Dr. John D. Cunningham/Visuals Unlimited. **34.10** (b) Siegel, R./Arco Images/Peter Arnold, Inc. (c, top) Richard H. Gross. (c, bottom) Andrew Syred/SPL/Photo Researchers, Inc. (d) Mark Rieger. (e) R. Carr. **34.12** (c) Herve Chaumeton/Agence Nature. **34.13** (c) Barry L. Runk/Grant Heilman, Inc. (c) James Mauseth. **34.14** Ed Reschke/Peter Arnold, Inc. **34.15** (left) R-R/S/Grant Heilman Photography, Inc. (center top and bottom; right) Professor Dr. Hans Hanks-Ulrich Koop. **Page 791** Courtesy of the Arabidopsis Information Resource, 2005. **34.16** (a) Kelly Yee and John J. Harada. (b) Damien Lovegrove/SPL/Photo Researchers, Inc. **Page 793** (left) Jonathan Plett and Sharon Regan. (center) Daniel Szymanski, *Plant Cell* 10:2047. (right) © Dr. Daniel Szymanski, Agronomy Department, Purdue University. **34.19** (all) S. M. Wick, *J Cell Biol*, 89:685, 1981, Rockefeller University Press. **34.21** (all) Jose Luis Riechmann. **34.22** (a) Roland R. Dute.

CHAPTER 35 **Page 801** © Garry Black/Masterfile. **35.1** Nigel Cattlin/Visuals Unlimited. **35.4** Kingsley R. Stern. **35.8** Sylvan H. Wittwer/Visuals Unlimited. **35.9** Sylvan Wittwer/Visuals Unlimited. **35.11** N.R. Lersten. **35.12** Larry D. Nooden. **35.13** Joanne Chory. **35.14** Amanda Darcy/Getty Images Inc. **35.15** David Cavagnaro/Peter Arnold, Inc. **35.16** Nigel Cattlin/Photo Researchers, Inc. **35.19** Cathlyn Melloan/Stone/Getty Images. **35.20** (both) Micrographs courtesy of Randy Moore, from "How Roots Respond to Gravity," M. L. Evans, R. Moore, and K. Hasenstein, *Scientific American,* December 1986. **35.21** (left) Michael Clayton, University of Wisconsin. (right) John Digby and Richard Firn. **35.23** Cary Mitchell. **35.24** (all) Frank B. Salisbury. **35.25** (a, b) David Sieren/Visuals Unlimited. **35.27** Dwight Kuhn. **35.28** Jan Zeevart. **35.29** (left) Clay Perry/Corbis. (right) Eric Chrichton/Corbis. **35.31** R. J. Downs. **35.32** (right) Eric Welzel/Fox Hill Nursery, Freeport, Maine. **Page 827** Tony Gibson at the National Science Foundation.

CHAPTER 36 **Page 831** Simon Fraser/SPL/Photo Researchers, Inc. **36.1** David Macdonald. **36.3** (b, left) Ray Simmons/Photo Researchers, Inc. (b, center) Ed Reschke/Peter Arnold, Inc. (b, right) Don Fawcett. **36.4** (a, top) Gregory Dimijian/Photo Researchers, Inc. **36.5** (a, top) Ed Reschke. (b, top) Ed Reschke. (c, top) Fred Hossler/Visuals Unlimited. (d, top) Ed Reschke. (e, top) Ed Reschke. (f, top) Ed Reschke. **36.6** (a, top) Ed Reschke. (b, top) Ed Reschke. (c, top) BioPhoto Associates/Photo Researchers, Inc. **36.7** Lennart Nilsson from *Behold Man,* © 1974 Albert Bonniers Forlag and Little, Brown and Company, Boston. **36.10** Fred Bruemmer.

CHAPTER 37 **Page 847** © C. J. Guerin, Ph.D., MRC Toxicology Unit/SPL/Photo Researchers, Inc. **37.1** © Gary Gerovac/Masterfile. **37.3** © Triarch/Visuals Unlimited **37.4** © Nancy Kedersha/UCLA/Photo Researchers, Inc. **37.5** © C. Raines/Visuals Unlimited. **37.13** © Dennis Kunkel/Visuals Unlimited. **37.15** E. R. Lewis, T. E. Everhart, Y. Y. Zevi/Visuals Unlimited.

CHAPTER 38 **Page 867** © Sovereign/ISM/SPL/Phototake, Inc. **38.7** Courtesy of Dr. Marcus Raichle, courtesy of Washington University School of Medicine, St. Louis.

CHAPTER 39 **Page 885** © Stephen Dalton/Animals, Animals—Earth Scenes. **39.3** (left) Herve Chaumeton/Agence Nature. **39.7** (bottom) © Andrew Syred/Photo Researchers, Inc. **39.10** (top) © E. R. Degginger. **39.11** © Chris Newbert. **39.18** (top) Carolina Biological Supply. (bottom) Dr. M. V. Parthasarathy/Cornell Integrated Microscopy Center. **39.19** (left) © A. Shay/OSF/Animals Animals—Earth Scenes. (right) Louisa Howard, Dartmouth College EM Facility. **39.22** David Hosking. **39.23** (top right) Kenneth Lohmann/University of North Carolina. **Page 908** Chase Smith.

CHAPTER 40 **Page 909** © Mark Wallner. **40.8** Syndication International Ltd., 1986. **40.9** (top) © John Shaw/Tom Stack & Associates, Inc. **40.14** (right) © Frans Lanting/Bruce Coleman Ltd.

CHAPTER 41 **Page 933** G. Delpho/Peter Arnold, Inc. **41.1** Kiisa Nishikawa/Northern Arizona University. **41.2** (both) © Don Fawcett/Visuals Unlimited. **41.8** Linda Pitkin/Planet Earth Pictures.

CHAPTER 42 **Page 949** GJLP/CNRI/SPL/Photo Researchers, Inc. **42.1** (a) From A. D. Waller, *Physiology, The Servant of Medicine,* Hitchcock Lectures, University of London Press, 1910. **42.6** (top) © National Cancer Institute/Photo Researchers, Inc. **42.8** © Professor P. Motta/Department of Anatomy/University La Sapienca, Rome/SPL/Photo Researchers, Inc. **42.13** (bottom) © Sheila Terry/SPL/Photo Researchers, Inc. **42.16** Lennart Nilsson from *Behold Man* ©1974 published by Albert Bonniers Forlag and Little, Brown and Company. **42.19** (a) Ed Reschke. (b) Biophoto Associates/Photo Researchers, Inc.

CHAPTER 43 **Page 971** © Dr. Andrejs Liepins/Science Photo Library/Photo Researchers, Inc. **43.1** Biology Media/Photo Researchers, Inc. **43.4** (right) From Harris, L. J.; Larson, S. B.; Hasel, K. W.; McPherson, A.; *Biochemistry* 36, p. 1581 (1997). Structure rendered with RIBBONS. **Page 979** © Peter Skinner/Photo Researchers, Inc. **43.13** Lennart Nilsson/Bonnier Fakta AB. **Page 991** (both) Z. Salahuddin, National Institutes of Health.

CHAPTER 44 **Page 997** © Steve Gschmeissner/Science Photo Library/Photo Researchers, Inc. **44.2** (a) Peter Parks/Oxford Scientific Films. (b) Jack Dermid/Visuals Unlimited. (c) © 2000 Photodisc, Inc. (with art by Lisa Starr). **44.3** (a) Alex Kirstitch. **44.5** (top) Ed Reschke. **44.9** (both) SIU/Visuals Unlimited.

CHAPTER 45 **Page 1015** Thomas Mangelsen/Minden Pictures. **45.1** David Shale/npl/Minden Pictures. **45.2** (a) Sanford/Angliolo/Corbis.

(b) James Watt/Animals, Animals—Earth Scenes. (c) Joe McDonald/Corbis. (d) Gunter Ziesler/Bruce Coleman, Inc. **45.7** Dr. Richard Kessel & Dr. Randy Kardon/Tissues & Organs/Visuals Unlimited. **45.11** (left) Mark Nielsen, University of Utah. (right) D.W. Fawcett/Photo Researchers, Inc. **45.18** (left) Fred Bruemmer. (right) Jane Burton/Bruce Coleman, Inc. **45.19** (left) D. Robert Franz/Plant Earth Pictures.

CHAPTER 46 **Page 1039** © Dennis Kunkel/Phototake, Inc. **46.10** (top) David Noble/FPG/Getty Images. (bottom) Claude Steelman/Tom Stack & Associates. **46.16** Rafael Winer/Corbis. **46.20** (right) John D. Cunningham/Visuals Unlimited. **46.21** (a) Joe McDonald/Corbis. (b) Fredrik Broman/Iconica/Getty Images, Inc. **46.22** (right) Inga Spence/Visuals Unlimited.

CHAPTER 47 **Page 1069** © Clouds Hill Imaging Ltd./Corbis. **47.1** Seapics. **47.2** Dr. Stanley Flegler/Visuals Unlimited. **47.4** (a) Dr. David M. Phillips/Visuals Unlimited. **47.5** Ryuzo Yanagimachi. **47.7** Hans Pfletschinger. **47.8** David M. Phillips/Visuals Unlimited. **47.9** John Cancalosi/Peter Arnold, Inc. **47.10** (a) Robin Chittenden/Frank Lane Picture Agency/Corbis. **47.12** Lennart Nilsson from *A Child Is Born,* © 1966, 1977 Dell Publishing Company, Inc. **47.15** (top right) Michael C. Webb/Visuals Unlimited. **47.17** (a) Lennart Nilsson From *A Child Is Born,* © 1966, 1977 Dell Publishing Company, Inc.

CHAPTER 48 **Page 1093** Daniel Sambraus/SPL/Photo Researchers, Inc. **48.2** (all) Carolina Biological Supply Company. **48.7** (all) Carolina Biological Supply Company. **48.8** Peter B. Armstrong, University of California, Davis. **48.10** Carolina Biological Supply Company. **48.11** (b) Carolina Biological Supply Company. **48.15** (all) Lennart Nilsson, *A Child Is Born,* © 1966, 1977, Dell Publishing Company, Inc. **Page 1117** David M. Parichy. **48.28** (left) Oliver Meckes/Nicole Ottawa/Photo Researchers, Inc. (right) UCSF Computer Graphics Laboratory, National Institutes, NCRR Grant 01081. **48.30** (a) Carolina Biological Supply Company. (b) Lennart Nilsson from *A Child Is Born,* © 1966, 1977 Dell Publishing Company, Inc.

Index

The letter i *designates illustration;* t *designates table;* **bold** *designates defined or introduced term.*